T0309900

Greenland
(Kalatdlit-Nunat)
Alaska
Yukon
N.W.T.
Nunavut
[Labrador]
B.C.
Alta.
Sask.
Man.
Ont.
Que.
Nfld.
St.Pierre
and Miquelon
P.E.I.
N.B.
N.S.
Wash.
Mont.
N.Dak.
Minn.
Vt.
Maine
N.H.
Mass.
Oreg.
Idaho
Wyo.
S.Dak.
Wis.
Mich.
N.Y.
R.I.
Conn.
Pa.
N.J.
Del.
Md.
Nev.
Utah
Colo.
Nebr.
Iowa
Ill.
Ind.
Ohio
W.Va.
Va.
Calif.
Kans.
Mo.
Ky.
N.C.
Tenn.
S.C.
Ariz.
N.Mex.
Okla.
Ark.
Ala.
Ga.
Miss.
La.
Fla.
Tex.
Mexico

Flora of North America

Contributors to Volume 10

J. Richard Abbott
Mitchell S. Alix
Fred R. Barrie
Paul E. Berry
David E. Boufford
Lyn A. Craven†
Shirley A. Graham
C. Barre Hellquist
Peter C. Hoch
Bruce K. Holst
Walter S. Judd
Leslie R. Landrum
Harlan Lewis†
Elizabeth McClintock†
Guy L. Nesom
James S. Pringle
Matt Ritter
Andrew Salywon
Robin W. Scribailo
Leila M. Shultz
Gordon C. Tucker
William A. Varga
Warren L. Wagner
Peter H. Weston

Editors for Volume 10

David E. Boufford
Taxon Editor for Combretaceae, Lythraceae, Melastomataceae, Myrtaceae, and Onagraceae

Luc Brouillet
Taxon Editor for Surianaceae

Tammy M. Charron
Managing Editor

Kanchi Gandhi
Nomenclatural and Etymological Editor

Martha J. Hill
Senior Technical Editor

Robert W. Kiger
Bibliographic Editor

Geoffrey A. Levin
Lead Editor (2019–2020) and Taxon Editor for Buxaceae

Jackie M. Poole
Taxon Editor for Polygalaceae

John L. Strother
Reviewing Editor

Leila M. Shultz
Taxon Editor for Gunneraceae, Haloragaceae, and Proteaceae

Gordon C. Tucker
Taxon Editor for Elaeagnaceae

James L. Zarucchi†
Editorial Director and Lead Editor (to 2019)

Volume 10 Composition

Kristin Pierce
Compositor and Editorial Assistant

Cuphea viscosissima

Flora of North America

North of Mexico

Edited by FLORA OF NORTH AMERICA EDITORIAL COMMITTEE

VOLUME 10

Magnoliophyta: Proteaceae to Elaeagnaceae

NEW YORK OXFORD • OXFORD UNIVERSITY PRESS • 2021

Oxford University Press is a department of the University of Oxford.
It furthers the University's objective of excellence in research,
scholarship, and education by publishing worldwide.

Oxford New York
Auckland Cape Town Dar es Salaam Hong Kong Karachi Kuala Lumpur
Madrid Melbourne Mexico City Nairobi New Delhi Shanghai Taipei Toronto

With offices in
Argentina Austria Brazil Chile Czech Republic France Greece Guatemala Hungary Italy
Japan Poland Portugal Singapore South Korea Switzerland Thailand Turkey Ukraine Vietnam

Published by Oxford University Press, Inc.
198 Madison Avenue, New York, New York 10016
www.oup.com

Library of Congress Cataloging-in-Publication Data
(Revised for Volume 10)
Flora of North America North of Mexico
edited by Flora of North America Editorial Committee.
Includes bibliographical references and indexes.
Contents: v. 1. Introduction—v. 2. Pteridophytes and gymnosperms—
v. 3. Magnoliophyta: Magnoliidae and Hamamelidae—
v. 22. Magnoliophyta: Alismatidae, Arecidae, Commelinidae (in part), and Zingiberidae—
v. 26. Magnoliophyta: Liliidae: Liliales and Orchidales—
v. 23. Magnoliophyta: Commelinidae (in part): Cyperaceae—
v. 25. Magnoliophyta: Commelinidae (in part): Poaceae, part 2—
v. 4. Magnoliophyta: Caryophyllidae (in part): part 1—
v. 5. Magnoliophyta: Caryophyllidae (in part): part 2—
v. 19, 20, 21. Magnoliophyta: Asteridae (in part): Asteraceae, parts 1–3—
v. 24. Magnoliophyta: Commelinidae (in part): Poaceae, part 1—
v. 27. Bryophyta, part 1—
v. 8. Magnoliophyta: Paeoniaceae to Ericaceae—
v. 7. Magnoliophyta: Salicaceae to Brassicaceae—
v. 28. Bryophyta, part 2—
v. 9. Magnoliophyta: Picramniaceae to Rosaceae—
v. 6. Magnoliophyta: Cucurbitaceae to Droseraceae—
v. 12. Magnoliophyta: Vitaceae to Garryaceae—
v. 17 Magnoliophyta: Tetrachondraceae to Orobanchaceae—
v. 10 Magnoliophyta: Proteaceae to Elaeagnaceae

ISBN: 9780197576076 (v. 10)
1. Botany—North America.
2. Botany—United States.
3. Botany—Canada.
I. Flora of North America Editorial Committee.
QK110.F55 2002 581.97 92-30459

1 2 3 4 5 6 7 8 9
Printed in the United States of America on acid-free paper

Contents

Walter Judd with *Ageratina illita* in the Dominican Republic, Sierra de Bahoruco.

This volume is dedicated to Walter Stephen Judd (b. 1951), eminent student of North American botany, educator about angiosperm systematics and phylogeny, and contributor to the Flora of North America North of Mexico.

FOUNDING MEMBER INSTITUTIONS

Flora of North America Association

Agriculture and Agri-Food Canada
Ottawa, Ontario

Arnold Arboretum
Jamaica Plain, Massachusetts

Canadian Museum of Nature
Ottawa, Ontario

Carnegie Museum of
Natural History
Pittsburgh, Pennsylvania

Field Museum of Natural History
Chicago, Illinois

Fish and Wildlife Service
United States Department of
the Interior
Washington, D.C.

Harvard University Herbaria
Cambridge, Massachusetts

Hunt Institute for Botanical
Documentation
Carnegie Mellon University
Pittsburgh, Pennsylvania

Jacksonville State University
Jacksonville, Alabama

Jardin Botanique de Montréal
Montréal, Québec

Kansas State University
Manhattan, Kansas

Missouri Botanical Garden
St. Louis, Missouri

New Mexico State University
Las Cruces, New Mexico

The New York Botanical Garden
Bronx, New York

New York State Museum
Albany, New York

Northern Kentucky University
Highland Heights, Kentucky

Université de Montréal
Montréal, Québec

University of Alaska
Fairbanks, Alaska

University of Alberta
Edmonton, Alberta

The University of British Columbia
Vancouver, British Columbia

University of California
Berkeley, California

University of California
Davis, California

University of Idaho
Moscow, Idaho

University of Illinois
Urbana-Champaign, Illinois

University of Iowa
Iowa City, Iowa

The University of Kansas
Lawrence, Kansas

University of Michigan
Ann Arbor, Michigan

University of Oklahoma
Norman, Oklahoma

University of Ottawa
Ottawa, Ontario

University of Louisiana
Lafayette, Louisiana

The University of Texas
Austin, Texas

University of Western Ontario
London, Ontario

University of Wyoming
Laramie, Wyoming

Utah State University
Logan, Utah

For their support of the preparation of this volume, we gratefully acknowledge and thank:

Fondation Franklinia

The Philecology Foundation

The Andrew W. Mellon Foundation

The David and Lucile Packard Foundation

an anonymous foundation

Chanticleer Foundation

The William and Flora Hewlett Foundation

The Stanley Smith Horticultural Trust

Hall Family Charitable Fund

WEM Foundation

William T. Kemper Foundation

For sponsorship of illustrations included in this volume, we express sincere appreciation to:

Mitchell S. Alix and Robin W. Scribailo, Westville Indiana
Myriophyllum alterniflorum, *M. aquaticum*, *M. farwellii*, *M. heterophyllum*, *M. hippuroides*, *M. humile*, *M. laxum*, *M. pinnatum*, *M. sibiricum*, *M. ussuriense*, *M. verticillatum*, *Proserpinaca palustris*, Haloragaceae

Alaska Native Plant Society, Anchorage Chapter
Chamaenerion angustifolium subsp. *angustifolium*, Onagraceae

Arizona Native Plant Society, Tucson, Arizona
Oenothera cespitosa subsp. *navajoensis*, Onagraceae

Julian P. Donahue, Tucson, Arizona
Monnina wrightii, Polygalaceae

Alan and Shirley Graham, St. Louis, Missouri
Volume 10 frontispiece, *Cuphea viscosissima*, Lythraceae

Nancy R. Morin, Point Arena, California
Ammannia grayi, Lythraceae, in honor of Shirley Graham
Oenothera organensis, Onagraceae, in honor of Warren Wagner

Jennifer H. Richards, Miami, Florida
Suriana maritima, Surianaceae

Clemens van Dijk, Eindhoven, The Netherlands
Oenothera canescens, Onagraceae

Flora of North America Editorial Committee

(as of October 2020)

Project Staff — past and present involved with the preparation of Volume 10

Barbara Alongi, *Illustrator*
Mike Blomberg, *Imaging*
Ariel S. Buback, *Assisting Technical Editor*
Trisha K. Distler, *GIS Analyst*
Linny Heagy, *Illustrator*
Rexford L. Hill, *Data Analyst*
Suzanne E. Hirth, *Editorial Assistant*
Cassandra L. Howard, *Senior Technical Editor*
Ruth T. King, *Editorial Assistant*
Marjorie C. Leggitt, *Illustrator*
John Myers, *Illustrator and Illustration Compositor*
Kristin Pierce, *Editorial Assistant and Compositor*
Andrew C. Pryor, *Assisting Technical Editor*
Heidi H. Schmidt, *Managing Editor (2007–2017)*
Yevonn Wilson-Ramsey, *Illustrator*

Contributors to Volume 10

J. Richard Abbott
University of Arkansas
Monticello, Arkansas

Mitchell S. Alix
Purdue University North Central
Westville, Indiana

Fred R. Barrie
Missouri Botanical Garden
St. Louis, Missouri

Paul E. Berry
University of Michigan
Ann Arbor, Michigan

David E. Boufford
Harvard University Herbaria
Cambridge, Massachusetts

Lyn A. Craven†
Australian National Herbarium
Canberra, Australia

Shirley A. Graham
Missouri Botanical Garden
St. Louis, Missouri

C. Barre Hellquist
Massachusetts College of Liberal Arts
North Adams, Massachusetts

Peter C. Hoch
Missouri Botanical Garden
St. Louis, Missouri

Bruce K. Holst
Marie Selby Botanical Gardens
Sarasota, Florida

Walter S. Judd
University of Florida
Gainesville, Florida

Leslie R. Landrum
Arizona State University
Tempe, Arizona

Harlan Lewis†
Pacific Palisades, California

Elizabeth McClintock†
San Francisco, California

Guy L. Nesom
Academy of Natural Sciences of Drexel University
Philadelphia, Pennsylvania

James S. Pringle
Royal Botanical Gardens
Hamilton, Ontario

Matt Ritter
California Polytechnic State University
San Luis Obispo, California

Andrew Salywon
Desert Botanical Garden
Phoenix, Arizona

Robin W. Scribailo
Purdue University North Central
Westville, Indiana

Leila M. Shultz
Utah State University
Logan, Utah

Gordon C. Tucker
Eastern Illinois University
Charleston, Illinois

William A. Varga
Utah State University
Logan, Utah

Warren L. Wagner
National Museum of Natural History
Smithsonian Institution
Washington, DC

Peter H. Weston
National Herbarium of New South Wales
The Royal Botanic Garden
Sydney, Australia

Volume 10 Taxonomic Reviewers

John Anderson
Bureau of Land Management
Wickenberg, Arizona

David E. Boufford
Harvard University Herbaria
Cambridge, Massachusetts

Glenn Dreyer
Connecticut College
New London, Connecticut

Alina Freire-Fierro
Universidad Técnica de Cotopaxi
Latacunga, Ecuador

Richard R. Halse
Oregon State University
Corvallis, Oregon

C. Barre Hellquist
Massachusetts College of Liberal Arts
North Adams, Massachusetts

Joseph H. Kirkbride Jr.
United States National Arboretum
Washington, DC

Aaron Liston
Oregon State University
Corvallis, Oregon

James B. Phipps
Western University
London, Ontario

Peter H. Raven
Missouri Botanical Garden
St. Louis, Missouri

Yuri Roskov
Illinois Natural History Survey
Champaign, Illinois

Marianne M. le Roux
University of Johannesburg
Johannesburg, South Africa

Richard W. Spellenberg
New Mexico State University
Las Cruces, New Mexico

Livia Wanntorp
Swedish Museum of Natural History
Stockholm, Sweden

Tom Wendt
The University of Texas at Austin
Austin, Texas

Dieter H. Wilken
Santa Barbara Botanic Garden
Santa Barbara, California

Martin F. Wojciechowski
Arizona State University
Tempe, Arizona

Regional Reviewers

ALASKA / YUKON

Bruce Bennett
Yukon Department of Environment
Whitehorse, Yukon

Justin Fulkerson
Alaska Center for Conservation Science
University of Alaska Anchorage
Anchorage, Alaska

Steffi M. Ickert-Bond, Regional Coordinator
University of Alaska, Museum of the North
Fairbanks, Alaska

Robert Lipkin
Alaska Natural Heritage Program
University of Alaska Anchorage
Anchorage, Alaska

David F. Murray
University of Alaska, Museum of the North
Fairbanks, Alaska

Carolyn Parker
University of Alaska, Museum of the North
Fairbanks, Alaska

Mary Stensvold
Sitka, Alaska

PACIFIC NORTHWEST

Edward R. Alverson
Lane County Parks Division
Eugene, Oregon

Curtis R. Björk
University of British Columbia
Vancouver, British Columbia

Mark Darrach
University of Washington
Seattle, Washington

Walter Fertig
Washington Natural Heritage Program
Olympia, Washington

David E. Giblin
University of Washington
Seattle, Washington

Richard R. Halse
Oregon State University
Corvallis, Oregon

Linda Jennings
University of British Columbia
Vancouver, British Columbia

Aaron Liston, Regional Coordinator
Oregon State University
Corvallis, Oregon

Frank Lomer
New Westminster, British Columbia

Kendrick L. Marr
Royal British Columbia Museum
Victoria, British Columbia

Jim Pojar
British Columbia Forest Service
Smithers, British Columbia

Peter F. Zika
University of Washington
Seattle, Washington

SOUTHWESTERN UNITED STATES

Tina J. Ayers
Northern Arizona University
Flagstaff, Arizona

Walter Fertig
Washington Natural Heritage Program
Olympia, Washington

H. David Hammond†
Northern Arizona University
Flagstaff, Arizona

G. F. Hrusa
Lone Mountain Institute
Brookings, Oregon

Max Licher
Northern Arizona University
Flagstaff, Arizona

Elizabeth Makings
Arizona State University
Tempe, Arizona

James D. Morefield
Nevada Natural Heritage Program
Carson City, Nevada

Nancy R. Morin, Regional Coordinator
Point Arena, California

Donald J. Pinkava†
Arizona State University
Tempe, Arizona

Jon P. Rebman
San Diego Natural History Museum
San Diego, California

Glenn Rink
Northern Arizona University
Flagstaff, Arizona

James P. Smith Jr.
Humboldt State University
Arcata, California

Gary D. Wallace
California Botanic Garden
Claremont, California

WESTERN CANADA

William J. Cody†
Agriculture and Agri-Food Canada
Ottawa, Ontario

Bruce A. Ford, Regional Coordinator
University of Manitoba
Winnipeg, Manitoba

Lynn Gillespie
Canadian Museum of Nature
Ottawa, Ontario

A. Joyce Gould
Alberta Environment and Parks
Edmonton, Alberta

Vernon L. Harms
University of Saskatchewan
Saskatoon, Saskatchewan

Elizabeth Punter
University of Manitoba
Winnipeg, Manitoba

ROCKY MOUNTAINS

Jennifer Ackerfield
Denver Botanic Gardens
Denver, Colorado

Walter Fertig
Washington Natural Heritage Program
Olympia, Washington

Ronald L. Hartman†
University of Wyoming
Laramie, Wyoming

Bonnie Heidel, Regional Coordinator
University of Wyoming
Laramie, Wyoming

Robert Johnson
Brigham Young University
Provo, Utah

Ben Legler
University of Idaho
Moscow, Idaho

Peter C. Lesica
University of Montana
Missoula, Montana

Donald H. Mansfield
The College of Idaho
Caldwell, Idaho

B. E. Nelson
University of Wyoming
Laramie, Wyoming

Leila M. Shultz
Utah State University
Logan, Utah

NORTH CENTRAL UNITED STATES

Anita F. Cholewa
University of Minnesota
St. Paul, Minnesota

Neil A. Harriman†
University of Wisconsin Oshkosh
Oshkosh, Wisconsin

Bruce W. Hoagland
University of Oklahoma
Norman, Oklahoma

Craig C. Freeman, Regional Coordinator
The University of Kansas
Lawrence, Kansas

Robert B. Kaul†
University of Nebraska
Lincoln, Nebraska

Gary E. Larson
South Dakota State University
Brookings, South Dakota

Deborah Q. Lewis
Iowa State University
Ames, Iowa

Ronald L. McGregor†
The University of Kansas
Lawrence, Kansas

Stephen G. Saupe
College of Saint Benedict/ St. John's University
Collegeville, Minnesota

Lawrence R. Stritch
Martinsburg, West Virginia

George Yatskievych
The Universtiy of Texas at Austin
Austin, Texas

SOUTH CENTRAL UNITED STATES

Jackie M. Poole, Regional Coordinator
Fort Davis, Texas

Robert C. Sivinski
University of New Mexico
Albuquerque, New Mexico

EASTERN CANADA

Sean Blaney
Atlantic Canada Conservation Data Centre
Sackville, New Brunswick

Luc Brouillet, Regional Coordinator
Institut de recherche en biologie végétale
Université de Montréal
Montréal, Québec

Jacques Cayouette
Agriculture and Agri-Food Canada
Ontario, Ottawa

Frédéric Coursol
Jardin botanique de Montréal
Montréal, Québec

William J. Crins
Ontario Ministry of Natural Resources
Peterborough, Ontario

Marian Munro
Nova Scotia Museum of Natural History
Halifax, Nova Scotia

Michael J. Oldham
Natural Heritage Information Centre
Peterborough, Ontario

NORTHEASTERN UNITED STATES

Ray Angelo
New England Botanical Club
Cambridge, Massachusetts

David E. Boufford, Regional Coordinator
Harvard University Herbaria
Cambridge, Massachusetts

Tom S. Cooperrider
Kent State University
Kent, Ohio

Allison Cusick
Carnegie Museum of Natural History
Pittsburgh, Pennsylvania

Arthur Haines
Canton, Maine

Michael A. Homoya
Indiana Department of Natural Resources
Indianapolis, Indiana

Robert F. C. Naczi
The New York Botanical Garden
Bronx, New York

Anton A. Reznicek
University of Michigan
Ann Arbor, Michigan

Edward G. Voss†
University of Michigan
Ann Arbor, Michigan

Kay Yatskievych
Austin, Texas

SOUTHEASTERN UNITED STATES

Mac H. Alford
University of Southern Mississippi
Hattiesburg, Mississippi

J. Richard Carter Jr.
Valdosta State University
Valdosta, Georgia

L. Dwayne Estes
Austin Peay State University
Clarksville, Tennessee

W. John Hayden
University of Richmond
Richmond, Virginia

Wesley Knapp
North Carolina Natural Heritage Program
Asheville, North Carolina

John B. Nelson
University of South Carolina
Columbia, South Carolina

Chris Reid
Louisiana State University
Baton Rouge, Louisiana

Bruce A. Sorrie
University of North Carolina
Chapel Hill, North Carolina

Dan Spaulding
Anniston Museum of Natural History
Anniston, Alabama

R. Dale Thomas
Seymour, Tennessee

Lowell E. Urbatsch
Louisiana State University
Baton Rouge, Louisiana

Alan S. Weakley, Regional Coordinator
University of North Carolina
Chapel Hill, North Carolina

Theo Witsell
Arkansas Natural Heritage Commission
Little Rock, Arkansas

B. Eugene Wofford
University of Tennessee
Knoxville, Tennessee

FLORIDA

Loran C. Anderson
Florida State University
Tallahassee, Florida

Bruce F. Hansen
University of South Florida
Tampa, Florida

Richard P. Wunderlin, Regional Coordinator
University of South Florida
Tampa, Florida

Preface for Volume 10

Since the publication of *Flora of North America* Volume 17 (the twenty-first volume in the *Flora* series) in 2019, the membership of the Flora of North America Association [FNAA] Board of Directors has changed. Thomas G. Lammers, Terry McIntosh, Richard K. Rabeler, Yuri Roskov, Dale H. Vitt, David H. Wagner, and James L. Zarucchi† have departed the board. New board members include Barbara Crandall-Stotler and Paul Davison. As a result of a reorganization finalized in 2003, the FNAA Board of Directors succeeded the former Editorial Committee; for the sake of continuity of citation, authorship of *Flora* volumes is to be cited as "Flora of North America Editorial Committee, eds."

Most of the editorial process for this volume was done at the Missouri Botanical Garden in St. Louis. Final processing and composition took place at the Missouri Botanical Garden; this included pre-press processing, typesetting and layout, plus coordination for all aspects of planning, executing, and scanning the illustrations. Other aspects of production, such as art panel composition plus labeling and occurrence map generation, were carried out elsewhere in North America.

Illustrations published in this volume were executed by four extremely talented artists: Barbara Alongi, Linny Heagy, Marjorie C. Leggitt, and Yevonn Wilson-Ramsey. Barbara Alongi prepared illustrations for Buxaceae, Combretaceae, Gunneraceae, Lythraceae, and Proteaceae; she also created the frontispiece depicting *Cuphea viscosissima* (Lythraceae). Linny Heagy illustrated species of Melastomataceae. Marjorie C. Leggitt illustrated species of Surianaceae. Yevonn Wilson-Ramsey prepared illustrations for Elaeagnaceae, Haloragaceae, Myrtaceae, Onagraceae, and Polygalaceae. John Myers composed and labeled all of the line drawings that appear in this volume.

Starting with Volume 8, published in 2009, the circumscription and ordering of some families within the *Flora* have been modified so that they mostly reflect that of the Angiosperm Phylogeny Group [APG] rather than the previously followed Cronquist organizational structure. The groups of families found in this and future volumes in the series are mostly ordered following E. M. Haston et al. (2007); since APG views of relationships and circumscriptions have evolved, and will change further through time, some discrepancies in organization will occur. Volume 30 of the *Flora of North America* will contain a comprehensive index to the published volumes. For practical publishing purposes, Surianaceae, Polygalaceae, and Elaeagnaceae were moved from Volume 11 to Volume 10.

Support from many institutions and by numerous individuals has enabled the *Flora* to be produced. Members of the Flora of North America Association remain deeply thankful to the many people who continue to help create, encourage, and sustain the *Flora*.

Introduction

Scope of the Work

Flora of North America North of Mexico is a synoptic account of the plants of North America north of Mexico: the continental United States of America (including the Florida Keys and Aleutian Islands), Canada, Greenland (Kalâtdlit-Nunât), and St. Pierre and Miquelon. The *Flora* is intended to serve both as a means of identifying plants within the region and as a systematic conspectus of the North American flora.

The *Flora* will be published in 30 volumes. Volume 1 contains background information that is useful for understanding patterns in the flora. Volume 2 contains treatments of ferns and gymnosperms. Families in volumes 3–26, the angiosperms, were first arranged according to the classification system of A. Cronquist (1981) with some modifications, and starting with Volume 8, the circumscriptions and ordering of families generally follow those of the Angiosperm Phylogeny Group [APG] (see E. Haston et al. 2007). Bryophytes are being covered in volumes 27–29. Volume 30 will contain the cumulative bibliography and index.

The first two volumes were published in 1993, Volume 3 in 1997, and Volumes 22, 23, and 26, the first three of five volumes covering the monocotyledons, appeared in 2000, 2002, and 2002, respectively. Volume 4, the first part of the Caryophyllales, was published in late 2003. Volume 25, the second part of the Poaceae, was published in mid 2003, and Volume 24, the first part, was published in January 2007. Volume 5, completing the Caryophyllales plus Polygonales and Plumbaginales, was published in early 2005. Volumes 19–21, treating Asteraceae, were published in early 2006. Volume 27, the first of two volumes treating mosses in North America, was published in late 2007. Volume 8, Paeoniaceae to Ericaceae, was published in September 2009, and Volume 7, Salicaceae to Brassicaceae, appeared in 2010. In 2014, Volume 28 was published, completing the treatment of mosses for the flora area, and at the end of 2014, Volume 9, Picramniaceae to Rosaceae was published. Volume 6, which covered Cucurbitaceae to Droseraceae, was published in 2015. Volume 12, Vitaceae to Garryaceae, was published in late 2016. Volume 17, Tetrachondraceae to Orobanchaceae, was published in 2019. The correct bibliographic citation for the *Flora* is: Flora of North America Editorial Committee, eds. 1993+. Flora of North America North of Mexico. 22+ vols. New York and Oxford.

Volume 10 treats 455 species in 65 genera contained in 12 families. For additional statistics please refer to Table 1 on p. xx.

Contents · General

The *Flora* includes accepted names, selected synonyms, literature citations, identification keys, descriptions, phenological information, summaries of habitats and geographic ranges, and other biological observations. Each volume contains a bibliography and an index to the taxa included in that volume. The treatments, written and reviewed by experts from throughout the systematic botanical community, are based on original observations of herbarium specimens and, whenever possible, on living plants. These observations are supplemented by critical reviews of the literature.

Table 1. *Statistics for Volume 10 of Flora of North America.*

Family	Total Genera	Endemic Genera	Introduced Genera	Total Species	Endemic Species	Introduced Species	Conservation Taxa
Proteaceae	1	0	1	1	0	1	0
Buxaceae	2	0	1	3	1	2	0
Gunneraceae	1	0	1	1	0	1	0
Haloragaceae	3	0	1	17	6	3	0
Combretaceae	5	0	2	8	0	5	0
Lythraceae	10	2	3	32	9	14	4
Onagraceae	17	3	1	277	173	6	46
Myrtaceae	13	0	9	38	0	30	2
Melastomataceae	3	0	1	15	14	1	0
Surianaceae	1	0	0	1	0	0	0
Polygalaceae	6	0	0	53	29	1	4
Elaeagnaceae	3	1	1	9	4	5	0
Totals	**65**	**6**	**21**	**455**	**236**	**69**	**56**

Italic = introduced

Basic Concepts

Our goal is to make the *Flora* as clear, concise, and informative as practicable so that it can be an important resource for both botanists and nonbotanists. To this end, we are attempting to be consistent in style and content from the first volume to the last. Readers may assume that a term has the same meaning each time it appears and that, within groups, descriptions may be compared directly with one another. Any departures from consistent usage will be explicitly noted in the treatments (see References).

Treatments are intended to reflect current knowledge of taxa throughout their ranges worldwide, and classifications are therefore based on all available evidence. Where notable differences of opinion about the classification of a group occur, appropriate references are mentioned in the discussion of the group.

Documentation and arguments supporting significantly revised classifications are published separately in botanical journals before publication of the pertinent volume of the *Flora*. Similarly, all new names and new combinations are published elsewhere prior to their use in the *Flora*. No nomenclatural innovations will be published intentionally in the *Flora*.

Taxa treated in full include extant and recently extinct or extirpated native species, named hybrids that are well established (or frequent), introduced plants that are naturalized, and cultivated plants that are found frequently outside cultivation. Taxa mentioned only in discussions include waifs known only from isolated old records and some non-native, economically important or extensively cultivated plants, particularly when they are relatives of native species. Excluded names and taxa are listed at the ends of appropriate sections, for example, species at the end of genus, genera at the end of family.

Treatments are intended to be succinct and diagnostic but adequately descriptive. Characters and character states used in the keys are repeated in the descriptions. Descriptions of related taxa at the same rank are directly comparable.

With few exceptions, taxa are presented in taxonomic sequence. If an author is unable to produce a classification, the taxa are arranged alphabetically and the reasons are given in the discussion.

Treatments of hybrids follow that of one of the putative parents. Hybrid complexes are treated at the ends of their genera, after the descriptions of species.

We have attempted to keep terminology as simple as accuracy permits. Common English equivalents usually have been used in place of Latin or Latinized terms or other specialized terminology, whenever the correct meaning could be conveyed in approximately the same space, for example, "pitted" rather than "foveolate," but "striate" rather than "with fine longitudinal lines." See *Categorical Glossary for the Flora of North America Project* (R. W. Kiger and D. M. Porter 2001; also available online at http://huntbot.andrew.cmu.edu) for standard definitions of generally used terms. Very specialized terms are defined, and sometimes illustrated, in the relevant family or generic treatments.

References

Authoritative general reference works used for style are *The Chicago Manual of Style,* ed. 14 (University of Chicago Press 1993); *Webster's New Geographical Dictionary* (Merriam-Webster 1988); and *The Random House Dictionary of the English Language*, ed. 2, unabridged (S. B. Flexner and L. C. Hauck 1987). *B-P-H/S. Botanico-Periodicum-Huntianum/Supplementum* (G. D. R. Bridson and E. R. Smith 1991), *BPH-2: Periodicals with Botanical Content* (Bridson 2004), and *BPH Online* [http://fmhibd.library.cmu.edu/HIBD-DB/bpho/findrecords.php] (Bridson and D. W. Brown) have been used for abbreviations of serial titles, and *Taxonomic Literature*, ed. 2 (F. A. Stafleu and R. S. Cowan 1976–1988) and its supplements by Stafleu et al. (1992–2009) have been used for abbreviations of book titles.

Graphic Elements

All genera and more than 25 percent of the species in this volume are illustrated. The illustrations may show diagnostic traits or complex structures. Most illustrations have been drawn from herbarium specimens selected by the authors. Data on specimens that were used and parts that were illustrated have been recorded. This information, together with the archivally preserved original drawings, is deposited in the Missouri Botanical Garden Library and is available for scholarly study.

Specific Information in Treatments

Keys

Dichotomous keys are included for all ranks below family if two or more taxa are treated. More than one key may be given to facilitate identification of sterile material or for flowering versus fruiting material.

Nomenclatural Information

Basionyms of accepted names, with author and bibliographic citations, are listed first in synonymy, followed by any other synonyms in common recent use, listed in alphabetical order, without bibliographic citations.

The last names of authors of taxonomic names have been spelled out. The conventions of *Authors of Plant Names* (R. K. Brummitt and C. E. Powell 1992) have been used as a guide for including first initials to discriminate individuals who share surnames.

If only one infraspecific taxon within a species occurs in the flora area, nomenclatural information (literature citation, basionym with literature citation, relevant other synonyms) is given for the species, as is information on the number of infraspecific taxa in the species and their distribution worldwide, if known. A description and detailed distributional information are given only for the infraspecific taxon.

Descriptions

Character states common to all taxa are noted in the description of the taxon at the next higher rank. For example, if sexual condition is dioecious for all species treated within a genus, that character state is given in the generic description. Characters used in keys are repeated in the descriptions. Characteristics are given as they occur in plants from the flora area. Characteristics that occur only in plants from outside the flora area may be given within square brackets, or instead may be noted in the discussion following the description. In families with one genus and one or more species, the family description is given as usual, the genus description is condensed, and the species are described as usual. Any special terms that may be used when describing members of a genus are presented and explained in the genus description or discussion.

In reading descriptions, the reader may assume, unless otherwise noted, that: the plants are green, photosynthetic, and reproductively mature; woody plants are perennial; stems are erect; roots are fibrous; leaves are simple and petiolate; flowers are bisexual, radially symmetric, and pediceled; perianth parts are hypogynous, distinct, and free; and ovaries are superior. Because measurements and elevations are almost always approximate, modifiers such as "about," "circa," or "±" are usually omitted.

Unless otherwise noted, dimensions are length × width. If only one dimension is given, it is length or height. All measurements are given in metric units. Measurements usually are based on dried specimens but these should not differ significantly from the measurements actually found in fresh or living material.

Chromosome numbers generally are given only if published and vouchered counts are available from North American material or from an adjacent region. No new counts are published intentionally in the *Flora*. Chromosome counts from nonsporophyte tissue have been converted to the $2n$ form. The base number ($x =$) is given for each genus. This represents the lowest known haploid count for the genus unless evidence is available that the base number differs.

Flowering time and often fruiting time are given by season, sometimes qualified by early, mid, or late, or by months. Elevations 100 m and under are rounded to the nearest 10 m, between 100 m and 200 m to the nearest 50 m, and over 200 m to the nearest 100 m. Mean sea level is shown as 0 m, with the understanding that this is approximate. Elevation often is omitted from herbarium specimen labels, particularly for collections made where the topography is not remarkable, and therefore precise elevation is sometimes not known for a given taxon.

The term "introduced" is defined broadly to refer to plants that were released deliberately or accidentally into the flora and that now are naturalized, that is, exist as wild plants in areas in which they were not recorded as native in the past. The distribution of introduced taxa are often poorly documented and changing, so the distribution statements for those taxa may not be fully accurate.

If a taxon is globally rare or if its continued existence is threatened in some way, the words "of conservation concern" appear before the statements of elevation and geographic range.

Criteria for taxa of conservation concern are based on NatureServe's (formerly The Nature Conservancy)—see http://www.natureserve.org—designations of global rank (G-rank) G1 and G2:

G1 Critically imperiled globally because of extreme rarity (5 or fewer occurrences or fewer than 1000 individuals or acres) or because of some factor(s) making it especially vulnerable to extinction.

G2 Imperiled globally because of rarity (5–20 occurrences or fewer than 3000 individuals or acres) or because of some factor(s) making it very vulnerable to extinction throughout its range.

The occurrence of species and infraspecific taxa within political subunits of the *Flora* area is depicted by dots placed on the outline map to indicate occurrence in a state or province. The Nunavut boundary on the maps has been provided by the GeoAccess Division, Canada Centre for Remote Sensing, Earth Science. Authors are expected to have seen at least one specimen documenting each geographic unit record (except in rare cases when undoubted literature reports may be used) and have been urged to examine as many specimens as possible from throughout the range of each taxon. Additional information about taxon distribution may be presented in the discussion.

Distributions are stated in the following order: Greenland; St. Pierre and Miquelon; Canada (provinces and territories in alphabetic order); United States (states in alphabetic order); Mexico (11 northern states may be listed specifically, in alphabetic order); West Indies; Bermuda; Central America (Belize, Costa Rica, El Salvador, Guatemala, Honduras, Nicaragua, Panama); South America; Europe, or Eurasia; Asia (including Indonesia); Africa; Atlantic Islands; Indian Ocean Islands; Pacific Islands; Australia; Antarctica.

Discussion

The discussion section may include information on taxonomic problems, distributional and ecological details, interesting biological phenomena, and economic uses.

Selected References

Major references used in preparation of a treatment or containing critical information about a taxon are cited following the discussion. These, and other works that are referred to in discussion or elsewhere, are included in Literature Cited at the end of the volume.

CAUTION

The Flora of North America Editorial Committee **does not encourage, recommend, promote, or endorse** any of the folk remedies, culinary practices, or various utilizations of any plant described within this volume. Information about medicinal practices and/or ingestion of plants, or of any part or preparation thereof, has been included only for historical background and as a matter of interest. Under no circumstances should the information contained in these volumes be used in connection with medical treatment. Readers are strongly cautioned to remember that many plants in the flora are toxic or can cause unpleasant or adverse reactions if used or encountered carelessly.

Key to boxed codes following accepted names:

[C] of conservation concern
[E] endemic to the flora area
[F] illustrated
[I] introduced to the flora area
[W] weedy applies to taxa listed as weeds on the State and Federal Composite List of All U.S. Noxious Weeds (https://plants.usda.gov/java/noxComposite) and the Composite List of Weeds from the Weed Science Society of America (https://wssa.net/wssa/weed/composite-list-of-weeds), with input from treatment authors and reviewers.

Flora of North America

PROTEACEAE Jussieu

• Protea Family

Peter H. Weston

Trees **[shrubs]**, terrestrial, unarmed, not clonal [clonal]; usually with 3-celled, biramose [simple] trichomes, sometimes also with glandular trichomes [glabrous]. **Branches** erect [prostrate]. **Leaves** persistent, alternate [opposite or whorled], simple [pinnate, bipinnate, or palmate]; stipules absent; petiolate [sessile]; blade papery [leathery], pinnatisect, bipinnatisect, or tripinnatisect, margins usually lobed [lobed and/or with marginal teeth]. **Inflorescences** terminal [axillary], simple or compound, racemelike or paniculate [racemose or capitate]; 1 bract present, scalelike, subtending paired flowers [absent]; bracteoles absent. **Flowers** usually bisexual [unisexual], zygomorphic [actinomorphic], pedicellate [sessile]; perianth and androecium hypogynous; tepals 4 [rarely 3 or 5], remaining loosely coherent except between dorsal tepals [distinct or connate], valvate; nectary usually present, 1[–4], fleshy [scalelike], crescentic [forming a ring, or distinct]; stamens 4 [rarely 3 or 5], opposite tepals, usually all fertile [1+ sterile]; filaments wholly [partly] adnate to tepals [free]; anthers basifixed, usually bilocular and tetrasporangiate [lateral anthers unilocular and bisporangiate]; pistil 1[or 2]-carpellate; ovary superior, 1-locular; placentation marginal [basal or apical]; style 1, distinct [vestigial], apex often functioning as pollen presenter; stigma 1; ovules [1 or]2[+], hemitropous [anatropous or orthotropous], bitegmic, crassinucellate. **Fruits** follicles [achenes, drupes, or drupelike], dehiscent [indehiscent]. **Seeds** [1 or]2[+], ovate-apiculate, winged [wingless].

Genera 81, species ca. 1700 (1 in the flora): introduced; Mexico, West Indies, Central America, South America, s, e Asia (including Malesia), s Africa, Indian Ocean Islands (Madagascar), Pacific Islands (Fiji, Micronesia, New Caledonia, New Guinea, New Zealand, Sulawesi, Vanuatu), Australia (including Tasmania).

Some species of Proteaceae produce edible seeds that have high protein and oil content, the most notable of which are macadamia nuts (*Macadamia integrifolia* Maiden & Betche and *M. tetraphylla* L. A. S. Johnson and their hybrids). Some rainforest species have been used as sources of timber for fine woodworking, including the one species naturalized in North America. The woods of these taxa have a distinctive, unusual grain due to the tall, wide rays, which are often lustrous and/or strikingly colored. Some Proteaceae are commercially exploited by the cut-flower industry, and the total area planted with these crops in 2002 has been estimated to be 10,000 hectares (P. W. Crous et al. 2004).

Molecular phylogenetic analyses have resolved the relationships of Proteaceae. Proteaceae are sister to Platanaceae and this clade is, in turn, the sister group of Nelumbonaceae (D. E. Soltis et al. 2000). These three families constitute Proteales as circumscribed by the Angiosperm Phylogeny Group (2016). Proteales are one of multiple lineages that branch near the base of the "eudicot" clade.

Prior to 1993, the affinities of Proteaceae were poorly understood, and the family was placed in a diverse range of taxa in different classifications of the angiosperms. None of these systems anticipated the close phylogenetic relationships of Proteaceae to Nelumbonaceae and Platanaceae, probably because the only morphological synapomorphy that any of them share is the distinctive hair base found in Platanaceae and Proteaceae (R. J. Carpenter et al. 2005).

1. GREVILLEA R. Brown ex Knight, Cult. Prot., xvii, 120. 1809 (as Grevillia), name and orthography conserved • [For Charles Francis Greville, 1749–1809, Fellow of the Royal Society] [I]

Inflorescences unbranched or 2–4[–20]-branched. **Flowers:** nectary solitary, ventral [surrounding pistil], crescent-shaped [to oblong or annular, rarely 4-lobed or absent]; pistil diagonally oriented; ovary stipitate [sessile]; style slightly sinuous [straight to curved, hooked, or bent], tip modified as ± radially symmetrical [to strongly oblique] pollen presenter; stigma terminal [ventral]. **Follicles** sessile. $x = 10$.

Species ca. 365 (1 in the flora): introduced; s Africa, Pacific Islands (New Caledonia, New Guinea, Sulawesi), Australia (including Tasmania); introduced also widely in tropical and subtropical areas.

1. Grevillea robusta A. Cunningham ex R. Brown, Suppl. Prodr. Fl. Nov. Holl., 24. 1830 • Silky oak [F] [I]

Trees 8–40 m, bark fissured. **Leaves** 10–34 × 9–15 cm; blade with 11–24(–31) primary lobes, margins of primary lobes entire or 2–5-lobed, sometimes with tertiary division; ultimate lobes oblong to elliptic or subtriangular, 0.5–5 × 0.2–1 cm, margins slightly recurved, surfaces subsericeous to subvillous abaxially, glabrous adaxially. **Inflorescences** erect, sometimes on relatively short, lateral leafy shoots; branches 24–154-flowered, secund, acropetal, 12–16 cm. **Pedicels** 7.5–16 mm. **Flowers** acroscopic; receptacle oblique; perianth usually golden yellow to orange, rarely reddish, sometimes with red blotches inside, glabrous; pistil 21–29 mm, glabrous; ovary stipitate; style yellow-orange; pollen presenter erect, conical. **Follicles** compressed, ellipsoidal to ovoid, 12–16 mm, glabrous. **Seeds** compressed, 8–12 × 4–6 mm.

Flowering Mar–Jul. Disturbed sites; 0–100 m; introduced; Calif., Fla.; Australia; introduced also in West Indies, s Africa, Pacific Islands.

Grevillea robusta is native to the coast and coastal ranges of southeastern Queensland and northeastern New South Wales, Australia. The plants usually grow in basaltic soils in three distinct habitats: riverine gallery rainforest and rainforest margins; riverine sclerophyll communities with *Casuarina cunninghamiana*; and *Araucaria* forest and vine thickets on higher slopes.

Grevillea robusta is cultivated as an ornamental for its fernlike foliage and spectacular, bird-attracting flowers, in agroforestry for timber and firewood, and as a shade tree in tea and coffee plantations (C. E. Harwood 1989, 1992). Some people have been reported to develop acute contact dermatitis after being exposed to sawdust of *G. robusta* (J. G. B. Derraik and M. Rademaker 2009).

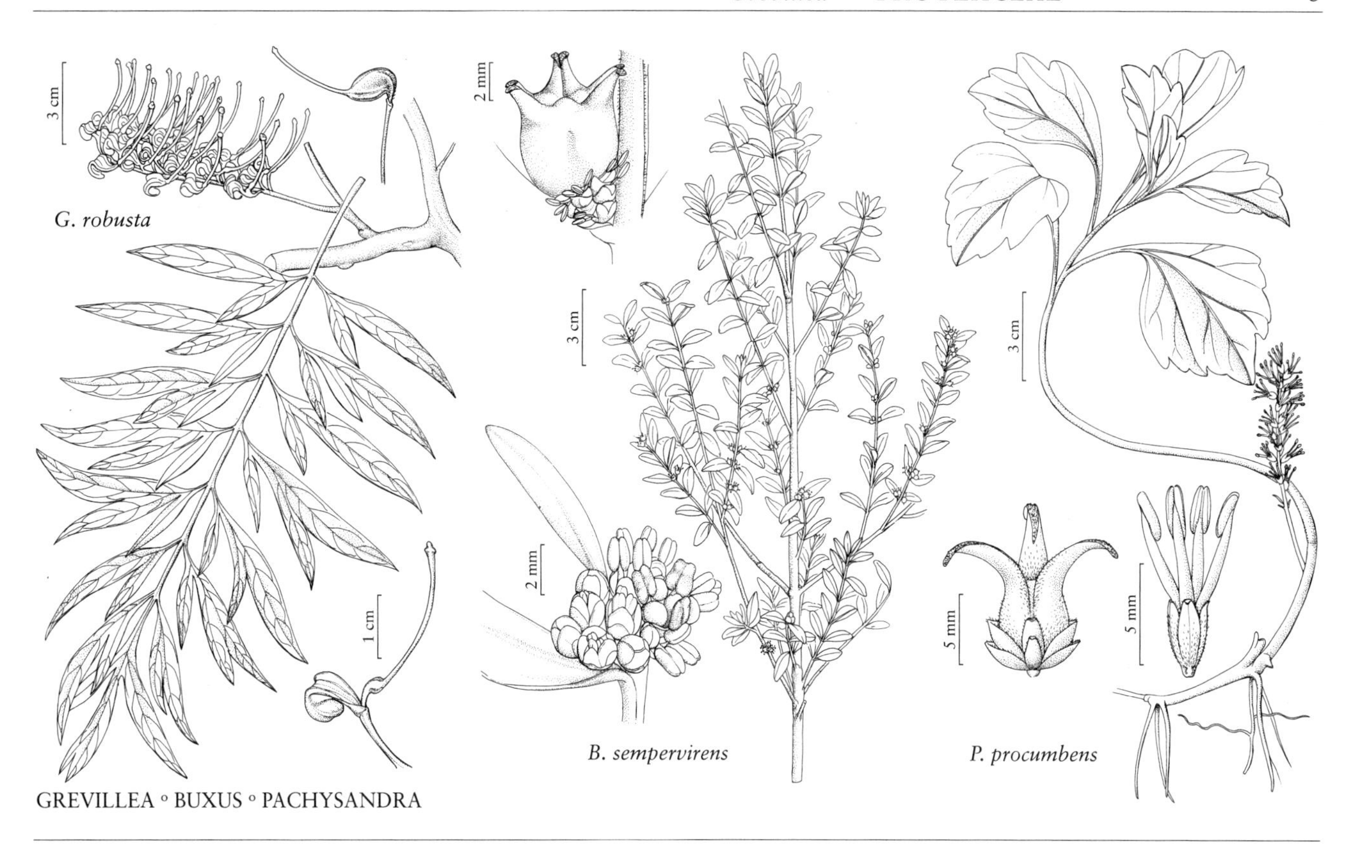

GREVILLEA ◦ BUXUS ◦ PACHYSANDRA

BUXACEAE Dumortier

• Boxwood Family

David E. Boufford

Herbs, perennial, shrubs, or subshrubs [trees], monoecious [dioecious or flowers bisexual], terrestrial, unarmed, clonal or not. **Stems** erect and ascending. **Leaves** persistent, opposite, subopposite, or alternate, simple; stipules absent; petiolate [sessile]; blade leathery, margins entire or dentate. **Inflorescences** terminal or axillary, spikes, racemes, or capitate clusters; bracts and/or bracteoles often present, subtending flowers. **Flowers** unisexual [bisexual], staminate and pistillate on same plant, sessile or pedicellate; perianth and androecium hypogynous; tepals 2, 4, 5, or 6, distinct; petals absent. **Staminate flowers:** stamens 4 [6], antitepalous, free, distinct; filaments thickened; anthers conspicuous, often red; pistillode present or absent. **Pistillate flowers:** pistils [1 or] 2- or 3(or 4)-carpellate, basally connate, apically distinct and grading into styles; ovary superior, locules 1 or 2 times carpels; placentation axile; styles equal to number of carpels, subulate, divergent to divaricate, often recurved in fruit, stigmatic along inner surface; interstylar nectaries present or absent; ovules 1 or 2 per locule, pendulous, anatropous, bitegmic, crassinucellate. **Fruits** capsules or berries, circumscissile near base or loculicidally dehiscent and forcibly ejecting seeds. **Seeds** brown or black, shiny, caruncular or not; endosperm fleshy.

Genera 7, species ca. 120 (2 genera, 3 species in the flora): North America, Mexico, West Indies, Central America, n South America, Europe, Asia, Africa; nearly worldwide.

SELECTED REFERENCE Channell, R. B. and C. E. Wood Jr. 1987. The Buxaceae in the southeastern United States. J. Arnold Arbor. 68: 241–257.

1. Shrubs, not clonal; leaves opposite or subopposite; tepals of pistillate flowers 5 or 6; seeds 2 per locule . . . . 1. *Buxus*, p. 7
1. Herbs or subshrubs, clonal; leaves alternate; tepals of pistillate flowers usually 4; seeds 1 per locule. . . . . 2. *Pachysandra*, p. 7

1. BUXUS Linnaeus, Sp. Pl. 2: 983. 1753; Gen. Pl. ed. 5, 423. 1754 • Box, boxwood [Latin *buxus*; from Greek *pyxos* (or *puxos*), name for box tree, of uncertain origin but supposedly from *pyknos*, dense, solid, alluding to excellent wood] I

Shrubs [trees], not clonal. **Leaves** opposite or subopposite; blade margins entire, surfaces shiny. **Inflorescences** axillary and terminal, spikes or capitate clusters [racemes]. **Staminate flowers** basal; tepals 4; pistillode present. **Pistillate flowers** 1 per inflorescence, apical; tepals 5 or 6; ovary 2- or 3-carpellate; locules 1 per carpel; stigma decurrent; interstylar nectaries or nectariferous tissue present; ovules 2 per locule. **Fruits** capsules, broadly oblong [globose or ovoid], loculicidally dehiscent, splitting into 2 or 3 valves, forcibly ejecting seeds at maturity, usually glabrous; style persistent. **Seeds** 2 per locule, black, oblong; with rudimentary aril. $x = 14$.

Species ca. 70 (1 in the flora): introduced; Mexico, West Indies, Central America, n South America, Europe, Asia, Africa.

The structures subtending the pistillate flowers are ambiguous; they may be tepals or bracts. The taxonomy and nomenclature of *Buxus* are unsettled. Species and cultivars other than *B. sempervirens* are likely to persist from cultivation.

1. **Buxus sempervirens** Linnaeus, Sp. Pl. 2: 983. 1753 F I

Shrubs 1–3(–8) m, glabrous, except puberulent on young shoots, petioles, and basal portion of leaves. **Leaves:** petiole to 2 mm; blade elliptic to widely elliptic, 1.5–3 × 0.7–1.7 cm, base cuneate, apex obtuse or, occasionally, retuse, surfaces darker green adaxially. **Capsules** 8 mm diam. **Seeds** 5–6 mm. $2n = 28$.

Flowering spring; fruiting late summer–fall. Old homesites, waste places; 0–1000 m; introduced; N.Y., N.C., Ohio, Pa., R.I., Tenn., Va., W.Va.; s, w Europe; sw Asia; nw Africa.

The hard, heavy wood of *Buxus sempervirens* is used for engraving, marquetry, turning, tool handles, mallet heads, and musical instruments. All parts of the plant are toxic if ingested; contact with the plant may cause dermatitis (W. H. Lewis and M. P. F. Elvin-Lewis 1977).

2. PACHYSANDRA Michaux, Fl. Bor.-Amer. 2: 177, plate 45. 1803 • [Greek *pachys*, stout, and *aner*, man, alluding to thickness of staminal filaments]

Herbs or subshrubs, clonal; rhizomes branched, creeping, long. **Leaves** alternate, often crowded or fasciculate; blades margins dentate, surfaces dull or slightly shiny. **Inflorescences** from leafless nodes near base of stem or terminal [axillary], rarely from rhizomes, dense spikes [racemes]. **Staminate flowers** distal and, sometimes, 1 or 2 proximally; tepals 2 or 4; stamens exserted; pistillode absent. **Pistillate flowers** 1–7 per inflorescence, proximal, subtended by 7 or more scalelike bracts; tepals usually 4, sometimes more, white; ovary 2- or 3(or 4)-carpellate, apically 2 or 3-lobed; locules 1(or 2) per carpel; stigma linear, apical along inner surface of style; interstylar nectaries or nectariferous tissue absent; ovules 1 or 2 per locule. **Fruits** capsules or berries, ovoid to subglobose, dehiscence basal, circumscissile, glabrous or densely hairy. **Seeds** 1 per locule, brown or black, ovoid, 3-sided, smooth; carunculate or ecarunculate. $x = 12$.

Species 3 (2 in the flora): e, s North America, e Asia.

The third species in the genus, *Pachysandra axillaris* Franchet, with two subspecies, *axillaris* and *stylosa* Boufford & Q. Y. Xiang, is intermediate between *P. procumbens* and *P. terminalis* in having axillary inflorescences. Similar to *P. procumbens*, *P. axillaris* has a 3-carpellate gynoecium; according to H. C. Robbins (1968), *P. procumbens* rarely has axillary inflorescences.

SELECTED REFERENCES Boufford, D. E. and Xiang Q. Y. 1992. *Pachysandra* reexamined. Bot. Bull. Acad. Sin. 33: 201–207. Robbins, H. C. 1968. The genus *Pachysandra* (Buxaceae). Sida 3: 211–248.

1. Inflorescences from leafless nodes near base of stem or rarely from rhizome; ovaries (2 or)3(or 4)-carpellate; fruits capsules . . . . 1. *Pachysandra procumbens*
1. Inflorescences terminal on aerial shoots; ovaries 2(or 3)-carpellate; fruits berries. . . 2. *Pachysandra terminalis*

1. Pachysandra procumbens Michaux, Fl. Bor.-Amer. 2: 178, plate 45. 1803 • Allegheny or mountain spurge E F

Herbs 10–40 cm, sparsely to densely pubescent, hairs short, curved. **Leaves** crowded distally on stem; petiole 1.5–8 cm; blade darker green adaxially, often with pale mottling along veins, elliptic to ovate, broadly ovate, rectangular, obovate, or nearly orbiculate, 3–11 × 2.5 8 cm, base cuneate to broadly cuneate or, sometimes, truncate and abruptly cuneate, margins coarsely dentate distal to middle, apex acute, surfaces dull, moderately to sparsely pubescent. **Inflorescences** 1–3(–10), from leafless nodes near base of stem, rarely from rhizome. **Staminate flowers** 18–38, sessile, each subtended by 1 bract; tepals 4, ovate to broadly ovate, 3–5 mm, margins with minute hairs, apex rounded to acute. **Pistillate flowers** 1–3(–7), sessile or pedicellate; tepals 3–5 mm, margins with minute hairs, apex rounded to acute; ovary (2 or)3(or 4)-carpellate, apical lobes (2 or) 3(or 4), locules 1 per carpel; styles 3; ovules 2 per locule. **Fruits** capsules, 12–16 mm, apex 3-lobed, densely and minutely hairy. **Seeds** 3–6, black, 3–4.5 × 2–2.5 mm; with small caruncule. $2n = 24$.

Flowering Feb–May; fruiting Jul–Aug. Rich, moist woods, near streams, on limestone soil; 50–1000 m; Ala., Fla., Ga., Ky., La., Md., Miss., Mo., N.C., Pa., S.C., Tenn., W.Va.

Pachysandra procumbens is native to the Gulf coastal plain and the Cumberland Plateau; it is believed to be introduced in Maryland, Missouri, and Pennsylvania.

2. Pachysandra terminalis Siebold & Zuccarini, Abh. Math.-Phys. Cl. Königl. Bayer. Akad. Wiss. 4(2): 142. 1845 • Japanese mountain spurge I

Pachysandra terminalis var. *variegata* Norton

Herbs 10–30 cm, glabrous or glabrate. **Leaves** crowded distally and in clusters at middle and often at proximal part of stem; petiole 1–3 cm; blade slightly darker green adaxially, without mottling along veins, elliptic to widely elliptic or ovate, broadly ovate, or obovate, 5–8 × 2–4 cm, base cuneate to broadly cuneate, margins coarsely dentate distal to middle, apex (terminal tooth) acute or obtuse, abaxial surface puberulent along veins, adaxial surface glabrous, shiny (not evident when dried). **Inflorescence** 1, terminal. **Staminate flowers** 15–20, sessile, each subtended by 1 bract and 2 sepal-like bracteoles; tepals 2, ovate, 2.5–4 mm, margins ciliate, apex rounded. **Pistillate flowers** 2–7, pedicellate; tepals 2.5–4 mm, margins ciliate, apex rounded; ovary 2(or 3)-carpellate, apical lobes 2(or 3), locules 1(or 2) per carpel; styles 2; ovules 1 or 2 per locule. **Fruits** berries, to 15 mm diam., apex 2-lobed, glabrous. **Seeds** 1–3, brown or black, 4–6 × 2–3 mm; ecarunculate. $2n = 48$.

Flowering Mar–Apr; fruiting Jul–Aug. Roadsides, railroad embankments, moist woods, along streams, near old homesites; 0–1000 m; introduced; Ont.; Conn., Del., Ill., Ind., Md., Mass., N.H., Ohio, Pa., R.I., Va., Wis.; e Asia.

Pachysandra terminalis, a native of China and Japan, is widely cultivated as an ornamental groundcover, usually in shaded situations, in temperate North America. The plants are more likely to spread vegetatively by rhizome pieces rather than by seeds. Many natural-appearing occurrences may be remnants of cultivation. The two sepal-like bracteoles of the staminate flowers are sometimes interpreted as tepals (R. B. Channell and C. E. Wood Jr. 1987).

GUNNERACEAE Meisner
• Chilean Rhubarb Family

Gordon C. Tucker

Herbs, perennial, monoecious or polygamomonoecious [dioecious], terrestrial or amphibious, acaulous, armed, clonal or not. **Roots** fibrous, rhizomes with [without] triangular scales (cataphylls). **Stems** absent. **Leaves** basal, alternate, simple; stipules scalelike or absent; petiolate; blade palmately lobed [toothed], venation prominent, margins irregularly toothed, abaxial surface with stiff, hard prickles. **Inflorescences** ± from rhizome, terminal [from axils of distal leaves], panicles of spikes; bracts absent. **Flowers** often pistillate proximally, staminate distally, bisexual between, sessile or pedicellate; perianth and androecium perigynous; sepals 2[0]; petals 0[2]; stamens 2[1] epipetalous; anthers basifixed, extrorse, pollen grains 3(–5)-aperturate, colpate, 2-nucleate; pistil 2-carpellate; ovary inferior, 1-locular; placentation apical; styles 2; stigmas 2, papillate; ovule 1, anatropous, bitegmic, crassinucellate. **Fruits** drupes. **Seeds** with oily endosperm, embryo straight.

Genus 1, species 35–50 (1 in the flora): introduced, California; Mexico, Central America, South America, w Europe, se Asia, s, e Africa, Pacific Islands (Hawaii), Australia.

The systematic position of Gunneraceae has been uncertain for a long time. *Gunnera* has sometimes been included in the mostly aquatic Haloragaceae, which also have reduced flowers with an inferior ovary. Molecular studies (combining *rbc*L, *atp*B, and 18S data) suggest that *Gunnera* occupies an isolated position among the core "eudicots." Some of these studies have also identified the African shrub *Myrothamnus* Welwitsch as the closest relative to *Gunnera*. Other work, based on leaf morphology and pollen, suggests a close relationship with Saxifragaceae (D. G. Fuller and L. J. Hickey 2005). *Gunnera* is remarkable for its symbiosis with cyanobacteria of the genus *Nostoc* (B. Osborne et al. 1991).

Species of *Gunnera* subg. *Panke* (Molina) Schindler are unique in having large, triangular scales between the leaves on the rhizomes. The morphological significance of these scales has been debated, and they have been interpreted as stipules, ligules, or cataphylls. Comparisons with features of the rhizomes in other subgenera support the interpretation of these structures as cataphylls (L. Wanntorp et al. 2003).

SELECTED REFERENCES Fuller, D. G. and L. J. Hickey. 2005. Systematics and leaf architecture of the Gunneraceae. Bot. Rev. (Lancaster) 71: 295–353. Osborne, B. et al. 1991. *Gunnera tinctoria*: An unusual nitrogen-fixing invader. BioScience 41: 224–234. Wanntorp, L., H.-E. Wanntorp, and M. Källersjö. 2002. Phylogenetic relationships of *Gunnera* based on nuclear ribosomal DNA ITS region, *rbc*L and *rps*16 intron sequences. Syst. Bot. 27: 512–521. Wilkinson, H. P. and L. Wanntorp. 2007. Gunneraceae. In: K. Kubitzki et al., eds. 1990+. The Families and Genera of Vascular Plants. 15+ vols. Berlin etc. Vol. 9, pp. 177–183.

1. GUNNERA Linnaeus, Syst. Nat. ed. 12, 2: 587, 597. 1767; Mant. Pl. 1: 16, 121. 1767 • Chilean rhubarb, gunnère du Chili [For John Ernest Gunner, 1718–1773, Norwegian bishop and botanist] [I]

Rhizomes ± horizontal; scales 10–20 cm, laciniate. **Leaves:** petiole attached to rhizome; blade ± orbiculate [reniform]. **Spikes** dense, stout, ± conic. **Flowers** sessile or pedicel as long as ovary; sepals green; stigmas linear, slender. **Drupes** ovate to oblong. x = 12, 17.

Species 35–50 (1 in the flora): introduced, North America; Mexico, Central America, South America; introduced also in w Europe, Pacific Islands.

In addition to *Gunnera tinctoria*, the smaller South American species *G. manicata* Linden ex Delchevalerie is cultivated along the Pacific Coast and has been noted as persisting in abandoned gardens in the Fraser River delta region, British Columbia; there is no evidence of self-sown populations (S. F. Lomer, pers. comm.).

1. Gunnera tinctoria (Molina) Mirbel in C. de Mirbel and N. Jolyclerc, Hist. Nat. Pl. 10: 141. 1805 • Giant rhubarb, dinosaur food, nalca, panque [F] [I]

Panke tinctoria Molina, Sag. Stor. Nat. Chili, 143, 351. 1782; *Gunnera chilensis* Lamarck

Herbs 1–2 m. **Leaves:** petiole 1–1.5 m × 20–45 mm; blade 1–2 m. **Peduncles** erect, 2–20 cm. **Inflorescences** 50–75[–100] × 10–25 cm; spikes many, 2–5 cm. **Flowers:** sepals 1 mm; filaments 0.5 mm; anthers broadly ellipsoid, 0.7–1 mm; styles glandular, 0.7–1.2 mm. **Drupes** red, 1.5–2[–3] × 1–2 mm. **Seeds** 1 per fruit. $2n$ = 34.

Flowering spring, fruiting summer. Disturbed, shaded, damp areas; 0–100 m; introduced; B.C.; Calif., Oreg., Wash.; s South America (Chile); introduced also in w Europe (Ireland), Pacific Islands (New Zealand).

Gunnera tinctoria is cultivated and escaped; it has become naturalized in coastal California (Marin and San Francisco counties), Oregon (Lincoln County), Washington (King County), and British Columbia (Vancouver Island, S. F. Lomer, pers. comm.). *Gunnera tinctoria* is suitable for a wetland or bog garden, and the rootstock is hardy to -10°C. The young leaf stalks can be peeled and cooked as a vegetable or eaten raw (U. P. Hedrick 1919). Some other species are occasionally cultivated in North America, including the superficially similar *G. manicata*; none is reported to escape. A good distinguishing character can be found in the large, laciniate, or jaggedly cut scales at the base of the leaves and inflorescences. In *G. manicata*, there is a prominent development of membranous webbing between the main lobes of the scale, whereas in *G. tinctoria* this is not well developed and the lobes are often nearly free to the main rachis of the scale (P. A. Williams et al. 2005). In *G. tinctoria*, the central part of the main inflorescence axis is 4–4.5 cm, versus 3–3.3 cm in *G. manicata*.

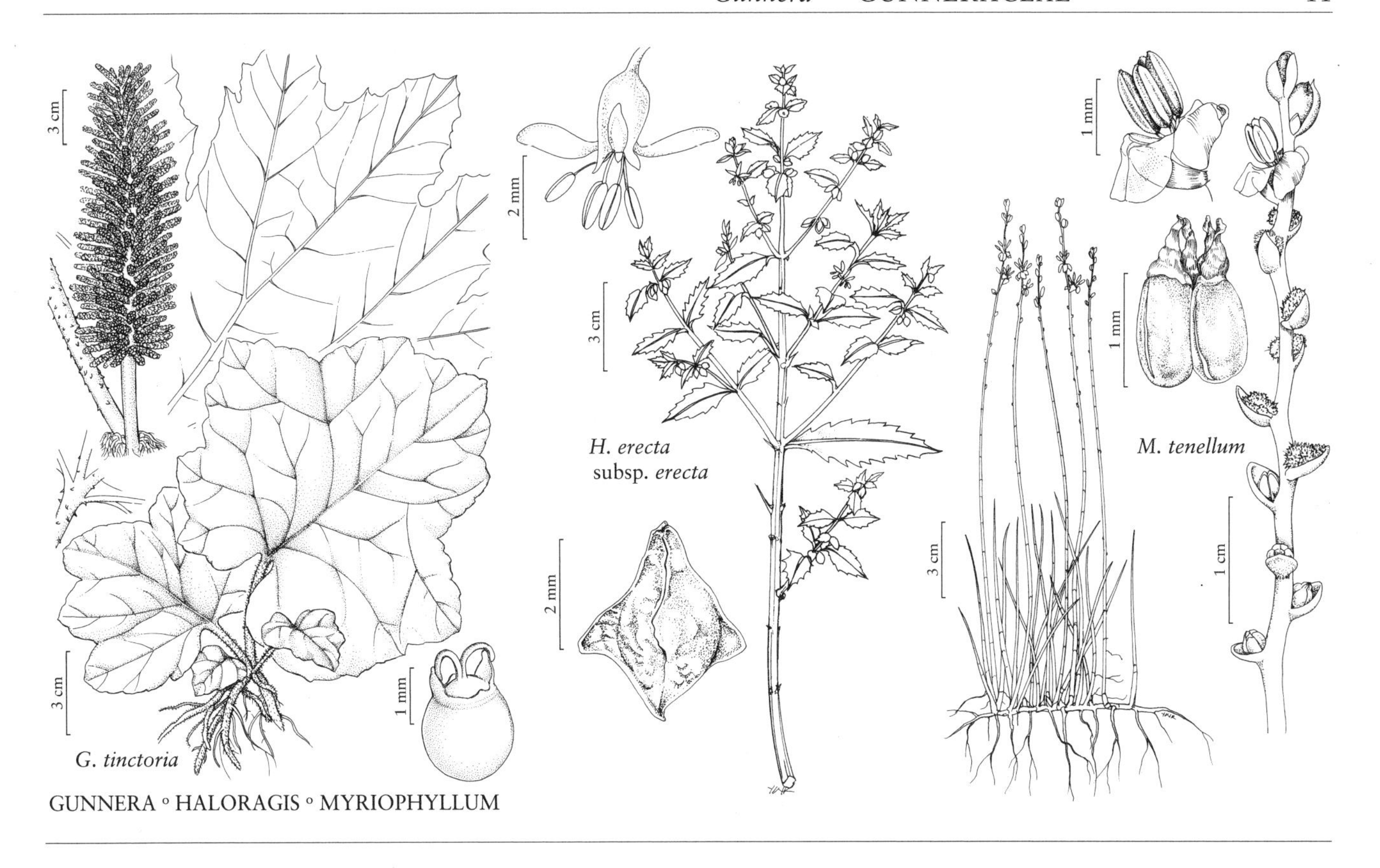

GUNNERA ∘ HALORAGIS ∘ MYRIOPHYLLUM

HALORAGACEAE R. Brown

• Water-milfoil Family

Robin W. Scribailo
Mitchell S. Alix

Herbs, perennial [**annual**], **or shrubs,** usually monoecious, rarely dioecious, usually aquatic to semiaquatic, sometimes terrestrial, unarmed, ± clonal. **Roots:** taproots or fibrous, and then often with adventitious nodal roots; rhizomes sometimes present [stolons in some *Haloragis*]. **Stems** erect, ascending, decumbent, or prostrate, cylindric to 4-ribbed, glabrous or scabrous to pubescent, hairs uniseriate and multiseriate, glands present or absent. **Turions** present or absent, lateral and/or terminal. **Leaves** opposite, alternate, or subverticillate to whorled, simple, often heteromorphic in *Myriophyllum* and *Proserpinaca*; stipules absent; sessile or petiolate; blade lobed, unlobed, or pinnatifid to pectinate, margins entire or serrate, surfaces glabrous or scabrous. **Inflorescences** terminal or lateral in axils of bracts or leaves, determinate or indeterminate, dichasia (*Haloragis* and *Proserpinaca*), or simple racemes (*Myriophyllum*); bracts and bracteoles present. **Flowers** bisexual or unisexual, staminate and pistillate usually on same plant, sessile or pedicellate (sometimes sessile in pistillate flowers of *Myriophyllum*); perianth and androecium epigynous; hypanthium subglobose; sepals persistent, (3 or)4, sometimes rudimentary (*Myriophyllum*), petals often caducous, sometimes persistent, (3 or)4, or 0 or rudimentary (*Proserpinaca*), keeled, cucullate, often distally cupulate; stamens 3–8 (1 or 2 times as many as sepals); anthers basifixed, dehiscing longitudinally; pistil 1, 3- or 4-carpellate; ovary 1, inferior, 1–4-locular; placentation axile; styles 1 per locule; stigmas 1 per locule, clavate, capitate, fimbriate; ovules 1(or 2, in *Haloragis* and *Proserpinaca*), anatropous, bitegmic, crassinucellate. **Fruit** a nutlet, indehiscent, or schizocarp, splitting into (2–)4 mericarps; exocarp glabrous, scabrous, rugose, tuberculate, or papillate, sometimes with ribs, ridges, or wings. **Seeds** 1 per locule; embryo straight, cylindric; endosperm ± copious and fleshy.

Genera 10, species ca. 120 (3 genera, 17 species in the flora): North America, Mexico, West Indies, Bermuda, Central America, South America, Eurasia, Africa, Indian Ocean Islands, Pacific Islands, Australia.

Morphologically, Haloragaceae are defined by the following floral characters: an epigynous ovary, usually 3- or 4-merous floral organization (always 3-merous in *Proserpinaca*), sometimes 2-merous, cucullate petals, and fruit a nutlet or schizocarp with 1 or 2 ovules per locule. In the aquatic members of the family, reliance on vegetative characters that are highly plastic and have evolved independently by convergent evolution has proven to be of limited usefulness for the delimitation of taxa (M. L. Moody and D. H. Les 2007).

J. Hutchinson (1959) suggested that Haloragaceae is closely allied to Onagraceae based on embryology, pollen morphology, and floral vasculature. A. Cronquist (1968) and A. L. Takhtajan (1969) believed Haloragaceae to be more closely allied to Podostemaceae. The work of A. E. Orchard (1975, 1985) has been important in circumscribing the family.

Molecular phylogenetic studies have placed Haloragaceae within the core eudicot order Saxifragales (D. R. Morgan and D. E. Soltis 1993; Soltis et al. 1997b). *Gunnera*, which had long been included within the family, has been moved to the monogeneric family Gunneraceae in Gunnerales (Angiosperm Phylogeny Group 2009).

SELECTED REFERENCE Moody, M. L. and D. H. Les. 2007. Phylogenetic systematics and character evolution in the angiosperm family Haloragaceae. Amer. J. Bot. 94: 2005–2025.

1. Flowers unisexual or bisexual, proximal pistillate, distal staminate, often with intermediate transitional zone of bisexual flowers; fruit a schizocarp, splitting into (2–)4 mericarps; plants aquatic or semiaquatic . . . . 2. *Myriophyllum*, p. 14
1. Flowers bisexual; fruit a nutlet, indehiscent; plants aquatic, semiaquatic, or terrestrial.
    2. Flowers 4-merous; plants shrubs or herbs, terrestrial . . . . 1. *Haloragis*, p. 13
    2. Flowers 3-merous; plants herbs, aquatic or semiaquatic . . . . 3. *Proserpinaca*, p. 29

1. HALORAGIS J. R. Forster & G. Forster, Char. Gen. Pl. ed. 2, 61, plate 31. 1776 • [Greek *halos*, sea, and *ragis*, grape-berry, alluding to maritime habitat and bunched fruits] I

Mitchell S. Alix

Robin W. Scribailo

Herbs or shrubs, monoecious, terrestrial; with taproot [stolons]. **Rhizomes** absent. **Stems** erect [creeping], 4-ribbed [smooth], scabrous to sparsely pubescent [glabrous]. **Turions** absent. **Leaves** opposite [proximally opposite, distally alternate], homomorphic; petiolate [sessile]; blade unlobed [pinnatifid], margins serrate [entire], surfaces glabrous or scabrous. **Inflorescences** dichasia, compound, 3- or 4[–7]-flowered, in axils of alternate bracts, bracts foliagelike proximally, highly reduced distally; bracteoles paired, opposite subtending bracts; flowers bisexual. **Flowers** [2–]4-merous; petals caducous; stamens 8; ovary 2- or 4-locular. **Fruit** a nutlet, silver-gray to dark green or red, 4-lobed to 4-angled, ridges often with wings and/or with protuberances opposite sepals or throughout, or tuberculate between ridges or wings, surfaces smooth to rugose, septa solid, endocarp woody, exocarp membranous or spongy. $x = 7$.

Species ca. 28 (1 in the flora): introduced, California; Pacific Islands (New Zealand), Australia.

SELECTED REFERENCE Forde, M. B. 1964. *Haloragis erecta*: A species complex in evolution. New Zealand J. Bot. 2: 425–453.

1. Haloragis erecta (Banks ex Murray) Oken, Allg. Naturgesch. 3: 1871. 1841 • Erect seaberry [F] [I]

Cercodia erecta Banks ex Murray, Commentat. Soc. Regiae Sci. Gott. 3: 3, plate 1. 1780

Subspecies 2 (1 in the flora): introduced, California; Pacific Islands (New Zealand).

A. E. Orchard (1975) recognized two subspecies of *Haloragis erecta* distinguished primarily on leaf characteristics. North American specimens have been called subsp. *erecta*, which has relatively thin, lanceolate to ovate leaf blades, and margins with 20–45 teeth. Subspecies *cartilaginea* (Cheeseman) Orchard is known from the North Cape Peninsula of the North Island of New Zealand and has relatively thick, orbiculate to broadly ovate leaf blades and margins with 10–15 teeth (Orchard).

1a. Haloragis erecta (Banks ex Murray) Oken subsp. **erecta** [F] [I]

Stems to 1 m, scabrous and tuberculate, with red glands and multiseriate hairs. **Leaves:** petiole (0.5–)0.8–1.7(–3) mm; blade [1.5–]1.8–5[–9] × (0.5–) 1–1.7[–3.5] cm, thin [cartilaginous], base attenuate, margins serrate, teeth 18–40, to 2 mm. **Inflorescences:** bracts petiolate, leaflike, lanceolate to elliptic, 0.5–2[–2.5] × 0.2–1.2 cm, margins serrate, teeth 22; primary bracteoles 0.3–1.3 × 0.1–0.3 mm; secondary bracteoles [0.3–] 0.5–0.9 × 0.1–0.2 mm. **Pedicels** 0.5–1.8 mm. **Flowers:** sepals red or green, deltate, 0.7–1.2 × 0.4–1 mm, glandular; petals cream or red, 1.5–1.8[–2.2] × 1.2–1.4 mm; keel with red tuberculate glands and multiseriate hairs; filaments to 0.4 mm; anthers yellow or red, linear-oblong, (1.2–)1.4–1.7 × 0.3–0.5 mm; ovary ovate, 0.9–1 × 0.7–1.2 mm, with 4 prominent, antisepalous ridges, locules each with 1 pendulous ovule; stigmas red, capitate, to 0.1 mm. **Fruits** obturbinate or pyriform to ovoid, 1.8–3 × 1.5–2.5(–4) mm, with 4 antisepalous, longitudinal ridges, ridges often with membranous, deltoid wings to 0.8 mm wide, surfaces smooth to rugose; stipes 0.8–1.3 mm; sepals erect, broadly deltoid. **2*n*** = 14.

Flowering and fruiting Jun–Sep. Open, disturbed areas; 0–100[–1000] m; introduced; Calif.; Pacific Islands (New Zealand).

Subspecies *erecta* is known from the San Francisco Bay area from Golden Gate Park and the San Francisco Botanical Garden, where it was intentionally planted. It is sold as the cultivar 'Wellington Bronze' or 'Toatoa.' The attractiveness and availability of this shrub are likely to contribute to its future spread within milder regions of the United States.

2. MYRIOPHYLLUM Linnaeus, Sp. Pl. 2: 992. 1753; Gen. Pl. ed. 5, 429. 1754 • Water-milfoil, myriophylle [Greek *myrios*, countless, and *phyllon*, leaf, alluding to capillary segments of lower and/or submersed leaves]

Robin W. Scribailo

Mitchell S. Alix

Herbs, usually monoecious (dioecious in *M. aquaticum* and *M. ussuriense*), aquatic to semiaquatic [semiterrestrial or terrestrial]; with fibrous and adventitious nodal roots. **Rhizomes** usually present, (absent in *M. farwellii*). **Stems** erect [prostrate], terete, glabrous, hydropoten (aggregates of transfer cells) sometimes present as scattered yellow, red, or brown splotches. **Turions** present or absent, lateral and/or terminal. **Leaves** submersed or emersed (not emersed in *M. farwellii*), usually whorled, sometimes alternate, opposite, subopposite, subverticillate, or irregular [scattered], usually heteromorphic (homomorphic in *M. aquaticum*, *M. farwellii*, and *M. tenellum*), emersed leaves usually becoming reduced floral bracts distally; sessile or petiolate; blade unlobed or lobed, pinnatifid, or pectinate, margins usually entire, sometimes serrate, surfaces glabrous, trichomes, when present, usually in axils of leaves and leaf segments, sometimes scattered. **Inflorescences** racemes, simple, 80+-flowered, or flowers borne singly (rarely dichasia in *M. humile*) in axils of emersed leaves (submersed in *M. farwellii*); bracteoles paired, alternate, opposite subtending leaf, apex often fringed with glandular, red trichomes, sometimes aristate; flowers unisexual or bisexual, proximally pistillate, distally staminate, often with intermediate transitional zone of bisexual flowers. **Flowers** 4-merous, staminate petals persistent or caducous, pistillate petals caducous or absent; stamens usually 4 or 8, sometimes 5–7 in bisexual flowers; ovary 4-locular. **Fruit** a schizocarp, light green, tan, olive-brown or -green, brown, red-brown, or purple, cylindric, ± ovoid to oblong, or ± globose, cruciate to ± 4-lobed, splitting into (2–)4 mericarps; mericarps compressed to ± flattened, concave, or ± rounded, adaxially rounded, sometimes with 1–4 abaxial ridges, ridges with or without wings, surfaces smooth to ± papillate to ± tuberculate, sometimes with red punctate glands. $x = 7$.

Species ca. 68 (14 in the flora): North America, Mexico, Central America, South America, Eurasia, n, nw Africa, Indian Ocean Islands, Pacific Islands, Australia.

M. L. Moody and D. H. Les (2010) realigned some species of *Myriophyllum* to conform to the results of their molecular analyses, formally recognizing three subgenera, five sections, and five subsections. Because of strong levels of independent convergence, the subgenera almost completely overlap in vegetative characters. Although we recognize the phylogenetic merits of the work of Moody and Les, use of their classification here would serve only to confuse and not clarify our attempts to present the most straightforward taxonomic arrangement for *Myriophyllum*.

Although *Myriophyllum* is easily recognized in the field, positive identification at the species level has been problematic. Much of this difficulty can be attributed to over-reliance on submersed vegetative material for identification. As noted by some authors (for example, C. D. Sculthorpe 1967; G. E. Hutchinson 1975), phenotypic plasticity can greatly alter leaf characters such as size and segment number of aquatic plants in different environments. In some cases, vegetative character states used in taxonomic treatments have been based largely on information repeated from regional floras. As a result, circumscriptions of species have often not encompassed the full extent of variability observed in some taxonomic characters. Reliance on these older treatments for identification, coupled with plasticity in vegetative characters, has led to misidentifications, particularly when flowering and fruiting materials are absent. Most misidentified herbarium specimens of *Myriophyllum* examined for this flora did not possess reproductive characters. Currently, the most effective method for identifying vegetative specimens of *Myriophyllum* is by analysis of ITS and cpDNA sequences (see discussion under 8. *M. spicatum*).

Floral structures and fruits offer good characters to distinguish *Myriophyllum* species. The inflorescences in *Myriophyllum* are described here as racemes because most flowers are short-pedicellate. The term spike often has been used in *Myriophyllum* implying that the flowers are sessile. Admittedly, the distinction is largely semantic, where the pedicels are so short that the flowers appear sessile. It is important to note that mericarp characters used to distinguish these species are not fully expressed morphologically until late in their development. As a result, mericarps that are farthest from the shoot apex typically best display diagnostic characters.

Many authors have mentioned that bisexual flowers often occur on the raceme where there is a transition from pistillate to staminate flowers. Bisexual flowers are more common in *Myriophyllum* than realized, and we found that many staminate flowers possess pistillodes at greater or lesser stages of development. This condition is most pronounced in *M. humile*, in which staminate flowers often have complete pistils rather than pistillodes.

Leaf characters in *Myriophyllum* often have been described using terms for compound leaves, such as pinnate in form, with a central rachis and leaf divisions referred to as pinnae. Because the leaves of *Myriophyllum* are simple, we use pectinate, central axis, and segments as descriptors.

Many *Myriophyllum* species will occasionally produce a small emergent form when plants become stranded along shorelines by wave action or when water levels decrease. Only those species that produce an emersed form as part of their normal life history are discussed in some detail.

Small ascidiate (flask-shaped) trichomes are found on the stems and in the axils, or at the bases of leaves and leaf segments. They have often been referred to as hydathodes (A. E. Orchard 1979; S. G. Aiken 1981), but they have been referred to also as enations, myriophyllin glands, pseudostipules, or scales (Orchard). Because the function of these structures is unknown and they do not appear to be secretory, the best approach would seem to be to refer to them as trichomes.

Correct identification of *Myriophyllum* species in North America is critical in the conservation and management of aquatic habitats because some species are introduced and highly invasive (for example, *M. aquaticum* and *M. spicatum*). This issue is further complicated by hybridization of *M. spicatum* with native *M. sibiricum* (see 8. *M. spicatum* discussion). The native *M. heterophyllum* is considered invasive in the northeastern United States and Pacific Northwest, where there is evidence of hybridization with other *Myriophyllum* species (see 13. *M. heterophyllum* discussion).

SELECTED REFERENCES Aiken, S. G. 1978. Pollen morphology in the genus *Myriophyllum* (Haloragaceae). Canad. J. Bot. 56: 976–982. Aiken, S. G. 1981. A conspectus of *Myriophyllum* (Haloragaceae) in North America. Brittonia 33: 57–69. Orchard, A. E. 1981. A revision of South American *Myriophyllum* (Haloragaceae), and its repercussions on some Australian and North American species. Brunonia 4: 27–65.

1. Leaves homomorphic, scalelike, submersed leaves 0.3–1(–1.5) mm . . . . 1. *Myriophyllum tenellum*
1. Leaves usually heteromorphic (homomorphic in *M. aquaticum* and *M. farwellii*), submersed leaves usually pectinate, rarely entire or lobed, 1.5+ mm.
    2. Flowers from axils of submersed leaves.
        3. Mericarps (1–)1.5–2.5 mm, abaxial surface 4-angled, smooth to sparsely tuberculate, winged, transversely hexagonal . . . . 2. *Myriophyllum farwellii*
        3. Mericarps (0.6–)0.8–1.2 mm, abaxial surface rounded, sparsely to densely tuberculate, not winged, transversely elliptic to ovate . . . . .11. *Myriophyllum humile* (in part)
    2. Flowers from axils of emersed leaves.
        4. Herbs dioecious.
            5. Emersed leaves in whorls of 4–6(–8), pectinate, (20–)25–70(–75) mm, with (14–)16–36(–40) filiform segments; turions absent . . . . .3. *Myriophyllum aquaticum*
            5. Emersed leaves opposite or in whorls of 3(or 4), usually linear, spatulate, or 2- or 3-lobed, sometimes pectinate, (1.7–)2.5–9(–10.5) mm, with (0–)2–8(–12) lobed to linear-filiform segments; turions present . . . . 4. *Myriophyllum ussuriense*
        4. Herbs monoecious.
            6. Stamens 8.
                7. Emersed leaves all pectinate; turions present . . . . 5. *Myriophyllum verticillatum*
                7. Emersed leaves transitional, proximally pectinate, distally pinnatifid to lobed, serrate, or entire; turions present or absent.
                    8. Distal flowers in axils of alternate leaves; mericarps 1.3–1.6 × 0.3–0.4 mm . . . . .6. *Myriophyllum alterniflorum*
                    8. Distal flowers in axils of whorled leaves; mericarps 1.5–2.7 × 0.6–1.6 mm.
                        9. Bracteoles deltate, margins dentate to serrate; emersed leaves pinnatisect to lobed or entire, ovate to oblong in outline, 2–9 × 1–6 mm, margins dentate to serrulate . . . . 7. *Myriophyllum quitense*
                        9. Bracteoles usually ovate to depressed-ovate or obovate, sometimes elliptic to triangular or rhombic, margins entire or serrate; emersed leaves proximally pectinate to pinnatifid, distally elliptic or obovate, sometimes spatulate in outline, 1–2.3 × 0.6–1(–1.5) mm, margins serrate to shallowly lobed or entire.

10. Submersed leaves with (20–)24–36(–42) segments, segments usually parallel and in one plane, forming angles less than 45° with central axis. . . . . 8. *Myriophyllum spicatum*
10. Submersed leaves with 6–18(–24) segments, segments often irregular in orientation, not parallel or in one plane, forming angles greater than 45° with central axis . . . . 9. *Myriophyllum sibiricum*

[6. Shifted to left margin.—Ed.]

6. Stamens 4.
   11. Emersed leaves (0.6–)0.7–2.3(–2.7) mm; mericarps densely tuberculate proximal to midpoint, tubercles crowded, large, often obscuring wings . . . . 10. *Myriophyllum laxum*
   11. Emersed leaves 3–17(–31) mm; mericarps papillate and/or sparsely to densely tuberculate, tubercles relatively small, shallow, not obscuring wings.
       12. Submersed leaves usually alternate or opposite, rarely in whorls of 3(or 4) . . . . 11. *Myriophyllum humile* (in part)
       12. Submersed leaves usually in whorls of (3 or)4–6, sometimes subverticillate, rarely alternate.
           13. Mericarps: abaxial surface sharply 2-angled, with 2 ridges, ridges with prominent membranous, undulating wings, wings erect to reflexed with 6–12 perpendicular ribs . . . . 12. *Myriophyllum pinnatum*
           13. Mericarps: abaxial surface bluntly 2- or 4-angled, with 2 or 4 ridges, ridges sometimes with inconspicuous wings proximally, ribs absent.
               14. Anthers 1.3–2.2 mm; staminate sepals 0.5–0.8(–0.9) mm; pistillate petals 1.5–3 mm; mericarps transversely orbiculate to widely elliptic, abaxial surface inconspicuously 4-angled. . . . . 13. *Myriophyllum heterophyllum*
               14. Anthers 0.5–0.9 mm; staminate sepals (0.1–)0.2–0.4(–0.5) mm; pistillate petals 0.7–1.3 mm; mericarps transversely elliptic to narrowly ovate, abaxial surface usually bluntly 2-angled, sometimes 4-angled . . . . 14. *Myriophyllum hippuroides*

1. Myriophyllum tenellum Bigelow, Fl. Boston. ed. 2, 346. 1824 • Dwarf or slender water-milfoil, myriophylle grêle [E] [F]

Herbs monoecious, aquatic or semiaquatic, often forming dense mats. **Stems** unbranched or 1-branched, to 0.7 m. **Turions** absent. **Leaves** alternate, homomorphic, scalelike; sessile; submersed leaves ovate, 0.3–1(–1.5) × 0.1–0.7 mm, margins entire; emersed leaves ovate to obovate, (0.5–)0.8–2.5(–3.3) × (0.2–)0.3–1.2(–3) mm, margins entire. **Inflorescences** to 9 cm; flowers proximally pistillate, medially bisexual, distally staminate; bracteoles cream, narrowly elliptic to ovate, (0.4–)0.5–0.8(–1) × (0.1–)0.2–0.4(–0.5) mm, margins usually entire, sometimes serrate. **Staminate flowers:** sepals green to cream, lanceolate, 0.1–0.5 × 0.1–0.2 mm; petals persistent, cream to pink, obovate, (0.6–)1–2(–2.3) × (0.3–)0.6–1.4(–1.6) mm; stamens 4, filaments to 1.2 mm, anthers yellow, narrowly elliptic, (0.3–)0.7–1.6 × 0.1–0.5(–0.7) mm. **Pistillate flowers:** sepals green to cream, lanceolate, 0.1–0.6(–1) × 0.1–0.2(–0.6) mm; petals often persistent, cream to pink, obovate, 0.5–1.7(–2.2) × (0.2–)0.4–1.2(–1.5) mm; pistils to 1.3 mm, stigmas white, to 0.7 mm. **Fruits** globose, distinctly 4-lobed. **Mericarps** olive-brown, cylindric to ± globose, 0.6–1.3 × (0.2–)0.4–1.4 mm, transversely ovate, abaxial surface rounded, smooth or minutely papillate, rarely with a single obscure, longitudinal ridge, wings absent. $2n = 14$.

Flowering and fruiting Jul–Oct. Oligotrophic to mesotrophic waters, lakes, sandy substrates; 0–900 m; St. Pierre and Miquelon; N.B., Nfld. and Labr., N.S., Ont., P.E.I., Que.; Conn., Ind., Maine, Md., Mass., Mich., Minn., N.H., N.J., N.Y., N.C., Pa., R.I., Vt., Va., Wis.

Myriophyllum tenellum is easily recognized; it is the only water-milfoil in the flora area with homomorphic scalelike leaves whether it is growing submersed or as a shoreline emergent. It is often associated with acidic to circumneutral waters. E. G. Voss (1972–1996, vol. 2) noted that it tends to be overlooked. Although it has been recently recorded from Indiana (R. W. Scribailo and M. S. Alix 2006), we have seen no specimens from Illinois or Ohio.

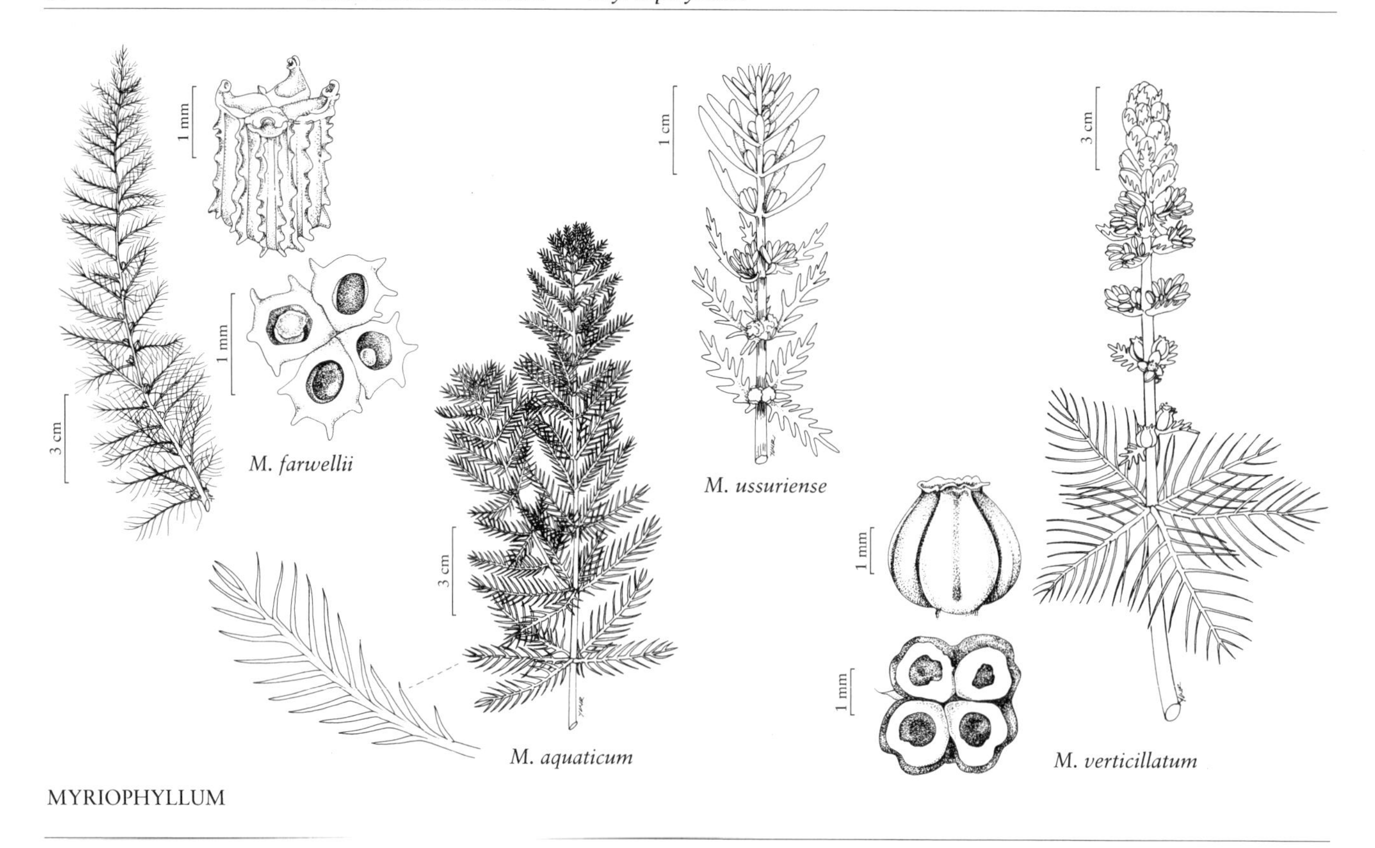

MYRIOPHYLLUM

2. Myriophyllum farwellii Morong, Bull. Torrey Bot. Club 18: 146. 1891 • Farwell's water-milfoil, myriophylle de Farwell E F

Herbs monoecious, aquatic, often forming dense stands. **Rhizomes** absent. **Stems** delicate, often branched, to 1 m, with numerous black, ascidiate trichomes. **Turions** present, green to brown, very narrowly cylindrical, with gradual and moderate transition from foliage leaves to slightly reduced turion leaves, to 10 cm, apex ± rounded; leaves pectinate and stiff, strongly appressed to axis, elliptic to narrowly obovate in outline, 11–23(–40) × 1.5–3(–4) mm; segments 6–12(–14), longest segment (1.5–)3–5.5 mm, basal segment less than or equal to ⅔ central axis of leaf, apex acute; brown, ascidiate trichomes scattered on surfaces and in axils. **Leaves** all submersed, usually in whorls of 3(or 4), sometimes alternate or subverticillate, homomorphic; petiole 0–3 mm; leaves pectinate, ovate to elliptic or obovate in outline, (6–)10–20(–25) × (4–)7–15 mm, segments (7–)10–14(–16), linear-filiform, longest segment (4–)8–15(–22) mm, with black ascidiate trichomes scattered on surfaces and in axils. **Inflorescences** submersed, to 30 cm; flowers bisexual; bracteoles cream, trullate, 0.2–0.5(–0.8) × (0.1–)0.3–0.5 mm, margins entire or irregularly lobed, lobes with glandular fringe. **Bisexual flowers:** sepals green to cream or ± purple, ovate to trullate, 0.1–0.5(–0.7) × 0.1–0.3(–0.5) mm; petals persistent, cream to purple, elliptic to obovate, 0.6–1.3 × 0.3–0.8 mm; stamens 4, filaments to 0.5 mm, anthers 0.2–0.4 × 0.1–0.3 mm; pistils 0.8–1.2 × 0.2–0.3 mm, stigmas red to ± purple, to 0.5 mm. **Fruits** cylindric to subglobose, deeply 4-lobed. **Mericarps** tan to brown, cylindric, (1–)1.5–2.5 × 0.5–0.9 mm, transversely hexagonal (rounded to obtusely angled adaxially), abaxial surface shallowly 4-angled, smooth or sparsely tuberculate, with 4 longitudinal ridges, ridges with irregular, shallow to pronounced, membranous wings, ribs absent. $2n = 14$.

Flowering and fruiting Jun–Aug. Oligotrophic to mesotrophic, often highly tannic waters, lakes, ponds, and marshes; 0–600; B.C., Man., N.B., Nfld. and Labr. (Nfld.), N.S., Ont., Que.; Alaska, Conn., Maine, Mass., Mich., Minn., N.H., N.Y., Pa., R.I., Vt., Wis.

Myriophyllum farwellii is distinguished from other water-milfoils, except *M. humile*, by submersed flowers and fruits in the axils of foliage leaves. It differs from *M. humile* by having distinctly larger, winged mericarps that are transversely hexagonal, versus smaller, wingless mericarps that are transversely elliptic to ovate. It produces elongate turions by midsummer with stiff, reduced leaves. These reduced leaves, which are dark green to black, are often visible at the base of new shoots in the next growing season, which can be an aid for identification. Its foliage leaves are mostly produced in whorls, contrary to what has been stated in the literature concerning an alternate or opposite arrangement.

SELECTED REFERENCE Aiken, S. G. 1976. Turion formation in watermilfoil, *Myriophyllum farwellii*. Michigan Bot. 15: 99–102.

3. Myriophyllum aquaticum (Vellozo) Verdcourt, Kew Bull. 28: 36. 1973 • Parrot- or water-feather F I W

Enydria aquatica Vellozo, Fl. Flumin., 57. 1829; *Myriophyllum brasiliense* Cambessèdes; *M. proserpinacoides* Gillies ex Hooker & Arnott

Herbs dioecious, pistillate, not staminate, in flora area, aquatic or semiaquatic, often forming dense stands. **Stems** branched or unbranched, to 5 m. **Turions** absent. **Leaves** in whorls of 4–6(–8), homomorphic; petiole to 9.6 mm; submersed leaves pectinate, oblanceolate to obovate in outline, (20–)25–70(–75) × (4–)5–26(–32) mm, segments (14–)16–36(–40), filiform, longest segment (2–)4–27(–33) mm; emersed leaves becoming unmodified floral bracts. **Inflorescences** to 20 cm; flowers unisexual; bracteoles cream to stramineous, (0.3–)0.5–1(–1.5) × 0.1–0.3(–0.5) mm, margins subulate to 3-fid. **Staminate flowers:** sepals cream, ovate to deltate, 0.7–0.8 × 0.3 mm; petals yellow, weakly cucullate, (2.3–)2.7–3.1 × 0.8–1.1 mm; stamens 8, filaments to 1.2 mm, anthers yellow, linear-oblong, (1.8–)2–2.7 × 0.2 mm. **Pistillate flowers:** sepals cream, lanceolate to deltate, 0.3–0.5 × 0.1–0.2(–0.4) mm; petals rudimentary or absent; pistils to 0.8 mm, stigmas white, to 0.3 mm. **Fruits** cylindric to ovoid, shallowly 4-lobed. **Mericarps** olive-green to brown, cylindric, 1.7 × 0.6–0.7 mm, narrowly obovate, abaxial surface rounded, ridges, wings and ribs absent.

Flowering and fruiting Apr–Sep. Lakes, canals, bays, ponds, slow moving ditches, creeks, rivers; 0–1500 m; introduced; B.C.; Ala., Ariz., Ark., Calif., Conn., Del., D.C., Fla., Ga., Idaho, Ill., Ind., Kans., Ky., La., Md., Mass., Minn., Miss., Mo., N.J., N.Mex., N.Y., N.C., Ohio, Okla., Oreg., Pa., R.I., S.C., Tenn., Tex., Va., Wash., W.Va., Wis.; South America; introduced also in Mexico, Central America, Eurasia, Africa, Indian Ocean Islands, Pacific Islands, Australia.

In the flora area, *Myriophyllum aquaticum* is an introduced invasive aquatic species, existing as pistillate populations throughout North America (R. Couch and E. Nelson 1992); it is native to the lowlands of South America (A. E. Orchard 1981). It has an unusual habit among North American species of *Myriophyllum*, where it is often observed as a robust emergent aquatic along shorelines. It can be found also growing to a depth of 5 m in lakes, with the largest submersed leaves recorded for any North American species of *Myriophyllum*. The leaves of *M. aquaticum* are very distinctive, being largely oblanceolate and two to three times as long as broad, with a large number of uniform, short-pinnate segments, often arranged in whorls of six or more.

Myriophyllum aquaticum has been reported from Iowa and Montana; no specimens have been seen that confirm these reports.

4. Myriophyllum ussuriense (Regel) Maximovicz, Bull. Acad. Imp. Sci. Saint-Pétersbourg 19: 182. 1873 • Ussurian water-milfoil F

Myriophyllum verticillatum Linnaeus var. *ussuriense* Regel, Mém. Acad. Imp. Sci. Saint Pétersbourg, ser. 7, 4(4): 60, plate 4, figs. 2–5. 1861

Herbs usually dioecious, rarely monoecious, aquatic or semiaquatic, usually not forming dense stands. **Stems** often branched, to 0.6 m. **Turions** present, ± brown, narrowly cylindrical, with gradual transition from foliage leaves to highly reduced turion leaves, (4–)7–12(–20) × 0.5–2(–3) mm, apex rounded to truncate; leaves often pectinate proximally and entire to 3-fid distally, strongly appressed to axis, lanceolate to narrowly elliptic or ovate in outline, (1.5–)2–4(–6.5) × (0.2–)0.3–2(–2.5) mm; segments 0–6(–10), longest segment 0.5–2 mm, basal segment less than or equal to ½ central axis of leaf, apex ± acute or rounded, brown, long-necked, ascidiate trichomes in axils present. **Leaves** opposite or in whorls of 3(or 4), heteromorphic; petiole 0–9 mm; submersed leaves usually pectinate, sometimes 2- or 3-lobed, ovate to widely ovate or trullate in outline, (1.3–)5–22(–26) × (0.3–)3–28(–35) mm, segments (0–)4–12(–14), distinctly alternate, lobed to linear-filiform, longest segment (0.5–)2–20(–25) mm; emersed leaves usually linear, spatulate, or 2- or 3-lobed, sometimes pectinate proximally, (1.7–)2.5–9(–10.5) × 0.3–3.5(–5) mm, segments (0–)2–8(–12), lobed to linear-filiform. **Inflorescences** to 12 cm; flowers usually unisexual, rarely bisexual; bracteoles cream to stramineous, lanceolate, elliptic, ovate, or obovate, (0.2–)0.3–0.7(–0.9) × (0.1–)0.2–0.4(–0.5) mm, margins entire, irregular, dentate, glandular, or lobed. **Staminate flowers:** sepals cream, elliptic to lanceolate, 0.5–0.7 × 0.2–0.5 mm; petals persistent, cream, sometimes apically suffused with purple, widely oblanceolate, 1.2–2.5 × 0.7–1.2 mm; stamens 8, filaments to 1.4 mm, anthers 0.9–1.8 × 0.2–0.4 mm. **Pistillate flowers:** sepals and petals rudimentary or absent; pistils to 0.7 mm, stigmas white, to 0.3 mm. **Fruits** subglobose, 4-lobed. **Mericarps** brown, obovate, 0.8 × 0.6 mm, abaxial surface rounded, minutely tuberculate, wings and ribs absent. $2n$ = [14] 21.

Flowering and fruiting Jul–Nov. Streams, rivers, muddy shorelines of ponds and lakes, intertidal wetlands; 0–600 m; B.C.; Oreg., Wash.; Eurasia.

Plants of *Myriophyllum ussuriense* typically grow in a semi-terrestrial habit in shallow water or on saturated sediments to a height of 20 cm. Shoots often have swollen stem bases that taper dramatically towards the apex. In some populations, extensive production of erect shoots from rhizomes produce dense stands. The floral bracts are distinctive, being opposite or alternate and elongate with usually 2–8 relatively short segments. Dimorphism in size between staminate and pistillate flowers of *M. ussuriense* is distinctive. Although most populations appear to be unisexual with staminate plants predominating and pistillate plants rare (O. Ceska et al. 1986), the latter are extremely small with a vestigial perianth and are easily overlooked, indicating that monoecy may be more common than thought in this species. S. Ueno and Y. Kadono (2001) reported that seven of 80 populations of *M. ussuriense* in Japan had some monoecious plants. No fruit was found despite an extensive examination of available material.

Submersed plants have pectinate leaves that are extremely delicate with usually fewer than 12 straight segments. A useful characteristic of some leaves is that the central axis terminates in a right-angled bifurcation.

5. **Myriophyllum verticillatum** Linnaeus, Sp. Pl. 2: 992. 1753 • Whorled water-milfoil, myriophylle verticillé F W

Myriophyllum verticillatum var. *cheneyi* Fassett; *M. verticillatum* var. *intermedium* W. D. J. Koch; *M. verticillatum* subsp. *pectinatum* (Wallroth) Piper & Beattie

Herbs monoecious, aquatic, sometimes forming dense stands. **Stems** branched or unbranched, to 3 m. **Turions** present, becoming brown to red-brown at maturity, clavate to obdeltoid, with abrupt transition from foliage leaves to reduced turion leaves, (6–)11–37(–52) × (3–)4–6 (–9) mm, apex ± rounded, lateral turions with several whorls of minute, brown prophylls, entire proximally and toothed to lobed distally, ovate to elliptic or lanceolate in outline; leaves pectinate, strongly appressed to axis throughout, narrowly flabelliform in outline, 4.5–7.5 × 1.2–1.8(–4) mm; segments 8–12(–18), flattened, linear-lanceolate, longest segment 1.5–6 mm, basal segment usually greater than or equal to ⅔ central axis of leaf, apex ± acute, trichomes usually absent. **Leaves** in whorls of (3 or)4, heteromorphic; petiole to 6 mm; submersed leaves pectinate, ovate to elliptic in outline, (7–)12–30(–46) × 9–24(–40) mm, segments (9–)12–22(–34), linear-filiform, longest segment (2–)6–19(–29) mm; emersed leaves pectinate, lanceolate to elliptic to ovate in outline, 2–5(–15) × 0.9–2.6 mm, segments (9–)12–20, greater than 0.5 mm. **Inflorescences** to 25 cm; flowers proximally pistillate, medially bisexual, distally staminate, in whorls of 4; bracteoles cream, ovate, 0.3–0.6(–1) × 0.1–0.6 (–1.3) mm, margins deeply dissected into irregular lobes. **Staminate flowers:** sepals cream, narrowly triangular to deltate, 0.5–0.7(–0.9) × 0.4–0.6(–0.8) mm; petals persistent, cream, sometimes apically suffused with purple, elliptic to obovate, 1.8–2.2(–2.4) × 0.7–1.5 mm; stamens 8, filaments to 2 mm, anthers 0.8–1.7 × 0.3–0.6 mm. **Pistillate flowers:** sepals greenish to cream, elliptic to triangular, 0.2–0.7 × 0.2–0.7 mm; petals often caducous, sometimes persistent, cream to purple, elliptic to obovate, 0.4–0.7(–0.9) × 0.3–0.5 (–0.8) mm; pistils (1.3–)1.8–2.7 mm, stigmas red to ± purple, to 0.8 mm. **Fruits** globose, shallowly 4-lobed. **Mericarps** olive-green to brown, subglobose to globose, 2–2.7(–3) × (0.9–)1.1–1.3(–1.7) mm, transversely widely obovate, abaxial surface broadly rounded to moderately flattened, smooth, often with 2 shallow longitudinal ridges, wings and ribs absent. $2n = 28$.

Flowering and fruiting Jul–Oct. Streams, rivers, ponds, lakes, sloughs, tannic waters; 0–2700 m; St. Pierre and Miquelon; Alta., B.C., Man., N.B., Nfld. and Labr., N.W.T., N.S., Nunavut, Ont., P.E.I., Que., Sask., Yukon; Alaska, Ariz., Calif., Colo., Conn., Del., Idaho, Ill., Ind., Iowa, Maine, Md., Mass., Mich., Minn., Mont., Nebr., Nev., N.H., N.J., N.Y., N.Dak., Ohio, Oreg., Pa., S.Dak., Tex., Utah, Vt., Wash., Wis., Wyo.; Eurasia; nw Africa (Algeria, Morocco).

Four varieties of *Myriophyllum verticillatum* have been proposed. Fassett based *M. verticillatum* var. *cheneyi* solely on the presence of four stamens. M. L. Fernald (1950) considered var. *cheneyi* conspecific with *M. hippuroides*. Some specimens labeled as var. *cheneyi* examined during this treatment were confirmed to represent *M. hippuroides* as noted by S. G. Aiken (1981). Fernald recognized vars. *intermedium*, *pectinatum* Wallroth, and *pinnatifidum* Wallroth based on differences in the length of floral bracts. All of these varieties can be found in the flora area though floral bract length is a very plastic character and there is no distinct separation among the forms. The presence and morphology of turions of *M. verticillatum* can be very helpful in the identification of vegetative material. The clavate to obdeltoid shape and reddish brown color of the turions in this species differ from the cylindrical, typically dark green turions of both *M. farwellii* and *M. sibiricum*.

Previous floristic studies have reported that the submersed leaves of *Myriophyllum verticillatum* have 18–34 segments. An examination of specimens unambiguously assignable to this species based on floral and fruit characters found the lower value of the range to be nine. The broad range for segment number and other leaf characters for *M. verticillatum* and overlap of these values with those for other *Myriophyllum* species underscores the importance of relying on floral and fruit characters for identifications.

Flowering plants of *Myriophyllum verticillatum* are distinguished from other water-milfoils by reduced, pectinate floral bracts. Many *Myriophyllum* species can be mistaken for *M. verticillatum* at early stages of flowering because the proximal floral bracts are transitional in form from submersed leaves and are typically pectinate. The bracteoles also are highly distinctive and are deeply divided into as many as seven narrow lobes. The schizocarps are unique among water-milfoils in the flora area because they have a relatively thick exocarp that typically impedes the splitting and separation of the schizocarp into mericarps at maturity. The most robust and diminutive plants of *M. verticillatum* can be found in Saskatchewan. No specimens identifiable as *M. verticillatum* have been seen from New Mexico though it has been reported from there (P. T. Adams 1998). It is likely to occur in New Mexico given that it is found in neighboring states except Oklahoma. The presence and morphology of turions of *M. verticillatum* can be very helpful in the identification of vegetative material. The clavate to obdeltoid shape and reddish brown color of the turions in this species distinguishes it from the cylindrical, typically dark green turions of *M. farwellii* and *M. sibiricum*.

SELECTED REFERENCES Weber, J. A. and L. D. Noodén. 1974. Turion formation and germination in *Myriophyllum verticillatum*: Phenology and its interpretation. Michigan Bot. 13: 151–158. Weber, J. A. and L. D. Noodén. 1976. Environmental and hormonal control of turion germination in *Myriophyllum verticillatum*. Amer. J. Bot. 63: 936–944.

6. Myriophyllum alterniflorum de Candolle in J. Lamarck and A. P. de Candolle, Fl. Franç., ed. 3, 6: 529. 1815 • Alternate-flowered water-milfoil, myriophylle à fleurs alternes [F]

Myriophyllum alterniflorum var. *americanum* Pugsley

Herbs monoecious, aquatic, often forming dense stands. **Stems** often branched, to 2.5 m. **Turions** absent. **Leaves** usually in whorls of 3 or 4(or 5), rarely opposite, distal emersed leaves alternate, heteromorphic; petiole 0–0.5 mm; submersed leaves pectinate, ovate to elliptic or obovate in outline, 3–16(–40) × (3–)4–10(–14) mm, segments 6–16(–20), linear-filiform, longest segment 2–11(–15) mm; emersed leaves elliptic to oblanceolate or spatulate proximally, pectinate to pinnatifid distally, 0.7–2 × 0.3–0.5 mm, segments 0–10(–12), margins ± serrate to shallowly lobed. **Inflorescences** to 12 cm; flowers proximally pistillate, in whorls of 4, often medially bisexual, distally staminate from alternate bracts; bracteoles cream to stramineous, with distinct brown to purple margins, ovate to depressed-ovate, 0.2–0.5 × 0.2–0.5 mm, margins serrate to irregularly fringed, apex often aristate. **Staminate flowers:** sepals cream to stramineous, narrowly triangular, 0.1–0.2 × 0.1 mm; petals caducous, cream, elliptic to obovate, (0.7–)1.3–1.8 × (0.4–)0.6–1 mm; stamens 8, filaments to 0.6 mm, anthers cream to purple, 1.1–1.5 × 0.2–0.4 mm. **Pistillate flowers:** sepals cream, lanceolate to triangular or deltate, 0.1–0.2 × 0.1 mm; petals usually caducous, rarely persistent, cream, widely ovate, 0.2–0.4 × 0.2–0.3 mm; pistils 0.7–1.1 mm, stigmas red to ± purple, pulvinate, to 0.3 mm. **Fruits** ovoid, 4-lobed. **Mericarps** olive-green to red-brown, cylindric to narrowly ovoid, 1.3–1.6 × 0.3–0.4 mm, transversely widely obovate, abaxial surface broadly rounded, irregularly to densely papillate, sometimes with orange to red, punctate trichomes, wings and ridges absent. **2*n*** = 14.

Flowering and fruiting Apr–Sep. Oligotrophic to mesotrophic waters, lakes, rivers, sandy substrates; 0–500 m; Greenland; St. Pierre and Miquelon; Man., N.B., Nfld. and Labr. (Nfld.), N.W.T., N.S., Ont., Que., Sask.; Alaska, Conn., Maine, Mass., Mich., Minn., N.H., N.Y., R.I., Vt., Wis.; Eurasia; n Africa.

Myriophyllum alterniflorum is distinguished in flower and fruit by minute, alternate, entire distal floral bracts, a characteristic shared only with *M. laxum*. Unlike the flowers of other *Myriophyllum* species in the flora, those of *M. alterniflorum* are widely scattered along the inflorescence and the tip of the spike is often nutant. The pistils are relatively small, each having a prominent, purple, pulvinate stigma that is as broad as or broader than the ovary and persistent in fruit. The pistil and mericarp surface has scattered orange to red punctate trichomes not seen on the pistils or mericarps of any other water-milfoil in the flora area. The cream, depressed-ovate, fringed, often aristate bracteoles with brown to red margins are also distinctive. Seed-set appears to be rare in *M. alterniflorum*. Specimens examined from Greenland far exceed other North American plants in the size of their leaves and staminate flowers. North American plants with relatively short leaves and a bushy habit have been referred to as var. *americanum*. Plants of this form can be confused with the relatively small *M. sibiricum* and *M. verticillatum* in northern latitudes but are distinguishable based on floral characters. Leaf size in *M. alterniflorum* is plastic and continuous in variation, thus varieties are not recognized here as separate taxa.

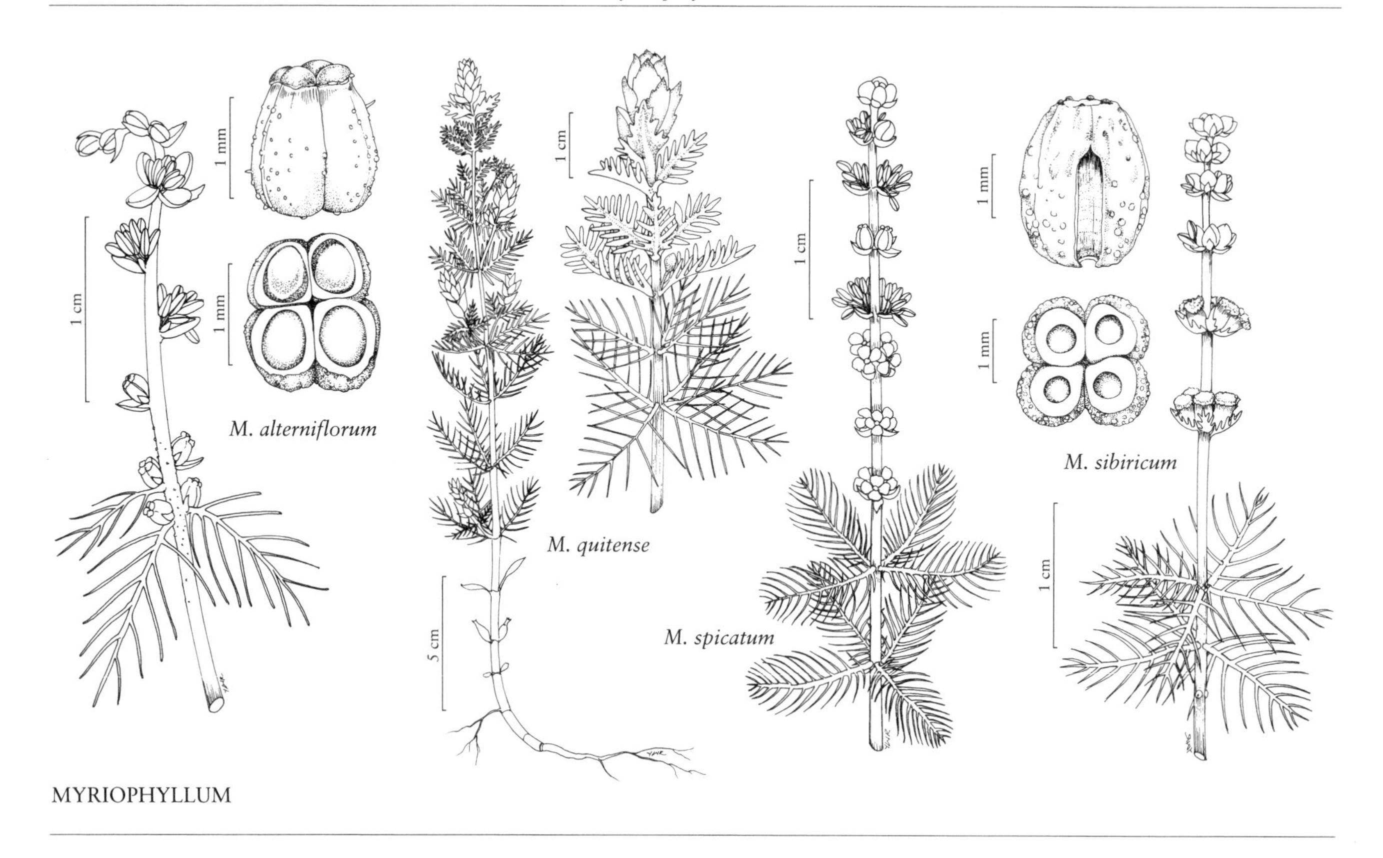

MYRIOPHYLLUM

7. **Myriophyllum quitense** Kunth in A. von Humboldt et al., Nov. Gen. Sp. Pl. 6(fol.): 71; 6(qto.): 89. 1823 • Andean or waterwort water-milfoil [F]

Myriophyllum elatinoides Gaudichaud-Beaupré

Herbs monoecious, aquatic or semiaquatic, often forming dense stands. **Stems** often branched, to 3 m. **Turions** absent. **Leaves** mostly in whorls of (3 or)4 (or 5), sometimes opposite to subopposite, heteromorphic; petiole 0–4 mm; submersed leaves pectinate to lobed (basalmost leaves opposite to subopposite, reduced, margins entire), ovate to obovate in outline, (3–)5–25(–35) × (2–)3–18(–20) mm, segments (2 or)3–9 (–11), linear, ± applanate, longest segment (7–)8–15 (–17) mm; emersed leaves pinnatisect to lobed or entire, ovate to oblong in outline, 2–9 × 1–6 mm, margins dentate to minutely serrate. **Inflorescences** to 8 cm; flowers proximally pistillate, medially bisexual, distally staminate; bracteoles cream, deltate, 0.5–1 × 0.2–0.6 mm, margins dentate to serrate, with glandular tip. **Staminate flowers:** sepals green to cream, ovate to deltate, (0.2–)0.3–0.5(–0.7) × (0.1–)0.2–0.4 (–0.5) mm; petals persistent, ± purple, oblong, 2–3 × 0.5–1.5 mm; stamens 8, filaments to 0.6 mm, anthers 1.8–2.5 × 0.2–0.6 mm. **Pistillate flowers:** sepals cream, deltate, 0.2–0.5 × 0.1–0.4 mm; petals ± persistent, cream, ± cucullate, elliptic, 0.1–0.5 × 0.2–0.3 mm; pistils 1.1–2.2 mm, stigmas cream to ± purple, to 0.6 mm. **Fruits** cylindric to oblong, 4-lobed. **Mericarps** tan to olive-brown, cylindric to ovoid, 1.5–1.8 × 0.6–0.8 mm, transversely elliptic, abaxial surface rounded, smooth, sometimes with a shallow, longitudinal ridge, wings and ribs absent. $2n = 42$.

Flowering Jun–Aug. Cold oligotrophic waters, lakes, rivers, streams; 0–2800 m; B.C., N.B., P.E.I.; Ariz., Calif., Idaho, Mont., Oreg., Utah, Wash., Wyo.; s Mexico; South America.

The most distinguishing feature of *Myriophyllum quitense* is the production of relatively large, ovate distal floral bracts with serrate margins. This characteristic is shared only with *M. heterophyllum*; however, *M. quitense* has eight stamens and the latter has four.

The proximal submersed leaves of most *Myriophyllum* species are uniformly pectinate, but those of *M. quitense* can range from entire or lobed to pectinate. This species often exhibits a pronounced transition from three or four proximal nodes of large, opposite, spatulate or lobed prophylls, to nodes of besomiform whorled leaves having obtriangular laminar surfaces and distal pinnatifid segments resembling those of pectinate leaves. In addition, the unusual grayish blue color of the foliage and whitish rhizomes are useful characteristics for distinguishing submersed vegetative specimens of *M. quitense* from similar species, such as *M. sibiricum*.

Myriophyllum quitense has a highly disjunct distribution in North America and South America (A. E. Orchard 1981; O. Ceska et al. 1986). S. G. Aiken (1981) and R. Couch and E. Nelson (1988) suggested that *M. quitense* was introduced into North America by migratory waterfowl. Both Ceska et al. and M. L. Moody and D. H. Les (2010) regarded it as native to North America. It has been reported from New Brunswick and Prince Edward Island (D. F. McAlpine et al. 2007) and further range extensions would seem likely given the level of disjunction in distribution.

8. **Myriophyllum spicatum** Linnaeus, Sp. Pl. 2: 992. 1753 • Eurasian water-milfoil, myriophylle en épi F I W

Herbs monoecious, aquatic, often forming dense stands. **Stems** often much-branched distally, to 6 m. **Turions** absent. **Leaves** in whorls of (3 or)4 (or 5), heteromorphic; petiole 0–0.4 mm; submersed leaves pectinate, obovate in outline, (14–)18–32(–36) × 10–20(–30) mm, segments (20–)24–36(–42), linear-filiform, longest segment 2–20(–26) mm, usually parallel and all in 1 plane, forming angles less than 45° with central axis; emersed leaves pectinate to pinnatifid proximally, with abrupt transition to obovate or elliptic, sometimes distally spatulate, in outline, margins of distal leaves entire to serrate to shallowly lobed, 1–2.3 × 0.6–1(–1.5) mm. **Inflorescences** to 15 cm; flowers proximally pistillate, medially bisexual, distally staminate; bracteoles cream to stramineous to purple, with distinct reddish or brown margins, usually ovate to depressed-ovate or obovate, sometimes elliptic to triangular or rhombic, 0.5–0.9 × 0.4–0.7 mm, margins entire or serrate, sometimes with distal, irregular, membranous fringe. **Staminate flowers:** sepals cream to stramineous, triangular, 0.3–0.4 × 0.2–0.3 mm; petals caducous, cream to red or dark purple, oblong to elliptic or obovate, 1.5–2.5(–3) × 0.8–1 mm; stamens 8, filaments to 1.2 mm, anthers greenish cream to yellow or purple, 1–2.2 × 0.4–0.8 mm. **Pistillate flowers:** sepals cream to green to purple, lanceolate to deltate or ovate, 0.1–0.3 × 0.1–0.2 mm; petals often persistent, cream, widely ovate, 0.6–0.8 × 0.3–0.4 mm; pistils 0.9–1.2 mm, stigmas white to red to ± purple, 0.2–0.7 mm. **Fruits** globose, 4-lobed. **Mericarps** olive-green to brown, cylindric to narrowly ovoid, 1.5–2.2 × 0.8–1.3 mm, transversely widely obovate, abaxial surface broadly rounded, sparsely and irregularly tuberculate, margins smooth to tuberculate, sometimes with 2 shallow, longitudinal ridges, wings and ribs absent. **2*n*** = 42.

Flowering and fruiting Apr–Oct. Oligotrophic to eutrophic waters, lakes, ponds, canals, streams; 0–1500 m; introduced; B.C., N.B., Ont., Que.; Ala., Ariz., Ark., Calif., Colo., Conn., Del., D.C., Fla., Ga., Idaho, Ill., Ind., Iowa, Kans., Ky., La., Md., Mass., Mich., Minn., Miss., Mo., Mont., Nebr., Nev., N.H., N.J., N.Y., N.C., N.Dak., Ohio, Okla., Oreg., Pa., R.I., S.C., Tenn., Tex., Utah, Vt., Va., Wash., W.Va., Wis.; Eurasia; n Africa.

Myriophyllum spicatum is considered one of the worst nuisance aquatic weeds in North America. Identification of this species is critical for management of lakes. Until the early 1900s, the widely accepted view was that *M. spicatum* was native to North America and was conspecific with European *M. spicatum*. M. L. Fernald (1919c) described *M. exalbescens* to distinguish all North American specimens from European *M. spicatum*, with the former name subsequently being changed to *M. sibiricum* due to nomenclatural precedence (S. G. Aiken and A. Cronquist 1988). The first to recognize the presence of both species in North America was apparently C. F. Reed (1970b). E. Hultén (1941–1950), B. C. Patten (1954), and S. A. Nichols (1975) proposed alternatively that *M. spicatum* and the native *M. sibiricum* form a continuum of variation, suggesting the two taxa may simply represent varieties or subspecies of a highly variable cosmopolitan species. Based on a study of herbarium collections, R. Couch and E. Nelson (1985, 1992) believed that *M. spicatum* was introduced from Europe in the 1940s and subsequently spread throughout the United States and Canada. A recent biogeographic study of cpDNA haplotypes indicates this species was introduced to North America from China and Korea (M. L. Moody et al. 2016).

Based upon examination of specimens for this treatment, and as pointed out by A. E. Orchard (1981), most of the characters initially proposed by M. L. Fernald (1919c) and expanded upon by S. G. Aiken (1981) that are thought to be reliable for distinguishing the two species, such as size and shape of floral bracts and bracteoles, anther length, swollen base of inflorescence, color of the stem in dried material, extent of branching, and differences in mericarps, break down when a wide range of North American herbarium material is examined. One of the few useful vegetative characters to distinguish these species in the northern regions of North America is that *M. sibiricum* often produces turions in the latter part of the growing season, whereas *M. spicatum* does not (E. Hultén 1947). The most commonly used vegetative character to distinguish the two species is the number of leaf segments in submersed leaves (Fernald). When attempting to distinguish plants of the latter two species, this is a reliable character, but only when specimens have low (6–18) or high (24–42) segment numbers; plants often have submersed leaves with intermediate segment numbers.

Molecular studies have shown that the overlap seen in morphological characters, such as leaf segment number, between *Myriophyllum sibiricum* and *M. spicatum* may be the result of frequent and widespread hybridization (M. L. Moody and D. H. Les 2002, 2007b; A. P. Sturtevant et al. 2009). Hybrids between these two species can have leaf segment numbers from 16–28 (Moody and Les 2007b), which overlaps with leaf segment numbers for both *M. sibiricum* (6–18) and *M. spicatum* (24–36). A reliable method to distinguish these taxa when there is overlap in this character state is DNA fingerprinting (Moody and Les 2002).

SELECTED REFERENCES Aiken, S. G. and R. R. Picard. 1980. The influence of substrate on the growth and morphology of *Myriophyllum exalbescens* and *Myriophyllum spicatum*. Canad. J. Bot. 58: 1111–1118. Aiken, S. G., P. R. Newroth, and I. Wile. 1979. Biology of Canadian weeds. 34. *Myriophyllum spicatum* Linnaeus. Canad. J. P1. Sci. 59: 201–215. Grace, J. B. and R. J. Wetzel. 1978. The production biology of Eurasian watermilfoil (*Myriophyllum spicatum* Linnaeus). A review. J. Aquatic Pl. Managem. 16: 1–11. Löve, A. 1961. Some notes on *Myriophyllum spicatum*. Rhodora 63: 139–145.

9. Myriophyllum sibiricum Komarov, Repert. Spec. Nov. Regni Veg. 13: 168. 1914 • Northern or short-spike water-milfoil, myriophylle de Sibérie [F]

Myriophyllum exalbescens Fernald; *M. magdalenense* Fernald; *M. spicatum* Linnaeus var. *capillaceum* Lange; *M. spicatum* subsp. *exalbescens* (Fernald) Hultén; *M. spicatum* var. *exalbescens* (Fernald) Jepson; *M. spicatum* var. *muricatum* Maximovicz

Herbs monoecious, aquatic, often forming dense stands. **Stems** usually unbranched, to 6 m. **Turions** present, ± dark green, cylindrical, with gradual transition from foliage leaves to reduced turion leaves, 12–40(–45) × (3–)5–12(–15) mm, apex ± rounded; leaves pectinate, stiff, strongly appressed to axis distally, not proximally, elliptic in outline, 5–15 × 1.4–5 mm, with clusters of brown, conical trichomes between leaf bases; segments 13–15(–17), elongate botuliform, longest segment 1.8–5.2(–6) mm, basal segment usually less than or equal to ½ central axis of leaf, apex apiculate, with single, brown, conical trichome in each axil. **Leaves** in whorls of (3 or)4, heteromorphic; petiole 0–4 mm; submersed leaves pectinate, usually obovate in outline, (2.8–)13–32(–44) × (2.1–)16–35 mm, segments 6–18 (–24), linear-filiform, often perpendicular to central axis, basal segments often as long as leaf axis, segments often irregular in orientation, not parallel and not in same plane, longest segment 2–20(–26) mm; emersed leaves, basal sometimes pectinate to pinnatifid proximally, with abrupt transition to obovate, elliptic, sometimes distally spatulate, in outline, margins of distal leaves entire to serrate to shallowly lobed, 1–2.3 × 0.6–1 (–1.5) mm. **Inflorescences** to 15 cm; flowers proximally pistillate, medially bisexual, distally staminate; bracteoles cream to stramineous or purple with distinct, reddish or brown margin, usually ovate to depressed-ovate, sometimes elliptic to triangular, (0.4–)0.6–1.3 × 0.3–0.7 mm, margins entire or serrate, sometimes with distal, irregular, membranous fringe. **Staminate flowers:** sepals cream to stramineous, usually depressed-ovate, sometimes ovate to triangular, 0.2–0.4 × 0.2–0.5 mm; petals caducous, cream to red or dark purple, oblong to elliptic or obovate, 1.7–2.3(–3) × 1–2 mm; stamens 8, filaments to 1.5 mm, anthers greenish cream to yellow or purple, 1–2.2 × 0.3–0.7 mm. **Pistillate flowers:** sepals cream to green to purple, lanceolate to deltate or ovate, 0.1–0.3 × 0.1–0.2 mm; petals often persistent, cream, widely ovate, 0.3–0.5 × 0.2–0.5 mm; pistils 1–2 mm, stigmas white to red or ± purple, ± pulvinate, 0.2–0.4 mm. **Fruits** globose, 4-lobed. **Mericarps** olive-green to brown, cylindric to narrowly ovoid, 1.5–2.7 × 1.2–1.6 mm, transversely widely obovate, abaxial surface broadly rounded, sparsely and irregularly tuberculate, margins smooth or tuberculate, sometimes with 2 shallow, partial, longitudinal ridges, wings and ribs absent. 2*n* = 42.

Flowering and fruiting May–Oct. Oligotrophic to eutrophic waters, lakes; 0–3300 m; Greenland; Alta., B.C., Man., N.B., Nfld. and Labr., N.W.T., N.S., Nunavut, Ont., P.E.I., Que., Sask., Yukon; Alaska, Ariz., Calif., Colo., Conn., Del., Idaho, Ill., Ind., Iowa, Kans., Maine, Md., Mass., Mich., Minn., Mo., Mont., Nebr., Nev., N.H., N.J., N.Mex., N.Y., N.Dak., Ohio, Oreg., Pa., R.I., S.Dak., Tex., Utah, Vt., Wash., Wis., Wyo.; Eurasia.

Myriophyllum exalbescens (*M. sibiricum*) was considered to be a North American endemic until the discovery of European specimens (S. G. Aiken and J. McNeill 1980). Since the taxonomic name of Russian material pre-dated that for North American specimens, all material of *M. exalbescens* was synonymized under the name *M. sibiricum* (A. Ceska and O. Ceska 1986; Aiken and A. Cronquist 1988). *Myriophyllum sibiricum* is widely recognized as circumpolar with an affinity for colder climates and is rarely found south of the 0°C January isotherm (Aiken 1981). *Myriophyllum sibiricum* is distinctive when growing with low leaf segment numbers. Hybridization with *M. spicatum* and subsequent introgression has apparently blurred the boundaries between these two taxa to the point that some specimens are not assignable to either species without molecular analysis (see 8. *M. spicatum*). There is concern that *M. sibiricum* is being rapidly outcompeted in lakes by either *M. spicatum* or its hybrid (M. L. Moody and D. H. Les 2007b; A. P. Sturtevant et al. 2009; E. A. LaRue et al. 2013). The dark green cylindric turions in *M. sibiricum*, which have reduced

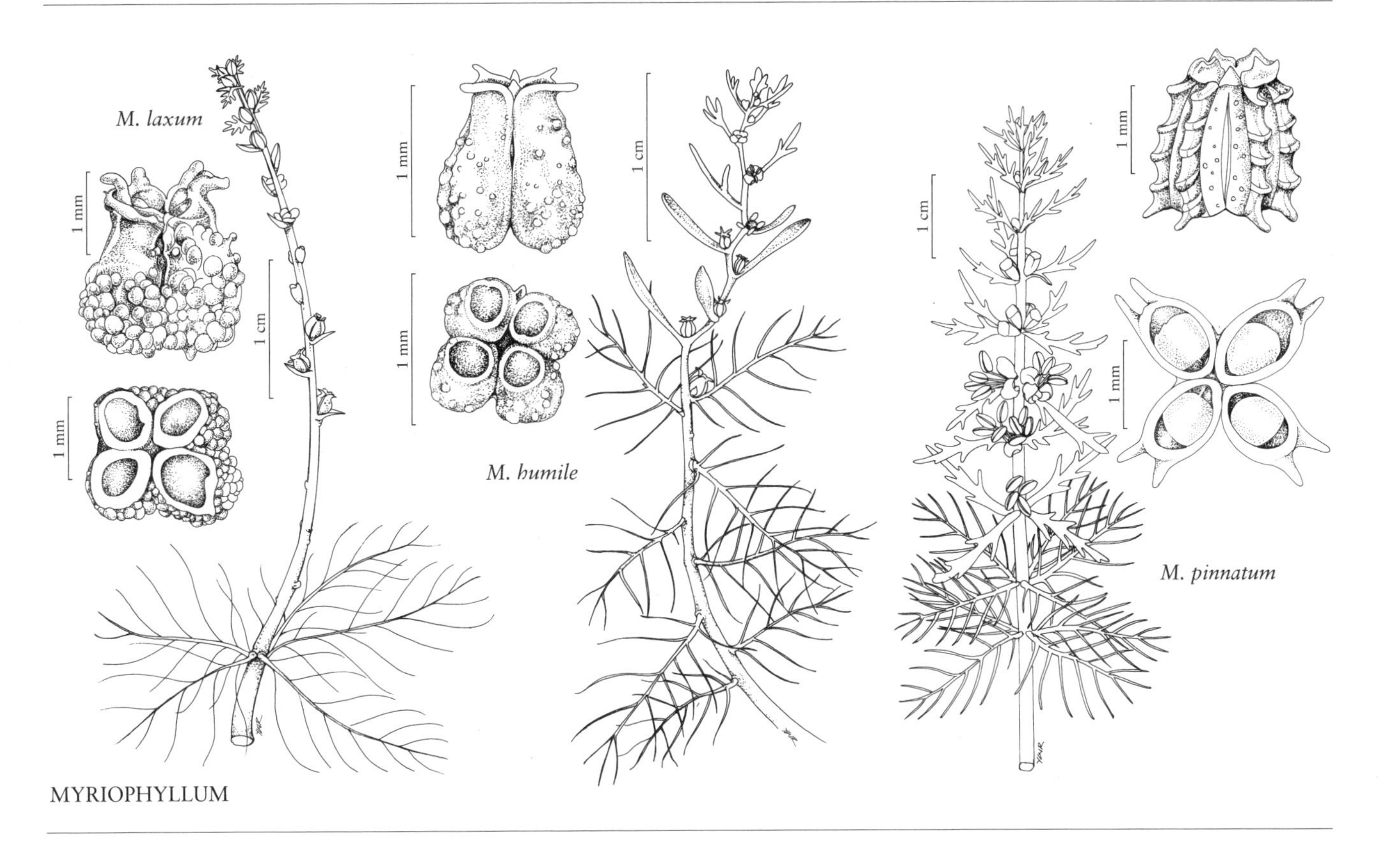

MYRIOPHYLLUM

and thickened storage leaves, are useful for identification. These reduced leaves are often blackened and visible at the base of new shoots in the next growing season, which can aid in the identification of vegetative material.

10. Myriophyllum laxum Shuttleworth ex Chapman, Fl. South. U.S. ed. 2, 143. 1883 • Piedmont water-milfoil [E] [F]

Herbs monoecious, aquatic, sometimes forming dense stands. **Stems** sometimes branched, to 1 m. **Turions** absent. **Leaves** usually alternate and/or in whorls of 3(or 4), sometimes opposite or irregular, heteromorphic; petiole 0–3 mm; submersed leaves pectinate, elliptic to obovate in outline, 9–27(–31) × 6–18(–22) mm, segments (6–)8–12(–16), linear-filiform, longest segment (4–)8–17(–21) mm, surfaces with numerous black, ascidiate trichomes; emersed leaves pectinate to pinnatifid proximally, elliptic to obovate, spatulate, or oblanceolate, (0.6–)0.7–2.3(–2.7) × (0.1–)0.2–0.7(–1) mm, with black, ascidiate trichomes scattered on surfaces and in axils. **Inflorescences** to 27 cm; flowers proximally pistillate, medially bisexual, distally staminate; bracteoles cream, lanceolate to triangular, 0.3–0.8 × 0.1–0.2(–0.3) mm, margins irregularly lobed, apex often narrowly apiculate. **Staminate flowers:** sepals cream, lanceolate to narrowly triangular, 0.1–0.3 × 0.1–0.2 mm; petals persistent, cream, suffused with ± red at tips, obovate to oblanceolate, 1.5–1.9 × 0.6–1.1 mm; stamens 4, filaments to 1.3 mm, anthers 1.2–1.5(–1.7) × 0.2–0.4(–0.6) mm. **Pistillate flowers:** sepals cream, lanceolate to ovate, 0.1–0.3 × 0.1–0.2 mm; petals caducous, cream, elliptic to obovate, 0.4–0.5 × 0.1–0.3 mm; pistils (0.5–)0.7–1.1(–1.3) mm, stigmas red to ± purple, to 0.5 mm. **Fruits** narrowly globose, 4-lobed. **Mericarps** brown to purple, cylindric to narrowly ovoid, 1–1.4 × (0.4–)0.6–0.8(–1.1) mm, transversely elliptic, abaxial surface rounded to shallowly 2-angled, densely tuberculate proximal to midpoint, tubercles crowded, relatively large, rarely with 2 shallow, partial longitudinal wings, tubercles often obscuring wings, ribs absent.

Flowering and fruiting Jun–Oct. Oligotrophic waters, lakes, ponds, streams; 0–150 m; Ala., Fla., Ga., La., Miss., N.C., S.C.

Myriophyllum laxum is a coastal plain species that has a very restricted range in the southeastern United States. It is most similar vegetatively to *M. humile*, with which it shares a delicate habit. The two have historically been reported to overlap in range in Virginia; no specimens of the former have been seen from that state. In *M. laxum* the submersed leaves are usually whorled but often irregular and sometimes alternate; in *M. humile* the leaves are usually opposite or alternate and almost never whorled. The proximal half of mericarps of

M. laxum is densely covered with large, mounded tubercles; mericarps of *M. humile* tend to be uniformly, sparsely to densely covered with smaller tubercles and have distinct tuberculate ridges typically running the entire length of the mericarp. The floral bracts are also very different in the two species and the flowers are much smaller in *M. laxum*.

Myriophyllum laxum has sometimes been confused with *M. heterophyllum*; however, they can be distinguished by the number of submersed leaf segments with *M. laxum* typically having 8–12 versus 12–20 for *M. heterophyllum*. They can also be easily distinguished by differences in floral and fruit characters. *Myriophyllum laxum* also interbreeds with *M. heterophyllum* (M. L. Moody and D. H. Les 2010; R. A. Thum et al. 2011), producing a hybrid that is often fertile, having floral and fruit characters very similar to those observed in *M. laxum*, but having highly variable leaf morphology typical of both parental species.

11. Myriophyllum humile (Rafinesque) Morong, Bull. Torrey Bot. Club 18: 242. 1891 • Western water-milfoil, myriophylle menu E F

Burshia humilis Rafinesque, Med. Repos., hexade 2, 5: 357. 1808; *Myriophyllum ambiguum* Nuttall var. *limosum* Nuttall; *M. procumbens* Bigelow

Herbs monoecious, aquatic or semiaquatic, usually not forming dense stands. **Stems** often branched, to 1 m. **Turions** absent. **Leaves** usually alternate or opposite, rarely in whorls of 3(or 4), heteromorphic; petiole to 4 mm; submersed leaves pectinate, ovate to elliptic in outline, (5.5–)10–27(–30) × (4.4–)6–22(–33) mm, segments (2–)4–13(–14), linear-filiform, longest segment (3–)8–17(–22.5) mm; emersed leaves usually pectinate to pinnatifid proximally, linear to spatulate or lobed distally, 5–9(–12.5) × 0.3–3(–6) mm, segments (0–)4–6(–9). **Inflorescences** to 35 cm, sometimes submersed with flowers in axils of unmodified, pectinate leaves; flowers proximally pistillate, distally staminate; bracteoles cream, oblong to elliptic to ovate or triangular, 0.3–0.7 × 0.1–0.4 mm, margins entire or irregularly lobed, apex often aristate. **Staminate flowers:** sepals cream to stramineous, triangular, 0.1–0.2 × 0.1–0.2 mm; petals caducous, purple, elliptic to obovate, 0.6–1.5 × 0.3–0.7 mm; stamens 4, filaments to 0.9 mm, anthers 0.3–0.8 × 0.1–0.3 mm. **Pistillate flowers:** sepals cream to stramineous, triangular, 0.1–0.2 × 0.1 mm; petals caducous, purple, elliptic to obovate, 0.3–0.5 × 0.2–0.3 mm; pistils 0.7–0.9 mm, stigmas red to ± purple, to 0.2 mm. **Fruits** cylindric, deeply 4-lobed. **Mericarps** tan to red-brown or purple, cylindric to narrowly ovoid, (0.6–)0.8–1.2 × 0.4–0.6 mm, transversely elliptic to ovate, abaxial surface rounded, sparsely to densely tuberculate, tubercles relatively small, shallow, wings and ridges absent.

Flowering and fruiting Jun–Oct. Oligotrophic waters, lakes, ponds, streams; 0–700 m; N.B., N.S., Que.; Conn., Del., Maine, Md., Mass., N.H., N.J., N.Y., Pa., R.I., Vt., Va.

Myriophyllum humile has a diminutive semiterrestrial growth form, referred to as var. *limosum* by T. Morong (1891), that has alternate and, typically, spatulate floral bracts that can be confused with the emergent form of *M. pinnatum*. In vegetative form, the leaves of *M. pinnatum* tend to be pectinate with a greater number of longer segments than those of *M. humile*, which are often linear, spatulate, or 4–6-lobed. When fruits are present, the two can be distinguished by the presence of winged ridges on the mericarps in *M. pinnatum*, which are absent in *M. humile*. Submersed forms of *M. humile* can be confused with *M. farwellii* because of the delicate nature of their leaves. *M. humile* has leaves mostly alternate and opposite; leaves in *M. farwellii* tend to be whorled or, sometimes, subverticillate, giving these plants a bushy appearance. *Myriophyllum humile* can be also confused with *M. laxum* (see 10. *M. laxum* discussion).

All specimens examined from Minnesota labeled *Myriophyllum humile* have been misidentified and are other species of the genus. Only one sterile herbarium specimen labeled as *M. humile* from Wisconsin has been seen, but its identity could not be confirmed.

12. Myriophyllum pinnatum (Walter) Britton, Sterns & Poggenburg, Prelim. Cat., 19. 1888 • Cut-leaf water-milfoil E F W

Potamogeton pinnatum Walter, Fl. Carol., 90. 1788; *Myriophyllum scabratum* Michaux

Herbs monoecious, aquatic or semiaquatic, sometimes forming dense stands. **Stems** often branched, to 1 m. **Turions** absent. **Leaves** usually in whorls of (3 or)4, often subverticillate or irregular, sometimes alternate, heteromorphic; petiole 0–5 mm; submersed leaves pectinate, ovate to obovate in outline, (9.5–)12–28(–33) × (7–)10–20(–25) mm, segments 6–12(–13), linear-filiform, longest segment (5–)8–15(–20) mm; emersed leaves linear-lanceolate, margins lobed to serrate, pectinate to pinnatifid, (4–)6–17(–19) × 1–6.5 mm, segments (0–)3–10. **Inflorescences** to 20 cm; flowers proximally pistillate, medially bisexual, distally staminate; bracteoles cream, ovate to triangular or deltate, 0.5–0.8 × 0.2–0.5 mm, margins serrate to irregularly lobed. **Staminate flowers:** sepals cream to stramineous, elliptic to linear-lanceolate or narrowly triangular, 0.1–0.3(–0.4) × 0.1–0.2 mm; petals persistent, cream to purple, elliptic to obovate, 0.7–1.7(–1.9) × (0.4–)0.5–0.8(–1) mm; stamens 4, filaments to 1.4 mm, anthers 0.7–1.5 × 0.2–0.5(–0.7) mm. **Pistillate flowers:** sepals cream, triangular, 0.1–0.2 × 0.1–0.2 mm; petals usually caducous, rarely persistent, cream, elliptic to obovate, 1–2 × 0.5–0.7 mm; pistils 0.8–1.2 mm, stigmas red to ± purple, to 0.5 mm. **Fruits** ovoid, cruciate. **Mericarps** tan to brown, cylindric to ovoid, (1–)1.2–1.8 × 0.6–0.8(–1) mm, transversely elliptic, flattened, abaxial surface sharply 2-angled, flattened to concave, papillate, sometimes minutely tuberculate, with 2 distinct longitudinal ridges, ridges with prominent, membranous, undulating wings, wings erect to reflexed, with 6–12 perpendicular ribs.

Flowering and fruiting Mar–Oct. Oligotrophic to mesotrophic waters, lakes, ponds, sloughs, mudflats; 0–700 m; B.C., N.B., Sask.; Ark., Conn., Del., Fla., Ga., Ill., Ind., Iowa, Kans., La., Md., Mass., Minn., Miss., Mo., Nebr., N.J., N.Y., N.C., Okla., R.I., S.C., S.Dak., Tenn., Tex., Va., W.Va.

Myriophyllum pinnatum is often confused with *M. heterophyllum* and/or *M. hippuroides*. The most distinctive characters to separate these species are their fruits and mericarps. In *M. pinnatum*, mericarps are cylindric with sharply angled faces and a prominent ribbed wing; both *M. heterophyllum* and *M. hippuroides* have globose to subglobose mericarps with inconspicuous ridges and prominent tubercles. *Myriophyllum pinnatum* has leaves with significantly fewer segments, and plants of *M. hippuroides* tend to be very delicate in appearance. Both *M. hippuroides* and *M. pinnatum* produce elongate, strap-shaped to linear-lanceolate to spatulate leaves distally, either when they grow submersed and produce emergent flowering racemes or when they grow as emergent plants stranded along shorelines. The former species typically produces pinnatifid emersed leaves with fewer segments. Although *M. heterophyllum* also produces emergent leaves in response to flowering, the leaves typically grade from pectinate to lobed to entire to ovate serrate leaves distally. *Myriophyllum heterophyllum* sometimes also produces a low growing semi-terrestrial form on mudflats having thickened and abbreviated pectinate leaves. The pectinate leaves produced under these conditions also have more segments than those seen in the other species. The emergent leaves in *M. pinnatum* have very shallow dentate margins compared to the lobed leaves of *M. hippuroides*. *Myriophyllum pinnatum* has been misidentified as *M. verticillatum* based on the presence of distally reduced pectinate leaves; in *M. verticillatum* the transition is almost always associated with flowering.

Records for Vermont and New Brunswick are range extensions for *Myriophyllum pinnatum*.

13. Myriophyllum heterophyllum Michaux, Fl. Bor.-Amer. 2: 191. 1803 • Two-leaf or various-leaved water-milfoil, myriophylle à feuilles variées [E] [F] [W]

Herbs monoecious, aquatic, often forming dense stands. **Stems** often branched, to 2.5 m. **Turions** absent. **Leaves** usually in whorls of 4(–6), often subverticillate, sometimes alternate, heteromorphic; petiole to 5 mm; submersed leaves pectinate, ovate to obovate in outline, (6–)12–29(–65) × (12–)14–18(–50) mm, segments (10–)12–20(–28), linear-filiform, longest segment (7–)9–25(–29) mm; emersed leaves pectinate to pinnatifid proximally, lanceolate to ovate or elliptic distally, 3–14(–31) × 1–5(–7) mm, margins serrate to lobed. **Inflorescences** to 60 cm; flowers proximally pistillate, medially bisexual, distally staminate; bracteoles cream, ovate to triangular or deltate, 0.6–1.1 × 0.3–0.9 mm, margins serrate to irregularly lobed. **Staminate flowers:** sepals cream, lanceolate to narrowly triangular, 0.5–0.8(–0.9) × 0.1–0.2 mm; petals persistent, cream, elliptic to obovate, 1.4–3 × 0.7–1.3 mm; stamens 4, filaments to 1.6 mm, anthers 1.3–2.2 × 0.3–0.7 mm. **Pistillate flowers:** sepals cream, triangular, (0.1–)0.2–0.6 × 0.1–0.3(–0.4) mm; petals caducous, cream, elliptic to obovate, 1.5–2(–3) × 0.8–1 mm; pistils 0.8–1.7 mm, stigmas red to ± purple, to 0.4 mm. **Fruits** ovoid to subglobose, deeply 4-lobed. **Mericarps** tan to red-brown, cylindric to narrowly ovoid, 1–1.5 × 0.5–0.8 mm, transversely orbiculate to widely elliptic, abaxial surface bluntly 4-angled, rounded to slightly flattened, densely papillate, with 4 shallow, longitudinal ridges, ridges sometimes with inconspicuous, shallow wings proximally, ribs absent.

Flowering and fruiting May–Oct. Oligotrophic to eutrophic waters, lakes, ponds; 0–600 m; B.C., N.B., Ont., P.E.I., Que.; Ala., Ark., Conn., Del., D.C., Fla., Ga., Ill., Ind., Iowa, Kans., Ky., La., Maine, Md., Mass., Mich., Minn., Miss., Mo., N.H., N.J., N.Y., N.C., Ohio, Okla., Pa., R.I., S.C., Tenn., Tex., Vt., Va., Wash., W.Va., Wis.

In flower, *Myriophyllum heterophyllum* is one of the more distinctive water-milfoils; it has relatively large, wide, ovate bracts and elongate spikes, on which leaves transition from pectinate to entire and often trail along the water surface. Plants are often very robust and bushy, with thickened red stems and highly crowded leaf whorls.

D. H. Les and L. J. Mehrhoff (1999) suggested that *Myriophyllum heterophyllum* is invasive in New England and progressively spread northward from a more southern native range. It is known to be introduced in British Columbia. R. A. Thum et al. (2011) provided genetic evidence that invasive

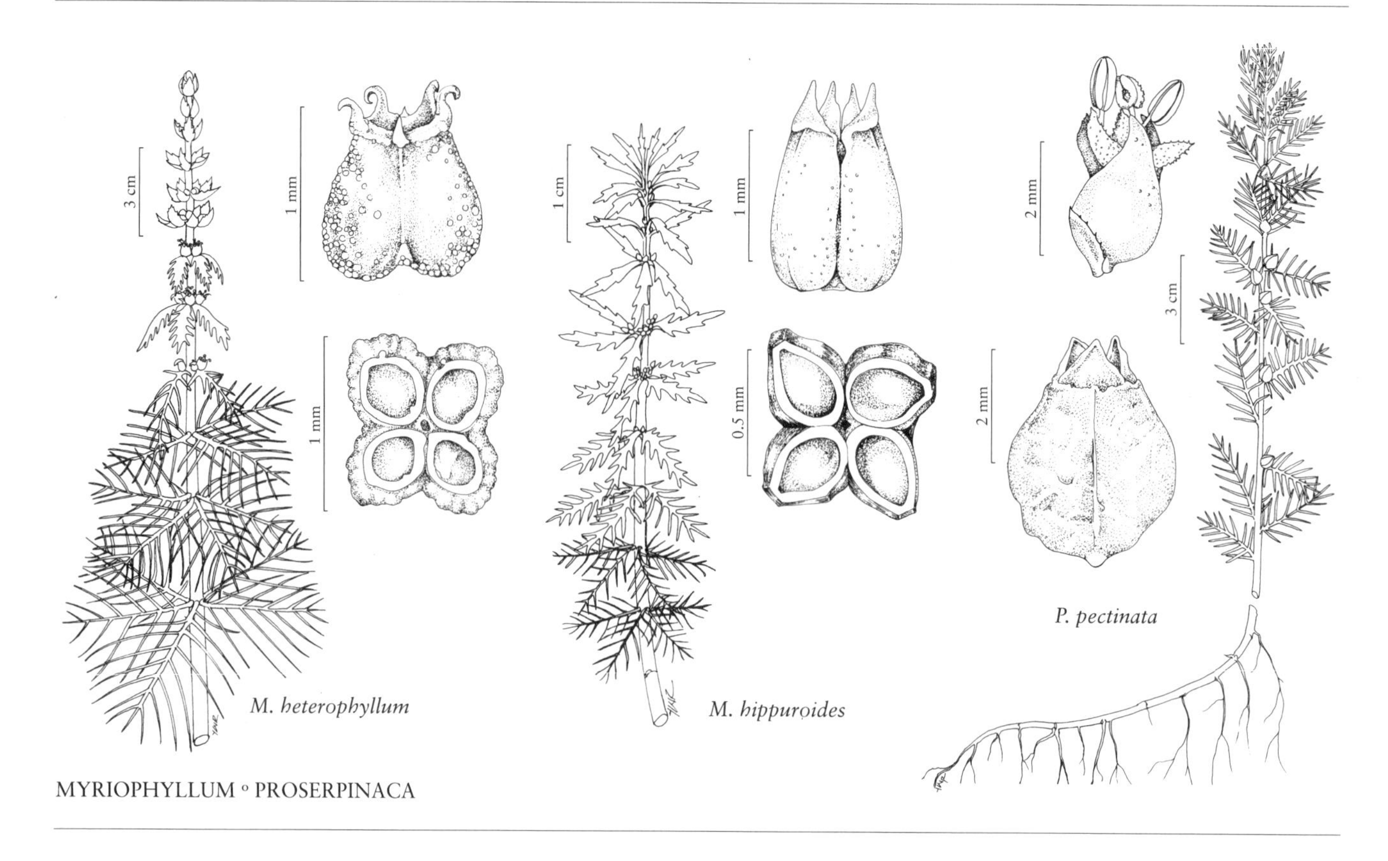

MYRIOPHYLLUM ∘ PROSERPINACA

populations of *M. heterophyllum* from New England, the Pacific Northwest, and California represent multiple introductions from the native Atlantic coastal plain and interior continental range.

S. G. Aiken (1981) suggested that *Myriophyllum heterophyllum* produces turions; we have seen no evidence of this, and it is likely that new shoots produced along rhizomes late in the growing season have been mistaken for turions.

14. Myriophyllum hippuroides Nuttall ex Torrey & A. Gray, Fl. N. Amer. 1: 530. 1840 • Western water-milfoil [F]

Herbs monoecious, aquatic, sometimes forming dense stands. **Stems** often branched, to 1.5 m. **Turions** absent. **Leaves** usually in whorls of 4(–6), sometimes subverticillate, heteromorphic; petiole to 3 mm; submersed leaves pectinate, obovate in outline, (12–)16–30(–41) × (10) 12–22(–30) mm, segments 12–24, linear-filiform, longest segment (6–)9–25(–36) mm; emersed leaves pectinate to pinnatifid proximally, linear to narrowly oblong or linear-lanceolate distally, (5–)7–16(–23) × (0.5–)1–2.2(–2.6) mm, often increasing in size distally, segments (5–)10–16(–22), 0–2.5 mm, margins serrate to shallowly lobed. **Inflorescences** to 35 cm; flowers proximally pistillate, medially bisexual, distally staminate; bracteoles cream, ovate to triangular, (0.4–) 0.7–1 × 0.2–0.7 mm, margins irregularly lobed. **Staminate flowers:** sepals cream, trullate to lanceolate, (0.1–)0.2–0.4(–0.5) × 0.1–0.4(–0.5) mm; petals persistent, cream to pink, elliptic to ovate, 1–1.5 × 0.5–0.8(–0.9) mm; stamens 4, filaments to 1 mm, anthers 0.5–0.9 × 0.2–0.4 mm. **Pistillate flowers:** sepals cream, trullate to lanceolate, 0.2–0.4 × 0.1–0.2 mm; petals persistent, cream to pink, elliptic to ovate, 0.7–1.3 × 0.4–0.7 mm; pistils 1.1–1.5 mm, stigmas red to ± purple, to 0.4 mm. **Fruits** cylindric to subglobose, cruciate to deeply 4-lobed. **Mericarps** tan to light green to red-brown or purple, cylindric to narrowly ovoid, 1.1–1.5 × 0.4–0.6 mm, transversely elliptic to narrowly ovate, compressed to flattened, abaxial surface bluntly 2-angled, flattened to slightly rounded, papillate to tuberculate, with 2(or 4) shallow, longitudinal ridges, wings and ribs absent.

Flowering and fruiting late Apr–Nov. Oligotrophic to mesotrophic waters, lakes, ponds, pools, riparian sloughs, small streams; 0–1800 m; B.C.; Calif., Ill., Oreg., Wash.; Mexico; Central America.

Myriophyllum hippuroides has a disjunct range in North America. It is native to the Pacific Northwest and California, where it can be confused with *M. pinnatum* and *M. heterophyllum* (see 12. *M. pinnatum*). *Myriophyllum hippuroides* often produces a terrestrial form consisting primarily of emergent leaves. The linear-lanceolate emergent leaves of *M. hippuroides* are at least three times as long as broad compared to the elliptic to ovate emergent leaves of *M. heterophyllum*. The most distinctive difference between these species is the morphology of their mericarps. The relatively narrow, transversely elliptic mericarps of *M. hippuroides* have a sharply 2-angled abaxial surface and are distinctly different from the subglobose and inconspicuously 4-angled mericarps of *M. heterophyllum* and the widely winged and ribbed mericarps of *M. pinnatum*.

3. PROSERPINACA Linnaeus, Sp. Pl. 1: 88. 1753; Gen. Pl. ed. 5, 38. 1754 • Mermaid-weed, proserpinie [Ancient name used by Pliny for a *Polygonum* taxon; derivation uncertain; probably for Proserpina, Roman goddess of spring, or Latin *proserpo*, creep, alluding to habit]

Mitchell S. Alix

Robin W. Scribailo

Herbs, monoecious, aquatic or semiaquatic [terrestrial]; with taproot or fibrous and, often, adventitious nodal roots. **Rhizomes** present. **Stems** creeping to erect [decumbent to ascending], terete, unbranched proximally, often branched distally, glabrous or ± punctate, often with scattered, black, ascidiate trichomes. **Turions** absent. **Leaves** submersed or emersed, alternate, homomorphic, dimorphic, or heteromorphic; sessile or petiolate; blade entire or lobed to pinnatifid or pectinate, margins entire or serrate; with trichomes in axils of leaves and scattered on surfaces. **Inflorescences** usually racemes or cymes, rarely dichasia, in axils of unreduced, emersed leaves; bracteoles paired, alternate, opposite subtending emersed leaf, secondary and tertiary bracteoles sometimes present; flowers bisexual. **Flowers** 3(or 4)-merous; petals caducous, rudimentary, sometimes absent; stamens 3, in 1 antisepalous whorl; ovary 3(or 4)-locular. **Fruit** a nutlet, tan to brown, ovoid, obturbinate, or pyramidal, transversely 3-angled, with acute angles or shallowly 3-lobed, faces concave, flat, or curvilinear, without ridges, surfaces smooth, rugose, papillate, or tuberculate. $x = 7$.

Species ca. 2 (2 in the flora): North America, Mexico, West Indies, Bermuda, Central America, South America.

SELECTED REFERENCES Catling, P. M. 1998. A synopsis of the genus *Proserpinaca* in the southeastern United States. Castanea 63: 404–414. Fassett, N. C. 1953c. *Proserpinaca*. Comun. Inst. Trop. Invest. Ci. Univ. El Salvador 2: 139–162.

1. Leaves homomorphic, pectinate, elliptic to ovate in outline . . . . 1. *Proserpinaca pectinata*
1. Leaves dimorphic to heteromorphic, submersed leaves pectinate, ovate, elliptic, obovate, or trullate in outline, emersed leaves pinnatifid to shallowly lobed to minutely serrate, lanceolate, narrowly elliptic, oblanceolate, spatulate, or obovate in outline . . . . 2. *Proserpinaca palustris*

1. Proserpinaca pectinata Lamarck in J. Lamarck and J. Poiret, Tabl. Encycl. 1: 214, plate 50, fig. 1. 1792 • Comb-leaved or coastal plain mermaid-weed [F]

Stems to 0.5 m. **Leaves** homomorphic, pectinate; petiole (0.4–)0.8–4(–6.5) mm; submersed leaves ovate in outline, 15–35 × 10–25 mm, segments 6–16, longest segment (7–)12–14(–18) mm; emersed leaves similar to submersed, transitional, elliptic to ovate in outline, (4–)10–20(–35) × (2–)5–15(–17) mm, segments (4–)8–18(–22), longest segment (1.4–)2–6(–8.5) mm. **Inflorescences:** bracteoles lanceolate to narrowly ovate, margins irregular, entire or broadly serrate to pinnatifid; primary bracteoles 1–3.8 × 0.3–0.9 mm, secondary bracteoles 0.7–1.4 × 0.2–0.7 mm. **Flowers:** sepals green to purple, shallowly triangular, 0.6–1(–1.5) × 0.5–1.2 mm; petals to 0.2 mm; filaments to 1.7 mm; anthers yellow, widely oblong, 0.4–1 × 0.2–0.6 mm; pistil 1.6–2.8 mm; style to 0.5 mm; stigma pink to purple, lanceolate, to 1 mm, fimbriate. **Fruits** obtusely 3-angled in cross-section, 2.2–3.2 × 2–3.8 mm, margins rounded to slightly winged, faces widely ovate to shallowly triangular, surface smooth to ± rugose or ± papillate to ± tuberculate; sepals strongly accrescent, ascending with appressed margins. $2n$ = 14.

Flowering and fruiting late Jun–Oct. Shores of lakes, ponds, rivers, streams, bogs, and marshes; 0–500 m; N.B., Nfld. and Labr. (Nfld.), N.S.; Ala., Del., D.C., Fla., Ga., Ky., La., Maine, Md., Mass., Mich., Miss., N.H., N.J., N.Y., N.C., Pa., R.I., S.C., Tenn., Tex., Va.; s Mexico; Central America.

Proserpinaca pectinata occurs in aquatic and semiaquatic habitats and displays little morphological variation across its range. It is easily distinguished from *P. palustris* by having flowers and fruits in the axils of pectinate leaves.

2. Proserpinaca palustris Linnaeus, Sp. Pl. 1: 88. 1753 • Marsh or common mermaid-weed, proserpinie des marais [F]

Proserpinaca palustris var. *amblyogona* Fernald; *P. palustris* var. *crebra* Fernald & Griscom

Stems to 0.5 m. **Leaves** dimorphic to heteromorphic, pectinate or lobed to serrate; petiole (0–)1–7(–9) mm; submersed leaves pectinate, ovate, elliptic, obovate, or trullate in outline, (6–)20–35(–66) × (4–)10–25(–42) mm, segments (8–)14–20(–27), longest segment (2–)7–21(–31) mm; emersed leaves transitional, pinnatifid to shallowly lobed to minutely serrate, lanceolate, narrowly elliptic, oblanceolate, spatulate, or obovate in outline, (6–)15–65(–100) × (1–)2–17(–23) mm. **Inflorescences:** bracteoles lanceolate to deltate, margins irregular, serrate to lobed; primary bracteoles 0.6–16 × 0.3–0.9 mm, secondary bracteoles 0.6–0.9 × 0.3–0.5 mm. **Flowers:** sepals green to purple, cucullate, shallowly triangular, (0.6–)0.8–1.3(–1.5) × 0.5–1.2 mm; petals to 0.1 mm; filaments to 2.5 mm; anthers yellow, oblong, 0.6–1.2 × 0.2–0.7 mm; pistil 1.5–3.3 mm; style to 0.5 mm; stigma pink to purple, lanceolate, to 1.5 mm, fimbriate. **Fruits** 3-angled or -lobed in cross-section, (2.3–)2.5–4(–4.5) × (1.5–)2.5–4(–5.7) mm, margins convex to strongly concave, often winged, faces widely ovate, cordiform, or shallowly triangular, surface smooth to ± rugose, sometimes with weak and shallow lateral ridges; sepals strongly accrescent, ascending with appressed margins. $2n$ = 14.

Flowering and fruiting May–Oct. Shores of lakes, ponds, rivers, streams, fens, and marshes; 0–700 m; N.B., N.S., Ont., Que.; Ala., Ark., Conn., Del., D.C., Fla., Ga., Ill., Ind., Iowa, Ky., La., Maine, Md., Mass., Mich., Minn., Miss., N.H., N.J., N.Y., N.C., Ohio, Okla., Pa., R.I., S.C., Tenn., Tex., Vt., Va., W.Va., Wis.; s Mexico; West Indies; Bermuda; Central America; South America.

Historically, *Proserpinaca palustris* has been split into varieties, primarily based on differences in fruit width-to-height ratios and surface morphology. In the flora area, three varieties (vars. *amblyogona*, *crebra*, and *palustris*) have been recognized (N. C. Fassett 1953c; P. M. Catling 1998). Distributions of the varieties overlap considerably and variation in fruit is continuous; only one taxon is recognized here.

The leaves of *Proserpinaca palustris* can be highly variable, transitioning from submersed and pectinate with a narrow central axis to emersed with broader blades with serrate margins (B. McCallum 1902; G. P. Burns 1904; N. C. Fassett 1953c; W. B. L. Schmidt and W. F. Millington 1968). Intermediate leaf forms often occur, which vary in central axis width and extent of blade division (Fassett). A reversion in the transition from submersed forms to emersed forms can occur later in the growing season (G. J. Davis 1967) or when the latter leaf types become inundated (Schmidt and Millington). The divergence in leaf form is influenced by photoperiod and temperature (Davis; Schmidt and Millington; A. Wallenstein and L. S. Albert 1963), hydration upon submergence (McCallum; Schmidt and Millington; Wallenstein and Albert), light intensity (Millington), and endogenous growth factors (Wallenstein and Albert; M. E. Kane and L. S. Albert 1989).

Plants with leaf characteristics intermediate between *Proserpinaca palustris* and *P. pectinata* were described as *P. intermedia* Mackenzie. N. C. Fassett (1953c) and P. M. Catling (1998) also considered *P. intermedia* to be distinct, possibly of hybrid origin between

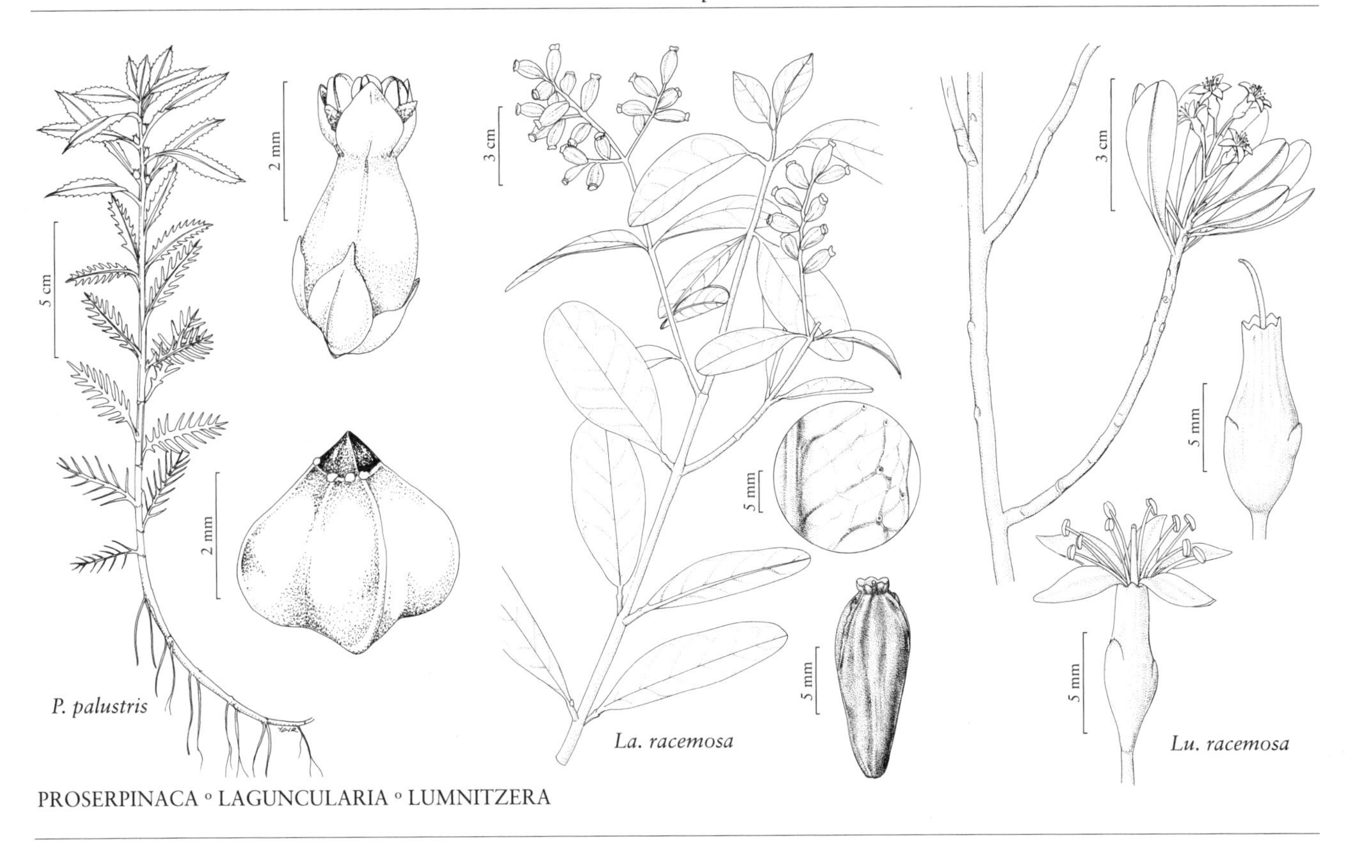

PROSERPINACA ◦ LAGUNCULARIA ◦ LUMNITZERA

P. palustris and *P. pectinata*. This intermediate form appears to arise sporadically; genetic evidence is needed to confirm the taxonomic status of this taxon. A wide range of leaf forms is also inducible in *P. intermedia* (M. E. Kane and L. S. Albert 1982) and these leaf forms completely overlap with those observed in *P. palustris*.

The suite of characters used by previous authors to distinguish *Proserpinaca intermedia* from *P. palustris* presents a continuum of variation. Therefore, in the absence of more definitive information, the former species is not recognized here as a distinct taxon.

COMBRETACEAE R. Brown • White Mangrove Family

Walter S. Judd

Trees, shrubs, or lianas, monoecious, dioecious, or polygamous, terrestrial, amphibious, or aquatic, sometimes with erect, monopodial trunk supporting a series of horizontal, sympodial branches, armed or unarmed, not clonal; mucilaginous cells or canals often present in parenchymatous tissues; hairs various, some long, straight, sharp-pointed, eglandular, unicellular, and very thick-walled, with conic internal compartment at base (combretaceous), sometimes hairs also short-stipitate, glandular; roots occasionally with pneumatophores. **Stems** erect, horizontal, spreading, or prostrate; bud scales absent. **Leaves** persistent or deciduous, alternate and spiral, or opposite or subopposite and decussate, simple; stipules usually absent, sometimes present and minute; petiolate, petiole or base of blade often with 2 circular to elliptic and flat structures or flask-shaped nectariferous cavities; blade papery to leathery, venation pinnate, secondary veins often forming loops or smoothly arching toward margin, margins entire, surfaces glabrous or hairy; domatia often present as pits, pouches, or hair tufts. **Inflorescences** terminal and/or axillary, spikes or panicles [racemes]; bracts present (each flower in axil of a usually caducous bract); bracteoles 0 or 2, adnate to hypanthium. **Flowers** bisexual or unisexual, staminate and pistillate on same or different plants, or bisexual and staminate on different plants, or all bisexual, usually actinomorphic; perianth and androecium borne on hypanthium; hypanthium cupulate, cylindric, or tubular, slightly to conspicuously prolonged beyond ovary and differentiated into 2 parts—proximally adnate to ovary, distally free; sepals 4 or 5, distinct or connate basally, imbricate to valvate; petals (0[4] or)5, distinct, imbricate or valvate; nectary present, usually a lobed or unlobed disc on top of ovary, often hairy; stamens [4 or]5–10 [–16], inflexed in bud; filaments distinct, included to long-exserted; anthers dorsifixed, dehiscent by 2 longitudinal slits; pollen grains 3-colporate, often also with poreless furrows; pistil 2–5-carpellate, carpels connate; ovary inferior [semi-inferior], 1-locular; placentation apical; style 1; stigma 1, punctate to capitate; ovules 2–5[or 6], pendulous on elongate funiculi from apex of locule, anatropous, bitegmic, crassinucellate. **Fruits** drupes, ribbed and/or winged, dry to spongy or fleshy. **Seed** 1 per fruit, relatively large, outer portion of seed coat fibrous; embryo with usually 2 folded or spirally twisted cotyledons; endosperm absent.

Genera 11, species ca. 500 (5 genera, 8 species in the flora): s United States, e Mexico, West Indies, Central America, South America, Asia, Africa, Australia; pantropical, warm temperate areas.

Combretaceae are clearly placed within Myrtales, a major angiosperm order whose monophyly is well supported by anatomy, embryology, and morphology (L. A. S. Johnson and B. G. Briggs 1984), as well as by plastid and nuclear DNA sequences (M. W. Chase et al. 1993; E. Conti et al. 1996, 1997; V. Savolainen et al. 2000, 2000b; D. E. Soltis et al. 2000, 2001; K. W. Hilu et al. 2003; Wang H. et al. 2009; O. Maurin et al. 2010, 2017). Morphological characteristics of Combretaceae supporting this placement include vessel elements with vestured pits, stems with internal phloem, stipules absent or present as very small lateral structures, flowers each with a short to elongate hypanthium, stamens incurved in bud, and a single style (W. S. Judd et al. 2008).

C. A. Stace (2007) recognized two subfamilies within Combretaceae: Strephonematoideae Engler & Diels (only *Strephonema* Hooker f. of western tropical Africa) and Combretoideae Burnett (all remaining genera); DNA-based phylogenetic analyses have supported this distinction (Tan F. X. et al. 2001, 2002; O. Maurin et al. 2010). Within the latter subfamily, two tribes are distinguished: Laguncularieae Engler & Diels (*Dansiea* Byrnes, *Laguncularia*, *Lumnitzera*, *Macropteranthes* F. Mueller) and Combreteae de Candolle [*Combretum*, *Conocarpus*, *Finetia* Gagnepain, *Getonia* Roxburgh, *Guiera* Adanson ex Jussieu, *Terminalia*]. Phylogenetic relationships within the family are imperfectly understood, and problems of generic circumscription exist (Stace).

Combretaceae are distributed pantropically, with extensions into warm temperate regions; within the geographical coverage of this flora, the species here treated are usually considered to be restricted to Florida; however, *Conocarpus erectus* and *Laguncularia racemosa* likely have become established on South Padre Island (Willacy County), Texas (B. L. Turner et al. 2003; Texas Non-Native Plants Group, http://www.texasnonnatives.org). The family is ecologically significant in coastal regions of southern and central Florida due to the presence of the mangrove species *Laguncularia racemosa* and the mangrove-associate *C. erectus*. Both occur in extensive coastal stands [along with *Avicennia germinans* and *Rhizophora mangle*]. *Lumnitzera racemosa* is also a mangrove species; it occurs in southern Florida only in a single naturalized population and shows potential to be an invasive species (J. W. Fourqurean et al. 2010). *Combretum indicum*, *Terminalia catappa*, *T. muelleri*, and, probably, *T. buceras* are also non-native; only the first two are considered invasive. Fruit dispersal in *Conocarpus*, *Laguncularia*, and *Lumnitzera* is by floating in water; the fruits of *Terminalia* are eaten by mammals and birds; they float and may be secondarily water-dispersed.

Combretaceae are not of great economic importance, although *Terminalia buceras*, *T. catappa*, and *T. molinetii* are widely used as shade trees and/or ornamentals in southern Florida due to their distinctive growth architecture, and cultivars of *Conocarpus erectus* with densely pubescent leaves are widely used as ornamental shrubs or small trees because of their distinctive coloration and salt tolerance. *Terminalia catappa* has fruits with edible kernels. *Combretum indicum* is often grown as an ornamental vine because of its showy flowers, which open white and change to pink and then red as the day progresses.

SELECTED REFERENCES Exell, A. W. 1931. The genera of Combretaceae. J. Bot. 69: 113–128. Exell, A. W. and C. A. Stace. 1966. Revision of the Combretaceae. Bol. Soc. Brot., ser. 2, 40: 5–25. Graham, S. A. 1964. The genera of Rhizophoraceae and Combretaceae in the southeastern United States. J. Arnold Arbor. 45: 285–301. Stace, C. A. 1965. The significance of the leaf epidermis in the taxonomy of the Combretaceae. Bot. J. Linn. Soc. 59: 229–252. Stace, C. A. 2004. Combretaceae. In: N. P. Smith et al., eds. 2004. Flowering Plants of the Neotropics. Princeton. Pp. 110–111. Stace, C. A. 2007. Combretaceae. In: K. Kubitzki et al., eds. 1990+. The Families and Genera of Vascular Plants. 15+ vols. Berlin etc. Vol. 9, pp. 67–82. Stace, C. A. 2010. Combretaceae. In: Organization for Flora Neotropica. 1968+. Flora Neotropica. 121+ nos. New York. No. 107.

1. Leaves opposite or subopposite and decussate.
   2. Trees or erect shrubs; petioles with conspicuous nectar glands; leaf blades with pit-domatia abaxially toward margins at junction of secondary and lower order veins, each containing a basal gland, and with scattered salt-excreting glandular hairs; flowers inconspicous, hypanthia free portions 1.3–2.5 mm, petals 1–1.3 mm, greenish white . . . 1. *Laguncularia*, p. 34
   2. Lianas or scandent shrubs; petioles without nectar glands; leaf blades with or without pouch-domatia abaxially, in axils of secondary and midveins, without salt-excreting glandular hairs; flowers conspicuous, hypanthia free portions 38–80 mm, petals 9–25 mm, white becoming pink then red . . . . 5. *Combretum*, p. 41
1. Leaves alternate and spiral (sometimes densely clustered distally on twigs, appearing verticillate).
   3. Stems with leaves clustered at ends of erect, short shoots, separated by longer, ± bare stem segments; trees with *Terminalia*-branching (branches in tiers, borne on central trunk, main axis erect and monopodial, lateral branches horizontal and sympodial) . . . . 4. *Terminalia*, p. 38
   3. Stems without clustered leaves on short shoots; trees or shrubs with branches ± similar, not with *Terminalia*-branching.
      4. Leaves with pit-domatia on distal margins and apex, blade apex rounded to retuse; petioles without nectar glands; inflorescences axillary, spikes, not in spherical or oblong heads; flowers bisexual, with sepals and petals . . . . .2. *Lumnitzera*, p. 35
      4. Leaves with pit-domatia abaxially at junction of veins with midveins, blade apex often acuminate to acute, occasionally obtuse, rarely rounded-mucronate; petioles with nectar glands at junction of petiole and blade base; inflorescences terminal and axillary, racemes or panicles of spherical or oblong heads; flowers unisexual (staminate and pistillate on different plants), with a single perianth whorl . . . . . 3. *Conocarpus*, p. 36

1. LAGUNCULARIA C. F. Gaertner, Suppl. Carp., 209, plate 217, fig. 3. 1807 • [Latin *laguncula*, flask or bottle, and *aria*, pertaining, alluding to fruit shape]

Shrubs or trees, often with ± erect pneumatophores. **Stems** erect, equal; twigs glabrous or sparsely hairy, hairs short, combretaceous. **Leaves** persistent, opposite and decussate; stipules absent; petiole not differentiated proximally and no part of it persistent, nectar glands conspicuous; blade fleshy-leathery, venation brochidodromous, apex obtuse, rounded, or retuse, surfaces often appearing glabrous, but with minute, scattered, salt-excreting glandular hairs, these sunken and similar in form to domatial glands; with pit-domatia abaxially at junction of secondary and lower order veins, each containing a basal gland. **Inflorescences** terminal and axillary, spikes or panicles; bracteoles present. **Flowers** inconspicuous, bisexual or staminate on different plants; hypanthium shallowly cupulate, free portion 1.3–2.5 mm, densely pubescent abaxially; sepals 5, green, triangular, pubescent abaxially; petals 5, greenish white, orbiculate, 1–1.3 mm, apex rounded to obtuse, pubescent (especially marginally); stamens 10, ± included; nectary disc atop ovary, pubescent; ovary somewhat flattened; style straight, free from hypanthium, with well-developed pistillode in staminate flowers; ovules 2. **Drupes** slightly flattened, oblong to obovoid, slightly ridged; with 2 major ridges, those subtending bracteoles better developed than others and forming spongy wings; hypanthium and calyx persistent.

Species 1: s United States, e Mexico, West Indies, Central America, South America, w Africa; tropical and subtropical mangrove habitats.

SELECTED REFERENCES Biehl, R. and H. Kinzel. 1965. Blattbau und Salzhaushalt von *Laguncularia racemosa* (L.) Gaertn. f. und anderer Mangrovebaüme auf Puerto Rico. Oesterr. Bot. Z. 112: 56–93. Landry, C. L., B. J. Rathcke, and L. B. Kass. 2009. Distribution of androdioecious and hermaphroditic populations of the mangrove *Laguncularia racemosa* (Combretaceae) in Florida and the Bahamas. J. Trop. Ecol. 25: 75–83.

1. Laguncularia racemosa (Linnaeus) C. F. Gaertner, Suppl. Carp., 209. 1807 • White mangrove [F]

Conocarpus racemosus Linnaeus, Syst. Nat. ed. 10, 2: 930. 1759 (as racemos.)

Shrubs or trees to 10(–20) m. **Leaves:** petiole 6–15 mm; blade ovate to obovate, oblong, or suborbiculate, 2–9.7 × 1.4–5 cm, base obtuse to rounded, folded longitudinally when young, with a faint longitudinal line each side of midvein in age. **Spikes** (or spicate units of panicles), 2–13 cm. **Flowers:** sepals 1 mm; stamens 1.5–2 mm; style 1–2 mm. **Drupes** greenish or gray-green [reddish green], 13–20 × 5–9 mm, pubescent [glabrous].

Flowering spring–early summer. Tidal swamps, mangrove communities; 0 m; Fla., Tex.; e Mexico; West Indies; Central America; South America; w Africa.

Laguncularia racemosa is an important component of mangrove swamps in central and southern Florida (extending northward to Levy County on the Gulf Coast and to Volusia County on the Atlantic Coast); it also recently has been reported from Willacy County, Texas. The species is occasionally cultivated as an ornamental tree or shrub in coastal situations. The flowers are quite fragrant and are pollinated by bees. The fruits are semiviviparous, the green embryo piercing the seed coat while the fruit is still on the tree. Some populations have some plants with bisexual flowers and others with staminate flowers, while in others all flowers are bisexual. The frequency of staminate plants in Florida is variable (0–68%).

2. LUMNITZERA Willdenow, Ges. Naturf. Freunde Berlin Neue Schriften 4: 186. 1803 • [For Istrán Lumnitzer, 1750–1806, Hungarian botanist] [I]

Shrubs or trees, rarely [sometimes] with pneumatophores, these looping upward then downward into substrate. **Stems** erect, equal; twigs glabrous or sparsely hairy, hairs short, combretaceous. **Leaves** persistent, alternate and spiral; stipules absent; petiole not differentiated proximally and no part of it persistent, nectar glands absent; blade fleshy-leathery, venation obscure, apex rounded to retuse; with pit-domatia on distal margins and at apex, each containing a basal gland. **Inflorescences** axillary [terminal], spikes [racemes]; bracteoles present. **Flowers** somewhat conspicuous, bisexual; hypanthium cylindrical, free portion 1.3–2.5 mm, slightly contracted beyond ovary, slightly conic distally, glabrous abaxially [pubescent]; sepals 5, green, triangular, glabrous [pubescent], margins ciliate, without [with] glands; petals 5, white or cream [yellow, red], ovate to elliptic, 4–5 mm, apex acute to obtuse, glabrous, margins sparsely ciliate; stamens [5–]10, exserted; nectary disc inconspicuous, from proximal portion of hypanthium adaxial surface, glabrous; ovary somewhat flattened; style straight to slightly curved, free from hypanthium; ovules 2–5. **Drupes** slightly flattened, ovoid, slightly ridged; with 2 major ridges, those subtending bracteoles better developed and expressed proximally, the remaining minor ridges only developed distally; hypanthium and calyx persistent.

Species 2 (1 in the flora): introduced, Florida; Asia, e Africa, Indian Ocean Islands (Madagascar), w Pacific Islands (Fiji, Tonga), n Australia.

1. Lumnitzera racemosa Willdenow, Ges. Naturf. Freunde Berlin Neue Schriften 4: 187. 1803 • Black mangrove [F] [I]

Shrubs or trees to 6 m (possibly greater). **Leaves:** petiole poorly developed, 1–7 mm; blade obovate to elliptic, 1.7–9 × 0.6–3 cm, base attenuate or cuneate, margins revolute when young, surrounding shoot apex, surfaces without scattered salt-excretory glands, glabrous [pubescent]. **Spikes** 1–3[–7] cm. **Flowers:** sepals 0.8–1.3 mm; stamens 4–5.5 mm; style 4–7.5 mm. **Drupes** green or brown, 10–20 × 3–8 mm, glabrous [pubescent]. $2n$ = 24.

Flowering spring–summer. Tidal swamps, mangrove communities; 0 m; introduced; Fla.; s Asia (Indochina), e Africa to w Pacific Ocean Islands (Fiji, Tonga), n Australia.

In the flora area, *Lumnitzera racemosa* is known from a single naturalized population in Miami-Dade County.

The open, white flowers of *Lumnitzera racemosa* are self-compatible and pollinated by bees, butterflies, moths, and wasps. The trees show high fruit set and the fruits, which have a corky outer layer, are water dispersed. The species shows invasive potential; eradication procedures have been attempted, and the cultivated specimens at Fairchild Tropical Botanic Garden have been removed (J. W. Fourqurean et al. 2010).

3. CONOCARPUS Linnaeus, Sp. Pl. 1: 176. 1753; Gen. Pl. ed. 5, 81. 1754 • [Greek *konos*, cone, and *karpos*, fruit, alluding to shape of densely clustered fruits]

Shrubs or trees, dioecious, without pneumatophores. **Stems** erect or prostrate, equal; twigs glabrous or hairy, hairs appressed to spreading, short to elongate, combretaceous. **Leaves** persistent, alternate and spiral; stipules absent; petiole not differentiated proximally and no part of it persistent, nectar glands glands conspicuous to inconspicuous, at junction of petiole and blade base; blade leathery, venation brochidodromous, apex often acuminate to acute, occasionally obtuse, rarely rounded-mucronate; pit-domatia conspicuous, at junction of abaxial veins with midvein. **Inflorescences** terminal and axillary, racemes or panicles of spherical or oblong heads, sometimes lateral ones reduced to single head; bracteoles absent. **Flowers** inconspicuous, each in axil of a bract, unisexual (staminate and pistillate on different plants); hypanthium shallowly cupulate, 0.3–0.5 mm, from constriction at apex of ovary, pubescent abaxially, often densely so; sepals 5, green to pale green, sometimes red-tinged, triangular, glabrate to pubescent abaxially; petals 0; stamens 5–10, long-exserted, with well-developed staminodia; nectary disc atop ovary, pubescent; ovary clearly flattened; style straight to slightly curved, free from hypanthium, with well-developed pistillode; ovules 2. **Drupes** somewhat flattened, ellipsoid to shortly obovoid, ridged or winged; ridges 2, forming shortly developed wings, slightly convex abaxially, concave adaxially; hypanthium and calyx persistent; fruits densely clustered in conelike heads, fragmenting in age.

Species 2 (1 in the flora): s United States, Mexico, West Indies, South America, sw Asia (Arabian Peninsula), w, ne Africa; tropical and subtropical areas.

Conocarpus occurs from Florida and extreme southern Texas through the West Indies to Mexico and coastal South America (to northern Peru on the Pacific Coast and to about the Tropic of Capricorn on the Atlantic Coast), and also in coastal western tropical Africa, northeastern Africa, and the Arabian Peninsula.

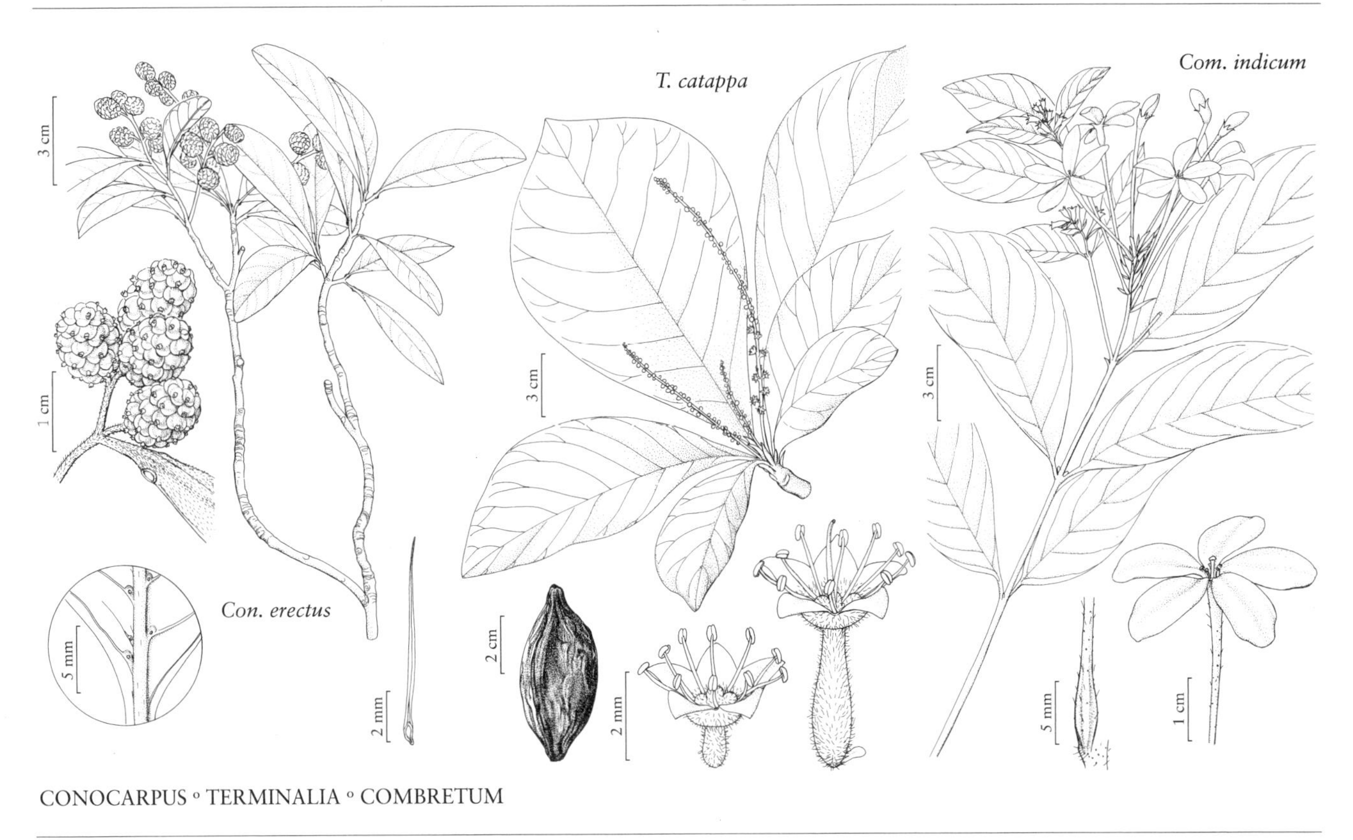

CONOCARPUS ◦ TERMINALIA ◦ COMBRETUM

1. **Conocarpus erectus** Linnaeus, Sp. Pl. 1: 176. 1753 (as erecta) • Buttonwood [F]

Conocarpus erectus var. *sericeus* Fors ex de Candolle

Shrubs or trees to 12(–20) m, sometimes ± prostrate. **Leaves:** petiole 1.5–16 mm; blade elliptic to obovate, 1.5–10 × 0.5–4 [–5] cm, smaller in terminal inflorescences, base attenuate to cuneate, surfaces glabrous or sparsely to densely pubescent, hairs appressed to ± erect, straight to crisped/tangled, combretaceous, some also minute, glandular. **Inflorescences** racemes or panicles of spherical or oblong heads (appearing paniculate), branches reproductive 3–25 cm distally, axis of lateral inflorescence 1–10 cm, heads conelike, 4–9 × 4–8 mm. **Flowers:** sepals 0.4–0.8 mm; stamens 1.8–3.5 mm; style 1–1.8 mm. **Drupes** greenish or brown, 2.5–4.5 × 2.7–5.5 mm, pubescent distally; clusters 6–12 mm. $2n = 24$.

Flowering spring–fall. Inland margins of tidal swamps, coastal habitats, open wetlands, hammocks, stream margins; 0–10 m; Fla., Tex.; Mexico; West Indies; South America; w Africa; in coastal and tropical areas.

In Florida, *Conocarpus erectus* is distributed from the Keys northward along the Gulf Coast to Levy County and along the Atlantic Coast to Volusia County. It has recently been reported from Willacy County, Texas.

Conocarpus erectus is commonly present on the landward margins of mangrove communities, as a mangrove associate. Although salt tolerant, it lacks obvious biological specializations of the mangrove community (for example, it is not viviparous and does not show root modifications). It is also common in open wetland communities and in tropical hammocks (tree islands). The densely packed flowers likely are generalist-insect pollinated; the fruits float and are dispersed by water. Some plants are consistently and densely pubescent, others are consistently glabrous, and still others show intermediate conditions such as a few slightly pubescent leaves on an otherwise densely pubescent plant, a few sparsely pubescent leaves on an otherwise glabrous plant, and plants producing both pubescent and glabrous leaves in a cyclic pattern. The densely pubescent form, "silver-leaved buttonwood," is a common ornamental in coastal regions, and its hairs give the leaves a gray to silver, occasionally even golden, coloration. *Conocarpus erectus* is extremely variable in stature; some plants form low-growing and almost prostrate shrubs, whereas others are erect trees.

4. TERMINALIA Linnaeus, Syst. Nat. ed. 12, 2: 665, 674. 1767, name conserved • [Latin *terminus*, terminal, and *alis*, pertaining, alluding to leaf clusters at branch tips]

Anogeissus (de Candolle) Wallich ex Guillemin & Perrottet, Fl. Seneg. Tent. 1(7): 279. 1832; *Buchenavia* Eichler; *Bucida* Linnaeus, name conserved; *Pteleopsis* Engler; *Ramatuela* Kunth; *Terminaliopsis* Danguy

Trees or shrubs, with Terminalia branching (branches in tiers, borne on central trunk, main axis erect and monopodial, lateral branches horizontal and sympodial), without pneumatophores. **Stems** erect and horizontal with leaves clustered at ends of erect, short shoots, separated by longer, ± bare stem segments. **Leaves** persistent or tardily deciduous (plants then briefly leafless), alternate and spiral; stipules present, appearing as fingerlike, multicellular, glandular hairs at petiole base; petiole not differentiated proximally and no part of it persistent, nectar glands absent [sometimes with sessile glands], sessile glands sometimes present at blade base; blade papery to leathery, venation brochidodromous [nearly eucamptodromous], apex acuminate, acute, mucronate, obtuse, retuse, or rounded; with or without hair-tuft or pit- [pocket-] domatia abaxially, at junction of secondary and tertiary veins with midvein, or junction of tertiary with secondary veins, sometimes also at other vein junctions, without glands. **Inflorescences** axillary, spikes, usually clustered on branches distally; bracteoles absent. **Flowers** somewhat conspicuous [inconspicuous], bisexual or bisexual and staminate on same plant, with staminate flowers distal [proximal] on spikes (or mixed with bisexual flowers); hypanthium cupulate, free portion 1–2 mm, from constriction at apex of ovary, pubescent abaxially throughout or pubescent proximally and glabrous distally [glabrous throughout]; sepals 4 or 5, pale green, yellow, or white, sometimes red-tinged, triangular to ovate-triangular, glabrous abaxially or pubescent; petals 0; stamens [4–]8–10, exserted; nectary disc lobate, atop ovary, pubescent; ovary terete, straight to slightly curved; style ± straight, free from hypanthium, pistillode absent in staminate flowers; ovules 2 or 3. **Drupes** radially symmetric in cross section to slightly [strongly] flattened, ovoid to ellipsoid; with 2 poorly developed ridges, 2 well-developed ridges or wings [2–5 well-developed wings], or 5 poorly developed and rounded lobes [terete and without lobes or wings]; hypanthium and calyx persistent or deciduous.

Species ca. 200 (4 in the flora): Florida; nearly worldwide in pantropics.

The flowers, with their exposed nectar glands, attract a variety of insects. *Terminalia* has a distinctive branching pattern, in which the main axis is erect and monopodial and the lateral branches are horizontal and sympodial. On the lateral branches, each segment originates as a lateral branch within the terminal leaf cluster of the previous segment. At first, each lateral branch grows horizontally, but eventually its apex becomes erect as a short shoot, producing a new terminal leaf cluster. Each sympodium as a whole branches where two lateral shoots continue its growth, substituting for the erect, terminal short shoot.

Terminalia buceras and *T. molinetii* often have been placed in *Bucida*, a genus traditionally separated from *Terminalia* by the retention of the free portion of the hypanthium and calyx on the fruit. The calyx is more or less deciduous in *T. molinetii* and it is persistent in three species of Madagascar and two of the Solomon Islands, which are phenetically quite different and likely not closely related to *T. buceras* or *T. molinetii*, calling this generic distinction into question (C. A. Stace 2007, 2010). A DNA-based phylogenetic analysis (O. Maurin et al. 2010) indicated that the recognition of *Bucida* renders *Terminalia* non-monophyletic. *Bucida*, therefore, has been placed in the synonymy of *Terminalia* (Stace 2007, 2010) as sect. *Bucida* (Linnaeus) Alwan & Stace.

SELECTED REFERENCES Fisher, J. B. and H. Honda. 1979. Branch geometry and effective leaf area: A study of *Terminalia*-branching pattern. 1. Theoretical trees. Amer. J. Bot. 66: 633–644. Fisher, J. B. and H. Honda. 1979b. Branch geometry and effective leaf area: A study of *Terminalia*-branching pattern. 2. Survey of real trees. Amer. J. Bot. 66: 645–655.

1. Flowers bisexual and staminate; spikes with numerous staminate flowers distally; drupes slightly flattened, with 2 poorly developed and rounded ridges, or 2 well-developed ridges or wings; hypanthia usually deciduous in age; leaf blades usually with pit- or hair-tuft domatia, also with nectar glands (near base).
    2. Drupes 35–70 mm, with 2 well-developed ridges or wings; leaf blades 6–35 cm, with pit-domatia . . . . 1. *Terminalia catappa*
    2. Drupes 12–20 mm, with 2 poorly developed and rounded ridges; leaf blades 4.2–19.5 cm, usually with hair-tuft domatia . . . . 2. *Terminalia muelleri*
1. Flowers bisexual; spikes with bisexual flowers distributed evenly or clustered distally; drupes radially symmetric in cross section, with 5 poorly developed and rounded lobes; hypanthia ± persistent or deciduous in age; leaf blades without domatia and nectar glands.
    3. Leaf blades 2–11 cm; branches without thorns in mature plants, with stout thorns in juvenile plants; spikes with flowers distributed along 1–6 cm; drupes (4–)5–8 mm . . . . 3. *Terminalia buceras*
    3. Leaf blades 0.4–2.8(–3.5) cm; branches with slender thorns; spikes with flowers clustered in distal 0.2–1 cm; drupes 3–6 mm . . . . 4. *Terminalia molinetii*

1. Terminalia catappa Linnaeus, Syst. Nat. ed. 12, 2: 674. 1767; Mant. Pl. 1: 128. 1767 • West Indian or Indian or tropical almond [F] [I]

Trees or shrubs to 20(–35) m; branches without thorns. **Leaves** persistent or tardily deciduous (then turning red and plants briefly leafless); petioles 5–28 mm; blade obovate, 6–35 × 2.6–16.5 cm, base narrowly cuneate to rounded or narrowly and obscurely cordate, apex acuminate or short-acuminate to obtuse or rounded, surfaces glabrate to moderately pubescent abaxially, midvein and secondary veins sparsely to densely pubescent, with nectar glands near base, glabrous or glabrate adaxially, midvein densely to sparsely pubescent, at least basally; with pit-domatia at junction of secondary and tertiary veins with midvein, or junction of tertiary with secondary veins, or other vein junctions. **Spikes** 5–25 cm, with bisexual flowers proximally, staminate flowers distally. **Flowers** 5-merous, bisexual and staminate; free portion of hypanthium 1–2 mm; sepals 1–2.8 mm; stamens 3–4.5 mm; style 3.5–4 mm. **Drupes** green or red, slightly flattened, ovoid to ellipsoid, 35–70 × 20–50 mm, sparsely pubescent or glabrous; with 2 well-developed ridges or wings; hypanthium and calyx deciduous in age.

Flowering spring–summer. Disturbed habitats, especially near coast; 0–10 m; introduced; Fla.; Asia; Pacific Islands; n Australia; introduced also widely in Neotropics.

Terminalia catappa is commonly used as an ornamental tree in southern Florida, and has naturalized in Brevard, Broward, and Miami-Dade counties.

2. Terminalia muelleri Bentham, Fl. Austral. 2: 500. 1864 • Australian almond [I]

Trees or shrubs to 10 m; branches without thorns. **Leaves** persistent or tardily deciduous (then turning red and plants briefly leafless); petiole 8–20 mm; blade obovate, 4.2–19.5 × 2–7.5 cm, base attenuate to cuneate, apex short-acuminate or mucronate, acute, obtuse, or rounded, surfaces glabrate to sparsely pubescent abaxially, midvein and secondary veins densely to sparsely pubescent or glabrous, when glabrous, elongate hairs usually present in axils, forming domatia, usually with nectar glands near base, glabrous or glabrate to sparsely pubescent adaxially, midvein densely to sparsely pubescent, at least basally; hair-tuft domatia usually present at junction of secondary veins and midvein, these sometimes poorly developed. **Spikes** 6.5–15 cm, with bisexual flowers proximally, staminate flowers distally. **Flowers** 4- or 5-merous, bisexual and staminate; free portion of hypanthium 1–2 mm; sepals 1.5–3 mm; stamens 2–4.3 mm; style 3.5–4 mm. **Drupes** green or red becoming blue or blue-black, slightly flattened, ovoid to ellipsoid, 12–20 × 8–15 mm, sparsely pubescent or glabrous; with 2 poorly developed and rounded ridges; hypanthium and calyx usually deciduous in age. $2n = 36$.

Flowering spring, summer. Mangrove swamps, disturbed hammocks, other disturbed habitats; 0–10 m; introduced; Fla.; ne Australia.

Terminalia muelleri is occasionally cultivated as an ornamental tree in southern and central Florida; it has naturalized coastally from Palm Beach County in the east to Manatee County in the west, and extending southward on the peninsula.

Terminalia muelleri is a triploid; polyploidy is common in the genus (D. Ohri 1996; C. A. Stace 2007).

3. **Terminalia buceras** (Linnaeus) C. Wright, Anales Real Acad. Ci. Méd. Fís. Nat. Habana Revista Ci. 5: 410. 1869 • Black olive [I]

Bucida buceras Linnaeus, Syst. Nat. ed. 10, 2: 1025. 1759

Trees or shrubs to 25 m; branches without thorns in mature plants, usually with stout thorns in juvenile plants. **Leaves** persistent; petiole 2–15[–26] mm; blade obovate to narrowly obovate, 2–11 × 0.8–4.5[–6.7] cm, base attenuate to cuneate, apex obtuse to rounded, occasionally very slightly retuse, surfaces glabrous or sparsely pubescent abaxially, midvein glabrous or moderately pubescent, without nectar glands near base, glabrous or sparsely pubescent adaxially, midvein glabrous or sparsely pubescent, at least basally, with 2–7 elongate, multicellular, and glandular hairs on adaxial surface of petiole near base; domatia absent. **Spikes** 3–19 cm, with flowers distributed along 1–6 cm. **Flowers** 5-merous, bisexual; free portion of hypanthium 1–2 mm; sepals 0.5–0.8 mm; stamens 3–4.5 mm; style 2.5–4 mm. **Drupes** green to brown, radially symmetrical in cross section, ovoid, (4–)5–8 × 2.5–4 mm, glabrate to densely pubescent; with 5 poorly developed and rounded lobes; hypanthium and calyx ± persistent in age.

Flowering late winter–summer. Hammocks; 0–10 m; introduced; Fla.; Mexico; West Indies; Central America.

Terminalia buceras is native to Mexico southward through Central America and throughout the Caribbean region; it is commonly used as a shade tree in southern Florida, and probably should be considered naturalized in Broward, Charlotte, Collier, and Miami-Dade counties, although some have considered the species to be native in extreme southern Florida (C. A. Stace 2010). Hybrids with *T. molinetii* are also available and used horticulturally. Frequently some of the fruits are infested with mites, causing them to form elongate twisted galls (to 16 cm). The generic name *Bucida*, here considered a synonym of *Terminalia*, is derived from the Latin for "horn of an ox," alluding to the resemblance of the galls to such horns.

4. **Terminalia molinetii** M. Gómez, Anales Soc. Esp. Hist. Nat. 19: 244. 1890 (as molineti) • Spiny black olive, dwarf geometry tree

Bucida molinetii (M. Gómez) Alwan & Stace; *B. spinosa* Jennings

Trees or shrubs to 8 m; branches with slender thorns. **Leaves** persistent; petiole 1–3 mm; blade obovate to narrowly obovate, 0.4–2.8(–3.5) × 0.2–1(–1.7) cm, base acute to cuneate, apex obtuse to retuse, surfaces glabrous or sparsely pubescent abaxially, midvein glabrous or sparsely pubescent, without nectar glands near base, glabrous adaxially, midvein glabrous or very sparsely pubescent, at least basally, with 1–4 elongate, multicellular, and glandular hairs on adaxial surface of petiole; domatia absent. **Spikes** 1–4 cm, with few flowers clustered in distal 0.2–1 cm. **Flowers** 5-merous, bisexual; free portion of hypanthium 1–1.5 mm; sepals 0.5 mm; stamens 3–4 mm; style 2.5–3 mm. **Drupes** green to brown, radially symmetrical in cross section, ovoid, 3–6 × 2–3.5 mm, glabrous or sparsely pubescent; with 5 poorly developed and rounded lobes; hypanthium and calyx deciduous in age.

Flowering late winter–summer. Low hammocks, pineland margins; 0–10 m; Fla.; s Mexico (Quintana Roo); West Indies (Bahamas, Cuba, Hispaniola, Puerto Rico, Virgin Islands); Central America (Belize).

In the flora area, *Terminalia molinetii* is rare and known only from Miami-Dade County. The species is often used as an ornamental shrub or small tree due to its graceful habit, especially in coastal regions, as are hybrids with *T. buceras*.

5. COMBRETUM Loefling, Iter Hispan., 308. 1758, name conserved • [Latin, derived from a name applied by Pliny the Elder to a climbing plant of uncertain identity] [I]

Cacoucia Aublet; *Calopyxis* Tulasne; *Meiostemon* Exell & Stace; *Quisqualis* Linnaeus; *Thiloa* Eichler

Scandent shrubs or lianas [shrubs, occasionally trees], without pneumatophores. **Stems** elongate, wandlike, equal; twigs hairy, hairs short to elongate, combretaceous [or absent], and short-stalked, glandular [peltate scales]. **Leaves** persistent, opposite or subopposite and decussate [whorled]; stipules absent; petiole persistent, differentiated [or not] proximally, becoming spinelike in age, nectar glands absent; blade papery [to leathery], venation eucamptodromous to brochidodromous, apex acuminate to long-acuminate; without salt-excreting glands; pouch-domatia present or absent abaxially in axils of secondary veins and midvein. **Inflorescences** terminal or axillary, spikes [racemes or panicles, sometimes nearly capitate]; bracteoles absent. **Flowers** conspicuous, each in axil of a bract, bisexual; hypanthium tubular and slightly broadened distally [to trumpet-shaped, narrowly campanulate, or cup-shaped], free portion 38–80 mm, from apex of ovary, with sparse, gland-tipped hairs abaxially, [or with peltate scales, and/or simple, nonglandular hairs]; sepals [4] 5, green or pale green, triangular [ovate, sometimes nearly obsolete], glandular-pubescent [with peltate scales] abaxially, and with simple, nonglandular hairs; petals [0, 4] 5, white becoming pink then red [greenish white, yellow, orange, or purple], ovate to slightly obovate, [sometimes narrowly so, or suborbiculate], 9–25 mm, apex obtuse or rounded [emarginate]; nonglandular-pubescent [glabrous or with peltate scales]; stamens [4, 8] 10, included [to exserted]; nectary on inner surface of hypanthium near base, glabrous [pubescent]; ovary ± terete; style straight, adnate to proximal portion of hypanthial tube [free]; ovules 2–4[–6]. **Drupes** terete, ellipsoid or ovoid [oblong], fleshy [dry], ridged or winged; ridges or wings [4] 5, poorly to well developed; hypanthium and calyx deciduous.

Species ca. 255 (1 in the flora): introduced, Florida; South America, Asia, Africa, n Australia; pantropical.

Combretum is naturalized in southern Florida; it is native pantropically but absent in the Pacific Islands.

1. **Combretum indicum** (Linnaeus) DeFilipps, Useful Pl. Dominica, 277. 1998 • Rangoon creeper [F] [I]

Quisqualis indica Linnaeus, Sp. Pl. ed. 2, 1: 556. 1762

Shrubs or lianas climbing, to 6 m. **Leaves:** petiole 5–20 mm; blade elliptic to oblong-elliptic to slightly obovate, 4–18.5 × 1.5–9 cm, base obtuse to slightly cordate, surfaces glabrate to densely pubescent, hairs combretaceous, and also scattered, inconspicuous, short-stalked, gland-tipped. **Spikes** 1.3–6[–12.5] cm. **Flowers:** sepals 1–3 mm; stamens 4–7.5 mm; style 10–20 mm (not including portion adnate to hypanthium). **Drupes** red, 25–40 × 7.5–14 mm, glabrous or sparsely glandular-pubescent. **2*n*** = 22, 24, 26.

Flowering spring–summer. Disturbed habitats; 0–20 m; introduced; Fla.; Asia (India to New Guinea); introduced also widely in Neotropics.

Combretum indicum has been found in Broward, Highlands, and Miami-Dade counties.

Combretum indicum is cultivated in warm regions for its beautiful and fragrant flowers, which have petals that are white when the flowers open and change to pink and then deep red as the day progresses. This species (and relatives) often have been placed in the genus *Quisqualis*, a name derived from the Latin "quis" (who?) and "qualis" (what?), alluding to the initial uncertain taxonomic placement of the genus, or possibly in astonishment at its variable habit or the distinctive color change. The approximately 17 species of the *Quisqualis* clade are closely related to species of *Combretum* subg. *Cacoucia* (Aublet) Exell & Stace, which have similar gland-tipped hairs (C. A. Stace 2007, 2010), and recognition of this group of species at the generic level would render *Combretum* non-monophyletic (O. Maurin et al. 2010).

LYTHRACEAE J. Saint-Hilaire • Loosestrife Family

Shirley A. Graham

Herbs, annual or perennial, shrubs, subshrubs, or trees, terrestrial, amphibious, or aquatic, usually unarmed (armed in *Punica*), rarely glaucous, clonal or not. **Stems** erect, decumbent, lax, spreading, creeping, trailing, floating, or submerged, young ones often 4-angled. **Leaves** deciduous, usually opposite, sometimes alternate, subalternate, or whorled, simple, rarely dimorphic, emergent and submerged leaves dissimilar; stipules usually absent; sessile, subsessile, or petiolate; blade membranous [leathery], venation brochidodromous, each marginal vein forming a series of loops, margins entire (except coarsely toothed in *Trapa*), trichomelike glands present in axil at petiole base. **Inflorescences** indeterminate or determinate, terminal or axillary, racemes, spikes, cymes, panicles, [thyrses], or flowers solitary; bracts absent (except present in *Cuphea aspera*); bracteoles present, paired on pedicels. **Flowers** bisexual [unisexual], actinomorphic or zygomorphic; perianth and androecium usually perigynous (semi-epigynous to epigynous in *Punica*, perigynous to semi-epigynous in *Trapa*), mono-, di-, or tristylous; floral tube or cup persistent, campanulate, turbinate, urceolate, cylindrical, or obconic, green (except red or yellow in some *Cuphea* and *Punica*), sometimes conspicuously ribbed; sepals persistent, 4–8, valvate, usually deltate, sometimes subulate in *Lythrum*, alternating with segments of epicalyx or epicalyx absent; petals usually caducous or deciduous, rarely persistent, 0–8, distinct, on interior margin of floral tube between sepals, crinkled, pinnately-veined; nectary present as distinct organ, as nectariferous tissue in wall of ovary or floral tube, or absent; stamens (1–) 4–42[+], usually equal to or 2 times number of petals, usually in 2 whorls of unequal length, sometimes 1 whorl; anthers versatile [or basifixed], introrse, 2-locular, longitudinally dehiscent; pistil 1; ovary usually superior (semi-inferior to inferior in *Punica* and *Trapa*), 2–6-locular (to 9 twisted locules in *Punica*); placentation axile, placenta elongate or globose; septa incomplete near ovary apex (reduced to thin threads in *Cuphea*); style 1; stigma 1, capitate to punctiform, dry (wet in *Heimia*, *Lagerstroemia*, and *Punica*); ovules 1–1400, anatropous, bitegmic. **Fruits** capsules, walls usually thin and dry (thick in *Decodon*, woody in *Lagerstroemia*), or leathery berries (*Punica*), or indurated, 2–4-horned drupes (*Trapa*), usually smooth (striate in *Rotala*), dehiscence loculicidal, septicidal, or septifragal, or indehiscent and splitting irregularly. **Seeds** 3–250(–1400), usually brown (wine in *Punica*) [black, red], narrowly obovoid to fusiform, oblong to pyramidal, obovoid-semiovoid, semiorbicular, suborbicular, or elliptic (in outline), usually dry, fleshy in *Punica*, winged or not; endosperm not developed; embryo straight, oily, cotyledons usually ± complanate, rarely rolled, often auriculate or cordate.

Genera 28, species ca. 600 (10 genera, 32 species in the flora): North America; nearly worldwide; mostly in tropical and subtropical areas.

Lythraceae are widely distributed in both hemispheres, generally in wet habitats, with limited temperate representation.

Taxa of Lythraceae in the flora area range in elevation from coastal plain to mid montane. Elevational ranges of the species are approximations based on regional topography taken from collection data where possible.

Lythraceae (order Myrtales) have recently been expanded to include the previously recognized families Duabangaceae, Punicaceae, Sonneratiaceae, and Trapaceae. Together they form a monophyletic family sister to Onagraceae (S. A. Graham et al. 2005). The traditional, more narrowly defined family (E. Koehne 1903) was divided into two tribes and four subtribes, a classification not supported by morphological or molecular phylogenetic evidence (Graham et al. 2011). Lythraceae are notably among only 28 angiosperm families that have evolved a heterostylous breeding system and one of three families in which the tristylous condition has evolved (S. C. H. Barrett et al. 2000). Six genera (*Adenaria*, *Ammannia*, *Decodon*, *Lythrum*, *Pemphis*, and *Rotala*) have distylous and/or tristylous species. Seed coats of many of the genera possess mucilaginous trichomes in the outermost cell layer that evaginate upon wetting, coming to resemble fungal growth. In *Cuphea* and in some genera outside the flora area, the trichomes reach half the length of the seeds and are spirally twisted. Floral merosity is highly variable in the family, and it is common to find a range of merosity in flowers of a single plant or within a population. The most frequent states are 4- and 6-merous. Only *Decodon* is commonly 5-merous. Submerged portions of the stems of the marsh-inhabiting genera *Ammannia*, *Decodon*, *Lythrum*, and *Trapa* can develop external spongy, aerenchymatous tissue that enhances oxygen exchange to the interior.

Some genera are widely cultivated: *Cuphea* and *Lagerstroemia* as landscape and garden plants, *Punica* for the pomegranate fruit, and *Lawsonia* for the leaves from which henna dye is obtained. *Trapa* and some species of *Lythrum* and *Rotala* can be invasive and pose significant problems for wetlands and river systems in North America.

SELECTED REFERENCES Graham, S. A. 1975. Taxonomy of the Lythraceae in the southeastern United States. Sida 6: 80–103. Graham, S. A. 1977. The American species of *Nesaea* (Lythraceae) and their relationship to *Heimia* and *Decodon*. Syst. Bot. 2: 61–71. Graham, S. A. 2007. Lythraceae. In: K. Kubitzki et al., eds. 1990+. The Families and Genera of Vascular Plants. 15+ vols. Berlin etc. Vol. 9, pp. 226–246. Graham, S. A. et al. 2005. Phylogenetic analysis of the Lythraceae based on four gene regions and morphology. Int. J. Pl. Sci. 166: 995–1017. Graham, S. A. et al. 2011. Relationships among the confounding genera *Ammannia*, *Hionanthera*, *Nesaea* and *Rotala* (Lythraceae). Bot. J. Linn. Soc. 166: 1–19. Graham, S. A. and A. Graham. 2014. Ovary, fruit, and seed morphology of the Lythraceae. Int. J. Pl. Sci. 175: 202–240. Koehne, E. 1903. Lythraceae. In: H. G. A. Engler, ed. 1900–1953. Das Pflanzenreich.... 107 vols. Berlin. Vol. 17[IV,216], pp. 1–326. Morris, J. A. 2007. A Molecular Phylogeny of the Lythraceae and Inference of the Evolution of Heterostyly. Ph.D. dissertation. Kent State University. Tobe, H., S. A. Graham, and P. H. Raven. 1998. Floral morphology and evolution in Lythraceae sensu lato. In: S. J. Owens and P. J. Rudall, eds. 1998. Reproductive Biology in Systematics, Conservation and Economic Botany. Kew. Pp. 329–334.

1. Leaf blade margins coarsely toothed in distal ½; fruits drupes, with 2–4 hardened, spiny horns; rosette-forming, rooted or floating aquatic herbs .... 10. *Trapa*, p. 65
1. Leaf blade margins entire; fruits capsules or berries, without spines or horns; aquatic, amphibious, or terrestrial herbs, shrubs, subshrubs, or trees.
    2. Branches often terminating as indurate thorns; floral tubes semi-epigynous to epigynous; fruits berries, leathery, crowned by persistent sepals and stamens; seeds fleshy.. 8. *Punica*, p. 62
    2. Branches without thorns; floral tubes perigynous; fruits capsules, enclosed by persistent floral tube; seeds dry.

[3. Shifted to left margin.—Ed.]

3. Stamens 36–42, 3+ times number of sepals; seeds unilaterally winged; shrubs or small trees . . . . 6. *Lagerstroemia*, p. 54
3. Stamens (1 or)2–12(–27) usually equal to or 2 times number of sepals; seeds not winged; herbs, shrubs, or subshrubs.
    4. Floral tubes cylindrical or obconic; flowers actinomorphic or zygomorphic.
        5. Flowers zygomorphic; capsule dehiscence by longitudinal complementary slits in wall and floral tube, placenta and seeds ultimately exserted . . . . 2. *Cuphea*, p. 48
        5. Flowers actinomorphic; capsule dehiscence septicidal or septifragal, placenta and seeds remaining within capsules . . . . 7. *Lythrum* (in part), p. 56
    4. Floral tubes campanulate, urceolate, or turbinate; flowers actinomorphic.
        6. Shrubs or subshrubs, (5–)10–30 dm; petals rose purple or bright yellow, 5–15 mm.
            7. Inflorescences simple or compound cymes, 3+-flowered; flowers pedicellate, petals rose purple, 8–15 mm . . . . 3. *Decodon*, p. 51
            7. Inflorescences leafy racemes; flowers sessile or subsessile, petals bright yellow, 5–10(–14) mm . . . . 5. *Heimia*, p. 53
        6. Herbs, 0.4–10 dm; petals deep to pale purple, rose purple, rose, pink, lavender, or white, 1–4(–7) mm, or petals absent.
            8. Capsules septicidally dehiscent, walls finely, transversely striate (10×); inflorescences terminal or axillary racemes, axillary spikes, or solitary flowers . . . . 9. *Rotala*, p. 63
            8. Capsules indehiscent, splitting irregularly or circumscissile then splitting irregularly, walls smooth; inflorescences axillary cymes, axillary or terminal racemes (sometimes spikelike in *Lythrum*), or solitary flowers.
                9. Leaf blade bases cordate or auriculate; floral tubes campanulate or urceolate, semiglobose to globose in fruit; sepals to 1/4 floral tube length . . . . 1. *Ammannia*, p. 44
                9. Leaf blade bases rounded, cordate, tapered, attenuate, or truncate; floral tubes broadly campanulate; sepals 1/3–1/2 floral tube length.
                    10. Sepals 4; petals 0; epicalyx segments absent . . . . 4. *Didiplis*, p. 53
                    10. Sepals 6; petals (0 or)6; epicalyx segments present, equal to 2 times longer than sepals . . . . 7. *Lythrum* (in part), p. 56

1. AMMANNIA Linnaeus, Sp. Pl. 1: 119. 1753; Gen. Pl. ed. 5, 55. 1754 • Toothcup, redstem [For Paul Ammann, 1634–1691, Professor of Botany at Leipzig]

Shirley A. Graham

Hionanthera A. Fernandes & Diniz; *Nesaea* Commerson ex Kunth

Herbs, annual or perennial, terrestrial or amphibious [aquatic], 1–10 dm, glabrous [puberulent]. **Stems** erect or trailing, glaucous or not throughout, branched or unbranched, anthocyanic in age, submerged stems sometimes thickened by aerenchymatous spongy tissue. **Leaves** opposite [subopposite or whorled]; sessile; blade linear, linear-lanceolate, oblong, linear-oblong, lanceolate, elliptic, or spatulate [ovate], base cordate or auriculate [attenuate]. **Inflorescences** indeterminate, axillary, simple or compound cymes or solitary flowers [racemes or capitula]. **Flowers** sessile, subsessile, or pedicellate, actinomorphic, monostylous [di- or tristylous]; floral tube perigynous, campanulate or urceolate, semiglobose to globose in fruit, often 8–12[–16]-ribbed; epicalyx segments shorter than or as long as sepals [longer than or absent]; sepals 4–6[–8], to 1/4 floral tube length; petals caducous [persistent], (0–)4–8, deep rose purple, pale pink, white, or pale lavender, sometimes with rose purple midvein or basal spot, obovate to suborbiculate; nectariferous tissue present at ovary-floral tube junction; stamens 4–12(–14)[–27]; ovary incompletely (1 or)2–5-locular; placenta globose [elongate]; style slender and well-exserted or stout and included; stigma capitate or punctiform. **Fruits** capsules, walls thin and dry, indehiscent, splitting irregularly at maturity or initially circumscissile then splitting irregularly. **Seeds** [2–]100+, obovoid to semiglobose, concave-convex, 0.3–0.4 × 0.3–0.5 mm; cotyledons ± complanate.

Species ca. 80 (5 in the flora): North America, Mexico, West Indies, Central America, South America, Europe, Asia, Africa, Pacific Islands, Australia; worldwide in temperate to subtropical and tropical areas.

Ammannia was expanded from ca. 25 species to ca. 80 with the inclusion of *Hionanthera* and *Nesaea* following multi-gene molecular phylogenetic evidence that the genera form a monophyly (S. A. Graham et al. 2011). Africa is the center of diversity for the genus.

Ammannia frequently grows with the closely similar *Rotala ramosior*. In the flora area, the auriculate leaf bases of *Ammannia* expediently separate it from *Rotala*; leaf bases are tapered in *R. ramosior*.

SELECTED REFERENCES Graham, S. A. 1985. A revision of *Ammannia* (Lythraceae) in the Western Hemisphere. J. Arnold Arbor. 66: 395–420. Howell, J. T. 1985. The genus *Ammannia* (Lythraceae) in western United States and Canada. Wasmann J. Biol. 43: 72–74.

1. Styles stout, 0.5 mm, included; petals 0–4(–6) . . . . 3. *Ammannia latifolia*
1. Styles slender, 1.5–9 mm, exserted; petals 4–6(–8).
   2. Inflorescences solitary flowers; peduncles 7–19 mm; petals 6, 4–7 mm; stigmas punctiform . . . . 4. *Ammannia grayi*
   2. Inflorescences simple or compound cymes (flowers sometimes solitary at proximalmost and distalmost nodes); peduncles 0–9 mm; petals 4(–8), 1.5–3 mm; stigmas capitate.
      3. Cymes 1–3(–5)-flowered mid stem; petals pale lavender, sometimes with rose purple midvein or rose purple basal spot; anthers light yellow . . . . 5. *Ammannia robusta*
      3. Cymes (1–)3–15-flowered mid stem; petals deep rose purple, sometimes with darker rose purple basal spot; anthers deep yellow.
         4. Peduncles slender (nearly filiform); cymes (3–)7–15-flowered mid stem; capsules (1–)1.5–2.5(–3.5) mm diam. . . . . 1. *Ammannia auriculata*
         4. Peduncles stout; cymes (1–)3–7(–15)-flowered mid stem; capsules 3.5–5 mm diam. . . . . 2. *Ammannia coccinea*

1. Ammannia auriculata Willdenow, Hort. Berol. 1: 7, plate 7. 1803 F W

Ammannia arenaria Kunth; *A. auriculata* var. *arenaria* (Kunth) Koehne; *A. wrightii* A. Gray

Herbs annual, slender, 1–4(–8) dm. **Stems** erect, usually much-branched distal to base. **Leaf blades** narrowly linear-lanceolate to linear-oblong, 20–80 × 2–15 mm. **Inflorescences** simple or compound cymes, (3–)7–15-flowered mid stem; peduncle slender (nearly filiform), 3–9 mm. **Pedicels** slender, 1–3 mm. **Floral tube** campanulate to urceolate, 1–3 mm; epicalyx segments shorter than sepals or absent; sepals 4; petals 4, deep rose purple, 1.5 × 1.5 mm; stamens 4(–8); anthers deep yellow; style slender, well-exserted, 1+ mm; stigma capitate. **Capsules** (1–)1.5–2.5(–3.5) mm diam., equal to or surpassing sepals, splitting irregularly. 2*n* = 30 (Egypt), 32 (Mexico, Tanzania).

Flowering late spring–late fall. Freshwater marshes, margins of pools and lakes, drying mud and sand flats, ditches, rice fields; 50–1500 m; Kans., Nebr., N.Mex., Okla., S.Dak., Tex.; Mexico; West Indies; Central America; South America; Asia; Africa; Australia.

Ammannia auriculata, the most widely distributed species in the genus, is common and probably native to Africa but sporadic in occurrence, depending on the availability of suitable habitats, which may change from year to year. The species in the flora area is most common in the central United States, where it often grows with *A. coccinea* or *A. robusta*.

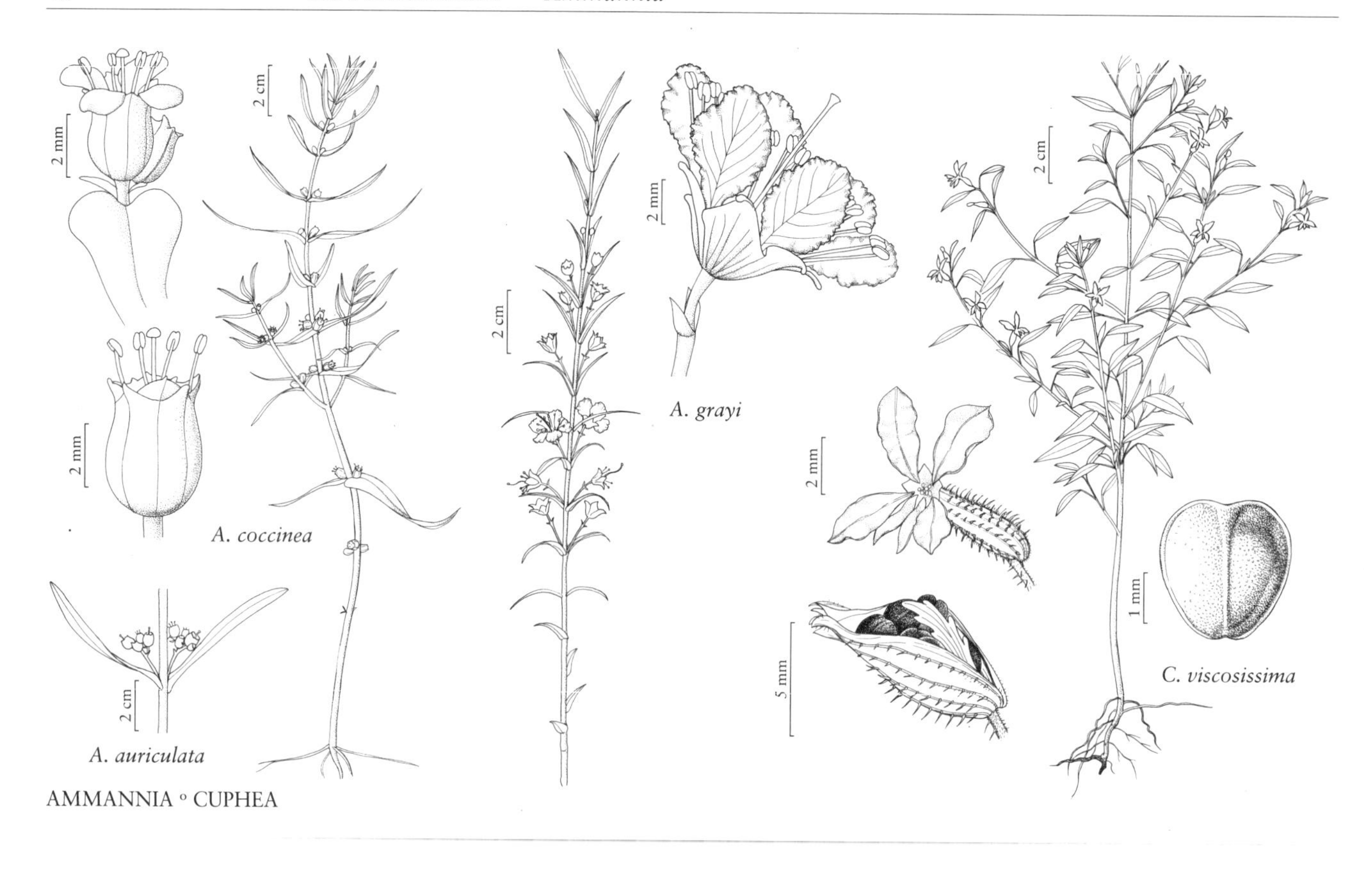

AMMANNIA ◦ CUPHEA

2. Ammannia coccinea Rottbøll, Pl. Horti Univ. Rar. Progr., 7. 1773 [F] [W]

Ammannia coccinea subsp. *purpurea* (Lamarck) Koehne; *A. latifolia* Linnaeus var. *octandra* A. Gray; *A. purpurea* Lamarck; *A. sanguinolenta* Swartz subsp. *purpurea* (Lamarck) Koehne; *A. stylosa* Fischer & C. A. Meyer; *A. teres* Rafinesque; *A. texana* Scheele

Herbs annual, robust, 2–10 dm. **Stems** erect, much-branched distal to base. **Leaf blades** usually lanceolate to linear-oblong, rarely elliptic to spatulate, 20–80 × 2–15 mm. **Inflorescences** simple or compound cymes, (1–)3–7(–15)-flowered mid stem; peduncle stout, (1–)3–9 mm. **Pedicels** stout, 1–2 mm. **Floral tube** campanulate to urceolate, (2.5–)3–5 mm; epicalyx segments as long as sepals; sepals 4(or 5); petals 4 (or 5), deep rose purple, sometimes with darker rose purple basal spot, 2 × 2 mm; stamens 4(–7); anthers deep yellow; style slender, well-exserted, 2–3.5 mm; stigma capitate. **Capsules** 3.5–5 mm diam., equal to or surpassing sepals, splitting irregularly. **2*n*** = 66.

Flowering late spring–late fall. Freshwater marshes, margins of pools and lakes, drying mud and sand flats, ditches, rice fields; 10–1500 m; B.C.; Ala., Ariz., Ark., Calif., Colo., Fla., Ga., Ill., Ind., Iowa, Kans., Ky., La., Md., Miss., Mo., Nebr., N.J., N.Mex., N.C., Ohio, Okla., Pa., S.C., S.Dak., Tenn., Tex., W.Va.; Mexico; West Indies; Central America; Europe (Italy, Spain); Asia (Afghanistan); Africa (Sudan); Pacific Islands (Guam, Hawaii); widely introduced in s Europe, elsewhere in Asia, Africa, other Pacific Islands, Australia.

Ammannia coccinea is considered an amphidiploid (n = 33) derived from *A. auriculata* (n = 16 in North America) and *A. robusta* (n = 17) (S. A. Graham 1979). The species displays a wide range of morphological variability, approaching at extremes each of the putative parents and making identification occasionally difficult, especially in recognizing co-occurring *A. coccinea* and *A. robusta*. Differences in petal and anther color are important distinctions, easily determined in the field but not always apparent on herbarium specimens. Occasional specimens of *A. coccinea* that approach *A. auriculata* in peduncle length or flower number are best recognized by the larger flowers of *A. coccinea*. Care must be taken in determining style length because the aging style of *A. coccinea* tends to break away leaving a short stump that resembles the true short style of *A. latifolia*.

Ammannia coccinea is native to North America but disjunctly distributed around the world as a result of introductions through rice culture.

SELECTED REFERENCE Graham, S. A. 1979. The origin of *Ammannia coccinea* Rottboell. Taxon 28: 169–178.

3. Ammannia latifolia Linnaeus, Sp. Pl. 1: 119. 1753 [W]

Ammannia koehnei Britton; *A. koehnei* var. *exauriculata* Fernald; *A. teres* Rafinesque var. *exauriculata* (Fernald) Fernald

Herbs annual, robust, 4–10 dm. **Stems** erect, sparsely branched, or branched proximally. **Leaf blades** lanceolate-linear to oblong, elliptic, or spatulate, 15–70(–100) × 4–15(–21) mm. **Inflorescences** simple cymes, (1–)3–10-flowered mid stem; peduncle stout, 3 mm. **Pedicels** 0–1 mm. **Floral tube** urceolate, 3–4 mm; epicalyx segments as long as or slightly longer than sepals; sepals 4(–6), shallowly deltate, with mucronate apex; petals 0–4(–6), pale pink to white, 1 × 1 mm; stamens 4(–8); anthers yellow; style stout, included, ca. 0.5 mm; stigma capitate. **Capsules** 3.5–5.5 mm diam., included to scarcely surpassing sepals, splitting irregularly. **2*n*** = 48.

Flowering late spring–late fall. Coastal plains in brackish to freshwater marshes, ditches; 0–200 m; Fla., Ga., La., Md., Miss., N.J., N.C., S.C., Tex., Va.; Mexico (Yucatán); West Indies; Central America (Panama); South America.

Ammannia latifolia flowers spring through fall in the northern part of the range; year-round from southern Florida southward.

Petals in *Ammannia latifolia* are consistently present from the Gulf Coast and from Georgia northward along the Atlantic Coast, and are commonly reduced in number or absent along the southern Atlantic and Gulf coasts and southward. Presence of petals is best determined in the bud.

4. Ammannia grayi S. A. Graham & Gandhi, Harvard Pap. Bot. 18: 75. 2013 • Stalkflower [C] [F]

Nesaea longipes A. Gray, Smithsonian Contr. Knowl. 3(5): 68. 1852; *Heimia longipes* (A. Gray) Cory

Herbs perennial, slender, 1–6 dm, glaucous throughout. **Stems** decumbent or trailing, branched or unbranched. **Leaf blades** linear to narrowly lanceolate, 10–50 × 1–5 mm. **Inflorescences** solitary flowers; peduncle slender, 7–19 mm. **Pedicels** slender, 0.5–1 mm. **Floral tube** campanulate, 3–6 mm; epicalyx segments shorter than sepals; sepals 6; petals 6, pale pink or lavender, 4–7 × 3–5 mm; stamens (10–)12(–14); anthers green; style slender, well-exserted, 6–9 mm; stigma punctiform. **Capsules** 3.5 mm diam., included, circumscissile then splitting irregularly. **2*n*** = ca. 50.

Flowering late spring–fall (May–Sep). Limestone seeps along creek banks, on highly alkaline soil around springs and moist areas; of conservation concern; 500–1500 m; N.Mex., Tex.; Mexico (Coahuila, Sonora).

5. Ammannia robusta Heer & Regel, Index Seminum (Zürich), adn. 1. 1842 [W]

Ammannia alcalina Blankinship; *A. coccinea* Rottbøll subsp. *robusta* (Heer & Regel) Koehne

Herbs annual, robust, 3–10 dm. **Stems:** proximal branches often decumbent, length often equaling main stem, unbranched, or branched near base. **Leaf blades** usually lanceolate-linear, rarely spatulate, 15–80 × 4–15 mm. **Inflorescences** simple cymes, 1–3(–5)-flowered mid stem; peduncle stout, 0–0.5 mm. **Pedicels** 0. **Floral tube** campanulate to urceolate, 3–5 mm; epicalyx segments as long as sepals; sepals 4(–8); petals 4(–8), pale lavender, sometimes with rose purple midvein or rose purple basal spot, 2–3 × 2.5–3.3 mm; stamens 4(–8); anthers light yellow; style slender, well-exserted, 2 mm; stigma capitate. **Capsules** 4–6 mm diam., equal to or surpassing sepals, splitting irregularly. **2*n*** = 34.

Flowering late spring–late fall. Freshwater marshes, margins of pools and lakes, drying mud and sand flats, ditches, rice fields; 0–2100 m; B.C., Ont.; Ariz., Ark., Calif., Colo., Idaho, Ill., Ind., Iowa, Kans., Ky., La., Mich., Minn., Mo., Mont., Nebr., Nev., N.Dak., Ohio, Okla., Oreg., S.Dak., Tenn., Tex., Utah, Wash., Wyo.; Mexico; West Indies; Central America; South America; Pacific Islands (Guam).

Ammannia robusta flowers spring through fall in the northern part of the range and year-round southward. The species has spread worldwide as a result of introduction through rice cultivation.

Ammannia robusta is documented for New Jersey by a collection from the mid 1800s (*P. Knieskern s.n.*, NY); there is no evidence that it became established in the state.

2. CUPHEA P. Browne, Civ. Nat. Hist. Jamaica, 216. 1756 • Waxweed, Mexican heather, cigar flower [Greek *kyphos*, humped, alluding to protruding base of floral tube]

Shirley A. Graham

Parsonsia P. Browne, name rejected

Herbs, annual or perennial, [subshrubs], terrestrial [amphibious], 1–10 dm, [glaucous], often glandular-viscid with 1+ types of glandular/eglandular indument, trichomes colorless, white, or red-purple [glabrous]; with fibrous roots or woody xylopodium. **Stems** erect, decumbent, or spreading, branched or unbranched. **Leaves** usually opposite, rarely 3- or 4-whorled, when whorled, opposite at proximal nodes; sessile, subsessile, or petiolate; blade ovate to lanceolate, oblong, elliptic, or linear, base attenuate or rounded [cuneate, cordate], surfaces finely scabrous. **Inflorescences** indeterminate, terminal or axillary, leafy or bracteate racemes [thyrses], 1 flower emerging between petioles at a node, others, when present, on axillary branchlets. **Flowers** sessile or pedicellate, zygomorphic, monostylous; floral tube perigynous, cylindrical, rounded, or spurred basally, conspicuously 12-ribbed, inner surface villous or glabrous; epicalyx segments shorter than sepals [to longer than sepals]; sepals 6, to 1/4 floral tube length; petals caducous [persistent], [0 or](2–)6, purple, rose purple, rose, or pink, subequal or unequal, sometimes 2 upper petals larger or of different color than others; nectary present at base of ovary; stamens (5–)11, deeply included or equal to surpassing sinus of sepals, 2 stamens usually shorter, more deeply inserted than others; ovary 2-locular; placenta elongate; septa reduced to thin threads, 1 locule reduced; style slender; stigma capitate to punctiform. **Fruits** capsules, walls thin and dry, dehiscence by longitudinal complementary slits in wall and floral tube, placenta and seeds ultimately exserted. **Seeds** 3–13(–20)[–100+], orbiculate, suborbiculate, oblong, or elliptic in outline; cotyledons ± complanate.

Species ca. 240 (6 in the flora): North America, Mexico, West Indies, Central America, South America, Pacific Islands, Australia.

The seeds of *Cuphea* store oils composed of medium-chain fatty acids that are widely used in the manufacture of soaps and detergents, in specialized food products, and in medicine. The endemic *C. viscosissima* is under development as a domestic source of the fatty acids that traditionally have been obtained from imported palm oils. Other species are cultivated as annual garden plants, and new hybrids and cultivars appear yearly in the nursery trade. Among the most popular cultivated species are: *C. calophylla* Chamisso & Schlechtendal and *C. hyssopifolia* Kunth, both sold under the name Mexican heather and identified by their small, tubular flowers with six purple petals; *C. ignea* A. de Candolle, the cigar flower or firecracker plant, a species with trailing stems and elongate, red flowers with black and white tips; and *C. llavea* Lexarza, marketed as 'Tiny Mice' or 'Bat-Faced Cuphea,' which has two large, red petals often with a black spot at the base, and purple trichomes filling the opening of the floral tube. The purple-petalled *C. procumbens* Ortega (firefly cuphea) and hybrids of *C. procumbens* with *C. llavea* are also popular flowering annuals. Garden escapes of *C. procumbens* account for reports in Florida, Georgia, Massachusetts, North Carolina, and South Carolina; the species does not appear to be naturalized anywhere in the flora area. All the cultivated species named above are native to Mexico.

SELECTED REFERENCES Barber, J. C., A. Ghebretinsae, and S. A. Graham. 2010. An expanded phylogeny of *Cuphea* (Lythraceae) and a North American monophyly. Pl. Syst. Evol. 289: 35–44. Graham, S. A. 1988. Revision of *Cuphea* section *Heterodon* (Lythraceae). Syst. Bot. Mongr. 20: 1–168. Graham, S. A., J. V. Freudenstein, and M. Luker. 2006. A phylogenetic study of *Cuphea* (Lythraceae) based on morphology and nuclear rDNA ITS sequences. Syst. Bot. 31: 764–778.

1. Leaves 3- or 4-whorled mid stem; flowers opposite or 3- or 4-whorled; pedicels 4–25 mm. . . . 1. *Cuphea aspera*
1. Leaves opposite; flowers alternate; pedicels 0–6 mm.
    2. Stamens deeply included, extending 2/3 distance to sinus of sepals; floral tube 4–6 mm. . . . . 2. *Cuphea carthagenensis*
    2. Stamens reaching or surpassing sinus of sepals; floral tube 5–12 mm.
        3. Sepals equal; leaves sessile or subsessile, petiole 0–2 mm; seeds 1.5–2 mm.
            4. Floral tube bases rounded; seed margins rounded . . . . 3. *Cuphea glutinosa*
            4. Floral tube bases descending spurs; seed margins flattened, thin. . . . . .4. *Cuphea strigulosa*
        3. Sepals unequal, adaxialmost longer; leaves petiolate (at least proximally), petiole (1–)2–15(–20) mm; seeds 2–2.8 mm.
            5. Leaf blade bases attenuate; seeds 7–10; upper petals 3–6 mm . . . . .5. *Cuphea viscosissima*
            5. Leaf blade bases rounded to cuneate; seeds 3–6; upper petals 1–2 mm. . . . . . . 6. *Cuphea wrightii*

1. Cuphea aspera Chapman, Fl. South. U.S., 135. 1860 C E

Herbs perennial, 2.5–5 dm, with woody xylopodium. **Stems** erect, usually unbranched, sometimes with 1 or 2 long branches from near base, minutely white-strigose and purple glandular-setose, especially viscid on youngest internodes. **Leaves** 3- or 4-whorled at mid stem, sessile; blade lanceolate, or linear distally, 10–25 × 1.5–5 mm, base rounded. **Racemes** bracteate, ± terminal. **Pedicels** 4–25 mm. **Flowers** opposite or 3- or 4-whorled; floral tube often fading abaxially, red-purple to rose adaxially, 7–9 × 1–2 mm, white-strigose, veins purple glandular-setose; base rounded or a descending spur, 1 mm; inner surface glabrous proximally, villous distal to stamens; epicalyx segments thick, often terminated by a bristle; sepals equal; petals 6, pale purple or pink, cuneate-oblong, subequal, 3–4.5 × 0.6–1 mm; stamens 11, reaching or surpassing sinus of sepals. **Seeds** 3, orbiculate in outline, 2.5 × 2.5 mm, margin rounded. **2*n*** = 48.

Flowering early–late summer. Pine flatwoods, sandy soil; of conservation concern; 0–200 m; Fla.

Cuphea aspera is known from Calhoun, Franklin, and Gulf counties. Its morphological relationships are with species of eastern Brazil. An ancestral form may have been carried northward from Brazil along the storm tracks that are noted for passage across the Apalachicola area where this species persists. It is listed in the Center for Plant Conservation's National Collection of Endangered Plants.

Parsonsia lythroides Small is an illegitimate name that pertains here.

2. Cuphea carthagenensis (Jacquin) J. F. Macbride, Publ. Field Mus. Nat. Hist., Bot. Ser. 8: 124. 1930 I W

Lythrum carthagenense Jacquin, Enum. Syst. Pl., 22. 1760; *Balsamona pinto* Vandelli; *Cuphea balsamona* Chamisso & Schlechtendal; *Parsonsia pinto* (Vandelli) A. Heller

Herbs annual, [**subshrubs**], 1–6 dm, with fibrous roots. **Stems** erect to decumbent and spreading, usually much-branched, hispid and setose, sometimes also puberulent. **Leaves** opposite, subsessile or sessile; petiole 0–2 mm; blade broadly elliptic to lanceolate, 12–55 × 5–25 mm, base attenuate. **Racemes** leafy. **Pedicels** 1–2 mm. **Flowers** alternate, 1 interpetiolar, with 1–3 flowers on axillary branchlets; floral tube purple adaxially and distally, or green throughout, 4–6 × 1–1.5 mm, glabrous except veins sparsely and coarsely setose; base rounded or a descending spur, 0.5 mm; inner surface glabrous; epicalyx segments thick, often terminated by a bristle; sepals equal; petals 6, deep purple or rose purple, subspatulate, subequal, 1.5–2.5 × 0.5–1 mm; stamens 11, extending 2/3 distance to sinus of sepals. **Seeds** (4–)6(–9), elliptic to suborbiculate in outline, 1.5–1.7 × 0.2–1.5 mm, margin narrow, flattened, thin. **2*n*** = 16.

Flowering late spring–fall. Atlantic and Gulf coastal plain, ditches, margins of moist woods, roadsides, moist open, disturbed areas; 0–200 m; introduced; Ala., Ark., Fla., Ga., La., Miss., N.C., S.C., Tenn., Tex.; Mexico; Central America; South America; introduced also in Pacific Islands (Fiji, Guam, Hawaii, Philippines), Australia.

The weedy, self-fertilizing *Cuphea carthagenensis* is the most widely distributed species of the genus and one of the more common in South America. It was first collected in the United States in Florida and North Carolina in the 1920s. Fossilized pollen very similar to pollen of *C. carthagenensis* and close relatives is known from the late Miocene of Alabama (S. A. Graham 2013). The species flowers year-round in subtropical and tropical regions.

3. Cuphea glutinosa Chamisso & Schlechtendal, Linnaea 2: 369. 1827 [I]

Parsonsia glutinosa (Chamisso & Schlechtendal) A. Heller

Herbs perennial, 1–4 dm, with fibrous roots. **Stems** often several from base, erect to decumbent, branched or unbranched, glandular-hispid and puberulent. **Leaves** opposite, sessile or subsessile; petiole 0–1 mm; blade ovate-lanceolate to oblong or elliptic, 5–15[–20] × 2–7[–10] mm, base cuneate to rounded. **Racemes** leafy. **Pedicels** 0–3 mm. **Flowers** alternate, solitary, interpetiolar; floral tube green abaxially, purple adaxially, 5.5–8(–9) × 2–2.5 mm, sparsely glandular-hispid; base rounded, 0.5 mm; inner surface glabrous proximally, villous distal to stamens; epicalyx segments thick, not terminated by a bristle; sepals equal; petals (2–)6, oblanceolate or oblong, unequal, 4 abaxial ones pale purple and 4–4.7 × 2.5–2.8 mm, 2 adaxial ones purple [deep purple or with deep purple midvein] and 4 × 1.5–1.9 mm; stamens 11, reaching or surpassing sinus of sepals. **Seeds** 8–13(–20), suborbiculate in outline, 1.5–2 × 1.5–1.7 mm, margin rounded. **2*n*** = 28, 32 (Bolivia), 34 (Paraguay).

Flowering spring–summer. Moist areas in open woods and pastures; 10–200 m; introduced; La., Tex.; South America.

First noted in the United States in 1884 in Vermilion Parish, Louisiana, *Cuphea glutinosa* is now more widespread in southern Louisiana and occurs in four counties in eastern Texas. The species is agamospermous in the United States, with sterile pollen but producing abundant seed. Sexually reproductive plants occur in eastern Brazil.

4. Cuphea strigulosa Kunth in A. von Humboldt et al., Nov. Gen. Sp. 6(fol.): 161; 6(qto.): 204. 1824 [I]

Cuphea strigulosa subsp. *opaca* Koehne

Herbs perennial, sometimes subshrubs, 2.5–10 dm, with fibrous roots. **Stems** erect to semi-decumbent, sparsely branched, puberulent and often sparsely red-purple glandular-setose. **Leaves** opposite, sessile to subsessile; petiole 0–2 mm; blade elliptic, 15–45 × 7–25 mm, base attenuate. **Racemes** leafy. **Pedicels** 1–2 mm. **Flowers** alternate, solitary, interpetiolar; floral tube green abaxially, purple or green adaxially, 6.5–7.5 × 1 mm, puberulent and sparsely glandular-setose; base a descending spur, 0.5 mm; inner surface glabrous proximally, glabrous or finely puberulent distal to stamens; epicalyx segments thick, often terminated by a bristle; sepals equal; petals 6, pale rose or pink, oblong, subequal, 2.5–5 × 1.5–2 mm; stamens 11, scarcely reaching sinus of sepals. **Seeds** 6–13, suborbiculate to oblong in outline, 1.5–1.8 × 1.3–1.5 mm, margin narrow, flattened, thin. **2*n*** = 16 (Brazil).

Flowering summer. Moist pastures, disturbed open, wet areas, roadsides, river margins; 0–50 m; introduced, Fla.; South America; introduced also in West Indies.

Cuphea strigulosa is widespread in Andean South America and in Brazil. It was first noted in Puerto Rico in 1964 and first collected in the Florida Everglades in 1995.

5. Cuphea viscosissima Jacquin, Hort. Bot. Vindob. 2: 83, plate 177. 1772 • Clammy cuphea, blue waxweed [E] [F] [W]

Lythrum petiolatum Linnaeus 1753, not *Cuphea petiolata* Pohl ex Koehne 1877; *Parsonsia petiolata* (Linnaeus) Rusby

Herbs annual, 1–6 dm, with fibrous roots. **Stems** erect to decumbent, often reddish, much-branched, purple-red glandular-setose, glandular-viscid. **Leaves** opposite, petiolate; petiole (2–)5–15(–20) mm; blade narrowly lanceolate to narrowly ovate, 20–50 × 6–20 mm, base attenuate. **Racemes** leafy. **Pedicels** 1–5 mm. **Flowers** alternate, solitary, interpetiolar, sometimes with 1 axillary; floral tube pale abaxially, deep purple-red adaxially, 8–12 × 1–2 mm, purple-red glandular-setose; base rounded or a descending spur, 0.5–1 mm; inner surface glabrous proximally, densely villous distal to stamens; epicalyx segments thick, 2 flanking the adaxialmost sepal terminated by a bristle; sepals unequal, adaxialmost longer; petals 6, purple, oblanceolate, unequal, 4 abaxial petals 2–5 × 0.4–0.6 mm, 2 upper petals 3–6 × 1.5–2.5 mm; stamens (5–)11, reaching or surpassing sinus of sepals. **Seeds** 7–10, oblong-elliptic in outline, 2.3–2.8 × 1.8–2.3 mm, margin rounded. **2*n*** = 12.

Flowering summer–fall. Weedy in pastures, roadsides, ditches, grassy borders, disturbed moist woods along trails; 0–900 m; Ont.; Ala., Ark., Conn., Del., D.C., Fla., Ga., Ill., Ind., Iowa, Kans., Ky., La., Md., Mass., Miss., Mo., Nebr., N.H., N.J., N.Y., N.C., Ohio, Okla., Pa., S.C., Tenn., Tex., Vt., Va., W.Va.

Cuphea viscosissima of the eastern and central United States is the most common and widespread species of *Cuphea* in the flora area; it is naturalized in Ontario. It is closely related to *C. lanceolata* W. T. Aiton of eastern and central Mexico, with which it shares the lowest known chromosome number in the genus, 2*n* = 12. The stamen number is typically 11, but varies in some populations (W. H. Duncan 1950).

Cuphea petiolata (Linnaeus) Koehne is an illegitimate name that pertains here.

6. Cuphea wrightii A. Gray, Smithsonian Contr. Knowl. 5(6): 56. 1853

Cuphea wrightii var. *nematopetala* Bacigalupi; *Parsonsia wrightii* (A. Gray) Kearney

Herbs annual, 1–4 dm, with fibrous roots. **Stems** erect, sparsely branched, dark purple-red-glandular-setose, glandular-viscid. **Leaves** opposite, petiolate proximally, sessile distally; petiole 1–10[–22] mm; blade ovate to lanceolate, 10–35 [–50] × 3–15[–20] mm, base rounded to cuneate. **Racemes** leafy. **Pedicels** (1–)4–6 mm. **Flowers** alternate, solitary and interpetiolar, (with 2 or 3 axillary); floral tube pale abaxially, purple-red to purple-black with dark veins adaxially, 5–11 × 1 mm, purple-red-glandular-setose; base rounded [or a descending spur], to 0.5 mm; inner surface glabrous proximally, lightly villous to glabrous distal to stamens; epicalyx segments thick, 2 flanking the adaxialmost sepal terminated by a bristle; sepals unequal, adaxialmost longer; petals 6, purple or rose [bicolor, with white abaxial petals], obovate to orbiculate, unequal, 4 abaxial 0.5–1[–2.5] × 0.2 mm, 2 adaxial 1–2[–5] × 0.7 mm; stamens 11, reaching or surpassing sinus of sepals. **Seeds** 3–6 [or 7], oblong-elliptic in outline, 2–2.5 × 1.8–2 mm, margin rounded. **2*n*** = 20 (Mexico), 44 (Mexico).

Flowering mid–late summer. Local in moist, open habitats, pastures, roadsides, rocky washes; 1000–1800 [–2900] m; Ariz.; Mexico; Central America.

Cuphea wrightii reaches its northernmost distribution in the southeastern corner of Arizona. Plants from Arizona with filiform petals (var. *nematopetala*) have been reported growing mixed with plants having normal obovate petals.

3. DECODON J. F. Gmelin, Syst. Nat. 1: 656, 677. 1791 • Swamp loosestrife, water willow, décodon verticillé [Greek *dekas*, ten, and *odon*, tooth, alluding to combination of five sepals and five alternating epicalyx segments] E

Shirley A. Graham

Shrubs or subshrubs, aquatic, (5–)10–30 dm, glabrous or velutinous. **Stems** erect or arching, branched or unbranched, spreading vegetatively by rooting from submerged tips of arching stems, submerged stems often thickened by external layers of spongy aerenchyma. **Leaves** opposite or 3-whorled; petiolate; blade lanceolate, base attenuate. **Inflorescences** indeterminate, axillary, simple or compound cymes, 3+-flowered. **Flowers** pedicellate, actinomorphic, tristylous; floral tube perigynous, campanulate; epicalyx segments ca. 2 times longer than sepals; sepals (4 or)5 (–7), to 1/3 floral tube length; petals caducous or deciduous, (4 or)5(–7), rose purple; nectariferous tissue present in lower ovary wall; stamens (8–)10; ovary 3(–5)-locular; placenta elongate; style slender; stigma capitate. **Fruits** capsules, walls moderately thick and dry, dehiscence loculicidal. **Seeds** 20–30, obpyramidal; cotyledons ± complanate.

Species 1: c, e North America.

Although *Decodon* is now monospecific and endemic to the central and eastern United States, it formerly had a nearly global distribution in the northern latitudes. Fossil seeds of *Decodon* from the Late Cretaceous (73.5 Ma) of northern Mexico represent the second oldest known occurrence of Lythraceae in the world, second only to pollen of *Lythrum* from the Late Cretaceous (82–81 Ma) of Wyoming. *Decodon* was widespread and diverse during the Miocene in both hemispheres; it became severely reduced as climates cooled in the Pliocene and thereafter, ultimately becoming restricted to *D. verticillatus* in eastern North America (S. A. Graham 2013).

SELECTED REFERENCES Dorken, M. E. and C. G. Eckert. 2001. Severely reduced sexual reproduction in northern populations of a clonal plant, *Decodon verticillatus* (Lythraceae). J. Ecol. 89: 339–350. Eckert, C. G. and S. C. H. Barrett. 1994. Post-pollination mechanisms and the maintenance of outcrossing in self-compatible, tristylous, *Decodon verticillatus* (Lythraceae). Heredity 72: 396–411.

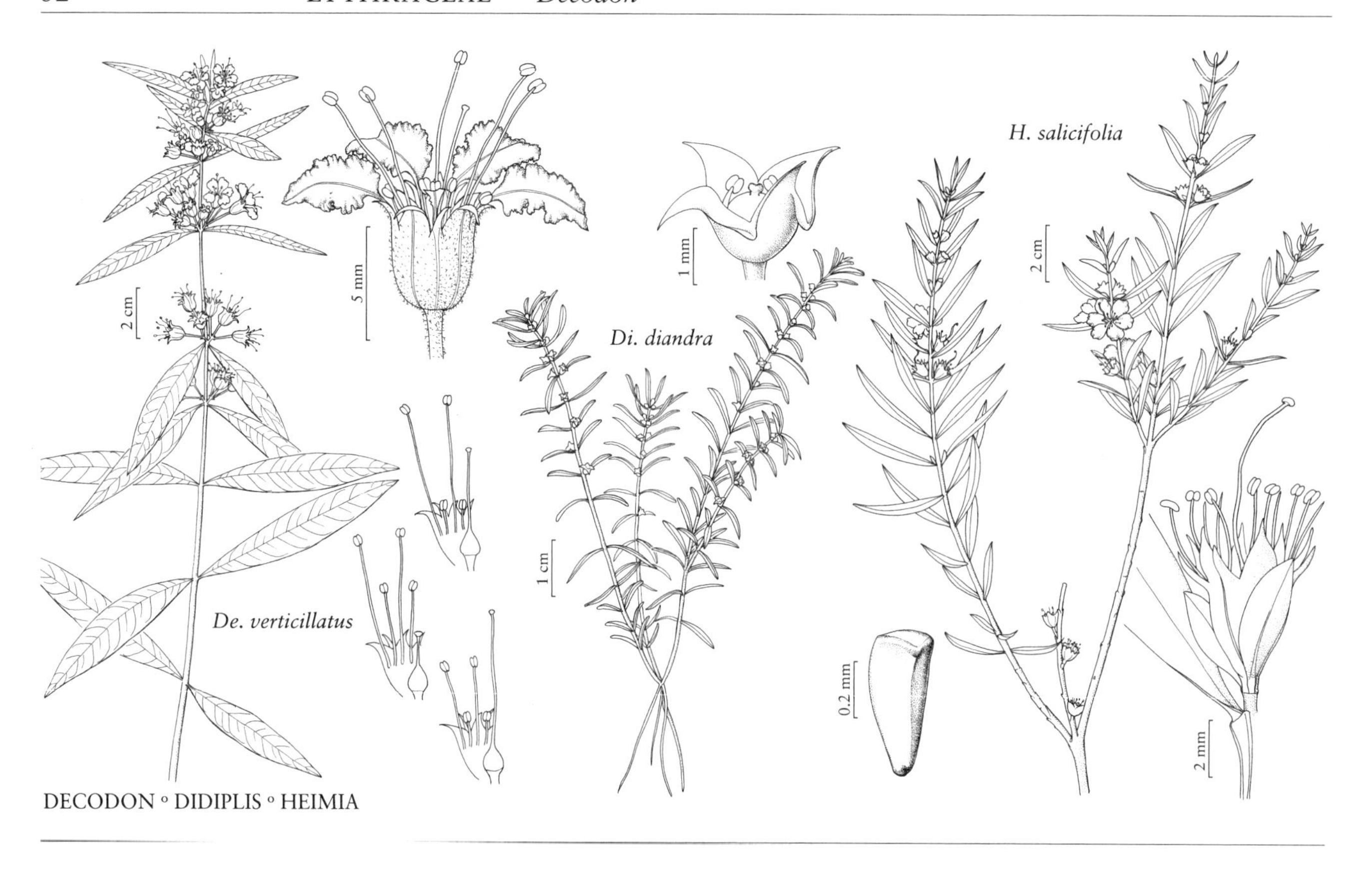

DECODON ∘ DIDIPLIS ∘ HEIMIA

1. Decodon verticillatus (Linnaeus) Elliott, Sketch Bot. S. Carolina 1: 544. 1821 (as verticillatum) E F

Lythrum verticillatum Linnaeus, Sp. Pl. 1: 446. 1753; *Decodon verticillatus* var. *laevigatus* Torrey & A. Gray; *D. verticillatus* var. *pubescens* Torrey & A. Gray

Leaves: petiole 3–12 mm; blade 30–200 × 5–50 mm, surfaces glabrous or puberulent abaxially, trichomes pale brown. **Inflorescences** 6–24-flowered. **Pedicels** 4–13 mm. **Floral tube** 5–8 × 3–5 mm, glabrous or puberulent; epicalyx segments linear, 2–3 mm; petals obovate or elliptic, 8–15 × 3.5–6 mm; stamens in 2 whorls, of 3 lengths, each flower with 2 of the 3 lengths. **Capsules** globose, loculicidally dehiscent. **Seeds** 1.5 × 1.5 mm; embryo surrounded by dense, small-celled spongy tissue. **2*n*** = 32.

Flowering summer–fall. Swamps, marshes, margins of lakes, ponds, or pools, margins of pools in fens; 0–500 m; N.B., N.S., Ont., P.E.I., Que.; Ala., Ark., Conn., Del., D.C., Fla., Ga., Ill., Ind., Ky., La., Maine, Md., Mass., Mich., Minn., Miss., Mo., N.H., N.J., N.Y., N.C., Ohio, Pa., R.I., S.C., Tenn., Tex., Vt., Va., W.Va., Wis.

The distinction made between glabrous inland and puberulent coastal plain populations of *Decodon verticillatus* has proven only partly accurate. Plants with some degree of indument are the most common form throughout the range; there are relatively few individuals that are entirely glabrous. There are no distinct geographical limits based on presence or absence of trichomes. Regional variability is seen in the extent to which the three style morphs are expressed; mid-style floral morphs are absent toward the northern limits of the range (M. E. Dorken and C. G. Eckert 2001).

4. DIDIPLIS Rafinesque, Atlantic J. 1: 177. 1833 • Water purslane or starwort [Derivation uncertain; Greek *dis*, twice, and *diploos*, double, possibly alluding to 2 stamens in 4-merous floral tube, or to 2-stamened *Didiplis*, segregated from 6-stamened *Peplis*] E

Shirley A. Graham

Herbs, annual or short-lived perennial, aquatic or amphibious, 0.5–4 dm, glabrous throughout. **Stems** erect, creeping, floating, or submerged, irregularly branched, frequently rooting at nodes when submerged. **Leaves** usually opposite, sometimes subalternate or whorled, dimorphic; sessile or subsessile; blade linear when submerged, narrowly elliptic to lanceolate when aerial, base truncate when submerged, tapered when aerial. **Inflorescences** indeterminate, flowers solitary, axillary, opposite. **Flowers** sessile, actinomorphic, monostylous; floral tube perigynous, broadly campanulate; epicalyx segments absent; sepals 4, broadly deltate, ½ floral tube length; petals 0; nectariferous tissue present at ovary-floral tube junction; stamens 2–4; ovary 2-locular; placenta globose; style sturdy, relatively short; stigma capitate. **Fruits** capsules, walls thin and dry, indehiscent, splitting irregularly. **Seeds** ca. 25, narrowly obovoid to fusiform, slightly convex-concave; cotyledons ± complanate.

Species 1: c, e United States.

Didiplis is easily overlooked due to its undistinguished aspect. In the past it has been included in *Ammannia*, *Lythrum*, and *Peplis*. Molecular evidence now strongly supports the sister relationship of *Didiplis* to *Rotala* (J. A. Morris et al. 2007).

1. Didiplis diandra (Nuttall ex de Candolle) Alph. Wood, Amer. Bot. Fl., 124. 1870 E F

Peplis diandra Nuttall ex de Candolle in A. P. de Candolle and A. L. P. P. de Candolle, Prodr. 3: 77. 1828

Leaves: blade 5–33(–40) × 0.5–5 (–10) mm. **Floral tube** 2 × 2 mm. **Capsules** globose, 1.5–2 × 1.5–2 mm. **Seeds** 0.6–0.7 × 0.2 mm. $2n$ = 32.

Flowering late spring–fall. Shallow streams, margins of swamps and ponds, temporary pools, ditches, drying mudflats; 50–500 m; Ala., Ark., Fla., Ga., Ill., Ind., Iowa, Kans., Ky., La., Minn., Miss., Mo., Nebr., N.C., Tenn., Tex., Va., Wis.

A Utah record is based on a specimen collected at Fish Lake, Sevier County, in 1894 (*M. E. Jones 5781*, BM) and on inclusion of the genus in the Utah flora without reference to a collection (I. Tidestrom 1925). *Didiplis* is not confirmed as a past or present member of that flora. Early reports of the species in Ohio by M. L. Fernald (1950) and others have not been verified (T. S. Cooperrider 1995). *Didiplis diandra* usually produces flowers and seeds underwater. Plants are often grown and sold internationally for the aquarium trade.

Ammannia nuttallii A. Gray, *Didiplis linearis* Rafinesque, *Hypobrichia nuttallii* Torrey & A. Gray, and *Ptilina aquatica* Nuttall ex Torrey & A. Gray are illegitimate names that pertain to *D. diandra*.

5. HEIMIA Link, Enum. Hort. Berol. Alt. 2: 3. 1822; Link in J. H. F. Link and C. F. Otto, Icon. Pl. Select. 5: 63, plate 28. 1822 • [For Ernst Ludwig Heim, 1747–1834, medical doctor in Berlin renowned for establishing sanitary health practices and said to have introduced Alexander von Humboldt to botany]

Shirley A. Graham

Shrubs or subshrubs, terrestrial, 5–30 dm, glabrous throughout. **Stems** erect, sparsely branched. **Leaves** opposite to subopposite; sessile or subsessile [petiolate]; blade lanceolate to linear-lanceolate, base narrowly attenuate. **Inflorescences** indeterminate, axillary, leafy racemes. **Flowers** sessile or subsessile, actinomorphic, monostylous; floral tube perigynous, campanulate to turbinate; epicalyx segments equal to or longer than sepals; sepals (5 or)6, ⅓–½ floral tube length; petals caducous, (5 or)6, bright yellow; nectary absent; stamens (10 or)12 (–24); ovary (3 or)4(–6)-locular; placenta globose; style slender; stigma capitate. **Fruits** capsules, walls thin and dry, dehiscence loculicidal. **Seeds** 200+, subpyramidal and elongate; cotyledons ± complanate.

Species 1–3 (1 in the flora): Texas, Mexico, South America; introduced in West Indies, Central America.

1. **Heimia salicifolia** Link, Enum. Hort. Berol. Alt. 2: 3. 1822; Link in J. H. F. Link and C. F. Otto, Icon. Pl. Select. 5: 63, plate 28. 1822 • Sinicuiche, Hachinal [F]

Heimia syphilitica de Candolle; *Nesaea salicifolia* Kunth

Stems narrowly 4-winged or 4-angled. **Leaves** 15–55[–87] × 2–10 mm. **Floral tube** 5–9 × 3–4 mm; epicalyx segments linear-corniform, 2–2.5 mm, apex often in-curved; sepals 1.5–2 × 1.5 mm; petals broadly obovate, 5–10(–14) × 3–5 mm. **Capsules** globose, 3–4.5 × 3.5–5.5 mm. **Seeds** 0.5–0.8 × 0.2–0.3 mm. 2*n* = 16 (Mexico, Bolivia).

Flowering spring–summer. Moist soil, streamsides in brushland; 0–1500 m; Tex.; Mexico; South America; introduced in West Indies, Central America.

Heimia salicifolia has an extensive native distribution from northern Argentina to southeastern Texas, with a gap in Central America. The species is employed in Mexico and parts of South America in traditional medicine and religious rites to induce mild auditory and visionary hallucinogenic effects. The array of medicinal applications also includes use as a strong anti-inflammatory, soporific, diuretic, and antisyphilitic, and for healing wounds. The effects obtained are due to the synergy of alkaloids present in the leaves (M. H. Malone and A. Rother 1994).

6. LAGERSTROEMIA Linnaeus, Syst. Nat. ed. 10, 2: 1068, 1076, 1372. 1759 • Crape or crepe myrtle [For Magnus Lagerstroem, 1696–1759, friend of Linnaeus and supporter of Uppsala University] [I]

Shirley A. Graham

Shrubs or small trees, terrestrial, 30–70 dm, outer bark whitish, smooth, flaking in thin plates, inner bark light brown or coral, glabrous or puberulent. **Stems** erect, branched. **Leaves** usually alternate or subalternate; subsessile [petiolate]; blade obovate or elliptic [lanceolate, oblong], base rounded or cuneate [attenuate]. **Inflorescences** determinate, terminal or axillary panicles. **Flowers** subsessile, actinomorphic, monostylous; floral tube perigynous, campanulate [semiglobose to turbinate], externally smooth [ridged]; epicalyx segments absent; sepals (5 or)6(or 7), ⅓–½ floral tube length; petals caducous or deciduous, (5 or)6(or 7)[–9], rose, purple, or white, obovate to orbiculate, often clawed; nectary absent; stamens 36–42[–200], dimorphic [uniform]; ovary (3–)6-locular; placenta elongate; style slender; stigma capitate to punctiform. **Fruits** capsules, woody, dehiscence loculicidal. **Seeds** ca. 20–50, obpyramidal, unilaterally winged from raphe; cotyledons rolled.

Species ca. 56 (1 in the flora); introduced; Asia, Pacific Islands, n Australia; introduced also widely in tropical and subtropical areas.

SELECTED REFERENCES Furtado, C. X. and S. Montien. 1969. A revision of *Lagerstroemia* Linnaeus (Lythraceae). Gard. Bull. Straits Settlem. 24: 185–334. Kim, S. C., S. A. Graham, and A. Graham. 1994. Palynology and pollen dimorphism in the genus *Lagerstroemia* (Lythraceae). Grana 33: 1–20.

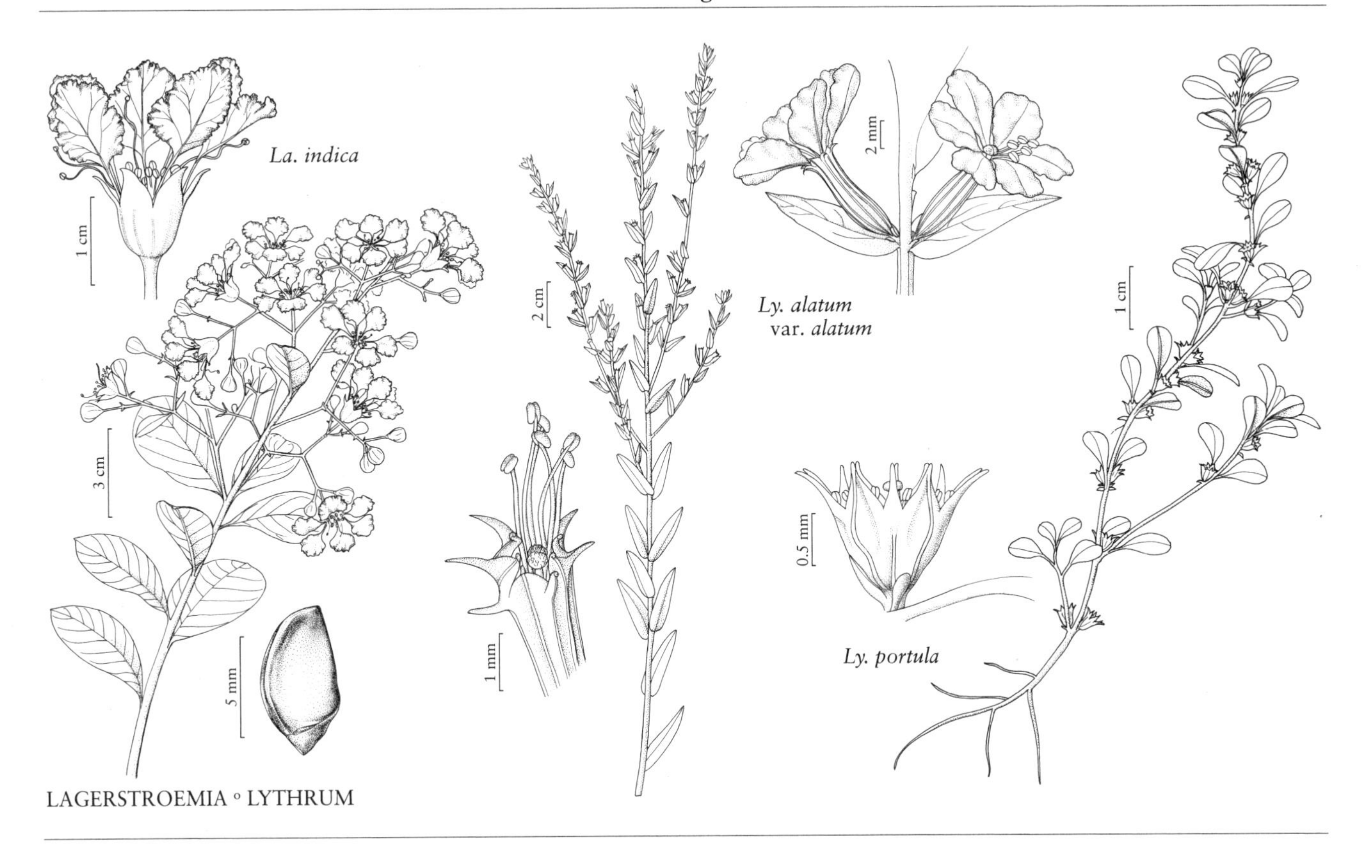

LAGERSTROEMIA ◦ LYTHRUM

1. Lagerstroemia indica Linnaeus, Syst. Nat. ed. 10, 2: 1076. 1759 [F] [I]

Leaves 20–70 × 15–30 mm. **Panicles** showy, 5–25 cm. **Flowers:** floral tube 7–10 × 6–10 mm; sepals erect; petals 10–12 × 7–15 mm, slender claw 7–8 mm; antisepalous stamens 6, stout, anthers green, relatively large, antipetalous stamens 30+, in clusters of 5 or 6, [rarely antipetalous stamens solitary], slender, anthers yellow, relatively small. **Capsules** globose to oblong, 9–15 × 8–11 mm. **Seeds** 7–9 mm, including unilateral wing. **2*n*** = 48, 50 (China, India).

Flowering summer–fall. Roadsides, secondary woodlands; 50–1000 m; introduced; Fla., Ga., La., Miss., S.C., Tex., Utah; Asia; introduced also widely in tropical and subtropical areas.

Lagerstroemia indica is popular in the southern and western United States as a landscape and street tree for its showy, densely flowered inflorescences. Numerous cultivars varying in habit and flower color have been developed. The dimorphic anthers of *L. indica* fill different reproductive roles. The larger, green anthers produce fertile pollen and the smaller, yellow anthers produce abundant, sterile pollen that attracts pollinators. A larger tree species, *L. speciosa* (Linnaeus) Persoon, is cultivated in southern Florida and southward; it has not become naturalized. It is distinguished from *L. indica* by oblong leaves to 28 cm and a yellow-brown puberulent, deeply ridged floral tube. Hybrid cultivars of the Japanese *L. subcostata* Koehne var. *fauriei* Hatusima × *L. indica* show a range of flower colors and habit and are increasingly popular in landscape use.

7. LYTHRUM Linnaeus, Sp. Pl. 1: 446. 1753; Gen. Pl. ed. 5, 205. 1754 • Loosestrife [Greek *lythron*, gore, alluding to use of *L. salicaria* in arresting hemorrhages]

Shirley A. Graham

Herbs, annual or perennial, or subshrubs, terrestrial or amphibious, 0.5–15(–30) dm, green, gray-green, or gray-white glaucous, usually glabrous, sometimes puberulent. **Stems** erect, weakly erect, or procumbent, usually branched, youngest growth narrowly 4-ridged or winged, submerged stems sometimes thickened by spongy tissue. **Leaves** opposite, subopposite, alternate, subalternate, or whorled; sessile or subsessile; blade ovate, obovate, oblong, oblong-lanceolate, lanceolate, oblanceolate, linear, linear-oblong, linear-lanceolate, orbiculate, or suborbiculate, base rounded, cordate, or attenuate. **Inflorescences** terminal and spikelike or racemes. **Flowers** sessile, subsessile, or short-pedicellate, actinomorphic, mono-, di-, or tristylous; floral tube perigynous, usually cylindrical or obconic (campanulate in *L. portula*), 8–12-ribbed; epicalyx segments shorter to or up to 2 times longer than sepals; sepals 6, narrowly deltate to subulate, obtuse and thickened in *L. tribracteatum*; petals caducous, (0 or)6, purple, lavender, rose, rose purple, pink, or white, sometimes with purple or red midvein; nectary encircling base of ovary or absent; stamens (2–)4–6(–12), usually 6 or 12, in 1 or 2 whorls, complementing style lengths; ovary 2-locular; placenta elongate; style slender, included or well exserted; stigma capitate. **Fruits** capsules, walls thin and dry, usually dehiscent, dehiscence septicidal or septifragal (indehiscent and splitting irregularly in *L. portula*, placenta and seeds remaining within capsules). **Seeds** 10–90(+), obovoid to fusiform or subglobose, to 1 mm; cotyledons ± complanate.

Species ca. 35 (12 in the flora): North America, Mexico, South America, Europe, Asia, Africa, Australia; introduced in West Indies.

Lythrum is represented in North America by equal numbers of native and introduced species. *Lythrum thymifolia* Linnaeus, native to the Mediterranean, was reported from Mobile, Alabama, on ballast in 1893 (C. T. Mohr 1901). It is similar to *L. hyssopifolia* but consistently has stamens reduced to two or three, leaves usually less than 2 mm wide, and is monostylous.

Lythrum is one of six genera in Lythraceae with a heterostylous breeding system and one of three in the family with tristylous species (also *Decodon* and the incipiently tristylous tropical genus *Adenaria* Kunth). *Lythrum* includes mono-, di-, and tristylous species. All native North American *Lythrum* species are distylous. Among the introduced species, *L. junceum*, *L. salicaria*, and *L. virgatum* are tristylous and *L. hyssopifolia*, *L. portula*, and *L. tribracteatum* are monostylous. The native North American species represent a single lineage corresponding taxonomically to subsect. *Pythagorea* Koehne. Some species of the subsection are taxonomically difficult and have shown little molecular divergence (J. A. Morris 2007). Hybridization and introgression are suspected where two native species co-occur, and also possibly between non-native species or native with non-native (B. Ertter and D. Gowen 2019).

SELECTED REFERENCE Shinners, L. H. 1953. Synopsis of the United States species of *Lythrum* (Lythraceae). Field & Lab. 21: 80–89.

1. Floral tubes broadly campanulate, 1 × 1.5 mm, widths greater than lengths; capsules indehiscent, splitting irregularly; stems mostly decumbent to creeping. . . . . .9. *Lythrum portula*
1. Floral tubes cylindrical or obconic, 3–7 × (0.4–)1–3 mm, lengths distinctly greater than widths; capsules septicidal or septifragal; stems erect, decumbent, or prostrate.
    2. Inflorescences spikelike, terminal (in *L. virgatum* racemose proximally, spikelike distally); flowers tristylous; stamens 12.
        3. Leaf blades lanceolate, bases cordate or rounded; plants usually gray-puberulent, sometimes glabrate . . . . 10. *Lythrum salicaria*
        3. Leaf blades lanceolate to narrowly linear, bases attenuate; plants glabrous. . . . 12. *Lythrum virgatum*
    2. Inflorescences racemose, diffuse, leafy; flowers mono-, di-, or tristylous; stamens (2–)4–12.
        4. Herbs or subshrubs 0.5–6 dm; stems decumbent or procumbent to erect, unbranched, branched from near base, or sparsely branched (much-branched distally in *L. ovalifolium*), sometimes with short accessory branches distally.
            5. Floral tubes obconic, with red spots on proximal half; epicalyx segments about equal to and more prominent than sepals; flowers tristylous; stamens 12 . . . . .6. *Lythrum junceum*
            5. Floral tubes obconic or cylindric, without red spots; epicalyx segments shorter than, equal to, or longer than sepals; flowers mono- or distylous; stamens (2–)4–6(–12).
                6. Floral tubes obcconic, lengths 8–10 times width at tube base; epicalyx segments as long as or longer than sepals; flowers monostylous . . . . .11. *Lythrum tribracteatum*
                6. Floral tubes cylindric or slightly obconic, lengths 5 times or less widths; epicalyx segments about 2 times longer than sepals; flowers mono- or distylous.
                    7. Leaves opposite throughout, equal to or shorter than internodes; flowers distylous . . . . 4. *Lythrum flagellare*
                    7. Leaves mostly alternate, sometimes opposite proximally, mostly longer than internodes and closely overlapping distally; flowers mono- or distylous.
                        8. Plants gray-green glaucous; leaf blade bases rounded; petals pink or rose, 1/2 length of floral tube; flowers monostylous . . . . 5. *Lythrum hyssopifolia*
                        8. Plants green to slightly gray glaucous; leaf blade bases attenuate; petals pale purple to purple, sometimes with red midvein, about as long as floral tube; flowers distylous. . . . . 8. *Lythrum ovalifolium*
        4. Herbs or subshrubs (3–)5–15 dm; stems erect, much-branched distally; flowers distylous.
            9. Leaves usually opposite or subopposite throughout, rarely alternate; floral nectaries absent . . . . 7. *Lythrum lineare*
            9. Leaves usually opposite to subopposite proximally, alternate distally or sometimes throughout; floral nectaries encircling bases of ovaries.
                10. Branch leaves abruptly and conspicuously smaller than those on main stems; floral tube obconic; pedicels slender; epicalyx segments and sepals about equal length . . . . 3. *Lythrum curtissii*
                10. Branch leaves gradually smaller than those on main stems; floral tube cylindrical; pedicels stout; epicalyx segments equal to or to 2 times length of sepals.
                    11. Herbs or subshrubs green or slightly gray glaucous; leaf blades ovate to oblong and bases subcordate to rounded, or lanceolate to linear-lanceolate and bases attenuate. . . . . .1. *Lythrum alatum*
                    11. Herbs whitish gray glaucous; leaf blades oblong-lanceolate proximally, mostly linear or linear-oblong distally, bases rounded . . . . 2. *Lythrum californicum*

1. Lythrum alatum Pursh, Fl. Amer. Sept. 1: 334. 1813
• Blue waxweed, winged loosestrife [E] [F] [W]

Herbs perennial, or subshrubs, slender or robust, 5–15 dm, green or slightly gray glaucous, glabrous. **Stems** erect, much-branched distally. **Leaves** opposite to subopposite proximally, alternate distally, branch leaves gradually smaller than those on main stem; sessile; blade ovate to oblong and base subcordate to rounded, or lanceolate to linear-lanceolate and base attenuate, 10–76 × 2–14 mm. **Inflorescences** racemes. **Flowers** alternate, subsessile, pedicel stout, distylous; floral tube cylindrical, 3–7 mm × 1 mm; epicalyx segments 2 times length of sepals; petals purple, obovate or oblong, 2–6.5 × 1.5–3 mm; nectary encircling base of ovary; stamens 6. **Capsules** septicidal or septifragal. **Seeds** ca. 30, obovoid to fusiform.

Varieties 2 (2 in the flora): North America; introduced in Mexico, West Indies.

Taxonomic judgement has varied as to limits of *Lythrum alatum* and closely related taxa centered around it (*L. californicum*, *L. flagellare*, and *L. ovalifolium*). Identification of species from sympatric areas is often difficult and morphology suggests that hybridization occurs where ranges overlap. *Lythrum californicum* apparently hybridizes with *L. alatum* var. *alatum* in Kansas and Oklahoma, and with *L. alatum* var. *lanceolatum* in Oklahoma and eastern Texas. Variety *lanceolatum* may also hybridize with *L. curtissii* in Calhoun County, Georgia, where they co-occur. A specimen combining features of *L. alatum* and *L. flagellare* is known from Hernando County, Florida.

1. Leaf blades ovate to oblong, bases subcordate to rounded; stems mostly less than 10 dm. . . . . 1a. *Lythrum alatum* var. *alatum*
1. Leaf blades lanceolate to linear-lanceolate, bases attenuate; stems mostly greater than 10 dm . . . . 1b. *Lythrum alatum* var. *lanceolatum*

1a. Lythrum alatum Pursh var. **alatum** [E] [F] [W]

Lythrum dacotanum Nieuwland

Stems mostly less than 10 dm, often 1+ stems closely arising from enlarged rootstock. **Leaf blades** ovate to oblong, base subcordate to rounded. **2*n*** = 20.

Flowering summer. Wet soil, low, open fields, prairies, woodland margins; 50–1500 m; Ont.; Ala., Ark., Colo., Conn., D.C., Fla., Ga., Ill., Ind., Iowa, Kans., Ky., Maine, Md., Mass., Mich., Minn., Miss., Mo., Nebr., N.H., N.J., N.Y., N.Dak., Ohio, Okla., Pa., R.I., S.Dak., Tenn., Vt., Va., W.Va., Wis., Wyo.; introduced in Mexico.

Lythrum cordifolium Nieuwland is a later homonym that pertains here.

1b. Lythrum alatum Pursh var. **lanceolatum** (Elliott) Rothrock, Rep. U.S. Geogr. Surv., Wheeler, 120. 1879 [E] [W]

Lythrum lanceolatum Elliott, Sketch Bot. S. Carolina 1: 544. 1821

Stems mostly greater than 10 dm, usually 1 main stem. **Leaf blades** lanceolate to linear-lanceolate, base attenuate. **2*n*** = 20.

Flowering summer. Wet soil, ditches, thickets, low fields; 50–1500 m; Ala., Ark., Fla., Ga., Ky., La., Miss., Mo., N.C., Okla., S.C., Tenn., Tex., Va.; introduced in West Indies (Cuba, Dominican Republic).

2. Lythrum californicum Torrey & A. Gray, Fl. N. Amer. 1: 482. 1840 [W]

Lythrum alatum Pursh var. *breviflorum* A. Gray; *L. alatum* var. *linearifolium* A. Gray; *L. breviflorum* (A. Gray) S. Watson; *L. linearifolium* (A. Gray) Small; *L. parvulum* Nieuwland; *L. sanfordii* Greene

Herbs perennial, slender, 5–10(–15) dm, whitish gray glaucous, glabrous. **Stems** erect, much-branched distally. **Leaves** mostly alternate, sometimes opposite proximally, branch leaves gradually smaller than those on main stem; sessile; blade oblong-lanceolate proximally, mostly linear or linear-oblong distally, 7–60 × 1–7 mm, base rounded. **Inflorescences** racemes. **Flowers** alternate, subsessile, pedicel stout, distylous; floral tube cylindrical, 5–7 × 1–1.5 mm; epicalyx segments equal or to 2 times length of sepals; petals bright rose purple, obovate, 4–6 × 2–4.5 mm; nectary encircling base of ovary; stamens 6. **Capsules** septicidal or septifragal. **Seeds** ca. 50, obovoid to fusiform. **2*n*** = 20.

Flowering spring–fall. Wet or moist soil, margins of ponds, streams, in ditches, on salt flats; 100–2200 m; Ariz., Calif., Kans., Nev., N.Mex., Okla., Utah, Tex.; Mexico (Baja California, Chihuahua, Coahuila, Nuevo León).

Lythrum californicum sometimes is difficult to distinguish from *L. alatum*; it generally has a more open vegetative habit with narrowly linear leaves. Problematic intermediates between *L. californicum* and *L. alatum* var. *alatum* occur in Kansas and Oklahoma, and between *L. alatum* var. *lanceolatum* and *L. californicum* in Oklahoma and eastern Texas. Prior to the recent recognition of *L. junceum* in California, older collections were identified as *L. californicum*.

3. Lythrum curtissii Fernald, Bot. Gaz. 33: 155. 1902 [C] [E]

Herbs perennial, slender, 5–10 dm, green, glabrous. **Stems** erect, much-branched distally. **Leaves** opposite proximally, alternate distally, branch leaves abruptly and conspicuously smaller than those on main stem; sessile or subsessile; blade (on main stem) broadly to narrowly lanceolate or oblong, 20–75 × 5–17 mm, (on branches) oblong to narrowly oblong, 3–15 × 1.5–3 mm, base narrowly attenuate. **Inflorescences** racemes. **Flowers** alternate, subsessile or pedicellate, pedicel slender, 1–1.5 mm, distylous; floral tube obconic, 3–6 × 1 mm; epicalyx segments about equal to length of sepals; petals deep to pale purple with dark central vein, oblanceolate or oblong, 1.5–2 × 0.5–1 mm; nectary encircling base of ovary; stamens 6. **Capsules** septicidal or septifragal. **Seeds** ca. 20, narrowly obovoid to fusiform. $2n = 20$.

Flowering summer–fall. Wet or moist shady woodlands, streamsides; of conservation concern; 0–100 m; Fla., Ga.

The shady habitat of *Lythrum curtissii*, a rare, delicate-flowered species, is unlike that of other native species of *Lythrum*, which tend toward more open, sunny areas. Putative hybrids of *L. alatum* var. *lanceolatum* and *L. curtissii* have been noted in the vicinity of Leary, Calhoun County, Georgia.

4. Lythrum flagellare Shuttleworth ex Chapman, Fl. South. U.S. ed. 2, 620. 1883 [C] [E]

Herbs perennial, slender, 1–4 dm, green, glabrous. **Stems** from creeping rhizome, rooting at nodes, weakly erect, unbranched or sparsely branched. **Leaves** opposite throughout, scarcely smaller distally, equal to or shorter than internodes; sessile or subsessile; blade oblong, 5–13 × 2–6 mm, base rounded. **Inflorescences** racemes, sparsely flowered. **Flowers** alternate or opposite, subsessile, distylous; floral tube without red spots, obconic, 4–5 × 1 mm; epicalyx segments 2 times length of sepals; petals pale purple to purple, obovate, 2.5–4 × 0.5–1 mm; nectary encircling base of ovary; stamens 6. **Capsules** septicidal or septifragal. **Seeds** undescribed. $2n = 20$.

Flowering spring–summer. Wet, springy areas; of conservation concern; 0–50 m; Fla.

Lythrum flagellare is relatively rare. A collection from Hernando County is possibly a hybrid of *L. alatum* and *L. flagellare*, as suggested by a more robust habit, leaves ultimately crowded and alternate on the stems, and less slender, more cylindrical floral tube than is typical for *L. flagellare*.

5. Lythrum hyssopifolia Linnaeus, Sp. Pl. 1: 447. 1753 • Hyssop-leaved loosestrife [I] [W]

Lythrum adsurgens Greene

Herbs annual or short-lived perennial, slender, 1.5–6 dm, gray-green glaucous, glabrous. **Stems** often from creeping rhizome, erect or weakly erect, sparsely branched distally. **Leaves** mostly alternate, sometimes opposite proximally, overlapping and scarcely smaller distally, equal to or longer than internodes; sessile; blade oblong to linear, 5–30 × 1–10 mm, base rounded. **Inflorescences** racemes. **Flowers** alternate, subsessile, monostylous; floral tube without red spots, obconic, becoming cylindrical, 4–6 × 0.5–1 mm; epicalyx segments 2 times length of sepals; petals pink or rose, oblong to obovate, 1.5–3(–5) × 0.7–1.3 mm, 1/2 floral tube length; nectary absent; stamens (2–)4–6(–12). **Capsules** septicidal or septifragal. **Seeds** ca. 20, obovoid, 0.8 × 0.5 mm. $2n = 20$.

Flowering summer–fall. Disturbed moist or seasonally flooded ground, drying pond margins, vernal pools, marshes; 50–1600 m; introduced; B.C., Ont.; Calif., Maine, Mass., Mich., N.H., N.J., N.Y., Ohio, Oreg., Pa., R.I., Wash.; Europe; Asia; introduced also in South America (Argentina, Chile), Africa, Pacific Islands (New Zealand), Australia.

Lythrum hyssopifolia has been present in the eastern United States since at least the early 1800s and now has a scattered and disjunct distribution in the eastern and western states and Ontario. Successful establishment is attributed in part to its self-compatible, monostylous breeding system. In Australia, *L. hyssopifolia* has been responsible for the poisoning death of young sheep that grazed on canola stubble contaminated with it (B. Crawford 2002).

6. Lythrum junceum Banks & Solander in A. Russell, Nat. Hist. Aleppo ed. 2, 2: 253. 1794 [I]

Herbs annual or short-lived perennial, slender 2–7 dm, green, glabrous. **Stems** sprawling or ascending, often branched from base, lax. **Leaves** mostly alternate, overlapping and smaller distally, subsessile; blade oblong at midstem to narrowly linear distally 8–38 × 1–11 mm, base obtuse to truncate. **Inflorescences** racemes. **Flowers** opposite to alternate, solitary in leaf axils, subsessile, tristylous; floral tube red-dotted on proximal half, narrowly obconic, 5–7 mm; epicalyx segments about equal to and more prominent than sepals; petals purple to rose, often with wihte base, obovate to oblong, 5–8 × 4–6 mm, stamens 12. **Capsules** septicidal or septifragal. **Seeds** many, oblong, 2 × 1 mm. **2*n*** = 10.

Flowering summer–fall. Moist or wet places, seasonal pools, lake margins, springs; introduced; 0–500 m; Calif.; s Europe; sw Asia; n Africa; Atlantic Islands (Macaronesia); introduced also in South America (Brazil), c Europe, Pacific Islands (New Zealand), Australia.

Lythrum junceum was present in Alameda County in 1905. It is now locally abundant there and also established in Santa Clara County (B. Ertter & D. Gowen 2019). It was also collected in Massachusetts in 1883, but did not persist there.

7. **Lythrum lineare** Linnaeus, Sp. Pl. 1: 447. 1753 [E]

Herbs perennial, or subshrubs, slender, 3–15 dm, green, glabrous. **Stems** erect, much-branched distally. **Leaves** usually opposite or subopposite throughout, rarely alternate, not crowded or overlapping distally; sessile; blade narrowly linear to narrowly lanceolate, 5–40 × 1–4 mm (mostly 5–10 mm distally in inflorescences), base attenuate. **Inflorescences** racemes. **Flowers** opposite, subsessile to pedicellate, pedicel slender, 0.5–1.5 mm, distylous; floral tube obconic becoming cylindrical, 3–5 × 1 mm; epicalyx segments equal to or slightly surpassing length of sepals; petals pale purple or white, oblong to obovate, 3–4 × 1 mm, mostly shorter than floral tube; nectary absent; stamens 6. **Capsules** septicidal or septifragal. **Seeds** ca. 30, obovoid to fusiform. **2*n*** = 20.

Flowering summer–fall. In and near brackish marshes of the Atlantic and Gulf coastal plains; 0–100 m; Ala., Conn., Del., Fla., Ga., La., Md., Miss., N.J., N.Y., N.C., S.C., Tex., Va.

8. **Lythrum ovalifolium** (A. Gray) Engelmann ex Koehne, Bot. Jahrb. Syst. 1: 321. 1881 [E]

Lythrum alatum Pursh var. *ovalifolium* A. Gray, Boston J. Nat. Hist. 6: 187. 1850; *L. alatum* var. *pumilum* A. Gray

Herbs perennial, slender, 1.5–4 dm, green to slightly gray glaucous, glabrous. **Stems** from creeping rhizome, erect or decumbent, much-branched distally. **Leaves** mostly alternate; sessile or subsessile; blade (on main stem) oblong to suborbiculate, 15–25 × 6–15 mm, mostly longer than internodes and closely overlapping distally, (on branches) oblong to linear, 6–10 × 1–2 mm, base attenuate. **Inflorescences** racemes. **Flowers** alternate, subsessile, pedicel slender, distylous; floral tube without red spots, obconic, becoming cylindrical, 3.5–6 × 1.3–3 mm; epicalyx segments 2 times length of sepals; petals pale purple to purple, sometimes with red midveins, obovate, 3–5.5 × 1.5–5.5 mm, about as long as floral tube; nectary encircling base of ovary, very narrow; stamens 6. **Capsules** septicidal or septifragal. **Seeds** ca. 25, obovoid to fusiform. **2*n*** = 20.

Flowering spring–fall. Wet areas; 200–1500 m; Tex.

Lythrum ovalifolium is known only from the Edwards Plateau. It is part of a complex of species related to *L. alatum* (J. A. Morris 2007).

9. **Lythrum portula** (Linnaeus) D. A. Webb, Feddes Repert. 74: 13. 1967 • Water purslane [F] [I] [W]

Peplis portula Linnaeus, Sp. Pl. 1: 332. 1753

Herbs annual, slender, delicate, 0.5–2.5 dm, green, glabrous. **Stems** frequently creeping and rooting at nodes, procumbent, decumbent, or weakly erect, often branched near base. **Leaves** opposite; sessile; blade spatulate or oblong to broadly obovate or orbiculate, 5–15 × 3–8 mm, base narrowly attenuate. **Inflorescences** spikelike. **Flowers** opposite or alternate, along most of stem, sessile to subsessile, monostylous; floral tube broadly campanulate, 1 × 1.5 mm; epicalyx segments equal to or to 2 times longer than sepals; sepals $1/3$–$1/2$ floral tube length, apex dark red; petals early caducous, 0 or 6, white to pink or rose, 1 × 0.7 mm; nectary absent; stamens 6. **Capsules** surpassing floral tube, indehiscent, splitting irregularly. **Seeds** ca. 10–25, subglobose. **2*n*** = 10 (Europe).

Flowering summer–fall. Drying ponds, lake margins, shallow water; 1000–2200 m; introduced; B.C.; Calif., Oreg., Wash.; Europe; n Africa.

Lythrum portula was long regarded as belonging to *Peplis* and is still accepted in that genus in some floras (D. A. Webb 1967). It is widespread in western Asia and Europe and has become established in the northwestern United States and adjacent Canada. It may be expected occasionally elsewhere in cool temperate regions in the flora area, as suggested by a 1999 introduction in Lake County, Ohio, presumably by seeds in soil accompanying plants purchased from a nursery on the West Coast. The Ohio population was recognized as non-native and destroyed (J. K. Bissell, pers. comm.).

10. Lythrum salicaria Linnaeus, Sp. Pl. 1: 446. 1753
• Purple loosestrife, salicaire commune [I] [W]

Lythrum salicaria var. *gracilior* Turczaninow; *L. salicaria* var. *tomentosum* (Miller) de Candolle

Herbs perennial, or subshrubs, robust, 5–15(–30) dm, green, gray-puberulent, sometimes glabrate. **Stems** erect, much-branched distally. **Leaves** opposite proximally, alternate distally, sometimes whorled, gradually smaller and overlapping in inflorescences; sessile; blade lanceolate, 2–10(–14) × 0.5–2 mm, base cordate or rounded. **Inflorescences** spikelike with flowers mostly in densely clustered, whorled cymes. **Flowers** whorled, sessile or subsessile, tristylous; floral tube cylindrical, 4–6 × 2 mm; epicalyx segments equal to or to 2 times length of sepals; petals rose purple, oblong to obovate, 6–14 × 2.5–4 mm; nectary encircling base of ovary; stamens 12, of 2 lengths. **Capsules** septicidal or septifragal. **Seeds** ca. 90, obovoid. **2*n*** = 30 (Asia), 60.

Flowering summer–fall. Wet areas, marshes, ditches, swamps, shores; 0–2200 m; introduced; St. Pierre and Miquelon; Alta., B.C., Man., N.B., Nfld. and Labr. (Nfld.), N.S., Ont., P.E.I., Que., Sask.; Ala., Alaska, Ark., Calif., Colo., Conn., Del., D.C., Idaho, Ill., Ind., Iowa, Kans., Ky., Maine, Md., Mass., Mich., Minn., Miss., Mo., Mont., Nebr., Nev., N.H., N.J., N.Mex., N.Y., N.C., N.Dak., Ohio, Okla., Oreg., Pa., R.I., S.Dak., Tenn., Tex., Utah, Vt., Va., Wash., W.Va., Wis., Wyo.; Eurasia; introduced also in temperate zones nearly worldwide.

Lythrum salicaria is one of the ten most frequently listed noxious weeds in North America. It forms extensive, showy monocultures that, although beautiful to the eye, crowd out native vegetation in wetlands, severely degrade wildlife habitats, and eliminate wildlife food sources. It was first reported in North America in 1814, possibly arriving in ship ballast, and was well-established along the New England coast by the 1830s (R. L. Stuckey 1980; D. Q. Thompson et al. 1987). It slowly spread throughout the northern United States and Canada, reaching the Pacific Northwest by 1940. In recent years, the spread has been exponential, aided by its wide tolerance of soil and climate, expansion of the highway system, increased availability of disturbed and degraded habitats, and horticultural escapes. Reproductive success is due to immense seed production, estimated at over 2.5 million seeds per year for a single mature plant. Seeds have been shown to be 80% viable after two years and to form extensive seed banks. The woody roots are difficult to eliminate and easily resprout if not removed completely. Biological control using host-specific beetles introduced from Europe (*Galerucella* and *Hylobius* species that attack leaves and roots, respectively) is showing some success, but whether these insects will effectively control *L. salicaria* throughout North America and remain specific to it remain to be fully determined. At least 33 U.S. states and many Canadian provinces have declared purple loosestrife a noxious weed, prohibiting sale or distribution of the species and its cultivars. Putative fertile hybrids with the closely related, possibly conspecific, *L. virgatum*, also a European introduction, has caused *L. virgatum* also to be placed on the prohibited list in several states. Results of genetic comparisons among *L. salicaria*, *L. virgatum*, their cultivars, and native *L. alatum* using isozymes indicate that *L. salicaria* and *L. virgatum* cannot be distinguished, whereas *L. alatum* is distinct from the introduced species. The F_1 generations of *L. salicaria* cultivars × native *L. alatum* have been found to be highly fertile (N. O. Anderson and P. D. Ascher 1993; M. S. Strefeler et al. 1996).

SELECTED REFERENCES Blossey, B., L. C. Skinner, and J. Taylor. 2001. Impact and management of purple loosestrife (*Lythrum salicaria*) in North America. Biodivers. & Conservation 10: 1787–1807. Brown, B. J., R. J. Mitchell, and S. A. Graham. 2002. Competition for pollination between an invasive species (purple loosestrife) and a native congener. Ecology 83: 2328–2336. Mal, T. K. et al. 1992. The biology of Canadian weeds. 100. *Lythrum salicaria*. Canad. J. Pl. Sci. 72: 1305–1330. Stuckey, R. L. 1980. Distributional history of *Lythrum salicaria* (purple loosestrife) in North America. Bartonia 47: 3–20. Thompson, D. Q., R. L. Stuckey, and E. B. Thompson. 1987. Spread, impact, and control of purple loosestrife (*Lythrum salicaria*) in North American wetlands. Fish Wildlife Res. 2: i–v, 1–55.

11. Lythrum tribracteatum Salzmann ex Sprengel, Syst. Veg. 4(2): 190. 1827 [I] [W]

Herbs annual, slender, 0.5–3 dm, green, glabrous or sparsely, minutely glandular-hispid. **Stems** prostrate to weakly erect, often branched near base, with relatively short accessory branches distally. **Leaves** opposite proximally, subalternate distally; sessile; blade oblong to oblanceolate, 3–25 × 0.1–0.5 mm, base attenuate. **Inflorescences** spikelike. **Flowers** alternate, often crowded on branches, sessile or subsessile, monostylous; floral tube without red spots, narrowly cylindrical, 4–6 × 0.4–0.5 mm; epicalyx segments equal to length of sepals; sepals obtuse, thick; petals lavender, oblong, 1–2(–3) × 0.5 mm; nectary absent; stamens 4–6. **Capsules** septicidal or septifragal. **Seeds** ca. 11, subglobose. **2*n*** = 10.

Flowering spring–fall. Seasonally wet areas, drying ponds, vernal pools, ditches; 0–1300 m; introduced; Calif., Idaho, Utah; s Europe.

First collected in Solano County, California, in 1930 (*J. T. Howell 5208*, UC), *Lythrum tribracteatum* has subsequently been documented in 12 counties in

California and in southwestern Idaho and Utah. In Utah the species may have been introduced through contamination of seed stock of the cultivated dye-plant indigo (*Indigofera*).

12. Lythrum virgatum Linnaeus, Sp. Pl. 1: 447. 1753 [I] [W]

Herbs perennial, slender, 5–12 dm, green, glabrous. **Stems** erect, sparsely branched proximally, branched distally. **Leaves** opposite proximally, subopposite to alternate distally, well separated on stems, not densely crowded distally; sessile; blade lanceolate to narrowly linear, 30–120 × 10–20 mm, narrower distally, base usually attenuate (or rounded in possible hybrids with *L. salicaria*). **Inflorescences** racemes at proximal nodes, spikelike distally with flowers in loose, ± whorled clusters. **Flowers** alternate, or loosely clustered and ± whorled at distal nodes, sessile or subsessile, tristylous; floral tube cylindrical, 4–6 × 2 mm, 3+ times longer than wide; epicalyx segments awl-shaped, equal to or shorter than length of sepals; petals rose purple, narrowly spatulate, 5–7 × 1.5–3 mm; nectariferous disc surrounding base of ovary; stamens usually 12, of 2 lengths. **Capsules** septicidal or septifragal. **Seeds** 75+, obovoid to fusiform. **2*n*** = 30 (Slovakia).

Flowering summer–fall. Marshes, ditches, wet areas; 100–600 m; introduced; Mass., Mont., N.H., Pa., Va., Wash.; Europe.

Lythrum virgatum is distinguished from *L. salicaria* in its native Europe by glabrous stems, less robust habit, narrower, more widely spaced leaves with attenuate leaf bases, and less floriferous, more open inflorescences. The species are said to hybridize in their native ranges as well as in North America when they come into contact under cultivation. In the flora area, *L. virgatum* escapes from cultivation and is apparently naturalized to an unknown extent. The ability to produce fertile hybrids is a major concern because of the invasive nature of *L. salicaria*. For this reason, the sale of *L. virgatum*, like that of *L. salicaria*, is illegal in some states.

8. PUNICA Linnaeus, Sp. Pl. 1: 472. 1753; Gen. Pl. ed. 5, 212. 1754 • Pomegranate [Greek *phoenikeos*, reddish purple, alluding to classical name *punicum malum*, apple of Carthage] [I]

Shirley A. Graham

Shrubs or small trees, terrestrial, 20–60 dm, glabrous throughout. **Stems** erect, much-branched, branches often terminating as indurate thorns. **Leaves** opposite to subopposite; subsessile to shortly petiolate; blade oblong to lanceolate-elliptic, base rounded to attenuate, apex often developing foliar nectary. **Inflorescences** determinate, usually terminal, sometimes axillary, in 1–5 clusters, racemes. **Flowers** sessile, actinomorphic, monostylous; floral tube semi-epigynous to epigynous, usually red, sometimes yellow, campanulate or urceolate-campanulate, thick, leathery, slightly constricted medially; epicalyx segments absent; sepals 5–8, erect to recurved, 1/4–1/3 floral tube length; petals deciduous, 5–8, usually bright red, white, or variegated, sometimes yellow, showy, crumpled; nectariferous tissue present at ovary-floral tube junction; stamens many, covering inner surface of floral tube distal to ovary; ovary semi- or fully inferior, carpels in 2(or 3) superposed, twisted layers [in 1 regular whorl]; placenta elongate; style slender; stigma capitate. **Fruits** berries, crowned by persistent sepals and stamens, walls thick and leathery, indehiscent, splitting irregularly. **Seeds** ca. 100–1400, reddish purple, oblong-pyramidal, outer layer fleshy, inner layer hardened; cotyledons rolled.

Species 2 (1 in the flora): introduced; Europe, Asia (India, Iran, Middle East), Africa; introduced also in warmer regions of the New World and Old World.

Punica protopunica Balfour f. is known only from the Indian Ocean island of Socotra in Yemen.

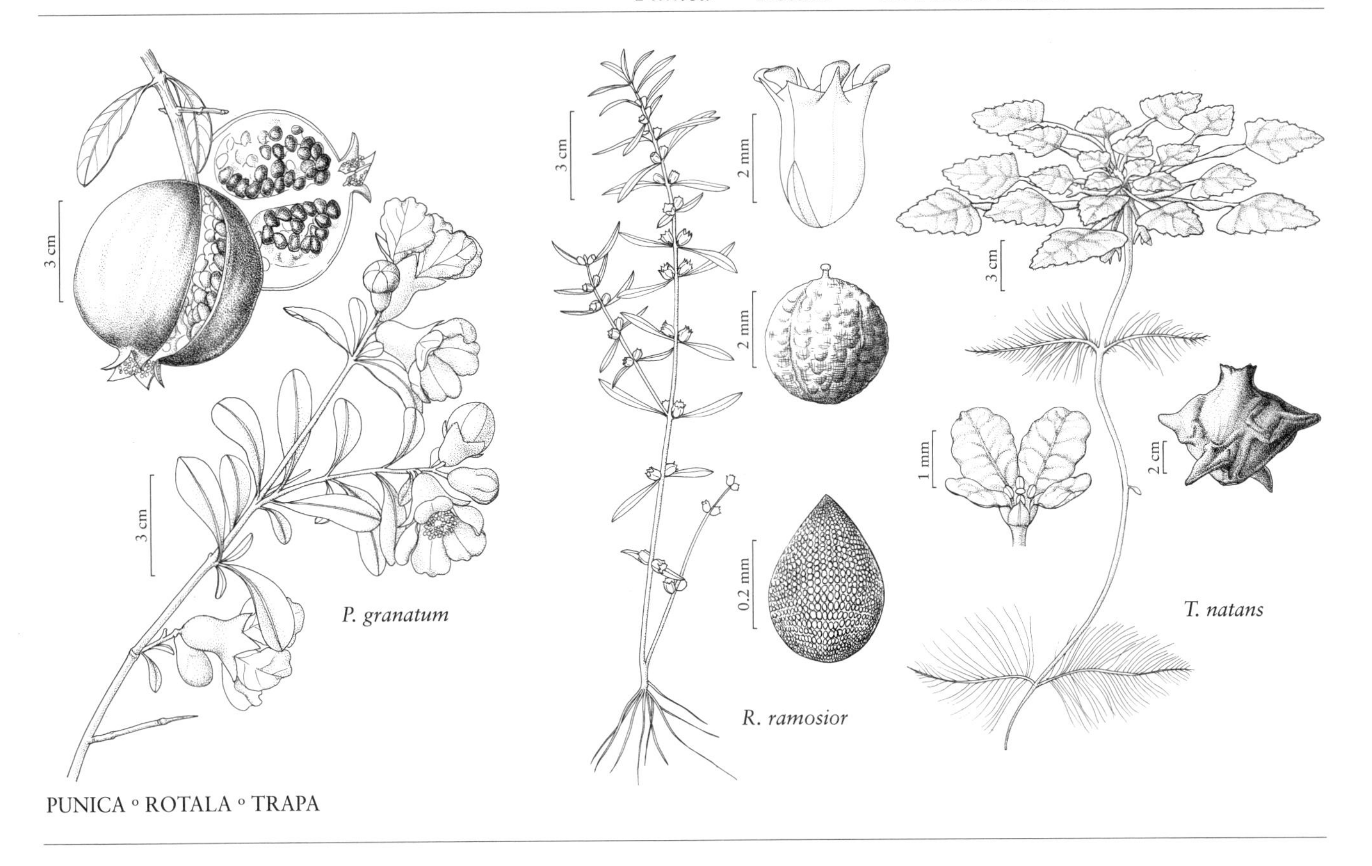

PUNICA ◦ ROTALA ◦ TRAPA

1. Punica granatum Linnaeus, Sp. Pl. 1: 472. 1753 [F] [I]

Leaves: petiole 2–5 mm; blade 25–65 × 10–20 mm, surfaces shiny. **Floral tube** 25–40 × 15–25 mm; sepals erect to recurved, thick, leathery; petals 15–25 × 10–20 mm; ovary proximal layer of 2 or 3 carpels with axile placentation, distal layer of ca. 5–7 carpels with parietal placentation. **Berries** reddish brown to yellow, globose, 5–12 cm diam. **Seeds** 7–12 mm. **2*n*** = 14, 16, 18, 19 (cultivated).

Flowering spring–fall. Old homesites; 400–1600 m; introduced; Calif., Tex.; s Europe; Asia; Africa; introduced also in Mexico, West Indies, Central America, South America.

Punica granatum has been widely planted as an ornamental in the western and southwestern United States and has naturalized around old homesites. It has been cultivated for centuries in warm regions of the world, probably having originated in the Middle East in the area of modern Iran, Afghanistan, and Iraq. The seeds are eaten in salads and other dishes, especially in Mediterranean cuisines. Juice expressed from the fruit rind and seeds is currently a popular beverage in North America, drunk for its antioxidant and anti-inflammatory properties. Dwarf ('nana') and double-flowered forms are also cultivated.

9. ROTALA Linnaeus, Mant. Pl. 2: 143, 175. 1771 • Toothcup [Latin *rota*, wheel, alluding to whorled leaves of *R. verticillaris* Linnaeus]

Shirley A. Graham

Herbs, annual or perennial, terrestrial, amphibious, or aquatic, often turning pink or red, 0.4–4 dm, glabrous throughout. **Stems** erect, decumbent, or creeping [floating], unbranched or branched, often rooting at nodes. **Leaves** usually opposite, rarely whorled [subalternate], dimorphic or uniform; sessile or subsessile; blade obovate, orbiculate, oblong-elliptic, oblanceolate, subspatulate, or linear [lanceolate, spatulate], base attenuate, obtuse, or rounded, margins membranous or cartilaginous. **Inflorescences** indeterminate, terminal or axillary racemes, axillary spikes, or solitary flowers (on main stem). **Flowers** sessile or subsessile, actinomorphic, monostylous [distylous]; floral tube perigynous, campanulate; epicalyx segments shorter to longer than sepals or absent; sepals [3 or] 4[–6], to ½ floral tube length; petals caducous or persistent, 0 or 4[–6], rose, pink, or white [purple]; nectariferous tissue variably present at base of inner wall of floral tube; stamens (1–)4[–6], in 1 whorl; ovary 2–4-locular; placenta globose, ultimately nearly free-central; style slender or stout; stigma capitate or thick-capitate. **Fruits** capsules, walls thin and dry, finely, transversely striate (10\x), 2–4-valved, dehiscence septicidal. **Seeds** 30–100+, obovoid-semiovoid, convex-concave, 0.3–0.5 mm; cotyledons ± complanate. $x = 8$.

Species ca. 50 (3 in the flora): North America, Mexico, West Indies, Central America, South America, Europe, s Asia, Africa.

The native *Rotala ramosior* and introduced *R. indica* and *R. rotundifolia* are weeds in North American rice fields, irrigation ditches and canals, and other disturbed wet areas. *Rotala ramosior* is increasingly found elsewhere in the world in rice-growing regions. The genus often grows with one or more species of the superficially similar *Ammannia* (S. A. Graham et al. 2011). Some species of *Rotala* are part of the international aquatic plant trade.

SELECTED REFERENCE Cook, C. D. K. 1979. A revision of the genus *Rotala* (Lythraceae). Boissiera 29: 1–156.

1. Inflorescences terminal racemes, simple or compound; leaves dimorphic, aerial blades obovate to orbiculate, submerged blades linear to orbiculate; style stout; stigmas thick-capitate . . . . 3. *Rotala rotundifolia*
1. Inflorescences terminal or axillary racemes, axillary spikes, or solitary flowers; leaves monomorphic, blades oblong-elliptic, oblanceolate, obovate to oblong, or subspatulate; style slender; stigmas capitate, not excessively thickened.
   2. Margins of leaves white-cartilaginous, blades obovate to oblong or subspatulate; flowers frequently in short axillary spikes or terminal or axillary racemes, sometimes solitary in axils of main stems; epicalyx segments absent . . . . 1. *Rotala indica*
   2. Margins of leaves membranous, blades oblong-elliptic to oblanceolate; flowers solitary in axils of main stems; epicalyx segments present . . . . 2. *Rotala ramosior*

1. Rotala indica (Willdenow) Koehne, Bot. Jahrb. Syst. 1: 172. 1880 • Indian toothcup [I] [W]

Peplis indica Willdenow, Sp. Pl. 2: 244. 1799

Herbs annual [perennial], terrestrial, amphibious, or aquatic. **Stems** erect or decumbent, unbranched or branched. **Leaves** opposite, monomorphic; blade obovate to oblong or subspatulate [ovate, lanceolate, linear], 4–20 × 3–8 mm, base attenuate to obtuse, margins white-cartilaginous, especially prominent when dry. **Inflorescences** terminal or axillary racemes, short axillary spikes, or solitary flowers. **Floral tube** 1.5–2.5 × 1–1.5 mm; epicalyx segments absent; sepal margins cartilaginous; petals persistent, 4, pink, linear to narrowly ovate, less than ½ length of sepals; stamens inserted about mid level in floral tube; style slender; stigma capitate. **Capsules** ca. 2 × 1.5 mm, 2-valved. $2n = 32$ (China, Japan).

Flowering late spring–summer. Rice fields, irrigation ditches; 50–100 m; introduced; Calif., La.; se Asia (to Japan), introduced also in Europe, Africa.

Rotala indica probably was introduced into the flora area as a contaminant of imported rice seed stock. It frequently occurs in rice fields together with the native *R. ramosior* and/or *Ammannia coccinea*.

2. Rotala ramosior (Linnaeus) Koehne in C. F. P. von Martius et al., Fl. Bras. 13(2): 194. 1877 [F] [W]

Ammannia ramosior Linnaeus, Sp. Pl. 1: 120. 1753; *A. dentifera* A. Gray; *A. humilis* Michaux; *Rotala dentifera* (A. Gray) Koehne; *R. ramosior* var. *dentifera* (A. Gray) Lundell; *R. ramosior* var. *interior* Fernald & Griscom

Herbs annual, terrestrial or amphibious. **Stems** erect, unbranched or branched. **Leaves** opposite, monomorphic; blade oblong-elliptic to oblanceolate, 10–50 × 2–12 mm, base attenuate, margins membranous. **Inflorescences** solitary flowers in axils of main stems. **Floral tube** 2–4 × 1–2 mm; epicalyx segments shorter to longer than sepals; sepals margins membranous; petals caducous, often 0, sometimes 4, white or pink, obovate, scarcely surpassing length of sepals; stamens inserted near base of floral tube; style slender; stigma capitate. **Capsules** 2–4.5 × 2–4.5 mm, (3 or)4-valved. $2n$ = 16 (Mexico), 32.

Flowering summer–fall. Low wet areas, marshes, temporary pools, rice fields, ditches; 0–1900 m; B.C., Ont.; Ala., Ariz., Ark., Calif., Colo., Conn., Del., D.C., Fla., Ga., Idaho, Ill., Ind., Iowa, Kans., Ky., La., Md., Mass., Mich., Minn., Miss., Mo., Mont., Nebr., N.H., N.J., N.Mex., N.Y., N.C., Ohio, Okla., Oreg., Pa., R.I., S.C., S.Dak., Tenn., Tex., Va., Wash., W.Va., Wis.; Mexico; West Indies; Central America; South America; introduced in Europe (Italy), Pacific Islands (Philippines).

Plants growing in the interior of the continent on rich soil are on average taller with wider leaves, longer bracteoles, and larger capsules than those growing on the sandy coastal plain. They were the basis for var. *interior*. Plants in the United States with relatively short bracteoles are apparently all diploids ($2n$ = 16). Plants with relatively long bracteoles and longer epicalyx segments, sometimes recognized as var. *dentifera*, occur in Mexico and southward; they are tetraploids ($2n$ = 32). The extensive distribution in the western United States is due to spread with rice cultivation and with the movement of cattle.

3. **Rotala rotundifolia** (Buchanan-Hamilton ex Roxburgh) Koehne, Bot. Jahrb. Syst. 1: 175. 1880 • Roundleaf toothcup [I]

Ammannia rotundifolia Buchanan-Hamilton ex Roxburgh, Fl. Ind. 1: 446. 1820

Herbs perennial [annual], amphibious or aquatic. **Stems** creeping, branched proximally, mostly unbranched distally (except in inflorescences). **Leaves** usually opposite, sometimes whorled, dimorphic; blade 3–20 × 2.5–14 mm, aerial blades obovate to orbiculate, submerged blades linear to orbiculate, (somewhat smaller than aerials), base rounded, margins membranous, abaxial surfaces of aerial leaves usually green, of submerged leaves often red or purple. **Inflorescences** terminal racemes, simple or compound. **Floral tube** 2 × 1.3 mm; epicalyx segments absent; sepals margins membranous; petals caducous, 4, bright rose, obovate, 2–3 times length of sepals; stamens inserted near base of floral tube; style stout; stigma thick-capitate. **Capsules** 1.5 × 1.5 mm, 4-valved. $2n$ = 16, 28, 30 (Asia).

Flowering spring–early summer. Rice fields in standing water, along drying shores and pond edges, open water; 0–100 m; introduced; Ala., Fla.; Asia.

Rotala rotundifolia, a plant of the international aquatic plant trade, was introduced in the southern United States through inappropriate disposal of unwanted aquarium plants. It was first documented in the flora area in 1996 in a canal in Broward County, Florida (C. C. Jacono and V. V. Vandiver 2007). It is currently known from two other counties in Florida and from Tuscaloosa, Alabama. The flowers are self-fertilizing and produce large numbers of minute, buoyant, viable seeds. Spread also occurs by the adventitious rooting of stem fragments. In Florida, extensive floating mats have become well-established in flood-control canals.

10. TRAPA Linnaeus, Sp. Pl. 1: 120. 1753; Gen. Pl. ed. 5, 56. 1754 • Water chestnut or caltrop, châtaigne d'eau [Greek *kalkitrapa*, ancient 4-spiked weapon, alluding to spiny fruit] [I]

C. Barre Hellquist

Shirley A. Graham

Herbs, annual, aquatic, rosette-forming, rooted or floating, to 50 dm (when rooted in deep water), glabrous or velutinous. **Stems** submerged, flexible, unbranched, submerged nodes often developing green, pinnately-dissected to filiform adventitious leaflike roots. **Leaves** dimorphic, flanked by deeply cleft stipules; floating leaves alternately or spirally arranged in terminal rosette, supported by slender to bulbous float at mid petiole, blade rhombic to triangular, base cuneate, margins coarsely toothed in distal ½; submerged leaves subopposite, sessile, blade linear, margins entire. **Inflorescences** indeterminate, emergent, solitary flowers in axils of floating leaves, pedicels elongating and declining after pollination, submerging developing fruit. **Flowers** pedicellate, actinomorphic, monostylous; floral tube perigynous to semi-epigynous, campanulate; epicalyx segments absent; sepals 4, 3 times floral tube length, persistent on fruit as hardened spines; petals caducous, 4, white or pale lavender; nectary development unknown; stamens 4, in 1 whorl; ovary 2-locular, surrounded by coronary disc, semi-inferior, fully inferior in fruit; placenta axile, abbreviated, at apex of ovary; style slender; stigma capitate; ovules 1 per locule, pendulous, 1 locule and ovule failing to develop after anthesis. **Fruits** drupes, top-shaped, woody, with 2–4 hardened, spiny horns, endocarp indurate, exocarp evanescent, with tubercles. **Seed** 1, oblong; cotyledons 2, unequal, 1 large, starchy, retained in fruit, the other scalelike, growing out of fruit apex, becoming photosynthetic.

Species 1–30+ (1 in the flora): introduced; Europe, Asia, Africa; introduced widely in subtropical and temperate regions.

Trapa is cultivated in China for the edible nut or enlarged cotyledon, which is referred to as water chestnut. The familiar cultivated Chinese water chestnut marketed in North America is usually another plant, the sedge *Eleocharis dulcis* (Burman f.) Trinius ex Henschel. Specialized features of *Trapa* (inflated leaf petioles, semi-inferior or fully inferior ovary, hardened horned fruits, unequal cotyledons) long made exact taxonomic placement difficult. Until recently, the genus was generally treated as the sole member of Trapaceae, with close relationship to Lythraceae or Onagraceae. Substantial molecular evidence indicates that *Trapa* is sister genus to *Sonneratia* Linnaeus f., a genus of mangrove trees in southeastern Asia. The two genera are well-nested within Lythraceae and closely related to two other Asian Lythraceae, *Duabanga* Buchanan-Hamilton, a genus of southeastern Asian trees, and *Lagerstroemia*, the crape myrtle (S. A. Graham et al. 2005). The great discrepancy in number of species attributed to the genus is primarily the result of extensive variability in the fruit shape and in the shape and number of horns and spines on the fruit. *Trapa* has become a notorious invader of rivers and lakes in the northeastern United States since its introduction in the nineteenth century. It was reported in southeastern Canada for the first time in 1998, extending the range northward from localities in the Lake Champlain watershed in the United States. *Trapa* fruits are well known from the Miocene, and younger, fossil deposits in the United States, Europe, and Asia.

SELECTED REFERENCES Couillault, J. 1973. Organisation de l'appareil conducteur de *Trapa natans* L. Bull. Soc. Bot. France 119: 177–198. Groth, A. T., L. Lovett-Doust, and J. Lovett-Doust. 1996. Population density and module demography in *Trapa natans* (Trapaceae), an annual, clonal aquatic macrophyte. Amer. J. Bot. 83: 1406–1415. Ram, M. 1956. Floral morphology and embryology of *Trapa bispinosa* Roxb. with a discussion on the systematic position of the genus. Phytomorphology 6: 312–323.

1. Trapa natans Linnaeus, Sp. Pl. 1: 120. 1753 F I W

Stems slender, young growth and flowering parts velutinous. **Leaves** of floating rosettes bearing successively longer petioles toward outer edges of rosette, to 20 cm; blade 20–40 × 25–60 mm, width greater than length, surfaces velutinous abaxially, glabrous adaxially. **Floral tube** 2 mm; sepals 4–7 mm, keeled; petals obovate, 8–15 mm. **Drupes** 20–25 mm diam., excluding spines; horns 2–4, to ca. 10 mm. **2*n*** = 48 (Poland, Japan), 96 (Japan).

Flowering summer–fall. Relatively neutral, nutrient-rich, flowing or still waters, rivers, ponds, lakes; 0–400 m; introduced; Que.; Conn., Del., D.C., Maine, Md., Mass., N.H., N.J., N.Y., Pa., R.I., Vt., Va.; Europe; Asia; Africa.

Trapa natans was first noted in Massachusetts in 1859 from an unknown origin and was recorded from the Charles River, Cambridge, in 1879. The hard, spiny fruits can cause severe puncture wounds and are slow to decay in lake and river bottoms. The species propagates by seed and by detached floating rosettes to form extensive floating mats that reduce oxygen, restrict light, crowd out native plants, and make navigation difficult. Populations grow rapidly and are difficult to eradicate except by sustained efforts over multiple seasons. Federal regulations now prohibit interstate sale and transport of *T. natans*. The species is considered rare and threatened in Europe.

ONAGRACEAE Jussieu

• Evening Primrose Family

Warren L. Wagner
Peter C. Hoch

Herbs, annual or perennial, shrubs, or subshrubs, [lianas or trees], terrestrial, amphibious, or aquatic, unarmed, not clonal; often with epidermal oil cells, usually with internal phloem, abundant raphides in vegetative cells. **Stems** erect to decumbent or prostrate. **Leaves** usually deciduous, usually alternate or opposite, sometimes whorled, simple, usually cauline, sometimes basal and forming rosettes; stipules present, intrapetiolar, usually caducous, relatively small, or absent (tribes Epilobieae and Onagreae); sessile or subsessile to petiolate; blade margins usually entire, toothed, or pinnately lobed, rarely bipinnately lobed. **Inflorescences** axillary, flowers solitary, leafy spikes, racemes, or panicles. **Flowers** usually bisexual, (protandrous in *Chamaenerion*, *Clarkia*, *Epilobium*, [and most species of *Lopezia*]; protogynous in *Circaea* and *Fuchsia*), sometimes unisexual (gynodioecious or dioecious, [subdioecious]), usually actinomorphic, sometimes zygomorphic, (2–)4(–7)-merous; perianth and androecium epigynous; sepals persistent after anthesis (in *Ludwigia*), or all flower parts deciduous after anthesis; floral tube present or absent in *Chamaenerion*, *Ludwigia*, [and most species of *Lopezia*]; sepals usually green or red, rarely pink or purple, valvate; petals present, rarely absent, often fading darker with age, imbricate or convolute, sometimes clawed; nectary present; stamens 2 times as many as sepals and in 2 series, antisepalous set usually longer, rarely all equal (*Chamaenerion*), or as many as sepals, [in *Lopezia* reduced to 2 or 1 plus 1 sterile staminode]; filaments distinct; anthers usually versatile, sometimes basifixed, dithecal, polysporangiate, with tapetal septa, sometimes also with parenchymatous septa, opening by longitudinal slits, pollen grains united by viscin threads, (2 or)3(–5)-aperturate, shed singly or in tetrads or polyads; ovary inferior, usually with as many carpels and locules as sepals, rarely 1 or 2 (*Circaea* and *Gayophytum*), septa sometimes thin or absent at maturity; placentation axile or parietal; style 1, stigma 1, with as many lobes as sepals or clavate to globose, papillate or not, and wet with free-running secretions to dry without the secretions; ovules 1 to numerous per locule, in 1 or several rows or clustered, anatropous, bitegmic. **Fruit** a loculicidal capsule or indehiscent berry or nutlike. **Seeds** smooth or sculptured, sometimes with a coma or wings, with straight, oily embryo, 4-nucleate embryo sac, endosperm absent. x = 7, 8, 10, 11, 15, 18.

Genera 22, species 664 (17 genera, 277 species in the flora): North America, Mexico, West Indies, Bermuda, Central America, South America, Eurasia, Africa, Atlantic Islands, Indian Ocean Islands, Pacific Islands, Australasia; nearly worldwide, primarily New World.

Members of the Onagraceae are especially richly represented in North America. The family comprises annual and perennial herbs, with some shrubs and a few small to medium-sized trees. Most species occur in open habitats, ranging from dry to wet, with a few species of *Ludwigia* aquatic, from the tropics to the deserts of western North America, temperate forests, and arctic tundra; some species of *Epilobium*, *Ludwigia*, and *Oenothera* can be weeds in disturbed habitats. Members of the family are characterized by 4-merous flowers (sometimes 2-, 5-, or 7-merous), an inferior ovary, a floral tube in most species, stamens usually two times as many as sepals, and pollen connected by viscin threads. Flowers are usually bisexual, sometimes unisexual, and plants are gynodioecious, matinal, diurnal, or vespertine, self-compatible or self-incompatible, often outcrossing and then pollinated by a wide variety of insects or birds, or autogamous (P. H. Raven 1979; W. L. Wagner et al. 2007).

Onagraceae are known in considerable systematic detail, and information is available on comparative breeding systems and pollination biology, on chromosome numbers and cytogenetic relations, often involving translocations, and on vegetative, floral, and seed anatomy, palynology, and embryology. The phylogeny of the family is known in reasonably good detail, with most parts of the trees generally well-supported.

The suprageneric and generic classification presented by W. L. Wagner et al. (2007) differs in a number of ways from the previous classification (P. H. Raven 1979, 1988). Onagraceae are divided into two subfamilies based on a fundamental basal split recognized in all phylogenetic studies (R. H. Eyde 1981; P. C. Hoch et al. 1993; R. A. Levin et al. 2003, 2004; V. S. Ford and L. D. Gottlieb 2007), with *Ludwigia* on one branch (as Ludwigioideae), and the rest of the family on a second branch (as Onagroideae). Onagroideae are subdivided into six tribes: Circaeeae (including Fuchsieae), Epilobieae, Gongylocarpeae, Hauyeae, Lopezieae, and Onagreae. The Epilobieae and Onagreae are diverse; together they constitute fully two-thirds of the species in the family and include 15 of the 22 genera. The classification following Wagner et al. can be viewed on the Onagraceae web site by Wagner and Hoch at http://botany.si.edu/Onagraceae.

SELECTED REFERENCES Bult, C. J. and E. A. Zimmer. 1993. Nuclear ribosomal RNA sequences for inferring tribal relationships within Onagraceae. Syst. Bot. 18: 48–63. Carlquist, S. 1975. Wood anatomy of Onagraceae, with notes on alternative modes of photosynthate movement in dicotyledon woods. Ann. Missouri Bot. Gard. 62: 386–424. Ford, V. S. and L. D. Gottlieb. 2007. Tribal relationships within Onagraceae inferred from *Pgi*C sequences. Syst. Bot. 32: 348–356. Hoch, P. C. et al. 1993. A cladistic analysis of the plant family Onagraceae. Syst. Bot. 18: 31–47. Kurabayashi, M., H. Lewis, and P. H. Raven. 1962. A comparative study of mitosis in the Onagraceae. Amer. J. Bot. 9: 1003–1026. Levin, R. A. et al. 2003. Family-level relationships of Onagraceae based on chloroplast *rbc*L and *ndh*F data. Amer. J. Bot. 90: 107–115. Levin, R. A. et al. 2004. Paraphyly in tribe Onagreae: Insights into phylogenetic relationships of Onagraceae based on nuclear and chloroplast sequence data. Syst. Bot. 29: 147–164. Lewis, H. and P. H. Raven. 1961. Phylogeny of the Onagraceae. Recent Advances Bot. 2: 1466–1469. Munz, P. A. 1965. Onagraceae. In: N. L. Britton et al. 1905+. North American Flora.... 47+ vols. New York. Ser. 2, part 5, pp. 1–278. Raven, P. H. 1979. A survey of reproductive biology in Onagraceae. New Zealand J. Bot. 17: 575–593. Raven, P. H. 1988. Onagraceae as a model of plant evolution. In: L. D. Gottlieb and S. K. Jain, eds. 1988. Plant Evolutionary Biology. A symposium Honoring G. Ledyard Stebbins. London. Pp. 85–107. Skvarla, J. J. et al. 1978. An ultrastructural study of viscin threads in Onagraceae pollen. Pollen & Spores 20: 5–143. Skvarla, J. J., P. H. Raven, and J. Praglowski. 1975. The evolution of pollen tetrads in Onagraceae. Amer. J. Bot. 62: 6–35. Skvarla, J. J., P. H. Raven, and J. Praglowski. 1976. Ultrastructural survey of Onagraceae pollen. In: I. K. Ferguson and J. Muller, eds. 1976. The Evolutionary Significance of the Exine. London. Pp. 447–479. Tobe, H. and P. H. Raven. 1985. The histogenesis and evolution of integuments in Onagraceae. Ann. Missouri Bot. Gard. 72: 451–468. Tobe, H. and P. H. Raven. 1996. Embryology of Onagraceae (Myrtales): Characteristics, variation and relationships. Telopea 6: 667–688. Wagner, W. L., P. C. Hoch, and P. H. Raven. 2007. Revised classification of the Onagraceae. Syst. Bot. Monogr. 83: 1–240.

1. Sepals persistent or tardily caducous after anthesis; flowers (3 or)4 or 5(–7)-merous; floral tube absent; petals yellow or white [a. Onagraceae subfam. Ludwigioideae, p. 70] . . . . 1. *Ludwigia*, p. 70
1. Sepals deciduous after anthesis (along with other flower parts); flowers (2–)4-merous; floral tube usually present, often elongate, if absent then petals usually rose purple or pink, rarely white [b. Onagraceae subfam. Onagroideae, p. 101].
    2. Stipules present and soon deciduous; fruit indehiscent (berry or burlike capsule with hooked hairs) [b1. Onagraceae subfam. Onagroideae tribe Circaeeae, p. 101].
        3. Fruit a berry; seeds few to ca. 500; flowers 4-merous . . . . .2. *Fuchsia*, p. 102
        3. Fruit a capsule, burlike, with stiff, hooked hairs; seeds 1 or 2; flowers 2-merous . . .3. *Circaea*, p. 104
    2. Stipules absent; fruit usually a capsule, sometimes indehiscent.
        4. Seeds usually comose, coma rarely secondarily lost; sepals erect or spreading; stigmas with dry multicellular papillae, entire or 4-lobed, lobes commissural; $x = 18$ [b2. Onagraceae subfam. Onagroideae tribe Epilobieae, p. 107].
            5. Floral tube absent; stamens subequal; style deflexed at anthesis, later erect, stamens initially erect, later deflexed; leaves usually alternate, rarely subopposite or subverticillate proximally . . . . 4. *Chamaenerion*, p. 108
            5. Floral tube present; stamens in 2 unequal whorls; style and stamens erect; leaves opposite, at least near base of stem . . . . 5. *Epilobium*, p. 112
        4. Seeds not comose; sepals reflexed; stigmas usually wet, non-papillate, and entire or (3 or)4-lobed (non-commissural), sometimes (*Clarkia*) lobes commissural and then with dry unicellular papillae; $x = 7$ [b3. Onagraceae subfam. Onagroideae tribe Onagreae, p. 159].

[6. Shifted to left margin.—Ed.]

6. Stigmas with commissural lobes and dry, unicellular papillae; flowers usually protandrous . . . . . 6. *Clarkia*, p. 160
6. Stigmas hemispherical to subglobose or subcapitate, peltate, or 4-lobed, not commissural, surface wet, non-papillate; flowers not protandrous.
    7. Ovaries 2-locular; stems usually delicate . . . . . 7. *Gayophytum*, p. 189
    7. Ovaries (3 or)4-locular; stems usually not especially delicate.
        8. Seeds with concave and convex sides, concave side with a thick wing, convex side covered with glasslike, clavate hairs; petals white with yellow basal area . . . . . 8. *Chylismiella*, p. 196
        8. Seeds not concave/convex and not with a wing and clavate hairs; petals yellow, purple, red, or white, if white, mostly without yellow base.
            9. Ovaries with a slender, sterile apical projection; plants usually acaulescent.
                10. Herbs perennial; sterile projection of ovary persistent with fertile part in fruit, projection without visible abscission lines at its junctures with floral tube or fertile part of ovary . . . . . 9. *Taraxia*, p. 197
                10. Herbs annual; sterile projection of ovary with visible abscission lines at its junctures with both short floral tube and fertile part of ovary . . . . 13. *Tetrapteron*, p. 222
            9. Ovaries without an apical projection; plants usually caulescent, sometimes acaulescent (in *Oenothera*).
                11. Styles with peltate indusium at base of stigma, at least at younger stages prior to anthesis; stigmas (3 or)4-lobed, receptive all around (or peltate to discoid or nearly square in sect. *Calylophus*) . . . . . 17. *Oenothera*, p. 243
                11. Styles without indusium; stigmas usually subglobose to globose, subcapitate, capitate, or cylindrical (*Eulobus*), rarely conical-peltate and ± 4-lobed.
                    12. Seeds in 2 rows per locule; capsules pedicellate; leaves mostly basal, blades often pinnately lobed, rarely bipinnately, sometimes unlobed, or lateral lobes greatly reduced or absent, terminal lobe usually large, abaxial surface of leaves or leaf margins with conspicuous, usually brown, oil cells . . . . . 16. *Chylismia*, p. 227
                    12. Seeds in 1 row per locule; capsules usually sessile, rarely very shortly pedicellate; leaves not predominately basal, blades not lobed or pinnatifid, leaves without oil cells.
                        13. Petals usually white, rarely red or tinged red; flowers vespertine . . . . . 11. *Eremothera*, p. 207
                        13. Petals yellow, often with red flecks or spots; flowers diurnal.
                            14. Flowering stems virgate; leaf blades pinnatifid to lobed; petals yellow with red flecks near base; seeds usually with purple spots; floral tube with a lobed disc . . . . . 15. *Eulobus*, p. 226
                            14. Flowering stems not virgate; leaf blades not pinnatifid, margins entire or toothed; petals yellow, sometimes with 1+ red spots at base; seeds without purple spots; floral tube without a lobed disc.
                                15. Stems densely leafy distally, nearly leafless proximally, with many slender, ascending branches from base; capsules strongly flattened, straight . . . . . 14. *Neoholmgrenia*, p. 224
                                15. Stems usually leafy throughout, branched throughout or with a few basal branches; capsules not flattened, subterete or 4-angled, often flexuous or curled, sometimes straight.
                                    16. Capsules subterete; flowers only from distal nodes; seeds appearing smooth, glossy, triangular in cross section . . . . . 10. *Camissonia*, p. 200
                                    16. Capsules 4-angled, at least when dry; flowers from basalmost to distal nodes; seeds dull, flattened . . . . . 12. *Camissoniopsis*, p. 214

a. ONAGRACEAE Jussieu subfam. LUDWIGIOIDEAE W. L. Wagner & Hoch, Syst. Bot. Monogr. 83: 20. 2007

Leaves: stipules present. **Flowers:** floral tube absent; sepals (3 or)4 or 5(–7), persistent at ovary apex after dehiscence of other floral parts; petals yellow or white, rarely absent. $x = 8$.

Genus 1, species 82 (31 in the flora): North America, Mexico, West Indies, Bermuda, Central America, South America, se Asia, Africa, Indian Ocean Islands, Pacific Islands, Australasia; introduced in Europe, w Asia.

Ludwigioideae were segregated as a distinct subfamily (W. L. Wagner et al. 2007) to reflect the phylogenetic relationship of *Ludwigia* as sister to other genera of Onagraceae in morphological and molecular analyses (see R. A. Levin et al. 2003, 2004). *Ludwigia* is distinguished by the absence of a floral tube, persistence of sepals on capsules after other floral parts dehisce, pollen shed in tetrads or polyads (or as monads in some sections, tetrads sometimes found elsewhere in Onagraceae), double ovule vascular supply, uniquely including a central supply (R. H. Eyde 1981), single-celled ovule archesporium (H. Tobe and P. H. Raven 1996), and a base chromosome number of $x = 8$.

1. LUDWIGIA Linnaeus, Sp. Pl. 1: 118 (as Ludvigia), [1204]. 1753; Gen. Pl. ed. 5, 55. 1754 • Water-primrose, primrose-willow, false loosestrife [For Christian Gottlieb Ludwig, 1709–1773, botanist and physician of Leipzig]

Peter C. Hoch

Herbs, usually perennial, rarely annual, or shrubs, [rarely trees], caulescent, usually glabrous, strigillose, villous, or hirtellous, rarely glandular-puberulent. **Stems** erect to spreading or prostrate and then often rooting at nodes, sometimes floating, submerged parts, when present, sometimes swollen with spongy aerenchyma or bearing inflated, white, spongy pneumatophores, usually branched. **Leaves** cauline, usually alternate, rarely opposite; stipules present, often deciduous, usually dark reddish green; usually petiolate, sometimes sessile; blade usually reduced distally, usually linear to lanceolate, oblong, or obovate, rarely deltate, with 1[or 2] ± conspicuous submarginal vein[s], margins entire, serrulate, or glandular-serrulate, usually without oil cells. **Inflorescences** spikes, racemes, or clusters, solitary or paired in leaf axils, erect or decumbent and ascending at tip; bracteoles usually 2 and conspicuous, black or dark red, often scalelike, at or near base of ovary, sometimes deciduous early, rarely absent. **Flowers** bisexual, actinomorphic, pedicellate or sessile; floral tube absent; sepals persistent after anthesis or tardily caducous, (3 or)4 or 5(–7), green, sometimes yellow or cream, often becoming flushed with red post-anthesis, spreading to suberect; petals caducous, usually (3 or)4 or 5(–7), sometimes 0, usually yellow, sometimes white, when yellow, then often ultraviolet-reflecting, margins entire; stamens as many as sepals in 1 series, or 2 times as many as sepals in 2 subequal or unequal series; anthers versatile, on smallest flowers appearing basifixed, pollen shed singly or in polyads or tetrads, 3(–5)-aperturate; ovary usually with as many locules as sepals, rarely more, apex flat or conical, often with raised or depressed nectary lobes surrounding base of each epipetalous stamen; style present [very rarely absent in sect. *Arborescentes*], usually glabrous; stigma entire or irregularly lobed, capitate or hemispherical, distal ½ receptive, surface wet and papillate. **Fruit** a capsule, spreading to erect, obconic, cylindric to clavate, turbinate, obpyramidal, or globose to cuboid, terete to sharply 4+-angled, straight to slightly curved, dehiscent irregularly or by a terminal pore, an apical ring, or flaps separating from valvelike apex, long-pedicellate to subsessile. **Seeds** 50–400, in 1–several rows per locule, usually free, sometimes embedded in endocarp, narrowly ovoid, smooth or finely pitted, raphe usually inconspicuous, sometimes expanded and nearly equal to seed [very rarely expanded into asymmetrical wing]. $x = 8$.

Species 82 (31 in the flora): North America, Mexico, West Indies, Bermuda, Central America, South America, se Asia, Africa, Indian Ocean Islands, Pacific Islands; introduced in Europe, w, s Asia, Australia.

Ludwigia is a pansubtropical genus currently divided into 22 sections (P. H. Raven [1963] 1964; T. P. Ramamoorthy and E. Zardini 1987; Zardini and Raven 1992; W. L. Wagner et al. 2007). The genus is well represented in North America and South America, and is frequent in Africa and south Asia. Raven provided a synopsis of *Ludwigia*, including in it all previously segregated genera (*Isnardia*, *Jussiaea*, *Ludwigiantha*, and *Oocarpon*), based in part on P. A. Munz (1942, 1944), H. Hara (1953), and others. H. E. Baillon (1866–1895, vol. 6) was the first author to merge *Isnardia* and *Jussiaea* under *Ludwigia*, and consequently *Ludwigia* is treated as having priority over the former two genera. Therefore, J. P. M. Brenan's (1953) subsequent merging of *Ludwigia* under *Jussiaea* is not acceptable (vide Shenzhen Code Art. 11.5). Hara (1953) referenced both Baillon and Brenan, and followed the then-practiced botanical code to accept *Ludwigia* over *Jussiaea*. Hara, and later Raven, made most of the new combinations needed in *Ludwigia*, and their works firmly established *Ludwigia*. Raven subdivided the genus into 17 sections using a combination of characters: sepal number, stamens as many or two times as many as sepals; pollen as monads or in tetrads (polyads were not distinguished until reported by J. Praglowski et al. 1983), capsule morphology, and seed morphology. The large sect. *Myrtocarpus*, primarily distributed in South America, was later subdivided into a total of eight sections (Ramamoorthy 1979; Ramamoorthy and Zardini; Zardini and Raven). Wagner et al. placed sect. *Oocarpon* into sect. *Oligospermum* (= sect. *Jussiaea*), and the present treatment, based on recent molecular analysis (Liu S. H. et al. 2017), combines formerly recognized sect. *Microcarpium* with sect. *Isnardia*. This reduces the number of sections to 22, 14 of which are monospecific; Liu et al. did not have sufficient resolution to evaluate classification of sect. *Myrtocarpus* and its segregates.

Since the synopsis by P. H. Raven ([1963]1964), data have become available for *Ludwigia* from cytology (M. Kurabayashi et al. 1962; Raven and W. Tai 1979; E. Zardini et al. 1991), palynology (J. J. Skvarla et al. 1975, 1976, 1978; J. Praglowski et al. 1983; V. C. Patel et al. 1984), embryology (H. Tobe and Raven 1983, 1985, 1986, 1986b, 1996), and anatomy (S. Carlquist 1975, 1977, 1982b; R. H. Eyde 1977, 1979, 1981, 1982; R. C. Keating 1982), as well as several published and unpublished revisions of sections. These data provide a rich source of potential characters for phylogenetic analysis. All recent analyses, whether morphological or molecular (see especially Eyde 1977, 1979; R. A. Levin et al. 2003, 2004; Liu S. H. et al. 2017), strongly support *Ludwigia* as monophyletic and sister to the rest of the family. Liu et al. also found strong support for a monophyletic sect. *Ludwigia*, a monophyletic sect. *Isnardia* that includes sect. *Microcarpium*, and a clade comprised of sects. *Isnardia*, *Ludwigia*, and *Miquelia* that is sister to the rest of the genus. Liu et al. found poor resolution in that second branch, due in part to inadequate sampling, but strong support for monophyletic sects. *Jussiaea* and *Macrocarpon*.

Ludwigia appears to have diverged from the common ancestor of the family between 80 and 93 Ma (E. Conti et al. 1997; K. J. Sytsma et al. 2004). The genus exhibits a complex biogeographic pattern, with ten sections endemic or centered in South America (39 species), two in North America (24 species), five in Africa (seven species), two in Asia (two species), and two not clearly centered in a single continent (10 species). *Ludwigia* has a base chromosome number of $x = 8$; aneuploidy is unknown, but polyploidy is extensive (P. H. Raven and W. Tai 1979; E. Zardini et al. 1991). S. A. Graham and T. B. Cavalcanti (2001) proposed that $x = 8$ is the base chromosome number for Lythraceae, which is sister to Onagraceae. This suggests that

$x = 8$ in *Ludwigia* is a plesiomorphy for Onagraceae, and that the chromosome number changed to $x = 11$ or $x = 10$ (in *Hauya*) on the branch leading to the rest of the family.

In the absence of a more thorough revision and formal phylogenetic analysis of *Ludwigia*, this treatment follows the most recent classification of the genus by W. L. Wagner et al. (2007), which is based primarily on P. H. Raven ([1963]1964) and supported by subsequent systematic and anatomical studies (especially Raven and W. Tai 1979; J. Praglowski et al. 1983; T. P. Ramamoorthy and E. Zardini 1987; Zardini and Raven 1992). Molecular analyses by Liu S. H. et al. (2017) support inclusion of sect. *Microcarpium* within sect. *Isnardia*, which involves more than half of the North American species. The sections are arranged using characters from Raven and elsewhere, as described in Wagner et al.

Species of *Ludwigia* characteristically grow in wet habitats, and some are nearly or fully aquatic. Those species often have adaptations for growing in water: aerenchyma—respiratory tissue with particularly large intercellular spaces—in the proximal stems, and/or pneumatophores, which are spongy, white roots arising from internodes on floating stems that facilitate aeration needed for root respiration in hydrophytic plants. Species in sect. *Ludwigia* have fusiform, tuberous roots that may also serve an adaptive function in wet habitats.

Seventy-five species of *Ludwigia* are self-compatible and seven (in sects. *Macrocarpon* and *Myrtocarpus*) are self-incompatible (P. H. Raven 1979). Flowers of *Ludwigia* are diurnal, remaining open for several days or, sometimes, for only one day (in small-flowered autogamous species); species may be outcrossing and pollinated by bees, flower-flies, or butterflies, or autogamous (Raven).

Several species of *Ludwigia* are cultivated as aquarium plants (for example, *L. repens*); others are grown in water gardens. Several species, especially in sect. *Jussiaea*, are considered noxious, invasive species (M. Wood 2006).

SELECTED REFERENCES Eyde, R. H. 1977. Reproductive structures and evolution in *Ludwigia* (Onagraceae). I. Androecium, placentation, meristem. Ann. Missouri Bot. Gard. 64: 644–655. Eyde, R. H. 1979. Reproductive structures and evolution in *Ludwigia* (Onagraceae). II. Fruit and seed. Ann. Missouri Bot. Gard. 66: 656–675. Eyde, R. H. 1981. Reproductive structures and evolution in *Ludwigia* (Onagraceae). III. Vasculature, nectaries, conclusions. Ann. Missouri Bot. Gard. 68: 470–503. Hara, H. 1953. *Ludwigia* versus *Jussiaea*. J. Jap. Bot. 28: 289–294. Liu, S. H. et al. 2017. Multi-locus phylogeny of *Ludwigia* (Onagraceae): Insights on infrageneric relationships and the current classification of the genus. Taxon 66: 1112–1127. Munz, P. A. 1942. Studies in Onagraceae XII. A revision of the New World species of *Jussiaea*. Darwiniana 4: 179–284. Munz, P. A. 1944. Studies in Onagraceae XIII. The American species of *Ludwigia*. Bull. Torrey Bot. Club 71: 152–165. Raven, P. H. 1963[1964]. The old world species of *Ludwigia* (including *Jussiaea*), with a synopsis of the genus (Onagraceae). Reinwardtia. 6: 327–427. Tobe, H. and P. H. Raven. 1986b. A comparative study of the embryology of *Ludwigia* (Onagraceae): Characteristics, variation, and relationships. Ann. Missouri Bot. Gard. 73: 768–787. Zardini, E. and P. H. Raven. 1992. A new section of *Ludwigia* (Onagraceae) with a key to the sections of the genus. Syst. Bot. 17: 481–485.

1. Stamens 2 times as many as sepals, in 2 series; seeds in 1 row per locule and embedded in endocarp, or 2 to several rows and free.
   2. Seeds in 1 row per locule, embedded in segment of endocarp, with inconspicuous raphe; capsules cylindric, subcylindric, or subclavate, terete, subterete, or obscurely angled.
      3. Stems usually erect or ascending, rarely floating or creeping; seeds loosely embedded in horseshoe-shaped segment of endocarp, easily detached; pollen in polyads; leaves alternate [1d. *Ludwigia* sect. *Seminudae*] . . . . 6. *Ludwigia leptocarpa*
      3. Stems floating or creeping and ascending to erect; seeds firmly embedded in woody segment of endocarp; pollen as monads; leaves alternate or fascicled [1e. *Ludwigia* sect. *Jussiaea*].
         4. Bracteoles deltate, 0.5–1 × 0.5–1 mm; leaf blades mostly oblong or elliptic, petioles 0.3–6 cm; capsules 10–14 mm, pedicels 7–60(–90) mm; seeds 1–1.5 × 0.9–1.3 mm; sepals 3–12 mm . . . . 9. *Ludwigia peploides*
         4. Bracteoles narrowly or broadly obovate, 1–1.8 × 0.7–0.8 mm; leaf blades mostly lanceolate or oblanceolate, petioles 0.1–2(–2.5) cm; capsules (11–)14–25(–30) mm, pedicels (9–)13–25(–85) mm; seeds 0.8–1 × 0.8–1 mm; sepals 6–19 mm.
            5. Emergent plant sparsely to densely villous; petioles 0.5–2(–2.5) cm; sepals (8–)12–19 mm; petals (15–)20–30 mm . . . . 7. *Ludwigia hexapetala*

5. Emergent plant generally villous and viscid; petioles 0.1–1.1 cm; sepals 6–12(–16) mm; petals (12–)16–20(–26) mm . . . . 8. *Ludwigia grandiflora*

2. Seeds in several rows per locule, free, raphe enlarged or inconspicuous; capsules obconic, obpyramidal, or cylindric, angled, winged, or terete.
   6. Seeds with enlarged raphe ± equal to seed body; capsules cylindric to clavate, subterete to ± angled; sepals 4 [1c. *Ludwigia* sect. *Macrocarpon*].
      7. Stems 60–250(–400) cm; sepals (6–)8–13 × 3–7 mm; petals (5–)10–20 × 5–20 mm; capsules (17–50 mm) usually exceeding pedicels (5–25 mm) . . . 4. *Ludwigia octovalvis*
      7. Stems 20–120 cm; sepals 10–20 × 7–12 mm; petals 20–35 × 10–30 mm; capsules (20–35 mm) rarely exceeding pedicels (10–40 mm) . . . . 5. *Ludwigia bonariensis*
   6. Seeds with inconspicuous raphe; capsules obconic or obpyramidal, oblong-obovoid, or subclavate to squarish cylindric, 4+-angled; sepals 4 or 5[–7].
      8. Stems terete or angled; capsules ± 4- or 5-angled; plants perennial herbs or shrubs, usually pubescent, rarely glabrous [1a. *Ludwigia* sect. *Myrtocarpus*] . . . . 1. *Ludwigia peruviana*
      8. Stems strongly 4-angled or -winged; capsules strongly 4+-angled or -winged; plants annual or short-lived perennial herbs, glabrous or sometimes strigillose on leaves and/or inflorescences [1b. *Ludwigia* sect. *Pterocaulon*].
         9. Leaves sessile; stems sharply 4-angled and -winged; sepals 7–12 × 1.5–4 mm; petals 10–20 × 10–18 mm . . . . 2. *Ludwigia decurrens*
         9. Leaves with petioles 0.2–2.2 cm; stems 4-angled, rarely -winged; sepals 3–6 × 1–2 mm; petals 3.5–5 × 2–2.5 mm. . . . . 3. *Ludwigia erecta*

1. Stamens as many as sepals, in 1 series; seeds in several rows per locule, free.
   10. Stems erect; roots often fusiform, fascicled; capsules globose, subcuboid, or ellipsoid, terete to 4-angled or 4-winged, with hard walls, dehiscing by terminal pore; petals present; leaves alternate [1f. *Ludwigia* sect. *Ludwigia*].
      11. Leaves: petioles 0.1–0.3(–0.7) cm, blades attenuate; nectary disc slightly elevated, rounded; pedicels 2–7 mm, shorter than or equaling capsule . . . . 10. *Ludwigia alternifolia*
      11. Leaves: sessile, blades cuneate to attenuate; nectary disc elevated, domed, (prominently 4-lobed); pedicels 3–17 mm, equal to or exceeding capsule.
         12. Stems usually densely erect-hirsute, sometimes glabrous, well branched distally; leaf blades lanceolate to ovate-oblong; bracteoles 2.5–7 mm; pedicels 3–10 mm . . . . 11. *Ludwigia hirtella*
         12. Stems strigillose to glabrate, simple or sparsely branched, often near base; leaf blades ovate or obovate proximally, lanceolate-linear or linear distally; bracteoles 0.7–3.2(–5) mm; pedicels 5–17 mm.
            13. Petals 9–12 mm; style 1.5–3.3 mm, stigma not exserted beyond anthers; capsules 4–7 × 4–5 mm, subcuboid to squarish globose, often 4-winged . . . . .12. *Ludwigia maritima*
            13. Petals 14–19 mm; style 5–9.5 mm, stigma as long as or exserted beyond anthers; capsules 2–6.8 × 1.6–3.3 mm, subglobose to ellipsoid, not winged. . . . . 13. *Ludwigia virgata*
   10. Stems erect, ascending, prostrate, or decumbent; roots not fusiform, often with stolons or rhizomes; capsules subcylindric to clavate, oblong-obovoid, obconic, broadly obpyramidal or subglobose, terete to sharply angled, with hard or thin walls, dehiscent by apical ring or lenticular slits along locule edges or indehiscent; petals present or absent; leaves alternate or opposite [1g. *Ludwigia* sect. *Isnardia*].
      14. Leaves opposite; stems prostrate or decumbent, erect at tips, sometimes ascending; plants often forming mats, without stolons or rhizomes.
         15. Petals 0; sepals 1–2 mm; anthers 0.2–0.4 mm; capsules (1.6–)2–5 mm, less than 2 times as long as broad, walls thin; pollen shed as monads.
            16. Plants nearly glabrous; petioles 0.1–2.5 cm; seeds yellowish brown. . . 16. *Ludwigia palustris*
            16. Plants densely strigillose; petioles 0.3–0.9 cm; seeds dark reddish brown . . . . .26. *Ludwigia spathulata*
         15. Petals 4, sometimes caducous; sepals 1.8–10 mm; anthers 0.4–2 mm; capsules 4–10.5 mm, generally more than 2 times as long as broad, walls hard; pollen usually shed in tetrads, rarely as monads.

17. Leaf blades narrowly elliptic to broadly lanceolate-elliptic or suborbiculate; petals 1.1–3 mm; sepals 1.8–5 mm, about as long as wide; mature pedicels 0.1–3 mm . . . . 30. *Ludwigia repens*
17. Leaf blades narrowly elliptic to oblanceolate-elliptic or narrowly oblanceolate to linear; petals 4.5–11 mm; sepals (3.5–)4–10 mm, about 2–3 times as long as wide; mature pedicels (4.5–)6–45 mm.
    18. Petals 7–11 mm; anthers 1.3–2 mm; mature pedicels (12–)17–45 mm, generally much longer than subtending leaves . . . . 25. *Ludwigia arcuata*
    18. Petals 4.5–5.5 mm; anthers 0.7–1 mm; mature pedicels (4.5–)6–15 (–20) mm, as long as or shorter than subtending leaves . . . . 29. *Ludwigia brevipes*

[14. Shifted to left margin.—Ed.}

14. Leaves alternate, rarely opposite proximally; stems erect or ascending, rarely prostrate; plants not forming mats, often with stolons, rarely with rhizomes (in *L. suffruticosa*).
    19. Capsules subcylindric to elongate-obpyramidal, 2–10(–12) mm, usually at least 2 times as long as broad.
        20. Petals 0; sepals 1.1–2.3 mm; stem leaves usually narrowly elliptic to elliptic, rarely sublinear, petioles 0–1.5 cm, blades 3–12 cm . . . . 18. *Ludwigia glandulosa*
        20. Petals 4; sepals 2.5–7 mm; stem leaves linear to elliptic-linear or linear-oblanceolate, sessile or subsessile, blades 1.5–6(–8.5) cm.
            21. Capsules subcylindric, irregularly dehiscent; seeds reddish brown; anthers 0.6–1.1 mm . . . . 14. *Ludwigia linifolia*
            21. Capsules elongate-obpyramidal, dehiscent by apical ring; seeds light brown; anthers 1–2 mm . . . . 15. *Ludwigia linearis*
    19. Capsules oblong-obovoid, obpyramidal, or obconic to subglobose, (1–)1.5–5(–7) mm, less than 2 times as long as broad.
        22. Flowers in densely clustered terminal racemes or spikes; stems unbranched or slightly branched; rhizomes often present . . . . 20. *Ludwigia suffruticosa*
        22. Flowers in elongate, loose, leafy axillary racemes or spikes; stems usually well branched, sometimes unbranched; rhizomes absent.
            23. Plants usually densely pubescent throughout.
                24. Plants densely strigillose; bracteoles 0.5–1.5 mm; mature pedicels 0.5–1.2(–2.3) mm . . . . 24. *Ludwigia sphaerocarpa* (in part)
                24. Plants densely hirtellous; bracteoles (1.5–)2–6.5(–7.2) mm; mature pedicels 0–1 mm.
                    25. Sepals 3.5–5.5(–6) mm, adaxial surface creamy white, tips reflexed; style 1–2 mm; seed surface cells nearly isodiametric . . . . 22. *Ludwigia pilosa*
                    25. Sepals 1.5–3 mm, adaxial surface green, tips not reflexed; style 0.3–0.5 mm; seed surface cells transversely elongate . . . . 23. *Ludwigia ravenii*
            23. Plants glabrous or nearly so, sometimes with hairs on raised lines decurrent from leaf axils.
                26. Capsules obpyramidal, sharply 4-angled and 4-winged, dehiscent by apical ring.
                    27. Stems subterete or slightly ridged; sepals pale green, 1.5–2.5 mm, about ½ as long as capsule; capsule wall not bulging; pollen shed in tetrads; seed surface cells nearly isodiametric . . . . 21. *Ludwigia lanceolata*
                    27. Stems slightly to distinctly winged; sepals creamy white adaxially, 2–4 mm, about as long as capsule; capsule wall somewhat bulging; pollen shed as monads; seed surface cells transversely elongate . . . . 27. *Ludwigia alata*
                26. Capsules obconic or oblong-obovoid, with 4 rounded angles or subterete, dehiscent by loculicidal slits, apical ring, or irregularly.
                    28. Nectary disc nearly flat at ovary apex; capsules 1–1.5 mm, thin- walled, dehiscent by apical ring; seeds dark reddish brown . . . . 17. *Ludwigia microcarpa*
                    28. Nectary disc raised 0.3–0.75 mm at ovary apex; capsules 1.5–7 mm, hard-walled, dehiscent irregularly or by loculicidal slits; seeds light brown to brown.

[29. Shifted to left margin.—Ed.]

29. Leaf blades elliptic, lanceolate, oblong-elliptic to very narrowly so; stolons present; capsules irregularly dehiscent; pollen shed in tetrads.
    30. Bracteoles 3.5–6.5(–8) mm; sepals pale green, apex elongate-acuminate, spreading or reflexed; capsules 4–7 mm, oblong-obovoid, pedicels 0.1–0.3 mm . . . . 19. *Ludwigia polycarpa*
    30. Bracteoles 0.5–1.5 mm; sepals yellow or cream adaxially, apex acuminate, ascending; capsules 2–4(–4.5) mm, subglobose, pedicels 0.5–1.2(–2.3) mm . . . . 24. *Ludwigia sphaerocarpa* (in part)
29. Leaf blades spatulate to oblanceolate to very narrowly so; stolons rarely present; capsules dehiscent by loculicidal slits; pollen shed as monads.
    31. Capsules 1.5–2.5 mm; sepals 1.2–1.8 mm; vestigial petals 0 or very rare; basal leaves sometimes opposite . . . . 28. *Ludwigia simpsonii*
    31. Capsules (2–)2.5–4(–4.7) mm; sepals 1.5–3 mm; 1–3 vestigial petals sometimes present; all leaves alternate . . . . 31. *Ludwigia curtissii*

1a. LUDWIGIA Linnaeus sect. MYRTOCARPUS (Munz) H. Hara, J. Jap. Bot. 28: 291. 1953

Jussiaea Linnaeus sect. *Myrtocarpus* Munz, Darwiniana 4: 184. 1942; *Corynostigma* C. Presl; *Ludwigia* sect. *Michelia* Ramamoorthy

Herbs, perennial, or shrubs. Stems usually erect, rarely subscandent, terete or angled. **Leaves** alternate. **Flowers** 4(or 5)[–7]-merous; petals present, yellow; stamens 2 times as many as sepals; pollen shed in polyads. **Capsules** usually obpyramidal to obconic, rarely subglobose, prominently 4+-angled or subterete, with thin walls, irregularly dehiscent. **Seeds** in several rows per locule, free, raphe inconspicuous. **2*n*** = 16, 32, 48, 64, 80, 96, 128.

Species 19 (1 in the flora): se United States, Mexico, West Indies, Central America, South America; introduced in Eurasia, Africa, Pacific Islands, Australasia.

Section *Myrtocarpus*, the second largest in the genus, has a center of distribution in southern Brazil, Argentina, and Paraguay; 14 species are restricted to that area and five extend to northern South America [*Ludwigia elegans* (Cambessèdes) H. Hara], to Central America and southern Mexico [*L. caparosa* (Cambessèdes) H. Hara, *L. nervosa* (Poiret) H. Hara, and *L. rigida* (Miquel) Sandwith], or to the Caribbean and the southern United States [*L. peruviana* (Linnaeus) H. Hara]. *Ludwigia peruviana* is naturalized in scattered localities in Asia and Australia (P. H. Raven 1963[1964]; T. P. Ramamoorthy and E. Zardini 1987). South America has been proposed to be the center of origin for Onagraceae and for *Ludwigia* (Raven 1963[1964], 1988). R. H. Eyde (1977, 1979, 1981) and Raven (1963[1964]) suggested that sect. *Myrtocarpus* includes species with the most generalized (plesiomorphic) morphology in *Ludwigia*; the section also includes at least five self-incompatible species (Ramamoorthy and Zardini). Most species outcross; only two are autogamous (*L. hassleriana*, *L. tomentosa*). Section *Myrtocarpus* comprises a polyploid complex consisting of one diploid (*L. nervosa*, $n = 8$), four tetraploids ($n = 16$), five hexaploids ($n = 24$), five octoploids ($n = 32$), one decaploid (*L. caparosa*, $n = 40$), and one multiploid (*L. peruviana*, $n = 32, 40, 48, 64$; Ramamoorthy and Zardini). The chromosome numbers of *L. burchellii* (Micheli) H. Hara and *L. rigida* are unknown.

P. A. Munz (1965) included nine species from sect. *Myrtocarpus* in his treatment for the North American Flora, but the circumscription of the section has changed (Munz 1944; T. P. Ramamoorthy and E. Zardini 1987) and his geographical definition included the West Indies and parts of Mexico/Central America; only *Ludwigia peruviana* is included in this treatment. Of the other species he included, *L. foliobracteolata* (Munz) H. Hara, *L. nervosa*, and *L. rigida* [as *L. lithospermifolia* (Kunth) H. Hara] remain in sect. *Myrtocarpus* and have not been reported in the flora area. The other five species are now placed in other sections: *L. latifolia* (Bentham) H. Hara (sect. *Tectiflora* Ramamoorthy), *L. erecta* (Linnaeus) H. Hara and *L. decurrens* Walter (sect. *Pterocaulon* Ramamoorthy), *L. inclinata* (Linnaeus f.) M. Gómez (sect. *Heterophylla* Ramamoorthy), and *L. sedoides* (Humboldt & Bonpland) H. Hara (sect. *Humboldtia* Ramamoorthy); of those, only *L. decurrens* and *L. erecta* occur in the flora area. The classification of this large *Myrtocarpus* complex was not supported by molecular analysis (Liu S. H. et al. 2017), but any revision must await more complete taxon sampling and better resolution.

SELECTED REFERENCES Ramamoorthy, T. P. 1979. A sectional revision of *Ludwigia* sect. *Myrtocarpus* sensu lato (Onagraceae). Ann. Missouri Bot. Gard. 66: 893–896. Ramamoorthy, T. P. and E. Zardini. 1987. The systematics and evolution of *Ludwigia* sect. *Myrtocarpus* sensu lato (Onagraceae). Monogr. Syst. Bot. Missouri Bot. Gard. 19: 1–120.

1. Ludwigia peruviana (Linnaeus) H. Hara, J. Jap. Bot. 28: 293. 1953 • Peruvian primrose-willow [F] [W]

Jussiaea peruviana Linnaeus, Sp. Pl. 1: 388. 1753; *J. hirsuta* Miller; *J. macrocarpa* Kunth; *J. mollis* Kunth; *J. peruviana* var. *glaberrima* Donnell Smith; *J. speciosa* Ridley; *J. sprengeri* hort. ex L. H. Bailey; *Ludwigia hirta* (Linnaeus) M. Gómez; *Oenothera hirta* Linnaeus

Herbs often woody at base, with peeling bark. **Stems** usually ridged, rarely succulent, profusely branched, 100–400 cm, usually villous, rarely glabrous, hairs deciduous in age, multicellular, usually tawny. **Leaves:** stipules deciduous, narrowly deltate, 1–1.5 × 0.3–0.5 mm, setaceous; petiole 0–1.5 cm; blade usually lanceolate, elliptic or broadly elliptic, sometimes ovate, obovate, or rounded, 2–45 × 1–10 cm, base obtuse or cuneate, rarely asymmetrical, margins entire or gland-toothed, apex usually acute or acuminate, rarely rounded and emarginate, mostly scabrid, membranous or papery, surfaces usually villous, sometimes glabrous; bracts usually not strongly reduced. **Inflorescences** leafy racemes, flowers solitary in distal axils; bracteoles deciduous, usually attached near base or on lower ½ of ovary, sometimes on upper pedicel, subtended by reduced, glandlike stipels, ovate or lanceolate to linear, 5–20 × 1–6 mm, apex acute or short-acuminate, surfaces villous. **Flowers:** sepals ovate or ovate-lanceolate, 10–23 × 4–9 mm, apex acute or short-acuminate, sometimes glandular-serrulate; petals bright yellow, orbiculate or obovate, 10–40 × 10–40 mm, apex rarely emarginate, short-clawed; stamens 8(or 10) in 2 unequal series, yellow, shorter filaments 1.5–4 mm, longer ones 3.5–5 mm, anthers oblong, 3–6 mm; ovary obconic, 4- or 5-angled, sometimes subterete, 5–20 × 3–7 mm, narrowed to pedicel, usually densely villous, sometimes glabrous; nectary disc elevated 0.3–3.2 mm on ovary apex, 4–6 mm diam., 4(or 5)-lobed, sunken, ringed by long, white hairs; style 1.5–3.5 mm, stigma globose, 1.6–3.5 × 1.8–3.5 mm, usually as long as stamens, rarely exserted beyond them. **Capsules** ± sharply 4- or 5-angled, 10–40 × 6–13 mm, thin-walled, irregularly dehiscent, pedicel 5–65 mm. **Seeds** brown or reddish brown, oblong, 0.6–0.9 × 0.3–0.4 mm, rounded at ends, with inconspicuous raphe. **2*n*** = 64, 80, 96, 128.

Flowering Jun–Aug(–Sep) (sometimes in any month). Wet places, ditches, drainage canals, sloughs, swales, marshy shores, wet clearings; 0–200[–2600] m; Ala., Fla., Ga., N.C., Tex.; Mexico (Chiapas, Durango, Oaxaca, Puebla, Tabasco, Veracruz); West Indies; Central America; South America; introduced in Eurasia (India, Indonesia, Singapore, Sri Lanka), Australia.

Ludwigia peruviana is sometimes cultivated and naturalized, which may account for occurrences in North America and Eurasia. Except in the Amazon basin, where it is known only from few collections in western Amazonia, and in northeastern Brazil, where it is scarce, *L. peruviana* is common throughout its range, and may behave as a weed, especially along slow-flowing canals and drainage ditches. *Ludwigia peruviana* is also naturalized at scattered localities in Asia (P. H. Raven 1963[1964]) and around Sydney, Australia. The earliest collection from Asia is labeled "ex horto bot. Bogoriensi Javae misit 1869" (Raven). It also occurs locally in the Nilgiri Mountains of southwest India and in Sri Lanka, as well as in scattered locations in Bangka, Java, Malaysia, and Sumatra from sea level to 1000 m (Raven 1963[1964], 1978).

Jussiaea grandiflora Ruíz & Pavon, a synonym for *J. peruviana* Linnaeus, appeared in 1830, not in 1802 (P. A. Munz 1942; P. H. Raven 1963[1964]); it is a later homonym of *J. grandiflora* Michaux (1803), as reported in W. Greuter and T. Raus (1987). *Jussiaea hirta* (Linnaeus) Swartz is an illegitimate homonym and *J. hirta* (Linnaeus) Vahl is an illegitimate isonym; both pertain here.

1b. Ludwigia Linnaeus sect. Pterocaulon Ramamoorthy, Ann. Missouri Bot. Gard. 66: 894. 1980

Herbs, annual or short-lived perennial. Stems erect or strongly ascending, 4-angled, sometimes 4-winged. **Leaves** alternate. **Flowers** 4[or 5]-merous; petals present, yellow [white]; stamens 2 times as many as sepals; pollen shed in polyads [rarely tetrads]. **Capsules** oblong-linear to subcylindric, subclavate to oblong-obovoid, or narrowly obpyramidal, prominently 4(or 5)-ribbed or -winged, with thin walls, irregularly dehiscent. **Seeds** in several rows per locule, free, raphe inconspicuous. **2*n*** = 16.

Species 5 (2 in the flora): c, e United States, s Mexico, West Indies, Bermuda, Central America, South America; introduced in Eurasia, Africa, Indian Ocean Islands, Pacific Islands.

Section *Pterocaulon* consists of five erect, mainly annual, diploid species (T. P. Ramamoorthy 1979; Ramamoorthy and E. Zardini 1987); three occur only in Argentina, Brazil, Paraguay, and Uruguay. *Ludwigia decurrens* and *L. erecta* occur more widely in South America and extend to the southeastern United States, Mexico, and the Caribbean; both also occur in Africa, and *L. decurrens* is naturalized in Asia and Europe (P. H. Raven 1963[1964]; Ramamoorthy and Zardini). Four species are autogamous, with the remaining one (*L. longifolia*) self-compatible but primarily outcrossing. Unlike species of sect. *Myrtocarpus*, in which section these species were treated prior to 1979, all are diploid annuals with mostly winged stems and very narrow, sublinear leaves. Sect. *Pterocaulon* appears to be polyphyletic in analyses by Liu S. H. et al. (2017), but not all species were sampled, and no change in classification is proposed here.

This section was delimited and named earlier as *Diplandra* Rafinesque 1840, but that name was pre-occupied by *Diplandra* Bertero 1830 and *Diplandra* Hooker & Arnott 1838, name conserved.

2. Ludwigia decurrens Walter, Fl. Carol., 89. 1788 • Wingleaf primrose-willow [F] [W]

Diplandra decurrens (Walter) Rafinesque; *D. montana* Rafinesque; *Jussiaea alata* C. Presl; *J. bertonii* H. Léveillé; *J. decurrens* (Walter) de Candolle; *J. palustris* G. Meyer; *J. pterophora* Miquel; *J. tenuifolia* Nuttall

Herbs annual or short-lived perennial, roots and lower stem sometimes inflated and spongy. **Stems** erect or strongly ascending, sharply 4-angled and 4-winged, 30–200 cm, simple to densely branched, glabrous. **Leaves:** stipules deltate, 0.4–0.5 × 0.2 mm; sessile and continuous with wings on stem; blade lanceolate or narrowly lanceolate to narrowly ovate or elliptic, 2–20 × 0.2–5 cm, base acute or rounded, margins entire, often minutely scabrid, apex acute or acuminate, membranous, surfaces glabrous or sometimes minutely puberulent on abaxial veins; bracts linear, mostly reduced. **Inflorescences** open, leafy racemes, flowers solitary in distal axils; bracteoles deciduous, usually attached near base of ovary, without subtending glands, lanceolate to subovate, 0.5–1 × 0.2–0.5 mm, apex acute. **Flowers:** sepals ovate to lanceolate, 7–12 × 1.5–4 mm, apex acute or short acuminate, membranous, margins usually minutely scabrid, surfaces usually glabrous, sometimes puberulent abaxially; petals orbiculate-obovate, 10–20 × 10–18 mm, short-clawed; stamens 8 in 2 subequal series, yellow, filaments 1.3–2.5 mm, anthers oblong, 1.3–1.6 × 0.5–0.6 mm; ovary obconic, sharply 4-angled and 4-winged, 6–10 × 2–4.5 mm, glabrous or minutely puberulent; nectary disc plane on ovary apex, 3–5 mm diam., 4-lobed, glabrate or ringed with short hairs; style 2.5–3.2 × 0.5–0.6 mm, stigma globose, 1–2 × 1–2 mm, not exserted beyond anthers and pollen shed directly on it. **Capsules** rarely sharply curved, subclavate to oblong-obovoid or narrowly obpyramidal, sharply 4-angled and 4-winged, 10–25 × 3–5 mm, glabrous, pedicel 2–15 mm. **Seeds** oblong or subcylindric, 0.5–0.6 × 0.2 mm, striate, raphe very narrow and inconspicuous. **2*n*** = 16.

Flowering summer–early fall. Moist or swampy habitats along sloughs, muddy stream banks, marshy shores of lakes and ponds, ditches, swamps; 0–300 [–600] m; Ala., Ark., D.C., Fla., Ga., Ill., Ky., La., Md., Miss., Mo., N.C., Ohio, Okla., Pa., S.C., Tenn., Tex., Va., W.Va., Wis.; Mexico (Campeche, Chiapas, Michoacán, Tabasco); West Indies; Bermuda; Central America; South America; introduced in Europe (France), e Asia (Japan), Africa (Cameroon), Pacific Islands (Philippines).

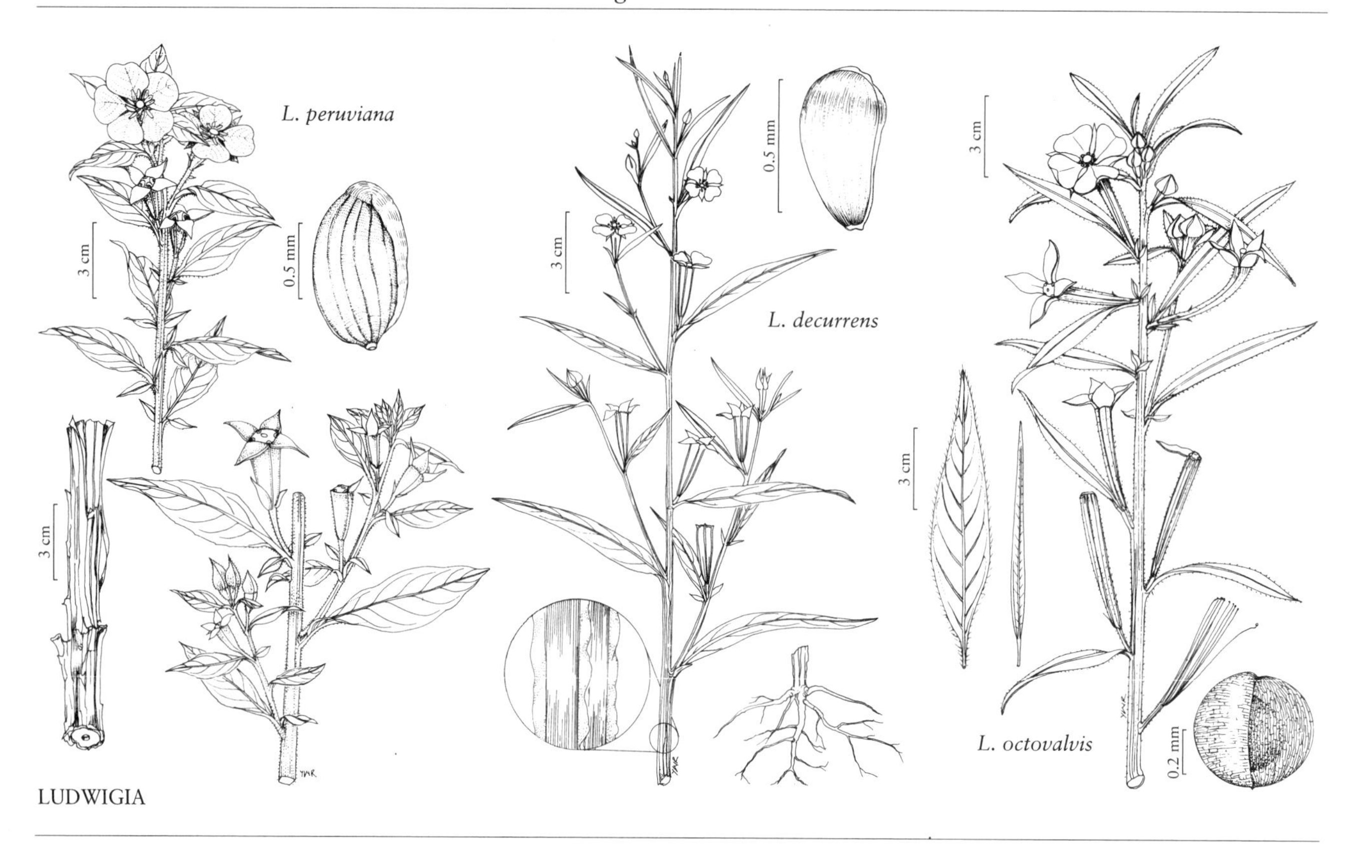

LUDWIGIA

Ludwigia decurrens appears to be most closely related to *L. erecta*, differing by having winged, not angled, capsules and larger flowers, with sepals 7–12 mm and petals 10–20 mm (in *L. erecta*, sepals 3–6 mm, petals 3.5–5 mm). They often grow in close proximity and may hybridize, but hybrids would be difficult to detect. *Ludwigia decurrens* is often self-pollinating, but larger flowers may promote outcrossing.

Ludwigia jussiaeoides Michaux is an illegitimate later homonym of *L. jussiaeoides* Desrousseaux and pertains here.

3. **Ludwigia erecta** (Linnaeus) H. Hara, J. Jap. Bot. 28: 292. 1953 • Yerba de jicotea

Jussiaea erecta Linnaeus, Sp. Pl. 1: 388. 1753; *Isnardia discolor* Klotzsch; *J. acuminata* Swartz; *J. acuminata* var. *latifolia* Grisebach; *J. acuminata* var. *longifolia* Grisebach; *J. altissima* Perrottet ex de Candolle; *J. declinata* Sessé & Mociño; *J. erecta* var. *plumeriana* de Candolle; *J. erecta* var. *sebana* de Candolle; *J. onagra* Miller; *J. plumeriana* (de Candolle) Bello; *J. ramosa* J. Jacquin ex Reichenbach; *Ludwigia acuminata* (Swartz) M. Gómez

Herbs annual, rarely persistent a second year from woody base. **Stems** erect, 4-angled, rarely 4-winged, sometimes basally terete, 40–280 cm, simple to densely branched, branches often ascending, glabrous. **Leaves:** stipules deltate, 0.2–0.3 × 0.15–0.2 mm; petiole 0.2–2.2 cm, somewhat flattened and continuous with ridges or wings on stem; blade elliptic to narrowly lanceolate, 2–20 × 0.2–4 cm, base cuneate, margins minutely scabrid, apex acute or acuminate, membranous, surfaces glabrous or sometimes minutely strigillose along abaxial veins; bracts often reduced. **Inflorescences** leafy spikes, flowers solitary in distal axils; bracteoles attached at base of ovary or on lower ½, without subtending glands, deltate, 0.3–0.5 × 0.2–0.3 mm, apex acute. **Flowers:** sepals ovate or lanceolate, 3–6 × 1–2 mm, apex acute or short-acuminate, surfaces usually glabrous, sometimes strigillose; petals obovate, 3.5–5 × 2–2.5 mm; stamens 8 in 2 subequal series, filaments 1.3–1.5 mm, anthers oblong, 0.6–1 × 0.4–0.5 mm; ovary obconic, 4-angled, 4–10 × 2–4 mm, usually glabrous, rarely strigillose; nectary disc plane on ovary apex, 3–4 mm diam., 4-lobed, glabrate; style 0.5–1.5 × 0.5–0.6 mm, stigma globose, 0.8–1 × 1–1.2 mm, not exserted beyond anthers and pollen shed directly on it. **Capsules** oblong-linear to squarish-cylindric, 4-angled, 10–22 × 2–4 mm, thin-walled, irregularly dehiscent, subsessile. **Seeds** elongate-ovoid, 0.3–0.5 × 0.2–0.3 mm, raphe very reduced and inconspicuous. $2n = 16$.

Flowering summer–early fall. Pond margins and depressions, wet sand ditches and prairies; 0–100 [–300] m; Ala., Ariz., Fla.; Mexico (Campeche, Chiapas, Jalisco, Michoacán, Nayarit, Oaxaca, Tabasco); West Indies; Central America; South America; Africa

(Nigeria, Tanzania); Indian Ocean Islands (Comoros Islands, Madagascar, Seychelles).

Ludwigia erecta, which is morphologically similar to *L. decurrens* and often growing with it, is modally self-pollinating and is usually easy to distinguish from that species.

Although *Ludwigia erecta* is widely distributed in warm temperate regions in the New World and Africa, it appears to be most closely related to species restricted to South America. Its appearance in a rather remote locality in Arizona in 2006 may be attributable to transport there in mud on migrating birds.

1c. LUDWIGIA Linnaeus sect. MACROCARPON (Micheli) H. Hara, J. Jap. Bot. 28: 291. 1953

Jussiaea Linnaeus sect. *Macrocarpon* Micheli, Flora 57: 302. 1874

Herbs, perennial, or shrubs. **Stems** erect or spreading, terete, subterete, or ridged. **Leaves** alternate. **Flowers** 4-merous; petals present, yellow; stamens 2 times as many as sepals; pollen shed in tetrads or polyads. **Capsules** cylindric to clavate-cylindric, ± angled to subterete, with thin walls, irregularly dehiscent. **Seeds** in several rows per locule, free, raphe enlarged, nearly equal to seed. **2*n*** = 16, 32, 48.

Species 4 (2 in the flora): se, s United States, Mexico, West Indies, Central America, South America, Eurasia, Africa, Indian Ocean Islands, Pacific Islands, Australasia.

Section *Macrocarpon* consists of four species, all of which occur in Argentina, Brazil, and Paraguay. *Ludwigia bonariensis* also occurs as a disjunct in Mexico and the southern United States; *L. octovalvis* is found worldwide in subtropical and tropical areas and is widely distributed in the southern United States. Section *Macrocarpon*, which is supported as monophyletic by molecular data (Liu S. H. et al. 2017), differs from sect. *Myrtocarpus* by having strictly 4-merous (versus 4+-merous) flowers, cylindric (versus obconic) capsules, and distinctive seeds with an enlarged raphe.

Ludwigia bonariensis and *L. lagunae* are diploid (*n* = 8) and self-incompatible; *L. neograndiflora* is tetraploid (*n* = 16; Liu S. H. et al. 2017), but compatibility is unknown. In contrast, most populations of the multiploid *L. octovalvis* are polyploid, with some reports of diploids in the New World (P. H. Raven and W. Tai 1979), and self-compatible. *Ludwigia octovalvis* is extremely variable in morphology and ploidy level, and the entire section is in need of taxonomic revision.

4. Ludwigia octovalvis (Jacquin) P. H. Raven, Kew Bull. 15: 476. 1962 • Mexican primrose-willow [F] [W]

Oenothera octovalvis Jacquin, Enum. Syst. Pl., 19. 1760; *Jussiaea calycina* C. Presl; *J. clavata* M. E. Jones; *J. frutescens* J. Jacquin ex de Candolle; *J. haenkeana* Steudel; *J. hirta* Lamarck; *J. ligustrifolia* Kunth; *J. occidentalis* Nuttall ex Torrey & A. Gray; *J. octofila* de Candolle; *J. octonervia* Lamarck; *J. octonervia* var. *sessiliflora* Micheli; *J. octovalvis* (Jacquin) Swartz; *J. parviflora* Cambessèdes; *J. peruviana* Linnaeus var. *octofila* (de Candolle) Bertoni; *J. pubescens* Linnaeus; *J. sagrana* A. Richard; *J. salicifolia* Kunth; *J. scabra* Willdenow; *J. suffruticosa* Linnaeus; *J. suffruticosa* var. *ligustrifolia* (Kunth) Grisebach; *J. suffruticosa* var. *linearifolia* Hassler; *J. suffruticosa* var. *octofila* (de Candolle) Munz; *J. suffruticosa* subsp. *octonervia* (Lamarck) Hassler; *J. suffruticosa* var. *octonervia* (Lamarck) Bertoni; *J. suffruticosa* var. *sessiliflora* (Micheli) Hassler; *J. suffruticosa* var. *sintenisii* Urban; *J. venosa* C. Presl; *J. villosa* Lamarck; *Ludwigia octovalvis* var. *ligustrifolia* (Kunth) Alain; *L. octovalvis* var. *octofila* (de Candolle) Alain; *L. octovalvis* subsp. *sessiliflora* (Micheli) P. H. Raven; *L. octovalvis* var. *sessiliflora* (Micheli) Shinners; *L. pubescens* (Linnaeus) H. Hara; *L. pubescens* var. *ligustrifolia* (Kunth) H. Hara; *L. pubescens* var. *linearifolia* (Hassler) A. Fernandes & R. Fernandes; *L. pubescens* var. *sessiliflora* (Micheli) H. Hara; *L. sagrana* (A. Richard) M. Gómez

Herbs (robust) or shrubs, herbs tap-rooted, often woody at base, with peeling bark. **Stems** erect to spreading, terete or sometimes ridged, 60–250(–400) cm, densely branched, densely villous to glabrate, especially near base. **Leaves:** stipules deltate, 0.5–0.8 × 0.6–0.7 mm, fleshy; petiole 0–1 cm; blade linear to oblong or oblanceolate, sometimes narrowly ovate, 0.7–14.5 × 0.1–4 cm, base tapered, margins entire, apex acute to acuminate, surfaces ± densely villous or strigillose; bracts not or scarcely reduced. **Inflorescences** open, leafy racemes, flowers solitary in axils, presentation often radial; bracteoles ovate, 3–8 × 1.4–4 mm, apex acuminate, attached near base of ovary. **Flowers:** sepals lanceolate to ovate, (6–)8–13 × 3–7 mm, apex acuminate, surfaces strigillose adaxially; petals bright yellow, fan-shaped, (5–)10–20 × 5–20 mm, apex sometimes shallowly notched; stamens 8 in 2 subequal series, yellowish white, filaments spreading, 4–6 mm, anthers oblong, 2.5–5 × 1–2 mm; pollen shed in tetrads or sometimes polyads; ovary cylindric, 4-angled, sometimes slightly twisted, 8–22 × 1–3 mm; nectary disc elevated 0.4–0.5 mm on ovary apex, 1–2.4 mm diam., with 4 white-pubescent sunken lobes opposite petals; style 2.5–3.5 mm, stigma capitate-globose, 1.8–3.5 × 2.5–3.5 mm, surrounded by anthers and pollen shed directly on it. **Capsules** cylindric to clavate-cylindric, subterete to ± 4-angled, 17–50 × 2.5–8 mm, thin-walled, irregularly dehiscent by 4–8 linear valves splitting from apex, short-villous, pedicel 5–25 mm. **Seeds** in several indistinct rows per locule, broad-cylindric with rounded ends, 0.6–0.9 × 0.5–0.8 mm, raphe inflated and nearly equal to seed body. **2*n*** = [16], 32, 48.

Flowering summer–early fall. Wet or moist places, along coasts, streams, ditches, swamps, often near disturbance or cultivation; 0–300[–2200] m; Ala., Fla., Ga., La., Miss., N.C., S.C., Tex.; Mexico; West Indies; Central America; South America; s, e Asia (Burma, China, India, Japan, Thailand, Vietnam); Africa; Indian Ocean Islands (Comoros Islands, Madagascar); Pacific Islands (New Caledonia, Papua New Guinea, Philippines).

Ludwigia octovalvis is perhaps the most widespread species of *Ludwigia* worldwide and exhibits a very complex pattern of morphological and ecological variation, correlated only in part with multiple ploidy levels; this complexity is reflected in its extensive synonymy. P. H. Raven (1963[1964]), P. A. Munz (1942, 1965), and others have proposed formal classifications to account for this variation, with mixed results and additional study using more powerful analytical tools is clearly needed in order to develop a more stable classification. In the absence of better understanding, and despite some correlated patterns of morphological and geographical variation on a global scale, this treatment does not recognize infraspecific taxa.

Jussiaea hirsuta Velloso, *J. suffruticosa* var. *angustifolia* Chodat & Hassler, *J. velutina* Kunze, and *Ludwigia suffruticosa* (Linnaeus) M. Gómez are later homonyms; these four names pertain here.

5. **Ludwigia bonariensis** (Micheli) H. Hara, J. Jap. Bot. 28: 291. 1953 • Carolina primrose-willow

Jussiaea bonariensis Micheli, Flora 57: 303. 1874; *J. neglecta* Small; *J. suffruticosa* Linnaeus var. *bonariensis* (Micheli) H. Léveillé

Herbs, from woody rootstock. **Stems** erect, subterete, 20–120 cm, branched, glabrate proximally, or strigillose, especially in distal parts, with raised strigillose lines decurrent from leaf axils mid stem. **Leaves:** stipules narrowly deltate, 0.3–0.5 × 0.1–0.2 mm; petiole winged, 0.1–0.7 cm; blade narrowly to broadly lanceolate, 4–15 × 0.3–1(–3) cm, base tapered, margins subentire to inconspicuously glandular-serrulate, apex acute to acuminate, surfaces finely strigillose, especially on abaxial veins, sometimes glabrate; bracts narrower, reduced in size. **Inflorescences** open, leafy racemes, flowers solitary in axils; bracteoles lanceolate-linear or setaceous, 2–6 × 0.3–0.8 mm, attached on pedicel just proximal to base of ovary. **Flowers:** sepals ovate-deltate, 10–20 × 7–12 mm, abruptly acuminate or acute, inconspicuously 5–7-nerved, surfaces strigillose; petals deep golden yellow, broadly obovate, 20–35 × 10–30 mm, apex shallowly emarginate, claw 1.5–3 mm; stamens 8 in 2 unequal series, filaments flattened and dilated near base, epipetalous set 3.4–4.5 mm, episepalous set 4.5–5.5 mm, anthers oblong, 4–5 mm; pollen shed in polyads; ovary subcylindric, slightly 4-angled, 8–12(–20) mm; nectary disc slightly elevated on ovary apex, 2–3 mm diam., 4-lobed, ringed by short hairs; style 3–3.5 mm, stigma clavate-capitate, 2.5–3 × 2–2.5 mm, often exserted beyond anthers. **Capsules** clavate-cylindric, subterete to obtusely 4-angled, 20–35 × 3.5–5 mm, thin walls, irregularly dehiscent, tapering to pedicel 10–40 mm. **Seeds** in several indistinct rows per locule, yellow-brown, oblong (appearing round), 0.5 mm, shiny, raphe ⅔ as wide as body. **2*n*** = 16.

Flowering summer. Wet places, mainly along coastal areas, especially ditches, banks near brackish water; 0–200[–2600] m; Ala., Fla., Miss., N.C., S.C., Tex.; Mexico (Chihuahua, Durango, Puebla, Quintana Roo, Tabasco, Tamaulipas, Veracruz); South America (Argentina, Bolivia, Brazil, Paraguay, Uruguay).

1d. Ludwigia Linnaeus sect. Seminudae P. H. Raven, Reinwardtia 6: 334. 1964

Herbs, annual or perennial, or shrubs, often forming pneumatophores when submerged. **Stems** erect or ascending, terete or angled. **Leaves** alternate. **Flowers** (4 or)5 or 6(or 7)-merous; petals present, orange-yellow; stamens 2 times as many as sepals; pollen shed in polyads. **Capsules** subcylindric [cylindric], angled or subterete [terete], with relatively thin walls, irregularly dehiscent. **Seeds** in 1 row per locule, embedded in horseshoe-shaped segment of endocarp from which it easily detaches, raphe inconspicuous. $2n$ = 32, 48, 64, 80.

Species 5 (1 in the flora): c, e United States, Mexico, West Indies, Central America, South America, Africa, Indian Ocean Islands (Madagascar).

Section *Seminudae* consists of a polyploid complex of five species (P. H. Raven and W. Tai 1979), one [hexaploid (n = 24) *L. africana* (Brenan) H. Hara] endemic to Africa, two [hexaploid (n = 24) *L. dodecandra* (de Candolle) Zardini & P. H. Raven and chromosomally unknown *L. quadrangularis* (Micheli) H. Hara] endemic to Central America and South America, and two widely distributed multiploids in the New World but present also in Africa. Of the latter, *L. affinis* (de Candolle) H. Hara (n = 16, 24) occurs only in a small area of West Africa, probably as a recent introduction from South America (Raven 1963[1964]), whereas *L. leptocarpa* (n = 32, 40) is widespread and possibly native in sub-Saharan Africa. This section is distinctive among the diplostemonous taxa in having uniseriate seeds embedded in loose segments of endocarp. However, sect. *Seminudae* so far is poorly resolved by molecular analysis (Liu S. H. et al. 2017).

6. Ludwigia leptocarpa (Nuttall) H. Hara, J. Jap. Bot. 28: 292. 1953 • Anglestem primrose-willow

Jussiaea leptocarpa Nuttall, Gen. N. Amer. Pl. 1: 279. 1818; *J. biacuminata* Rusby; *J. foliosa* C. Wright ex Grisebach; *J. leptocarpa* subsp. *angustissima* (Helwig) Acevedo-Rodríguez; *J. leptocarpa* var. *angustissima* Helwig; *J. leptocarpa* var. *meyeriana* (Kuntze) Munz; *J. pilosa* Kunth; *J. pilosa* var. *robustior* Donnell Smith; *J. schottii* Micheli; *J. surinamensis* Miquel; *J. variabilis* G. Meyer; *J. variabilis* var. *meyeriana* Kuntze; *J. variabilis* var. *pilosa* (Kunth) Kuntze; *Ludwigia leptocarpa* var. *angustissima* (Helwig) Alain; *L. leptocarpa* var. *meyeriana* (Kuntze) Alain

Herbs, perennial or (robust) annual, or shrubs, often with woody base, when aquatic, forming white pneumatophores from nodes. **Stems** usually erect or strongly ascending, rarely floating or creeping, terete to somewhat angled on young branches, 30–250 cm, well branched to sparsely branched or simple, usually villous, often also strigillose, rarely glabrous, with raised lines decurrent from leaf axils. **Leaves:** stipules narrowly deltate, 0.1–0.2 × 0.1–0.2 mm; petiole 0.2–3.5 cm; blade broadly lanceolate, 3.5–18 × 1–4 cm, base narrowly cuneate, margins subentire, apex acuminate, surfaces hirsute or villous; bracts slightly to much reduced. **Inflorescences** leafy racemes, flowers solitary in axils; bracteoles often absent, when present, narrowly deltate, 2–3 × 1.2–2.4 mm, attached near ovary base. **Flowers:** sepals ovate-deltate, 5.5–11 × 1.5–3 mm, margins entire, apex acuminate, surfaces villous; petals orange-yellow, obovate, 5–11 × 4–8 mm; stamens (8 or) 10 or 12(or 14), in 2 unequal series, longer filaments 2.5–4.5 mm, shorter ones 1.5–2.5 mm, anthers oblong, 1.2–1.6 × 0.7–1 mm, extrorse; ovary cylindric, subterete, 10–16 × 2–3 mm, glabrate to strigillose or villous; nectary disc slightly elevated at ovary apex, 2–4 mm diam., lobed, depressed, surrounded by densely matted white hairs; style 3–4.5 mm, glabrous, stigma capitate-globose, 1–1.5 × 2–2.5 mm, ± exserted beyond anthers. **Capsules** obscurely [4 or]5 or 6[or 7]-angled or subterete, straight or curved, 15–50 × 2.5–4 mm, relatively thin-walled, seeds visible as bumps, tardily and irregularly loculicidal, villous, pedicel 2–20 mm. **Seeds** in 1 row per locule, horizontal and loosely embedded in an easily detached horseshoe-shaped segment of firm endocarp, pale brown, obovoid, 1–1.2 mm, shiny, finely pitted, raphe much narrower than seed body. $2n$ = 32, 48.

Flowering summer. Wet places, mainly along coastal areas, especially ditches, banks near brackish water; 0–200[–1300] m; Ala., Ark., Del., Fla., Ga., Ill., Ind., Ky., La., Md., Miss., Mo., N.C., Ohio, Okla., Pa., S.C., Tenn., Tex., Va., W.Va.; Mexico (Campeche, Chiapas, Jalisco, Nayarit, Oaxaca, San Luis Potosí, Tabasco, Tamaulipas, Veracruz); West Indies (Cuba, Hispaniola, Jamaica, Puerto Rico); Central America; South America; Africa; Indian Ocean Islands (Madagascar).

Ludwigia leptocarpa is a globally widespread and morphologically variable species; in the flora area it is distributed widely in wet areas of the southeastern United States. Both tetraploid and hexaploid plants are known, but it is not clear if or how ploidy level is related to the considerable morphological variation, especially in pubescence type and pattern.

1e. Ludwigia Linnaeus sect. Jussiaea (Linnaeus) Baillon, Hist. Pl. 6: 463. 1876

Jussiaea Linnaeus, Sp. Pl. 1: 388. 1753; Gen. Pl. ed. 5, 183. 1754; *Adenola* Rafinesque; *Cubospermum* Loureiro; *Jussiaea* sect. *Oligospermum* Micheli; *Ludwigia* sect. *Oligospermum* (Micheli) H. Hara; *Ludwigia* sect. *Oocarpon* (Micheli) P. H. Raven; *Oocarpon* Micheli

Herbs, perennial, subshrubs, or emergent aquatics, creeping, floating, or emergent and ascending, rooting at nodes, when floating often forming spongy, white pneumatophores at nodes, when erect with spongy base. **Stems** decumbent to erect or ascending, terete, sometimes angled distally. **Leaves** alternate or fascicled. **Flowers** 5(or 6)-merous; petals present, yellow [white]; stamens 2 times as many as sepals, [rarely as many as]; pollen shed as monads. **Capsules** cylindric, subcylindric, or subclavate, terete, subterete, or obscurely angled, often up-curved, with thick, woody walls, irregularly and tardily dehiscent. **Seeds** in 1 row per locule, pendulous and firmly embedded in woody, coherent segment of endocarp, raphe inconspicuous. **2*n*** = 16, 32, 48, 80, 96.

Species 9 (3 in the flora): United States, Mexico, West Indies, Central America, South America, s Asia, Africa; introduced in Europe, Pacific Islands, Australia.

This cosmopolitan polyploid section of nine variable species (11 taxa) includes three diploid species ($n = 8$), four tetraploids ($n = 16$), one hexaploid [*Ludwigia grandiflora*, $n = 24$; P. H. Raven and W. Tai (1979) reported one anomalous count of $n = 48$], and one decaploid (*L. hexapetala*, $n = 40$; see also G. L. Nesom and J. T. Kartesz 2000). One persistent triploid ($2n = 24$) hybrid, described as *L.* ×*taiwanensis* C. I. Peng [*L. adscendens* ($n = 16$) × *L. peploides* subsp. *stipulacea* ($n = 8$); Peng 1990] is widespread in southern China and Taiwan. E. Zardini et al. (1991) reported several other natural hybrids.

Most species of sect. *Jussiaea* have non-naturalized distributions restricted to the New World. Section *Jussiaea* differs from most diplostemonous sections by releasing its pollen as monads and having woody, subcylindric capsules with uniseriate, firmly embedded seeds. Most species in sect. *Jussiaea* are vigorously aquatic, and several (including *L. peploides*, *L. hexapetala*, and *L. grandiflora*) can be invasive weeds in wetlands and wet agricultural areas; the latter two, or all three, species have recently become major invasive species in California, particularly in the Russian and Sacramento river drainages and in the San Diego region, in Arizona (especially Gila and Salt rivers), and in Washington (M. Wood 2006; P. C. Hoch and B. J. Grewell 2012).

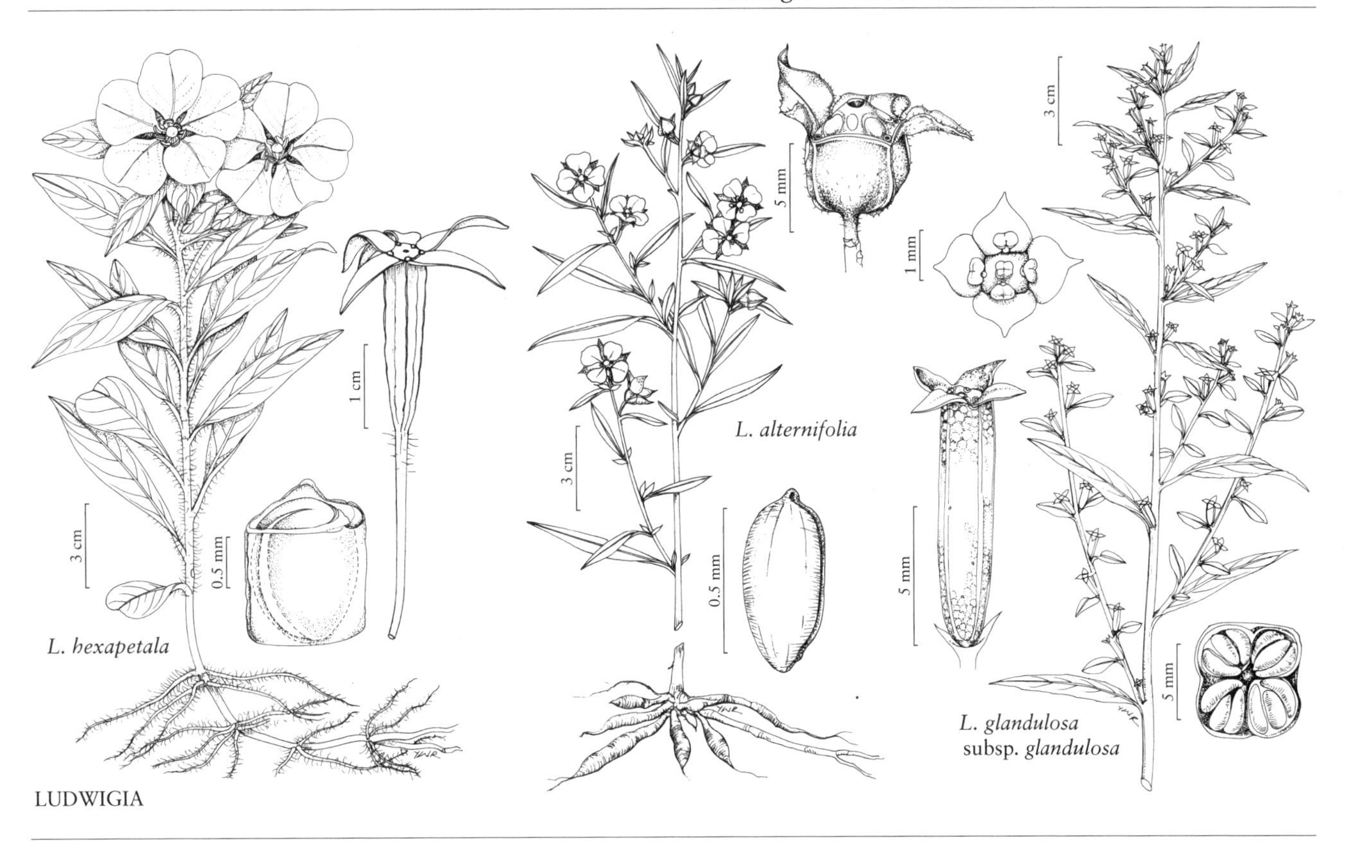

LUDWIGIA

7. **Ludwigia hexapetala** (Hooker & Arnott) Zardini, H. Y. Gu & P. H. Raven, Syst. Bot. 16: 243. 1991

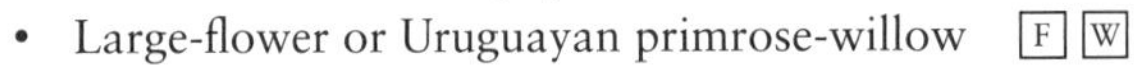
• Large-flower or Uruguayan primrose-willow [F] [W]

Jussiaea hexapetala Hooker & Arnott, Bot. Misc. 3: 312. 1833 (as Jussieua); *J. repens* Linnaeus var. *major* Hassler; *Ludwigia grandiflora* (Michaux) Greuter & Burdet subsp. *hexapetala* (Hooker & Arnott) G. L. Nesom & Kartesz; *L. grandiflora* var. *hexapetala* (Hooker & Arnott) D. B. Ward; *L. uruguayensis* (Cambessèdes) H. Hara var. *major* (Hassler) Munz

Herbs, subshrubs, or emergent aquatics, adventitious roots sometimes forming a thick mass 10–23 cm at submerged nodes, sometimes woody at base, white pneumatophores 5–10 cm often on submerged stems. **Stems** floating or creeping and ascending to erect, terete, 20–200(–400) cm, simple to densely branched apically, glabrous (floating) or sparsely to densely villous (emergent), sometimes villous only on inflorescence. **Leaves:** stipules ovate or deltate, 0.7–2 × 0.5–1.1 mm, not succulent, apex subacute, mucronate; petiole flattened, 0.5–2(–2.5) cm; blade narrowly oblanceolate, narrowly elliptic, or lanceolate to obovate or spatulate, (1.5–)4.2–10.7(–13.5) × (0.5–)0.8–3 cm, chartaceous, base cuneate or attenuate, margins entire, apex acute to obtuse, rounded or truncate, sometimes mucronate, surfaces not shiny, usually glabrous, sometimes villous on petiole and veins or throughout; bracts not reduced. **Inflorescences:** emergent stems sometimes in leafy racemes, sometimes reflexed, flowers solitary in leaf axils; bracteoles obovate to narrowly obovate, 1–1.8 × 0.7–0.8 mm, apex acute or acuminate, attached on distal ½ of pedicel or at ovary base. **Flowers:** sepals ovate-deltate or lanceolate-deltate, (8–)12–19 × 2–5 mm, chartaceous, margins entire, apex acuminate, surfaces ± densely villous; petals bright yellow, sometimes with orange base, fan-shaped, (15–)20–30 × (12–)16–25 mm, apex emarginate or mucronate; stamens 10(or 12), in 2 unequal series, yellow, filaments recurved, shorter ones (1.6–)2.3–5.2 mm, longer ones (3.1–)3.6–7.5 mm, anthers oblong, (1.2–)1.7–4 × 1–1.5 mm; ovary subcylindric, terete, 10–18 × 2–3 mm, apex ± broadened, glabrous or sparsely to densely villous; nectary disc slightly raised on ovary apex, yellowish green, 2–4 mm diam., lobed, glabrous or ringed with white hairs; style yellow, 6–10 mm, glabrous, stigma subcapitate-globose, 0.5–1.5 × 1.5–2.5 mm, often exserted beyond anthers. **Capsules** cylindric or subclavate, terete, sometimes curved, (12–)16–24(–30) × 2.5–4 mm, with thick woody walls, irregularly and tardily dehiscent, pedicel (9–)13–25(–85) mm. **Seeds** embedded in wedge-shaped piece of endocarp, 0.8–1 × 0.8–1 mm. **2*n*** = 80.

Flowering spring–late fall. Wet places, along slow-moving rivers, streams, canals, ditches, often growing into main channel as aquatic weed; 0–200[–2600] m;

Ala., Ark., Calif., Fla., Ga., Ky., La., Miss., N.Y., N.C., Oreg., Pa., S.C., Tenn., Wash.; Central America (Costa Rica); South America (Argentina, Bolivia, Brazil, Chile, Colombia, Ecuador, Paraguay, Peru, Uruguay); introduced in w Europe (Belgium, France, Spain).

Ludwigia hexapetala ($2n = 80$) was formerly included with *L. grandiflora* ($2n = 48$) in *L. uruguayensis* (Cambessèdes) H. Hara, and some authors (G. L. Nesom and J. T. Kartesz 2000) still consider them to be a single species. The small but consistent morphological differences and different ploidy levels argue for keeping them distinct at the species level.

Fernald described *Jussiaea michauxiana* (1944), since he thought that *J. grandiflora* Michaux (1803) was a homonym (not *J. grandiflora* Ruíz & Pavon). However, it was later determined that the volume containing the Ruíz & Pavon name was published in 1830 (not 1802) making the name by Michaux valid and legitimate, and the name by Fernald an illegitimate substitution. Plants now known as *Ludwigia hexapetala* were included in the circumscription of *L. uruguayensis* (Cambessèdes) H. Hara (based on *J. uruguayensis* Cambessèdes) by P. H. Raven (1963[1964]) and P. A. Munz (1965).

8. **Ludwigia grandiflora** (Michaux) Greuter & Burdet, Willdenowia 16: 448. 1987 • Large-flower or Uruguayan primrose-willow

Jussiaea grandiflora Michaux, Fl. Bor. Amer. 1: 267. 1803; *J. repens* var. *grandiflora* (Michaux) Micheli; *J. repens* var. *hispida* Hauman; *J. stenophylla* Gillies ex Hooker & Arnott; *J. stuckertii* H. Léveillé; *J. uruguayensis* Cambessèdes; *Ludwigia clavellina* M. Gómez & Molinet var. *grandiflora* (Michaux) M. Gómez; *L. uruguayensis* (Cambessèdes) H. Hara

Herbs, subshrubs, or emergent aquatics, rooting at lower nodes, sometimes woody at base, white pneumatophores 8–10 cm often on submerged stems. **Stems** erect or ascending to creeping or floating, terete or sometimes angled distally, 20–300(–450) cm, usually densely branched, sometimes simple, glabrous if floating, or densely villous and viscid throughout, or rarely just on inflorescence. **Leaves:** stipules (rarely in clusters of 3), ovate-deltate, 0.6–2 × 0.6–1.5 mm, fleshy, apex subacute, often mucronate; petiole 0.1–1.1 cm; blade usually lanceolate to (narrowly) elliptic or oblanceolate, rarely narrowly obovate, (1.7–)3.1–8(–10.5) × 0.5–2(–2.5) cm, chartaceous, viscid, base cuneate or attenuate, margins entire, apex obtuse or acute, always glandular-mucronate, surfaces densely villous, sometimes less dense adaxially, distal leaves more pubescent than proximal ones; bracts scarcely reduced. **Inflorescences** on emergent stems sometimes in leafy racemes, flowers solitary in leaf axils; bracteoles narrowly to broadly obovate, 1–1.2 × 0.7–0.8 mm, succulent, apex acute, oppositely attached at ovary base. **Flowers:** sepals usually deciduous, not persistent on capsule, lanceolate, 6–12(–16) × 2–4 mm, chartaceous, apex acute, surfaces densely villous; petals yellow, fan-shaped, (12–)16–20(–26) × 11–16(–21) mm, apex rounded, usually emarginate, rarely mucronate; stamens 10(or 12), in 2 unequal series, yellow, filaments reflexed, shorter ones (2.8–)3.8–5.3 mm, longer ones (3.7–)6–6.5 mm, anthers oblong, 1–2.5 × (0.6–)0.8–1.2 mm; ovary subcylindric, terete, 6–12 × 1.5–2.5 mm, apex thickened, densely villous; nectary disc slightly raised on ovary apex, yellow, 1.5–2.5 mm diam., lobed, ringed with villous hairs; style yellow, 4.7–6.7(–8) mm, glabrous or sparsely pubescent near base, stigma subcapitate-globose, 1–1.3 × 1.6–2.5 mm, usually exserted beyond anthers. **Capsules** subcylindric, terete, straight or curved, (11–)14–25 × 3–4 mm, with thick woody walls, irregularly and tardily dehiscent, villous-viscid, pedicel 13–25(–27) mm. **Seeds** embedded in wedge-shaped piece of endocarp, 0.8–1 × 0.8–0.9 mm. $2n = 48$.

Flowering summer. Wet places, along slow-moving rivers, streams, canals, ditches, often growing into main channel as aquatic weed; 0–200[–1200] m; Ala., Ark., Calif., Fla., Ga., Ky., La., Miss., Mo., N.J., N.Y., N.C., Okla., Oreg., Pa., S.C., Tenn., Tex., Va., Wash., W.Va.; Central America (Guatemala); South America (Argentina, Bolivia, Brazil, Paraguay, Uruguay).

Ludwigia grandiflora occurs in two disjunct areas: the southeastern United States on the coastal plain of southern South Carolina, Georgia, northern Florida, Louisiana, west to central Texas, and recently in southern California (P. C. Hoch and B. J. Grewell 2012) and Oregon; and central South America from south of the Amazon basin of Brazil and Bolivia where it is very scattered, to Uruguay, northeastern Argentina, and Paraguay where it is very frequent. It has been collected three times in Guatemala and twice in Missouri, although it is not clearly established in either region. It usually grows below 200 m elevation, but in Guatemala and in Santa Catarina, Brazil (*Smith & Klein 13383*, MO), it has been collected as high as 1200 m elevation. Populations of *L. grandiflora* in the United States are fairly variable, although not as much as in South American populations.

As noted by Greuter and Burdet, the publication of *Jussiaea grandiflora* Ruíz & Pavon, which was a synonym of *J. peruviana*, occurred in 1830, not in 1802 as reported (P. A. Munz 1942; P. H. Raven 1963 [1964]). Therefore, *J. grandiflora* Michaux in 1803 is legitimate, and *J. grandiflora* Ruíz & Pavon is an illegitimate homonym.

9. Ludwigia peploides (Kunth) P. H. Raven, Reinwardtia 6: 393. 1964 • Floating primrose-willow W

Jussiaea peploides Kunth in A. von Humboldt et al., Nov. Gen. Sp. 6(fol.): 77; 6(qto.): 97. 1823; *J. repens* Linnaeus var. *peploides* (Kunth) Grisebach; *Ludwigia adscendens* (Linnaeus) H. Hara var. *peploides* (Kunth) H. Hara

Herbs or emergent aquatics, rooting at nodes, sometimes with fleshy, white pneumatophores at submerged nodes. **Stems** floating or creeping and ascending to erect, terete, 10–100(–300) cm, simple or branched, glabrous or sparsely to densely villous, hairs sometimes viscid on emergent distal stem. **Leaves:** stipules broadly ovate-deltate, 0.6–1.6 × 0.4–1 mm, succulent, apex acute or obtuse, gland-tipped, rarely divided into 3 parts; petiole flattened or narrowly winged, 0.2–6 cm; blade narrowly oblong or elliptic to ovate, broadly obovate, or orbiculate, (0.4–)1–10 × 0.4–4 cm, base narrowly cuneate or attenuate, margins entire, apex obtuse or rounded to acute, sometimes mucronate or glandular-mucronate, surfaces of floating leaves glabrous, those of emergent leaves glabrous to sparsely or densely strigillose at least adaxially; bracts scarcely reduced. **Inflorescences** on emergent stems sometimes in leafy racemes, flowers solitary in leaf axils; bracteoles (rarely absent), deltate, squamate, 0.5–1 × 0.5–1 mm, apex acute, attached near base or on lower ½ of ovary. **Flowers:** sepals narrowly deltate or lanceolate, 3–12 × 1.5–4 mm, apex acute or acuminate, surfaces glabrous or sparsely to densely hirtellous; petals yellow, obpyramidal, 7–24 × 4–13 mm, apex mucronate or emarginate, up-curved; stamens 10(or 12), in 2 unequal series, bright yellow, filaments suberect or reflexed, shorter ones 1.4–4.2 mm, longer ones (1.9–)3.3–6 mm, anthers oblong, 0.5–2.2 mm; ovary subcylindric or truncate, 6–20 × 1.5–3 mm, apex somewhat broader, glabrous or sparsely to densely hirtellous; nectary disc slightly raised on ovary apex, 2–2.5 mm diam., lobed, glabrous or fringed with long hairs; style (1.9–)2.4–7.3 mm, glabrous or sparsely to densely hirtellous on proximal ½, stigma flattened-globose, 0.9–1.2 × 1–2.5 mm, sometimes shallowly or deeply 5-lobed, as long as or exserted beyond anthers. **Capsules** cylindric, subterete to obscurely 5-angled, straight or curved, 10–40 × 2–4 mm, with thick woody walls, irregularly and tardily dehiscent, pedicel 7–60(–90) mm. **Seeds** embedded in elongated piece of endocarp, 1–1.5 × 0.9–1.3 mm. **2*n*** = 16.

Subspecies 4 (3 in the flora): United States, Mexico, West Indies (Cuba), Central America, South America, Asia (China), Pacific Islands (Galapagos Islands); introduced in Europe (France), elsewhere in the Pacific Islands (New Zealand, Society Islands), Australia.

Ludwigia peploides consists of four subspecies more or less well defined geographically and morphologically, with three present in the flora area: subsp. *glabrescens*, subsp. *montevidensis*, and subsp. *peploides* (P. H. Raven 1963[1964]); these subspecies have ranges that are mostly distinct. Subspecies *peploides* has a wide distribution in the New World, from the southern United States south to Argentina. Subspecies *glabrescens* is widespread in eastern United States. Subspecies *montevidensis* occurs primarily in southern South America and scattered (probably introduced) in the southern United States, Australia, France, and New Zealand. Subspecies *peploides* and *montevidensis* occur together locally in California and Louisiana, where subsp. *montevidensis* is introduced. The ranges of subsp. *glabrescens* and *peploides* come together in Texas.

Subspecies *stipulacea* (Ohwi) P. H. Raven is known from eastern Asia (e China).

1. Stems usually densely villous, rarely sparsely so; leaf blades not shiny, apices glandular-mucronate; capsules (20–)24–32 mm. . . . . 9c. *Ludwigia peploides* subsp. *montevidensis*
1. Stems glabrous or sparely villous; leaf blades shiny, apices usually eglandular-mucronate; capsules 10–40 mm.
   2. Principal leaf blades 0.8–4(–8.5) cm; petioles 0.2–2.5 cm; pedicels 10–35 mm; capsules 10–17(–25) mm; seeds 7–14 per locule . . . . .9a. *Ludwigia peploides* subsp. *peploides*
   2. Principal leaf blades (2–)4–10 cm; petioles 0.7–6 cm; pedicels 35–90 mm; capsule 25–40 mm; seeds 16–18 per locule . . . . 9b. *Ludwigia peploides* subsp. *glabrescens*

9a. Ludwigia peploides (Kunth) P. H. Raven subsp. **peploides**

Jussiaea californica (S. Watson) Jepson; *J. diffusa* Forsskål var. *californica* (S. Watson) Greene ex Jepson; *J. repens* Linnaeus var. *californica* S. Watson

Stems glabrous or sparsely villous. **Leaves** alternate or fascicled; stipules symmetrical; petioles of basal leaves 0.3–0.8 cm, those of distal leaves 0.2–2.5 cm; blade 0.8–4(–8.5) cm, apex sometimes glandular-mucronate, surfaces shiny, glabrous, rarely ciliate or minutely pellucid-punctate. **Flowers:** anthers on short filaments 0.5–1.1 mm, those on long filaments 0.8–1.2 mm; ovary 6–14 mm, apex truncate, glabrous or with scattered hairs, stigma usually as long as anthers, rarely exserted beyond them. **Capsules** 10–17(–25) × 2–3 mm, pedicel 10–35 mm. **Seeds** 7–14 per locule. **2*n*** = 16.

Flowering summer–early fall. Wet places, along slow-moving rivers, streams, canals, ditches, often growing into main channels as aquatic weeds; 0–900[–3000] m; Ariz., Calif., N.Mex., Oreg., Tex.; Mexico; West Indies (Cuba); Central America; South America; Pacific Islands (Galapagos Islands); introduced elsewhere in Pacific Islands (Society Islands), Australia.

9b. Ludwigia peploides (Kunth) P. H. Raven subsp. **glabrescens** (Kuntze) P. H. Raven, Reinwardtia 6: 394. 1964

Jussiaea repens Linnaeus var. *glabrescens* Kuntze, Revis. Gen. Pl. 1: 251. 1891 (as Jussieua); *J. boydiana* Featherman; *Ludwigia peploides* var. *glabrescens* (Kuntze) Shinners

Stems glabrous. **Leaves** alternate, usually not fascicled; stipules symmetrical; petioles of basal leaves 0.7–2.5 cm, those of distal leaves 0.7–6 cm; blade (2–)4–10 cm, apex eglandular-mucronate, surfaces shiny, glabrous. **Flowers:** anthers subequal, 1.2–1.4 mm; ovary 14–20 mm, apex somewhat broader, glabrous or with scattered hairs, stigma usually exserted beyond anthers, rarely as long as anthers. **Capsules** 25–40 × 3–4 mm, pedicel 35–90 mm. **Seeds** 16–18 per locule. 2*n* = 16.

Flowering summer-early fall. Wet places, along slow-moving rivers, streams, canals, ditches, often growing into main channels as aquatic weeds; 0–900[–3000] m; Ala., Ark., Del., D.C., Fla., Ga., Ill., Ind., Iowa, Kans., Ky., La., Md., Miss., Mo., Nebr., N.J., N.Y., N.C., Ohio, Okla., Pa., S.C., Tenn., Tex., Va., W.Va.; South America (Venezuela).

9c. Ludwigia peploides (Kunth) P. H. Raven subsp. **montevidensis** (Sprengel) P. H. Raven, Reinwardtia 6: 395. 1964 [I]

Jussiaea montevidensis Sprengel, Syst. Veg. 2: 232. 1825 (as Jussieva); *J. repens* Linnaeus var. *montevidensis* (Sprengel) Munz; *Ludwigia adscendens* (Linnaeus) H. Hara var. *montevidensis* (Sprengel) H. Hara; *L. peploides* var. *montevidensis* (Sprengel) Shinners

Stems usually densely villous, rarely sparsely so, hairs often viscid when fresh, or glabrate on submerged stems. **Leaves** alternate, sometimes fascicled; stipules often asymmetrical; petioles of basal leaves (0.5–)0.8–1.6 cm, those of distal leaves 0.5–2.8 cm; blade (0.4–)1–6 (–9.5) cm, apex glandular-mucronate, surfaces not shiny, usually densely hirtellous, rarely glabrous abaxially. **Flowers:** anthers on short filaments (0.7–)0.9–1.8 mm, those on long filaments (0.8–)1.1–2.2 mm; ovary 6–10 mm, apex truncate, densely hirtellous, sometimes only on apical ½, stigma usually as long as anthers, rarely exserted beyond them. **Capsules** (20–)24–32 × 2–4 mm, pedicel 7–38(–60) mm. **Seeds** 10–15 per locule. 2*n* = 16 (32).

Flowering summer-early fall. Wet places, along slow-moving rivers, streams, canals, ditches, often growing into main channels as aquatic weeds; 0–500[–2000] m; introduced; Calif., La.; South America (Argentina, Brazil, Chile, Peru, Uruguay); introduced also in Europe (France), Pacific Islands (New Zealand), Australia.

In the flora area, subsp. *montevidensis* is introduced in California (P. H. Raven 1963c), where it was first collected in 1906 (El Dorado County, *Rixford* s.n., CAS), and in Louisiana. Subspecies *montevidensis* occasionally forms masses of vegetation that can obstruct water flow and navigation in California and elsewhere.

1f. LUDWIGIA Linnaeus sect. LUDWIGIA

Isnardia subg. *Ludwigiaria* (de Candolle) Reichenbach

Herbs, perennial, roots often fusiform, thickened, and fascicled. **Stems** erect, subterete to angled. **Leaves** alternate. **Flowers** 4-merous; petals present, yellow; stamens as many as sepals; pollen usually shed in tetrads. **Capsules** globose to cuboid, 4-angled or terete, with hard walls, dehiscent by an apical pore. **Seeds** in several rows per locule, free, raphe an inconspicuous, longitudinal ridge. 2*n* = 16.

Species 4 (4 in the flora): c, e North America.

Species of sect. *Ludwigia* are found mainly along the coastal plain of southeastern United States.

Section *Ludwigia* consists of four perennial diploid species native to the southeastern United States; *L. hirtella* extends north to Rhode Island and *L. alternifolia* to Ontario and into the Great Plains of central North America (P. A. Munz 1965). *Ludwigia virgata* is outcrossing, the other three are autogamous. This strongly supported monophyletic section (P. H. Raven 1963[1964]; Liu S. H. et al. 2017) differs from the other sections found in North America by having fusiform, tuberous roots that may have an adaptive function in wet habitats, and capsules regularly dehiscing by an apical pore. The free seeds are loose in the globose capsules until they dehisce through the pore, giving rise to the common names seedbox or rattlebox. All species of sect. *Ludwigia* consistently shed their pollen in tetrads; J. Praglowski et al. (1983) reported for all taxa in this section that sometimes the tetrads appeared to be linked, suggesting polyads.

10. Ludwigia alternifolia Linnaeus, Sp. Pl. 1: 118. 1753 • Seedbox, rattlebox, ludwigie à feuilles alternes [E] [F] [W]

Isnardia alternifolia (Linnaeus) de Candolle; *I. alternifolia* var. *salicifolia* (Poiret) de Candolle; *I. alternifolia* var. *uniflora* (Rafinesque) de Candolle; *I. aurantiaca* (Rafinesque) de Candolle; *Ludwigia alternifolia* var. *linearifolia* Britton; *L. alternifolia* var. *pubescens* E. J. Palmer & Steyermark; *L. angustifolia* Michaux var. *ramosissima* (Walter) Poiret; *L. aurantiaca* Rafinesque; *L. ramosissima* Walter; *L. salicifolia* Poiret; *L. uniflora* Rafinesque

Roots fusiform, fascicled, thickened, epidermis splitting or peeling near base. **Stems** subterete or somewhat angled, with narrow raised lines or wings decurrent from leaf axils, 40–150 cm, well branched in distal ½, glabrous or sparsely to densely strigillose. **Leaves:** stipules narrowly deltate, 0.05–0.2 × 0.05–0.1 mm; petiole 0.1–0.3(–0.7) cm; blade lanceolate-elliptic, (0.6–)2–12 × (0.3–)1–1.5(–2.5) cm, base attenuate, margins entire, apex acute, surfaces strigillose throughout or glabrate with strigillose veins; bracts often reduced and more linear. **Inflorescences** leafy racemes, flowers solitary in leaf axils; bracteoles linear-lanceolate, 1–2.5 mm, margins entire, apex acute or subacuminate, glabrous or with scattered hairs, attached near base of ovary. **Flowers:** sepals narrowly ovate-deltate, (6–)6.5–9.5 × 4–6.5 mm, margins entire, apex acute to obtuse, surfaces strigillose, sometimes mixed with villous hairs, or glabrate; petals cordate, 10–14 × 8–12 mm, base attenuate, apex emarginate; filaments opaque white, awl-shaped, 1–3 mm, anthers 1–1.7 × 0.6–0.8 mm; ovary subcuboid to globose, 2.5–4 × 2.5–3.5 mm; nectary disc slightly elevated on ovary apex, 0.8–1.5 mm diam., 4-lobed, ringed with soft, curly hairs or glabrous; style 1.5–2.4 mm, glabrous, stigma light yellow, capitate to hemispherical, 1–1.3 × 1.4–2 mm, shallowly 4-lobed, not or scarcely exserted beyond anthers. **Capsules** subcuboid to squarish globose, 4–6(–7) × 4–6 mm, 4-angled and 4-winged, wings 0.3–1.5 mm wide, pedicel 2–7 mm. **Seeds** light brown, oblong to reniform, 0.5–0.8 × 0.2–0.4 mm, surface cells elongate transversely to seed length. **2*n*** = 16.

Flowering summer–early fall. Swamps, damp, peaty places, roadside ditches, margins of cultivated fields; 0–1800 m; Ont., Que.; Ala., Ark., Colo., Conn., Del., D.C., Fla., Ga., Ill., Ind., Iowa, Kans., Ky., La., Md., Mass., Mich., Miss., Mo., Nebr., N.H., N.J., N.Y., N.C., Ohio, Okla., Pa., R.I., S.C., Tenn., Tex., Vt., Va., W.Va., Wis.

Ludwigia alternifolia is widespread and common in the eastern half of the flora area, as far west as Ontario in Canada, eastern Colorado, and Texas.

Ludwigia macrocarpa Michaux 1803 is a superfluous, illegitimate name for *L. alternifolia*. *Rhexia linearifolia* Poiret was originally described as a species of Melastomataceae, but the description is superficially similar to *Ludwigia* and has often been included in synonymy with *L. alternifolia*. The type of *R. linearifolia* has not been located or studied.

11. Ludwigia hirtella Rafinesque, Med. Repos., hexade 2, 5: 358. 1808 • Spindleroot, hairy or Rafinesque's seedbox [E]

Isnardia hirsuta (Desrousseaux) Roemer & Schultes var. *permollis* (Barton) de Candolle; *I. hirtella* (Rafinesque) Kuntze; *Ludwigia permollis* Barton

Roots fusiform, somewhat thickened, often fascicled. **Stems** subterete to angled, with raised lines or narrow wings decurrent from leaf axils, (30–)50–120 cm, well branched in distal ½, densely erect-hirsute to sometimes glabrous. **Leaves:** stipules narrowly deltate, 0.05–0.15 × 0.05–0.1 mm; sessile; blade closely appressed to stem, lanceolate to ovate-oblong, 1.4–5.5 × 0.4–1.2 cm, base attenuate, margins entire, apex acute to rounded, surfaces glabrous or densely erect-hirsute; bracts usually very reduced, sublinear. **Inflorescences** leafy racemes, flowers solitary in leaf axils; bracteoles attached in subopposite pairs just proximal to ovary base, lanceolate-linear, 2.5–7 × 0.5–1.5 mm, margins entire, apex acute, surfaces hirtellous. **Flowers:** sepals narrowly ovate-deltate, (5.4–)6–12 × 2.2–4.5(–5) mm, margins entire, apex acute, surfaces hirtellous or, sometimes, glabrous; petals cordate, 12–14 × 13–14 mm, base attenuate, apex emarginate; filaments opaque white, awl-shaped, 2–5.5 mm, anthers 1.4–2.6 × 0.4–0.8 mm; ovary subcuboid or globose, 3–4 × 3–4 mm; nectary disc elevated, domed, 1–1.4 mm diam., prominently 4-lobed, ringed with stiff, spreading hairs or glabrous; style 1.3–4.5 (–5.5) mm, glabrous, stigma capitate to hemispherical, 0.9–1.5 × 0.9–2(–2.5) mm, often shallowly 4-lobed, as long as stamens, sometimes exserted beyond them. **Capsules** subcuboid to squarish globose, 4–6 × 4–5 mm, 4-angled, often also 4-winged, wings 0.3–1.1 mm wide, pedicel 3–10 mm. **Seeds** light brown, oblong to reniform, 0.6–0.7 × 0.3–0.4 mm, surface cells elongate transversely to seed length. $2n$ = 16.

Flowering summer–early fall. Boggy depressions, roadside ditches, margins of boggy streams; 0–300 m; Ala., Ark., Del., D.C., Fla., Ga., Ky., La., Md., Miss., N.J., N.C., Okla., S.C., Tenn., Tex., Va.

A name often found in synonymy with *Ludwigia hirtella* is a later homonym: *L. hirsuta* Pursh 1813, not Desrousseaux 1789.

12. Ludwigia maritima R. M. Harper, Torreya 4: 163, fig. 2. 1904 • Seaside primrose-willow [E]

Roots fibrous or fusiform, sometimes fascicled. **Stems** subterete to scarcely angled, with narrow raised lines or wings decurrent from leaf axils, 30–90 cm, simple or sparsely branched distally, strigillose to sometimes glabrate. **Leaves:** stipules narrowly deltate, 0.05–0.2 × 0.05–0.1 mm; sessile; blade ovate proximally, lanceolate to lanceolate-linear distally, (2–) 3–8 × 0.3–1.5 cm, base cuneate, margins entire, apex acute, surfaces glabrate to strigillose or hirsute; bracts usually much reduced, sublinear. **Inflorescences** sparse racemes, flowers solitary in leaf axils; bracteoles attached in subopposite pairs on distal 1/3 of pedicel, lanceolate-linear, 0.7–3.2(–5) × 0.2–0.5 mm, margins entire, apex acute, surfaces strigillose. **Flowers:** sepals often spreading, ovate-deltate, (4.5–)5.5–8(–9) × 3–5 mm, margins entire, apex acute to obtuse, surfaces strigillose; petals cordate, 9–12 × 8–10 mm, base attenuate, apex emarginate; filaments yellow, awl-shaped, 1.9–3.2 mm, anthers 1.2–2.5 × 0.4–0.7 mm; ovary subcuboid or globose, 3–4.5 × 3–4 mm; nectary disc elevated, domed, 0.9–1.3 mm diam., prominently 4-lobed, ringed with sparse, spreading hairs; style 1.5–3.3 mm, glabrous, stigma capitate to hemispherical, 0.6–1.2 × 1.4–1.9 mm, shallowly 4-lobed, not exserted beyond anthers. **Capsules** subcuboid to squarish globose, 4–7 ×4–5 mm, 4-angled, often also 4-winged, wings 0.3–1.2 mm wide, pedicel 5–17 mm. **Seeds** light brown, oblong to reniform, 0.4–0.6 × 0.2–0.4 mm, surface cells elongate transversely to seed length, except may be parallel to seed length near raphe. $2n$ = 16.

Flowering summer. Damp, sandy, or peaty habitats, roadside ditches, margins of bogs or fields, usually within 75 miles of sea coast; 0–200 m; Ala., Fla., Ga., La., Miss., N.C., S.C.

13. Ludwigia virgata Michaux, Fl. Bor.-Amer. 1: 89. 1803 • Savannah primrose-willow [E]

Isnardia virgata (Michaux) de Candolle

Roots fascicled, often fusiform, or spreading horizontally. **Stems** subterete to scarcely angled, with narrow raised lines or wings decurrent from leaf axils, 45–85 cm, branched mainly near base, glabrate with strigillose raised lines or strigillose. **Leaves:** stipules narrowly deltate, 0.1–0.2 × 0.1–0.15 mm; sessile; blade ovate to obovate proximally, lanceolate-linear to linear distally, 2–7 × 0.2–1(–1.5) cm, base cuneate, margins entire, apex acute to rounded, surfaces strigillose, densely so particularly along veins; bracts usually very reduced in size, sublinear. **Inflorescences** sparse racemes, flowers solitary in leaf axils; bracteoles attached in subopposite pairs on distal 1/3 of pedicel, lanceolate-linear, 0.7–3.2(–5) × 0.2–0.5 mm, margins entire, apex acute, surfaces strigillose. **Flowers:** sepals strongly reflexed, narrowly ovate-deltate, 6–10 × 2.5–4.5 mm, margins entire, apex acute, surfaces finely strigillose to glabrate; petals cordate, 14–19 × 13–15 mm, base attenuate, apex emarginate; filaments opaque white, awl-shaped, 1.9–4.2 mm, anthers 2–4 × 0.6–1 mm; ovary subcuboid to globose, 3–4 × 3–4 mm; nectary disc elevated, domed, 0.8–1.4 mm diam., prominently 4-lobed, ringed with silky-curly hairs; style 5–9.5 mm, glabrous, stigma capitate to hemispherical, 0.6–1.3 × 1.3–2.6 mm, shallowly 4-lobed, as long as or exserted beyond anthers. **Capsules** subglobose to ellipsoid, 3.5–6.8 × 3.3–4.3 mm, 4-angled, angles not developed into wings, pedicel 6.5–17 mm. **Seeds** light brown, elliptic-oblong to reniform, 0.5–0.7 × 0.3–0.4 mm, surface cells elongate transversely to seed length or elongate parallel to length near raphe. $2n$ = 16.

Flowering summer. Sandy savannas, pinelands, damp roadside ditches, margins of ponds, bogs, irrigated fields, usually within 75 miles of sea coast; 0–300 m; Ala., Fla., Ga., Miss., N.C., S.C., Va.

1g. Ludwigia Linnaeus sect. Isnardia (Linnaeus) W. L. Wagner & Hoch, Syst. Bot. Monogr. 83: 36. 2007

Isnardia Linnaeus, Sp. Pl. 1: 120. 1753; Gen. Pl. ed. 5, 56. 1754; *Ludwigia* [unranked] *Isnardia* (Linnaeus) Torrey & A. Gray; *Ludwigia* [unranked] *Ludwigiantha* Torrey & A. Gray; *Ludwigia* sect. *Microcarpium* Munz; *Ludwigiantha* (Torrey & A. Gray) Small

Herbs, perennial, with stolons or rhizomes, or creeping and rooting at nodes, rarely floating. **Stems** usually erect or ascending, sometimes prostrate, decumbent, or sprawling, terete, subterete, or slightly ridged, rarely winged. **Leaves** alternate or opposite. **Flowers** 4-merous; petals absent or present, yellow; stamens as many as sepals; pollen shed in tetrads or as monads. **Capsules** subcylindric to clavate, oblong-obovoid, obconic, broadly obpyramidal or subglobose, terete to sharply 4-angled, sometimes 4-winged, with hard or thin walls, irregularly dehiscent or dehiscent by an apical ring or lenticular slits along locule edges, rarely dispersing as unit. **Seeds** in several rows per locule, free, raphe inconspicuous. **2*n*** = 16, 32, 48, 64.

Species 19 (18 in the flora): North America, Mexico, West Indies, Central America, n South America, e Asia; introduced in Eurasia, Africa, Pacific Islands, Australasia.

Based on recent molecular and genomic analyses of the monophyletic group of North Temperate haplostemonous species (Liu S. H. et al. 2017), sect. *Isnardia* (previously referred to as sect. *Dantia* Baillon; C. I. Peng et al. 2005) has been expanded to include the species treated previously as sect. *Microcarpium* (Peng 1988, 1989). Section *Isnardia* now comprises a polyploid complex of 19 species with a center of distribution in the southeastern United States. Most species are restricted to the Gulf and/or southeastern Coastal Plains, usually from Florida to Texas and to New Jersey. Several species extend farther north (*L. glandulosa* and *L. sphaerocarpa*) or only occur farther north (*L. polycarpa*); *L. palustris* occurs widely across eastern North America from Newfoundland, Quebec, Minnesota, and Kansas to the Gulf and east coasts, is well established on the west coast from California to British Columbia, ranges south to the West Indies, Mexico, Central America, and northern South America, and also occurs in Europe. *Ludwigia repens* also extends to the West Indies and Mexico, with disjunct occurrences in Arizona, California, and Oregon; several other species also extend to the West Indies, and *L. stricta* (C. Wright ex Grisebach) C. Wright is endemic to Cuba (Peng and H. Tobe 1987; Peng 1988, 1989).

Section *Isnardia* now consists of five diploids (2*n* = 16), nine tetraploids (2*n* = 32), four hexaploids (2*n* = 48), and one octoploid (2*n* = 64); one diploid, *Ludwigia stricta*, does not occur in North America. Species in this treatment are arranged by ploidy level, then by presumed genomic and phylogenetic order (H. Tobe et al. 1988; C. I. Peng 1988, 1989). Despite the different ploidy levels, natural hybridization among the species of this section is relatively common, and all species above the diploid level are allopolyploid (Peng 1988, 1989; Peng et al. 2005). Most hybrids are sterile, but may persist vegetatively. One sterile hybrid, *L.* ×*lacustris* Eames (*L. brevipes* × *L. palustris*), has persisted in the same area in Connecticut and Rhode Island for at least 70 years through vegetative reproduction; one of its parents, *L. brevipes*, no longer occurs in this region (Peng et al. 2005).

Most species in sect. *Isnardia* have stolons and fibrous roots, and the more aquatic species have aerenchyma near the base of their stems. Five species (the former sect. *Dantia*, in the strict sense) differ from all others in *Ludwigia* by having opposite rather than alternate leaves. Most species (14 of 19 in the section) have no petals (some occasionally have one to four vestigial petals). Eight species shed their pollen as monads, most others as tetrads. Some of the apetalous species have showy sepals that are creamy white, light yellow, or light green, and insect visitors have been observed visiting them (C. I. Peng 1989; Peng et al. 2005); all species in the section that have been tested are self-compatible, and only *Ludwigia arcuata* has large enough flowers, with the stigma regularly exserted beyond the anthers, that it regularly outcrosses.

Several species in sect. *Isnardia* are popular aquarium plants due to their aquatic nature and small stature.

Several superfluous generic names have been proposed for this section, including *Dantia* Boehmer (and several combinations based on the genus at the sectional level).

SELECTED REFERENCES Peng, C. I. 1988. The biosystematics of *Ludwigia* sect. *Microcarpium* (Onagraceae). Ann. Missouri Bot. Gard. 75: 970–1009. Peng, C. I. 1989. The systematics and evolution of *Ludwigia* sect. *Microcarpium* (Onagraceae). Ann. Missouri Bot. Gard. 76: 221–302. Peng, C. I. et al. 2005. Systematics and evolution of *Ludwigia* section *Dantia* (Onagraceae). Ann. Missouri Bot. Gard. 92: 307–359. Peng, C. I. and H. Tobe. 1987. Capsule wall anatomy in relation to capsular dehiscence in *Ludwigia* sect. *Microcarpium* (Onagraceae). Amer. J. Bot. 74: 1102–1110. Tobe, H., P. H. Raven, and C. I. Peng. 1988. Seed coat anatomy and relationships of *Ludwigia* sects. *Microcarpium*, *Dantia*, and *Miquelia* (Onagraceae) and notes on fossil seeds of *Ludwigia* from Europe. Bot. Gaz. 149: 450–457.

14. **Ludwigia linifolia** Poiret in J. Lamarck et al., Encycl., Suppl. 3: 513. 1814 • Southeastern primrose-willow, flaxleaf seedbox

Isnardia linifolia (Poiret) Kuntze

Herbs slender, rarely with aerenchyma, forming slender stolons 4–15(–30) cm, 0.7–1(–1.5) mm thick. **Stems** erect or ascending, slightly ridged, usually well branched, 12–55(–62) cm, glabrous. **Leaves** alternate; stipules narrowly ovate to narrowly lanceolate, 0.2–0.3 × 0.1–0.2 mm; stolons: petiole narrowly attenuate, 0.05–0.5 cm, blade narrowly obovate or oblanceolate to spatulate, 0.5–2 × 0.1–0.6 cm; stems: sessile, blade linear to linear-oblanceolate, 1.5–4 × 0.1–0.4(–0.6) cm, base very narrowly cuneate, margins entire with obscure hydathodal glands, apex acuminate to acute; bracts linear, reduced. **Inflorescences** leafy spikes, flowers solitary in leaf axils; bracteoles attached 0–1.5 mm distal to base of ovary, very narrowly oblanceolate to linear, (1.5–)2.5–9(–13) × 0.2–0.8 mm, margins entire, apex acute. **Flowers:** sepals ascending, green, narrowly lanceolate-deltate, (3–)4–7 × 1.1–1.7 mm, margins entire, apex narrowly acute, surfaces glabrous or minutely papillose; petals narrowly obovate-elliptic, 4–6 × 2–4 mm, base obtuse, apex obtuse or rounded; filaments pale yellow, (1.3–)1.5–2.5(–3) mm, anthers oblong, 0.6–1.1 × 0.5–0.8 mm; pollen shed in tetrads; ovary subcylindric, 3–4 × 1–1.5 mm; nectary disc elevated 0.3–0.7 mm on ovary apex, bright yellow, 0.8–1.5 mm diam., prominently 4-lobed, minutely papillose; style yellow, 1.25–2.5 mm, glabrous, stigma subcapitate, 0.3–0.6 × 0.6–0.8 mm, shallowly 4-lobed, not exserted beyond anthers. **Capsules** subcylindric, terete or slightly angled, 5–10(–12) × 1.3–2(–2.2) mm, hard-walled, irregularly dehiscent, pedicel 0 mm. **Seeds** reddish brown, oblong-elliptic, 0.6–0.7 × 0.2–0.3 mm, surface cells nearly isodiametric. $2n = 16$.

Flowering late Jun–Oct. Drainage ditches, margins of creeks or swamps, open edges of cypress swamps, moist pinelands, edges of brackish lakes; 0–300 m; Ala., Fla., Ga., Miss., N.C., S.C.; Mexico (Tabasco).

Ludwigia linifolia is primarily a coastal species that extends farther inland in Georgia and the Carolinas. Being one of five diploids in sect. *Isnardia*, it has particularly prominent nectary lobes and appears to be modally outcrossing (C. I. Peng 1989). It also is known from a disjunct population on the Yucatán Peninsula in Tabasco, Mexico.

15. **Ludwigia linearis** Walter, Fl. Carol., 89. 1788 • Narrowleaf primrose-willow [E]

Isnardia linearis (Walter) de Candolle; *Ludwigia angustifolia* Michaux; *L. linearis* var. *puberula* Engelmann & A. Gray

Herbs slender, usually with aerenchyma near base, forming stolons 10–30 cm, 0.8–2.5 mm thick. **Stems** erect, slightly ridged, often well branched, (22–)50–100(–145) cm, glabrous or sparsely to densely, minutely strigillose. **Leaves** alternate; stipules narrowly ovate or lanceolate, 0.15–0.3 × 0.05–0.15 mm; stolons: petiole attenuate, 0.2–0.5 cm, blade narrowly to very narrowly elliptic, 1–2.5 × 0.3–0.9 cm, surfaces glabrous or minutely strigillose; stems: subsessile, blade linear to elliptic-linear, 1.6–6(–8.5) × 0.1–0.4(–0.6) cm, base very narrowly cuneate, margins entire with obscure hydathodal glands, apex very narrowly acute, surfaces glabrous or sparsely to densely, minutely strigillose or puberulent; bracts not much reduced. **Inflorescences** leafy racemes or spikes, flowers solitary in leaf axils; bracteoles deciduous, attached on pedicel near ovary base or to 4.5 mm distal to base, linear, 0.4–4(–7.5) × 0.1–0.3 mm. **Flowers:** sepals ascending, green, lanceolate-deltate to narrowly so, 2.5–5(–5.5) × 1–3(–3.5) mm, margins entire, apex acuminate or elongate-acuminate to cuspidate, surfaces sparsely to densely strigillose; petals obovate to suborbiculate, 3–6 × (2–)2.5–5 mm, base attenuate, margins entire, apex obtuse; filaments white or cream, 1.1–2.2 mm, anthers lanceolate-oblong, 1–2 × 0.6–1 mm; pollen shed in tetrads; ovary cylindric (wider at apex), 2.5–4.5 × 1–2.5 mm, strigillose; nectary disc elevated (0.2–) 0.3–0.6 mm on ovary apex, yellow, 1.3–2.5 mm diam., 4-lobed, margins glabrous or minutely strigillose; style yellowish green, (0.4–)0.7–1.5 mm, glabrous or densely strigillose on proximal part, stigma clavate to subcapitate, (0.6–)1–1.9 × 0.6–0.9 mm, shallowly 4-lobed, not exserted beyond anthers. **Capsules** elongate-obpyramidal, obscurely 4-angled, often with central, longitudinal groove on each side, 5–10(–12) × 2–5.5 mm, hard-walled, dehiscent by apical ring, pedicel 0–3.5(–5) mm. **Seeds** light brown, oblong-elliptic, 0.5–0.7 × 0.2–0.3 mm, surface cells oblong, elongate either parallel or transversely to seed length. 2*n* = 16.

Flowering late Jun–Sep. Drainage ditches, along river or stream banks, swales, edges of pocosins, sandy soil in wet meadows, brackish marshes, disturbed ground; 0–300 m; Ala., Ark., Del., Fla., Ga., La., Md., Miss., Mo., N.J., N.C., Okla., S.C., Tenn., Tex., Va.

Ludwigia linearis is widespread in the southeastern United States, with a complex pattern of morphological variation, especially in stem pubescence, ranging from glabrous to densely strigillose, but without strong geographical separation (C. I. Peng 1989).

16. Ludwigia palustris (Linnaeus) Elliott, Sketch Bot. S. Carolina 1: 211. 1817 • Marsh seedbox, water purslane, ludwigie palustre [W]

Isnardia palustris Linnaeus, Sp. Pl. 1: 120. 1753; *I. ascendens* J. Hall; *I. nitida* (Michaux) Poiret; *I. palustris* var. *americana* de Candolle; *Ludwigia apetala* Walter; *L. nitida* Michaux; *L. palustris* var. *americana* (de Candolle) Fernald & Griscom; *L. palustris* var. *inundata* Svenson; *L. palustris* var. *liebmannii* H. Léveillé; *L. palustris* var. *nana* Fernald & Griscom; *L. palustris* var. *pacifica* Fernald & Griscom

Herbs often creeping, rooting at nodes, forming mats. **Stems** prostrate or decumbent and ascending at tips, subterete or with raised lines decurrent from leaf axils, well branched, 10–50(–70) cm, glabrous or, sometimes, minutely strigillose on leaf margins and inflorescence. **Leaves** opposite; stipules narrowly deltate, 0.05–0.1 × 0.05–0.1 mm; petiole narrowly winged, 0.1–2.5 cm, blade narrowly to broadly elliptic or ovate-elliptic, 0.5–4.5 × 0.3–2.3 cm, base abruptly attenuate, margins entire and minutely strigillose, apex subacute, surfaces glabrous; bracts not reduced. **Inflorescences** leafy spikes or racemes, flowers usually paired in leaf axils of prostrate stems; bracteoles attached at base or to 2.5 mm distal to base of ovary, sublinear, 0.3–1(–1.8) × 0.1–0.8 mm. **Flowers:** sepals ascending, green, ovate-deltate, 1.1–2 × 1–2.1 mm, margins finely serrulate with minute hairs, apex acuminate, sometimes with blunt tip, surfaces glabrous; petals 0; filaments translucent, 0.4–0.6 mm, anthers 0.2–0.4 × 0.3–0.6 mm; pollen shed singly; ovary oblong, 1.5–3.5 × 1–2 mm, glabrate; nectary disc elevated 0.15–0.3 mm on ovary apex, green, 1–2 mm diam., 4-lobed, glabrous; style pale green, 0.3–0.7 mm, glabrous, stigma subglobose or capitate, 0.4–0.6 × 0.2–0.4 mm, not exserted beyond anthers. **Capsules** oblong obovoid, 4-angled, (1.6–)2–5 × 1.5–3(–3.5) mm, thin-walled, irregularly dehiscent or dispersing as unit, pedicel 0–0.5 mm. **Seeds** yellowish brown, ellipsoid, 0.5–0.7 × 0.3–0.4 mm, surface cells transversely elongate. 2*n* = 16.

Flowering Feb–Oct. Roadside ditches, wet meadows, dried pond bottoms, margins of ponds, swamps, rivers, alluvial sand bars; 0–1000[–2700] m; B.C., N.B., Nfld. and Labr. (Nfld.), N.S., Ont., Que.; Ala., Ariz., Ark., Calif., Conn., Del., D.C., Fla., Ga., Idaho, Ill., Ind., Iowa, Kans., Ky., La., Maine, Md., Mass., Mich., Minn., Miss., Mo., Nebr., N.H., N.J., N.Mex., N.Y., N.C., Ohio, Okla., Oreg., Pa., R.I., S.C., Tenn., Tex., Vt., Va., Wash., W.Va., Wis.; Mexico (Chiapas, Chihuahua, Guerrero, Hidalgo, Jalisco, México, Michoacán, Morelos, Nuevo León, Oaxaca, San Luis Potosí, Sinaloa, Sonora, Tamaulipas, Veracruz); West Indies (Cuba, Hispaniola, Jamaica, Puerto Rico); Bermuda; Central America (Costa Rica, Guatemala, Panama); South America (Colombia, Peru, Venezuela); Europe; sw Asia; Africa; introduced in Pacific Islands (Hawaii, New Zealand), Australia.

Ludwigia palustris is a common diploid and most widely distributed species in sect. *Isnardia*. It is particularly widespread in temperate North America and in Europe [the type is thought to be from Europe (P. H. Raven 1963[1964]; C. I. Peng et al. 2005)], more sporadically in Africa and sw Asia, and introduced in Australasia, and Hawaii. The close sister relationship of sect. *Isnardia* with sect. *Miquelia* (*L. ovalis* Miquel only, endemic to eastern Asia) suggests that this clade may have had a history connected with the evolution of the Arcto-Tertiary Geoflora (P. H. Raven and D. I. Axelrod 1974; Peng et al.).

Ludwigia palustris is known to hybridize with *L. brevipes* producing the sterile *L.* ×*lacustris* Eames.

17. Ludwigia microcarpa Michaux, Fl. Bor.-Amer. 1: 88. 1803

Isnardia microcarpa (Michaux) Poiret

Herbs slender, sometimes suffrutescent from woody base, often with aerenchyma, rarely creeping and rooting at nodes, often forming slender stolons 4–15(–25) cm, 0.4–0.8 mm thick. **Stems** usually erect or ascending, rarely prostrate, slightly to distinctly winged, wings to 1.8 mm wide, unbranched to densely branched, 5–60 cm, glabrous. **Leaves** alternate; stipules lanceolate-deltate, 0.13–0.15 × 0.1–0.13 mm; stolons: petiole attenuate, 0.1–0.5 cm, blade broadly elliptic to suborbiculate, 0.2–0.7 × 0.2–0.5 cm; stems: petiole winged, 0.1–0.5 cm, blade obovate-spatulate or oblanceolate, sometimes narrowly oblanceolate-elliptic, 0.4–1.7 × 0.15–1 cm, base attenuate, margins subentire or often with hydathodal glands forming minute teeth, or minutely papillose-strigillose, apex acute or mucronate, surfaces glabrous; leaves on side branches much reduced, glabrous; bracts near apex and on branches reduced. **Inflorescences** leafy spikes or racemes, flowers solitary in axils, usually not crowded; bracteoles attached at base of ovary, sublinear or narrowly oblong, 0.4–1.2(–1.5) × 0.1–0.4 mm, usually with swollen base. **Flowers:** sepals ascending or spreading, pale green to cream adaxially, ovate-deltate, 0.9–2 × 1–1.9 mm, margins minutely papillose-strigillose or entire, apex acuminate, surfaces glabrous; petals 0; filaments translucent, 0.4–0.55 mm, anthers 0.1–0.2 × 0.2–0.3 mm; pollen shed singly; ovary pale green, obovoid-subglobose, 0.8–1 × 0.8–1.2 mm, glabrate; nectary disc nearly flat on ovary apex, light green, 0.5–1.2 mm diam., 4-lobed, glabrous; style light green, 0.3–0.6 mm, glabrous, stigma subcapitate, 0.15–0.3 × 0.05–0.15 mm, not exserted beyond anthers. **Capsules** obconic, subterete, 1–1.5 × 1.4–1.9 mm, thin-walled, seeds often visible as bumps, dehiscent by apical ring, pedicel 0–0.2 mm. **Seeds** dark reddish brown, oblong-ovoid, 0.5–0.6 × 0.3–0.4 mm, surface cells transversely elongate, glabrous or, sometimes, densely covered by waxy hairs. $2n = 16$.

Flowering Mar–Nov (year-round). Roadside ditches, marshes, borders of ponds and streams, low meadows, low areas in open woods, edges of swamp forests, brackish marshes, hammocks, solution pits of limestone on marl prairies; 0–400 m; Ala., Ark., Fla., Ga., La., Miss., Mo., N.C., S.C., Tenn., Tex., Va.; West Indies (Bahamas, Cuba, Jamaica).

The diploid *Ludwigia microcarpa* has the smallest stature, leaves, flowers, fruits, and fewest seeds (ca. 10–20) per capsule of any species in sect. *Isnardia* (C. I. Peng 1989). Most plants start to flower when young.

18. Ludwigia glandulosa Walter, Fl. Carol., 88. 1788 [E] [F]

Herbs slender, forming stolons 5–20 cm, 0.4–0.8 mm thick. **Stems** erect, slightly ridged, usually well branched, 10–80 (–100) cm, glabrate or often with strigillose raised lines decurrent from leaf axils. **Leaves** alternate; stipules ovate-triangular, 0.15–0.35 × 0.05–0.25 mm, succulent; stolons: petiole attenuate, 0.3–1 cm, blade narrowly elliptic, 1.5–3.5(–5.5) × 0.5–1.3 (–2) cm; main stem: petiole 0–1.5 cm, blade usually narrowly elliptic to elliptic, sometimes linear, 3–12 × 0.3–2.1 cm, base attenuate, margins subentire with hydathodal glands often visible, apex acute to very narrowly acute, surfaces densely papillose-strigillose, abaxial veins glabrous or sparingly, minutely strigillose; leaves on side branches usually reduced, 0.8–4.5 × 0.2–1 cm; bracts much reduced. **Inflorescences** open, leafy racemes or spikes, flowers solitary in axils, often congested, especially on branches; bracteoles attached on pedicel at base of ovary or to 2 mm distal to base, narrowly lanceolate to sublinear, 0.4–1 × 0.1–0.4 mm, apex acuminate, surfaces glabrate. **Flowers:** sepals ascending, light green, ovate-deltate, 1.1–2.3 × 1–1.8 mm, margins entire, fringed with minute, strigillose hairs, apex short-acuminate or acute, surfaces glabrous; petals 0; filaments nearly translucent, 0.6–1.1 mm, anthers 0.3–0.5 × 0.3–0.6 mm; pollen shed in tetrads; ovary subcylindric, 2–5 × 0.8–1.9 mm; nectary disc elevated 0.3–0.4 mm on ovary apex, light green, 0.6–1.8 mm diam., 4-lobed, glabrous or minutely papillose; style pale green, 0.3–0.8 mm, glabrous, stigma broadly clavate to subglobose, 0.2–0.5 × 0.2–0.5 mm, not exserted beyond anthers. **Capsules** subcylindric, subterete to obscurely 4-angled with 4 shallow grooves, 2–8(–9) × 1.3–2(–3) mm, hard-walled, irregularly dehiscent, pedicel 0–0.3(–0.5) mm. **Seeds** light brown, kidney-shaped with slightly pointed ends, 0.5–0.8 × 0.3–0.4 mm, surface cells columnar, elongate either parallel or transversely to seed length. $2n = 32$.

Subspecies 2 (2 in the flora): c, e United States.

Ludwigia glandulosa consists of two subspecies: subsp. *glandulosa* is very common and widespread throughout the Atlantic and Gulf coastal plains and the Mississippi Embayment, westward to eastern Texas and southeastern Oklahoma; subsp. *brachycarpa* grows only in the western portion of the range of subsp. *glandulosa*, extending farther west in Texas and Oklahoma. The two taxa grow in similar habitats, but subsp. *glandulosa* prefers drier habitats farther south and west. The general distinctiveness of these subspecies is probably maintained by their modal autogamy; vegetative reproduction by means of stolons may likewise play a role in preserving favored genotypes (C. I. Peng 1989).

1. Capsules (4–)5–8(–9) mm; cauline leaf blades 3.2–12 × 0.4–2.1 cm; seeds: surface cells elongate parallel to seed length . . . . 18a. *Ludwigia glandulosa* subsp. *glandulosa*
1. Capsules 2–5 mm; cauline leaf blades 3–5(–7) × 0.3–0.5(–1) cm; seeds: surface cells elongate transversely to seed length. . . . . 18b. *Ludwigia glandulosa* subsp. *brachycarpa*

18a. Ludwigia glandulosa Walter subsp. **glandulosa** [E] [F]

Jussiaea brachycarpa Lamarck; *Ludwigia cylindrica* Elliott; *L. cylindrica* var. *brachycarpa* (Lamarck) Torrey & A. Gray; *L. heterophylla* Poiret

Stems usually reddish green, (20–)40–80(–100) cm. **Leaves:** petiole 0.1–1.5 cm, blade elliptic to very narrowly elliptic, those on main axis 3.2–12 × 0.4–2.1 cm, those on branches 1–4.5 × 0.3–1 cm. **Inflorescences:** bracteoles attached at base of ovary or to 2 mm distal to base, rarely on pedicel, 0.5–1 × 0.2–0.4 mm. **Flowers:** sepals 1.3–2.3 × 1.2–1.7 mm, apex acuminate; nectary disc glabrous; style 0.3–0.5 mm, stigma 0.3–0.5 mm diam. **Capsules** subterete, (4–)5–8(–9) × 1.6–2(–3) mm, pedicel 0–0.4 (–0.5) mm. **Seeds** 0.5–0.7 × 0.3–0.4 mm, surface cells elongate parallel to seed length.

Flowering Jun–Sep. Roadside ditches, marshes, pond borders, wet meadows, swales, alluvial floodplains, peaty bogs, moist pinelands, swampy woodlands, waste ground; 0–300 m; Ala., Ark., Fla., Ga., Ill., Ind., Ky., La., Md., Miss., Mo., N.C., Okla., S.C., Tenn., Tex., Va.

The distribution of subsp. *glandulosa* is relatively continuous along the Atlantic and Gulf coastal plains, extending inland to Tennessee, extreme western Kentucky, extreme southern Indiana, Illinois, and Missouri, to southeastern Oklahoma and eastern Texas, with disjunct populations in Maryland and in north-central Missouri. In Florida, it occurs only in the panhandle region.

18b. Ludwigia glandulosa Walter subsp. **brachycarpa** C. I. Peng, PhytoKeys 145: 58. 2020 [E]

Stems rarely reddish green, 10–55(–90) cm. **Leaves:** petiole 0–1 cm, blades linear-elliptic to linear, sometimes very narrowly elliptic, those on main axis 3–5(–7) × 0.3–0.5(–1) cm, those on branches 0.8–3.6 × 0.2–0.3 (–0.8) cm. **Inflorescences:** bracteoles attached at base of ovary, 0.4–0.8 × 0.1–0.2 mm. **Flowers:** sepals 1.1–1.9 × 1–1.8 mm, apex acute or short-acuminate; nectary disc obscurely, minutely papillose; style 0.4–0.8 mm, stigma 0.2–0.3 mm diam. **Capsules** obscurely 4-angled, 2–5 × 1.3–2 mm, pedicel 0–0.2 mm. **Seeds** 0.6–0.8 × 0.3–0.4 mm, surface cells elongate transversely to seed length.

Flowering Apr–Nov. Ditches, low meadows, coastal prairies, seeps in sandy woods, moist sinkholes in granite outcrops, old clay fields; 0–200 m; La., Okla., Tex.

Subspecies *brachycarpa* grows along the Gulf Coast from southwestern Louisiana to Nueces County, Texas, and more sporadically northward in eastern Texas to south-central Oklahoma.

Subspecies *brachycarpa* was published initially by Peng as a new combination based on *Ludwigia cylindrica* var. *brachycarpa* Torrey & A. Gray, not realizing that the variety was based on *Jussiaea brachycarpa* Lamarck, which Peng considered to be a synonym of *L. glandulosa* subsp. *glandulosa*. Therefore, subsp. *brachycarpa* is a new subspecies but was invalid when published by Peng since it was not accompanied by a Latin description in 1986; that situation has since been remedied.

19. Ludwigia polycarpa Short & R. Peter, Transylvania J. Med. Assoc. Sci. 8: 581. 1835 • Manyfruit primrose-willow, false loosestrife [E]

Isnardia polycarpa (Short & R. Peter) Kuntze

Herbs slender, with well-developed aerenchyma on submerged stems, forming stolons 2.5–15(–22) cm, 1–2.3 mm thick, well branched. **Stems** erect or ascending, slightly ridged, well branched, (10–) 25–60(–85) cm, glabrate with raised ± strigillose lines decurrent from leaf axils. **Leaves** alternate; stipules narrowly to broadly ovate, 0.1–0.4 × 0.1–0.3 mm; stolons: leaves often clustered near apex of stolon, petiole 0–0.5 cm, blade narrowly elliptic or oblanceolate, 0.8–2(–3.2) × 0.2–0.8(–1.2) cm, base attenuate, margins entire or remotely denticulate, apex acute, surfaces glabrous; stems: petiole winged, 0.1–1 cm, blade very narrowly oblong-elliptic, 3.5–11 × 0.4–1(–1.7) cm, base very narrowly cuneate or long-attenuate, margins entire and densely, minutely papillose-serrulate with obscure hydathodal glands, apex narrowly acute or acuminate, surfaces glabrous; bracts not much reduced. **Inflorescences** elongated, leafy spikes, flowers solitary in leaf axils, sometimes borne almost to base of stems; bracteoles attached 0.5–2.5(–3) mm distal to base of ovary, linear-lanceolate, 3.5–6.5(–8) × 0. 4–1(–1.3) mm, with a swollen base, margins minutely papillose-serrulate. **Flowers:** sepals spreading horizontally with reflexed tips, pale green, narrowly ovate-deltate, 2.5–4.5 × 1.5–3.2 mm, margins entire, minutely papillose-serrulate, apex elongate-acuminate, surfaces glabrous;

petals 0; filaments yellowish green, 0.7–1.5 mm, base dilated, anthers 0.5–0.9 × 0.5–0.7 mm; pollen shed in tetrads; ovary oblong, barely 4-angled, 3–4.5 × 2–3.5 mm; nectary disc elevated 0.5–0.8 mm on ovary apex, yellowish green, 1.8–3 mm diam., 4-lobed, glabrous; style yellowish green, 0.5–0.8 mm, glabrous, stigma broadly clavate to subglobose, 0.4–0.8 × 0.3–0.6 mm, usually 4-lobed, not exserted beyond anthers. **Capsules** oblong-obovoid, obscurely 4-angled, 4–7 × 2.5–5 mm, hard-walled, irregularly dehiscent, pedicel 0.1–0.3 mm. **Seeds** light brown, narrowly oblong with curved ends, 0.5–0.6 × 0.2–0.3 mm, surface cells elongate parallel to seed length. 2*n* = 32.

Flowering Jun–Sep. Ditches, moist prairies, alluvial ground of ponds, lakes, and rivers, marshes, swales, edges of lagoons, low fallow fields; 100–300 m; Ont.; Conn., Idaho, Ill., Ind., Iowa, Kans., Ky., Mass., Mich., Minn., Mo., Nebr., Ohio, Pa., Va., W.Va., Wis.

Ludwigia polycarpa, unlike all other species in sect. *Isnardia*, is distributed primarily in the central Midwest and Great Lakes regions, with one highly disjunct population recorded from Kootenai County, Idaho, which is presumably introduced. This species has also been found scattered as far east as Connecticut and Massachusetts, and reports of it from Arkansas, Maine, Tennessee, and Vermont cannot be confirmed. As indicated by C. I. Peng (1989), a report of this species from Alabama involved a natural hybrid between *L. glandulosa* and *L. pilosa*.

The basal stolons formed by *Ludwigia polycarpa* tend to be shorter, more condensed, and more branched than those found in other species, and may be a morphological adaptation to perennial survival in the colder areas in which it grows.

20. Ludwigia suffruticosa Walter, Fl. Carol., 90. 1788 • Shrubby primrose-willow or seedbox [E]

Isnardia suffruticosa (Walter) Kuntze

Herbs with 1–3 rhizomes 0.6–5.5 × 0.2–0.5 mm, often branched, glabrous or densely hirtellous, sometimes also forming stolons 8–80 cm, 1.1–2.2 mm thick, branched, glabrous or sparsely to densely hirtellous. **Stems** erect, slightly ridged, unbranched or slightly branched, (16–)30–90 cm, glabrous or strigillose to hirtellous, especially on distal parts. **Leaves** alternate; stipules deltate, 0.25–0.45 × 0.15–0.4 mm; rhizomes: sessile, blades minute, appressed, and scalelike, oblate or suborbiculate, 0.3–0.6 × 0.6–0.9 mm; stolons: petiole 0.1–0.6 cm, blade oblong or oblanceolate-elliptic to spatulate, 0.4–3.5 × 0.2–1.5 cm; main stem: sessile, blade lanceolate-elliptic or lanceolate-linear to linear, 2.5–9.5 × (0.1–)0.3–0.9 cm, proximal ones shorter and often oblong or oblong-lanceolate, base rounded or obtuse, margins entire with obscure hydathodal glands, apex acuminate to acute, surfaces glabrous or, sometimes, pilose on proximal blades; bracts very reduced. **Inflorescences** densely clustered, terminal racemes or spikes, 1–5(–12) cm; bracteoles attached at base of ovary or on pedicel distally, narrowly lanceolate, 3.5–5(–6) × (1.2–)1.4–2 mm, surfaces usually glabrous, sometimes pilose abaxially. **Flowers:** sepals ascending, pale green or white adaxially, broadly ovate-deltate, 2.3–3.5(–4) × 2.3–3.2(–3.8) mm, margins entire, apex acuminate, surfaces glabrous; petals 0; filaments yellow or cream, 1.2–2 mm, distinctly dilated toward base, anthers 0.7–1(–1.3) × 0.5–0.7 mm; pollen shed in tetrads; ovary broadly obovoid or cup-shaped, 2.2–3 × 2.3–3.3 mm; nectary disc elevated 0.5–0.6 mm on ovary apex, pale yellow, 1.8–3.1 mm diam., obscurely 4-lobed, glabrous; style pale yellow, 0.9–1.7 mm, glabrous, stigma pale green to white, globose to capitate, 0.4–0.8 × 0.5–0.8 mm, distinctly 4-lobed, not exserted beyond anthers. **Capsules** broadly obpyramidal, angles rounded, sometimes subspherical, 2.5–4.3 × 2.5–4.5(–5) mm, hard-walled, dehiscent by apical ring, pedicel 0.5–1.5(–2) mm. **Seeds** brown, elliptic-oblong, curved on both ends, 0.5–0.6 × 0.2–0.3 mm, surface cells ± isodiametric. 2*n* = 32.

Flowering May–Sep. Sandy ditches, marshes, wet meadows, limestone sinks, cypress swamps, moist pinelands; 0–150 m; Ala., Fla., Ga., N.C., S.C.

Ludwigia suffruticosa is distinctive by virtue of its highly condensed inflorescence and sessile leaves. It also differs from other species in sect. *Isnardia* in that it perennates mainly by underground rhizomes. This apetalous species has showy bracts and attracts many insects, including bumblebees, wasps, and honeybees (C. I. Peng 1989). Its center of distribution is in Florida, extending along the coastal plain barely to Alabama on the west and barely to southern North Carolina on the northeast. Recent reports of this species from Mississippi and from Mexico (Chiapas and Oaxaca) have not been confirmed.

21. Ludwigia lanceolata Elliott, Sketch Bot. S. Carolina 1: 213. 1817 • Lanceleaf primrose-willow [E]

Isnardia lanceolata (Elliott) de Candolle

Herbs slender, with well-developed aerenchyma when base submerged, often forming stolons 10–40 cm, 2–3 mm thick, stolons with widely spaced leaves. **Stems** erect, subterete or slightly ridged, well branched distally, 45–100 cm, glabrous, with raised lines decurrent from leaf axils. **Leaves** alternate; stipules ovate to very broadly ovate, 0.2–0.5 ×

0.1–0.3 mm, succulent; stolons: petiole winged, 0.2–1 cm, blade orbiculate or elliptic to broadly elliptic, 0.5–2.7 × 0.6–1.2 cm, base attenuate, apex rounded to acute; main stem: petiole winged, 0.1–0.5 cm, blade elliptic, oblanceolate, or narrowly oblanceolate to sublinear, 2–7.5 × 0.2–0.8(–1.4) cm, base narrowly cuneate, sometimes attenuate, margins entire with minute hydathodal glands, apex acute to narrowly acute, leaves on side branches much reduced; bracts sublinear, moderately reduced. **Inflorescences** open spikes, flowers solitary in distal leaf axils; bracteoles attached in opposite pairs at or slightly distal to base of ovary, ovate-elliptic to very narrowly elliptic, 1.5–4.3 × 0.4–1.4 mm, often with a swollen base, margins minutely papillose, apex subacute. **Flowers:** sepals ascending, pale green, broadly ovate-deltate, 1.5–2.5 × 1.8–3.3 mm, margins minutely papillose, apex usually acute, rarely acuminate, surfaces glabrous; petals 0; filaments nearly translucent, 1–1.4 mm, base dilated, anthers 0.4–0.6(–0.8) × 0.5–0.6 mm; pollen shed in tetrads; ovary broadly obovoid or cup-shaped, 2.5–3.5 × 2.2–3.2 mm; nectary disc elevated 0.4–0.6 mm on ovary apex, yellowish green, 1.8–2.6 mm diam., 4-lobed, glabrous; style yellowish green, 0.5–0.7 mm, glabrous, stigma yellowish green, broadly capitate to subglobose, 0.3–0.5 × 0.7–1 mm, not exserted beyond anthers. **Capsules** obpyramidal, sharply 4-angled and 4-winged, wings 0.3–0.7 mm wide, 3.5–5 × 2.5–4.5 mm, hard-walled, dehiscent by apical ring, pedicel 0.1–0.5 mm. **Seeds** light brown, narrowly oblong with constricted ends, 0.6–0.8 × 0.2–0.3 mm, surface cells nearly isodiametric. $2n = 32$.

Flowering Jun–Oct. Ditches, low meadows, cypress swamps, moist pinelands, edges of pocosins, sandy peaty soil; 0–100 m; Fla., Ga., N.C., S.C.

Ludwigia lanceolata is fairly uncommon, with scattered populations occurring along the Atlantic coast of southern North Carolina, South Carolina, eastern and southern Georgia, and peninsular Florida. It reaches its western limit in the central panhandle of Florida. C. I. Peng (1988, 1989) found this tetraploid species to be interfertile with other tetraploid species in the section, but few natural hybrids are found, perhaps due to persistent autogamy and habitat specialization.

22. Ludwigia pilosa Walter, Fl. Carol., 89. 1788 • Hairy primrose-willow [E]

Isnardia mollis (Michaux) Poiret; *I. pilosa* (Walter) Kuntze; *Ludwigia hirsuta* Desrousseaux; *L. mollis* Michaux; *L. rudis* Walter

Herbs often with prominent aerenchyma when base submerged, forming stolons 30–250 cm, 2–4 mm thick, creeping in mud or floating in water, sometimes bearing flowers and fruits. **Stems** erect, subterete, densely branched, 40–120 cm, densely hirtellous. **Leaves** alternate; stipules ovate to lanceolate, 0.2–0.25 × 0.1–0.15 mm, usually obscured by pubescence; stolons: petiole 0.2–0.7 cm, blade obovate or elliptic to orbiculate, 0.6–2 × 0.5–1.1 cm, margins with distinct hydathodal teeth, base attenuate, surfaces densely hirtellous to glabrate; stems: petiole 0–0.2(–1) cm, blade elliptic or lanceolate-elliptic to very narrowly elliptic, 1.5–8(–10) × 0.3–1.2(–1.4) cm, base cuneate or attenuate, margins entire with obscure hydathodal glands, apex acute or narrowly acute, surfaces ± densely hirtellous, leaves on branches much reduced; bracts much reduced. **Inflorescences** usually congested, leafy spikes or racemes, flowers solitary in distal leaf axils; bracteoles attached 1–2.2 mm distal to base of ovary, linear-lanceolate or narrowly elliptic, 3–6.5(–7.2) × 0.3–1.5(–1.7) mm, apex acuminate, surfaces hirtellous. **Flowers:** sepals ascending with reflexed tips, pale green abaxially, creamy white adaxially, often tinged with pink or red, ovate-deltate, 3.5–5.5(–6) × 2–4 mm, margins entire, apex elongate-acuminate to subcuspidate, surfaces densely hirtellous; petals 0; filaments yellowish, 1.5–2.5 mm, base dilated, anthers 0.6–0.9 (–1.3) × 0.5–0.7 mm; pollen shed in tetrads; ovary obovoid to cup-shaped, 2.5–4 × 2.5–4 mm; nectary disc elevated 0.3–0.7 mm on ovary apex, bright yellow, turning black upon drying, 2–3.6 mm diam., indistinctly 4-lobed, densely hirtellous around style base and between lobes; style cream, 1–2 mm, sparsely to densely hirtellous, especially proximally, stigma capitate, 0.3–0.6 × 0.3–0.6 mm, not exserted beyond anthers. **Capsules** subglobose or, sometimes, oblong-obovoid, subterete or with 4 rounded corners, 3–5 × 3–4.5 mm, hard-walled, irregularly dehiscent, pedicel 0–1 mm. **Seeds** brown, elliptic-oblong or oblong-ovoid, slightly curved on both ends, 0.5–0.7 × 0.3–0.4 mm, surface cells ± isodiametric. $2n = 32$.

Flowering Jun–Sep. Roadside ditches, marshes, swales in sandy pine flats, edges of pocosins, peaty bogs, low grassy savannas, swamp forests; 0–300 m; Ala., Fla., Ga., La., Miss., N.C., S.C., Tex., Va.

The distribution of *Ludwigia pilosa* is nearly continuous along the Atlantic and Gulf coastal plains, from extreme southeastern Virginia to northern Florida, and west to Louisiana and southeastern Texas. Disjunct populations occur in northern Alabama and central North Carolina.

Ludwigia pilosa is easily distinguished from most others in sect. *Isnardia* by being densely hirtellous throughout. Its showy sepals and nectary disc attract multiple insect visitors including ants, bumblebees, honeybees, moths, and wasps (C. I. Peng 1989).

23. Ludwigia ravenii C. I. Peng, Syst. Bot. 9: 129, fig. 1. 1984 • Raven's primrose-willow [C] [E]

Herbs slender, rarely with aerenchyma when base submerged, forming slender, glabrate stolons 10–18 cm, 0.6–1.5 mm thick, occasionally bearing flowers and fruits. **Stems** erect, slightly ridged, usually well branched, (15–)35–90 cm, densely hirtellous. **Leaves** alternate; stipules lanceolate to broadly deltate, 0.2–0.5 × 0.1–0.3 mm; stolons: petiole 0.1–0.3 cm, blade elliptic to orbiculate, 1–1.8 × 0.6–1.4 cm, base attenuate, apex rounded to acute; stems: petiole narrowly winged, 0.1–0.8 cm, blade narrowly lanceolate-elliptic, 1.3–6.5 × 0.4–1.5 cm, base attenuate, margins entire with minute hydathodal glands, apex acute, surfaces densely hirtellous; bracts not much reduced. **Inflorescences** leafy racemes, flowers solitary in leaf axils; bracteoles attached near base of ovary, lanceolate or elliptic to narrowly so, (1.5–)2–4.3 × 0.3–0.9 mm, apex acuminate, surfaces hirtellous. **Flowers:** sepals ascending-spreading, green, broadly ovate-deltate, 1.5–3 × 1.4–2.1 mm, margins entire, apex acuminate, surfaces densely hirtellous; petals 0; filaments light green, 0.7–1.1 mm, anthers 0.3–0.4 × 0.4–0.5 mm; pollen shed in tetrads; ovary obovoid to obconic, 2.8–3.5 × 2–3 mm; nectary disc elevated 0.3–0.4 mm on ovary apex, light green, 1.4–2.5 mm diam., 4-lobed, glabrous; style light green, 0.3–0.5 mm, glabrous, stigma clavate to subcapitate, 0.4–0.6 × 0.3–0.5 mm, not exserted beyond anthers. **Capsules** oblong-obovoid, subterete to scarcely 4-angled, (3–)4–5(–5.3) × 2.5–3.5(–4) mm, hard-walled, irregularly dehiscent, pedicel 0.2–0.5 mm. **Seeds** light brown, elliptic-oblong with slightly curved ends, 0.5–0.7 × 0.3–0.4 mm, surface cells elongate transversely to seed length. $2n = 32$.

Flowering Jul–Sep. Wet, peaty habitats, ditches, margins of ponds, bogs, swamps; of conservation concern; 0–100 m; Fla., N.C., S.C., Va.

Ludwigia ravenii is an uncommon species occurring in scattered populations in coastal southeastern Virginia and eastern North Carolina, with single disjunct populations in southeastern South Carolina and northeastern Florida. C. I. Peng (1989) observed its similarity to *L. pilosa* by virtue of its dense, hirtellous pubescence, but noted its smaller, more consistently autogamous flowers. It is in the Center for Plant Conservation's National Collection of Endangered Plants.

24. Ludwigia sphaerocarpa Elliott, Sketch Bot. S. Carolina 1: 213. 1817 • Globefruit primrose-willow [E]

Isnardia sphaerocarpa (Elliott) de Candolle; *Ludwigia sphaerocarpa* var. *deamii* Fernald & Griscom; *L. sphaerocarpa* var. *jungens* Fernald & Griscom; *L. sphaerocarpa* var. *macrocarpa* Fernald & Griscom

Herbs often with prominent aerenchyma when base submerged, forming stolons 20–90 cm, 2–3.5 mm thick, floating, sometimes branched. **Stems** erect, slightly ridged, well branched, (40–)60–110 cm, densely strigillose or glabrous. **Leaves** alternate; stipules lanceolate-deltate, 0.1–0.4 × 0.1–0.2 mm; stolons: petiole ± winged, 0.1–0.3 cm, blade narrowly elliptic to oblanceolate or spatulate, 0.9–3 × 0.4–0.8(–1.3) cm, base attenuate, margins subentire with hydathodal glands, apex acute or obtuse; stems: petiole 0.1–0.4(–1) cm, blade narrowly elliptic or lanceolate to sublinear, on main stem (2.6–)6–10 × 0.5–1.1(–1.6) cm, on branches 2–5(–6) × 0.3–0.5(–0.6) cm, base attenuate or narrowly cuneate, margins entire with hydathodal glands mainly on primary cauline leaves, apex acute to very narrowly acute, surfaces glabrous or densely strigillose; bracts not much reduced. **Inflorescences** open, leafy racemes, more congested on branches, flowers solitary in leaf axils; bracteoles attached in subopposite pairs near base of ovary, usually linear to very narrowly lanceolate, rarely lanceolate, 0.5–1.5 × 0.1–0.3 mm, apex acuminate. **Flowers:** sepals ascending, yellow or cream adaxially, ovate-deltate, 2–3.5(–4) × 1.6–3(–3.3) mm, margins entire, apex acuminate, surfaces glabrous or densely strigillose; petals 0; filaments yellow, 1–1.7 mm, slightly dilated toward base, anthers 0.5–0.8 × 0.4–0.7 mm; pollen shed in tetrads; ovary broadly obovoid or cup-shaped, 1.5–3.5 × 2–3 mm; nectary disc elevated 0.4–0.6 mm on ovary apex, bright yellow, 1.5–3 mm diam., 4-lobed, glabrous or short-hirtellous between lobes; style yellow, 0.6–1(–1.3) mm, glabrous or strigillose proximally, stigma yellow, capitate to subglobose, 0.3–0.5 × 0.4–0.7 mm, not exserted beyond anthers. **Capsules** sometimes tinged pink, subglobose, subterete, 2–4(–4.5) × 2–4 mm, hard-walled, irregularly dehiscent, pedicel 0.5–1.2(–2.3) mm. **Seeds** brown to light brown, elliptic, 0.4–0.7 × 0.3–0.4 mm, surface cells transversely elongate to seed length, sometimes oblique. $2n = 32$.

Flowering Jun–Sep. Drainage ditches, shores of slow-moving streams or ponds, marshes, swales, swamp forests, edges of limestone sinks, peaty bogs in pastures, interdunal marshes; 0–300 m; Ala., Conn., Del., Fla., Ga., Ill., Ind., La., Md., Mass., Mich., Miss., N.J., N.Y., N.C., Pa., R.I., S.C., Tenn., Tex., Va.

Ludwigia sphaerocarpa has its primary distribution along the Atlantic coastal plain, from Massachusetts to north-central Florida, and west along the Gulf coastal plain sporadically to southeastern Texas. Disjunct populations occur in south-central Tennessee, extreme southwestern Indiana, along Lake Michigan in northeastern Illinois and northwestern Indiana, and in west-central New York. In Michigan, *L. sphaerocarpa* is known from Allegan and Berrien counties, as reported by Reznicek and Voss in the Michigan Flora (https://michiganflora.net/species.aspx?id=1757).

25. Ludwigia arcuata Walter, Fl. Carol., 89. 1788 • Piedmont primrose-willow [E]

Isnardia arcuata (Walter) Kuntze; *I. pedunculosa* (Michaux) de Candolle; *Ludwigia pedunculosa* Michaux; *Ludwigiantha arcuata* (Walter) Small

Herbs usually creeping and rooting at nodes, forming mats. **Stems** prostrate or decumbent and ascending at tips, slightly ridged, often well branched, 5–70 cm, glabrate to sparsely strigillose, denser on distal parts. **Leaves** opposite; stipules narrowly deltate or ovate, 0.05–0.15 × 0.05–0.1 mm; submerged stems: petiole 0–0.2 cm, blade narrowly linear, 1.9–4 × 0.1–0.25 cm; emergent stems: petiole 0–0.2 cm, blade narrowly elliptic or narrowly oblanceolate-elliptic to linear, 0.6–1.8 × 0.2–0.5 cm, base narrowly cuneate, margins entire, apex acute, surfaces glabrous or sparingly strigillose on margins and abaxial midveins; bracts reduced. **Inflorescences** in racemes or spikes, well-formed on ascending stems, not on prostrate stems; bracteoles attached at base of ovary or 1.5–8 mm proximally on pedicels, sublinear to very narrowly elliptic, 1.4–5 × 0.2–0.8 mm, apex acute, surfaces minutely strigillose. **Flowers:** sepals reflexed or spreading, green, lanceolate-deltate, 5.2–10 × 1.5–2.7 mm, with 3 prominent parallel veins, margins entire and minutely strigillose, apex acute or elongate-acuminate, surfaces minutely strigillose abaxially; petals rarely caducous, elliptic-obovate to spatulate-obovate, 7–11 × 4.5–8 mm, base attenuate, apex rounded; filaments initially spreading, becoming erect, yellow, 2.5–4.5 mm, anthers 1.3–2 × 0.7–1.1 mm; pollen shed in tight tetrads; ovary cylindric to funnelform, 4–5.5 × 1.5–2.8 mm; nectary disc elevated 0.6–1 mm on ovary apex, 1.5–2.6 mm diam., bright yellow, with 4 distinct domed lobes, minutely strigillose between lobes or glabrous; style yellow, 2.3–4(–4.8) mm, glabrous, stigma yellow, broadly capitate, 0.3–0.6 × 0.6–1.8 mm, as long as or exserted beyond anthers. **Capsules** clavate, subterete, sometimes slightly curved, 5.5–10 × 2.3–4 mm, hard-walled, irregularly dehiscent, pedicel (12–)17–45 mm. **Seeds** light to dark brown, elliptic-oblong, 0.5–0.7 × 0.3–0.4 mm, surface cells transversely elongate. **2*n*** = 32.

Flowering Mar–Aug. Roadside ditches, edges of lakes or ponds, swampy prairies, springs, mucky or sandy beach strands; 0–150 m; Ala., Fla., Ga., S.C.

Ludwigia arcuata is common in its range, but geographically restricted to central and western parts of peninsular Florida and adjacent Georgia, extending to southern South Carolina. Disjunct populations have been collected in Bibb County in central Georgia and Mobile County, Alabama.

The tetraploid *Ludwigia arcuata* has the largest flowers in sect. *Isnardia* and is the most consistently outcrossing species; C. I. Peng (1989) reported abundant insect visitors on this species. It is morphologically most similar to the hexaploid *L. brevipes*, with which it shares two genomes (Peng).

26. Ludwigia spathulata Torrey & A. Gray, Fl. N. Amer. 1: 526. 1840 • Spoon primrose-willow, southern water purslane [C] [E]

Isnardia spathulata (Torrey & A. Gray) Kuntze

Herbs creeping and rooting at nodes, often forming mats. **Stems** prostrate or decumbent and ascending distally, slightly ridged, well branched, 10–40 cm, densely strigillose throughout. **Leaves** opposite; stipules narrowly deltate or ovate, 0.05–0.15 × 0.05–0.1 mm; petiole very narrowly winged, 0.3–0.9 cm, blade elliptic-spatulate or narrowly so, 0.9–1.7 × 0.3–0.9 cm, base attenuate, margins entire, apex acute, surfaces strigillose; bracts not reduced except at branch tips. **Inflorescences** leafy spikes or racemes, flowers usually paired in leaf axils; bracteoles attached at base of ovary or on short pedicel, narrowly oblong or oblanceolate, 0.2–0.8 × 0.05–0.2 mm, apex acute, often obscured by hairs. **Flowers:** sepals ascending, pale green, broadly ovate-deltate, 1–1.7 × 1.1–1.7 mm, margins entire, apex acuminate, surfaces densely strigillose; petals 0; filaments translucent, 0.5–0.8 mm, anthers 0.2–0.4 × 0.3–0.5 mm; pollen shed singly; ovary oblong-obovoid, 4-angled to subterete, 1.5–2.5 × 1–1.5 mm; nectary disc elevated 0.1–0.2 mm on ovary apex, yellowish green, 0.7–0.9 mm diam., 4-lobed, glabrous; style yellowish green, 0.3–0.5 mm, glabrous, stigma pale yellow, capitate, 0.2–0.3 × 0.2–0.3 mm, not exserted beyond anthers. **Capsules** oblong-obovoid, subterete, 2.5–4(–4.5) × 1.5–2.5 mm, thin-walled, seeds often visible on exocarp as small bumps, irregularly dehiscent or dispersing as unit, pedicel 0–0.5 mm. **Seeds** dark reddish brown, ellipsoid, 0.5–0.7 × 0.4–0.5 mm, surface cells transversely elongate. **2*n*** = 32.

Flowers May–Sep. Ditches, swales, edges of ponds, lakes, sinks, swamps, sandy river bars, dried seasonal ponds, disturbed low savannas; of conservation concern; 0–200 m; Ala., Fla., Ga., S.C.

The tetraploid *Ludwigia spathulata* is relatively uncommon and occurs primarily on the Gulf Coastal Plain in the panhandle of Florida, southern Alabama, and southwestern Georgia. Outlying populations have also been collected in transitional areas between the Coastal Plain and the Piedmont in South Carolina and Georgia. With its small apetalous flowers, *L. spathulata* is modally autogamous and shows low morphological variability. Its strongest affinities appear to be with *L. palustris*, with which it shares a genome (C. I. Peng 1988, 1989).

27. Ludwigia alata Elliott, Sketch Bot. S. Carolina 1: 212. 1817 • Winged primrose-willow

Isnardia alata (Elliott) de Candolle

Herbs with aerenchyma when base submerged, forming stolons from lower nodes, 8–65(–95) cm, 0.7–2.5 mm thick. **Stems** erect or somewhat sprawling, slightly to distinctly winged (wings to 1.8 mm wide), branched distally, 40–120(–160) cm, glabrous. **Leaves** alternate; stipules ovate-deltate, often narrowly so, 0.2–0.4 × 0.1–0.3 mm, succulent; stolons: petiole 0.15–1 cm, blades orbiculate to oblanceolate or broadly elliptic, 0.4–2.6 × 0.4–1.5 cm, base attenuate, apex rounded to subacute; stems: petiole 0–0.3 cm, blade lanceolate-elliptic or very narrowly elliptic to linear, sometimes to oblanceolate or oblanceolate-elliptic near base, 1.8–10 × 0.2–1.3(–2) cm, base narrowly cuneate or attenuate, margins subentire with remote hydathodal glands, rarely minutely papillose-serrulate near apex, apex acute to narrowly acute, leaves on side branches much reduced; bracts much reduced. **Inflorescences** sometimes congested, leafy spikes or racemes, flowers solitary in distal leaf axils; bracteoles attached near base of ovary, lanceolate-elliptic or narrowly so, 2.4–4.7 × 0.6–1.5 mm, margins minutely papillose or smooth, apex acute, surfaces glabrous. **Flowers:** sepals spreading to reflexed, creamy white adaxially, broadly ovate-deltate, 2–4 × 1.6–4 mm, margins smooth or minutely papillose-serrulate, apex acute or acuminate, surfaces glabrous; petals 0; filaments nearly translucent, 1.1–1.7 mm, slightly dilated near base, anthers 0.5–0.9 × 0.4–0.7 mm; pollen shed singly; ovary obpyramidal, sharply 4-angled, 2–3.8 × 2–3.5 mm; nectary disc elevated 0.5–0.8 mm on ovary apex, bright yellow, square with rounded corners, 2–3.3 mm diam., prominently 4-lobed, glabrous; style pale green, 0.8–1.3 mm, glabrous, stigma pale yellow, subglobose, 0.3–0.6 × 0.3–0.7 mm, shallowly 4-lobed on top, not exserted beyond anthers. **Capsules** obpyramidal, sharply 4-angled and 4-winged, 3–5 × 2.8–4.5 mm, with hard walls somewhat bulging, dehiscent by apical ring, pedicel 0–0.8 mm. **Seeds** light brown, ellipsoid, slightly curved on both ends, 0.6–0.7 × 0.3–0.4 mm, surface cells elongate transversely to seed length. **2*n*** = 48.

Flowering Jun–Oct. Ditches, edges of ponds and lagoons, peaty or sandy swales, open cypress swamps, sandy borrow pits in open pine woods, swampy, flat outcrops of oolitic rocks, wet savannas, tidal flats, brackish marshes, sandy beach strands and hammocks; 0–50 m; Ala., Fla., Ga., La., Miss., N.C., S.C., Va.; West Indies (Jamaica).

Ludwigia alata occurs only at very low elevations along the Atlantic and Gulf coastal plains from Virginia to the tip of Florida, and west to southwestern Louisiana, with disjunct populations on Jamaica (C. I. Peng 1989). This hexaploid species is often confused with *L. lanceolata*, with which it shares two genomes (Peng 1988, 1989) and with which it is frequently sympatric. The showy petals of *L. alata* suggest a higher level of outcrossing, and numerous natural hybrids have been documented (Peng 1988, 1989).

28. Ludwigia simpsonii Chapman, Fl. South. U.S. ed. 2 repr. 2, 685. 1892 (as simpsoni) • Simpson's primrose-willow

Ludwigia cubensis Helwig; *L. curtissii* Chapman var. *simpsonii* (Chapman) D. B. Ward

Herbs sometimes creeping and rooting at nodes, new shoots arising from trailing stems or main caudex, rarely forming stolons. **Stems** erect, ascending, decumbent, or prostrate, slightly ridged, well branched, 10–60(–75) cm, glabrous, with raised lines decurrent from leaf axils. **Leaves** alternate or proximal pairs opposite; stipules narrowly ovate-deltate, 0.1–0.3 × 0.1–0.2 mm, succulent; petiole winged, 0.2–1 cm, blade spatulate or oblanceolate to very narrowly oblanceolate or sublinear, 0.6–1.5(–2) × (0.1–)0.3–0.7(–1.1) cm, base attenuate, margins subentire with hydathodal glands, apex acute or mucronate; bracts not much reduced. **Inflorescences** open, leafy spikes or racemes, flowers solitary in leaf axils; bracteoles attached in opposite pairs near base of ovary, lanceolate-elliptic, 0.9–1.5(–2.5) × 0.4–0.9 mm, swollen at base, apex acuminate. **Flowers:** sepals ascending, creamy white near base adaxially, ovate-deltate, 1.2–1.8 × 1–2 mm, margins entire, apex narrowly acute or acuminate, surfaces glabrous; petals 0 or very rare; filaments nearly translucent, 0.5–0.8 mm, anthers 0.2–0.4 × 0.3–0.4 mm; pollen shed singly;

ovary obconic, subterete or scarcely 4-angled, 1.2–1.8 × 1.2–2 mm; nectary disc elevated 0.3–0.4 mm on ovary apex, green, 0.9–1.3 mm diam., distinctly 4-lobed, glabrous; style pale green, 0.4–0.6 mm, stigma pale yellow, subglobose, 0.2–0.3 × 0.2–0.3 mm, not exserted beyond anthers. **Capsules** obconic, obscurely 4-angled, 1.5–2.5 × 1.5–3 mm, hard-walled, dehiscent by loculicidal slits, pedicel 0–0.4 mm. **Seeds** light brown or brown, ellipsoid, 0.5–0.6 × 0.3–0.4 mm, surface cells transversely elongate, glabrous, occasionally covered by minute waxy hairs. **2*n*** = 48.

Flowering Apr–Nov (year-round). Sandy, peaty ditches, open pineland swamps, edges of cypress swamps, tidal flats and nearby marshes, limestone sinks; 0–50 m; Fla., Miss.; West Indies (Cuba, Jamaica).

Ludwigia simpsonii is a hexaploid species occurring primarily in Florida, with outlier populations in southern Mississippi, western Cuba, and Jamaica. The species grows frequently in close proximity to *L. curtissii*; it grows mainly along roadside ditches with other weeds, whereas *L. curtissii* grows in less disturbed habitats, and the two seldom occur side by side (C. I. Peng 1989). Peng (1988, 1989) suggested that the hexaploid (2*n* = 48) *L. simpsonii* and the diploid (2*n* = 16) *L. microcarpa* gave rise to *L. curtissii*, the only octoploid (2*n* = 64) in sect. *Isnardia*.

29. Ludwigia brevipes (Long) Eames, Rhodora 35: 228. 1933 (as Ludvigia) • Long Beach primrose-willow [E]

Ludwigiantha brevipes Long in N. L. Britton and A. Brown, Ill. Fl. N. U.S. ed. 2, 2: 586, fig. 3015. 1913

Herbs creeping and rooting at nodes, sometimes forming large mats. **Stems** prostrate, ascending or erect at tips, terete, well branched, 20–70 cm, glabrous or, sometimes, minutely strigillose on leaf margins and inflorescence. **Leaves** opposite; stipules narrowly deltate, 0.05–0.15 × 0.05–0.1 mm; petiole narrowly winged, 0.2–0.8 cm, blades on submerged stems linear, 3.2–4.7 × 0.2–0.3 cm, those on emergent ones oblanceolate-elliptic to narrowly oblanceolate, (0.7–)1–1.7(–2) × 0.2–0.7(–1.1) cm, base very narrowly cuneate or attenuate, margins entire, apex acute; bracts not reduced. **Inflorescences** sometimes few-flowered, erect racemes, flowers paired in leaf axils of prostrate stems; bracteoles attached in opposite pairs at base of ovary or on pedicel distally, linear, 1–3(–4.5) × 0.1–0.7 mm, apex acuminate. **Flowers:** sepals slightly reflexed at anthesis, ascending in fruit, light green, ovate-deltate or narrowly so, (3.5–)4–5(–6) × 1.7–3 mm, with 3 parallel veins, margins strigillose and finely serrulate, apex narrowly acute to elongate-acuminate, surfaces glabrous; petals sometimes caducous, oblong-spatulate, 4.5–5.5 × 1.5–3 mm, base attenuate, apex obtuse; filaments spreading, pale cream, 1.8–2.5 mm, anthers 0.7–1 × 0.5–0.7 mm; pollen shed in very loose tetrads; ovary obconic-cylindric, subterete or scarcely 4-angled, 3–5 × 2–2.5 mm; nectary disc elevated 0.5–0.7 mm on ovary apex, bright yellow, 1.7–2.3 mm diam., 4-lobed, glabrate; style cream, 1.1–1.7 mm, stigma cream, broadly capitate, 0.4–0.6 × 0.5–0.8 mm, often exserted beyond spreading stamens. **Capsules** clavate, subterete to obscurely 4-angled, sometimes slightly curved, 6–10.5 × 2.5–4 mm, hard-walled, irregularly dehiscent, pedicel (4.5–)6–15(–20) mm. **Seeds** light to dark brown, ellipsoid, 0.6–0.7 × 0.3–0.5 mm, surface cells transversely elongate. **2*n*** = 48.

Flowering Jun–Sep. Wet soil or sand along edges of ponds, lakes, marshes, or rivers, moist dune hollows, seasonal ponds; 0–200 m; Fla., N.J., N.C., S.C., Va.

The hexaploid *Ludwigia brevipes* is mainly restricted to the Atlantic coastal plain from central and eastern South Carolina to eastern North Carolina and extreme southeastern Virginia. The type collection of *L. brevipes* from middle New Jersey remains the only disjunct population north of the main range of this species more than 100 years after it was found. In 1988, an isolated population was found in the panhandle of Florida (Escambia County, *Burkhalter 11065*, MO) far to the southwest of the main range of *L. brevipes*; other reports of the species from Florida were erroneous (C. I. Peng 1989).

Ludwigia brevipes is known to hybridize with *L. palustris* producing the sterile *L.* ×*lacustris* Eames.

30. Ludwigia repens J. R. Forster, Fl. Amer. Sept., 6. 1771 (as Ludvigia), name conserved • Creeping primrose-willow, red ludwigia [W]

Isnardia intermedia Small & Alexander; *I. natans* (Elliott) Kuntze; *I. repens* de Candolle; *I. repens* var. *rotundata* Grisebach; *Ludwigia fluitans* Scheele; *L. natans* Elliott; *L. natans* var. *rotundata* (Grisebach) Fernald & Griscom; *L. natans* var. *stipitata* Fernald & Griscom; *L. repens* Swartz var. *rotundata* (Grisebach) M. Gómez; *L. repens* var. *stipitata* (Fernald & Griscom) Munz

Herbs creeping and rooting at nodes, often forming loose mats. **Stems** prostrate, ascending to suberect at tips, terete, sparsely branched, 30–80 cm, glabrous or, sometimes, minutely strigillose on leaf margins and inflorescence. **Leaves** opposite; stipules narrowly deltate, 0.05–0.1 × 0.05–0.1 mm; petiole narrowly winged, 0.3–2.3 cm, blade narrowly elliptic to broadly lanceolate-elliptic or suborbiculate, 0.8–4.5 × 0.4–2.7 cm, base attenuate, margins entire or sometimes with hydathodal glands, apex acute or apiculate, rarely obtuse, surfaces lustrous, subglabrous or sparingly to densely papillose strigillose; bracts not much reduced. **Inflorescences** sometimes few-flowered, erect racemes, flowers paired in leaf axils of prostrate stems; bracteoles attached in opposite pairs to pedicel 1–5 mm proximal to base of ovary, lanceolate to narrowly oblong-lanceolate or sublinear, 1–5(–8) × 0.2–1 mm, apex acute, surfaces sparingly minutely strigillose. **Flowers:** sepals ascending, light green, ovate deltate to narrowly so, 1.8–5 × 1.5–3.5 mm, margins minutely strigillose, apex acuminate to elongate-acuminate, surfaces subglabrous; petals caducous, oblanceolate to elliptic-oblong, 1.1–3 × 0.4–1.4 mm, base attenuate, apex obtuse, often variable in size and shape in same flower; filaments pale yellow, 0.5–1.5 mm, slightly inflated near base, anthers 0.4–0.9 × 0.3–0.8 mm; pollen shed singly or in tetrads; ovary obconic-cylindric, barely 4-angled to subterete, 2–6 × 2.5–3.5 mm; nectary disc elevated 0.3–0.8 mm on ovary apex, yellow, 1.1–3 mm diam., 4-lobed, glabrous; style pale yellow, 0.6–0.9 mm, glabrous, stigma pale yellow, broadly capitate, 0.3–0.5 × 0.3–0.8 mm, usually not exserted beyond anthers. **Capsules** elongate-obpyramidal, 4-angled, corners sometimes rounded, 4–10 × 2.5–4 mm, hard-walled, irregularly dehiscent, pedicel 0.1–3 mm. **Seeds** yellowish brown, ellipsoid, 0.6–0.8 × 0.3–0.5 mm, surface cells transversely elongate. **2*n*** = 48.

Flowers Mar–Nov (year-round). Muddy or damp, sandy edges of pools, lakes, swamps, creeks, and roadside ditches, moist soil in solution pits in limerock and hammock clearings in Florida Everglades, shade or sun; 0–1200[–1600] m; Ala., Ariz., Calif., Fla., Ga., Kans., La., Miss., Nev., N.J., N.Mex., N.C., Okla., Oreg., S.C., Tenn., Tex., Va.; Mexico (Chihuahua, Coahuila, México, Morelos, Nuevo León, Puebla, San Luis Potosí, Sonora); West Indies (Bahamas, Cuba, Hispaniola, Jamaica); Bermuda; introduced in Asia (Bangladesh, Japan).

Ludwigia repens occurs primarily on the Atlantic and Gulf coastal plains of the United States from North Carolina to Texas, with more scattered distribution into west Texas, New Mexico, Oklahoma, and Kansas, and disjunct populations in New Jersey, Tennessee, southern Arizona, Nevada, California, and western Oregon (Marion County); it also occurs in northern and central Mexico and the Caribbean region. In some parts of this wide range it is considered an aquatic weed.

Ludwigia repens is one of the most popular species of *Ludwigia* used in the aquarium trade; this may help to account for its wide distribution.

Much like the related diploid *Ludwigia palustris*, the hexaploid *L. repens* is widespread and morphologically variable, and infraspecific taxa have been proposed to describe this variation. Given the tendency of species in sect. *Isnardia* to hybridize, and the lack of a geographical basis for much of the variation, C. I. Peng (1989) declined to adopt any infraspecific classification, and this treatment follows Peng.

As described in C. I. Peng et al. (2005), *Ludwigia repens* J. R. Forster was conserved with a new type (the original type selected was from Virginia, where the species does not occur); *L. repens* Swartz (1797) was a later homonym described from Jamaica and pertains here.

31. Ludwigia curtissii Chapman, Fl. South. U.S. ed. 2, 621. 1883 • Curtiss's primrose-willow

Ludwigia spathulifolia Small

Herbs rarely creeping and rooting at nodes, stolons usually absent. **Stems** erect or ascending at base, very rarely prostrate, unbranched to well branched, branches sometimes very slender, 15–75 cm, glabrous, with slightly raised lines decurrent from leaf axils. **Leaves** alternate; stipules reddish purple, narrowly ovate, 0.2–0.3 × 0.1–0.3 mm, succulent; petiole winged, 0.3–1.2 cm, blade usually oblanceolate-spatulate to spatulate or oblanceolate, rarely sublinear, 1–2.5(–3) × 0.1–0.8 cm, base attenuate, margins subentire with hydathodal glands, apex acute or mucronate, surfaces glabrous; bracts not reduced. **Inflorescences** usually not congested, leafy racemes or spikes, flowers solitary in leaf axils; bracteoles attached in opposite pairs near base of ovary, narrowly lanceolate, lanceolate-elliptic, or oblong-linear, 1.5–3.5 (–4) × 0.4–0.8 mm, swollen at base, apex acuminate. **Flowers:** sepals ascending, green fading to white near base, ovate-deltate, 1.5–3 × 1.2–2 mm, margins entire, apex narrowly acute or acuminate, surfaces glabrous; petals 0(–3), narrowly elliptic or spatulate, 1–2.5 × 0.5–1 mm, base attenuate, apex obtuse; filaments pale yellow, 0.8–1(–1.3) mm, anthers 0.3–0.6 × 0.3–0.5 mm; pollen shed singly; ovary obovate-obpyramidal, 2–2.5 × 1.8–2.3 mm, glabrous; nectary disc elevated 0.3–0.4 mm on ovary apex, green, 0.9–1.6 mm diam., prominently 4-lobed, glabrous; style pale green, 0.4–0.7 mm, glabrous, stigma pale yellow, subglobose, 0.3–0.4 × 0.2–0.4 mm, not exserted beyond anthers. **Capsules** obconic, obscurely 4-angled, (2–)2.5–4(–4.7) × 2–3(–3.5) mm, hard-walled, dehiscent by loculicidal slits, pedicel 0.1–0.5 mm. **Seeds** light brown, ellipsoid, 0.4–0.6 × 0.3–0.4 mm, surface cells transversely elongate, glabrous or, sometimes, with surface wax that mimics appressed hairs. **2*n*** = 64.

Flowering Mar–Nov (year-round). Pine savannas and flatwoods, marshes, edges of ponds and streams, sandy or peaty swales, limestone prairies, solution pits on limestone; 0–50 m; Fla.; West Indies (Bahamas).

Ludwigia curtissii is the only octoploid (n = 32) in sect. *Isnardia*, and is restricted to peninsular Florida and the Bahamas. It and *L. simpsonii* they are the only members of the section that do not form true stolons; instead they simply sprout new shoots from the base. The two species are similar in many ways and appear to share three genomes (C. I. Peng 1988).

b. ONAGRACEAE Jussieu subfam. ONAGROIDEAE W. L. Wagner & Hoch, Syst. Bot. Monogr. 83: 41. 2007

Leaves: stipules present or absent. **Flowers:** floral tube present or, rarely, absent; sepals 2 or 4 (very rarely 3), deciduous with floral tube, petals, and stamens; petals yellow, white, pink, red, rarely in combination. x = 7, 10, 11, 15, 18.

Genera 21, species 582 (16 genera, 246 species in the flora): North America, Mexico, West Indies, Central America, South America, Eurasia, Pacific Islands (New Zealand, Society Islands), Australia.

Onagroideae encompass the main lineage of the family, after the early branching of *Ludwigia* (R. A. Levin et al. 2003, 2004). This large and diverse lineage is distinguished by the presence of a floral tube beyond the apex of the ovary; sepals deciduous with the floral tube, petals, and stamens; pollen shed in monads (or tetrads in *Chylismia* sect. *Lignothera* and all but one species of *Epilobium*); ovular vascular system exclusively transseptal (R. H. Eyde 1981); ovule archesporium multicellular (H. Tobe and P. H. Raven 1996); and change in base chromosome number from x = 8 in *Ludwigia* to x = 10 or x = 11 at the base of Onagroideae (Raven 1979; Levin et al. 2003). Molecular work (Levin et al. 2003, 2004) substantially supports the traditional tribal classification (P. A. Munz 1965; Raven 1979, 1988); tribes are recognized to delimit major branches within the phylogeny of Onagroideae, where the branches comprise strongly supported monophyletic groups of one or more genera.

b1. ONAGRACEAE Jussieu (subfam. ONAGROIDEAE) tribe CIRCAEEAE Dumortier, Fl. Belg., 88. 1827

Fuchsieae de Candolle

Herbs, perennial, or shrubs, [epiphytes, lianas, or trees]. Leaves opposite or whorled, [alternate]; stipules present. **Flowers** primarily protogynous, actinomorphic and 4-merous, or zygomorphic and 2-merous; stamens 2 times as many, or as many, as sepals; pollen shed in monads. **Fruits** indehiscent, either a fleshy berry or a dry capsule, covered with stiff, hooked hairs. **Seeds** 1–500, without hairs or wings.

Genera 2, species 117 (2 genera, 4 species, including 1 hybrid, in the flora): North America, Mexico, West Indies (Hispaniola), Central America, South America, Eurasia, n Africa, Pacific Islands (New Zealand, Society Islands).

All previous classification systems have placed *Circaea* and *Fuchsia* into different tribes, based on their morphological and geographical differences. Molecular analyses place these genera into a single clade (C. J. Bult and E. A. Zimmer 1993; E. Conti et al. 1993; R. A. Levin et al. 2003, 2004; V. S. Ford and L. D. Gottlieb 2007) that is as or more strongly supported than are other clades. The two genera share the feature of indehiscent fruits, expressed in *Fuchsia* as fleshy berries and in *Circaea* as dry fruits covered with hooklike hairs; nonhomologous indehiscent fruits also occur in Onagreae. The only occurrences of protogyny in the family occur in these two genera (not in all species of either, P. H. Raven 1979).

2. FUCHSIA Linnaeus, Sp. Pl. 2: 1191. 1753; Gen. Pl. ed. 5, 498. 1754 • [For Leonhart Fuchs, 1501–1566, German physician, herbalist, and professor at Tübingen] [I]

Paul E. Berry

Warren L. Wagner

Shrubs, [lianas, epiphytes, or trees]. Stems scandent, [erect, ascending, pendant, or decumbent], branched. **Leaves** cauline, opposite or whorled [alternate]; stipules present, usually deciduous; petiolate; blade margins serrulate to crenate-dentate [entire]. **Inflorescences** solitary flowers in leaf axils [racemes, panicles, or involucrate when flowers numerous]. **Flowers** bisexual and protogynous [or unisexual (and plants gynodioecious, dioecious, or subdioecious)], actinomorphic, buds pendent; floral tube deciduous (with sepals, petals, and stamens) after anthesis, well developed, cylindrical to obconical, with basal nectary, adnate or mostly free, nectary unlobed or shallowly 4–8-lobed; sepals 4, reflexed or spreading singly in anthesis; petals 4, [minute or absent], dark purple [shades of purple, red, or orange, rarely lavender, green, or rose-pink]; stamens 8, in 2 unequal series, [in equal series, or shorter whorl reflexed into floral tube], anthers basifixed, pollen shed singly; ovary 4-locular, stigma subentire or 4-lobed, clavate [capitate or globose], surface wet and non-papillate. **Fruit** a fleshy berry, oblong [to ellipsoid or subglobose]; pedicellate. **Seeds** few to ca. 500, in 2 to several rows per locule, rarely in 1 row, embedded in pulp. $x = 11$.

Species 107 (1 in the flora): introduced; Mexico, West Indies (Hispaniola), Central America, South America, Pacific Islands (New Zealand, Society Islands); introduced also in w Europe (Ireland), Asia (India), e, s Africa, Pacific Islands (Hawaiian Islands, New Zealand), Australia.

Nearly three-fourths of the species of *Fuchsia* occur in the northern Andes of South America, where the largest section (sect. *Fuchsia*) is centered. Other species occur in New Zealand (3 spp.), the Society Islands (1 sp.), Mexico, Hispaniola, Central America, and the southern Andes. All species of *Fuchsia* share the strong morphological synapomorphy of a fleshy berry, a fruit type unique in the family. Most species of *Fuchsia* characteristically have 2-aperturate pollen [3-aperturate pollen in the polyploid sections *Kierschlegeria* (Spach) Munz and *Quelusia* (Vandelli) de Candolle], are usually shrubs or subshrubs, sometimes lianas or trees, and have prolonged floral tubes on flowers that are mostly purple, red, or orange, and generally with petaloid sepals. Within the Circaeeae, P. E. Berry et al. (2004) found strong support for a monophyletic *Fuchsia*.

SELECTED REFERENCES Berry, P. E. 1989. A systematic revision of *Fuchsia* sect. *Quelusia* (Onagraceae). Ann. Missouri Bot. Gard. 76: 532–584. Berry, P. E. et al. 2004. Phylogenetic relationships and biogeography of *Fuchsia* (Onagraceae) based on noncoding nuclear and chloroplast DNA data. Amer. J. Bot. 91: 601–614.

1. **Fuchsia magellanica** Lamarck in J. Lamarck et al., Encycl. 2: 565. 1788 • Hardy fuchsia, earring flower [F] [I]

Shrubs glabrous or sparsely strigillose. **Stems** arcuate, 1–3(–5) m. **Leaves** in whorls of 3 or 4 per node, sometimes opposite; petiole 0.5–1(–2) cm; blade elliptic-ovate or ovate to lanceolate, 1.5–6(–7) × 1–2(–4) cm. **Flowers** 1 or 2 per node of distal leaves, pendent; floral tube crimson, 7–15 mm; sepals crimson, (15–)17–25(–30) mm; petals convolute after anthesis, (8–)11–20 mm; stamens exserted; filaments 18–35 mm; stigma clavate. **Berry** oblong, 10–22 mm; pedicel (10–)20–55 mm. $2n = 44$.

Flowering May–Nov. Sea bluffs, ditches, creek banks, damp thickets in partial shade or full sun; 10–200 m; introduced; Calif., Oreg.; South America; introduced also in w Europe (Ireland), Asia (India), e Africa, Pacific Islands (Hawaiian Islands, New Zealand), Australia.

Fuchsia magellanica is known in the flora area from Alameda, Contra Costa, Humboldt, Mendocino, Monterey, San Francisco, San Luis Obispo, San Mateo, and Sonoma counties in California, and Coos, Curry, Lane, and Lincoln counties in Oregon; it is native to the southern Andes in South America (Argentina, Chile).

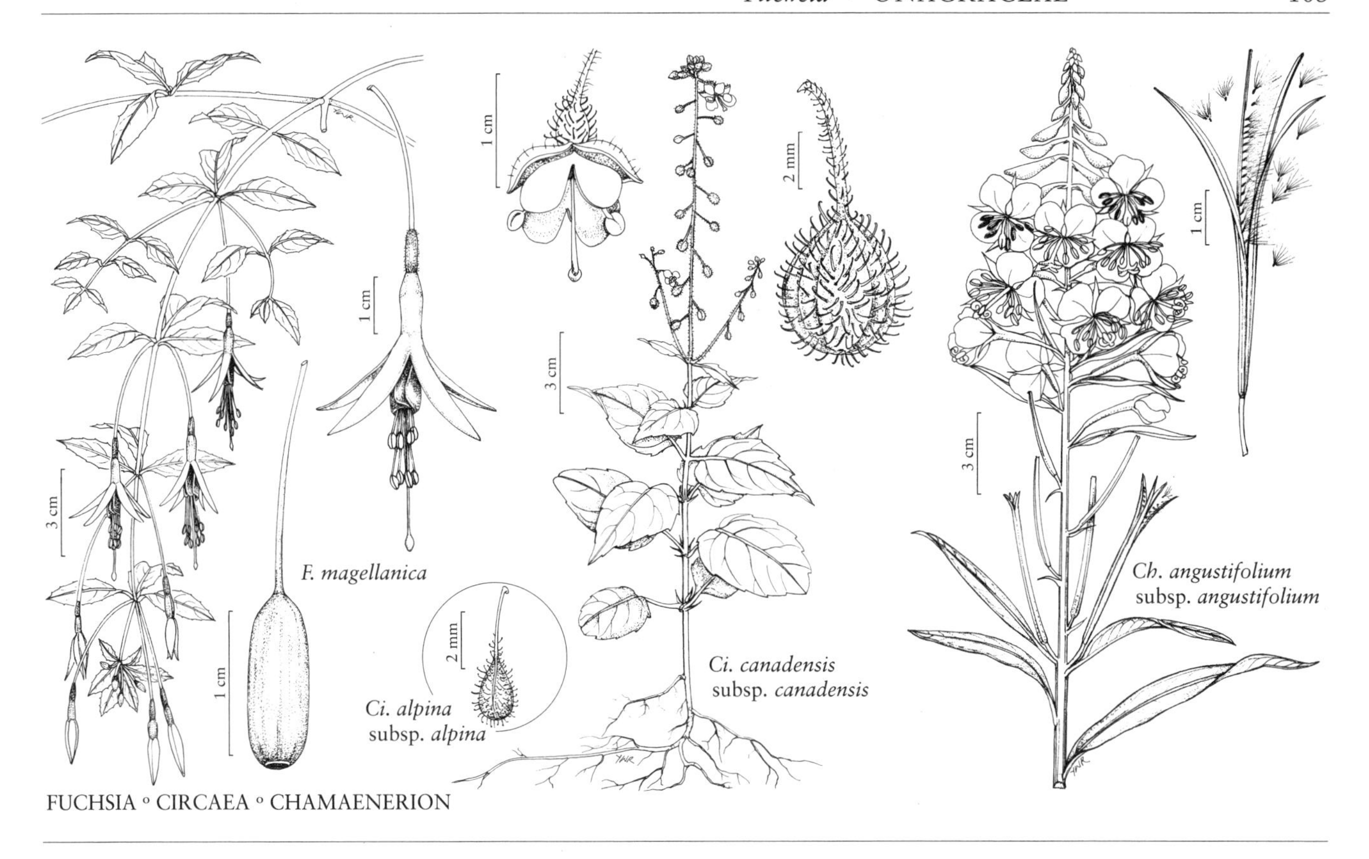

FUCHSIA ◦ CIRCAEA ◦ CHAMAENERION

Fuchsia magellanica is a member of sect. *Quelusia* (Vandelli) de Candolle, characterized by its shrubby-lianoid habit, opposite or whorled leaves, and distinctive floral pattern associated with hummingbird pollination, including violet, convolute petals, strongly exserted stamens, and partially connate sepals that are longer than the floral tubes. In the flora area, it is visited by Anna's Hummingbird (*Calypte anna*).

In Oregon, besides the normal red and purple flowered plants, there are also naturalized populations of a much paler flowered form of *Fuchsia magellanica* in which the floral tubes and sepals are pale white or tinged with pink, and the petals are only slightly purple. Similar populations can be found in the wild in Patagonia, and similar color variants are naturalized in New Zealand. One particular naturalized population near Brookings, Curry County, Oregon, is different from the other naturalized *F. magellanica* populations, and it appears to have originated from escaped cultivars involving a cross between *F. magellanica* and *F. coccinea* Aiton, a Brazilian species also in sect. *Quelusia*. Another naturalized population at the northern end of Bodega Bay, Sonoma County, California, is adjacent to banks full of *F. magellanica*, but it is different and appears to be intermediate between *F. regia* Vellozo and *F. coccinea*. In California, additional species of *Fuchsia* have been reported by collections or observations. *Fuchsia hybrida* hort. ex Siebold & Voss, which can distinguished from *F. magellanica* by the longer floral tube, 10–20 mm, and sepals 25–30 mm, is a variable garden hybrid involving *F. magellanica* or *F. regia* as one of the parents. Most of the observations or collection occurrence outside of cultivation of *F. hybrida* usually are near buildings or gardens or on a university campus. *Fuchsia boliviana* Carrière, which can be separated from other others by its floral tube (25–)30–60(–70) mm, scarlet petals, and flowers in pendent racemes or few-branched panicles, was reported as naturalized based on an erroneous observation.

3. CIRCAEA Linnaeus, Sp. Pl. 1: 8. 1753; Gen. Pl. ed. 5, 10. 1754 • [Greek *kirkaia*, a poetic name, alluding to mythical enchantress Circe's usage of an unknown plant as a charm]

David E. Boufford

Herbs, perennial, caulescent, colonial; stolons numerous. **Stems** erect, unbranched or sparsely branched. **Leaves** cauline, opposite; stipules present, soon deciduous; petiolate; blade margins dentate to prominently dentate. **Inflorescences** simple or branched racemes, terminal on main stem or also at apex of branches, erect. **Flowers** bisexual, zygomorphic, buds erect; floral tube inconspicuous, deciduous (with sepals, petals, and stamens) after anthesis, with a nectary wholly within and filling proximal portion of floral tube or elongated and projecting above opening of floral tube as a fleshy, cylindrical or ringlike disc; sepals 2, reflexed to spreading; petals 2, alternate sepals, white or pink, without spots, clawed, apex notched; stamens 2, anthers basifixed, pollen shed singly; ovary 1- or 2-locular, stigma bilobed or obpyramidal, surface wet, minutely papillate. **Fruit** a capsule, spreading or slightly reflexed, globose to clavoid or obovoid, indehiscent, surface smooth or with prominent longitudinal grooves (sulci) and rounded ridges, burlike, with stiff, hooked hairs; pedicellate, deciduous at maturity. **Seeds** 1 or 2, ellipsoid, glabrous, without appendages. $x = 11$.

Species 8 (3, including 1 hybrid, in the flora): North America, Europe, Asia, n Africa.

Circaea occurs throughout the temperate and boreal northern hemisphere, but is most diverse in eastern Asia, where all but one species occur. Reproductive features include: self-compatible; flowers diurnal, outcrossing, and pollinated by syrphid flies and small bees, or, sometimes, autogamous. It is found in rich, moist soils in deciduous forests and thickets, forest margins, and in moss or soil in mixed, coniferous-broadleaved deciduous, boreal forests. *Circaea alpina* subsp. *alpina* and *C. canadensis* subsp. *canadensis* often grow in close proximity and hybridize in eastern North America to produce *C.* ×*sterilis*. The unilocular *C. alpina*, with petals less than 2 mm, is self-pollinating under adverse weather conditions, but outcrosses on warm, sunny days. Because of its shorter style and much smaller pollen grains, it is probably the pollen recipient during hybridization events. Artificial hybridization experiments in England using *C. alpina* as the pollen donor and *C. lutetiana* as the pollen recipient failed to result in offspring, although hybrids were easily produced in the other direction (P. M. Benoit 1966). Recent molecular phylogenetic analysis supported the separation of the *C. canadensis* complex into two species; *C. alpina* subsp. *pacifica* was found to be sister to the remainder of the genus rather than being nested with other members of *C. alpina* (Xie L. et al. 2009). Thus, despite the strong morphological similarities of taxa within the *C. canadensis* and *C. alpina* complexes, these North American taxa may be better treated as separate species. Further detailed molecular studies are underway to examine this in more detail (Xie et al., unpubl.).

SELECTED REFERENCES Boufford, D. E. 1982b. The systematics and evolution of *Circaea* (Onagraceae). Ann. Missouri Bot. Gard. 69: 804–994. Boufford, D. E. et al. 1990. A cladistic analysis of *Circaea* (Onagraceae). Cladistics 6: 171–182. Xie, L. et al. 2009. Molecular phylogeny, divergence time estimates, and historical biogeography of *Circaea* (Onagraceae). Molec. Phylogen. Evol. 53: 995–1009.

1. Flowers opening before elongation of raceme axis, clustered and corymbiform at apex of raceme, on ascending to erect pedicels; capsules clavoid, without corky ribs or grooves; stolons terminated by a tuber . . . . 2. *Circaea alpina*
1. Flowers opening after elongation of raceme axis, more or less loosely spaced, borne on spreading pedicels; capsules usually obovoid to pyriform or subglobose, rarely clavoid, with corky, thickened ribs with deep grooves, or fruit sterile and aborting shortly after anthesis; stolons without or with a terminal tuber.

[2. Shifted to left margin.—Ed.]

2. Ovaries all or nearly all developing to maturity; capsules with corky thickened ribs separated by deep grooves; pollen highly fertile (greater than 80%); stolons without a tuber . . . . . 1. *Circaea canadensis*
2. Ovaries aborting shortly after anthesis, very rarely a few persistent, but easily detached, after anthesis; capsules, when somewhat persistent, smooth or with only low ribs and with shallow grooves; pollen highly sterile (less than 2% fertile); stolons terminated by a tuber or, more commonly, apex sligltly dilated . . . . . 3. *Circaea* ×*sterilis*

1. Circaea canadensis (Linnaeus) Hill, Veg. Syst. 10: 21. 1765 [F]

Circaea lutetiana Linnaeus var. *canadensis* Linnaeus, Sp. Pl. 1: 9. 1753; C. *quadrisulcata* (Maximowicz) Franchet & Savatier subsp. *canadensis* (Linnaeus) Á. Löve & D. Löve; *C. quadrisulcata* var. *canadensis* (Linnaeus) H. Hara

Subspecies 2 (1 in the flora): North America, e Asia.

1a. Circaea canadensis (Linnaeus) Hill subsp. **canadensis** • Enchanter's nightshade, circée du Canada [E] [F]

Circaea canadensis var. *virginiana* Fernald

Herbs glabrous, glandular puberulent distally; stolons not tipped by tubers. **Stems** 20–90 cm. **Leaves:** petiole (1.3–)2.5–5.5 cm; blade narrowly to broadly ovate or oblong-ovate, 5–16 × 2.5–8.5 cm, base cordate to rounded, margins denticulate, apex acute to slightly acuminate. **Inflorescences** 2.5–30 cm. **Flowers** opening after elongation of axis, ± loosely spaced; pedicels spreading at anthesis, 2.5–6.5 mm, with a minute, setaceous bracteole at base; floral tube (0.4–)0.7–1.2 mm, funnelform, nectary projecting 0.2–0.7 mm beyond opening of floral tube; sepals green or purple, broadly elliptic or oblong to oblong-ovate, 1.9–3.8 × 1.2–2.4 mm; petals usually white, rarely pink, broadly deltate to broadly obovate, obcordate, or broadly obovate, (1.3–)1.6–2.9 × (1.5–)2.2–4 mm; apical notch ⅓ to slightly more than ½ length of petal; filaments 1.2–2.8 mm, pollen highly fertile (greater than 80%); style 2.5–5.5 mm, stigma surrounded by anthers at anthesis. **Capsules** pyriform to subglobose, rounded, usually obliquely, to pedicel, with corky, prominent ribs and deep grooves, 2.8–4.5 × 1.9–3.6 mm; pedicel and mature fruit combined length 6.3–11.2 mm. $2n = 22$.

Flowering summer. Cool, temperate deciduous and mixed forests, forest margins, along streams; 0–600 m; Man., N.B., N.S., Ont., P.E.I., Que.; Ala., Ark., Conn., Del., D.C., Ga., Ill., Ind., Iowa, Kans., Ky., La., Maine, Md., Mass., Mich., Minn., Miss., Mo., Nebr., N.H., N.J., N.Y., N.C., N.Dak., Ohio, Okla., Pa., R.I., S.C., S.Dak., Tenn., Vt., Va., W.Va., Wis.

Subspecies *quadrisulcata* (Maximowicz) Boufford, which closely resembles subsp. *canadensis*, is in eastern Asia, extending from Japan across Russia to the vicinity of Moscow (D. E. Boufford 1982b). The presence of a minute bracteole at the base of the pedicel in subsp. *canadensis* is the only consistent feature that separates it from subsp. *quadrisulcata*, although subsp. *canadensis* tends to be somewhat larger overall.

2. Circaea alpina Linnaeus, Sp. Pl. 1: 9. 1753 [F]

Herbs glabrous or pubescent with at least a few recurved, falcate hairs, glabrous or glandular puberulent distally; stolons with apical tuber. **Stems** 3–50 cm. **Leaves:** petiole 0.3–5 cm; blade usually ovate to broadly ovate, rarely suborbiculate, 1.5–7.5(–11) × 1.5–5.5(–8) cm. **Inflorescences** 0.7–12(–17) cm. **Flowers** opening before elongation of axis, corymbiform; pedicels erect or ascending at anthesis, 0.7–3.5 mm, with or without a minute, setaceous bracteole at base; floral tube a mere constriction to 0.6 mm, funnelform to very broadly so, nectary wholly within floral tube; sepals white or pink, sometimes purple tinged apically, oblong or ovate to broadly ovate, 0.8–1.8(–2.2) × 0.6–1.3 mm; petals white, obtriangular or obdeltate to obovate or broadly obovate, 0.6–2 × 0.6–1.8 mm; apical notch to ½ length of petal; filaments 0.7–2.2 mm; style 0.6–2.3 mm. **Capsules** clavoid, tapering smoothly to pedicel, without ribs or grooves, 1.6–2.6 × 0.5–1.2 mm, 1-locular, 1-seeded; pedicel and mature fruit combined length 3.5–7.8 mm.

Subspecies 6 (2 in the flora): North America, Europe, Asia.

Circaea alpina inhabits moist places, and is also found on moss covered rocks and logs in cold temperate and boreal forests at high altitudes and latitudes throughout the northern hemisphere and in the tropics and subtropics at high elevations in southern and southeastern Asia, at elevations 0–5000 m.

1. Stems glabrous; leaf blade margins conspicuously dentate, base usually cordate to subcordate, rarely truncate or rounded . . . 2a. *Circaea alpina* subsp. *alpina*
1. Stems with at least a few recurved, falcate hairs; leaf blade margins subentire to minutely denticulate, base usually rounded to subcordate, rarely cordate . . . . 2b. *Circaea alpina* subsp. *pacifica*

2a. Circaea alpina Linnaeus subsp. **alpina** E F

Stems soft, flattened after pressing and appearing winged, glabrous. **Leaves:** petiole 0.3–4 cm; blade 1.5–7.5 × 1.5–5.5 cm, margins conspicuously dentate, base usually cordate to subcordate, rarely truncate or rounded, apex short acuminate to acute. **Inflorescences** glabrous or sparsely to densely glandular puberulent. **Flowers** clustered at apex of raceme, opening before elongation of raceme axis; floral tube a mere constriction between ovary and base of sepals to 0.5 mm; apical notch of petal ¼–½ length of petal; pedicel and mature fruit combined length 3.7–6.5 mm. **2*n*** = 22.

Flowering summer. Moist to wet places, on moss covered rocks and logs, cool, temperate and boreal forests; 0–2500 m; St. Pierre and Miquelon; Alta., B.C., Man., N.B., Nfld. and Labr. (Nfld.), N.W.T., N.S., Ont., P.E.I., Que., Sask., Yukon; Ariz., Colo., Conn., D.C., Idaho, Ill., Ind., Iowa, Ky., Maine, Md., Mass., Mich., Minn., Miss., Mo., Mont., N.H., N.J., N.Mex., N.Y., N.C., N.Dak., Ohio, Pa., R.I., S.C., S.Dak., Tenn., Vt., Va., Wash., W.Va., Wis., Wyo.

2b. Circaea alpina Linnaeus subsp. **pacifica** (Ascherson & Magnus) P. H. Raven, Canad. J. Bot. 43: 1396. 1965 E

Circaea pacifica Ascherson & Magnus, Bot. Zeitung (Berlin) 29: 392. 1871; *C. alpina* var. *pacifica* (Ascherson & Magnus) M. E. Jones

Stems firm, terete, remaining mostly unflattened after pressing, pubescent with at least a few recurved, falcate hairs. **Leaves:** petiole 1.5–5 cm; blade 3–7.5(–11) × 2.5–5.5(–8) cm, margins subentire to minutely denticulate, base usually rounded to subcordate, rarely cordate, apex acute to short acuminate. **Inflorescences** usually densely, less often sparsely, glandular puberulent. **Flowers** clustered at apex of raceme, opening before elongation of raceme axis; floral tube 0.3–0.6 mm; apical notch of petal ¼–⅓ length of petal; pedicel and mature fruit combined length 3.5–6.5 mm. **2*n*** = 22.

Flowering summer. Cool, temperate deciduous and mixed forests, forest margins, along streams; (0–)200–2700(–2900) m; Alta., B.C.; Alaska, Ariz., Calif., Colo., Idaho, Mont., Nev., N.Mex., Oreg., Utah, Wash., Wyo.

Subspecies *pacifica* and subsp. *alpina* are easily separated by the stems with at least a few hairs and by the minutely denticulate to nearly entire leaf blade margins in subsp. *pacifica*. The stems of subsp. *pacifica* remain unflattened in herbarium specimens, whereas the stems of subsp. *alpina* are flattened and appear to be very narrowly winged after pressing. Subspecies *pacifica* is usually larger and more robust than subsp. *alpina* and is similar to some of the Asian subspecies of *Circaea alpina* in its somewhat thicker, deeper green leaves and less delicate stems. Plants in the areas where the two subspecies are sympatric are sometimes difficult to assign to one subspecies or the other, the few minute hairs on the stem being the most reliable distinguishing feature.

3. Circaea ×sterilis Boufford, Harvard Pap. Bot. 9: 256. 2005 E

Herbs glabrous or pubescent with at least a few hairs at nodes; stolons tipped by apical tuber or, more commonly, apex slightly dilated. **Stems** 7–70 cm. **Leaves:** petiole 0.8–5.5 cm; blade narrowly to broadly ovate, 2–7(–11) × 1.3–4.2(–7) cm, margins denticulate to prominently serrate, base cordate, apex short-acuminate to acute. **Inflorescences** 1.5–18 cm, sparsely to densely stipitate glandular. **Flowers** opening after elongation of axis, ± loosely spaced; pedicels ascending to spreading at anthesis, 1.8–5.5 mm, with a minute, setaceous bracteole at base; floral tube 0.4–1.2 mm, funnelform to narrowly so; nectary projecting beyond opening of floral tube, 0.1–0.4 mm; sepals pink or pale green with apex commonly purple tinged, narrowly oblong to oblong-obovate or ovate, 1.6–3.5 × 0.9–2 mm; petals white or pink, narrowly obtriangular to very broadly obovate, base cuneate to rounded, 1–3.6 × (0.6–)1.5–3.6 mm; apical notch (⅓–)½–¾ length of petal; filaments 1.3–3.7 mm, pollen highly sterile (less than 2% fertile); style 2.7–4.7 mm. **Capsules** sterile, usually aborting before maturity, rarely developing to 3 × 1.5 mm, clavoid to obovoid, somewhat rounded to smooth or with low ribs and shallow grooves, tapering to pedicel, 2-locular, often with one locule larger, each with infertile seed. **2*n*** = 22 (9 bivalents plus a ring or chain of 4, or 11 bivalents at meiosis).

Flowering summer. Moist places in deciduous, mixed, or coniferous forests, especially in naturally disturbed areas along streams; 0–1000 m; N.B., N.S., Ont., P.E.I., Que.; Conn., Maine, Mass., Mich., Minn., N.H., N.J., N.Y., N.C., Ohio, Pa., S.Dak., Vt., Va., W.Va., Wis.

Circaea ×*sterilis* is a sterile hybrid intermediate in morphology between *C. alpina* × *C. canadensis*. Although sexually sterile, like other hybrids in *Circaea*, it reproduces more vigorously vegetatively than either of the parents. Its occurrence in naturally disturbed places along streams no doubt results in portions of stolons being broken off and carried away during high water to establish new colonies some distance away; colonies often extend for great distances along streams. The hybrids persist many years and colonies discovered 100 or more years ago can still be found in the same locality, often in the absence of one or both parents. The hybrid also occurs outside the range of *C. canadensis* subsp. *canadensis*, most conspicuously on the Gaspé Peninsula, Quebec, and in northern New Brunswick, but also elsewhere. Those hybrids may indicate a once more northerly range for *C. canadensis* subsp. *canadensis*, or that seeds resulting from hybridization events can be transported outside the range of the parents by birds or mammals.

M. L. Fernald (1950) misapplied the name *Circaea canadensis* Hill to plants now known to be hybrids (*C.* ×*sterilis*) between *C. canadensis* and *C. alpina* in North America and placed *C. intermedia* Ehrhart, a name published later, in its synonymy. *Circaea* ×*intermedia* is the hybrid between the European *C. lutetiana* Linnaeus and *C. alpina* Linnaeus. Hill used the name *C. canadensis* for plants in North America, to distinguish them from the European *C. lutetiana*.

b2. ONAGRACEAE Jussieu (subfam. ONAGROIDEAE) tribe EPILOBIEAE Endlicher, Fl. Poson., 366. 1830

Epilobioideae Alph. Wood; *Boisduvaliinae* Raimann; *Epilobiinae* Torrey & A. Gray

Herbs, annual or perennial, sometimes suffrutescent. Leaves opposite at least near base or throughout, alternate distally, or sometimes alternate throughout; stipules absent. **Flowers** actinomorphic or slightly zygomorphic, 4-merous; sepals erect or spreading; stamens 2 times as many as sepals; pollen shed in tetrads or monads. **Fruit** a slender, cylindrical, loculicidal capsule. **Seeds** (1 or) many per locule, with tuft of hairs (coma) at chalazal end, sometimes without coma.

Genera 2, species 173 (2 genera, 43 species in the flora): North America, Mexico, West Indies, Central America, South America, Eurasia, Africa, Atlantic Islands, Australasia; introduced in Pacific Islands.

Epilobium and its close relatives have been recognized historically either as part of Onagreae (A. P. de Candolle 1828b), sometimes as a subtribe (É. Spach 1834–1848, vol. 4; J. Torrey and A. Gray 1838–1843, vol. 1), or as the distinct Epilobieae (R. Raimann 1893; P. A. Munz 1965; P. H. Raven 1976). The synapomorphies for Epilobieae as currently delimited include its highly condensed, heteropycnotic chromosomes with a base chromosome number of $x = 18$, sepals held erect or spreading (not reflexed) throughout anthesis, and the presence of a coma of hairs on the seeds (secondarily lost in some species). Molecular support for the tribe is strong (97–100% BOOTSTRAP support; D. A. Baum et al. 1994; R. A. Levin et al. 2004).

Endlicher established Epilobieae and included *Oenothera* within it; it is unclear how or whether his concept differed from Onagreae of A. P. de Candolle (1828b). É. Spach (1834–1848, vol. 4) recognized these genera as Onagreae and differentiated so-called sect. *Oenotherinae* from sect. *Epilobieae*, placing in the latter not only *Epilobium* and related groups, but also *Clarkia* and its segregates. J. Torrey and A. Gray (1840) excluded *Clarkia* from their Epilobiinae and also excluded *Boisduvalia*, a delimitation also followed by R. Raimann (1893) for his Epilobieae. Epilobieae did not assume its current delimitation, including only *Epilobium* and its close relatives, until the works of P. A. Munz (1941) and P. H. Raven (1964).

G. L. Stebbins (1971) and P. H. Raven (1976) considered the diverse chromosome numbers in Epilobieae and proposed that the species of *Boisduvalia* (now a section of *Epilobium*) with $n = 9$ or 10 represented the original base chromosome number for the tribe, and that these numbers were derived from $x = 11$ which is found in Circaeeae, Gongylocarpeae, and Lopezieae (W. L. Wagner et al. 2007). They proposed a series of aneuploid reductions from $n = 9$ or 10 to $n = 6$, followed by polyploidy to produce the array of numbers in *Epilobium* ($n = 12, 13, 15, 16, 18, 30$) and *Boisduvalia* ($n = 9, 10, 15, 19$). D. A. Baum et al. (1994) demonstrated that molecular data did not support that hypothesis and found that a monophyletic *Chamaenerion* ($n = 18, 36, 54$) forms a strongly distinct sister branch to *Epilobium*, within which sect. *Epilobium* ($n = 18$) is monophyletic and sister to the rest of *Epilobium*, including the former segregates *Boisduvalia* and *Zauschneria*. The data from Baum et al. suggested that Epilobieae are primitively polyploid, with numbers based on $x = 18$, which is unique in the family. Using comparable sampling and some additional genes, R. A. Levin et al. (2004) found strong support for the phylogeny proposed by Baum et al.

SELECTED REFERENCES Raven, P. H. 1976. Generic and sectional delimitation in Onagraceae, tribe Epilobieae. Ann. Missouri Bot. Gard. 63: 326–340. Seavey, S. R. et al. 1977. Evolution of seed size, shape, and surface architecture in the tribe Epilobieae (Onagraceae). Ann. Missouri Bot. Gard. 64: 18–47.

4. CHAMAENERION Séguier, Pl. Veron. 3: 168. 1754 • Fireweed, rosebay willowherb [Greek *chamae-*, on the ground or dwarf, and *nērion*, oleander, alluding to resemblance of flower color and foliage]

Peter C. Hoch

Chamerion (Rafinesque) Rafinesque ex Holub; *Epilobium* Linnaeus sect. *Chamaenerion* (Séguier) Tausch; *Epilobium* subg. *Chamerion* Rafinesque

Herbs, perennial, often clumped, caulescent; from woody caudex or forming shoots from spreading rhizomes. **Stems** erect, usually unbranched, rarely sparsely branched, strigillose or glabrous. **Leaves** cauline, usually alternate, rarely subopposite or subverticillate in proximal pairs; subsessile; blades scalelike and minute below ground, proximal blades small, coriaceous to submembranous, triangular-ovate to lanceolate, distal blades usually linear to lanceolate or elliptic, rarely ovate, often subcoriaceous. **Inflorescences** terminal, racemes [spikes], erect or suberect. **Flowers** bisexual, protandrous, slightly zygomorphic, opening on axis nearly perpendicular to stem axis, deciduous from ovary apex; floral tube absent; sepals green or reddish green, spreading; petals spreading, usually rose purple to pink, rarely white, margins entire; stamens subequal; filament bases slightly bulged proximally to form a chamber around nectary disc, initially erect, later deflexed; anthers versatile, pollen bluish gray [yellow], shed in monads, 3(–5)-aperturate; ovary 4-locular, nectary disc raised on ovary apex; style initially deflexed, becoming erect 2–3 days after onset of anthesis, stigma deeply 4-lobed, lobes spreading open to revolute as style becomes erect, commissural, receptive only on inner surfaces, surface dry, with multicellular papillae. **Fruit** a capsule, loculicidal, spreading to erect, narrowly cylindrical, terete to quadrangular, splitting to base with intact central column; pedicellate. **Seeds** 200–500, in 1 row per locule, gray-brown, narrowly clavate, with ± persistent coma at chalazal end. $x = 18$.

Species 8 (2 in the flora): North America, Mexico, Eurasia, nw Africa, Atlantic Islands (Iceland).

J. Holub (1972b) argued that the correct name for this group, when segregated from *Epilobium*, should be *Chamerion* (Rafinesque) Rafinesque ex Holub. However, A. N. Sennikov (2011), using the more recent and clarified ICBN (Melbourne Code), correctly argued that the name of the segregated genus should be *Chamaenerion*, and that opinion is followed here.

All species of *Chamaenerion* tested to date have been self-compatible, although the strong protandry can result in low fertilization rates if there is a shortage of pollinators. Flowering is diurnal, with each flower generally remaining open for three to five or more days.

All eight species (nine taxa) of *Chamaenerion* are restricted to the northern hemisphere, and six of eight species occur only in Eurasia; *Circaea* is the only other genus of Onagraceae in which most species occur outside of the western hemisphere (D. E. Boufford 1982b). *Chamaenerion* is generally divided into two sections, each with four species (T. Tacik 1959; and, as *Chamerion*, J. Holub 1972b; W. L. Wagner et al. 2007). P. H. Raven (1976) and most others who treated this group as a section of *Epilobium* divided it into two corresponding subsections. Both species of *Chamaenerion* in North America are placed in sect. *Chamaenerion* (not treated formerly here), along with two species endemic to the Himalayan region, *C. conspersum* (Haussknecht) Kitamura and *C. speciosum* (Decaisne) Hoch & Gandhi (Chen C. J. et al. 1992); the other four species, all endemic in Eurasia, comprise sect. *Rosmarinifolium* Tacik.

Chamaenerion differs from *Epilobium* in leaves nearly always alternate, rarely subopposite or verticillate near stem base (versus opposite at least on proximal stem); absence of a floral tube (versus a distinct floral tube); slightly zygomorphic flowers with subequal stamens that are erect, later deflexed, and a style that is deflexed, later erect (versus actinomorphic flowers with erect stamens in two unequal series and erect style); and pollen shed in monads (versus pollen shed in tetrads, or monads in one species). Molecular analyses (D. A. Baum et al. 1994; R. A. Levin et al. 2004) provided strong support for *Chamaenerion* as a monophyletic group separate from *Epilobium*, and for sect. *Chamaenerion* (only *C. angustifolium* and *C. latifolium* sampled) as monophyletic.

The superfluous and illegitimate name *Pyrogennema* Lunell pertains here.

SELECTED REFERENCE Holub, J. 1972b. Taxonomic and nomenclatural remarks on *Chamaenerion* auct. Folia Geobot. Phytotax. 7: 81–90.

1. Leaf blades elliptic or ovate to lanceolate-elliptic, 2–5(–8) × 0.6–1.7(–2) cm, without submarginal vein, petioles 0–2 mm; bracts ca. ½ as long as cauline leaves; seeds fusiform, 1.2–2.1 × 0.4–0.6 mm, with distinct chalazal collar (0.1–0.2 mm); styles glabrous, shorter than stamens . . . . 1. *Chamaenerion latifolium*
1. Leaf blades lanceolate to linear, proximally oblong to obovate, (3–)5–23 × 0.3–3.4 cm, with ± prominent submarginal vein, petioles 0–7 mm; bracts much smaller than cauline leaves; seeds narrowly obovoid to oblong, 0.9–1.3 × 0.3–0.4 mm, with inconspicuous chalazal collar (to 0.1 mm); styles pubescent proximally, longer than stamens . . . . 2. *Chamaenerion angustifolium*

1. Chamaenerion latifolium (Linnaeus) Sweet, Hort. Brit. ed. 2, 198. 1830 • Alpine or dwarf fireweed, river-beauty, épilobe à feuilles larges

Epilobium latifolium Linnaeus, Sp. Pl. 1: 347. 1753 (as latifolia); *Chamaenerion latifolium* var. *grandiflorum* (Britton) Rydberg; *C. latifolium* var. *megalobum* Nieuwland; *C. latifolium* var. *parviflorum* Hartz; *C. subdentatum* Rydberg; *Chamerion latifolium* (Linnaeus) Holub; *C. subdentatum* (Rydberg) Á. Löve & D. Löve; *E. latifolium* var. *albiflorum* A. E. Porsild; *E. latifolium* var. *album* I. W. Hutchison; *E. latifolium* var. *grandiflorum* Britton; *E. latifolium* subsp. *leucanthum* Ulke; *E. latifolium* var. *tetrapetalum* Pallas; *E. latifolium* var. *venustum* Douglas ex G. Don; *E. opacum* Lehmann

Herbs with ± woody caudex 4–10 mm diam., fleshy shoots from caudex, roots wiry. **Stems** usually clumped, terete, unbranched or scarcely branched, 12–35 cm, glabrous proximally, subglabrous or sparsely to, rarely, densely strigillose distally. **Leaves:** petiole 0–2 mm; basal blades often brown, triangular-ovate, 0.7–1 cm, cauline blades elliptic or ovate to lanceolate-elliptic, 2–5(–8) × 0.6–1.7(–2) cm, base cuneate to subobtuse, margins subentire or remotely punctate-denticulate (with 4–7 teeth), apex obtuse or acuminate, lateral veins obscure, 3 or 4 on each side of midrib, submarginal vein absent, surfaces subglabrous or strigillose; bracts ca. ½ as long as blades. **Inflorescences** erect racemes, sparsely to densely strigillose. **Flowers** erect in bud, nodding at anthesis, opening laterally; buds oblong-obovoid, 8–17 × 3–6 mm; sepals often flushed purplish red, oblong-lanceolate, 10–16 × 1.5–3.5 mm, base attenuate or

± short-clawed, apex acute, usually subglabrous; petals rose purple to pink, obovate or oblong-obovate, 10–24(–32) × 7–15(–21) mm, sometimes slightly unequal, lower pair slightly shorter and narrower, apex rounded or retuse; filaments white or pink, 6–11 mm; anthers dark red, oblong or elliptic-oblong, 1.2–4 × 0.7–1.5 mm; ovary often flushed purplish red, 1.5–3 cm, usually gray-canescent; nectary disc raised 0.5–1 mm on ovary apex, 3–4 mm diam., glabrous; style light pink to rose purple, 3.5–8 mm, glabrous; stigma lobes 2–3.5 mm, recurved at maturity, surface white, papillose. **Capsules** 2.5–8 cm, strigillose-canescent; pedicels 1.2–2.5 cm. **Seeds** fusiform, 1.2–2.1 × 0.4–0.6 mm, acuminate at micropylar end, chalazal collar 0.1–0.2 mm, surface irregularly low-reticulate; coma tawny or dingy white, 9–15 mm, not readily deciduous. $2n$ = 36, 72.

Flowering Jun–Aug(–Sep). Consistently moist, gravelly or sandy places along rivers and creeks, near base of talus slopes in arctic and alpine regions; 0–2000(–4500) m; Greenland; Alta., B.C., Man., Nfld. and Labr. (Nfld.), N.W.T., Nunavut, Ont., Que., Yukon; Alaska, Calif., Colo., Idaho, Mont., Nev., Oreg., Utah, Wash., Wyo.; Eurasia; Atlantic Islands (Iceland).

Chamaenerion latifolium is a widespread arctic-alpine species found abundantly across arctic Alaska, Canada, and Greenland, as well as in comparable arctic regions in Eurasia. It also occurs farther south along the cordilleras at high alpine elevations (to 4500 meters) (Chen C. J. et al. 1992; P. C. Hoch 2012 [as *Chamerion*]).

Both diploid (n = 18) and tetraploid (n = 36) plants of *Chamaenerion latifolium* have been documented, but variation in ploidy shows no obvious association with any morphological features, except that diploids have mostly three-pored pollen and tetraploids a larger proportion of four-pored grains (E. Small 1968). Using a few meiotic counts and pollen pore number on herbarium specimens, Small found diploids in Alaska and western North America, and tetraploids in eastern Canada, Greenland, and Iceland, but the ranges overlap (T. Mosquin and E. Small 1971). The lack of correlation of geography with any other variable characters led Small to oppose any infraspecific classification, a decision followed in this treatment. The mechanism of polyploidization is probably autopolyploidy, based in part on the high rate of quadrivalent formation in tetraploid meiosis (Small).

In 1813, M. Wormskjöld described *Epilobium intermedium* Wormskjöld [recombined variously as *Chamaenerion angustifolium* var. *intermedium* (Wormskjöld) Lange; *E. angustifolium* var. *intermedium* (Wormskjöld) Fernald; and *Chamerion angustifolium* subsp. *intermedium* (Wormskjöld) Á. Löve] from Greenland, noting that it was intermediate between *E. angustifolium* and *E. latifolium*. From the description by Wormskjöld, it is not clear whether his plant was actually a hybrid, or simply an unusual variant of one or the other species. Hybrids between these two species have been reported in transitional habitats (T. W. Böcher 1962), but they are surprisingly infrequent, given the huge region of sympatric occurrence (T. Mosquin and E. Small 1971).

White-petaled individuals or populations occur at low frequency and have sometimes been given taxonomic status, but flower color differences do not correlate with other morphological or geographical variation.

Chamaenerion halimifolium Salisbury is illegitimate and pertains here.

2. Chamaenerion angustifolium (Linnaeus) Scopoli, Fl. Carniol. ed. 2, 1: 271. 1771 • Fireweed, rosebay willowherb, épilobe à feuilles étroites [F] [W]

Epilobium angustifolium Linnaeus, Sp. Pl. 1: 347. 1753 (as angustifolia); *Chamerion angustifolium* (Linnaeus) Holub; *Pyrogennema angustifolium* (Linnaeus) Lunell

Herbs robust, caudex woody, often forming large clones by horizontal rhizomes 0.5–2 cm diam., extending 0.5–5 m, forming shoots from caudex and along rhizomes. **Stems** erect, terete, usually unbranched, rarely branched distally, 20–200 cm, glabrous or sparsely to densely strigillose, often exfoliating proximally. **Leaves:** usually spirally arranged, very rarely subopposite proximally; petiole 0–7 mm; blade lanceolate to linear or proximally oblong to obovate, (3–)5–23 × 0.3–3.4 cm, proximal ones much reduced, base obtuse or cuneate to attenuate, margins entire or scarcely denticulate, apex attenuate-acute, lateral veins 10–25 on each side of midrib, diverging nearly at right angles, confluent into ± prominent submarginal vein, surfaces glabrous or strigillose on abaxial midrib; bracts much reduced and linear. **Inflorescences** erect racemes, 5–50 cm, glabrous or strigillose. **Flowers** nodding in bud, erect at anthesis, opening laterally; buds oblong to oblanceoloid, 7–14 × 3–35 mm; sepals purplish green, oblong-lanceolate or upper pair somewhat oblanceolate, 8–19 × 1.5–3 mm, base usually attenuate, rarely ± clawed, apex acuminate or attenuate, canescent abaxially, glabrous adaxially; petals usually rose purple to pale pink, very rarely white, obovate to narrowly obovate or nearly rotund, 9–25 × 3–15 mm, upper pair often ± longer and broader, base attenuate, apex entire or shallowly emarginate; filaments pink, 7–15 mm, subequal; anthers red to rose purple, oblong, 2–3 × 1–1.7 mm; ovary 6–25 mm, surface densely canescent; nectary disc raised 0.5–1 mm on ovary apex, 2–4 mm diam., smooth or slightly 4-lobed, glabrous; style sharply deflexed initially, later erect, white or flushed pink, 8–16 mm, proximally villous; stigma spreading to revolute, lobes 3–6 × 0.6–1 mm, surfaces white, densely dry-papillose. **Capsules** 4–9.5 cm, densely appressed-canescent; pedicel 0.5–3 cm. **Seeds** narrowly obovoid to oblong, 0.9–1.3 × 0.3–0.4 mm, chalazal end tapering abruptly to very short neck (to 0.1 mm), surface appearing smooth and often shiny, but irregularly reticulate; coma 10–17 mm, white or dingy, dense, not easily detached. **2*n*** = 36, 72, 108.

Subspecies 2 (2 in the flora): North America, n Mexico, Europe, Asia.

Chamaenerion angustifolium, which in North America is known mostly as fireweed, is a widespread circumpolar, circumboreal species that can be locally dominant, often in disturbed habitats, particularly following fires (T. Mosquin 1966, 1967; G. Henderson et al. 1979). In addition to producing numerous, highly vagile, comose seeds, aggressive rhizomatous growth enables it to survive clonally and spread rapidly after fires (L. E. Vodolazskij 1979). As succession proceeds, production of the familiar spikes of magenta flowers is inhibited, but fireweed often persists in non-flowering condition even in relatively mature forests, where it is able to sprout and spread quickly following disturbance. The species is one of the few within the genus and tribe to form true rhizomes (R. C. Keating et al. 1982), enabling it to colonize large areas very rapidly.

Polyploidy is well documented in *Chamaenerion angustifolium*, but unlike the similar situation in *C. latifolium* (T. Mosquin and E. Small 1971), diploid and tetraploid populations differ morphologically and have partially overlapping, but distinct, geographical ranges (Mosquin 1966; Chen C. J. et al. 1992). The diploid subsp. *angustifolium* occupies the northern part of its range, in North America across Canada and interior Alaska, and in Asia across Siberia and northern Europe, ranging southward at higher elevations. Farther south in Eurasia and North America, it is replaced by the tetraploid subsp. *circumvagum* (Mosquin 1966). Hexaploid (*n* = 54) populations have been detected only in Japan, and cannot be distinguished morphologically from tetraploids. B. C. Husband and D. W. Schemske (1998), Husband and H. A. Sabara (2004), and Sabara et al. (2013) have documented population segregation and mixing in contact areas between the two ploidy levels, including populations with a single cytotype, a high proportion of mixed populations, and presence of triploids (*n* = 27), but at low levels, indicating strong reproductive isolation between cytotypes.

The floral phenology of *Chamaenerion angustifolium* was described in 1793 by C. K. Sprengel (P. Knuth 1906–1909, vol. 2) as a classic example of protandry. The flowers are presented in tall spikes, opening from the base of the spike, initially with stamens erect and style sharply reflexed. By the second or third day after opening, the stamens reflex as the style straightens into the floral axis, the four lobes of the stigma spread open, and nectaries commence production. Bees start at the lowest flowers in search of nectar, moving up the spike until there is no more nectar, by which time they are well dusted with pollen, which they bring to the lower functionally (female) flowers of the next spike.

1. Leaves subsessile, leaf blades glabrous on abaxial midrib, (3–)7–14(–18.5) × (0.3–)0.7–1.3(–2.5) cm, base rounded to cuneate, margins usually entire, rarely low-denticulate; stems subglabrous; petals 9–15(–19) × 3–9(–11) mm; pollen usually 3-porate, to 75 µm diam. . . . . 2a. *Chamaenerion angustifolium* subsp. *angustifolium*
1. Leaves with petioles 2–7 mm, leaf blades usually strigillose, rarely glabrous on abaxial midrib, (6–)9–23 × (0.7–)1.5–3.4 cm, base subcuneate to attenuate, margins ± denticulate; stems strigillose, usually glabrous proximally; petals 14–25 × 7–14 mm; pollen mixed 3- and 4-porate (rarely 5-porate), more than 75 µm diam. . . . . 2b. *Chamaenerion angustifolium* subsp. *circumvagum*

2a. Chamaenerion angustifolium (Linnaeus) Scopoli subsp. **angustifolium** F W

Chamaenerion angustifolium var. *spectabile* Simmons; *C. spicatum* (Lamarck) Gray; *Epilobium angustifolium* Linnaeus var. *albiflorum* Dumortier; *E. angustifolium* var. *album* G. Don; *E. angustifolium* var. *pygmaeum* Jepson; *E. spicatum* Lamarck; *E. spicatum* var. *leucanthum* Wender

Stems 20–130 cm, subglabrous. **Leaves** subsessile; cauline blades linear-lanceolate to narrowly lanceolate, (3–)7–14(–18.5) × (0.3–)0.7–1.3(–2.5) cm, base rounded to cuneate, margins usually entire, rarely low-denticulate, surfaces and veins glabrous. **Inflorescences** subglabrous; bracts 5–20 mm. **Flowers:** sepals 6–15 × 1.5–2.5 mm, acute or acuminate; petals obovate to narrowly obovate, 9–15(–19) × 3–9(–11) mm, entire or scarcely emarginate; filaments 7–14 mm; anthers 2–2.5 mm; pollen 3-porate (rarely 4-porate), 60–75 µm diam.; ovaries 6–20 mm, style 8–14 mm. **Capsules** 4–8 cm; pedicels 0.5–2 cm. **Seeds** 0.9–1 × 0.3–0.4 mm. **2*n*** = 36.

Flowering Jun–Sep. Moist, often disturbed, places in mountains or in lower areas, frequent especially after fires; 0–3800 m; Greenland; Alta., B.C., Man., Nfld. and Labr., N.W.T., N.S., Nunavut, Ont., Que., Sask., Yukon; Alaska, Idaho, Minn., Mont., Pa., Utah, Wash., Wyo.; Eurasia.

Plants with white corollas, only a minor variant, sometimes occur. Because of reduced competition in Arctic regions after fires, large populations of subsp. *angustifolium* may persist for many years.

2b. Chamaenerion angustifolium (Linnaeus) Scopoli subsp. **circumvagum** (Mosquin) Moldenke, Phytologia 27: 289. 1973

Epilobium angustifolium Linnaeus subsp. *circumvagum* Mosquin, Brittonia 18: 169. 1966, based on *Chamaenerion angustifolium* var. *platyphyllum* Daniels, Univ. Missouri Stud., Sci. Ser. 2: 176. 1911; *C. angustifolium* var. *abbreviatum* Lunell; *C. angustifolium* var. *canescens* (Alph. Wood) Britton; *C. angustifolium* subsp. *macrophyllum* (Haussknecht) Czerepanov; *C. danielsii* (D. Löve) Czerepanov; *C. exaltatum* Rydberg; *Chamerion angustifolium* (Linnaeus) Holub var. *canescens* (Alph. Wood) N. H. Holmgren & P. K. Holmgren; *C. angustifolium* subsp. *circumvagum* (Mosquin) Hoch; *C. danielsii* (D. Löve) Czerepanov; *C. platyphyllum* (Daniels) Á. Löve & D. Löve; *E. angustifolium* var. *abbreviatum* (Lunell) Munz; *E. angustifolium* var. *canescens* Alph. Wood; *E. angustifolium* subsp. *macrophyllum* (Haussknecht) Hultén; *E. angustifolium* var. *macrophyllum* (Haussknecht) Fernald; *E. angustifolium* var. *platyphyllum* (Daniels) Fernald; *E. danielsii* D. Löve; *Pyrogennema angustifolium* (Linnaeus) Lunell var. *abbreviatum* (Lunell) Lunell

Stems 30–200 cm, sparsely strigillose, usually glabrous proximally. **Leaves:** petioles 2–7 mm; cauline blades oblong- to elliptic-lanceolate, (6–)9–23 × (0.7–)1.5–3.4 cm, base subcuneate to attenuate, margins ± denticulate, surface strigillose on adaxial side and on abaxial midrib, or, rarely, glabrous. **Inflorescences** strigillose; bracts 7–25 mm. **Flowers:** sepals 9–19 × 1.5–3 mm, acuminate; petals obovate to suborbicular, 14–25 × 7–14 mm, entire or scarcely emarginate; filaments 8–17 mm; anthers 2.5–3 mm; pollen mixed 3- and 4-porate (rarely 5-porate), 70–95 µm diam.; ovaries 8–25 mm, style 9–18 mm. **Capsules** 5–9.5 cm; pedicels 1–3 cm. **Seeds** 1–1.3 × 0.3–0.4 mm. $2n$ = 72, 108.

Flowering Jun–Sep. Moist, often disturbed places in montane and boreal forest regions, frequent after fires; 0–3800 m; St. Pierre and Miquelon; Alta., B.C., Man., N.B., Nfld. and Labr., N.W.T., N.S., Ont., P.E.I., Que., Sask., Yukon; Alaska, Ariz., Ark., Calif., Colo., Conn., Del., Idaho, Ill., Ind., Iowa, Maine, Md., Mass., Mich., Minn., Mont., Nebr., Nev., N.H., N.J., N.Mex., N.Y., N.C., N.Dak., Ohio, Oreg., Pa., R.I., S.Dak., Tenn., Utah, Vt., Va., Wash., W.Va., Wis., Wyo.; Mexico (Coahuila, Nuevo León); Eurasia.

5. EPILOBIUM Linnaeus, Sp. Pl. 1: 347. 1753; Gen. Pl. ed. 5, 164. 1754 • Willowherb [Greek *epi*, on, *lobos*, pod or capsule, and *ιον* (ion), violet, alluding to violet flower at apex of fruit]

Peter C. Hoch

Herbs, annual or perennial, sometimes suffrutescent, caulescent, often with basal rosettes, fleshy decussate turions, soboles, stolons, which may be tipped with turions, or rarely buds (gemmae) in leaf axils; with woody base or caudex, or with taproots. **Stems** erect to ascending or decumbent, simple to well-branched distally or sometimes from base, in some species proximal epidermis exfoliating, strigillose, glandular-puberulent, villous, often mixed, or glabrous, often with raised hairy lines decurrent from leaf axils. **Leaves** cauline, sometimes also basal, opposite and decussate proximal to inflorescence or only in proximal pairs, or rarely throughout, alternate or, sometimes, fasciculate distally; stipules absent; subsessile to petiolate; blade margins entire or toothed; bracts usually reduced distally. **Inflorescences** terminal, usually racemes or spikes, rarely panicles, flowers solitary in leaf axils, erect or ascending, sometimes also flowering from proximal nodes; bracteoles absent. **Flowers** bisexual, usually actinomorphic, rarely zygomorphic, buds often erect, sometimes recurved; floral tube deciduous (with sepals, petals, and stamens) after anthesis, obconic or cylindrical, sometimes with bulbous base, with scales or ring of hairs or raised ring of tissue near mouth inside, nectary at base of tube; sepals 4, spreading individually, usually green or flushed with red or cream, rarely same color as petals, lanceolate; petals 4, usually rose-purple to white, rarely cream-yellow or orange-red, usually obcordate or broadly obovate, emarginate; stamens 8, in 2 unequal series with episepalous longer or rarely subequal, erect, anthers versatile, on smallest flowers appearing basifixed, pollen usually shed in tetrads, rarely singly; ovary 4-locular, style erect, stigma entire and clavate to capitate, or deeply 4-lobed, commissural, receptive only on inner surfaces, surface dry with multicellular papillae. **Fruit** a capsule, straight or slightly curved, narrowly cylindrical to fusiform or rarely narrowly ellipsoidal, usually terete, rarely ± 4-angled, loculicidally dehiscent, usually splitting to base with intact central column, rarely splitting only on distal ⅓ with central column disintegrating; pedicellate or sessile. **Seeds** usually numerous (1–100+ per locule), usually in 1, rarely 2, rows per locule, very rarely 1 row per capsule by dissolution of septa and central column, surface papillose to finely reticulate or longitudinally ridged, sometimes constricted near micropylar end, rarely with inflated rim around body of seed on adaxial side, with persistent coma at chalazal end or coma absent. x = 9, 10, 12, 13, 15, 16, 18, 19.

Species ca. 165 (41 in the flora): North America, Mexico, West Indies, Central America, South America, Eurasia, Africa, Pacific Islands (New Zealand), Australasia.

Epilobium is the largest genus in Onagraceae, distributed mainly in cool temperate, montane, boreal, and arctic areas on all continents except Antarctica. P. H. Raven (1976) proposed many elements of the current classification of Epilobieae, recognizing six sections in *Epilobium*, including for the first time the genus *Zauschneria* as a separate section, based on patterns of variation in chromosome number (H. Lewis and Raven 1961; M. Kurabayashi et al. 1962), wood anatomy (S. Carlquist 1975), seed morphology (S. R. Seavey et al. 1977), and pollen wall (J. J. Skvarla et al. 1976) and viscin thread morphology (Skvarla et al. 1978). Raven excluded *Boisduvalia* from *Epilobium*, though it later was included (P. C. Hoch and Raven 1992) as two distinct sections, and Raven included *Chamaenerion* as a section, which was later excluded (Hoch and Raven 1999; W. L. Wagner et al. 2007). Both of these changes from the 1976 classification were based in large part on relationships supported by molecular data (D. A. Baum et al. 1994). The monophyly of this re-defined *Epilobium* was very strongly supported in analyses of ITS alone (Baum et al.) and ITS plus *trn*L-F (R. A. Levin et al. 2004), as was the segregation of *Chamaenerion* from *Epilobium*. In addition, both analyses strongly supported a clade that includes the former segregated genera *Boisduvalia* and *Zauschneria* as well as the sections of *Epilobium* with chromosome numbers other than n = 18. As treated here, *Epilobium* is divided into eight sections, one of which has two subsections (Wagner et al.).

The phylogeny of *Epilobium* as elucidated by D. A. Baum et al. (1994) and later R. A. Levin et al. (2004) suggests complex chromosomal evolution in the genus. Since the outgroup (*Chamaenerion*) and one major clade (sect. *Epilobium*) share the base number x = 18, the derivation of the diverse other gametic numbers (n = 9, 10, 12, 13, 15, 16, and 19) appears to have involved aneuploid reduction from x = 18. One explanation of the relationships proposed by the molecular data suggests the early evolution of a lineage with n = 15 (sects. *Cordylophorum*, *Epilobiopsis*, and *Zauschneria*), from which arose a branch with n = 12 (sect. *Xerolobium*), a separate branch with n = 13 and n = 16 (sect. *Crossostigma*), and a third branch with n = 9, 10, and 19 (sect. *Pachydium*). This hypothesis requires the fewest aneuploidy changes, but confirming it will require additional molecular and cytological analysis.

Almost all *Epilobium* species tested are self-compatible, but at least *E. obcordatum* is apparently self-incompatible (S. R. Seavey and S. K. Carter 1994, 1996). All species have hermaphroditic, diurnal flowers that usually remain open for more than one day. Many species are modally autogamous, a few primarily cleistogamous, but others, including 15 species in North America, are partly or wholly outcrossing. In the latter group, flowers are often markedly protandrous, with a 4-lobed stigma exserted beyond anthers. Primary pollinators include bees, flower flies, butterflies, moths, and sometimes hummingbirds (sect. *Zauschneria*). In general, flowering commences in the first year of growth.

Only the nine species (eight in the flora area) of sects. *Pachydium*, *Crossostigma*, *Epilobiopsis*, and *Xerolobium* are annuals; all other taxa of *Epilobium* (sects. *Cordylophorum*, *Epilobium*, *Macrocarpa*, and *Zauschneria*) are perennials. Annual species have taproots and proximal stems usually with peeling epidermis. Perennial species of *Epilobium*, which have more fibrous root systems and rarely peeling epidermis, vary greatly in the habit and stature, degree of clumping, degree of branching, and mode of perennation.

The various types of perennating structures in *Epilobium* (R. C. Keating et al. 1982; Chen C. J. et al. 1992) are extremely useful for identification of many taxa; unfortunately, many plants are collected without the structures. Unless care is taken to dig up the bases of the plants, some of the most useful diagnostic features such as turions are lost, making identification more difficult. Perennating structures are less useful for phylogenetic analysis due to the difficulty in determining whether the common possession of a particular type represents an analogous or homologous pattern. However, one type of structure (soboles) is found in at least some taxa of all perennial sections and may be the most generalized type in the genus.

Soboles are here defined as simple, scaly, underground shoots with somewhat elongated internodes and arising from the caudex. Soboles generally give rise to more or less clumped plants with ascending stems. In certain species, including *Epilobium rigidum* (sect. *Macrocarpa*), all taxa of sects. *Cordylophorum* and *Zauschneria*, and some in sect. *Epilobium*, soboles arise from more or less woody caudices, a character state that may be plesiomorphic in the genus; all other types of perennating structures are found only in sect. *Epilobium* and are discussed there.

In addition to habit and perennating structures, certain other morphological features particularly useful for diagnosing plants include: leaf size, shape, attachment, and arrangement on stem; seed surface (papillose, reticulate, or ridged), size, and shape (S. R. Seavey et al. 1977); pubescence type (mainly strigillose: short incurved hairs; villous: long, thin, spreading hairs; and/or glandular-puberulent) and pattern on stem (all around or restricted to raised lines decurrent from base of petioles) and leaves; and some floral features including size, shape, and color. In fact, floral features are diagnostic for only certain sections or species; most species have very similar flowers.

SELECTED REFERENCES Baum, D. A., K. J. Sytsma, and P. C. Hoch. 1994. The phylogeny of *Epilobium* (Onagraceae) based on nuclear ribosomal DNA sequences. Syst. Bot. 19: 363–388. Keating, R. C., K. J. Sytsma, and P. C. Hoch. 1982. Perennation in *Epilobium* (Onagraceae) and its relation to classification and ecology. Syst. Bot. 7: 379–404. Munz, P. A. 1941. A revision of the genus *Boisduvalia* (Onagraceae). Darwiniana 5: 124–153. Raven, P. H. and D. M. Moore. 1965. A revision of *Boisduvalia* (Onagraceae). Brittonia 17: 238–254. Trelease, W. 1891. A revision of the American species of *Epilobium* occurring north of Mexico. Rep. (Annual) Missouri Bot. Gard. 2: 69–117.

1. Herbs annual, with taproot, stem epidermis peeling proximally; leaves opposite in proximal pairs, alternate or fasciculate distally.
    2. Seed coma usually present, very rarely absent, often easily or readily detached.
        3. Herbs 15–200 cm; floral tubes 1–16 mm, sepals 1–8.5 mm, petals 1.5–15(–20) mm; seeds 1.5–2.7 mm; pollen usually shed singly, rarely in tetrads [5b. *Epilobium* sect. *Xerolobium*] . . . . 2. *Epilobium brachycarpum*
        3. Herbs 5–45 cm; floral tubes 0.4–1.5 mm, sepals 0.5–2.5 mm, petals 1.4–5 mm; seeds 0.6–1.2 mm; pollen shed in tetrads [5c. *Epilobium* sect. *Crossostigma*].
            4. Leaves usually folded along midrib, often fasciculate distally; inflorescences congested; petals 1.4–2.5 mm; seeds 0.6–0.9 mm, surface papillose . . . . . . 3. *Epilobium foliosum*
            4. Leaves flat, not fasciculate; inflorescences not congested; petals 2–5 mm; seeds 0.9–1.2 mm, surface reticulate . . . . 4. *Epilobium minutum*
    2. Seed coma absent.
        5. Capsule walls tough, usually splitting only in distal 1/3; seeds in 2 rows per locule; often with decumbent proximal branches; flowers usually cleistogamous [5e. *Epilobium* sect. *Epilobiopsis*].

6. Capsules 8–12 mm, sharply 4-angled; sepals 1.5–3 mm; seeds 1.2–1.5 mm . . . . 8. *Epilobium cleistogamum*
6. Capsules 4.5–8 mm, subterete; sepals 0.7–1.9 mm; seeds 1–1.3 mm . . . . 9. *Epilobium campestre*

5. Capsule walls flexible, splitting to base; seeds in 1 row per locule or compressed to 1 row per capsule; proximal branches ascending, erect, or absent; flowers chasmogamous or cleistogamous [5f. *Epilobium* sect. *Pachydium*].
7. Capsules with beak 0–0.5 mm, septifragal, septa adherent to intact central axis; bracts wider than cauline leaves . . . . 10. *Epilobium densiflorum*
7. Capsules with beak 2–5 mm, loculicidal, septa adhering to valves, central axis disintegrating; bracts not wider than cauline leaves.
8. Floral tubes 1.5–3 mm, sepals (1.5–)2–6 mm, petals (2.8–)3.5–10 mm; capsules (10–)14–21 mm; seeds in 1 staggered row distally, 1.8–2.3 mm; flowers ± chasmogamous . . . . 11. *Epilobium pallidum*
8. Floral tubes 0.4–1 mm, sepals 0.7–2 mm, petals 1.2–3.2 mm; capsules (6–)8–14 mm; seeds in 4 rows (1 row per locule), 0.9–1.6 mm; flowers ± cleistogamous . . . . 12. *Epilobium torreyi*

1. Herbs perennial, sometimes woody, ± without taproot, stem epidermis usually not peeling, sometimes peeling proximally; leaves usually opposite proximal to inflorescences, alternate distally, not fasciculate.
9. Floral tubes 16–32 mm, bulbous near base, irregular scales at mouth inside; flowers: floral tube, sepals, and petals usually orange-red, very rarely white, slightly zygomorphic, upper petals ± flared at right angle to calyx tube [5g. *Epilobium* sect. *Zauschneria*].
10. Herbs clumped perennials to subshrubs, 10–120 cm; leaves not dimorphic, usually mixed pubescent, rarely glabrate, not white-canescent . . . . 13. *Epilobium canum*
10. Herbs matted perennials, 5–25 cm; leaves dimorphic, proximal ones densely white-canescent and eglandular, distal ones glandular puberulent and scattered villous, not canescent . . . . 14. *Epilobium septentrionale*
9. Floral tubes 0.3–9.5 mm, not or very rarely bulbous near base, scales absent; flowers: floral tube and sepals green or flushed reddish green, petals usually pink, white, or rose-purple, rarely cream, usually actinomorphic, very rarely slightly zygomorphic.
11. Herbs suffrutescent or rhizomatous, rarely herbaceous; proximal stem epidermis usually peeling; stigmas 4-lobed; seeds 1.5–3.4 mm, ± prominently constricted near micropylar end.
12. Leaf blades 20–45 mm, not folded on midrib; sepals 9.5–14.5 mm; petals 16–20 mm; floral tubes 1–1.8 mm [5a. *Epilobium* sect. *Macrocarpa*]. . . . . . 1. *Epilobium rigidum*
12. Leaf blades 9–25 mm, folded on midrib or not; sepals 2.6–6.5 mm; petals 5–9.5 mm; floral tubes 1.8–9.5 mm [5d. *Epilobium* sect. *Cordylophorum*].
13. Leaf blades not folded on midrib; petals cream to light yellow; flowers slightly zygomorphic (upper petals slightly longer, styles declined); sepals 4–6.5 mm; capsules 10–30 mm; seeds 4–8+ seeds per locule; ovaries eglandular . . . . 7. *Epilobium suffruticosum*
13. Leaf blades ± folded on midrib; petals rose-purple; flowers not zygomorphic; sepals 2.6–4.4 mm; capsules 8–16 mm; seeds 1–3 per locule; ovaries glandular pubescent.
14. Leaf blade margins denticulate, surfaces usually glabrescent, rarely strigillose-villous; floral tubes 2.7–3.2(–5) mm; seeds 2.1–2.9 mm . . . . 5. *Epilobium nevadense*
14. Leaf blade margins subentire or low-denticulate, surfaces densely spreading-hairy; floral tubes 5.2–9.5 mm; seeds 1.5–2.4 mm. . . . . . 6. *Epilobium nivium*
11. Herbs herbaceous or, sometimes, ± suffrutescent; proximal stem epidermis not peeling; stigmas entire or 4-lobed; seeds 0.8–2.2 mm, not constricted near micropylar end [5h. *Epilobium* sect. *Epilobium*].
15. Stigmas 4-lobed, exserted beyond or rarely surrounded by anthers; floral tubes 1–5.5 mm; sepals (2.5–)5–14 mm; petals (4–)8–26 mm.
16. Petals cream-yellow; seed coma tawny, persistent . . . . 29. *Epilobium luteum*
16. Petals pink to rose-purple; seed coma white, dull white, or dingy, sometimes tawny, ± easily or readily detached.

17. Herbs ± suffrutescent, forming shoots from woody caudex; stems 5–25 cm, clumped to cespitose; floral tubes 2.1–5.5 mm.
    18. Herbs ± glabrous and glaucous proximal to inflorescences; leaf blades 6–18(–24) mm; floral tubes 3.2–5.5 mm, slightly raised ring of tissue inside . . . . 15. *Epilobium obcordatum*
    18. Herbs pubescent throughout, not glaucous; leaf blades 13–26 mm; floral tubes 2.1–4 mm, prominent raised ring of tissue inside . . . . 16. *Epilobium siskiyouense*
17. Herbs herbaceous, forming basal rosettes, stolons, turions, or soboles; stems (5–)15–120(–250) cm, erect, sometimes clumped; floral tubes 1–3 mm.
    19. Stems densely villous, often mixed glandular puberulent proximal to inflorescences, 18–120(–250) cm.
        20. Leaves sessile, clasping; sepals 6–12 mm; petals 9–20 mm; herbs with thick stolons sometimes terminating in rosette . . . . 26. *Epilobium hirsutum*
        20. Leaves with petioles 1–3 mm proximally, subsessile distally, not clasping; sepals 2.5–6 mm; petals 4–8.5 mm; herbs with short-stalked leafy rosettes . . . . 27. *Epilobium parviflorum*
    19. Stems densely strigillose to ± glabrous proximal to inflorescences, not villous, (5–)15–75 cm.
        21. Stems densely strigillose proximal to inflorescences; seeds densely papillose; sepals 5–6.5 mm; petals 7.5–10 mm. . . . . 28. *Epilobium montanum*
        21. Stems ± glabrous and glaucous proximal to inflorescences; seeds ridged; sepals 6–10 mm; petals (6–)10–15 mm . . . . 38. *Epilobium oreganum*

[15. Shifted to left margin.—Ed.}

15. Stigmas entire, clavate or capitate, surrounded by or, rarely, exserted beyond anthers; floral tubes 0.3–2.6 mm; sepals 1.1–7.5 mm; petals 1.6–10(–14) mm.
    22. Herbs with ± threadlike stolons, with or without terminal turions; stems usually erect from base, rarely ascending.
        23. Stolons terminating in fleshy turions; stems (5–)15–95 cm, simple to well branched.
            24. Stems densely villous, without raised lines from margins of petioles. . . . . . 19. *Epilobium densum*
            24. Stems strigillose to subglabrous, not villous, sometimes with faint raised strigillose lines from margins of petioles.
                25. Leaf blade surfaces strigillose abaxially, subglabrous adaxially, with strigillose margins and veins; inflorescences nodding in bud . . . . 17. *Epilobium palustre*
                25. Leaf blade surfaces densely strigillose on both sides; inflorescences erect in bud . . . . 18. *Epilobium leptophyllum*
        23. Stolons not terminating in turions; stems 5–30 (–40) cm, usually simple, rarely branched, rarely stems 20–80 cm, well branched.
            26. Leaf blades 1.5–10 cm; stems 20–80 cm, well branched, not clumped or matted . . . . 25. *Epilobium obscurum*
            26. Leaf blades 0.4–2.1 cm; stems 5–30(–40) cm, not or rarely branched, ± loosely clumped or matted.
                27. Stems subglabrous, often matted, 8–30(–40) cm; leaf blades broadly elliptic proximally, narrowly elliptic or lanceolate to sublinear distally . . . . 22. *Epilobium oregonense*
                27. Stems densely glandular puberulent, loosely clumped, 5–20 cm; leaf blades broadly obovate to orbiculate proximally to ovate or lanceolate distally . . . . 37. *Epilobium howellii*
    22. Herbs with rosettes, turions, fleshy soboles or shoots; stem bases erect or ascending.

[28. Shifted to left margin.—Ed.]

28. Herbs with small-leafed epigeous or scaly hypogeous soboles; stems usually ± ascending, rarely erect, ± clumped or matted, rarely with basal scales.
   29. Stems 8–75(–85) cm, glabrous and ± glaucous proximal to inflorescences, often without raised lines; leaves subsessile, ± clasping; seeds 0.7–1(–1.3) mm, surfaces densely papillose . . . . 40. *Epilobium glaberrimum*
   29. Stems 3–50 cm, subglabrous or strigillose, not glaucous, proximal to inflorescences, with distinct raised strigillose lines decurrent from margins of petioles; leaves subsessile or with petiole 1–12 mm, not clasping; seeds 0.7–2.1 mm, surfaces reticulate to densely papillose.
      30. Stems strigillose proximal to inflorescences, densely glandular puberulent distally . . . . 34. *Epilobium smithii*
      30. Stems subglabrous with raised strigillose lines proximal to inflorescences, mixed strigillose and glandular puberulent or subglabrous distally.
         31. Stems 10–50 cm; leaf blades 1.5–6.2 × 0.7–2.9 cm, margins denticulate; capsules 35–100 mm.
            32. Petals usually pink to rose-purple, rarely white; pedicels 5–15(–25) mm in fruit; capsules 35–65 mm; petioles 3–9 mm proximally, to 0 mm distally, not winged; seed surfaces usually papillose, sometimes reticulate . . . . 30. *Epilobium hornemannii*
            32. Petals white, rarely red-veined or fading pink; pedicels 15–45 mm in fruit; capsules 50–100 mm; petioles 3–12 mm, often winged; seed surfaces reticulate or, sometimes, barely rugose. . . . . 31. *Epilobium lactiflorum*
         31. Stems 3–20(–25) cm; leaf blades (0.5–)0.7–2.8 × 0.3–1.6 cm, margins subentire to sparsely denticulate; capsules 17–40(–55) mm.
            33. Herbs with thin leafy soboles, no woody rootstock, densely clumped or matted, stems nodding in bud; pedicels 5–35(–68) mm in fruit; seeds 0.7–1.4 mm; stems subglabrous, rarely mixed strigillose and sparsely glandular puberulent distally . . . . 32. *Epilobium anagallidifolium*
            33. Herbs with wiry, scaly soboles and extended semi-woody rootstock, clumped, stems ± erect; pedicels 2–21 mm in fruit; seeds (1.3–)1.5–2.1 mm; stems subglabrous proximally, ± densely strigillose and mixed glandular-puberulent distally . . . . 33. *Epilobium clavatum*

28. Herbs usually with leafy rosettes or fleshy hypogeous turions, rarely axillary bulblets; stems ± erect, usually not clumped, usually with dark basal scales.
   34. Herbs with basal sessile or, rarely, short-stalked, leafy epigeous rosettes.
      35. Stems (2–)5–40(–45) cm, simple; leaf blades 0.8-4.5 cm, margins usually subentire, rarely denticulate, subsessile; inflorescences racemes, nodding in bud.
         36. Stems 10–40(–45) cm; leaf blades (1.2–)2–4.5 cm, narrowly oblong to narrowly lanceolate proximally; seed surfaces papillate . . . . .20. *Epilobium davuricum*
         36. Stems (2–)5–18 cm; leaf blades 0.8–2.1 cm, obovate to narrowly elliptic proximally; seed surfaces reticulate . . . . 21. *Epilobium arcticum*
      35. Stems (3–)10–85(–190) cm, ± branched; leaf blades (1–)3–10(–16) cm, margins denticulate to densely serrulate, subsessile or petiole to 10 mm; inflorescences erect racemes, panicles, or corymbs.
         37. Seed coma tawny; seed surfaces papillose; leaf blade margins sharply serrulate, 30–75 teeth per side; petioles 4–10 mm; inflorescences eglandular; petals 2.5–5.5 mm, white . . . . 24. *Epilobium coloratum*
         37. Seed coma white or dingy; seed surfaces ridged; leaf blade margins serrulate, (8–)15–40 teeth per side; petioles 0–5(–10) mm; inflorescences usually mixed glandular pubescent; petals 2–14 mm, pink to rose-purple or white. . . . . 39. *Epilobium ciliatum* (in part)
   34. Herbs with fleshy underground turions, rarely also with bulblets in proximal or mid-cauline nodes.

[38. Shifted to left margin.—Ed.]

38. Herbs ± eglandular; stems 7–30 cm.
    39. Pedicels 5–16 mm; seeds 1.7–2.2 mm, surfaces reticulate to low papillose; stems only with hypogeous turions . . . . 35. *Epilobium mirabile*
    39. Pedicels 15–38 mm; seeds 0.8–1.2 mm, surfaces papillose; stems with turions and, usually, bulblets in proximal or mid-cauline nodes . . . . 36. *Epilobium leptocarpum*
38. Herbs mixed glandular puberulent, at least on inflorescences; stems 2–120(–190) cm, usually more than 30 cm.
    40. Seed surfaces ridged; stems (3–)10–120(–190) cm; leaf blades (1–)3–12(–16) cm; petals 2–14 mm, white, pink, or rose purple . . . . 39. *Epilobium ciliatum* (in part)
    40. Seed surfaces papillose or reticulate; stems 2–55(–60) cm; leaf blades 0.5–5.5(–6.5) cm; petals 1.6–5.5(–7) mm, white, fading or rarely pink.
        41. Pedicels 8–40 mm; capsules ascending, spreading; leaves not clasping, veins inconspicuous; inflorescences ± nodding in bud . . . . 23. *Epilobium hallianum*
        41. Pedicels 0–5 mm; capsules erect, appressed to stem; leaves clasping, veins conspicuous; inflorescences ± erect . . . . 41. *Epilobium saximontanum*

5a. Epilobium Linnaeus sect. Macrocarpa Hoch & W. L. Wagner, Syst. Bot. Monogr. 83: 82. 2007 [E]

Herbs perennial, suffrutescent, with basal shoots from caudex. **Stems** ± woody at base, epidermis peeling proximally. **Leaves** opposite proximal to inflorescence, alternate distally. **Flowers** actinomorphic; floral tube extremely short for size of flower, without bulbous base and scales inside; petals pink to rose-purple; pollen shed in tetrads; stigma deeply 4-lobed. **Capsules** narrowly fusiform to cylindrical, splitting to base, central column persistent, pedicellate. **Seeds** numerous, in 1 row per locule, narrowly obovoid, prominently constricted near micropylar end, coma present.

Species 1: w United States.

Section *Macrocarpa* consists of only *Epilobium rigidum*, an uncommon species endemic to the Klamath-Siskiyou region along the California-Oregon border in western United States. Since its publication in the worldwide monograph of *Epilobium* by C. Haussknecht (1884; see also P. C. Hoch and P. H. Raven 1990), this species has been included consistently in sect. *Epilobium* in the strict sense (Raven 1976), despite having a number of unusual morphological features. It has the largest seeds in the genus (2.5–3.4 × 0.9–1.4 mm), with a prominent constriction near the micropylar end (S. R. Seavey et al. 1977), in both features resembling seeds of sects. *Cordylophorum*, *Crossostigma* (only *E. foliosum*), *Xerolobium*, and *Zauschneria* but unlike any taxa in sect. *Epilobium*. Despite having some of the largest petals in the genus (16–20 mm), *E. rigidum* has an extremely short floral tube (1–1.8 mm), and this combination of character states is unique. It shares with sect. *Epilobium* the chromosome number $n = 18$, and can form hybrids with some species in that section; however, all hybrids are completely sterile, and no bivalents form at meiotic metaphase I (Seavey and Raven 1977b, 1978). Analysis of ITS sequence data (D. A. Baum et al. 1994) placed *E. rigidum* as the sister branch to sect. *Epilobium* but with essentially no support, even considering the sparse sampling of the section (five species). R. A. Levin et al. (2004), using both ITS and *trn*L-F sequence data, found very strong support for both a sparsely sampled sect. *Epilobium* and for a clade of all sections with chromosome numbers other than $n = 18$; *E. rigidum* is weakly supported as the basally diverging branch on the non-sect. *Epilobium* clade. The current best molecular evidence places *E. rigidum* near the base of the genus, possibly on a branch with all other species having $n = 18$ (sect. *Epilobium*), or more likely (with weak support) on the branch with all non-$n = 18$ species, with which it shares similarities in seed and pollen morphology. This apparent basal position and the unique short floral tube (which might even be seen as transitional to the character state in the tubeless sister genus, *Chamaenerion*) suggest that *E. rigidum* is best treated as a separate section, positioned near the base of the two well-supported clades in *Epilobium*.

1. Epilobium rigidum Haussknecht, Oesterr. Bot. Z. 29: 51. 1879 • Stiff willowherb [E]

Epilobium rigidum var. *canescens* Trelease

Herbs from woody caudex forming hypogeal shoots with barklike periderm. **Stems** several to many, suberect or ascending, terete, 10–40 cm, simple or sparsely branched, usually glabrous and ± glaucous proximal to inflorescence, strigillose distally, sometimes densely strigillose throughout. **Leaves** crowded distally, petiole 2–6 mm, blade narrowly ovate to ovate or broadly elliptic, often obovate in proximal pairs; cauline 2–4.5 × 0.8–2 cm, base rounded to attenuate, margins subentire or finely denticulate, 8–12 teeth per side, lateral veins inconspicuous, 3–5 per side, apex obtuse proximally to subacute distally, surfaces glaucous and subglabrous to densely strigillose; bracts narrower and much smaller. **Inflorescences** erect racemes, simple, ± densely strigillose, rarely mixed sparsely glandular puberulent. **Flowers** erect; buds 6–11 × 4–5 mm, apiculate; pedicels 4–8 mm; floral tube 1–1.8 × 2–3 mm, with raised ring of tissue edged with spreading hairs at mouth inside; sepals often reddish green, lanceolate, 9.5–14.5 × 2.5–3.5 mm, apex acuminate, abaxial surface densely strigillose; petals pink to rose-purple, obcordate, 16–20 × 13–16 mm, apical notch 3.4–5.5 mm; filaments light pink, those of longer stamens 9–14 mm, those of shorter ones 6.5–10 mm; anthers cream, 1.8–3.5 × 1–1.9 mm; ovary 6–12 mm, densely strigillose; style cream to light pink, 14.5–18.5 mm, stigma broadly 4-lobed, 1–1.5 × 3–3.5 mm, exserted beyond anthers. **Capsules** 20–35 mm, surfaces strigillose; pedicel 9–13 mm, bracts often attached 2–3 mm from base. **Seeds** narrowly obovoid, constriction 0.6–0.8 mm from micropylar end, 2.5–3.4 × 0.9–1.4 mm, chalazal collar obscure, light brown, surface papillose; coma easily detached, white, 6–8 mm. $2n = 36$.

Flowering Jul–Sep. Dry rocky or sandy benches, rocky hillsides, dry streambeds in coniferous forests, on seasonally moist serpentine slopes, rarely along disturbed roadsides; 100–1200(–1500) m; Calif., Oreg.

Within its range in northwestern California and southwestern Oregon, *Epilobium rigidum* is restricted to unusually dry habitats compared to most species in sect. *Epilobium*, but is not unlike taxa in the non-n = 18 clade, which are both perennial and annual. It is self-compatible, but with strongly protandrous flowers and an exserted stigma, and is modally outcrossing, pollinated by bees and flies.

Plant vestiture varies from subglabrous to densely strigillose throughout (var. *canescens*), but plants with these differences can be in the same population, and no other morphological differences between them have been found.

5b. Epilobium Linnaeus sect. Xerolobium P. H. Raven, Ann. Missouri Bot. Gard. 63: 334. 1977 [E]

Herbs annual, with taproot. **Stems:** epidermis peeling proximally. **Leaves** opposite and early deciduous in proximalmost pairs, alternate and sometimes fasciculate distally. **Flowers** actinomorphic; floral tube without bulbous base or scales inside; petals rose-purple to pink or white; pollen usually shed singly, rarely in tetrads; stigma entire and clavate to deeply 4-lobed. **Capsules** narrowly cylindrical or fusiform, splitting to base, central column persistent, pedicellate. **Seeds** numerous, in 1 row per locule, obovoid to broadly obovoid, prominently constricted near micropylar end, coma present.

Species 1: w North America; introduced in sw South America, Europe.

Section *Xerolobium* consists only of the annual species *Epilobium brachycarpum*, with $n = 12$, a unique chromosome number apparently derived by aneuploidy from the tribal base number of $x = 18$. It differs from the other annual taxa in sects. *Epilobiopsis* ($n = 15$) and *Pachydium* ($n = 9, 10, 19$) not only in chromosome number, but also by flowering in summer, whereas the other taxa are primarily spring-flowering.

Section *Xerolobium* shares with sects. *Cordylophorum* and *Zauschneria* the apomorphic features of the so-called incised compound pollen viscin threads (J. J. Skvarla et al. 1978; also found in sect. *Pachydium* and in *Epilobium minutum* of sect. *Crossostigma*) and in having seeds prominently constricted at the micropylar end (S. R. Seavey et al. 1977; also found in *E. rigidum* of sect. *Macrocarpa* and *E. foliosum* of sect. *Crossostigma*). In addition to its autapomorphic chromosome number, *E. brachycarpum* is marked as phylogenetically distinct by virtue of shedding its pollen in monads (Skvarla et al. 1975). All but a few exceptionally large-flowered populations in a small area of northern California (sometimes designated as *E. jucundum*) have monads, a feature otherwise found in Epilobieae only in *Chamaenerion*. The feature of shedding pollen as monads was clearly derived independently within *E. brachycarpum*.

2. **Epilobium brachycarpum** C. Presl, Reliq. Haenk. 2: 30. 1831 • Tall annual willowherb, épilobe d'automne [E] [F] [W]

Epilobium adenocladum (Haussknecht) Rydberg (as adenocladon); *E. altissimum* Suksdorf; *E. apricum* Suksdorf; *E. fasciculatum* Suksdorf; *E. hammondii* Howell; *E. jucundum* A. Gray; *E. jucundum* var. *viridifolium* Suksdorf; *E. laevicaule* Rydberg; *E. paniculatum* Nuttall ex Torrey & A. Gray; *E. paniculatum* var. *hammondii* (Howell) M. Peck; *E. paniculatum* var. *jucundum* (A. Gray) Trelease; *E. paniculatum* var. *laevicaule* (Rydberg) Munz; *E. paniculatum* var. *subulatum* (Haussknecht) Fernald; *E. paniculatum* var. *tracyi* (Rydberg) Munz; *E. subulatum* (Haussknecht) Rydberg; *E. tracyi* Rydberg

Herbs slender. **Stems** erect, terete, 15–200 cm, simple to paniculate-branched (especially in larger plants), glabrous proximally, strigillose distally, sometimes mixed glandular puberulent. **Leaves** subsessile or petiole 1–4 mm, blade linear to linear-lanceolate or narrowly elliptical, often folded along midrib, 1–5.5(–7) × 0.1–0.8 cm, usually shorter than internodes, base tapered or cuneate, margins remotely denticulate, 2–10 teeth per side, lateral veins obscure, 2–5 per side, apex acute or acuminate, surfaces subglabrous and sometimes glaucous to strigillose; bracts very reduced, sometimes attached to pedicel. **Inflorescences** erect, open panicles with filiform branches or simple racemes, glabrous or strigillose, often mixed glandular puberulent. **Flowers** usually erect; buds 1–12 × 1–3.5 mm; floral tube obconic to funnel-form, 1–16 × 0.8–2.9 mm, with ring of spreading hairs near mouth inside, in larger flowers, ring swollen; sepals green to reddish green, 1–8.5 × 0.8–2.1 mm, apex acute, abaxial surface strigillose and glandular puberulent to subglabrous; petals white to pink or deep rose-purple, 1.5–15(–20) × 1–7.5 mm, apical notch 0.5–6.5 mm; filaments usually cream-white, rarely pink, those of longer stamens 1–9.5 mm, those of shorter ones 0.5–6.3 mm; anthers cream, 0.5–4 × 0.3–2 mm; ovary 2–16 mm, strigillose, often mixed glandular puberulent, to subglabrous; style cream, 2–18 mm, stigma clavate to subcapitate, entire to deeply 4-lobed, 0.5–1.9 × 0.3–2.5 mm, surrounded by stamens or (in some larger flowers) exserted beyond anthers. **Capsules** erect or ascending, 15–32 mm, surfaces strigillose and glandular puberulent or glabrous; pedicel 1–17 mm. **Seeds** obovoid to broadly obovoid, with constriction 0.3–0.7 mm from micropylar end, 1.5–2.7 × 0.8–1.3 mm, chalazal collar inconspicuous, brown or gray, often flecked with darker spots, surface low papillose; coma easily detached, white to dingy white, 5–10 mm. $2n = 24$.

Flowering Jun–Sep. Open, dry or seasonally moist, often disturbed ground in open woods, meadows, prairies, roadsides and stream banks; 0–3000 m; Alta., B.C., Man., Ont., Que., Sask.; Ariz., Calif., Colo., Idaho, Ky., Mont., Nev., N.Mex., N.Dak., Oreg., S.Dak., Utah, Wash., Wis., Wyo.; introduced in South America (Argentina), Europe (Germany, Spain).

The presence of *Epilobium brachycarpum* in Kentucky and Wisconsin is in railroad yards, so possibly ephemeral. This species occurs as an adventive in Argentina, possibly by natural long-distance dispersal (J. C. Solomon 1982), and in Spain (J. Izco 1983) and Germany (T. Gregor et al. 2013), probably from human introduction, since it frequently occurs as a weed on margins of cultivated fields (L. H. Shinners 1941).

Most populations of this self-compatible species have small autogamous flowers, but some populations with larger outcrossing flowers that exhibit marked protandry and herkogamy are pollinated by bees, butterflies, and occasionally hummingbirds.

Mature plants of *Epilobium brachycarpum* sometimes have only alternate or fasciculate leaves. However, seedlings always have opposite leaves in first pairs, but these are often deciduous.

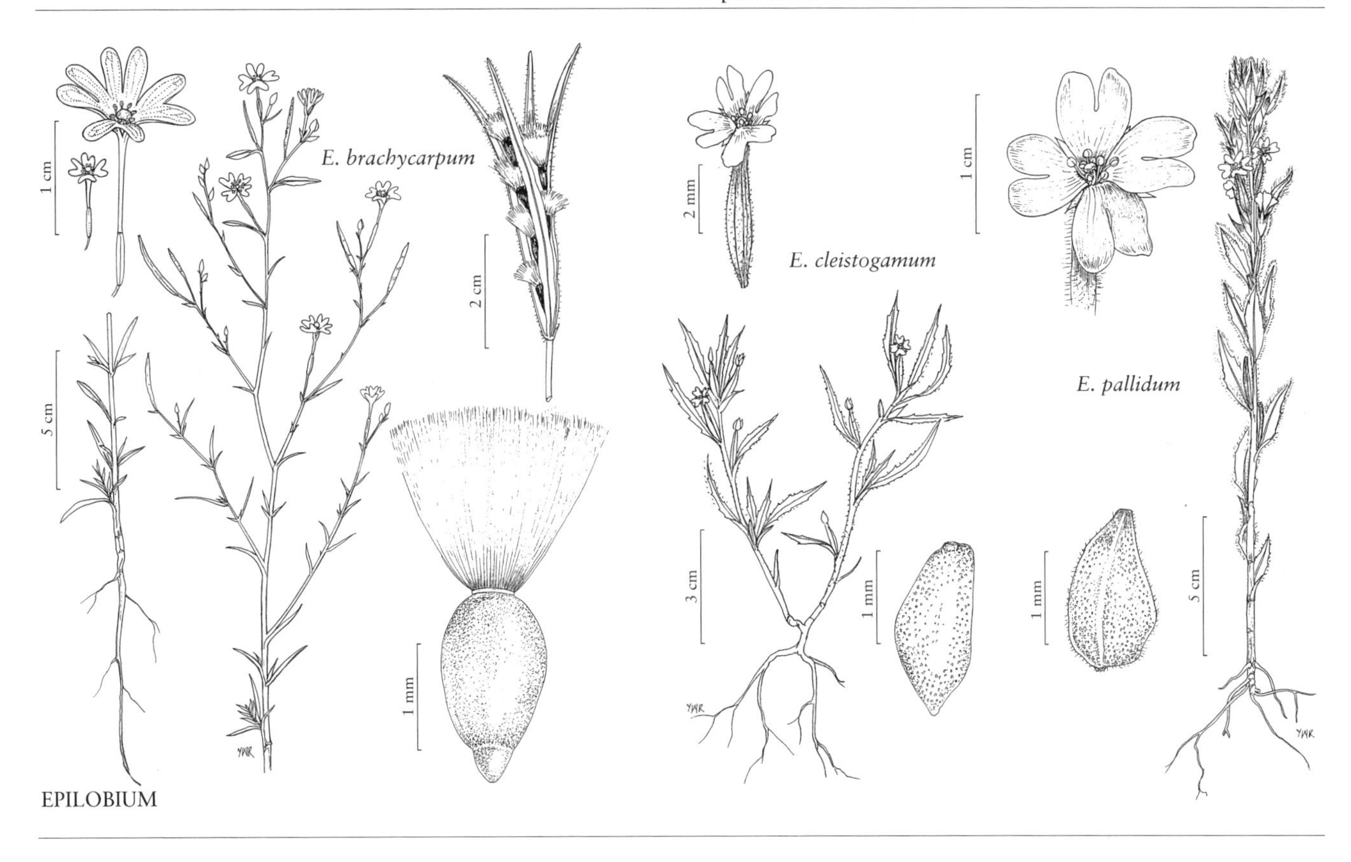

EPILOBIUM

5c. Epilobium Linnaeus sect. Crossostigma (Spach) P. H. Raven, Ann. Missouri Bot. Gard. 63: 336. 1977

Crossostigma Spach, Ann. Sci. Nat., Bot., sér. 2, 4: 174. 1835

Herbs annual, with taproot. **Stems:** epidermis often peeling proximally. **Leaves** opposite in proximal pairs, alternate or fasciculate distally. **Flowers** actinomorphic; floral tube without bulbous base or scales inside; petals white to pale rose-purple; pollen shed in tetrads; stigma usually entire, rarely ± 4-lobed. **Capsules** narrowly cylindrical, splitting to base, central column persistent, pedicellate to subsessile. **Seeds** numerous, in 1 row per locule, obovoid to oblanceoloid, entire or constricted near micropylar end, coma present.

Species 2 (2 in the flora): w North America, nw Mexico.

Section *Crossostigma* consists of two annual species endemic to western North America from California to British Columbia and Montana (*Epilobium minutum*) or from California to Idaho and Arizona, rarely to northwestern Mexico (*E. foliosum*). The two species are similar morphologically, but *E. foliosum* (n = 16) consistently has leaves folded along the midrib and densely clustered (fascicled) on the upper stem, more crowded inflorescences, smaller petals (1.8–3 mm), and smaller (0.6–0.9 mm) papillate seeds compared to *E. minutum* (n = 13), which has flat leaves, non-clustered open inflorescences, generally larger petals (2–5 mm), and larger (0.9–1.2 mm) reticulate seeds. Despite repeated efforts to cross them, no experimental hybrids resulted (S. R. Seavey et al. 1977b), nor are there any known natural hybrids, although such hybrids might be difficult to detect. In the only molecular analysis that included both species, they consistently formed a well-supported (92% BS) monophyletic group (D. A. Baum et al. 1994).

Among the few annual species of *Epilobium*, these two species differ from those of sects. *Pachydium* (n = 9, 10, 19) and *Epilobiopsis* (n = 15) by having comose seeds, and from *E. brachycarpum* (sect. *Xerolobium*, n = 12) by their smaller stature and numerous leaf, flower, and seed characters.

3. **Epilobium foliosum** (Torrey & A. Gray) Suksdorf, Deutsche Bot. Monatsschr. 18: 87. 1900 • Leafy willowherb

Epilobium minutum Lindley var. *foliosum* Torrey & A. Gray, Fl. N. Amer. 1: 490. 1840; *E. foliosum* var. *glabrum* Suksdorf; *E. minutum* var. *biolettii* Greene

Herbs slender. **Stems** strict, terete, 5–45 cm, simple to freely branched, subglabrous to strigillose proximally, strigillose and often villous and/or glandular puberulent distally. **Leaves** mostly alternate, often fasciculate distally, petiole 1–12 mm, blade spatulate proximally to narrowly lanceolate or linear distally, usually folded along midrib, 0.5–3 × 0.1–0.7 cm, base long-attenuate, margins subentire or scarcely serrulate, 2–4 remote teeth per side, lateral veins inconspicuous, apex blunt proximally to acute distally, surfaces subglabrous or with scattered short hairs on margins; bracts much reduced, sometimes attached to pedicel. **Inflorescences** suberect panicles or racemes, congested distally, with few thin branches, densely strigillose, often sparsely mixed villous and glandular puberulent. **Flowers** erect, often cleistogamous; buds 1.1–1.9 × 0.8–1.1 mm, often apiculate; floral tube 0.4–0.8 × 0.5–0.8 mm, with ring of short hairs at mouth inside; sepals often reddish green, 1.3–2.5 × 0.5–0.7 mm, apex subacute; petals white,1.4–2.5(–3) × 1–1.8 mm, apical notch 0.5–0.7 mm; filaments white, those of longer stamens 0.9–1.3 mm, those of shorter ones 0.5–0.7 mm; anthers 0.3–0.5 × 0.2–0.4 mm, apiculate; ovary 4–6 mm, strigillose; style white or cream, 1.2–1.7 mm, stigma subcapitate to obscurely 4-lobed, 0.3–0.4 × 0.4–0.6 mm, surrounded by longer anthers. **Capsules** 12–20 mm, surfaces sparsely hairy; pedicel 2–5 mm. **Seeds** obovoid, with slight constriction 0.2–0.3 mm from micropylar end, 0.6–0.9 × 0.3–0.5 mm, short chalazal collar 0.2–0.3 mm wide, grayish brown, surface low-papillose in ± irregular rows; coma easily detached, dingy white, 2–2.5 mm. $2n$ = 32.

Flowering Apr–Aug. Dry, rocky slopes, roadsides, disturbed dry areas in mountains; 50–2300 m; B.C.; Ariz., Calif., Idaho, Oreg., Wash.; Mexico (Baja California).

Epilobium foliosum is an autogamous self-compatible species, frequently with cleistogamous flowers, and even when the flowers are somewhat larger and chasmogamous, they rarely have insect visitors. S. R. Seavey et al. (1977b) tentatively determined a specimen from Guadalupe Island, 280 km off the coast of Baja California, Mexico (*Palmer 4217* in 1875) as *Epilobium foliosum*, and an additional sheet of the same collection (GH, as *Palmer 31*) had mature seeds (0.75–0.8 mm, low papillose) that verify that determination. No additional collections of this affinity since that by Palmer in 1875 have been found at this locality nor elsewhere in Mexico. Another disjunct occurrence of this species, at least 600 km east of California populations in Gila County, Arizona, is equally difficult to explain, since the collections, made between 1935 and the present, are from scattered localities in the region and do not seem obviously associated with introduction from human activity.

4. **Epilobium minutum** Lindley in W. J. Hooker, Fl. Bor.-Amer. 1: 207. 1832 • Chaparral willowherb [E]

Epilobium minutum var. *canescens* Suksdorf

Herbs slender. **Stems** strict, erect, sometimes reddish green, terete, 3.5–40 cm, simple or freely branched, subglabrous proximally to strigillose and glandular puberulent distally. **Leaves** alternate distally, not fasciculate, petiole 0–2 mm, blade subspatulate proximally to lanceolate, oblanceolate, or narrowly elliptical distally, not folded along midrib, 0.9–2.5 × 0.2–0.6 cm, shorter than internodes, base tapered, margins entire or scarcely denticulate, 1–4 teeth per side, lateral veins obscure, apex subacute or often blunt proximally, surfaces subglabrous or with scattered hairs along margins; bracts much reduced, sometimes attached to pedicel. **Inflorescences** erect racemes or open panicles, relatively loose and uncrowded, branches thin, mixed strigillose and glandular puberulent. **Flowers** erect or, sometimes, nodding in bud; buds broadly ovoid, 1.2–2.5 × 1–1.5 mm; floral tube 1.1–1.5 × 1–1.4 mm, usually with ring of spreading hairs at mouth inside; sepals 0.5–2.5 × 0.4–1.3 mm, apex acute, abaxial surface strigillose, sometimes mixed glandular puberulent; petals white to pink, 2–5 × 1.5–3 mm, apical notch 0.2–1.9 mm; filaments white, those of longer stamens 0.5–3 mm, those of shorter ones 0.3–2 mm; anthers 0.6–1 × 0.5–0.8 mm; ovary 4–9 mm, mixed strigillose and glandular puberulent; style light pink, 1–3.5 mm, stigma subclavate to obscurely 4-lobed, 0.4–0.6 × 0.4–0.5 mm, surrounded by longer anthers.

Capsules 9–28 mm, surfaces strigillose and glandular puberulent; pedicel 3–10 mm. **Seeds** obovoid, without constriction, 0.9–1.2 × 0.4–0.6 mm, low chalazal collar 0.1–0.2 mm wide, brown, surface reticulate; coma easily detached, white, 2.5–3 mm. **2*n*** = 26.

Flowering Apr–Sep. Open, dry places, along roads, disturbed areas; 90–1900 m; B.C.; Calif., Idaho, Mont., Nev., Oreg., Wash.

Epilobium minutum, like the similar *E. foliosum*, also occasionally produces cleistogamous flowers, and is modally autogamous in any event. S. R. Seavey et al. (1977b) observed that *E. minutum* is less common than *E. foliosum* in the southern part of their overlapping ranges and more common in the north. Several sheets (for example, *Lawler 3276*, California, Butte Co. [MO]; *Nelson & Gordon 5573*, California, Trinity Co. [MO]) mention that the plants were growing on serpentine soil. The earliest collection of this species appears to be one made by Archibald Menzies in 1792–1794 under the name *E. palustre* (BM).

Crossostigma lindleyi Spach (a substitute name for *Epilobium minutum*) and *E. lindleyi* (Spach) Rydberg are illegitimate names that pertain here.

5d. Epilobium Linnaeus sect. Cordylophorum (Nuttall ex Torrey & A. Gray) P. H. Raven, Ann. Missouri Bot. Gard. 63: 333. 1977 [E]

Epilobium [unranked] *Cordylophorum* Nuttall ex Torrey & A. Gray, Fl. N. Amer. 1: 488. 1840; *Cordylophorum* (Nuttall ex Torrey & A. Gray) Rydberg

Herbs perennial, suffrutescent or rhizomatous. **Stems** ± woody at base, epidermis usually peeling proximally. **Leaves** opposite proximal to inflorescence, alternate distally. **Flowers** actinomorphic or slightly zygomorphic; floral tube sometimes with bulbous base, without scales inside; petals rose-purple or cream-yellow; pollen in tetrads; stigma 4-lobed. **Capsules** fusiform to subcylindrical, splitting to base, central column persistent, pedicellate to subsessile. **Seeds** 1–8+ per locule in 1 row, obovoid to broadly obovoid, prominently constricted near micropylar end, coma present.

Species 3 (3 in the flora): w United States.

Section *Cordylophorum* consists of three perennial self-compatible but outcrossing species, each with an unusual and restricted distribution in western North America. All three species, which P. H. Raven (1976) placed in two subsections, share a chromosome number of *n* = 15 (found also in sects. *Zauschneria* and *Epilobiopsis*) and several unusual pollen, seed, and leaf character states. However, as Raven noted, this section can be delimited only by a combination of characteristics, with no unequivocal synapomorphy. S. R. Seavey and Raven (1977c) reported hybrids between *E. suffruticosum* and *E. nevadense* (only 24% pollen-fertile), and between *E. nevadense* and *E. nivium* (99% fertile). Attempts to cross any of these species to taxa of sect. *Zauschneria* (*n* = 15), sect. *Xerolobium* (*n* = 12), and sect. *Epilobium* (*n* = 18) failed in all cases. D. A. Baum et al. (1994) found only weak support for the monophyly of this section. Pending additional data, the similarities among these species are emphasized and the current classification is here retained.

5d.1. Epilobium Linnaeus (sect. Cordylophorum) subsect. Petrolobium P. H. Raven, Ann. Missouri Bot. Gard. 63: 334. 1977 [E]

Herbs suffrutescent, clumped. **Leaves** opposite only in proximal pairs, alternate or sometimes with fascicles of small leaves distally, caducous apical mucro prominent. **Flowers** actinomorphic; petals rose-purple. **Seeds** 1 or 2 per locule, obovoid to broadly obovoid.

Species 2: w United States.

P. H. Raven (1976), P. A. Munz (1965), and others noted the presence of an apiculus of brown oil cells at the leaf apex and suggested that this feature characterized sects. *Cordylophorum* and *Xerolobium*; however, further investigation of this character indicated that, while this feature is perhaps most prominent in these sections, similar apiculi occur in at least some taxa of all sections of *Epilobium*. Compared with the apiculi in subsect. *Nuttalia*, those in subsect. *Petrolobium* are particularly prominent and are a useful diagnostic character.

The two species of subsect. *Petrolobium* also characteristically have only four to eight seeds per capsule, which is fewer than the number found in any other species of the genus.

5. **Epilobium nevadense** Munz, Bull. Torrey Bot. Club 56: 166. 1929 • Nevada willowherb [E]

Herbs with many shoots from thick, woody caudex. **Stems** erect or ascending, terete, 10–50 cm, branched at base and apically, densely strigillose throughout, sometimes mixed villous distally. **Leaves:** proximal pairs often early-deciduous, petiole 1–4 mm, blade lanceolate-elliptic to narrowly so, ± folded along midrib, 0.9–1.7 × 0.2–0.6 cm, shorter than internodes, base attenuate or narrowly cuneate, margins denticulate, 6–10 low teeth per side, lateral veins inconspicuous or absent, apex acute with deciduous, rigid mucronate gland, surfaces usually glabrescent with scattered hairs on abaxial midrib, rarely strigillose-villous throughout; bracts much reduced, sublinear, often attached to pedicel. **Inflorescences** erect, open racemes or panicles, strigillose, often mixed glandular puberulent. **Flowers** erect to ± nodding; buds rounded-obovoid, 5–6 × 3–4 mm; floral tube with slight constriction 2–3 mm distal to base, 2.7–3.2(–5) × 1.8–2.5(–3.1) mm, without ring or scales inside, glabrous; sepals erect or sometimes deflexed in late anthesis, green or reddish green, lanceolate, 2.6–4.2 × 0.9–1.3 mm, apex acute; petals deep rose-purple, obcordate, 5–7.2 × 3.2–4.1 mm, apical notch 2–3 mm; filaments cream or white, those of longer stamens 5–7.5 mm, those of shorter ones 3.5–5.5 mm; anthers cream, 1–1.8 × 0.5–0.8 mm, scarcely apiculate; ovary 2.5–3.8 mm, densely strigillose and/or glandular puberulent; style cream, 6–9.5 mm, glabrous, stigma 4-lobed, 0.8–1.2 × 1–1.5 mm, lobes reflexed or sometimes incompletely spread, then forming cup-like structure, exserted beyond longer anthers. **Capsules** erect, subfusiform, 8–12 mm, surfaces strigillose and/or glandular puberulent; pedicel 1–1.8 mm. **Seeds** obovoid, with constriction 0.6–1 mm from micropylar end, 2.1–2.9 × 1.2–1.5 mm, very inconspicuous chalazal collar 0.05–0.06 mm wide, dark brown, surface low papillose, papillae often with central pit; coma easily detached, white, 6–7.5 mm. $2n = 30$.

Flowering Jul–Sep. Loose scree slopes, limestone talus, sandy soils at base of steep rock faces in pinyon pine-juniper-mountain brush communities; 1800–2800 m; Ariz., Nev., Utah.

In his description of *Epilobium nevadense*, Munz clearly recognized its affinity to *E. nivium* and suggested a close relationship between these two species and *E. brachycarpum*, based on similarities in seed and floral morphology. S. R. Seavey and P. H. Raven (1977c) demonstrated the close affinity between *E. nivium* and *E. nevadense* by forming fully fertile (99%) hybrids. However, compared to *E. nivium*, *E. nevadense* has denticulate, subglabrous leaves (versus subentire, densely pubescent leaves) and shorter floral tube [2.7–3.2(–5) mm] versus longer (5.2–9.5 mm) in *E. nivium*; furthermore, the two have completely non-overlapping geographical ranges. In overall morphology and cytology, these two species (and the somewhat more distantly related *E. suffruticosum*) are quite distinct from the rest of the genus.

Originally known only from the Charleston Mountains in southern Nevada, *Epilobium nevadense* has since been collected in northern Arizona, Eureka and Lincoln counties in Nevada, and in three counties of southwestern Utah. It may be more widespread in this region, much of which (especially in southern Nevada) consists of military reserves that are inaccessible to collectors. Although it was at one time considered endangered (S. D. Ripley 1975) due to the relatively low number of collections and threats from increased recreational use in its area of occurrence, it is no longer considered a candidate for listing (http://endangered.fws.gov). Several collections of this species show evidence of seed predation, apparently by moth larvae (H. N. Mozingo and Margaret Williams 1980), and S. R. Seavey and P. H. Raven (1977c) reported that larvae found in capsules from the locality in the Charleston Mountains were identified as *Mompha* (Momphidae, Gelechioidea).

6. Epilobium nivium Brandegee, Zoë 3: 242, plate 24. 1892 • Snow Mountain willowherb [E]

Herbs with many shoots from thick, woody caudex 5–12 mm diam.. **Stems** erect or ascending, terete, 10–25 cm, sparsely branched distally, densely grayish white-strigillose. **Leaves** subsessile or petioles 0.5–2.5 mm, blade elliptic or narrowly so to lanceolate, often folded along midrib, 0.9–1.8 × 0.3–0.7 cm, usually longer than internodes, base rounded to cuneate, margins subentire or low denticulate, 1–3 low teeth per side, lateral veins inconspicuous, 1–3 per side, apex blunt to acute with conspicuous dark brown mucronate tip, surfaces densely spreading-hairy; bracts very reduced, attached to pedicel 1–2 mm from base. **Inflorescences** erect open racemes or panicles, densely spreading-hairy. **Flowers** erect; buds 6–8 × 2.5–3.2 mm; floral tube 5.2–9.5 × 2.4–3.2 mm, constriction 4–6 mm distal to base, base ± bulbous, spreading-hairy from mouth nearly to base inside; sepals 2.7–4.2 × 1.6–2 mm, abaxial surface densely villous and glandular puberulent; petals rose-purple, 6–9.5 × 3.8–6.2 mm, apical notch 1.5–2.5 mm; filaments light pink, those of longer stamens 5–6.5 mm, those of shorter ones 3–3.5 mm; anthers 1.3–2.1 × 0.6–0.9 mm, apiculate; ovary 2.5–4.5 mm, densely villous and glandular puberulent; style pinkish cream, 11–17 mm, glabrous, stigma 4-lobed, 0.8–1 × 1.1–2.1 mm, lobes often not spread and then cuplike, usually exserted beyond anthers. **Capsules** fusiform, 8–16 mm, surfaces glandular puberulent; pedicel 2–5 mm. **Seeds** obovoid to broadly so, with slight constriction 0.4–0.6 mm from micropylar end, 1.5–2.4 × 0.8–1.3 mm, inconspicuous chalazal collar, dark brown, surface papillose; coma easily detached, dingy white, 6.5–7.5 mm. $2n = 30$.

Flowering late Jul–Sep. Crevices in rocky outcrops, shale or talus slopes, with scrub oak (*Quercus*), *Abies concolor*, and *Pinus jeffreyi*; 1600–2400 m; Calif.

Epilobium nivium has an extremely restricted range, mainly in the Snow Mountain region of Colusa and Lake counties, but recent collections from Mendocino and southern Trinity counties have extended its range several hundred km to the north. Many collections, notably including the type gathering, have strikingly woody bases, suggesting that these are long-lived plants. Like *E. nevadense* and some other species in the genus that characteristically grow on scree slopes, the lower part of the stems often lack leaves, which may be abraded by movement of the rocky substrate.

As reported by S. R. Seavey and P. H. Raven (1977c) and also noted on some herbarium labels, capsules of *Epilobium nivium* sometimes show signs of possible seed predation by moth larvae as reported for *E. nevadense*.

5d.2. Epilobium Linnaeus (sect. Cordylophorum) subsect. Nuttalia P. H. Raven, Ann. Missouri Bot. Gard. 63: 334. 1977 [E]

Herbs rhizomatous. **Leaves** opposite proximal to inflorescence, alternate distally, not fasciculate, caducous apical mucro inconspicuous or absent. **Flowers** slightly zygomorphic; petals cream-yellow. **Seeds** 4–8+ per locule, narrowly obovoid.

Species 1: w United States.

The single species of subsect. *Nuttalia*, *Epilobium suffruticosum*, shares several distinctive seed and pollen character states with *E. nevadense* and *E. nivium* (subsect. *Petrolobium*) and *E. brachycarpum* (sect. *Xerolobium*), as well as having the bract attached to the pedicel of each flower, not the stem (P. H. Raven 1976). In addition, *E. suffruticosum* can hybridize with *E. nevadense* and *E. nivium*, although the hybrids have sharply reduced fertility (S. R. Seavey and Raven 1977c). As noted by Raven in separating the subsections, *E. suffruticosum* has a very divergent range from the other two, and differs in having creamy, not rose-purple, petals, zygomorphic flowers, and many-seeded capsules.

7. **Epilobium suffruticosum** Nuttall in J. Torrey and A. Gray, Fl. N. Amer. 1: 488. 1840 • Shrubby willowherb [E]

Cordylophorum suffruticosum (Nuttall) Rydberg

Herbs with short, fleshy shoots from woody caudex, often extending 20+ cm underground; proximal epidermis peeling. **Stems** several–many, ascending to erect, terete, 10–25 cm, simple or well-branched, ± densely strigillose. **Leaves** often crowded, opposite and sometimes with fascicles of very small leaves at proximal nodes, subsessile or attenuate to broad petiole 0.5–1.5 mm, blade light grayish green, narrowly lanceolate to elliptic, 1–2.5 × 0.2–0.7 cm, often exceeding internodes, base cuneate to attenuate, margins entire or ± denticulate, 4–6 low teeth per side, lateral veins inconspicuous, apex blunt proximally to subacute, surfaces ± densely short-strigillose; bracts not much reduced in size. **Inflorescences** erect racemes or panicles, ± densely strigillose. **Flowers** slightly nodding; buds 4–8 × 1.5–3.5 mm, apiculate; floral tube funnelform to obconic, 1.8–3 × 1.9–2.6 mm, ring of spreading hairs 1–2.5 mm from base inside; sepals 3–6.5 × 1–2.6 mm, often apiculate, abaxial surface densely strigillose; petals cream to light yellow, obcordate, 5–9.3 × 2–3.8 mm, slightly unequal with upper 2 longer, apical notch 1–2.3 mm; filaments cream, slightly inflated at base, those of longer stamens 6–10 mm, those of shorter ones 4.5–8 mm; anthers cream-yellow, 1.4–2.2 × 0.6–1.1 mm; ovary 4–9 mm, densely white-canescent; style declined below main plane of flower, cream, 7.8–14.5 mm, glabrous, stigma deeply 4-lobed, 0.8–1.2 × 1.8–2.8 mm, lobes spreading-recurved 0.9–1.2 mm, exserted beyond longer anthers, often prematurely exserted and protogynous. **Capsules** often curved, fusiform-clavate, 10–30 mm, surfaces finely strigillose; pedicel 4.5–13 mm. **Seeds** narrowly obovoid to oblanceoloid, with constriction 0.7–1.3 mm from micropylar end, 2.1–3 × 0.7–1.1 mm, very inconspicuous chalazal collar, light brown, surface low-papillose; coma easily detached, tawny, 7–9.5 mm, with unusually dense hairs. $2n = 30$.

Flowering (Jun–)Jul–Aug. Gravel bars along rivers and streams, moist stabilized talus, moraines, other rocky places; 700–3000 m; Idaho, Mont., Utah, Wyo.

Epilobium suffruticosum shares its unusual cream-yellow flower color only with *E. luteum*, a distantly related species in sect. *Epilobium*. Both species have relatively large flowers with 4-lobed stigmas and are visited quite intensively by bees and other insect pollinators. Nevertheless, these species differ dramatically in habit, leaves, seeds, and many other characters, do not overlap at all in distribution, and are never confused with one another; the similar floral features must have been derived independently.

The flowers of *Epilobium suffruticosum* are also slightly zygomorphic, which is relatively rare in the genus. In the field and on many herbarium specimens of *E. suffruticosum*, the stigmas are clearly exserted even before the flowers are fully open. The label for *Raven 26451* (Wyoming, Park County, MO) notes: "protogynous; in late bloom, most flowers male-sterile." Several flowers from this collection have undeveloped anthers, suggesting that the flowers are functionally pistillate. However, these plants are not sterile since they have apparently fertile capsules with fully developed seeds.

The distribution of *Epilobium suffruticosum* consists of two clusters of fairly common occurrence—in northwestern Wyoming around Yellowstone and Teton national parks, and in south-central Idaho mainly in the drainages of the Boise and Payette rivers—with more scattered collections in western Montana north to Flathead County, and a single collection to the south in Weber County, northern Utah. There are no obvious morphological discontinuities among these specimens, nor any obvious explanation for the gaps in distribution; it may be due to collecting bias. This species is commonly found on gravel/sand bars of cold montane streams and rivers, in a stable association despite the apparent ephemeral nature of these habitats. It would appear that the plants have deep, woody roots by which they anchor themselves; in the spring flood stages of these rivers, they must experience complete inundation and considerable scouring, yet persist, often in moderately large colonies.

The exact locality of the type collection (streams east of Wallawallah, plains of the Upper Columbia River, Oregon) is problematic, since the closest known localities are at least 250 km southeast of the town of Walla Walla, Washington. Whether this is a matter of the historical accuracy of the locality by Nuttall or of the local extinction of this species from a locality in eastern Oregon cannot be determined at present. A collection by Hayden in 1859 (Powder River, Wyoming) is far outside the range of *E. suffruticosum* and may have been mislabeled.

5e. Epilobium Linnaeus sect. Epilobiopsis (Spegazzini) Lievens, Hoch & P. H. Raven, Syst. Bot. Monogr. 83: 89. 2007

Oenothera Linnaeus sect. *Epilobiopsis* Spegazzini, Anales Soc. Ci. Argent. 48: 46. 1899; *Boisduvalia* Spach sect. *Currania* Munz; *Epilobium* sect. *Currania* (Munz) Hoch & P. H. Raven

Herbs annual, with taproot. **Stems** often with decumbent proximal branches, rarely simple, epidermis peeling proximally. **Leaves** opposite and early deciduous in proximal pairs, alternate distally, often crowded and exceeding internodes, caducous apical mucro often prominent. **Flowers** actinomorphic, usually cleistogamous; floral tube not bulbous, without scales inside; petals pink or white; pollen in tetrads; stigma clavate, subentire to ± shallowly 4-lobed. **Capsules** fusiform to subclavate, sharply 4-angled or terete, walls tough, usually splitting only in distal ⅓, central column disintegrating, sessile. **Seeds** 7–14 per locule in 2 rows, irregularly angular-fusiform; coma absent.

Species 2 (2 in the flora): w North America, nw Mexico, s South America.

Section *Epilobiopsis*, treated formerly as *Boisduvalia* sect. *Currania* (P. A. Munz 1941, 1965; P. H. Raven 1976), consists of two self-compatible, sometimes cleistogamous, annual species with $n = 15$ that occur primarily in western North America; *Epilobium campestre* also occurs disjunctly in southern Argentina. The species in this section are distinctive by lacking a seed coma and by being spring-blooming annuals (both features shared with sect. *Pachydium*), and uniquely by having tough, only partially dehiscent, capsules with seeds in two rows per locule (Raven and D. M. Moore 1965). Although the species of sect. *Epilobiopsis* share the gametic chromosome number $n = 15$ with those of sects. *Cordylophorum* and *Zauschneria*, all attempts have failed to form inter-sectional hybrids with those groups, as well as with species of sect. *Pachydium* ($n = 9$, 10, 19) with which they were formerly grouped (Munz 1965; S. R. Seavey 1992). Unexpectedly, molecular analyses (D. A. Baum et al. 1994; R. A. Levin et al. 2004) placed sect. *Epilobiopsis* as sister to a clade of sects. *Pachydium* and *Zauschneria* with strong support. Baum et al. also found weak support (81%) for the monophyly of sect. *Epilobiopsis*.

8. Epilobium cleistogamum (Curran) Hoch & P. H. Raven, Phytologia 73: 458. 1993 • Selfing willowherb [E] [F]

Boisduvalia cleistogama Curran, Bull. Calif. Acad. Sci. 1: 12. 1884; *Oenothera cleistogama* (Curran) H. Léveillé

Herbs from slender taproot. **Stems** terete, 1.5–32 cm, simple or often with sprawling, stout, prostrate proximal branches, proximally glabrous, often distally spreading-hairy and ± glandular puberulent. **Leaves** subsessile, blade grayish green, linear to narrowly elliptic, proximally broader and surfaces subglabrous, distally narrower and surfaces densely villous, especially on margins and midrib, often folded along midrib, usually early-withering, 1.5–5.5 × 0.2–0.6 cm, base cuneate, margins serrulate, 5–18 low teeth per side, lateral veins obscure, 1–4 per side, apex acute; bracts scarcely reduced. **Inflorescences** erect spikes, leafy, densely villous and glandular puberulent, first flowers at most proximal nodes. **Flowers** ± cleistogamous, suberect, often hidden by subtending bracts; buds 2–4 × 1–1.5 mm, apiculate; floral tube 0.5–1 × 0.4–1 mm, raised ring of lax hairs near mouth inside; sepals pale green or reddish green, not keeled, 1.5–3 × 0.6–1.2 mm, apex acute, abaxial surface villous and glandular puberulent; petals white to pale pink, 2–5.8 × 0.8–1.8 mm, apical notch 0.5–1.5 mm; filaments light pink, those of longer stamens 0.6–1.6 mm, those of shorter ones 0.5–0.8 mm; anthers light yellow, 0.4–0.5 × 0.3–0.5 mm; ovary 8–11 mm, densely villous and glandular puberulent; style light pink, 1.4–2.4 mm, stigma capitate, ± 4-lobed to subentire, 0.5–0.9 × 0.4–0.8 mm, surrounded by longer anthers. **Capsules** narrowly cylindrical, often curved-ascending, sharply 4-angled with 4 strong ribs, 8–12 mm, beak 1.5–3 mm, tardily dehiscent on distal ⅓, central axis disintegrating, sparsely villous and glandular puberulent; sessile. **Seeds** 10–14 per tightly packed row, irregularly angular to fusiform, 1.2–1.5 × 0.4–0.6 mm, chalazal collar absent, surface irregularly reticulate. $2n = 30$.

Flowering May–Jul. Primarily around vernal pools, clay flats, other seasonally moist habitats, usually in heavy clay soil; 20–300(–1600) m; Calif.

Epilobium cleistogamum is an annual species endemic to heavy clay soil in the Central Valley of California and surrounding foothills, from southern Tehama County to northern Tulare County and into the Sacramento River delta in Contra Costa and Solano counties, and barely to San Luis Obispo County in the southern Coast Range. Flowering often commences at the first or second proximal node, and flowers are frequently cleistogamous. The seeds are arranged nearly horizontally and are irregularly angular due to tight packing in the rigid capsules. Plants characteristically have decumbent branches and tardily dehiscent capsules that shed their seeds only following rains, often many months after fruits matured and plants were green (P. H. Raven 1969).

9. **Epilobium campestre** (Jepson) Hoch & W. L. Wagner, Syst. Bot. Monogr. 83: 208. 2007 • Smooth spike-primrose

Boisduvalia campestris Jepson, Fl. W. Calif., 330. 1901; *B. glabella* (Nuttall) Walpers; *B. glabella* var. *campestris* (Jepson) Jepson; *B. pygmaea* Munz; *Epilobium pygmaeum* (Munz) Hoch & P. H. Raven; *Oenothera glabella* Nuttall 1840, not *E. glabellum* G. Forster 1786

Herbs with 1 or more unbranched taproots. **Stems** usually suberect, rarely matted, terete, 1.5–50 cm, often with sprawling, decumbent proximal branches, rarely simple, glabrous proximally or throughout, sometimes ± densely strigillose and/or villous distally. **Leaves** crowded, subsessile, blade lanceolate to narrowly lanceolate or oblong, 0.8–3.5 × 0.2–0.6(–1) cm, longer than subtending internodes, base cuneate, margins evenly serrulate, 4–7 teeth per side, lateral veins obscure, 2–5 per side, apex acute, surfaces strigillose and ± villous, at least along veins and margins; bracts not much reduced. **Inflorescences** erect spikes, congested, unbranched, densely strigillose and ± villous or subglabrous. **Flowers** erect, often hidden by subtending bracts, often cleistogamous; buds 1.2–2 × 0.7–1.1 mm; floral tube 0.3–1.1 × 0.2–0.8 mm, raised ciliate ring proximal to mouth inside; sepals reddish green, 0.7–1.9 × 0.6–1.2 mm; petals pale pink, fading purplish rose, 0.9–3.5 × 0.7–0.9 mm, apical notch 0.3–1.3 mm; filaments light pink, those of longer stamens 0.5–1.5 mm, those of shorter ones 0.4–0.9 mm; anthers pale yellow, 0.4–0.8 × 0.3–0.5 mm; ovary 3–5 mm, usually densely villous; style pale pink, 0.6–1.8 mm, stigma clavate, irregularly 4-lobed to subentire, 0.5–1 × 0.2–0.6 mm, surrounded by longer anthers. **Capsules** cylindrical to subfusiform, ± terete, 4.5–8 mm, beak 0.8–1 mm, usually dehiscing on distal ⅓, sometimes tardily splitting to base, central axis prematurely disintegrating, villous; subsessile. **Seeds** 7–14 per tightly packed row, irregularly angular-fusiform, 1–1.3 × 0.4–0.6 mm, chalazal collar absent, surface irregularly reticulate. $2n = 30$.

Flowering May–Sep. Vernally moist flats, depressions, shores, and open fields, usually clay soils; 30–3000 m; Alta., B.C., Sask.; Ariz., Calif., Idaho, Mont., Nev., N.Mex., N.Dak., Oreg., S.Dak., Utah, Wash., Wyo.; Mexico (Baja California); South America (Argentina).

Epilobium campestre is widespread in temperate western North America. Like *E. cleistogamum*, it also grows in habitats that are only moist early in the growing season, or otherwise ephemeral moist places, like shores of reservoirs with fluctuating water levels (P. H. Raven and D. M. Moore 1965), and consequently flowers earlier than most species of *Epilobium*.

The occurrence of this species in Chubut Province, Argentina, appears to be the result of natural long-distance dispersal, probably by birds.

Seeds of *Epilobium campestre* are inclined about 20° from vertical, which while unique and characteristic is a difficult character to observe. Seeds are tightly packed in rigid capsules, as described under *E. cleistogamum*.

Oenothera pygmaea Spegazzini 1899, an illegitimate name (not Douglas 1832), pertains here.

5f. Epilobium Linnaeus sect. Pachydium (Fischer & C. A. Meyer) Hoch & K. Gandhi, PhytoKeys, 145: 60. 2020

Oenothera sect. *Pachydium* Fischer & C. A. Meyer, Index Seminum (St. Petersburg) 2: 45. 1836; *Boisduvalia* Spach, Hist. Nat. Vég. 4: 383. 1835; *Boisduvalia* [unranked] *Dictyopetalum* (Fischer & C. A. Meyer) Endlicher; *Boisduvalia* [unranked] *Pachydium* (Fischer & C. A. Meyer) Endlicher; *Boisduvalia* subg. *Pachydium* (Fischer & C. A. Meyer) Reichenbach; *Cratericarpium* Spach; *Epilobium* sect. *Boisduvalia* (Spach) Hoch & P. H. Raven; *Dictyopetalum* (Fischer & C. A. Meyer) Baillon; *Oenothera* [unranked] *Boisduvalia* (Spach) Torrey & A. Gray; *Oenothera* sect. *Dictyopetalum* Fischer & C. A. Meyer

Herbs annual, with taproot. **Stems:** epidermis peeling proximally. **Leaves** opposite and early deciduous in proximal pairs, alternate distally. **Flowers** actinomorphic, chasmogamous or rarely cleistogamous; floral tube not bulbous, without scales inside; petals rose-purple to white; pollen in tetrads; stigma clavate, entire or sometimes obscurely 4-lobed. **Capsules** fusiform-clavate to lanceolate-linear, sometimes torulose, splitting to base, central column persistent or not, sessile. **Seeds** 2–8 per locule in 1 row, or 8–12 per capsule, pushed into a single row by disintegration of central column and septa, irregularly angular-fusiform; coma absent.

Species 4 (3 in the flora): w North America, nw Mexico, w South America.

Section *Pachydium* consists of four self-compatible, often cleistogamous, annual species, two of which are endemic to and widespread in western North America, and a third that also reaches northern Mexico. *Epilobium subdentatum* (Meyen) Lievens & Hoch, occurs only in Chile (widespread) and Argentina (only in Chubut Province), and has larger, often outcrossing flowers pollinated by bees (P. H. Raven and D. M. Moore 1965). The species in sect. *Pachydium* are diverse cytologically: *E. torreyi* has the lowest chromosome number in Epilobieae ($n = 9$), *E. densiflorum* and *E. pallidum* have $n = 10$, and *E. subdentatum* has $n = 19$. Contrary to suggestions by G. L. Stebbins (1971) and Raven (1976), the numbers $n = 9$ and 10 apparently were derived as direct or stepwise aneuploid reductions from the paleotetraploid $x = 18$ base number for Epilobieae (D. A. Baum et al. 1994). Experimental hybrids were formed in all combinations among the four species, although most had notably aberrant chromosome pairing and very low pollen fertility (S. R. Seavey 1992). The so-called tetraploid *E. subdentatum* ($n = 19$) apparently arose following hybridization between *E. torreyi* ($n = 9$) and *E. densiflorum* or a close relative with $n = 10$ (Seavey), followed by long-distance dispersal to South America (Raven and Moore). The monophyly of the section is scarcely supported by analysis of ITS sequence (Baum et al.), and the exact relationship of these species is not well-resolved.

10. Epilobium densiflorum (Lindley) Hoch & P. H. Raven, Phytologia 73: 457. 1993 • Denseflower willowherb

Oenothera densiflora Lindley, Edwards's Bot. Reg. 19: plate 1593. 1833; *Boisduvalia bipartita* Greene; *B. densiflora* (Lindley) Bartling; *B. densiflora* var. *bipartita* (Greene) Jepson; *B. densiflora* var. *imbricata* Greene; *B. densiflora* var. *montana* Jepson; *B. densiflora* var. *pallescens* Suksdorf; *B. densiflora* var. *salicina* (Rydberg) Munz; *B. imbricata* (Greene) A. Heller; *B. salicina* Rydberg; *B. sparsiflora* A. Heller; *B. sparsifolia* A. Nelson & P. B. Kennedy; *O. densiflora* var. *imbricata* (Greene) H. Léveillé

Herbs usually with taproot, sometimes with loose network of roots. **Stems** erect or ascending, terete, 4–150 cm, simple or branched with strong central axis, proximal branches ascending or suberect, villous or strigillose, often mixed glandular puberulent distally. **Leaves** opposite and often early-deciduous proximally, alternate and crowded distally, usually subsessile, rarely petiole 1–2 mm, blade usually narrowly lanceolate to sublinear, rarely to lanceolate, 1.4–7.5(–9.2) × 0.5–1.4 cm, base cuneate to attenuate, margins remotely to sharply serrulate, 5–12 teeth per side, lateral veins obscure, 2–5 per side, apex acute, surfaces densely villous and/or strigillose; bracts broader than cauline leaves, broadly lanceolate to ovate or subrotund, 0.5–2.5 × 0.3–1.8 cm, long-acuminate, sometimes folded on midrib. **Inflorescences** erect spikes, congested, simple, densely villous and strigillose, sometimes mixed glandular puberulent. **Flowers** erect, often hidden within subtending bracts, usually chasmogamous; buds sessile, narrowly elongate, 2–4 mm; floral tube 1.3–3.8 × 1–2.2 mm, ring of hairs 0.6–2 mm distal to base inside; sepals 2–7.5 × 0.5–2.2 mm, apex acute; petals rose-purple, magenta, pink, or white, 3–9.5(–11.5) × 1.2–5(–6.2) mm, apical notch 0.8–3.8 mm; filaments dark pink, those of longer stamens 1.5–4.5 mm, those of shorter ones 0.5–1.9 mm; anthers yellow, 0.5–1.2 × 0.3–0.7 mm; ovary 2–5 mm, densely villous, often mixed glandular puberulent; style white, 2.2–5.5 (–7.5) mm, glabrous, stigma subcapitate to irregularly 4-lobed, 0.3–0.8 × 0.3–1 mm, surrounded by longer anthers. **Capsules** cylindrical to subfusiform, 4–11 mm, beak to 0.5 mm, central column persistent, surfaces densely villous; subsessile or pedicel 1–2.5 mm. **Seeds** 3–8 per locule, irregularly angular-fusiform, 1.2–1.6(–1.9) × 0.4–1 mm, without a chalazal collar, light brown, surface irregularly reticulate with raised cells. $2n = 20$.

Flowering May–Oct. Vernally wet places, moist pastures, woodlands, meadows, along streams and ditches, alluvial valleys, often on low ground in volcanic or sandy soils; 0–2600 m; B.C.; Ariz., Calif., Idaho, Mont., Nev., Oreg., Utah, Wash.; Mexico (Baja California).

Epilobium densiflorum is an extremely variable species that changes its aspect through the flowering season. Collections made very early in the season may include only well-spaced, narrowly lanceolate leaves, the proximal ones usually opposite, and a short, sparse, somewhat open inflorescence. A late-season collection, even from the same population, may entirely lack cauline leaves, and consist instead of bare, peeling stems topped by dense, tightly imbricate-bracted inflorescences, with each broad bract enclosing a capsule or flower.

Boisduvalia douglasii Spach is an illegitimate substitute for *Oenothera densiflora* Lindley and pertains here.

11. Epilobium pallidum (Eastwood) Hoch & P. H. Raven, Phytologia 73: 458. 1993 • Largeflower spike-primrose [E] [F]

Boisduvalia pallida Eastwood, Leafl. W. Bot. 2: 54. 1937; *B. macrantha* A. Heller 1905, not *Epilobium macranthum* Hooker 1840

Herbs with taproot. **Stems** terete, 4.5–55 cm, simple or with proximal ascending branches and/or distal branches, ± densely villous and/or strigillose, mixed glandular puberulent distally or glabrescent. **Leaves** opposite only in proximal pairs, alternate and crowded distally, subsessile, blade narrowly elliptic or lanceolate, 1.2–5 × 0.2–1.2 cm, usually longer than internodes, base cuneate to attenuate, margins subentire or sparsely serrulate, 3–5 teeth per side, veins inconspicuous, 2–7 per side, apex acute, surfaces villous; bracts same size and shape as leaves. **Inflorescences** erect spikes, often congested proximally, sometimes nodding in bud, simple, or sometimes sparsely branched, ± densely mixed villous, strigillose, and glandular puberulent. **Flowers** erect, mostly chasmogamous, sometimes initiating at second or third proximal node; buds 3–5 × 2.5–3.5 mm, apex blunt; floral tube 1.5–3 × 1.1–2.8 mm, ring of lax hairs near base inside; sepals erect, (1.5–)2–6.2 × (0.5–)1–2.2 mm, densely villous and glandular pubescent abaxially; petals rose-purple, (2.8–)4.2–9.8 × (2.2–)3.5–5.3 mm, apical notch 1.2–3.5 mm; filaments rose-purple, those of longer stamens 2.4–6.5 mm, those of shorter ones 0.9–4.3 mm; anthers 0.6–1.6 × 0.2–0.8 mm; ovary 3.5–9 mm, densely pubescent; style pink, 3.3–9.2 mm, stigma 4-lobed, ± subentire in smaller flowers, 0.7–1.4 × 0.5–1.3 mm, usually surrounded by, rarely exserted beyond, longer anthers. **Capsules** narrowly fusiform, (10–)14–21 × 1–2.1 mm, beak 2–5 mm, ribs usually broad and prominent, septa and central column disintegrating with age, surfaces ± densely villous; subsessile. **Seeds** 8–12 per capsule, initially in 4 locules, at maturity pushed into 2 rows proximally, 1 overlapping row distally, irregularly angular-fusiform, 1.8–2.3 × 0.7–0.9 mm, chalazal collar absent, surface irregularly reticulate. **2*n*** = 20.

Flowering May–Aug(–Sep). Stream banks and washes in vernally moist areas in shrubland or lower forested regions; 60–2100 m; Calif., Idaho, Oreg.

P. H. Raven and D. M. Moore (1965) observed that smaller-flowered individuals of this species were sometimes difficult to separate from *Epilobium torreyi* (reported as *Boisduvalia stricta*), especially in the absence of mature fruits. The capsules of *E. pallidum* are usually longer and thinner than those of related species and the internal structure is highly distinctive. The stigmas of larger-flowered plants often are exserted beyond the anthers, increasing the chances for outcrossing.

12. Epilobium torreyi (S. Watson) Hoch & P. H. Raven, Phytologia 73: 458. 1993 • Torrey's willowherb [E]

Oenothera torreyi S. Watson, Proc. Amer. Acad. Arts 8: 600. 1873; *Boisduvalia parviflora* A. Heller; *B. stricta* (A. Gray) Greene; *B. torreyi* (S. Watson) S. Watson; *Gayophytum strictum* A. Gray 1868, not *Epilobium strictum* Muhlenberg ex Sprengel 1825; *Oenothera densiflora* Lindley var. *tenella* A. Gray

Herbs with taproot. **Stems** erect, terete, (1–)4–65 cm, usually with several erect or ascending virgate branches proximally or simple, densely villous, sometimes mixed strigillose, or sometimes subglabrous. **Leaves** opposite only in proximal pairs, distally alternate, subsessile, blade linear-lanceolate to very narrowly elliptic, 0.5–4.5 × 0.2–0.3(–0.6) cm, usually longer than internodes, base cuneate, margins subentire to sparsely serrulate, 2–5 low teeth per side, lateral veins obscure, 2–5 per side, apex acute, surfaces subglabrous proximally to densely villous and/or strigillose distally; bracts slightly reduced. **Inflorescences** erect spikes, simple or sparsely branched, ± densely villous and/or strigillose. **Flowers** erect, often initiating in most proximal nodes, usually cleistogamous; buds 0.8–1.5 × 0.5–1 mm; floral tube 0.4–1 × 0.6–1.2 mm, ring of lax hairs near base inside; sepals 0.7–2 × 0.4–0.8 mm; petals pink with darker veins or white, 1.2–3.2 × 0.9–1.8 mm, apical notch 0.4–1 mm; filaments pale pink, those of longer stamens 0.4–1.8

mm, those of shorter ones 0.3–1 mm; anthers 0.3–0.6 × 0.3–0.4 mm, apiculate; ovary 3–6 mm, densely pubescent; style pale pink, 1.2–2.8 mm, stigma clavate to subcapitate or irregularly 4-lobed, 0.5–1.2 × 0.3–0.6 mm, surrounded by longer anthers. **Capsules** cylindrical to subfusiform, terete to slightly 4-angled, (6–)8–14 × 1.1–2 mm, beak 2–3 mm, central column disintegrating, villous; sessile. **Seeds** 3–6 per locule, irregularly angular-oblong or fusiform, 0.9–1.6 × 0.4–0.9 mm, chalazal collar absent, brown, surface irregularly reticulate. **2*n*** = 18.

Flowering May–Aug. Moist places along stream banks, seasonal streambeds, seeps, and roadside ditches, often in gravelly red or granite soil; 0–2600 m; B.C.; Calif., Idaho, Nev., Oreg., Wash.

Epilobium torreyi is an annual species that has a unique gametic chromosome number of *n* = 9, the lowest found in the genus. Its morphological similarity to *Epilobium pallidum* (*n* = 10) suggested to P. H. Raven and D. M. Moore (1965) that *E. torreyi* may be an aneuploid derivative of *E. pallidum* or a close relative.

No natural hybrids among species of sect. *Pachydium* have been reported, but in fact such plants might be difficult to detect. S. R. Seavey (1992) was able to form hybrids, but they were highly sterile. Nevertheless, hybridization between species with *n* = 9 and *n* = 10, followed by polyploidization, may well have given rise to the South American species of this section, *Epilobium subdentatum* (*n* = 19).

5g. EPILOBIUM Linnaeus sect. ZAUSCHNERIA (C. Presl) P. H. Raven, Ann. Missouri Bot. Gard. 63: 335. 1977

Zauschneria C. Presl, Reliq. Haenk. 2: 28, plate 52. 1831

Herbs perennial, herbaceous or suffrutescent, with shoots from caudex. **Stems** ± woody at base, epidermis not peeling proximally. **Leaves** opposite proximal to inflorescence or throughout, alternate and/or fasciculate distally. **Flowers** slightly zygomorphic (upper petals ± flared at right angle to calyx tube, lower ones parallel with it), spread laterally to suberect; floral tube elongate, bulbous near base, with irregular scales inside at base of staminal filaments; floral tube, sepals, and petals usually orange-red, very rarely white; pollen in tetrads; stigma deeply 4-lobed. **Capsules** subclavate to fusiform, splitting to base, central column persistent, subsessile to short-pedicellate. **Seeds** numerous, in 1 row per locule, ± broadly obovoid, prominently constricted near micropylar end; coma present.

Species 2 (2 in the flora): w North America, n Mexico.

Section *Zauschneria*, with its relatively large, tubular, red-orange flowers commonly pollinated by hummingbirds, differs sharply in aspect from the rest of *Epilobium*, and until recently (P. H. Raven 1976) was treated as a separate but related genus (for example, P. A. Munz 1965). The section consists of two species (four taxa) endemic to western North America, one of which is diploid (*E. septentrionale*, *n* = 15) and the other polyploid (*E. canum*, *n* = 15, 30). The latter is the only species in *Epilobium*, other than the allopolyploid *E. subdentatum* (sect. *Pachydium*), with polyploidy above the paleotetraploid level of *x* = 18. Despite their distinctive flowers, these taxa fit well within *Epilobium*, possessing seed comas and numerous similarities in pollen and seed morphology, especially to taxa in sections *Cordylophorum* and *Xerolobium* (Raven). Molecular analyses (D. A. Baum et al. 1994; R. A. Levin et al. 2004) placed sect. *Zauschneria* (*x* = 15) in a well-supported clade with sects. *Epilobiopsis* (*n* = 15) and *Pachydium* (*n* = 9, 10, 19), which together were previously segregated as *Boisduvalia*. Within this clade, Levin et al. found strong support for sect. *Zauschneria* as sister to sect. *Pachydium*, a surprising relationship based on morphology and cytology.

Section *Zauschneria* occurs primarily in the California Floristic Province (P. H. Raven and D. I. Axelrod 1978), except for *Epilobium canum* subsp. *garrettii*, which is confined primarily to the Great Basin Province (R. F. Thorne 1993d). In addition, *E. canum* subsp. *latifolium* extends into the desert ranges of the Sonoran Province (Thorne).

Section *Zauschneria* is sufficiently variable especially in vestiture and leaf size and shape that it was once considered to include 20 species (G. L. Moxley 1920). J. D. Clausen et al. (1940) treated the group as three diploid species and one tetraploid species with three subspecies; P. A. Munz (1965) recognized the same six taxa. P. H. Raven (1976) further reduced these to six subspecies of one variable species, one of which (the diploid *Epilobium septentrionale*) subsequently reinstated at the species level (R. N. Bowman and Hoch 1979).

13. Epilobium canum (Greene) P. H. Raven, Ann. Missouri Bot. Gard. 63: 335. 1977 [F]

Zauschneria cana Greene, Pittonia 1: 28. 1887

Herbs suffruticose or not, with basal shoots from ± woody caudex, often decussate scales at base. **Stems** erect to ascending, often clumped but not matted, green or gray-green, terete, 10–110(–120) cm, usually well-branched throughout, sometimes simple, strigillose and/or long-villous, usually mixed glandular puberulent distally, rarely glabrate. **Leaves** ± densely spaced, alternate and often fasciculate distally, subsessile, blade grayish green or green to silvery-canescent, usually narrowly linear to lanceolate or elliptic to ovate, rarely orbiculate, 0.6–5(–6) × 0.1–2.5 cm, base cuneate to attenuate, margins subentire to sharply toothed, 3–15 teeth per side, veins inconspicuous or prominent, 3–7 per side, apex acute, sometimes with caducous dark mucro, surfaces usually ± densely strigillose, sometimes mixed villous and/or glandular puberulent, rarely glabrate; bracts much smaller and narrower. **Inflorescences** erect spikes or racemes, loose to congested, often branched, glandular puberulent and sometimes mixed strigillose or villous. **Flowers:** buds 11–18 × 4–6 mm, subsessile or pedicels 1–2 mm; floral tube same color as petals, 16–32 × 5–8 mm, base slightly bulbous, ring of 8 irregular scales at base of filaments 4–6.5 mm from base inside; sepals same color as petals, 7–15 × 3.5–5 mm, abaxial surface densely pubescent; petals usually orange-red, very rarely white, obcordate, 8–17 × 5–9.5 mm, apical notch 2–3 mm; filaments light orange-red to white, those of longer stamens 12.5–32 mm, those of shorter ones 10–25 mm; anthers 2.7–4 × 0.8–1.2 mm, apiculate; ovary 8–15 mm, glandular puberulent, often mixed villous; style light orange-red, 42–65 mm, glabrous, stigma 4-lobed, 1–1.4 × 2.4–3 mm, exserted 8–15 mm beyond anthers. **Capsules** straight or ± curved-ascending, 15–35 mm, sometimes beaked, surfaces glandular puberulent and strigillose; subsessile or pedicel 0–3 mm. **Seeds** broadly to narrowly obovoid, with constriction 0.6–0.8 mm from micropylar end, 1.5–2.6 × 0.9–1.3 mm, chalazal collar inconspicuous, light brown, surface low papillose; coma easily detached, dingy white, 5.5–7 mm.

Subspecies 3 (3 in the flora): sw United States, nw Mexico.

This treatment recognizes three self-compatible but highly outcrossing subspecies marked by distinct but sometimes intergrading morphology and overlapping geographical ranges. R. N. Bowman and P. C. Hoch (1979) agreed with the treatment of *Epilobium canum* subsp. *garrettii* (n = 15) and subsp. *latifolium* (n = 30) by P. H. Raven (1976), but considering the complex intergrading patterns of variation involving the rest of this species, they combined the two remaining tetraploid subspecies recognized by Raven (subspp. *angustifolium* and *mexicanum*) with the remaining diploid subspecies into a single polyploid subsp. *canum*.

1. Leaf blades linear to narrowly lanceolate or elliptic, rarely to narrowly ovate, 0.6–4.5 × 0.1–0.6(–0.8) cm, usually less than 0.6 cm wide, often fascicled distally, herbaceous to suffrutescent; stems 20–110(–120) cm . . . . 13a. *Epilobium canum* subsp. *canum*
1. Leaf blades lanceolate to broadly ovate, 0.8–5(–6) × 0.4–2.3 cm, usually more than 0.6 cm wide, not fascicled; herbaceous; stems 10–50(–70) cm.
    2. Leaf blades usually lanceolate to ovate or broadly elliptical, rarely orbiculate, not coriaceous, margins subentire to distinctly denticulate, lateral veins obscure to conspicuous . . . . 13b. *Epilobium canum* subsp. *latifolium*
    2. Leaf blades ovate to broadly elliptical, coriaceous, margins prominently denticulate, lateral veins conspicuous . . . . 13c. *Epilobium canum* subsp. *garrettii*

13a. Epilobium canum (Greene) P. H. Raven subsp. **canum** • Hummingbird trumpet [F]

Epilobium canum subsp. *angustifolium* (D. D. Keck) P. H. Raven; *E. canum* subsp. *mexicanum* (C. Presl) P. H. Raven; *Zauschneria californica* C. Presl 1831, not *Epilobium californicum* Haussknecht 1884; *Z. californica* subsp. *angustifolia* D. D. Keck; *Z. californica* subsp. *mexicana* (C. Presl) P. H. Raven; *Z. californica* var. *microphylla* A. Gray ex W. H. Brewer & S. Watson; *Z. californica* var. *villosa* (Greene) Jepson; *Z. eastwoodiae* Eastwood & Moxley; *Z. mexicana* C. Presl 1831, not

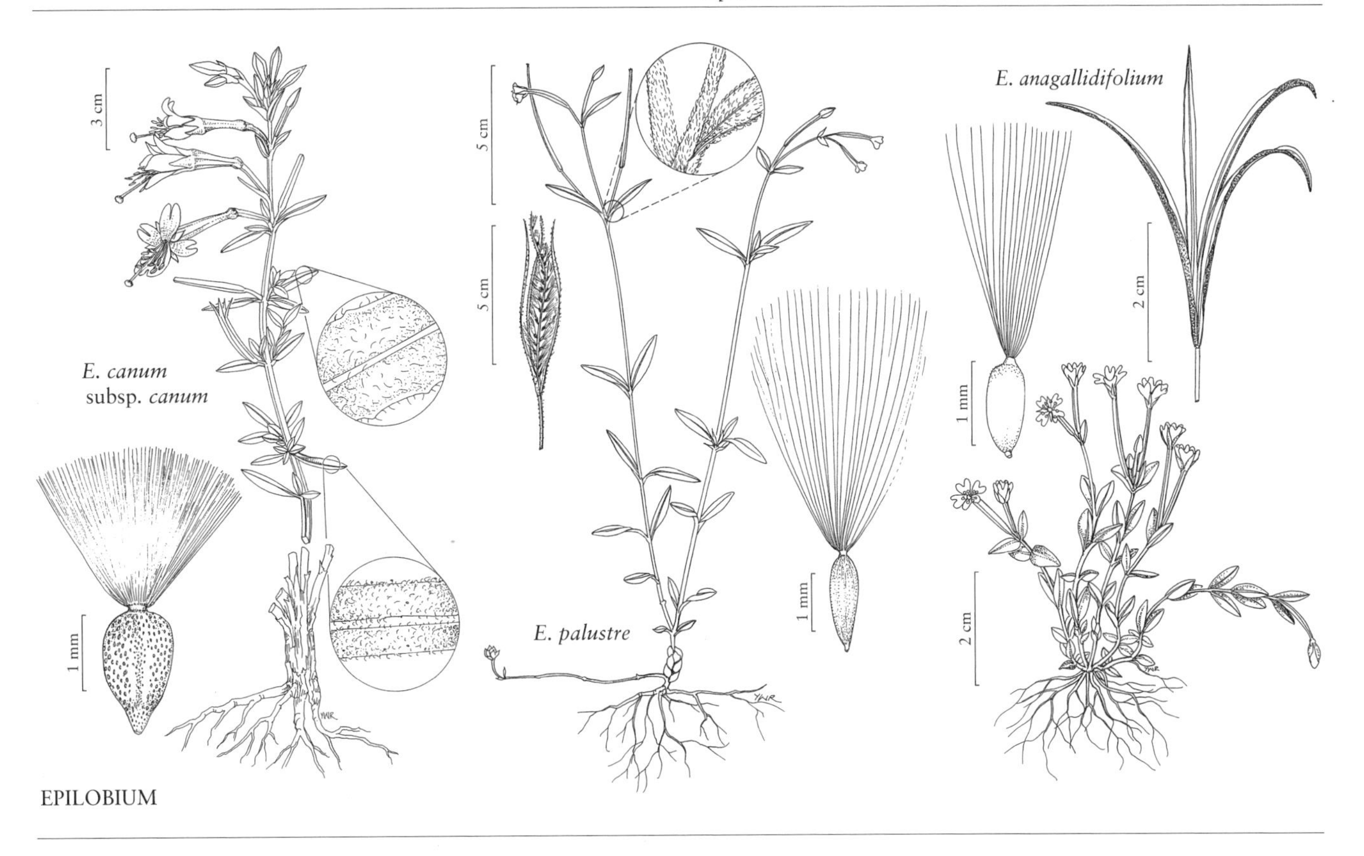

EPILOBIUM

E. mexicanum Seringe 1828; *Z. microphylla* (A. Gray ex W. H. Brewer & S. Watson) Moxley; *Z. velutina* Eastwood; *Z. villosa* Greene

Herbs suffruticose or not. **Stems** 20–110(–120) cm, ± densely strigillose, densely villous, or both, often mixed short glandular puberulent especially distally, sometimes predominantly short glandular puberulent, rarely glabrate. **Leaves** alternate and often fasciculate distally; blade usually green to grayish green, rarely silvery-canescent, usually narrowly linear to narrowly lanceolate or elliptic, rarely to narrowly ovate, 0.6–4.5 × 0.1–0.6(–0.8) cm, margins subentire to remotely serrulate, 4–10 teeth per side, veins usually inconspicuous, surfaces usually moderately to densely villous and/or strigillose, rarely glabrate. **Flowers**: floral tube 17–29 mm; sepals 7–10 mm; petals 8–15 mm. **Capsules** 18–30 mm; subsessile or pedicel 1–3 mm. **Seeds** 1.5–2.2 × 0.9–1.2 mm. **2*n*** = 30, 60.

Flowering (Apr–)Jun–Dec. Dry rocky or stony slopes and ridges, disturbed chaparral, along coastal bluffs, offshore islands, scattered in Sierra Nevada foothills; 0–1000(–2700) m; Calif.; Mexico (Baja California).

The extremely variable subsp. *canum* has the lowest elevational range, from sea level to 1000 m through most of its range, and to 2700 m in a few places in the southern Sierra Nevada. It occurs almost exclusively in the Californian Province as defined by R. F. Thorne (1993d), and in the Great Central Valley, Central Western California, and Southwestern California subdivisions of the Jepson Manual (B. G. Baldwin et al. 2012). Specifically, it occurs from north-central California (Lake, Sonoma, and Yolo counties) through the Coast Ranges and along the Pacific Coast south to Baja California, extending to the foothills of the Sierra Nevada (Fresno, Tulare, and Tuolumne counties), with one questionably native disjunct locality in Humboldt County.

In areas where subsp. *canum* overlaps in range with subsp. *latifolium*, especially in mountainous areas of southern California (notably Kern, Los Angeles, Monterey, Riverside, and San Bernardino counties), specimens of intermediate morphology occur. Most populations of subsp. *canum* occur at lower elevations, but where the taxa are sympatric, intermediates can be found. Some specimens from the New York, Panamint, and other mountain ranges in the Mojave Desert of San Bernardino County, which extend into Arizona, are treated here as subsp. *latifolium*, but these have leaves somewhat narrower than typical for that subspecies and approach those of subsp. *canum*.

Distinctive populations of subsp. *canum*, consisting of large, shrubby, densely pubescent plants with relatively narrow, fascicled leaves, occur particularly near the coast in southern California. Farther inland and at higher elevations, populations with variable combinations of characters led to a proliferation of names. Although some local series of populations appear quite distinctive, over the entire range there are no combinations of characters that consistently delimit morphological units, nor do any patterns differentiate diploids from tetraploids.

13b. Epilobium canum (Greene) P. H. Raven subsp. **latifolium** (Hooker) P. H. Raven, Ann. Missouri Bot. Gard. 63: 335. 1977 • Hummingbird trumpet

Zauschneria californica C. Presl var. *latifolia* Hooker, Bot. Mag. 76: plate 4493. 1850; *Z. argentea* A. Nelson; *Z. arizonica* Davidson; *Z. californica* subsp. *latifolia* (Hooker) D. D. Keck; *Z. canescens* Eastwood ex Moxley; *Z. crassifolia* Rydberg; *Z. elegans* Eastwood ex Moxley; *Z. hallii* Moxley; *Z. latifolia* (Hooker) Greene; *Z. latifolia* var. *arizonica* (Davidson) Hilend; *Z. latifolia* var. *johnstonii* Hilend; *Z. latifolia* var. *tomentella* (Greene) Jepson; *Z. latifolia* var. *viscosa* (Moxley) Jepson; *Z. pringlei* Eastwood ex Moxley; *Z. pulchella* Moxley; *Z. pulchella* var. *adpressa* Moxley; *Z. tomentella* Greene; *Z. viscosa* Moxley

Herbs not suffruticose. **Stems** 10–50(–70) cm, villous and short glandular-puberulent, mixed especially distally, sometimes predominantly glandular puberulent, rarely glabrate. **Leaves** not fasciculate; blade green to grayish green, usually lanceolate to ovate or broadly elliptical, rarely orbiculate, 0.8–5(–6) × 0.4–1.8(–2.2) cm, margins subentire to distinctly denticulate, 4–10 teeth per side, veins obscure to pronounced, 3–6 per side, surfaces villous and glandular puberulent; bracts sometimes overlapping pedicels of flowers (and fruits). **Flowers:** floral tube 16–24 mm; sepals 6–9 mm; petals 7–9 mm. **Capsules** 8–22 mm; subsessile or pedicel 1–2 mm. **Seeds** 1.2–1.6 × 0.8–1.1 mm. $2n$ = 30, 60.

Flowering Jun–Dec. Sandy or rocky soils in woodland and montane areas, stabilized talus slopes, disturbed ravines, roadsides, granite cliffs, stream banks, stabilized gravel bars; (100–)1000–3200 m; Ariz., Calif., N.Mex., Oreg.; Mexico (Baja California, Chihuahua, Sonora).

As delimited by J. D. Clausen et al. (1940) and P. H. Raven (1976), subsp. *latifolium* was considered to be strictly tetraploid. However, at least one population from Arizona (Pima County, *Yatskievych s.n.* in 1978, MO, count by S. Seavey) was diploid.

Although subsp. *latifolium* occurs primarily in the Sierra Nevada and Northwestern California subdivisions (B. G. Baldwin et al. 2012) from 1500–3200 m, or in the southern Vancouverian Province as defined by R. F. Thorne (1993d), it is also found at higher elevation in the Pacific Coast Ranges (Central Western and Southwestern California subdivisions). It extends eastward into the Sonoran Province into central and southeastern Arizona, southwestern New Mexico, and Sonora and western Chihuahua, Mexico. In most areas where subsp. *latifolium* occurs in proximity to subsp. *canum*, the former occurs at higher elevations.

13c. Epilobium canum (Greene) P. H. Raven subsp. **garrettii** (A. Nelson) P. H. Raven, Ann. Missouri Bot. Gard. 63: 335. 1977 • Garrett's firechalice [E]

Zauschneria garrettii A. Nelson, Proc. Biol. Soc. Wash. 20: 36. 1907 (as Zaushneria); *Epilobium canum* var. *garrettii* (A. Nelson) N. H. Holmgren & P. K. Holmgren; *Z. latifolia* (Hooker) Greene var. *garrettii* (A. Nelson) Hilend; *Z. orbiculata* Moxley

Herbs not suffruticose. **Stems** 15–40 cm, ± densely spreading villous and glandular puberulent, denser proximally. **Leaves** opposite throughout or distally alternate, not fasciculate; blade green, ovate to broadly elliptical, 2–4.5 × 0.5–2.3 mm, coriaceous, margins prominently denticulate, 5–15 teeth per side, veins prominent, surfaces villous ± mixed glandular puberulent. **Flowers:** floral tube 19–24 mm; sepals 4–8 mm; petals 8–13 mm. **Capsules** 15–20 mm; sessile or pedicel 1–2 mm. **Seeds** 0.9–1.5 × 0.6–1 mm. $2n$ = 30.

Flowering Jun–Oct. Sandy or rocky soils on steep limestone slopes, rocky hillsides, roadsides, dry streambeds; 1200–3400 m; Calif., Idaho, Nev., Utah, Wyo.

Subspecies *garrettii* is a diploid taxon occurring mainly in the Great Basin from northwestern Wyoming (Lincoln, Park, and Teton counties) to southeastern Idaho (Bannock, Bonneville, and Franklin counties); northern Nevada (Eureka County); extensively in montane areas of Utah from the north-central to the southwestern part of the state, with a disjunct locality in the Abajo Mountains in southeastern Utah (San Juan County); and barely to the eastern slope of the Sierra Nevada in California (Inyo County) and scattered in the desert ranges of the Mojave Desert in eastern San Bernardino County. In those areas in California, it overlaps and intergrades with subsp. *latifolium*.

14. Epilobium septentrionale (D. D. Keck) R. N. Bowman & Hoch, Ann. Missouri Bot. Gard. 66: 897. 1980 • Northern willowherb [E]

Zauschneria septentrionalis D. D. Keck, Publ. Carnegie Inst. Wash. 520: 219, figs. 84 [top center], 85 [upper right]. 1940; *Epilobium canum* (Greene) P. H. Raven subsp. *septentrionale* (D. D. Keck) P. H. Raven

Herbs usually not suffruticose, with basal shoots from thickened caudex, often decussate scales at base. **Stems** decumbent, often matted, grayish green, terete, 5–25 cm, well-branched throughout, densely white-canescent proximally, mixed strigillose, villous, and glandular puberulent distally. **Leaves** densely spaced, opposite proximal to inflorescence, alternate or fasciculate distally, petiole 0–2 mm, blade grayish green to green, lanceolate or elliptic to narrowly ovate, 1–3.5(–4) × 0.4–1.1 cm, base cuneate or attenuate, margins subentire to low-denticulate, 5–8 teeth per side, veins inconspicuous, 3–5 per side, apex acute, sometimes with caducous dark mucro, surfaces densely white-canescent and eglandular proximally, changing abruptly on inflorescence to glandular puberulent mixed with scattered villous; bracts somewhat reduced. **Inflorescences** ascending spikes or racemes, densely glandular puberulent. **Flowers:** buds 12–15 × 4–5.5 mm, subsessile or pedicels 1–2 mm; floral tube same color as petals, 17–23 × 3.5–5 mm, with ring of scales 4–5 mm from base inside; sepals same color as petals, 7–12 × 2.5–3.5 mm; petals red-orange, 8–14 × 5–6.5 mm, apical notch 1.8–2.4 mm; filaments orange or red, those of longer stamens 15–17 mm, those of shorter ones 13–15 mm; anthers 2.8–3.2 × 0.7–1 mm; ovary 5–11 mm, glandular puberulent; style light orange, 40–45 mm, stigma 4-lobed, 0.9–1.1 × 2.2–2.6 mm, exserted 8–10 mm beyond anthers. **Capsules** ± straight, 20–26 mm, ± beaked, surfaces glandular puberulent; sessile or pedicel 1–3 mm. **Seeds** broadly to narrowly obovoid, with constriction 0.7–0.8 mm from micropylar end, 1.8–2.4 × 1.1–1.3 mm, chalazal collar inconspicuous, light brown, surface low-papillose; coma easily detached, dingy white, 6–7 mm. $2n = 30$.

Flowering Aug–Sep. Rocky ledges and serpentine slopes along rivers; 10–1900 m; Calif.

Epilobium septentrionale is endemic to northern California, found only in the drainages of the Eel, Mattole, and Trinity rivers in Humboldt, Mendocino, and Trinity counties. The dimorphic pattern of vestiture on this species (eglandular, white-canescent lower leaves versus glandular puberulent upper leaves and bracts) is highly distinctive and not found in any specimens of *E. canum*. Most collections are relatively uniform in aspect, leaves, flowers, fruits, and seeds, despite the relative isolation from one another in the different river drainages where they occur.

S. R. Seavey and P. H. Raven (1977c) reported an experimental hybrid between *Epilobium septentrionale* and diploid *E. canum* subsp. *canum*; although it had normal chromosome pairing, pollen fertility was reduced (51%).

5h. Epilobium Linnaeus sect. Epilobium

Epilobium [unranked] *Chrysonerion* Torrey & A. Gray; *Epilobium* sect. *Lysimachion* Tausch

Herbs perennial, rarely suffrutescent, with basal rosettes, turions, soboles, stolons, sometimes tipped with turions, or rarely bulblets (gemmae) in leaf axils, sometimes cespitose. **Stems** not woody, epidermis not peeling. **Leaves** opposite proximal to inflorescence or throughout, usually alternate distally. **Flowers** actinomorphic; floral tube not bulbous, without scales inside; petals usually rose-purple to pink or white, very rarely cream-yellow (*E. luteum*); pollen in tetrads; stigma entire or 4-lobed. **Capsules** narrowly subcylindric to narrowly clavate, splitting to base, central column persistent, pedicellate or sessile. **Seeds** many, in 1 row per locule, narrowly obovoid or fusiform to narrowly ellipsoid, [rarely with inflated rim around perimeter on adaxial side]; coma usually present, very rarely absent.

Species ca. 150 (27 in the flora): North America, Mexico, West Indies (Hispaniola), Central America, South America, Eurasia, Africa, Pacific Islands, Australasia.

Section *Epilobium* is the largest in the genus, comprising about 150 species (168 taxa) distributed in cool montane, alpine, boreal, or arctic habitats on all continents except Antarctica, and extending to high elevations in mountains of the tropics. No other section of *Epilobium* occurs native outside of North America except for one species each of sects. *Pachydium* and *Epilobiopsis* in South America; *E. brachycarpum* (sect. *Xerolobium*) is adventive in South America and Europe. All species are diploid ($n = 18$) perennials, and the flowers in most species are protandrous and shed pollen directly onto the stigma when it becomes receptive, sometimes in bud. In all autogamous species the stigmas are undivided. Some species, 13 in North America, have deeply four-lobed stigmas that are often exserted beyond the anthers, the stigma lobes spreading after the pollen has begun to be shed; as a result, these plants are predominantly outcrossed. However, almost all are self-compatible and capable of self-pollination, except

E. obcordatum, which appears to be self-incompatible (S. R. Seavey and K. S. Bawa 1986; Seavey and S. K. Carter 1994, 1996). Primary pollinators of this group are bees, flies, and butterflies.

Perennating structures are diverse in sect. *Epilobium* and are important diagnostic characters for many species. As noted elsewhere, soboles from ± woody caudices may be the most generalized type of structure, from which have evolved several other major types (R. C. Keating et al. 1982), defined as follows.

Stolons, either hypogeous or epigeous, are found in many species of sect. *Epilobium*. Stolons may be filiform, with small decussate leaves or scales and a fleshy or leafy bud at the tip, or thicker and ropelike with larger leaves, or tough and wiry with scales. Species that produce stolons often grow in damp habitats, including boggy areas, wet scree slopes, and damp swales. New stems arise from the tip of the stolons, in clumps with ascending bases, or some distance from the parental stem.

Rosettes, clusters of leaves scarcely distinguishable from cauline leaves, are common in sect. *Epilobium*, notably including the very widespread *E. ciliatum*. In general, plants producing rosettes grow in less harsh habitats.

The other common type of perennating structure in sect. *Epilobium* is the sessile turion, which generally forms underground (to 5 cm), with tightly decussate, fleshy scales, a distinct round or strobiloid form, and little or no internode elongation. Turions may also form at the tips of stolons (as in *E. palustre* and relatives). A rare variant of these overwintering buds are bulblets (gemmae) that form in distal leaf axils. Stems arising from turions or rosettes are not or only loosely clumped, and generally strict, not ascending as found in most soboliferous or stoloniferous plants.

Species of sect. *Epilobium* have solid endexine in the distal pollen walls, unlike the rest of the genus, which have large endexine channels in the distal walls. The pollen viscin threads of species in this section are tightly compound (J. Praglowski et al. 1994), unlike those of other sections. These characters suggest that sect. *Epilobium* is sister to the remaining sections, as supported by recent molecular studies (D. A. Baum et al. 1994; R. A. Levin et al. 2004).

Based on extensive crossing studies, it appears that virtually all species of sect. *Epilobium* can hybridize with most or all other species, resulting in more or less fertile offspring (S. R. Seavey and P. H. Raven 1977, 1977b, 1977c, 1978). Natural hybridization occurs fairly frequently where two or more species occur sympatrically in nature (Raven and T. E. Raven 1976). Analysis of experimental hybrids revealed the presence of reciprocal chromosome translocation differences within this section; species or groups of species have been found to differ from one another by one or more reciprocal translocations, resulting in rings or chains of chromosomes, rather than bivalents, in hybrids between the groups (Raven and Raven; Seavey and Raven 1977, 1977b, 1977d, 1978). Most or all North American species tested belong to one of three groups, designated AA, BB, and CC by Seavey and Raven; there are additional arrangements that do not occur in North America. The BB arrangement is found in many species in North America, including all of those with most generalized morphology and distribution, most species in South America and Eurasia, and all species in Australasia, and is the presumed original arrangement. It differs from the AA (primarily including *E. ciliatum* and relatives) and CC (mainly the north temperate so-called Alpinae) groups by one reciprocal translocation each, and AA differs by two translocations from CC. These and other studies found no evidence that the type of permanent translocation heterozygosity found in Onagreae occurs in any species of *Epilobium*. As treated here, species are arranged according to their known or presumed chromosome group designation. Because the BB group includes not only the majority of species of sect. *Epilobium* worldwide, but also several North American species that appear to be more generalized than any others, the sequence of groups here is BB, CC, then AA.

15. Epilobium obcordatum A. Gray, Proc. Amer. Acad. Arts 6: 532. 1865 • Rockfringe [E]

Epilobium obcordatum var. *puberulum* Jepson

Herbs ± suffruticose, wiry shoots from woody caudex with barklike periderm extending to 25 cm below ground, shoots with scaly bases. **Stems** many, decumbent to ascending, clumped or cespitose, green to grayish green, terete, 5–15 cm, branched mainly proximally, subglabrous and ± glaucous proximal to inflorescence, ± canescent distally or throughout. **Leaves** opposite proximal to inflorescence, alternate distally, usually crowded and exceeding internodes, subsessile or petiole 1–2 mm; blade green or grayish green, usually broadly lanceolate-elliptic to ovate or obovate, rarely suborbiculate, 0.6–2.4 × 0.4–1.3 (–1.9) cm, base rounded to subcordate, margins low denticulate, 4–9 teeth per side, veins indistinct, 4–7 per side, apex obtuse proximally to acute distally, surfaces usually subglabrous, rarely canescent, especially on margins and veins; bracts much reduced. **Inflorescences** ascending to erect, sparse racemes or loose panicles, ± densely canescent and glandular puberulent. **Flowers** erect; buds 7–13 × 3–5 mm, apex acute, sometimes with stigma exserted; pedicel 3–10 mm; floral tube 3.2–5.5 × 2.2–4.2 mm, slightly raised ring of spreading hairs 0.4–1 mm from base inside; sepals (5–)8.5–14 × 1.8–2.9 mm; petals pink to rose-purple, obcordate, (12–)15–26 × (7–)9–14.6 mm, apical notch 2.5–7.2 mm; filaments cream to pink, those of longer stamens 8.5–16 mm, those of shorter ones 5.5–11 mm; anthers cream-yellow, 1.6–2.9 × 0.6–1.3 mm; ovary 9–22 mm, usually canescent and glandular puberulent, rarely subglabrous; style cream to light pink, 11–23 mm, glabrous, stigma deeply 4-lobed, 1–1.5 × 2.2–4.5 mm, exserted beyond anthers. **Capsules** straight, subclavate, 16–40 mm, surfaces canescent and glandular puberulent; pedicel 5–15 mm. **Seeds** narrowly obovoid, 1.4–2.1 × 0.6–0.9 mm, with low chalazal collar 0.4–0.5 mm wide, light or grayish brown, surface papillose; coma easily detached, tawny, 5–9 mm. $2n = 36$.

Flowering Jul–Sep. Dry, rocky montane or alpine ridges, basaltic cliffs, along edges of talus or gravel slopes; 1900–4000 m; Calif., Idaho, Nev., Oreg.

Epilobium obcordatum is an uncommon but relatively widespread and very characteristic species of the high Sierra Nevada, extending to scattered high ranges in northeastern Nevada, Idaho, and southeastern Oregon (Steens Mountains). Its low, clumped habit, dense green and often glaucous foliage, and large flowers make it one of the more attractive species of the genus, with considerable potential as a cultivated plant in rock gardens. Although it bears some general morphological similarities with two species in western North America, *E. rigidum* and *E. siskiyouense*, as discussed under those taxa, *E. obcordatum* also bears close resemblance to *E. nankotaizanense* Yamamoto, an alpine endemic from Taiwan, China (Chen C. J. et al. 1992). It is uncertain whether they are actually related or have evolved similar morphologies independently in similar high montane habitats on either side of the north Pacific.

Little has been reported on the pollination biology of *Epilobium obcordatum*, but its large flowers with marked protandry and herkogamy strongly suggest that the plants are outcrossing, probably pollinated by large bees.

Epilobium obcordatum shows considerable morphological variation, especially in leaf shape (ranging from narrowly ovate to orbiculate) and pubescence pattern. In the latter, plants mainly in the Sierra Nevada have stems glabrous and often glaucous below the inflorescence and mixed canescent and glandular puberulent distally. Plants mainly in Idaho and Nevada have stems sparsely to moderately canescent and inflorescences densely mixed canescent and glandular puberulent. But some collections, including the type of *E. obcordatum* var. *puberulum*, are mixed, and these pubescence differences do not correlate with other morphological or eco-geographical characters.

16. Epilobium siskiyouense (Munz) Hoch & P. H. Raven, Madroño 27: 146. 1980 • Siskiyou willowherb [E]

Epilobium obcordatum A. Gray subsp. *siskiyouense* Munz in N. L. Britton et al., N. Amer. Fl., ser. 2, 5: 205. 1965; *E. obcordatum* var. *laxum* Dempster

Herbs ± suffruticose, shoots from woody caudex with barklike periderm extending to 40 cm below ground, shoots with scaly bases. **Stems** several to many, erect to ascending, loosely clumped, terete, 10–25 cm, rarely branched distal to base, usually short-villous and strigillose throughout, mixed sparsely glandular puberulent distally, rarely subglabrous proximal to inflorescence. **Leaves** opposite proximal to inflorescence, alternate and usually crowded distally, sessile; blade gray-green, narrowly to broadly ovate, 1.3–2.6 × 0.8–2 cm, base rounded to subcordate, margins usually serrulate, 6–12 teeth per side, rarely subentire, veins inconspicuous, 3–5 per side, apex rounded proximally to acute distally, surfaces sparsely short-villous to subglabrous and glaucous; bracts much reduced. **Inflorescences** erect, compact racemes, densely canescent and glandular puberulent, or subglabrous, only ovaries pubescent.

Flowers erect; buds often purplish green, 9–11 × 3.5–5 mm, blunt; pedicel 6–12 mm; floral tube 2.1–4 × 2.9–5 mm, prominent ring of tissue 0.3–0.6 mm wide, edged by spreading hairs, 0.9–1.8 mm from base of tube inside; sepals purplish green, 5–10.5 × 2–3.5 mm, apex acute; petals pink to rose-purple, obcordate, 10–22.5 × 9.5–15.5 mm, apical notch 2–6.5 mm; filaments cream, those of longer stamens 6.5–14 mm, those of shorter ones 3.5–11 mm; anthers cream 1.9–3.3 × 0.7–1.2 mm; ovary 12–22 mm, ± densely canescent and glandular puberulent; style white to light pink, 10.5–18 mm, sparsely villous just proximal to stigma, stigma broadly 4-lobed, 1–1.8 × 2.4–4.2 mm, exserted beyond anthers. **Capsules** 25–45 mm, surfaces canescent and glandular puberulent; pedicel 6–25 mm. **Seeds** narrowly obovoid, 1.4–1.9 × 0.6–0.7 mm, with low, obscure chalazal collar, light brown, surface papillose; coma easily detached, somewhat tawny, 4–8 mm. $2n = 36$.

Flowering Jul–Aug. Stream banks, moist, rocky slopes, montane ridges, sometimes on serpentine areas; 1600–2500 m; Calif., Oreg.

Epilobium siskiyouense is endemic to the Klamath region in southwestern Oregon (Jackson County) and north-central California in the Salmon, Scott Bar, and Siskiyou mountains of Siskiyou and Trinity counties. As noted by Hoch and Raven, this geographical range and several morphological features appear to be intermediate between those of *E. obcordatum* and *E. rigidum*. All three species have unusually large flowers (12–26 mm) with four-lobed stigmas, and as a group are quite distinct from their congeners in the region. Despite these similarities, the three taxa differ substantially in details of floral structure, especially regarding the dimensions of the floral tube. Specifically, *E. rigidum* has mean petal length 18.2 mm, floral tubes 1–1.6 × 2.5–3.6 mm; *E. siskiyouense* mean petal length 17.1 mm, floral tubes 2.1–4 × 2.9–5 mm; and *E. obcordatum* mean petal length 18.6 mm, floral tubes 3.2–5.2 × 2.2–3.6 mm. Thus, in flowers that are similar in overall size and aspect, *E. rigidum* has a very short, broad floral tube, *E. obcordatum* has a relatively long, narrow tube, and *E. siskiyouense* has a tube intermediate in size and shape. In terms of the ratio of tube length to width, the three taxa do not overlap. Although these characters are difficult to include in a key (since they require floral dissection and/or precise measurements), they are diagnostic for these species.

Epilobium siskiyouense has an additional diagnostic floral character that is unique in the genus. Whereas most other species of *Epilobium* have a simple ring of spreading hairs, sometimes with a low ridge of tissue near the mouth of the floral tube, *E. siskiyouense* has a relatively broad ring of tissue (0.3–0.6 mm wide), shaped like a washer, from which spreading hairs arise; this feature may provide protection for the nectar.

Epilobium siskiyouense has two distinct patterns of vestiture on the stems. In some specimens, the lower stems are mixed canescent and glandular puberulent and the inflorescence only glandular puberulent. In other specimens, the stems are subglabrous below a sparsely canescent and glandular puberulent inflorescence. There is no obvious correlation of this difference with any other morphological, ecological, or geographical factors.

17. Epilobium palustre Linnaeus, Sp. Pl. 1: 348. 1753 • Marsh willowherb, épilobe palustre [F]

Chamaenerion palustre (Linnaeus) Schreber; *Epilobium molle* Torrey var. *sabulonense* Fernald; *E. nesophilum* (Fernald) Fernald var. *lupulinum* Hodgdon & R. B. Pike; *E. nesophilum* var. *sabulonense* (Fernald) Fernald; *E. oliganthum* Michaux; *E. palustre* var. *albescens* Wahlenberg; *E. palustre* var. *grammadophyllum* (Haussknecht) Munz; *E. palustre* var. *lapponicum* Wahlenburg; *E. palustre* var. *longirameum* Fernald & Wiegand; *E. palustre* var. *oliganthum* (Michaux) Fernald; *E. palustre* var. *sabulonense* (Fernald) B. Boivin; *E. pylaieanum* Fernald; *E. wyomingense* A. Nelson

Herbs with multiple filiform epigeous stolons with widely spaced, small leaves, terminating in condensed, dark, small turions 3–7 × 2–3 mm. **Stems** erect, loosely clustered, terete, 5–80 cm, simple to well branched, subglabrous on proximal internodes, sometimes with faint strigillose lines decurrent from margins of petioles, densely strigillose distally. **Leaves** opposite proximal to inflorescence, alternate distally, subsessile; blade lanceolate or narrowly elliptic to sublinear, 1.5–7 × 0.2–1.9 cm, base cuneate, margins entire or inconspicuously denticulate, 2–6 teeth per side, veins inconspicuous, apex acute or acuminate, surfaces strigillose abaxially, subglabrous adaxially or strigillose only on margins and veins; bracts smaller and narrower. **Inflorescences** nodding in bud, later erect, racemes, not branched, densely strigillose, rarely mixed sparsely glandular puberulent. **Flowers** erect to spreading; buds 2–5 × 1.5–2.5 mm; pedicel 8–12 mm; floral tube 0.6–1.8 × 1.3–2.2 mm, ring of spreading hairs at mouth inside; sepals 1.4–4.5 × 0.8–1.5 mm, abaxial surface strigillose; petals usually white, rarely pink, 2–9 × 1.8–5 mm, obcordate, apical notch 0.6–1.6 mm; filaments cream or white, those of longer stamens 0.8–3.5 mm, those of shorter ones 0.4–2.1 mm; anthers 0.4–1 × 0.3–0.7 mm; ovary 12–35 mm, strigillose; style cream, 1.5–4.5 mm, stigma clavate to cylindrical, entire, 0.8–2 × 0.5–1.4 mm, usually surrounded by, rarely exserted beyond, anthers. **Capsules** straight or upcurved, 25–90 mm, surfaces strigillose; pedicel 15–35(–60) mm. **Seeds**

elliptic, attenuate to narrowly fusiform, 1.4–2.2 × 0.4–0.5 mm, chalazal collar 0.1–0.25 mm, conspicuous, surface finely papillose; coma persistent, white, 5–7 mm. **2*n*** = 36.

Flowering Jul–Sep. Low, boggy areas, swamps, saturated stream banks, mossy meadows; 0–1000 (–2600) m; Greenland; St. Pierre and Miquelon; Alta., B.C., Man., N.B., Nfld. and Labr., N.W.T., N.S., Nunavut, Ont., P.E.I., Que., Sask., Yukon; Alaska, Calif., Colo., Conn., Maine, Mass., Mich., Minn., Mont., N.H., N.J., N.Y., Oreg., Pa., R.I., S.Dak., Utah, Vt., Wis., Wyo.; n, c Eurasia.

Epilobium palustre is an extremely widespread circumboreal species that is relatively common from Alaska across Canada, especially around the Great Lakes to the Maritime provinces (mostly below 1000 m). This species is somewhat less common in the northeastern United States, and only scattered in the mountains of the western United States, south to California and Colorado (to 3000 m). It also occurs in ice-free regions of Greenland south of 70° north latitude, in northern and central Europe, across most of subarctic Russia to the Caucasus, the Himalaya complex, and eastern Asia, including Japan and the Russian Far East.

Epilobium palustre shows considerable variation across its very wide distribution and, not surprisingly, distinctive local races, which sometimes have been named formally, including several from northeastern North America. Within series *Palustriformes* by C. Haussknecht (1884), M. L. Fernald (1944d) clarified much confusion in names at that time, clearly delineating and establishing the nomenclature of *E. davuricum*, *E. leptophyllum*, *E. palustre*, and *E. densum*. Fernald also recognized *E. nesophilum* (see discussion under 18. *E. leptophyllum*) and *E. pylaieanum*, the latter based on small plants of *E. palustre* from southern Newfoundland.

Epilobium lineare Muhlenberg and *E. palustre* var. *albiflorum* Lehmann are illegitimate names that pertain here.

18. Epilobium leptophyllum Rafinesque, Précis Découv. Somiol., 41. 1814 • Bog willowherb, épilobe leptophylle [E]

Epilobium densum Rafinesque var. *nesophilum* Fernald; *E. nesophilum* (Fernald) Fernald; *E. oliganthum* Michaux var. *gracile* Farwell; *E. palustre* Linnaeus var. *gracile* (Farwell) Dorn; *E. squamatum* Nuttall; *E. tenellum* Rafinesque

Herbs with threadlike, nearly leafless epigeous stolons terminating in compact, fleshy turions 3–8 × 2–4 mm. **Stems** erect, simple to loosely clustered, terete, 15–95 cm, simple to well branched, densely strigillose, often mixed glandular puberulent on inflorescence, rarely with faint strigillose lines decurrent from margins of petioles. **Leaves** opposite proximally, usually alternate, rarely fasciculate distally, subsessile; blade linear to very narrowly elliptic or sublanceolate, 2–7.5 × 0.1–0.7 cm, usually longer than internodes, base rounded to subcuneate, margins subentire, 4–7 inconspicuous teeth per side, sometimes revolute, lateral veins inconspicuous, apex obtuse proximally to acute distally, both surfaces densely strigillose, increasing distally; bracts not much reduced. **Inflorescences** erect racemes, densely strigillose, often mixed sparsely glandular puberulent. **Flowers** erect; buds 3–5 × 1.5–2.5 mm; pedicel 5–12 mm; floral tube 0.8–1.5 × 1.2–1.8 mm, ring of spreading hairs at mouth inside; sepals 2.5–4.5 × 0.9–1.3 mm, abaxial surface strigillose; petals obcordate, white to light pink, 3.5–7 × 1.6–4 mm, apical notch 1–1.8 mm; filaments white or cream, those of longer stamens 0.8–3.5 mm, those of shorter ones 0.6–2.5 mm; anthers cream, 0.5–0.9 × 0.4–0.6 mm; ovary 12–18 mm, densely strigillose, sometimes mixed glandular puberulent; style cream, 2–3.8 mm, stigma narrowly clavate, entire, 1–1.8 × 0.5–1.2 mm, usually surrounded by, rarely exserted beyond, anthers. **Capsules** straight, narrowly cylindrical, 35–80 mm, surfaces densely strigillose; pedicel 10–35 mm. **Seeds** narrowly fusiform to narrowly obovoid, 1.5–2.2 × 0.5–0.7 mm, chalazal collar 0.1–0.2 mm, ± pronounced, surface papillose; coma persistent, dingy white, 6–8 mm. **2*n*** = 36.

Flowering Jun–Sep. Marshy ground, bogs, fens, low thickets, seepage areas, damp pastures; 0–1000 (–2900) m; St. Pierre and Miquelon; Alta., B.C., Man., N.B., Nfld. and Labr. (Nfld.), N.W.T., N.S., Ont., P.E.I., Que., Sask.; Alaska, Calif., Colo., Conn., Idaho, Ill., Ind., Iowa, Kans., Ky., Maine, Md., Mass., Mich., Minn., Mo., Mont., Nebr., Nev., N.H., N.J., N.Mex., N.Y., N.C., N.Dak., Ohio, Okla., Oreg., Pa., R.I., S.Dak., Tenn., Tex., Utah, Vt., Va., Wash., W.Va., Wis., Wyo.

The range of *Epilobium leptophyllum* overlaps with that of the related *E. palustre*, but the former is less common to the north and more common south into the midwestern United States, and absent only from most of the southern tier of states. It is also relatively uncommon in the western United States and Canada. Judging by the number of herbarium sheets that also include *E. palustre*, *E. densum*, and even *E. coloratum*, it sometimes occurs in sympatry with those species and may rarely hybridize with them, based on plants with intermediate morphology and/or sterile fruits.

Fernald described *Epilobium nesophilum* from the Magdalen Islands (Quebec), and especially Newfoundland, first as a variety of *E. densum* (1918), then as a separate species (1925).

Epilobium rosmarinifolium Pursh 1813, an illegitimate name (not Haenke 1788), pertains here.

19. Epilobium densum Rafinesque, Précis Découv. Somiol. 42. 1814 • Downy willowherb, épilobe dressé [E]

Epilobium strictum Muhlenberg ex Sprengel

Herbs with threadlike, sparsely leaved epigeous stolons terminating with compact fleshy turions 4–8 × 3–4 mm. **Stems** erect, simple or loosely clustered, terete, 15–95 cm, often well branched distally, densely villous throughout, sometimes mixed glandular puberulent distally, decurrent lines absent. **Leaves** opposite proximally, alternate and often fasciculate distally, subsessile; blade oblong-lanceolate to sublinear, 2–4.5 × 0.3–0.9 cm, ± exceeding internodes, base cuneate, margins entire or denticulate, 3–6 inconspicuous teeth per side, sometimes revolute, veins apparent on abaxial side, 3–5 per side, apex acute, surfaces villous, especially distally; bracts somewhat reduced. **Inflorescences** erect racemes, branched or not, densely villous, sometimes mixed glandular puberulent. **Flowers** erect; buds 2.5–5 × 1.5–1.5 mm; pedicel 3–8 mm; floral tube 1–1.8 × 1–2 mm, ring of spreading hairs at mouth inside; sepals 2–4.5 × 1–1.3 mm, abaxial surface villous and glandular puberulent; petals light to dark pink, obcordate, 4–6 × 2–3 mm, apical notch 1–1.5 mm; filaments pink, those of longer stamens 2–3.5 mm, those of shorter ones 1–2 mm; anthers cream, 0.5–0.8 × 0.4–0.6 mm; ovary 12–20 mm, densely villous, often mixed glandular puberulent; style cream, 2–3.5 mm, stigma narrowly clavate, entire, 1–1.6 × 0.5–1 mm, surrounded by anthers. **Capsules** 35–68 mm, short-beaked, surfaces densely villous, sometimes mixed glandular puberulent; pedicel 5–15 mm. **Seeds** narrowly fusiform to narrowly obovoid, 1.5–2 × 0.5–0.6 mm, chalazal collar inconspicuous, 0.1 mm, surface low papillose; coma persistent, dingy white, 6–8 mm. $2n = 36$.

Flowering Jul–Sep. Sphagnum and peat bogs, marshes, seeps, damp pastures; 0–600 m; N.B., N.S., Ont., P.E.I., Que.; Conn., Ill., Ind., Iowa, Maine, Md., Mass., Mich., Minn., N.H., N.J., N.Y., Ohio, Pa., R.I., Vt., Va., Wis.

Epilobium densum is relatively uncommon but widely distributed within the Great Lakes-St. Lawrence River region in boggy or marshy areas. Within that region, it sometimes grows in close proximity with *E. leptophyllum* and *E. palustre*; it is densely pubescent like *E. leptophyllum*, and although the type of hairs differs markedly on the two species, they sometimes are found mixed on herbarium sheets. Because it is less pubescent, *E. palustre* is rarely mixed or confused with *E. densum*. Hybrids among these species occur but are uncommon and generally marked by slightly to moderately reduced seed fertility.

M. L. Fernald (1944d), among others, argued that names published in the *Catalogus* by Muhlenberg were not validly published. In the case of *Epilobium strictum*, K. Sprengel considered that Muhlenberg provided enough of a description (upright, soft) to validate the name in 1825. However, Rafinesque had validly published the name *E. densum* for the same taxon in 1814.

Epilobium molle Torrey is an illegitimate name that pertains here.

20. Epilobium davuricum Fischer ex Sprengel, Novi Provent., 44. 1818 • Dahurian willowherb, épilobe de Daourie

Epilobium palustre Linnaeus var. *davuricum* (Fischer ex Sprengel) S. L. Welsh

Herbs with basal rosettes of linear leaves 12–40 × 2–5 mm. **Stems** erect, rarely clumped, terete, 10–40(–45) cm, usually simple, rarely branched, glabrous proximal to inflorescence with sparsely strigillose raised lines decurrent from margins of petioles, mixed strigillose and glandular puberulent distally. **Leaves** opposite and crowded proximal to inflorescence, alternate and scattered distally, subsessile; blade narrowly oblong or narrowly lanceolate to linear, (1.2–)2–4.5 × 0.1–0.5 cm, base attenuate, margins irregularly denticulate, 2–4 teeth per side, veins inconspicuous, 3 or 4 per side, apex subacute to obtuse or ± truncate, surfaces glabrous with sparsely strigillose margins and adaxial midrib; bracts much reduced and narrower. **Inflorescences** nodding in bud, suberect later, racemes, simple, mixed strigillose and glandular puberulent, sometimes subglabrous. **Flowers** erect, sometimes starting in third most-proximal node; buds 3–6.5 × 1–2 mm; pedicel 4–15 mm; floral tube 0.7–1.6 × 1–2.2 mm, with or often without ring of spreading hairs at mouth inside; sepals 1.2–3.5 × 0.7–1.2 mm, apex subacute; petals white, 2–5.5 × 1.2–3.1 mm, apical notch 0.3–1 mm; filaments white or cream, those of longer stamens 1–2.5 mm, those of shorter ones 0.5–1.4 mm; anthers cream, 0.4–0.6 × 0.3–0.5 mm; ovary usually purplish red, 10–20 mm, sparsely mixed strigillose and glandular puberulent or subglabrous; style cream or white, 0.7–2.5 mm, stigma cylindrical to clavate, entire, 0.8–2 × 0.2–1.2 mm, surrounded by anthers. **Capsules** slender, 30–55 mm, surfaces sparsely strigillose and glandular puberulent; pedicel 8–38 mm. **Seeds** narrowly attenuate, 1.4–2 × 0.3–0.5 mm, chalazal collar conspicuous, 0.1–0.3 mm, light brown or blond, surface papillose; coma persistent, white, 3–5 mm. $2n = 36$.

Flowering Jul–Aug. Subarctic open balsam poplar and spruce forests, taiga, wet meadows, boggy coastal areas, limestone barrens, wet marly soils; 0–1500 (–2000) m; Alta., B.C., Man., Nfld. and Labr. (Nfld.), N.W.T., Nunavut, Ont., Que., Sask., Yukon; Alaska; Europe (Russia).

Even though *Epilobium davuricum* has not yet been characterized by chromosome group, its morphology and general distribution suggest that it (plus *E. arcticum*) is related to *E. palustre* as a member of the Palustriformes group. It does not have the turion-tipped stolons that characterize *E. palustre* and relatives but has similarly large seeds with a distinct chalazal collar. Some specimens with aberrant combinations of characters suggest that *E. davuricum* may hybridize with *E. palustre* and possibly *E. arcticum* in areas where their distributions overlap.

21. Epilobium arcticum Samuelsson, Bot. Not. 1922: 160, fig. 1. 1922 • Arctic willowherb, épilobe arctique

Epilobium davuricum Fischer ex Sprengel subsp. *arcticum* (Samuelsson) P. H. Raven; *E. davuricum* var. *arcticum* (Samuelsson) Polunin

Herbs with sessile, basal rosettes of broadly ovate to spatulate leaves 0.5–1.8 × 0.3–0.9 cm. **Stems** single or many, suberect or nodding in bud, often clumped, terete, (2–)5–18 cm, rarely branched, subglabrous proximal to inflorescence with raised strigillose lines decurrent from margins of petioles, strigillose distally. **Leaves** opposite proximally, alternate on inflorescence, subsessile; blade obovate to narrowly elliptic to distally sublinear, 0.8–2.1 × 0.2–0.5 cm, base cuneate to attenuate, margins subentire to minutely denticulate, 2–5 low teeth per side, veins inconspicuous, apex obtuse to truncate, surfaces glabrous or sparsely strigillose on abaxial midrib; bracts reduced and narrower. **Inflorescences** often nodding in bud, erect later, few-flowered racemes, subglabrous or sparsely strigillose. **Flowers** nodding to suberect; buds 2–5 × 1–2.5 mm; pedicel 6–12(–18) mm, exceeding subtending bracts; floral tube 0.5–1.1 × 0.6–1.3 mm, with or without sparse ring of hairs at mouth inside; sepals green or flushed purple, 1.1–1.8 × 0.6–1.2 mm; petals white, sometimes flushed pink, 2.2–4.5 × 1.4–2.5 mm, apical notch 0.5–0.7 mm; filaments white or light pink, those of longer stamens 4–5 mm, those of shorter ones 2–3 mm; anthers cream, 0.3–0.4 × 0.2–0.3 mm; ovary 8–18 mm, subglabrous to strigillose; style white, 2.5–3.5mm, stigma clavate, 1–1.8 × 0.5–1 mm, surrounded by anthers. **Capsules** erect, often reddish purple, 20–42 mm, surfaces sparsely strigillose; pedicel 25–40 mm. **Seeds** narrowly fusiform to narrowly obovoid, 1.1–1.7 × 0.3–0.5 mm, with distinct chalazal collar 0.1–0.2 mm, light brown, surface rugose or reticulate; coma persistent, dull white, 5–7 mm. $2n = 36$.

Flowering Jul–Aug. Boggy, wet meadows, along streams, seepage slopes, depressions of low-center polygons; 50–500 m; Greenland; N.W.T., Nunavut, Que., Yukon; Alaska; Europe (n Russia).

Epilobium arcticum has the most northern distribution in the genus, occurring almost exclusively above the Arctic Circle (about 66°N) in Alaska, Canada, coastal Greenland north of 69°N, and the Russian Federation, although it apparently is absent from northern Europe (P. H. Raven 1968).

Epilobium arcticum has often been combined or confused with *E. davuricum*, but they differ in size, leaf shape and size, and seed size, as well as in distribution (S. G. Aiken et al., http://nature.ca/aaflora/data). Most populations of *E. arcticum* occur at (62–)67–80°N, in the Tundra zone, often on the islands of the Arctic Archipelago (especially Axel Heiberg, Baffin, and Ellesmere islands). Most populations of *E. davuricum*, on the other hand, occur south of the Arctic Circle in the Boreal and Taiga zones, and rarely, if ever, on those islands.

Because *Epilobium arcticum* grows at latitudes with extremely short growing seasons, plants often commence flowering at the second or third most-proximal node.

22. Epilobium oregonense Haussknecht, Monogr. Epilobium, 276, plate 14, fig. 66. 1884 • Oregon willowherb [E]

Herbs with slender stolons to 18 cm with minute, rounded leaves. **Stems** erect or ascending, often loosely matted, often flushed purple distally, terete, 8–30 (–40) cm, simple or sparsely branched from base, subglabrous. **Leaves** opposite proximal to inflorescence, alternate distally, subsessile; blade broadly elliptic proximally, narrowly elliptic or lanceolate to sublinear distally, 5–25 × 1–7 mm, longer than internodes proximally to much shorter distally, base cuneate to rounded, margins subentire, veins extremely faint, 3–5 per side, apex obtuse, surfaces subglabrous; bracts extremely reduced and linear. **Inflorescences** usually erect, sometimes nodding in bud, racemes, open, unbranched, sparsely strigillose and glandular puberulent. **Flowers** suberect or nodding; buds 2–3.5 × 1–1.5 mm, apex blunt; pedicel 2–7 mm; floral tube 0.8–1.8 × 1–2.1 mm, with faint ring of hairs at mouth inside; sepals often flushed purple, 2.5–4.5 × 1–1.6 mm; petals white to pink, 5–8 × 2.8–4 mm, apical notch 0.8–1.5 mm; filaments white, those of longer stamens 2.8–4.5 mm, those of shorter ones 2–3.8 mm; anthers yellow-cream, 0.8–1.2 × 0.4–0.5 mm; ovary green to purple, 8–14 mm, sparsely strigillose and glandular puberulent; style white, 3.8–4.8 mm, glabrous, stigma subcapitate, 1–1.4 × 1–1.2 mm, surrounded by longer anthers. **Capsules** slender, often purplish green, 21–40(–52) mm, surfaces subglabrous; pedicel 20–65 mm. **Seeds** narrowly oblanceoloid or subfusiform, 1–1.4 × 0.4–0.6 mm, chalazal collar 0.1–0.2 mm, light brown, surface low papillose; coma persistent, whitish, 3–4 mm. **2*n*** = 36.

Flowering Jul–Aug. Montane to subalpine boggy or mossy areas, wet meadows, protected, semi-shaded stream banks; 1200–3000(–3500) m; B.C.; Ariz., Calif., Colo., Idaho, Mont., Nev., N.Mex., Oreg., Utah, Wash., Wyo.

Epilobium oregonense is a distinctive western North American endemic, found primarily throughout the Cascade–Sierra mountain complex barely into the Transverse Ranges of southern California, and very scattered through the Rocky Mountains. It is exceedingly rare in Arizona, Colorado, and New Mexico.

Even though *Epilobium oregonense* bears some similarity to *E. anagallidifolium* and other members of the Alpinae group, and often grows in close proximity to them, this species does not share the derived CC chromosomal arrangement with that group, instead having the more globally widespread BB arrangement. The similarities with *E. anagallidifolium* include the small stature, small, obtuse, and subentire leaves, and long pedicels in fruit; however, *E. oregonense* differs by its long, threadlike stolons, distal leaves extremely narrow and reduced in size relative to the internodes, and near complete absence of pubescence on the plant, including a lack of raised lines of hairs on the stems.

Another species with which *Epilobium oregonense* has been confused is *E. hallianum*, but that species always forms condensed basal turions, is more strictly erect, and generally has larger and more denticulate leaves. The distinctive and diagnostic stolons of *E. oregonense* are similar to those found in *E. palustre* and related species (all of which also have the BB chromosome arrangement), except that those of *E. oregonense* never terminate in a condensed turion, as found in those other species. The exact affinities of *E. oregonense* remain uncertain, but it appears to be most closely related to the *E. palustre* complex.

Some specimens of *Epilobium oregonense* grow as floating mats in cold streams; these specimens are notably large, with particularly strong development of basal stolons and larger, more lanceolate leaves. As evidenced by mixed herbarium collections, *E. oregonense* grows sympatrically with several congeners, including *E. anagallidifolium*, *E. ciliatum* subspp. *ciliatum* and *glandulosum*, *E. hallianum*, and *E. hornemannii*, and hybridizes occasionally, at least with *E. ciliatum* subsp. *ciliatum* and *E. hornemannii*.

Epilobium oregonense var. *gracillimum* Trelease, which pertains here, was not validly published, and other names based on it are also invalid.

23. Epilobium hallianum Haussknecht, Monogr. Epilobium, 261. 1884 (as halleanum) • Glandular willowherb [E]

Epilobium brevistylum Barbey var. *pringleanum* (Haussknecht) Jepson; *E. brevistylum* var. *subfalcatum* (Trelease) Munz; *E. brevistylum* var. *tenue* (Trelease) Jepson; *E. brevistylum* var. *ursinum* (Parish ex Trelease) Jepson; *E. delicatum* Trelease var. *tenue* Trelease; *E. glandulosum* Lehmann var. *tenue* (Trelease) C. L. Hitchcock; *E. pringleanum* Haussknecht; *E. pringleanum* var. *tenue* (Trelease) Munz; *E. ursinum* Parish ex Trelease; *E. ursinum* var. *subfalcatum* Trelease

Herbs with small, 3–6 mm, round or oblong, compact turions 1–5 cm below ground. **Stems** strict, erect, terete, 2–50(–60) cm, rarely branched only in larger plants, subglabrous proximal to inflorescence except for raised strigillose lines decurrent from margins of petioles, or sometimes ± densely long-villous throughout with inconspicuous decurrent lines. **Leaves** opposite proximally, alternate on inflorescence, subsessile or proximally with petioles 1–1.5 mm; blade ovate proximally to lanceolate or narrowly elliptic distally, 0.5–4.7 × 0.2–1.4 cm, base rounded to cuneate, margins subentire proximally to denticulate distally, 8–20 teeth per side, veins inconspicuous, 3–6 per side, apex obtuse to subacute, surfaces mostly glabrous with strigillose margins; bracts much reduced. **Inflorescences** usually nodding in bud, erect later, open racemes, sometimes congested, usually mixed strigillose and glandular puberulent, rarely also mixed villous, or rarely subglabrous. **Flowers** erect; buds 2–5 × 1–2 mm; pedicel 3–8 mm; floral tube 0.5–1.7 × 0.8–1.6 mm, with slightly raised ring of spreading hairs at mouth inside; sepals green, 1.2–2.8 × 0.5–1 mm, abaxial surface subglabrous or sparsely glandular puberulent; petals white, often fading pink, 1.6–5.5 × 1.2–3 mm, apical notch 0.3–1.2 mm; filaments white or cream, those of longer stamens 0.6–2.5 mm, those of shorter ones 0.4–1.5 mm; anthers cream, 0.2–0.9 × 0.2–0.5 mm; ovary 10–14 mm, strigillose and glandular puberulent or subglabrous; style cream, 0.8–5 mm, stigma clavate, entire, 0.4–1.2 × 0.3–0.7 mm, usually surrounded by, rarely exserted beyond, anthers. **Capsules** very narrowly cylindrical, (15–)24–60 mm, surfaces usually subglabrous to mixed strigillose and glandular puberulent, rarely sparsely villous; pedicel 8–40 mm. **Seeds** narrowly fusiform to narrowly obovoid, 1.1–1.6 × 0.4–0.6 mm, chalazal collar ± conspicuous, 0.05–0.2 mm, light brown, surface papillose; coma easily detached, white, 3–6 mm. **2*n*** = 36.

Flowering Jun–Sep. Semi-shaded stream banks, wet grassy slopes or meadows, bogs, seasonally wet sites, vernal pools; 100–3700 m; Alta., B.C., Sask.; Ariz., Calif., Colo., Idaho, Mont., Nev., N.Mex., Oreg., S.Dak., Utah, Wash., Wyo.

Epilobium hallianum has condensed fleshy turions and generally strict habit, suggesting an affinity with *E. ciliatum* or *E. saximontanum*, both with the AA arrangement. However, *E. hallianum* has the BB arrangement (S. R. Seavey and P. H. Raven 1978) and apparently more distant relationship to those other species.

In different parts of its geographical range *Epilobium hallianum* shows considerable morphological variability, especially in leaf shape and margins, and in type and pattern of stem vestiture, including plants in the southern part of its range with densely villous stems that have been treated as *E. ursinum*. In part because of this variability and in part because the very characteristic condensed turions are easily lost during collection and/or often overlooked, *E. hallianum* is frequently misidentified.

24. Epilobium coloratum Biehler, Pl. Nov. Herb. Spreng., 18. 1807

Epilobium coloratum var. *tenuifolium* H. H. Eaton; *E. domingense* Urban

Herbs often robust and rank, with sessile or short-stalked leafy basal rosettes and dense fibrous roots. **Stems** erect, subterete, (20–)40–85(–120) cm, freely branched distally, subglabrous proximal to inflorescence, often with raised strigillose lines decurrent from petioles, densely strigillose distally. **Leaves** opposite proximal to inflorescence, alternate distally, petiole 4–10 mm; blade narrowly lanceolate to lanceolate or narrowly elliptic, 3–10(–15) × 0.5–3 cm, subequal to internodes, base rounded to cuneate, margins sharply and irregularly serrulate, 30–75 teeth per side, veins prominent, often raised abaxially, 10–25 per side, apex acute or acuminate, surfaces subglabrous with strigillose margins and abaxial veins; bracts abruptly reduced. **Inflorescences** usually upright panicles, sometimes corymbiform, rarely racemes, densely strigillose. **Flowers** erect; buds 2–3 × 1.5–2.5 mm; pedicel 5–10 mm; floral tube 0.3–0.6 × 0.3–0.5 mm, raised ring of sparse spreading hairs at mouth inside; sepals 1.3–3.2 × 0.5–1.5 mm, abaxial surface strigillose; petals white, 2.5–5.5 × 2–3.8 mm, apical notch 0.5–1 mm; filaments white, those of longer stamens 1.8–2.5 mm, those of shorter ones 1.5–2 mm; anthers pale yellow, 0.3–0.4 × 0.2–0.4 mm; ovary 15–30 mm, ± densely strigillose; style erect, white, 1.5–2.8 mm, glabrous, stigma cylindrical to subcapitate, entire, 1–1.5 × 0.8–1.2 mm, surrounded by longer anthers. **Capsules** 40–65 mm, surfaces strigillose; pedicel 8–12 mm. **Seeds** narrowly oblanceoloid, 1.2–1.7 × 0.3–0.5 mm, abruptly rounded with very short chalazal neck, brownish gray, surface evenly papillose; coma not easily detached, cinnamon red, 8–12 mm. $2n = 36$.

Flowering Jul–Sep. Saturated swampy areas, stream banks in lowland forests, wet ditches, open, disturbed wetlands, secondary floodplain forests; 0–500(–1500) m; N.B., Nfld. and Labr. (Nfld.), N.S., Ont., Que.; Ala., Ariz., Ark., Conn., Del., D.C., Ga., Ill., Ind., Iowa, Kans., Ky., La., Maine, Md., Mass., Mich., Minn., Mo., Nebr., N.H., N.J., N.Y., N.C., N.Dak., Ohio, Okla., Pa., R.I., S.C., S.Dak., Tenn., Tex., Vt., Va., W.Va., Wis.; West Indies (Dominican Republic, Haiti).

Epilobium coloratum is highly unusual in having a distribution primarily in southeastern North America. In its overall morphology, particularly leaves and flowers, it more closely resembles European species such as *E. obscurum* Schreber or *E. tetragonum* Linnaeus than any other North American species, and shares with the European species the widespread BB chromosome arrangement. There is a general resemblance between *E. coloratum* and the very widespread *E. ciliatum* (AA chromosome arrangement); hybrids between these species, known as *E.* ×*wisconsinense* Ugent, are highly sterile due to different chromosome arrangements in the parental species. The true affinities of *E. coloratum* are uncertain. Its occurrence on Hispaniola is also unique and possibly due to a recent introduction.

Prior to 1950, all major treatments of *Epilobium* attributed the name of *E. coloratum* to Muhlenberg (as *E. coloratum* Muhlenberg ex Wildenow), including C. Haussknecht (1884), W. Trelease (1891), and others. However, M. L. Fernald (1945d) noted that *Index Kewensis* had overlooked the 1807 publication by Biehler and virtually all treatments since that time follow the interpretation by Fernald.

25. Epilobium obscurum Schreber, Spic. Fl. Lips., 147, [155]. 1771 • Dwarf willowherb [I]

Herbs with elongated, leafy epigeal stolons. **Stems** erect or ascending, subterete, 20–80 cm, often well branched from base, sometimes also distally, subglabrous proximal to inflorescence with raised strigillose lines decurrent from margins of petioles, strigillose distally. **Leaves** opposite proximal to inflorescence, alternate distally, petiole 0–2 mm; blade green or slightly bluish green, lanceolate to narrowly ovate, 1.5–10 × 0.4–1.8 cm, ± shorter than internodes, base rounded to attenuate, margins denticulate with 15–40 evenly spaced teeth per side, veins prominent, 3–7 per side, apex subacute, surfaces sparsely strigillose, mainly on margins and veins; bracts gradually reduced. **Inflorescences** erect racemes or sparse panicles, strigillose. **Flowers** erect; buds 2–5 × 1–2.5 mm; pedicel 2–14 mm; floral tube 0.8–1 × 1.1–1.5 mm, conspicuous ring of spreading hairs at mouth inside, mixed strigillose and sparse glandular puberulent externally; sepals lanceolate, somewhat keeled, 2.5–4 × 1–1.3 mm, abaxial surface strigillose; petals rose-purple, 3.5–6 × 1.8–3 mm, apical notch 0.8–1.4 mm; filaments pale pink, those of longer stamens 2–2.2 mm, those of shorter ones 0.8–1.3 mm; anthers yellow, 0.7–0.8 × 0.4–0.5 mm; ovary 12–38 mm, strigillose; style white, 2.5–3.5 mm, glabrous, stigma clavate, 1.5–2 × 0.6–0.8 mm, surrounded by longer anthers. **Capsules** 40–70 mm, surfaces strigillose; pedicel 4–16 mm. **Seeds** narrowly obovoid, 0.9–1 × 0.3–0.4 mm, chalazal collar inconspicuous, brown, surface coarsely papillose; coma readily detached, dull white, 4–5 mm. $2n = 36$.

Flowering Jul–Aug. Ruderal areas, banks of ditches, streams, edges of swampy areas; 0–200[–500] m; introduced; B.C.; Mich., Wash.; Europe; introduced also in South America (Chile), Africa (Morocco), Pacific Islands (New Zealand), Australia (Tasmania).

Epilobium obscurum, native throughout Europe and the European part of Russia except the far north, to Turkey and the Azores, is one of several Eurasian species that has naturalized in North America, following multiple early introductions around east coast port cities, and later around the Great Lakes. E. G. Voss (1972–1996, vol. 2) reported a collection of *E. obscurum* made in 1927 in Michigan near Detroit, but despite efforts by Voss and others, no additional collections of this species have been detected in that area, suggesting that it failed to become naturalized there.

In the Pacific Northwest, W. Suksdorf grew and collected at least six European taxa in his garden in Bingen (Klickitat County, Washington); this included *E. obscurum* (as early as 1922) but none of those taxa became naturalized. However, one or more independent new introductions in the Seattle (P. Zika, pers. comm.) and Vancouver (F. Lomer, pers. comm.) regions appear to be more persistent and the species should be considered naturalized there.

26. Epilobium hirsutum Linnaeus, Sp. Pl. 1: 347. 1753
• Great or great hairy willowherb, codlins and cream, épilobe hirsute [I] [W]

Chamaenerion hirsutum (Linnaeus) Scopoli; *Epilobium amplexicaule* Lamarck; *E. aquaticum* Thuillier; *E. hirsutum* var. *villosum* (Thunberg) H. Hara; *E. villosum* Thunberg

Herbs usually robust and rank, sometimes woody near base, with thick, ropelike stolons to 1 m with scattered cataphylls and, often, terminal leafy rosette. **Stems** erect to ascending, often clumped, terete, 25–120(–250) cm, unusually thick, 3–9 mm diam., well branched mainly in distal ½, densely long-villous throughout, usually mixed glandular puberulent distally, rarely sparsely villous or densely white-tomentose. **Leaves** opposite proximal to inflorescence, alternate distally, sessile and ± clasping stem; blade elliptic-lanceolate to narrowly obovate or elliptic, 4–12(–23) × 0.3–4(–5) cm, base cuneate to attenuate, margins serrulate, 15–50 teeth per side, veins 6–9 per side, apex acute to acuminate or obtuse proximally, surfaces ± densely villous; bracts moderately reduced. **Inflorescences** erect racemes or panicles, usually densely villous and glandular puberulent, rarely tomentose. **Flowers** erect; buds 5–9 × 1.8–4.5 mm, sometimes beaked; pedicel 3–11 mm; floral tube 1.3–2.9 × 2.2–4 mm, conspicuous ring of spreading hairs near mouth inside; sepals oblong-linear, often keeled, 6–12 × 1–3 mm, abaxial surface densely pubescent; petals bright pink to rose-purple, rarely white, broadly obcordate, 9–20 × 7–15 mm, apical notch 1–3 mm; filaments white or pink, those of longer stamens 5–10 mm, those of shorter ones 2.5–6 mm; anthers cream, 1.5–3 × 0.6–1.2 mm; ovary 15–34 mm, densely villous and glandular puberulent; style white or pink, 5–12 mm, usually glabrous, stigma deeply 4-lobed, 1.8–2.2 × 3–5.5 mm, lobes recurved or spreading, exserted beyond anthers. **Capsules** often flushed purple, 25–90 mm, surfaces usually densely villous and glandular puberulent, rarely glabrescent; pedicel 5–20 mm. **Seeds** narrowly obovoid, 0.8–1.2 × 0.3–0.6 mm, chalazal collar inconspicuous, dark brown, surface coarsely papillose; coma easily detached, tawny or dull white, 7–10 mm. $2n = 36$.

Flowering Jun–Sep. Low wet areas along streams, rivers, ponds, and lakes, roadside ditches, along railroad tracks, marshes and swampy areas; 0–150[–3000] m; introduced; B.C., N.B., N.S., Ont., P.E.I., Que.; Colo., Conn., Ill., Ind., Ky., Maine, Md., Mass., Mich., N.H., N.J., N.Y., Ohio, Oreg., Pa., R.I., Utah, Vt., Wash., W.Va., Wis.; Eurasia; Africa.

Epilobium hirsutum is very widespread in cool temperate Eurasia and montane regions, occurring throughout Europe except in the far north (P. H. Raven 1968), through the Caucasus and central Asia (E. I. Steinberg 1949) to Nepal (Raven 1962), China (Chen C. J. et al. 1992), and Japan (A. W. Lievens and P. C. Hoch 1999). It occurs as well along the Mediterranean coast of Africa, through East Africa to southern Africa, and in the Canary and Cape Verde Islands (Raven 1967).

Epilobium hirsutum exceeds almost all other species of the genus in stature, so its size, very large flowers, and densely villous aspect make it easy to identify. R. L. Stuckey (1970) provided a detailed account of the introduction and spread of *E. hirsutum* in North America, noting the earliest known collection (July 1829) was from Newport, Rhode Island. Most early collections appeared in waste areas, particularly near harbor ballast piles, although some may have been grown in gardens. By the 1890s this species was well established along the Atlantic coastal region from New Jersey and Philadelphia through New England, and around Niagara Falls in the Great Lakes region. During the twentieth century, *E. hirsutum* spread extensively in southern Ontario and Quebec, south along the Atlantic coast to Maryland, and to all of the states along the southern shores of the Great Lakes, most recently including Wisconsin (1970), and Indiana (1972). It occurs in much the same habitat as that of another, more widely publicized invader, *Lythrum salicaria*, and sometimes is recorded as a companion species. The earliest known collection in western North America was made in 1933 in Bingen (Klickitat County), Washington. Whether from that introduction or others, *E. hirsutum* is now naturalized and widespread in the Pacific Northwest. It also was reported recently from the Denver region in Colorado and near Midway in Utah.

Epilobium grandiflorum F. H. Wiggers and *E. grandiflorum* Allioni are illegitimate names that pertain here.

SELECTED REFERENCE Stuckey, R. L. 1970. Distributional history of *Epilobium hirsutum* (great hairy willow-herb) in North America. Rhodora 72: 164–181.

27. Epilobium parviflorum Schreber, Spic. Fl. Lips., 146, [155]. 1771 • Smaller hairy willowherb [I] [W]

Herbs often robust and rank, with short-stalked leafy basal rosettes. **Stems** erect, terete, 18–100(–160) cm, well branched distally, densely gray-villous proximally, mixed villous and glandular puberulent distally, often with raised strigillose lines decurrent from margins of petioles. **Leaves** opposite proximal to inflorescence, alternate distally, petioles 1–3 mm proximally, sessile distally; blade narrowly lanceolate or oblong-lanceolate, 3–12 × 0.5–2.5 cm, often exceeding internodes, base rounded to broadly cuneate, margins serrulate, with 15–60 teeth per side, veins 4–8 per side, apex subacute, surfaces ± densely villous, hairs sometimes appressed; bracts usually much reduced. **Inflorescences** erect racemes or often leafy panicles. **Flowers** erect; buds 3.5–5.5 × 1.8–3 mm; pedicel 3–10 mm; floral tube 1–1.9 × 1.3–2.5 mm, a ring of spreading hairs at mouth within, densely villous and glandular puberulent abaxially; sepals narrowly lanceolate, often keeled, 2.5–6 × 1–1.5 mm; petals usually pink to rose-purple, rarely white, broadly obovate, 4–8.5 × 3–4.5 mm, apical notch 1–4 mm; filaments cream to light purple, those of longer stamens 2–6 mm, those of shorter ones 1–3.5 mm; anthers oblong, 0.8–1.3 × 0.4–0.6 mm; ovary 10–30 mm, mixed villous and glandular puberulent; style white to pink, 2.2–6 mm, glabrous, stigma deeply 4-lobed, 1–1.5 × 2.2–4 mm, lobes 1–1.8 mm, initially erect, later recurved, surrounded by or barely exserted beyond anthers. **Capsules** 30–70 mm, surfaces usually glandular puberulent, often mixed villous, rarely glabrescent; pedicel 5–18 mm. **Seeds** obovoid, 0.8–1.1 × 0.4–0.5 mm, chalazal collar inconspicuous, brown, surface coarsely papillose; coma easily detached, dingy white, 5–9 mm. $2n = 36$.

Flowering Jun–Sep. Disturbed, wet areas near streams, bogs, rivers, and lakes, often calcareous; 0–150[–1800] m; introduced; B.C., Ont.; Mich., N.J., N.Y., Ohio, Pa., Vt., Wash.; Eurasia; n Africa; introduced also in Pacific Islands (New Zealand).

Epilobium parviflorum is widespread in Eurasia, from Europe through the Caucasus and southern Asia to eastern China (Chen C. J. et al. 1992), and in northwestern Africa and the Canary Islands (P. H. Raven 1967). Prior to the report of naturalized populations of *E. parviflorum* in Ontario by N. J. Purcell (1976), the species was considered an ephemeral adventive in New Jersey, New York, and Pennsylvania, where collections were made on ballast heaps in 1877–1880 (W. Trelease 1891; H. A. Gleason 1952, vol. 2) but not subsequently. However, recent collections indicate well-established populations scattered widely across the Great Lakes region (Purcell; E. G. Voss 1972–1996, vol. 2; T. S. Cooperrider and B. K. Andreas 1991) and more recently in the Pacific Northwest. *Epilobium parviflorum* is clearly naturalized and can be expected to spread farther, given its weedy nature (Raven and T. E. Raven 1976).

Epilobium parviflorum most closely resembles *E. hirsutum*, sharing the otherwise unique combination of densely villous pubescence and 4-lobed stigmas, but differs by having smaller flowers, leaves not clasping and/or decurrent on stems, and perennating by rosettes rather than by thick ropy stolons. The two species co-occur throughout most of their range in Eurasia, and although their adventive ranges in North America are quite similar, *E. hirsutum* has spread much more widely and rapidly.

28. Epilobium montanum Linnaeus, Sp. Pl. 1: 348. 1753 • Willowherb [I]

Chamaenerion montanum (Linnaeus) Scopoli

Herbs with short, basal stolons terminating in fleshy or leafy rosettes at, or just below, ground level. **Stems** erect, terete, (5–)20–95 cm, often well branched, especially distally, densely strigillose proximal to inflorescence, without decurrent lines, mixed strigillose and glandular puberulent distally. **Leaves** opposite proximal to inflorescence, alternate distally, petiole 1–6 mm; blade dark green, narrowly ovate to ovate, 3–8 × 1.5–3.4 cm, subequal to internodes, base broadly cuneate to truncate, margins irregularly serrulate with 20–30 teeth per side, veins prominent, 3–5 per side, apex acuminate to acute, surfaces strigillose, especially along veins and margins; bracts much reduced. **Inflorescences** nodding to erect, racemes or loose panicles, strigillose and glandular puberulent. **Flowers** erect; buds 4–5 × 1.5–3 mm; pedicel 2–18 mm; floral tube 1.2–2 × 1.5–2.5 mm, conspicuous ring of spreading hairs at mouth inside; sepals often flushed red, keeled, 5–6.5 × 1.2–1.7 mm, abaxial surface strigillose and glandular puberulent; petals rose-purple, obcordate, 7.5–10 × 4–5.5 mm, apical notch 2.7–4.2 mm; filaments light pink, those of longer stamens 3.5–5.5 mm, those of shorter ones 1.5–2.4 mm; anthers yellow, 1–1.2 × 0.5–0.6 mm; ovary 20–30 mm, strigillose and glandular puberulent; style rose-purple to white, 3.5–7.5 mm, glabrous or with scattered long hairs near base, stigma deeply 4-lobed, 1.5–2 × 2.5–4 mm, lobes 1.2–1.8 mm, exserted beyond or, sometimes, surrounded by anthers. **Capsules** 40–80 mm, surfaces strigillose and glandular puberulent; pedicel 8–20 mm. **Seeds** obovoid, 1–1.2 × 0.4–0.5 mm, chalazal collar inconspicuous, 0.6–0.8 mm wide, brown, surface coarsely papillose; coma readily detached, dull white, 5–6 mm. **2*n*** = 36.

Flowering Jul–Aug. Moist rocky banks, gravelly slopes, open woods, disturbed or waste areas; 0–150 [–1500] m; introduced; B.C., Nfld. and Labr. (Nfld.); Maine; Europe; Asia; introduced also in Pacific Islands (New Zealand).

Epilobium montanum is native and widely distributed throughout Europe and in Asia at least as far east as the Ural Mountains and the Caucasus, and in Japan (A. W. Lievens and P. C. Hoch 1999). It also is questionably naturalized on Campbell Island, New Zealand (P. H. Raven and T. E. Raven 1976). Although *E. montanum* was first collected in 1894 in the vicinity of St. John's, Newfoundland, and clearly naturalized in that area, it does not appear to have spread appreciably in eastern North America, unlike European adventives such as *E. hirsutum* or *E. parviflorum*. The collection by F. C. Seymour in 1971 in Maine may be an independent introduction. Collections of *E. montanum* in the University of British Columbia Botanical Garden in Vancouver appeared to have been ephemeral, but more recent collections in ruderal areas in that region suggest that it has naturalized there. Like other European introductions, *E. montanum* has the BB chromosome arrangement (S. R. Seavey and P. H. Raven 1977).

29. Epilobium luteum Pursh, Fl. Amer. Sept. 1: 259. 1813 • Yellow willowherb [E]

Herbs usually with scaly, subterranean soboles, rarely condensed basal turions. **Stems** several, ascending to suberect, loosely clumped, subterete, 15–75 cm, simple or sparsely branched distally, subglabrous proximal to inflorescence with densely strigillose lines decurrent from margins of petioles, mixed strigillose and glandular puberulent distally. **Leaves** opposite proximal to inflorescence, alternate distally, petioles 1–3 mm proximally, subsessile distally; blade ovate or elliptic, 2.5–7.8 × 1.2–3.5 cm, base attenuate, margins denticulate with 8–20 low, ± pellucid teeth per side, 4–7 prominent veins per side, apex acute to acuminate, surfaces subglabrous with scattered strigillose hairs on margin and abaxial midrib; bracts much reduced, narrower, more acuminate and denticulate. **Inflorescences** nodding, later erect, racemes, congested, simple to sparsely branched, densely mixed strigillose and glandular puberulent. **Flowers** erect; buds 6–10 × 4–5 mm; pedicel 5–8 mm; floral tube 1.2–3 × 3–4.6 mm, ring of spreading hairs at mouth inside; sepals greenish cream, lanceolate, 10–12 × 3–3.5 mm, abaxial surface densely glandular puberulent; petals cream to pale yellow, 12–22 × 9–13 mm, broad apical notch 1.2–2.4 mm; filaments cream, those of longer stamens 13–16.5 mm, those of shorter ones 10–15 mm; anthers yellow, 2.2–3 × 0.7–1.3 mm; ovary 20–35 mm, densely glandular puberulent; style cream, 15–22 mm, stigma 4-lobed, 1–2 × 2.8–4.5 mm, lobes spreading to recurved, strongly exserted beyond anthers. **Capsules** erect, 35–75 mm, surfaces sparsely glandular puberulent; pedicel 10–22 mm. **Seeds** narrowly fusiform or oblanceoloid, 1–1.2 × 0.4–0.5 mm, chalazal collar 0.06–0.1 mm, surface reticulate; coma persistent, tawny, 6.5–8 mm. **2*n*** = 36.

Flowering Jul–Sep. Moist rocky slopes, seeps, banks of lakes, streams, springs, and gravel bars along coastal (boreal) to montane or subalpine areas near snowfields; 0–2200 m; Alta., B.C.; Alaska, Calif., Oreg., Wash.

Epilobium luteum is one of the most distinctive species in the genus due to its large creamy flowers, a color otherwise known only in the distantly related *E. suffruticosum*. It appears that *E. luteum* is most closely related to species of the Alpinae group, sharing not only similarities in perennating habit and structures, but also the derived CC chromosome arrangement (S. R. Seavey and P. H. Raven 1977, 1978).

Because *Epilobium luteum* is modally outcrossing and commonly pollinated by bees, it sometimes hybridizes with several other species of *Epilobium* when they grow sympatrically. One recurrent hybrid combination is *E. luteum* × *E. ciliatum* subsp. *glandulosum*, first described and named *E.* ×*treleasianum* H. Léveillé, and later *E. luteum* var. *lilacinum* L. F. Henderson. Following a suggestion by P. A. Munz (1965) that *E.* ×*treleasianum* was a hybrid of that particular parentage, S. R. Seavey (1993) verified the relationship in a series of crossing experiments. *Epilobium* ×*treleasianum* occurs over a wide geographical range in Alaska, British Columbia, and Washington, forming repeatedly when the parental species co-occur, and often backcrossing with one or both parents, forming hybrid swarms with varying morphological combinations. It also grows vigorously vegetatively.

Another less common hybrid of *Epilobium luteum* was described as *E.* ×*pulchrum* Suksdorf, and a minor variant as *E.* ×*pulchrum* var. *albiflorum* Suksdorf (S. R. Seavey 1993). The second parent of these hybrids is less obvious, but based on morphological features, it is most likely *E. hornemannii*, which also grows frequently in sympatry with *E. luteum*.

30. Epilobium hornemannii Reichenbach, Iconogr. Bot. Pl. Crit. 2: 73. 1824 (as hornemanni) • Hornemann's willowherb, épilobe de Hornemann

Epilobium nutans Hornemann in G. C. Oeder et al., Fl. Dan. 8: plate 1387. 1810, not F. W. Schmidt 1794

Herbs with short, scaly hypogeous or leafy epigeous soboles. **Stems** ascending to erect, clumped, terete, 10–45 cm, usually simple, rarely branched proximally, subglabrous proximal to inflorescence with sparsely strigillose lines decurrent from margins of petioles, ± sparsely mixed strigillose and glandular puberulent distally. **Leaves** opposite proximal to inflorescence, usually alternate distally, petioles 3–9 mm proximally to subsessile distally; blade broadly elliptic to spatulate proximally, ovate to lanceolate distally, ± coriaceous or not, 1.5–6.2 × 0.7–2.9 cm, base attenuate to cuneate or rounded, margins subentire proximally, denticulate distally with 10–25 teeth per side, veins often inconspicuous, 4–7 per side, apex obtuse to subacute, surfaces glabrous or, sometimes, strigillose along margins; bracts reduced. **Inflorescences** erect or nodding, open racemes, mixed strigillose and glandular puberulent. **Flowers** erect; buds 2–5.5 × 2–4 mm; pedicel 2–5 mm; floral tube 1–2.2 × 1.3–2.8 mm, sparse ring of hairs at mouth inside or ring absent; sepals sometimes red-tipped or bright red, 2–7 × 1–2.2 mm, abaxial surface sparsely strigillose and glandular puberulent; petals usually rose-purple or magenta to light pink, rarely white, 3–10(–11) × 2–6 mm, apical notch 0.7–2.4 mm; filaments cream to light pink, those of longer stamens 1.4–5(–6) mm, those of shorter ones 1.2–4 mm; anthers light yellow, 0.4–1.2 × 0.3–0.6 mm; ovary 15–25 mm, glandular puberulent, sometimes mixed strigillose; style white or cream, 2–8 mm, stigma cream, clavate or cylindrical, entire, 1.2–3 × 0.5–1 mm, usually surrounded by, rarely exserted beyond, anthers. **Capsules** 35–65 mm, surfaces glandular puberulent, sometimes mixed strigillose; pedicel 5–15(–25) mm. **Seeds** narrowly fusiform or oblanceoloid, 0.9–1.6 × 0.3–0.5 mm, chalazal collar short, 0.05–0.1 mm, blond to brown, surface distinctly papillose or reticulate/smooth; coma readily detached, dingy white, 6–11 mm.

Subspecies 2 (2 in the flora): North America, Eurasia.

Epilobium hornemannii occurs widely in montane and boreal regions in North America and western Eurasia, and also in Japan and the Russian Far East. It is characterized by having the CC chromosome arrangement and is included in the Alpinae alliance with *E. anagallidifolium*, *E. lactiflorum*, and others (I. Kytövuori 1972).

W. Trelease (1891) discussed eastern and western forms of *Epilobium hornemannii*, the latter divided into two variations; however, he did not formally recognize any of these variants.

P. A. Munz (1965) included the Eurasian *Epilobium alsinifolium* Villars in his North American treatment, noting that it occurred in Greenland. However, B. Fredskild (1984) suggested that, for the most part, these determinations represent misidentifications of *E. hornemannii*.

The two subspecies recognized here intergrade throughout much of their shared range, but whereas subsp. *hornemannii* is commonly found in high montane to alpine regions, in the northern part of its range it grows at much lower elevations, and in maritime areas is replaced by coriaceous-leaved forms here designated as subsp. *behringianum*. The situation is rather analogous to the pattern seen in *E. ciliatum* in which subsp. *ciliatum* is wide-ranging and variable, but replaced in Pacific maritime areas by subsp. *watsonii*, from which it differs consistently in most specimens, but sometimes intergrades.

1. Leaves not coriaceous, petioles 3–7 mm proximally; sepals 2–4.5 mm; petals 3–9 mm; capsules 40–65 mm; seeds 0.9–1.2 mm, surface papillose . . . . 30a. *Epilobium hornemannii* subsp. *hornemannii*
1. Leaves ± coriaceous, petioles 4–9 mm proximally; sepals 5–7 mm; petals 8–10(–11) mm; capsules 35–55 mm; seeds 0.9–1.6 mm, surface reticulate . . . . . 30b. *Epilobium hornemannii* subsp. *behringianum*

30a. Epilobium hornemannii Reichenbach subsp. **hornemannii**

Epilobium cupreum Lange; *E. foucaudianum* H. Léveillé; *E. steckerianum* Fernald

Leaves not coriaceous, petiole 3–7 mm proximally; cauline blade 1.5–5.5 × 0.7–2.9 cm. **Flowers:** sepals sometimes red-tipped, 2–4.5 × 1–2.2 mm; petals usually rose-purple to light pink, rarely white, 3–9 × 2–5.5 mm. **Capsules** 40–65 mm. **Seeds** 0.9–1.2 × 0.3–0.5 mm, surfaces distinctly papillose. **2*n*** = 36.

Flowering (May–)Jun–Aug. Banks of montane to alpine streams and lakes, open tussock meadows, willow swales, gravelly ridges, stabilized scree slopes, roadside ditches; 50–3700 m; Greenland; Alta., B.C., N.B., Nfld. and Labr. (Nfld.), N.W.T., N.S., Nunavut, Ont., Que., Yukon; Alaska, Ariz., Calif., Colo., Idaho, Maine, Mont., Nev., N.H., N.Mex., N.Y., Oreg., S.Dak., Utah, Wash., Wyo.; n Eurasia.

Subspecies *hornemannii* is one of the more common and widely distributed members of the Alpinae alliance and carries the CC chromosomal arrangement. It grows from near sea level in the north (Alaska, Northwest Territories) to high alpine (to 3700 m) in the southern Rockies and Sierra Nevada; it also occurs in northern Europe, subarctic and Russian Far East, and Japan. In Nunuvut, it is found only on Charlton Island in James Bay. It often grows sympatrically with *E. ciliatum* and *Epilobium saximontanum* (both AA arrangement); occasional hybrids are found, but they often have much reduced seed fertility, probably due to chromosomal irregularities.

Subspecies *hornemannii* occurs occasionally in spring pools, forming distinctive dense floating mats of ascending stems. Specimens growing in very moist habitats often have pale, thin epigeal stolons with scales, rather than typical leafy soboles.

30b. Epilobium hornemannii Reichenbach subsp. **behringianum** (Haussknecht) Hoch & P. H. Raven, Ann. Missouri Bot. Gard. 64: 136. 1977

Epilobium behringianum Haussknecht, Monogr. Epilobium, 277. 1884; *E. alpinum* Linnaeus var. *behringianum* (Haussknecht) S. L. Welsh; *E. bongardii* Haussknecht

Leaves ± coriaceous, petiole 4–9 mm proximally; cauline blade 1.5–6.2 × 0.7–2.2 cm. **Flowers:** sepals bright red, 5–7 × 2–2.5 mm; petals magenta to light pink, 8–10(–11) × 3–6 mm. **Capsules** 35–55 mm. **Seeds** 0.9–1.6 × 0.4–0.5 mm, surfaces reticulate (smooth). **2*n*** = 36.

Flowering Jun–Aug. Moist meadows, seeps, stream banks, mossy or rocky crevices, usually coastal; 0–500 m; B.C., Yukon; Alaska; Asia (Russian Far East).

Subspecies *behringianum* is a very distinctive coastal variant of the species, essentially replacing it in maritime areas from northern British Columbia, including Haida Gwaii (the Queen Charlotte Islands), through Pacific coastal Alaska, and in similar habitats along the northwestern Pacific coast in Russia.

31. Epilobium lactiflorum Haussknecht, Oesterr. Bot. Z. 29: 89. 1879 • Milkflower willowherb, épilobe à fleurs blanches

Epilobium alpinum Linnaeus var. *lactiflorum* (Haussknecht) C. L. Hitchcock; *E. canadense* H. Léveillé; *E. canadense* var. *albescens* H. Léveillé; *E. hornemannii* Reichenbach var. *lactiflorum* (Haussknecht) D. Löve

Herbs with short, leafy epigeal soboles. **Stems** ascending to suberect, often clumped, terete, 15–50 cm, usually simple, rarely branched proximally, subglabrous proximal to inflorescence except for raised densely strigillose lines decurrent from margins of petioles, usually mixed strigillose and glandular puberulent distally. **Leaves** opposite proximal to inflorescence or just proximal ⅓, alternate distally, petiole 3–12 mm, ± winged; blade broadly spatulate to ovate proximally, narrowly ovate to narrowly lanceolate distally, 2–5.5 × 0.8–2.4 cm, base attenuate to cuneate, margins subentire proximally to denticulate distally with 7–16 teeth per side, more marked distally, lateral veins inconspicuous, 4–8 per side, apex obtuse proximally to subacute distally, surfaces glabrous except for strigillose margins; bracts reduced and narrower. **Inflorescences** nodding in bud, later erect, ± open racemes, mixed strigillose and glandular puberulent. **Flowers** suberect; buds 2–5 × 1.5–3.5 mm; pedicel 5–15 mm; floral tube 1–2.2 × 1–3 mm, inner surface glabrous without ring; sepals often flushed purplish red, frequently keeled, (2–)3–5.5 × 0.9–1.8 mm, abaxial surface sparsely glandular puberulent, sometimes mixed strigillose; petals white, rarely with red veins or flushed light pink, 3–8.5 × 1.6–4.5 mm, apical notch 0.7–1.4 mm; filaments white to cream, those of longer stamens 1.4–4 mm, those of shorter ones 1.1–3 mm; anthers light yellow, 0.4–0.9 × 0.3–0.6 mm; ovary 20–40 mm, glandular puberulent; style cream or white, 1.4–4.6 mm, stigma clavate or rarely subcapitate and indented apically, entire, 1.2–2.5 × 0.4–1.6 mm, surrounded by anthers. **Capsules** slender, sometimes flushed reddish green, ± ascending, 50–100 mm, surfaces sparsely glandular puberulent; pedicel 15–45 mm. **Seeds** narrowly obovoid, 1.1–1.7 × 0.4–0.6 mm, chalazal collar 0.05–0.1 mm, blond or light brown, surface reticulate or sometimes barely rugose; coma easily detached, white, 7–14 mm. $2n = 36$.

Flowering Jun–Sep. Montane stream banks, moist crevices and ledges, gravelly roadsides, burned-over woodlands, sandy moraines, subalpine forests, alpine meadows; 50–3800 m; Greenland; Alta., B.C., N.B., Nfld. and Labr., N.W.T., N.S., Ont., Que., Yukon; Alaska, Ariz., Calif., Colo., Idaho, Maine, Mont., Nev., N.H., N.Mex., Oreg., Utah, Vt., Wash., Wyo.; Eurasia.

Epilobium lactiflorum has a nearly circum-subarctic distribution in North America (including coastal Greenland) and Eurasia, extending south into alpine and cool montane habitats along mountain axes. This distribution is similar to that of *E. anagallidifolium* and *E. hornemannii* (all with CC chromosomal arrangement), and these species commonly grow in similar habitats as well.

Petal color can be variable in many *Epilobium* species, but *E. lactiflorum* (white flowers) differs quite consistently from *E. hornemannii* (rose-purple to light pink or rarely white) in that feature. Mature fruits and pedicels are also fairly longer in *E. lactiflorum*. Despite their morphological similarities and broadly overlapping ranges and habitats, *E. lactiflorum* and *E. hornemannii* subsp. *hornemannii* do not appear to hybridize with much frequency, although intermediates, with only moderately reduced seed fertility, might be difficult to detect.

32. Epilobium anagallidifolium Lamarck in J. Lamarck et al., Encycl. 2: 376. 1786 • Alpine or pimpernel willowherb, épilobe à feuilles de mouron [F]

Epilobium alpinum Linnaeus, name rejected; *E. pseudoscaposum* Haussknecht

Herbs with spreading thin, small-leafed epigeous soboles to 5 cm. **Stems** many, ascending, often sigmoidally bent, nodding distally, later erect, clumped or mat-forming, terete, 3–20(–25) cm, simple, subglabrous, sometimes with faint raised strigillose lines decurrent from margins of petioles, rarely mixed strigillose and sparsely glandular puberulent distally. **Leaves** opposite and crowded proximal to inflorescence, alternate distally, petioles 1–6 mm, rarely subsessile distally; blade spatulate to oblong proximally, elliptic to narrowly lanceolate or sublinear distally, (0.5–)0.8–2.5 × 0.3–1 cm, base attenuate to cuneate, margins subentire proximally, sparsely denticulate distally with 2–5 low teeth per side, veins obscure, 2–4 per side, apex obtuse or rounded proximally to subacute distally, surfaces subglabrous; bracts reduced, usually much narrower. **Inflorescences** nodding in bud, later suberect, few-flowered racemes, subglabrous to sparsely strigillose and/or glandular puberulent. **Flowers** suberect; buds 2–5 × 1–2 mm; pedicel 1–6(–15) mm; floral tube 0.6–1.2 × 0.8–1.8 mm, slightly raised subglabrous ring at mouth inside; sepals green to reddish purple, 1.5–5 × 0.6–1.5 mm, abaxial surface subglabrous to sparsely glandular; petals usually pink to rose-purple, rarely white, narrowly obcordate, (1.7–)2.5–6.5(–8) × 1.6–3.5 mm, apical notch 0.5–1.2 mm; filaments cream to light pink, those of longer stamens 1.4–3.2 mm, those of shorter ones 0.7–2 mm; anthers 0.3–0.6 × 0.2–0.4 mm; ovary often reddish purple, 6–20 mm, subglabrous or sparsely strigillose and glandular puberulent; style white, 1.2–2.5 mm, glabrous, stigma broadly clavate to subcapitate, entire, 0.9–1.5 × 0.4–0.7 mm, surrounded by longer anthers. **Capsules** slender, often reddish purple, 17–40(–55) mm, surfaces subglabrous or with scattered hairs; pedicel 5–35(–68) mm. **Seeds** narrowly obovoid, 0.7–1.4 × 0.3–0.5 mm, inconspicuous chalazal collar 0.1–0.2 mm wide, light brown, surface reticulate (smooth); coma persistent, dull white, 2–4 mm. $2n = 36$.

Flowering Jun–Sep. Moist flats, stream banks, subarctic coastal marsh edges, high montane and alpine meadows and seeps; 0–4500 m; Greenland; Alta., B.C., Nfld. and Labr., N.W.T., Nunavut, Que., Yukon; Alaska, Calif., Colo., Idaho, Maine, Mont., Nev., N.H., Oreg., Utah, Wash., Wyo.; Eurasia.

Epilobium anagallidifolium is widely but sparsely distributed in high montane-alpine and subarctic Eurasia, including Europe, Russia, China, and Japan.

Epilobium anagallidifolium usually forms low clumps or mats, with stems nodding in bud and usually subglabrous below the inflorescence. Many collections of *E. anagallidifolium* from eastern Canada and Greenland tend to be unusually tall (to 25 cm) and robust for the species, with somewhat larger, thicker leaves, and longer pedicels (to 60 mm). Similarly large and robust specimens occur scattered in Yukon and Washington, and may result from occasional hybridization and introgression with sympatric species such as *E. hornemannii* or *E. lactiflorum*, which also have the CC chromosomal arrangement. In an analysis of Fennoscandian populations of the Alpinae group, I. Kytövuori (1972) found a similar pattern of mostly smaller, sigmoidal plants of *E. anagallidifolium* with a small proportion of larger ones, and he also suggested the possibility of hybridization and/or introgression.

Plants of *Epilobium anagallidifolium*, and indeed of the whole Alpinae group, from Haida Gwaii (the Queen Charlotte Islands) of British Columbia (J. A. Calder and R. L. Taylor 1968), are particularly distinctive compared to those on the mainland, and difficult to interpret. The observed differences may be the result of hybridization with other sympatric species or a response to unique ecological conditions on the islands, reinforced by relative isolation from mainland British Columbia.

The Linnaean name *Epilobium alpinum* has long been a source of nomenclatural confusion and instability, since it circumscribed at least four distinct species, especially *E. anagallidifolium*. A proposal by P. C. Hoch et al. (1995) to permanently reject the name *E. alpinum* Linnaeus was approved.

33. Epilobium clavatum Trelease, Rep. (Annual) Missouri Bot. Gard. 2: 111, plate 48. 1891 • Talus or clavatefruit willowherb [E]

Epilobium alpinum Linnaeus var. *clavatum* (Trelease) C. L. Hitchcock

Herbs with wiry, scaly soboles just below ground level, often with extended semi-woody rootstock. **Stems** numerous, ascending, clumped, terete, 5–15(–22) cm, usually simple, rarely branched, subglabrous, with raised strigillose lines decurrent from petioles, ± densely strigillose and often mixed glandular puberulent distally. **Leaves** crowded and opposite proximal to inflorescence, alternate distally, petiole 0–3 mm; blade obovate proximally to ovate or elliptic distally, 0.5–2.8 × 0.6–1.6 cm, base attenuate proximally to obtuse distally, margins subentire to sparsely denticulate, 3–6 teeth per side, veins obscure, 2–4 per side, apex obtuse proximally to subacute distally, surfaces subglabrous or sparsely strigillose margins and abaxial midrib, sometimes subglaucous; bracts not much reduced, sessile. **Inflorescences** usually erect, rarely slightly nodding, racemes, strigillose and glandular puberulent, sometimes sparsely so. **Flowers** erect; buds often purplish green, 3–4.5 × 1.4–2.2 mm; pedicel 3–9 mm; floral tube 0.6–2 × 1–2 mm, glabrous or with a raised ring of sparse hairs at mouth inside; sepals often purplish green, 2.5–4.2 × 1–2 mm, abaxial surface sparsely glandular puberulent to subglabrous; petals rose-purple to pale pink, obcordate, 3.5–6(–7) × 2–4 mm, apical notch 0.5–1 mm; filaments cream, those of longer stamens 1.8–4 mm, those of shorter ones 1–3 mm; anthers light yellow, 0.4–0.9 × 0.25–0.5 mm; ovary often reddish purple, 8–20 mm, densely glandular puberulent, often mixed strigillose, rarely subglabrous; style white or pale pink, 1.4–3.2 mm, glabrous, stigma cream, narrowly clavate to subcapitate, 0.8–1.4 × 0.3–0.8 mm, surrounded by at least longer anthers. **Capsules** often purplish red, 20–42 mm, surfaces sparsely pubescent or subglabrous; pedicel 2–21 mm. **Seeds** narrowly obovoid or fusiform, (1.3–)1.5–2.1 × 0.4–0.7 mm, chalazal collar conspicuous, 0.04–0.16 × 0.2–0.4 mm, blond, surface finely reticulate; coma easily detached, white, 5–8 mm. **2*n*** = 36.

Flowering May–Sep. Rocky crevices, scree slopes, ledges, stream banks, often near snow banks or moraines in upper montane to alpine zones; 800–4200 m; Alta., B.C., N.W.T., Yukon; Alaska, Calif., Colo., Idaho, Mont., Nev., Oreg., Utah, Wash., Wyo.

Epilobium clavatum shares a clumped habit and the CC chromosomal arrangement with related species in the Alpinae group, but differs from them by its dense, wiry mass of basal soboles arising from an extended and somewhat woody caudex and relatively thick capsules and large seeds (1.3–2 mm). This unusual habit morphology may be the result of growing on unstable, shifting scree slopes. Like *E. anagallidifolium*, with which it often grows in near-sympatry in alpine areas, it is of notably low stature, often less than 15 cm, and has subentire leaves and capsules rarely exceeding 4 cm. However, *E. clavatum* does not nod in bud, and generally is more robust than *E. anagallidifolium*, and it has a much smaller range, being endemic only to the western North American cordilleran region, whereas *E. anagallidifolium* has a discontinuous circumboreal range.

34. Epilobium smithii H. Léveillé, Repert. Spec. Nov. Regni Veg. 5: 8. 1908 [E]

Epilobium clavatum Trelease var. *glareosum* (G. N. Jones) Munz; *E. glareosum* G. N. Jones

Herbs with sprawling, wiry underground soboles with brownish tan scalelike leaves, arising from semi-woody extended caudex. **Stems** 10–30+, ascending to erect, clumped, terete, 6–35 cm, usually simple, rarely slightly branched distally, strigillose throughout, especially on raised lines decurrent from margins of petioles, densely glandular puberulent distally. **Leaves** opposite proximal to inflorescence, alternate distally, petiole 0–5 mm; blade dark or grayish green, lanceolate to subovate, 1–3.8 × 0.3–1.5 cm, base attenuate proximally to rounded distally, margins low-denticulate with 4–15 teeth per side, lateral veins usually indistinct, 2–5 per side, apex subacute to blunt, surfaces sparsely glandular puberulent on margins and veins; bracts scarcely reduced. **Inflorescences** erect or sometimes nodding in bud, short racemes, glandular puberulent. **Flowers** few, erect or sometimes slightly nodding in bud; buds 3–4.5 × 2.5–3 mm; pedicel 5–10 mm; floral tube 1–2.2 × 1.2–2.2 mm, with raised ring of sparse hairs at mouth inside; sepals often red along margins, lanceolate, 3–4.8 × 1–2.1 mm, abaxial surface scattered mixed glandular puberulent and strigillose; petals dark pink to rose-purple, obcordate, (3–)5–7 × (2–)3–4.5 mm, apical notch 1–2.5 mm; filaments pale pink, those of longer stamens 2.4–4.2 mm, those of shorter ones 1.2–2.6 mm; anthers pale yellow, 0.5–1.1 × 0.3–0.7 mm; ovary 15–22(–26) mm, densely glandular puberulent; style cream, 2.5–3.5 mm, often with scattered hairs near base, stigma clavate to subcapitate, 0.8–1.5 × 0.5–1 mm, surrounded by longer anthers. **Capsules** 24–65 mm, surfaces glandular puberulent; pedicel 10–30 mm. **Seeds** narrowly obovoid, (1.2–)1.4–1.7 × 0.4–0.7 mm, chalazal collar conspicuous, 0.08–0.15 × 0.15–0.25 mm, light brown, surface densely irregular papillose; coma persistent, dull white, 6–12 mm. $2n$ = 36.

Flowering Jul–Sep. Moist talus or scree slopes, crevices of rocky outcrops, often on south-facing subalpine to alpine slopes; (1000–)1500–3000 m; Alta., B.C.; Mont., Utah, Wash.

Epilobium smithii has a restricted distribution, relatively abundant on the Olympic Peninsula (Washington) and Vancouver Island (British Columbia) and more scattered across northern Washington to the Waterton-Glacier International Peace Park in Montana and adjacent Alberta. A single collection from the Uinta Mountains in Utah suggests that the range may be larger.

Although *Epilobium smithii* has been generally ignored, it differs strikingly from most other species of *Epilobium* by virtue of being densely glandular puberulent all around the upper stems. It is most similar to *E. clavatum*, with which some authors combined it and with which it may be closely related in the CC chromosome group.

Collections of *Epilobium smithii* are often mixed and include other species such as *E. anagallifolium*, *E. clavatum*, and *E. lactiflorum*, and less often *E. leptocarpum* and *E. mirabile*, the range of which all overlap with that of *E. smithii*. Despite the observed sympatry of these species, their similarity in floral features, and their capacity to hybridize (S. R. Seavey and P. H. Raven 1978), few obvious hybrids have been found.

35. Epilobium mirabile Trelease, Contr. U.S. Natl. Herb. 11: 404. 1906 • Olympic Mountain willowherb E

Herbs with sessile, compact, fleshy turions that leave dark basal scales. **Stems** erect, loosely or not clumped, terete, 7–30 cm, usually simple, rarely branched, subglabrous with raised strigillose lines decurrent from margins of petioles, or densely strigillose and without raised lines. **Leaves** opposite proximal to inflorescence, alternate distally, petioles 1–3 mm proximally, subsessile distally; blade ovate to narrowly ovate, coriaceous, 1.5–3 × 0.7–1.2 cm, base rounded to cuneate, margins denticulate, 8–12 teeth per side, veins indistinct, 4–9 per side, apex obtuse proximally to acute distally, surfaces sparsely strigillose, mainly on margins and midrib; bracts not much reduced. **Inflorescences** erect racemes, rarely branched, densely strigillose. **Flowers** erect; buds 3–4 × 1.5–2.2 mm; pedicel 4–5 mm; floral tube 1.5–2 × 1.6–2.2 mm, sparsely glandular puberulent, sometimes mixed strigillose; sepals often purplish red, 2–3.2 × 1.5–2.4 mm; petals white, often red-tinged at apex, 3.8–5 × 2–3 mm, apical notch 0.4–0.8 mm; filaments cream, those of longer stamens 1.4–2.3 mm, those of shorter ones 0.8–1.4 mm; anthers 0.4–0.6 × 0.3–0.5 mm; ovary 10–18 mm, densely strigillose and glandular puberulent; style yellow or light pink, 2–2.3 mm, stigma broadly clavate, 0.8–1 × 0.6–0.7 mm, surrounded by longer anthers. **Capsules** 30–45 mm, relatively thick (2–3 mm), surfaces ± sparsely glandular puberulent and mixed strigillose; pedicel 5–16 mm. **Seeds** narrowly obovoid, 1.7–2.2 × 0.6–0.8 mm, chalazal collar inconspicuous, gray to light brown, surface low papillose or reticulate; coma readily detached, white, very full, 10–15 mm. $2n$ = 36.

Flowering Jul–Aug. Subalpine scree slopes, gravelly tussock meadows; 1500–2600 m; Alta., B.C.; Mont., Wash.

Epilobium mirabile also has the CC chromosomal arrangement and is one of the least common species of *Epilobium* in North America; fewer than 20 collections are known, even though its range is quite large. Most collections are from the Olympic Peninsula in Washington or Waterton-Glacier International Peace Park in Alberta and adjacent Montana. However, one collection is known from Powell County in central Montana, and one from Manning Provincial Park in British Columbia. The species may be more widespread but under-collected due to its restricted habitat, mainly on subalpine south-facing scree slopes.

Specimens of *Epilobium mirabile* from the northern Rocky Mountains (Alberta and Montana) have subglabrous stems with strong, raised, strigillose lines and seeds with low papillose surfaces, whereas specimens from the northern Cascades (British Columbia) and Olympic Mountains (Washington) have densely strigillose stems with no raised lines and seeds with reticulate surfaces. The plants otherwise have very similar and distinctive morphology and ecology.

36. Epilobium leptocarpum Haussknecht, Monogr. Epilobium, 258, plate 14, fig. 67. 1884 • Slenderfruit willowherb [E]

Epilobium paddoense H. Léveillé

Herbs with numerous small, (3–5 × 2–3 mm), fleshy, sessile turions at or just below ground level, often also with bulblets in proximal to mid cauline leaf axils. **Stems** erect or ascending, often clumped, often flushed red, terete, 8–30 cm, simple or branched, glabrous proximal to inflorescence except for elevated strigillose lines decurrent from margins of petiole, strigillose throughout distally. **Leaves** opposite proximal to inflorescence, alternate distally, petiole broad, 3–5 mm proximally to subsessile distally; blade lanceolate or elliptic to narrowly lanceolate distally, 0.8–4 × 0.4–1.3 cm, base attenuate to cuneate, margins subentire to denticulate distally, 5–9 teeth per side, veins 3 or 4 per side, apex obtuse to distally acute, surfaces glabrous with sparsely strigillose margins and abaxial midrib; bracts reduced and narrower. **Inflorescences** nodding to suberect, racemes, strigillose. **Flowers** erect; buds 2–4 × 1.5–2.5 mm; pedicel 7–12 mm; floral tube 1–2 × 1–2 mm, with ring of spreading hairs at mouth inside; sepals green, narrowly lanceolate, 3–4 × 0.8–1.2 mm; petals white fading to pink, 4–6.5 × 2.3–4 mm, apical notch 0.8–1.6 mm; filaments white, those of longer stamens 2.2–3 mm, those of shorter ones 1.2–1.6 mm; anthers 0.4–0.8 × 0.4–0.6 mm; ovary 12–18 mm, densely strigillose; style white, 3.2–4 mm, stigma cream, broadly clavate, 0.6–1.5 × 0.6–1.2 mm, surrounded by at least longer anthers. **Capsules** often reddish green, 25–55 mm, thin, surfaces sparsely strigillose; pedicel 15–38 mm. **Seeds** subfusiform, 0.8–1.2 × 0.3–0.4 mm, chalazal collar inconspicuous, 0.02–0.06 mm, light brown, surface papillose; coma persistent, tawny, 3–6 mm. $2n = 36$.

Flowering Jun–Aug. Boreal/montane moist meadows, stream banks, moist bluffs, sun to part shade, gravelly or sandy soils, mossy ground; 0–2400 m; Alta., B.C.; Alaska, Idaho, Mont., Oreg., Wash.

Epilobium leptocarpum often occurs sympatrically with other species of the Alpinae alliance within its Pacific Northwest range, and shares with them the CC chromosomal arrangement. However, it has small compact basal turions and, almost uniquely in *Epilobium*, bulblets in the leaf axils. The only other species with bulblets is *E. fauriei* H. Léveillé, an endemic in Japan that has strikingly similar overall morphology to that *E. leptocarpum*, and shares with it the CC chromosome arrangement.

37. Epilobium howellii Hoch, Phytologia 73: 460. 1993 • Yuba Pass willowherb [E]

Herbs with short, threadlike stolons with scattered, minute leaves. **Stems** delicate, erect, loosely clumped, terete, 5–20 cm, simple or branched from base, densely glandular puberulent, without decurrent lines. **Leaves** opposite proximal to inflorescence, alternate distally, subsessile; blade broadly obovate to orbiculate proximally to ovate or lanceolate distally, 0.4–2 × 0.3–1.2 cm, base cuneate to rounded, margins finely denticulate, 4–6 low teeth per side, veins indistinct, 2–6 per side, apex obtuse to subacute distally, surfaces subglabrous with strigillose margins and veins or both surfaces sparsely strigillose distally; bracts much reduced and narrower. **Inflorescences** erect, open racemes, glandular puberulent. **Flowers** sometimes cleistogamous; buds 2–3.5 × 1–1.5 mm, often nodding; pedicel 8–16 mm; floral tube 0.4–0.8 × 0.5–1 mm, ring of sparse hairs at mouth inside, or absent; sepals 1.5–2 × 0.8–1 mm; petals white, 2–3 × 1.5–2 mm, apical notch 0.4–0.5 mm; filaments white, those of longer stamens 1.8–3 mm, those of shorter ones 1–1.5 mm; anthers 0.3–0.4 × 0.3–0.4 mm; ovary 9–12 mm, sparsely glandular puberulent; style white, 2–3 mm, stigma white, narrowly clavate, entire, 0.5–0.6 × 0.4–0.5 mm, surrounded by anthers. **Capsules** 35–45 mm, surfaces subglabrous to sparsely glandular puberulent; pedicel 25–40 mm. **Seeds** narrowly oblanceoloid, 0.8–1.1 × 0.3–0.4 mm, chalazal collar inconspicuous, surface low papillose; coma easily detached, dingy white, 3–6 mm. $2n = 36$.

Flowering Jul–Aug. Mossy seeps, semi-shaded swales, grassy montane meadows; 2000–2700 m; Calif.

Epilobium howellii is an enigmatic species in that it was apparently overlooked until relatively recently by numerous botanists in the Sierra Nevada in Fresno, Mono, and Sierra counties. Its similarity to many of the smaller species of *Epilobium*, especially in the Alpinae group, may have caused this oversight; it is very clearly distinguished from similar species by its exclusively glandular puberulent stems. Recent focused collecting efforts mainly by United States Forest Service personnel have shown that it is much more widespread than originally thought, although still uncommon, and its geographical and ecological range is still quite restricted compared to most other species of *Epilobium*.

38. Epilobium oreganum Greene, Pittonia 1: 225. 1887 • Grant's Pass willowherb [C] [E]

Epilobium glaucum Howell, Howell's Pacific Coast Pl., no. 1139. 1887, not Philippi 1864; *E. brevistylum* Barbey var. *exaltatum* (Drew) Jepson; *E. californicum* Haussknecht var. *exaltatum* (Drew) Jepson; *E. exaltatum* Drew; *E. glandulosum* Lehmann var. *exaltatum* (Drew) Munz; *E. subcaesium* Greene

Herbs with leafy basal rosettes or short shoots. **Stems** several to many, ascending or erect, terete, loosely clumped, 40–100 cm, usually well-branched apically, ± glabrous and glaucous proximal to inflorescence, without distinct raised lines, sparsely mixed strigillose and glandular pubescent proximally. **Leaves** opposite proximal to inflorescence, alternate distally, petiole broad, 1–3 mm; blade narrowly lanceolate to narrowly ovate, 3–9 × 0.7–2.5 cm, base rounded to cuneate, margins finely serrulate, 20–40 teeth per side, veins reddish green, conspicuous, 6–10 per side, apex acute, surfaces glabrous and often glaucous, crowded proximally; bracts much reduced and narrower. **Inflorescences** erect racemes or open panicles, often branched, glandular puberulent, sometimes mixed strigillose. **Flowers** erect; buds 5–8 × 2–3.5 mm, often with stigma exserted; pedicel 2–4 mm; floral tube 2–3 × 1.8–3 mm, with ring of spreading hairs near base of tube inside; sepals often flushed red, 6–10 × 2.1–2.8 mm, abaxial surface mixed strigillose and glandular puberulent; petals dark pink to rose-purple, (6–)10–15 × 4.5–6 mm, apical notch 2.6–3 mm; filaments cream or light pink, those of longer stamens 6–8 mm, those of shorter ones 3.5–4.5 mm; anthers cream or yellow, 1–1.2 × 0.5–0.6 mm; ovary 20–25 mm, densely glandular puberulent and mixed strigillose; style cream or yellow, 9–13 mm, stigma broadly and sometimes irregularly 4-lobed, 1–1.5 × 2.1–2.9 mm, exserted beyond anthers. **Capsules** 25–45 mm, surfaces mixed sparsely glandular puberulent and strigillose, often with reduced fertile seed set; pedicel 3–6 mm. **Seeds** narrowly obovoid, 0.9–1.3 × 0.4–0.5 mm, chalazal collar 0.1–0.15 × 0.2–0.25 mm, gray-brown, surface with conspicuous parallel longitudinal ridges of laterally flattened papillae; coma readily detached, white, 4–6 mm. $2n = 36$.

Flowering Jul–Aug(–Sep). Damp seeps, swampy areas, stream banks; of conservation concern; 200–500 m; Calif., Oreg.

Epilobium oreganum is endemic to a small region of southern Oregon (Douglas and Josephine counties, mainly from Grants Pass south along the Illinois River) and northern California (Del Norte, Humboldt, Siskiyou, Tehama, and Trinity counties, especially along the South Fork of the Trinity and Klamath rivers).

Epilobium oreganum is the only other species that shares the distinctive ridged seeds found also in *E. ciliatum*, and looks quite similar to that species; both also have the AA chromosomal arrangement. However, it differs from *E. ciliatum* in being generally glabrous and glaucous, and by having exserted 4-lobed stigmas. W. Trelease (1891) and later P. A. Munz (1965) considered *E. oreganum* to be of hybrid origin, the presumptive parents being *E. glaberrimum* (glabrous) and so called *E. adenocaulon* (= *E. ciliatum*; ridged seeds). Some specimens have notably reduced seed set; whether that is the result of a hybrid origin, a failure to outcross in a plant with a very exserted stigma, or to another cause is not clear. The exact affinities of *E. oreganum* are uncertain, but it occupies a restricted and distinctive ecogeographical range and has a unique combination of morphological features.

39. Epilobium ciliatum Rafinesque, Med. Repos., hexade 2, 5: 361. 1808 • Fringed willowherb, épilobe cilié [F] [W]

Herbs with leafy basal rosettes or large, fleshy, condensed underground turions, or sometimes shoots from caudex. **Stems** erect, green or tan to reddish green, terete, (3–) 10–120(–190) cm, often thick, well branched or simple, subglabrous proximal to inflorescence with raised strigillose lines decurrent from margins of petioles, ± densely mixed strigillose and glandular puberulent distally, rarely densely strigillose or densely villous throughout. **Leaves** opposite proximal to inflorescence, alternate distally, petiole 0–5(–10) mm, often subsessile distally, sometimes clasping; blade narrowly obovate, obovate, broadly elliptic, or spatulate proximally, to very narrowly lanceolate to ovate or broadly elliptic distally, (1–)3–12(–16) × (0.2–) 0.6–5.5 cm, base rounded to cuneate or short-attenuate, margins serrulate, (8–)15–40 irregular teeth per side, veins prominent, 4–10 per side, apex obtuse to acute or subacuminate, surfaces usually subglabrous with strigillose margins, rarely densely strigillose or villous; bracts scarcely reduced to very reduced and narrower. **Inflorescences** usually erect, rarely nodding, racemes or panicles, well branched and open, to simple and congested, ± densely strigillose and glandular puberulent. **Flowers** erect; buds 1.5–7 × 1–3 mm; pedicel 2–14(–20) mm; floral tube 0.5–2.6 × 0.9–3.5 mm, ring of spreading hairs at mouth inside; sepals often reddish green, lanceolate, sometimes keeled, 2–7.5 × 0.7–2.5 mm; petals white or pink to rose-purple, obovate, 2–14 × 1.3–6.3 mm, apical notch 0.4–2.5 mm; filaments white to dark pink, those of longer stamens 1.4–7 mm, those of shorter ones 0.6–5.2 mm; anthers light yellow to cream, 0.5–1.8 × 0.3–0.9 mm; ovary

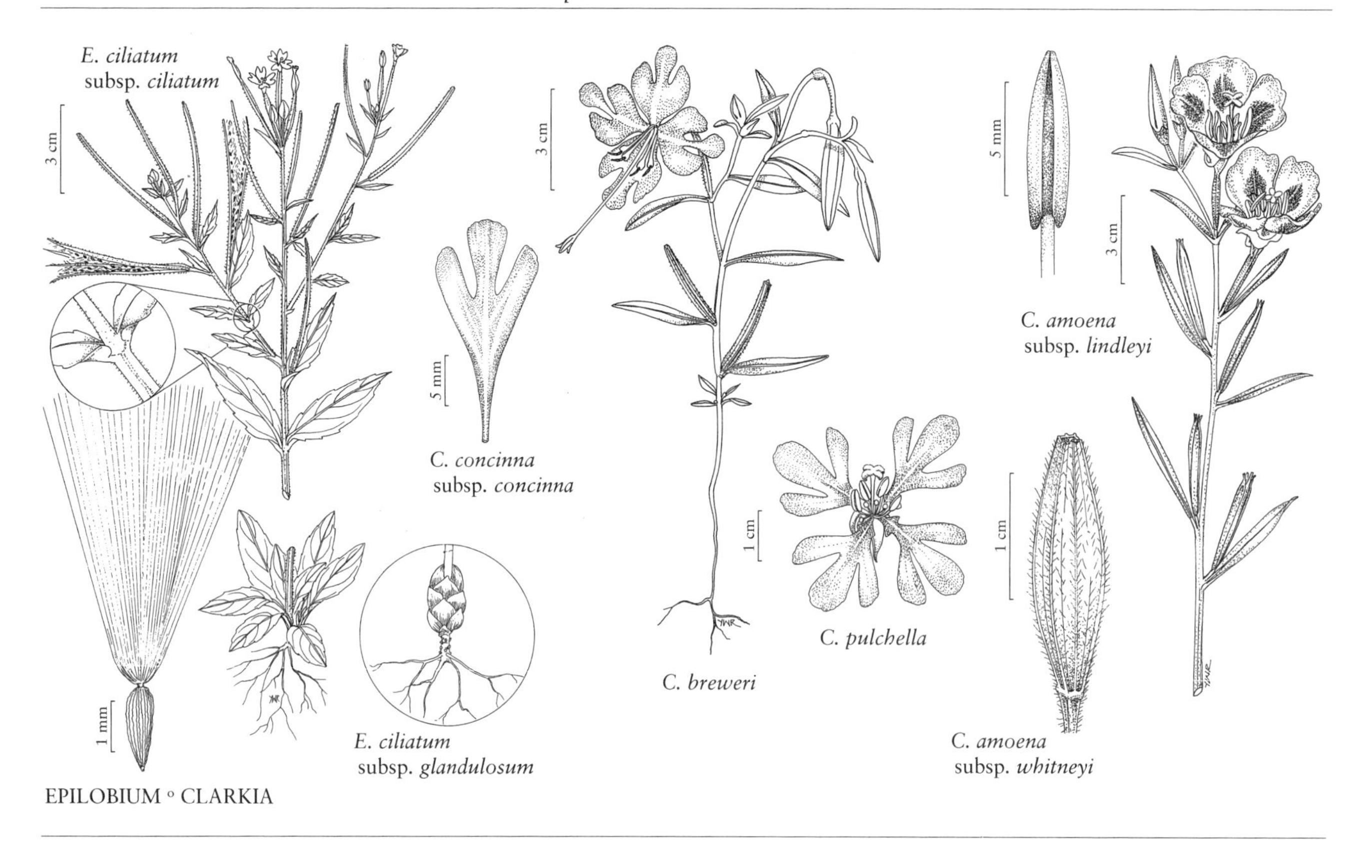

EPILOBIUM ◦ CLARKIA

often reddish green, 8–40 mm, ± densely mixed strigillose and glandular pubescent; style cream to light yellow, 1.1–8.5 mm, stigma cream to orange-yellow, narrowly to broadly clavate or subcapitate, 0.8–2.8 × 0.4–1.2 mm, rarely indented apically, usually surrounded by, rarely exserted beyond, anthers. **Capsules** erect, (15–) 30–100 mm, surfaces usually strigillose and glandular puberulent, rarely glabrescent; pedicel 2–15(–40) mm, rarely subsessile. **Seeds** narrowly obovoid or subfusiform, (0.6–)0.8–1.6(–1.9) × 0.3–0.6 mm, chalazal collar ± conspicuous, 0.1–0.3 × 0.2–0.4 mm, grayish tan to brown, surface with conspicuous parallel longitudinal ridges of laterally flattened papillae; coma readily detached, white or dingy white, 2–8 mm, very rarely absent. **2*n*** = 36.

Subspecies 3 (3 in the flora): North America, Mexico, Central America, South America, e Asia; introduced in Europe, Pacific Islands, Australia.

Epilobium ciliatum, which has the AA chromosomal arrangement, shows extraordinary variation in morphology. It has the largest geographical range among North American *Epilobium* species, and has spread invasively outside of its native range. Although almost certainly originating in North America, *E. ciliatum* is also considered native in South America (J. C. Solomon 1982) and East Asia (Chen C. J. et al. 1992), but adventive in Europe and western Russia (P. H. Raven 1968), Pacific Islands, especially New Zealand, and Australia (Raven and T. E. Raven 1976). Its chromosomal affinities and morphological similarities to a small group of species in western North America strongly suggest that that region is its center of origin.

Within the enormous variation displayed by *Epilobium ciliatum*, three broadly defined entities can be recognized: subsp. *watsonii*, characteristically with bracts scarcely reduced on an extended, crowded corymbose inflorescence, found only along the Pacific coast, usually within sight of the ocean; subsp. *glandulosum*, generally large, few-branched plants with condensed turions just below ground and crowded inflorescences of relatively large rose-purple flowers, found mainly in damp, cool, and relatively undisturbed habitats; and subsp. *ciliatum*, which range from small and simple to large and well-branched, usually with leafy basal rosettes and open inflorescences, relatively narrow leaves and small white flowers, found most often in disturbed damp to dry habitats throughout the entire range of the species. These subspecies often intergrade in regions where their ranges overlap, resulting in populations with diverse mixtures of intermediate characters, yet the subspecies consistently retain their main morphological characteristics in populations throughout most of their respective ranges. Each shows some degree of endogenous variability, most notably in the very widespread subsp. *ciliatum*.

1. Leaf blades very narrowly lanceolate to narrowly ovate or elliptic, proximally narrowly obovate to spatulate; bracts very reduced on open inflorescence; petals 2–6(–9) mm, white or sometimes pink; herbs usually with rosettes, rarely fleshy turions. . . . . 39a. *Epilobium ciliatum* subsp. *ciliatum*
1. Leaf blades narrowly ovate to ovate to broadly elliptic, sometimes lanceolate, proximally obovate to broadly elliptic; bracts little reduced on crowded inflorescence; petals 4.5–12(–15) mm, usually rose-purple to pink, rarely white; herbs usually with fleshy turions or rosettes, rarely fleshy shoots.
   2. Herbs usually with large, condensed subsessile turions 1–10 cm below ground, leaving dark scales, rarely with rosettes of fleshy leaves; inflorescences simple or branched, not corymbose . . . . 39b. *Epilobium ciliatum* subsp. *glandulosum*
   2. Herbs with leafy basal rosettes, sometimes fleshy shoots from woody caudex; inflorescences ± simple, subcorymbose. . . . . 39c. *Epilobium ciliatum* subsp. *watsonii*

39a. Epilobium ciliatum Rafinesque subsp. **ciliatum** F W

Epilobium adenocaulon Haussknecht; *E. adenocaulon* subsp. *americanum* (Haussknecht) Á. Löve & D. Löve; *E. adenocaulon* var. *ecomosum* (Fassett) Munz; *E. adenocaulon* var. *holosericeum* (Trelease) Munz; *E. adenocaulon* var. *occidentale* Trelease; *E. adenocaulon* var. *parishii* (Trelease) Munz; *E. adenocaulon* var. *perplexans* Trelease; *E. adenocaulon* var. *pseudocoloratum* Lunell; *E. americanum* Haussknecht; *E. californicum* Haussknecht; *E. californicum* var. *holosericeum* (Trelease) Jepson; *E. californicum* var. *occidentale* (Trelease) Jepson; *E. californicum* var. *parishii (*Trelease) Jepson; *E. ciliatum* var. *ecomosum* (Fassett) B. Boivin; *E. doriphyllum* Haussknecht; *E. ecomosum* (Fassett) Fernald; *E. fendleri* Haussknecht; *E. glandulosum* Lehmann var. *adenocaulon* (Haussknecht) Fernald; *E. glandulosum* var. *ecomosum* Fassett; *E. glandulosum* var. *macounii* (Trelease) C. L. Hitchcock; *E. glandulosum* var. *occidentale* (Trelease) Fernald; *E. glandulosum* var. *perplexans* (Trelease) Fernald; *E. griseum* Suksdorf; *E. holosericeum* Trelease; *E. leptocarpum* Haussknecht var. *macounii* Trelease; *E. macdougalii* Rydberg; *E. mexicanum* Seringe; *E. montezumae* Samuelsson; *E. novomexicanum* Haussknecht; *E. occidentale* (Trelease) Rydberg; *E. ostenfeldii* H. Léveillé; *E. ovale* Takeda; *E. palmeri* Rydberg; *E. parishii* Trelease; *E. perplexans* (Trelease) Trelease ex J. M. Coulter & A. Nelson; *E. praecox* Suksdorf; *E. repens* Schlechtendal; *E. watsonii* Barbey var. *occidentale* (Trelease) C. L. Hitchcock; *E. watsonii* var. *parishii* (Trelease) C. L. Hitchcock

Herbs usually with leafy basal rosettes, with subovate, obtuse, subentire leaves 15–35 mm, rarely fleshy turions. **Stems** (3–)10–120(–190) cm, usually well branched especially distally, rarely simple, subglabrous proximal to inflorescence with raised strigillose lines from margins of petioles, ± densely mixed strigillose and glandular puberulent distally, rarely densely villous throughout. **Leaves:** petiole 0–5(–8) mm; blade narrowly obovate to spatulate proximally, very narrowly lanceolate to narrowly ovate or elliptic distally, (1–)3–12 × (0.3–)0.6–3.7 cm, apex obtuse proximally to acute distally; bracts very reduced and narrower. **Inflorescences** usually erect, rarely nodding, racemes or panicles, open, not congested. **Flowers:** floral tube 0.5–1.8 × 0.9–3 mm; sepals 2–6 × 0.7–1.6 mm; petals white or sometimes pink, 2–6(–9) × 1.3–4 mm; filaments white or pink, those of longer stamens 1.4–4.8 mm, those of shorter ones 0.6–2.6 mm; style 1.1–6.5 mm, stigma 0.8–2.2 × 0.4–0.9 mm. **Capsules** (15–) 30–100 mm, surfaces usually strigillose and glandular puberulent, rarely glabrescent; pedicel 2–15(–40) mm, rarely subsessile. **Seeds** (0.6–)0.8–1.2(–1.4) × 0.3–0.5 mm, chalazal collar 0.02–0.2 × 0.1–0.3 mm; coma very rarely absent. $2n = 36$.

Flowering Jun–Oct(–Nov). Disturbed, open, mesic areas, along roadsides, stream banks, lake margins, seeps, coastal bluffs; 0–4200 m; Alta., B.C., Man., N.B., Nfld. and Labr., N.W.T., N.S., Ont., P.E.I., Que., Sask., Yukon; Alaska, Ariz., Calif., Colo., Conn., Del., Idaho, Ill., Ind., Iowa, Kans., Ky., Maine, Md., Mass., Mich., Minn., Mo., Mont., Nebr., Nev., N.H., N.J., N.Mex., N.Y., N.C., N.Dak., Ohio, Okla., Oreg., Pa., R.I., S.C., S.Dak., Tenn., Tex., Utah, Vt., Va., Wash., W.Va., Wis., Wyo.; Mexico; Central America (Guatemala); South America (Argentina, Chile); Asia (China, Japan, Korea, Russian Far East); introduced in Europe, Pacific Islands (Hawaii, New Zealand), Australia.

Subspecies *ciliatum* is the most widely distributed and common taxon in the genus, both in areas where it is native, and in those where it has become naturalized. The extensive synonymy of this taxon reflects its abundance in a wide ecogeographical range, and its extreme variability. Compared to most other taxa of *Epilobium*, subsp. *ciliatum* shows a very high degree of phenotypic plasticity, especially in plant size and branching, leaf size and shape, patterns of vestiture, and flower size and color. This plasticity, coupled with its persistent autogamy and high seed set, produces many diverse local and/or regional phenotypes, some of which have been segregated and named in the past. However, many intermediates also occur, and growing even very distinctive phenotypes in a common garden produces nearly indistinguishable plants with morphology typical for the subspecies. Some notable regional variants treated here in synonymy include *E. holosericeum* in southern California and occasionally elsewhere, which has densely sericeous pubescence throughout an otherwise typical plant of subsp. *ciliatum*, and *E. ecomosum*, found in southern Quebec in Canada along the St. Lawrence River, comprised of plants typical of subsp. *ciliatum* save their ecomose seeds. Ecomose seeds also characterize the annual sects. *Pachydium* and *Epilobiopsis*, but are known in sect. *Epilobium* only in *E. curtisiae* P. H. Raven from New Zealand. This treatment adopts a broad delimitation of taxa, despite such apparently distinctive variants, in part because plants with similar but slightly differing character combinations occur scattered through the wide range of the subspecies, and in part because a narrow species definition would require recognition of many additional regional taxa with slightly differing character combinations that would be very difficult to define and distinguish, and would obscure the overriding integrity of subsp. *ciliatum*.

Subspecies *ciliatum* occurs sympatrically with many other species of *Epilobium* and hybridizes with most, as evidenced by the low but persistent occurrence of apparent hybrids on herbarium sheets. Experimental hybrids between subsp. *ciliatum* and numerous other species (P. H. Raven and T. E. Raven 1976; S. R. Seavey and P. H. Raven 1977, 1977b, 1978) produce vigorous plants, often with reduced seed fertility. Even low levels of introgression following hybridization may account for some of the variation found in this taxon.

Epilobium palmeri H. Léveillé is an illegitimate name that pertains here.

39b. Epilobium ciliatum Rafinesque subsp. **glandulosum** (Lehmann) Hoch & P. H. Raven, Ann. Missouri Bot. Gard. 64: 136. 1977 [F]

Epilobium glandulosum Lehmann, Nov. Stirp. Pug. 2: 14. 1830; *E. affine* Bongard; *E. boreale* Haussknecht; *E. brevistylum* Barbey; *E. ciliatum* var. *glandulosum* (Lehmann) Dorn; *E. cinerascens* Piper; *E. delicatum* Trelease; *E. glandulosum* var. *asiaticum* H. Hara; *E. glandulosum* var. *brionense* Fernald; *E. glandulosum* var. *cardiophyllum* Fernald; *E. glandulosum* var. *cinerascens* (Piper) M. Peck; *E. glandulosum* var. *kurilense* (Nakai) H. Hara; *E. kurilense* Nakai; *E. maximowiczii* Haussknecht; *E. sandbergii* Rydberg; *E. tetragonum* Linnaeus var. *glandulosum* (Lehmann) Torrey & A. Gray

Herbs usually with large (5–12 mm), condensed, subsessile turions 1–10 cm below ground, leaving dark scales, rarely with rosettes of fleshy leaves. **Stems** 20–110(–170) cm, thick, simple or sparsely branched distally, subglabrous proximal to inflorescence with raised strigillose lines from margins of petioles, densely mixed strigillose and glandular puberulent distally, or rarely villous throughout. **Leaves:** petiole 0–4(–10) mm; blade obovate to broadly elliptic proximally, usually narrowly ovate to ovate or broadly elliptic, rarely lanceolate distally, 3–10.5(–16) × 1–4.5(–5.5) cm, apex obtuse to subacute; bracts usually not much reduced. **Inflorescences** erect racemes, simple or sometimes branched, congested, 5–60(–85) cm. **Flowers:** floral tube 1–2.6 × 1.4–3.5 mm; sepals 4.5–7.5 × 1.2–1.8 mm; petals usually rose-purple to pink, rarely white, 4.5–12(–14) × 2.5–6.3 mm; filaments white or cream to purple, those of longer stamens 3–6.5 mm, those of shorter ones 1.4–4.3 mm; style 2.4–8.2 mm, stigma 1.2–2.3 × 0.6–1 mm, sometimes exserted beyond anthers. **Capsules** 40–85 mm, mixed strigillose and glandular puberulent; pedicel 5–25 mm, rarely subsessile. **Seeds** 1.1–1.6(–1.9) × 0.4–0.6 mm, chalazal collar 0.05–0.3 × 0.2–0.4 mm. $2n$ = 36.

Flowering May–Sep. Damp banks of streams and lakes, seeps, wet meadows in montane, subalpine and maritime areas; 0–3400 m; St. Pierre and Miquelon; Alta., B.C., Man., N.B., Nfld. and Labr., N.W.T., N.S., Nunavut, Ont., P.E.I., Que., Sask., Yukon; Alaska, Calif., Colo., Idaho, Maine, Mass., Minn., Mont., Nev., N.H., N.Mex., Oreg., Utah, Vt., Wash., Wis., Wyo.; e Asia (Russian Far East).

Subspecies *glandulosum* has a geographical range almost entirely within the larger range of subsp. *ciliatum*. Despite this overlap, the two taxa tend to grow in different habitats and are only rarely functionally sympatric, with subsp. *glandulosum* in cooler, more stable habitats, often at higher elevations, and subsp. *ciliatum* in drier, more disturbed and open habitats. Where they occur in close sympatry, they sometimes intergrade, broadening each other's overall variability. In northern Canada and Alaska, subsp. *glandulosum* tends to replace subsp. *ciliatum*, notably along coastal Alaska, where it also replaces subsp. *watsonii*. In Nunavut, subsp. *glandulosum* is known only from Akamiski Island in James Bay.

As discussed under 30. *Epilobium luteum*, subsp. *glandulosum* hybridizes with that species—both have relatively large flowers that promote outcrossing—resulting in the distinctive hybrid, *E.* ×*treleasianum*, which has very notably large, pink flowers. Hybrids formed with other sympatric taxa are less obvious, but can be detected by both reduced seed set and intermediate morphology.

Epilobium adenocaulon Haussknecht var. *cinerascens* (Piper) M. Peck is an illegitimate name that pertains here.

39c. Epilobium ciliatum Rafinesque subsp. **watsonii** (Barbey) Hoch & P. H. Raven, Ann. Missouri Bot. Gard. 64: 136. 1977 [E]

Epilobium watsonii Barbey in W. H. Brewer et al., Bot. California 1: 219. 1876 (as watsoni); *E. congdonii* H. Léveillé; *E. franciscanum* Barbey; *E. watsonii* var. *franciscanum* (Barbey) Jepson

Herbs usually with leafy basal rosettes, sometimes fleshy shoots from woody caudex, often forming dense thickets. **Stems** 25–150(–180) cm, thick, often well branched proximally, rarely distally, subglabrous proximal to inflorescence with raised strigillose lines from margins of petioles, densely mixed strigillose and glandular puberulent distally, or densely strigillose throughout. **Leaves:** petiole 1–5 mm; blade ovate or broadly elliptic to lanceolate, 3.5–9.6 × 1.6–3.5 cm, apex acute or subacuminate; bracts only slightly reduced in size. **Inflorescences** erect racemes, congested, simple, subcorymbose. **Flowers:** floral tube 0.9–2.4 × 1.4–3.2 mm; sepals 3.5–7.5 × 1–2.5 mm; petals rose-purple to bright pink, 5–12(–15)× 2.3–5.5 mm; filaments pink or purple to white, those of longer stamens 2.8–7 mm, those of shorter ones 1–5.2 mm; style 2.3–7.5 mm, stigma 1.1–2.8 × 0.5–1.2 mm, sometimes exserted beyond anthers. **Capsules** 30–85 mm, mixed strigillose and glandular puberulent; pedicel 3–10 mm, rarely subsessile. **Seeds** 0.9–1.3 × 0.4–0.5 mm, chalazal collar 0.02–0.1 × 0.12–0.2 mm. $2n = 36$.

Flowering May–Oct. Stream banks, seepages, roadside ditches, open places along coast; 0–200 m; B.C.; Calif., Oreg., Wash.

Subspecies *watsonii* is a strongly maritime variant of the species that shows very consistent morphological integrity throughout its very long but narrow range along the Pacific Coast of North America from Point Conception, Santa Barbara County, California, to Vancouver Island, British Columbia. Virtually throughout its range it grows with or close to subsp. *ciliatum*, although the latter rarely grows in the cool maritime strip next to the coast. In areas where there is more extensive contact, such as the marshes of Sonoma County, California, the two taxa intergrade completely, from pure subsp. *watsonii* at Bodega Bay, to nearly pure subsp. *ciliatum* at Laguna de Santa Rosa.

On the Olympic Peninsula in Washington and extending north into British Columbia, subsp. *watsonii* intergrades not only with subsp. *ciliatum*, but also with subsp. *glandulosum*, which in the south grows at higher elevations but in the north eventually replaces subsp. *watsonii* along the coast into Alaska.

40. Epilobium glaberrimum Barbey in W. H. Brewer et al., Bot. California 1: 220. 1876 • Glaucous willowherb

Epilobium fastigiatum (Nuttall) Piper var. *glaberrimum* (Barbey) Piper

Herbs with branching, wiry basal shoots with crowded, often dark, scales. **Stems** many, erect or ascending, often densely clumped, slender, terete, 5–75(–85) cm, simple or distally branched, glabrous and glaucous, sometimes strigillose and/or mixed glandular puberulent distally, rarely with faint raised strigillose lines decurrent from margins of distal petioles. **Leaves** opposite and crowded proximal to inflorescence, alternate and widely spaced distally, subsessile and ± clasping; blade light green or blue-green, very narrowly lanceolate to narrowly ovate or elliptic, (0.5–)1–7.2 × (0.3–)0.5–1.8 cm, base cuneate, margins subentire or faintly denticulate, 8–15 teeth per side, veins inconspicuous, 3–6 per side, apex obtuse proximally to acute distally, glabrous and glaucous; bracts much reduced and narrower. **Inflorescences** erect racemes, glabrous or ± sparsely glandular puberulent, sometimes mixed strigillose. **Flowers** erect; buds 2.5–5 × 2–3.5 mm; pedicel 2–11(–20) mm; floral tube 0.7–2.3 × 1–2.5 mm, ring of spreading hairs at mouth inside or without a ring; sepals often red-tipped, sometimes keeled, 2.5–7.5 × 0.8–2 mm, abaxial surface glabrous or strigillose and/or glandular puberulent; petals usually light pink to rose-purple, rarely white, 3.4–10(–12) × 1.8–5 mm, apical notch 0.5–3 mm; filaments purple to white or cream, those of longer stamens 2.4–5(–6) mm, those of shorter ones 1.4–4 mm; anthers light yellow, 0.5–1.2(–1.5) × 0.3–0.7 mm; ovary 10–42 mm, subglabrous to sparsely mixed strigillose and glandular puberulent; style light yellow or cream, 1.7–5(–7) mm, stigma clavate to subcapitate, 0.8–3 × 0.5–1.2 mm, usually surrounded by, rarely exserted beyond, longer anthers. **Capsules** 15–75 mm, with small protuberance at base, surfaces subglabrous to sparsely mixed strigillose and glandular puberulent; pedicel 5–18(–25) mm. **Seeds** narrowly obovoid, 0.7–1(–1.3) × 0.3–0.5 mm, chalazal collar inconspicuous, light brown or gray, surface coarsely papillose in longitudinal rows; coma readily detached, white, 4–9 mm.

Subspecies 2 (2 in the flora): w North America, nw Mexico.

Epilobium glaberrimum is restricted mainly to the Sierra Nevada-Cascade region, but also (subsp. *fastigiatum*) in the northern Rocky Mountains. It is one of the more easily identified species due to its consistently glabrous and glaucous aspect. Although the distributions and elevational ranges of the two subspecies appear to overlap extensively, they are largely allopatric with subsp. *glaberrimum* occurring in lower to mid montane zones and subsp. *fastigiatum* in upper montane to subalpine zones in the southern part of its range, and in lower zones farther north.

1. Stems 20–75(–85) cm; leaf blades (2–)3–7.2 cm, very narrowly lanceolate to lanceolate or narrowly elliptic (length/width ratio more than 4); capsules 45–75 mm; petals (4–)5–10(–12) mm . . . . 40a. *Epilobium glaberrimum* subsp. *glaberrimum*
1. Stems 5–35 cm; leaf blades (0.5–)1–3.4 cm, lanceolate to narrowly ovate or elliptic (length/width ratio less than 4); capsules 15–45 (–55) mm; petals 3.4–8 mm . . . . .40b. *Epilobium glaberrimum* subsp. *fastigiatum*

40a. Epilobium glaberrimum Barbey subsp. glaberrimum

Epilobium madrense S. Watson; *E. pruinosum* Haussknecht

Stems 20–75(–85) cm, sometimes sparsely strigillose distally, without lines. **Leaf blades** usually light green, very narrowly lanceolate to lanceolate or narrowly elliptic, (2–)3–7.2 × 0.5–1.6 cm, length/width ratio more than 4. **Flowers:** floral tube 0.7–2.3 × 1.4–2.5 mm; sepals 3–7.5 × 1–1.7 mm; petals usually pink to rose-purple, rarely white, (4–)5–10 (–12) × 2–4.5(–5) mm; filaments of longer stamens 3–5.5(–6) mm, those of shorter ones 1.4–4 mm; ovary 19–42 mm, style 1.7–5(–7) mm. **Capsules** 45–75 mm. **2*n*** = 36.

Flowering Jun–Oct. Well-drained, gravelly soils, seepage areas, damp meadows, stream banks, roadside ditches, mainly lower to mid montane; (150–)300–2800(–3300) m; Ariz., Calif., Idaho, Nev., Oreg., Utah, Wash.; Mexico (Baja California, Chihuahua).

Subspecies *glaberrimum* extends into mountainous areas of Baja California and Chihuahua, Mexico, much farther south than subsp. *fastigiatum*, which in turn occurs much farther north.

40b. Epilobium glaberrimum Barbey subsp. fastigiatum (Nuttall) Hoch & P. H. Raven, Ann. Missouri Bot. Gard. 71: 342. 1984 [E]

Epilobium affine Bongard var. *fastigiatum* Nuttall in J. Torrey and A. Gray, Fl. N. Amer. 1: 489. 1840; *E. concinnum* Congdon; *E. fastigiatum* (Nuttall) Piper; *E. glaberrimum* var. *fastigiatum* (Nuttall) Trelease ex Jepson; *E. glaberrimum* var. *latifolium* Barbey; *E. platyphyllum* Rydberg

Stems 5–35 cm, sometimes sparsely strigillose and glandular puberulent distally, rarely with faint raised strigillose lines decurrent from margins of distal petioles. **Leaf blades** usually blue-green, lanceolate to narrowly ovate or elliptic, (0.5–)1–3.4 × (0.3–)0.7–1.8 cm, length/width ratio less than 4. **Flowers:** floral tube 0.9–1.4 × 1–2.2 mm; sepals 2.5–5.2 × 0.8–1.6 mm; petals rose-purple to pink, 3.4–8 × 1.8–3.9 mm; filaments of longer stamens 2.4–5 mm, those of shorter ones 1.6–3 mm; ovary 10–34 mm, style 2.2–3.6 mm. **Capsules** 15–45(–55) mm. **2*n*** = 36.

Flowering Jun–Sep. Rocky, well drained soil, moist alpine slopes, seepage, talus slope, stream banks, bogs, mainly upper montane to alpine; (1000–)1200–3500 m; Alta., B.C.; Calif., Idaho, Mont., Nev., Oreg., Utah, Wash., Wyo.

Subspecies *fastigiatum* occurs primarily along the Sierra Nevada-Cascade axis, but also ranges through the Basin Ranges to the Rocky Mountains from northern Wyoming to southern Alberta and British Columbia.

41. Epilobium saximontanum Haussknecht, Oesterr. Bot. Z. 29: 119. 1879 • Rocky Mountain willowherb, épilobe des Rocheuses [E]

Epilobium adenocaulon Haussknecht subsp. *rubescens* (Rydberg) Hiitonen; *E. drummondii* Haussknecht; *E. drummondii* var. *latiusculum* Rydberg; *E. latiusculum* (Rydberg) Rydberg; *E. ovatifolium* Rydberg; *E. rubescens* Rydberg; *E. scalare* Fernald; *E. stramineum* Rydberg

Herbs usually with sessile, fleshy, underground turions, or sometimes thick, elongated shoots with dark, decussate scales. **Stems** erect, strict, terete, 4–55 cm, simple or well branched in age, subglabrous proximally to mixed strigillose and glandular puberulent distally, with raised strigillose lines decurrent from margins of petioles. **Leaves** opposite proximal to inflorescence, alternate and reduced distally, often ± appressed, usually subsessile, rarely petiole 1–3 mm, often clasping; blade obovate proximally to ovate, lanceolate, or narrowly elliptic distally, 1–5.5(–6.5) × 0.4–2(–2.4) cm, base rounded or obtuse, margins low denticulate, 9–30 teeth per side, veins ± conspicuous, 3–6 per side, apex subacute, surfaces subglabrous with strigillose margins; bracts much reduced. **Inflorescences** erect, sometimes nodding in bud, racemes, sometimes sparsely branched. **Flowers** erect; buds 2–3.5 × 1.8–2.5 mm; pedicel 0–1 mm; floral tube 0.8–1.4 × 0.8–1.9 mm, ring of sparse spreading hairs at mouth inside; sepals sometimes flushed red, 1.2–3.5 × 0.6–1.4 mm, abaxial surface strigillose and sometimes mixed glandular puberulent; petals usually white, infrequently pink, 2.2–5(–7) × 1.7–3.2 mm, apical notch 0.4–1.5 mm; filaments usually cream, rarely light pink, those of longer stamens 2–3.5 mm, those of shorter ones 1–2 mm; anthers cream to light yellow, 0.3–0.8 × 0.3–0.5 mm; ovary 9–30 mm, densely strigillose and glandular puberulent; style cream or yellow, 1.6–2.8 mm, stigma usually narrowly to broadly clavate, rarely subcapitate, 1–3 × 0.8–2 mm, surrounded by at least longer anthers. **Capsules** 30–55(–70) mm, surfaces mixed strigillose and glandular puberulent; usually subsessile, rarely pedicel 1–5 mm, often appressed to stem. **Seeds** very narrowly obovoid, 1–1.6(–1.8) × 0.4–0.6 mm, chalazal collar 0.1–0.2 mm, light brown or gray, surface rugose to papillose; coma usually readily detached, white, 3–9 mm. **2*n*** = 36.

Flowering Jul–Sep. Montane semi-shaded stream banks, damp meadows, mossy seeps, wet slatey cliffs, disturbed or seasonally damp areas; 0–3700 m; Alta., B.C., Man., Nfld. and Labr. (Nfld.), Ont., Que.; Ariz., Calif., Colo., Idaho, Mont., Nev., N.Mex., Oreg., S.Dak., Utah, Wyo.

Epilobium saximontanum is morphologically similar to *E. ciliatum* (especially subsp. *glandulosum*) with which it also shares the AA chromosome arrangement. However, in addition to its fleshy compact turions, it very characteristically has notably appressed capsules, unlike most other species in the genus, and a notably strict habit.

The distribution of *Epilobium saximontanum* is unusual; it includes the Rocky Mountain region, only barely reaching the high southern Sierra Nevada, disjunct to the Black Hills of South Dakota, and more widely in eastern Canada, from the shores of Hudson Bay to Newfoundland. Specimens are fairly uniform across this wide and rather discontinuous range, although locally they show some variability, possibly due to hybridization with any of several species that may be sympatric with it. H. Lewis and D. M. Moore (1962) reported hybrids between *E. saximontanum* (cited as *E. brevistylum*) and *E. ciliatum* subsp. *ciliatum* (cited as *E. adenocaulon*) from Colorado, and herbarium specimens with *E. saximontanum* and apparent hybrids are not uncommon.

b3. ONAGRACEAE Jussieu (subfam. ONAGROIDEAE) tribe ONAGREAE Dumortier, Fl. Belg., 89. 1827

Herbs (annual or perennial), [shrubs]. Leaves alternate or basal; stipules absent. **Flowers** usually actinomorphic, rarely slightly zygomorphic (in *Oenothera*), (3 or)4-merous; stamens 2 times as many, or rarely as many, as sepals; pollen usually shed in monads, rarely tetrads (*Chylismia* sect. *Lignothera*). **Fruit** a dry capsule, usually dehiscent, sometimes indehiscent. **Seeds** few to numerous, without hairs or wings, [very rarely with asymmetrical dry wing (*Xylonagra*)], or with dry (*Oenothera*), erose or smooth wing, or with thick, papillate wings (*Chylismiella*).

Genera 13, species 265 (12 genera, 199 species in the flora): North America, Mexico, West Indies, Central America, South America.

Onagreae account for more than half the total genera in Onagraceae and diversified from a center in southwestern North America (L. Katinas et al. 2004). Delimitation of the tribe by W. L. Wagner et al. (2007) differs from previous ones by the exclusion of *Gongylocarpus*, now in its own tribe, by the segregation of eight genera (*Camissoniopsis*, *Chylismia*, *Chylismiella*, *Eremothera*, *Eulobus*, *Neoholmgrenia*, *Taraxia*, and *Tetrapteron*) from *Camissonia*, and by the inclusion of three previously separate genera (*Calylophus*, *Gaura*, and *Stenosiphon*) in *Oenothera*. Within the branch of the family that lacks stipules (Gongylocarpeae, Epilobieae, and Onagreae), the last two tribes form a clade that has very strong molecular support (R. A. Levin et al. 2003, 2004), but no obvious morphological synapomorphy. The clade may be defined by a cytogenetic change from the base chromosome number of $x = 11$ found in Circaeeae, Gongylocarpeae, and Lopezieae, to $x = 18$ in Epilobieae, and $x = 7$ in Onagreae; however, these changes could also have occurred independently. Other than the new chromosome number $x = 7$, the only apparent morphological synapomorphy for Onagreae alone is pollen with prominent apertural protrusions (J. Praglowski et al. 1987, 1989), a character state also found in Circaeeae (Praglowski et al. 1994). The monophyly of Onagreae has moderate (Levin et al. 2004) to strong support (V. S. Ford and L. D. Gottlieb 2007).

SELECTED REFERENCE Raven, P. H. 1964. The generic subdivision of Onagraceae, tribe Onagreae. Brittonia 16: 276–288.

6. CLARKIA Pursh, Fl. Amer. Sept. 1: 256, 260 [as Clarckia], plate 11. 1813 • [For Captain William Clark, 1770–1838, of the Lewis and Clark Expedition]

Harlan Lewis†

Peter C. Hoch

Herbs, annual, caulescent. **Stems** slender to stout, erect to prostrate or decumbent, unbranched to sparsely branched. **Leaves** cauline, alternate; stipules absent; sessile or petiolate, petiole usually shorter than blade; blade margins entire or denticulate. **Inflorescences** usually racemes or spikes, rarely panicles; buds erect or pendent. **Flowers** bisexual, usually actinomorphic, often protandrous; floral tube deciduous (with sepals, petals, and stamens) after anthesis, usually with basal nectary and a ring of hairs inside; sepals 4, often pink to purplish red, usually connate to tip in bud, reflexed singly, in pairs, or all together to 1 side at anthesis; petals 4, usually lavender or pink to dark reddish purple, pale yellow or white, [rarely blue (*C. tenella*)], often spotted, flecked or streaked with red, purple, or white; stamens 8, in 2 equal or unequal series, or 4 in 1 series, filaments filiform or expanded distally, sometimes subtended by hairy scales, anthers basifixed, often with short acute sterile tip, pollen cream, yellow, blue-gray, lavender, or red, shed singly; ovary 4-locular, stigma 4-lobed, commissural, lobes receptive only on dry, unicellular-papillose inner surfaces. **Fruit** a capsule, elongate, straight or curved, cylindrical, fusiform, or subclavate, often 4-angled (shallowly to deeply 4- or 8-grooved) or terete, loculicidal, often tardily dehiscent, sometimes with sterile beak, rarely short, indehiscent and nutlike (*C. heterandra*); sessile or pedicellate. **Seeds** usually numerous, rarely (*C. heterandra*) 1 or 2, usually angled, cubic, or elongate to spindle-shaped, often with crest of elongated cells, scaly or minutely tuberculate. $x = 7$.

Species 42 (41 in the flora): w North America, w South America.

Molecular phylogenetic analysis places a strongly monophyletic *Clarkia* as sister to *Gayophytum* and *Chylismiella* (R. A. Levin et al. 2004). All but one species of *Clarkia* are endemic to western North America; the only exception is *C. tenella* (Cavanilles) H. Lewis & M. E. Lewis, which occurs in Mediterranean-climate regions of Argentina and Chile. Several species of *Clarkia* are grown ornamentally, especially *C. amoena*, the so-called Godetia of horticulture, and *C. unguiculata*, the common garden clarkia. Both species and cultivars developed from them are used in annual border plantings or in hanging baskets. *Clarkia pulchella* is also commonly cultivated, especially in Europe. *Clarkia* has been the subject of detailed systematic and evolutionary studies for more than half a century, and there are many reports of these studies in the literature (H. Lewis and M. E. Lewis 1955; V. S. Ford and L. D. Gottlieb 2003; V. M. Eckhart et al 2004).

SELECTED REFERENCES Gottlieb, L. D. and V. S. Ford. 1996. Phylogenetic relationships among the sections of *Clarkia* (Onagraceae) inferred from the nucleotide sequences of *Pgi*C. Syst. Bot. 21: 45–62. Lewis, H. and M. E. Lewis. 1955. The genus *Clarkia*. Univ. Calif. Publ. Bot. 20: 241–392. Sytsma, K. J., J. F. Smith, and L. D. Gottlieb. 1990. Phylogenetics in *Clarkia* (Onagraceae): Restriction site mapping of chloroplast DNA. Syst. Bot. 15: 280–295.

1. Capsules indehiscent, nutlike, 2–3 mm, 1 or 2 seeded [6h.7. sect. *Phaeostoma* subsect. *Heterogaura*] . . . . 41. *Clarkia heterandra*
1. Capsules loculicidal, 10–70 mm, many-seeded.
    2. Stamens 4; floral tube slender, 13–35 mm; petals conspicuously 3-lobed [6a. sect. *Eucharidium*].
        3. Petal length 2 times width, lobes ± equal or middle lobe wider . . . . 1. *Clarkia concinna*
        3. Petal length equal to width, middle lobe longer and much narrower. . . . . 2. *Clarkia breweri*
    2. Stamens 8; floral tube obconic to campanulate or funnelform, 1–10(–15) mm; petals lobed or not.

[4. Shifted to left margin.—Ed.]

4. Inner anthers sterile; petals with 3 subequal lobes [6b. sect. *Clarkia*]. . . . . 3. *Clarkia pulchella*
4. Inner anthers fertile, anthers similar or inner much smaller, paler; petals not 3-lobed.
   5. Inflorescence axis recurved at tip in bud; buds pendent.
      6. Petal claw broad with pair of lateral basal lobes; stamens subtended by ciliate scales [6d. sect. *Myxocarpa*].
         7. Stigma not or rarely exserted beyond anthers; petals 6–12(–14) mm.
            8. Petals not spotted; pollen yellow . . . . 12. *Clarkia stellata*
            8. Petals spotted or mottled or with darker flecks; pollen blue-gray.
               9. Inflorescence axis recurved only at tip in bud, straight 4+ nodes distal to open flowers . . . . 14. *Clarkia virgata* (in part)
               9. Inflorescence axis in bud recurved 1–3 nodes distal to open flowers . . . . 16. *Clarkia rhomboidea*
         7. Stigma exserted beyond anthers (rarely so in *C. virgata*); petals (7–)12–20 mm.
            10. Inflorescence axis recurved in bud, straight 1–3 nodes distal to open flowers; petal length 1.4–1.6 times width. . . . . 11. *Clarkia mildrediae*
            10. Inflorescence axis recurved only at tip in bud, straight 4+ nodes distal to open flowers; petal length 1.5–3 times width.
               11. Flower buds fusiform, tip acute. . . . . 10. *Clarkia borealis*
               11. Flower buds narrowly obovoid, tip obtuse.
                  12. Petal length 1.5–2 times width . . . . .13. *Clarkia mosquinii*
                  12. Petal length 1.9–3 times width.
                     13. Leaf blades elliptic to ovate . . . . 14. *Clarkia virgata* (in part)
                     13. Leaf blades linear to lanceolate. . . . . .15. *Clarkia australis*
      6. Petals not clawed or claw not greater than 2 mm, without lateral lobes; stamens not subtended by scales.
         14. Petals shallowly to deeply 2-lobed [6h.1. sect. *Phaeostoma* subsect. *Lautiflorae*] . . . . 27. *Clarkia biloba*
         14. Petals not 2-lobed, sometimes emarginate.
            15. Stamens subequal, anthers of similar size and color.
               16. Ovary 4-grooved; capsules usually wider distally [6c.3. sect. *Rhodanthos* subsect. *Jugales*] . . . . .9. *Clarkia gracilis*
               16. Ovary conspicuously 8-grooved; capsules not wider distally [6c.2. sect. *Rhodanthos* subsect. *Flexicaules*].
                  17. Stigma not exserted beyond anthers; pedicel in fruit 0–3 mm . . . . 7. *Clarkia lassenensis*
                  17. Stigma exserted beyond anthers; pedicel in fruit 5–15 mm . . . . . . 8. *Clarkia arcuata*
            15. Inner stamens shorter, inner anthers much smaller, paler.
               18. Stigma not exserted beyond anthers; petals 5–12 mm.
                  19. Petals white to pale cream, not flecked, fading pink [6h.3. sect. *Phaeostoma* subsect. *Micranthae*]. . . . . 32. *Clarkia epilobioides*
                  19. Petals pale to dark pink, usually darker flecked.
                     20. Petals pink, usually darker flecked [6h.1. sect. *Phaeostoma* subsect. *Lautiflorae*] . . . . .30. *Clarkia modesta*
                     20. Petals pale pink, shading nearly white near base, purple-flecked [6h.2. sect. *Phaeostoma* subsect. *Prognatae*]. . . . . 31. *Clarkia similis*
               18. Stigma exserted beyond anthers; petals 10–35 mm.
                  21. Corollas rotate; petals oblanceolate [6h.1. sect. *Phaeostoma* subsect. *Lautiflorae*] . . . . 28. *Clarkia lingulata*
                  21. Corollas bowl-shaped; petals fan-shaped.
                     22. Ovary 8-grooved [6h.1. sect. *Phaeostoma* subsect. *Lautiflorae*] . . . . .29. *Clarkia dudleyana*
                     22. Ovary 4-grooved [6h.5. sect. *Phaeostoma* subsect. *Sympherica*].
                        23. Width of outer filaments about 2 times inner; floral tube 2–7 mm ring of hairs in floral tube below rim . . . . .35. *Clarkia cylindrica*
                        23. Width of all filaments about equal or inner slightly thinner; floral tube 1.5–4 mm ring of hairs in floral tube

24. Capsule beak 0–3 mm . . . . 34. *Clarkia lewisii*
24. Capsule beak 7–15 mm . . . . 36. *Clarkia rostrata*

[5. Shifted to left margin.—Ed.]

5. Inflorescence axis straight or erect; buds erect or pendent.
  25. Buds pendent; corollas rotate or bowl-shaped; inner stamens shorter, inner anthers smaller, paler.
    26. Petals 2-lobed, with a slender central tooth [6g. sect. *Fibula*] . . . . 26. *Clarkia xantiana*
    26. Petals not lobed.
      27. Corollas bowl-shaped; petals not clawed [6g. sect. *Fibula*].
        28. Seeds brown. . . . . 24. *Clarkia bottae*
        28. Seeds gray . . . . 25. *Clarkia jolonensis*
      27. Corollas rotate; petals clawed.
        29. Petal claw shorter than blade [6h.4. sect. *Phaeostoma* subsect. *Connubium*] . . . . 33. *Clarkia delicata*
        29. Petal claw equal to or longer than blade [6h.6. sect. *Phaeostoma* subsect. *Phaeostoma*].
          30. Sepals and ovary puberulent, mixed with longer, straight spreading hairs to 3 mm . . . . 37. *Clarkia unguiculata*
          30. Sepals and ovary sparsely to densely puberulent, without longer, straight spreading hairs.
            31. Leaf blades not glaucous, bright green . . . . 39. *Clarkia exilis*
            31. Leaf blades glaucous, gray-green or reddish green.
              32. Sepals usually dark red-purple; stigma exserted beyond anthers . . . . 38. *Clarkia springvillensis*
              32. Sepals green, red-tinged or not; stigma exserted or not beyond anthers. . . . . 40. *Clarkia tembloriensis*
  25. Buds erect; corollas bowl-shaped; stamens subequal, anthers of similar size and color.
    33. Sepals remaining connate along edges and reflexed together to 1 side.
      34. Ovary 8-grooved.
        35. Petals 30–60 mm; ovary fusiform [6c.1. sect. *Rhodanthos* subsect. *Primigenia*] . . . . 4. *Clarkia amoena* (in part)
        35. Petals 5–15 mm; ovary cylindrical [6f. sect. *Biortis*]. . . . . 23. *Clarkia affinis*
      34. Ovary 4-grooved [6c.1. sect. *Rhodanthos* subsect. *Primigenia*].
        36. Petals with distinct red spot or mark near middle . . . . 4. *Clarkia amoena* (in part)
        36. Petals red or reddish purple at base.
          37. Petals 10–30 mm; stigma exserted beyond anthers . . . . 5. *Clarkia rubicunda*
          37. Petals 5–13 mm; stigma not exserted beyond anthers . . . . 6. *Clarkia franciscana*
    33. Sepals reflexed individually or in pairs [6e. sect. *Godetia*].
      38. Stigma not exserted beyond anthers; petals 5–15 mm.
        39. Stems erect; leaf blades usually linear to lanceolate, apex acute . . . . 22. *Clarkia purpurea* (in part)
        39. Stems prostrate to decumbent; leaf blades elliptic to oblanceolate, apex usually obtuse.
          40. Petals 5–11 mm, without spot . . . . 20. *Clarkia davyi*
          40. Petals 10–15 mm, with red spot above base . . . . 21. *Clarkia prostrata*
      38. Stigma exserted beyond anthers; petals 10–30 mm.
        41. Inflorescences dense; ovary at anthesis longer than adjacent internode.
          42. Floral tube conspicuously veined; petals with large, wedge-shaped purplish red spot near apex . . . . 17. *Clarkia imbricata*
          42. Floral tube not conspicuously veined; petals with red spot near or proximal to middle . . . . 19. *Clarkia speciosa* (in part)
        41. Inflorescences open; ovary at anthesis shorter than adjacent internode.
          43. Buds mucronate, sepal tips distinct in bud. . . . . 18. *Clarkia williamsonii*
          43. Buds not mucronate, sepal tips connate to tip.
            44. Petals without red or red-purple spot . . . . 19. *Clarkia speciosa* (in part)
            44. Petals with conspicuous red or purple spot.
              45. Petal spot at or proximal to middle . . . . 19. *Clarkia speciosa* (in part)
              45. Petal spot distal to middle . . . . 22. *Clarkia purpurea* (in part)

6a. Clarkia Pursh sect. Eucharidium (Fischer & C. A. Meyer) H. Lewis & M. E. Lewis, Univ. Calif. Publl. Bot. 20: 359. 1955 [E]

Eucharidium Fischer & C. A. Meyer, Index Seminum (St. Petersburg) 2: 36. 1836; *Clarkia* subg. *Eucharidium* (Fischer & C. A. Meyer) Jepson

Inflorescences: axis suberect or slightly recurved; buds pendent. **Flowers:** floral tube narrowly tubular, 13–35 mm; sepals reflexed together to 1 side; petals pink, sometimes white-streaked, fan-shaped, conspicuously 3-lobed, middle lobe often longer than laterals, claw slender, not lobed; stamens 4. **Capsules** subterete; sessile or subsessile.

Species 2 (2 in the flora): California.

Section *Eucharidium* includes two species characterized by large tri-lobed petals, four rather than eight stamens, and a long floral tube that adapts them to pollination by long-tongued Lepidoptera or Diptera (G. A. Allen et al. 1990).

1. Clarkia concinna (Fischer & C. A. Meyer) Greene, Pittonia 1: 140. 1887 • Red ribbons [E] [F]

Eucharidium concinnum Fischer & C. A. Meyer, Index Seminum (St. Petersburg) 2: 37. 1836

Stems erect, to 40 cm, glabrous or puberulent. **Leaves:** petiole 5–25 mm; blade lanceolate to elliptic or ovate, 1–4.5 cm. **Inflorescences** racemes, axis suberect or slightly recurved; buds pendent. **Flowers:** floral tube 13–25 mm; sepals reflexed together to 1 side, sometimes nearly separating, petal-like, pink or red; corolla rotate, petals bright pink, usually white-streaked, narrowly fan-shaped, 10–30 mm, length 2 times width, tapered to a ± distinct claw, lobes ± equal or middle lobe wider, usually oblanceolate; stamens 4, filaments not wider distally; ovary 8-grooved; stigma exserted or not beyond anthers. **Capsules** 15–20 mm; sessile. **Seeds** reddish brown, 2–3 mm, scaly, crest to 0.8 mm, conspicuous.

Subspecies 3 (3 in the flora): California.

The two self-pollinating subspecies with smaller flowers probably arose independently from the outcrossing subsp. *concinna*.

1. Stigmas exserted beyond anthers; petals 15–30 mm. . . . . . . 1a. *Clarkia concinna* subsp. *concinna*
1. Stigmas not exserted beyond anthers; petals 10–20 mm.
    2. Sepals remaining connate only near tip; petal lobes prominent, blades gradually tapered to distinct claw. . . 1b. *Clarkia concinna* subsp. *automixa*
    2. Sepals remaining connate in distal ⅔; petal lobes short, blades abruptly tapered to inconspicuous claw . . . . 1c. *Clarkia concinna* subsp. *raichei*

1a. Clarkia concinna (Fischer & C. A. Meyer) Greene subsp. **concinna** [E] [F]

Flowers: sepals remaining connate only near tip, usually reflexed in pairs or singly; petals 15–30 mm, blades gradually tapered to distinct claw, lobes prominent, separated by deep sinuses; stigma exserted beyond anthers. **2*n*** = 14.

Flowering Apr–Jul. Mixed evergreen forests, woodlands, coastal scrub; 0–1500 m; Calif.

Subspecies *concinna* occurs from the San Francisco Bay area through much of northwestern California to Humboldt and Siskiyou counties, and in the Cascade Range foothills in Butte and Tehama counties.

1b. Clarkia concinna (Fischer & C. A. Meyer) Greene subsp. **automixa** R. N. Bowman, Madroño 34: 41, figs. 1 [bottom], 4, 5. 1987 • Santa Clara red ribbons [E]

Flowers: sepals remaining connate only near tip, usually reflexed in pairs or singly; petals 10–20 mm, blades gradually tapered to distinct claw, lobes prominent, separated by deep sinuses; stigma not exserted beyond anthers. **2*n*** = 14.

Flowering Apr–Jun. Woodlands; 0–1500 m; Calif.

Subspecies *automixa* is known from Alameda, Santa Clara, and barely Santa Cruz counties in the foothills around the Santa Clara Valley, and is designated as rare by the California Native Plant Society.

1c. **Clarkia concinna** (Fischer & C. A. Meyer) Greene subsp. **raichei** G. A. Allen, V. S. Ford & Gottlieb, Madroño 37: 309, fig. 1 [right]. 1991 • Raiche's clarkia or red ribbons [C] [E]

Flowers: sepals remaining connate in distal ⅔, usually reflexed together, sometimes in pairs; petals 10–20 mm, blades abruptly tapered to inconspicuous claw, lobes short, not separated by deep sinuses; stigma not exserted beyond anthers. **2*n*** = 14.

Flowering Apr–May. Exposed sites; of conservation concern; 20 m; Calif.

Subspecies *raichei* is known only from a small area near the type locality just south of Tomales, Marin County, and is listed as rare by the California Native Plant Society.

2. **Clarkia breweri** (A. Gray) Greene, Pittonia 1: 141. 1887 • Brewer's clarkia, fairy fans [E] [F]

Eucharidium breweri A. Gray, Proc. Amer. Acad. Arts 6: 532. 1865

Stems erect or decumbent, to 20 cm, glabrous or sparsely puberulent. **Leaves:** petiole to 20 mm; blade linear to lanceolate, 2–5 cm. **Inflorescences** racemes, axis straight or recurved; buds pendent. **Flowers:** floral tube 20–35 mm; sepals reflexed together to 1 side, not petal-like, green to magenta; corolla rotate, petals pink, broadly fan-shaped, 15–25 mm, length equal to width, without claw, 3-lobed, middle lobe longer and much narrower, linear to oblanceolate; stamens 4, filaments wider distally; ovary inconspicuously grooved; stigma exserted beyond anthers. **Capsules** 15–40 mm; subsessile. **Seeds** reddish brown, 2–3 mm, scaly-tuberculate, crest to 0.8 mm, conspicuous. **2*n*** = 14.

Flowering Apr–Jun. Woodlands, chaparral; 0–1000 m; Calif.

Clarkia breweri is restricted to dry woodlands and chaparral west of the Central Valley from the San Francisco Bay area into the southern Coast Ranges in Fresno, Monterey, and San Benito counties.

6b. Clarkia Pursh sect. Clarkia [E]

Inflorescences: axis recurved or suberect; buds pendent. **Flowers:** floral tube campanulate, 3–5 mm; sepals reflexed together to 1 side; petals bright pink to lavender, sometimes white- or purple-veined, fan-shaped, 3-lobed, claw with pair of prominent, lateral lobes; stamens 8, unequal, 4 fertile, 4 sterile and reduced. **Capsules** angled, 8-grooved; pedicellate.

Species 1: w North America.

3. **Clarkia pulchella** Pursh, Fl. Amer. Sept. 1: 260, plate 11. 1813 (as Clarckia) • Pinkfairies [E] [F]

Stems erect, to 50 cm, glabrous or puberulent. **Leaves:** petiole 0–10 mm; blade linear to lanceolate, 2–8 cm. **Inflorescences** racemes, axis straight or recurved; buds pendent. **Flowers:** floral tube minutely strigillose in distal ½ inside; sepals reflexed together to 1 side; corolla rotate, petals very broadly fan-shaped, 10–30 mm, lateral lobes 1–5 mm; stamens 8, unequal, 4 fertile, 4 sterile and reduced, subtended by puberulent scales, outer anthers lavender to white, inner much smaller, sterile; ovary shallowly 8-grooved; stigma exserted beyond anthers. **Capsules** 10–30 mm; pedicel 3–10 mm. **Seeds** dark brown, 1 mm, scaly, crest to 0.1 mm, inconspicuous. **2*n*** = 24.

Flowering May–Jul. Openings in sagebrush and coniferous forests; 500–2200 m; B.C.; Idaho, Mont., Oreg., S.Dak., Wash., Wyo.

Clarkia pulchella is the only North American species in the genus that does not occur in California; instead it is found throughout most of eastern Oregon and Washington, western Idaho, and northwestern Montana, to southern British Columbia, with disjunct occurrences in Bannock County in Idaho, Teton County in Wyoming, and Meade County in South Dakota. It was first discovered in 1806 by Meriwether Lewis during the Lewis and Clark expedition, and was the first species named in the new genus *Clarkia*.

Clarkia pulchella is an allopolyploid that combines morphological characteristics of sect. *Myxocarpa* (*C. borealis* and relatives), which includes two species with $2n = 10$, and sect. *Eucharidium* (*C. concinna* and *C. breweri*) with $2n = 14$. Molecular data support a relationship with sect. *Eucharidium* but at present show no direct association with sect. *Myxocarpa*.

Clarkia elegans Poiret is an illegitimate name that pertains here.

6c. Clarkia Pursh sect. Rhodanthos (Fischer & C. A. Meyer) P. H. Raven, Brittonia 16: 287. 1964 [E]

Oenothera Linnaeus sect. *Rhodanthos* Fischer & C. A. Meyer, Index Seminum (St. Petersburg) 2: 45. 1836

Inflorescences: axis recurved or straight and erect; buds pendent or erect. **Flowers:** floral tube obconic, 1–10 mm; sepals reflexed together to 1 side; petals usually pink to lavender or light purple, rarely white, usually with red spot near middle or a red zone at base, obovate to fan-shaped, unlobed, claw inconspicuous or absent; stamens 8, subequal. **Capsules** 4-angled, 4-grooved or 8-ribbed; sessile or pedicellate.

Species 6 (6 in the flora): w North America.

6c.1. Clarkia Pursh (sect. Rhodanthos) subsect. Primigenia H. Lewis & M. E. Lewis, Univ. Calif. Publ. Bot. 20: 261. 1955 [E]

Inflorescences: axis straight, erect; buds erect. **Flowers:** ovary usually 4-grooved, rarely 8-grooved (*C. amoena* subsp. *whitneyi*), not wider distally.

Species 3 (3 in the flora): w North America.

4. Clarkia amoena (Lehmann) A. Nelson & J. F. Macbride, Bot. Gaz. 65: 62. 1918 • Farewell-to-spring [E] [F]

Oenothera amoena Lehmann, Index Seminum (Hamburg) 1820: 8. 1821; *Godetia amoena* (Lehmann) G. Don

Stems erect to decumbent, 20–200 cm, puberulent. **Leaves:** petiole to 10 mm; blade linear to lanceolate, 1–6 cm. **Inflorescences** open or dense spikes or racemes, axis straight; buds erect. **Flowers:** floral tube 3–10 mm; sepals usually reflexed together to one side, or rarely in pairs or singly; corolla bowl-shaped, petals pale pink to lavender, usually with red spot or mark near middle, obovate to fan-shaped, 15–60 mm, not lobed, apex sometimes shallowly notched or erose; stamens 8, in 2 subequal sets; ovary cylindrical and 4-grooved or fusiform and 8-grooved, puberulent; stigma exserted or not beyond anthers. **Capsules** 15–40 mm, sometimes broader distally; pedicel 0–13 mm. **Seeds** brown to grayish brown, 1–1.5 mm, scaly, crest 0.1 mm.

Subspecies 5 (5 in the flora): w North America.

Clarkia amoena is closely related to *C. rubicunda*, which differs morphologically mainly in the color pattern of the petals. Petals of *C. amoena* have a conspicuous red spot or group of small red spots or marks near the middle, whereas those of *C. rubicunda* have a red area at the base and are not spotted near the middle. The areas of distribution of the two species barely overlap in California just north of San Francisco Bay, and *C. rubicunda* could be considered a southern geographical race or subspecies of *C. amoena* were it not that their readily formed hybrids are sterile due to chromosomal rearrangement. *Clarkia amoena* is one of the parent species of the allotetraploid *C. gracilis*. Intermediates between subspecies are frequent.

1. Stigmas not exserted beyond anthers; petals less than 20 mm; coastal British Columbia, Oregon, Washington. . . . . 4b. *Clarkia amoena* subsp. *caurina*
1. Stigmas exserted beyond anthers; petals 15–60 mm.
    2. Ovaries 8-grooved, broadly fusiform, 8–12 mm wide . . . . . . 4e. *Clarkia amoena* subsp. *whitneyi*
    2. Ovaries 4-grooved, cylindrical to subclavate, 2–6 mm wide.

[3. Shifted to left margin.—Ed.]

3. Inflorescences open, internodes longer than subtending flowers, bracts sublinear . . . . 4c. *Clarkia amoena* subsp. *huntiana*
3. Inflorescences congested, internodes shorter than subtending flowers, bracts narrowly lanceolate or wider.
   4. Stems decumbent to suberect, to 100 cm; petals 20–35 mm, usually with red spot mid blade; coastal California . . . . 4a. *Clarkia amoena* subsp. *amoena*
   4. Stems erect, to 200 cm; petals 30–40 mm, usually without red spot or with very small spot or streak mid blade; inland Oregon, Washington . . . . . 4d. *Clarkia amoena* subsp. *lindleyi*

4a. Clarkia amoena (Lehmann) A. Nelson & J. F. Macbride subsp. **amoena** [E]

Stems decumbent to suberect, to 100 cm. **Inflorescences** congested spikes; bracts narrowly lanceolate; internodes shorter than subtending flowers. **Flowers:** petals usually with bright red spot mid blade, 20–35 mm; ovary cylindrical, 2–5 mm wide, 4-grooved; stigma exserted beyond anthers. $2n = 14$.

Flowering Jun–Aug. Open slopes and bluffs near coast; 0–100 m; Calif.

Subsp. *amoena* is found primarily in the North Coast and North Coast Ranges, and in the San Francisco Bay and Central Coast regions south to Monterey County.

4b. Clarkia amoena (Lehmann) A. Nelson & J. F. Macbride subsp. **caurina** (Abrams ex Piper) H. Lewis & M. E. Lewis, Univ. Calif. Publ. Bot. 20: 268. 1955 [E]

Godetia caurina Abrams ex Piper, Contr. U.S. Natl. Herb. 11: 410. 1906; *Clarkia amoena* var. *caurina* (Abrams ex Piper) C. L. Hitchcock; *C. amoena* var. *pacifica* (M. Peck) C. L. Hitchcock; *G. pacifica* M. Peck

Stems erect, to 50 cm. **Inflorescences** open racemes; bracts ± narrowly lanceolate; internodes as long as or longer than subtending flowers. **Flowers:** petals usually with bright red spot mid blade or on distal part, less than 20 mm; ovary cylindrical, 2–3 mm wide, 4-grooved, grooves sometimes inconspicuous; stigma not exserted beyond anthers. $2n = 14$.

Flowering Jun–Aug. Sea bluffs, coastal slopes, along Columbia River; 0–100 m; B.C.; Oreg., Wash.

Subspecies *caurina*, the only subspecies that does not occur in California, is found along the coast and along the Columbia River in Oregon and Washington to British Columbia (only on Vancouver Island).

4c. Clarkia amoena (Lehmann) A. Nelson & J. F. Macbride subsp. **huntiana** (Jepson) H. Lewis & M. E. Lewis, Univ. Calif. Publ. Bot. 20: 264. 1955 [E]

Godetia amoena (Lehmann) G. Don forma *huntiana* Jepson, Univ. Calif. Publ. Bot. 2: 329. 1907; *G. amoena* var. *huntiana* (Jepson) Jepson

Stems erect, to 100 cm. **Inflorescences** open racemes; bracts sublinear; internodes longer than subtending flowers. **Flowers:** petals usually with bright red spot mid blade, 15–30 mm; ovary cylindrical to subclavate, 2–4 mm wide, 4-grooved, grooves sometimes inconspicuous; stigma exserted beyond anthers. $2n = 14$.

Flowering Jun–Aug. Openings in forests and woodlands, often near but rarely along immediate coast; 0–700 m; Calif., Oreg.

Subspecies *huntiana* is found in open sites within woodland regions in northwestern California and into southwestern Oregon, and scattered near the coast in west-central California.

4d. Clarkia amoena (Lehmann) A. Nelson & J. F. Macbride subsp. **lindleyi** (Douglas) H. Lewis & M. E. Lewis, Univ. Calif. Publ. Bot. 20: 267. 1955 • Lindley's clarkia [E] [F]

Oenothera lindleyi Douglas, Bot. Mag. 55: plate 2832. 1828 (as lindleyii); *Clarkia amoena* var. *lindleyi* (Douglas) C. L. Hitchcock; *Godetia amoena* (Lehmann) G. Don var. *lindleyi* (Douglas) Jepson

Stems erect, to 200 cm. **Inflorescences** congested racemes; bracts narrowly lanceolate to lanceolate; internodes shorter than subtending flowers. **Flowers:** petals without red spot near middle of blade or with very small spot or streak, 30–40 mm; ovary cylindrical, 2–3 mm wide, 4-grooved; stigma exserted beyond anthers. $2n = 14$.

Flowering Jul–Aug. Woodlands, margins of fields, along railroad tracks; 0–300 m; Oreg., Wash.

Subspecies *lindleyi* is found in west-central Oregon and southwestern Washington in the Coast Ranges.

4e. Clarkia amoena (Lehmann) A. Nelson & J. F. Macbride subsp. **whitneyi** (A. Gray) H. Lewis & M. E. Lewis, Univ. Calif. Publ. Bot. 20: 265. 1955 • Whitney's clarkia [C] [E] [F]

Oenothera whitneyi A. Gray, Proc. Amer. Acad. Arts 7: 340, 400. 1868; *Godetia whitneyi* (A. Gray) T. Moore

Stems decumbent, to 100 cm. **Inflorescences** congested racemes; bracts lanceolate; internodes shorter than subtending flowers. **Flowers:** petals with bright red spot mid blade, 30–60 mm; ovary broadly fusiform, 8–12 mm wide, 8-grooved, 4 deeper; stigma exserted beyond anthers. $2n = 14$.

Flowering Jul–Aug. Coastal scrub; of conservation concern; 0–100 m; Calif.

Subspecies *whitneyi* is common in cultivation, mostly as hybrid derivatives with other subspecies, but is scarce in the wild, found only rarely along the coast in Humboldt and Mendocino counties; it is listed as rare by the California Native Plant Society.

5. Clarkia rubicunda (Lindley) H. Lewis & M. E. Lewis, Madroño 12: 34. 1953 • Farewell-to-spring, ruby chalice clarkia [E]

Godetia rubicunda Lindley, Edwards's Bot. Reg. 22: plate 1856. 1836; *Clarkia rubicunda* subsp. *blasdalei* (Jepson) H. Lewis & M. E. Lewis; *G. blasdalei* Jepson

Stems erect or decumbent, to 150 cm, puberulent; buds erect. **Leaves:** petiole to 10 mm; blade lanceolate to elliptic, 1–4 cm. **Inflorescences** open or dense spikes or racemes, axis straight; buds erect. **Flowers:** floral tube 4–10 mm; sepals reflexed together to one side; corolla bowl-shaped, petals pink to lavender, base red or purplish red, fan-shaped, 10–30 mm, apex erose; stamens 8, subequal; ovary cylindrical, 4-grooved, puberulent; stigma exserted beyond anthers. **Capsules** 20–40 mm; pedicel 0–25(–40) mm. **Seeds** brown to grayish brown, 1.2–1.5 mm, scaly, crest 0.2 mm. $2n = 14$.

Flowering May–Aug. Openings in woodlands, forests, chaparral, coastal scrub; 0–500 m; Calif.

Clarkia rubicunda is known from the central coast of California, from Contra Costa and Marin counties south along the coast and foothills to northern San Luis Obispo County.

Clarkia rubicunda is probably a derivative of *C. amoena* and may be ancestral to *C. franciscana*. *Clarkia rubicunda* is distinguishable from some populations of *C. amoena* only by the absence of a red spot or group of spots near the middle of the petal and the presence of a red area at the base of the petal. *Clarkia rubicunda* can be distinguished from *C. franciscana* by the position of the stigma and size and shape of the petals. All three species differ in chromosome arrangement and hybrids are highly sterile.

6. Clarkia franciscana H. Lewis & P. H. Raven, Brittonia 10: 7, fig. 1a, b, d. 1958 • Presidio clarkia [C] [E]

Stems erect, to 40 cm, strigillose. **Leaves:** petiole 0–5 mm; blade narrowly lanceolate, 1–5.5 cm. **Inflorescences** racemes, axis straight; buds erect. **Flowers:** floral tube 1–3 mm; sepals reflexed together to one side; corolla bowl-shaped, petals lavender-pink shading white near middle, base bright reddish purple, fan-shaped, 5–13 mm, apex erose; stamens 8, subequal; ovary cylindrical, 4-grooved, puberulent; stigma not exserted beyond anthers. **Capsules** 20–40 mm; pedicel 0–15 mm. **Seeds** gray, 1.2–1.5 mm, scaly, crest 0.2 mm. $2n = 14$.

Flowering May–Jun. Serpentine soil; of conservation concern; 50 m; Calif.

Clarkia franciscana is an endangered species (designated rare by the California Native Plant Society), restricted to serpentine soils in coastal grass and shrub communities. The only known localities for it are the Presidio in San Francisco County, and the Oakland Hills in Alameda and Contra Costa counties.

Geographic distribution and petal color pattern suggest that *Clarkia franciscana* may be a self-pollinating derivative of *C. rubicunda*. If true, enzyme studies indicate that the origin is not recent.

Clarkia franciscana is in the Center for Plant Conservation's National Collection of Endangered Plants.

6c.2. Clarkia Pursh (sect. Rhodanthos) subsect. Flexicaules H. Lewis & M. E. Lewis, Univ. Calif. Publ. Bot. 20: 273. 1955 [E]

Inflorescences: axis recurved at tip in bud; buds pendent. **Flowers:** ovary 8-grooved or 8-ribbed, not wider distally.

Species 2 (2 in the flora): w United States.

7. **Clarkia lassenensis** (Eastwood) H. Lewis & M. E. Lewis, Madroño 12: 33. 1953 • Lassen clarkia or godetia [E]

Godetia lassenensis Eastwood, Leafl. W. Bot. 2: 281. 1940

Stems erect, to 90 cm, puberulent. **Leaves:** petiole to 10 mm; blade linear to narrow lanceolate, 2–5 cm. **Inflorescences** racemes, axis recurved at tip in bud; buds pendent. **Flowers:** floral tube 3–5 mm; corolla bowl-shaped, petals pinkish lavender shading lighter proximally, with reddish purple base, broadly obovate, 8–16 mm; stamens 8, subequal; ovary 8-grooved, densely puberulent; stigma not exserted beyond anthers. **Capsules** 25–40 mm; pedicel 0–3 mm. **Seeds** light brown or mottled with dark spots, 1.5 mm, minutely tuberculate, crest 0.2 mm. 2*n* = 14.

Flowering May–Jun. Woodlands, conifer forests; 500–2000 m; Calif., Nev., Oreg.

Clarkia lassenensis is found primarily in the northeastern counties of California, south-central counties of Oregon, and northwestern counties of Nevada, with a few outlier collections from Alpine and Placer counties in the Sierra Nevada, and from northern Mendocino and Glenn counties in the North Coast Ranges.

Clarkia lassenensis is morphologically most similar to *C. arcuata* and *C. gracilis*. At one time, *C. lassenensis* was considered a probable parent of the tetraploid *C. gracilis* but cytogenetic and molecular sequence data indicate that an unknown related species, presumably now extinct, was involved. *Clarkia lassenensis* can be distinguished readily from *C. arcuata* by flower size, position of the stigma, and pubescence of the immature capsule, and from *C. gracilis* by having immature capsules with eight grooves rather than four.

8. **Clarkia arcuata** (Kellogg) A. Nelson & J. F. Macbride, Bot. Gaz. 65: 62. 1918 • Glandular clarkia or fairyfan [E] [F]

Oenothera arcuata Kellogg, Proc. Calif. Acad. Sci. 1: 58. 1855; *Godetia hispidula* (S. Watson) S. Watson; *O. hispidula* S. Watson

Stems erect, to 8 cm, glabrous or puberulent. **Leaves** sessile; blade linear to narrowly lanceolate or oblanceolate, 1.5–6 cm, surfaces glabrate. **Inflorescences** racemes, axis recurved at tip in bud; buds pendent. **Flowers:** floral tube 3–7 mm, puberulent with spreading hairs and shorter glandular hairs; corolla bowl-shaped, petals pinkish lavender, lighter proximally, often with dark reddish spot at base, broadly obovate or obdeltate, 10–30 mm; stamens 8, subequal; ovary 8-grooved, sparsely puberulent, hairs mostly glandular; stigma exserted beyond anthers. **Capsules** 10–35 mm; pedicel 5–15 mm. **Seeds** brown, 2 mm, minutely scaly, crest 0.5 mm, prominent. 2*n* = 14.

Flowering Apr–Jun. Openings in woodlands and chaparral, serpentine soil; 0–1600 m; Calif.

Clarkia arcuata is primarily distributed in the foothills of the Sierra Nevada from Mariposa to Tehama counties, sparsely south to Kern County, and with one disjunct occurrence in Shasta County; it has also been reported from Napa County.

Clarkia arcuata is morphologically most similar to *C. lassenensis*, from which it differs in fruit characteristics. The two species have distinct areas of distribution and attempts to produce hybrids have not been successful.

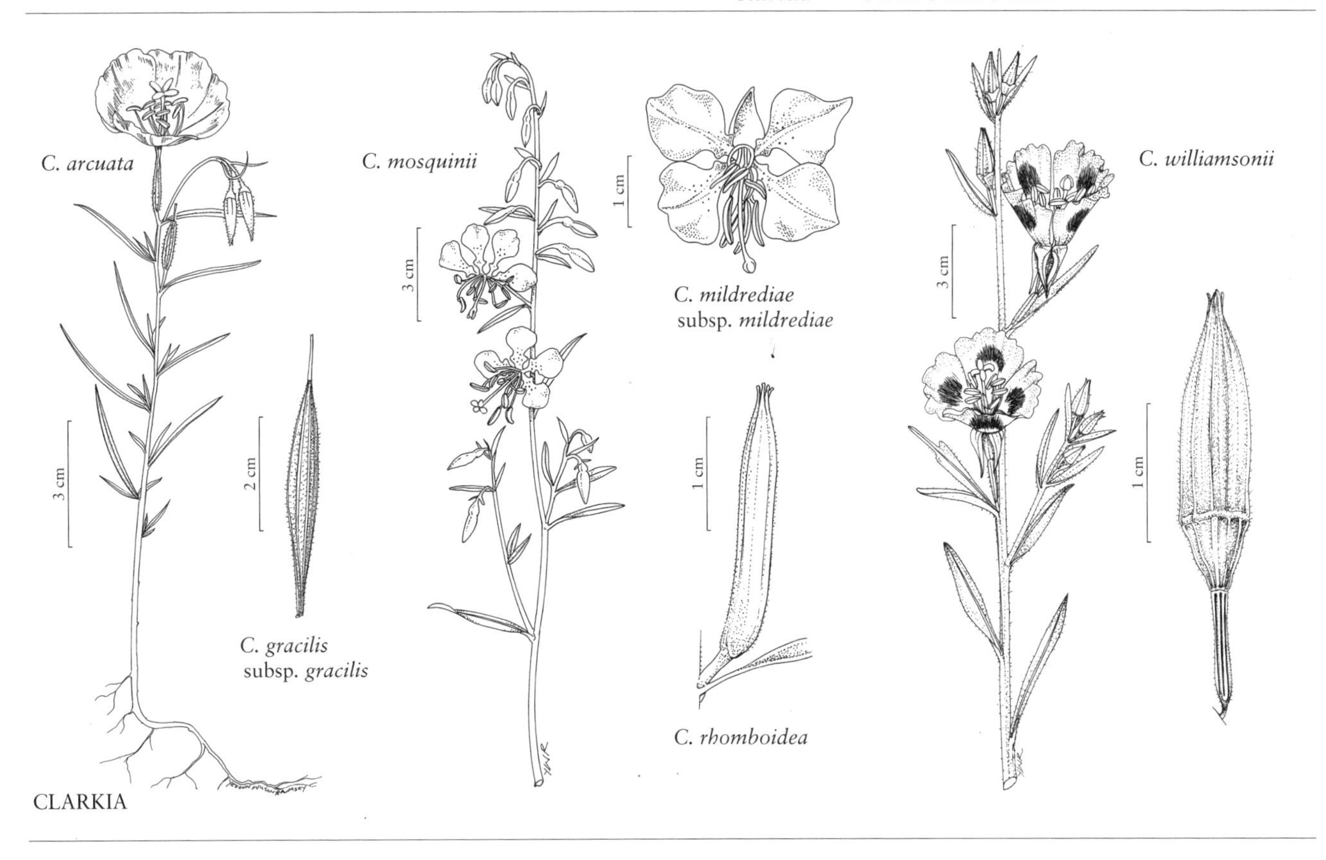

CLARKIA

6c.3. Clarkia Pursh (sect. Rhodanthos) subsect. Jugales H. Lewis & M. E. Lewis, Univ. Calif. Publ. Bot. 20: 277. 1955 [E]

Inflorescences: axis recurved at tip in bud; buds pendent. **Flowers:** ovary 4-grooved, wider distally.

Species 1: w United States.

9. Clarkia gracilis (Piper) A. Nelson & J. F. Macbride, Bot. Gaz. 65: 63. 1918 • Slender clarkia [E] [F]

Godetia gracilis Piper in C. V. Piper and R. K. Beattie, Fl. N.W. Coast, 251. 1915; *G. amoena* (Lehmann) G. Don var. *gracilis* (Piper) C. L. Hitchcock

Stems erect, to 90 cm, glabrous or puberulent. **Leaves:** petiole to 11 mm; blade linear to narrowly lanceolate, 2–7 cm. **Inflorescences** racemes, axis recurved at tip in bud; buds pendent. **Flowers:** floral tube 1.5–10 mm; corolla bowl-shaped, petals pink to lavender or light purple, often lighter toward base, unspotted or with red spot near middle or base, obovate or obdeltate, 6–40 mm; stamens 8, subequal; ovary 4-grooved, puberulent; stigma exserted or not beyond anthers. **Capsules** 25–50 mm, usually wider distally; pedicel 0–10 mm. **Seeds** brown or mottled with dark spots, 1.5–2 mm, scaly, crest 0.1 mm.

Subspecies 4 (4 in the flora): w United States.

Clarkia gracilis is a tetraploid derived from *C. amoena* and an unknown, presumably extinct, species related to *C. arcuata* and *C. lassenensis*. *Clarkia gracilis* is readily distinguished from *C. amoena* by the axis of the inflorescence in bud, which is not recurved in *C. amoena*; it is distinguished from *C. arcuata* and *C. lassenensis* by its four rather than eight-grooved ovary.

1. Stigmas not exserted beyond anthers; petals 6–23 mm. . . . . 9a. *Clarkia gracilis* subsp. *gracilis*
1. Stigmas exserted beyond anthers; petals 20–40 mm.
    2. Petals with red spot near mid blade, base of blades white . . . 9b. *Clarkia gracilis* subsp. *sonomensis*
    2. Petals unspotted, base of blades red or bright red.
        3. Petals 20–30 mm . . . . 9c. *Clarkia gracilis* subsp. *tracyi*
        3. Petals 30–40 mm . . . . 9d. *Clarkia gracilis* subsp. *albicaulis*

9a. Clarkia gracilis (Piper) A. Nelson & J. F. Macbride subsp. **gracilis** [E] [F]

Godetia amoena (Lehmann) G. Don var. *concolor* Jepson

Flowers: petals pink to lavender, unspotted, 6–23 mm, base darker or not; stigma not exserted beyond anthers. **2*n*** = 28.

Flowering Apr–Jul. Woodlands, forest openings; 0–1500 m; Calif., Oreg., Wash.

Subspecies *gracilis* is found throughout most of northern California below 1500 m, and is more scattered in western Oregon and Washington.

9b. Clarkia gracilis (Piper) A. Nelson & J. F. Macbride subsp. **sonomensis** (C. L. Hitchcock) H. Lewis & M. E. Lewis, Madroño 12: 33. 1953 • Sonoma clarkia [E]

Godetia amoena (Lehmann) G. Don var. *sonomensis* C. L. Hitchcock, Bot. Gaz. 89: 338. 1930

Flowers: petals pinkish lavender shading to white proximally with red spot near middle, 20–40 mm; stigma exserted beyond anthers. **2*n*** = 28.

Flowering May–Jun. Woodlands, forest openings; 0–500 m; Calif., Oreg.

Subspecies *sonomensis* is restricted to the North Coast Ranges of California in Colusa, Humboldt, Lake, Marin, Mendocino, Napa, and Sonoma counties, and of southern Oregon in Douglas and Josephine counties.

9c. Clarkia gracilis (Piper) A. Nelson & J. F. Macbride subsp. **tracyi** (Jepson) Abdel-Hameed & R. Snow, Amer. J. Bot. 55: 1048. 1968 • Tracy's clarkia [E]

Godetia cylindrica Jepson var. *tracyi* Jepson, Fl. Calif. 2: 584. 1936

Flowers: petals pinkish lavender, base red, unspotted, 20–30 mm; stigma exserted beyond anthers. **2*n*** = 28.

Flowering May–Jul. Woodlands, serpentine soil; 100–500 m; Calif.

Subspecies *tracyi* is known from the Inner North Coast Ranges, mainly in Colusa, Lake, and Napa counties, less commonly in Humboldt, Mendocino, Sonoma, Tehama, and Yolo counties.

9d. Clarkia gracilis (Piper) A. Nelson & J. F. Macbride subsp. **albicaulis** (Jepson) H. Lewis & M. E. Lewis, Madroño 12: 33. 1953 • Whitestem clarkia [C] [E]

Godetia amoena (Lehmann) G. Don var. *albicaulis* Jepson, Univ. Calif. Publ. Bot. 2: 329. 1907

Flowers: petals pinkish lavender to light purple shading to white near middle, base bright red, 30–40 mm; stigma exserted beyond anthers. **2*n*** = 28.

Flowering May–Jul. Woodlands; of conservation concern; 500 m; Calif.

Subspecies *albicaulis* is found in foothill woodlands around the upper Sacramento Valley, mainly in Butte, Colusa, Lake, and Tehama counties, less commonly in Lassen, Mariposa, Modoc, Shasta, and Solano counties.

6d. Clarkia Pursh sect. Myxocarpa H. Lewis & M. E. Lewis, Univ. Calif. Publ. Bot. 20: 347. 1955

Inflorescences: axis recurved at tip in bud, becoming erect; buds pendent. **Flowers:** floral tube campanulate, 1–5 mm; sepals reflexed individually or, sometimes, all together to 1 side; petals lavender-purple or -pink, reddish purple, or pinkish lavender, often dark-flecked or -spotted, obovate, suborbiculate, or obdeltate to rhombic, rarely ± 3-lobed, constricted into ± broad claw with pair of lateral basal lobes; stamens 8, subequal; ovary subterete or 4-grooved, often 8-striate. **Capsules** 4-angled; sessile or pedicellate.

Species 7 (7 in the flora): nw United States, w Mexico.

SELECTED REFERENCE Small, E. 1971. The systematics of *Clarkia*, section *Myxocarpa*. Canad. J. Bot. 49: 1211–1217.

10. Clarkia borealis E. Small, Canad. J. Bot. 49: 1215, figs. 2B, 3A,B. 1971 • Northern clarkia [C] [E]

Stems erect, to 100 cm, puberulent. **Leaves:** petiole 15–40 mm; blade elliptic to ovate, 2–6 cm. **Inflorescences** open racemes, axis recurved only at tip in bud, straight 4+ nodes distal to open flowers; buds pendent, fusiform, base slightly swollen, tip acute. **Flowers:** floral tube 2–4 mm; sepals reflexed individually; corolla rotate, petals lavender-pink, often dark-flecked, obdeltate to suborbiculate, unlobed, 13–19 × 7–12 mm, length 1.6–2 times width; stamens 8, subequal, subtended by ciliate scales, pollen blue-gray; ovary shallowly 4-grooved, puberulent; stigma exserted beyond anthers. **Capsules** 20–30 mm; pedicel 0–3 mm. **Seeds** light brown or mottled with dark spots, 1.5–2.5 mm, minutely tuberculate, crest 0.2 mm.

Subspecies 2 (2 in the flora): California.

Clarkia borealis is closely related, and possibly ancestral, to *C. mildrediae.* The two species can be distinguished most readily by the degree of curvature of the inflorescence and the petal color.

1. Seeds 1.5–1.8 mm . . . . 10a. *Clarkia borealis* subsp. *borealis*
1. Seeds 1.8–2.5 mm . . . . 10b. *Clarkia borealis* subsp. *arida*

10a. Clarkia borealis E. Small subsp. **borealis** [C] [E]

Seeds 1.5–1.8 mm. **2*n*** = 14.

Flowering Jun–Jul. Foothill woodlands, forest margins; of conservation concern; 300–800 m; Calif.

Subspecies *borealis* is known only from western Shasta and eastern Trinity counties and is listed as rare by the California Native Plant Society.

10b. Clarkia borealis E. Small subsp. **arida** E. Small, Canad. J. Bot. 49: 1215, figs. 2B, 3B. 1971 • Shasta clarkia [C] [E]

Seeds 1.8–2.5 mm. **2*n*** = 14.

Flowering Jun–Aug. Foothill woodlands; of conservation concern; 500 m; Calif.

Subspecies *arida* is known only from south-central Shasta County, and barely into adjacent northern Tehama County and is designated rare by the California Native Plant Society.

11. Clarkia mildrediae (A. Heller) H. Lewis & M. E. Lewis, Madroño 12: 34. 1953 • Mildred's clarkia [E] [F]

Phaeostoma mildrediae A. Heller, Leafl. W. Bot. 2: 221. 1940 (as mildredae)

Stems erect, to 100 cm, puberulent. **Leaves:** petiole 15–40 mm; blade elliptic to ovate, 3–6 cm. **Inflorescences** open racemes, axis recurved in bud, straight only 1–3 nodes distal to open flowers; buds pendent, tip acute. **Flowers:** floral tube 2–3 mm; sepals usually reflexed individually; corolla rotate, petals reddish purple, often darker flecked or spotted, narrowly obdeltate to suborbiculate, unlobed, 11–25 × 7–18 mm, length 1.4–1.6 times width; stamens 8, subequal, subtended by ciliate scales, anthers magenta or orange-red to yellow, pollen blue-gray or bright yellow to tan; ovary shallowly 4-grooved, puberulent; stigma exserted beyond anthers. **Capsules** 20–30 mm; pedicel 0–5 mm. **Seeds** brown or gray, 1.5–1.8 mm, scaly-echinate, crest 0.1 mm.

Subspecies 2 (2 in the flora): California.

Clarkia mildrediae is closely related to, and may be derived from, *C. borealis. Clarkia mildrediae* subsp. *lutescens* is probably the direct ancestor of *C. stellata. Clarkia mildrediae* differs from *C. borealis* in petal color and inflorescence habit and from *C. stellata* in flower size and position of the stigma. On the basis of morphology, chromosome number and pairing, *C. mildrediae* appears to be one of the parents of the tetraploid species *C. rhomboidea.*

1. Petals 12–18 mm wide; anthers magenta, fresh pollen blue-gray . . . . 11a. *Clarkia mildrediae* subsp. *mildrediae*
1. Petals 7–16 mm wide; anthers yellow to orange-red, fresh pollen bright yellow to tan . . . . 11b. *Clarkia mildrediae* subsp. *lutescens*

11a. Clarkia mildrediae (A. Heller) H. Lewis & M. E. Lewis subsp. **mildrediae** [E] [F]

Flowers: petals 12–18 mm wide; anthers magenta, fresh pollen blue-gray. **2*n*** = 14.

Flowering Jun–Jul. Yellow-pine forests; 400–1700 m; Calif.

Subspecies *mildrediae* is known only from the Feather River drainage in the foothills of the southern Cascade–northern Sierra Nevada ranges in Butte, Plumas, Yuba, and barely Sierra counties, but is relatively abundant within that area. Nevertheless, it is listed as rare by the California Native Plant Society.

11b. Clarkia mildrediae (A. Heller) H. Lewis & M. E. Lewis subsp. **lutescens** Gottlieb & Janeway, Madroño 44: 250, fig. 1 [top]. 1998 [E]

Flowers: petals 7–16 mm wide; anthers yellow to orange-red, fresh pollen bright yellow to tan. $2n = 14$.

Flowering Jun–Jul. Yellow-pine forests; 400–1700 m; Calif.

The distribution of subsp. *lutescens* overlaps with that of subsp. *mildrediae* in Butte and Plumas counties, but extends farther to the southeast into Sierra and Yuba counties.

12. Clarkia stellata Mosquin, Leafl. W. Bot. 9: 215. 1962 • Lake Amador clarkia [E]

Stems erect, to 100 cm, puberulent. **Leaves:** petiole 5–30 mm; blade lanceolate to elliptic or ovate, 1–5 cm. **Inflorescences** open racemes, axis in bud recurved 1–3 nodes distal to open flowers; buds pendent, narrowly obovoid, tip acute. **Flowers:** floral tube 1.5–2 mm; sepals reflexed individually; corolla rotate, petals lavender-purple, not dark-flecked or spotted, obovate, 6–8 × 3–5 mm, inconspicuously 3-lobed; stamens 8, subequal, subtended by ciliate scales, pollen yellow; ovary shallowly 4-grooved, puberulent; stigma not exserted beyond anthers. **Capsules** 20–25 mm; pedicel 1–3 mm. **Seeds** unknown. $2n = 14$.

Flowering Jun–Jul. Open coniferous forests; 1000–1500 m; Calif.

Clarkia stellata is known from the southern Cascade–northern Sierra Nevada region, including Lassen, Nevada, Plumas, Shasta, Sierra, and Tehama counties (with unverified reports from Butte and Modoc counties).

Clarkia stellata is probably a self-pollinating derivative of *C. mildrediae* subsp. *lutescens*, to judge from pollen color. The two species are readily distinguished by the much smaller flowers of *C. stellata* and the position of the stigma. Hybrids have low fertility due to chromosomal rearrangement. *Clarkia stellata* is morphologically very similar to the self-pollinating tetraploid *C. rhomboidea* but can be distinguished from it by yellow pollen and shallowly lobed, unspotted petals.

13. Clarkia mosquinii E. Small, Canad. J. Bot. 49: 1216, fig. 4A,B. 1971 • Mosquin's clarkia [C] [E] [F]

Clarkia mosquinii subsp. *xerophylla* E. Small

Stems erect, to 100 cm, puberulent. **Leaves:** petiole 10–30 mm; blade linear-lanceolate to ovate or elliptic, 2–5 cm. **Inflorescences** open racemes, axis recurved only at tip in bud, straight 4+ nodes distal to open flowers; buds pendent, narrowly obovoid, tip obtuse. **Flowers:** floral tube 2–5 mm; sepals reflexed individually; corolla rotate, petals lavender-purple, often with darker spots, ± rhombic, unlobed, 10–20 × 6–13 mm, length 1.5–2 times width; stamens 8, subequal, subtended by ciliate scales, pollen blue-gray; ovary shallowly 4-grooved; stigma exserted beyond anthers. **Capsules** 15–25 mm; pedicel 0–3 mm. **Seeds** brown or gray, 0.9–1.2 mm, scaly. $2n = 12$.

Flowering Jun–Jul. Yellow-pine forests; of conservation concern; 200–300 m; Calif.

Clarkia mosquinii is known only from a small area in the Feather River drainage at the northern limits of the Sierra Nevada range in Butte and (barely) Plumas counties; it is listed as rare by the California Native Plant Society.

Clarkia mosquinii is closely related to *C. borealis* and may be a derivative of that species with a reduced chromosome number. In addition to chromosome number, they differ in geographical distribution and shape of the buds, which are blunt at the tip in *C. mosquinii* and acute or acuminate in *C. borealis*. *Clarkia mosquinii* is also closely related, and probably ancestral, to two species with $2n = 10$, *C. australis* and *C. virgata*, which have more southern distributions.

SELECTED REFERENCE Gottlieb, L. D. and L. P. Janeway. 1995. The status of *Clarkia mosquinii* (Onagraceae). Madroño 42: 79–82.

14. Clarkia virgata Greene, Erythea 3: 123. 1895 • Sierra clarkia [E]

Stems erect, to 100 cm, puberulent. **Leaves:** petiole 15–50 mm; blade elliptic to ovate, 2–5 cm. **Inflorescences** open racemes, axis recurved only at tip in bud, straight 4+ nodes distal to open flowers; buds pendent, narrowly obovoid, tip obtuse. **Flowers:** floral tube 2–4 mm; sepals reflexed individually; corolla rotate, petals lavender-purple, mottled or spotted with reddish purple, ± rhombic, unlobed, 7–14 × 3–7 mm, length 1.9–3 times width; stamens 8, subequal, subtended by ciliate scales, pollen blue-gray; ovary shallowly

4-grooved; stigma not or rarely exserted beyond anthers. **Capsules** 10–20 mm; pedicel 1–4 mm. **Seeds** brown or gray, 1–1.5 mm, scaly-echinate, crest 0.1 mm. $2n = 10$.

Flowering Jun–Jul. Yellow-pine forests, foothill woodlands; 400–1100 m; Calif.

Clarkia virgata is known primarily from El Dorado to Tuolumne counties in the north-central Sierra Nevada range, with scattered collections to Mariposa and Yuba counties.

Clarkia virgata is very similar to *C. mosquinii* and *C. australis* and is probably derived from the former through chromosome reduction in number and rearrangement and may be ancestral to the latter, which differs in chromosome arrangement. Experimental hybrids in all combinations have very low fertility. The three species are difficult to distinguish morphologically but replace one another ecogeographically with *C. australis* in the south and *C. virgata* in the middle with non-overlapping distributions. Other than geographical distribution, *C. virgata* is usually distinguishable from *C. mosquinii* by having narrower petal blades and from *C. australis* by having broader leaves.

15. Clarkia australis E. Small, Canad. J. Bot. 49: 1216, fig. 4D. 1971 • Small's southern clarkia [C] [E]

Clarkia virgata Greene var. *australis* (E. Small) D. W. Taylor

Stems erect, to 100 cm, puberulent. **Leaves:** petiole 10–30 mm; blade linear to lanceolate, 2–5 cm. **Inflorescences** open racemes, axis recurved only at tip in bud, straight 4+ nodes distal to open flowers; buds pendent, narrowly obovoid, tip obtuse. **Flowers:** floral tube 2–4 mm; sepals reflexed individually; corolla rotate, petals lavender-purple, mottled or spotted reddish purple, ± rhombic, unlobed, 6–12(–14) × 3–7 mm, length 2.2–3 times width; stamens 8, subequal, subtended by ciliate scales, pollen blue-gray; ovary shallowly 4-grooved; stigma exserted beyond anthers. **Capsules** 10–20 mm; pedicel 0–4 mm. **Seeds** brown, 1–1.5 mm, scaly. $2n = 10$.

Flowering Jun–Jul. Yellow-pine forests; of conservation concern; 800–1500 m; Calif.

Clarkia australis is found in the foothills of the central Sierra Nevada range, from Calaveras, Madera, Mariposa, and Tuolumne counties, and has been designated as rare by the California Native Plant Society.

Clarkia australis is morphologically very similar to *C. virgata* and, based on its more southern distribution, may be derived from it. They are most readily distinguished morphologically by the narrower leaves of *C. australis*.

16. Clarkia rhomboidea Douglas in W. J. Hooker, Fl. Bor.-Amer. 1: 214. 1832 • Diamond or tongue clarkia, diamond fairyfan, rhomboid farewell-to-spring [F]

Stems erect, to 100 cm, puberulent. **Leaves:** petiole 5–25 mm; blade lanceolate to elliptic or ovate, 1–6 cm. **Inflorescences** open racemes, axis in bud recurved 1–3 nodes distal to open flowers; buds pendent, narrowly obovoid, tip acute to obtuse, often curved to one side. **Flowers:** floral tube 1–3 mm; sepals reflexed individually; corolla rotate, petals pinkish lavender, often with darker flecks, narrowly to broadly obovate or rhombic, sometimes ± 3-lobed, 6–12(–14) × 3–7 mm; stamens 8, subequal, subtended by ciliate scales, pollen blue-gray; ovary shallowly 4-grooved; stigma not or rarely exserted beyond anthers. **Capsules** 10–25 mm; pedicel 1–4 mm. **Seeds** brown, gray, or mottled, 1–1.5 mm, scaly-echinate, crest 0.1 mm, inconspicuous. $2n = 24$.

Flowering May–Sep. Yellow-pine forests, woodlands; 0–3000 m; Ariz., Calif., Idaho, Mont., Nev., Oreg., Utah, Wash.; Mexico (Baja California).

Clarkia rhomboidea is a tetraploid derived from *C. mildrediae* and *C. virgata* or a closely related species. The six diploid species in sect. *Myxocarpa* closely related to *C. rhomboidea* (*C. australis*, *C. borealis*, *C. mildrediae*, *C. mosquinii*, *C. stellata*, and *C. virgata*) are California endemics with relatively small areas of distribution, whereas *C. rhomboidea* occurs throughout much of the western United States and is morphologically much more variable. *Clarkia rhomboidea* characteristically has relatively small, self-pollinating flowers with the stigma in contact with the anthers. Among the diploid species only *C. stellata* has similar small, self-pollinating flowers but is distinguished by yellow pollen and petals with a shallowly 3-lobed blade that is not flecked. Rare populations of *C. rhomboidea* have relatively large flowers with the stigma exserted beyond the anthers. When they occur within the geographical range of the outcrossing diploid species, they may be difficult to distinguish without determining chromosome number.

6e. Clarkia Pursh sect. Godetia (Spach) H. Lewis & M. E. Lewis, Univ. Calif. Publ. Bot. 20: 283. 1955

Godetia Spach, Hist. Nat. Vég. 4: 386. 1835; *Clarkia* subgen. *Godetia* (Spach) Tzvelev; *Oenotheridium* Reiche

Inflorescences: axis erect, or prostrate to decumbent (*C. davyi*, *C. prostrata*); buds erect. **Flowers:** floral tube funnelform to obconical, 2–15 mm; sepals reflexed individually or in pairs; petals lavender-pink to dark wine-red, shading white or yellow near middle or base, usually purplish red-spotted, obovate to obdeltate, unlobed, claw inconspicuous or absent; stamens 8, subequal. **Capsules** conspicuously 8-ribbed; subsessile.

Species 7 (6 in the flora): w North America, nw Mexico, w South America.

Clarkia tenella (Cavanilles) H. Lewis & M. E. Lewis, with three subspecies, is known from Argentina and Chile, and is the only species of *Clarkia* not known to occur naturally in North America.

Some of the common cultivars of *Clarkia* in the horticulture trade are members of sect. *Godetia*, with their relatively large bowl-shaped flowers, and some cultivars (particularly of *C. speciosa*) are still on the market as *Godetia*.

17. Clarkia imbricata H. Lewis & M. E. Lewis, Madroño 12: 38. 1953 • Vine Hill clarkia [C] [F]

Stems erect, to 60 cm, glabrous or sparsely puberulent. **Leaves:** petiole 0–2 mm; blade lanceolate, 2–2.5 cm. **Inflorescences** dense racemes, axis straight; buds erect. **Flowers:** floral tube 10–15 mm, conspicuously veined, lavender striate within; sepals reflexed individually; corolla bowl-shaped, petals lavender shading to white proximally, with large, wedge-shaped purplish red spot near apex, 20–25 mm; stamens 8, subequal; ovary 8-grooved, longer than adjacent internode; stigma exserted beyond anthers. **Capsules** 10–15 mm. **Seeds** brown or gray, 2 mm, scaly, crest 0.2 mm. **2*n*** = 16.

Flowering Jun–Jul. Clearings, roadsides, chaparral; of conservation concern; 50 m; Calif.

Clarkia imbricata, known from only one small area of Sonoma County, is designated as rare by the California Native Plant Society, and is in the Center for Plant Conservation's National Collection of Endangered Plants; it is a relict of a lineage with 2*n* = 16 that presumably contributed a genome to the tetraploid species *C. davyi* and the South American *C. tenella*.

Morphologically, *Clarkia imbricata* is most similar to *C. speciosa*, *C. williamsonii*, and some populations of *C. purpurea*. *Clarkia imbricata* can be distinguished from *C. speciosa* by the color pattern of the petals and from *C. williamsonii* and populations of *C. purpurea* with similar flower size and color pattern by its broader, ascending, overlapping leaves.

18. Clarkia williamsonii (Durand & Hilgard) H. Lewis & M. E. Lewis, Madroño 12: 34. 1953 • Fort Miller clarkia or fairyfan [E] [F]

Godetia williamsonii Durand & Hilgard in War Department [U.S.], Pacif. Railr. Rep. 5(3): 7, plate 5. 1857 (as williamsoni)

Stems erect, to 100 cm, puberulent. **Leaves:** petiole 0–10 mm; blade linear to narrowly lanceolate, 2–7 cm. **Inflorescences** open racemes, axis straight; buds erect, mucronate. **Flowers:** floral tube 7–13 mm; sepals reflexed individually or in pairs, tips distinct in bud; corolla bowl-shaped, petals usually lavender, white near middle with purple spot distally, rarely uniformly wine-red, 10–30 mm; stamens 8, subequal; ovary 8-grooved, shorter than adjacent internode; stigma usually exserted beyond anthers. **Capsules** 10–30 mm. **Seeds** brown or gray, 1–1.5 mm, scaly, crest 0.1 mm. **2*n*** = 18.

Flowering Apr–Sep. Foothill woodlands, yellow-pine forests; 400–2000 m; Calif.

Clarkia williamsonii occurs widely along the western slope of the Sierra Nevada from Nevada to Kern counties, and the Tehachapi Mountains barely to Los Angeles and Santa Barbara counties (one collection each). There are unverified reports from Riverside and Shasta counties.

Clarkia williamsonii is similar to *C. speciosa* and some populations of the hexaploid *C. purpurea* but can be distinguished from the former by petal color pattern and from both by having sepals that have distinctly free tips in bud, a trait most obvious in pressed specimens when the tips tend to spread apart.

19. Clarkia speciosa H. Lewis & M. E. Lewis, Madroño 12: 34. 1953 • Red spotted or redspot clarkia [E]

Oenothera viminea Douglas var. *parviflora* Hooker & Arnott, Bot. Beechey Voy., 342. 1839, not *Clarkia parviflora* Eastwood 1903; *Godetia parviflora* (Hooker & Arnott) Jepson

Stems erect or decumbent, to 60 cm, puberulent. **Leaves:** petiole 0–5 mm; blade linear to lanceolate, 1–6 cm. **Inflorescences** open or dense racemes or panicles, axis straight; buds erect. **Flowers:** floral tube 5–15 mm; sepals usually reflexed individually or in pairs, or rarely together to one side, tips not distinct in bud; corolla bowl-shaped, petals lavender to wine-red, often shading white or pale yellow toward base, unspotted or with bright red or purplish red spot near or proximal to middle, 10–25 mm; stamens 8, subequal; ovary 8-grooved; stigma exserted beyond anthers. **Capsules** 15–20 mm. **Seeds** brown or gray, 0.7–1 mm, scaly, crest 0.1 mm.

Subspecies 4 (4 in the flora): California.

1. Petals without red or purplish red spot.
   2. Stems erect to decumbent; petals dark purplish red to lavender, ± white or pale yellow near base, rarely pale yellow throughout . . . . 19a. *Clarkia speciosa* subsp. *speciosa* (in part)
   2. Stems decumbent; petals lavender to pink shading pale yellow or white in proximal ½ . . . . 19b. *Clarkia speciosa* subsp. *immaculata*
1. Petals with bright red or purplish red spot near or proximal to middle.
   3. Inflorescences dense racemes . . . . 19c. *Clarkia speciosa* subsp. *nitens*
   3. Inflorescences open racemes or panicles.
      4. Petals usually with bright red spot; branches on well-developed plants many, few-flowered. . . . . 19a. *Clarkia speciosa* subsp. *speciosa* (in part)
      4. Petals with purplish red spot; branches on well-developed plants virgate, few, many-flowered . . . 19d. *Clarkia speciosa* subsp. *polyantha*

19a. Clarkia speciosa H. Lewis & M. E. Lewis subsp. **speciosa** [E]

Godetia parviflora (Hooker & Arnott) Jepson var. *luteola* C. L. Hitchcock; *G. parviflora* var. *margaritae* (Jepson) C. L. Hitchcock; *G. viminea* (Douglas) Spach var. *margaritae* Jepson

Stems erect to decumbent, branches on well-developed plants many, few-flowered. **Leaf blades** linear to narrowly lanceolate. **Inflorescences** open racemes or few-branched panicles. **Petals** dark purplish red to lavender, ± white or pale yellow toward base, rarely pale yellow throughout, usually with bright red spot near middle. $2n = 18$.

Flowering May–Jul. Woodlands; 0–500 m; Calif.

Subspecies *speciosa* is known from the South Coast Ranges in Monterey, San Benito, San Luis Obispo, Santa Barbara, and scarcely to Ventura counties, and very sporadically in the Tehachapi and southern Sierra Nevada Mountains in Kern and Tulare counties.

19b. Clarkia speciosa H. Lewis & M. E. Lewis subsp. **immaculata** H. Lewis & M. E. Lewis, Univ. Calif. Publ. Bot. 20: 291. 1955 • Pismo clarkia [C] [E]

Stems decumbent, branches on well-developed plants few-flowered. **Leaf blades** linear to narrowly lanceolate. **Inflorescences** open racemes. **Petals** lavender to pink shading pale yellow or white in proximal ½, unspotted. $2n = 18$.

Flowering May–Jul. Sandy hills near coast; of conservation concern; 0–100 m; Calif.

Subspecies *immaculata*, known only from a small portion of San Luis Obispo County, is designated as rare by the California Native Plant Society.

19c. Clarkia speciosa H. Lewis & M. E. Lewis subsp. **nitens** (H. Lewis & M. E. Lewis) H. Lewis & P. H. Raven, Madroño 39: 166. 1992 [E]

Clarkia nitens H. Lewis & M. E. Lewis, Univ. Calif. Publ. Bot. 20: 287, fig. 8a,b. 1955

Stems erect. **Leaf blades** lanceolate. **Inflorescences** dense racemes. **Petals** lavender shading pale yellow proximally or pale yellow throughout, with red spot near or proximal to middle. $2n = 18$.

Flowering May–Jul. Woodlands; 0–700 m; Calif.

Subspecies *nitens* is known from the Central Valley and foothills of the southern Sierra Nevada from San Joaquin to Tulare counties, with a few collections reported in Kern and San Bernardino counties; it intergrades particularly with subsp. *polyantha* in the southern Sierra Nevada foothills.

19d. Clarkia speciosa H. Lewis & M. E. Lewis subsp. **polyantha** H. Lewis & M. E. Lewis, Univ. Calif. Publ. Bot. 20: 291. 1955 [E]

Stems erect, branches on well-developed plants few, virgate, many-flowered. **Leaf blades** linear to narrowly lanceolate. **Inflorescences** open racemes or panicles. **Petals** purple or lavender, usually lighter toward base, with a purplish red spot near middle. $2n = 18$.

Flowering May–Jul. Woodlands; 0–100 m; Calif.

Subspecies *polyantha* is relatively common in Fresno, Kern, Madera, and Tulare counties in the foothills of the southern Sierra Nevada, and is uncommon in the Tehachapi Mountains in Kern and Los Angeles counties. A report from Riverside County (cited in Calflora) is erroneous.

20. Clarkia davyi (Jepson) H. Lewis & M. E. Lewis, Madroño 12: 33. 1953 • Davyi's clarkia [E] [F]

Godetia quadrivulnera (Douglas ex Lindley) Spach var. *davyi* Jepson, Univ. Calif. Publ. Bot. 2: 341. 1907

Stems prostrate or decumbent, to 90 cm, sparsely puberulent. **Leaves:** sessile or subsessile; blade oblanceolate to broadly elliptic or obovate, 1–2.5 cm, apex usually obtuse. **Inflorescences** open racemes, axis straight; buds erect. **Flowers:** floral tube 2–5 mm; sepals reflexed in pairs or individually; corolla bowl-shaped, petals lavender-pink shading white or pale yellow basally, unspotted, 5–11 mm; stamens 8, subequal; ovary 8-grooved; stigma not exserted beyond anthers. **Capsules** 15–25 mm. **Seeds** brown or gray, 1 mm, scaly, crest inconspicuous. $2n = 34$.

Flowering Apr–Jun. Grasslands, low sea bluffs; 0–100 m; Calif.

Clarkia davyi is ecologically restricted, mainly growing along the Pacific coast on bluffs and grassy stabilized sand dunes, rarely farther inland. Its range extends from Humboldt and (barely) Del Norte counties in the north through all coastal counties to Santa Barbara County in the south, including Santa Rosa Island.

Clarkia davyi is morphologically similar to the polytypic South American tetraploid *C. tenella* and appears to be one of the parental species of the hexaploid *C. prostrata*.

21. Clarkia prostrata H. Lewis & M. E. Lewis, Madroño 12: 36. 1953 • Prostrate clarkia [E]

Stems prostrate or decumbent, to 50 cm, sparsely puberulent. **Leaves:** sessile or subsessile; blade oblanceolate to elliptic, 1–2.5 cm, apex usually obtuse. **Inflorescences** prostrate, dense racemes, axis straight; buds erect. **Flowers:** floral tube 4–7 mm; sepals usually reflexed in pairs; corolla bowl-shaped, petals lavender-pink shading pale yellow basally, with reddish purple spot above base, 10–15 mm; stamens 8, subequal; ovary 8-grooved; stigma not exserted beyond anthers. **Capsules** 20–30 mm. **Seeds** brown or gray, 1–1.5 mm, scaly, crest 0.2 mm. $2n = 52$.

Flowering May–Jul. Coastal bluffs in grasslands and closed-cone pine forests; 0–100 m; Calif.

Clarkia prostrata, like *C. davyi*, occurs only on coastal bluffs and adjacent low elevation pine forests along the Pacific coast, and in this case only in the California Central Coast Subregion in Monterey, Santa Barbara, Santa Cruz, and San Luis Obispo counties.

Clarkia prostrata is a hexaploid that combines the tetraploid genome of *C. davyi* and the diploid genome of *C. speciosa*. *Clarkia prostrata* is morphologically and ecologically very similar to *C. davyi* but can usually be distinguished by its larger flowers with a spot on each petal. It differs from *C. speciosa* by having smaller flowers with the stigma not exserted beyond the anthers.

22. Clarkia purpurea (Curtis) A. Nelson & J. F. Macbride, Bot. Gaz. 65: 64. 1918 • Purple or winecup clarkia

Oenothera purpurea Curtis, Bot. Mag. 10: plate 352. 1796; *Godetia purpurea* (Curtis) G. Don

Stems erect or rarely decumbent, to 100 cm, glabrous and sometimes glaucous or sparsely to densely puberulent, sometimes mixed with longer, spreading hairs. **Leaves:** petiole 0–2 mm; blade linear or narrowly lanceolate to elliptic or ovate, 1.5–7 cm. **Inflorescences** open or dense racemes, axis straight; buds erect. **Flowers:** floral tube 2–10 mm; sepals reflexed individually or in pairs; corolla bowl-shaped, petals lavender to purple, purplish red, or dark wine-red, often with red or purple spot near middle, tip, or base, 9–25 mm; stamens 8, subequal; ovary 8-grooved, length less than 8 times width; stigma as long as or exserted beyond anthers. **Capsules** 10–30 mm, beak 0–2 mm. **Seeds** brown or gray, 1–2 mm, scaly, crest 0.2 mm.

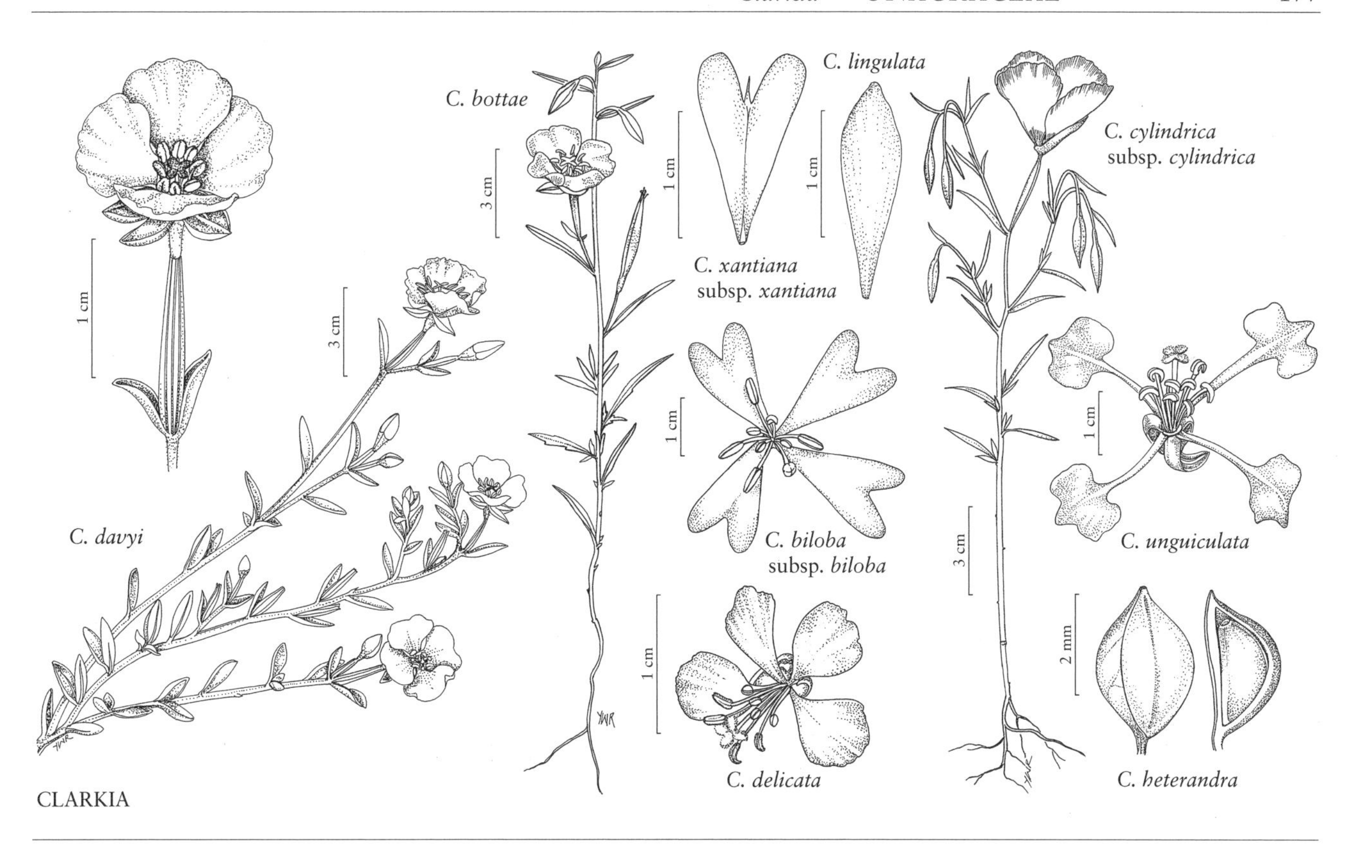

CLARKIA

Subspecies 3 (3 in the flora): w North America, nw Mexico.

Clarkia purpurea consists of a diverse assemblage of hexaploid populations and is almost certainly derived from multiple origins followed by hybridization and, perhaps, backcrossing to parental species. Three morphological forms are recognized as subspecies; intergrades are frequent.

1. Leaf blades broadly lanceolate to elliptic or ovate; inflorescences dense racemes . . . . 22a. *Clarkia purpurea* subsp. *purpurea*
1. Leaf blades linear to lanceolate or narrowly lanceolate; inflorescences usually open racemes.
   2. Petals 15–25 mm; stigmas exserted beyond anthers. . . . . . .22b. *Clarkia purpurea* subsp. *viminea*
   2. Petals 9–14 mm; stigmas not exserted beyond anthers. . . . . 22c. *Clarkia purpurea* subsp. *quadrivulnera*

22a. Clarkia purpurea (Curtis) A. Nelson & J. F. Macbride subsp. **purpurea** E

Leaf blades broadly lanceolate to elliptic or ovate, 1.5–4.5 cm, length usually less than 5 times width. **Inflorescences** dense racemes. **Flowers:** petals lavender to purple or purplish red, often with darker spot near tip, 10–25 mm; stigma exserted beyond anthers. **2*n*** = 52.

Flowering Apr–Jun. Grasslands, often in moist conditions; 0–100 m; Calif., Oreg.

Subspecies *purpurea* is widely distributed but uncommon in California and southern Oregon.

Collections matching the original description and illustration are rare, probably because the grassland habitat in and around the Central Valley where it grew is very desirable for development and, therefore, much altered. Intermediates with the two other subspecies are now more frequent than the typical subspecies.

22b. Clarkia purpurea (Curtis) A. Nelson & J. F. Macbride subsp. **viminea** (Douglas) H. Lewis & M. E. Lewis, Univ. Calif. Publ. Bot. 20: 303. 1955 • Large godetia E

Oenothera viminea Douglas, Bot. Mag. 55: plate 2873. 1828; *Clarkia viminea* (Douglas) A. Nelson & J. F. Macbride; *Godetia viminea* (Douglas) Spach

Leaf blades linear to narrowly lanceolate, 3–7 cm. **Inflorescences** open racemes. **Flowers:** petals lavender to purple, darker proximally or not, usually with darker spot distally, 15–25 mm; stigma exserted beyond anthers. **2*n*** = 52.

Flowering May–Jul. Open, grassy or shrubby places; 0–500 m; Calif., Oreg.

Subspecies *viminea* is widespread at low to middle elevations in California and Oregon. It is a morphologically heterogeneous assemblage of populations, all of which are characterized by relatively large flowers with the stigma exserted beyond the anthers.

22c. Clarkia purpurea (Curtis) A. Nelson & J. F. Macbride subsp. **quadrivulnera** (Douglas ex Lindley) H. Lewis & M. E. Lewis, Univ. Calif. Publ. Bot. 20: 305. 1955 • Purple godetia

Oenothera quadrivulnera Douglas ex Lindley, Bot. Reg. 13: plate 1119. 1828; *Clarkia quadrivulnera* (Douglas ex Lindley) A. Nelson & J. F. Macbride; *Godetia purpurea* (Curtis) G. Don var. *parviflora* (S. Watson) C. L. Hitchcock; *G. quadrivulnera* (Douglas ex Lindley) Spach; *G. quadrivulnera* var. *vacensis* Jepson

Leaf blades linear to lanceolate, 1.5–5 cm. **Inflorescences** usually open racemes. **Flowers:** petals lavender to purple or dark wine-red, often with purple spot near middle or distally, 9–14 mm; stigma not exserted beyond anthers. $2n = 52$.

Flowering Apr–Aug. Open grassy or shrubby places; 0–500 m; B.C.; Ariz., Calif., Oreg., Wash.; Mexico (Baja California).

Subspecies *quadrivulnera* is extremely widespread from Baja California in Mexico north through California, Oregon, and Washington to the islands of extreme southwestern British Columbia.

Subspecies *quadrivulnera* is diverse in petal color and color pattern but all of the populations have relatively small flowers that are primarily self-pollinated, usually with pollen in contact with the stigma at the time the flowers open.

6f. CLARKIA Pursh sect. BIORTIS H. Lewis & M. E. Lewis, Univ. Calif. Publ. Bot. 20: 309. 1955 [E]

Inflorescences: axis erect; buds erect. **Flowers:** floral tube obconic, 1.5–4 mm; sepals reflexed together to 1 side; petals pale pink to dark wine-red, often purple-flecked or -marked, obovate, unlobed, claw inconspicuous or absent; stamens 8, subequal. **Capsules** slender, conspicuously 8-ribbed; sessile or pedicellate.

Species 1: California.

23. Clarkia affinis H. Lewis & M. E. Lewis, Madroño 12: 34. 1953 • Chaparral clarkia or fairyfan [E]

Stems erect, to 80 cm, puberulent. **Leaves:** petiole 0–3 mm; blade linear to narrowly lanceolate, 1.5–7 cm. **Inflorescences** dense spikes, axis straight; buds erect. **Flowers:** floral tube 1.5–4 mm; sepals reflexed together to 1 side; corolla bowl-shaped, petals 5–15 mm; stamens 8, subequal; ovary cylindrical, 8-grooved, length at least 10 times width; stigma not exserted beyond anthers. **Capsules** 15–30 mm, beak 3–7 mm; pedicel 0–5 mm. **Seeds** brown or gray, 1–1.5 mm, scaly, crest 0.1 mm. $2n = 52$.

Flowering Apr–Jun. Openings in woodlands and chaparral; 0–500 m; Calif.

Clarkia affinis is known primarily from west-central California and the North Coast Ranges, and more scattered in the Sierra Nevada Foothills and Western Transverse Ranges.

Clarkia affinis is a hexaploid most closely related to *C. purpurea*; both have $2n = 52$. Chromosome pairing in hybrids between them, as well as morphology, suggest that they have a tetraploid ($2n = 34$) genome in common. The two species are most readily distinguished by their immature capsules, which in *C. affinis* are slender, at least ten times longer than wide, beaked, and shallowly grooved, whereas those of *C. purpurea* are stout, not more than eight times longer than wide, not prominently beaked, and deeply grooved; the sepals of the former are generally reflexed together in fours whereas those of the latter are reflexed individually or in twos. Based on morphology and molecular data, the diploid genome probably came from *C. cylindrica* or a related species.

6g. Clarkia Pursh sect. Fibula H. Lewis & M. E. Lewis, Univ. Calif. Publ. Bot. 20: 333. 1955 [E]

Clarkia subsect. *Xantianae* K. E. Holsinger

Inflorescences: axis erect; buds pendent. **Flowers:** floral tube obconic to campanulate, 2–5 mm; sepals reflexed together to 1 side; petals pale to pink-lavender or reddish purple, white near base and red-flecked or upper petals with a white-bordered dark spot (*C. xantiana*), fan-shaped, unlobed or shallowly 2-lobed, with subulate tooth in sinus, claw inconspicuous (conspicuous in *C. xantiana*); stamens 8, in 2 unequal sets, outer anthers with lavender to purple pollen, inner shorter, with cream pollen. **Capsules** terete, 4-grooved, or conspicuously 8-ribbed (*C. xantiana*); sessile or subsessile to long-pedicellate.

Species 3 (3 in the flora): California.

SELECTED REFERENCE Sytsma, K. J. and L. D. Gottlieb. 1986. Chloroplast DNA evolution and phylogenetic relationships in *Clarkia* sect. *Peripetasma* (Onagraceae). Evolution 40: 1248–1261.

24. Clarkia bottae (Spach) H. Lewis & M. E. Lewis, Madroño 12: 33. 1953 • Botta's clarkia, punch bowl godetia [E] [F]

Godetia bottae Spach, Nouv. Ann. Mus. Hist. Nat. 4: 393. 1836; *Clarkia deflexa* (Jepson) H. Lewis & M. E. Lewis; *G. bottae* var. *deflexa* (Jepson) C. L. Hitchcock; *G. deflexa* Jepson

Stems erect, to 100 cm, glabrous, glaucous, rarely puberulent. **Leaves:** petiole to 5 mm; blade narrowly lanceolate to lanceolate, 3–10 cm. **Inflorescences** open racemes, axis straight in bud; buds pendent. **Flowers:** floral tube 2–3 mm; sepals reflexed together to 1 side; corolla bowl-shaped, petals pale lavender to pinkish lavender, often white toward base, usually red-flecked, unlobed, 15–30 mm; stamens 8, unequal, outer anthers lavender, inner smaller, paler; ovary obscurely 4-grooved; stigma exserted beyond anthers. **Capsules** 30–40 mm; pedicel 0–30 mm. **Seeds** brown or gray, 1.2–1.8 mm, ± papillose, crest 0.2 mm. $2n = 18$.

Flowering Apr–Jul. Chaparral, woodlands, coastal scrub; 0–1000 m; Calif.

Clarkia bottae is known from the Outer South Coast Ranges from Monterey and San Benito to Santa Barbara counties, throughout southwestern California, and more sporadically in the southern Sierra Nevada in Fresno and Kern counties. Reports from Del Norte, Madera, and Napa counties have not been verified.

Clarkia bottae is most closely related to *C. jolonensis* and may be closely related to *C. xantiana*; all three have $2n = 18$, and spontaneous hybrids were formed when *C. bottae* was grown adjacent to a natural population of *C. xantiana*.

25. Clarkia jolonensis D. R. Parnell, Madroño 20: 322. 1970 • Jolon clarkia [C] [E]

Stems erect, to 60 cm, glabrous, glaucous. **Leaves:** petiole to 10 mm; blade narrowly lanceolate to lanceolate, 3–5 cm. **Inflorescences** open racemes, axis straight; buds pendent. **Flowers:** floral tube 2–3 mm; sepals reflexed together to 1 side; corolla bowl-shaped, petals pale lavender to pinkish lavender, usually red-flecked, unlobed, 10–20 mm; stamens 8, unequal, outer anthers lavender, inner smaller, paler; ovary obscurely 4-grooved; stigma exserted beyond anthers. **Capsules** 30–40 mm; pedicel 0–10 mm. **Seeds** dark gray, 1.2–1.4 mm, scaly, crest 0.2 mm. $2n = 18$.

Flowering Apr–Jun. Dry woodlands; of conservation concern; 200–700 m; Calif.

Clarkia jolonensis is known only from Monterey County, mainly in the Outer South Coast Ranges. It is scarcely distinguishable morphologically from *C. bottae* except for seed color, but attempts to produce hybrids have been unsuccessful. Although in describing *C. jolonensis* Parnell suggested that its range did not overlap with that of *C. bottae*, the ranges in fact overlap extensively in Monterey County.

26. Clarkia xantiana A. Gray, Proc. Boston Soc. Nat. Hist. 7: 146. 1859 • Gunsight or Xantus's clarkia [E] [F]

Stems erect, to 80 cm, glabrous, glaucous. **Leaves:** petiole 0–2 mm; blade linear to lanceolate, 2–6 cm. **Inflorescences** open racemes, axis straight; buds pendent. **Flowers:** floral tube 2–5 mm; sepals reflexed together to 1 side; corolla rotate, petals lavender to reddish purple, lavender-pink, or white, often with white-surrounded dark reddish purple spot distally, clawed, 2-lobed, with slender central tooth, 1–3 mm, 6–20 mm; stamens 8, in 2 unequal sets, outer anthers lavender to purple, inner smaller, paler; ovary 8-grooved; stigma exserted or not beyond anthers. **Capsules** 15–25 mm; pedicel 0–5 mm. **Seeds** brown, 1.3–1.5 mm, tuberculate, crest minute.

Subspecies 2 (2 in the flora): California.

Originally placed in the group now delimited as subsect. *Phaeostoma* by H. Lewis and M. E. Lewis (1955), *Clarkia xantiana* was treated as a monotypic subsect. *Xantianae* within sect. *Phaeostoma* by K. E. Holsinger (1985), based mainly on its unusual 2-lobed petals with a tooth in the sinus. Molecular data (R. A. Levin et al. 2004) placed *C. xantiana* close to *C. bottae*, and both of them close to but not within sect. *Phaeostoma*. Both species share the chromosome number $2n = 18$ with *C. jolonensis*.

SELECTED REFERENCE Eckhart, V. M., M. A. Geber, and C. McGuire. 2004. Experimental studies of selection and adaptation in *Clarkia xantiana* (Onagraceae). I. Sources of phenotypic variation across a subspecies border. Evolution 58: 59–70.

1. Petals 12–20 mm; stigmas exserted beyond anthers . . . . 26a. *Clarkia xantiana* subsp. *xantiana*
1. Petals 6–12 mm; stigmas not exserted beyond anthers . . . . 26b. *Clarkia xantiana* subsp. *parviflora*

26a. Clarkia xantiana A. Gray subsp. **xantiana** [E] [F]

Flowers: petals lavender to reddish purple, 12–20 mm; stigma exserted beyond anthers. $2n = 18$.

Flowering May–Jun. Dry slopes, woodlands, forest margins; 500–2000 m; Calif.

Subspecies *xantiana* is known from the southern Sierra Nevada (especially Kern River drainage) and Tehachapi Mountain area in Kern and Tulare counties, and the Western Transverse Ranges in Los Angeles County.

26b. Clarkia xantiana A. Gray subsp. **parviflora** (Eastwood) H. Lewis & P. H. Raven, Madroño 39: 168. 1992 • Kern Canyon clarkia [C] [E]

Clarkia parviflora Eastwood, Bull. Torrey Bot. Club 30: 492. 1903

Flowers: petals lavender-pink or white, 6–12 mm; stigma not exserted beyond anthers. $2n = 18$.

Flowering May–Jun. Foothill woodlands; of conservation concern; 1000–1500 m; Calif.

Subspecies *parviflora* has a range similar to that of subsp. *xantiana* in the southern Sierra Nevada (especially Kern River drainage) and Tehachapi Mountain area in Inyo, Kern, and Tulare counties, and especially the Western Transverse Ranges in Los Angeles and barely San Bernardino counties.

6h. CLARKIA Pursh sect. PHAEOSTOMA (Spach) H. Lewis & M. E. Lewis, Univ. Calif. Publ. Bot. 20: 338. 1955

Phaeostoma Spach, Hist. Nat. Vég. 4: 392. 1835; *Clarkia* subgen. *Phaeostoma* (Spach) Munz & C. L. Hitchcock

Inflorescences: axis recurved or erect; buds pendent or erect. **Flowers:** floral tube obconic or campanulate, 1–5(–7) mm; sepals reflexed together to 1 side (sometimes in pairs in *C. epilobioides*); petals lavender, pink purplish, or white, obovate or spatulate to fan-shaped, unlobed or rarely 2-lobed, tapering to claw; stamens 8, unequal, outer anthers darker pink or purple, inner shorter and pale pink or cream, rarely sterile (*C. heterandra*); ovary 4- or 8-grooved or ribbed, or smooth (*C. heterandra*). **Capsules** narrowly cylindrical or subfusiform, or rarely broadly subclavate (indehiscent, nutlike capsule in *C. heterandra*), 4 or 8-grooved or smooth; sessile or subsessile to long-pedicellate.

Species 15 (15 in the flora): California, nw Mexico.

Section *Phaeostoma* is the largest section in *Clarkia*, with the 15 species classified in seven subsections. For the most part, these correspond with groups delineated originally by H. Lewis and M. E. Lewis (1955). Within the part of the genus with dimorphic stamens (8 in 2 dissimilar sets that vary in size and color), the species of sect. *Phaeostoma* have entire petals that taper to a distinct claw. Even though molecular support for sect. *Phaeostoma* is not strong (R. A. Levin et al. 2004), at this time it seems preferable to stress similarities and maintain this variable group in one section.

6h.1. CLARKIA Pursh (sect. PHAEOSTOMA) subsect. LAUTIFLORAE H. Lewis & M. E. Lewis, Univ. Calif. Publ. Bot. 20: 319. 1955 [E]

Inflorescences open racemes, axis recurved at tip in bud; buds pendent. **Flowers** protandrous; petals lavender to pink or magenta, without zones or flecks of color, 8–30 mm, sometimes 2-lobed; ovary conspicuously 8-grooved; stigma exserted beyond anthers.

Species 4 (4 in the flora): California.

27. Clarkia biloba (Durand) A. Nelson & J. F. Macbride, Bot. Gaz. 65: 60. 1918 • Two lobed clarkia [E] [F]

Oenothera biloba Durand, Pl. Pratten. Calif., 87. 1855; *Godetia biloba* (Durand) S. Watson

Stems erect, 30–100 cm, strigillose. **Leaves:** petiole to 15 mm; blade linear to lanceolate, 2–8 cm. **Inflorescences** open racemes, axis recurved at tip in bud; buds pendent. **Flowers:** floral tube 1–4 mm; sepals reflexed together to 1 side; corolla rotate to bowl-shaped, petals purplish to pale pink, lavender, or bright pink to magenta, often red-flecked, broadly to narrowly fan-shaped, 10–25 mm, shallowly to deeply 2-lobed; stamens 8, unequal, outer anthers lavender, inner ones smaller, paler. **Capsules** 10–25 mm. **Seeds** brown, 1 mm, minutely scaly to puberulent, crest inconspicuous.

Subspecies 3 (3 in the flora): California.

Clarkia biloba is most closely related to *C. lingulata*, which is derived from *C. biloba* subsp. *australis*. Some populations of *C. biloba* subsp. *brandegeeae* (originally described as a form of *C. dudleyana*) are morphologically very similar to some individuals of *C. dudleyana* but the two taxa are separated geographically, have different chromosome numbers, and hybrids between them are sterile.

1. Petals bright pink to magenta, narrowly fan-shaped, length greater than 1.5 times width ....27c. *Clarkia biloba* subsp. *australis*
1. Petals lavender to pale or purplish pink, broadly fan-shaped, length not greater than 1.5 times width.
   2. Petals purplish to pale pink, deeply 2-lobed, lobes usually $\frac{1}{5}$–$\frac{1}{2}$ petal length .... 27a. *Clarkia biloba* subsp. *biloba*
   2. Petals lavender, shallowly 2-lobed, lobes usually less than $\frac{1}{5}$ petal length, sometimes obscure .... 27b. *Clarkia biloba* subsp. *brandegeeae*

27a. Clarkia biloba (Durand) A. Nelson & J. F. Macbride subsp. **biloba** [E] [F]

Leaf blades narrowly lanceolate, 5–8 mm wide. **Flowers:** corolla rotate, petals purplish to pale pink, broadly fan-shaped, length not greater than 1.5 times width, deeply 2-lobed, lobes usually $\frac{1}{5}$–$\frac{1}{2}$ petal length. $2n = 16$.

Flowering May–Aug. Foothill woodlands; 0–1000 m; Calif.

Subspecies *biloba* is known from the central and northern Sierra Nevada Foothills, especially in Calaveras, El Dorado, and Tuolumne counties, and from the eastern San Francisco Bay area in Contra Costa County.

27b. Clarkia biloba (Durand) A. Nelson & J. F. Macbride subsp. **brandegeeae** (Jepson) H. Lewis & M. E. Lewis, Univ. Calif. Publ. Bot. 20: 323. 1955 (as brandegeae) • Brandegee's clarkia [E]

Godetia dudleyana Abrams forma *brandegeeae* Jepson, Univ. Calif. Publ. Bot. 2: 334. 1907 (as brandegeae)

Leaf blades lanceolate, 5–8 mm wide. **Flowers:** corolla bowl-shaped to rotate, petals lavender, broadly fan-shaped, length not greater than 1.5 times width, shallowly 2-lobed, lobes usually less than 1/5 petal length, sometimes obscure. **2*n*** = 16.

Flowering Jun–Jul. Foothill woodlands; 0–500 m; Calif.

Subspecies *brandegeeae* is known from the northern Sierra Nevada Foothills, from Amador to Butte counties, including El Dorado, Nevada, Placer, Sacramento, and Yuba counties.

27c. Clarkia biloba (Durand) A. Nelson & J. F. Macbride subsp. **australis** H. Lewis & M. E. Lewis, Univ. Calif. Publ. Bot. 20: 322, fig. 17b. 1955 • Mariposa clarkia [C] [E]

Leaf blades linear to narrowly lanceolate, 2–6 mm wide. **Flowers:** corolla rotate, petals bright pink to magenta, narrowly fan-shaped, length greater than 1.5 times width, 2-lobed, lobes usually 1/5–1/3 petal length. **2*n*** = 16.

Flowering May–Jun. Chaparral, woodlands; of conservation concern; 300–500 m; Calif.

Subspecies *australis* is known from the foothills of the south-central Sierra Nevada, mainly in Mariposa and Tuolumne counties, less commonly in Calaveras and El Dorado counties.

28. Clarkia lingulata H. Lewis & M. E. Lewis, Madroño 12: 35. 1953 • Merced clarkia [C] [E] [F]

Stems erect, to 60 cm, puberulent. **Leaves:** petiole to 15 mm; blade linear to narrowly lanceolate, 2–6 cm. **Inflorescences** open racemes, axis recurved at tip in bud; buds pendent. **Flowers:** floral tube 1–4 mm; sepals reflexed together to 1 side; corolla rotate, petals bright pink, red-flecked or not, oblanceolate, 10–20 mm, apex subentire or minutely notched; stamens 8, unequal, outer anthers lavender, inner smaller, paler. **Capsules** 10–20 mm. **Seeds** brown, 1 mm, minutely scaly to puberulent, crest inconspicuous. **2*n*** = 18.

Flowering May–Jun. Open chaparral; of conservation concern; 400–500 m; Calif.

Clarkia lingulata is listed as endangered by the State of California, known from only a few populations in Merced River Canyon, Mariposa County. It is derived from *C. biloba* subsp. *australis*, from which it can be distinguished morphologically by its narrower, unlobed petals; the two taxa also differ in chromosome number, and form only highly sterile hybrids.

29. Clarkia dudleyana (Abrams) J. F. Macbride, Contr. Gray Herb. 56: 54. 1918 • Dudley's clarkia [E]

Godetia dudleyana Abrams, Fl. Los Angeles, 267. 1904

Stems erect, to 70 cm, puberulent. **Leaves:** petiole 3–10 mm; blade narrowly lanceolate, 1.5–7 cm. **Inflorescences** open racemes, axis recurved at tip in bud; buds pendent. **Flowers:** floral tube 1–3 mm; sepals reflexed together to 1 side; corolla bowl-shaped, petals lavender-pink, usually white-streaked, often red-flecked, broadly fan-shaped, 10–30 mm, apex subentire to crenulate; stamens 8, unequal, outer anthers lavender, inner smaller, paler. **Capsules** 10–30 mm. **Seeds** brown, 1 mm, minutely scaly to puberulent, crest inconspicuous. **2*n*** = 18.

Flowering May–Jul. Openings in woodlands, chaparral, yellow-pine forests, coastal sage; 0–1500 m; Calif.

Clarkia dudleyana is a rather widespread species in California, known primarily from the central and southern Sierra Nevada foothills, the Tehachapi Mountain area, the Transverse Ranges, and the Peninsular Ranges, ranging from Tuolumne to Riverside counties, sporadically in the north to Nevada County and in the south to San Diego County

Clarkia dudleyana is morphologically most similar to *C. biloba* and *C. modesta*, but molecular data suggest that the relationship is not close. On the basis of chloroplast DNA sequence, *C. dudleyana* and *C. heterandra* are closely related.

30. Clarkia modesta Jepson, Man. Fl. Pl. Calif., 673. 1925 • Waltham Creek clarkia [E]

Stems erect, 20–70 cm, puberulent. **Leaves:** petiole 5–15 mm; blade linear to narrowly lanceolate or elliptic, 2–4 cm. **Inflorescences** open racemes, axis recurved at tip in bud; buds pendent. **Flowers:** floral tube 1–3 mm; sepals reflexed together to 1 side; corolla generally rotate, petals usually arranged in lateral pairs, pink, usually darker flecked, oblanceolate to diamond-shaped, scarcely clawed, 8–12 mm; stamens 8, unequal, outer anthers lavender, inner smaller, paler. **Capsules** 15–30 mm. **Seeds** brown, 0.8–1 mm, tuberculate, crest inconspicuous. $2n = 16$.

Flowering Apr–Jun. Sandy places in woodlands; 0–1000 m; Calif.

Clarkia modesta occurs mainly in the Inner North Coast Ranges, the San Francisco Bay area, and the South Coast Ranges, from Trinity to Santa Barbara counties, and in the central and southern Sierra Nevada Foothills, from Mariposa to Tulare counties.

Clarkia modesta is one of the parents of the tetraploid species *C. similis*, from which it differs by having darker pink petals.

6h.2. Clarkia Pursh (sect. Phaeostoma) subsect. Prognatae H. Lewis & M. E. Lewis, Univ. Calif. Publ. Bot. 20: 332. 1955

Inflorescences open racemes, axis recurved at tip in bud; buds pendent. **Flowers** not protandrous; petals pink to white, without zones of color, purple-flecked in proximal ½, 6–10 mm; ovary 8-grooved; stigma not exserted beyond anthers.

Species 1: California, nw Mexico.

31. Clarkia similis H. Lewis & W. R. Ernst, Madroño 12: 89. 1953 • Ramona clarkia

Stems erect, 30–90 cm, puberulent. **Leaves:** petiole to 8 mm; blade narrowly lanceolate to elliptic, 2–4 cm. **Inflorescences** open racemes, axis recurved at tip in bud; buds pendent. **Flowers:** floral tube 1.5–2 mm; sepals reflexed together to 1 side; corolla rotate to bowl-shaped, petals pale pink shading nearly white near base, purple-flecked, fading pink, oblanceolate or obovate to diamond-shaped, 6–10 mm; stamens 8, unequal, outer anthers dark pink, inner smaller, paler; ovary shallowly 8-grooved. **Capsules** 15–30 mm; subsessile. **Seeds** brown, 1 mm, tuberculate, crest inconspicuous. $2n = 34$.

Flowering Apr–Jun. Shady sites, oak woodlands, chaparral; 0–1000 m; Calif.; Mexico (Baja California).

Clarkia similis is known from southwestern California and northern Baja California, Mexico, mainly in the South Coast, Transverse, and Peninsular ranges, from Monterey and San Benito counties in the north to San Diego County in the south. Reports from the Southern Sierra Nevada Foothills in Kern County and the Sacramento Valley in Tehama County are not confirmed.

Clarkia similis is a tetraploid species derived through polyploidization following hybridization between *C. epilobioides* and *C. modesta*.

6h.3. Clarkia Pursh (sect. Phaeostoma) subsect. Micranthae K. E. Holsinger & H. Lewis, Ann. Missouri Bot. Gard. 73: 493. 1986

Inflorescences open racemes, axis recurved at tip in bud; buds pendent. **Flowers** autogamous, not protandrous; petals white to pale cream, without zones or flecks of color, 5–10 mm; ovary subterete, not obviously grooved or ribbed; stigma not exserted beyond anthers.

Species 1: sw United States, nw Mexico.

32. Clarkia epilobioides (Nuttall ex Torrey & A. Gray) A. Nelson & J. F. Macbride, Bot. Gaz. 65: 60. 1918 • Canyon or willowherb clarkia

Oenothera epilobioides Nuttall ex Torrey & A. Gray, Fl. N. Amer. 1: 511. 1840; *Godetia epilobioides* (Nuttall ex Torrey & A. Gray) S. Watson

Stems erect, 20–70 cm, sparsely puberulent. **Leaves:** petiole to 7 mm; blade linear to narrowly lanceolate or oblanceolate, 1.5–2.5 cm. **Inflorescences** open racemes, sometimes few-branched, axis recurved at tip in bud; buds pendent. **Flowers** usually cleistogamous; floral tube 1–3 mm; sepals reflexed together to 1 side or in pairs; corolla bowl-shaped, petals fading pink, obovate; stamens 8, unequal, anthers white or cream, outer ones larger than inner. **Capsules** 10–30 mm; pedicel 5–11 mm. **Seeds** brown, 0.5–1 mm, scaly, crest inconspicuous. $2n = 18$.

Flowering Apr–May. Shady sites, woodlands, chaparral; 0–1000 m; Ariz., Calif.; Mexico (Baja California).

Clarkia epilobioides is known from south-central Arizona in Gila, Maricopa, Pima, and Pinal counties, and widely in west-central and southwestern California and adjacent Baja California, Mexico. In California, it occurs from Contra Costa and San Mateo counties in the San Francisco Bay area to San Diego County in the south, including most of the Channel Islands.

Clarkia epilobioides is modally self-pollinating, and up to half of its flowers do not open, yet set a full complement of seeds (H. Lewis and M. E. Lewis 1955). However, outcrossing does occur, and *C. epilobioides* is one of the parents of the tetraploid species *C. similis*, from which it differs by having white, unflecked petals; it is also one of the parents of the tetraploid *C. delicata*.

6h.4. Clarkia Pursh (sect. Phaeostoma) subsect. Connubium (H. Lewis & M. E. Lewis) W. L. Wagner & Hoch, Syst. Bot. Monogr. 83: 111. 2007

Clarkia sect. *Connubium* H. Lewis & M. E. Lewis, Univ. Calif. Publ. Bot. 20: 337. 1955

Inflorescences open racemes, axis erect; buds pendent. **Flowers** not protandrous; petals rose-lavender to pale pink, without zones or flecks of color, 8–12 mm; ovary 8-grooved; stigma not exserted beyond anthers.

Species 1: California, nw Mexico.

33. Clarkia delicata (Abrams) A. Nelson & J. F. Macbride, Bot. Gaz. 65: 60. 1905 • Campo or delicate clarkia [C] [F]

Godetia delicata Abrams, Bull. Torrey Bot. Club 32: 539. 1905

Stems erect, 20–70 cm, glabrous and glaucous distally, usually puberulent basally. **Leaves:** petiole to 10 mm; blade lanceolate to elliptic or ovate, 1.5–4 cm. **Inflorescences** open racemes, sometimes branched, axis straight; buds pendent. **Flowers:** floral tube 2 mm; sepals reflexed together to 1 side; corolla rotate, petals oblanceolate to obovate, 8–12 mm, claw tapered, shorter than blade, apex entire; stamens 8, unequal, outer anthers orange-red, inner smaller, paler. **Capsules** 15–35 mm; subsessile. **Seeds** brown, 1–1.5 mm, tuberculate (especially on raphe), crest inconspicuous. $2n = 36$.

Flowering Apr–May. Oak woodlands, chaparral; of conservation concern; 0–1000 m; Calif.; Mexico (Baja California).

Clarkia delicata is known in California only from the Peninsular Ranges, mainly in San Diego County with outliers in Riverside and San Bernardino counties, and in northern Baja California, Mexico. Because of its limited range, it is listed as rare by the California Native Plant Society. It is a tetraploid derived from hybridization between *C. epilobioides* and *C. unguiculata*.

6h.5. Clarkia Pursh (sect. Phaeostoma) subsect. Sympherica K. E. Holsinger & H. Lewis, Ann. Missouri Bot. Gard. 73: 493. 1986 [E]

Inflorescences open racemes, axis recurved at tip in bud; buds pendent. **Flowers** protandrous; petals pink or purplish red fading white in middle, base purplish red, sometimes reddish or purple-flecked, 10–35 mm; ovary 4-grooved; stigma exserted beyond anthers.

Species 3 (3 in the flora): California.

SELECTED REFERENCE Ford, V. S. and L. D. Gottlieb. 2003. Reassessment of phylogenetic relationships in *Clarkia* sect. *Sympherica*. Amer. J. Bot. 90: 167–174.

34. **Clarkia lewisii** P. H. Raven & D. R. Parnell, Ann. Missouri Bot. Gard. 64: 642. 1978 • Lewis's clarkia [E]

Stems erect, to 50 cm, puberulent to glabrate. **Leaves:** petiole to 7 mm; blade narrowly lanceolate to lanceolate, 2–5 cm. **Inflorescences** open racemes, axis recurved at tip in bud; buds pendent. **Flowers:** floral tube 1.5–4 mm, with ring of hairs at distal margin inside; sepals reflexed together to 1 side; corolla bowl-shaped, petals pinkish lavender shading white near middle, base purplish red or with red line, sometimes reddish purple-flecked, 10–30 mm; stamens 8, unequal, width of all filaments subequal or inner slightly thinner, outer anthers lavender, inner smaller, paler. **Capsules** 15–70 mm, beak 0–3 mm. **Seeds** brown, 1 mm, scaly to puberulent, crest inconspicuous. $2n = 18$.

Flowering Jun. Coastal scrub, woodlands, chaparral; 0–300 m; Calif.

Clarkia lewisii is known primarily from Monterey County, sparsely in San Benito County, barely reaching Santa Clara County, and is listed as rare by the California Native Plant Society. It is most closely related and morphologically similar to *C. cylindrica*, from which it can be distinguished by having all filaments about equally wide and a ring of hairs at the rim of the floral tube; outer filaments of *C. cylindrica* are two times as wide as the inner ones, and the ring of hairs is within the tube below the rim. *Clarkia lewisii* is also closely related to *C. rostrata*, from which it differs conspicuously by having a much shorter capsule beak.

Clarkia lewisii is a new name applied to the species known until 1978 as *C. bottae*, following examination and reinterpretation of the type of *Godetia bottae* Spach by P. H. Raven and D. R. Parnell (1978). They determined that the type specimens of *G. bottae* actually referred to the species then known as *C. deflexa* (Jepson) H. Lewis & M. E. Lewis, and reapplied the name *C. bottae* to that species in sect. *Fibula*.

35. **Clarkia cylindrica** (Jepson) H. Lewis & M. E. Lewis, Madroño 12: 33. 1953 (as cyclindrica) • Speckled clarkia [E] [F]

Godetia bottae Spach var. *cylindrica* Jepson, Univ. Calif. Publ. Bot. 2: 332. 1907

Stems erect, to 60 cm, puberulent or glabrous. **Leaves:** petiole to 5 mm; blade linear to narrowly lanceolate, 1–6 cm. **Inflorescences** open racemes, axis recurved at tip in bud; buds pendent. **Flowers:** floral tube 2–7 mm, with ring of hairs proximal to distal margin inside; sepals reflexed together to 1 side; corolla bowl-shaped, petals purple to pinkish lavender shading white near middle, often reddish purple-flecked, base bright purplish red, 10–35 mm; stamens 8, unequal, width of outer filaments about 2 times inner, outer anthers lavender, inner smaller, paler. **Capsules** 20–50 mm, beak 3–5 mm. **Seeds** brown, 1–1.5 mm, minutely scaly to puberulent, crest 0.1 mm.

Subspecies 2 (2 in the flora): California.

As defined by Davis, the subspecies of *Clarkia cylindrica* have distinct but partly overlapping geographical ranges; subsp. *cylindrica* mainly in the South Coast and Transverse Ranges to the Tehachapi Mountain area, and subsp. *clavicarpa* mainly in the central and southern Sierra Nevada Foothills to the Tehachapi Mountain area. More recent collections suggest more substantial geographical overlap. Morphological variation correlates with geographical distribution, with the most consistent difference in ovary and capsule shape. According to Davis, the taxa are moderately interfertile, less so for more distantly separated individuals.

1. Ovaries and capsules cylindrical . . . . 35a. *Clarkia cylindrica* subsp. *cylindrica*
1. Ovaries and capsules subclavate, enlarged distally . . . . . . 35b. *Clarkia cylindrica* subsp. *clavicarpa*

35a. Clarkia cylindrica (Jepson) H. Lewis & M. E. Lewis subsp. **cylindrica** [E] [F]

Flowers: floral tube length ± equal width; ovary and capsule cylindrical. **2*n*** = 18.

Flowering Apr–Jun. Grasslands, woodlands, chaparral; 0–1000 m; Calif.

Subspecies *cylindrica* is known from the South Coast and Transverse Ranges from Monterey and San Benito to Los Angeles counties, and from the Tehachapi Mountain area more sporadically along the southern and central Sierra Nevada Foothills from Kern to Stanislaus counties.

35b. Clarkia cylindrica (Jepson) H. Lewis & M. E. Lewis subsp. **clavicarpa** W. S. Davis, Brittonia 22: 283. 1970 [E]

Flowers: floral tube length at least 2 times width; ovary and capsule subclavate, enlarged distally. **2*n*** = 18.

Flowering Apr–Jun. Grasslands, woodlands, chaparral; 0–1000 m; Calif.

Subspecies *clavicarpa* is known primarily from the central and southern Sierra Nevada Foothills to the Tehachapi Mountain area from Tuolumne to Kern counties, with a few scattered occurrences in the Inner South Coast Ranges to Fresno County, and in the Transverse Ranges to San Bernardino County.

36. Clarkia rostrata W. S. Davis, Brittonia 22: 281. 1970 • Beaked clarkia [C] [E]

Stems erect, to 60 cm, puberulent. **Leaves:** petiole to 10 mm; blade lanceolate, 1–6 cm. **Inflorescences** open racemes, axis recurved at tip in bud; buds pendent. **Flowers:** floral tube 1.5–2.5 mm, with ring of hairs at distal margin inside; sepals reflexed together to 1 side; corolla bowl-shaped, petals pinkish lavender shading white near middle, often flecked reddish purple, base reddish purple, 10–25 mm; stamens 8, unequal, width of all filaments equal or inner slightly thinner, outer anthers lavender, inner smaller, paler. **Capsules** 10–30 mm, beak 7–15 mm. **Seeds** unknown. **2*n*** = 18.

Flowering Apr–May. Oak-pine woodlands; of conservation concern; 500 m; Calif.

Clarkia rostrata is known only from the Merced River drainage in the central Sierra Nevada Foothills, including Mariposa, Merced, Stanislaus, and (barely) Tuolumne counties. Because of its very limited distribution, *C. rostrata* is listed as rare by the California Native Plant Society.

Clarkia rostrata is closely related to *C. cylindrica* and *C. lewisii* but can be distinguished readily from both by the conspicuous beak of the capsule.

6h.6. Clarkia Pursh (sect. Phaeostoma) subsect. Phaeostoma (Spach) K. E. Holsinger, Taxon 34: 705. 1985 [E]

Phaeostoma Spach, Hist. Nat Vég. 4: 392. 1835; *Gauropsis* C. Presl

Inflorescences open racemes, axis erect; buds pendent. **Flowers** protandrous or not; petals lavender-pink to salmon, dark reddish purple, or white, generally with purplish red zone at base, 5–25 mm; ovary 4-grooved; stigma exserted beyond or subequal to anthers.

Species 4 (4 in the flora): California.

Subsection *Phaeostoma* forms a distinctive group by virtue of having petal claws as long as or longer than the petal blades.

37. Clarkia unguiculata Lindley, Edwards's Bot. Reg. 23: sub plate 1981. 1837 • Elegant or woodland clarkia, mountain garland [E] [F]

Stems erect, 30–100 cm, glabrous, glaucous. **Leaves:** petiole 0–10 mm; blade lanceolate to elliptic or ovate, 1–6 cm. **Inflorescences** open racemes, sometimes branched, axis erect; buds pendent. **Flowers:** floral tube 2–5 mm; sepals reflexed together to 1 side, green to dark red, sparsely to densely puberulent abaxially, with longer, straight, spreading hairs to 3 mm; corolla rotate, petals lavender-pink to salmon or dark reddish purple, triangular or diamond-shaped to suborbiculate, 10–25 mm, claw slender, equal to or longer than blade, entire, rarely somewhat expanded at base; stamens 8, unequal, outer anthers red, inner smaller, paler; ovary with hairs as on sepals; stigma exserted beyond anthers. **Capsules** 15–30 mm. **Seeds** brown, 1–1.5 mm, tuberculate, crest inconspicuous. $2n = 18$.

Flowering Apr–Sep. Woodlands; 0–1500 m; Calif.

Clarkia unguiculata is a widely distributed species in California, and occurs throughout much of the southern two-thirds of the state in appropriate woodland habitats.

Clarkia unguiculata is ancestral to *C. exilis*, *C. springvillensis*, and *C. tembloriensis*. It is one of the parents of the tetraploid species *C. delicata* and may have been involved in the origin of *C. heterandra*.

38. Clarkia springvillensis Vasek, Madroño 17: 220. 1964 • Springville clarkia [C] [E]

Stems erect, 30–90 cm, glabrous, glaucous. **Leaves:** petiole 0–5 mm; blade lanceolate, 2–9 cm, surfaces glaucous, glabrous. **Inflorescences** open racemes, axis erect; buds pendent. **Flowers:** floral tube 3–4 mm; sepals reflexed together to 1 side, usually dark red-purple, sparsely to densely puberulent abaxially, without long, spreading hairs; corolla rotate, petals lavender-pink, usually with dark purplish spot near base, ± diamond-shaped, 13–15 mm, claw slender, equal to or longer than blade, entire; stamens 8, unequal, outer anthers red, inner smaller, paler; ovary with hairs as on sepals; stigma exserted beyond anthers. **Capsules** 15–30 mm. **Seeds** unknown. $2n = 18$.

Flowering May. Woodlands; of conservation concern; 500 m; Calif.

Clarkia springvillensis is a rare species known primarily from the vicinity of Springville in Tulare County, with one ambiguous collection from Kern County. Due to its very limited distribution, *C. springvillensis* is listed as rare by the California Native Plant Society. It is derived from *C. unguiculata* and is closely related to *C. exilis* and *C. tembloriensis*.

39. Clarkia exilis H. Lewis & Vasek, Madroño 12: 211. 1954 • Kern River or slender clarkia [E]

Stems erect, 30–100 cm, glabrous, glaucous. **Leaves:** petiole 0–5 mm; blade bright green, lanceolate to narrowly elliptic, 1–6 cm, surfaces not glaucous, glabrous. **Inflorescences** open racemes, axis erect; buds pendent. **Flowers:** floral tube 1–3 mm; sepals reflexed together to 1 side, usually green, sparsely to densely puberulent inside, without longer, spreading hairs; corolla rotate, petals lavender-pink or white, often with dark purplish spot, usually diamond-shaped, 5–15 mm, claw slender, equal to or longer than blade, entire; stamens 8, unequal, outer anthers red, inner smaller, paler; ovary with hairs as on sepals; stigma subequal to anthers. **Capsules** 10–30 mm. **Seeds** brown, 1 mm, tuberculate, crest inconspicuous. $2n = 18$.

Flowering Apr–May. Woodlands; 0–1000 m; Calif.

Clarkia exilis is of limited distribution, known primarily from the southern Sierra Nevada Foothills and Tehachapi Mountain area in Kern and Tulare counties, with unverified reports from Fresno, San Luis Obispo, and Ventura counties. It is listed as rare by the California Native Plant Society.

Clarkia exilis is derived from *C. unguiculata* and is closely related to *C. springvillensis* and *C. tembloriensis*.

40. Clarkia tembloriensis Vasek, Madroño 17: 220. 1964 • Temblor Range clarkia [E]

Stems erect, to 80 cm, glabrous, glaucous. **Leaves:** petiole 0–5 mm; blade gray-green, lanceolate, 2–7 cm, surfaces glaucous. **Inflorescences** open racemes, axis erect; buds pendent. **Flowers:** floral tube 2–3 mm; sepals reflexed together to 1 side, green, red-tinged or not, sparsely to densely puberulent abaxially, without longer, spreading hairs; corolla rotate, petals lavender-pink, spot purplish or absent, ± diamond-shaped, 10–25 mm, claw slender, equal to or longer than blade, entire; stamens 8, unequal, outer anthers lavender to red, inner smaller, paler; ovary with hairs as on sepals; stigma exserted or not beyond anthers. **Capsules** 15–30 mm. **Seeds** unknown.

Subspecies 2 (2 in the flora): California.

Clarkia tembloriensis is derived from *C. unguiculata* and is closely related to *C. exilis* and *C. springvillensis*. Hybrids between the subspecies of *C. tembloriensis* have low fertility and the two taxa are rarely found together.

1. Petals usually less than 10 mm wide; stigmas exserted beyond anthers or not . . . . 40a. *Clarkia tembloriensis* subsp. *tembloriensis*
1 Petals about 10 mm wide; stigmas not exserted beyond anthers . . . . 40b. *Clarkia tembloriensis* subsp. *calientensis*

40a. Clarkia tembloriensis Vasek subsp. **tembloriensis** E

Clarkia tembloriensis subsp. *longistyla* Vasek

Flowers: petals usually less than 10 mm wide; stigma exserted beyond anthers or not. $2n$ = 18.

Flowering Apr–May. Grasslands, scrub; 100–500 m; Calif.

Subspecies *tembloriensis* is known from the eastern San Francisco Bay area and Inner South Coast Ranges, along the western edge of the Great Valley, from San Joaquin and Alameda counties to Ventura County, with scattered reports from the southern Sierra Nevada Foothills in Kern County.

40b. Clarkia tembloriensis Vasek subsp. **calientensis** (Vasek) K. E. Holsinger, Syst. Bot. 10: 157. 1985 • Vasek's clarkia C E

Clarkia calientensis Vasek, Syst. Bot. 2: 252. 1978

Flowers: petals 10 mm wide; stigma not exserted beyond anthers. $2n$ = 18.

Flowering Apr. Grasslands; of conservation concern; 500 m; Calif.

Subspecies *calientensis* is known only from the vicinity of the type locality in central Kern County and is possibly extinct.

6h.7. Clarkia Pursh (sect. Phaeostoma) subsect. Heterogaura (Rothrock) W. L. Wagner & Hoch, Syst. Bot. Monogr. 83: 112. 2007 E

Heterogaura Rothrock, Proc. Amer. Acad. Arts 6: 350, 354. 1864; *Clarkia* sect. *Heterogaura* (Rothrock) H. Lewis & P. H. Raven

Inflorescences open racemes or panicles, axis erect; buds erect. **Flowers** not protandrous; petals pink darkening to lavender, without markings, 3–5 mm; stamens 8, 4 fertile, 4 sterile and reduced, 3–5 mm; ovary not grooved; stigma not exserted beyond anthers. **Fruit** an indehiscent capsule, subterete, 2–3 mm, nutlike, 1 or 2 seeded; pedicel 0–2 mm.

Species 1: w United States.

SELECTED REFERENCE Sytsma, K. J. and L. D. Gottlieb. 1986b. Chloroplast DNA evidence for the origin of the genus *Heterogaura* from a species of *Clarkia* (Onagraceae). Proc. Natl. Acad. Sci. U.S.A. 83: 5554–5557.

41. Clarkia heterandra (Torrey) H. Lewis & P. H. Raven, Madroño 39: 163. 1992 • California gaura, heterogaura, mountain clarkia E F

Gaura heterandra Torrey in War Department [U.S.], Pacif. Railr. Rep. 5(4): 87. 1857; *Heterogaura heterandra* (Torrey) Coville

Stems erect, to 60 cm, glandular puberulent. **Leaves:** petiole 5–20 mm; blade lanceolate to ovate, 2–8 cm. **Inflorescences** open racemes or panicles, axis straight; buds erect. **Flowers** sometimes autogamous; floral tube 1–2 mm; sepals reflexed together to 1 side; corolla rotate, petals elliptic to obovate, tapered to claw; stamens 8, unequal, anthers cream or light pink, inner ones smaller, sterile; ovary subglobose, grooves obscure; stigma not exserted beyond anthers. **Capsules** 2–3 mm, indehiscent, nutlike; sessile or pedicellate (to 2 mm). **Seeds** 1 or 2. **2*n*** = 18.

Flowering May–Jul. Shady sites, woodlands, yellow-pine forests; 500–2000 m; Calif., Oreg.

Clarkia heterandra occurs in southern Oregon and California, where it is known from the Klamath Ranges in Trinity County, widely in the Sierra Nevada and Tehachapi Mountain area from Nevada to Kern counties, and in the South Coast and Transverse Ranges from San Luis Obispo to Riverside and San Bernardino counties.

Prior to 1986, *Clarkia heterandra* was treated as the monotypic genus *Heterogaura*, thought to be possibly related to the genus *Gaura* (now a section of *Oenothera*) due to its similar indehiscent fruits. However, molecular data indicate that *C. heterandra* is clearly within *Clarkia*, most closely related to *C. dudleyana* and *C. unguiculata*, which suggests a possible hybrid origin at the diploid level (K. J. Sytsma and L. D. Gottlieb 1986, 1986b; W. L. Wagner et al. 2007). *Clarkia dudleyana* and *C. unguiculata* produce spontaneous hybrids when grown adjacent to each other if they come from allopatric populations.

7. GAYOPHYTUM A. Jussieu, Ann. Sci. Nat. (Paris) 25: 18, plate 4. 1832 • Groundsmoke [For Claude Gay, 1800–1873, French author of Flora of Chile, and Greek *phyton*, plant]

Harlan Lewis†

Warren L. Wagner

Herbs, annual, caulescent; taproot simple or branched. **Stems** usually delicate, erect or spreading, densely branched or unbranched, epidermis usually exfoliating near base. **Leaves** cauline, usually alternate, sometimes subopposite proximally; stipules absent; sessile or petiolate; blade margins entire. **Inflorescences** panicles or racemes, erect, usually flowering at distal nodes only, sometimes from basal nodes. **Flowers** bisexual, actinomorphic, buds erect; floral tube inconspicuous, deciduous (with sepals, petals, and stamens) after anthesis, nectary unknown, presumably basal; sepals 4, reflexed singly or in pairs; petals 4, white with 1 or 2 yellow or greenish yellow areas at base, fading pink or red; stamens 8, in 2 unequal series; anthers ± basifixed, pollen shed singly; ovary 2-locular; stigma entire, subglobose to hemispheric, surface wet and non-papillate. **Fruit** a capsule, straight or slightly curved, flattened or subterete and often constricted between seeds, loculicidally dehiscent, valves all free or 2 free and 2 remaining attached to septum after dehiscence; usually pedicellate. **Seeds** 4–50+, in 1 row per locule, ovoid. $x = 7$.

Species 9 (8 in the flora): w North America, nw Mexico, w South America.

Gayophytum consists of nine annual species that usually occur in gravelly or sandy soil in open coniferous forests, among sagebrush, at the drying edges of meadows, or along roadsides at 100–4200 m. Six species are diploid ($n = 7$) and three are tetraploid ($n = 14$, *G. diffusum*, *G. micranthum*, and *G. racemosum*). Seven species occur exclusively in western North America, with some species reaching Baja California, Mexico, or southern Canada; two species are endemic to California (*G. oligospermum* in the Transverse and Peninsular ranges of southern California, and *G. eriospermum* in the central and southern Sierra Nevada). One species, *G. micranthum* Hooker & Arnott, is endemic to southern South America in Argentina and Chile, and one species, *G. humile*, occurs on both continents, with a wide distribution in the western United States and a limited one in Argentina and Chile. Two or more species are often found growing together.

Reproductive features include: self-compatible; flowers diurnal, fading mid day; floating chromosomal reciprocal translocations in at least one species (*Gayophytum eriospermum*), and one species (*G. heterozygum*) is a permanent translocation heterozygote (PTH); most species are autogamous (or even cleistogamous); some species with larger flowers and the stigma elevated above the anthers are outcrossing and pollinated by syrphid flies or bees.

Gayophytum is the only genus in Onagreae that has a 2-locular capsule with one row of seeds in each locule. In addition, seeds maturing in the capsules exhibit a variety of arrangements that are useful in the taxonomy of the genus. The terminology that has been used in describing these patterns has been somewhat confusing. The seed arrangement patterns are described here so that minimal and directly comparable terminology can be used in the key and descriptions. In the basic arrangement the septum is straight and the seeds in each locule are arranged obliquely to the septum and subopposite to seeds in the adjacent locule, forming two even rows in the capsule. This condition occurs in two species (*G. humile*, *G. racemosum*). In these species the capsule walls are smooth and flat or slightly constricted between the seeds. The remaining species have seeds arranged more or less parallel to the septum, and have either a straight septum or a somewhat to strongly sinuous septum. In *G. decipiens* the septum is straight and the seeds are in two subopposite rows with the capsule walls smooth and flat or slightly constricted between the seeds. All of the remaining species have a somewhat to strongly sinuous septum with the seeds arranged in an alternating pattern between the locules, and the adjacent seeds are either strongly or slightly overlapping or sometimes not overlapping and well spaced from each other. In these species the capsules walls are somewhat to strongly constricted between the seeds giving a torulose appearance to the capsule, sometimes described in the literature as lumpy, especially in the most extreme case in the capsules of the PTH *G. heterozygum* where roughly half of the seeds abort before maturity giving a very irregular, lumpy appearance to the capsule. In two of these species (*G. ramosissimum* and some populations of *G. diffusum* subsp. *parviflorum*) the seeds are crowded and strongly overlapping, so much so that they appear to form two irregular rows in each locule. In all other species with alternate seeds there is no or little overlap and overall they form a single row in the capsule.

SELECTED REFERENCES Lewis, H. and J. Szweykowski. 1964. The genus *Gayophytum* (Onagraceae). Brittonia 16: 343–391. Thien, L. B. 1969. Translocations in *Gayophytum* (Onagraceae). Evolution 23: 456–465.

1. Petals 3–8 mm; sepals 2–6 mm; stigmas usually exserted beyond anthers of longer stamens at anthesis.
   2. Petals 4–8 mm; sepals 3–6 mm; seeds 1.2–2.3 mm; Sierra Nevada and Greenhorn mountains, California . . . . .5. *Gayophytum eriospermum*
   2. Petals 3–5(–7) mm; sepals 2–3(–5) mm; seeds 1–1.6 mm; widely distributed in w United States but absent from Sierra Nevada and Greenhorn mountains . . . . . 8. *Gayophytum diffusum* (in part)
1. Petals 0.5–3 mm; sepals 0.4–2 mm; stigmas surrounded by anthers at anthesis.
   3. Capsules flattened, not constricted between seeds; stems usually branched only near base, secondary branches few or none; seeds 0.7–1.1 mm, arranged obliquely to septum, subopposite to seeds in adjacent locule forming 2 even rows.

4. Two valves of capsule remaining attached to septum after dehiscence; pedicels 0–0.5 mm; seeds 24–50 per capsule . . . . 1. *Gayophytum humile*
4. All valves of capsule free from septum after dehiscence; pedicels 0.4–2 mm; seeds (10–)14–34 per capsule. . . . . 2. *Gayophytum racemosum*

[3. Shifted to left margin.—Ed.]

3. Capsules terete to ± flattened, somewhat to conspicuously constricted between seeds; stems branched only distally or throughout, secondary branches usually many; seeds (0.8–)1–2 mm, arranged ± parallel to septum, and either subopposite forming 2 even rows with a straight septum, or arranged in alternating pattern with sinuous septum.
   5. Seeds ca. ½ aborted (pollen ca. 50% fertile); capsules with conspicuous irregular constrictions between seeds (due to aborted seeds) . . . . 6. *Gayophytum heterozygum*
   5. Seeds all developing (pollen 90–100% fertile); capsules with somewhat to conspicuous irregular constrictions between seeds.
      6. Seeds in each locule crowded and overlapping.
         7. Petals 0.7–1.2(–1.5) mm; pedicels longer than capsules, (3–)5–12 mm; seeds 10–30 per capsule . . . . 4. *Gayophytum ramosissimum*
         7. Petals 1.2–3 mm; pedicels usually shorter than capsules, 2–10(–15) mm; seeds (3–)6–18 per capsule . . . . 8. *Gayophytum diffusum* (in part)
      6. Seeds in each locule not crowded and overlapping.
         8. Seeds mostly more than 9 per capsule, subopposite or alternate; capsules with inconspicuous constrictions between seeds, septum nearly straight.
            9. Stems branched throughout, usually with 2–8 nodes between branches; petals 1.1–1.8 mm; petioles 0–5 mm . . . . 3. *Gayophytum decipiens*
            9. Stems branched or unbranched at base, much branched distally, usually with 1 or 2 nodes between branches; petals 1.2–3 mm; petioles 0–10 mm . . . . 8. *Gayophytum diffusum* (in part)
         8. Seeds mostly 9 or fewer per capsule, usually alternate; capsules with conspicuous constrictions between seeds, septum sinuous.
            10. Seeds (1–)3–5 per capsule; pedicels about equal to capsules, 3–11 mm; s California . . . . 7. *Gayophytum oligospermum*
            10. Seeds (3–)6–18 per capsule; pedicels usually shorter than capsules, 2–10 (–15) mm; widely distributed, British Columbia, w United States . . . . 8. *Gayophytum diffusum* (in part)

1. Gayophytum humile A. Jussieu, Ann. Sci. Nat. (Paris) 25: 18, plate 4. 1832 [F]

Gayophytum nuttallii Torrey & A. Gray

Herbs glabrous or sparsely and minutely glandular distally. **Stems** erect, unbranched or spreading and branched near base, secondary branches few or none, branching not dichotomous, 5–20(–30) cm. **Leaves** little reduced distally, crowded and often exceeding subtending internode, 10–25 × 1–3 mm; petiole 0–10 mm; blade very narrowly elliptic or lanceolate to sublinear. **Inflorescences** with flowers arising as proximally as first 1–3 nodes from base. **Flowers:** sepals 0.6–1.3 mm, reflexed singly; petals 0.8–1.5 mm; pollen 90–100% fertile; stigma subglobose, surrounded by anthers at anthesis. **Capsules** ascending, flattened, 8–17 × 1–2 mm, not constricted between seeds, valve margins entire or minutely undulate, 2 valves remaining attached to septum after dehiscence, septum straight; pedicel 0–0.5 mm. **Seeds** 24–50, all developing, arranged obliquely to septum and subopposite seeds in adjacent locule, forming 2 even rows in capsule, light brown, 0.7–1.1 × 0.3–0.4 mm, glabrous. $2n = 14$.

Flowering May–Sep. Drying margins of meadows, streams, lakes and pools; 800–3000 m; Calif., Idaho, Mont., Nev., Oreg., Wash., Wyo.; South America (Argentina, Chile).

Gayophytum humile is most similar to the allotetraploid *G. racemosum* and, on the basis of morphology, may be one of its parents. The two species are most distinct in their mature capsules; in *G. humile* the two lateral valves remain attached to the septum at dehiscence, whereas in *G. racemosum* all four valves separate from the septum at dehiscence.

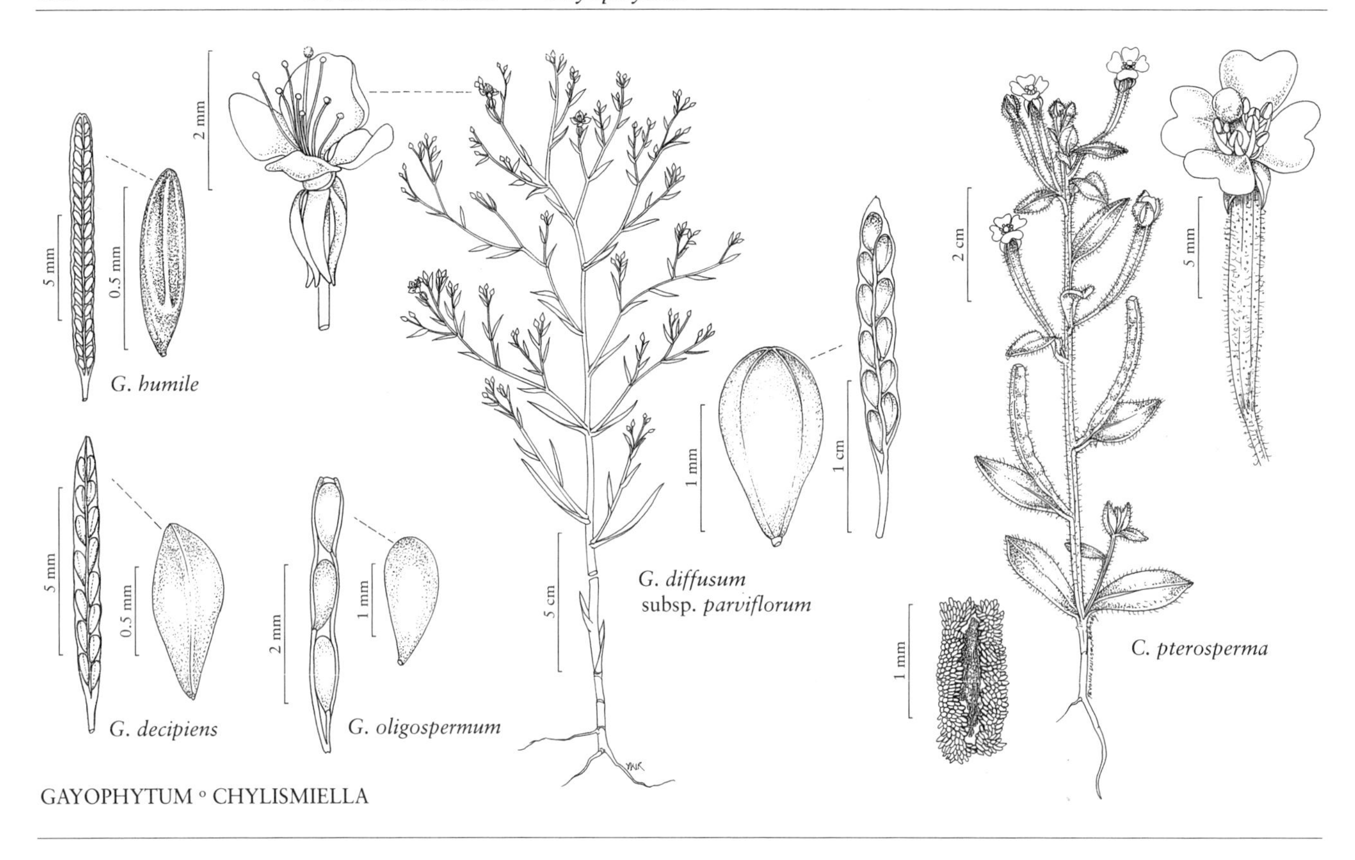

GAYOPHYTUM ◦ CHYLISMIELLA

2. Gayophytum racemosum Torrey & A. Gray, Fl. N. Amer. 1: 514. 1840 [E]

Gayophytum caesium Torrey & A. Gray; *G. helleri* Rydberg; *G. helleri* var. *glabrum* Munz; *G. humile* A. Jussieu var. *hirtellum* Munz; *G. racemosum* var. *caesium* (Torrey & A. Gray) Munz; *G. ramosissimum* Torrey & A. Gray var. *pygmeum* Jepson

Herbs glabrous or strigillose, rarely villous. **Stems** erect or spreading, branched mostly proximally, secondary branches few or none, branching not dichotomous, (5–)10–40 cm. **Leaves** little reduced distally, crowded, 10–25 × 1–3 mm; petiole 0–5 mm; blade narrowly lanceolate or elliptic to sublinear. **Inflorescences** with flowers arising as proximally as first 1–3 nodes from base. **Flowers:** sepals 0.8–1.4 mm, reflexed singly or in pairs; petals 1.3–1.8 mm; pollen 90–100% fertile; stigma subglobose, surrounded by anthers at anthesis. **Capsules** ascending, flattened, 10–15 × 0.8–1.4 mm, not constricted between seeds, valve margins entire or slightly undulate, all valves free from septum after dehiscence, septum straight; pedicel 0.4–2 mm. **Seeds** (10–)14–34, all developing, arranged obliquely to septum and subopposite seeds in adjacent locule, forming 2 even rows in capsule, light to dark brown, 0.8–1 × 0.3–0.5 mm, glabrous or densely puberulent. $2n$ = 28.

Flowering May–Oct. Drying margins of meadows and pools; 1000–4000 m; Alta.; Ariz., Calif., Colo., Idaho, Mont., Nev., Oreg., S.Dak., Utah, Wash., Wyo.

Gayophytum racemosum, on the basis of morphological and ecological observation, appears to be an allotetraploid derived from hybridization between *G. decipiens* and *G. humile* or a related species that is presumably extinct. *Gayophytum racem*osum is most readily distinguished from *G. humile* by the dehiscence of the capsules and from *G. decipiens* by branching habit and by the capsules, which are not as conspicuously flat in *G. decipiens*.

3. Gayophytum decipiens H. Lewis & Szweykowski, Brittonia 16: 368, figs. 5H, 6C. 1964 [E] [F]

Herbs glabrous or villous to strigillose. **Stems** erect or proximalmost branches decumbent, branched throughout, usually with 2–8 nodes between branches, branching usually not dichotomous, 5–50 cm. **Leaves** somewhat reduced distally, 10–32 × 1–4 mm; petiole 0–5 mm; blade narrowly lanceolate to sublinear. **Inflorescences** with flowers arising as proximally as first 1–5 nodes from base. **Flowers:** sepals 0.7–1.1 mm, reflexed singly; petals 1.1–1.8 mm; pollen 90–100% fertile; stigma subglobose, surrounded by anthers at anthesis. **Capsules**

usually ascending, rarely reflexed, not conspicuously flattened, 6–15 × 0.6–1 mm, with inconspicuous constrictions between seeds, valve margins slightly undulate, all valves free from septum after dehiscence, septum straight; pedicel 0–5 mm. **Seeds** 10–25, all developing, arranged ± parallel to septum and subopposite seeds in adjacent locule, forming 2 even rows in capsule, light brown, sometimes mottled with dark brown, 0.8–1.8 × 0.3–0.7 mm, glabrous or densely puberulent. **2*n*** = 14.

Flowering May–Sep. Sandy or gravely soil, in pinyon-juniper woodlands and pine forests, desert ranges, mountains bordering desert areas; 1800–4200 m; Ariz., Calif., Colo., Idaho, Mont., Nev., N.Mex., Oreg., Utah, Wash., Wyo.

Gayophytum decipiens may be one of the parents of the allotetraploid species *G. racemosum*. Since some collections of the tetraploid *G. diffusum* subsp. *parviflorum* closely resemble *G. decipiens*, the latter has probably contributed to the extensive variation in the *G. diffusum* polyploid complex. Among the diploid species, *G. decipiens* is most similar to *G. ramosissimum* but can be distinguished by its branching habit and capsules.

4. Gayophytum ramosissimum Torrey & A. Gray, Fl. N. Amer. 1: 513. 1840 E

Gayophytum ramosissimum var. *deflexum* Hooker; *G. ramosissimum* var. *obtusum* Jepson

Herbs glabrous or sparsely strigillose distally. **Stems** erect, profusely branched throughout, usually at every other node, branching dichotomous except near base, 10–50 cm. **Leaves** reduced distally, 10–40 × 1–5 mm; petiole 0–3(–10) mm; blade very narrowly lanceolate to sublinear. **Inflorescences** with flowers arising as proximally as first 5–15 nodes from base. **Flowers:** sepals 0.4–0.8 mm, reflexed singly; petals 0.7–1.2(–1.5) mm; pollen 90–100% fertile; stigma subglobose, surrounded by anthers at anthesis. **Capsules** ascending to reflexed, subterete, 3–9 × 0.8–1.2 mm, with inconspicuous constrictions between seeds, valve margins entire or weakly undulate, all valves free from septum after dehiscence, septum straight; pedicel (3–)5–12 mm. **Seeds** 10–30, all developing, arranged ± parallel to septum and in alternating pattern between locules, crowded, overlapping, often appearing to form 2 irregular rows in each locule, brown or gray mottled with brown, 1–1.5 × 0.5–0.7 mm, glabrous. **2*n*** = 14.

Flowering May–Sep. Sagebrush communities; 500–3000 m; Ariz., Calif., Colo., Idaho, Mont., Nev., N.Mex., Oreg., Utah, Wash., Wyo.

Gayophytum ramosissimum is similar to some individuals of *G. diffusum* subsp. *parviflorum* and has probably contributed to the extensive variation of the *G. diffusum* tetraploid complex. *Gayophytum ramosissimum* is most readily distinguished from *G. diffusum* subsp. *parviflorum* by its capsule being shorter than the pedicel.

5. Gayophytum eriospermum Coville, Contr. U.S. Natl. Herb. 4: 103. 1893 E

Gayophytum lasiospermum Greene var. *eriospermum* (Coville) Jepson

Herbs usually glabrous or strigillose, rarely villous. **Stems** erect, usually unbranched near base, branched at each of several nodes proximal to first flower, less branched distally, branching subdichotomous, 15–100 cm. **Leaves** much reduced distally, 20–75 × 1–6 mm; petiole 0–10 mm; blade narrowly lanceolate. **Inflorescences** with flowers arising as proximally as first 10–20 nodes from base. **Flowers:** sepals 3–6 mm, reflexed singly or in pairs; petals 4–8 mm; pollen 90–100% fertile; stigma hemispheric, usually exserted beyond anthers of longer stamens at anthesis. **Capsules** ascending or reflexed, subterete, 4–16 × 1–1.5 mm, with constrictions between seeds, valve margins undulate, all valves free from septum after dehiscence, septum sinuous; pedicel 3–12 mm. **Seeds** 4–10(–13), all developing, arranged ± parallel to septum and in alternating pattern between locules, adjacent seeds not overlapping, well spaced from each, forming a single row in capsule, brown or gray mottled with brown, 1.2–2.3 × 0.6–1 mm, glabrous or puberulent. **2*n*** = 14.

Flowering Jun–Oct. Open montane forests; 1000–3000 m; Calif.

Gayophytum eriospermum is the only regularly outcrossing diploid species in the genus. The species is known from southwestern Sierra Nevada and Greenhorn mountains (south of Placer County). Except in having larger flowers and generally larger capsules with more seeds, *G. eriospermum* is very similar to *G. oligospermum*, suggesting that the latter may be a self-pollinating derivative. *Gayophytum eriospermum* appears to have been involved in the origin of the *G. diffusum* polyploid complex. Some populations of *G. diffusum* subsp. *diffusum* are so similar to *G. eriospermum* that it has been necessary to determine chromosome number to make a positive identification. However, *G. diffusum* subsp. *diffusum* is not known to occur within the area of distribution of *G. eriospermum*.

6. Gayophytum heterozygum H. Lewis & Szweykowski, Brittonia 16: 377, figs. 5C, 5K, 13B. 1964 E

Gayophytum diffusum Torrey & A. Gray var. *villosum* Munz

Herbs usually glabrous or strigillose, rarely villous. **Stems** erect, usually unbranched near base, branched at each of several nodes proximal to first flower, less branched distally, branching dichotomous, 15–80 cm. **Leaves** much reduced distally, 15–60 × 1–5 mm; petiole 0–10 mm; blade narrowly lanceolate to sublinear. **Inflorescences** with flowers arising usually as proximally as first 10–20 nodes from base. **Flowers:** sepals 1–1.8 mm, reflexed singly or in pairs; petals 2–3 (–4) mm; pollen ca. 50% fertile; stigma hemispheric to subglobose, surrounded by anthers at anthesis. **Capsules** ascending or reflexed, terete to ± flattened, 6–15 × 0.8–1.1 mm, with conspicuous irregular constrictions between seeds, valve margins undulate, all valves free from septum after dehiscence, septum sinuous; pedicel (2–)3–12 mm. **Seeds** 2–10, ca. ½ aborted, arranged ± parallel to septum and in alternating pattern between locules, adjacent seeds not overlapping, irregularly well spaced from each, forming a single row in capsule, brown or gray mottled with brown, 1.4–2 × 0.6–0.8 mm, glabrous or puberulent. $2n$ = 14.

Flowering Jun–Oct. Open montane forests; 800–3000 m; Calif., Nev., Oreg., Wash.

Gayophytum heterozygum is the only species of Onagraceae outside of *Oenothera* that is a permanent translocation heterozygote (PTH) (W. L. Wagner et al. 2007), in which the chromosomes form a ring of 14 at meiosis rather than the seven pairs characteristic of all other self-pollinating diploid species of the genus. As is true for other PTH species, *G. heterozygum* produces only about fifty percent fertile pollen and a correspondingly reduced number of fertile seeds. All other species have less than ten percent of the pollen sterile.

It is not clear whether one or two species were involved in the parentage of *Gayophytum heterozygum* (H. Lewis and J. Szweykowski 1964; L. B. Thien 1969). This species is morphologically intermediate between *G. eriospermum* and *G. oligospermum* and has been suggested by Lewis and Szweykowski to be a PTH species derived by hybridization between them, much like the PTH species in 17. *Oenothera*.

7. Gayophytum oligospermum H. Lewis & Szweykowski, Brittonia 16: 375, figs. 5J, 13A. 1964 E F

Herbs glabrous or very sparsely strigillose distally. **Stems** erect, usually unbranched near base, branched at each of several nodes proximal to first flower, less branched distally, branching dichotomous, 20–70 cm. **Leaves** much reduced distally, 10–60 × 1–4 mm; petiole 0–10 mm; blade very narrowly lanceolate to sublinear. **Inflorescences** with flowers arising usually as proximally as first 10–20 nodes from base. **Flowers:** sepals 1–1.4 mm, reflexed singly or in pairs; petals 1.5–2.5 mm; pollen 90–100% fertile; stigma subglobose, surrounded by anthers at anthesis. **Capsules** ascending to strongly reflexed, subterete, 4–9 × 0.9–1 mm, with conspicuous constrictions between seeds, valve margins undulate, all valves free from septum after dehiscence, septum sinuous; pedicel 3–11 mm, usually about equal to capsule. **Seeds** usually 3–6, all developing, arranged ± parallel to septum and in alternating pattern between locules, adjacent seeds not overlapping, well spaced from each, forming a single row in capsule, brown, sometimes mottled with darker spots, 1.3–1.7 × 0.6–0.8 mm, glabrous. $2n$ = 14.

Flowering Jun–Sep. Open pine forests; 1300–2800 m; Calif.

Gayophytum oligospermum is known only from Los Angeles, Riverside, San Bernardino, and San Diego counties in southern California.

Gayophytum oligospermum is morphologically similar to *G. heterozygum* and has been shown to have chromosomes that pair completely with one of the two arrangements that make up the genome of the complex heterozygote *G. heterozygum*, indicating that *G. oligospermum* is one of the parents of *G. heterozygum* or that it is derived from *G. heterozygum* (L. B. Thien 1969).

Gayophytum oligospermum has been reported from the summit of Breckenridge Mountain, Kern County, from a single collection (*Charlton 3750*, UCR). This record needs to be verified.

8. Gayophytum diffusum Torrey & A. Gray, Fl. N. Amer. 1: 513. 1840 [F]

Herbs usually glabrous to strigillose, sometimes villous. **Stems** erect, branched or unbranched near base, much branched distally, usually with 1 or 2 nodes between branches, distal branching dichotomous or lateral branches shortened, 5–60 cm. **Leaves** reduced distally, 10–60 × 1–5 mm; petiole 0–10 mm; blade very narrowly lanceolate. **Inflorescences** with flowers arising usually as proximally as first 1–20 nodes from base. **Flowers:** sepals 0.9–3(–5) mm, reflexed singly or in pairs; petals 1.2–5(–7) mm; pollen 90–100% fertile; stigma hemispheric to subglobose, exserted beyond anthers of longer stamens or surrounded by them at anthesis. **Capsules** ascending to reflexed, subterete, 3–15 × 1–1.5 mm, with inconspicuous or conspicuous constrictions between seeds, valve margins somewhat undulate, all valves free from septum after dehiscence, septum straight or sinuous; pedicel 2–10(–15) mm, usually shorter than capsule. **Seeds** (3–)6–18, all or most developing, arranged ± parallel to septum and in alternating pattern between locules, crowded, overlapping, often appearing to form 2 irregular rows in each locule, or well spaced, forming a single row in capsule, brown, sometimes mottled with gray, 1–1.6 × 0.5–0.8 mm, glabrous or puberulent. **2*n*** = 28.

Subspecies 2 (2 in the flora): w North America, n Mexico.

Gayophytum diffusum consists of a diverse assemblage of tetraploid populations, some of which are similar to every known diploid species except *G. humile*. The combination of characteristics of at least five diploid species in various ways suggests that the complex is derived from several independently formed allopolyploids that subsequently hybridized and segregated to produce the observed diversity.

Populations of *Gayophytum diffusum* differ in breeding behavior. Populations with relatively large flowers and stigmas that extend beyond the anthers are obviously outcrossing, whereas most populations are small-flowered and modally self-pollinated. It is among the latter that the greatest morphological diversity is found. Often two or more morphologically different, apparently true-breeding strains can be found growing together. In such a variable complex, recognition of infraspecific taxa becomes arbitrary. In this treatment the striking morphological differences associated with breeding behavior have been used as a basis for subspecies recognition. At some localities the two subspecies intergrade.

1. Petals 3–5(–7) mm; sepals 2–3(–5) mm; stigma usually exserted beyond anthers of longer stamens at anthesis. . .8a. *Gayophytum diffusum* subsp. *diffusum*
1. Petals 1.2–3 mm; sepals 0.9–2 mm; stigma surrounded by anthers at anthesis. 8b. *Gayophytum diffusum* subsp. *parviflorum*

8a. Gayophytum diffusum Torrey & A. Gray subsp. **diffusum** [E]

Flowers: sepals 2–3(–5) mm; petals 3–5(–7) mm; stigma hemispheric, usually exserted beyond anthers of longer stamens at anthesis. **2*n*** = 28.

Flowering May–Sep. Open pine forests, sagebrush slopes, dry margins of meadows; 500–2500 m; Calif., Idaho, Mont., Nev., Oreg., Wash., Wyo.

Subspecies *diffusum* is very similar to *Gayophytum eriospermum* but is not known from the area of distribution of the latter.

8b. Gayophytum diffusum Torrey & A. Gray subsp. **parviflorum** H. Lewis & Szweykowski, Brittonia 16: 386, figs. 5B, 17B. 1964 [F]

Gayophytum diffusum var. *strictipes* (Hooker) Dorn; *G. lasiospermum* Greene; *G. lasiospermum* var. *hoffmannii* Munz; *G. nuttallii* Torrey & A. Gray var. *abramsii* Munz; *G. nuttallii* var. *intermedium* (Rydberg) Munz; *G. ramosissimum* Torrey & A. Gray var. *strictipes* Hooker

Flowers: sepals 0.9–2 mm; petals 1.2–3 mm; stigma subglobose, surrounded by anthers at anthesis. **2*n*** = 28.

Flowering Jun–Oct. Open pine forests, sagebrush slopes, dry margins of meadows; 800–3700 m; B.C.; Ariz., Calif., Colo., Idaho, Mont., Nev., N.Mex., Oreg., S.Dak., Utah, Wash., Wyo.; Mexico (Baja California).

Subspecies *parviflorum* is extremely variable and may sometimes closely resemble diploid species *Gayophytum decipiens*, *G. heterozygum*, *G. oligospermum*, and *G. ramosissimum*, the first of which are hypothesized to have been involved in its origin and variation (H. Lewis and J. Szweykowski 1964).

8. CHYLISMIELLA (Munz) W. L. Wagner & Hoch, Syst. Bot. Monogr. 83: 115. 2007 • [Genus *Chylismia* and Latin *-ella*, diminutive, alluding to flower size] E

Warren L. Wagner

Oenothera Linnaeus sect. *Chylismiella* Munz, Amer. J. Bot. 15: 224. 1928; *Camissonia* Link sect. *Chylismiella* (Munz) P. H. Raven

Herbs, annual, caulescent; from a taproot. **Stems** erect, branched. **Leaves** cauline, alternate; stipules absent; petiolate; blade margins entire. **Inflorescences** racemes, nodding at anthesis, erect in fruit. **Flowers** bisexual, actinomorphic, buds erect; floral tube deciduous (with sepals, petals, and stamens) after anthesis, nectary unknown, presumably at base of tube; sepals 4, reflexed singly or in pairs; petals 4, white, yellow at base, fading purple; stamens 8 in 2 subequal series, anthers basifixed, pollen shed singly; ovary 4-locular, stigma subentire or slightly lobed, subcapitate, surface unknown, probably wet and non-papillate. **Fruit** a capsule, straight or slightly curved, terete, loculicidally dehiscent; pedicellate. **Seeds** numerous, in 2 rows per locule, (appearing as 1 by crowding), with thick wing on concave side, wing and convex side covered with glassy, clavate hairs. $x = 7$.

Species 1: w United States.

Chylismiella was formerly included as a section in *Oenothera* (P. A. Munz 1928; P. H. Raven 1962, 1964) or *Camissonia* (Raven 1969), and elevated to generic rank based on molecular analyses of the Onagreae. *Chylismiella* was placed by R. A. Levin et al. (2004) sister to *Gayophytum* rather than with other groups placed by Raven (1964, 1969) in *Camissonia*. *Chylismiella* plus *Gayophytum* together have a sister relationship to *Clarkia* in the Levin et al. analyses. Raven (1969), in noting that *Chylismiella pterosperma* (as *Camissonia*) is extremely distinctive and not closely allied to other groups in *Camissonia*, suggested the possibility of a close relationship between *Chylismiella* and *Gayophytum* due to the shared character of white petals with a yellow band near the base, but did not suggest a close relationship with *Clarkia*. *Chylismiella* is distinguished by seeds with thick, papillate wings. Reproductive features include: self-compatible and flowers diurnal, mainly autogamous (Raven 1962, 1969).

1. Chylismiella pterosperma (S. Watson) W. L. Wagner & Hoch, Syst. Bot. Monogr. 83: 208. 2007 E F

Oenothera pterosperma S. Watson, Botany (Fortieth Parallel), 112, plate 14, figs. 4–7. 1871; *Camissonia pterosperma* (S. Watson) P. H. Raven; *Chylismia pterosperma* (S. Watson) Small; *Sphaerostigma pterospermum* (S. Watson) A. Nelson

Herbs sparsely to moderately spreading hirtellous, becoming glandular puberulent distally. **Stems** slender, ascending, branching, 2–15 cm. **Leaves** 0.3–3 × 0.1–0.6 cm; petiole 0.2–0.7 cm; blade sometimes with purple spots, narrowly lanceolate to oblanceolate. **Flowers** opening at sunrise; floral tube 1–2 mm; sepals often reddish green, 1.5–2.5 mm; petals white with yellow area at base, broadly obovate, apex notched, 1.5–3 mm; episepalous staminal filaments 1–1.7 mm, epipetalous ones 0.5–1 mm, anthers 0.3–0.4 mm; style 2.2–4 mm, stigma 1–1.5 mm diam., surrounded by anthers at anthesis. **Capsules** ascending or spreading, 12–18 × 1.2–1.6 mm; pedicel 4–8 mm. **Seeds** 1–1.5 × 0.6–0.8 mm, broader at 1 end, ends truncate (where contacting next seed in row), clavate hairs longer at one end, shortest in middle. $2n = 14$.

Flowering Apr–Jun. Well-drained slopes, often of volcanic origin, with sagebrush and pinyon-juniper; 700–2600 m; Ariz., Calif., Idaho, Nev., Oreg., Utah.

The range of *Chylismiella pterosperma*, an inconspicuous and uncommon plant, centers across Nevada and extends to southeast Oregon (Lake and Malheur counties), western Utah, northwest Arizona (Coconino and Mohave counties), eastern California (Inyo County), and southern Idaho (Butte County).

9. TARAXIA (Torrey & A. Gray) Nuttall ex Raimann in H. G. A. Engler and K. Prantl, Nat. Pflanzenfam. 96[III,7]: 216. 1893 • [Species *Leontodon taraxacoides*, alluding to similar leaves] [E]

Warren L. Wagner

Oenothera Linnaeus [unranked] *Taraxia* Torrey & A. Gray, Fl. N. Amer. 1: 506. 1840; *Oenothera* sect. *Heterostemon* (Nuttall) Munz; *Oenothera* subg. *Heterostemon* Nuttall; *Oenothera* [unranked] *Primulopsis* Torrey & A. Gray; *Oenothera* subg. *Taraxia* (Torrey & A. Gray) Jepson

Herbs, fleshy perennial, acaulescent; with thick or slender, sometimes woody taproot, sometimes branched and then usually producing new rosettes. **Leaves** in a basal rosette; stipules absent; petiolate; blade margins subentire to deeply sinuate or pinnatifid. **Inflorescences** solitary flowers in leaf axils. **Flowers** bisexual, actinomorphic, buds erect; floral tube deciduous (with sepals, petals, and stamens) after anthesis, with fleshy basal nectary; sepals 4, reflexed separately; petals 4, usually yellow, rarely white, without spots, usually fading orange, strongly ultraviolet reflective, or sometimes not reflective near base; stamens 8, in 2 unequal series, anthers basifixed, pollen shed singly; ovary 4-locular, with a long, slender, sterile apical projection proximal to opening of floral tube, projection without visible abscission lines at its junctures with floral tube or fertile part of ovary, stigma entire or irregularly lobed, globose, surface unknown, probably wet and non-papillate. **Fruit** a capsule, straight or slightly irregularly curved, subterete to 4-angled, cylindric-lanceoloid or -ovoid, or oblong-ellipsoid, irregularly loculicidal, gradually tapering into a slender, sterile portion (4–)15–180 mm, sometimes persistent 1+ years, often blackened, thin- or thick-walled; subsessile. **Seeds** numerous, in 2 rows per locule, pitted or coarsely papillose. $x = 7$.

Species 4 (4 in the flora): w North America.

Taraxia is known from the western United States and southwestern Canada in open, moist clay or sandy sites, usually at low to middle elevations. *Taraxia* is characterized by its acaulescent habit, seeds in two rows per locule in unwinged, irregularly dehiscent capsules, and notably by having a relatively long, slender, sterile projection at the apex of the ovary that persists on the mature capsule after the floral tube and perianth detach.

This distinctive group of species has been treated variously as a subgenus or section of *Oenothera* (J. Torrey and A. Gray 1838–1843, vol. 1; P. A. Munz 1965), as a section of *Camissonia* (P. H. Raven 1969), or as a separate genus (J. K. Small 1896). Traditionally, the two acaulescent annual species now viewed as composing the genus *Tetrapteron*, which share with *Taraxia* a sterile apical projection on the ovary, have been included in this group (R. Raimann 1893; Raven 1969). Munz (1965) included the six species in his *Oenothera* subg. *Heterostemon*, but separated the four perennials (as sect. *Heterostemon*) from the two annuals (as sect. *Tetrapteron*). On the basis of additional information, W. L. Wagner et al. (2007) recognized the two annual species as the genus *Tetrapteron*. R. A. Levin et al. (2004) found strong molecular support for *Taraxia* on a weakly supported branch sister to *Clarkia* + *Gayophytum* + *Chylismiella*, whereas the two annual species are strongly monophyletic on a weakly supported branch with *Camissoniopsis* and *Neoholmgrenia*. Even though the molecular support for the clade of *Clarkia* + *Gayophytum* + *Chylismiella* + *Taraxia* is weak, this group of genera shares the feature of basifixed anthers, unlike the versatile anthers of all other genera of tribe Onagreae. P. H. Raven (1964) first pointed out that the basifixed anthers in *Taraxia* are similar to those found in *Clarkia*. Species of *Taraxia* are sometimes grown as ornamentals in rock gardens. Reproductive features include: self-incompatible, flowers diurnal, outcrossing, and pollinated by small bees [*T. ovata* (E. G. Linsley et al. 1973), *T. tanacetifolia* (Linsley et al. 1963b)] or facultatively autogamous [*T breviflora T subacaulis* (Raven 1969)].

1. Leaf blade margins usually subentire to sinuate or crisped, sometimes irregularly sinuate-lobed toward base, rarely pinnatifid; herbs sparsely to densely short-hirsute or strigillose, especially on leaf blade margins and ± veins.
   2. Leaf blades usually densely short-hirsute on margins and ± veins, sometimes sparsely so; capsules subterete, walls much distended by seeds . . . . 1. *Taraxia ovata*
   2. Leaf blades glabrate, veins and margins rarely sparsely strigillose; capsules 4-angled, walls nearly flat, not noticeably distended by seeds . . . . 2. *Taraxia subacaulis*
1. Leaf blade margins pinnatifid; herbs usually sparsely to densely strigillose or short-hirtellous, hairs spreading or appressed.
   3. Styles 9.5–20(–25) mm; stigmas exserted beyond anthers at anthesis; petals (8–)10–23 mm; herbs densely or, sometimes, sparsely short-hirtellous and/or strigillose . . . . 3. *Taraxia tanacetifolia*
   3. Styles 3–6.5 mm; stigmas surrounded by anthers at anthesis; petals 5–7(–9) mm; herbs densely to sparsely strigillose, sometimes also appressed-hirtellous, hairs spreading or appressed . . . . 4. *Taraxia breviflora*

1. Taraxia ovata (Nuttall) Small, Bull. Torrey Bot. Club 23: 185. 1896 E F

Oenothera ovata Nuttall in J. Torrey and A. Gray, Fl. N. Amer. 1: 507. 1840; *Camissonia ovata* (Nuttall) P. H. Raven

Herbs densely, sometimes sparsely, short-hirsute, especially on leaf blade margins and ± veins; taproot thick, often branched in age, producing new rosettes. **Leaves** 3–15 × 1.6–5 cm; petiole narrowly winged, 0.8–15 cm, base slightly dilated; blade ovate to very narrowly elliptic, base attenuate, margins usually subentire to shallowly sinuate or crisped, rarely deeply sinuate, apex acute to acuminate. **Flowers** opening near sunrise; floral tube 2–3 mm, with short, matted hairs inside near base; sepals 11–19 mm; petals usually yellow, rarely white, 8–23 mm; episepalous staminal filaments 3.5–8 mm, epipetalous ones 2–6 mm, anthers 3–5 mm; sterile prolongation of ovary 25–180 mm, style 4.5–11 mm, shortly pubescent near base, stigma exserted slightly beyond anthers at anthesis. **Capsules** subterete, cylindric-lanceoloid, 11–30 × 3–5 mm, walls thin, much distended by seeds; rarely with pedicel to 0.4 mm. **Seeds** uniformly brown, elongate-ovoid, 1.8–2.2 × 1.2–1.4 mm, densely and coarsely papillose. 2*n* = 14.

Flowering Feb–May. Grassy fields, clay soil, usually near coast; 0–500 m; Calif., Oreg.

Taraxia ovata occurs in counties near the coast and is found in Humboldt, Lake, and Mendocino counties south to the vicinity of Monterey Bay, Monterey County, and again south of the Santa Lucia Mountains in northern San Luis Obispo County, California, and Douglas and Josephine counties in Oregon. It has no close relatives in the genus (P. H. Raven 1969), an assertion supported by its early branching in molecular analyses (R. A. Levin et al. 2004; W. L. Wagner et al. 2007). The species is self-incompatible and pollinated by the oligolectic bee *Andrena* (*Diandrena*) *chalybea* (Cresson) (Raven).

Oenothera primuloidea H. Léveillé is an illegitimate, superfluous name that pertains here.

2. Taraxia subacaulis (Pursh) Rydberg, Mem. New York Bot. Gard. 1: 281. 1900 E F

Jussiaea subacaulis Pursh, Fl. Amer. Sept. 1: 304. 1813 (as Jussieua); *Camissonia subacaulis* (Pursh) P. H. Raven; *Oenothera heterantha* Nuttall; *O. heterantha* var. *taraxacifolia* S. Watson; *O. subacaulis* (Pursh) Garrett; *O. subacaulis* var. *taraxacifolia* (S. Watson) Jepson; *Taraxia heterantha* (Nuttall) Small; *T. taraxacifolia* (S. Watson) A. Heller

Herbs usually glabrate, rarely sparsely strigillose, especially on leaf blade veins and margins; taproot thick, deep, sometimes branched in age, producing new rosettes. **Leaves** 2–22 × 0.7–4.2 cm; petiole narrowly winged, 1–12 cm; blade lanceolate to narrowly elliptic, base attenuate, margins usually subentire to sinuate, sometimes deeply and irregularly pinnatifid, apex acuminate. **Flowers** opening near sunrise; floral tube 1.5–3 mm, with short, matted hairs inside near base; sepals 4–15 mm, mostly glabrous, sometimes very minutely strigillose; petals yellow, 5–16 mm, often apiculate; episepalous staminal filaments 1.8–6.5 mm, epipetalous ones 0.5–2.5 mm, anthers 0.9–2 mm; sterile prolongation of ovary 15–80 mm, style 4–8.5(–11) mm, glabrous or sparsely pubescent near base, stigma usually surrounded by longer anthers at anthesis, rarely exserted beyond anthers. **Capsules** 4-angled, oblong-ellipsoid, 11–28 × 5–8 mm, walls thick, scarcely distended by seeds, becoming blackened and persistent on plants for 1+ years after shedding seeds; rarely with pedicel to 10 mm. **Seeds** uniformly tan to light brown, ellipsoid-oblanceoloid, 1.3–1.9 × 0.6–1 mm, coarsely pitted. 2*n* = 14.

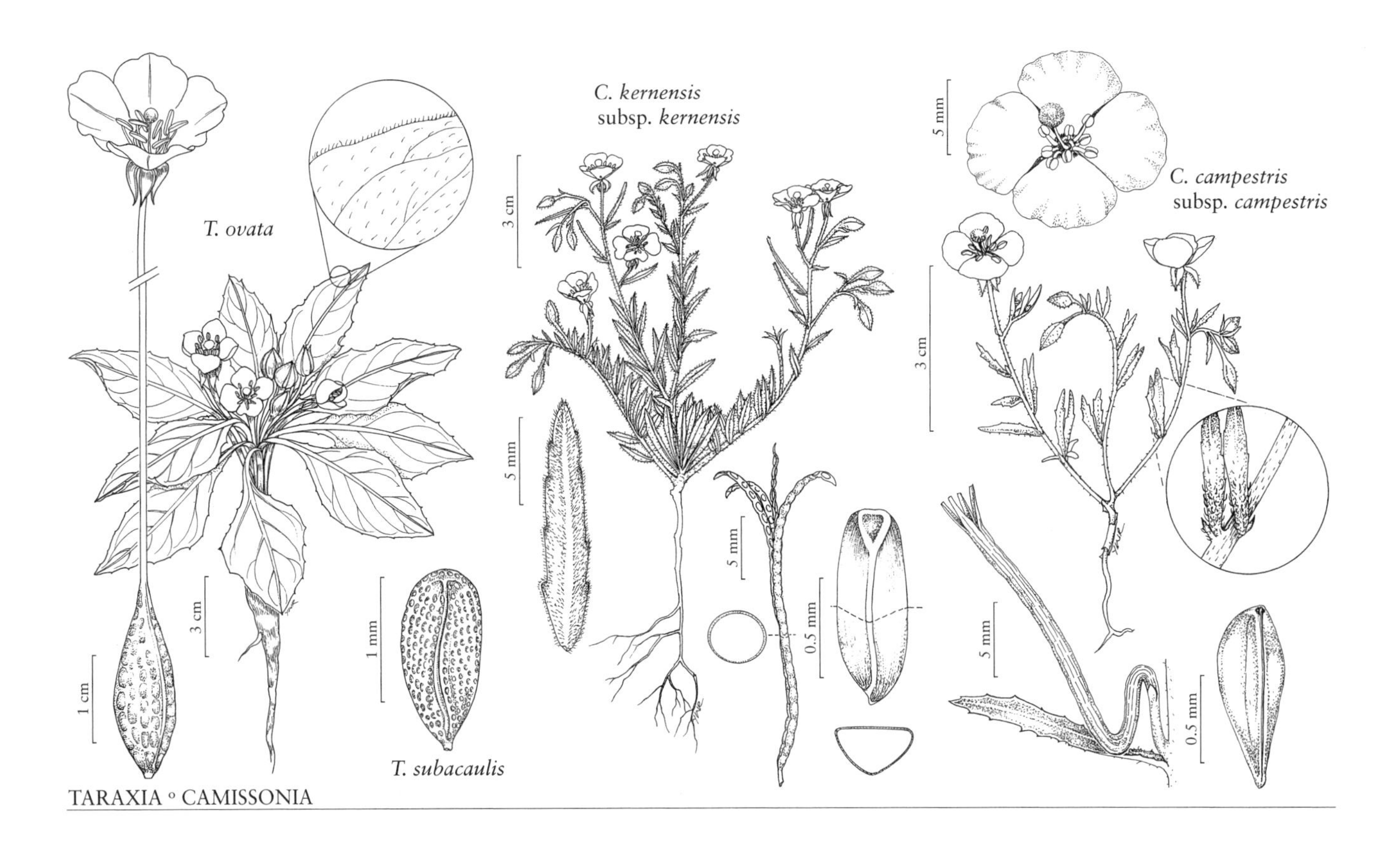

TARAXIA ◦ CAMISSONIA

Flowering late Mar–Jul. Meadows, seasonally moist open places, from foothills to mountains; 400–2900 m; Calif., Colo., Idaho, Mont., Nev., Oreg., Utah, Wash., Wyo.

Taraxia subacaulis occurs well inland from Pacific coastal areas. Its only occurrence in Colorado is in Moffat and Routt counties. P. H. Raven (1969) determined *Taraxia subacaulis* to be self-compatible and, usually, facultatively autogamous. The species is distinctive, but is sister to *T. tanacetifolia* in molecular analyses (R. A. Levin et al. 2004; W. L. Wagner et al. 2007).

3. Taraxia tanacetifolia (Torrey & A. Gray) Piper, Contr. U.S. Natl. Herb. 11: 405. 1906 [E]

Oenothera tanacetifolia Torrey & A. Gray in War Department [U.S.], Pacif. Railr. Rep. 2(1): 121, plate 4. 1857; *Camissonia tanacetifolia* (Torrey & A. Gray) P. H. Raven; *C. tanacetifolia* subsp. *quadriperforata* P. H. Raven; *Taraxia longiflora* Nuttall ex Raimann; *T. tikurana* A. Nelson

Herbs usually densely, sometimes sparsely, short-hirtellous, hairs spreading or appressed, and/or sparsely to densely strigillose; taproot deep, woody, with numerous slender branches in age, producing new rosettes. **Leaves** 6.5–32 × 0.7–3.3 cm; petiole not winged, 1–8 cm; blade very narrowly elliptic, base attenuate to narrowly cuneate, margins deeply and irregularly pinnatifid, apex acute to long-acuminate. **Flowers** opening near sunrise; floral tube 4–6.5(–8.5) mm, soft-pilose in proximal ½ inside; sepals 5.5–13 mm; petals yellow, (8–)10–23 mm; episepalous staminal filaments 5.5–12 mm, epipetalous ones 2.5–8 mm, anthers (2.3–) 2.8–3.5 mm; sterile prolongation of ovary 14–55 mm, style 9.5–20(–25) mm, pilose near base, stigma exserted beyond anthers at anthesis. **Capsules** subterete, cylindric-ovoid, 7–25 × 3–5 mm, walls thick, evidently distended by seeds; sessile. **Seeds** uniformly tan to brown, curved-cylindric, 1.5–2 × 0.6–0.8 mm, pitted in rows. **2*n*** = 14, 28, 42.

Flowering May–Jul. Open areas on clay soil, moist swales and meadows, dry streambeds, edges of drying ponds; 400–2500 m; Calif., Idaho, Nev., Oreg., Wash.

P. H. Raven (1969) determined that the majority of plants examined cytologically were tetraploid (2*n* = 28), making one diploid (2*n* = 14) determination. Raven also found hexaploid plants in a restricted area from Sierra Valley in Plumas and immediately adjacent Lassen and Sierra counties, California. They do not differ morphologically from diploid and tetraploid plants other than having greater than ten percent of pollen grains 4-pored, but there was no known diploid population that could have combined with tetraploids to give rise to this hexaploid. Raven named it *Camissonia tanacetifolia* subsp. *quadriperforata*; however, it is not currently recognized due to the lack of morphological differentiation. Raven also found *Taraxia tanacetifolia* to be primarily self-incompatible with possibly some self-compatible plants occurring.

Oenothera nuttallii Torrey & A. Gray 1840 (not Sweet 1830) pertains here.

4. **Taraxia breviflora** (Torrey & A. Gray) Nuttall ex Small, Bull. Torrey Club. 23: 185. 1896 [E]

Oenothera breviflora Torrey & A. Gray, Fl. N. Amer. 1: 506. 1840; *Camissonia breviflora* (Torrey & A. Gray) P. H. Raven

Herbs densely to sparsely strigillose, sometimes also short-hirtellous, the hairs spreading or appressed; taproot deep, woody with numerous slender branches in age, producing multiple rosettes. **Leaves** 1.7–12 × 0.3–3.3 cm; petiole not winged, 1.5–3 cm; blade very narrowly elliptic, base attenuate, margins irregularly pinnatifid, apex acute to long-acuminate. **Flowers** opening near sunrise; floral tube 1.8–2(–2.5) mm, glabrate to very sparsely pilose inside near base; sepals 3–6(–7.5) mm; petals yellow, 5–7(–9) mm; episepalous staminal filaments 2.8–4(–5) mm, epipetalous ones 1.5–1.8(–2) mm, anthers 0.9–1.3 mm; sterile prolongation of ovary 4–15 mm, style 3–6.5 mm, sparsely pilose near base to glabrate, stigma surrounded by anthers of longer stamens at anthesis. **Capsules** subterete, cylindric-ovoid, often slightly asymmetric, 8–17 × 2–5 mm, walls thick, evidently distended by seeds; sessile. **Seeds** tan to brown, curved-cylindric, 1.3–1.8 × 0.7–0.9 mm, pitted in rows. $2n = 14$.

Flowering Jun–Aug. Seasonally moist meadows, dry streambeds, edges of drying ponds, mountains and upper foothills; 100–3100 m; Alta., B.C., Sask.; Colo., Idaho, Mont., Nev., Utah, Wyo.

P. H. Raven (1969) determined that *Taraxia breviflora* is self-compatible and primarily autogamous.

10. CAMISSONIA Link, Jahrb. Gewächsk. 1(1): 186. 1818 • Suncups [For Ludolf Karl Adelbert von Chamisso, 1781–1838, French-born German botanist]

Warren L. Wagner

Camissonia sect. *Sphaerostigma* (Seringe) P. H. Raven; *Oenothera* Linnaeus sect. *Sphaerostigma* Seringe; *Oenothera* [unranked] *Sphaerostigma* (Seringe) Torrey & A. Gray; *Oenothera* subg. *Sphaerostigma* (Seringe) Jepson; *Sphaerostigma* (Seringe) Fischer & C. A. Meyer

Herbs, annual, caulescent; with a taproot. **Stems** erect, decumbent, or ascending, usually branched from base and distally, epidermis white or reddish brown, often exfoliating. **Leaves** cauline, proximalmost often clustered near base, alternate; stipules absent; subsessile; blade margins entire, serrulate, or serrate. **Inflorescences** usually leafy spikes, sometimes racemes, nodding at anthesis, erect in fruit. **Flowers** bisexual, actinomorphic, buds erect; floral tube deciduous (with sepals, petals, and stamens) after anthesis, with basal nectary; sepals 4, reflexed separately or in pairs; petals 4, yellow, fading red, often with red dots basally; stamens 8, in 2 unequal series, anthers versatile, pollen shed singly; ovary 4-locular, without apical projection, stigma subentire, subcapitate to subglobose, surface unknown, probably wet and non-papillate. **Fruit** a capsule, straight to flexuous, cylindrical, subterete, regularly, sometimes tardily, loculicidally dehiscent; usually sessile, sometimes pedicellate. **Seeds** numerous, in 1 row per locule, narrowly obovoid to narrowly oblanceoloid, triangular in cross-section, appearing smooth, glossy. $x = 7$.

Species 12 (11 in the flora): w North America, nw Mexico, w South America.

Species of *Camissonia* occur in desert scrub, grasslands, or pinyon-juniper woodlands, on brushy or open slopes and flats, washes, and, sometimes, on serpentine barrens, at elevations 0 to 2300 meters. *Camissonia campestris* and *C. kernensis* are self-incompatible diploids; *C. pusilla* and *C. sierrae* are self-compatible diploids; *C. contorta* is an autogamous hexaploid; the other species are autogamous tetraploids. Identification of the polyploid species of *Camissonia* is aided by their pollen having a high proportion of grains with more pores, usually 4 or 5, than typical 3-pored pollen in Onagraceae. This can be observed under low magnification (for example, 10\×) since 3-pored pollen is triangular, while 4-pored is quadrangular, and 5-pored is pentangular.

P. H. Raven (1969) delineated a group of four closely related species (*Camissonia kernensis*, *C. parvula*, *C. pubens*, and *C. pusilla*), marked by having sepals reflexed separately (rather than in pairs), which occur mainly in the Great Basin. Of the remaining species of *Camissonia* in the flora area, several (*C. benitensis*, *C. campestris*, *C. integrifolia*, *C. lacustris*, and *C. sierrae*) have more or less restricted ranges within California, or (*C. strigulosa*) extend also to Baja California, Mexico, or (*C. contorta*) to Washington, Idaho, and disjunctly to Vancouver Island, British Columbia. *Camissonia dentata* (Cavanilles) Reiche is the twelfth species in the genus, disjunct in western South America from southern Peru southward into Chile and Argentina.

W. L. Wagner et al. (2007) departed significantly from the most recent monograph by P. H. Raven (1969) in the delimitation of *Camissonia* based on the molecular analysis (R. A. Levin et al. 2004); they recognized eight genera in addition to the much reduced *Camissonia*: *Camissoniopsis*, *Chylismia*, *Chylismiella*, *Eremothera*, *Eulobus*, *Neoholmgrenia*, *Taraxia*, and *Tetrapteron*. Raven noted that *Camissonia* was the most heterogeneous genus in Onagreae, consisting of sharply distinct sections. He further noted that the capitate or subglobose stigma found in *Camissonia*, by which he distinguished the genus from the broadly circumscribed *Oenothera* of P. A. Munz (1965), was also found in *Gayophytum*, *Gongylocarpus* Schlechtendal & Chamisso, and *Xylonagra* Donnell Smith & Rose, thus the primary defining character state for the genus at that time is a plesiomorphy (P. C. Hoch et al. 1993). The redefined *Camissonia* is morphologically delimited by having subterete capsules that are more or less swollen by seeds, linear to narrowly elliptic leaves, and glossy seeds that are triangular in cross-section and mostly smaller than 1 mm in length; and flowering only at the distal, not basal, nodes; and is without ultraviolet reflectance pattern on petals (Raven). Reproductive features include: self-incompatible (*C. campestris*, *C. kernensis)* or self-compatible; flowers diurnal; outcrossing and pollinated by bees (E. G. Linsley et al. 1963, 1963b, 1964, 1973), or autogamous, rarely cleistogamous (Raven).

1. Sepals reflexed separately.
    2. Petals 8–15(–18) mm; sepals 5–9(–11) mm; stigma exserted beyond anthers at anthesis ..... 1. *Camissonia kernensis*
    2. Petals 1.5–4 mm; sepals 1.2–3.8 mm; stigma surrounded by anthers at anthesis.
        3. Leaves cauline, none clustered near base, blade margins entire or subentire; plants usually glabrous or densely strigillose, rarely villous proximally, sometimes sparsely glandular puberulent distally ..... 4. *Camissonia parvula*
        3. Leaves cauline with some clustered near base, blade margins serrulate or undulate-serrulate; plants moderately to densely villous, usually also glandular puberulent, especially distally.
            4. Leaf blades 0.04–0.2 cm wide; floral tubes 0.8–1.6 mm; sepals 1.2–2 mm; capsules 18–32 mm ..... 2. *Camissonia pusilla*
            4. Leaf blades 0.2–0.6 cm wide; floral tubes 1.3–3 mm; sepals 2.2–3.8 mm; capsules (18–)26–50 mm ..... 3. *Camissonia pubens*
1. Sepals reflexed in pairs.
    5. Stigma exserted beyond anthers at anthesis; sepals 3.5–8(–12) mm; petals (3.5–)5–15.5 mm ..... 5. *Camissonia campestris*
    5. Stigma surrounded by anthers at anthesis (except sometimes slightly exserted beyond anthers in *C. sierrae*); sepals (1.2–)1.6–4(–5.5) mm; petals 2–7 mm.
        6. Leaf blade margins usually entire, rarely with 1 or 2 small teeth; plants densely strigillose distally ..... 10. *Camissonia integrifolia*
        6. Leaf blade margins usually sparsely serrulate (*C. sierrae* sometimes with 1–several small teeth); plants glandular puberulent, usually also villous (except also strigillose in *C. strigulosa*) distally.

[7. Shifted to left margin.—Ed]

7. Leaf blades usually lanceolate to narrowly ovate, sometimes elliptic, base rounded. . . . . 6. *Camissonia sierrae*
7. Leaf blades linear to narrowly elliptic, base cuneate or attenuate.
   8. Plants usually densely strigillose, often also glandular puberulent, especially distally, sometimes glandular puberulent only and glabrate, sometimes also villous near base; sepals 1.6–4 mm; less than 10% of pollen grains 4-pored . . . . .7. *Camissonia strigulosa*
   8. Plants villous, often also glandular puberulent distally, rarely strigillose and glandular puberulent, if so, more than 30% of pollen grains 4 or 5-pored; sepals 1.6–5.5 mm; less than 10% or more than 30% of pollen grains 4 or 5-pored.
      9. Sepals (3–)3.8–5.5 mm; petals (4–)4.5–7 mm . . . . .9. *Camissonia lacustris*
      9. Sepals 1.6–4 mm; petals 2.5–4(–5) mm.
         10. Anthers with less than 10% of pollen grains 4-pored; leaves green or slightly bluish green; floral tubes 1.2 mm; plants villous and also glandular puberulent distally; San Benito County and adjacent Fresno and Monterey counties, California. . . . . 8. *Camissonia benitensis*
         10. Anthers usually with more than 30% of pollen grains 4 or 5-pored; leaves usually bluish green; floral tubes 1.6–2.7 mm; plants usually villous throughout, often also glandular puberulent distally, rarely entirely strigillose and glandular puberulent; British Columbia, California, Idaho, Nevada, Oregon, Washington . . . . 11. *Camissonia contorta*

1. Camissonia kernensis (Munz) P. H. Raven, Brittonia 16: 284. 1964 [E] [F]

Oenothera kernensis Munz, Amer. J. Bot. 18: 737. 1931

Herbs sparsely or densely villous and glandular puberulent, especially distally, sometimes glabrate or sparsely glandular puberulent. **Stems** erect, often many-branched, 5–30 cm. **Leaves:** proximalmost sometimes clustered near base; blade usually very narrowly elliptic to narrowly so, rarely lanceolate, 1–3.8(–5.5) × 0.2–0.5 cm, base narrowly cuneate, margins sparsely serrate, apex acuminate. **Flowers** opening near sunrise; floral tube 2.2–3.8(–5.5) mm, villous inside; sepals 5–9(–11) mm, reflexed separately; petals 8–15(–18) mm, each with 2 large red dots basally; episepalous filaments 3.5–5.5(–7) mm, epipetalous filaments 1.3–2 (–4.5) mm, anthers 1.8–2(–3) mm, pollen with less than 5% of grains 4- or 5-pored; style 7–10(–14) mm, stigma well exserted beyond anthers at anthesis. **Capsules** 22–37 × 1.5–1.7 mm; pedicel 0–15 mm. **Seeds** 1.1–1.2 × 0.5–0.6 mm.

Subspecies 2 (2 in the flora): sw United States.

Camissonia kernensis occurs in sagebrush scrub, and Joshua-tree and pinyon-juniper woodlands at elevations of 700–1900 m in southern and central California and southern Nevada. The species is self-incompatible and is apparently pollinated by oligolectic bees of *Andrena* subg. *Onagrandrena* (P. H. Raven 1969); Raven subdivided the species into two intergrading subspecies.

1. Herbs compact, sparsely glandular puberulent and villous throughout; leaves: proximalmost often clustered near base; pedicels 3–15 mm in fruit. . . . . 1a. *Camissonia kernensis* subsp. *kernensis*
1. Herbs with open habit, glandular puberulent throughout, usually also sparsely villous, or glabrate, with few glandular hairs; leaves: none clustered near base; pedicels 0–5(–15) mm in fruit. . . . . 1b. *Camissonia kernensis* subsp. *gilmanii*

1a. Camissonia kernensis (Munz) P. H. Raven subsp. **kernensis** [E] [F]

Herbs compact, villous throughout and sparsely glandular puberulent. **Stems** 5–15(–22) cm. **Leaves:** proximalmost often clustered near base. **Capsules:** pedicel 3–15 mm. **2*n*** = 14.

Flowering Mar–Jul. Sandy slopes, flats, often with *Artemisia tridentata* and *Yucca brevifolia*; 700–2100 m; Calif.

Subspecies *kernensis* is known only from northeastern Kern County where it is sometimes locally abundant.

1b. Camissonia kernensis (Munz) P. H. Raven subsp. **gilmanii** (Munz) P. H. Raven, Contr. U.S. Natl. Herb. 37: 310. 1969 E

Oenothera dentata Cavanilles var. *gilmanii* Munz, Leafl. W. Bot. 2: 87. 1938; *Camissonia kernensis* var. *gilmanii* (Munz) Cronquist; *O. kernensis* Munz subsp. *gilmanii* (Munz) Munz; *O. kernensis* subsp. *mojavensis* Munz

Herbs with open habit, glandular puberulent throughout, usually also sparsely villous, or glabrate with few glandular hairs. **Stems** to 30 cm. **Leaves:** none clustered near base. **Capsules:** pedicel 0–5(–15) mm. **2*n*** = 14.

Flowering Mar–Jul. Desert washes, slopes; 500–1900 m; Calif., Nev.

Subspecies *gilmanii* is known from south-central to southeastern California and southern Nevada.

2. Camissonia pusilla P. H. Raven, Contr. U.S. Natl. Herb. 37: 312; fig. 53. 1969 E

Herbs glandular puberulent and villous. **Stems** erect, slender, often branched, 2–22 cm. **Leaves:** proximalmost usually clustered near base; blade sometimes with purple splotches or dots, linear, 1–3 × 0.04–0.2 cm, base narrowly cuneate, margins serrulate, apex acuminate. **Flowers** opening near sunrise; floral tube 0.8–1.6 mm, glabrous inside; sepals 1.2–2 mm, reflexed separately; petals 1.8–3.1 mm, each with 2 red dots basally; episepalous filaments 0.8–2 mm, epipetalous filaments 0.4–0.9 mm, anthers 0.3–0.4 mm, pollen with less than 5% of grains 4- or 5-pored; style 1.6–3.2 mm, stigma surrounded by anthers at anthesis. **Capsules** 18–32 × 0.6–1 mm; pedicel 0–2 mm. **Seeds** 0.7–0.8 × 0.4 mm. **2*n*** = 14.

Flowering Apr–Jun. Sandy soils on open or brushy slopes, usually with sagebrush scrub; 100–3000 m; Calif., Idaho, Nev., Oreg., Utah, Wash.

P. H. Raven (1969) determined that *Camissonia pusilla* is a self-compatible diploid and autogamous; it is closely related to *C. kernensis*, *C. parvula*, and *C. pubens*.

3. Camissonia pubens (S. Watson) P. H. Raven, Ann. Missouri Bot. Gard. 69: 995. 1983 E

Oenothera strigulosa (Fischer & C. A. Meyer) Torrey & A. Gray var. *pubens* S. Watson, Proc. Amer. Acad. Arts 8: 594. 1873; *Camissonia contorta* (Douglas) Kearney var. *pubens* (S. Watson) Kearney; *O. pubens* (S. Watson) Munz; *Sphaerostigma contortum* (Douglas) Walpers var. *pubens* (S. Watson) Small; *S. orthocarpum* A. Nelson & P. B. Kennedy; *S. pubens* (S. Watson) Rydberg

Herbs moderately to densely villous throughout, also glandular puberulent, especially distally. **Stems** erect, moderately robust, often branched at base, 5–20(–38) cm. **Leaves:** proximalmost usually clustered near base; blade narrowly lanceolate, 1.5–3(–4.5) × 0.2–0.6 cm, base narrowly cuneate, margins undulate-serrate, apex acuminate. **Flowers** opening near sunrise; floral tube 1.3–3 mm, usually pubescent on proximal ½ inside; sepals 2.2–3.8 mm, reflexed separately; petals (2.2–)3–4 mm, each with 1–several red dots basally; filaments 0.5–1 mm, anthers 0.4–0.5 mm, pollen with less than 5% of grains 4- or 5-pored; style 3.2–4.1 mm, stigma surrounded by anthers at anthesis. **Capsules** (18–)26–50 × 0.8–1.2 mm; pedicel 0–2 mm. **Seeds** 1–1.1 × 0.4 mm. **2*n*** = 28.

Flowering Apr–Jun. Sandy soil on open or brushy slopes and flats, usually sagebrush scrub or pinyon-juniper woodlands; 900–3000 m; Calif., Nev.

Camissonia pubens occurs in eastern California and west-central Nevada.

P. H. Raven (1969) determined that *Camissonia pubens* is self-compatible and autogamous. The species is closely related to *C. kernensis*, *C. parvula*, and *C. pusilla*.

4. Camissonia parvula (Nuttall ex Torrey & A. Gray) P. H. Raven, Brittonia 16: 284. 1964 E

Oenothera parvula Nuttall ex Torrey & A. Gray, Fl. N. Amer. 1: 511. 1840; *O. contorta* Douglas var. *flexuosa* (A. Nelson) Munz; *Sphaerostigma contortum* (Douglas) Walpers var. *flexuosum* A. Nelson; *S. filiforme* A. Nelson; *S. flexuosum* (A. Nelson) Rydberg; *S. parvulum* (Nuttall ex Torrey & A. Gray) Walpers

Herbs usually glabrous or densely strigillose, rarely villous (mostly proximally), also often sparsely glandular puberulent, especially distally. **Stems** erect, slender, wiry, often branched, 2–15 cm. **Leaves:** proximalmost not clustered near base; blade linear or linear-filiform,

1–3 × 0.04–0.1 cm, base attenuate, margins subentire, apex acute. **Flowers** opening near sunrise; floral tube 1.3–2 mm, glabrate; sepals 1.5–2.5 mm, reflexed separately; petals 1.5–3.6 mm, without red dots at base; filaments 0.5–1 mm, anthers 0.3–0.6 mm, pollen with less than 5% of grains 4- or 5-pored; style 1.5–3 mm, stigma surrounded by anthers at anthesis. **Capsules** 15–28 × 0.6–1 mm; pedicel 0–2 mm. **Seeds** 0.7–0.8 × 0.4 mm. **2*n*** = 28.

Flowering Apr–Jun. Sandy soils, usually with sagebrush scrub; 100–2700 m; Ariz., Calif., Colo., Idaho, Mont., Nev., N.Mex., Oreg., Utah, Wash., Wyo.

P. H. Raven (1969) determined that *Camissonia parvula* is a self-compatible tetraploid and autogamous. The species is closely related to *C. kernensis* and *C. pubens*.

5. Camissonia campestris (Greene) P. H. Raven, Brittonia 16: 284. 1964 [E] [F]

Oenothera campestris Greene, Fl. Francisc., 216. 1891; *O. dentata* Cavanilles var. *campestris* (Greene) Jepson; *Sphaerostigma campestre* (Greene) Small; *S. dentatum* (Cavanilles) Walpers subsp. *campestre* (Greene) Johansen

Herbs glabrous, villous, strigillose, or glandular puberulent, especially distally, sometimes glabrous distally. **Stems** erect or decumbent, slender, wiry, usually well-branched, 5–25(–50) cm. **Leaves:** proximalmost not clustered near base; blade linear to narrowly elliptic or narrowly oblanceolate, 0.5–2.5(–3) × 0.1–0.15 (–0.5) cm, base attenuate, margins sparsely serrulate to coarsely serrate, apex acuminate. **Flowers** opening near sunrise; floral tube 1.5–5.5 mm, ± densely villous on proximal ½ inside; sepals 3.5–8(–12) mm, reflexed in pairs; petals (3.5–)5–15.5 mm, each usually with 1 or 2 red dots basally; episepalous filaments (1.4–) 2.1–5.5 mm, epipetalous filaments (0.7–)1.2–3.2 mm, anthers 1–2.4 mm, pollen with less than 5% of grains 4- or 5-pored; style (3.2–)4–12(–15) mm, stigma well exserted beyond anthers at anthesis. **Capsules** 20–43 × 0.7–1.5(–2) mm; subsessile. **Seeds** 0.8–1.6 × 0.4–0.6 mm.

Subspecies 2 (2 in the flora): California.

P. H. Raven (1969) determined that *Camissonia campestris* is self-incompatible.

1. Stems usually erect; leaf blade margins sparsely serrulate . . . 5a. *Camissonia campestris* subsp. *campestris*
1. Stems usually decumbent; leaf blade margins coarsely serrate . . . . 5b. *Camissonia campestris* subsp. *obispoensis*

5a. Camissonia campestris (Greene) P. H. Raven subsp. **campestris** [E] [F]

Oenothera campestris Greene subsp. *parishii* (Abrams) Munz; *O. dentata* Cavanilles var. *johnstonii* Munz; *O. dentata* var. *parishii* (Abrams) Munz; *Sphaerostigma campestre* (Greene) Small var. *helianthemiflorum* (H. Léveillé) A. Nelson; *S. campestre* var. *parishii* Abrams

Stems usually erect. **Leaves:** blade linear to narrowly elliptic or narrowly oblanceolate, margins sparsely serrulate. **2*n*** = 14.

Flowering May–Jun. Open, sandy flats, desert scrub, non-coastal grasslands; 0–2000 m; Calif.

Subspecies *campestris* is known from central and southern California; it sometimes occurs sympatrically with the hexaploid *Camissonia contorta*, forming usually sterile hybrids.

5b. Camissonia campestris (Greene) P. H. Raven subsp. **obispoensis** P. H. Raven, Contr. U.S. Natl. Herb. 37: 325. 1969 [E]

Stems usually decumbent. **Leaves:** blade narrowly elliptic, margins coarsely serrate. **2*n*** = 14.

Flowering May–Jun. Marine sand deposits in openings in chaparral and oak woodlands; 100–500 m; Calif.

Subspecies *obispoensis* is known from the coastal and somewhat inland areas of central California from southern Monterey County to northwestern Santa Barbara County; the subspecies intergrades with subsp. *campestris* and sometimes occurs sympatrically with *C. contorta* and *C. strigulosa*, forming usually sterile hybrids. Plants of subsp. *obispoensis* in the western part of its range have narrower leaves with smaller serrations on the margin and are less pubescent than those in the eastern part of the range.

6. Camissonia sierrae P. H. Raven, Contr. U.S. Natl. Herb. 37: 326, figs. 58, 59. 1969 [E]

Herbs glabrous, villous, and glandular puberulent distally. **Stems** erect or ascending, slender, wiry, usually many-branched, 5–15 cm. **Leaves:** proximalmost not clustered near base; blade usually lanceolate to narrowly ovate, sometimes elliptic, 0.5–1.8 × 0.2–0.5 cm, base rounded, margins inconspicuously serrulate or with 1–several small teeth, apex acute. **Flowers** opening near sunrise; floral tube 1–2.2 mm, villous on proximal ½ inside; sepals 1.2–4.2 mm, reflexed in pairs; petals 2.2–7 mm, each usually with 0 or 2 red dots basally; episepalous filaments 2.4–3.2 mm, epipetalous filaments 1.2–2 mm, anthers 0.6–1.2 mm, pollen with less than 5% of grains 4- or 5-pored; style 2.8–7 mm, stigma 0.6–0.8 mm diam., surround by, or slightly exserted beyond, anthers at anthesis. **Capsules** 20–30 × 0.5–0.7 mm; subsessile. **Seeds** 0.8–1.6 × 0.4–0.6 mm.

Subspecies 2 (2 in the flora): California.

P. H. Raven (1969) determined that *Camissonia sierrae* is self-compatible and outcrossing or autogamous; it is closely related to *C. campestris*.

1. Sepals 3–4.2 mm; base of petals with 2 red dots; styles 3–7 mm . . . . 6a. *Camissonia sierrae* subsp. *sierrae*
1. Sepals 1.2–3 mm; base of petals without red dots; styles 2.8–5 mm . . . . 6b. *Camissonia sierrae* subsp. *alticola*

6a. Camissonia sierrae P. H. Raven subsp. **sierrae** E

Flowers: floral tube 1.3–2.2 mm; sepals 3–4.2 mm; petals 4–7 mm, with 2 red dots basally; style 3–7 mm, stigma surrounded by or slightly elevated beyond anthers at anthesis. $2n = 14$.

Flowering May–Jul. Granite outcrops, *Pinus ponderosa* forests, upper limits of *Pinus sabiniana-Quercus douglasii* forests; 500–1200 m; Calif.

Subspecies *sierrae* is known from the Sierra Nevada foothills of Madera and central Mariposa counties.

6b. Camissonia sierrae P. H. Raven subsp. **alticola** P. H. Raven, Contr. U.S. Natl. Herb. 37: 329, fig. 59. 1969 E

Flowers: floral tube 1–2.2 mm; sepals 1.2–3 mm; petals 2.2–4 mm, without red dots at base; style 2.8–5 mm, stigma surrounded by anthers at anthesis. $2n = 14$.

Flowering May–Jul. Shallow soil on granite outcrops, *Pinus ponderosa* forests; 1800–2400 m; Calif.

Subspecies *alticola* is known from northeastern Fresno County and, possibly, the Merced River region in Mariposa County.

7. Camissonia strigulosa (Fischer & C. A. Meyer) P. H. Raven, Contr. U.S. Natl. Herb. 37: 333. 1969

Sphaerostigma strigulosum Fischer & C. A. Meyer, Index Seminum (St. Petersburg) 2: 50. 1836; *Oenothera contorta* Douglas var. *epilobioides* (Greene) Munz; *O. contorta* var. *strigulosa* (Fischer & C. A. Meyer) Munz; *O. strigulosa* (Fischer & C. A. Meyer) Torrey & A. Gray; *O. strigulosa* var. *epilobioides* Greene

Herbs densely strigillose, often also glandular puberulent, especially distally, or sometimes glandular puberulent only and then glabrate, sometimes also villous near base. **Stems** usually erect, sometimes decumbent, slender, wiry, usually many-branched, to 50 cm. **Leaves:** proximalmost not clustered near base; blade linear to very narrowly elliptic, 0.8–3.5 × 0.1–0.3 cm, base cuneate or attenuate, margins sparsely serrulate, apex acute. **Flowers** opening near sunrise; floral tube 1.6–2.7 mm, usually moderately to sparsely pubescent inside on proximal ½, rarely glabrous; sepals 1.6–4 mm, reflexed in pairs; petals 2.1–4.2(–4.5) mm, each ± with 2 red dots basally; episepalous filaments 0.9–2(–2.2) mm, epipetalous filaments 0.5–1.3 mm, anthers 0.3–0.6 mm, pollen with usually less than 10% of grains 4-pored; style 2.3–4.8 mm, stigma surrounded by anthers at anthesis. **Capsules** 15–45 × 0.8–1.3 mm; subsessile. **Seeds** 0.6–0.8 × 0.3–0.4 mm. $2n = 28$.

Flowering Mar–Aug. Open, sandy soils of dunes, grasslands, desert scrub; 0–2100 m; Calif.; Mexico (Baja California).

Camissonia strigulosa is known in the flora area from central to southern California, west of the Sierra Nevada.

P. H. Raven (1969) determined that *Camissonia strigulosa* is a self-compatible tetraploid and autogamous; it is closely related to *C. benitensis*, *C. contorta*, *C. integrifolia*, and *C. lacustris*.

8. Camissonia benitensis P. H. Raven, Contr. U.S. Natl. Herb. 37: 332, fig. 60. 1969 • San Benito suncup C E

Herbs villous and also glandular puberulent distally. **Stems** erect or decumbent, slender, wiry, usually branched, 3–20 cm. **Leaves:** proximalmost not clustered near base, green or slightly bluish green; blade very narrowly elliptic, 0.7–2 × 0.1–0.3 cm, base cuneate or attenuate, margins sparsely serrulate, apex acute. **Flowers** opening near sunrise; floral tube ca. 1.2 mm, moderately to very sparsely pubescent inside on proximal ½; sepals 3.2–3.5 mm, reflexed in pairs; petals 3.5–4 mm, each ± with 2 red dots basally; episepalous filaments 2 mm, epipetalous filaments 1.2 mm, anthers 0.3–0.6 mm, pollen with usually less than 10% of grains 4-pored; style 2.1–2.5 mm, stigma surrounded by anthers at anthesis. **Capsules** 15–45 × 0.8–1.3 mm; subsessile. **Seeds** 0.6–0.8 × 0.3–0.4 mm. $2n = 28$.

Flowering Apr–Jul. Sandy or gravelly serpentine soil on alluvial terraces and sandy or gravelly serpentine soil in upland areas in geologic interfaces between serpentine and non-serpentine rock types in *Quercus douglasii* and *Juniperus californicus* woodlands; of conservation concern; 600–1400 m; Calif.

Camissonia benitensis is known from New Idria and nearby serpentine areas, lower Clear Creek drainage and San Carlos Creek, San Benito County, and is reported from adjacent Fresno and Monterey counties. More than 50,000 individuals are known, but their habitat is threatened by off-road vehicles.

P. H. Raven (1969) determined that *Camissonia benitensis* is a self-compatible tetraploid and autogamous, also stating that *C. benitensis* is most likely closely related to *C. strigulosa*.

9. Camissonia lacustris P. H. Raven, Contr. U.S. Natl. Herb. 37: 329, fig. 61. 1969 [E]

Herbs densely villous, usually also glandular puberulent distally. **Stems** usually erect, sometimes decumbent, slender, wiry, usually many-branched, to 50 cm. **Leaves:** proximalmost not clustered near base; blade linear to very narrowly elliptic, 0.8–3.5 × 0.1–0.3 cm, base cuneate or attenuate, margins sparsely serrulate, apex acute. **Flowers** opening near sunrise; floral tube 1.6–2.7 mm, usually moderately to very sparsely pubescent inside on proximal ½, rarely glabrous; sepals (3–)3.8–5.5 mm, reflexed in pairs; petals (4–)4.5–7 mm, each with 2 red dots basally; episepalous filaments 2.5–3.5 mm, epipetalous filaments 1.7–2.5 mm, anthers 0.8–1.3 mm, pollen with usually less than 10% of grains 4-pored; style (3.5–)4–7 mm, stigma surrounded by anthers at anthesis. **Capsules** 15–45 × 0.8–1.3 mm; subsessile. **Seeds** 0.6–0.8 × 0.3–0.4 mm. $2n = 28$.

Flowering Mar–Aug. Open grasslands; 200–1600 m; Calif.

Camissonia lacustris is known from two disjunct areas: serpentine soil in Lake County and the Sierra Nevada foothills from El Dorado to Fresno counties.

P. H. Raven (1969) determined that *Camissonia lacustris* is a self-compatible tetraploid and autogamous; it is closely related to *C. strigulosa*.

10. Camissonia integrifolia P. H. Raven, Contr. U.S. Natl. Herb. 37: 344, fig. 62. 1969

Herbs sparsely strigillose or glabrate, more densely so distally. **Stems** usually erect, sometimes decumbent, slender, wiry, usually many-branched, to 30 cm. **Leaves:** proximalmost not clustered near base; blade linear, 1–3 × 0.1–0.3 cm, base cuneate or attenuate, margins usually entire, rarely with 1 or 2 small teeth, apex acute. **Flowers** opening near sunrise; floral tube 1.5–2.5 mm, moderately to sparsely pubescent inside on proximal ½; sepals 1.6–4 mm, reflexed in pairs; petals 2–4.2 mm, each ± with 2 red dots basally; episepalous filaments 0.9–2.1 mm, epipetalous filaments 0.5–1.4 mm, anthers 0.3–0.6 mm, pollen with usually less than 10% of grains 4-pored; style 2.3–4.8 mm, stigma surrounded by anthers at anthesis. **Capsules** 45–60 × 0.8–1.3 mm; subsessile. **Seeds** 1–2 × 0.4–0.5 mm. $2n = 28$.

Flowering Apr–May. Sagebrush slopes; 700–1000 m; Calif.; Mexico (Baja California).

Camissonia integrifolia is known in the flora area from central to southern California, west of the Sierra Nevada.

P. H. Raven (1969) determined that *Camissonia integrifolia* is a self-compatible tetraploid and autogamous. The species forms sterile natural hybrids with *C. strigulosa*, to which it is presumably most closely related.

11. Camissonia contorta (Douglas) Kearney, Trans. New York Acad. Sci. 14: 37. 1895 [E] [W]

Oenothera contorta Douglas in W. J. Hooker, Fl. Bor.-Amer. 1: 214. 1832; *Sphaerostigma contortum* (Douglas) Walpers

Herbs usually villous throughout, often also glandular puberulent distally, or, rarely, entirely strigillose and glandular puberulent throughout. **Stems** usually erect, sometimes decumbent, slender, wiry, usually many-branched, to 50 cm. **Leaves:** proximalmost not clustered near base, usually bluish green; blade linear to narrowly elliptic, 1–3.5 × 0.1–0.5 cm, base cuneate or attenuate, margins sparsely serrulate, apex acute. **Flowers** opening near sunrise; floral tube 1.6–2.7 mm, usually moderately to very sparsely pubescent inside on proximal ½, rarely glabrous; sepals 1.6–4 mm, reflexed in pairs; petals 2.5–5 mm, each ± with 2 red dots basally; episepalous filaments 1–2.6 mm, epipetalous filaments 0.5–1.5 mm, anthers 0.3–0.6 mm, pollen with usually more than 30% of grains 4- or 5-pored; style 2.5–5.1 mm, stigma surrounded by anthers at anthesis. **Capsules** 15–45 × 0.8–1.3 mm; subsessile. **Seeds** 0.7–0.9 × 0.3–0.4 mm. $2n = 42$.

Flowering Mar–Jul. Sandy soil, slopes, flats, disturbed areas, grasslands, chaparral, pinyon-juniper woodlands; 0–2300(–2700) m; B.C.; Calif., Idaho, Nev., Oreg., Wash.

Camissonia contorta is known from south Vancouver Island in British Columbia to south San Joaquin Valley and bordering foothills in Kern County, California, Ada and Adams counties in Idaho, western Nevada, east-central and southwest Oregon, and in Washington from San Juan and Whidbey islands, and Klickitat and Walla Walla counties.

P. H. Raven (1969) determined that *Camissonia contorta* is a self-compatible hexaploid and autogamous. The species probably arose, at least in part, following hybridization between the diploid *C. campestris* subsp. *campestris* and the tetraploid *C. strigulosa*, but some populations referred to as this species may also have originated following the functioning of an unreduced gamete in a tetraploid plant.

Although W. L. Wagner and P. C. Hoch (2009) came to a different conclusion for the valid publication of *Camissonia contorta*, the phrase "*Camissonia contorta pubens*" used by Kearney should be accepted as the telescoped representation of two different names: varietal and specific (K. N. Gandhi, pers. comm.).

11. EREMOTHERA (P. H. Raven) W. L. Wagner & Hoch, Syst. Bot. Monogr. 83: 125. 2007 • [Greek *eremia*, desert, and *thera* from genus *Oenothera*, probably alluding to habitat and likeness]

Warren L. Wagner

Camissonia Link sect. *Eremothera* P. H. Raven, Brittonia 16: 285. 1964; *Oenothera* Linnaeus sect. *Eremothera* (P. H. Raven) Munz

Herbs, annual, caulescent; with a taproot. **Stems** usually erect, sometimes ascending, usually well-branched from base, sometimes also distally, with white or reddish green exfoliating epidermis. **Leaves** cauline, proximal ones often clustered near base, alternate; stipules absent; petiolate, often subsessile distally; blade margins denticulate, crenate-dentate, serrulate, sinuate-toothed, or entire. **Inflorescences** spikes, erect or nodding at anthesis, or flowers also in proximal leaf axils in some taxa. **Flowers** bisexual, actinomorphic, buds erect; floral tube deciduous (with sepals, petals, and stamens) after anthesis, with basal nectary; sepals 4, reflexed singly or in pairs; petals 4, usually white, rarely red or tinged red, without spots, fading pink or red; stamens 8 in 2 unequal series, episepalous ones rarely abortive (*E. minor*), anthers versatile, pollen shed singly; ovary 4-locular, without apical projection, style villous near base, strigillose, or glabrous, stigma entire, subglobose, surface unknown, probably wet and non-papillate. **Fruit** a capsule, straight or much contorted, narrowly cylindrical throughout or thickened proximally, terete or 4-angled, regularly but tardily loculicidal; sessile. **Seeds** numerous, in 1 row per locule, usually monomorphic and narrowly obovoid to oblanceoloid, sometimes dimorphic, with seeds near base of capsule sharply angular and truncate-ellipsoid, finely reticulate, or seeds near base of capsule coarsely papillose. $x = 7$.

Species 7 (7 in the flora): w, sc United States, nw Mexico.

Species of *Eremothera* are found mainly in the interior deserts and bordering areas of the western United States.

R. A. Levin et al. (2004) found strong molecular support for paraphyly in the broadly delimited *Camissonia* of P. H. Raven (1969). There was some support for a clade of *Camissonia* and *Eremothera* and another clade of *Camissoniopsis*, *Neoholmgrenia*, and *Tetrapteron* (Levin et al.), but without morphological features linking the members of these two clades. The monophyletic subclades of these two clades were recognized as genera by W. L. Wagner et al. (2007) whereas they were all treated by Raven as clearly distinguishable sections. Raven recognized four distinct groups within *Eremothera* (as *Camissonia* sect. *Eremothera*): *E. refracta* and its autogamous derivative, *E. chamaenerioides*; the very diverse *E. boothii* (with six subspecies) and two rare autogamous derivatives, *E. gouldii* and *E. pygmaea*; the local clay endemic *E. nevadensis*; and the widespread autogamous and often cleistogamous *E. minor*. Levin et al. included one species from each of these four groups in their molecular analyses and found strong support for *Eremothera* as circumscribed by Raven and maintained by Wagner et al. *Eremothera* is well defined by white petals that open in the evening and an entire, subglobose stigma; some species are visited by moths at anthesis and by bees the following morning (Raven). Reproductive features include: self-incompatible (*E. boothii*, *E. refracta*, and, possibly, *E. nevadensis*) or self-compatible; flowers vespertine; outcrossing and pollinated in the evening by small moths and the following morning by bees, in *E. boothii* subsp. *decorticans* by large oligolectic andrenid bees (E. G. Linsley et al. 1963, 1964, 1973), or autogamous, rarely cleistogamous (Raven).

1. Capsules narrowly cylindrical; stems flowering distally.
   2. Sepals 4–6 mm; petals 3.5–10 mm; floral tube 4–7 mm; stigmas exserted beyond anthers at anthesis . . . . 1. *Eremothera refracta*
   2. Sepals 1.5–2.5 mm; petals 1.8–3 mm; floral tube 1.5–3 mm; stigmas surrounded by anthers at anthesis . . . . 2. *Eremothera chamaenerioides*
1. Capsules cylindrical, proximally thickened; stems flowering distally and proximally.
   3. Stigma exserted beyond anthers at anthesis; sepals (2.7–)3.2–8 mm; petals (3–)3.5–9 mm.
      4. Primary stem short, lateral stems decumbent; leaves usually in a tuft distally; inflorescences erect; petals 3–5 mm; style 6–7 mm, glabrous. . . . . 3. *Eremothera nevadensis*
      4. Stems usually well branched at base and distally; leaves well distributed or basally clustered; inflorescences nodding; petals (3–)3.5–9 mm; style (6.5–)8.2–13.5(–15) mm, proximally villous. . . . . 4. *Eremothera boothii*
   3. Stigma surrounded by anthers at anthesis; sepals 0.8–2.6 mm; petals 0.8–2.5 mm.
      5. Stems flowering from base; herbs densely strigillose, often also glandular puberulent in inflorescences . . . . 7. *Eremothera minor*
      5. Stems usually not flowering near base; herbs glandular puberulent, usually also villous.
         6. Seeds dimorphic, basal ones coarsely papillose; capsules 8–20 mm; leaves 1.5–6.5 × 0.5–2 cm . . . . 5. *Eremothera pygmaea*
         6. Seeds monomorphic, all appearing smooth, finely reticulate; capsules 8–12 mm; leaves 0.5–3.5 × 0.5–1 cm . . . . 6. *Eremothera gouldii*

1. Eremothera refracta (S. Watson) W. L. Wagner & Hoch, Syst. Bot. Monogr. 83: 210. 2007 [E]

Oenothera refracta S. Watson, Proc. Amer. Acad. Arts 17: 373. 1882; *Camissonia refracta* (S. Watson) P. H. Raven; *O. deserti* M. E. Jones; *Sphaerostigma deserti* (M. E. Jones) A. Heller; *S. refractum* (S. Watson) Small

Herbs sparsely strigillose, sometimes also glandular puberulent, especially in inflorescence. **Stems** usually well branched from base and distally, 6–45 cm, flowering only distally. **Leaves** cauline, with lower ones clustered near base and these often withered by flowering, 2–6(–8) × 0.1–0.8 cm; petiole 0–2 cm; blade narrowly lanceolate to narrowly elliptic-lanceolate or narrowly oblanceolate, those distally on stems usually linear to linear-lanceolate, margins usually sparsely and weakly denticulate, sometimes sinuate-toothed. **Inflorescences** nodding. **Flowers** opening at sunset; floral tube 4–7 mm, villous in proximal ½ inside; sepals 4–6 mm; petals white, fading pinkish, 3.5–10 mm; episepalous filaments 2–4.5 mm, epipetalous filaments slightly shorter, anthers 1.5–2.5 mm; style 9–13 mm, villous proximally, stigma 1–1.5 mm diam., exserted beyond anthers at anthesis. **Capsules** narrowly cylindrical throughout, spreading or reflexed, straight to ± contorted, terete, 20–50 × 0.7–1 mm, regularly but tardily dehiscent. **Seeds** monomorphic, gray, 0.9–1.5 × 0.4–0.5 mm, finely reticulate. $2n = 14$.

Flowering Mar–May. Sandy desert slopes and flats; -30–1700 m; Ariz., Calif., Nev., N.Mex., Utah.

Eremothera refracta is known from Esmeralda, southern Nye, and Clark counties in Nevada, Washington County in Utah, south throughout the Mojave and Colorado deserts of Inyo, San Bernardino, Imperial, central and eastern Riverside, and eastern Kern and San Diego counties in California, Mohave, Yuma, and western Pima counties in Arizona, and a single collection well east of normal range has been seen from Hidalgo County in New Mexico (east of Lordsburg, *Jones* in 1930, POM). P. H. Raven (1969) determined *E. refracta* to be self-incompatible.

2. **Eremothera chamaenerioides** (A. Gray) W. L. Wagner & Hoch, Syst. Bot. Monogr. 83: 209. 2007

Oenothera chamaenerioides A. Gray, Smithsonian Contr. Knowl. 5(6): 58. 1853; *Camissonia chamaenerioides* (A. Gray) P. H. Raven; *O. erythra* (Davidson) J. F. Macbride; *Sphaerostigma chamaenerioides* (A. Gray) Small; *S. erythrum* Davidson

Herbs glandular puberulent and sparsely strigillose distally, especially in inflorescence. **Stems** usually well branched from base, 8–50 cm, flowering only distally. **Leaves** cauline, with lower ones clustered near base, (0.7–)2–8(–10) × 0.1–2.5 cm; petiole 0.1–3.5 cm; blade very narrowly elliptic to narrowly elliptic, margins entire or sparsely denticulate. **Inflorescences** nodding. **Flowers** opening at sunset; floral tube 1.5–3 mm, villous in proximal ½ inside; sepals 1.5–2.5 mm; petals white, fading pinkish, 1.8–3 mm; episepalous filaments 0.7–1.5 mm, epipetalous filaments slightly shorter, anthers 0.5–1.1 mm; style 2.3–4.5 mm, villous proximally, stigma 0.7–1 mm diam., surrounded by anthers at anthesis. **Capsules** narrowly cylindrical throughout, spreading, straight, terete, 35–60 × 0.7–1 mm, regularly but tardily dehiscent. **Seeds** monomorphic, gray, 0.9–1 × 0.3–0.4 mm, finely reticulate. $2n = 14$.

Flowering (Jan–)Feb–Jun. Sandy desert slopes and flats; -50–1700 m; Ariz., Calif., Nev., N.Mex., Tex., Utah; Mexico (Baja California, Sonora).

Eremothera chamaenerioides occurs in sub-Mogollon Arizona, southeastern California, southern Nevada, southern New Mexico, trans-Pecos Texas, and Kane, Millard, Tooele, and Washington counties, Utah. P. H. Raven (1969) determined *Eremothera chamaenerioides* to be self-compatible and autogamous.

3. **Eremothera nevadensis** (Kellogg) W. L. Wagner & Hoch, Syst. Bot. Monogr. 83: 210. 2007 [E]

Oenothera nevadensis Kellogg, Proc. Calif. Acad. Sci. 2: 224, fig. 70. 1863; *Camissonia nevadensis* (Kellogg) P. H. Raven; *O. gauriflora* Torrey & A. Gray var. *caput-medusae* H. Léveillé; *O. gauriflora* var. *vermiculata* M. E. Jones; *Sphaerostigma nevadense* (Kellogg) A. Heller; *S. tortuosum* A. Nelson

Herbs glabrate to strigillose in distal and younger parts and capsules. **Stems** branched from base, primary stem short, lateral stems decumbent, usually with leaves mostly in a tuft toward apex, 1–5(–18) cm, flowering proximally and distally. **Leaves** mostly cauline, clustered toward ends of branches, 1–4.5 × 0.2–0.8 cm; petiole 1–3 cm; blade oblanceolate or narrowly oblanceolate, margins entire. **Inflorescences** erect. **Flowers** opening at sunset; floral tube 2.2–3.5 mm, glabrous inside; sepals 3.2–4 mm; petals white, fading pinkish, 3–5 mm; episepalous filaments 4.5–4.8 mm, epipetalous filaments 3–4 mm, anthers 0.4–1.5 mm; style 6–7 mm, glabrous, stigma 0.5–0.8 mm diam., exserted beyond anthers at anthesis. **Capsules** cylindrical and thickened proximally, spreading and highly contorted, 4-angled, 8–14 × 1–2 mm, regularly but tardily dehiscent. **Seeds** monomorphic, gray, 1.2–1.5 × 0.3–0.4 mm, finely reticulate. $2n = 14$.

Flowering Apr–May(–early Jun). Local and colonial on clay, sandy, or gravelly soils, often vernally wet sites, somewhat tolerant of alkali soils; 1200–1700 m; Nev.

Eremothera nevadensis is known only from west-central Nevada in western Churchill, Douglas, northern Lyon, Ormsby, Pershing, Storey, and southern Washoe counties. P. H. Raven (1969) presumed *Eremothera nevadensis* to be self-compatible.

4. **Eremothera boothii** (Douglas) W. L. Wagner & Hoch, Syst. Bot. Monogr. 83: 209. 2007 [F]

Oenothera boothii Douglas in W. J. Hooker, Fl. Bor.-Amer. 1: 213. 1832; *Camissonia boothii* (Douglas) P. H. Raven; *Sphaerostigma boothii* (Douglas) Walpers

Herbs slender or stout, glabrate to strigillose or villous, and/or glandular puberulent, especially in inflorescence. **Stems** usually well branched at base and distally, 3–65 cm, usually flowering only distally. **Leaves** cauline, sometimes with lower ones clustered near base, these often withered by flowering, 1–11(–13) × 0.2–2.2(–3) cm; petiole 0–6 cm; blade very narrowly elliptic to narrowly ovate, lanceolate, or oblanceolate, margins denticulate, serrulate, sinuate-toothed, or subentire. **Inflorescences** nodding. **Flowers** opening at sunset; floral tube (2–)3–8 mm, villous in proximal ½ inside; sepals (2.7–)4–8 mm; petals usually white fading pink, rarely red and fading red, (3–)3.5–9 mm; episepalous filaments (1.5–)2–5.8 mm, epipetalous filaments slightly shorter, anthers (1–)1.8–2.3 mm; style (6.5–)8.2–13.5(–15) mm, villous near base, stigma 1.2–2 mm diam., exserted beyond anthers at anthesis. **Capsules** cylindrical and thickened proximally, spreading to curved downward, contorted to straight, terete or 4-angled, 8–35 × 0.9–3.8 mm, held on dried plants and regularly but tardily dehiscent. **Seeds** usually dimorphic, rarely monomorphic, those with relatively smooth surfaces light brown, 1.4–2.1 × 0.5–0.7 mm, those of lower portion dark brown, 1.4–2.1 × 0.6–0.9 mm, coarsely papillose. **2*n*** = 14.

Subspecies 6 (6 in the flora): w United States, nw Mexico.

P. H. Raven (1969) included within *Eremothera boothii* a complex of intergrading entities that have been variously treated, ranging from three species with infraspecific taxa to one species with infraspecific taxa. Raven determined *E. boothii* to be self-incompatible.

1. Cluster of basal leaves prominent at time of flowering; plants blooming February through June (or August), strigillose and/or glandular puberulent.
   2. Capsules 2–3.8 mm diam. near base, thickened and indurate along angles, curved outward but not downward; inflorescences very dense . . . . 4c. *Eremothera boothii* subsp. *condensata*
   2. Capsules 1–2.3 mm diam. near base, not thickened and indurate, curved outward or downward; inflorescences ± open.
      3. Capsules 1.7–2.3 mm diam. near base, curved outward . . . . 4d. *Eremothera boothii* subsp. *decorticans*
      3. Capsules 1–1.6 mm diam. near base, apex often curved downward . . . . 4e. *Eremothera boothii* subsp. *desertorum*
1. Cluster of basal leaves rarely prominent at time of flowering, leaves evenly distributed; plants blooming May through August (or September), villous and/or glandular puberulent, rarely (subsp. *alyssoides*) densely strigillose.
   4. Herbs usually strigillose, often densely so, rarely villous and/or glandular puberulent; plants flowering May through June (or August) . . . 4b. *Eremothera boothii* subsp. *alyssoides*
   4. Herbs villous and glandular puberulent; plants flowering (May to) June through August (or September).

[5. Shifted to left margin—Ed.]

5. Herbs villous, also glandular puberulent, especially in inflorescences; stems usually 15–60 cm; leaf blades narrowly lanceolate, narrowly ovate, or ovate to elliptic, 0.6–2.2 cm wide, margins coarsely serrulate to sinuate-toothed . . . . 4a. *Eremothera boothii* subsp. *boothii*
5. Herbs densely villous, also densely glandular puberulent, especially in inflorescences; stems 5–20 cm; leaf blades narrowly elliptic to narrowly lanceolate or lanceolate, sometimes oblanceolate proximally, 0.3–1.5 cm wide, margins sparsely serrulate to sometimes sinuate-toothed . . . . 4f. *Eremothera boothii* subsp. *intermedia*

4a. Eremothera boothii (Douglas) W. L. Wagner & Hoch subsp. **boothii** [E]

Sphaerostigma lemmonii A. Nelson; *S. senex* A. Nelson

Herbs villous and glandular puberulent, especially in inflorescence. **Stems** usually 15–60 cm. **Leaves** not especially clustered toward base, mostly evenly distributed, 1–6 × 0.6–2.2 cm; petiole 0.1–2.6 cm; blade narrowly lanceolate, narrowly ovate, or ovate to elliptic, margins coarsely serrulate to sinuate-toothed. **Inflorescences** ± open. **Flowers:** floral tube 4–8 mm; petals white, 3.5–9 mm. **Capsules** slightly curved outward to contorted, 1.3–1.9 mm diam. near base. **Seeds** dimorphic.

Flowering Jun–Aug. Sandy flats, steep loose slopes, sagebrush shrublands, Joshua tree and pinyon-juniper woodlands; 900–2400 m; Ariz., Calif., Idaho, Nev., Oreg., Wash.

Subspecies *boothii* is known from northwest Arizona; east-central California and west Nevada; south Idaho, east Oregon, south and east Washington.

Subspecies *boothii* intergrades broadly with subspp. *alyssoides* and *intermedia* in the Nevada portion of its range.

4b. Eremothera boothii (Douglas) W. L. Wagner & Hoch subsp. **alyssoides** (Hooker & Arnott) W. L. Wagner & Hoch, Syst. Bot. Monogr. 83: 209. 2007 [E] [F]

Oenothera alyssoides Hooker & Arnott, Bot. Beechey Voy., 340. 1839; *Camissonia boothii* (Douglas) P. H. Raven subsp. *alyssoides* (Hooker & Arnott) P. H. Raven; *C. boothii* var. *alyssoides* (Hooker & Arnott) N. H. Holmgren & P. K. Holmgren; *C. boothii* var. *villosa* (S. Watson) Cronquist; *Holostigma alyssoides* (Hooker & Arnott) Hooker; *O. alyssoides* var. *villosa* S. Watson;

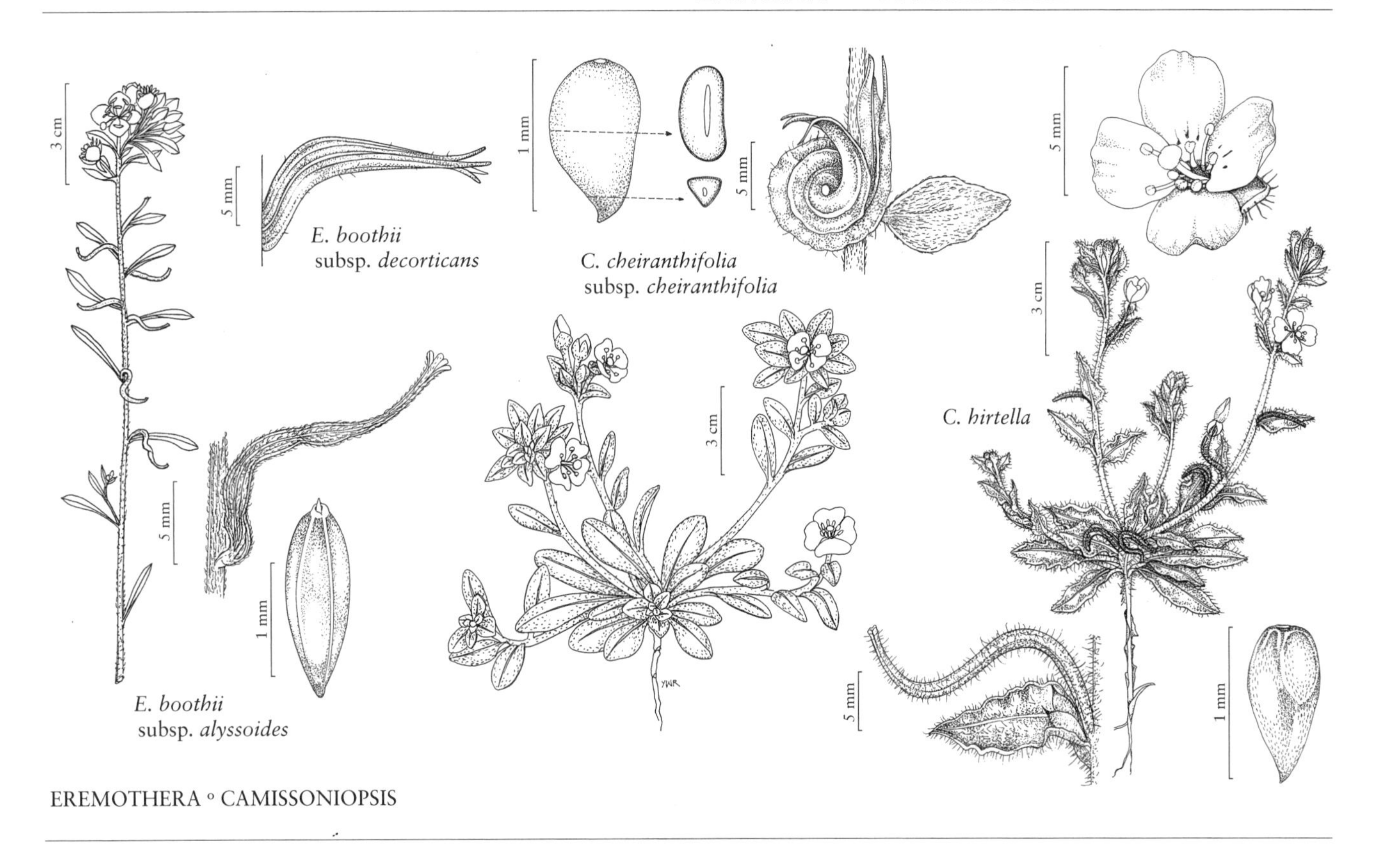

EREMOTHERA ◦ CAMISSONIOPSIS

O. boothii subsp. *alyssoides* (Hooker & Arnott) Munz; *O. gauriflora* Torrey & A. Gray var. *hitchcockii* H. Léveillé; *O. utahensis* (Small) Garrett; *Sphaerostigma alyssoides* (Hooker & Arnott) Walpers; *S. hitchcockii* (H. Léveillé) A. Nelson; *S. implexum* A. Nelson; *S. macrophyllum* Rydberg; *S. utahense* Small

Herbs usually strigillose, often densely so, especially in inflorescence, rarely villous and/or glandular puberulent. **Stems** 3–35 cm. **Leaves** not especially clustered toward base, mostly evenly distributed, 1–5.5(–7.5) × 0.3–1.6 cm; petiole 0–6 cm; blade very narrowly elliptic to narrowly ovate or ovate, margins sparsely denticulate. **Inflorescences** ± open. **Flowers:** floral tube 3–8 mm; petals white, 4–5 mm. **Capsules** usually very contorted, 1–1.5 mm diam. near base. **Seeds** monomorphic.

Flowering May–Jun(–Aug). Sandy slopes and flats, sagebrush, shadscale, rabbitbrush shrublands or pinyon-juniper woodlands; 600–2000 m; Calif., Idaho, Nev., Oreg., Utah.

Subspecies *alyssoides* intergrades broadly with subsp. *intermedia* in Nevada and is very similar to subsp. *desertorum*. *Sphaerostigma alyssoides* var. *macrophyllum* Small is an illegitimate name that applies here.

4c. Eremothera boothii (Douglas) W. L. Wagner & Hoch subsp. **condensata** (Munz) W. L. Wagner & Hoch, Syst. Bot. Monogr. 83: 209. 2007

Oenothera decorticans (Hooker & Arnott) Greene var. *condensata* Munz, Bot. Gaz. 85: 247. 1928; *Camissonia boothii* (Douglas) P. H. Raven subsp. *condensata* (Munz) P. H. Raven; *C. boothii* var. *condensata* (Munz) Cronquist; *O. boothii* Douglas subsp. *condensata* (Munz) Munz

Herbs glabrate to strigillose, sometimes also glandular puberulent in inflorescence. **Stems** 5–20(–30) cm. **Leaves** primarily clustered toward base, 2.5–10(–13) × 0.3–2(–2.5) cm; petiole 0–6 cm; blade usually lanceolate to oblanceolate, margins subentire to denticulate. **Inflorescences** very dense. **Flowers:** floral tube 3.5–8 mm; petals white, 4–5 mm. **Capsules** curved outward, not downward, sometimes ± contorted, tapering abruptly distally, 4-angled, much thickened along angles, 2–3.8 mm diam. near base. **Seeds** dimorphic.

Flowering Feb–May. Sandy slopes and washes, desert scrublands; 80–1300 m; Ariz., Calif., Nev., Utah; Mexico (Baja California, Sonora).

Subspecies *condensata* intergrades with subsp. *desertorum*.

4d. Eremothera boothii (Douglas) W. L. Wagner & Hoch subsp. **decorticans** (Hooker & Arnott) W. L. Wagner & Hoch, Syst. Bot. Monogr. 83: 209. 2007 [E] [F]

Gaura decorticans Hooker & Arnott, Bot. Beechey Voy., 343. 1839; *Camissonia boothii* (Douglas) P. H. Raven subsp. *decorticans* (Hooker & Arnott) P. H. Raven; *C. boothii* var. *decorticans* (Hooker & Arnott) Cronquist; *C. boothii* subsp. *rutila* (Davidson) Munz; *Oenothera alyssoides* Hooker & Arnott var. *decorticans* (Hooker & Arnott) Jepson; *O. boothii* Douglas subsp. *decorticans* (Hooker & Arnott) Munz; *O. boothii* subsp. *rutila* (Davidson) Munz; *O. decorticans* (Hooker & Arnott) Greene; *O. decorticans* var. *rutila* (Davidson) Munz; *O. rutila* Davidson; *Sphaerostigma decorticans* (Hooker & Arnott) Small; *S. rutilum* (Davidson) Parish

Herbs glabrate to strigillose, sometimes also glandular puberulent in inflorescence. **Stems** stout, hollow, 12–65 cm. **Leaves** primarily clustered toward base, 2–11 × 0.3–2(–3) cm; petiole 0–4.2 cm; blade lanceolate, sometimes narrowly ovate or elliptic proximally, margins entire or denticulate. **Inflorescences** ± open. **Flowers:** floral tube 4–6 mm; petals usually white, rarely red, 3.5–5 mm. **Capsules** nearly straight, curved outward, tapering abruptly distally, 1.7–2.3 mm diam. near base. **Seeds** dimorphic.

Flowering Feb–Jun(–Jul). Open, usually steep slopes, shale or other loose rocky sites; 0–1900 m; Calif.

Subspecies *decorticans* is known from the southern Sierra Nevada Foothills, Tehachapi Mountain area, southern San Joaquin Valley, San Francisco Bay area, Inner South Coast Ranges, and Western Transverse Ranges.

Subspecies *decorticans* intergrades with subsp. *desertorum*. Occasional populations of subsp. *decorticans* at relatively high elevations in the southern part of its range have somewhat smaller flowers and red petals; they have been distinguished as *Oenothera rutila* or, more recently, *O. boothii* subsp. *rutila*. P. H. Raven (1969) saw no evidence that the various populations with this combination of characteristics had a common origin and, even if they did, they do not appear to constitute a major geographical race comparable with the other subspecies of *Eremothera boothii*.

4e. Eremothera boothii (Douglas) W. L. Wagner & Hoch subsp. **desertorum** (Munz) W. L. Wagner & Hoch, Syst. Bot. Monogr. 83: 209. 2007 [E]

Oenothera decorticans (Hooker & Arnott) Greene var. *desertorum* Munz, Bot. Gaz. 85: 246. 1928; *Camissonia boothii* (Douglas) P. H. Raven subsp. *desertorum* (Munz) P. H. Raven; *C. boothii* var. *desertorum* (Munz) Cronquist; *C. boothii* subsp. *inyoensis* (Munz) Munz; *O. boothii* Douglas subsp. *desertorum* (Munz) Munz; *O. boothii* subsp. *inyoensis* Munz

Herbs strigillose and/or glandular puberulent, especially in inflorescence. **Stems** 10–35 cm. **Leaves** often clustered toward base, 1–9 × 0.2–1.4 cm; petiole 0–4.5 cm; blade narrowly elliptic to narrowly ovate or oblanceolate, margins subentire to sparsely denticulate. **Inflorescences** relatively leafless. **Flowers:** floral tube (2–)4–8 mm; petals white, 3–7 mm. **Capsules** flexuous-contorted, apex often curved downward, 1–1.6 mm diam. near base. **Seeds** dimorphic.

Flowering Apr–Aug. Sandy or gravelly slopes and washes, creosote desert shrublands, pinyon-juniper woodlands; 400–2400 m; Calif.

Subspecies *desertorum* is intermediate geographically and morphologically between subspp. *condensata* and *decorticans* (P. H. Raven 1969). Populations from Inyo County, where they grow on limestone, have relatively small flowers and lax inflorescences, which gives them a distinctive appearance; these were named *Oenothera boothii* subsp. *inyoensis*. Very extensive intergradation between them and more typical subsp. *desertorum* caused Raven to consider them part of subsp. *desertorum*.

4f. Eremothera boothii (Douglas) W. L. Wagner & Hoch subsp. **intermedia** (Munz) W. L. Wagner & Hoch, Syst. Bot. Monogr. 83: 209. 2007 [E]

Oenothera boothii Douglas subsp. *intermedia* Munz in N. L. Britton et al., N. Amer. Fl., ser. 2, 5: 152. 1965; *Camissonia boothii* (Douglas) P. H. Raven subsp. *intermedia* (Munz) P. H. Raven

Herbs densely villous, also densely glandular puberulent, especially in inflorescence. **Stems** 5–20 cm. **Leaves** not especially clustered toward base, mostly evenly distributed, 2–5 × 0.3–1.5 cm; petiole 0–1.5 cm; blade narrowly elliptic to narrowly lanceolate or lanceolate, sometimes oblanceolate proximally, margins sparsely serrulate to sometimes sinuate-toothed. **Inflorescences** leafy. **Flowers:** floral tube 3–5 mm; petals white, 4–5 mm. **Capsules** usually curved outward or slightly contorted, 0.9–1.4 mm diam. near base. **Seeds** dimorphic.

Flowering late May–Aug(–Sep). Sandy or gravelly slopes and flats, sagebrush, shadscale, and rabbitbrush shrublands, pinyon-juniper woodlands; 1200–2000 m; Calif., Nev.

Subspecies *intermedia* is somewhat invariant; it bridges the morphological gap between subspp. *alyssoides* and *boothii*. It occurs in Nevada from southeastern Churchill and southern Lander counties south throughout Nye and Esmeralda counties and in southern Mineral County, in northeastern Inyo County, California, and in the Kingston Range, northeastern San Bernardino County, California.

5. **Eremothera pygmaea** (Douglas) W. L. Wagner & Hoch, Syst. Bot. Monogr. 83: 210. 2007 [E]

Oenothera pygmaea Douglas in W. J. Hooker, Fl. Bor.-Amer. 1: 213. 1832; *Camissonia pygmaea* (Douglas) P. H. Raven; *O. boothii* Douglas var. *pygmaea* (Douglas) Torrey & A. Gray; *Sphaerostigma boothii* (Douglas) Walpers var. *pygmaeum* (Douglas) Walpers

Herbs glandular puberulent, also moderately villous, sometimes sparsely so. **Stems** simple or loosely branched, (4–)12–35 cm, usually flowering only distally. **Leaves** cauline, sometimes with lower ones clustered near base,1.5–6.5 × 0.5–2 cm; petiole 0–3.5 cm; blade lanceolate to ovate or elliptic to subrhombic, margins crenate-dentate or serrulate. **Inflorescences** nodding. **Flowers** opening at sunset; floral tube 1.7–2.2(–4) mm, villous in proximal ½ inside; sepals 1.7–2.6 mm; petals white, fading pinkish, 1.5–2.5 mm; episepalous filaments 1–2.2 mm, epipetalous filaments slightly shorter, anthers 0.4–0.9 mm; style 3.2–4 mm, villous near base, stigma 0.5–0.8 mm diam., surrounded by anthers at anthesis. **Capsules** cylindrical and thickened proximally, spreading, straight to arcuate or loosely sigmoid, terete, 8–20 × 2–3 mm, regularly but tardily dehiscent. **Seeds** dimorphic, light brown, ca. 1 mm, those at base of capsule coarsely papillose, those of upper portion finely reticulate. **2*n*** = 14, 28.

Flowering late May–Aug. Steep, loose slopes, in scree, on gravelly flats or washes; 150–1500 m; Idaho, Oreg., Wash.

P. H. Raven (1969) determined *Eremothera pygmaea* to be self-compatible and autogamous. It is rare and local at scattered localities in eastern Washington (Douglas, Grant, and Kittitas counties), eastern Oregon (Gilliam, Grant, Harney, and Wheeler counties), and at one locality in adjacent southern Idaho (Jerome County).

6. **Eremothera gouldii** (P. H. Raven) W. L. Wagner & Hoch, Syst. Bot. Monogr. 83: 210. 2007 [E]

Camissonia gouldii P. H. Raven, Contr. U.S. Natl. Herb. 37: 368, fig. 70. 1969; *Oenothera gouldii* (P. H. Raven) S. L. Welsh & N. D. Atwood

Herbs glandular puberulent, sometimes also moderately or sparsely villous. **Stems** simple or loosely branched, 6–20 cm, usually flowering only distally. **Leaves** cauline, sometimes with lower ones clustered near base, 0.5–3.5 × 0.5–1 cm; petiole 0–0.6 cm; blade elliptic-lanceolate to elliptic, margins crenate-dentate or -serrulate. **Inflorescences** only at terminal nodes, nodding at anthesis. **Flowers** opening at sunset; floral tube 1.5–3 mm, villous in proximal ½ inside; sepals 1–1.5 mm; petals white, fading pinkish, 1.5–2.5 mm; episepalous filaments 1–2 mm, epipetalous filaments slightly shorter, anthers 0.4–0.9 mm; style 3–4.5 mm, villous near base, stigma 0.5–0.8 mm diam., surrounded by anthers at anthesis. **Capsules** cylindrical and thickened proximally, spreading, straight to arcuate or weakly sigmoid, terete, thickened near base, tapering distally, 8–12 × 2–3 mm, regularly but tardily dehiscent. **Seeds** monomorphic, light brown, ca. 1 mm, finely reticulate.

Flowering May–Oct. Volcanic scree or cinder flats; 1000–2000 m; Ariz., Utah.

Eremothera gouldii is known from Coconino and Mohave counties in Arizona and Washington County in Utah.

P. H. Raven (1969) determined *Eremothera gouldii* to be self-compatible and autogamous.

7. **Eremothera minor** (A. Nelson) W. L. Wagner & Hoch, Syst. Bot. Monogr. 83: 210. 2007 [E]

Sphaerostigma minus A. Nelson, Bull. Torrey Bot. Club 26: 130. 1899 (as minor); *Camissonia minor* (A. Nelson) P. H. Raven; *Oenothera alyssoides* Hooker & Arnott var. *minutiflora* S. Watson; *O. chamaenerioides* A. Gray var. *torta* H. Léveillé; *O. minor* (A. Nelson) Munz; *O. minor* var. *cusickii* Munz; *S. alyssoides* (Hooker & Arnott) Walpers var. *minutiflorum* (S. Watson) Small; *S. tortum* (H. Léveillé) A. Nelson; *S. tortum* var. *eastwoodiae* A. Nelson

Herbs densely strigillose, inflorescence often also glandular puberulent. **Stems** usually well branched from base, 3–30 cm, usually flowering proximally and distally. **Leaves** cauline, mostly clustered near base, reduced distally, 0.5–2.5 × 0.3–1.5 cm; petiole 0.5–2 cm; blade oblanceolate or narrowly oblanceolate, margins entire or sparsely denticulate. **Inflorescences** erect. **Flowers** opening at sunset; floral tube 0.5–1.9 mm, strigillose in proximal ½ inside; sepals 0.8–1.8 mm; petals white, fading pinkish, 0.8–1.3 mm; filaments 0.3–1.3 mm, epipetalous filaments shorter than episepalous, sometimes apparently abortive, anthers 0.5–0.8 mm; style 1.2–3.2 mm, sparsely short-villous near base, stigma 0.5–0.6 mm diam., surrounded by anthers at anthesis. **Capsules** cylindrical and thickened proximally, spreading, contorted, subterete, 10–25 × 0.8–1.2 mm, regularly but tardily dehiscent. **Seeds** monomorphic, gray, 1.1–1.2 × 0.4 mm, finely reticulate. $2n = 14$.

Flowering May–early Jun. Clay or sandy soils, slopes, flats, sagebrush, rabbitbrush, bitterbrush or saltbush shrublands; 700–1800 m; Calif., Colo., Idaho, Nev., Oreg., Utah, Wash., Wyo.

P. H. Raven (1969) determined *Eremothera minor* to be self-compatible and autogamous. The name *Sphaerostigma nelsonii* A. Heller is superfluous and pertains here.

12. CAMISSONIOPSIS W. L. Wagner & Hoch, Syst. Bot. Monogr. 83: 123. 2007 • [Genus *Camissonia* and Greek *-opsis*, resemblance]

Warren L. Wagner

Agassizia Spach, Hist. Nat Vég. 4: 347. 1835, not Chavennes 1833 [Plantaginaceae]; *Camissonia* Link sect. *Holostigma* P. H. Raven; *Holostigma* Spach 1835, not G. Don 1834 [Campanulaceae]

Herbs, usually annual, rarely short-lived perennial, caulescent. **Stems** prostrate to ascending or erect, often with reddish brown or white exfoliating epidermis. **Leaves** cauline and often in a basal rosette, alternate; stipules absent; sessile or petiolate; blade margins dentate, denticulate, or serrulate. **Inflorescences** spikes, erect or nodding at anthesis. **Flowers** bisexual, actinomorphic, buds erect; floral tube deciduous (with sepals, petals, and stamens) after anthesis, with basal nectary; sepals 4, usually reflexed in pairs, sometimes separately; petals 4, yellow, fading red, with 1+ red dots basally; stamens 8, in 2 unequal series, anthers versatile, pollen shed singly; ovary 4-locular, without apical projection, style glabrous or pubescent distally, stigma entire, subcapitate to subglobose, surface unknown, probably wet and non-papillate. **Fruit** a capsule, contorted or curled 1 to 5 times, or straight, narrowly cylindrical and thickened proximally, 4-angled (at least when dry), regularly but tardily loculicidally dehiscent, not swollen by seeds; sessile. **Seeds** numerous, in 1 row per locule, flattened, narrowly obovoid, dull. $x = 7$.

Species 14 (13 in the flora): w United States, nw Mexico.

Camissoniopsis proavita (P. H. Raven) W. L. Wagner & Hoch is known from northern Baja California, Mexico. It is a diploid, closely related to *C. micrantha* but differing in having numerous flowers in the basal rosette, which is densely leafy.

All species of *Camissoniopsis* occur near coasts or on dry slopes or desert flats inland from 0–2500 m. R. A. Levin et al. (2004) found strong molecular support for *Camissoniopsis* in a clade with *Neoholmgrenia* and *Tetrapteron*. *Camissoniopsis* was segregated from *Camissonia* as delimited by P. H. Raven (1969). *Camissoniopsis* is distinguished by having 4-angled fruits, at least when dry, and not swollen by seeds, dull seeds usually smaller than 1 mm, and by flowering from both basal and distal nodes (Raven). Relationships within *Camissoniopsis* are complex and reticulate. Several diploids (especially *C. hirtella*) appear to have contributed to the formation of the tetraploids and, in turn, the hexaploids (Raven), and, as a result, are very similar morphologically to each other. Identification of the polyploid species of *Camissoniopsis* is aided by their pollen having a high proportion of grains with higher number of pores than typical Onagraceae 3-pored pollen, usually 4- or 5-pored. This can be observed under low magnification (for example, 10\x) since the 3-pored pollen is triangular while the 4-pored is quadrangular and 5-pored is pentangular. Raven proposed *Camissonia* sect. *Holostigma* as a new combination based on Spach's generic name. He was unaware that *Holostigma* Spach, like *Agassizia* Spach, is a later homonym and thus illegitimate; however, he satisfied all requirements for valid publication of a new sectional name in *Camissonia*. Reproductive features include: self-incompatible (*C. cheiranthifolia* and *C. bistorta*) or self-compatible; flowers diurnal; outcrossing and pollinated by bees (E. G. Linsley et al. 1963, 1964, 1973) or autogamous (Raven).

1. Herbs perennial; coastal habitats . . . . 1. *Camissoniopsis cheiranthifolia*
1. Herbs usually annual, rarely short-lived perennial (in *C. bistorta*); primarily inland habitats.
    2. Stigma exserted beyond anthers at anthesis; sepals (2.3–)5–8(–11) mm; petals (4.2–)7–15 mm . . . . 2. *Camissoniopsis bistorta*
    2. Stigma surrounded by all anthers, or at least those of longer filaments, at anthesis; sepals 1–6(–8.5) mm; petals 1.5–10.5(–13) mm.
        3. Capsules 2.8–3.5 mm diam. near base, straight or slightly curved outward, deeply grooved along lines of dehiscence . . . . 4. *Camissoniopsis guadalupensis*
        3. Capsules 0.7–2.2 mm diam. near base, straight or curved into 1+-coiled spirals, not deeply grooved.
            4. Pollen with 25–100% of grains 4- or 5-pored.
                5. Inflorescences exclusively villous; 25–60% of pollen grains 4- or 5-pored . . . . 12. *Camissoniopsis luciae*
                5. Inflorescences villous and glandular puberulent; 70–100 % of pollen grains 4- or 5-pored.
                    6. Capsules 1.3–1.6 mm diam., subterete in living material (obscurely 4-angled when dry); southernmost Monterey County to central San Luis Obispo County, California . . . . 11. *Camissoniopsis hardhamiae*
                    6. Capsules 1.5–2 mm diam., 4-angled in living material; San Diego County, California, adjacent Baja California, and offshore islands . . . . 13. *Camissoniopsis robusta*
            4. Pollen with less than 5% of grains 4-pored (rarely more in *C. intermedia*).
                7. Capsules 1.8–2.2 mm diam., conspicuously 4-angled in living material . . . . 3. *Camissoniopsis lewisii*
                7. Capsules 0.7–1.2(–1.8) mm diam., terete, subterete, or obscurely 4-angled, at least in living material.
                    8. Distal leaves petiolate, blade base attenuate; capsules usually much contorted, irregularly to 5-coiled; herbs moderately to sparsely strigillose, sometimes also sparsely villous . . . . 5. *Camissoniopsis ignota*
                    8. Distal leaves usually subsessile, blade base rounded, cuneate, or truncate; capsules straight to 1–2-coiled; herbs strigillose to villous.
                        9. Herbs conspicuously grayish in appearance, densely strigillose; lateral stems usually decumbent; plants of the deserts . . . . . 8. *Camissoniopsis pallida*
                        9. Herbs not conspicuously gray in appearance, mostly villous; lateral stems erect to decumbent; plants not of deserts or only at desert margins (except *C. confusa* in central Arizona).
                            10. Capsules 0.7–0.9 mm diam.; distal leaf blades elliptic-ovate or ovate; stems ascending to erect . . . . 7. *Camissoniopsis hirtella*
                            10. Capsules 0.9–1.2(–1.8) mm diam.; distal leaf blades narrowly lanceolate to narrowly ovate; stems decumbent or erect.
                                11. Stems decumbent; inflorescences usually densely villous, rarely also glandular puberulent . . . . 6. *Camissoniopsis micrantha*
                                11. Stems erect; inflorescences usually moderately to densely villous, also glandular puberulent.

[12. Shifted to left margin.—Ed.]

12. Floral tube (1.8–)2–3.8 mm; petals (2.5–)5–10.5 mm; styles (2.5–)4.5–7.5 mm; herbs densely villous, often also stigillose . . . . 9. *Camissoniopsis confusa*
12. Floral tube 1.2–2 mm; petals 1.5–3.5(–4.5) mm; styles 2–3.5 mm; herbs moderately villous . . . . 10. *Camissoniopsis intermedia*

1. Camissoniopsis cheiranthifolia (Hornemann ex Sprengel) W. L. Wagner & Hoch, Syst. Bot. Monogr. 83: 204. 2007 [F]

Oenothera cheiranthifolia Hornemann ex Sprengel, Syst. Veg. 2: 228. 1825; *Agassizia cheiranthifolia* (Hornemann ex Sprengel) Spach; *Camissonia cheiranthifolia* (Hornemann ex Sprengel) Raimann; *Holostigma cheiranthifolium* (Hornemann ex Sprengel) Spach; *Sphaerostigma cheiranthifolium* (Hornemann ex Sprengel) Fischer & C. A. Meyer

Herbs short-lived perennial, sometimes woody at base, usually densely strigillose throughout, rarely glabrous, also villous distally. **Stems** prostrate, decumbent, or ascending from base, to 60(–130) cm. **Leaves** 0.5–5 × 0.3–2.2 cm; petiole 0–1.5(–2.5) cm, distal ones to 1 cm; blade narrowly ovate, base attenuate, cuneate, or cordate, margins sparsely serrulate, apex acute. **Flowers** opening near sunrise; floral tube 2.1–8.5 mm; sepals 4–11.5 mm; petals yellow, often red-dotted near base, 6–20 mm; episepalous filaments 2.8–8 mm, epipetalous filaments 1.5–6 mm, anthers 1–3 mm, less than 5% of pollen grains 4- or 5-pored; style 6–23 mm, stigma surrounded by or exserted beyond anthers at anthesis. **Capsules** often coiled in 1–2 spirals, 4-angled, 10–25 × 2–2.5 mm. **Seeds** 1.2–1.3 mm.

Subspecies 2 (2 in the flora): w United States, nw Mexico.

Camissoniopsis cheiranthifolia occurs on slopes and dunes along the immediate coast and on islands from Coos Bay, Curry County, Oregon, to the vicinity of San Quintín, Baja California; it is also known from the east shore of San Francisco Bay and locally on sand dunes along the lower Sacramento River, California, 0–100 m. P. H. Raven (1969) determined *C. cheiranthifolia* to be self-incompatible (some populations in subsp. *suffruticosa*) or self-compatible (both subspecies) and apparently pollinated by oligolectic bees of *Andrena* subg. *Onagrandrena* (Raven); Raven subdivided the species into two intergrading subspecies.

1. Plants primarily herbaceous; petals 6–11 mm; style 6–9 mm; stigma usually surrounded by anthers . . . . 1a. *Camissoniopsis cheiranthifolia* subsp. *cheiranthifolia*
1. Plants woody at base; petals (10–)12–20 mm; style 13–23 mm; stigma exserted beyond anthers at anthesis . . . . 1b. *Camissoniopsis cheiranthifolia* subsp. *suffruticosa*

1a. Camissoniopsis cheiranthifolia (Hornemann ex Sprengel) W. L. Wagner & Hoch subsp. **cheiranthifolia** [E] [F]

Holostigma spirale (Hooker) Spach; *Oenothera cheiranthifolia* Hornemann ex Sprengel var. *nitida* (Greene) Munz; *O. nitida* Greene; *O. spiralis* Hooker; *O. spiralis* var. *nitida* (Greene) Jepson; *Sphaerostigma nitidum* (Greene) Small; *S. spirale* (Hooker) Fischer & C. A. Meyer; *S. spirale* var. *clypeatum* (H. Léveillé) A. Nelson

Stems primarily herbaceous, usually prostrate or decumbent, moderately strigillose, rarely densely silvery strigillose. **Flowers:** floral tube 2.1–4.2(–4.8) mm; sepals 4–5.6(–6.7) mm; petals 6–11 mm; episepalous filaments 2.8–4.5 mm, epipetalous filaments 1.5–3 mm, anthers 1–1.5 mm; style 6–9 mm, stigma usually surrounded by anthers at anthesis, rarely exserted beyond anthers. $2n = 14$.

Flowering Apr–Aug. Coastal sandy slopes and flats; 0–100 m; Calif., Oreg.

Subspecies *cheiranthifolia* occurs from Coos Bay, Curry County, Oregon, south to Point Conception, Santa Barbara County, California; on the east shore of San Francisco Bay in Alameda, Contra Costa, and Sacramento counties, California, and on San Miguel, Santa Rosa, Santa Cruz, San Nicolas, and San Clemente islands (P. H. Raven 1969).

1b. Camissoniopsis cheiranthifolia (Hornemann ex Sprengel) W. L. Wagner & Hoch subsp. **suffruticosa** (S. Watson) W. L. Wagner & Hoch, Syst. Bot. Monogr. 83: 204. 2007

Oenothera cheiranthifolia Hornemann ex Sprengel var. *suffruticosa* S. Watson, Proc. Amer. Acad. Arts 8: 592. 1873; *Camissonia cheiranthifolia* (Hornemann ex Sprengel) Raimann subsp. *suffruticosa* (S. Watson) P. H. Raven; *O. spiralis* Hooker var. *linearis* Jepson; *O. spiralis* var. *viridescens* (Hooker) Jepson; *Sphaerostigma spirale* (Hooker) Fischer & C. A. Meyer var. *viridescens* (Hooker) A. Nelson; *S. viridescens* (Hooker) Walpers

Stems woody at least at base, ascending, usually densely silvery strigillose. **Flowers:** floral tube 5–8.5 mm; sepals 6–11.5 mm; petals (10–)12–20 mm; episepalous filaments 5–8 mm, epipetalous filaments 3–6 mm, anthers 2.2–3 mm; style 13–23 mm, stigma exserted beyond anthers at anthesis. **2*n*** = 14.

Flowering Jan–Aug. Coastal sandy slopes, dunes, and flats; 0–100 m; Calif.; Mexico (Baja California).

Subspecies *suffruticosa* occurs from near Goleta, Santa Barbara County, California, and San Nicolas Island, south to near San Quintín and Isla San Martín, Baja California, Mexico (P. H. Raven 1969). The subspecies hybridizes with *C. bistorta* in areas of near sympatry.

2. Camissoniopsis bistorta (Nuttall ex Torrey & A. Gray) W. L. Wagner & Hoch, Syst. Bot. Monogr. 83: 204. 2007

Oenothera bistorta Nuttall ex Torrey & A. Gray, Fl. N. Amer. 1: 508. 1840; *Camissonia bistorta* (Nuttall ex Torrey & A. Gray) P. H. Raven; *O. bistorta* var. *veitchiana* Hooker; *Sphaerostigma bistortum* (Nuttall ex Torrey & A. Gray) Walpers; *S. bistortum* var. *veitchianum* (Hooker) A. Nelson; *S. veitchianum* (Hooker) Small

Herbs annual, rarely short-lived perennial, usually villous, sometimes strigillose. **Stems** 1–several from base, ascending or decumbent, to 80 cm. **Leaves** 1.2–12 × 0.2–1.5 cm; petiole 0–4 cm, distal ones 0–0.3 cm; blade (basal) narrowly elliptic or (cauline) usually narrowly lanceolate or lanceolate, rarely linear, base (basal) narrowly cuneate, (cauline) cuneate or subcordate, margins usually sparsely and inconspicuously denticulate, apex acute. **Flowers** opening near sunrise; floral tube 2–5(–7.5) mm; sepals (2.3–) 5–8(–11) mm; petals yellow, each usually with 1 bright red dot, rarely 2, near base, (4.2–)7–15 mm; episepalous filaments (1–)1.5–3.5 mm, epipetalous filaments (0.5–)1–2.5 mm, anthers (0.5–)1.3–2(–2.5) mm, less than 5% of pollen grains 4- or 5-pored; style (5.5–)7–12 mm, stigma exserted beyond anthers at anthesis. **Capsules** straight or somewhat contorted, weakly 4-angled, 12–40 × 1.5–2.5 mm. **Seeds** 0.9–1 mm. **2*n*** = 14.

Flowering Mar–Jun. Sandy or clayey soils, coastal strands, grasslands, coastal sage scrub, chaparral, oak woodlands, margins of Sonoran and Mojave deserts, rarely higher elevation meadows; 0–1600(–2600) m; Calif.; Mexico (Baja California).

Camissoniopsis bistorta occurs in California from Ventura County south and east through the counties of southern Los Angeles, southwestern San Bernardino, Orange, western Riverside, and the western two-thirds of San Diego, reaching the margins of the desert in San Bernardino and San Diego counties, and southward in cismontane Baja California to Ojos Negros and San Vicente. The species occurs at exceptionally high elevations in the Santa Ana drainage of the San Bernardino Mountains. P. H. Raven (1969) indicated that there were occasional apparent hybrids between *C. cheiranthifolia* subsp. *suffruticosa* and *C. bistorta* occurring in intermediate habitats in areas where the two species co-occur. He determined that *C. bistorta* is self-incompatible.

Camissoniopsis bistorta was apparently introduced with stream gravel in 1959 in Goleta Marsh, Santa Barbara, California, and on ballast heaps at Nanaimo, Vancouver Island, British Columbia, in 1893. It has apparently not persisted at either site.

Oenothera heterophylla Nuttall ex Hooker & Arnott (1839), not Spach (1836), is an illegitimate name that pertains to *Camissoniopsis bistorta*.

3. Camissoniopsis lewisii (P. H. Raven) W. L. Wagner & Hoch, Syst. Bot. Monogr. 83: 205. 2007

Camissonia lewisii P. H. Raven, Contr. U.S. Natl. Herb. 37: 275, fig. 32. 1969

Herbs annual, villous, also glandular puberulent distally. **Stems** usually several, decumbent, rarely 1 erect stem, 30–60 cm. **Leaves** 1–8 × 0.2–1.1 cm; petiole 0–3 cm; blade narrowly lanceolate-elliptic, base cuneate or subcordate, margins denticulate, apex acute. **Flowers** opening near sunrise; floral tube 1.5–4 m; sepals 1.7–3.4 mm; petals yellow, with 1 or 2 red dots basally, 2.5–5.5 mm; episepalous filaments 2–2.8 mm, epipetalous filaments 1–1.7 mm, anthers 0.7–1.2 mm, less than 5% of pollen grains 4- or 5-pored; style 2.8–4.5 mm, stigma surrounded by anthers at anthesis. **Capsules** usually loosely 1-coiled, conspicuously 4-angled in living material, 13–20 × 1.8–2.2 mm. **Seeds** 0.7–0.8 mm. **2*n*** = 14.

Flowering Mar–May(–Sep). Open sandy and clayey grasslands, coastal dunes and beaches; 0–300 m; Calif.; Mexico (Baja California).

Camissoniopsis lewisii occurs from Point Dume and the Los Angeles Basin, Los Angeles County, south to Cardon Grande at the northern edge of Baja California Sur. P. H. Raven (1969) determined *C. lewisii* to be self-compatible and primarily autogamous, and suggested that this coastal *Camissoniopsis* may have been derived more or less directly from coastal populations of *C. bistorta*.

4. Camissoniopsis guadalupensis (S. Watson) W. L. Wagner & Hoch, Syst. Bot. Monogr. 83: 204. 2007

Oenothera guadalupensis S. Watson, Proc. Amer. Acad. Arts 11: 115, 137. 1876; *Camissonia guadalupensis* (S. Watson) P. H. Raven

Subspecies 2 (1 in the flora): California, nw Mexico.

Camissoniopsis guadalupensis is known from San Clemente Island, Los Angeles County, California (subsp. *clementiana*), and Isla Guadalupe, Baja California (subsp. *guadalupensis*). P. H. Raven (1969) determined *C. guadalupensis* to be self-compatible and primarily autogamous.

4a. Camissoniopsis guadalupensis (S. Watson) W. L. Wagner & Hoch subsp. **clementiana** (P. H. Raven) W. L. Wagner & Hoch, Syst. Bot. Monogr. 83: 204. 2007 [E]

Oenothera guadalupensis S. Watson subsp. *clementina* P. H. Raven, Aliso 5: 332. 1963; *Camissonia guadalupensis* (S. Watson) P. H. Raven subsp. *clementina* (P. H. Raven) P. H. Raven

Herbs annual, subsucculent and heavy-set, densely villous, also glandular puberulent distally. **Stems** erect, with branches usually arising near base, 2–18(–35) cm. **Leaves** 1.2–3.8(–9.5) × 0.5–1.2(–1.8) cm; petiole 0–3 cm; blade (basal and proximal cauline) narrowly elliptic, (cauline) narrowly ovate, base attenuate, margins sparsely and inconspicuously denticulate, apex (basal) acute, (cauline) obtuse, rounded, or truncate. **Flowers** opening near sunrise; floral tube 1.6–2.4 mm; sepals 1.9–3.2 mm; petals yellow, each with 1 red dot near base, 2.8–4.2 mm; episepalous filaments 1.3–2.3 mm, epipetalous filaments 0.4–1.6 mm, anthers 0.4–0.8 mm, less than 5% of pollen grains 4- or 5-pored; style 3.2–4.5 mm, stigma surrounded by anthers at anthesis. **Capsules** stout, straight or slightly curved outward, 4-angled, 10–18 × 2.8–3.5 mm, deeply grooved along lines of dehiscence. **Seeds** 0.7–0.9 mm. **2*n*** = 14.

Flowering Mar–Jun(–Sep). Sandy flats, dunes; 0–60 m; Calif.

Subspecies *clementiana* is restricted to San Clemente Island, Los Angeles County, where it is common on dunes around the north end and down the west shore, perhaps to the south end.

5. Camissoniopsis ignota (Jepson) W. L. Wagner & Hoch, Syst. Bot. Monogr. 83: 205. 2007

Oenothera micrantha Hornemann ex Sprengel var. *ignota* Jepson, Man. Fl. Pl. Calif., 684. 1925; *Camissonia ignota* (Jepson) P. H. Raven; *O. hirta* Link var. *ignota* (Jepson) Munz; *O. ignota* (Jepson) Munz

Herbs annual, strigillose, usually also sparsely villous, often also glandular puberulent distally. **Stems** arising from base, usually decumbent, rarely with only 1, erect stem, 10–55 cm. **Leaves** 1.5–7 × 0.3–1.3 cm; petiole (0–)0.2–2.5 cm, petiolate distally; blade narrowly lanceolate, lanceolate, or narrowly elliptic, base attenuate, margins serrulate, apex acute. **Flowers** opening near sunrise; floral tube (1.1–)1.8–3 mm; sepals 2.6–5.5 mm; petals yellow, sometimes red-dotted near base, (3–)4–8 mm; episepalous filaments (1.2–)2.5–3.6 mm, epipetalous filaments (1–)1.3–2 mm, anthers (0.6–)0.8–1.6 mm, less than 5% of pollen grains 4- or 5-pored; style (3–)4.5–7 mm, stigma surrounded by anthers at anthesis. **Capsules** very slender, usually much contorted, irregularly to 5-coiled, rarely simply flexuous, terete in living material, 4-angled when dry, 20–30 × 0.8–1 mm. **Seeds** 1.2–1.3 mm. **2*n*** = 14.

Flowering (Jan–)Mar–Apr(–Aug). Clay or sandy soils, flats and slopes in coastal sage scrub or chaparral, sandy soils in mountains; 100–1100(–1500) m; Calif.; Mexico (Baja California).

Camissoniopsis ignota is most common in clay fields and slopes at low elevations, but occasional on sandy soil and higher in the mountains in the Coast Ranges and bordering valleys from Yolo County, California, south to the southern end of the Sierra San Miguel, in Baja California, usually away from the immediate coast and barely reaching the margins of the desert. P. H. Raven (1969) determined *C. ignota* to be self-compatible and primarily autogamous.

6. Camissoniopsis micrantha (Hornemmann ex Sprengel) W. L. Wagner & Hoch, Syst. Bot. Monogr. 83: 205. 2007 [E]

Oenothera micrantha Hornemann ex Sprengel, Syst. Veg. 2: 228. 1825; *Camissonia micrantha* (Hornemann ex Sprengel) P. H. Raven; *Holostigma micranthum* (Hornemann ex Sprengel) Spach; *S. micranthum* (Hornemann ex Sprengel) Walpers

Herbs annual, densely villous, more densely so distally, also rarely glandular puberulent distally. **Stems** arising from base, usually decumbent, rarely with 1 erect, 15–60 cm. **Leaves**

1–12 × 0.2–1.7 cm; petiole 0–2 cm, distal ones 0–0.5 cm; blade (basal) narrowly elliptic, (cauline) narrowly elliptic-lanceolate to lanceolate, base (basal) narrowly cuneate, (cauline) rounded, margins denticulate, apex acute. **Flowers** opening near sunrise; floral tube 1.2–2 mm; sepals 1–2.2(–2.5) mm; petals yellow, sometimes with 1 or 2 red dots near base, 1.5–3.5(–4.5) mm; episepalous filaments 0.8–1.5 mm, epipetalous filaments 0.5–0.8(–1) mm, anthers 0.4–0.6 mm, less than 5% of pollen grains 4- or 5-pored; style 2–3.5 mm, stigma surrounded by anthers at anthesis. **Capsules** straight or curved, equal to or slightly more than 1 complete spiral, subterete in living material, 4-angled when dry, 13–20(–25) × 1.1–1.2(–1.8) mm. **Seeds** 0.7–1.1 mm. **2*n*** = 14.

Flowering (Jan–)Mar–Jun(–Sep). Coastal strand, coastal sage scrub, chaparral; 0–300(–800) m; Calif.

Camissoniopsis micrantha occurs from the vicinity of Bodega Bay, Sonoma County, near Lower Lake, Lake County, and near Rio Vista, Sacramento County, south in the Coast Ranges to the Los Angeles Basin and the northern edge of San Diego County; also on San Miguel, Santa Rosa, Santa Cruz, and Santa Catalina islands. The species was introduced, apparently on ballast heaps, at Nanaimo, Vancouver Island, British Columbia (*Macoun s.n.* in 1893, NMC). It has apparently not persisted in this area. P. H. Raven (1969) determined *C. micrantha* to be self-compatible and primarily autogamous. Excluded populations are now recognized as *C. hirtella*, *C. ignota*, *C. lewisii*, and *C. pallida*.

Oenothera hirta Link (1821), not Linnaeus (1759), is an illegitimate name that pertains to *Camissoniopsis micrantha*.

7. **Camissoniopsis hirtella** (Greene) W. L. Wagner & Hoch, Syst. Bot. Monogr. 83: 204. 2007 [F]

Oenothera hirtella Greene, Fl. Francisc., 215. 1891; *Camissonia hirtella* (Greene) P. H. Raven; *O. hirta* Link var. *jonesii* H. Léveillé; *O. micrantha* Hornemann ex Sprengel var. *hirtella* (Greene) Jepson; *O. micrantha* var. *jonesii* (H. Léveillé) Munz; *O. micrantha* var. *reedii* (Parish) Jepson; *Sphaerostigma arenicola* A. Nelson; *S. bistortum* (Nuttall ex Torrey & A. Gray) Walpers var. *reedii* Parish; *S. hirtellum* (Greene) Small; *S. hirtellum* var. *montanum* Davidson; *S. micranthum* (Hornemann ex Sprengel) Walpers var. *jonesii* (H. Léveillé) A. Nelson

Herbs annual, densely villous throughout, also glandular puberulent distally. **Stems** erect, with 1 or more ascending branches from near base, to 60 cm. **Leaves** 1–11 × 0.3–2.1 cm; petiole 0–5 cm, distal ones 0–0.5 cm; blade lanceolate to ovate, sometimes elliptic-ovate or ovate distally, base cordate to truncate, sometimes cuneate or attenuate, margins dentate, apex acute. **Flowers** opening near sunrise; floral tube 1–3 mm; sepals 2.5–6 mm; petals yellow, sometimes red-dotted near base, 2–9 mm, sometimes with a tooth arising from emarginate apex; episepalous filaments 1.2–6 mm, epipetalous filaments 0.5–3 mm, anthers 0.4–1 mm, less than 5% of pollen grains 4- or 5-pored; style 2–8 mm, stigma surrounded by anthers at anthesis. **Capsules** 1–2-coiled spiral, subterete in living material, 4-angled when dry, 13–20(–25) × 0.7–0.9 mm. **Seeds** 1–1.2 mm. **2*n*** = 14.

Flowering (Jan–)Mar–Jul(–Nov). Brushy hills and slopes, on burns; 0–2300 m; Calif.; Mexico (Baja California).

Camissoniopsis hirtella occurs from Amador and Trinity counties southward in the Coast Ranges and Sierra Nevada of California to the Sierra de Juárez and Sierra San Pedro Mártir, Baja California, usually away from the immediate coast and barely reaching the margins of the desert. P. H. Raven (1969) determined *C. hirtella* to be self-compatible and primarily autogamous. The species occasionally hybridizes with *C. ignota* (Raven).

8. **Camissoniopsis pallida** (Abrams) W. L. Wagner & Hoch, Syst. Bot. Monogr. 83: 205. 2007

Sphaerostigma pallidum Abrams, Bull. Torrey Bot. Club 32: 539. 1905; *Camissonia pallida* (Abrams) P. H. Raven; *Oenothera abramsii* J. F. Macbride; *O. micrantha* Hornemann ex Sprengel var. *abramsii* (J. F. Macbride) Jepson

Herbs annual, appearing conspicuously grayish, densely strigillose, sometimes also glandular puberulent distally. **Stems** usually with decumbent lateral branches from basal rosette, 5–60 cm. **Leaves** 1–5(–11) × 0.2–0.7 (–1.4) cm; petiole 0–0.2(–0.4) cm, distal ones sessile; blade lanceolate to narrowly ovate, base often cuneate to truncate, sometimes attenuate, margins sparsely denticulate, apex acute to obtuse. **Flowers** opening near sunrise; floral tube 1–4.2 mm; sepals (1.5–)2.5–8 mm; petals yellow, sometimes with 1–3 red dots basally, (2–)3.5–13 mm; episepalous filaments (0.5–)1.5–6.5 mm, epipetalous filaments (0.2–)0.5–3.8 mm, anthers (0.4–)0.8–2.2 mm, less than 5% of pollen grains 4- or 5-pored; style (2.1–)3–10.5 mm, stigma surrounded by at least anthers of longer stamens, often by both sets, at anthesis. **Capsules** usually 1–3-coiled spiral, subterete in living material, 4-angled when dry, 13–24 × 0.7–1.2 mm. **Seeds** 1–1.5 mm.

Subspecies 2 (2 in the flora): w United States, nw Mexico.

P. H. Raven (1969) determined *Camissoniopsis pallida* to be self-compatible and primarily autogamous.

1. Floral tube 1–3 mm; petals (2–)3.5–6(–8) mm; styles (2.1–)3–6.5 mm. . . . . 8a. *Camissoniopsis pallida* subsp. *pallida*
1. Floral tube 3.8–4.2 mm; petals 6.5–13 mm; styles 6.5–10.5 mm. . . 8b. *Camissoniopsis pallida* subsp. *hallii*

8a. Camissoniopsis pallida (Abrams) W. L. Wagner & Hoch subsp. **pallida**

Oenothera hirta Link var. *exfoliata* (A. Nelson) Munz; *O. micrantha* Hornemann ex Sprengel var. *exfoliata* (A. Nelson) Munz; *Sphaerostigma micranthum* (Hornemann ex Sprengel) Walpers var. *exfoliatum* A. Nelson

Flowers: floral tube 1–3 mm; sepals (1.5–)2.5–5.5 mm; petals rarely with red dots basally, (2–)3.5–6(–8) mm; episepalous filaments (0.5–)1.5–4 mm, epipetalous filaments (0.2–)0.5–2.2 mm, anthers (0.4–)0.8–1.2 mm; style (2.1–)3–6.5 mm. **2*n*** = 14.

Flowering (Jan–)Mar–Aug(–Nov). Desert slopes and flats, along washes, creosote bush scrub, Joshua tree woodlands; 30–1900 m; Ariz., Calif., Nev.; Mexico (Baja California).

Subspecies *pallida* occurs from the head of the San Joaquin Valley in Ventura and Kern counties, California, across the Colorado and Mojave deserts, north to the vicinity of Independence, Inyo County, California, and Esmeralda County, Nevada (the only member of the genus in Nevada), east to Mohave and Yavapai counties, Arizona, and also in the vicinity of Tucson, Pima County; also south along the eastern side of Baja California to the Sierra de San Borjas, Mexico.

8b. Camissoniopsis pallida (Abrams) W. L. Wagner & Hoch subsp. **hallii** (Davidson) W. L. Wagner & Hoch, Syst. Bot. Monogr. 83: 204. 2007 [C] [E]

Sphaerostigma hallii Davidson, Muhlenbergia 3: 107. 1907; *Camissonia pallida* (Abrams) P. H. Raven subsp. *hallii* (Davidson) P. H. Raven; *Oenothera bistorta* Nuttall ex Torrey & A. Gray var. *hallii* (Davidson) Jepson; *O. hallii* (Davidson) Munz

Flowers: floral tube 3.8–4.2 mm; sepals 4.8–8 mm; petals with 1–3 red dots basally, 6.5–13 mm; episepalous filaments 3–6.5 mm, epipetalous filaments 1.8–3.8 mm, anthers 1.5–2.2 mm; style 6.5–10.5 mm. **2*n*** = 14.

Flowering Mar–May(–Jun). Sandy washes; of conservation concern; 30–1400 m; Calif.

Subspecies *hallii* occurs in Riverside and San Bernardino counties, from Banning east throughout the Little San Bernardino Mountains and their northern slopes and southeast to the vicinity of Mecca and Box Canyon.

9. Camissoniopsis confusa (P. H. Raven) W. L. Wagner & Hoch, Syst. Bot. Monogr. 83: 204. 2007 [E]

Camissonia confusa P. H. Raven, Contr. U.S. Natl. Herb. 37: 298, fig. 51. 1969

Herbs annual, densely villous, often also strigillose, at least sparsely villous and glandular puberulent on stems distally and on inflorescences. **Stems** erect, with multiple branches, rarely with 1 stem, to 70 cm. **Leaves** 1–6 × 0.4–2 cm; petiole 0–3 cm, distal ones sessile; blade lanceolate or narrowly ovate, base round or truncate, margins sparsely denticulate, apex long-acuminate. **Flowers** opening near sunrise; floral tube (1.8–)2–3.8 mm; sepals (1.5–)3.2–8.5 mm; petals yellow, usually with 1 or 2 red dots basally, (2.5–)5–10.5 mm; episepalous filaments (1.2–)2.5–4.5 mm, epipetalous filaments (0.8–)1.5–2.5 mm, anthers (0.4–)0.8–1.5 mm, less than 5% of pollen grains 4- or 5-pored; style (2.5–)4.5–7.5 mm, stigma surrounded by anthers at anthesis. **Capsules** straight or 1–2-coiled spiral, subterete in living material, 4-angled when dry, 13–23 × 0.9–1.2 mm. **Seeds** 0.7–1.1 mm. **2*n*** = 14.

Flowering Mar–Jun(–Jul). Dry inland slopes, chaparral; 300–2000 m; Ariz., Calif.

Camissoniopsis confusa occurs in California from the La Panza Range of central San Luis Obispo County south through the Coast Ranges to the San Bernardino Mountains and southern San Diego County; also in central Arizona (westernmost Gila, Maricopa, and northern Pinal counties). P. H. Raven (1969) determined *C. confusa* to be self-compatible and primarily autogamous. The species apparently is a tetraploid derived via hybridization between two diploid (2*n* = 14) species, *C. hirtella* and *C. pallida*.

10. Camissoniopsis intermedia (P. H. Raven) W. L. Wagner & Hoch, Syst. Bot. Monogr. 83: 205. 2007

Camissonia intermedia P. H. Raven, Contr. U.S. Natl. Herb. 37: 295, fig. 48. 1969

Herbs annual, appearing greenish, moderately villous, often also glandular puberulent on stems distally and on inflorescences. **Stems** erect, usually with 1+ ascending branches from basal rosette, 30–60 cm. **Leaves** 1–12 × 0.2–1.7 cm; petiole 0–1 cm, distal ones sessile; blade lanceolate to narrowly ovate, base cuneate to truncate, basal and proximal cauline often attenuate, margins denticulate, apex acute. **Flowers** opening near sunrise; floral tube

1.2–2 mm; sepals 1–2.5 mm; petals yellow, with 1 or 2 red dots basally, 1.5–3.5(–4.5) mm; episepalous filaments 0.8–1.5 mm, epipetalous filaments 0.5–0.9 mm, anthers 0.4–0.5 mm, less than 5% of pollen grains 4- or 5-pored; style 2–3.5 mm, stigma surrounded by anthers at anthesis. **Capsules** straight or 1-coiled spiral, subterete in living material, 4-angled when dry, 13–25 × 1.1–1.2 mm. **Seeds** 0.7–1.1 mm. **2*n*** = 28.

Flowering Mar–Jun(–Sep). Disturbed brushy slopes, on burns; (150–)300–800 m; Calif.; Mexico (Baja California).

Camissoniopsis intermedia occurs from Lake and Yolo counties (where rare) south in the Coast Ranges of California to the western San Gabriel Mountains, western Riverside and San Diego counties, and south in Baja California to the south end of the Sierra San Miguel; also on Santa Catalina and Santa Cruz islands. P. H. Raven (1969) determined *C. intermedia* to be self-compatible and primarily autogamous. The species is apparently a tetraploid derived via hybridization between two diploid (2*n* = 14) species, *C. hirtella* and *C. micrantha*.

11. Camissoniopsis hardhamiae (P. H. Raven) W. L. Wagner & Hoch, Syst. Bot. Monogr. 83: 204. 2007 [C] [E]

Camissonia hardhamiae P. H. Raven, Contr. U.S. Natl. Herb. 37: 301, fig. 47. 1969

Herbs annual, villous, also glandular puberulent distally. **Stems** erect, with 1 or more branches from basal rosette, to 60 cm. **Leaves** 1–12 × 0.4–1.8 cm; subsessile; blade lanceolate, narrowly elliptic, or narrowly ovate, base truncate, margins dentate, apex acute. **Flowers** opening near sunrise; floral tube 1.7–2 mm; sepals 1.8–3.2 mm; petals yellow, immaculate, 2–4 mm; episepalous filaments 1.5–2 mm, epipetalous filaments 1–1.5 mm, anthers 0.7 mm, 70–100% of pollen grains 4- or 5-pored; style 3–4 mm, stigma surrounded by anthers at anthesis. **Capsules** straight or 1-coiled, subterete in living material, obscurely 4-angled when dry, 13–25 × 1.3–1.6 mm. **Seeds** 0.7–1.1 mm. **2*n*** = 42.

Flowering Mar–May. Sandy soils, limestone, disturbed oak woodlands; of conservation concern; 150–1000 m; Calif.

Camissoniopsis hardhamiae is narrowly endemic to the Outer South Coast Ranges. Populations are very local, known only from a few localities in sandy soil in disturbed oak woodland, southernmost Monterey to central San Luis Obispo County. P. H. Raven (1969) determined *C. hardhamiae* to be self-compatible and primarily autogamous. The species is apparently a hexaploid derived via hybridization between the tetraploid *C. intermedia* (2*n* = 28) and the diploid *C. micrantha* (2*n* = 14).

12. Camissoniopsis luciae (P. H. Raven) W. L. Wagner & Hoch, Syst. Bot. Monogr. 83: 205. 2007 [E]

Camissonia luciae P. H. Raven, Contr. U.S. Natl. Herb. 37: 302, fig. 49. 1969

Herbs annual, villous throughout. **Stems** erect or ascending, 20–50 cm. **Leaves** 1.3–5.5 × 1.2–2.5 cm; sessile; blade lanceolate to narrowly oblong, base rounded or truncate, sometimes cuneate, margins sparsely denticulate, apex acuminate to, sometimes, rounded. **Flowers** opening near sunrise; floral tube 2–3 mm; sepals 2.5–4.5 mm; petals yellow, with 1 red dot basally, 4–7 mm, sometimes with a tooth arising from emarginate apex; episepalous filaments 2–6 mm, epipetalous filaments 0.8–1.6 mm, anthers 0.4–1 mm, 25–60% of pollen grains 4-pored; style 3–6 mm, stigma surrounded by anthers at anthesis. **Capsules** straight or 1.5–2+-coiled spiral, subterete in living material, obscurely 4-angled when dry, 15–20 × 1.3–2 mm. **Seeds** 1.3–1.5 mm. **2*n*** = 42.

Flowering Apr–May(–Jul). Openings in chaparral; 300–1400 m; Calif.

Camissoniopsis luciae is known from the Santa Lucia Mountains, Monterey County, and scattered southward to San Benito, San Luis Obispo, and Santa Barbara counties. P. H. Raven (1969) determined *C. luciae* to be self-compatible and primarily autogamous. The species is a hexaploid that parallels the widespread diploid *C. hirtella* in the variable notching of its petals. Presumably, it has been derived from the tetraploid *C. intermedia* (2*n* = 28) and the diploid *C. hirtella* (2*n* = 14), but it is rather easily separated from both by the absence of glandular hairs in the inflorescence, relatively large flowers, and pollen characteristics.

13. Camissoniopsis robusta (P. H. Raven) W. L. Wagner & Hoch, Syst. Bot. Monogr. 83: 205. 2007

Camissonia robusta P. H. Raven, Contr. U.S. Natl. Herb. 37: 304, fig. 41. 1969

Herbs annual, villous, usually also glandular puberulent distally. **Stems** erect, with 1 or more ascending branches from base, to 60 cm. **Leaves** 1–8 × 1.5–2 cm; subsessile; blade narrowly lanceolate-elliptic, base cuneate to truncate, basal often attenuate, margins denticulate, apex acute. **Flowers** opening near sunrise; floral tube 1.8–3.7 mm; sepals 2.6–4.2 mm; petals yellow, usually with 1 or 2 red dots basally, 3.2–7 mm; episepalous filaments 1.8–3 mm, epipetalous filaments 1–1.5 mm, anthers 0.8–3 mm, 70–100% of pollen grains 4-pored; style 3–6.2 mm, stigma surrounded by anthers at anthesis. **Capsules** usually 1-coiled spiral, 4-angled, 14–25 × 1.5–2 mm. **Seeds** 0.9–1.2 mm. **2*n*** = 42.

Flowering (Jan–)Mar–Jun(–Sep). Coastal sage, chaparral, disturbed or open places; 0–600(–800) m; Calif.; Mexico (Baja California).

Camissoniopsis robusta occurs in coastal San Diego County, California, and coastal northwestern Baja California, south to the vicinity of El Rosario; also on Guadalupe, San Clemente, Santa Catalina, and Santa Cruz (rare) islands. P. H. Raven (1969) determined *C. robusta* to be self-compatible and primarily autogamous. Based on the intermediate morphology of this hexaploid, Raven suggested that it was derived from two species with which it occurs nearly throughout its rather limited range, the tetraploid *C. intermedia* (2*n* = 28) and the diploid *C. lewisii* (2*n* = 14).

13. TETRAPTERON (Munz) W. L. Wagner & Hoch, Syst. Bot. Monogr. 83: 129. 2007 • [Greek *tettares*, four, and *pteron*, wing, alluding to fruit appearance]

Warren L. Wagner

Oenothera Linnaeus sect. *Tetrapteron* Munz, Amer. J. Bot. 16: 247. 1929; *Camissonia* Link sect. *Tetrapteron* (Munz) P. H. Raven

Herbs, annual, usually acaulescent, sometimes with very short lateral stems; with slender taproot. **Stems** (when present) thickened, becoming tough in age, with loose, white, exfoliating epidermis. **Leaves** in a basal rosette; stipules absent; sessile; blade margins entire or sparsely serrulate. **Inflorescences** solitary flowers in leaf axils. **Flowers** bisexual, actinomorphic, buds nodding, becoming erect; floral tube deciduous (with sepals, petals, and stamens) after anthesis, with fleshy nectary disc near base; sepals 4, reflexed in pairs; petals 4, yellow, unspotted or with red basal spot, strongly ultraviolet reflective; stamens 8 in 2 unequal series, anthers subbasifixed, pollen shed singly; ovary 4-locular, with slender, tubular, sterile apical projection 6–45 mm proximal to floral tube, with visible abscission lines at its junctures with both fertile part of ovary and short floral tube, style glabrous or shortly pubescent, stigma entire, globose, surface unknown, probably wet and non-papillate. **Fruit** a capsule, obovoid, with pointed wing near center distal portion of each valve, very tardily dehiscent in distal ½ only and persistent on plant, often blackened, apical sterile projection often breaking off; subsessile. **Seeds** few to numerous, in 2 crowded rows per locule, obovoid or narrowly obovoid, finely papillose, tan with dark splotches or brown. ***x*** = 7.

Species 2 (2 in the flora): w United States, nw Mexico.

P. H. Raven (1969), like R. Raimann (1893) and others, placed together all species of *Camissonia* with a sterile apical projection on the ovary; however, it is clear from molecular evidence (R. A. Levin et al. 2004; W. L. Wagner et al. 2007) that the perennial species with this feature (*Taraxia*) are distinct from annual species (*Tetrapteron*) and do not share an immediate common ancestor with them. *Tetrapteron* has distinctive capsules that differ from those of other genera in its irregularly obovoid shape, woody walls, and especially the pointed wings on the distal half of each valve. In addition, species of *Tetrapteron* are acaulescent or nearly so and have anthers intermediate between the basifixed anthers of *Taraxia* and its close relatives, and the versatile anthers of other species in the former *Camissonia*. The two species of *Tetrapteron* often occur on clay soil and retain their seeds very late, long after the plant has otherwise died and shriveled, which probably accounts for their unique capsular attributes. Steve Boyd (unpubl.) has observed that *T. graciliflorum* has hygrochastic capsules and suspects that *T. palmeri* also does. Reproductive features include: self-compatible; flowers diurnal; autogamous and, sometimes, cleistogamous, or in *T. graciliflorum* rarely outcrossing and pollinated by small bees or flies (Raven). The unusual sterile projections on the ovary in *Taraxia* and in *Tetrapteron* are presumed to have arisen independently. The species of both are acaulescent, so that the projection on the ovary raises the flowers above the leaves and, presumably, makes them accessible to pollinators (R. A. Raguso et al. 2007).

1. Herbs densely pilose; leaf blades linear to narrowly lanceolate; petals 5–18 mm; floral tubes 1.6–3.2 mm . . . . 1. *Tetrapteron graciliflorum*
1. Herbs sparsely to moderately strigose and sometimes also sparsely pilose; leaf blades narrowly oblanceolate; petals 2–5 mm; floral tubes 0.8–1.3 mm . . . . 2. *Tetrapteron palmeri*

1. Tetrapteron graciliflorum (Hooker & Arnott) W. L. Wagner & Hoch, Syst. Bot. Monogr. 83: 214. 2007 • Hill suncup [F]

Oenothera graciliflora Hooker & Arnott, Bot. Beechey Voy., 341. 1839; *Camissonia graciliflora* (Hooker & Arnott) P. H. Raven; *Taraxia graciliflora* (Hooker & Arnott) Raimann

Herbs densely pilose. **Stems** rarely with ascending lateral branches to 2.5 cm. **Leaves:** blade linear to very narrowly lanceolate, 1–9.8 × 0.1–0.9 cm, dilated at base, margins entire or very sparsely serrulate. **Flowers** opening near sunrise; floral tube 1.6–3.2 mm; sepals 4.5–8 mm; petals 5–18 mm; episepalous filaments 1.8–3.2 mm, epipetalous filaments 0.8–1.6 mm; sterile projection of ovary 6–45 mm; style 3–5.5 mm, short-hairy near base; stigma 1–1.6 mm diam., surrounded by anthers of longer stamens at anthesis. **Capsules** irregularly obovoid, sharply 4-angled, thick-walled, somewhat woody, with pointed wing near center-top of each valve, 4–8 × 2.6–4.8 mm, tardily dehiscent in distal ⅓. **Seeds** tan with dark splotches, obovoid, 1.2–2 mm. $2n = 14$.

Flowering Mar–May. Colonial on open or brushy slopes, on clay soil, grasslands, *Yucca* or juniper and oak shrublands; 0–800 m; Calif., Oreg.; Mexico (Baja California).

Tetrapteron graciliflorum is rare in Oregon, known only from a few collections in Jackson and Josephine counties. In Baja California, Mexico, it is known only from Rancho Aguajito.

2. Tetrapteron palmeri (S. Watson) W. L. Wagner & Hoch, Syst. Bot. Monogr. 83: 214. 2007 [E]

Oenothera palmeri S. Watson, Proc. Amer. Acad. Arts 12: 251. 1877; *Camissonia palmeri* (S. Watson) P. H. Raven; *Taraxia palmeri* (S. Watson) Small

Herbs sparsely to moderately strigose and sometimes also sparsely pilose. **Stems** rarely present, swollen, ascending to 2 cm. **Leaves:** blade narrowly oblanceolate, 1.5–5.5 × 0.2–0.7 cm, dilated at base, margins sparsely and evenly serrulate. **Flowers** opening near sunrise; floral tube 0.8–1.3 mm; sepals 1.6–2.8 mm; petals 2–5 mm; episepalous filaments 0.8–1 mm, epipetalous filaments 0.2 mm; sterile projection of ovary 5.5–12 mm; style 1–2.2 mm, glabrous; stigma 0.3–0.6 mm diam., surrounded by anthers of long and short stamens. **Capsules** irregularly obovoid, thick-walled, somewhat woody, sharply 4-angled, with pointed wing near center-top of each valve, 5–7 × 4.5–7 mm, tardily dehiscent in distal ½ only. **Seeds** brown, narrowly obovoid, 1.2–2 mm. $2n = 14$.

Flowering Mar–May. Desert habitats, on clay or sandy soil, creosote to sagebrush-juniper woodlands; 600–1400 m; Ariz., Calif., Nev., Oreg.

Tetrapteron palmeri has a disjunct distribution, occurring in four distinct areas: Arizona, represented only by the type specimens collected in the Colorado River valley in 1876; near Harper and Vale, Malheur County, Oregon; north of Winnemucca, Humboldt County, and Empire City, Ormsby County, Nevada; in California it is fairly common from southern Inyo County to the southwestern border of the Mojave Desert, west to the vicinity of Tejon Pass and southeastern San Luis Obispo County in the inner South Coast Ranges, and also east of Jacumba on the road to Mountain Springs in San Diego County.

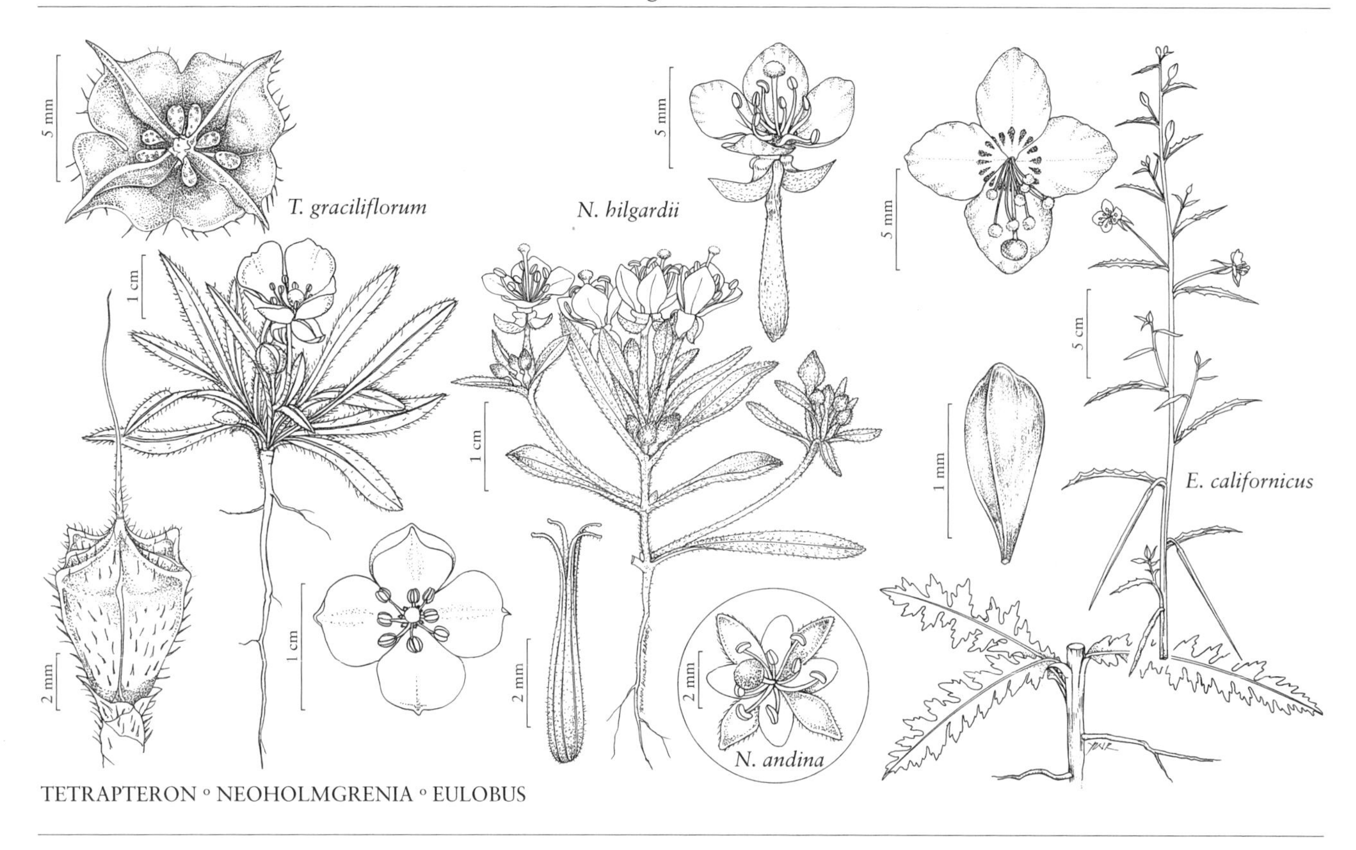

TETRAPTERON ◦ NEOHOLMGRENIA ◦ EULOBUS

14. NEOHOLMGRENIA W. L. Wagner & Hoch, Novon 19: 131. 2009 • [For the Holmgren family: Arthur Hermann Holmgren, 1912–1992, Noel Herman Holmgren, b. 1937, and Patricia Kern Holmgren, b. 1940] E

Warren L. Wagner

Holmgrenia W. L. Wagner & Hoch, Syst. Bot. Monogr. 83: 127. 2007, not Lindberg 1863 [Bryophyta], based on *Camissonia* Link sect. *Nematocaulis* P. H. Raven, Brittonia 16: 285. 1964; *Oenothera* Linnaeus sect. *Nematocaulis* (P. H. Raven) Munz

Herbs, annual, caulescent; with slender taproot. **Stems** densely leafy distally, nearly leafless proximally, with many slender, ascending branches from base. **Leaves** cauline, alternate, often appearing subverticillate and densely tufted; stipules absent; sessile; blade margins entire. **Inflorescences** spikes, densely leafy. **Flowers** bisexual, actinomorphic, buds nodding, becoming erect; floral tube deciduous (with sepals, petals, and stamens) after anthesis, relatively narrow, with basal nectary; sepals 4, reflexed singly or in pairs; petals 4, yellow, fading pale yellowish orange, without spots; stamens 8 in 2 unequal series, epipetalous ones sometimes very reduced, or 4 in 1 series, epipetalous ones absent, anthers versatile, pollen shed singly; ovary 4-locular, stigma entire, subglobose, surface probably wet and non-papillate. **Fruit** a capsule, straight, strongly flattened, somewhat torulose, regularly loculicidal, dehiscent nearly throughout length; sessile. **Seeds** numerous, in 1 row per locule, narrowly obovoid, smooth, shiny, without dots or blotches. $x = 7$.

Species 2 (2 in the flora): w North America.

R. A. Levin et al. (2004) found strong molecular support for *Neoholmgrenia* in a clade with *Camissoniopsis* and *Tetrapteron*. P. H. Raven (1969) found *Neoholmgrenia* (as *Camissonia* sect. *Nematocaulis*) to be one of the most distinctive groups in *Camissonia* by virtue of its relatively short (less than 10 mm), flattened capsules and densely clustered leaves near the tips of otherwise leafless stems. Reproductive features include: self-compatible; flowers diurnal; outcrossing and pollinated by small bees or flies, or autogamous and occasionally cleistogamous (Raven).

1. Petals 2.5–5 mm; styles 4.5–6 mm . . . . 1. *Neoholmgrenia hilgardii*
1. Petals 0.8–2.3 mm; styles 1.7–3 mm . . . . 2. *Neoholmgrenia andina*

1. Neoholmgrenia hilgardii (Greene) W. L. Wagner & Hoch, Novon 19: 132. 2009 [E] [F]

Oenothera hilgardii Greene, Bull. Torrey Bot. Club 10: 41. 1883 (as hilgardi); *Camissonia hilgardii* (Greene) P. H. Raven; *Holmgrenia hilgardii* (Greene) W. L. Wagner & Hoch; *O. andina* Nuttall var. *hilgardii* (Greene) Munz; *Sphaerostigma andinum* (Nuttall) Walpers var. *hilgardii* (Greene) A. Nelson; *S. hilgardii* (Greene) Small

Herbs finely strigillose, more densely so distally, especially on ovary. **Stems** erect to ascending, capillary, 1–15 cm. **Leaves** linear to linear-oblanceolate, 1–3 × 0.1–0.3 cm. **Flowers** opening near sunrise; floral tube 1.5–2 mm; sepals 2–3 mm; petals 2.5–5 mm; episepalous filaments 2–4 mm, epipetalous filaments 1.5–2.8 mm, sometimes epipetalous stamens absent, anthers of longer stamens 0.7–0.8 mm, those of shorter ones 0.5–0.6 mm; style 4.5–6 mm, sparsely pubescent near base, stigma 0.7–0.8 mm diam., surrounded by anthers at anthesis. **Capsules** ascending, strongly flattened from unequal width of valves, 5–10 × 1–1.3 mm. **Seeds** 0.7–1.3 × 0.3–0.4 mm. $2n = 14$.

Flowering May–Jul. Sandy or clay soil, in sagebrush scrub; 0–500 m; Oreg., Wash.

Neoholmgrenia hilgardii is probably self-compatible but outcrossing. The species occasionally grows sympatrically with *N. andina* (P. H. Raven 1969). It is known from Chelan, Douglas, Grant, Kittitas, Okanagan, and Yakima counties, Washington. The species was collected twice along the lower Columbia River, at Bingen, Klickitat County, Washington, and Hayden Island, Multnomah County, Oregon.

2. Neoholmgrenia andina (Nuttall) W. L. Wagner & Hoch, Novon 19: 131. 2009 [E] [F]

Oenothera andina Nuttall in J. Torrey and A. Gray, Fl. N. Amer. 1: 512. 1840; *Camissonia andina* (Nuttall) P. H. Raven; *Holmgrenia andina* (Nuttall) W. L. Wagner & Hoch; *O. andina* var. *anomala* M. Peck; *Sphaerostigma andinum* (Nuttall) Walpers

Herbs finely strigillose, more densely so distally, especially on ovary. **Stems** erect to ascending, capillary, 1–15 cm. **Leaves** linear to linear-oblanceolate. 1–3 × 0.1–0.3 cm. **Flowers** opening near sunrise; floral tube 0.8–2 mm; sepals 0.8–2.5 mm; petals 0.8–2.3 mm; episepalous filaments 0.5–2.2 mm, epipetalous filaments 0.1–0.5 mm, sometimes epipetalous stamens absent, anthers of longer stamens 0.2–0.5 mm, those of shorter ones 0.1–0.5 mm; style 1.7–3 mm, usually glabrous, rarely sparsely pubescent near base, stigma 0.4–0.6 mm diam., surrounded by anthers at anthesis. **Capsules** ascending, strongly flattened from unequal width of valves, (5–)8–10 × 1–1.3 mm. **Seeds** 0.7–1.3 × 0.3–0.4 mm. $2n = 28, 42$.

Flowering May–Jul. Open places, clay or sandy soil, swales or drying meadows, playa bottoms, gravelly slopes, sagebrush scrub, pinyon-juniper woodlands; 500–2000 m; Alta., B.C.; Calif., Idaho, Mont., Nev., Oreg., Utah, Wash., Wyo.

Some individuals of *Neoholmgrenia andina* have 3-merous flowers and some have stamens reduced to one whorl of four; both character states are possibly related to its predominant autogamous, and sometimes cleistogamous, habit (P. H. Raven 1969).

15. EULOBUS Nuttall ex Torrey & A. Gray, Fl. N. Amer. 1: 514. 1840 • [Greek *eu*, good or well, and *lobos*, capsule, alluding to long, linear pods]

Warren L. Wagner

Camissonia Link sect. *Eulobus* (Nuttall ex Torrey & A. Gray) P. H. Raven; *Oenothera* Linnaeus sect. *Eulobus* (Nuttall ex Torrey & A. Gray) Baillon; *Oenothera* subg. *Eulobus* (Nuttall ex Torrey & A. Gray) Munz

Herbs, annual, [perennial, subshrubs], caulescent; with a taproot. **Stems** erect [to prostrate], unbranched or branched. **Leaves** cauline, sometimes also in basal rosette, alternate; stipules absent; sessile or petiolate; blade pinnatifid to lobed, [subentire]. **Inflorescences** open spikes, erect. **Flowers** bisexual, actinomorphic, buds erect; floral tube relatively short, deciduous (with sepals, petals, and stamens) after anthesis, lined with a lobed, fleshy nectary disc [or with rounded, fleshy disc at base of style]; sepals 4, reflexed separately, in pairs, or as a unit, (rarely spreading); petals 4, yellow, fading orangish red, usually finely flecked with red near base; stamens 8, in 2 subequal series or epipetalous series shorter, anthers versatile, pollen shed singly; ovary 4-locular, stigma globose to cylindrical, surface wet and non-papillate. **Fruit** a capsule, straight or slightly curved [contorted], somewhat torulose or subterete, loculicidally dehiscent, midrib of each valve prominent; sessile. **Seeds** numerous, in 1 row per locule, narrowly obovoid, ± triangular in cross section, finely papillose. $x = 7$.

Species 4 (1 in the flora): sw United States, nw Mexico.

Eulobus is characterized by olive brown seeds with distinct purple spots, yellow petals with maroon flecks near the base, the distal part ultraviolet-reflective, leaves mostly lobed or pinnatifid, and capsules somewhat contorted or straight, and often sharply reflexed at maturity. R. A. Levin et al. (2004) included *E. californicus* and *E. crassifolius* (Greene) W. L. Wagner & Hoch (a species of Baja California, Mexico) in their molecular analysis and found *Eulobus* to be strongly supported as monophyletic; they also found that *Eulobus* plus (*Chylismia* plus *Oenothera*) formed a weakly supported clade. Reproductive features include: self-incompatible [*E. crassifolius*, *E. sceptrostigma* (Brandegee) W. L. Wagner & Hoch] or self-compatible [*E. angelorum* (S. Watson) W. L. Wagner & Hoch, *E. californicus*], flowers diurnal, outcrossing and pollinated mainly by small oligolectic bees (E. G. Linsley 1963b, 1973) or autogamous (P. H. Raven 1969).

1. Eulobus californicus Nuttall ex Torrey & A. Gray, Fl. N. Amer. 1: 515. 1840 [F]

Camissonia californica (Nuttall ex Torrey & A. Gray) P. H. Raven; *Oenothera leptocarpa* Greene

Herbs sparsely strigillose, sometimes also sparsely glandular puberulent, especially on leaves and inflorescences. **Stems** erect, virgate, bluish green and often glaucescent, or bright green, thick, hollow, fleshy, 2–180 cm. **Leaves** in a well-defined basal rosette, usually withered by anthesis, and also cauline; basal 2.5–10(–30) × 0.6–2.5(–6.5) cm, petiole 0–6 cm, blade narrowly elliptic, margins irregularly pinnatifid; cauline very much reduced distally, 1–8 × (0.1–)0.3–1(–1.6) cm, petiole 0–3 cm, blade narrowly elliptic, margins irregularly pinnatifid. **Flowers** opening at sunrise; floral tube 0.6–1.5 mm, closed by a conspicuous rounded, fleshy, reddish brown disc; sepals reddish green, 3.9–8 mm; petals ± with fine, red flecking at base, 6–14 mm; episepalous filaments 3–9 mm, epipetalous filaments 2–5 mm, anthers of longer stamens 1–2.5 mm, those of shorter stamens 0.5–1.2 mm; style 4–10 mm, stigma globose, 0.8–2 mm diam., surrounded by anthers of longer stamens at anthesis. **Capsules** sharply reflexed at maturity, cylindrical fresh, 4-angled dry, (45–)60–110 × 1–1.2 mm. **Seeds** olive brown, often flecked with purple dots, 1.3–1.6 × 0.5–0.7 mm. $2n = 14, 28$.

Flowering Dec–May. Open places in coastal sage scrub, chaparral, desert scrub, valley grasslands, foothill woodlands, washes, flats, loose soils; 0–1300 m; Ariz., Calif.; Mexico (Baja California, Sonora).

Eulobus californicus is known from western and southern Arizona, central and southern California, as well as from adjacent areas of northwestern Mexico.

P. H. Raven (1969) determined that *Eulobus californicus* is self-compatible and primarily autogamous. Known diploid ($2n$ = 14) populations are all in California.

16. CHYLISMIA (Torrey & A. Gray) Nuttall ex Raimann in H. G. A. Engler and K. Prantl, Nat. Pflanzenfam. 96[III,7]: 217. 1893 • [Greek *chylos*, juice or succulence, and *-isma*, condition, alluding to fleshy leaves of *C. scapoidea*, the type species]

Warren L. Wagner

Oenothera Linnaeus [unranked] *Chylismia* Torrey & A. Gray, Fl. N. Amer. 1: 506. 1840; *Camissonia* Link sect. *Chylismia* (Torrey & A. Gray) P. H. Raven; *Oenothera* sect. *Chylismia* (Torrey & A. Gray) P. H. Raven; *Oenothera* subg. *Chylismia* (Torrey & A. Gray) Jepson

Herbs, usually annual, sometimes perennial, rarely biennial, usually caulescent. **Stems** ascending to erect, usually branched. **Leaves** basal and cauline, cauline often reduced, basal often forming well-developed rosette, alternate; stipules absent; long-petiolate; blade often pinnately (rarely bipinnately) lobed, sometimes unlobed, or lateral lobes greatly reduced or absent, terminal lobe usually large, margins usually regularly or irregularly dentate to serrate, sometimes denticulate, serrulate, or entire, abaxial surface or margin with ± conspicuous, usually brown, oil cells. **Inflorescences** racemes, erect or nodding. **Flowers** bisexual, actinomorphic, buds usually erect, sometimes reflexed; floral tube deciduous (with sepals, petals, and stamens after anthesis), with basal nectary; sepals 4, reflexed singly; petals 4, usually yellow or white, often fading orange-red, sometimes lavender or purple, rarely cream, often with 1+ red dots near base; stamens usually 8, in 2 subequal series, rarely 4 in 1 series (usually in *C. exilis*), anthers versatile, pollen shed singly or in tetrads; ovary 4-locular, stigma usually entire and capitate, rarely conical-peltate and ± 4-lobed, surface unknown, probably wet and non-papillate. **Fruit** a capsule, straight or slightly curved, subterete and clavate or oblong-cylindrical, regularly loculicidal; pedicellate. **Seeds** numerous, in 2 rows per locule, lenticular to narrowly ovoid to narrowly obovoid, finely pitted, with ± pronounced membranous margin when immature. x = 7.

Species 16 (16 in the flora): w United States, nw Mexico.

Chylismia is distinguished from other genera formerly included in *Camissonia* by straight to arcuate (never twisted or curled) capsules on distinct pedicels and seeds in 2 rows per locule. R. A. Levin et al. (2004) included only one species each from *Camissonia* sects. *Chylismia* and *Lignothera*; the two formed a moderately supported branch, which led W. L. Wagner et al. (2007) to recognize *Chylismia* as a distinct genus. *Chylismia* is strongly supported in a sister relationship to the realigned *Oenothera*. This clade is in turn sister to *Eulobus*. Reproductive features include: self-incompatible (*C. brevipes*, *C. claviformis*, *C. multijuga*, *C. munzii*, and probably *C. confertiflora*, *C. eastwoodiae*, and *C. parryi*; P. H. Raven 1962, 1969) or self-compatible; flowers diurnal, outcrossing and pollinated by mostly oligolectic bees or autogamous, or opening one to two hours before sunset (in one subspecies of *C. claviformis* and the two species of sect. *Lignothera*); the evening-opening subspecies of *C. claviformis* pollinated mostly by oligolectic bees and moths, *C. cardiophylla* mainly by small moths, and *C. arenaria*, with its long floral tubes, by hawkmoths (E. G. Linsley et al. 1963, 1963b, 1964). Most species are diploid ($2n$ = 14) but there are occasional tetraploids ($2n$ = 28); floating translocations are relatively common (Raven 1962, 1969).

SELECTED REFERENCE Raven, P. H. 1962. The systematics of *Oenothera* subgenus *Chylismia*. Univ. Calif. Publ. Bot. 34: 1–122.

1. Floral tubes 0.4–9 mm; pollen shed singly; leaves basal and cauline, usually with well-developed basal rosettes, blades usually pinnately or bipinnately lobed, lateral lobes sometimes greatly reduced or absent; plants usually annual, sometimes perennial, rarely biennial . . . . 16a. *Chylismia* sect. *Chylismia*, p. 228
1. Floral tubes 4.5–40 mm; pollen shed in tetrads; leaves cauline, blades unlobed; plants usually perennial, sometimes annual . . . . 16b. *Chylismia* sect. *Lignothera*, p. 242

16a. CHYLISMIA (Torrey & A. Gray) Nuttall ex Raimann sect. CHYLISMIA

Camissonia Link sect. *Tetranthera* (P. H. Raven) P. H. Raven; *Oenothera* Linnaeus sect. *Tetranthera* P. H. Raven

Herbs usually annual, sometimes perennial, rarely biennial. Leaves basal and cauline, usually with well-developed basal rosette; blade usually pinnately or bipinnately lobed, sometimes with scattered, irregular lobes, sometimes lateral lobes greatly reduced or absent, terminal lobe elliptic, narrowly to broadly ovate to oblong, lanceolate, oblanceolate, cordate, or subcordate. **Flowers** usually opening at sunrise, rarely at sunset; floral tube 0.4–9 mm; petals usually bright yellow, rarely white or cream, usually with red dots basally, or lavender to purple with white or yellow basally, sometimes with darker flecks near base, fading yellow, orange, reddish, or lavender; pollen shed singly.

Species 14 (14 in the flora): w United States, nw Mexico.

Section *Chylismia* consists of 10 diploid ($2n = 14$) species, and two that are partly polyploid ($2n = 14, 28$) [*C. scapoidea* subsp. *scapoidea* and *C. walkeri* subsp. *walkeri* (only one tetraploid population)]; no chromosome counts are available for the remaining two species, *C. atwoodii* and *C. confertiflora* (P. H. Raven 1962, 1969). Species of sect. *Chylismia* usually occur on sandy desert slopes, flats, and washes, often in sagebrush shrubland in the northern part of its range, or on rock slides or cliffs, mainly in the Mojave and northwestern Sonoran deserts, the Great Basin, and the lower elevations of the surrounding Sierra Nevada and Rocky Mountains. The limits of this range stretch from southeastern Oregon, central and southern Idaho, and central Wyoming, south through Nevada and Utah to eastern and southeastern California, northern Baja California and northwestern Sonora, Mexico, Arizona, northwestern New Mexico, and western Colorado. Several species are rare; *C. confertiflora* and *C. specicola* are known only from the Grand Canyon in northwestern Arizona, *C. megalantha* only from the vicinity of the type locality in Nye County, Nevada, and *C. atwoodii* also only from a narrow area around its type locality in Kane County, Utah. Others are widespread, especially the very diverse *C. claviformis* (11 subspp.), *C. scapoidea* (4 subspp.), *C. walkeri* (2 subspp.), and *C. brevipes* (3 subspp.). *Chylismia scapoidea* is the only species in the genus to occur east of the continental divide, both in Colorado on the upper Arkansas River in Fremont and Pueblo counties, and much more widely in Wyoming. *Chylismia* does not occur west of the Cascade-Sierra Nevada axis. Because R. A. Levin et al. (2004) included only *C. claviformis* in their analysis, they did not test the monophyly of sect. *Chylismia*; however, this section is both geographically distinct and morphologically set apart by the characteristic pinnate leaves (modified in some species, which have retained the entire apical lobe but do not have the smaller lateral lobes). Most species have bright yellow petals with red dots proximally and ultraviolet reflectance distally; some subspecies of *C. claviformis* have white petals; three species (*C. atwoodii*, *C. heterochroma*, and *C. megalantha*) have lavender or purple petals, often with lavender or purple flecks toward base, and white or yellow at the base and no reflectance, clearly a derived condition within the section (Raven 1962, 1969).

1. Stamens usually 4, rarely 8, then antipetalous anthers abortive, smaller .... 11. *Chylismia exilis*
1. Stamens 8, anthers all fertile.
   2. Petals lavender or purple, often with purple or lavender flecks near base.
      3. Stigma surrounded by anthers at anthesis, plants autogamous; petals 2–6 mm... ....14. *Chylismia heterochroma*
      3. Stigma exserted beyond anthers at anthesis, plants outcrossing; petals 7–14 mm.
         4. Capsules 8–14 mm, erect or ascending; floral tubes 4–9 mm .... 12. *Chylismia megalantha*
         4. Capsules 11–25 mm, spreading to reflexed; floral tubes 0.6–1 mm...... 13. *Chylismia atwoodii*
   2. Petals yellow or white, often with red dots at base.
      5. Capsules distinctly clavate.
         6. Stigma surrounded by anthers at anthesis; petals 1.5–5.5(–8) mm ....9. *Chylismia scapoidea*
         6. Stigma exserted beyond anthers at anthesis; petals 2–10 mm.
            7. Mature capsules sharply reflexed; petals bright yellow............... 7. *Chylismia munzii*
            7. Mature capsules ascending, spreading, erect, or slightly reflexed; petals yellow or white.
               8. Raceme branches intricate and filiform; leaves in poorly defined basal rosette; capsules 4–10 mm.................................10. *Chylismia parryi*
               8. Raceme branches not filiform; leaves primarily in well-defined basal rosette; capsules 8–40 mm.
                  9. Stamens subequal; leaves usually pinnately lobed, with well-developed lateral lobes, sometimes these reduced or absent, margins dentate, sinuate-dentate, or serrate ....6. *Chylismia claviformis*
                  9. Stamens unequal, differentiated into 2 sets; leaves usually unlobed, rarely pinnately lobed with reduced lateral lobes, margins entire or sparsely denticulate ....8. *Chylismia eastwoodiae*
      5. Capsules oblong-cylindrical.
         10. Stigma surrounded by anthers at anthesis; petals 1–6 mm; styles 1.5–6(–7) mm; racemes erect in bud, mature buds individually reflexed.
             11. Plants perennial, base sometimes woody; leaf blades glabrous or sparsely villous; anthers glabrous .... 4. *Chylismia specicola*
             11. Plants annual or short-lived perennial, base not woody; leaf blades moderately to densely villous; anthers glabrous or sparsely ciliate..... 5. *Chylismia walkeri*
         10. Stigma exserted beyond anthers at anthesis; petals 3–18 mm; styles 7–18 mm; racemes erect or nodding in bud, mature buds individually reflexed or not.
             12. Racemes erect to nodding, elongating in bud, mature buds individually reflexed; plants virgate; floral tubes 1–3 mm ....3. *Chylismia multijuga*
             12. Racemes nodding, mostly elongating after flowers open; plants not virgate; floral tubes 3–8 mm.
                 13. Buds without free tips or with subapical free tips 1–2 mm, or with minute apical free tips less than 1 mm, and then bud pubescent but not glandular puberulent; plants branched or unbranched distally; stamens subequal....................................... 1. *Chylismia brevipes*
                 13. Buds with subapical free tips 1–2 mm, glandular puberulent; plants well branched; stamens differentiated into 2 sets............2. *Chylismia confertiflora*

1. Chylismia brevipes (A. Gray) Small, Bull. Torrey Bot. Club 23: 194. 1896 (as Chylisma) [E]

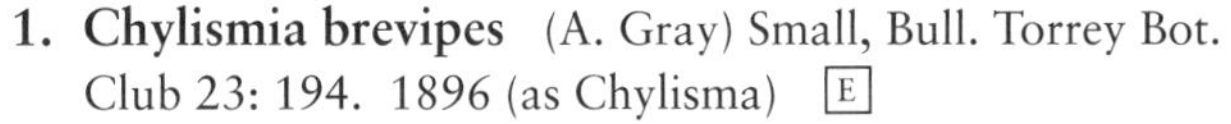

Oenothera brevipes A. Gray in War Department [U.S.], Pacif. Rail. Rep. 4(5): 87. 1857; *Camissonia brevipes* (A. Gray) P. H. Raven

Herbs annual, moderately to densely villous, sometimes strigillose. **Stems** branched, 3–75 cm. **Leaves** primarily in basal rosette, cauline greatly reduced when present, 6–14 × 1.5–3.5 cm; petiole 1.5–4(–11) cm; blade pinnately lobed or lateral lobes greatly reduced or absent, often mixed on same plant, terminal lobe usually ovate, rarely elliptic, 2.5–6.9 × 1.5–7 cm, margins irregularly dentate, oil cells on abaxial surface inconspicuous. **Racemes** nodding, mostly elongating after flowers. **Flowers** opening at sunrise; buds sometimes individually reflexed, without free tips or with subapical free tips 1–2 mm, or with minute, apical free tips less than 1 mm; floral tube 3–8 mm, densely short-villous inside proximally; sepals 5–9 mm; petals bright yellow, sometimes with red dots at base, fading yellow to orange or reddish, 3–18 mm; stamens subequal, filaments 3–6 mm, anthers 2.5–6 mm, ciliate;

style 10–18 mm, stigma exserted beyond anthers at anthesis. **Capsules** ascending or spreading, oblong-cylindrical, 18–92 mm; pedicel 2–20 mm. **Seeds** 1–1.5 mm.

Subspecies 3 (3 in the flora): sw United States.

P. H. Raven (1962, 1969) determined this species to be self-incompatible.

1. Flower buds individually reflexed; petals often fading reddish, 3–8 mm . . . . 1c. *Chylismia brevipes* subsp. *arizonica*
1. Flower buds not individually reflexed; petals fading yellow to orange, 6–18 mm.
   2. Plants stout, villous; buds with subapical free tips 1–2 mm; petals usually without red dots at base . . . . . . 1a. *Chylismia brevipes* subsp. *brevipes*
   2. Plants slender, usually strigillose, sometimes also villous proximally; buds with apical free tips 0–1 mm; petals often with red dots near base . . . . .1b. *Chylismia brevipes* subsp. *pallidula*

1a. Chylismia brevipes (A. Gray) Small subsp. **brevipes** E

Oenothera divaricata Greene

Herbs stout, somewhat succulent, usually unbranched distally, usually densely villous, usually also strigillose. **Flowers:** buds not individually reflexed, with subapical free tips 1–2 mm; floral tube yellow inside, 4–8 mm; petals usually without red dots at base, fading yellow to orange, 6–18 mm; anthers 4–6 mm. **Capsules** 20–92 mm; pedicel 5–20 mm. $2n = 14$.

Flowering Feb–May. Hillsides, alluvial fans, sandy slopes; -70–2100 m; Ariz., Calif., Nev., Utah.

Subspecies *brevipes* occurs in the Sonoran and Mojave deserts from southern Esmeralda County, Nevada, and southwestern Washington County, Utah, south to northern Yuma County, Arizona, and northern Imperial County, California. It intergrades with subsp. *pallidula* and hybridizes with *Chylismia claviformis* subspp. *claviformis* and *integrior*, as well as *C. multijuga* and *C. munzii*.

1b. Chylismia brevipes (A. Gray) Small subsp. **pallidula** (Munz) W. L. Wagner & Hoch, Syst. Bot. Monogr. 83: 206. 2007 E

Oenothera brevipes A. Gray var. *pallidula* Munz, Amer. J. Bot. 15: 229. 1928; *Camissonia brevipes* (A. Gray) P. H. Raven subsp. *pallidula* (Munz) P. H. Raven; *O. brevipes* subsp. *pallidula* (Munz) P. H. Raven; *O. pallidula* (Munz) Munz

Herbs slender, usually branched distally, usually strigillose, sometimes also villous proximally. **Flowers:** buds not individually reflexed, without free tips or with apical free tips less than 1 mm; floral tube yellow inside, 4–5 mm; petals often with red dots near base, fading yellow to orange, 7–12 mm; anthers 4–6 mm. **Capsules** 20–42 mm; pedicel 2–10 mm. $2n = 14$.

Flowering Mar–May. Dry flats, desert pavement, with *Ambrosia dumosa* and *Larrea*; 70–1100 m; Ariz., Calif., Nev., Utah.

Subspecies *pallidula* occurs in the Sonoran and Mojave deserts from Washington County, Utah, to southeastern Inyo County, California, south to Riverside County, California, Mohave County, Arizona, and Clark County, Nevada. It hybridizes with *Chylismia claviformis* subspp. *aurantiaca* and *peeblesii*, as well as with *C. multijuga*.

1c. Chylismia brevipes (A. Gray) subsp. **arizonica** (P. H. Raven) W. L. Wagner & Hoch, Syst. Bot. Monogr. 83: 205. 2007 C E

Oenothera brevipes A. Gray subsp. *arizonica* P. H. Raven, Univ. Calif. Publ. Bot. 34: 84. 1962; *Camissonia brevipes* (A. Gray) P. H. Raven subsp. *arizonica* (P. H. Raven) P. H. Raven

Herbs slender, often branched distally, villous, sometimes also strigillose, at least distally. **Flowers:** buds individually reflexed, without free tips or with apical free tips less than 1 mm; floral tube orange-brown inside, 3–5 mm; petals often with red dots at base, often fading reddish, 3–8 mm; anthers 2.5–5 mm. **Capsules** 18–60 mm; pedicel 2.5–5 mm. $2n = 14$.

Flowering Feb–Apr. Rocky slopes, flats; of conservation concern; 70–300 m; Ariz., Calif.

Subspecies *arizonica* is known from Yuma County, Arizona, and eastern Imperial County, California. It hybridizes with *Chylismia claviformis* subspp. *peeblesii*, *rubescens*, and *yumae*.

2. Chylismia confertiflora (P. H. Raven) W. L. Wagner & Hoch, Syst. Bot. Monogr. 83: 207. 2007 [C] [E]

Oenothera confertiflora P. H. Raven, Univ. Calif. Publ. Bot. 34: 80. 1962; *Camissonia confertiflora* (P. H. Raven) P. H. Raven

Herbs annual, densely villous and strigillose, glandular puberulent on distal parts. **Stems** well branched, 15–50 cm. **Leaves** in well-developed basal rosette and also cauline, 7–20 × 1.5–2.5 cm; petiole 1–4(–8) cm; blade pinnately lobed, terminal lobe oblanceolate to narrowly ovate, 2.5–5 × 1–2.5 cm, margins irregularly dentate, oil cells on abaxial surface inconspicuous. **Racemes** nodding, dense, mostly elongating after flowers open. **Flowers** opening at sunrise; buds with conspicuous, subapical free tips 1–2 mm; floral tube 3–5 mm, short-villous inside proximally; sepals 9–12 mm; petals bright yellow, with red dots at base, fading lavender, 12–18 mm; stamens unequal, filaments of antisepalous stamens 6–8 mm, those of antipetalous ones 4–5 mm, anthers 4–6 mm, ciliate; style 11–18 mm, stigma exserted beyond anthers at anthesis. **Capsules** ascending or spreading, oblong-cylindrical, immature capsule to 35 mm; pedicel 5–15 mm. **Seeds** not known.

Flowering Apr–May. Cinder soil; of conservation concern; 1300–1400 m; Ariz.

Chylismia confertiflora is known only from the type locality on the east side and base of Vulcan's Throne, Toroweap Valley, Grand Canyon National Monument in Mohave County. P. H. Raven (1962, 1969) assumed this species to be self-incompatible, based on the large flowers with the stigma elevated above the anthers.

A. Cronquist et al. (1997c) treated *Chylismia confertiflora* as part of *C. brevipes*, with a comment about the capsule dimensions, and indicated they consider the differences to be one end of the spectrum of a variation within *C. brevipes*. Although known from very few collections and sparse field data, P. H. Raven (1962, 1969) considered it to be distinct and to be most closely related to *C. brevipes* and to *C. multijuga*. He distinguished it from the latter by its larger flowers, nodding inflorescences, and large buds; and from the former by its glandular puberulent sepals, unequal stamens, and uniformly branched habit. In addition, the very restricted range of *C. confertiflora* is outside (to the east) of the range of *C. brevipes*. *Chylismia multijuga* grows within a few miles of the only know locality, but in this area *C. confertiflora* is very distinct from that species. Further collections and more detailed study of the overall morphological patterns as well as, perhaps, molecular data may clarify whether this is best considered to be an extremely restricted distinct species or a somewhat distinct outlier of the variable *C. brevipes*.

3. Chylismia multijuga (S. Watson) Small, Bull. Torrey Bot. Club 23: 193. 1896 (as Chylisma) [E]

Oenothera multijuga S. Watson, Amer. Naturalist 7: 300. 1873; *Camissonia multijuga* (S. Watson) P. H. Raven; *Chylisma hirta* A. Nelson; *C. parviflora* (S. Watson) Rydberg; *C. venosa* A. Nelson & P. B. Kennedy; *O. brevipes* A. Gray var. *multijuga* (S. Watson) Jepson; *O. brevipes* var. *parviflora* S. Watson; *O. multijuga* var. *parviflora* (S. Watson) Munz; *O. phlebophylla* Tidestrom; *O. watsonii* Tidestrom

Herbs annual or biennial, villous, at least proximally, glabrous or even glaucous distally, rarely glandular puberulent. **Stems** virgate with numerous divergent branches, 20–150 cm. **Leaves** primarily in well-developed basal rosette, cauline reduced or absent, 6–30 × 1.4–6.5 cm; petiole 0.3–6 cm; blade pinnately or bipinnately lobed, terminal lobe ovate to elliptic, 2.5–6.5 × 1.5–3 cm, margins irregularly serrate, dark brown oil cells prominently lining veins abaxially. **Racemes** erect to nodding, elongating in bud. **Flowers** opening at sunrise; buds individually reflexed, with apical or slightly subapical free tips less than 1 mm; floral tube 1–3 mm, glabrous or villous inside proximally; sepals 3–8 mm; petals usually bright yellow, rarely cream, fading yellow to lavender, 4–9 mm; stamens unequal, filaments of antisepalous stamens 2.5–4 mm, those of antipetalous ones 1.3–3 mm, anthers 2–4 mm, ciliate; style 7–11 mm, stigma exserted beyond anthers at anthesis. **Capsules** usually spreading, rarely slightly reflexed, oblong-cylindrical, 10–52 mm; pedicel 7–20 mm. **Seeds** 1–1.3 mm. $2n = 14$.

Flowering Mar–Jun(–Sep). Forming colonies on rocky slopes and banks of eroded sedimentaries, on gypsum or limestone, on conglomerates, often with *Juniperus* and *Pinus edulis*, with *Encelia farinosa* and *Larrea*; 300–1100 m; Ariz., Nev., Utah.

Chylismia multijuga is known from Washington County, Utah, and southern Lincoln County, Nevada, to northern Mohave County, Arizona. P. H. Raven (1962, 1969) determined this species to be self-incompatible. It hybridizes with *C. brevipes* subspp. *brevipes* and *pallidula*.

4. Chylismia specicola (P. H. Raven) W. L. Wagner & Hoch, Syst. Bot. Monogr. 83: 208. 2007 [C] [E]

Oenothera specicola P. H. Raven, Univ. Calif. Publ. Bot. 34: 87. 1962 (as specuicola); *Camissonia specicola* (P. H. Raven) P. H. Raven

Herbs perennial, base sometimes woody, glabrous or sparsely villous proximally. **Stems** with several divergent branches from base, 10–50 cm. **Leaves** primarily in basal rosette and cauline, 3–20 × 0.7–2.5 cm; petiole 0.5–4 cm; blade pinnately or bipinnately lobed, terminal lobe ovate to elliptic, 0.4–2.5 × 0.3–1.5 cm, margins irregularly serrate, dark brown oil cells prominently lining veins abaxially. **Racemes** erect, elongating after anthesis. **Flowers** opening at sunrise; buds individually reflexed, with apical free tips less than 1 mm; floral tube 1.5–2 mm, glabrous inside; sepals 2–5 mm; petals bright yellow, with red dots near base, fading pale lavender, 2–6 mm; stamens unequal, filaments of antisepalous stamens 1.5–3 mm, those of antipetalous ones 1–2 mm, anthers 1.2–2 mm, glabrous; style 4–7 mm, stigma surrounded by anthers at anthesis. **Capsules** spreading to ascending, oblong-cylindrical, 8–20 mm; pedicel 6–10 mm. **Seeds** 0.6–1 mm.

Subspecies 2 (2 in the flora): Arizona.

P. H. Raven (1962, 1969) thought that this species is most likely self-compatible but primarily outcrossing.

1. Leaf blades and styles glabrous; herbs ± woody at base. . . . . 4a. *Chylismia specicola* subsp. *specicola*
1. Leaf blades and styles villous, at least at base; herbs not woody . . . . 4b. *Chylismia specicola* subsp. *hesperia*

4a. Chylismia specicola (P. H. Raven) W. L. Wagner & Hoch subsp. **specicola** [C] [E]

Herbs somewhat woody at base. **Leaf blades** glabrous. **Styles** 4–5 mm, glabrous. **2*n*** = 14.

Flowering Jun–Sep. Crevices of broken limestone; of conservation concern; 1300–1700 m; Ariz.

Subspecies *specicola* is known from the south rim of the Grand Canyon in Coconino County.

4b. Chylismia specicola (P. H. Raven) W. L. Wagner & Hoch subsp. **hesperia** (P. H. Raven) W. L. Wagner & Hoch, Syst. Bot. Monogr. 83: 208. 2007 [C] [E]

Oenothera specicola P. H. Raven subsp. *hesperia* P. H. Raven, Univ. Calif. Publ. Bot. 34: 88. 1962 (as specuicola); *Camissonia specicola* (P. H. Raven) P. H. Raven subsp. *hesperia* (P. H. Raven) P. H. Raven

Herbs not woody at base. **Leaf blades** villous. **Styles** 5–7 mm, villous at base. **2*n*** = 14.

Flowering May–Oct. Washes, dry stream beds, on limestone; of conservation concern; 300–1100 m; Ariz.

Subspecies *hesperia* is known from two disjunct areas along the Colorado River: Havasu and Hualapai canyons in Coconino County and from Separation to Spencer canyons in Mohave County.

5. Chylismia walkeri A. Nelson, Bot. Gaz. 56: 66. 1913 (as Chylisma) [E]

Camissonia walkeri (A. Nelson) P. H. Raven; *Oenothera walkeri* (A. Nelson) P. H. Raven

Herbs annual or short-lived perennial, villous, usually densely so proximally, less dense to glabrate distally, sometimes hairs somewhat appressed and shorter on leaves, also sometimes glandular puberulent on distal parts. **Stems** slender, unbranched or branched from base, 10–60 cm. **Leaves** in basal rosette and/or cauline, often purple-dotted, 2–22 × 0.4–3.5 cm; petiole 0.4–8 cm; blade pinnately lobed, sometimes lateral lobes greatly reduced or absent and blade reduced to terminal lobe only, terminal lobe oblong or cordate to ovate, 1–5 × 0.5–3.2 cm, margins serrate, brown oil cells prominently lining veins abaxially. **Racemes** erect, elongating after anthesis. **Flowers** opening at sunrise; buds individually reflexed, with apical free tips less than 1 mm; floral tube 0.5–1.5 mm, glabrous or sparsely villous inside; sepals 1.5–5 mm; petals bright yellow, fading pale orange or lavender, 1–6 mm; stamens unequal, filaments of antisepalous stamens 1–3 mm, those of antipetalous ones 0.3–2 mm, anthers 0.5–2 mm, glabrous or sparsely ciliate; style 1.5–6 mm, stigma surrounded by anthers at anthesis. **Capsules** spreading or ascending, oblong-cylindrical, 11–45 mm; pedicel 5–30 mm. **Seeds** 0.6–1.2 mm.

Subspecies 2 (2 in the flora): sw United States.

P. H. Raven (1962, 1969) determined this species to be self-incompatible and primarily autogamous.

1. Leaves primarily cauline, rarely forming inconspicuous basal rosette, lateral lobes usually greatly reduced or absent; petals 1–3 mm; anthers 0.5–0.8 mm. . . . . . 5a. *Chylismia walkeri* subsp. *walkeri*
1. Leaves primarily basal, forming conspicuous rosette, cauline reduced or absent, lateral lobes usually well developed; petals 2.8–6 mm; anthers 1–2 mm. . . . . 5b. *Chylismia walkeri* subsp. *tortilis*

5a. Chylismia walkeri A. Nelson subsp. **walkeri** [E]

Camissonia bolanderi N. D. Atwood & S. L. Welsh; *C. dominguezescalanteorum* N. D. Atwood, L. C. Higgins & S. L. Welsh; *Oenothera multijuga* S. Watson var. *orientalis* Munz

Leaves primarily cauline, rarely forming basal rosette; blade with only terminal lobe well developed, sometimes pinnately lobed, lateral lobes usually greatly reduced or absent. **Flowers:** floral tube 0.5–1.3 mm; sepals 1.5–2 mm; petals 1–3 mm; anthers 0.5–0.8 mm, glabrous; style 1.5–4 mm. **Capsules:** pedicel 5–15 mm. **2*n*** = 14, 28.

Flowering Apr–Jun. Loose slides of limestone and other sedimentaries, sandy washes; 900–1800 m; Ariz., Colo., N.Mex., Utah.

Subspecies *walkeri* is nearly confined to the Colorado Plateau, known from Mesa County, Colorado, and Emery County, Utah, south to Montezuma County, Colorado, San Juan County, New Mexico, and central Coconino and northeastern Mohave counties, Arizona.

5b. Chylismia walkeri A. Nelson subsp. **tortilis** (Jepson) W. L. Wagner & Hoch, Syst. Bot. Monogr. 83: 208. 2007 [E]

Oenothera scapoidea Torrey & A. Gray var. *tortilis* Jepson, Man. Fl. Pl. Calif., 687. 1925; *Camissonia walkeri* (A. Nelson) P. H. Raven subsp. *tortilis* (Jepson) P. H. Raven; *C. walkeri* var. *tortilis* (Jepson) S. L. Welsh; *O. walkeri* (A. Nelson) P. H. Raven subsp. *tortilis* (Jepson) P. H. Raven

Leaves primarily basal, forming conspicuous rosette, cauline reduced or absent; blade usually pinnately lobed, rarely bipinnately, lateral lobes usually well developed. **Flowers:** floral tube 1–1.5 mm; sepals 1.5–5 mm; petals 2.8–6 mm; anthers 1–2 mm, glabrous or sparsely ciliate; style 4–6 mm. **Capsules:** pedicel 5–30 mm. **2*n*** = 14.

Flowering Apr–Jun. Colonial in rocky debris near cliffs, along ephemeral streams, on limestone; 600–1900 m; Calif., Nev., Utah.

Subspecies *tortilis* is known from Millard and Washington counties, Utah, west to southern Elko and Mineral counties, Nevada, southwest to Inyo and northeastern San Bernardino counties, California, eastward to Clark County, Nevada.

6. Chylismia claviformis (Torrey & Frémont) A. Heller, Muhlenbergia 2: 105. 1906 (as Chylisma clavaeformis) [F]

Oenothera claviformis Torrey & Frémont in J. C. Frémont, Rep. Exped. Rocky Mts., 314. 1845 (as clavaeformis); *Camissonia claviformis* (Torrey & Frémont) P. H. Raven; *Chylismia scapoidea* (Torrey & A. Gray) Nuttall ex Raimann var. *claviformis* (Torrey & Frémont) Small; *O. scapoidea* Torrey & A. Gray var. *claviformis* (Torrey & Frémont) S. Watson

Herbs annual, glabrous, strigillose, glandular puberulent, or, sometimes, villous. **Stems** branched mostly from base, 3–70 cm. **Leaves** primarily in basal rosette, cauline reduced or absent, 1.5–20 × 0.3–3.5 cm; petiole 0.7–12 cm; blade usually pinnately lobed, sometimes lateral lobes poorly developed or absent, terminal lobe usually narrowly ovate to lanceolate, sometimes cordate or subcordate, 0.8–9 × 0.2–4.5 cm, margins dentate, sinuate-dentate, or serrate, brown oil cells conspicuously lining veins abaxially. **Racemes** nodding, elongating after anthesis. **Flowers** opening at sunset or sunrise; buds with or without subapical or apical free tips; floral tube 2–6.5 mm, villous inside proximally; sepals 2–8 mm; petals pale to bright yellow or white, sometimes red- or purple-dotted near base, fading purple, sometimes red or orange, or not changing color, 1.5–8 mm; stamens subequal, filaments 1.5–5.5 mm, anthers 1.5–6 mm, ciliate; style 5–16 mm, stigma exserted beyond anthers at anthesis. **Capsules** ascending to spreading, clavate, 8–40 mm; pedicel 4–40 mm. **Seeds** 0.6–1.5 mm.

Subspecies 11 (10 in the flora): w United States, nw Mexico.

P. H. Raven (1962) subdivided this species into 12 subspecies and, subsequently (1969), he combined two of them. The latter approach is used here. Only subsp. *wigginsii* P. H. Raven does not occur in the United States; its narrow range is restricted to northern Baja California. Raven (1962, 1969) determined this species to be self-incompatible.

Chylismia claviformis is the most complex and, along with *C. scapoidea*, the most widely distributed species of the genus. The central part of its geographical range is occupied by five closely related white-petaled subspecies (*aurantiaca*, *claviformis*, *funerea*, *integrior*, and *peeblesii*) that are very similar morphologically.

South of this area four additional subspecies occur, all yellow-petaled (*peirsonii*, *rubescens*, *wigginsii*, and *yumae*). These four subspecies have sepals and petal color similar to those of *C. brevipes*, and P. H. Raven (1962, 1969) thought it likely that they were derived following hybridization between that species and one of the white-petaled populations of *C. claviformis*. North of the range of the white-petaled subspecies are found two additional yellow-petaled subspecies (*cruciformis* and *lancifolia*). Most populations of subsp. *cruciformis* consist of plants in which the flowers open in the early morning; in all other subspecies the flowers open in the late afternoon (Raven 1962, 1969). The following key will separate them, but there are many intergrades among the subspecies so that not all specimens will be easily identified.

1. Herbs villous proximally; buds with subapical free tips; petals usually yellow, rarely white.
    2. Petals not changing color in fading, 4.5–7 mm; lateral lobes of leaf blades well developed . . . . 6h. *Chylismia claviformis* subsp. *peirsonii*
    2. Petals fading brick red, 3–5 mm; lateral lobes of leaf blades poorly developed, small or absent . . . 6i. *Chylismia claviformis* subsp. *rubescens*
1. Herbs strigillose, glandular puberulent, or glabrous proximally; buds with or without subapical or apical free tips; petals yellow or white.
    3. Petals yellow.
        4. Herbs strigillose, sometimes also glandular puberulent distally . . . . .6j. *Chylismia* claviformis subsp. *yumae*
        4. Herbs glabrous or glandular puberulent distally.
            5. Leaf blades with narrowly ovate to subcordate terminal lobes, at least some lateral lobes developed; flowers usually opening at dawn . . . 6c. *Chylismia claviformis* subsp. *cruciformis*
            5. Leaf blades with lanceolate terminal lobes, lateral lobes usually greatly reduced or absent; flowers usually opening at dusk. . . . 6f. *Chylismia claviformis* subsp. *lancifolia*
    3. Petals usually white, rarely pale yellow (in subsp. *claviformis*).
        6. Herbs usually glabrous distally, rarely sparsely strigillose or glandular puberulent; lateral lobes of leaf blades usually well developed . . . . .6a. *Chylismia claviformis* subsp. *claviformis*
        6. Herbs usually strigillose and/or glandular puberulent, rarely glabrate; lateral lobes of leaf blades reduced, absent, or well developed.

[7. Shifted to left margin.—Ed.]

7. Lateral lobes of leaf blades well developed.
    8. Herbs strigillose, sometimes glabrate distally . . . . 6b. *Chylismia claviformis* subsp. *aurantiaca*
    8. Herbs glandular puberulent and strigillose . . . . .6g. *Chylismia claviformis* subsp. *peeblesii*
7. Lateral lobes of leaf blades poorly developed, small, or absent.
    9. Leaf blades usually with at least some poorly developed lateral lobes; plants strigillose . . . . 6d. *Chylismia claviformis* subsp. *funerea*
    9. Leaf blades often with only terminal lobe developed; plants strigillose proximally, strigillose and glandular puberulent or glabrate distally. . . . 6e. *Chylismia claviformis* subsp. *integrior*

6a. Chylismia claviformis (Torrey & Frémont) A. Heller subsp. **claviformis** [E] [F]

Herbs glabrous or strigillose proximally, sometimes also glandular puberulent, usually glabrous distally, rarely very sparsely strigillose or glandular puberulent. **Stems** 6–55 cm. **Leaves:** blade lateral lobes usually well developed, terminal lobe narrowly ovate, to 6 × 3.5 cm, margins irregularly sinuate-dentate. **Flowers** opening at sunset; buds with apical free tips less than 1 mm; floral tube orange-brown inside, 3–5.5 mm; petals usually white, very rarely pale yellow, sometimes purple-dotted near base, often fading purple, 3.5–8 mm. $2n = 14$.

Flowering Mar–May. Alluvial slopes and flats, with *Ambrosia dumosa* and *Larrea*; 800–1700 m; Calif.

Subspecies *claviformis* is known from western Inyo, eastern Kern, northern Los Angeles, western San Bernardino, and northern Riverside counties, almost entirely in the Mojave Desert. It intergrades widely and gradually with subspp. *aurantiaca* and *funerea*, and hybridizes with *Chylismia brevipes* subsp. *brevipes*.

6b. Chylismia claviformis (Torrey & Frémont) A. Heller subsp. **aurantiaca** (Munz) W. L. Wagner & Hoch, Syst. Bot. Monogr. 83: 206. 2007 [F]

Oenothera claviformis Torrey & Frémont var. *aurantiaca* Munz, Amer. J. Bot. 15: 237. 1928 (as clavaeformis); *Camissonia claviformis* (Torrey & Frémont) P. H. Raven subsp. *aurantiaca* (Munz) P. H. Raven; *C. claviformis* var. *aurantiaca* (Munz) Cronquist; *Chylismia aurantiaca* (Munz) D. A. Johansen; *O. claviformis* subsp. *aurantiaca* (Munz) P. H. Raven

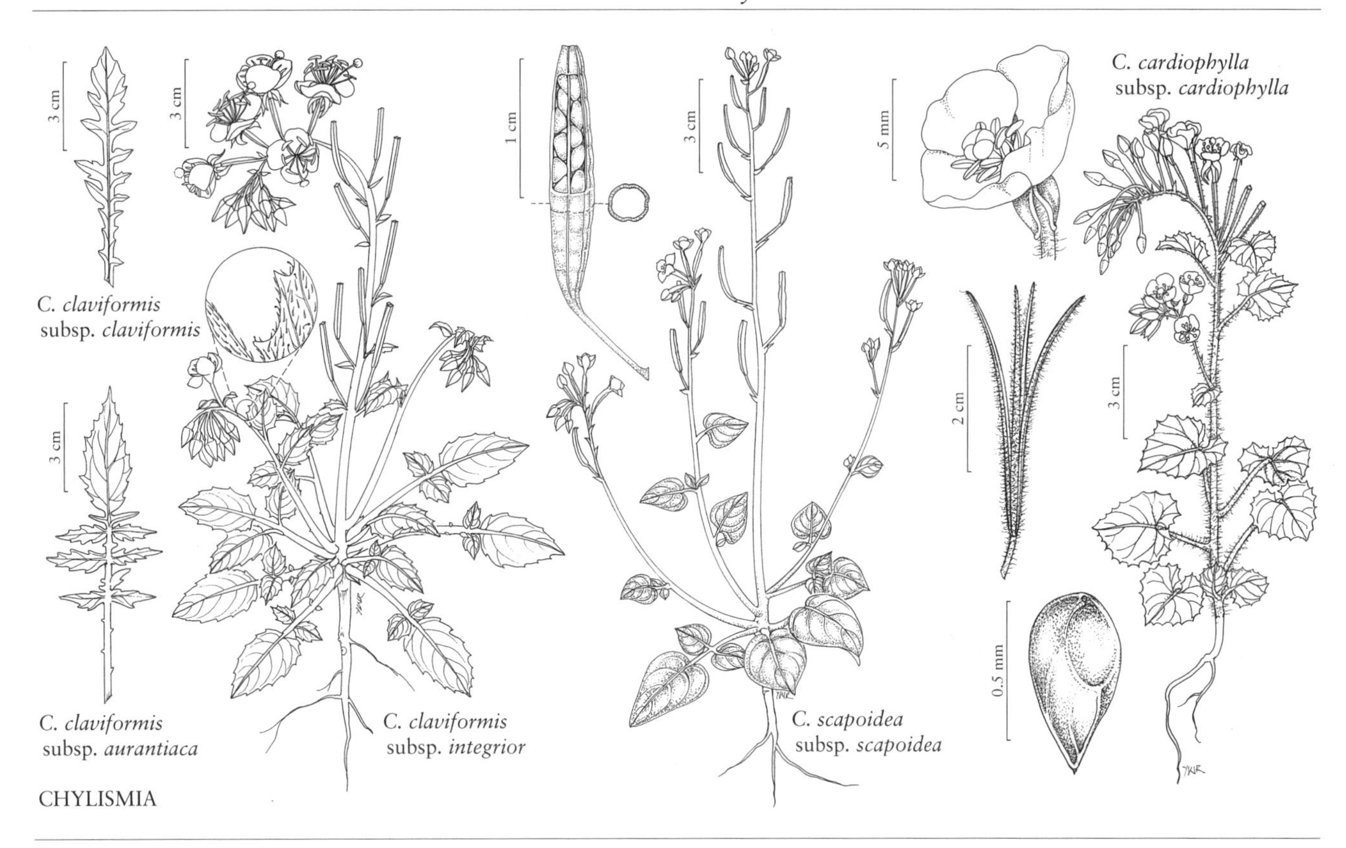

CHYLISMIA

Herbs strigillose, especially proximally, sometimes glabrate distally. **Stems** 5–50 cm. **Leaves:** blade sometimes purple-dotted, lateral lobes irregular, well developed, terminal lobe narrowly ovate, to 3 × 1.5 cm, margins irregularly sinuate-dentate. **Flowers** opening at sunset; buds usually without free tips, rarely with apical free tips less than 1 mm; floral tube orange-brown inside, 3–5 mm; petals white, rarely purple-dotted at base, often fading purple, rarely orange, 2.5–8 mm. **2*n*** = 14.

Flowering (Dec–)Feb–Jun. Sandy flats, washes, with *Ambrosia dumosa*, *Fouquieria splendens*, and *Larrea*; -70–1100 m; Ariz., Calif., Nev.; Mexico (Baja California).

Subspecies *aurantiaca* is known from Lincoln County, Nevada, south through southeastern California to northeasternmost Baja California, and in Arizona only in westernmost Mohave and Yuma counties. It intergrades extensively with subspp. *peirsonii* and *yumae*, and those with white petals (subspp. *claviformis*, *funerea*, *integrior*, and *peeblesii*); it sometimes hybridizes with *Chylismia brevipes* subspp. *brevipes* and *pallidula*, and with *C. munzii*.

The name *Oenothera scapoidea* var. *aurantiaca* S. Watson is superfluous because Watson included *O. claviformis* Torrey & Frémont as a synonym, which is the basionym of *O. scapoidea* var. *claviformis* (Torrey & Frémont) S. Watson. Likewise, *Chylismia scapoidea* (Torrey & A. Gray) Nuttall ex Raimann var. *aurantiaca* Davidson & Moxley is also an illegitimate name that pertains here.

6c. **Chylismia claviformis** (Torrey & Frémont) A. Heller subsp. **cruciformis** (Kellogg) W. L. Wagner & Hoch, Syst. Bot. Monogr. 83: 206. 2007 [E]

Oenothera cruciformis Kellogg, Proc. Calif. Acad. Sci. 2: 227, fig. 71. 1863; *Camissonia claviformis* (Torrey & Frémont) P. H. Raven subsp. *cruciformis* (Kellogg) P. H. Raven; *C. claviformis* var. *cruciformis* (Kellogg) Cronquist; *Chylisma cruciformis* (Kellogg) Howell; *C. scapoidea* (Torrey & A. Gray) Nuttall ex Raimann var. *cruciformis* (Kellogg) Small; *O. claviformis* Torrey & Frémont subsp. *citrina* P. H. Raven; *O. claviformis* subsp. *cruciformis* (Kellogg) P. H. Raven; *O. claviformis* var. *cruciformis* (Kellogg) Munz

Herbs strigillose or glandular puberulent proximally, glandular puberulent or glabrous distally. **Stems** 3–55 cm. **Leaves:** blade lateral lobes well developed, few to numerous, terminal lobe narrowly ovate to subcordate, to 8 × to 4 cm, margins serrate-dentate. **Flowers** opening at sunrise; buds without free tips, sometimes with apical free tips less than 1 mm; floral tube yellow or orange-brown inside, 2–6.5 mm; petals bright yellow, sometimes red-dotted in proximal ½, often fading purple, 2.5–8 mm. **2*n*** = 14.

Flowering Mar–May. Sandy or clay flats and slopes, with *Artemisia tridentata*, *Ericameria*, *Grayia spinosa*, or *Purshia tridentata*; 600–1500 m; Calif., Idaho, Nev., Oreg.

Subspecies *cruciformis* is known from Lassen County, California, western Canyon and Owyhee counties, Idaho, Harney, Lake, and Malheur counties, Oregon, and central and southern Washoe County, Nevada. It intergrades with subsp. *integrior.*

6d. Chylismia claviformis (Torrey & Frémont) A. Heller subsp. **funerea** (P. H. Raven) W. L. Wagner & Hoch, Syst. Bot. Monogr. 83: 206. 2007 [E]

Oenothera claviformis Torrey & Frémont subsp. *funerea* P. H. Raven, Univ. Calif. Publ. Bot. 34: 106. 1962 (as clavaeformis); *Camissonia claviformis* (Torrey & Frémont) P. H. Raven subsp. *funerea* (P. H. Raven) P. H. Raven; *C. claviformis* var. *funerea* (P. H. Raven) Cronquist

Herbs strigillose proximally, often densely so. **Stems** 6–60 cm. **Leaves:** blade lateral lobes usually present, but often poorly developed, sometimes absent, terminal lobe ovate to cordate, to 8 × 4.5 cm, margins dentate. **Flowers** opening at sunset; buds usually with conspicuous, subapical free tips; floral tube orange-brown inside, 3–5.5 mm; petals white, often fading purple, 3.5–7.5 mm. **2*n*** = 14.

Flowering Feb–May. Dry slopes, flats, with *Ambrosia dumosa* and *Larrea tridentata*; -80–900 m; Calif.

Subspecies *funerea* is known from Eureka and Saline valleys, and the region of Death Valley, Inyo County, and northernmost San Bernardino County. It intergrades with subspp. *aurantiaca* and *claviformis*; sometimes it hybridizes with *Chylismia brevipes* subsp. *brevipes* and *C. munzii*.

6e. Chylismia claviformis (Torrey & Frémont) A. Heller subsp. **integrior** (P. H. Raven) W. L. Wagner & Hoch, Syst. Bot. Monogr. 83: 206. 2007 [E] [F]

Oenothera claviformis Torrey & Frémont subsp. *integrior* Raven, Univ. Calif. Publ. Bot. 34: 106. 1962 (as clavaeformis); *Camissonia claviformis* (Torrey & Frémont) P. H. Raven subsp. *integrior* (P. H. Raven) P. H. Raven; *C. claviformis* var. *purpurascens* (S. Watson) Cronquist; *O. claviformis* var. *purpurascens* (S. Watson) Munz; *O. scapoidea* Torrey & A. Gray var. *purpurascens* S. Watson

Herbs strigillose proximally, usually also glandular puberulent distally, or glabrate. **Stems** 6–70 cm. **Leaves:** blade sometimes purple-dotted, lateral lobes often absent, sometimes present but poorly developed, terminal lobe narrowly ovate to ovate or subcordate, to 7 × 3 cm, margins serrate. **Flowers** opening at sunset; buds usually without free tips, rarely with apical free tips less than 1 mm; floral tube orange-brown inside, 3–6 mm; petals white, sometimes purple-dotted near base, fading purple, 4.5–8 mm. **2*n*** = 14.

Flowering Apr–Jun. Dry flats, with *Artemisia tridentata*, *Ericameria*, or *Juniperus*, on banks and flats; 1100–2200 m; Calif., Nev., Oreg.

Subspecies *integrior* is known from southern Harney County, Oregon, southward nearly throughout Nevada to Esmeralda and Lincoln counties and to Mono and northeastern Inyo counties, California, and often forms large colonies on banks and flats in Great Basin Desert. It intergrades with subspp. *aurantiaca* and *cruciformis*, and hybridizes with *Chylismia brevipes* subsp. *brevipes*.

6f. Chylismia claviformis (Torrey & Frémont) A. Heller subsp. **lancifolia** (A. Heller) W. L. Wagner & Hoch, Syst. Bot. Monogr. 83: 206. 2007 [E]

Chylismia lancifolia A. Heller, Muhlenbergia 2: 226. 1906 (as Chylisma); *Camissonia claviformis* (Torrey & Frémont) P. H. Raven subsp. *lancifolia* (A. Heller) P. H. Raven; *C. claviformis* var. *lancifolia* (A. Heller) Cronquist; *Oenothera claviformis* Torrey & Frémont subsp. *lancifolia* (A. Heller) P. H. Raven

Herbs strigillose proximally, glabrous and often glaucous distally. **Stems** 5–60 cm. **Leaves:** blade lateral lobes usually poorly developed or absent, terminal lobe lanceolate, to 5 × 2.8 cm, margins irregularly serrate-dentate. **Flowers** usually opening at sunset; buds usually without free tips, sometimes with apical free tips less than 1 mm; floral tube orange-brown inside, 3.5–6 mm; petals bright yellow, sometimes red-dotted in proximal ½, fading pale orange, 3.5–7 mm. **2*n*** = 14.

Flowering Apr–Jul. Sandy slopes and flats, often with *Artemisia tridentata* or *Ericameria*; 600–1700(–1900) m; Calif., Nev.

Subspecies *lancifolia* is known from east of the Sierra Nevada in southern Mono and Inyo counties, California, and adjacent Mineral County, Nevada.

6g. Chylismia claviformis (Torrey & Frémont) A. Heller subsp. **peeblesii** (Munz) W. L. Wagner & Hoch, Syst. Bot. Monogr. 83: 206. 2007

Oenothera claviformis Torrey & Frémont var. *peeblesii* Munz, Leafl. W. Bot. 2: 158. 1939 (as clavaeformis); *Camissonia claviformis* (Torrey & Frémont) P. H. Raven subsp. *peeblesii* (Munz) P. H. Raven; *O. claviformis* subsp. *peeblesii* (Munz) P. H. Raven

Herbs glandular puberulent and strigillose. **Stems** 5–60 cm. **Leaves:** blade lateral lobes irregular, well developed, terminal lobe narrowly ovate, to 7 × 3 cm, margins irregularly sinuate-dentate. **Flowers** opening at sunset; buds without free tips; floral tube orange-brown inside, 3–5.5 mm; petals white, often fading purple, 3–7.5 mm. **2*n*** = 14.

Flowering (Dec–)Jan–Apr. Flat, sandy plains, washes, with *Ambrosia dumosa*, *Carnegiea gigantea*, *Larrea tridentata*, and *Prosopis*; 100–700 m; Ariz., N.Mex.; Mexico (Sonora).

Subspecies *peeblesii* is known throughout almost all the southwestern half of Arizona and locally in northwesternmost Sonora, and was recently collected in Grant and Hildago counties in New Mexico. It intergrades with subspp. *aurantiaca* and *rubescens*, and hybridizes with all subspecies of *Chylismia brevipes*.

6h. Chylismia claviformis (Torrey & Frémont) A. Heller subsp. **peirsonii** (Munz) W. L. Wagner & Hoch, Syst. Bot. Monogr. 83: 207. 2007

Oenothera claviformis Torrey & Frémont var. *peirsonii* Munz, Amer. J. Bot. 15: 238. 1928 (as clavaeformis); *Camissonia claviformis* (Torrey & Frémont) P. H. Raven subsp. *peirsonii* (Munz) P. H. Raven; *Chylismia peirsonii* (Munz) D. A. Johansen; *O. claviformis* subsp. *peirsonii* (Munz) P. H. Raven

Herbs villous, at least proximally, rarely strigillose or glandular puberulent. **Stems** 5–60 cm. **Leaves:** blade lateral lobes usually well developed, terminal lobe narrowly ovate, to 9 × 3.5 cm, margins irregularly sinuate-dentate. **Flowers** opening at sunset; buds with conspicuous, subapical free tips; floral tube orange-brown inside, 2.5–4.5 mm; petals usually yellow, rarely white, not changing color in fading, 4.5–7 mm. **2*n*** = 14.

Flowering Feb–Apr. Sandy flats or washes, with *Ambrosia dumosa*, *Fouquieria splendens*, and *Larrea*; -70–400 m; Calif.; Mexico (Baja California).

Subspecies *peirsonii* is known from east of the Salton Sea and Imperial Valley in Imperial and San Diego counties, California, and in northeastern Baja California to the Gulf of California. It intergrades with subsp. *aurantiaca*.

6i. Chylismia claviformis (Torrey & Frémont) A. Heller subsp. **rubescens** (P. H. Raven) W. L. Wagner & Hoch, Syst. Bot. Monogr. 83: 207. 2007

Oenothera claviformis Torrey & Frémont subsp. *rubescens* P. H. Raven, Univ. Calif. Publ. Bot. 34: 103. 1962 (as clavaeformis); *Camissonia claviformis* (Torrey & Frémont) P. H. Raven subsp. *rubescens* (P. H. Raven) P. H. Raven

Herbs villous, at least proximally, also glandular puberulent. **Stems** 5–40 cm. **Leaves:** blade lateral lobes poorly developed, small or absent, terminal lobe lanceolate, to 4 × 2 cm, margins irregularly dentate. **Flowers** opening at sunset; buds with conspicuous, subapical free tips; floral tube orange-brown inside, 3.5–5 mm; petals yellow, fading brick red, 3–5 mm. **2*n*** = 14.

Flowering Mar–Apr. Sandy flats, with *Ambrosia dumosa*, *Carnegiea gigantea*, and *Larrea*; 60–400 m; Ariz.; Mexico (Sonora).

Subspecies *rubescens* is known from western Maricopa, eastern Pima, and southeastern Yuma counties, Arizona, and northwestern Sonora. It intergrades with subsp. *peeblesii* and hybridizes with *Chylismia brevipes* subsp. *arizonica*.

6j. Chylismia claviformis (Torrey & Frémont) A. Heller subsp. **yumae** (P. H. Raven) W. L. Wagner & Hoch, Syst. Bot. Monogr. 83: 207. 2007

Oenothera claviformis Torrey & Frémont subsp. *yumae* P. H. Raven, Univ. Calif. Publ. Bot. 34: 104. 1962 (as clavaeformis); *Camissonia claviformis* (Torrey & Frémont) P. H. Raven subsp. *yumae* (P. H. Raven) P. H. Raven

Herbs strigillose, often densely so, sometimes also glandular puberulent distally. **Stems** 5–40 cm. **Leaves:** blade lateral lobes poorly or well developed, terminal lobe lanceolate, to 6.5 × 2 cm, margins irregularly sinuate-dentate. **Flowers** opening at sunset; buds usually without free tips, sometimes with apical free tips less than 1 mm; floral tube orange-brown inside, 2.5–4 mm; petals pale yellow, fading reddish or not changing color, 3–5 mm. **2*n*** = 14.

Flowering Feb–May. Very arid dunes and sandy flats, with *Ambrosia dumosa* and *Larrea tridentata*; 0–300 m; Ariz., Calif.; Mexico (Baja California, Sonora).

Subspecies *yumae* is known from southeastern Imperial County, California, Yuma Desert, Arizona, and from El Gran Desierto to Puerto Peñasco in northwestern Sonora, and in northeastern Baja California. The subspecies is probably from hybridization between subspp. *aurantiaca* and *peirsonii*; it intergrades with subsp. *aurantiaca* and rarely hybridizes with *Chylismia brevipes* subsp. *arizonica*.

7. **Chylismia munzii** (P. H. Raven) W. L. Wagner & Hoch, Syst. Bot. Monogr. 83: 207. 2007 E

Oenothera munzii P. H. Raven, Univ. Calif. Publ. Bot. 34: 91. 1962; *Camissonia munzii* (P. H. Raven) P. H. Raven

Herbs annual, strigillose, often densely so. **Stems** several, 8–50 cm. **Leaves** primarily in basal rosette and also cauline, 1.5–20 × 0.5–3 cm; petiole 0.5–5 cm; blade pinnately lobed, terminal lobe ovate to narrowly ovate, 1.3–6 × 0.6–3 cm, margins denticulate, brownish oil cells lining veins abaxially. **Racemes** nodding, not congested, elongating in mature bud. **Flowers** opening at sunrise; buds with or without subapical free tips; floral tube orange-brown inside, 2–3 mm, villous inside; sepals 4–7 mm; petals bright yellow, with red dots near base, fading pale yellow or yellowish orange, 3–10 mm; stamens subequal, filaments 4–8 mm, anthers 3–6 mm, ciliate; style 8–18 mm, stigma exserted beyond anthers at anthesis. **Capsules** widely spreading, becoming sharply reflexed, clavate, 8–24 mm; pedicel 8–28 mm. **Seeds** 0.8–1.6 mm. $2n = 14$.

Flowering Mar–Jun. Mesic slopes, washes; 600–1600 m; Calif., Nev.

Chylismia munzii is known from middle elevations in the mountains at the north end, eastward from, and south of Death Valley, from Saline Valley and the Grapevine Mountains, Inyo County, California, and Yucca Flat, Nye County, Nevada, southward to the Kingston Range, San Bernardino County, California. P. H. Raven (1962, 1969) determined this species to be self-incompatible. It sometimes hybridizes with *C. brevipes* subsp. *brevipes* and *C. claviformis* subsp. *aurantiaca*.

8. **Chylismia eastwoodiae** (Munz) W. L. Wagner & Hoch, Syst. Bot. Monogr. 83: 207. 2007 C E

Oenothera scapoidea Torrey & A. Gray var. *eastwoodiae* Munz, Amer. J. Bot. 15: 234. 1928 (as eastwoodae); *Camissonia eastwoodiae* (Munz) P. H. Raven; *O. eastwoodiae* (Munz) P. H. Raven

Herbs annual, succulent, glabrous, glandular puberulent, or villous proximally. **Stems** unbranched or branched from base, 3–30 cm. **Leaves** primarily in basal rosette and also cauline; petiole 0.5–8 cm; blade usually not pinnately lobed or, if so, then lateral lobes greatly reduced, terminal lobe oblanceolate to cordate, 0.8–7.5 × 0.4–3 cm, margins entire or sparsely denticulate, pale brown oil cells lining veins abaxially. **Racemes** nodding, elongating after anthesis. **Flowers** opening at sunrise; buds without subapical free tips; floral tube 2–4.5 mm, villous inside proximally; sepals 3–8 mm; petals bright yellow, with red dots near base, fading pale yellow or yellowish orange, 5.5–9 mm; stamens unequal, filaments of antisepalous stamens 3–8 mm, those of antipetalous ones 2.8–5.5 mm, anthers 2–4 mm, ciliate; style 10–17 mm, stigma exserted beyond anthers at anthesis. **Capsules** erect, spreading, or slightly reflexed, clavate, 18–40 mm; pedicel 4–28 mm. **Seeds** 1.2–1.7 mm. $2n = 14$.

Flowering Apr–Jun. Clay flats, on gray, alkaline, marine-deposited gumbo, sandy draws; of conservation concern; 1200–1800 m; Colo., Utah.

Chylismia eastwoodiae is known from Mesa County, Colorado, and from Emery County south to San Juan County, Utah. P. H. Raven (1962, 1969) suspected this species to be self-incompatible, but did not have data to make the determination.

9. **Chylismia scapoidea** (Torrey & A. Gray) Nuttall ex Raimann in H. G. A. Engler and K. Prantl, Nat. Pflanzenfam. 96[III,7]: 217. 1893 E F

Oenothera scapoidea Torrey & A. Gray, Fl. N. Amer. 1: 506. 1840; *Camissonia scapoidea* (Torrey & A. Gray) P. H. Raven; *O. brevipes* A. Gray var. *scapoidea* (Torrey & A. Gray) H. Léveillé

Herbs annual, strigillose, villous, or glandular puberulent. **Stems** usually unbranched, sometimes branched from base, 3–45 cm. **Leaves** primarily in basal rosette, cauline poorly developed or absent, 1–18 × 0.5–3.5 cm; petiole 0.5–6.5 cm; blade pinnately lobed or lateral lobes greatly reduced or absent, sometimes mixed on same plant, terminal lobe narrowly ovate to ovate or elliptic, 1–6.5 × 0.5–3.3 cm, margins irregularly dentate to subentire, oil cells on abaxial surface inconspicuous or conspicuous, pale yellowish brown or dark brown. **Racemes** nodding, elongating in fruit. **Flowers** opening at sunrise; buds with or without subapical free tips less than 1 mm; floral tube 1–4 mm, sparsely villous or glabrous inside; sepals 1.2–5 mm; petals bright yellow, often with red dots near base, fading pale yellow or yellowish orange, 1.5–5.5(–8) mm; stamens unequal, filaments of antisepalous stamens 1.2–6 mm, those of antipetalous ones 0.5–4 mm, anthers 1–2.5 mm, ciliate or glabrous; style 3–11 mm, stigma surrounded by anthers at anthesis. **Capsules** ascending, clavate, 8–50 mm; pedicel 4–20 mm. **Seeds** 1–2 mm.

Subspecies 4 (4 in the flora): w United States.

P. H. Raven (1962, 1969) determined this species to be self-compatible and primarily autogamous.

1. Oil cells lining veins of leaf blades abaxially usually pale yellowish brown, inconspicuous, rarely darker; leaf blades usually unlobed, rarely pinnately lobed with few, small lateral lobes, margins subentire.
   2. Capsules (10–)15–30 mm; petals 1.7–5 mm . . . . 9a. *Chylismia scapoidea* subsp. *scapoidea*
   2. Capsules 25–50 mm; petals 1.5–2 mm. . . . . 9b. *Chylismia scapoidea* subsp. *macrocarpa*
1. Oil cells lining veins of leaf blades abaxially dark brown, conspicuous; leaf blades pinnately lobed, lobes well developed or with few, small lateral lobes or unlobed, margins dentate.
   3. Petals 1.8–4 mm; lateral lobes of leaves often few or absent, sometimes well developed; capsules 8–20(–28) mm . . . . 9c. *Chylismia scapoidea* subsp. *brachycarpa*
   3. Petals 4–5.5(–8) mm; lateral lobes of leaves usually well developed, sometimes poorly developed; capsules 16–38 mm . . . . 9d. *Chylismia scapoidea* subsp. *utahensis*

9a. Chylismia scapoidea (Torrey & A. Gray) Nuttall ex Raimann subsp. **scapoidea** E F

Chylismia scapoidea var. *seorsa* A. Nelson; *Oenothera scapoidea* Torrey & A. Gray var. *seorsa* (A. Nelson) Munz

Leaves: blade usually unlobed, rarely pinnately lobed with few, small lateral lobes, terminal lobe narrowly ovate to ovate or, sometimes, elliptic, 1–5.5 × 0.5–3 cm, margins subentire, oil cells on abaxial surface inconspicuous, usually pale yellowish brown, rarely darker. **Flowers:** petals 1.7–5 mm. **Capsules** (10–)15–30 mm. **2*n*** = 14, 28.

Flowering Mar–May. Sandy or clay flats; 1200–2000(–2500) m; Ariz., Colo., Nev., N.Mex., Oreg., Utah, Wyo.

Subspecies *scapoidea* is known from western Wyoming, west to Lake County, Oregon, and Elko and White Pine counties, Nevada, south to Beaver County, Utah, northern Coconino, Navajo, and Apache counties, Arizona, Sandoval and San Juan counties, New Mexico, and western Colorado; also on the headwaters of the Arkansas River in eastern Colorado (Fremont and Pueblo counties). It is the only member of sect. *Chylismia* that occurs east of the Continental Divide.

9b. Chylismia scapoidea (Torrey & A. Gray) Nuttall ex Raimann subsp. **macrocarpa** (P. H. Raven) W. L. Wagner & Hoch, Syst. Bot. Monogr. 83: 208. 2007 C E

Oenothera scapoidea Torrey & A. Gray subsp. *macrocarpa* P. H. Raven, Univ. Calif. Publ. Bot. 34: 95. 1962; *Camissonia bairdii* S. L. Welsh; *C. scapoidea* (Torrey & A. Gray) P. H. Raven subsp. *macrocarpa* (P. H. Raven) P. H. Raven; *C. scapoidea* var. *macrocarpa* (P. H. Raven) Cronquist

Leaves: blade usually unlobed, rarely pinnately lobed with few, small lateral lobes, terminal lobe ovate, 1.7–3.5 × 1–1.5 cm, margins subentire, oil cells on abaxial surface inconspicuous, usually pale yellowish brown, rarely darker. **Flowers:** petals 1.5–2 mm. **Capsules** 25–50 mm.

Flowering May–Jun. On detrital clay knobs, gravelly flats; of conservation concern; 1500–2000 m; Ariz.

Subspecies *macrocarpa* is known from northern Apache and Navajo counties and northeastern Coconino County.

9c. Chylismia scapoidea (Torrey & A. Gray) Nuttall ex Raimann subsp. **brachycarpa** (P. H. Raven) W. L. Wagner & Hoch, Syst. Bot. Monogr. 83: 208. 2007 E

Oenothera scapoidea Torrey & A. Gray subsp. *brachycarpa* P. H. Raven, Univ. Calif. Publ. Bot. 34: 95. 1962; *Camissonia scapoidea* (Torrey & A. Gray) P. H. Raven subsp. *brachycarpa* (P. H. Raven) P. H. Raven

Leaves: blade with lateral lobes few or absent, sometimes well developed, terminal lobe narrowly ovate to ovate or, sometimes, elliptic, 1–6.5 × 0.7–3.3 cm, margins irregularly dentate, oil cells on abaxial surface conspicuous, dark brown. **Flowers:** petals 1.8–4 mm. **Capsules** 8–20(–28) mm. **2*n*** = 14.

Flowering May–Jun. Sandy slopes and flats, with *Artemisia*, *Grayia*, or *Juniperus*; 700–2000 m; Idaho, Nev., Oreg., Utah.

Subspecies *brachycarpa* is known from southeastern Oregon, southwestern Idaho, northwestern Utah, and northeastern Nevada.

9d. Chylismia scapoidea (Torrey & A. Gray) Nuttall ex Raimann subsp. **utahensis** (P. H. Raven) W. L. Wagner & Hoch, Syst. Bot. Monogr. 83: 208. 2007 [E]

Oenothera scapoidea Torrey & A. Gray subsp. *utahensis* P. H. Raven, Univ. Calif. Publ. Bot. 34: 96. 1962; *Camissonia scapoidea* (Torrey & A. Gray) P. H. Raven subsp. *utahensis* (P. H. Raven) P. H. Raven; *C. scapoidea* var. *utahensis* (P. H. Raven) S. L. Welsh

Leaves: blade usually pinnately lobed, terminal lobe narrowly ovate to ovate or, sometimes, elliptic, 2–5 × 1–2.8 cm, margins dentate, oil cells on abaxial surface conspicuous, usually dark brown. **Flowers:** petals 4–5.5(–8) mm. **Capsules** 16–38 mm.

Flowering May–Jun. Dry rocky slopes, flats, with *Atriplex* and *Ericameria*; 1200–1700 m; Nev., Utah.

Subspecies *utahensis* is known from extreme northeastern Nevada (Elko County) and western Utah. P. H. Raven (1962, 1969) described this subspecies for populations that are somewhat intermediate between *Chylismia walkeri* subsp. *tortilis* and *C. scapoidea* subsp. *brachycarpa*, suggesting that this entity perhaps developed via hybridization between them. The ranges of the two putative parents heavily overlap that of subsp. *utahensis*. A. Cronquist et al. (1997c) combined subsp. *brachycarpa* and subsp. *utahensis*.

10. Chylismia parryi (S. Watson) Small, Bull. Torrey Bot. Club 23: 193. 1896 (as Chylisma) [E]

Oenothera parryi S. Watson, Amer. Naturalist 9: 270. 1875; *Camissonia parryi* (S. Watson) P. H. Raven; *Chylismia tenuissima* (M. E. Jones) Rydberg; *O. scapoidea* Torrey & A. Gray var. *parryi* (S. Watson) M. E. Jones; *O. tenuissima* M. E. Jones

Herbs annual, sparsely to densely villous throughout or, sometimes, glabrate distally. **Stems** often intricately branched, 5–80 cm. **Leaves** in poorly defined basal rosette and also cauline; petiole 0.3–3.8 cm; blade usually unlobed, very rarely pinnately lobed with few, small lateral lobes, ovate to elliptic, margins sparsely denticulate to subentire, pale or dark brown oil cells lining veins abaxially. **Racemes** nodding, with intricate, filiform branches, elongating in fruit. **Flowers** opening at sunrise; buds without free tips; floral tube 0.5–2 mm, glabrous or villous inside; sepals 1.5–4 mm; petals bright yellow, often with red dots near base, fading pale yellow or yellowish orange, 2–7 mm; stamens unequal, filaments of antisepalous stamens 1.7–3.5 mm, those of antipetalous ones 1.2–2.5 mm, anthers 0.9–1.2 mm, glabrous; style 4–9 mm, stigma exserted beyond anthers at anthesis. **Capsules** erect or ascending, clavate, 4–10 mm; pedicel 4–20 mm. **Seeds** 0.7–1.2 mm. $2n = 14$.

Flowering May–Jun(–Sep). Red clay and sand slopes weathered from red (freshwater-deposited) sandstone cliffs, with *Juniperus* or *Larrea tridentata*; 800–1300 m; Ariz., Utah.

Chylismia parryi is known from northwestern Arizona (Coconino to Mohave counties) and southwestern Utah (Beaver to Washington counties), and is apparently disjunct to San Juan County, Utah. It is outcrossing and, perhaps, self-incompatible (P. H. Raven 1962, 1969). There are two morphological forms of this species. Raven (1962) noted that a later flowering form has narrower, smaller leaves, and less overall pubescence. It is not clear what these represent, but Raven (1962) made the combination *Oenothera parryi* forma *tenuissima* (M. E. Jones) P. H. Raven for it. He later (Raven 1969) noted that these plants did not seem to merit formal recognition, without any discussion.

11. Chylismia exilis (P. H. Raven) W. L. Wagner & Hoch, Syst. Bot. Monogr. 83: 207. 2007 [C] [E]

Oenothera exilis P. H. Raven, Univ. Calif. Publ. Bot. 34: 114. 1962; *Camissonia exilis* (P. H. Raven) P. H. Raven

Herbs annual, glandular puberulent and sparsely villous. **Stems** slender, unbranched or branched, 10–20 cm. **Leaves** primarily cauline; petiole 0.3–1.8 cm; blade unlobed, narrowly ovate to elliptic, 0.3–2 × 0.3–1 cm, margins entire or inconspicuously denticulate, brownish oil cells lining veins abaxially. **Racemes** erect, elongating in fruit. **Flowers** opening at sunrise; buds without free tips; floral tube 0.4–0.5 mm, glabrous inside; sepals 1–1.2 mm; petals yellow, fading pale lavender, 1–1.5 mm; stamens 4 (or 8), antisepalous, filaments 0.5 mm, anthers 0.5–0.7 mm, glabrous, when 8, then antipetalous ones smaller and abortive; style 1.5 mm, stigma surrounded by anthers at anthesis. **Capsules** spreading or reflexed, clavate, 4–10 mm; pedicel 3–9 mm. **Seeds** 0.8 mm. $2n = 14$.

Flowering Apr–Jun. Calcareous sand, gypseous clay flats, juniper woodlands; of conservation concern; 1000–1900 m; Ariz., Utah.

Chylismia exilis, known from Kane and San Juan counties in Utah and northern Coconino and Mohave counties in Arizona, is cryptic due to its small size. It may not be as rare as assumed, since it is difficult to spot in the field. P. H. Raven (1962, 1969) determined this species to be self-compatible and autogamous.

12. Chylismia megalantha (Munz) W. L. Wagner & Hoch, Syst. Bot. Monogr. 83: 207. 2007 [E]

Oenothera heterochroma S. Watson var. *megalantha* Munz, Leafl. W. Bot. 3: 52. 1941; *Camissonia megalantha* (Munz) P. H. Raven; *O. megalantha* (Munz) P. H. Raven

Herbs annual, glandular pubescent throughout. **Stems** several, 10–200 cm. **Leaves** in poorly defined basal rosette and cauline; petiole 1.8–5.5 cm; blade unlobed, broadly cordate to ovate, 2.4–8 × 7 cm, margins sinuate-dentate, yellowish oil cells prominently lining veins abaxially. **Racemes** erect, elongating in flower. **Flowers** opening at sunrise; buds without free tips; floral tube 4–9 mm, with matted, villous hairs inside; sepals 4.5–9 mm; petals pale to dark lavender, diffusely purplish-flecked near base, white at very base, fading darker lavender, 9–14 mm; stamens unequal, filaments of antisepalous stamens 6–12 mm, of antipetalous ones 3.5–8 mm, anthers 2 mm, glabrous; style 14–22.5 mm, stigma exserted beyond anthers at anthesis. **Capsules** erect or ascending, clavate, 8–14 mm; pedicel 2–3.5 mm. **Seeds** 1–1.3 mm. $2n = 14$.

Flowering Jun–Oct. Rubble derived from volcanic tuff, partly on moist soil along springs; 1200–1400 m; Nev.

Chylismia megalantha is known from around Frenchman Drainage to French Peak and Skull Mountain in southern Nye County. P. H. Raven (1962, 1969) determined this species to be self-compatible, but outcrossing.

13. Chylismia atwoodii (Cronquist) W. L. Wagner & Hoch, Syst. Bot. Monogr. 83: 205. 2007 [C] [E]

Camissonia atwoodii Cronquist, Great Basin Naturalist 46: 258. 1986

Herbs annual, glandular puberulent. **Stems** several, 5–150 cm. **Leaves** in poorly defined basal rosette and cauline; petiole 0.7–3.4 cm; blade unlobed, broadly ovate to oblong-ovate, elliptic, or subcordate, 1.2–7.6 × 0.8–5.5 cm, margins serrulate to serrate-denticulate, brown oil cells prominently lining veins abaxially. **Racemes** erect, elongating in flower. **Flowers** opening at sunrise; buds without free tips; floral tube 0.6–1 mm; sepals 5–7 mm; petals purple, fading darker purple, 7–14 mm; stamens 4 + 4, unequal, anthers 1.5–2 mm, glabrous, stigma exserted beyond anthers at anthesis. **Capsules** spreading to reflexed, clavate, 11–25 mm; pedicel 3–5 mm. **Seeds** 1.5–1.8 mm.

Flowering Aug–Nov. Open slopes in desert shrub communities, on clay soil; of conservation concern; 1100–1600 m; Utah.

Chylismia atwoodii is known only from eastern Kane County, and only from a few collections, so is still poorly characterized morphologically, but clearly distinct among the purple-petaled species.

14. Chylismia heterochroma (S. Watson) Small, Bull. Torrey Bot. Club 23: 193. 1896 (as Chylisma) [E]

Oenothera heterochroma S. Watson, Proc. Amer. Acad. Arts 17: 373. 1882; *Camissonia heterochroma* (S. Watson) P. H. Raven; *C. heterochroma* var. *monoensis* (Munz) Cronquist; *O. heterochroma* subsp. *monoensis* (Munz) P. H. Raven; *O. heterochroma* var. *monoensis* Munz

Herbs annual, glandular puberulent throughout, or glabrate and glaucous distally. **Stems** several, 10–100 cm. **Leaves** primarily in poorly defined basal rosette, cauline greatly reduced when present; petiole 0.4–8 cm; blade unlobed, ovate to cordate, 2–11.5 × 1.4–5 cm, margins sinuate-dentate, brown oil cells prominently lining veins abaxially. **Racemes** erect, elongating in anthesis. **Flowers** opening at sunrise; buds without free tips; floral tube 2–5 mm, villous inside; sepals 1.5–3.5 mm; petals lavender, paler and often with flecks toward base, often yellow at very base, fading darker lavender, 2–6 mm; stamens unequal, filaments of antisepalous ones 1.8–3 mm, of antipetalous ones 1–2.5 mm, anthers 0.6–1 mm, glabrous or sparsely ciliate; style 4–7 mm, stigma surrounded by anthers at anthesis. **Capsules** erect, clavate, 7–13 mm; pedicel 2–5 mm. **Seeds** 1–1.2 mm. $2n = 14$.

Flowering May–Jun. Alluvial and rocky slopes; 600–2200 m; Calif., Nev.

Chylismia heterochroma is known from Churchill and Lander counties, Nevada, south to Lincoln and southern Nye counties, Nevada, to adjacent California (Mono Lake, Mono County, and central Inyo counties). P. H. Raven (1962, 1969) determined this species to be self-compatible and autogamous.

16b. Chylismia (Torrey & A. Gray) Nuttall ex Raimann sect. Lignothera (P. H. Raven) W. L. Wagner & Hoch, Syst. Bot. Monogr. 83: 136. 2007

Oenothera Linnaeus sect. *Lignothera* P. H. Raven, Univ. Calif. Publ. Bot. 34: 76. 1962; *Camissonia* Link sect. *Lignothera* (P. H. Raven) P. H. Raven

Herbs annual or perennial. Leaves cauline; blade unlobed, cordate-orbicular or -deltate. **Flowers** opening at sunset; floral tube 4.5–40 mm; petals yellow, without dots or flecks, fading brick red or orange; pollen shed in tetrads.

Species 2 (2 in the flora): sw United States, nw Mexico.

Section *Lignothera* consists of two diploid ($2n = 14$) species (four taxa) that occur on rocky slopes and in washes in the Mojave and western Sonoran Deserts. *Chylismia arenaria* occurs from southeastern California into adjacent southwestern Arizona and barely to northern Sonora, Mexico; the more widespread *C. cardiophylla* occurs in that same region but also reaches to south-central Baja California, Mexico, farther east in Arizona, and north to the western and southern margins of Death Valley in Inyo County, California. P. H. Raven (1962) considered this group to be an early evolutionary offshoot within *Camissonia*. He revised his position (Raven 1969) to regard the late afternoon-opening flowers, pollen shed in tetrads, and semi-woody habit as specializations within Onagreae and in *Camissonia*, and, consequently, to regard sect. *Lignothera* as a derivative of sect. *Chylismia* and its long floral tubes an adaptation for hawkmoth pollination.

1. Floral tubes 4.5–14 mm; racemes congested; sepals 3–9 mm . . . . 15. *Chylismia cardiophylla*
1. Floral tubes 18–40 mm; racemes open; sepals 8–15 mm . . . . 16. *Chylismia arenaria*

15. Chylismia cardiophylla (Torrey) Small, Bull. Torrey Bot. Club 23: 193. 1896 (as Chylisma) [F]

Oenothera cardiophylla Torrey in War Department [U.S.], Pacif. Rail. Rep. 5(2): 360. 1857; *Camissonia cardiophylla* (Torrey) P. H. Raven

Herbs annual or perennial, villous and glandular puberulent. **Stems** usually well branched, forming bushy habit, 20–100 cm. **Leaves** cauline, mostly toward base; petiole (0.7–)2.5–7.5 cm; blade cordate-ovate to -orbiculate, 2.5–7.5 × 2.3–5.5 cm, smaller distally, margins erose-dentate. **Racemes** nodding, congested. **Flowers:** floral tube 4.5–14 mm, villous inside; sepals 3–9 mm; petals yellow, 3–12 mm; filaments 1–3 mm, anthers 2–4 mm; style 8–23 mm, stigma surrounded by or exserted just beyond anthers at anthesis. **Capsules** ascending, cylindrical, 20–55 mm; pedicel 1–18 mm. **Seeds** 0.5–0.7 mm.

Subspecies 3 (2 in the flora): sw United States, nw Mexico.

P. H. Raven (1962, 1969) determined this species to be self-compatible, but primarily outcrossing. Subspecies *cedrosensis* (Greene) W. L. Wagner & Hoch, occurs in Baja California and adjacent Sonora, Mexico.

1. Herbs villous, sometimes also glandular puberulent; floral tubes 4.5–12 mm . . . . 15a. *Chylismia cardiophylla* subsp. *cardiophylla*
1. Herbs glandular puberulent throughout, often also sparsely villous; floral tubes 9–14 mm . . . . 15b. *Chylismia cardiophylla* subsp. *robusta*

15a. Chylismia cardiophylla (Torrey) Small subsp. **cardiophylla** [F]

Oenothera cardiophylla Torrey var. *petiolaris* M. E. Jones

Herbs villous, sometimes also glandular puberulent. **Flowers:** floral tube 4.5–12 mm; petals 3–12 mm; style 8–23 mm, stigma often exserted beyond anthers at anthesis. $2n = 14$.

Flowering Mar–May. Rocky walls, sandy alluvial flats, with *Ambrosia dumosa*, *Hyptis emoryi*, and *Larrea tridentata*; 0–700 m; Ariz., Calif.; Mexico (Baja California, Sonora).

Subspecies *cardiophylla* is known from southern San Bernardino County south to eastern San Diego County, California, and Yuma County (and possibly western Pinal County), Arizona, and south in northeastern and central Baja California, Mexico; it is also found on Isla Ángel de la Guarda, Isla San Esteban, Isla San Luis, Isla San Marcos, and Isla San Pedro Mártir in the Gulf of California. Isla San Esteban is the only locality for this subspecies in Sonora.

15b. Chylismia cardiophylla (Torrey) Small subsp. **robusta** (P. H. Raven) W. L. Wagner & Hoch, Syst. Bot. Monogr. 83: 206. 2007 E

Oenothera cardiophylla Torrey subsp. *robusta* P. H. Raven, Univ. Calif. Publ. Bot. 34: 79. 1962; *Camissonia cardiophylla* (Torrey) P. H. Raven subsp. *robusta* (P. H. Raven) P. H. Raven

Herbs glandular puberulent throughout, often also sparsely villous. **Flowers:** floral tube 9–14 mm; petals 7–11 mm; style 14–20 mm, stigma usually exserted beyond anthers at anthesis. $2n = 14$.

Flowering Mar–May. Rocky borders of washes and hillsides, with *Ambrosia dumosa* and *Larrea tridentata*; 600–1400 m; Calif.

Subspecies *robusta* is known from the western and southern margins of Death Valley, Inyo County.

16. Chylismia arenaria A. Nelson, Amer. J. Bot. 21: 575. 1934 (as Chylisma)

Camissonia arenaria (A. Nelson) P. H. Raven; *Oenothera arenaria* (A. Nelson) P. H. Raven; *O. cardiophylla* Torrey var. *longituba* Jepson; *O. cardiophylla* var. *splendens* Munz & I. M. Johnston

Herbs perennial, sometimes facultative annual, villous, sometimes also sparsely glandular puberulent in inflorescences. **Stems** well branched, 25–180 cm. **Leaves** cauline, often mostly toward base; petiole 3–6 cm; blade cordate-deltate, 2.5–4(–6) × 2.5–4(–6) cm, smaller distally, margins coarsely dentate. **Racemes** nodding, open. **Flowers:** floral tube 18–40 mm, finely pubescent inside; sepals 8–15 mm; petals bright to pale yellow, 8–20 mm; filaments 5–9 mm, anthers 5–8 mm; style 30–58 mm, stigma exserted beyond anthers at anthesis. **Capsules** ascending, cylindrical, 30–44 mm; pedicel 2–5 mm. **Seeds** 0.5–0.7 mm. $2n = 14$.

Flowering Mar–Apr. Sandy washes, rocky slopes, desert scrub in Sonoran Desert shrublands, usually with *Ambrosia dumosa*, *Carnegiea*, *Larrea tridentata*, and *Prosopis*; -50–500 m; Ariz., Calif.; Mexico (Sonora).

Chylismia arenaria is known from the foot of the Needles in Mohave County, Arizona, and from the north end of the Salton Sea, Riverside County, California, southeastward to the Tinajas Atlas Range, Arizona, and Sonora, Mexico. P. H. Raven (1962, 1969) determined *C. arenaria* to be self-compatible, but primarily outcrossing.

17. OENOTHERA Linnaeus, Sp. Pl. 1: 346. 1753; Gen. Pl. ed. 5, 163. 1754 • Evening primrose [Greek *oinos*, wine, and *thera*, seeking or catching, alluding to roots of some unknown plants possessing perfume of wine, perhaps misapplied by Linnaeus]

Warren L. Wagner

Herbs, annual, biennial, or perennial, sometimes suffrutescent, caulescent or acaulescent; usually with taproots, sometimes fibrous roots, sometimes with shoots arising from spreading lateral roots, or with rhizomes. **Stems** erect or ascending, sometimes decumbent, and then sometimes rooting at nodes, epidermis green or whitish and exfoliating. **Leaves** in basal rosette and cauline, basal leaves present before flowering, often absent later, alternate; stipules absent; sessile or petiolate; blade margins toothed to pinnatifid, often irregularly so, sometimes subentire. **Inflorescences:** flowers solitary in leaf axils, when terminal, often forming leafy spikes, rarely racemes, erect or nodding. **Flowers** bisexual, usually actinomorphic, sometimes zygomorphic and petals positioned in upper ½ of flower (in sect. *Gaura*), buds erect or recurved; floral tube deciduous (with sepals, petals, and stamens) after anthesis, usually glabrous, sometimes lanate or densely hispid, with short, interlocking hairs within, with basal nectary; sepals (3 or)4, reflexed individually, in pairs, or as a unit and reflexed to one side at anthesis; petals (3 or)4, usually yellow, purple, or white, rarely pink or red, sometimes base pale green to yellow, usually fading orange, purple, pale yellow, reddish, or whitish; stamens (6 or)8, subequal or in 2 unequal series, anthers versatile, filaments usually without basal scale, sometimes with basal scale (sect. *Gaura*), 0.3–0.5 mm, scales nearly closing mouth of floral tube, pollen shed singly; ovary (3 or)4-locular or septa incomplete (sect. *Gaura*) and 1-locular, ovules numerous, or 1–8 (sect. *Gaura*), style glabrous or pubescent, stigma usually deeply divided into (3 or)4 linear lobes, sometimes peltate, discoid to nearly quadrangular or obscurely and shallowly 4-lobed (sect. *Calylophus*), entire surface of lobes receptive, surface probably wet and non-papillate, subtended by ± conspicuous peltate indusium in early development, persisting to anthesis, often obscured by developing stigma. **Fruit** a capsule, straight or curved, lanceoloid or ovoid, ellipsoid to cylindrical, rhombic-obovoid, or globose, sometimes clavate or ellipsoid with proximal part sterile and cylindrical, tapering to a pedicel-like base (stipe), terete or (3 or)4-angled or -winged, usually loculicidally dehiscent, sometimes tardily so, sometimes an indehiscent, nutlike capsule with hard, woody walls (sects. *Gaura*, *Gauropsis*), usually (3 or)4-locular, sometimes septa incomplete and fragile and then ovary 1-locular (sect. *Gaura*); usually sessile, sometimes pedicellate. **Seeds** usually numerous (1–160+), in 1 or 2(–4) rows or clustered in each locule, or sometimes reduced to 1–8 (sect. *Gaura*), surface papillose, beaded, rugose, furrowed, reticulate, or smooth. $x = 7$.

Species ca. 150 (93 in the flora): North America, Mexico, West Indies, Bermuda, Central America, South America; introduced in temperate to subtropical areas nearly worldwide.

Oenothera is distributed widely in temperate to subtropical areas of North America and South America, usually in open, often disturbed habitats, from sea level to nearly 5000 m elevation; several species are widely naturalized worldwide. The center of diversity for *Oenothera* is in southwestern North America, but farther east than other genera of the Onagreae. *Oenothera* is delimited here more broadly than in the past, when as many as 15 genera were recognized, which broadly correspond to sections recognized here, and as few as three (P. A. Munz 1965) or four genera (P. H. Raven 1964). *Oenothera*, as it has been delimited in recent decades, lacked a clear generic synapomorphy, but was considered distinctive by virtue of having a 4-lobed stigma receptive over its entire surface; however, *Gaura*, *Stenosiphon* (short lobes with a distinctive disk at the base), and to some degree *Calylophus* (peltate, discoid to quadrangular, sometimes shallowly 4-lobed) each have similar but slightly different variations on the basic lobed stigma. *Calylophus* and *Gaura* have been considered to be closely related and distinct from *Oenothera* and *Stenosiphon* by having anthers with both tapetal and parenchymatous septa (Raven). Molecular studies with broad sampling in Onagreae (R. A. Levin et al. 2004) showed high support for a monophyletic *Oenothera* only when *Calylophus*, *Gaura*, and *Stenosiphon* are included within it. Subsequently, it was recognized that while stigma morphology varies among these four groups from deeply divided into long or short linear lobes to peltate, discoid, nearly square or obscurely and shallowly 4-lobed, the stigmas of all four groups are subtended by a more or less conspicuous peltate indusium (W. L. Wagner et al. 2007). This structure, long known in *Gaura* and *Stenosiphon* because it is conspicuous throughout anthesis, had been overlooked in *Oenothera*, apparently because it usually is evident only in early development; in some species it becomes less conspicuous and even hidden by the stigma. Molecular analyses (Levin et al.; G. D. Hoggard et al. 2004) and the consistent synapomorphy of the indusiate style lead to a broadened concept of *Oenothera* (Wagner et al.) by including in it the previously separate genera *Gaura* and *Stenosiphon* for the first time. *Calylophus* is also included in the expanded *Oenothera* as was done by Munz and others, but not by Raven and H. F. Towner (1977).

Permanent translocation heterozygosity (PTH) appears to have been a major element in the evolution of species of *Oenothera* (R. E. Cleland 1972; P. H. Raven 1979; K. E. Holsinger and N. C. Ellstrand 1984; C. Harte 1994; W. Dietrich et al. 1997; M. T. Johnson et al. 2009, 2011), and otherwise occurs in the Onagraceae only in *Gayophytum heterozygum*. Permanent translocation heterozygosity has been recorded in only 57 species in seven plant families (for example, Holsinger and Ellstrand), including Onagraceae (47 spp.), Campanulaceae (2 spp.), Commelinaceae (2 spp.), Clusiaceae (2 spp.), Iridaceae (3 spp.), Paeoniaceae (2 spp.), and Papaveraceae (1 sp.). The taxonomic distribution of PTH in the Onagraceae now appears to be even more concentrated in *Oenothera* than when Raven reviewed PTH in the family; in his treatment, PTH occurred in *Calylophus* (1 sp.), *Gaura* (2 spp.), *Gayophytum* (1 sp.), and *Oenothera* (43 spp., including 37 in sect. *Oenothera*, and the other six in sections now included in one of the two primary *Oenothera* lineages [including *Hartmannia* (1 sp.), *Kneiffia* (1 sp.), *Lavauxia* (1 sp.), *Leucocoryne* (2 spp.), and *Xanthocoryne* (1 sp.)]. It appears that the specific chromosomal structure in *Oenothera*—metacentric chromosomes with pycnotic, condensed proximal regions—enables reciprocal translocations, resulting in the regular occurrence of rings of chromosomes at meiosis and ultimately the specialized system of PTH, in which all seven pairs of chromosomes exchange arms and segregate as a unit (M. Kurabayashi et al. 1962; Cleland; Raven). The best known species possessing this system are the members of subsect. *Oenothera*, in which the structure and mechanisms of PTH were elucidated (Cleland; Harte; Dietrich et al.). In addition to the translocations, the system requires balanced lethals, which prevent the formation of the homozygous combinations (PTH species have ca. 50% infertile pollen (sterile pollen smaller than fertile under minimal magnification, but fertility can range from 30% to 85%), facultative autogamy, and alternate disjunction of the chromosomes during meiosis. In PTH species, a ring of 14 chromosomes or occasionally 12 + 1 pair chromosomes is formed at meiotic metaphase. A PTH species in the sense employed here is an aggregation of true-breeding populations having similar morphological and genetic attributes.

P. H. Raven (1979) pointed out that most PTH species are annuals or biennials; only about 10% of the known PTH species in Onagraceae are perennial. Most of the species of *Oenothera* that have become naturalized outside their natural range are PTH, and all of the species that have achieved a wide naturalized distribution are PTH. Most of the PTH species in *Oenothera* seem to have originated within the limits of a taxonomic species (Raven). In a few instances, notably *O. triangulata* (Raven and D. P. Gregory 1972[1973]), four species of subsect. *Oenothera* (Raven et al. 1979; W. Dietrich et al. 1997), and several species of subsect. *Munzia* (Dietrich), the PTH taxa seem to have originated after hybridization between species. In all instances, PTH seems to have arisen as a way to limit recombination (Raven). In *Oenothera*, floating translocations are common (that is, a ring of four or a ring of six) and 46 species (30% of the species in the genus) are PTH with 1II + a ring of 12 chromosomes or a ring of 14.

Evening primrose oil (EPO), found mainly in seeds of sect. *Oenothera*, contains a rare omega-6 essential fatty acid, gamma-linolenic acid (GLA), which is considered to be the active therapeutic ingredient with value as a pharmaceutical and nutritional supplement. EPO is used for the treatment of a variety of disorders, particularly those affected by metabolic products of essential fatty acids; convincing evidence for its efficacy in treating most disorders is still lacking. The most promising uses are in the treatment of eczema and other skin irritations, multiple sclerosis, and diabetes.

Many species of *Oenothera* are cultivated and are among the most popular ornamentals in the family. The species most commonly cultivated and used for creating horticultural cultivars are *O. capillifolia* (sect. *Calylophus*), *O. lindheimeri* (sect. *Gaura*), *O. speciosa* (sect. *Hartmannia*), *O. fruticosa* (sect. *Kneiffia*), *O. acaulis* and *O. flava* (sect. *Lavauxia*), *O. macrocarpa* (sect. *Megapterium*), *O. glazioviana* (sect. *Oenothera*), and *O. cespitosa* (sect. *Pachylophus*).

Reproductive features include: self-incompatible or self-compatible; flowers vespertine or diurnal, usually lasting less than one day, but sometimes lasting two to several days (sects. *Kneiffia*, *Megapterium*); outcrossing species with diurnal flowers pollinated by bees [especially Halictidae (halictids), Anthophoridae (anthophorids), and *Bombus*], small moths, butterflies, syrphid flies, or hummingbirds (two spp.), and outcrossing species with vespertine flowers pollinated by hawkmoths or, sometimes, other small moths, rarely wasps or antlions (in *O. cinerea* of sect. *Gaura*), or autogamous, occasionally cleistogamous.

Seed numbers are not known precisely for all sections of *Oenothera*; here, seeds numerous indicates usually 30+ per capsule; for some taxa, seed number ranges from one to eight per capsule; when eight or fewer, the actual range is usually given.

Several sections (for example, *Gaura*, *Hartmannia*, *Kneiffia*, *Leucocoryne*, *Peniophyllum*) of *Oenothera* have capsules much wider in the distal, fertile part than in the proximal, sterile part, which tapers to a pedicel-like stipe. In other sections (for example, *Megapterium*, *Pachylophus*) the proximal part of the capsule is abruptly constricted to what has traditionally been considered to represent a pedicel. That terminology has been continued here, but anatomically it is not clear whether the stalk in these sections is developmentally the same as those with a stipe. The origin of pedicel-like structures should be studied in depth across the family.

SELECTED REFERENCES Cleland, R. E. 1972. *Oenothera*: Cytogenetics and Evolution. London. Evans, M. E. K. et al. 2005. Climate and life-history evolution in evening primroses (*Oenothera*, Onagraceae): A phylogenetic comparative analysis. Evolution 59: 1914–1927. Evans M. E. K. et al. 2009. Climate, niche evolution, and diversification of the "bird-cage" evening primroses (*Oenothera*, sections *Anogra* and *Kleinia*). Amer. Naturalist 173: 225–240. Gregory, D. P. 1963. Hawkmoth pollination in the genus *Oenothera*. Aliso 5: 375–384. Gregory, D. P. 1964. Hawkmoth pollination in the genus *Oenothera*. Aliso 5: 385–419. Harte, C. 1994. *Oenothera*: Contributions of a Plant to Biology. Berlin. Stubbe, W. 1964. The role of the plastome in the evolution of the genus *Oenothera*. Genetica 35: 28–33. Tobe, H., W. L. Wagner, and Chin H. C. 1987. Systematic and evolutionary studies of *Oenothera* (Onagraceae): Seed-coat anatomy. Bot. Gaz. 148: 235–257. Wagner, W. L. 2005. Systematics of *Oenothera* sections *Contortae*, *Eremia*, *Pachylophus*, and *Ravenia* (Onagraceae). Syst. Bot. 30: 332–355.

1. Stigmas peltate, discoid to quadrangular or obscurely and shallowly 4-lobed.
   2. Seeds in 2 rows per locule; capsules cylindrical .... 17b. *Oenothera* sect. *Calylophus*, p. 254
   2. Seeds clustered in each locule; capsules ellipsoid-rhombic to subglobose.......... ....17i. *Oenothera* sect. *Peniophyllum*, p. 277
1. Stigmas deeply divided into (3 or) 4 linear lobes.
   3. Capsules cylindrical to lanceoloid or ovoid, without ridges or wings, sometimes angled.
      4. Petals white.
         5. Seeds in 2 rows per locule, ellipsoid to subglobose, surface regularly pitted, pits in longitudinal lines; capsules straight or sometimes curved upward, dehiscent ½ their length.................................. 17n. *Oenothera* sect. *Kleinia*, p. 302
         5. Seeds in 1 row per locule, obovoid, surface minutely alveolate but appearing smooth; capsules straight, curved upward, or contorted, dehiscent ½ to nearly throughout their length .... 17o. *Oenothera* sect. *Anogra*, p. 304
      4. Petals yellow.
         6. Seeds prismatic and angled, usually ellipsoid to subglobose, rarely obovoid and obtusely angled, surfaces reticulate and regularly or irregularly pitted .... 17p. *Oenothera* sect. *Oenothera*, p. 313
         6. Seeds obovoid to oblanceoloid, surfaces coarsely rugose, also with turgid and collapsed papillae.
            7. Herbs winter-annuals; capsules not twisted, surfaces not wrinkled; seeds with conspicuous raphial groove and a pore at distal end of raphial face ....17l. *Oenothera* sect. *Eremia*, p. 300
            7. Herbs perennials; capsules twisted, surfaces conspicuously wrinkled; seeds lacking a raphial groove and pore.......... 17m. *Oenothera* sect. *Contortae*, p. 301
   3. Capsules usually ellipsoid to oblong, ovoid, or obovoid, sometimes lanceoloid, cylindrical, pyramidal, fusiform, or globose, angled or winged, or valve margins with ridges or tubercles.

[8. Shifted to left margin.—Ed]

8. Petals white, pink, or rose purple, rarely streaked or flecked with red.
   9. Capsules cylindrical to obtusely angled, with tubercles or an undulate ridge along valve margins . . . . 17a. *Oenothera* sect. *Pachylophus*, p. 247
   9. Capsules angled or winged, without tubercles or ridge along valve margins.
      10. Capsules indehiscent; seeds in 2–4 rows in locules or reduced to 1–8.
         11. Petals usually pink, rarely white, streaked or flecked with red; flowers actinomorphic; seeds in 2–4 rows per locule . . . . 17d. *Oenothera* sect. *Gauropsis*, p. 265
         11. Petals white, not streaked or flecked; flowers usually zygomorphic; seeds not in rows, reduced to 1–8 . . . . 17k. *Oenothera* sect. *Gaura*, p. 283
      10. Capsules dehiscent, at least in distal portion; seeds clustered in each locule.
         12. Petals pink or rose purple, rarely white (*O. speciosa*); capsules usually angled, rarely with a narrow wing to 0.5 mm wide, apex attenuate to a beak . . . . .17e. *Oenothera* sect. *Hartmannia*, p. 267
         12. Petals white; capsules with wings 0.5–4 mm wide, apex rounded, obtuse or bluntly acuminate. . . . . 17f. *Oenothera* sect. *Leucocoryne*, p. 269
8. Petals yellow.
   13. Seeds ovoid, 0.7–1 mm, clustered in each locule . . . . 17j. *Oenothera* sect. *Kneiffia*, p. 278
   13. Seeds obovoid, subcuboid, cuneiform, or rhombic, (1.8–)2–8 mm, in 1 or 2 rows per locule.
      14. Capsules with wings 10–32 mm wide throughout capsule length, or capsule walls with corky thickening and wings not developed (sometimes in *O. brachycarpa*), then capsule appearing only 4-angled; seeds usually with erose wing distally, surfaces coarsely rugose . . . . 17h. *Oenothera* sect. *Megapterium*, p. 272
      14. Capsules angled and valves with a prominent median ridge or, if winged, then wing oblong to triangular, confined to distal ½–⅔, (1–)3–5(–10) mm wide; seeds sometimes with a small wing at distal end and with a ridge or small wing along one adaxial margin, surfaces beaded.
         15. Petals obovate or obcordate; capsules winged, wings oblong to triangular, confined to distal ½–⅔ of capsule. .17c. *Oenothera* sect. *Lavauxia*, p. 262
         15. Petals usually elliptic, sometimes oblanceolate; capsules angled and valves with a prominent, broad median ridge . . . . .17g. *Oenothera* sect. *Paradoxus*, p. 271

17a. OENOTHERA Linnaeus sect. PACHYLOPHUS (Spach) Walpers, Repert. Bot. Syst. 2: 83. 1843 (as Pachylophis)

Pachylophus Spach, Hist. Nat. Vég. 4: 365. 1835; *Oenothera* [unranked] *Pachylophus* (Spach) Endlicher; *Oenothera* subg. *Pachylophus* (Spach) Reichenbach

Herbs annual or perennial, acaulescent or caulescent; from a usually stout taproot, sometimes lateral roots producing adventitious shoots. **Stems** (when present) usually ascending, sometimes erect or decumbent, branched or unbranched. **Leaves** in a basal rosette, sometimes also cauline, (0.5–)1.7–26(–36) cm; blade margins usually coarsely dentate to pinnatifid, sometimes serrate or subentire. **Inflorescences** solitary flowers from rosette or in axils of distal leaves. **Flowers** opening near sunset with a sweet scent or nearly unscented; buds erect or nodding by recurved floral tube, quadrangular, without free tips; floral tube (2–)3–140(–165) mm; sepals separating individually or in pairs; petals white, fading rose purple to pink, obovate or obcordate; pollen 90–100% fertile; stigma deeply divided into 4 linear lobes. **Capsules** thick-walled and woody, straight to falcate or sigmoid, lanceoloid or ellipsoid-ovoid to cylindrical, or sometimes obtusely 4-angled, tapering to a sterile beak, valve margins with a row of tubercles or a thickened, undulate ridge, dehiscent ⅓–⅞ their length; sessile or pedicellate. **Seeds** usually numerous, in (1 or) 2 rows per locule, obovoid to oblong, sometimes suborbicular or triangular, adaxial face with hollow chamber (seed collar) or, rarely (in *O. brandegeei*), filled with large, spongy cells, area above raphe a translucent membrane, surface papillose, reticulate, or irregularly roughened. **2*n*** = 14, 28.

Species 5 (4 in the flora): w North America, nw Mexico.

Members of sect. *Pachylophus* occur from southern Canadian prairies through the western United States and northern Mexico (northern Chihuahua and Sonora); *O. brandegeei* (Munz) P. H. Raven is disjunct in central Baja California. The center of diversity of sect. *Pachylophus* is in the Great Basin region, especially in Colorado (five taxa) and Utah (six taxa). The section is characterized by white petals, capsule valve margins tuberculate or ridged, and seeds with an unusual hollow seed collar, and rarely (only *O. psammophila*) a stem epidermis that produces viscid exudates. Two species of the section were included in a molecular analysis showing 100% strong support for the section (R. A. Levin et al. 2004). The position of sect. *Pachylophus* was not supported as a member of the two main lineages within the genus; it was sister to sect. *Calylophus* at the base of the phylogenetic tree. Reproductive features include: self-incompatible (3 spp.) or self-compatible (2 spp.); flowers vespertine, fragrance sweet or like rubber; large-flowered species outcrossing and pollinated by hawkmoths (*Hyles*, *Manduca*, and *Sphinx*) or Noctuidae (noctuids), with pollen-gathering bees sometimes effecting pollination (E. G. Linsley et al. 1963b; D. P. Gregory 1964; W. L. Wagner et al. 1985; D. Artz et al. 2010), and small-flowered species (*O. brandegeei* and *O. cavernae*) largely autogamous. Wagner et al. reported that for *O. psammophila* noctuids were the primary pollinators and hawkmoths secondary; recent study of populations by R. Raguso (unpubl.) indicates predominant hawkmoth pollination.

SELECTED REFERENCE Wagner, W. L., R. Stockhouse, and W. M. Klein. 1985. A systematic and evolutionary study of the *Oenothera cespitosa* species complex (Onagraceae). Monogr. Syst. Bot. Missouri Bot. Gard. 12: 1–103.

1. Petals (6.5–)8–20(–25) mm; herbs winter or spring annuals, stems 2–4 cm; stigmas surrounded by anthers at anthesis . . . . 4. *Oenothera cavernae*
1. Petals (16–)20–50(–60) mm; herbs perennial or, sometimes, annual; stems 10–40 cm; stigmas exserted beyond anthers at anthesis.
    2. Plants glabrous, with resinous exudate, especially on younger leaves; capsules somewhat curved and often somewhat twisted, valve margins with irregular, wavy ridges . . . . 2. *Oenothera psammophila*
    2. Plants usually pubescent, sometimes glabrous, without resinous exudate; capsules not twisted, valve margins tuberculate or ridged.
        3. Herbs perennial, acaulescent or caulescent; stems, when present, usually ascending, sometimes decumbent; capsule valve margins with tubercles or ridges; flowers: 1–4(–6) per stem opening per day . . . . 1. *Oenothera cespitosa*
        3. Herbs robust spring annuals, rarely overwintering for a 2nd year, caulescent; stems densely leafy, ascending to erect; capsule valve margins with conspicuous tubercles; flowers: usually 5–10 per stem opening per day . . . . 3. *Oenothera harringtonii*

1. Oenothera cespitosa Nuttall, Cat. Pl. Upper Louisiana, no. 53. 1813 F

Pachylophus cespitosus (Nuttall) Raimann

Herbs perennial, acaulescent or caulescent, usually hirsute or villous, usually also glandular puberulent, or exclusively strigillose, rarely glabrous; from stout taproot, sometimes lateral roots producing adventitious shoots. **Stems** (when present), usually ascending or decumbent, unbranched or branched from near base, 0–40 cm. **Leaves** 1.7–26(–36) × (0.3–)0.5–4.5(–6.5) cm; petiole (0.2–)1.7–11(–14) cm; blade usually oblanceolate to rhombic or spatulate, rarely elliptic, obovate, lanceolate, or linear-oblanceolate, margins irregularly sinuate-dentate, serrate, pinnatifid, lobed, or subentire, apex usually acute to rounded, rarely acuminate. **Flowers** 1–4(–6) per stem opening per day near sunset, with moderate to strong sweet scent with a rubbery background scent; buds usually erect, rarely recurved (during early development); floral tube (20–)40–140(–165) mm; sepals (15–)18–45(–54) mm; petals white, fading rose or rose pink to dark or deep rose purple, or pink to pale or light rose, or lavender, obovate or obcordate, (16–)20–50(–60) mm; filaments (6–)10–30(–35) mm, anthers (6–)9–17(–20) mm; style (45–)60–180(–185) mm, stigma exserted beyond anthers at anthesis. **Capsules** straight, curved, falcate, or sigmoid, usually cylindrical to lanceoloid or ellipsoid, sometimes ovoid, usually obtusely 4-angled, (10–)13–50(–68) × 4–9 mm, tapering to a sterile beak 6–8 mm, valve margins with rows of distinct tubercles to sinuate or nearly smooth ridges, dehiscent ⅓–⅞ their length; pedicel (0–)1–40(–55) mm. **Seeds** numerous in 1 or 2 rows per locule, usually obovoid, oblong, or triangular, rarely suborbicular, 2.1–3.9 × 1–2.6 mm, embryo ⅕–⅔ of seed volume, surface papillose, reticulate or rarely irregularly roughened; seed collar sealed by a thin membrane, this flat or depressed into raphial cavity, when depressed often splitting, becoming separated from seed collar. **2*n*** = 14, 28.

Subspecies 5 (5 in the flora): w North America, nw Mexico.

Oenothera cespitosa occurs in a wide array of habitats, from grassland, desert scrub, pinyon-juniper woodland, or Arizona chaparral to montane conifer forests, rarely at timberline, at elevations from (450–)800–3370 m. *Oenothera cespitosa* is self-incompatible (W. L. Wagner et al. 1985; Wagner 2005).

Pachylophus nuttallii Spach is an illegitimate name that pertains here.

1. Plants glabrous.
    2. Floral tubes (28–)35–60(–85) mm; petals fading rose pink to dark rose purple; capsules falcate or sigmoid, valve margins tuberculate .... 1a. *Oenothera cespitosa* subsp. *cespitosa* (in part)
    2. Floral tubes (45–)75–110(–153) mm; petals fading pink or rarely pale rose; capsules somewhat curved, valve margins with smooth to irregular, undulate ridges .... 1c. *Oenothera cespitosa* subsp. *macroglottis* (in part)
1. Plants hirsute, villous, glandular puberulent, or strigillose.
    3. Plants strigillose, rarely glandular puberulent; petals fading rose pink to dark rose purple .... 1a. *Oenothera cespitosa* subsp. *cespitosa* (in part)
    3. Plants hirsute or villous, usually also glandular puberulent, rarely only glandular puberulent; petals fading pink to light or pale rose or lavender-rose, sometimes deep rose purple.
        4. Stems unbranched to many-branched, sometimes producing dense clumps 5–50 cm diam.; petals fading rose; seed collar sinuate distally .... 1b. *Oenothera cespitosa* subsp. *crinita* (in part)
        4. Stems unbranched to several-branched, not forming clumps; petals fading rose purple or pink to pale rose or lavender; seed collar various.

[5. Shifted to left margin.—Ed.]

5. Petals fading rose or sometimes deep rose purple; capsules ellipsoid to lanceoloid-ellipsoid, falcate or sigmoid; pedicels 0.5–1 mm; seed collar membrane depressed and often splitting at maturity, margin conspicuously sinuate throughout .... 1b. *Oenothera cespitosa* subsp. *crinita* (in part)
5. Petals fading pink to pale rose or lavender; capsules lanceoloid to cylindrical, straight or somewhat curved; pedicels usually (0–)1–40(–55) mm; seed collar membrane neither depressed nor splitting at maturity, margin not sinuate, sometimes somewhat so distally.
    6. Capsules oblong-lanceoloid; buds often recurved when young; floral tube (35–)40–70(–80) mm; plants shaggy-villous, sometimes densely so .... 1e. *Oenothera cespitosa* subsp. *navajoensis*
    6. Capsules cylindrical to lanceoloid-cylindrical; buds erect; floral tube (41–)75–140-(–165) mm; plants hirsute.
        7. Capsules somewhat curved, valve margins with nearly smooth to irregular, undulate ridges; leaf blades oblanceolate to spatulate, margins dentate .... 1c. *Oenothera cespitosa* subsp. *macroglottis* (in part)
        7. Capsules straight, valve margins with minute to conspicuous tubercles, these sometimes coalesced into a sinuate ridge; leaf blades usually oblanceolate to narrowly elliptic, rarely lanceolate, margins usually pinnately lobed to dentate, rarely serrate ...... 1d. *Oenothera cespitosa* subsp. *marginata*

1a. Oenothera cespitosa Nuttall subsp. cespitosa [E]

Oenothera cespitosa subsp. *montana* (Nuttall) Munz; *O. cespitosa* var. *montana* (Nuttall) Durand; *O. cespitosa* subsp. *purpurea* (S. Watson) Munz; *O. cespitosa* var. *purpurea* (S. Watson) Munz; *O. marginata* Nuttall ex Hooker & Arnott var. *purpurea* S. Watson; *O. montana* Nuttall; *O. scapigera* Pursh; *Pachylophus canescens* Piper; *P. glaber* A. Nelson; *P. montanus* (Nuttall) A. Nelson

Herbs acaulescent or short-caulescent, glabrous or densely strigillose or hairs sometimes ± spreading, rarely sparsely glandular puberulent on flower parts. **Stems** (if present) usually unbranched, rarely with 1–several short laterals, 0–6(–21) cm. **Leaves** (2.8–)7–16(–21) × (0.3–)1–3(–5) cm; petiole (1–)3–7(–10) cm; blade obovate to linear-oblanceolate, margins coarsely and irregularly serrate or dentate, sometimes pinnately lobed or subentire. **Flowers:** floral tube (20–)40–60(–85) mm; sepals (15–)24–35(–40) mm; petals fading rose pink to dark rose purple, (16–)25–40(–48) mm; filaments (12–)15–24(–26) mm, anthers 9–12 mm; style (45–)60–120 mm. **Capsules** falcate or sigmoid, becoming nearly straight at maturity, asymmetrical and often somewhat flattened, lanceoloid to ovoid, (10–)20–40(–50) × 4–6 mm, valve margins with prominent, sinuate ridge with 5–10 peaks, or nearly distinct tubercles; pedicel 0.5–3 mm. **Seeds** narrowly obovoid, 2.5–3.9 × 1.2–1.7 mm embryo ½–⅔ of seed volume, surface papillose; seed collar oblong, membrane depressed deeply into raphial cavity, margin usually sinuate only distally, sometimes sinuate throughout. 2*n* = 14.

Flowering May–Jul(–Aug). Scattered or forming colonies in open sites, loose to hard, compacted clay, sandy soil, rocky slopes of shale, volcanic, or fine sandstone, gumbo flats, badlands, bluffs, exposed rocky ridges, roadcuts, grasslands, sagebrush, shadscale scrub, exposed sites in montane conifer forests; 800–3100 m; Alta., Man., Sask.; Colo., Idaho, Mont., Nebr., Nev., N.Dak., Oreg., S.Dak., Utah, Wash., Wyo.

1b. Oenothera cespitosa Nuttall subsp. **crinita** (Rydberg) Munz in N. L. Britton et al., N. Amer. Fl., ser. 2, 5: 100. 1965 (as caespitosa) E

Pachylophus crinitus Rydberg, Fl. Rocky Mts., 598, 1064. 1917; *Oenothera cespitosa* var. *crinita* (Rydberg) Munz; *O. cespitosa* subsp. *jonesii* (Munz) Munz; *O. cespitosa* var. *jonesii* Munz; *O. cespitosa* var. *stellae* S. L. Welsh

Herbs acaulescent or caulescent, densely hirsute, also sparsely glandular puberulent. **Stems** unbranched to many-branched, and then sometimes producing dense clumps 5–50 cm diam., 2–14 cm. **Leaves** 1.7–10(–18) × (0.3–)0.5–2.5(–3.4) cm; petiole (0.2–)3–5(–8) cm; blade usually oblanceolate to linear-oblanceolate, rarely obovate, margins subentire, sinuate, or dentate to pinnatifid. **Flowers:** floral tube (28–)35–75(–85) mm; sepals (15–)18–25(–27) mm; petals fading rose or sometimes deep rose purple, (16–)20–30(–35) mm; filaments (6–)10–17(–20) mm, anthers 6–8(–10) mm; style (45–)60–90(–105) mm. **Capsules** usually falcate or sigmoid, especially when young, also somewhat flattened, ellipsoid-ovoid to lanceoloid, 10–31(–34) × 4–9 mm, valve margins with 8–15 tubercles or these coalesced into a sinuate ridge; pedicel 0.5–1 mm. **Seeds** obovoid, oblong, or ± triangular, 2.9–3.5 × 1.1–2 mm, embryo ½–⅔ of seed volume, surface papillose, reticulate or very minutely roughened; seed collar membrane depressed and often splitting, becoming separated from collar at maturity, margin conspicuously sinuate throughout, surface often ribbed, ribs forming partial or complete vertical partitions in collar. 2*n* = 14, 28.

Flowering Apr–Jul. Open sites, compacted or loose soil derived from dolomite, limestone, tufa, or marble, exposed knolls, gravelly benches, steep slopes, scree, rocky mesas, rocky arroyos, from mountain summits in alpine or subalpine communities with *Pinus longaeva* and *P. flexilis* or pinyon-juniper woodlands to Great Basin or Mojave Desert shrub communities dominated by *Artemisia*, *Atriplex confertifolia*, *Coleogyne*, *Hilaria*, *Lycium*; 1100–3400 m; Ariz., Calif., Nev., Utah.

Subspecies *crinita* is the most polymorphic subspecies of *Oenothera cespitosa*; it is also the least understood. W. L. Wagner et al. (1985) grouped two series of populations that appear to intergrade together within the limits of this subspecies. One population is a morphologically relatively uniform form characterized by a many-branched habit, which may form dense clumps to 50 cm diameter, leaves that are 2–7 cm, floral tubes 25–60 mm and petals that fade to a rose color, and it occurs at high elevations on rocky, limestone sites or at lower elevations on extreme, chalky, white limestone and dolomite substrates or sometimes scree slopes. A more common form occurs at low to mid elevations in pinyon-juniper woodlands to Great Basin or Mojave Desert scrub on rocky slopes, talus, or along arroyos that is much less compact with one to several clustered rosettes, rarely more, with leaves 8–16 cm, floral tubes 45–75 mm, and petals that fade rose purple. The common form also grows on limestone and dolomite but, unlike the clumped form, it does not seem to be restricted to it. To compound the problem, many foothill and valley populations of subsp. *crinita* intergrade extensively with subspp. *cespitosa* and *marginata*.

1c. Oenothera cespitosa Nuttall subsp. **macroglottis** (Rydberg) W. L. Wagner, Stockhouse & W. M. Klein, Ann. Missouri Bot. Gard. 70: 195. 1983 (as caespitosa) E

Pachylophus macroglottis Rydberg, Bull. Torrey Bot. Club 30: 259. 1903; *Oenothera cespitosa* var. *macroglottis* (Rydberg) Cronquist; *P. hirsutus* Rydberg

Herbs acaulescent or short-caulescent, hirsute and glandular puberulent, or glabrous. **Stems** (if present), usually unbranched, rarely with 1–several short laterals, 4–8 cm. **Leaves** (6.8–)9.5–23(–32) × (1.3–)2.4–4.5(–6.5) cm; petiole (3–)4–11(–14) cm; blade oblanceolate to spatulate, margins often undulate, usually regularly to irregularly dentate, rarely coarsely and irregularly pinnately lobed. **Flowers:** floral tube (45–)75–110(–153) mm; sepals (22–)30–45(–50) mm; petals fading pink to pale rose, (21–)35–43(–50) mm; filaments (16–)19–28(–35) mm, anthers (10–)12–17 mm; style 85–180 mm. **Capsules** somewhat curved, lanceoloid-cylindrical to cylindrical, symmetrical throughout, sometimes slightly flattened on one side at base, (17–)25–45(–56) × 5–8 mm, valve margins with conspicuous, nearly smooth to irregular undulate ridge; pedicel 2–7 mm. **Seeds** narrowly obovoid, 2.5–3 × 1–1.4 mm, embryo ½ of seed volume, surface minutely papillose to reticulate; seed collar forming narrow slit above raphe with a slightly sunken membrane, margin entire or obscurely sinuate distally. $2n$ = 14, 28.

Flowering May–Jul(–Sep). Open, igneous rocky slopes, talus, roadcuts, open or shaded and sandy or gravelly sites along streams, rarely on shale, in upper pinyon-juniper woodlands, Gambel oak scrub, ponderosa pine forests, ponderosa pine-Douglas fir forests, spruce-fir-lodgepole pine forests; 2000–3100 m; Colo., N.Mex., Utah, Wyo.

1d. Oenothera cespitosa Nuttall subsp. marginata (Nuttall ex Hooker & Arnott) Munz in N. L. Britton et al., N. Amer. Fl., ser. 2, 5: 101. 1965 (as caespitosa)

Oenothera marginata Nuttall ex Hooker & Arnott, Bot. Beechey Voy., 342. 1839; *Anogra longiflora* A. Heller; *O. cespitosa* subsp. *eximia* (A. Gray) Munz; *O. cespitosa* var. *eximia* (A. Gray) Munz; *O. cespitosa* var. *longiflora* (A. Heller) Munz; *O. cespitosa* var. *marginata* (Nuttall ex Hooker & Arnott) Munz; *O. eximia* A. Gray; *O. idahoensis* Mulford; *Pachylophus cylindrocarpus* A. Nelson; *P. exiguus* (A. Gray) Rydberg; *P. eximius* (A. Gray) Rydberg; *P. longiflorus* (A. Heller) A. Heller; *P. marginatus* (Nuttall ex Hooker & Arnott) Rydberg; *P. prolatus* A. Nelson

Herbs caulescent or acaulescent, usually moderately to densely hirsute and glandular puberulent, rarely exclusively glandular puberulent. **Stems** usually unbranched, rarely with 1–several branches from near base, 10–40 cm. **Leaves** (2.8–)10–26(–36) × (0.6–)1–3(–4.5) cm; petiole (3–)4–11(–14) cm; blade usually oblanceolate to narrowly elliptic, rarely lanceolate, margins coarsely and irregularly pinnately lobed to dentate or, rarely, serrate. **Flowers:** floral tube (41–)80–140(–165) mm; sepals (22–)34–45(–54) mm; petals fading pink to lavender, (24–)35–50(–60) mm; filaments (16–)20–30(–35) mm, anthers (10–)12–17(–20) mm; style (78–)100–150(–185) mm. **Capsules** straight, cylindrical to, sometimes, lanceoloid-cylindrical, slightly asymmetrical, (21–)25–50(–68) × 6–8 mm, valve margins with minute to conspicuous tubercles, these sometimes coalesced into a sinuate ridge; pedicel (0–)1–40(–55) mm. **Seeds** usually narrowly to broadly obovoid, rarely suborbicular, 2.2–3.4 × 1.1–2.6 mm, embryo ⅕–½ of seed volume, surface appearing longitudinally striate, reticulate under magnification; seed collar large, usually appearing inflated, sealed by a flat membrane which is often slightly depressed into seed collar cavity, margin entire or obscurely sinuate distally. $2n$ = 14.

Flowering Mar–Aug. Rocky slopes, cracks in rocks, talus, along gravelly creek beds and arroyos, roadcuts in loose to somewhat compacted soil derived from granite, sandstone, limestone, volcanic cinder, rarely shale, mostly in foothill communities of pinyon-juniper woodlands, big sagebrush scrub, chaparral, grasslands, openings in ponderosa pine forests; (400–)1200–2300(–3100) m; Ariz., Calif., Colo., Idaho, Nev., N.Mex., Oreg., Tex., Utah, Wash., Wyo.; Mexico (Chihuahua, Sonora).

1e. Oenothera cespitosa Nuttall subsp. navajoensis W. L. Wagner, Stockhouse & W. M. Klein, Monogr. Syst. Bot. Missouri Bot. Gard. 12: 66, fig. 104. 1985 (as caespitosa) [E] [F]

Oenothera cespitosa var. *navajoensis* (W. L. Wagner, Stockhouse & W. M. Klein) Cronquist; *Pachylophus caulescens* Rydberg

Herbs caulescent or acaulescent, moderately to densely crinkly-villous (often appearing shaggy), and glandular puberulent. **Stems** (if present) unbranched to few-branched, (0–)10–25 cm. **Leaves** (3.5–)4–13(–16) × (0.7–)1–3.2 cm; petiole (1.3–)1.7–10(–12) cm; blade oblanceolate to rhombic-obovate, margins often coarsely and irregularly dentate or serrate, sometimes pinnately lobed, often with several larger lobes near blade base. **Flowers:** buds often recurved when young; floral tube (35–)40–70(–80) mm; sepals 22–27(–32) mm; petals fading pink to light rose, (25–)28–32(–34) mm; filaments 11–15(–17) mm, anthers 9–12 mm; style (50–)59–85(–96) mm. **Capsules** straight, oblong-lanceoloid, base asymmetrical, 13–35(–40) × 5–6 mm; valve margins with a low sinuate ridge to 8–15 small, nearly distinct tubercles; pedicel 1–3 mm. **Seeds** narrowly obovoid, 2.1–2.6 × 1.1–1.3 mm, embryo ½ of seed volume, surface minutely papillose; seed collar forming a narrow slit above raphe with a slightly sunken membrane, margin entire or obscurely sinuate distally. $2n$ = 14, 28.

Flowering (Apr–)May–Jun. Colorado Plateau region, forming small colonies on loose or compacted soil derived from clay, shale, fine-textured sandstone, or gypsum, on slopes and along small drainage patterns, often around harvester ant mounds, arroyos in somewhat sandy or gravelly soil, in shrubby communities dominated by *Atriplex confertifolia*, *A. corrugata*, *A. cuneata*, *Artemisia spinescens*, *Coleogyne ramosissima*, *Frankenia jamesii*, *Hilaria jamesii*, with big sagebrush scrub or sage-grasslands, rarely in lower pinyon-juniper woodlands; 1100–1900(–2100) m; Ariz., Colo., N.Mex., Utah.

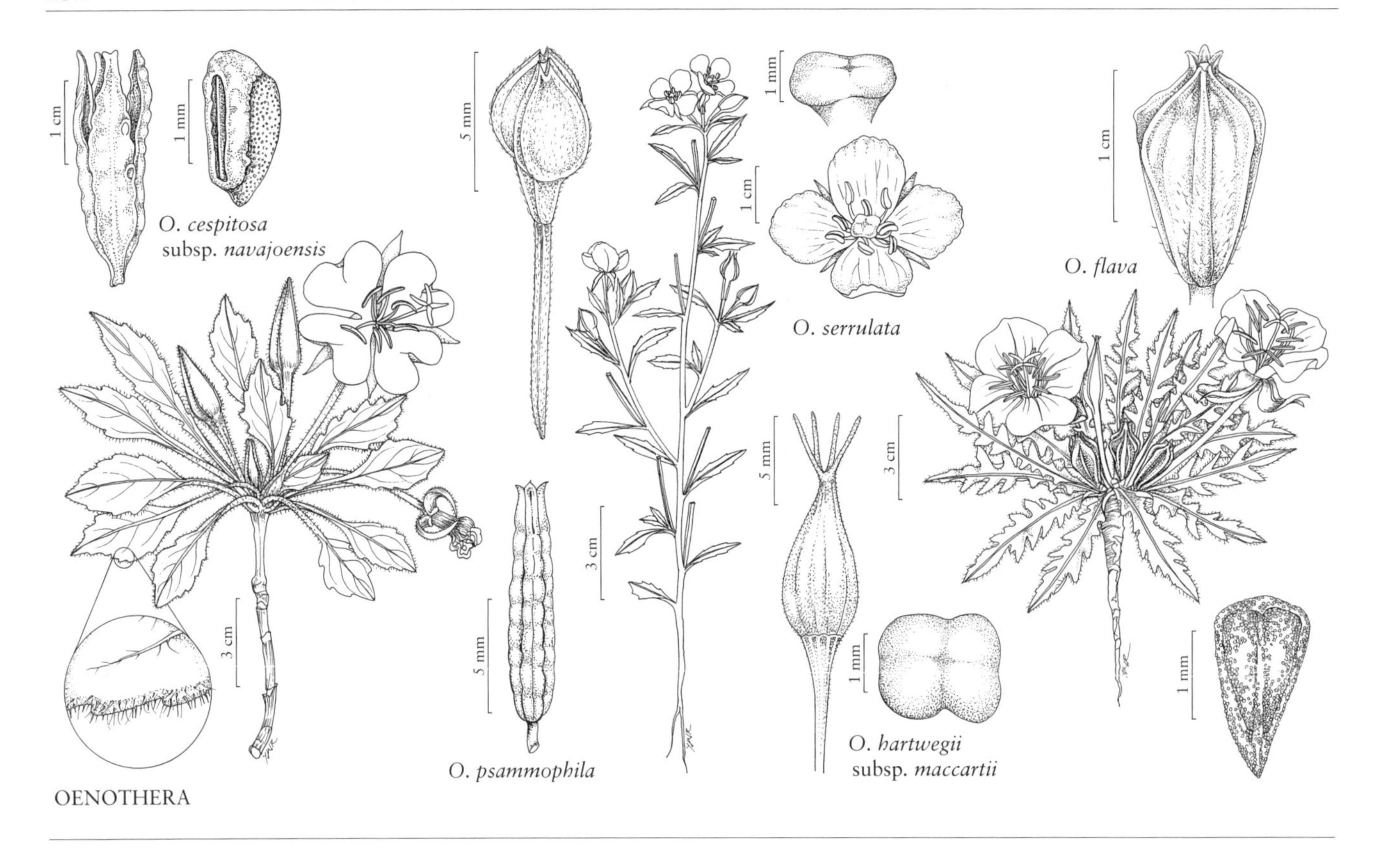

OENOTHERA

2. Oenothera psammophila (A. Nelson & J. F. Macbride) W. L. Wagner, Stockhouse & W. M. Klein, Monogr. Syst. Bot. Missouri Bot. Gard. 12: 84. 1985 • St. Anthony Dunes evening primrose [C] [E] [F]

Pachylophus psammophilus A. Nelson & J. F. Macbride, Bot. Gaz. 61: 32. 1916; *Oenothera cespitosa* Nuttall var. *psammophila* (A. Nelson & J. F. Macbride) Munz

Herbs perennial, caulescent, glabrous, also with resinous exudate, especially on younger leaves; from woody taproot. **Stems** decumbent, usually branched, 10–30 cm, becoming woody and buried in sand. **Leaves** (6–)8–9(–14.2) × (0.7–)1.5–2(–3.1) cm; petiole 3–9 cm; blade narrowly oblanceolate to oblanceolate, margins usually subentire or repand or remotely dentate, rarely serrate, apex acute. **Flowers** 1–3 per stem opening per day near sunset, with sweet scent; buds erect; floral tube 42–60 mm; sepals 22–28 mm; petals white, fading rose pink to rose, 23–40 mm; filaments 17–19 mm, anthers 13–16 mm; style 60–72(–88) mm, stigma exserted beyond anthers at anthesis. **Capsules** somewhat curved and often somewhat twisted, lanceoloid-cylindrical, nearly cylindrical, (20–)30–47 × 7–8 mm, gradually tapering to apex, 6–8 mm, dehiscent nearly throughout their length, valve margins with a conspicuous, irregular, wavy ridge; pedicel 1–5 mm. **Seeds** numerous, in 2 distinct rows per locule, narrowly obovoid, 2.5–3 × 1.2–1.4 mm, embryo ⅞ of seed volume, surface finely reticulate; seed collar with a broad membrane sealing cavity, margin entire. $2n$ = 14.

Flowering May–Jul. On barren areas of drifting sand at interface between outcrops of lava and sand dunes; of conservation concern; 1500–1700 m; Idaho.

Oenothera psammophila is known only from the dune area north and west of St. Anthony, Fremont County. It is unique in *Oenothera* because of the exudate produced on the leaves to which sand particles adhere, forming a sand sheath, presumably for protection from the constantly blowing sand particles. It is self-incompatible (W. L. Wagner et al. 1985; Wagner 2005).

3. Oenothera harringtonii W. L. Wagner, Stockhouse & W. M. Klein, Ann. Missouri Bot. Gard. 70: 195. 1983 [C] [E]

Herbs robust spring annual or, rarely, surviving a second year, caulescent, hirtellous, also glandular puberulent; from stout taproot. **Stems** ascending to erect, stout, unbranched or with lateral stems from basal rosette, densely leafy, 15–30 cm. **Leaves** 10–14(–14.5) × 1.5–2.3 (–3) cm; petiole 4.3–6.6 cm; blade narrowly oblanceolate, margins irregularly and coarsely dentate, apex acute. **Flowers** usually 5–10 per stem opening per day near sunset, with heavy, sweet scent; buds erect; floral tube 31–60 mm; sepals 17–26 mm; petals white, fading pale pink, 20–26 mm; filaments 11–16 mm, anthers 8–11 mm; style 65–96 mm, stigma exserted beyond anthers at anthesis. **Capsules** straight, lanceoloid, obtusely 4-angled, (21–)25–30(–35) × (5–)6–8 mm, tapering to a sterile beak 6–8 mm, dehiscent ½–⅔ their length, valve margins with 5–8 conspicuous, irregular tubercles, sometimes 2 or more coalesced into a sinuate ridge, also with conspicuous medial ridge throughout; pedicel 0.5–1 mm. **Seeds** numerous, usually in 2 distinct rows per locule, sometimes rows partially overlapping, narrowly obovoid, 2.1–2.3 × 1–1.3 mm, embryo slightly less than ½ seed volume, surface appearing finely striate but papillose under magnification; seed collar with membrane intact at maturity, membrane rarely splitting and separating from collar, margin entire. $2n$ = 14.

Flowering May–Jun. On compacted, silty clay to looser rocky and sandy soil in open grassland; of conservation concern; 1400–1900 m; Colo.

Oenothera harringtonii is known only from southeastern Colorado from western El Paso and eastern Fremont counties, southeast through Pueblo to Otero counties, and south to Las Animas County; it may also occur in adjacent Colfax and Union counties in New Mexico but has not been collected there. *Oenothera harringtonii* is self-incompatible (W. L. Wagner et al. 1985; Wagner 2005).

4. Oenothera cavernae Munz, Leafl. W. Bot. 3: 50. 1941 [C] [E]

Herbs winter or spring annual, acaulescent or short-caulescent, glandular puberulent, sometimes also sparsely hirsute; from a taproot. **Stems** (when present) 1–several, ascending, usually unbranched, 2–4 cm. **Leaves** primarily in a basal rosette, sometimes also cauline, (0.5–)2.5–13(–19.5) × (0.2–)0.6–2.3(–2.7) cm; petiole 0.5–5.2 cm; blade oblanceolate to elliptic-oblanceolate (in some exceptionally large leaves), margins lyrate-pinnatifid to subentire (in very small ones), apex usually rounded, rarely acute. **Flowers** 1–3(–10) per stem opening per day near sunset, without noticeable scent; buds sometimes ± recurved before anthesis; floral tube (20–)30–37(–47) mm; sepals 4.5–12 mm; petals white, fading pale pink, (6.5–)8–20(–25) mm; filaments 5.2–7.5(–12) mm, anthers (1.4–)3–4.5(–6) mm; style (24–)35–45(–56) mm, stigma surrounded by anthers at anthesis. **Capsules** falcate (especially before maturity), ellipsoid-ovoid to ovoid, obtusely 4-angled, 12–38 × 6–14 mm, tapering to a sterile beak 2–8 mm, dehiscent to ½ their length, valve margins with a sinuate ridge or 8–20 nearly distinct tubercles; pedicel 0–10 mm. **Seeds** usually numerous, sometimes as few as 5, in 2 adjacent rows per locule, obovoid, 2.5–3.1 × 1.1–1.4 mm, embryo ½ of seed volume, surface minutely papillose to reticulate; seed collar without membrane, producing a large empty cavity, margin irregularly sinuate. $2n$ = 14.

Flowering Mar–May. Exposed calcareous slopes, crevices in limestone, dolomite, or loose talus, sandy arroyos, sandstone, granitic crevices, volcanic cinders in Mojave Desert or Great Basin scrub communities, rarely in arid juniper woodlands; of conservation concern; 400–1700 m; Ariz., Calif., Nev.

Oenothera cavernae is known from the Arrow Canyon, Las Vegas, and Sheep ranges and the low hills near Arden and Sloan in Clark County, Nevada, eastward along the Grand Canyon to the vicinity of Page, Arizona, and perhaps Washington County, Utah and formerly in Glenn Canyon, and more recently collected in eastern San Bernardino County, California (eastern Clark Mountain Range, and the base of range in Ivanpah Valley). W. L. Wagner et al. (1985) determined *O. cavernae* to be self-compatible and autogamous.

17b. Oenothera Linnaeus sect. Calylophus (Spach) W. L. Wagner & Hoch, Syst. Bot. Monogr. 83: 147. 2007

Calylophus Spach, Hist. Nat. Vég. 4: 349. 1835; *Oenothera* [unranked] *Calylophus* (Spach) Torrey & A. Gray

Herbs perennial, rarely annual, sometimes suffrutescent, caulescent; from a stout to slender taproot. **Stems** decumbent to ascending or erect, branched or unbranched, epidermis gray to brown, ± exfoliating. **Leaves** cauline, 0.3–9 cm, fascicles of small leaves often present in larger leaf axils; blade margins entire or subentire, serrate, or serrulate, sometimes spinulose. **Inflorescences** solitary flowers in axils of distal leaves. **Flowers** opening near sunset or sunrise, sometimes in afternoon, usually with a sweet scent; buds erect, terete or quadrangular, with free tips; floral tube 2–60(–70) mm; sepals flat or with keeled midribs, reflexed individually; petals yellow, usually fading dark yellow, orange, pale pink, or pale purple, suborbiculate to rhombic or obcordate; stigma usually yellow to yellow-green, blue-black in *O. capillifolia* subsp. *capillifolia*, peltate, discoid to quadrangular, sometimes shallowly 4-lobed. **Capsules** woody and hard to thin and ± papery, straight, cylindrical to obtusely 4-angled, often tapering at each end, dehiscent ½ to throughout their length; sessile. **Seeds** usually numerous, in 2 rows per locule, usually obovoid and somewhat angled, rarely oblanceoloid, surface smooth. **2*n*** = 14 (28).

Species 7 (7 in thc flora): w, c North America, Mexico.

Section *Calylophus* consists of 7 species (13 taxa) classified in two subsections distributed throughout the Great Plains to Arizona and south to central Mexico with a center of diversity in Texas (H. F. Towner 1977). P. H. Raven (1964) separated *Calylophus* from *Oenothera* as treated by P. A. Munz (1965) based on its peltate stigma and unusual sporogenous tissue (Towner). The peltate stigma can now be more properly interpreted as a variation from the typical *Oenothera* stigma, with the indusium enlarged and the lobes reduced (W. L. Wagner et al. 2007). Molecular studies (R. A. Levin et al. 2004) strongly support both the inclusion of *Calylophus* within *Oenothera* and the monophyly of the section by inclusion of a species from each of the subsections. Subsequent detailed analyses (B. Cooper, unpubl.) indicate that subsect. *Salpingia* is not monophyletic with subsect. *Calylophus* nested within it, indicating that subsections may not be justifiable as currently defined. All species, except *O. serrulata*, are self-incompatible and outcrossing; flowers diurnal to vespertine, opening in the early morning or from midafternoon to near sunset, wilting in one and one-half to two days; those species with diurnal flowers are pollinated by bees [especially Halictidae (halictids) and Anthophoridae (anthophorids), often oligolectic species], those with larger vespertine flowers are pollinated by hawkmoths (Towner). The self-compatible *O. serrulata* is autogamous and a PTH species. Recent work (M. Moore, pers. comm.) suggests that there is additional edaphic endemism within sect. *Calylophus*, which is being investigated using morphology and molecular analyses (B. Cooper, unpubl.), and will likely result in detection of previously unknown species in the section.

Meriolix Rafinesque ex Endlicher 1840 is a superfluous name that pertains here.

SELECTED REFERENCE Towner, H. F. 1977. The biosystematics of *Calylophus* (Onagraceae). Ann. Missouri Bot. Gard. 64: 48–120.

1. Sepals with conspicuously keeled midribs; stamens in 2 unequal series, antisepalous filaments 2 times as long as antipetalous filaments. . . . . 17b.1. *Oenothera* subsect. *Calylophus*, p. 255
1. Sepals without keeled midribs; stamens in subequal series . . . . . . .17b.2. *Oenothera* subsect. *Salpingia*, p. 258

17b.1. Oenothera Linnaeus (sect. Calylophus) subsect. Calylophus (Spach) W. L. Wagner & Hoch, Syst. Bot. Monogr. 83: 147. 2007

Calylophus Spach, Hist. Nat. Vég. 4: 349. 1835

Herbs perennial, short-lived or suffrutescent, or annual. Leaves: blade margins subentire or serrulate or spinulose-serrate. **Flowers** usually crowded near stem apex; buds quadrangular; floral tube 2–20 mm; sepals with keeled midrib; stamens in 2 unequal series, antisepalous set conspicuously longer, pollen 30–100% fertile. **Capsules** hard, straight or, sometimes, slightly recurved, tardily dehiscent ½ to throughout their length. **Seeds** obovoid or oblanceoloid, sometimes sharply angled. **2*n*** = 14.

Species 3 (3 in the flora): w, c North America, n Mexico.

Subsection *Calylophus* consists of three species (four taxa), one of which (*Oenothera capillifolia*) occurs primarily in Texas, but extends to southeastern Colorado, southern Kansas, southern Louisiana, and northeastern Mexico, and is disjunct in northwestern New Mexico; *O. serrulata* occurs throughout the Great Plains to Alberta, Manitoba, and Saskatchewan east to Wisconsin, west to east-central Arizona, with one occurrence in Chihuahua, Mexico, at elevations 0–1800(–2100) m. Species of subsect. *Calylophus* have diurnal flowers.

1. Petals 5–12(–20) mm; stigmas surrounded by anthers at anthesis; pollen 30–60(–80)% fertile . . . . 7. *Oenothera serrulata*
1. Petals 6–25 mm; stigmas exserted beyond anthers at anthesis; pollen 90–100% fertile.
    2. Leaves (0.1–)0.3–1 cm wide; stems (10–)25–80 cm, 1–many, unbranched to moderately branched, weakly decumbent to erect; sepals 4–12 mm; floral tube 5–20 mm; petals 6–25 mm . . . . 5. *Oenothera capillifolia*
    2. Leaves 0.1–0.2 cm wide; stems 15–30(–40) cm, many, branched from base, ascending to erect; sepals 4–6 mm; floral tube 7 mm; petals 15–20 mm . . . . 6. *Oenothera gayleana*

5. Oenothera capillifolia Scheele, Linnaea 21: 576. 1848

Meriolix capillifolia (Scheele) Small

Herbs perennial (short-lived or, sometimes, suffrutescent) or annual, glabrous or strigillose; from a stout taproot. **Stems** 1–many, weakly decumbent to ascending or erect, unbranched to moderately branched, (10–)25–80 cm. **Leaves** 1–9 × (0.1–)0.3–1 cm, sometimes fascicles of small leaves to 2 cm present in non-flowering axils; petiole 0–0.6 cm; blade linear to narrowly lanceolate or oblanceolate, often folded lengthwise, usually not much reduced distally, proximalmost leaves sometimes spatulate, base attenuate, margins subentire or serrulate or spinulose-serrate, apex acute. **Flowers** opening at sunrise; buds with free tips 0–4 mm; floral tube 5–20 mm; sepals 4–12 mm, midribs keeled; petals yellow, fading orangish to purplish, 6–25 mm; antisepalous filaments 2–8 mm, antipetalous filaments 1–4 mm, anthers 2–7 mm, pollen 90–100% fertile; style 9–30 mm, stigma sometimes blue-black, discoid to quadrangular, exserted beyond anthers. **Capsules** 10–35 × 1–2 mm, hard, dehiscent ½ their length, often tardily dehiscent throughout their length. **Seeds** obovoid, 1–1.8 mm, sharply angled, apex truncate.

Subspecies 2 (2 in the flora): c, sc United States, n Mexico.

Oenothera capillifolia is self-incompatible (H. F. Towner 1977).

Oenothera berlandieri (Spach) Steudel 1841, not D. Dietrich 1840, is superfluous and cannot be used in *Oenothera* when transferred from *Calylophus*, and pertains here.

1. Stems 1–several, ascending to erect, 30–80 cm; leaves 2.5–9 cm . . . . 5a. *Oenothera capillifolia* subsp. *capillifolia*
1. Stems several–many, decumbent to ascending, (10–)25–40 cm; leaves 1–4 cm . . . . 5b. *Oenothera capillifolia* subsp. *berlandieri*

5a. Oenothera capillifolia Scheele subsp. **capillifolia** [E]

Calylophus berlandieri Spach subsp. *pinifolius* (Engelmann ex A. Gray) Towner; *Meriolix hillii* Small; *M. melanoglottis* Rydberg; *M. serrulata* (Nuttall) Walpers var. *pinifolia* (Engelmann ex A. Gray) Small; *Oenothera berlandieri* (Spach) Steudel subsp. *pinifolia* (Engelmann ex A. Gray) W. L. Wagner & Hoch; *O. serrulata* Nuttall var. *maculata* H. Léveillé; *O. serrulata* subsp. *pinifolia* (Engelmann ex A. Gray) Munz; *O. serrulata* Nuttall var. *pinifolia* Engelmann ex A. Gray

Herbs annual or short-lived perennial. Stems 1–several, unbranched or sparsely branched, ascending to erect, 30–80 cm. **Leaves** 2.5–9 × 0.2–1 cm; blade margins remotely serrulate to spinulose-serrate. **Flowers:** buds with free tips 0.5–4 mm; sepals with conspicuously keeled midribs. **2*n*** = 14.

Flowering Mar–Sep. Prairies, open places in oak savannas on rocky, clay, or sandy soil, often calcareous; 0–900 m; La., Okla., Tex.

Subspecies *capillifolia* occurs from Blaine and Lincoln counties, Oklahoma, south through a narrow portion of north-central Texas to central Texas, where it is widely distributed, especially on the Edwards Plateau; also occurring locally in western and southern Louisiana. The floral tubes and/or stigma of some populations, especially from Bexar, Blanco, Comal, Gillespie, Hays, Kendall, Kerr, and Travis counties, Texas, are a deep blackish purple, while others are yellow, which is typical of the section.

5b. Oenothera capillifolia Scheele subsp. **berlandieri** (Spach) W. L. Wagner & Hoch, PhytoKeys 28: 71. 2013

Calylophus berlandieri Spach, Ann. Sci. Nat., Bot., sér. 2, 4: 273. 1835 (as Calylophis); *C. drummondianus* Spach subsp. *berlandieri* (Spach) Towner & P. H. Raven; *Meriolix berlandieri* (Spach) Walpers

Herbs perennial. Stems several–many, decumbent to ascending, moderately branched, (10–)25–40 cm. **Leaves** 1–4 × (0.1–)0.3–0.6 cm; blade margins subentire or serrate, sometimes weakly undulate. **Flowers:** buds with free tips 0–2 mm; sepals often with slightly keeled midribs. **2*n*** = 14.

Flowering Mar–Sep. Grassy prairies, plains, low hills, sandy, gravelly, and limestone soil, relatively dry areas, in vegetation dominated by mesquite, oaks, and *Opuntia*; 0–1200 m; Kans., N.Mex., Okla., Tex.; Mexico (Coahuila, Nuevo León).

Subspecies *berlandieri* occurs in Meade, Reno, and Seward counties in Kansas, south through eastern New Mexico, the Texas Panhandle, and western Oklahoma to Crane, Culberson, and Ward counties in Texas, southeastward to near the Pecos and Rio Grande rivers to the Gulf Coast, becoming widespread on the Coastal Plain north to Milam County in Texas; it also occurs in the Santa Rosa Mountains of northern Coahuila, Mexico.

6. Oenothera gayleana B. L. Turner & M. J. Moore, Phytologia 96: 200, figs. 1, 2. 2014 [E]

Herbs perennial, sometimes suffrutescent, usually strigillose, sometimes glabrous; from a stout taproot. **Stems** many, ascending to erect, branched from base, 15–30(–40) cm. **Leaves** 2.5–3.5 × 0.1–0.2 cm, rarely fascicles of small leaves present in non-flowering axils; petiole 0–0.1 cm; blade linear to narrowly linear-lanceolate, folded lengthwise, base long-attenuate, margins subentire or serrulate, apex acute. **Flowers** opening near sunrise; buds with free tips 0–0.5 mm; floral tube 7 mm; sepals 4–6 mm, midribs keeled; petals yellow, fading yellow to orange, 15–20 mm; antisepalous filaments 5 mm, antipetalous filaments 2 mm, anthers 3–4 mm, pollen 90–100% fertile; style 10 mm, stigma discoid to quadrangular, exserted beyond anthers at anthesis. **Capsules** 18–20 × 2 mm, hard, dehiscent ½ their length, often tardily dehiscent throughout their length. **Seeds** oblanceoloid, 1–1.8 mm, sharply angled, apex truncate. **2*n*** = 14.

Flowering May–Sep. Gypsum outcrops; 500–1400 m; N.Mex., Tex.

Oenothera gayleana is a recently discovered gypsum endemic known only from scattered outcrops from De Baca and Eddy counties in New Mexico, and Culberson County in Texas. When published, the delimitation of *O. gayleana* included populations in Collinsworth and Dickens counties in the Texas panhandle, and adjacent Harmon County in Oklahoma. Subsequent study (B. Cooper et al., unpubl.) has determined they are actually *O. serrulata*.

7. Oenothera serrulata Nuttall, Gen. N. Amer. Pl. 1: 246. 1818 F

Calylophus australis Towner & P. H. Raven; *C. drummondianus* Spach; *C. serrulatus* (Nuttall) P. H. Raven; *C. serrulatus* var. *arizonicus* Shinners; *C. serrulatus* var. *spinulosus* (Torrey & A. Gray) Shinners; *Meriolix drummondiana* (Spach) Small; *M. intermedia* Rydberg; *M. oblanceolata* Rydberg; *M. serrulata* (Nuttall) Walpers; *M. serrulata* var. *drummondii* (Torrey & A. Gray) Walpers; *M. serrulata* var. *spinulosa* (Torrey & A. Gray) Walpers; *M. spinulosa* (Torrey & A. Gray) A. Heller; *Oenothera leucocarpa* Comien; *O. serrulata* var. *douglasii* Torrey & A. Gray; *O. serrulata* subsp. *drummondii* (Torrey & A. Gray) Munz; *O. serrulata* var. *drummondii* Torrey & A. Gray; *O. serrulata* var. *integrifolia* H. Léveillé; *O. serrulata* var. *spinulosa* Torrey & A. Gray

Herbs perennial, glabrous or strigillose; from a stout taproot. **Stems** 1–many, weakly decumbent to erect, unbranched to moderately branched, 10–60(–80) cm. **Leaves** 1–9 × 0.1–1 cm, sometimes fascicles of small leaves to 2 cm present in non-flowering axils; petiole 0–0.6 cm; blade linear to narrowly lanceolate or oblanceolate, often folded lengthwise, usually not much reduced distally, proximalmost stem leaves often narrowly oblanceolate to oblanceolate, sometimes spatulate, base attenuate, margins subentire or spinulose-serrate, apex acute. **Flowers** opening near sunrise; buds with free tips 0–4 mm; floral tube 2–12(–16) mm; sepals 1.5–9 mm, midribs keeled; petals yellow, fading dark yellow to orange, 5–12(–20) mm; antisepalous filaments 1–5(–7) mm, antipetalous filaments 0.5–3 mm, anthers 1.5–4(–6) mm, pollen 30–60(–80)% fertile; style 2–15 (–20) mm, stigma discoid to quadrangular, surrounded by anthers at anthesis. **Capsules** 6–25 × 1–3 mm, hard, dehiscent ½ their length, often tardily dehiscent through their length. **Seeds** obovoid, 1–1.8 mm, sharply angled, apex truncate. **2*n*** = 14.

Flowering Mar–Aug. Prairies, in grassy, open areas in woods, rarely in mountains, usually sandy or rocky soil; 0–2100 m; Alta., Man., Ont., Sask.; Ariz., Colo., Ill., Ind., Iowa, Kans., Mich., Minn., Mo., Mont., Nebr., N.Mex., N.Dak., Okla., S.Dak., Tex., Wis., Wyo.; Mexico (Chihuahua).

Oenothera serrulata occurs from southern Alberta, southern Saskatchewan, and southern Manitoba to eastern New Mexico, the Texas Panhandle, and the Gulf Coast of Texas, including eastern Montana, eastern Wyoming, eastern Colorado, North Dakota, South Dakota, Nebraska, Kansas, western and central Oklahoma, western and southern Minnesota, Iowa, northwestern Missouri, and with outlying populations in central Illinois, northern Indiana, southeastern Wisconsin, northwestern peninsular Michigan, east-central Arizona, and west-central Chihuahua, Mexico; it is naturalized in Ontario. It was documented in 1909 as a non-native in Vermont and has not been collected since. *Oenothera serrulata* is a PTH species and forms a ring of 12 + 1II or a ring of 14 chromosomes in meiosis, and is self-compatible and autogamous (H. F. Towner 1977).

Calylophus nuttallii Spach is a superfluous name that pertains here. *Oenothera spachiana* Steudel August 1840 (not Torrey & A. Gray June 1840) is an illegitimate later homonym and also pertains here.

H. F. Towner (1977) is followed here in recognition of a broadly delimited *Oenothera serrulata* as a complex assemblage of populations that are all primarily autogamous and are PTH. These populations consist of wide morphological diversity involving leaf size and shape, stature, pubescence, and flower size. Some of these variants may have evolved independently from *O. capillifolia*. Flower size is variable throughout the geographical range, and some of the largest flowered forms occur near large-flowered populations of *O. capillifolia* subsp. *capillifolia* in central Oklahoma. Most populations occurring west of approximately 98°W longitude comprise well-branched, short-leaved, and relatively low-statured plants, while those east of that line are less branched, taller and more erect, long-leaved, and densely strigillose. Populations along the Texas Gulf Coast described as *Calylophus australis* are rather distinctive and are separated from the remainder of the populations of *O. serrulata* primarily in less dense pubescence, shorter, coarsely serrate leaves, and more erect stems. They may have been independently derived from *O. capillifolia*. In his revision, Towner did not continue to recognize them because there were no data available on the phylogeny of other populations of *O. serrulata*. Subsequent detailed analyses (B. Cooper, unpubl.) indicate that the Texas coastal populations described as *C. australis* arose independently from other populations of *O. serrulata*, but *O. serrulata* also has multiple apparent origins from *O. capillifolia*.

17b.2. Oenothera Linnaeus (sect. Calylophus) subsect. Salpingia (Torrey & A. Gray) W. L. Wagner & Hoch, Syst. Bot. Monogr. 83: 148. 2007

Oenothera [unranked] *Salpingia* Torrey & A. Gray, Fl. N. Amer. 1: 501. 1840; *Galpinsia* Britton; *Oenothera* subg. *Salpingia* (Torrey & A. Gray) Munz

Herbs perennial, sometimes suffrutescent, or annual. Leaves: blade margins entire, serrulate, or serrate. **Flowers** usually 1 to several per stem opening per day in afternoon or near sunset or sunrise; buds terete; floral tube 5–60(–70) mm; sepals flat, midrib not keeled; stamens in 2 subequal series, pollen 85–100% fertile. **Capsules** hard or papery, straight or slightly curved, promptly dehiscent ½ to throughout their length. **Seeds** obovoid, rounded or sharply angled. **2*n*** = 14 (28).

Species 4 (4 in the flora): w United States, Mexico.

Subsection *Salpingia* consists of four species (nine taxa) ranging from eastern Nevada and Utah to western Nebraska and adjacent Wyoming, south in Mexico to Chihuahua, northeastern Sonora, Nuevo León, southwestern Tamaulipas, and San Luis Potosí, at elevations 30–2600 m. Flowers of subsect. *Salpingia* are diurnal or vespertine (and opening in afternoon or evening). *Salpingia* (Torrey & A. Gray) Raimann (1893), not *Salpinga* Martius ex de Candolle (1828), is a later homonym and pertains here.

1. Floral tubes 5–25(–33) mm, funnelform in distal ½ or more; flowers diurnal, opening near sunrise . . . . . 11. *Oenothera tubicula*
1. Floral tubes (15–)16–60(–70) mm, funnelform in distal ½ or less; flowers vespertine opening in afternoon or near sunset.
    2. Leaf axils with conspicuous fascicles of small leaves, these 0.2–2.5 cm; flower buds with free tips 2–9(–12) mm; capsules somewhat papery and dehiscent in distal ½. . . . . 10. *Oenothera toumeyi*
    2. Leaf axils sometimes without axillary fascicles of leaves, when present, these 0.1–1.5 cm; flower buds with free tips 0.3–6 mm; capsules hard, dehiscent throughout.
        3. Plants not cespitose, stems erect to ascending, 4–60 cm, strigillose, glandular puberulent, glabrous, hirtellous or short-pilose. . . . . . 8. *Oenothera hartwegii*
        3. Plants low, often cespitose, stems spreading to decumbent or ascending, 4–20(–30) cm, densely strigillose throughout, sometimes glandular puberulent distally. . . . . 9. *Oenothera lavandulifolia*

8. Oenothera hartwegii Bentham, Pl. Hartw., 5. 1839 (as hartwegi) [F]

Calylophus hartwegii (Bentham) P. H. Raven; *Galpinsia hartwegii* (Bentham) Britton; *Salpingia hartwegii* (Bentham) Raimann

Herbs perennial, sometimes suffrutescent, strigillose, glandular puberulent, glabrous, hirtellous, or short-pilose; from a stout taproot. **Stems** 1–many, erect to ascending, unbranched to densely branched, 4–60 cm. **Leaves** 0.3–6.5 × 0.04–1.2 cm, sometimes fascicles of small leaves 0.1–1.5 cm present in non-flowering axils; petiole 0–0.2 cm; blade elliptic, lanceolate, linear, or filiform to ovate or oblanceolate, usually not much reduced distally, proximalmost leaves sometimes obovate to spatulate, base attenuate to obtuse, truncate, or subcordate, sometimes clasping, margins entire or serrate, often undulate, apex acute. **Flowers** usually 1 per stem opening per day in afternoon or near sunset; buds with free tips 0.5–6 mm; floral tube 16–50(–60) mm, funnelform in distal ½ or less; sepals 7–28 mm; petals yellow, fading pale pinkish or pale purple, 10–35 mm; filaments 4–13 mm, anthers 5–13 mm, pollen 85–100% fertile; style 25–65(–75) mm, stigma yellow, quadrangular, usually exserted beyond anthers. **Capsules** 6–40 × 2–4 mm, hard, promptly dehiscent throughout their length. **Seeds** obovoid, 1–2.5 mm. **2*n*** = 14, 28.

Subspecies 5 (5 in the flora): sw, c United States, n Mexico.

Oenothera hartwegii consists of five intergrading subspecies, which are generally locally common on rocky, sandy, gypsum, or limestone soil in arid to relatively mesic open areas, in southeastern Colorado, southwestern Kansas, western Oklahoma, Texas (except eastern part), New Mexico, southeastern and east-central Arizona, and in Mexico from Chihuahua, northern Coahuila, and northwestern Tamaulipas south to Aguascalientes. H. F. Towner (1977) found that *O. hartwegii* is self-incompatible and usually vespertine; two of the subspecies (*filifolia* and *maccartii*) open early in the afternoon and are pollinated both day and evening.

1. Leaves (except proximalmost): blade base truncate or subcordate and clasping; plants densely pubescent with mixture of hair types, but always short-pilose and usually also hirtellous, sometimes also strigillose, especially on leaves, or glandular puberulent distally. . . . . 8e. *Oenothera hartwegii* subsp. *pubescens*
1. Leaves: blade base attenuate or obtuse; plants glabrous, sparsely strigillose, or glandular puberulent.
   2. Plants usually glabrous throughout, sometimes glandular puberulent on distal parts, especially on ovary . . . 8d. *Oenothera hartwegii* subsp. *fendleri*
   2. Plants strigillose and/or glandular puberulent, especially on distal parts.
      3. Plants glandular puberulent throughout, more densely so on distal parts, sometimes also sparsely strigillose on ovary and leaves; leaf blades filiform to narrowly lanceolate . . . . .8c. *Oenothera hartwegii* subsp. *filifolia*
      3. Plants usually strigillose, rarely glandular puberulent; leaf blade narrowly lanceolate to oblanceolate, sometimes linear.
         4. Leaf blades narrowly lanceolate, sometimes linear, margins entire or shallowly and sparsely serrulate, sometimes undulate; plants sparsely to densely strigillose throughout . . . . .8a. *Oenothera hartwegii* subsp. *hartwegii*
         4. Leaf blades usually narrowly lanceolate to lanceolate or oblanceolate, rarely linear, margins subentire or serrulate, usually crinkled-undulate; plants usually sparsely strigillose, sometimes glandular puberulent . . . . .8b. *Oenothera hartwegii* subsp. *maccartii*

8a. Oenothera hartwegii Bentham subsp. **hartwegii**

Oenothera greggii A. Gray var. *pringlei* Munz; *O. pringlei* (Munz) Munz

Herbs sparsely to densely strigillose throughout, more densely so on distal parts. **Leaves** 1–3.5 × 0.05–0.4 cm, fascicles of small leaves 0.2–1.5 cm usually present in axils; blade narrowly lanceolate, sometimes linear, base attenuate, margins entire or shallowly and sparsely serrulate, sometimes undulate. **Flowers:** buds with free tips (1–)2–6 mm; floral tube (18–)30–50(–60) mm; sepals 8–20 mm; petals 13–30 mm; filaments 5–10 mm, anthers 5–9 mm; style 30–65(–75) mm. $2n$ = 14, 28.

Flowering Feb–Oct. Rocky or gravelly soil, sometimes limestone, grasslands, conifer woodlands; 900–2300 m; Tex.; Mexico (Aguascalientes, Chihuahua, Coahuila, Durango, Nuevo León, San Luis Potosí, Zacatecas).

Subspecies *hartwegii* is the most southerly distributed among the taxa in sect. *Calylophus*, occurring widely from western Texas south into northern Mexico. It is often found in canyons and high plains in the northern part of its range, and reaching pine forests at its southern limits. It is weakly distinct from subsp. *maccartii*.

8b. Oenothera hartwegii Bentham subsp. **maccartii** (Shinners) W. L. Wagner & Hoch, Syst. Bot. Monogr. 83: 212. 2007 [F]

Calylophus hartwegii (Bentham) P. H. Raven var. *maccartii* Shinners, Sida 1: 343. 1964

Herbs usually sparsely strigillose, sometimes glandular puberulent. **Leaves** 0.6–3.5 × 0.1–0.6 cm, fascicles of small leaves to 1.5 cm usually present in axils; blade usually narrowly lanceolate to lanceolate or oblanceolate, rarely linear, base attenuate, margins subentire or serrulate, usually crinkled-undulate or undulate. **Flowers:** buds with free tips 1–6 mm; floral tube 17–45 mm; sepals 11–27 mm; petals 10–30 mm; filaments 6–12 mm, anthers 5–9 mm; style 25–60 mm. $2n$ = 14, 28.

Flowering Mar–Sep. Grasslands, sandy to gravelly soil, limestone, with *Acacia*, *Larrea*, *Opuntia*, *Prosopis*, and *Yucca*; 30–1500 m; Tex.; Mexico (Coahuila, Nuevo León, Tamaulipas).

Subspecies *maccartii* occurs on the south Texas Plains and along the Rio Grande from Kinney, Milam, Uvalde, and Val Verde counties south to southeastern Coahuila, central Nuevo León, and northwestern Tamaulipas.

8c. Oenothera hartwegii Bentham subsp. **filifolia** (Eastwood) W. L. Wagner & Hoch, Syst. Bot. Monogr. 83: 212. 2007

Oenothera tubicula A. Gray var. *filifolia* Eastwood, Proc. Calif. Acad. Sci., ser. 3, 1: 72, plate 6, fig. 1. 1897; *Calylophus hartwegii* (Bentham) P. H. Raven subsp. *filifolius* (Eastwood) Towner & P. H. Raven; *C. hartwegii* var. *filifolius* (Eastwood) Shinners; *Galpinsia filifolia* (Eastwood) A. Heller; *O. filifolia* (Eastwood) Tidestrom; *O. hartwegii* var. *filifolia* (Eastwood) Munz

Herbs glandular puberulent throughout, more densely so on distal parts, sometimes also sparsely strigillose on ovaries and leaves. **Leaves** 0.3–4 × 0.04–0.3(–0.4) cm, fascicles of small leaves to 1.5 cm often present in axils; blade filiform to narrowly lanceolate, base attenuate, margins entire or remotely serrulate, sometimes undulate. **Flowers:** buds with free tips 0.5–4 mm; floral tube 16–50 mm; sepals 7–17 mm; petals 12–23 mm; filaments 6–13 mm, anthers 6–11 mm; style 26–60 mm. **2*n*** = 14.

Flowering May–Oct. Highly local, often abundant, almost always on semiarid gypsum flats, dunes, or outcrops, with *Juniperus*, *Larrea*, and *Yucca*; 600–1900 m; N.Mex., Tex.; Mexico (Chihuahua, Coahuila, Zacatecas).

Subspecies *filifolia* occurs in Otero and Torrance counties in southeastern New Mexico, south and east through the Trans-Pecos and southern Panhandle of Texas to Cottle County in Texas, and southward from widely scattered localities in central Chihuahua, Coahuila, and Zacatecas, Mexico.

8d. Oenothera hartwegii Bentham subsp. **fendleri** (A. Gray) W. L. Wagner & Hoch, Syst. Bot. Monogr. 83: 212. 2007

Oenothera fendleri A. Gray, Mem. Amer. Acad. Arts, n. s. 4: 45. 1849; *Calylophus hartwegii* (Bentham) P. H. Raven subsp. *fendleri* (A. Gray) Towner & P. H. Raven; *Galpinsia fendleri* (A. Gray) A. Heller; *G. hartwegii* (Bentham) Britton var. *fendleri* (A. Gray) Small; *O. hartwegii* var. *fendleri* (A. Gray) A. Gray

Herbs usually glabrous throughout, sometimes glandular puberulent on distal parts, especially on ovaries. **Leaves** 1–5 × 0.15–1 cm, fascicles of small leaves to 1 cm (when present); blade linear to oblanceolate or lanceolate, base attenuate to obtuse, rarely nearly clasping, margins entire or subentire, rarely undulate. **Flowers:** buds with free tips 0.5–3 mm; floral tube 30–50 mm; sepals 9–28 mm; petals 10–30 mm; filaments 5–12 mm, anthers 5–13 mm; style 40–75 mm. **2*n*** = 14.

Flowering Apr–Oct. In scattered populations on clay or gravelly soil, sometimes calcareous, in grasslands, often with *Juniperus* and *Prosopis*, to woodlands with *Juniperus*, *Pinus edulis*, sometimes *Pinus ponderosa*; 300–2200 m; Ariz., Kans., N.Mex., Okla., Tex.; Mexico (Chihuahua).

Subspecies *fendleri* is known from Barber, Comanche, and Morton counties, Kansas, south through western Oklahoma and scattered sites in the Texas Panhandle to eastern Chihuahua, central trans-Pecos Texas, central and western New Mexico, and east-central Arizona. It is the most distinctive subspecies in the complex.

8e. Oenothera hartwegii Bentham subsp. **pubescens** (A. Gray) W. L. Wagner & Hoch, Syst. Bot. Monogr. 83: 212. 2007

Oenothera greggii A. Gray var. *pubescens* A. Gray, Smithsonian Contr. Knowl. 3(5): 72. 1852; *Calylophus hartwegii* (Bentham) P. H. Raven subsp. *pubescens* (A. Gray) Towner & P. H. Raven; *C. hartwegii* var. *pubescens* (A. Gray) Shinners; *Galpinsia camporum* Wooton & Standley; *G. greggii* (A. Gray) Small; *G. interior* Small; *G. lampasana* (Buckley) Wooton & Standley; *O. camporum* (Wooton & Standley) Tidestrom; *O. greggii* A. Gray; *O. greggii* var. *lampasana* (Buckley) Munz; *O. interior* (Small) G. W. Stevens; *O. lampasana* Buckley

Herbs densely pubescent with mixture of hair types, always short-pilose, especially on ovary and stem, usually also hirtellous, especially on stem and distal parts, sometimes also strigillose, especially on leaves, or glandular puberulent distally. **Leaves** 0.6–4 × 0.15–1.2 cm, fascicles of small leaves often absent or much reduced, sometimes to 1.5 cm; blade very narrowly elliptic or narrowly lanceolate to ovate, base truncate or subcordate and clasping, rarely nearly clasping, margins entire or sparsely serrulate, rarely crinkled-undulate. **Flowers:** buds with free tips 0.5–3 mm; floral tube 20–50 mm; sepals 9–26 mm; petals 12–35 mm; filaments 5–12 mm, anthers 4–13 mm; style 25–70 mm. **2*n*** = 14, 28.

Flowering Mar–Oct. Colonial in moderately dry, open places, plains, hills, sandy to gravelly soil, limestone or gypsum, grasslands with *Juniperus* and *Prosopis*; 200–2100 m; Ariz., Colo., Kans., N.Mex., Okla., Tex.; Mexico (Coahuila, Durango).

Subspecies *pubescens* occurs from Baca and Las Animas counties, Colorado, Clark, Meade, Morton, and Seward counties, Kansas, to western Oklahoma and the Texas Panhandle, throughout central and trans-Pecos Texas, west through eastern and southern New Mexico to central and southeastern Arizona, and also very locally in northern Mexico.

9. Oenothera lavandulifolia Torrey & A. Gray, Fl. N. Amer. 1: 501. 1840 (as lavandulaefolia)

Calylophus hartwegii (Bentham) P. H. Raven subsp. *lavandulifolius* (Torrey & A. Gray) Towner & P. H. Raven; *C. hartwegii* var. *lavandulifolius* (Torrey & A. Gray) Shinners; *C. lavandulifolius* (Torrey & A. Gray) P. H. Raven; *Galpinsia lavandulifolia* (Torrey & A. Gray) Small; *G. lavandulifolia* var. *glandulosa* (Munz) Moldenke; *O. hartwegii* Bentham var. *lavandulifolia* (Torrey & A. Gray) S. Watson; *O. lavandulifolia* var. *glandulosa* Munz

Herbs perennial, densely strigillose throughout, sometimes glandular puberulent distally; from a stout taproot. **Stems** several to many, decumbent to ascending, branched, 4–20(–30) cm. **Leaves** 0.6–5 × 0.08–0.6 cm, fascicles of small leaves 0.2–1 cm often present in nonflowering axils; petiole 0 cm; blade narrowly lanceolate or narrowly oblanceolate, base attenuate to truncate, sometimes clasping, margins entire or subentire, sometimes revolute, sometimes weakly undulate, apex acute to obtuse. **Flowers** usually 1 per stem opening per day near sunset; buds with free tips 0.3–3 mm; floral tube 25–60 mm, funnelform in distal ½ or less; sepals 8–20 mm; petals yellow, fading pale pink or pale purple, 12–28 mm; filaments 6–12 mm, anthers 5–11 mm, pollen 85–100% fertile; style 30–75 mm, stigma yellow, quadrangular, usually exserted beyond anthers. **Capsules** 6–25 × 1–3 mm, hard, promptly dehiscent throughout their length. **Seeds** obovoid, 1.5–2.5 mm. **2*n*** = 14.

Flowering Apr–Aug. Local and sparse, on sandy and rocky, calcareous soil, high plains, mountains, often with *Artemisia tridentata*, *Cercocarpus*, *Juniperus*, *Pinus edulis*, or *P. monophylla*, sometimes in lower zones with *Larrea*, or in higher zones with *P. ponderosa*; 600–2800 m; Ariz., Colo., Kans., Nebr., N.Mex., Okla., S.Dak., Tex., Wyo.; Mexico (Nuevo León).

Oenothera lavandulifolia is known from southern Fall River County, South Dakota, southeastern Wyoming, and far western Nebraska, through western Kansas, Colorado, eastern and southern Utah, northwestern Oklahoma, and the Texas Panhandle to trans-Pecos Texas, central New Mexico, northern and central Arizona, and eastern Nevada. It also occurs in Nuevo León, Mexico, and may be more widespread in northern Mexico. H. F. Towner (1977) found that *O. lavandulifolia* is self-incompatible and vespertine.

10. Oenothera toumeyi (Small) Tidestrom, Proc. Biol. Soc. Wash. 48: 41. 1935

Galpinsia toumeyi Small, Bull. Torrey Bot. Club 25: 317. 1898; *Calylophus hartwegii* (Bentham) P. H. Raven subsp. *toumeyi* (Small) Towner & P. H. Raven; *C. hartwegii* var. *toumeyi* (Small) Shinners; *C. toumeyi* (Small) Towner; *Oenothera hartwegii* Bentham var. *toumeyi* (Small) Munz

Herbs perennial or sometimes annual, glabrate to strigillose throughout; from a stout taproot. **Stems** 1–several, ascending to erect, unbranched to densely branched, 15–70 cm. **Leaves** 1–3.5 × 0.1–0.7 cm, fascicles of small leaves 0.2–2.5 cm present in nonflowering axils; petiole 0 cm; blade narrowly lanceolate, base acute-attenuate, margins entire or obscurely and sparsely serrulate, not undulate, apex acute. **Flowers** usually 1 per stem opening per day at sunset; buds with free tips 2–9(–12) mm; floral tube (15–)30–60 (–70) mm, funnelform in distal ½ or less; sepals 10–25 mm; petals yellow, fading pale pink or pale purple, 10–20 mm; filaments 4–12 mm, anthers 6–10 mm, pollen 85–100% fertile; style 35–70(–80) mm, stigma yellow, quadrangular, usually exserted beyond anthers. **Capsules** 10–50 × 1.5–4 mm, somewhat papery, promptly dehiscent in distal ½. **Seeds** obovoid, 2–3 mm. **2*n*** = 14.

Flowering (May–)Jul–Oct. Local and uncommon on shaded, rocky slopes or disturbed areas, pine-oak forests; 1500–2600 m; Ariz., N.Mex.; Mexico (Chihuahua, Sonora).

Oenothera toumeyi occurs locally from the Chiricahua, Huachuca, and Santa Rita mountains in Cochise and Santa Cruz counties, Arizona, and the Mogollon Mountains in southern Catron County, New Mexico, south through northeastern Sonora in the Sierra Madre Occidental to west-central Chihuahua. H. F. Towner (1977) found that *O. toumeyi* is self-incompatible and vespertine.

11. Oenothera tubicula A. Gray, Smithsonian Contr. Knowl. 3(5): 71. 1852

Calylophus tubiculus (A. Gray) P. H. Raven; *Galpinsia tubicula* (A. Gray) Small; *Oenothera hartwegii* Bentham var. *tubicula* (A. Gray) H. Léveillé

Subspecies 2 (1 in the flora): sw, sc United States, n Mexico.

H. F. Towner (1977) found that *Oenothera tubicula* is self-incompatible and diurnal with opening times just prior to sunrise. It occurs primarily on limestone soil in arid lowlands, but occasionally in montane areas, from Guadalupe County, New Mexico, south to western Texas, northeast to Howard County, Texas, and south to northern Zacatecas, south-central Nuevo León, and southwestern Tamaulipas, 600–1800 m. Subspecies *strigulosa* (Towner) W. L. Wagner & Hoch is known only from rocky, open sites and canyons in relatively montane areas, sometimes in pine forests in southernmost Coahuila, south-central Nuevo León, and southeastern Tamaulipas, from 1500 to 2300 m. It differs in being strigillose on the ovary and distally on stems, leaves linear to narrowly lanceolate, and the petals fading red or purple.

11a. Oenothera tubicula A. Gray subsp. **tubicula**

Galpinsia carlsbadiana A. Nelson; *Oenothera tubicula* var. *demissa* A. Gray; O. ×*serrulatoides* H. Léveillé

Herbs short-lived perennial, glandular puberulent; from a stout taproot. **Stems** 1–many, unbranched to densely branched, decumbent to erect, 4–53 cm. **Leaves** 0.7–4.6 × 0.1–1.2 cm, sometimes fascicles of small leaves 0.2–1.5 cm present in non-flowering axils; petiole 0–0.2 cm; blade linear to ovate or obovate, base attenuate, margins entire, apex acute. **Flowers** usually several per stem opening per day near sunrise; buds with free tips 0.5–2 mm; floral tube 5–25(–33) mm, funnelform in distal ½ or more; sepals 3–13 mm; petals yellow, fading pale pink or pale purple, 5–20(–25) mm; filaments 1–6 mm, anthers 2–7 mm, pollen 85–100% fertile; style 9–30(–40) mm, stigma yellow, quadrangular, usually exserted beyond anthers. **Capsules** 8–20 × 1.5–2.5 mm, hard, promptly dehiscent throughout their length. **Seeds** obovoid, 1–1.4 mm. $2n = 14$.

Flowering Apr–Aug. Colonial, primarily on limestone soil, in flat arid grasslands, with *Larrea* and *Yucca*; 600–1400 m; N.Mex., Tex.; Mexico (Coahuila, Nuevo León, Tamaulipas, Zacatecas).

Subspecies *tubicula* is known from Guadalupe County, New Mexico, south in the western side of the Pecos River drainage to western Texas, where it occurs from Culberson County east to Howard County, thence south through Brewster, Presidio, and Terrell counties, and probably most of central Coahuila, to northern Zacatecas, southwestern Nuevo León, and southwestern Tamaulipas.

17c. OENOTHERA Linnaeus sect. LAVAUXIA (Spach) Walpers, Repert. Bot. Syst. 2: 83. 1843

Lavauxia Spach, Hist. Nat. Vég. 4: 366. 1835; *Oenothera* [unranked] *Lavauxia* (Spach) Endlicher; *Oenothera* subg. *Lavauxia* (Spach) Reichenbach

Herbs usually perennial, sometimes annual or biennial, usually acaulescent or short-caulescent; from a taproot, sometimes lateral roots producing adventitious shoots. **Stems** (when present) ascending [or decumbent]. **Leaves** in a basal rosette, sometimes also cauline, (4–)8–36 cm; blade margins coarsely pinnatifid or lobed to dentate or subentire. **Inflorescences** solitary flowers in axils of leaves. **Flowers** opening near sunset, with a slightly sweet fragrance and a spermaceous background odor, or nearly unscented; buds erect, obtusely quadrangular, with unequal or subequal free tips; floral tube (20–)40–200(–265) mm; sepals splitting along one suture, remaining coherent and reflexed as a unit at anthesis or separating in pairs; petals yellow [white], fading orange to lavender, drying purple to lavender or purplish brown, obovate with acute apex or obcordate; stigma deeply divided into 4 linear lobes. **Capsules** leathery or woody in age, usually narrowly ovoid or ellipsoid to rhombic-obovoid, sometimes ovoid or lanceoloid, winged, wings oblong to triangular in distal [½–]⅔ of capsule, constricted to a short beak, dehiscent ¼–½ their length; sessile. **Seeds** numerous, usually in 2 rows per locule, rarely 3 adjacent and distinct rows (*O. triloba*), asymmetrically cuneiform, narrowly winged distally, also along one adaxial margin, surface minutely beaded. $2n = 14$.

Species 5 (3 in the flora): w, c North America, n, c Mexico, s South America.

Section *Lavauxia* consists of five diploid ($n = 7$) species, divided into two subsections, which are locally common in seasonally wet depressions, flats, meadows, stream banks, or disturbed sites from southern Canada through the western and east-central United States to central Mexico, disjunct in Baja California, and two species in southern South America, at 0–3200 m elevation. The monophyly of sect. *Lavauxia* is strongly supported, but its relationships to the rest of the genus are obscure. No member of the South American subsect. *Australis* has been included in molecular analyses, but it shares unique morphological features with the North American subsect. *Lavauxia*; the monophyly of this section is not in doubt, based on the shared dandelionlike acaulescent habit and unique capsule morphology. Species are self-compatible or, rarely, apparently self-incompatible (some plants of *O. acutissima*); flowers vespertine, fragrance pungent, often slightly sweet with a spermaceous background scent in *O. acutissima* and outcrossing populations of *O. flava*, or without noticeable scent in autogamous populations of *O. flava* and in *O. triloba*. Species are outcrossing and pollinated by hawkmoths (*O. acutissima* and large-flowered populations of *O. flava*) or autogamous, occasionally cleistogamous in *O. flava* (W. L. Wagner et al 2007; R. A. Raguso et al. 2007; H. E. Summers et al. 2015).

SELECTED REFERENCE Raguso, R. A. et al. 2007. Floral biology of North American *Oenothera* sect. *Lavauxia*: Advertisements, rewards and extreme variation in the floral depth. Ann. Missouri Bot. Gard. 94: 236–257.

17c.1. OENOTHERA Linnaeus (sect. LAVAUXIA) subsect. LAVAUXIA (Spach) W. L. Wagner & Hoch, Syst. Bot. Monogr. 83: 150. 2007

Lavauxia Spach, Hist. Nat. Vég. 4: 366. 1835

Flowers petals yellow, fading orange, drying purple to lavender. **Capsules** narrowly winged, wings oblong or triangular (*O. triloba*) in distal ⅔. **2*n*** = 14.

Species 3 (3 in the flora): w North America, n, c Mexico.

Subsection *Lavauxia* consists of three yellow-flowered species that are native from southern Alberta and southwestern Saskatchewan, scattered in the western United States and east through Kansas, Missouri south of the Missouri River, northern Arkansas, Texas, northern Alabama, and Kentucky, south into Mexico from the Sierra Madre Occidental to the Trans-Mexican Volcanic Belt in Guanajuato and Hidalgo, and disjunct in the mountains of Baja California, at 300–3200 m elevation. One species (*O. triloba*) is introduced in the eastern United States.

1. Petals pale yellow; capsules woody, with broad, triangular wings 5–10 mm wide, these often terminating in a hooked tooth; herbs annual or, sometimes, biennial .... 14. *Oenothera triloba*
1. Petals bright yellow; capsules leathery, with narrowly oblong wings 1–5(–6) mm wide, not with a hooked tooth; herbs perennial, rarely short-lived.
   2. Leaves moderately stiff, blades linear to narrowly elliptic, margins irregularly and coarsely dentate; petals fading deep reddish orange, drying purplish brown; lateral roots often producing adventitious shoots; capsules 14–18(–22) mm, valve wings 1–2(–4) mm wide .... 12. *Oenothera acutissima*
   2. Leaves flexible, sometimes ± fleshy, blades oblanceolate to linear-oblong, margins usually irregularly and coarsely pinnately lobed, rarely subentire; petals fading pale orange, drying purple; lateral roots not producing adventitious shoots; capsules (10–)20–35(–43) mm, wings (2–)3–5(–6) mm wide.......................... 13. *Oenothera flava*

12. Oenothera acutissima W. L. Wagner, Syst. Bot. 6: 153, fig. 1. 1981 C E

Oenothera flava (A. Nelson) Garrett var. *acutissima* (W. L. Wagner) S. L. Welsh

Herbs perennial, subacaulescent or very short-caulescent, strigillose mostly along leaf margins and flower parts, also sparsely glandular puberulent, sometimes also sparsely hirsute distally; from a stout taproot, usually with several long, lateral roots often producing adventitious shoots. **Stems** (when present) ascending, (1–)several–10, densely leafy, 1–2 cm. **Leaves** primarily in a basal rosette, 7–14(–18) × (0.3–)0.5–1(–1.5) cm, moderately thick and stiff; petiole (1.2–)3–5 cm; blade linear to very narrowly elliptic, margins irregularly and coarsely dentate or pinnately lobed, apex long-attenuate. **Flowers** 1–3 opening per day near sunset; buds with unequal free tips 1–3 mm; floral tube (53–)60–100 mm; sepals 26–50 mm; petals bright yellow, fading deep reddish orange, drying purplish brown, 28–50 mm; filaments 21–35 mm, anthers 9–11 mm; style 75–143 mm, stigma exserted beyond anthers. **Capsules** leathery in age, oblong-oblanceoloid, narrowly winged, wings oblong, 1–2(–4) mm wide, broadest near apex, 14–18(–22) × 7–8 mm (excluding wings), apex abruptly constricted, dehiscent ¼–⅓ their length, valve surface with inconspicuous veins; sessile. **Seeds** asymmetrically cuneiform, 2–2.5 mm. **2*n*** = 14.

Flowering May–Jun. Restricted to sandy and gravelly, reddish, soil in seasonally wet sites, meadows, depressions, along arroyos, among rocks, in mixed conifer forests, sagebrush scrub; of conservation concern; 1800–2400(–2600) m; Colo., Utah.

Oenothera acutissima is known only from the vicinity of Manila, eastern Uinta Mountains, Daggett and Duchesne counties, Utah, east to areas in and near the foothills of the Douglas and Blue mountains, in Uinta County, Utah, and Moffat County, Colorado.

13. Oenothera flava (A. Nelson) Garrett, Spring Fl. Wasatch ed. 4, 106. 1927 [F]

Lavauxia flava A. Nelson, Bull. Torrey Bot. Club 31: 243. 1904; *L. palustris* Rose; *L. taraxacoides* Wooton & Standley; *Oenothera flava* subsp. *taraxacoides* (Wooton & Standley) W. L. Wagner; *O. murdockii* S. L. Welsh & N. D. Atwood; *O. taraxacoides* (Wooton & Standley) Munz; *O. triloba* Nuttall var. *ecristata* M. E. Jones

Herbs perennial, rarely short-lived, acaulescent or very short-caulescent, glabrate to moderately strigillose, usually also glandular puberulent, sometimes sparsely hirsute distally; from a taproot. **Stems** (when present) ascending, 1–several, usually densely leafy, 0–2 cm. **Leaves** primarily in a basal rosette, (3.4–)6–30(–36) × (0.5–)1.5–5(–7) cm, flexible, sometimes ± fleshy; petiole (0.2–)2–7(–10) cm; blade oblanceolate to linear, margins usually irregularly and coarsely pinnately lobed, rarely subentire, apex acute. **Flowers** 1–4 opening per day near sunset; buds with free tips (1–)2–10 (–12) mm; floral tube (24–)40–200(–265) mm; sepals (8–)11–40(–42) mm; petals bright yellow, sometimes paler (in smaller-flowered plants), fading pale orange, drying purple, (7–)10–45(–50) mm; filaments (5–)8–23(–26) mm, anthers (2–)3–13(–16) mm; style (40–)50–250(–290) mm, stigma exserted beyond or surrounded by ring of anthers. **Capsules** leathery in age, surface usually conspicuously reticulate, usually narrowly ovoid or ellipsoid, sometimes ovoid or lanceoloid, winged, wings narrowly oblong, (2–)3–5 (–6) mm wide, confined to distal ⅔ of capsule, (10–) 20–35(–43) × 4–8 mm (excluding wings), gradually constricted to a short beak, dehiscent ¼–½ their length, valve surface usually conspicuously reticulate; sessile. **Seeds** asymmetrically cuneiform, 1.8–2.2(–2.6) mm. **2*n*** = 14.

Flowering Mar–Aug(–Oct). Local and colonial, sometimes abundant in wet (at least seasonally moist) clay to gravelly sand of swales, desiccating flats and ponds, montane meadows, margins of permanent or seasonal watercourses, open sites; 300–3200 m; Alta., Man., Sask.; Ariz., Calif., Colo., Idaho, Mont., Nebr., Nev., N.Mex., N.Dak., Oreg., S.Dak., Utah, Wash., Wyo.; Mexico (Chihuahua, Durango, Guanajuato, Hidalgo, Jalisco, Sonora).

Petals in *Oenothera flava* typically range from 7–32 mm with floral tubes 24–100 mm; however, plants from three disjunct areas: the Mogollon Plateau in Arizona to Catron County, New Mexico; Sacramento Mountains and Sierra Blanca, Lincoln and Otero counties, New Mexico; and the Sierra Madre Occidental from northern Chihuahua south to Durango, have much larger petals (30–55 mm) and longer floral tubes (80–265 mm). They were originally recognized as a distinct species or most recently as a subspecies (*O. flava* subsp. *taraxacoides*), but detailed study of the variation pattern suggests that the larger flowers occur in areas of high hawkmoth species diversity and higher rates of outcrossing, similar to the pattern discussed in detail by D. P. Gregory (1963, 1964). R. A. Raguso et al. (2007) and H. E. Summers et al. (2015) came to the same conclusion in an independent study of floral biology of sect. *Lavauxia*. Because populations from the three disjunct areas appear to have diverged independently from lower-elevation source populations, it seems best to treat the complex as one variable species without any formal subdivision.

14. Oenothera triloba Nuttall, J. Acad. Nat. Sci. Philadelphia 2: 118. 1821

Lavauxia hamata Wooton & Standley; *L. triloba* (Nuttall) Spach; *L. watsonii* Small; *Oenothera hamata* (Wooton & Standley) Tidestrom; *O. rhizocarpa* Sprengel; *O. roemeriana* Scheele; *O. triloba* var. *parviflora* S. Watson; *O. triloba* [unranked] *watsonii* (Small) F. C. Gates

Herbs winter-annual, sometimes biennial, acaulescent or very short-caulescent, sparsely to moderately strigillose and glandular puberulent, sometimes one hair type predominant, rarely glabrate, sometimes also very sparsely hirsute, especially on leaf veins; from a slender or, sometimes, stout taproot. **Stems** (when present) ascending, 1–several, densely leafy, 0–20 cm. **Leaves** in a basal rosette, sometimes also cauline, (2.5–)6–25(–32) × (0.6–)1.5–4(–5) cm, thin; petiole (0.5–)1–8 cm; blade oblanceolate to elliptic, margins irregularly pinnatifid, sometimes subentire, apex acute to obtuse or rounded. **Flowers** 1–4 opening per day near sunset, without noticeable scent; buds with subequal free tips 2–7 mm; floral tube (20–)28–95(–138) mm; sepals (6–)10–30(–35) mm; petals pale yellow, fading pale orange, drying lavender, (10–)12–30(–38) mm; filaments (5–)8–15(–18) mm, anthers (3.5–)4–11 mm; style (3.4–)4.2–11.5(–16.3) mm, stigma usually surrounded by anthers, sometimes (especially in some Texas populations) exserted beyond anthers. **Capsules**

woody in age, rhombic-obovoid, winged, wings broadly triangular, 5–10 mm wide, often terminating in a hooked tooth, (10–)15–25(–28) × 4–8 mm (excluding wings), valve surface reticulate, dehiscent 1/8–1/3 their length. **Seeds** asymmetrically cuneiform, (2.1–)2.5–3(–3.3) mm. $2n = 14$.

Flowering (Feb–)Mar–May(–Jul). Scattered to common in clay, sandy or rocky soil, playas, floodplains, creek beds, slopes and flats, moist sites, disturbed sites, roadsides, old fields, in *Larrea* deserts, prairies, glades; 300–1900 m; Ala., Ark., Colo., D.C., Ill., Ind., Kans., Ky., Md., Mo., N.Mex., Ohio, Okla., Pa., Tenn., Tex., Va.; Mexico (Baja California, Chihuahua, Nuevo León).

Oenothera triloba is primarily a species of the high plains from eastern Socorro County, New Mexico, east through all but eastern Texas, Oklahoma, to southern Kansas, east of Meade and Pawnee counties and south of Douglas and Saline counties. It becomes more sporadic eastward into Missouri south of the Missouri River, northwestern and north-central Arkansas, central and eastern Tennessee, northern Alabama, and Logan and Warren counties, Kentucky; also known from disjunct sites in northern Mexico from Nuevo León, Chihuahua, and Baja California, Mexico; and, introduced in Illinois, Indiana, Ohio, Kentucky (Campbell and Fayette counties), Pennsylvania, Virginia, Maryland, and the District of Columbia. Areas where it was introduced are represented by old collections; no current information indicates their continued presence in any of these areas. It was recently collected in Baca County, Colorado.

Capsules of dead plants sometimes form pineconelike clusters of ten to 100 or more capsules.

The illegitimate names *Lavauxia nuttalliana* Spach and *L. triloba* (Nuttall) Spach var. *watsonii* Britton pertain here.

17d. Oenothera Linnaeus sect. Gauropsis (Torrey & Frémont) W. L. Wagner, Ann. Missouri Bot. Gard. 71: 1124. 1985 [E]

Oenothera subsect. *Gauropsis* Torrey & Frémont in J. C. Frémont, Rep. Exped. Rocky Mts., 315. 1845; *Gaurella* Small; *Oenothera* subg. *Gauropsis* (Torrey & Frémont) Munz

Herbs perennial, caulescent; from a taproot, lateral roots producing adventitious shoots. **Stems** decumbent to ascending, branched. **Leaves** cauline (basal rosette not observed), (0.3–)0.6–1.5(–2.5) cm; blade margins sinuate-denticulate to subentire. **Inflorescences** solitary flowers in axils of distal leaves. **Flowers** opening near sunset, with a sweet scent or nearly unscented; buds erect, terete, without free tips or free tips minute; floral tube (8–)10–15(–17) mm; sepals usually splitting along one suture, remaining coherent and reflexed as a unit at anthesis, rarely separating in pairs; petals usually pink, rarely white, streaked or flecked with red, fading bright purple, obovate; stigma deeply divided into 4 linear lobes. **Capsules** woody, ovoid, narrowly winged, abruptly constricted to a conspicuous, sterile beak, indehiscent; sessile. **Seeds** numerous, in 2–4 irregular rows per locule, asymmetrically cuneiform or oblanceoloid and weakly angled (probably resulting from compression from adjacent seeds during development), surface glossy, obscurely reticulate, appearing finely granular. $2n = 14$.

Species 1: w United States.

Section *Gauropsis* consists of one distinctive species with pink petals, flecked with red, and ovoid, indehiscent, narrowly winged capsules and an apex abruptly constricted to a conspicuous sterile beak. It is self-compatible with flowers vespertine, outcrossing, and pollinated by noctuid moths and occasionally hawkmoths, such as *Hyles lineata* (W. L. Wagner 1984). Wagner indicated that *O. canescens* forms adventitious shoots from lateral roots, but this feature should be re-examined, since the closely related *O. speciosa* (sect. *Hartmannia*) reproduces by rhizomes.

One illegitimate later homonym has been published for this taxon: *Gauropsis* (Torrey & Frémont) Cockerell (1900) not C. Presl (1851).

SELECTED REFERENCE Wagner, W. L. 1984. Reconsideration of *Oenothera* subg. *Gauropsis* (Onagraceae). Ann. Missouri Bot. Gard. 71: 1114–1127.

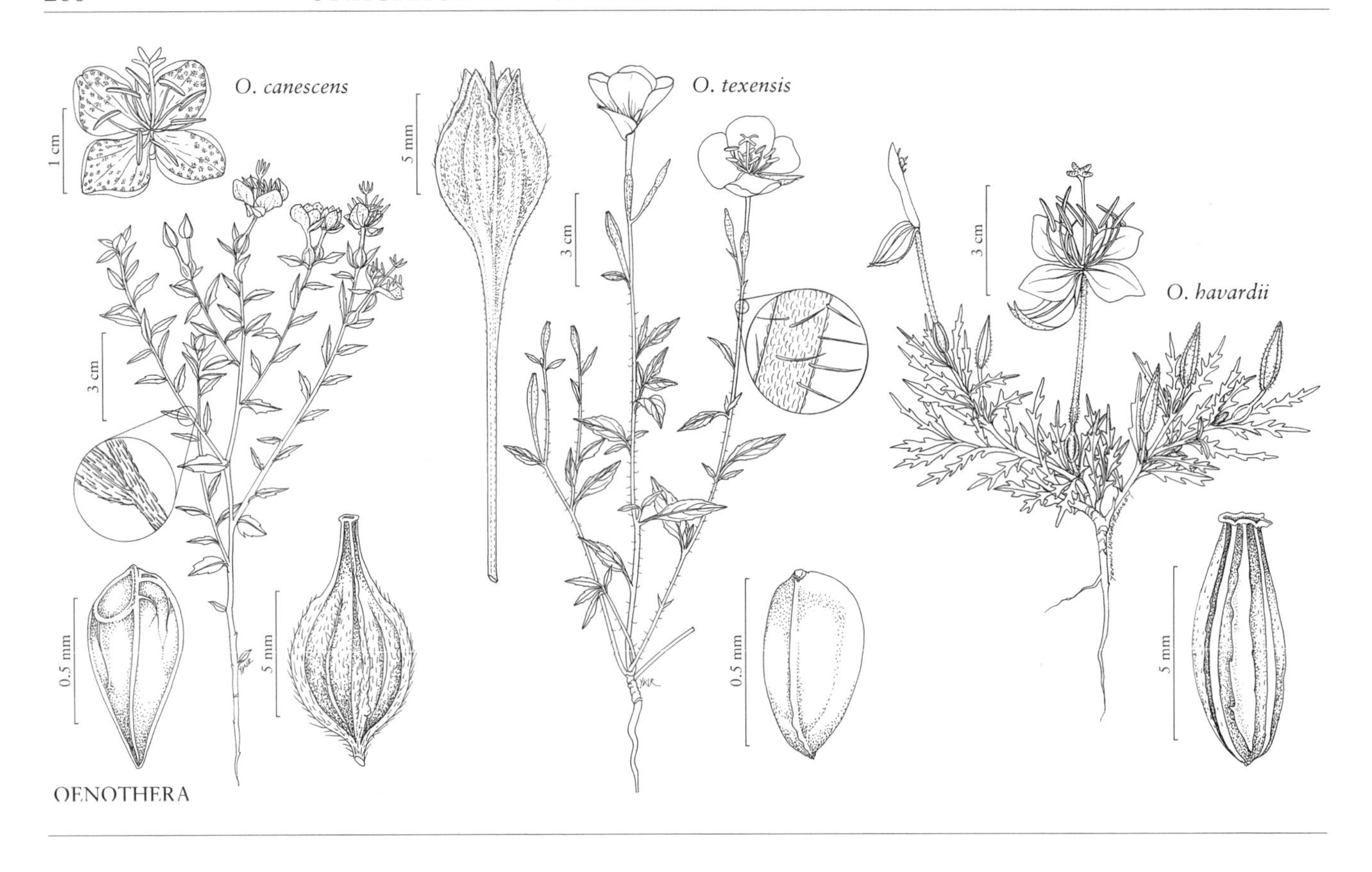

OENOTHERA

15. Oenothera canescens Torrey & Frémont in J. C. Fremont, Rep. Exped. Rocky Mts., 315. 1845 [E] [F]

Gaurella canescens (Torrey & Frémont) A. Nelson; *Megapterium canescens* (Torrey & Frémont) Britton; *Oenothera guttulata* Geyer ex Hooker

Herbs low, forming clumps 10–50 cm diam., densely strigillose throughout; from a taproot, lateral roots producing adventitious shoots. **Stems** many-branched from base, leafy, (10–)15–25(–38) cm. **Leaves** cauline, (0.3–)0.6–1.5(–2.5) × (0.05–)0.15–0.4(–0.6) cm, fascicles of small leaves 0.2–0.6 cm often present in non-flowering axils; petiole 0–0.1 cm; blade lanceolate to linear, base cuneate, apex acute. **Flowers** several opening per day near sunset; buds usually without free tips, rarely free tips 0.2–0.3 mm; sepals (7–)8–12 mm; petals pink, rarely white, streaked or flecked with red, fading bright purple, (8–)10–17 mm; filaments 6–8 mm, anthers often with red longitudinal stripe, 3–6 mm; style (16–)22–27 mm, stigma exserted beyond anthers at anthesis. **Capsules** woody, ovoid, narrowly winged, wings 0.8–1.5 mm wide, (7–)9–12(–14) × 2–4 mm (excluding wings), abruptly constricted to a conspicuous, sterile beak, (2–)3–4.5 mm, indehiscent; sessile. **Seeds** asymmetrically cuneiform or oblanceoloid, 1.2–1.5 × 0.4–0.5 mm. **2*n*** = 14.

Flowering May–Aug. Prairie depressions, playas, margins of ditches, temporary wet areas; (400–)700–1800 m; Colo., Kans., Nebr., N.Mex., Okla., Tex., Wyo.

Oenothera canescens is restricted to prairie depressions, playas, ditch margins, and other places of temporary water in the High Plains of the western United States from Goshen County, Wyoming, southeast to Hayes County, Nebraska, south through eastern Colorado, the eastern tier of counties in New Mexico, western Kansas, and to Garza and Dawson counties in the Texas Panhandle; also disjunct populations from Chautauqua, Sedgwick, and Stafford counties, Kansas.

The illegitimate names *Gaurella guttulata* (Geyer ex Hooker) Small, *G. canescens* (Torrey & Frémont) Cockerell, and *Gauropsis guttulata* (Geyer ex Hooker) Cockerell pertain here.

17e. Oenothera Linnaeus sect. Hartmannia (Spach) Walpers, Repert. Bot. Syst. 2: 84. 1843

Hartmannia Spach, Hist. Nat. Vég. 4: 370. 1835; *Oenothera* [unranked] *Hartmannia* (Spach) Endlicher; *Oenothera* subg. *Hartmannia* (Spach) Reichenbach; *Oenothera* [unranked] *Xylopleurum* (Spach) Endlicher; *Oenothera* sect. *Xylopleurum* (Spach) Walpers; *Oenothera* subg. *Xylopleurum* (Spach) Reichenbach; *Xylopleurum* Spach

Herbs perennial, caulescent; from a taproot, sometimes producing new shoots from rhizomes. **Stems** decumbent to ascending or erect, usually branched, sometimes unbranched. **Leaves** in a basal rosette and cauline, cauline 1–6(–10) cm; blade margins subentire, weakly serrate to weakly sinuate-toothed, sometimes sinuate-pinnatifid. **Inflorescences** solitary flowers in axils of distal leaves. **Flowers** opening near sunrise or sunset; buds erect, terete, without free tips or free tips minute (except to 4 mm in *O. speciosa*); floral tube 4–26 mm; sepals splitting along one suture, remaining coherent and reflexed as a unit at anthesis or, rarely, separating in pairs; petals usually pink or rose purple, fading darker, rarely white, fading pink, obovate to obcordate; stigma deeply divided into 4 linear lobes. **Capsules** hard and leathery, straight, clavate or narrowly obovoid to rhombic-ellipsoid, angled or narrowly winged (to 0.5 mm, apex attenuate to a sterile beak, proximal part tapering to a sterile, pedicel-like base (stipe), valve midrib raised (prominent in *O. speciosa*), dehiscent at apex or nearly throughout fertile part; sessile. **Seeds** numerous, clustered in each locule, narrowly obovoid, surface glossy, appearing granular, but minutely papillose under magnification. $2n$ = 14, 28, 42.

Species 5 (4 in the flora): w, c United States, Mexico, West Indies, Bermuda, Central America, South America; introduced widely in tropical and subtropical regions.

Section *Hartmannia* consists of five species, mostly diploid ($2n$ = 14), but *Oenothera speciosa* is diploid and also has polyploid populations ($2n$ = 28, 42). All of the species are morphologically very similar, characterized by purple to pink (or white in *O. speciosa*) petals, leaves often lobed toward the base or distally, capsules often straight, with the fertile portion of the capsule relatively short, angled or narrowly winged (less than 5 mm), apex attenuate, tapering to an acute beak. All species form a definite rosette, which persists at least until the onset of flowering, and ascending to decumbent stems. Most of the species occur in an area from Arizona and Texas south into Mexico, but *O. speciosa* extends to the Central Plains in the United States, and *O. rosea* also ranges farther to Central America, the Caribbean (Bermuda, Hispaniola, Jamaica), and northern South America to central Chile. Only *Oenothera deserticola* (Loesener) Munz occurs outside the area and is restricted to high elevations in the Trans-Volcanic Belt of Mexico. *Oenothera rosea* is widely naturalized worldwide in tropical and subtropical areas. Four species form bivalents in meiosis, and are self-compatible, except for *O. speciosa*, which is self-incompatible. *Oenothera rosea* is a PTH species and forms a ring of 14 chromosomes in meiosis, and is self-compatible and autogamous. *Oenothera speciosa* has mostly diploid ($2n$ = 14) populations with vespertine or diurnal white flowers in the northern part of its range; many of those from central Texas southward are morning-opening, pink-flowered, and mostly tetraploid ($2n$ = 28) or hexaploid ($2n$ = 42). Pollinators are primarily bees, noctuids, and very few hawkmoths (in *O. speciosa*), and sometimes secondarily skippers and other butterflies (summarized by W. L. Wagner et al. 2007).

1. Inflorescences sharply nodding; capsule stipes cylindrical .... 16. *Oenothera speciosa*
1. Inflorescences erect; capsule stipes gradually tapering to base.
    2. Floral tubes 4–8 mm; sepals 6–12 mm; petals 4–12 mm; pollen 35–65% fertile .....19. *Oenothera rosea*
    2. Floral tubes 9–26 mm; sepals 7.5–23 mm; petals 8–25(–30) mm; pollen 85–100% fertile.
        3. Floral tubes 9–14 mm; sepals 7.5–12 mm; petals 8–15 mm; styles 12–19 mm; plants strigillose, often densely so ....17. *Oenothera platanorum*
        3. Floral tubes 15–26 mm; sepals 15–23 mm; petals 12–25(–30) mm; styles 26–36 mm; plants strigillose, also sparsely hirsute ....18. *Oenothera texensis*

16. Oenothera speciosa Nuttall, J. Acad. Nat. Sci. Philadelphia 2: 119. 1821 W

Hartmannia berlandieri (Spach) Rose; *H. reverchonii* Rose; *Oenothera berlandieri* (Spach) Spach ex D. Dietrich; *O. delessertiana* Steudel; *O. hirsuta* (Spach) D. Dietrich; *O. obtusifolia* (Spach) D. Dietrich; *O. shimekii* H. Léveillé & Guffroy; *O. spachii* D. Dietrich; *O. speciosa* var. *berlandieri* (Spach) Munz; *O. speciosa* var. *childsii* (L. H. Bailey) Munz; *O. tetraptera* Cavanilles var. *childsii* L. H. Bailey; *Xylopleurum berlandieri* Spach; *X. drummondii* Spach; *X. hirsutum* Spach; *X. obtusifolium* Spach; *X. speciosum* (Nuttall) Raimann

Herbs perennial, caulescent, glabrate to strigillose, usually also sparsely hirsute; from slender taproot and spreading by rhizomes. **Stems** many, erect, 4–60 cm. **Leaves** in a basal rosette and cauline, basal 2–9 × 0.3–3.2 cm, blade oblanceolate to obovate, margins subentire or sinuate-pinnatifid; cauline 1–10 × 0.3–3.5 cm, blade narrowly elliptic to ovate, margins subentire or serrulate to sinuate-pinnatifid. **Inflorescences** sharply nodding. **Flowers** 1–3 opening per day near sunrise in some populations, near sunset in others; buds with free tips 0–4 mm; floral tube 12–25 mm; sepals 15–50 mm; petals pink to rose, fading darker, or white, fading pink, 15–45 mm; filaments 10–22 mm, anthers 6–16 mm, pollen 85–100% fertile; style 20–55 mm, stigma exserted beyond anthers at anthesis. **Capsules** narrowly obovoid to narrowly rhombic-ellipsoid, angled, 10–25 × 3.5–6 mm, apex attenuate to a sterile beak, valve midrib prominent, proximal stipe cylindrical, not tapering to base, (4–)8–15 mm; sessile. **Seeds** narrowly obovoid, 1–1.2 × 0.5–0.6 mm. $2n$ = 14, 28, 42.

Flowering (Feb–)Apr–Jul(–Oct). Grasslands, glades, open woodlands, disturbed places, pastures, railroads, roadsides, loamy or sandy soil, sometimes clay; 10–900 m; Ark., Ill., Ind., Iowa, Kans., La., Mo., Nebr., N.Mex., Okla., Tex.; Mexico (Chihuahua, Coahuila, Durango, Nuevo León, San Luis Potosí, Tamaulipas).

Oenothera speciosa is widely cultivated worldwide for its showy flowers and easy maintenance. It is not known to be definitely naturalized, but tends to persist or become adventive due to its aggressive vegetative reproduction.

17. Oenothera platanorum P. H. Raven & D. R. Parnell, Madroño 20: 246. 1970

Herbs perennial, caulescent, strigillose, often densely so; from slender taproot. **Stems** 1–several, ascending, 5–60 cm. **Leaves** in a basal rosette and cauline, basal 2–7 × 0.3–1.4 cm, blade narrowly elliptic to narrowly ovate, margins weakly serrulate to sinuate-pinnatifid; cauline 1.2–6 × 0.3–1 cm, blade narrowly elliptic to elliptic or ovate, proximal ones sinuate-pinnatifid, margins subentire or weakly serrulate. **Inflorescences** erect. **Flowers** 1–3 opening per day near sunrise; buds with free tips 0–0.1 mm; floral tube 9–14 mm; sepals 7.5–13 mm; petals rose purple, fading darker, 8–15 mm; filaments 4–9 mm, anthers 2.5–4 mm, pollen 85–100% fertile; style 12–19 mm, stigma surrounded by anthers at anthesis. **Capsules** clavate or narrowly obovoid, 9–14 × 3–4 mm, apex attenuate to a sterile beak, valve midrib prominent in distal part, proximal stipe 4–15 mm, gradually tapering to base; sessile. **Seeds** narrowly obovoid, 0.7–0.9 × 0.3–0.5 mm. $2n$ = 14.

Flowering Mar–Aug. Streambeds and near springs; 700–1900 m; Ariz.; Mexico (Sonora).

Oenothera platanorum is known only from the southeastern counties of Cochise, Pinal, and Santa Cruz in Arizona. It was recently collected in Sonora, Mexico. The species is very similar to both *O. texensis*, from which it differs in its smaller flowers, and the widespread *O. rosea*, from which it differs in the somewhat larger flowers and in forming seven bivalents in meiosis and fully fertile pollen, whereas *O. rosea* is a PTH species.

18. Oenothera texensis P. H. Raven & D. R. Parnell, Madroño 20: 247. 1970 [F]

Herbs perennial, caulescent, strigillose and also sparsely hirsute; from slender taproot. **Stems** several–many, ascending, unbranched or branched, 25–50 cm. **Leaves** in a basal rosette and cauline, basal (1–)2.5–6.5 × 0.6–2.3 cm, blade narrowly elliptic to narrowly ovate or ovate, margins weakly serrulate to sinuate-pinnatifid; cauline 1–5.5 × 0.6–2 cm, blade narrowly elliptic to narrowly ovate, margins weakly serrulate. **Inflorescences** erect. **Flowers** 1–3 opening per day near sunrise; buds with free tips 0–1 mm; floral tube 15–26 mm; sepals 15–23 mm; petals rose purple, fading darker, 12–25(–30) mm; filaments 9–13 mm, anthers 3.5–6 mm, pollen 85–100% fertile; style 26–36 mm, stigma exserted beyond anthers at anthesis. **Capsules** clavate or narrowly obovoid, 9–15 × 3.5–6 mm, apex attenuate to a sterile beak, valve midrib prominent in distal part, proximal stipe 7–12(–28) mm, gradually tapering to base; sessile. **Seeds** narrowly obovoid, 0.8–1 × 0.2–0.4 mm. **2*n*** = 14.

Flowering May–Sep. Sandy and gravel bars of streambeds and along streams; 900–2500 m; Tex.; Mexico (Coahuila, Tamaulipas).

In the flora area, *Oenothera texensis* is known only from Jeff Davis County.

19. Oenothera rosea L'Héritier ex Aiton, Hort. Kew. 2: 3. 1789

Gaura epilobia Seringe; *Godetia heuckii* Philippi; *Hartmannia rosea* (L'Héritier ex Aiton) G. Don; *H. rosea* var. *parvifolia* (J. M. Coulter) Small; *H. virgata* (Ruíz & Pavon) Spach; *Oenothera psycrophila* Ball; *O. rosea* var. *parvifolia* J. M. Coulter; *O. rubra* Cavanilles; *O. virgata* Ruíz & Pavon; *Xylopleurum roseum* (L'Héritier ex Aiton) Raimann

Herbs perennial, caulescent, strigillose and often also sparsely hirsute; from slender taproot. **Stems** 1–several, ascending to decumbent, 7–65 cm. **Leaves** in a basal rosette and cauline, basal 1–6 × 0.3–2 cm, blade narrowly elliptic to elliptic or ovate, margins subentire, weakly serrulate, or sinuate-pinnatifid; cauline 1–6 × 0.3–2 cm, blade narrowly elliptic to narrowly ovate, margins subentire or weakly serrulate, proximal ones sinuate-pinnatifid. **Inflorescences** erect. **Flowers** 1–3 opening per day near sunrise; buds with free tips 0.1–1 mm; floral tube 4–8 mm; sepals 6–12 mm; petals rose purple, fading darker, 4–12 mm; filaments 4–6 mm, anthers 2–3.5 mm, pollen 35–65% fertile; style 7–13.5 mm, stigma surrounded by anthers at anthesis. **Capsules** narrowly obovoid, 4–12 × 2–4 mm, apex attenuate to a sterile beak, proximal stipe 5–20 mm, gradually tapering to base, valve midrib prominent in distal part; sessile. **Seeds** narrowly obovoid, 0.5–0.9 × 0.3–0.5 mm. **2*n*** = 14.

Flowering Mar–Sep. Disturbed habitats, along creeks, low, weedy places; 10–600 m; Ariz., Calif., Tex.; Mexico; West Indies; Central America; introduced in South America (Argentina), Europe, Asia, s Africa, Atlantic Islands (Azores, Canary Islands); tropical areas.

Oenothera rosea is a PTH species, forming a ring of 14 chromosomes in meiosis, and is self-compatible and autogamous. In the flora area, it is known from Cochise, Pima, and Santa Cruz counties in Arizona, Alameda, Los Angeles, Riverside, San Bernardino, San Diego, San Francisco, and Santa Barbara counties in California (primarily in urban areas), and from southern Texas. It is clearly of North American origin, since all of its close relatives are confined to North America, and has spread south along the Andes. It occurs at 500–3700 m in South America but generally at lower elevations in most areas.

The name *Hartmannia affinis* Spach is illegitimate, being based on *Oenothera virgata*; *H. gauroides* Spach is also illegitimate, being based on *O. rosea*; *O. purpurea* Lamarck is a later homonym; these three names pertain here.

17f. Oenothera Linnaeus sect. Leucocoryne W. L. Wagner & Hoch, Syst. Bot. Monogr. 83: 157. 2007

Herbs annual or perennial, caulescent; from a slender taproot, [sometimes lateral roots producing adventitious shoots]. **Stems** several(–many), decumbent to ascending, usually branched. **Leaves** in a basal rosette and cauline, cauline 1.8–7 cm; blade margins weakly serrulate to subentire, sinuate-pinnatifid, or coarsely sinuate-dentate. **Inflorescences** solitary flowers in axils of distal leaves. **Flowers** opening near sunset, scent unknown or unscented; buds erect, terete, with free tips; floral tube 5–30 mm; sepals usually splitting along one suture, remaining coherent and reflexed as a unit at anthesis, rarely separating in pairs; petals white, fading lavender to pink [or deep purple], obovate to obcordate; stigma deeply divided into 4 linear lobes. **Capsules** hard and leathery, straight, clavate or obovoid, winged, wings 0.5–4 mm, apex rounded, obtuse or bluntly acuminate, proximal part a sterile pedicel-like base (stipe), gradually tapering to base, valve midrib raised, dehiscent at apex or nearly throughout body; sessile. **Seeds** numerous, clustered in each locule, narrowly obovoid, surface glossy, appearing granular, but minutely papillose under magnification. $2n = 14$.

Species 5 (2 in the flora): Texas, Mexico, West Indies, Central America, South America; introduced in Europe, Asia, s Africa, Pacific Islands (Hawaii), Australia.

Section *Leucocoryne* consists of five species that occur from southern Texas, through northern Mexico to the Trans-Volcanic Belt of central Mexico, southward to Guatemala, Nicaragua, and Costa Rica, in openings in pine-oak or conifer forests on slopes or along streams (or arroyos), sometimes along roadsides or other weedy habitats, at 1400–2800(–3250) m elevation.

Species of section *Leucocoryne* share white petals, have similar capsule shape with a rounded to bluntly acuminate apex, and buds with free tips. All but *Oenothera dissecta* A. Gray ex S. Watson were formerly included in a more broadly defined sect. *Hartmannia*. *Oenothera dissecta*, previously included in sect. *Gauropsis* (P. A. Munz 1965; W. L. Wagner 1984), was transferred to sect. *Leucocoryne*, because it has capsules most similar to those of other white-flowered species grouped here (Wagner et al. 2007). All sect. *Leucocoryne* species are self-compatible (unknown in *O. luciae-julianiae* W. L. Wagner), the flowers vespertine, pollinated by hawkmoths or autogamous (*O. kunthiana* and presumably *O. luciae-julianiae*). *Oenothera kunthiana* is a PTH species and forms a ring of 14 chromosomes in meiosis; *O. luciae-julianiae* also appears to be a PTH species, indicated by lowered pollen fertility (Wagner 2004). Section *Leucocoryne* species are diploid ($n = 7$), except *O. dissecta*, which is known exclusively as a tetraploid ($n = 14$); the chromosomes of *O. luciae-julianiae* have not been examined.

Oenothera dissecta and *O. luciae-julianiae* both occur in Mexico. *Oenothera dissecta* occurs in flats in Durango, Jalisco, San Luis Potosí, and Zacatecas; *O. luciae-julianiae* occurs in pine-oak forest habitats in the Sierra Madre Occidental from Chihuahua south through Durango, Guanajuato, Nayarit, Jalisco, Michoacán, and Querétaro.

1. Petals 20–43 mm; pollen 85–100% fertile; stigmas exserted beyond anthers at anthesis; floral tubes 10–30 mm . . . . 20. *Oenothera tetraptera*
1. Petals 8–25 mm; pollen 35–65% fertile; stigmas surrounded by anthers at anthesis; floral tubes 5–31 mm . . . . 21. *Oenothera kunthiana*

20. Oenothera tetraptera Cavanilles, Icon. 3: 40, plate 279. 1796

Hartmannia latiflora (Seringe) Rose; *H. tetraptera* (Cavanilles) Small; *Oenothera latiflora* Seringe; *O. tetraptera* var. *immutabilis* H. Léveillé; *Xylopleurum tetrapterum* (Cavanilles) Raimann

Herbs annual or perennial, strigillose and also hirsute; from a slender taproot. **Stems** 15–50 cm. **Leaves** 2.5–10 × 0.6–2.2 cm; petiole 0.2–2.2 cm; blade usually lanceolate to oblanceolate, sometimes elliptic, margins weakly serrate to sinuate-pinnatifid. **Flowers** 1–3 opening per day near sunset; buds with free tips 0.5–3 mm; floral tube 10–30 mm; sepals 20–40 mm; petals white, fading pink, 20–43 mm; filaments 12–18 mm, anthers 5–10 mm, pollen 85–100% fertile; style 19–67 mm, stigma exserted beyond anthers at anthesis. **Capsules** broadly clavate or obovoid, 20–51 × 5–7 mm, winged, wings 2–4 mm, valve surface with prominent midrib, proximal stipe 8–45 mm; sessile. **Seeds** narrowly obovoid, 1–1.5 × 0.5–0.7 mm. $2n = 14$.

Flowering Feb–May. Alluvial flats, open areas, sandy soil, weedy sites; 10–300[–2000] m; Tex.; Mexico; West Indies (Jamaica); Central America; n South America; introduced widely in temperate Europe, Asia, s Africa, Australia.

In the flora area, *Oenothera tetraptera* is known only from southern Texas. *Oenothera tetraptera* presumably has become naturalized in South America (Argentina, Bolivia, Colombia, Ecuador, Venezuela), West Indies (Jamaica), Europe, Asia, South Africa, and Australia.

Oenothera candida Dumont Courset is a superfluous name, as is *O. candida* Bellardi ex Colla, and they both pertain here. The name *Hartmannia macrantha* Spach is illegitimate and pertains here.

21. Oenothera kunthiana (Spach) Munz, Amer. J. Bot. 19: 759. 1932

Hartmannia kunthiana Spach, Nouv. Ann. Mus. Hist. Nat. 4: 362. 1836; *H. domingensis* Urban & Ekman; *H. parviflora* Spach; *Oenothera domingensis* (Urban & Ekman) Munz; *O. fissifolia* Steudel; *O. walpersii* Donnell Smith

Herbs annual, strigillose and also hirsute; from a slender taproot. **Stems** 5–40 cm. **Leaves** 1–6 × 0.3–3 cm; petiole 0.1–1.1 cm; blade usually lanceolate to oblanceolate, sometimes elliptic, margins weakly serrate to sinuate-pinnatifid. **Flowers** 1–3 opening per day near sunset; buds with free tips 0–0.5 mm; floral tube 5–31 mm; sepals 10–27 mm; petals white, fading pink, 8–25 mm; filaments 6–12 mm, anthers 3–5 mm, pollen 35–65% fertile; style 12–30 mm, stigma surrounded by anthers at anthesis. **Capsules** broadly clavate or obovoid, 7–31 × 3–5 mm, winged, wings 0.5–1.5 mm, valve surface with prominent midrib, proximal stipe 3–17 mm; sessile. **Seeds** narrowly obovoid, 0.9–1.2 × 0.4–0.5 mm. $2n = 14$.

Flowering Feb–May. Alluvial flats, open areas, sandy soil, weedy sites; 10–1300[–2000] m; Tex.; Mexico; Central America; n South America; introduced widely in temperate Europe, Asia, Pacific Islands (Hawaii), Australia.

Oenothera kunthiana is a PTH species and forms a ring of 14 chromosomes in meiosis, and is self-compatible and autogamous, common and widespread from sea level to middle elevations in the mountains from southern Texas south throughout Mexico except for Baja California and the tropical lowlands southward to Guatemala, Nicaragua, and Costa Rica; it was once collected in Minas Gerais, Brazil. *Oenothera kunthiana* was recently found to be naturalized in the Hawaiian Islands.

Oenothera pinnatifida Kunth is a later homonym of *O. pinnatifida* Nuttall and another later homonym is *O. micrantha* Walpers, not Hornemann ex Sprengel (1825); they both pertain here.

17g. Oenothera Linnaeus sect. Paradoxus W. L. Wagner, Ann. Missouri Bot. Gard. 71: 1122. 1985

Herbs perennial, caulescent; from a taproot, lateral roots producing adventitious shoots. **Stems** weakly erect becoming decumbent, branched or unbranched. **Leaves** in a basal rosette and cauline, 1–5 cm; blade margins pinnately lobed to sinuate-toothed or dentate. **Inflorescences** solitary flowers in axils of distal leaves. **Flowers** opening near sunset; buds erect, terete, with free tips coherent; floral tube (37–)45–60(–65) mm; sepals splitting along one suture, reflexed as a unit to one side; petals lemon-yellow, fading orange-red to reddish purple, usually elliptic, sometimes oblanceolate; stigma deeply divided into 4 linear lobes. **Capsules** hard and woody, narrowly ovoid to ovoid, 4-angled, apex tapering to a short, sterile beak, valves with prominent midrib and capsule appearing 8-ribbed, tardily dehiscent ⅓ their length; sessile. **Seeds** numerous, in 1 or 2 partially overlapping rows per locule, asymmetrically cuneiform to rhombic, irregularly angled, surface minutely beaded. $2n = 14, 28$.

Species 1: sw, sc United States, n Mexico.

Section *Paradoxus* consists of a single species, restricted to the Chihuahuan Desert.

22. Oenothera havardii S. Watson, Proc. Amer. Acad. Arts 20: 366. 1885 (as havardi) [F]

Hartmannia havardii (S. Watson) Rose; *H. palmeri* Rose

Herbs compact to sprawling, strigillose; from a taproot, lateral roots producing adventitious shoots. **Stems** usually many-branched, sometimes unbranched, often twining among vegetation, sometimes rooting at nodes, 5–25(–70) cm. **Leaves** in a basal rosette and cauline, basal usually quickly deciduous, (1–)2–5 × (0.2–)0.5–1.5 cm; petiole 0–0.6 cm; blade oblanceolate, linear-lanceolate to linear distally, margins few toothed to pinnately lobed to sinuate-dentate distally. **Flowers** 1–few opening per day near sunset; buds often twisted, free tips coherent; floral tube (37–)45–60(–65) mm; sepals (16–)18–26(–30) mm; petals lemon-yellow, fading orange-red to reddish purple, usually elliptic, sometimes oblanceolate, (18–)21–30(–32) mm; filaments 15–18 (–22) mm, anthers red, 6–13 mm; style (55–)65–86 (–94) mm, stigma exserted beyond anthers at anthesis. **Capsules** woody, narrowly ovoid to ovoid, 4-angled, 8–13(–16) × 3–4 mm, apex tapering to a short sterile beak 2–3 mm, valves with a prominent, broad midrib and capsule appearing 8-ribbed, tardily dehiscent ca. ⅓ capsule length. **Seeds** 2–2.5(–3.3) × 1.2–1.5 mm, sometimes with a small wing at distal end or a raised ridge along one longitudinal margin. **2*n*** = 14, 28.

Flowering Apr–Oct. In depressions, seasonally wet flats, stream banks, margins of irrigated fields, sandy or clay soil, among tufted grasses like *Sporobolus wrightii*, primarily in Chihuahuan Desert; 1300–2000 m; Ariz., Tex.; Mexico (Chihuahua, Durango, Sonora, Zacatecas).

Oenothera havardii ranges from Brewster and Presidio counties, Texas, and Cochise County, Arizona, south to Durango and Zacatecas, Mexico. W. L. Wagner (1984) found that *O. havardii* is self-incompatible and vespertine.

17h. OENOTHERA Linnaeus sect. MEGAPTERIUM (Spach) Walpers, Repert. Bot. Syst. 2: 82. 1843

Megapterium Spach, Hist. Nat. Vég. 4: 363. 1835; *Oenothera* [unranked] *Megapterium* (Spach) Endlicher; *Oenothera* subg. *Megapterium* (Spach) Reichenbach

Herbs perennial, acaulescent or caulescent; from a stout, woody taproot, sometimes (*O. brachycarpa*, *O. howardii*) lateral roots producing adventitious shoots. **Stems** ascending or becoming decumbent, usually unbranched. **Leaves** in a basal rosette, often also cauline, (2.8–)5–21(–34) cm; blade margins entire, dentate, or pinnatifid. **Inflorescences** solitary flowers in axils of distal leaves. **Flowers** opening near sunset, with sweet scent or nearly unscented; buds erect, quadrangular, with free tips; floral tube (21–)35–210(–220) mm; sepals splitting along one suture, remaining coherent and reflexed as a unit at anthesis; petals yellow, fading yellow, orange, pink, or deep red, obovate to rhombic-obovate; stigma deeply divided into 4 linear lobes. **Capsules** papery, leathery, or corky in age, ovoid, narrowly lanceoloid to broadly ellipsoid, or globose, winged, wings 10–32 mm wide throughout, or capsule walls with corky thickening and wings not developed (sometimes in *O. brachycarpa*), then capsule appearing only 4-angled, apex truncate to cuneate, dehiscent ¼–⅓ their length; pedicellate, sometimes disarticulating from plant at maturity. **Seeds** numerous, in 1 or 2 rows per locule, grayish to yellowish brown, brown, or dark purplish brown, obovoid or subcuboid, angled or rounded, usually with an erose wing distally, surface coarsely rugose and reticulate, thickened, especially at distal end, this area with an internal cavity adjacent to embryo. **2*n*** = 14, 28, 42, 56.

Species 4 (4 in the flora): w, c United States, n Mexico.

Section *Megapterium* consists of four species (eight taxa); two (*Oenothera brachycarpa*, *O. macrocarpa*) are diploid (2*n* = 14), one (*O. coryi*) is hexaploid (2*n* = 42), and one (*O. howardii*) has tetraploid, hexaploid, and octoploid populations (2*n* = 28, 42, 56) (W. L. Wagner et al. 2007). The species usually occur on xeric rocky sites of limestone, sandstone, shale, or gypsum, rarely (*O. brachycarpa*) on volcanic soil, from eastern Nevada, Utah, and eastern Colorado east to the Mississippi River in Missouri, and south through northern Arkansas and Texas, to Coahuila, Durango, and Nuevo León, Mexico; there are only two isolated records (*O. macrocarpa* subsp. *macrocarpa* from St. Clair County, Illinois, and Rutherford County, Tennessee) from east of the Mississippi River, at 130–3000 m elevation. All species are self-incompatible and vespertine, the flowers fading the following morning, or sometimes remaining open for a second day in *O. macrocarpa*, pollinated by hawkmoths including *Hyles*, *Manduca*, and *Sphinx* (see Wagner et al. for summary).

1. Leaves (2.8–)3.7–12.5(–17) cm, blade margins entire or denticulate or serrulate; plants caulescent; stems (1–)4–40(–60) cm . . . . .26. *Oenothera macrocarpa*
1. Leaves (3.1–)5–21(–34) cm, blade margins usually pinnately lobed, sometimes lobed only proximally; plants acaulescent or sometimes caulescent; stems 0–20(–36) cm.
   2. Petals pale yellow to yellow, drying lavender to purple, usually broadly rhombic-obovate, rarely obovate, usually appearing crumpled; floral tubes (90–)120–210(–220) mm; leaf blade usually with a large terminal lobe, margins usually irregularly pinnatifid with some of the sinuses extending nearly to midrib, margins usually erose; capsule wings 0–3(–5) mm wide . . . . 23. *Oenothera brachycarpa*
   2. Petals lemon-yellow to brilliant yellow, drying purple to reddish brown, rarely lavender, broadly obovate, not crumpled; floral tubes (43–)60–110(–125) mm; leaf blade without a large terminal lobe, margins entire or irregularly pinnately lobed, sinuses usually extending less than ½ to midrib, margins not erose; capsule wings (2–)4–7(–11) mm wide.
      3. Petals brilliant yellow, (30–)40–60(–73) mm, drying deep reddish purple to reddish brown; flowers with strong, sweet scent; buds with free tips 1–3(–4) mm; leaves (0.5–)1–2(–3) cm wide, blades usually oblanceolate, elliptic to narrowly elliptic, rarely lanceolate . . . . .24. *Oenothera howardii*
      3. Petals lemon-yellow, 35–43 mm, drying lavender to purple; flowers with weak scent; buds with free tips 0.7–1.2 mm; leaves (0.2–)0.3–0.5(–0.7) cm wide, blades linear to narrowly lanceolate . . . . .25. *Oenothera coryi*

23. Oenothera brachycarpa A. Gray, Smithsonian Contr. Knowl. 3(5): 70. 1852 F

Lavauxia brachycarpa (A. Gray) Britton; *L. graminifolia* (H. Léveillé) Rose; *L. wrightii* (A. Gray) Small; *Megapterium brachycarpum* (A. Gray) Rydberg; *M. brachycarpum* var. *wrightii* (A. Gray) Moldenke; *Oenothera australis* (Wooton & Standley) Tidestrom; *O. brachycarpa* var. *wrightii* (A. Gray) H. Léveillé; *O. cespitosa* Nuttall subsp. *australis* (Wooton & Standley) Munz; *O. cespitosa* var. *australis* (Wooton & Standley) Munz; *O. graminifolia* H. Léveillé; *O. wrightii* A. Gray; *Pachylophus australis* Wooton & Standley

Herbs acaulescent or sometimes caulescent, strigillose, also hirsute, hairs often with reddish purple pustulate base, glandular puberulent distally; from a woody taproot, sometimes lateral roots producing adventitious shoots. **Stems** (when present) ascending, longer ones becoming decumbent, usually densely leafy, 0–20(–36) cm. **Leaves** in a basal rosette, sometimes also cauline, (3.1–)5–21(–34) × (0.3–)1.5–3.5(–5.3) cm; petiole (0.8–)2.5–11(–15) cm; blade usually lanceolate to elliptic, rhombic-obovate, sometimes suborbicular or linear, usually irregularly pinnatifid, some sinuses extending nearly to midrib, usually with a large terminal lobe (0.1–)1.5–2(–2.4) cm, margins erose, apex acute to obtuse or rounded. **Flowers** usually 1–3, rarely more, opening per day near sunset, weakly scented; buds with unequal free tips 1–7 mm; floral tube (90–)120–210(–220) mm; sepals 38–55 mm; petals pale yellow to yellow, fading pale orange to pink, drying lavender to purple, usually broadly rhombic-obovate, sometimes obovate, (38–)45–58(–62) mm, distal margin usually erose; filaments (16–)20–32 mm, anthers (8–)13–21 mm; style (123–)155–240(–255) mm, stigma exserted beyond anthers at anthesis. **Capsules** leathery or corky, ovoid to narrowly ellipsoid, ± winged, wings 0–3(–5) mm wide, sometimes capsule with corky thickening between wings, then capsule only 4-angled, body (12–)18–40 × 6–10 mm, dehiscent ¼ their length; pedicel 0–3 mm. **Seeds** usually numerous, in 1 or 2 rows per locule, obovoid to subcuboid, 3–5 × 1.8–2.2 mm. $2n = 14$.

OENOTHERA

Flowering Mar–Aug. Rocky sites, usually on limestone, shale, or gypsum, on igneous substrates from canyons and slopes in Chihuahuan Desert scrub, grasslands, oak-pine-juniper woodlands, open sites in ponderosa pine-Douglas fir forests; 1000–2700 m; Ariz., N.Mex., Tex.; Mexico (Chihuahua, Coahuila, Durango, Sonora).

Oenothera brachycarpa occurs from southeastern Arizona in southern Navajo, southeastern Pima, Graham, Santa Cruz, and Cochise counties, east across southern New Mexico to Val Verde and Pecos counties in trans-Pecos Texas.

24. Oenothera howardii (A. Nelson) M. E. Jones ex Prain in B. D. Jackson et al., Index Kew., suppl. 3: 121. 1908 (as howardi) E

Lavauxia howardii A. Nelson, Bot. Gaz. 34: 368. 1902 (as howardi)

Herbs acaulescent or sometimes caulescent, moderately to densely strigillose and glandular puberulent, sometimes also sparsely to moderately hirsute; from a taproot, sometimes lateral roots producing adventitious shoots. **Stems** (when present) ascending, longer ones becoming decumbent, leafy, sometimes densely so, 0–10(–30) cm. **Leaves** in a basal rosette, sometimes also cauline, (6–)8.5–17(–23) × (0.5–)1–2(–3) cm; petiole 2–7.5 cm; blade usually oblanceolate, elliptic to narrowly oblong, rarely lanceolate, margins often undulate, entire or remotely and irregularly pinnately lobed mostly in proximal 1/2, rarely more regularly pinnately lobed and lobing extending to distal 1/2, sinuses usually extending less than 1/2 to midrib, lobes triangular to oblong or linear, (1–)4–9(–13) mm, apex acute to obtuse. **Flowers** usually 1 or 2, rarely more, opening per day near sunset, strongly and sweetly scented; buds with unequal free tips 1–3(–4) mm; floral tube (43–)60–110(–125) mm; sepals (30–)35–60(–80) mm; petals brilliant yellow, fading deep red, drying deep reddish purple to reddish brown, usually broadly obovate, rarely subrhombic, (30–)40–60(–73) mm, sometimes with a terminal tooth; filaments (19–)25–38 mm, anthers 10–17 mm; style (90–)110–145(–165) mm, stigma exserted beyond anthers at anthesis. **Capsules** leathery, ovoid, narrowly ovoid, or narrowly lanceoloid to broadly ellipsoid, winged, wings (2–)4–7(–11) mm wide, body (20–)25–50(–80) × 4–6 mm, dehiscent 1/4–1/3 their length; pedicel 2–6 mm. **Seeds** numerous, usually in 1 row per locule, rarely in 2 rows toward base, obovoid to subcuboid, 3–8 × 2.5–3.5 mm. **2*n*** = 28, 42, 56.

Flowering May–Jul. Open or rocky areas, in shale, fine-textured sandstones, clays, gypsum, or limestone from High Plains grasslands, open sites in pinyon-juniper woodlands, ponderosa pine-Douglas fir forests; (1000–)1500–2300(–3000) m; Colo., Kans., Nev., Utah, Wyo.

Oenothera howardii is known from three disjunct areas: three collections on the High Plains (Baca and Otero counties, Colorado, and Hamilton County, Kansas); open yucca-juniper grassland, rocky slopes or disturbed areas on shale substrates along the Colorado counties of Boulder, Denver, Jefferson, and Larimer, and just over the state line in Wyoming; and, common to scattered, mostly on rocky slopes but also in shaded canyon sites on fine-textured red sandstones, clays, gypsum, chalky white degraded limestone or limestone in pinyon-juniper woodland to ponderosa pine-Douglas fir forest in southern Utah and eastern Nevada.

A. Nelson intended to publish *Lavauxia howardii* as a new combination and base it on *Oenothera howardii* M. E. Jones (1893), which was not validly published at the time, but inadvertently published *L. howardii* as a new species. *Oenothera howardii* (A. Nelson) W. L. Wagner is an isonym.

25. Oenothera coryi W. L. Wagner, Ann. Missouri Bot. Gard. 73: 475. 1986 [E]

Herbs acaulescent or caulescent, densely strigillose and glandular puberulent distally; from a taproot. **Stems** densely leafy, 4–20 cm. **Leaves** in a basal rosette, sometimes also cauline, 5–16 × (0.2–)0.3–0.5(–0.7) cm; petiole 0.6–3.5 cm; blade linear to narrowly lanceolate, margins entire or sometimes proximal ½ of blade remotely lobed, apex long-attenuate, acute to rounded. **Flowers** usually 1–3, rarely more, opening per day near sunset, weakly scented; buds with unequal free tips 0.7–1.2 mm; floral tube (55–)75–100(–125) mm; sepals 34–40 mm; petals lemon-yellow, fading orange, drying lavender to purple, broadly obovate, 35–43 mm, sometimes with terminal tooth; filaments 17–25 mm, anthers 14–17 mm; style (85–)105–135(–143) mm, stigma exserted beyond anthers at anthesis. **Capsules** leathery, lanceoloid to ovoid, winged, wings 4–6 mm wide, body 25–30 × 8 mm, dehiscent ¼–⅓ their length; pedicel 1–2(–3) mm. **Seeds** numerous, usually in 2 distinct rows per locule, often reduced to 1 row near apex, rarely 1 row throughout, obovoid to subcuboid, 2.5–4 × 2.5–3.5 mm. **2*n*** = 42.

Flowering Apr–May. Open grasslands, disturbed areas; 300–1000 m; Tex.

Oenothera coryi is known only from Baylor, Callahan, Knox, Nolan, Taylor, and Throckmorton counties in north-central Texas and Crosby and Garza counties in the Texas Panhandle.

26. Oenothera macrocarpa Nuttall, Cat. Pl. Upper Louisiana, no. 56. 1813

Megapterium macrocarpum (Nuttall) R. R. Gates; *M. nuttallianum* Spach

Herbs caulescent, strigillose or glabrous, sometimes glandular puberulent distally; from a stout taproot, sometimes lateral roots producing adventitious shoots. **Stems** moderately leafy, (1–)4–40(–60) cm. **Leaves** cauline, (2.8–)3.7–12.5(–17) × (0.1–)0.4–3(–4.5) cm; petiole (0.4–)1–4(–6) cm; blade linear, lanceolate-elliptic, elliptic to oblanceolate or suborbiculate, margins entire or conspicuously or inconspicuously denticulate or serrulate, sometimes undulate, apex usually acute, sometimes obtuse or retuse (subsp. *incana*). **Flowers** usually 1 or 2, rarely more, opening per day near sunset, fading next morning, sometimes (subspp. *macrocarpa* and *oklahomensis*) lasting for 2 days, weakly scented; buds with unequal free tips 1–11(–15) mm; floral tube (21–)35–140(–160) mm; sepals (20–)25–65(–75) mm; petals bright yellow, fading orange, reddish orange or mostly unchanged, obovate to very broadly obovate, (17–)25–65(–68) mm, usually with terminal notch and/or tooth, margin sometimes erose; filaments 13–40(–44) mm, anthers 10–24(–25) mm; style (45–)55–192 mm, stigma usually exserted beyond anthers at anthesis. **Capsules** papery in age, narrowly ellipsoid to lanceoloid, sometimes twisted (subsp. *fremontii*), winged, wings (2–)10–28(–34) mm wide, body (13–)25–70(–115) × 2–9 mm, dehiscent ¼–⅓ their length; pedicel 1–12(–25) mm. **Seeds** numerous, rarely as few as 8, in 1 row per locule, obovoid, (2–)3–5 × 1–2.3 mm.

Subspecies 5 (4 in the flora): c, s United States, n Mexico.

Oenothera macrocarpa is variable and has differentiated extensively in the Great Plains region. Each of the five distinctive subspecies occupies a different geographical and ecological situation. Only subsp. *mexicana* W. L. Wagner from Coahuila, Mexico, occurs outside of the flora area. In general, the subspecies are sharply distinct and each is characterized by a number of features, including pubescence, leaf features, flower and floral tube size, and size and morphology of the capsules and seeds. The five entities are treated as subspecies primarily because of their complete interfertility and extensive intergradation in any area of marginal contact. Intermediates are known between subsp. *macrocarpa* and subspp. *fremontii* and *oklahomensis* and between subspp. *incana* and *oklahomensis*. There is also some evidence that suggests past hybridization between subspp. *fremontii* and *incana* although there is no present contact between them. All subspecies are self-incompatible.

Oenothera alata Nuttall (1818) is an illegitimate name based on *O. macrocarpa* and pertains here.

1. Herbs glabrous.
   2. Leaf blades usually very broadly elliptic to suborbiculate, rarely oblanceolate or elliptic, 2–4.3 cm wide, margins usually entire, sometimes inconspicuously denticulate, usually flat, sometimes undulate .... 26b. *Oenothera macrocarpa* subsp. *incana* (in part)
   2. Leaf blades usually elliptic to lanceolate-elliptic, sometimes linear, (0.3–)0.8–2(–3) cm wide, margins usually conspicuously denticulate to serrulate, rarely subentire, usually undulate....... 26c. *Oenothera macrocarpa* subsp. *oklahomensis*
1. Herbs strigillose.
   3. Bud free tips 1–2(–5) mm; petals (17–)25–33 (–37) mm; floral tubes (21–)35–65(–80) mm; capsules often twisted, 13–30(–65) mm, wings 2–5(–9) mm wide...... 26d. *Oenothera macrocarpa* subsp. *fremontii*
   3. Bud free tips (4–)5–11(–12) mm; petals (25–) 31–65(–68) mm; floral tubes (50–)70–140 (–160) mm; capsules not twisted, (28–)30–70 (–115) mm, wings 10–28(–34) mm wide.
      4. Leaves green or younger ones grayish green, moderately strigillose, rarely more densely strigillose, blades often lanceolate-elliptic to broadly elliptic, sometimes linear or lanceolate, 0.4–2.3(–3) cm wide; capsule wings (14–)18–28(–34) mm wide .... 26a. *Oenothera macrocarpa* subsp. *macrocarpa*
      4. Leaves gray, densely strigillose, blades very broadly elliptic to suborbiculate, rarely oblanceolate or elliptic, 2–4.3 cm wide; capsule wings 10–15(–24) mm wide..... .... 26b. *Oenothera macrocarpa* subsp. *incana* (in part)

26a. Oenothera macrocarpa Nuttall subsp. **macrocarpa** [E]

Megapterium missourense (Sims) Spach; *Oenothera macrocarpa* var. *missourensis* (Sims) Carrière; *O. missourensis* Sims; *O. missourensis* var. *latifolia* A. Gray

Herbs strigillose and glandular puberulent distally. **Stems** several, unbranched, sometimes with shorter secondary branches, 10–40(–60) cm. **Leaves** green, younger ones grayish green, (6–)8–12(–14.5) × 0.4–2.3(–3) cm; blade often lanceolate-elliptic to broadly elliptic, sometimes linear or lanceolate, margins usually flat, sometimes undulate, entire or inconspicuously denticulate, apex acute. **Flowers:** buds with unequal free tips (4–)8–10(–12) mm; floral tube (78–)95–115(–140) mm; sepals (45–)50–65(–75) mm; petals (40–)50–65 (–68) mm; filaments (25–)30–40(–44) mm, anthers (15–)17–24 mm; style (120–)135–160(–190) mm. **Capsules** ovoid, narrowly ovoid, narrowly lanceoloid to broadly ellipsoid, or subglobose, not twisted, wings (14–)18–34 mm wide, body 52–70(–115) × 7–8 mm. **2*n*** = 14.

Flowering Apr–Jun(Sep). Rocky, clay, alkaline soil, unglaciated prairies, glades, bluffs, open prairie hillsides, disturbed sites, limestone or dolomite; 100–500 m; Ark., Ill., Kans., Mo., Nebr., Okla., Tenn., Tex.

Subspecies *macrocarpa* is known from three disjunct areas: southeastern Nebraska, eastern half of Kansas, Craig and Washington counties, Oklahoma, east to Missouri south of the Missouri River and St. Clair County, Illinois, and to the northern tier of counties in Arkansas; glades near Murfreesboro, Rutherford County, Tennessee; and Blackland Prairies, Cross Timbers and eastern Edwards Plateau, from Bryan, Johnston, and Pontotoc counties, Oklahoma, southwest to Bexar, Coke, Kerr, and McCulloch counties, Texas.

26b. Oenothera macrocarpa Nuttall subsp. **incana** (A. Gray) W. L. Wagner, Ann. Missouri Bot. Gard. 70: 194. 1983 [E]

Oenothera missourensis Sims var. *incana* A. Gray, Boston J. Nat. Hist. 6: 189. 1850; *Megapterium argyrophyllum* R. R. Gates; *O. macrocarpa* var. *incana* (A. Gray) Reveal

Herbs strigillose, usually densely so, rarely glabrous, and sometimes glandular puberulent distally. **Stems** several, unbranched, sometimes with shorter secondary branches, 1–20(–30) cm. **Leaves** usually gray, rarely green, (5–)6.2–12.5(–17) × 2–4.3 cm; blade usually very broadly elliptic to suborbiculate, rarely oblanceolate or elliptic, margins usually flat, sometimes undulate, usually entire, sometimes inconspicuously denticulate, apex usually acute to obtuse, sometimes retuse. **Flowers:** buds with unequal free tips 5–11 mm; floral tube (50–)70–140 (–160) mm; sepals (25–)35–50 mm; petals (25–)31–50 (–52) mm; filaments (13–)15–25(–28) mm, anthers (10–)14–20 mm; style (75–)100–192 mm. **Capsules** broadly ellipsoid to globose, not twisted, wings 10–15 (–24) mm wide, body 28–48(–74) × 6–8 mm. **2*n*** = 14.

Flowering Apr–Jun(–Aug). Rocky, clay soil, grasslands, disturbed sites, limestone, gypsum, rarely igneous soil; (500–)600–1200 m; Kans., Okla., Tex.

Subspecies *incana* occurs on the high plains in Clark, Comanche, Kiowa, and Meade counties, Kansas, south across Oklahoma as far east as Comanche and Harper counties to the Texas Panhandle to Garza and Knox counties; one collection is known from Taylor County, Texas.

26c. Oenothera macrocarpa Nuttall subsp. **oklahomensis** (Norton) W. L. Wagner, Ann. Missouri Bot. Gard. 70: 194. 1983 [E]

Megapterium oklahomense Norton, Rep. (Annual) Missouri Bot. Gard. 9: 153, plate 47, figs. 1–3. 1898; *Oenothera macrocarpa* var. *oklahomensis* (Norton) Reveal; *O. missourensis* Sims var. *oklahomensis* (Norton) Munz

Herbs glabrous. **Stems** several, unbranched, sometimes with shorter secondary branches, 5–30(–42) cm. **Leaves** green, usually flecked with reddish purple splotches or reddish tinged, (5.5–)7–11(–12.5) × (0.3–)0.8–2(–3) cm; blade usually elliptic to lanceolate-elliptic, sometimes linear, moderately thick, fleshy, margins usually undulate, usually conspicuously denticulate to serrulate, rarely subentire, apex acute. **Flowers:** buds with unequal free tips (5–)6–10(–15) mm; floral tube (82–)90–140(–147) mm; sepals 35–60 mm; petals (38–)42–55(–65) mm; filaments 24–34 mm, anthers (15–)19–21(–25) mm; style (120–)140–190 mm. **Capsules** globose to narrowly ellipsoid, not twisted, wings (10–)15–20(–25) mm wide, body 35–55(–75) × 7–9 mm. $2n$ = 14.

Flowering Apr–Jun(–Sep). Rocky, clay soil, open sites of fine-textured, red sandstone usually mixed with gypsum, on pure gypsum, on limestone; 200–500(–800) m; Kans., Okla., Tex.

Subspecies *oklahomensis* is known from Barber, Harper, Meade, and Montgomery counties in extreme southern Kansas south across central Oklahoma to Harmon County in southwestern Oklahoma, and Cooke and Knox counties in adjacent Texas.

26d. Oenothera macrocarpa Nuttall subsp. **fremontii** (S. Watson) W. L. Wagner, Ann. Missouri Bot. Gard. 70: 194. 1983 [E]

Oenothera fremontii S. Watson, Proc. Amer. Acad. Arts 8: 587. 1873; *Megapterium fremontii* (S. Watson) Britton

Herbs densely strigillose. **Stems** numerous, with numerous short secondary branches, 3–30 cm. **Leaves** gray, (2.8–)3.7–11 × 0.1–0.6(–1.5) cm; blade linear to narrowly elliptic to narrowly elliptic-lanceolate or narrowly oblanceolate, margins flat, entire or inconspicuously denticulate, apex acute. **Flowers:** buds with unequal free tips 1–2(–5) mm; floral tube (21–)35–65(–80) mm; sepals (20–)25–30(–37) mm; petals (17–)25–33(–37) mm; filaments 13–18 mm, anthers 10–12 mm; style (45–)55–80(–98) mm. **Capsules** ellipsoid to narrowly ellipsoid, often twisted, wings 2–5(–9) mm wide, body 13–30(–65) × 2–6 mm. $2n$ = 14.

Flowering May–Aug. Rocky soil derived from fine-textured sandstone, shale or chalk on rocky hillsides, bluffs, badlands; 400–900 m; Kans., Nebr.

Subspecies *fremontii* occurs from Franklin and Webster counties in south-central Nebraska south into Kansas to Ellsworth, Hodgeman, and Logan counties; also with disjunct locations in Antelope and Cedar counties in northeastern Nebraska, and Barber County in south-central Kansas. Some specimens from the eastern part of the range, where subsp. *fremontii* and subsp. *macrocarpa* are sympatric, appear intermediate between the two subspecies and are difficult to assign.

17i. Oenothera Linnaeus sect. Peniophyllum (Pennell) Munz, Bull. Torrey Bot. Club 64: 288. 1937 [E]

Peniophyllum Pennell, Bull. Torrey Bot. Club 46: 373. 1919; *Oenothera* subsect. *Peniophyllum* (Pennell) Straley

Herbs annual, caulescent; from a taproot. **Stems** erect, unbranched or branched. **Leaves** in a basal rosette and cauline, cauline 1–2(–)4 cm; blade margins subentire or remotely dentate. **Inflorescences** solitary flowers in axils of distal leaves. **Flowers** opening near sunrise, nearly unscented; buds erect, terete, without free tips; floral tube 1–2 mm; sepals separating in pairs; petals bright yellow, fading pink, obcordate to obovate; stigma shallowly divided into 4 linear lobes. **Capsules** hard and leathery, ellipsoid-rhombic to subglobose, 4-angled, apex rounded or obtuse, proximal part tapering to a sterile, pedicel-like base (stipe), valve midrib raised at distal end, indehiscent or tardily dehiscent distally; sessile. **Seeds** usually numerous, clustered in each locule, ovoid, surface minutely papillose. $2n = 14$.

Species 1: se, sc United States.

Section *Peniophyllum* consists of a single diploid species. It has been treated as the sole member of subg. *Kneiffia* sect. *Peniophyllum* (P. A. Munz 1937, 1965) or sect. *Kneiffia* subsect. *Peniophyllum* (G. B. Straley 1977). It was separated from sect. *Kneiffia* by W. L. Wagner et al. (2007) because molecular data (R. A. Levin et al. 2004) failed to place these two groups in a single clade. Instead, *Oenothera linifolia* forms a very weakly supported clade with *O. havardii* (sect. *Paradoxus*), with which it shares no known morphological similarity other than characteristics of subclade B. The most distinctive characteristics of sect. *Peniophyllum* are the heteromorphic rosette versus cauline leaves, the cauline leaves crowded and linear, capsules ellipsoid-rhombic to subglobose, indehiscent or tardily dehiscent only at distal end, and the entire capsule falling from the plant prior to senescence. In most sections of *Oenothera*, the capsules are persistent on the stem, but in sect. *Peniophyllum* and two other sections (*Gaura* and *Megapterium*) of subclade B, the capsules often disarticulate from the plant. The flowers are self-compatible, diurnal, and autogamous or cleistogamous.

27. Oenothera linifolia Nuttall, J. Acad. Nat. Sci. Philadelphia 2: 120. 1821 [E]

Kneiffia linifolia (Nuttall) Spach; *Oenothera linifolia* var. *glandulosa* Munz; *Peniophyllum linifolium* (Nuttall) Pennell

Herbs annual, caulescent, strigillose or glabrous, also often glandular puberulent, especially distally; from a sparsely branched taproot. **Stems** unbranched or with many ascending branches arising near base, erect, 10–50 cm. **Leaves** in a basal rosette and cauline, basal 1–2(–4) × 0.2–0.6 cm, petiole 0.2–1(–1.5) cm, blade ovate to obovate or narrowly elliptic; cauline 1–4 × less than 0.1 cm, sessile, blade linear or filiform. **Flowers** usually 1–3 opening per day near sunrise; buds without free tips; sepals 1.5–2 mm; petals bright yellow, fading pink, obcordate to obovate, 3–5(–7) mm; filaments 1–2 mm, anthers 0.5–1 mm; style 1–2 mm, stigma surrounded by anthers at anthesis. **Capsules** ellipsoid-rhombic to subglobose, 4-angled, 4–6(–10) × 1.5–3 mm, stipe 0–4 mm, valve midrib raised at distal end, indehiscent or tardily dehiscent distally; sessile. **Seeds** clustered in each locule, ovoid, surface minutely papillose, 1 × 0.5 mm. $2n = 14$.

Flowering Apr–Jun(–Aug). Prairies, open woodlands, open rocky and sandy sites, roadsides; 0–400 m; Ala., Ark., Fla., Ga., Ill., Kans., Ky., La., Miss., Mo., N.C., Okla., S.C., Tenn., Tex.

Kneiffia linearifolia Spach (1835) is an illegitimate name based on *Oenothera linifolia*.

17j. Oenothera Linnaeus sect. Kneiffia (Spach) Walpers, Repert. Bot. Syst. 2: 84. 1843 [E]

Kneiffia Spach, Hist. Nat. Vég. 4: 373. 1835; *Blennoderma* Spach; *Oenothera* [unranked] *Blennoderma* (Spach) Endlicher; *Oenothera* sect. *Blennoderma* (Spach) Walpers; *Oenothera* subg. *Blennoderma* (Spach) Reichenbach; *Oenothera* [unranked] *Kneiffia* (Spach) Endlicher; *Oenothera* subg. *Kneiffia* (Spach) Munz

Herbs annual or perennial, caulescent; from fibrous roots or a taproot, sometimes somewhat fleshy, or sometimes producing rhizomes. **Stems** usually erect or ascending, sometimes decumbent, branched or unbranched. **Leaves** in a basal rosette and cauline, cauline 3–10(–13) cm; blade margins entire, subentire, denticulate, or coarsely dentate. **Inflorescences** solitary flowers in axils of distal leaves. **Flowers** opening near sunrise, faintly scented; buds erect, terete, with free tips; floral tube 3–25 mm; sepals splitting along one suture, remaining coherent and reflexed as a unit at anthesis, or separating in pairs, or all separating individually; petals yellow, fading pale pink or lavender, or orangish pink or yellow, or not changing color, obcordate to obovate; stigma deeply divided into 4 linear lobes. **Capsules** leathery, straight, usually clavate or oblong, sometimes ellipsoid, angled or winged, apex rounded to truncate or weakly emarginate, tapering to a sterile, pedicel-like base (stipe), valve midrib raised, initially apically dehiscent, eventually dehiscent nearly throughout; sessile. **Seeds** usually numerous, clustered in each locule, ovoid, surface minutely papillose. $2n$ = 14, 28, 42, 56.

Species 6 (6 in the flora): North America.

Section *Kneiffia* consists of six species (seven taxa) widely distributed in the eastern half of the United States and adjacent Canada, at 0–1900 m elevation (G. B. Straley 1977). *Oenothera spachiana* is the only annual species and occurs in open fields, prairies, rocky or sandy sites, and along roadsides; the remaining species are all perennial and occupy diverse habitats, including fresh and brackish swampy areas, wood margins, meadows, prairies, and sandy sites. Three species are self-incompatible (*O. fruticosa*, *O. pilosella*, and *O. riparia*) and the other three are self-compatible (*O. perennis*, *O. sessilis*, and *O. spachiana*). The flowers are diurnal, opening near sunrise and closing near sunset; in some populations of outcrossing species they may reopen for several days. In the outcrossing taxa, the flowers are pollinated by bees (Halictidae and *Bombus*). *Oenothera spachiana* and *O. perennis* are autogamous, the former often cleistogamous and the latter a PTH species. Taxonomy of the section here differs somewhat from Straley in that molecular data (K. N. Krakos et al. 2014) indicate that *O. sessilis* should be recognized at the species level since the closest relative is *O. fruticosa*, not *O. pilosella* as previously thought. Similarly, *O. riparia* is separated from *O. fruticosa* since it apparently is most closely related to *O. perennis* despite the morphological resemblance to *O. fruticosa*.

SELECTED REFERENCES Krakos, K. N., J. S. Reece, and P. H. Raven. 2014. Molecular phylogenetics and reproductive biology of *Oenothera* section *Kneiffia* (Onagraceae). Syst. Bot. 39: 523–532. Munz, P. A. 1937. Studies in Onagraceae X. The subgenus *Kneiffia*. Bull. Torrey Bot. Club 64: 287–306. Straley, G. B. 1977. Systematics of *Oenothera* sect. *Kneiffia* (Onagraceae). Ann. Missouri Bot. Gard. 64: 381–424.

1. Herbs annual, from a taproot; flowers in leaf axils in distal ½ of plant . . . . 28. *Oenothera spachiana*
1. Herbs perennial, from fibrous roots, sometimes producing rhizomes; flowers in axils of distalmost few nodes of plant.
   2. Stigmas surrounded by anthers at anthesis; petals 5–10 mm; inflorescences nodding; pollen 40–70% fertile . . . . 33. *Oenothera perennis*
   2. Stigmas exserted beyond anthers at anthesis; petals (8–)15–30 mm; inflorescences usually erect, sometimes nodding; pollen 85–100% fertile.
      3. Plants usually exclusively hirsute, rarely glabrous, from fibrous roots and producing rhizomes . . . . 32. *Oenothera pilosella*
      3. Plants usually strigillose, glandular puberulent, glabrate, or glabrous, sometimes villous, usually with fibrous or fleshy roots, not or rarely producing rhizomes.
         4. Capsules ellipsoid, 8–10 mm; plants densely strigillose, glabrate proximally . . . . .29. *Oenothera sessilis*
         4. Capsules clavate to oblong-clavate or oblong-ellipsoid, (5–)10–17(–20) mm; plants usually strigillose, and/or glandular puberulent, or glabrous, sometimes villous.
            5. Plants usually moderately to densely strigillose throughout, sometimes and/or glandular puberulent or villous; cauline leaf blades 0.1–1(–1.7) cm wide, margins entire or weakly and remotely denticulate. . . 30. *Oenothera fruticosa* (in part)
            5. Plants glabrous or sparsely strigillose, sometimes also glandular puberulent distally; cauline leaf blades (0.2–)0.8–3(–5) cm wide, margins dentate to remotely denticulate.

[6. Shifted to left margin.—Ed.]

6. Roots fibrous; leaf blade margins dentate to remotely denticulate; petioles 0.1–2(–6) cm; inland upland meadows, stream margins, edges of woods . . . . 30. *Oenothera fruticosa* (in part)
6. Roots somewhat fleshy, with adventitious roots where submerged; petioles 0–1.5 cm; leaf blade margins remotely denticulate; coastal marshes, slow-running rivers . . . . 31. *Oenothera riparia*

28. Oenothera spachiana Torrey & A. Gray, Fl. N. Amer. 1: 498. 1840 [E]

Blennoderma drummondii Spach; *Kneiffia spachiana* (Torrey & A. Gray) Small; *Oenothera uncinata* Scheele

Herbs annual, densely strigillose; from a sparsely branched taproot. **Stems** erect, usually unbranched or with few ascending branches, 10–30(–45) cm. **Leaves** in a basal rosette and cauline, basal 2–5 × 0.5–1.5 cm, petiole 0.5–2 cm, blade oblanceolate to elliptic, margins subentire; cauline 3–6 × 0.2–0.6 cm, petiole 0.2–0.6(–1.5) cm, blade narrowly lanceolate to linear, margins subentire. **Inflorescences** erect, flowers in leaf axils in distal ½ of plant. **Flowers** opening near sunrise; buds with free tips to 1 mm, erect to spreading; floral tube 4–10 mm; sepals 4–8 mm; petals pale yellow, fading pale pink, 5–14 mm; filaments 3–7 mm, anthers 2–4 mm, pollen 85–100% fertile; style 3–7 mm, stigma surrounded by anthers. **Capsules** broadly clavate, 4-angled, 5–15 × 3–5 mm, stipe 2–5 mm; sessile. **Seeds** 1 × 0.5 mm. **2*n*** = 14.

Flowering Mar–May. Prairies, open roadsides, sandy places; 10–200 m; Ala., Ark., La., Miss., Okla., Tex.

G. B. Straley (1977) determined *Oenothera spachiana* to be self-compatible and autogamous. Collections outside the native range of *O. spachiana* have been made as a ballast weed in Camden County, New Jersey.

Oenothera drummondii (Spach) Walpers (1843), not Hooker (1834) is a later homonym and pertains here.

29. Oenothera sessilis (Pennell) Munz, Bull. Torrey Bot. Club 64: 291. 1937 [C] [E]

Kneiffia sessilis Pennell, Bull. Torrey Bot. Club 46: 366. 1919; *Oenothera pilosella* Rafinesque subsp. *sessilis* (Pennell) Straley; *O. pilosella* var. *sessilis* (Pennell) B. L. Turner

Herbs perennial, densely strigillose, glabrate proximally; from fibrous roots. **Stems** ascending, unbranched to few-branched, 30–65 cm. **Leaves** in a basal rosette and cauline, basal 2.5–7 × 0.7–2.3 cm, petiole 1–1.5 cm, blade oblanceolate, margins subentire, undulate; cauline (3–)6–7(–9) × (0.3–)0.6–0.8(–1.1) cm, sessile, blade lanceolate to narrowly lanceolate, margins subentire. **Inflorescences** nodding, flowers in axils of distalmost few nodes. **Flowers** opening near sunrise; buds with free tips 1–2 mm, connivent to spreading; floral tube 10–15(–20) mm; sepals 10–18 mm; petals bright yellow, fading pale pink, 15–25 mm; filaments 7–9 mm, anthers 5–8 mm, pollen 85–100% fertile; style 10–12 mm, stigma exserted beyond anthers at anthesis. **Capsules** ellipsoid, 4-angled, 8–10 × 3–4 mm, stipe 0–2 mm; sessile. **Seeds** 1 × 0.5 mm. **2*n*** = 56.

Flowering May–Jun. Moist remnant prairies in sandy or silty soil; of conservation concern; 0–100 m; Ark., La., Tex.

Oenothera sessilis is relatively rare within its range and has a narrow overall distribution, occurring in Ashley, Phillips, Prairie, and St. Francis counties in Arkansas, Allen, Claiborne, and Tensas parishes in Louisiana, and Galveston County in Texas. *Oenothera sessilis* appears to be relatively rare and may no longer occur in Texas; it was last collected there in the 1840s by Lindheimer on Galveston Island. It is also rare in Louisiana but has been collected in recent decades. It is currently most common in Arkansas. P. A. Munz (1937, 1965) treated this taxon as *O. sessilis*, but G. B. Straley (1977) in his revision of sect. *Kneiffia* placed it as a subspecies of *O. pilosella* based on a common octoploid (2*n* = 56) chromosome number, morphology, and field studies. K. N. Krakos et al. (2014), based on new field studies, controlled greenhouse breeding experiments, and phylogenetic data found that this taxon differs morphologically from *O. pilosella* by having consistently shorter stature and smaller flowers, is self-compatible, and does not form a monophyletic group with *O. pilosella* in molecular analyses, and is here reinstated as a distinct species.

Oenothera sessilis is in the Center for Plant Conservation's National Collection of Endangered Plants as *O. pilosella* subsp. *sessilis*.

30. Oenothera fruticosa Linnaeus, Sp. Pl. 1: 346. 1753 [E]

Kneiffia fruticosa (Linnaeus) Spach ex Raimann

Herbs perennial, moderately to densely strigillose and/or villous, glandular puberulent, or glabrous; from fibrous roots, not or rarely producing rhizomes. **Stems** erect to decumbent, branched or unbranched, (10–)30–80 cm. **Leaves** in a basal rosette and cauline, rosette usually withered by anthesis, surfaces sometimes glaucous, especially abaxially, basal 3–12 × 0.5–3 cm, petiole 1–4 cm, blade oblanceolate to obovate, margins subentire, dentate, or denticulate, sometimes undulate; cauline 2–6(–11) × (0.1–)0.5–2(–5) cm, petiole 0.1–2 (–6) cm, blade linear, lanceolate to oblanceolate, narrowly elliptic, or ovate, margins subentire or dentate or denticulate, sometimes undulate. **Inflorescences** usually erect, rarely nodding, flowers in axils of distalmost few nodes. **Flowers** opening near sunrise; buds with free tips 0.5–8(–13) mm, connivent, sometimes spreading; floral tube 5–20 mm; sepals 5–20 mm; petals pale to bright yellow, fading pale pink, orangish pink, or yellow, (8–)15–25(–30) mm; filaments 5–15 mm, anthers 4–7 mm, pollen 85–100% fertile; style 12–20 mm, stigma exserted beyond anthers at anthesis. **Capsules** clavate to oblong-clavate or oblong-ellipsoid, 4-angled to 4-winged, (5–)10–17(–20) × (2–)3–4(–6) mm, stipe 0.1–10 mm; sessile. **Seeds** 1 × 0.5 mm.

Subspecies 2 (2 in the flora): c, e North America.

Oenothera fruticosa as delimited here is a polymorphic species. Previous classification of this group has undergone numerous reorganizations due to the difficulties in separating populations into discrete morphological taxa. In the past it has most frequently been treated as two species, *O. fruticosa* and *O. tetragona* Roth, often with a dozen or more infraspecific taxa recognized. The broad delimitation of G. B. Straley (1977) is followed here with one species consisting of two subspecies that appear to intergrade extensively across a wide area of overlap. Straley determined *O. fruticosa* to be self-incompatible and polyploid.

1. Capsules clavate to oblong-clavate, widest above middle; plants moderately to densely strigillose, and/or sometimes glandular puberulent or villous; leaf blade margins subentire or weakly and remotely denticulate . . . . 30a. *Oenothera fruticosa* subsp. *fruticosa*
1. Capsules oblong to oblong-ellipsoid, widest at middle; plants glandular puberulent or glabrous, sometimes villous or strigillose; leaf blade margins dentate to remotely denticulate . . . . 30b. *Oenothera fruticosa* subsp. *tetragona*

30a. Oenothera fruticosa Linnaeus subsp. fruticosa [E]

Kneiffia allenii Small; *K. arenicola* Small; *K. brevistipata* Pennell; *K. charlesii* Lahman; *K. fruticosa* (Linnaeus) Spach ex Raimann var. *humifusa* (Allen) Pennell; *K. fruticosa* var. *unguiculata* (Fernald) Moldenke; *K. linearis* (Michaux) Spach; *K. longipedicellata* Small; *K. semiglandulosa* Pennell; *K. subglobosa* Small; *K. tetragona* (Roth) Pennell var. *longistipata* Pennell; *K. velutina* Pennell; *O. fruticosa* var. *eamesii* (B. L. Robinson) S. F. Blake; *O. fruticosa* var. *goodmanii* Munz; *O. fruticosa* var. *humifusa* Allen; *O. fruticosa* var. *linearifolia* Hooker; *O. fruticosa* var. *linearis* (Michaux) S. Watson; *O. fruticosa* var. *microcarpa* Fernald; *O. fruticosa* var. *subglobosa* (Small) Munz; *O. fruticosa* var. *unguiculata* Fernald; *O. fruticosa* var. *vera* Hooker; *O. linearis* Michaux; *O. linearis* var. *eamesii* B. L. Robinson; *O. longipedicellata* (Small) B. L. Robinson; *O. subglobosa* (Small) Weatherby & Griscom; *O. subglobosa* var. *arenicola* (Small) Weatherby & Griscom; *O. tetragona* Roth var. *brevistipata* (Pennell) Munz; *O. tetragona* var. *longistipata* (Pennell) Munz; *O. tetragona* var. *sharpii* Munz; *O. tetragona* var. *velutina* (Pennell) Munz; *O. unguiculata* (Fernald) Sorrie, LaBlond & Weakley

Herbs moderately to densely strigillose, and/or sometimes glandular puberulent or villous; from fibrous or, sometimes, slightly fleshy rootstock or producing rhizomes. **Leaves:** basal 3–10 × 0.5–2 cm, petiole 1–4 cm, blade oblanceolate to obovate, margins subentire or weakly and remotely denticulate; cauline 2–6(–8) × 0.1–1(–1.7) cm, petiole 0.2–2(–4) cm, blade linear, lanceolate to oblanceolate, narrowly elliptic, or narrowly ovate, margins subentire or weakly and remotely denticulate. **Flowers:** buds with free tips 0.5–1(–6) mm, usually connivent, sometimes spreading; floral tube 5–15 mm; sepals 5–20 mm; petals (8–)15–25 mm. **Capsules** clavate to oblong-clavate, widest distal to middle, angled, rarely narrowly winged, (5–)10–17(–20) × (2–)3–4 mm, stipe 3–10 mm; sessile. $2n$ = 28, 42.

Flowering Apr–Aug(–Sep). Open habitats, meadows, stream margins, edges of woods, open woods, margins of freshwater or saltwater marshes, stabilized sand dunes, roadsides, partially disturbed habitats; 0–500(–1100) m; Ala., Ark., Conn., Del., D.C., Fla., Ga., Ill., Ind., Ky., La., Md., Mass., Miss., Mo., N.Y., N.C., Ohio, Okla., Pa., S.C., Tenn., Va., W.Va.

Kneiffia suffruticosa Spach (1835) is an illegitimate substitution for *Oenothera fruticosa*, as is *K. angustifolia* Spach (1835) one for *O. linearis* Michaux, and *O. linearis* var. *allenii* Britton (1894) is one based on *O. fruticosa* var. *humifusa*.

30b. Oenothera fruticosa Linnaeus subsp. tetragona (Roth) W. L. Wagner, PhytoKeys 34: 16. 2014 [E]

Oenothera tetragona Roth, Catal. Bot. 2: 39. 1800; *Kneiffia fraseri* (Pursh) Spach; *K. fruticosa* (Linnaeus) Spach ex Raimann var. *differta* (Millspaugh) Millspaugh; *K. glauca* (Michaux) Spach; *K. latifolia* Rydberg; *K. serotina* (Lehmann) Bartling; *K. tetragona* (Roth) Pennell; *K. tetragona* var. *hybrida* (Michaux) Pennell; *O. ambigua* (Nuttall) Sprengel; *O. canadensis* Goldie; *O. fraseri* Pursh; *O. fruticosa* var. *ambigua* Nuttall; *O. fruticosa* var. *differta* Millspaugh; *O. fruticosa* var. *fraseri* (Pursh) Hooker; *O. fruticosa* subsp.

glauca (Michaux) Straley; *O. fruticosa* var. *glauca* (Michaux) H. Léveillé; *O. fruticosa* var. *incana* (Nuttall) Hooker; *O. fruticosa* var. *maculata* H. Léveillé; *O. glauca* Michaux; *O. glauca* var. *fraseri* (Pursh) Torrey & A. Gray; *O. hybrida* Michaux; *O. hybrida* var. *ambigua* (Nuttall) S. F. Blake; *O. incana* Nuttall; *O. serotina* Lehmann; *O. tetragona* var. *fraseri* (Pursh) Munz; *O. tetragona* subsp. *glauca* (Michaux) Munz; *O. tetragona* var. *hybrida* (Michaux) Fernald; *O. tetragona* var. *latifolia* (Rydberg) Fernald

Herbs mostly glandular puberulent or glabrous, sometimes strigillose or villous on stems; from fibrous rootstock. **Leaves** in a basal rosette and cauline, basal 3–12 × 0.5–3 cm, petiole 1–3 cm, blade oblanceolate to obovate, margins dentate to remotely denticulate; cauline 2–6(–11) × (0.5–)1–2(–5) cm, sometimes glaucous; petiole 0.1–2(–6) cm; blade narrowly elliptic to broadly ovate, margins dentate to remotely denticulate. **Flowers:** buds with free tips (3–)4–8(–13) mm, connivent to spreading; floral tube 5–20 mm; sepals 8–22 mm; petals (8–)15–20(–30) mm. **Capsules** oblong to oblong-ellipsoid, widest at middle, (5–)10–17(–20) × (2–)3–4(–6) mm, 4-winged, sometimes 4-angled, stipe 0.1–3(–7) mm; sessile. **2*n*** = 28.

Flowering Apr–Aug(–Sep). Open meadows, stream margins, edges of woods, open woods; (100–)300–1700 m; N.S., Ont., Que.; Ala., Conn., Del., D.C., Ga., Ill., Ind., Ky., La., Maine, Md., Mass., Mich., Mo., N.H., N.Y., N.C., Ohio, Pa., R.I., S.C., Tenn., Va., W.Va.

Oenothera serotina Sweet (1826) is a homonym, not Lehmann (1826), while *Kneiffia floribunda* Spach (1835) is an illegitimate substitution based on *O. tetragona*, and both names pertain here.

31. Oenothera riparia Nuttall, Gen. N. Amer. Pl. 1: 247. 1818 [C] [E]

Kneiffia riparia (Nuttall) Small; *Oenothera tetragona* Roth var. *riparia* (Nuttall) Munz

Herbs perennial, sparsely strigillose, becoming glabrate distally, usually also glandular puberulent distally; from a somewhat fleshy rootstock, forming adventitious roots where submerged. **Stems** erect or ascending, usually many-branched throughout, proximal branches often somewhat spongy, 50–120 cm. **Leaves** in a basal rosette and cauline, softly succulent, 4–13.5 × 0.8–2.1 cm, petiole 0–1.5 cm, blade lanceolate to elliptic-lanceolate, margins remotely denticulate. **Inflorescences** erect, flowers in axils of distalmost few nodes. **Flowers** opening near sunrise; buds with free tips 0.5–2(–5) mm, erect to spreading; floral tube 14–20 mm; sepals 20–30 mm; petals bright yellow, fading pale pink or lavender, 16–27 mm; filaments 10–15 mm, anthers 7–8 mm, pollen 85–100% fertile; style 15–30 mm, stigma exserted beyond anthers at anthesis. **Capsules** oblong-clavate to oblong-ellipsoid, 4-angled or 4-winged, and wings 0.1–0.2 mm, 7–15 × 4–6 mm, stipe 2–5 mm; sessile. **Seeds** 0.8 × 0.4 mm. **2*n*** = 56.

Flowering May–Jul. Isolated colonies in or at edge of water in marshes or slow-running rivers, apparently with at least some tidal influence; of conservation concern; 0–10 m; Ga., N.C., S.C., Va.

Oenothera riparia has not been accepted as a distinct species for a long time. The conservative approach used by G. B. Straley (1977) is here followed in recognizing a broadly delimited *O. fruticosa*, except that the very distinctive coastal tidal-freshwater semi-aquatic octoploid populations (Straley 1982) are recognized as *O. riparia*. Plants of these coastal populations, which occur from southern Virginia to North Carolina, are more robust, more branched, and less pubescent than those of the two subspecies of *O. fruticosa*, and have slightly succulent leaves and more prominent adventitious roots (Straley 1982; D. Boufford, pers. comm.). Further field studies and cytological analyses are needed to fully understand the geographical and ecological limits of *O. riparia* and confirm that it is strictly an octoploid species. K. N. Krakos et al. (2014) determined that *Oenothera riparia* is self-incompatible and is pollinated by bees (*Bombus*, *Lassioglossum*, *Megachile*, *Parallelia*, *Xylocopa*, and *Zale*).

32. Oenothera pilosella Rafinesque, Ann. Nat. 1: 15. 1820 [E] [F]

Kneiffia fruticosa (Linnaeus) Spach ex Raimann var. *pilosella* (Rafinesque) Britton; *K. pilosella* (Rafinesque) A. Heller; *K. pratensis* Small; *K. sumstinei* Jennings; *Oenothera fruticosa* Linnaeus var. *hirsuta* Nuttall ex Torrey & A. Gray; *O. fruticosa* var. *pilosella* (Rafinesque) Small & A. Heller; *O. pratensis* (Small) B. L. Robinson

Herbs perennial, usually densely to sparsely hirsute, rarely glabrous; from a thickened base, rhizomatous. **Stems** spreading or ascending, unbranched or few-branched distally, 20–80 cm. **Leaves** in a basal rosette and cauline, basal 4–8 × 2–5 cm, petiole (0.5–)1–3 (–4) cm, blade oblanceolate to ovate, margins entire; cauline 3–10(–13) × 1–2(–4) cm, petiole 0–0.5(–2) cm, blade lanceolate to ovate, abruptly narrowed to base, margins subentire or coarsely dentate. **Inflorescences** erect, flowers in axils of distalmost few nodes. **Flowers** opening near sunrise; buds with free tips 1–3 mm, spreading; floral tube 10–25 mm; sepals 10–20 mm; petals bright yellow, fading pale pink or pale yellow, 15–30 mm; filaments 7–15 mm, anthers 4–8 mm, pollen

85–100% fertile; style 10–20 mm, stigma exserted beyond anthers at anthesis. **Capsules** usually oblong-clavate to oblong-ellipsoid or ellipsoid, 4-angled or weakly 4-winged, (5–)10–15(–28) × 2–4(–5) mm, stipe (1–)3–5(–9) mm; sessile. **Seeds** 1 × 0.5 mm. **2*n*** = 56.

Flowering May–Jul(–Aug). Open fields, edge of woods, marshes and bottomland prairies, open disturbed sites, ditches, old fields, railroads, roadsides; 100–600 m; Ont.; Ala., Ark., Conn., Ill., Ind., Iowa, Ky., La., Maine, Mass., Mich., Miss., Mo., N.H., N.Y., Ohio, Okla., Pa., R.I., Vt., Va., W.Va., Wis.

Oenothera pilosella is widespread in cultivation in gardens and frequently escapes and becomes naturalized; the northern and eastern natural limits of *O. pilosella* are not clear. According to G. B. Straley (1977) the natural limits are from Wayne County, West Virginia, along the Ohio River and Erie County, New York, for the eastern limits, and Tuscola County, Michigan, and Manitowoc County, Wisconsin, for the northern limits. K. N. Krakos (2014), based on new field studies and phylogenetic data, found that *O. pilosella* does not form a monophyletic group with plants previously treated by Straley as *O. pilosella* subsp. *sessilis* in molecular analyses, and thus is here reinstated as the distinct species *O. sessilis*. Straley determined that *O. pilosella* is self-incompatible and an octoploid, one of the few in the genus.

33. **Oenothera perennis** Linnaeus, Syst. Nat. ed. 10, 2: 998. 1759 [E] [W]

Kneiffia chrysantha (Michaux) Spach; *K. depauperata* Jennings; *K. perennis* (Linnaeus) Pennell; *Oenothera chrysantha* Michaux; *O. perennis* var. *rectipilis* (S. F. Blake) S. F. Blake; *O. pumila* Linnaeus var. *chrysantha* (Michaux) Gordinier & Howe; *O. pumila* var. *minima* Lehmann; *O. pumila* var. *pusilla* (Michaux) Torrey & A. Gray; *O. pumila* var. *rectipilis* S. F. Blake; *O. pusilla* Michaux

Herbs perennial, sparsely to moderately strigillose, glandular puberulent distally; from fibrous roots. **Stems** usually erect to slightly decumbent, unbranched to few-branched distally, (3–)15–30(–75) cm. **Leaves** in a basal rosette and cauline, basal 2–4 × 0.2–1.2 cm, petiole (0.2–)0.5–1.2(–2.5) cm, blade oblanceolate to obovate; cauline 3–7 × 0.2–1.2 cm, petiole 0.1–1 cm, blade oblanceolate to obovate, margins entire or weakly and remotely denticulate. **Inflorescences** nodding, flowers in axils of distalmost few nodes. **Flowers** opening near sunrise, nearly unscented; buds with free tips to 1 mm, connivent; floral tube 3–10 mm; sepals 2–4 mm; petals bright yellow, fading pale yellow, or orangish yellow to pale pink, 5–10 mm; filaments 3–4 mm, anthers 1–2 mm, pollen 40–70% fertile; style 3–4 mm, stigma surrounded by anthers at anthesis. **Capsules** clavate, 4-angled or narrowly 4-winged, 5–10 × 2–3 mm, stipe 1–2 mm; sessile. **Seeds** 0.7–0.8 × 0.2–0.3 mm. **2*n*** = 14.

Flowering May–Jul(–Aug). Fields, open woods, boggy areas; (0–)150–900(–1400) m; St. Pierre and Miquelon; B.C., Man., N.B., Nfld. and Labr. (Nfld.), N.S., Ont., P.E.I., Que.; Conn., Del., D.C., Ill., Ind., Iowa, Ky., Maine, Md., Mass., Mich., Minn., Mo., Nebr., N.H., N.Y., N.C., Ohio, Pa., R.I., S.C., Tenn., Vt., Va., W.Va., Wis.

Oenothera perennis, a PTH species that forms a ring of 14 chromosomes in meiosis, is self-compatible and autogamous (G. B. Straley 1977). It is disjunct in Nebraska from the rest of its range in eastern North America, occurring in Garfield, Holt, and Rock counties (R. Kaul, pers. comm.). It is introduced in British Columbia.

Oenothera pumila Linnaeus is an illegitimate substitution based on *O. perennis* Linnaeus, while *Kneiffia michauxii* Spach is an illegitimate substitution based *O. pumila*, as is *K. pumila* Spach, and the three pertain here.

17k. OENOTHERA Linnaeus sect. GAURA (Linnaeus) W. L. Wagner & Hoch, Syst. Bot. Monogr. 83: 165. 2007

Gaura Linnaeus, Sp. Pl. 1: 347. 1753

Herbs annual, biennial, or perennial, caulescent; from a taproot, sometimes woody or producing rhizomes. **Stems** usually erect or ascending, sometimes decumbent, branched or unbranched. **Leaves** in a basal rosette and cauline (sometimes not present at flowering), (0.5–)2–8(–13) cm; blade margins sinuate-dentate to denticulate, serrate, lobed, or entire. **Inflorescences** solitary flowers in axils of distal leaves, forming a spike, erect or nodding. **Flowers** opening near sunset or sunrise; buds erect, terete, without free tips; floral tube 1.5–20[–42] mm, usually lanate in distal ½ within; sepals splitting along one suture, remaining coherent and reflexed as a unit at anthesis, or separating in pairs or sometimes individually; petals white [rarely yellow], fading pink to red, purple, or off-white, spatulate to elliptic, rhombic, or, sometimes, oblanceolate, usually clawed; filaments with basal scale 0.3–0.5 mm, these nearly closing mouth of floral tube, or sometimes reduced or absent; stigma deeply divided into (3 or) 4 linear lobes. **Capsules** woody and nutlike, ovoid, fusiform, lanceoloid, ellipsoid, obovoid, or pyramidal, (3- or)4-angled, sometimes weakly so, or (3- or)4-winged, apex acute to attenuate or, sometimes, rounded, indehiscent, septa fragile, not evident at maturity; sessile, sometimes disarticulating from plant at maturity. **Seeds** reduced to 1–4(–8), usually ovoid, rarely oblanceoloid (*O. glaucifolia*), surface smooth. $2n$ = 14, 28, 42, 56.

Species 25 (23 in the flora): North America, Mexico, Central America (Guatemala); introduced in South America, Europe, Asia, s Africa, Australia.

Section *Gaura* consists of 25 species (26 taxa) that are subdivided into eight subsections [seven in the flora area; subsect. *Gauridium* (Spach) W. L. Wagner & Hoch is is found only in Mexico]. All species have indehiscent capsules, a feature otherwise found in *Oenothera* only in *O. canescens* (sect. *Gauropsis*). *Oenothera havardii* (sect. *Paradoxus*) and *O. linifolia* (sect. *Peniophyllum*) have tardily and only partially dehiscent capsules. Twenty-one species in four subsections (*Campogaura*, *Gaura*, *Stipogaura*, and *Xenogaura*) have zygomorphic flowers; the other four subsections (*Gauridium*, *Schizocarya*, *Stenosiphon*, and *Xerogaura*), each with a single species, have actinomorphic, or nearly actinomorphic, flowers. Since Linnaeus described *Gaura*, it has been maintained as distinct at the generic level and, at various times, even at the tribal level. Its distinct status rested on several characteristic features including a scale at the base of the filaments; a peltate indusium at the base of the stigma; indehiscent, nutlike capsules; and seeds reduced to 1–4(–8) (P. H. Raven and D. P. Gregory 1972[1973]; W. L. Wagner et al. 2007). The most recent molecular studies (G. D. Hoggard et al. 2004; R. A. Levin et al. 2004) place *Gaura* strongly within the *Oenothera* clade and equally strongly in a clade with other sections possessing winged/angled capsules that are sometimes indehiscent or nearly so. Hoggard et al. found strong support for the inclusion of the monotypic *Stenosiphon* within *Gaura*; Levin et al. concurred, finding strong support for the monophyly of the *Gaura* lineage, but placed it unequivocally within *Oenothera*. Reconsidering the distinctive features of *Gaura*, Wagner et al. found that the indusium characterizes the whole genus *Oenothera*, and the indehiscent fruits seem to characterize a larger clade in the genus. The reduction in seed number appears to be a strong synapomorphy for the reconstituted *Oenothera* sect. *Gaura*. Wagner et al. recognized eight subsections within sect. *Gaura* that also includes *Stenosiphon*. The subsections are arranged according to the synthesis of morphological characters, crossing analyses, and molecular data (Raven and Gregory; Wagner et al.). *Oenothera anomala* Curtis and *O. hexandra* (Ortega) W. L. Wagner & Hoch are Mexican species that occur well south of the flora area.

SELECTED REFERENCES Hoggard, G. D. et al. 2004. The phylogeny of *Gaura* (Onagraceae) based on ITS, ETS and *trn*L-F sequence data. Amer. J. Bot. 91: 139–148. Raven, P. H. and D. P. Gregory. 1972[1973]. A revision of the genus *Gaura* (Onagraceae). Mem. Torrey Bot. Club. 23(1): 1–96.

1. Filaments without basal scales or with minute scale; flowers nearly actinomorphic.
    2. Floral tubes 6–17 mm; herbs probably biennial, glaucous at least in proximal part . . . . 17k.1. *Oenothera* subsect. *Stenosiphon*, p. 285
    2. Floral tubes 1.5–5 mm; herbs annual, not glaucous. . . . . . 17k.2. *Oenothera* subsect. *Schizocarya*, p. 286
1. Filaments with basal scales; flowers zygomorphic or sometimes nearly actinomorphic.
    3. Capsules with a slender stipe (0.5–)2–10 mm . . . . 17k.5. *Oenothera* subsect. *Stipogaura*, p. 289
    3. Capsules usually with a stipe 0.2–2.2 mm.

[4. Shifted to left margin.—Ed]

4. Capsules pyramidal in distal ½, abruptly constricted to a cylindrical proximal part.
   5. Capsules not conspicuously bulging at base of distal pyramidal ½; plants not rhizomatous . . . . .17k.4. *Oenothera* subsect. *Campogaura* (in part), p. 288
   5. Capsule conspicuously bulging at base of the distal pyramidal ½; plants rhizomatous . . . . 17k.6. *Oenothera* subsect. *Xenogaura*, p. 293
4. Capsules fusiform, ellipsoid, ovoid, or obovoid and then abruptly constricted or cuneate to base.
   6. Capsules ellipsoid, ovoid, or obovoid . . . . .17k.7. *Oenothera* subsect. *Gaura*, p. 294
   6. Capsules fusiform.
      7. Floral tubes 9–13 mm; flowers nearly actinomorphic . . 17k.3. *Oenothera* subsect. *Xerogaura*, p. 287
      7. Floral tubes 3–11(–13) mm; flowers zygomorphic . . . . .17k.4. *Oenothera* subsect. *Campogaura* (in part), p. 288

17k.1. OENOTHERA Linnaeus (sect. GAURA) subsect. STENOSIPHON (Spach) W. L. Wagner & Hoch, Syst. Bot. Monogr. 83: 167. 2007

Stenosiphon Spach, Ann. Sci. Nat., Bot., sér. 2, 4: 170. 1835

Herbs probably biennial; from stout roots. **Stems** erect, branched or unbranched. **Inflorescences** wandlike, nodding. **Flowers** 4-merous, nearly actinomorphic, opening near sunrise; floral tube 6–17 mm; petals white, slightly unequal; filaments without basal scales. **Capsules** erect in age, ovoid, somewhat flattened, 4-angled, valves with raised midrib and conspicuous lateral veins; sessile. $2n = 14$.

Species 1: c United States.

34. Oenothera glaucifolia W. L. Wagner & Hoch, Syst. Bot. Monogr. 83: 212. 2007 [E] [F]

Gaura linifolia Nuttall ex E. James, Account Exped. Pittsburgh 2: 100. 1822, not *Oenothera linifolia* Nuttall 1821; *Stenosiphon linifolius* (Nuttall ex E. James) Heynhold

Herbs probably biennial, glabrous, becoming sparsely to densely glandular puberulent and short-villous distally, glaucous at least in proximal parts; from stout roots. **Stems** erect, branched or unbranched, 30–300 cm. **Leaves** in a basal rosette and cauline, basal 3–7 × 0.5–2 cm, sessile, blade oblong to oblong-lanceolate, base usually ± auriculate, margins entire; cauline 3–8(–10) × 0.4–1.8 cm, blade lanceolate to oblong-lanceolate, gradually smaller, becoming linear-subulate distally. **Inflorescences** long, wandlike, unbranched or branched. **Flowers** 4-merous, nearly actinomorphic, opening near sunrise; floral tube 6–17 mm; sepals 4–6 mm; petals white, fading off-white or tinged pink, slightly unequal, rhombic, 4–6 mm, abruptly clawed; filaments 5–8 mm, anthers 1.5–2 mm, pollen 85–100% fertile; style 6–10, stigma exserted beyond anthers at anthesis. **Capsules** ovoid, 4-angled, somewhat flattened, 3–4 × 1.5–2.3 mm, valves with raised midrib and conspicuous lateral veins; sessile. **Seeds** 1, pale yellow, oblanceoloid, 2.4–2.6 × 1–1.5 mm. $2n = 14$.

Flowering May–Oct(–Nov). Rocky prairie slopes and outcrops or bluffs, along streams, roadsides, usually on limestone; 200–1300 m; Ark., Colo., Kans., Mo., Nebr., N.Mex., Okla., Tex., Wyo.

Oenothera glaucifolia is self-incompatible, the flowers diurnal, pollinated primarily by wasps (R. Clinebell, unpubl.), as well as bees, flies, butterflies, and occasionally beetles (summarized by W. L. Wagner et al. 2007). It was collected once in 1988 in Indiana at Miller Woods Visitor Center (Lake County), *Dritz 596* (MOR); it seems likely that it was introduced, and has not been collected there since.

Stenosiphon virgatus Spach is a superfluous name and pertains here.

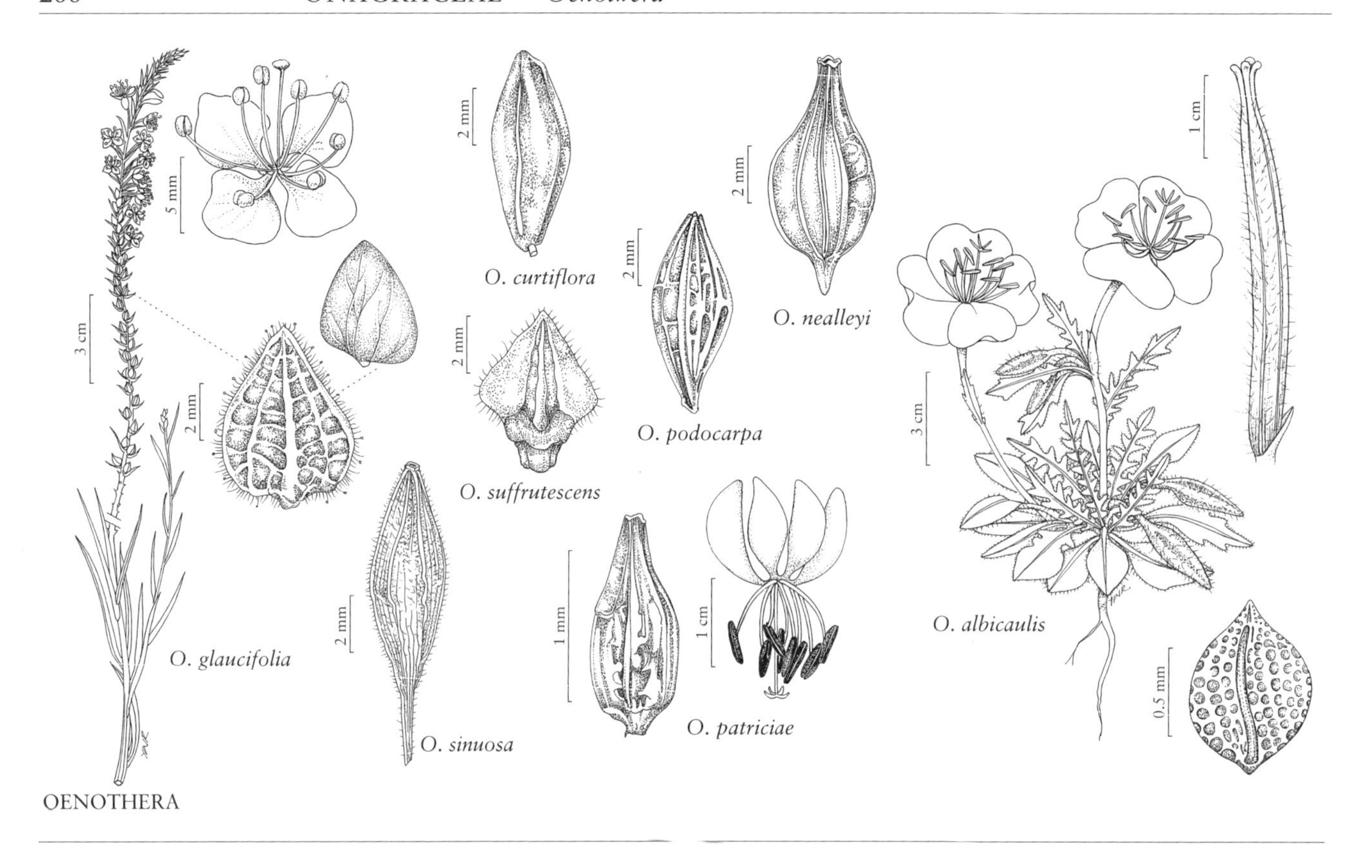

OENOTHERA

17k.2. Oenothera Linnaeus (sect. Gaura) subsect. Schizocarya (Spach) W. L. Wagner & Hoch, Syst. Bot. Monogr. 83: 168. 2007

Schizocarya Spach, Ann. Sci. Nat., Bot., sér. 2, 4: 170. 1835; *Gaura* [unranked] *Schizocarya* (Spach) Endlicher; *Gaura* sect. *Schizocarya* (Spach) P. H. Raven & D. P. Gregory

Herbs annual; from enlarged taproot. **Stems** usually unbranched, sometimes branched. **Inflorescences** slender, nodding. **Flowers** 4-merous, nearly actinomorphic, opening near sunset; floral tube 1.5–5 mm; petals white, slightly unequal; filaments with minute basal scale. **Capsules** reflexed in age, fusiform, terete, weakly 4-angled in distal ⅓, angles becoming broad and rounded in proximal part, tapering abruptly toward base; sessile. **2*n*** = 14.

Species 1: w, c United States, n, c Mexico; introduced in s South America, Asia (China, Japan), Australia.

35. Oenothera curtiflora W. L. Wagner & Hoch, Syst. Bot. Monogr. 83: 211. 2007 F

Gaura parviflora Douglas ex Lehmann, Nov. Stirp. Pug. 2: 15. 1830, not *Oenothera parviflora* Linnaeus 1759; *G. australis* Grisebach; *G. hirsuta* Scheele; *G. micrantha* (Spach) D. Dietrich; *G. parviflora* var. *lachnocarpa* Weatherby; *Schizocarya micrantha* Spach 1835, not *O. micrantha* Hornemann ex Sprengel 1825

Herbs annual, strigillose, glandular puberulent, and long-villous; from heavy taproot, 2–4 cm diam. **Stems** erect, unbranched or many-branched distally, (20–) 30–200(–300) cm. **Leaves** in a basal rosette and cauline, basal 4–15 × 1.5–3 cm, petiole 0–1.8 cm, blade broadly oblanceolate, margins sinuate-dentate to dentate; cauline 2–13 × 0.5–5 cm, petiole 0–2 cm, blade narrowly elliptic to narrowly ovate, margins sinuate-dentate to dentate. **Inflorescences** relatively long, dense. **Flowers** 4-merous, nearly actinomorphic, opening near sunset; floral tube 1.5–5 mm; sepals 2–3.5 mm; petals white, fading pale to dark pink, slightly unequal, oblong-obovate to elliptic-oblanceolate, 1.5–3 mm, abruptly clawed; filaments 1.5–3 mm, anthers 0.5–1 mm, pollen 85–100% fertile; style 3–9 mm, stigma surrounded by anthers at anthesis. **Capsules** fusiform, terete, weakly angled in distal ⅓, angles becoming broad and rounded in proximal part, 5–11 × 1.5–3 mm, tapering abruptly toward base; sessile. **Seeds** 3 or 4, reddish brown, 2–3 × 1–1.5 mm. **2*n*** = 14.

Flowering (Feb–)Apr–Oct. Rocky prairie slopes, woodlands, along streams, roadsides, disturbed areas; 10–2800 m; Ala., Ariz., Ark., Calif., Colo., Fla., Ga., Idaho, Ill., Ind., Iowa, Kans., Ky., La., Md., Mass., Minn., Mo., Mont., Nebr., Nev., N.Mex., Okla., Oreg., S.C., S.Dak., Tenn., Tex., Utah, Va., Wash., Wyo.; Mexico (Baja California, Chihuahua, Coahuila, Durango, Nuevo León, Sinaloa, Zacatecas); introduced in South America (Argentina), Asia (China, Japan), Australia.

Oenothera curtiflora is self-compatible and autogamous (P. H. Raven and D. P. Gregory 1972[1973]). Sometimes it is apparently a biennial. The species is native to grassland regions and open areas across much of interior North America. The full extent of its indigenous range is not clear and collections from the eastern half of the United States (Alabama, Florida, Georgia, Indiana, Massachusetts, and Tennessee) and California may be more recent introductions. *Gaura mollis* Nuttall ex Torrey 1827 is an isonym of *G. mollis* E. James 1822, a suppressed name.

17k.3. OENOTHERA Linnaeus (sect. GAURA) subsect. XEROGAURA (P. H. Raven & D. P. Gregory) W. L. Wagner & Hoch, Syst. Bot. Monogr. 83: 168. 2007

Gaura Linnaeus sect. *Xerogaura* P. H. Raven & D. P. Gregory, Mem. Torrey Bot. Club 23(1): 19. 1973

Herbs perennial, clumped; from stout roots. **Stems** usually branched from base, sometimes branched distally. **Inflorescences** slender, nodding. **Flowers** 4-merous, nearly actinomorphic, opening near sunset; floral tube 9–13 mm; petals white, slightly unequal; filaments with basal scales. **Capsules** erect, fusiform, often slightly curved, weakly 4-angled, valve midrib inconspicuous and narrow; sessile. **2*n*** = 14.

Species 1: Texas, n Mexico.

Subsection *Xerogaura* consists of a single diploid (2*n* = 14) species with a restricted distribution from two disjunct areas: Davis Mountains of west Texas, and Chihuahua, Mexico, near Chihuahua and Gallego. The white, nearly actinomorphic flowers and elongate capsules of *Oenothera arida* were considered plesiomorphic features by P. H. Raven and D. P. Gregory (1972[1973]). In molecular phylogenetic (G. D. Hoggard et al. 2004; R. A. Levin et al. 2004) analyses, *O. arida* did not group with *O. boquillensis*, the other species placed in subsect. *Xerogaura* by Raven and Gregory; instead *O. boquillensis* paired with *O. suffrutescens* in subsect. *Campogaura* (Hoggard et al.) and was transferred to subsect. *Campogaura* by Wagner et al.

36. Oenothera arida W. L. Wagner & Hoch, Syst. Bot. Monogr. 83: 211. 2007

Gaura macrocarpa Rothrock, Proc. Amer. Acad. Arts 6: 353. 1864, not *Oenothera macrocarpa* Nuttall 1813

Herbs perennial, clumped, strigillose and glandular puberulent throughout, also sparsely villous; from stout roots. **Stems** erect, usually branched several cm belowground or from near base, sometimes also branched distally, 20–60(–100) cm. **Leaves** in a basal rosette and cauline, basal 2–4 × 0.4–0.8 cm, petiole 0–0.4 cm, blade narrowly oblanceolate to narrowly spatulate; cauline 0.5–5 × 0.1–0.8 cm, petiole 0–0.3 cm, blade narrowly lanceolate or very narrowly elliptic, margins subentire or sinuate-denticulate. **Flowers** 4-merous, nearly actinomorphic, opening near sunset; floral tube 9–13 mm; sepals 7–9 mm; petals white, fading pink to pale red, slightly unequal, rhombic, 7–8 mm, short-clawed; filaments 3–4 mm, anthers 4–5 mm, pollen 85–100% fertile; style 18–22 mm, stigma exserted beyond anthers at anthesis. **Capsules** erect, fusiform, often slightly curved, weakly 4-angled, (9–)13–17 × 2–3 mm, valves with inconspicuous raised midrib; sessile. **Seeds** (1–)3 or 4, yellowish or light brown, 2–3.5 × 1–2 mm. **2*n*** = 14.

Flowering Apr–Aug. Sandy flats and washes; 1300–1800 m; Tex.; Mexico (Chihuahua).

Oenothera arida is known only from several areas in the foothills of the Davis Mountains in eastern Jeff Davis County, northeastern Presidio County, and northern Brewster County, and from areas near Gallego and Chihuahua in Chihuahua, Mexico. P. H. Raven and D. P. Gregory (1972[1973]) determined *O. arida* to be self-incompatible.

17k.4. OENOTHERA Linnaeus (sect. GAURA) subsect. CAMPOGAURA (P. H. Raven & D. P. Gregory) W. L. Wagner & Hoch, Syst. Bot. Monogr. 83: 169. 2007

Gaura Linnaeus sect. *Campogaura* P. H. Raven & D. P. Gregory, Mem. Torrey Bot. Club 23(1): 27. 1973

Herbs perennial, clumped; from taproot. **Stems** ascending to erect, few-branched to well-branched from near base, sometimes branched distally. **Inflorescences** nodding. **Flowers** 4-merous, slightly to strongly zygomorphic, opening near sunset; floral tube 3–11(–13) mm; petals white, slightly unequal; filaments with basal scales. **Capsules** erect, sometimes reflexed in age, fusiform and inconspicuously ribbed, or distal ½ pyramidal and abruptly constricted to terete proximal part, weakly or strongly angled in distal ½, ribs inconspicuous; sessile. $2n$ = 14, 28, 42, 56.

Species 2 (2 in the flora): w North America, n Mexico.

Subsection *Campogaura* consists of two species, the diploid ($2n$ = 14) *Oenothera boquillensis*, which occurs in washes and sandy slopes in dry mountains near the Rio Grande River in Brewster County, Texas, south to central Chihuahua, Coahuila, and Nuevo León, Mexico, and the diploid and polyploid ($2n$ = 14, 28, 42, 56) *O. suffrutescens* (= *Gaura coccinea*), which occurs widely in western North America from southern British Columbia to Ontario, Canada, southward through the central plains states to the Trans-Volcanic Belt in Oaxaca and Puebla, Mexico.

1. Capsules 5.5–13 mm, fusiform . . . . 37. *Oenothera boquillensis*
1. Capsules 4–9 mm, distal ½ pyramidal and abruptly constricted to terete proximal part, weakly or strongly angled in distal ½. . . . . .38. *Oenothera suffrutescens*

37. Oenothera boquillensis (P. H. Raven & D. P. Gregory) W. L. Wagner & Hoch, Syst. Bot. Monogr. 83: 211. 2007

Gaura boquillensis P. H. Raven & D. P. Gregory, Mem. Torrey Bot. Club 23(1): 21, figs. 3, 33. 1973

Herbs perennial, strigillose and glandular puberulent; from a narrow taproot. **Stems** erect, usually branched several cm belowground or near base, sometimes also branched distally, 25–100 cm. **Leaves** in a basal rosette and cauline, basal 1–6.5(–13) × 0.4–1.5 cm, blade narrowly oblanceolate; cauline 0.5–3(–6.5) × 0.1–1.1 cm, blade very narrowly elliptic, narrowly lanceolate, or linear, margins sinuate-dentate to subentire. **Flowers** 4-merous, slightly zygomorphic, opening near sunset; floral tube 3–8.5 mm; sepals 3–9 mm; petals white, fading pink to red, slightly unequal, elliptic-obovate, 4–10 mm, clawed; filaments 2–4.5 mm, anthers 2–4 mm, pollen 85–100% fertile; style 6.5–15 mm, stigma exserted beyond anthers at anthesis. **Capsules** erect, sometimes reflexed in age, fusiform, sometimes slightly narrowed in proximal ⅓, 5.5–13 × 1–2.5 mm, valves with inconspicuous raised midrib; sessile. **Seeds** (1 or)2–4, yellowish or light brown, 1.5–2.5 × 0.8–1.3 mm. $2n$ = 14.

Flowering May–Aug. Washes, sandy canyon sides; 600–1400 m; Tex.; Mexico (Chihuahua, Coahuila, Nuevo León).

Oenothera boquillensis has a narrow distribution in canyons from near the Rio Grande River in southern Brewster County southward into Mexico from central Chihuahua, Coahuila, and western Nuevo León. P. H. Raven and D. P. Gregory (1972[1973]) determined *O. boquillensis* to be self-incompatible.

38. Oenothera suffrutescens (Seringe) W. L. Wagner & Hoch, Syst. Bot. Monogr. 83: 214. 2007 F

Gaura suffrutescens Seringe in A. P. de Candolle and A. L. P. P. de Candolle, Prodr. 3: 45. 1828; *G. bracteata* Seringe; *G. coccinea* Pursh 1813, not *Oenothera coccinea* Britton 1890; *G. coccinea* var. *arizonica* Munz; *G. coccinea* var. *epilobioides* (Kunth) Munz; *G. coccinea* var. *glabra* (Lehmann) Munz; *G. coccinea* var. *integerrima* Torrey; *G. coccinea* var. *parvifolia* (Torrey) Rickett; *G. epilobioides* Kunth 1823, not *O. epilobioides* Nuttall ex Torrey & A. Gray 1840; *G. glabra* Lehmann; *G. induta* Wooton & Standley; *G. linearis* Wooton & Standley; *G. marginata* Lehmann; *G. multicaulis* Rafinesque; *G. odorata* Sessé ex Lagasca 1816, not *O. odorata* Jacquin 1794; *G. parvifolia* Torrey; *G. spicata* Sessé & Mociño

Herbs perennial, densely strigillose, sometimes also long-villous proximally, sometimes glabrate; from a deep, thick taproot, often with branching underground stems, or branching only at surface, these often becoming horizontal or nearly so and giving rise to new plants. **Stems** erect or ascending, usually many-branched, 10–120 cm. **Leaves** in a basal rosette (but not present at flowering) and cauline, 0.7–6.5 × 0.1–1.5 cm, blade linear to narrowly elliptic, margins entire or remotely and coarsely serrate. **Flowers** 4-merous, zygomorphic, opening near sunset; floral tube 4–11(–13) mm; sepals 5–9(–10) mm; petals white, fading salmon pink to scarlet-red, slightly unequal, obovate to elliptic-obovate or elliptic, 3–7(–8) mm, abruptly clawed; filaments 3–6.5(–7) mm, anthers (2.5–)3–5(–5.5) mm; style 10–22 mm, stigma exserted beyond anthers at anthesis. **Capsules** erect, pyramidal in distal ½ and abruptly constricted to terete proximal part, pyramidal part weakly or strongly angled, not conspicuously bulging at base, 4–9 × (1–)1.5–3 mm; sessile. **Seeds** (1–)3 or 4, light to reddish brown, 1.5–3 × 1–1.5 mm. $2n$ = 14, 28, 42, 56.

Flowering Apr–Aug(–Nov). Sandy or clay soil, often calcareous, desert shrublands to pinyon-juniper or oak woodlands, grasslands, disturbed areas; 150–2000 (–3000) m; Alta., B.C., Man., Ont., Sask.; Ariz., Calif., Colo., Ill., Ind., Iowa, Kans., Mich., Minn., Mo., Mont., Nebr., Nev., N.Mex., N.Y., N.Dak., Okla., S.Dak., Tex., Utah, Wis., Wyo.; Mexico (Aguascalientes, Chihuahua, Coahuila, Durango, Guanajuato, Hidalgo, Jalisco, México, Nuevo León, Oaxaca, Puebla, San Luis Potosí, Sonora, Tamaulipas, Veracruz, Zacatecas); introduced in South America (Brazil), Europe (Wales).

Oenothera suffrutescens is naturalized sporadically in southern California (Los Angeles, San Diego, Santa Barbara, and Ventura counties; although native in eastern part of the state), Illinois, Indiana, Michigan, New York, southern Ontario, and Wisconsin.

P. H. Raven and D. P. Gregory (1972[1973]) determined *Oenothera suffrutescens* to be self-incompatible and polyploid. It is known to form hybrids with *O. calcicola* and *O. hispida*.

Schizocarya kunthii Spach is an illegitimate name based on *Gaura epilobioides* that pertains here.

17k.5. Oenothera Linnaeus (sect. Gaura) subsect. Stipogaura (P. H. Raven & D. P. Gregory) W. L. Wagner & Hoch, Syst. Bot. Monogr. 83: 169. 2007

Gaura Linnaeus sect. *Stipogaura* P. H. Raven & D. P. Gregory, Mem. Torrey Bot. Club 23(1): 34. 1973

Herbs perennial, sometimes suffrutescent, clumped or spreading by rhizomes; from woody, often twisted roots. **Stems** ascending to erect, several from near base, usually also branched from shortened internodes (forming a whorl of branches just proximal to inflorescence). **Inflorescences** wandlike, erect. **Flowers** 4-merous, strongly zygomorphic, opening near sunset; floral tube 1.5–9 mm; petals white, slightly unequal; filaments with basal scales. **Capsules** reflexed, lanceoloid to ovoid, narrowly 4-winged or, sometimes, 4-angled, abruptly constricted or tapered to a slender, sterile stipe; sessile. $2n$ = 14, 28.

Species 5 (5 in the flora): sw, e United States, ne Mexico.

Species of subsect. *Stipogaura* are distributed in sandy or rocky open sites from eastern New Mexico, western Texas, and northeastern Mexico to the eastern United States in southern Indiana, Kentucky, and South Carolina, with partially overlapping ranges, replacing one another geographically (P. H. Raven and D. P. Gregory 1972[1973]). All of the species of the section are self-incompatible, the flowers vespertine, pollinated by a wide variety of insects (Raven and Gregory); one species (*Oenothera cinerea*) is visited by at least 32 species of insect pollen carriers, the most important of which are antlions in the genus *Scotoleon*, two species of noctuid moths, and two species of the nocturnal, oligolectic halictid bee *Sphecodogastra* (R. R. Clinebell et al. 2004).

1. Herbs suffrutescent, densely soft-villous, hairs mostly appressed, also strigillose and, rarely, glandular puberulent, plant parts grayish green . . . . 41. *Oenothera cinerea*
1. Herbs not suffrutescent, strigillose and/or villous with spreading hairs, sometimes also glandular puberulent, plant parts green.
   2. Cauline leaves 1–3(–7) cm, blade margins conspicuously sinuate-dentate; stems ascending, 30–70(–120) cm; plants strigillose and villous, hairs 2–4 mm, sometimes glabrate or glandular puberulent distally. . . . . .39. *Oenothera mckelveyae*
   2. Cauline leaves (1–)2.5–12 cm, blade margins subentire or sinuate-dentate; stems erect, 40–250(–300) cm; plants glabrous or strigillose, sometimes also sparsely villous, hairs 1–2 mm, sometimes glandular puberulent distally.
      3. Herbs rhizomatous perennials (usually forming extensive colonies), usually glabrous, sometimes strigillose . . . . .43. *Oenothera sinuosa*
      3. Herbs clumped perennials, usually sparsely to densely strigillose or glabrous, sometimes glabrate or sparsely villous, sometimes sparsely glandular puberulent distally.
         4. Stamens presented at anthesis in lower ½ of flower; capsules lanceoloid to narrowly ovoid; plants usually sparsely strigillose, sometimes sparsely glandular puberulent distally, rarely glabrate, or sparsely villous . . . . 40. *Oenothera calcicola*
         4. Stamens presented at anthesis evenly around flower parts; capsules ovoid; plants sparsely to densely strigillose, inflorescence usually glabrous or glandular puberulent, sometimes proximalmost parts villous. 42. *Oenothera filipes*

39. Oenothera mckelveyae (Munz) W. L. Wagner & Hoch, Syst. Bot. Monogr. 83: 213. 2007

Gaura villosa Torrey var. *mckelveyae* Munz, Bull. Torrey Bot. Club 65: 214. 1938; *Gaura mckelveyae* (Munz) P. H. Raven & D. P. Gregory

Herbs perennial, clumped, long-villous, more sparsely so distally, hairs erect, 2–4 mm, also strigillose, sometimes glabrate distally or also sparsely glandular puberulent; from twisted, woody rootstock. **Stems** ascending, branched below or just above ground, branched also proximal to inflorescences, 30–70(–120) cm. **Leaves** in a basal rosette and cauline, basal 3–17 × 0.8–2 cm, blade oblanceolate, cauline 1–6.5 × 0.1–1.5 cm, sessile, blade narrowly oblanceolate to elliptic, margins conspicuously sinuate-dentate, often undulate. **Inflorescences** slender. **Flowers** 4-merous, zygomorphic, opening near sunset; floral tube 2–3.5 mm; sepals 6–12 mm; petals white, fading dark pink to red, slightly unequal, elliptic-obovate, 7–11 mm, long-clawed; stamens presented in lower ½ of flower, filaments 5–9 mm, lanate at very base, anthers 2–4 mm, pollen 90–100% fertile; style 9–16 mm, stigma exserted beyond anthers at anthesis. **Capsules** reflexed, lanceoloid to narrowly ovoid, narrowly 4-winged, 8–19 × 1.5–2 mm, tapering to a sterile stipe 3–9 mm. **Seeds** (1 or)2–4, 2–3 × 1 mm, yellowish to reddish brown. $2n = 14$.

Flowering Mar–Jun. Sandy soil; 0–300 m; Tex.; Mexico (Nuevo León, Tamaulipas).

Oenothera mckelveyae, on the Rio Grande Plain, is found in an area bounded by from Dimmit and LaSalle counties east to Karnes and Refugio counties in the north, southward through south Texas, extending to northeastern Tamaulipas and adjacent Nuevo León. P. H. Raven and D. P. Gregory (1972[1973]) found *Oenothera mckelveyae* to be self-incompatible.

40. Oenothera calcicola (P. H. Raven & D. P. Gregory) W. L. Wagner & Hoch, Syst. Bot. Monogr. 83: 211. 2007

Gaura calcicola P. H. Raven & D. P. Gregory, Mem. Torrey Bot. Club 23(1): 40, figs. 8, 43. 1973

Herbs perennial, clumped, usually sparsely strigillose, rarely glabrate or sparsely villous, hairs erect, sometimes sparsely glandular puberulent distally; from twisted, woody rootstock. **Stems** erect, branched below and just above ground, branched also proximal to inflorescences,

(40–)60–250 cm. **Leaves** in a basal rosette and cauline, basal 3–13 × 0.6–2.5 cm, blade spatulate to oblanceolate, cauline (1–)2.5–12 × 0.1–1 cm, blade linear to narrowly oblanceolate, margins slightly to conspicuously sinuate-dentate. **Inflorescences** slender. **Flowers** 4-merous, zygomorphic, opening near sunset; floral tube 3–9 mm; sepals 6–12 mm; petals white, fading dark pink to red, slightly unequal, elliptic-obovate, 7–11 mm; stamens presented in lower ½ of flower, filaments 3–7 mm, anthers 2.5–5 mm, pollen 90–100% fertile; style 9.5–19 mm, stigma exserted beyond anthers at anthesis. **Capsules** lanceoloid to narrowly ovoid, narrowly 4-winged, 7–12 × 1.5–2.5 mm, tapered to a sterile stipe 2–5 mm. **Seeds** (2 or)3 or 4(or 5), light brown or reddish brown, 1.5–2.5 × 0.8–1.3 mm. **2*n*** = 14.

Flowering Apr–May. Dry limestone, gypsum, or caliche soil, slopes; 400–1800(–2100) m; Tex.; Mexico (Coahuila, Durango, Nuevo León, Tamaulipas).

Oenothera calcicola occurs at mostly higher elevations and more montane areas than other species of subsect. *Stipogaura*, from the southern Trans-Pecos and Edwards Plateau regions of Texas southward into northern Mexico. P. H. Raven and D. P. Gregory (1972[1973]) determined *O. calcicola* to be self-incompatible. It is known to form hybrids with *O. suffrutescens*.

41. Oenothera cinerea (Wooton & Standley) W. L. Wagner & Hoch, Syst. Bot. Monogr. 83: 211. 2007 E

Gaura cinerea Wooton & Standley, Contr. U.S. Natl. Herb. 16: 152. 1913

Herbs suffrutescent, densely soft-villous, hairs mostly appressed, 2–3 mm, becoming less villous distally, also strigillose, rarely glandular puberulent or hispidulous, plant parts grayish green; from deep, twisted, woody rootstock. **Stems** erect, several-branched near ground, also branched proximal to inflorescences, 60–280 cm. **Leaves** in a basal rosette and cauline, 0.5–8 × 0.15–2 cm, sessile, blade narrowly lanceolate or oblanceolate to very narrowly elliptic or linear, margins usually subentire or shallowly sinuate-dentate, sometimes deeply sinuate-dentate, often undulate. **Inflorescences** slender. **Flowers** 4-merous, zygomorphic, opening near sunset; floral tube 1.5–5 mm; sepals 6–14 mm; petals white, fading pink to red, slightly unequal, elliptic, 7–13 mm, clawed; stamens presented in lower ½ of flower, filaments 4.5–11 mm, anthers 2–4.5 mm, pollen 90–100% fertile; style 9–19 mm, stigma exserted beyond anthers at anthesis. **Capsules** lanceoloid to narrowly ovoid, 4-winged, 9–19 × 1–3.5 mm, abruptly constricted to a long, sterile stipe 2–10 mm. **Seeds** (1 or)2–4, 2–3(–4) × 0.8–1.3 mm, yellowish to light brown or rarely reddish brown.

Subspecies 2 (2 in the flora): sc United States.

P. H. Raven and D. P. Gregory (1972[1973]) determined *Oenothera cinerea* to be self-incompatible. The two subspecies recognized here have disjunct distributions but are very similar morphologically.

1. Herbs soft-villous, also strigillose or glandular puberulent. . . . . . 41a. *Oenothera cinerea* subsp. *cinerea*
1. Herbs soft-villous, also hispidulous. . . . . 41b. *Oenothera cinerea* subsp. *parksii*

41a. Oenothera cinerea (Wooton & Standley) W. L. Wagner & Hoch subsp. **cinerea** E

Gaura villosa Torrey 1827, not *Oenothera villosa* Thunberg 1794; *G. villosa* var. *arenicola* Munz

Herbs densely soft-villous, hairs mostly appressed, also strigillose, rarely glandular puberulent, branches of inflorescences glabrous or sparsely glandular puberulent. **Leaves:** blade narrowly lanceolate to very narrowly elliptic or linear, sometimes narrowly elliptic to lanceolate or narrowly oblanceolate proximally, margins subentire or shallowly sinuate-dentate, sometimes markedly so. **Flowers:** floral tube 2–5 mm; petals 8.5–13 mm; style 10–19 mm. **2*n*** = 14.

Flowering May–Aug. Sandy flats and dunes on high plains and rolling plains; 700–1700 m; Colo., Kans., N.J., N.Mex., Okla., Tex.

Subspecies *cinerea* is locally escaped in New Jersey. It occurs in northwestern Texas, eastern New Mexico, southeastern-most Colorado, southwestern Kansas (one station farther north, in Ellis County), and the western half of Oklahoma.

41b. Oenothera cinerea (Wooton & Standley) W. L. Wagner & Hoch subsp. **parksii** (Munz) W. L. Wagner & Hoch, Syst. Bot. Monogr. 83: 211. 2007 E

Gaura villosa Torrey var. *parksii* Munz, Bull. Torrey Bot. Club 65: 215. 1938; *G. villosa* subsp. *parksii* (Munz) P. H. Raven & D. P. Gregory

Herbs densely soft-villous, hairs mostly appressed, also hispidulous, branches of inflorescences glabrous. **Leaves:** blade narrowly oblanceolate or oblanceolate to narrowly elliptic, margins shallowly sinuate-dentate. **Flowers:** floral tube 1.5–4 mm; petals 7–12 mm; style 9–16 mm. **2*n*** = 14.

Flowering (Apr–)May–Aug(–Oct). Flats and hills of red sand local on Rio Grande Plain; 100–200 m; Tex.

Subspecies *parksii* is known from a narrow area of south Texas, including Atascosa, Bexar, Dimmit, Frio, Guadalupe, Jim Hogg, Maverick, Medina, Wilson, Zapata, and Zavala counties.

42. Oenothera filipes (Spach) W. L. Wagner & Hoch, Syst. Bot. Monogr. 83: 212. 2007 [E]

Gaura filipes Spach, Nouv. Ann. Mus. Hist. Nat. 4: 379. 1836; *G. filipes* var. *major* Torrey & A. Gray; *G. michauxii* Spach

Herbs perennial, clumped, sparsely to densely strigillose, inflorescence usually glabrous or glandular puberulent, sometimes proximalmost parts villous, hairs erect, 1–2 mm; from heavy, twisted, woody rootstock. **Stems** erect, branched below and just above ground, branched also proximal to inflorescences, 60–250(–300) cm. **Leaves** in a basal rosette and cauline, (1–)3–9 × (0.1–)0.5–1.3 cm, blade linear or narrowly lanceolate to narrowly oblanceolate, margins slightly to coarsely sinuate-dentate. **Inflorescences** slender, often well-branched, buds small and well-spaced. **Flowers** 4-merous, zygomorphic, opening near sunset; floral tube 2.5–6 mm; sepals 5–12.5 mm; petals white, fading pink to red, slightly unequal, elliptic, 5–10 mm, clawed; stamens presented evenly around flower parts, filaments 3–8.5 mm, anthers 1.4–4 mm, pollen 90–100% fertile; style 8.5–19 mm, stigma exserted beyond anthers at anthesis. **Capsules** ovoid, narrowly 4-winged or 4-angled, 5–10 × 1.5–2 mm, abruptly constricted to a sterile stipe 0.5–4.5 mm. **Seeds** 1 or 2, yellowish to reddish brown, 1.5–3.5 × 1–1.5 mm. **2*n*** = 14.

Flowering May–Sep(–Oct). Sandy hills and flats, open woods; 0–300 m; Ala., Fla., Ga., Ill., Ind., Ky., La., Miss., Mo., S.C., Tenn.

Oenothera filipes occurs marginally in several states, including: southernmost Illinois and southern Indiana; northern Florida; southeastern Mississippi; and, Washington Parish, Louisiana. P. H. Raven and D. P. Gregory (1972[1973]) determined *O. filipes* to be self-incompatible.

43. Oenothera sinuosa W. L. Wagner & Hoch, Syst. Bot. Monogr. 83: 214. 2007 [E] [F]

Gaura sinuata Nuttall ex Seringe in A. P. de Candolle and A. L. P. P. de Candolle, Prodr. 3: 44. 1828, not *Oenothera sinuata* Linnaeus 1771

Herbs perennial, usually glabrous, sometimes strigillose and villous, hairs erect; from a woody taproot but spreading by rhizomes (forming extensive colonies). **Stems** erect, branched below and just above ground, branched also proximal to inflorescences, 40–120(–250) cm. **Leaves** in a basal rosette and cauline, (1–)3–11 × (0.1–)0.5–2 cm, blade linear to narrowly oblanceolate, margins usually sparsely sinuate-dentate, rarely subentire, often undulate. **Inflorescences** stout. **Flowers** 4-merous, zygomorphic, opening near sunset; floral tube 2.5–5 mm; sepals 7–14 mm; petals white, fading pink to red, slightly unequal, elliptic, 7–15 mm; stamens presented in lower ½ of flower, filaments 5–11 mm, lanate at very base, anthers 3–5 mm, pollen 90–100% fertile; style 12–19 mm, stigma exserted beyond anthers at anthesis. **Capsules** narrowly ovoid, narrowly 4-winged or 4-angled, 8–15 × 1.5–3.5 mm, abruptly constricted to a long, sterile stipe 2–8 mm. **Seeds** (1 or)2–4, light to reddish brown, 2–3 × 1–1.5 mm. **2*n*** = 28.

Flowering Apr–Aug. Flats and washes in light sandy loam; 0–300(–1300) m; Ala., Ark., Calif., Fla., Ga., Mo., N.Y., Okla., Tex.; introduced in Europe (Italy), s Africa.

P. H. Raven and D. P. Gregory (1972[1973]) determined *Oenothera sinuosa* to be self-incompatible.

Oenothera sinuosa is endemic to Oklahoma and Texas and is escaped or naturalized in Alabama, Arkansas, California (where found to 1300 m), Florida, Georgia, Missouri, and New York.

Oenothera sinuosa is potentially a noxious weed due to the aggressive rhizomatous habit, but is somewhat limited by its self-incompatibility. Molecular data (G. D. Hoggard et al. 2004) are consistent with the hypothesis that the allotetraploid (2*n* = 28) *O. sinuosa* arose by interspecific hybridization of two species within subsect. *Stipogaura* as suggested by P. H. Raven and D. P. Gregory (1972[1973]). The molecular data indicate that the pistillate parent came from *O. calcicola* or a close relative, while the staminate parent originated from the lineage that gave rise to *O. cinerea* and *O. filipes*.

17k.6. Oenothera Linnaeus (sect. Gaura) subsect. Xenogaura (P. H. Raven & D. P. Gregory) W. L. Wagner & Hoch, Syst. Bot. Monogr. 83: 170. 2007

Gaura Linnaeus sect. *Xenogaura* P. H. Raven & D. P. Gregory, Mem. Torrey Bot. Club 23(1): 50. 1973

Herbs perennial, colonial; from woody taproot, spreading by rhizomes. **Stems** ascending to decumbent, several-branched from base, usually also irregularly branched distally, sometimes with a single, unbranched stem. **Inflorescences** slightly nodding. **Flowers** 4-merous, slightly to strongly zygomorphic, opening near sunset; floral tube 4–14 mm; petals white, slightly unequal; filaments with basal scales. **Capsules** erect, pyramidal in distal ½, conspicuously bulging at base of pyramidal part, strongly 4-angled, abruptly constricted to terete proximal part; sessile. **2*n*** = 28.

Species 1: s, w United States, n, c Mexico.

Subsection *Xenogaura* consists of the allotetraploid (2*n* = 28) *Oenothera hispida* (= *Gaura drummondii*) from Texas to central Mexico. P. H. Raven and D. P. Gregory (1972[1972]) suggested that *O. hispida* arose following interspecific hybridization between *O. suffrutescens* (subsect. *Campogaura*) and a species in subsect. *Stipogaura*, possibly near *O. mckelveyae*. G. D. Hoggard et al. (2004) found that the pistillate parent of *O. hispida* was indeed *O. mckelveyae* or a close relative, but that the staminate parent probably came from a lineage related to *O. dodgeniana* or *O. lindheimeri* in subsect. *Gaura*. *Oenothera hispida* is not easily distinguished morphologically from *O. suffrutescens* (subsect. *Campogaura*), with which it shares similar fruit characters; *O. hispida* is an aggressively rhizomatous perennial with fruits conspicuously bulging on the distal half (Raven and Gregory). The rhizomatous habit makes this species potentially invasive, despite its self-incompatibility, but so far it has established itself most heavily in coastal southern California (W. L. Wagner et al. 2007).

44. Oenothera hispida (Bentham) W. L. Wagner, Hoch & Zarucchi, PhytoKeys 50: 26. 2015

Gaura hispida Bentham, Pl. Hartw., 288. 1849; *G. crispa* (Spach) D. Dietrich; *G. drummondii* (Spach) Torrey & A. Gray; *G. roemeriana* Scheele 1848, not *Oenothera roemeriana* Scheele 1849; *O. xenogaura* W. L. Wagner & Hoch; *Schizocarya crispa* Spach 1839, not *O. crispa* Schultes 1809; *S. drummondii* Spach 1836, not *O. drummondii* Hooker 1834

Herbs perennial, spreading by rhizomes (forming colonies), strigillose, often also villous; from taproot. **Stems** ascending to decumbent, several-branched from base, usually also irregularly branched distally, sometimes with a single, unbranched stem, 20–60(–120) cm. **Leaves** in a basal rosette and cauline, 0.5–7.5(–9.5) × 0.1–2.2 cm, blade narrowly lanceolate to elliptic, margins subentire or shallowly sinuate-dentate. **Flowers** 4-merous, zygomorphic, opening near sunset; floral tube 4–14 mm; sepals 7–11(–14) mm; petals white, fading red, slightly unequal, elliptic, 6–10 mm, clawed; filaments 4–8.5 mm, anthers 3–6 mm, pollen 90–100% fertile; style 12–26 mm, stigma exserted beyond anthers at anthesis. **Capsules** erect, pyramidal in distal ½, conspicuously bulging at base of distal pyramidal part, strongly 4-angled, conspicuously bulging at base, abruptly constricted to terete proximal part, 7–13 × 3–5 mm; sessile. **Seeds** (2 or)3 or 4(–8), reddish brown, 2–2.5 × 1–1.3 mm. **2*n*** = 28.

Flowering May–Jul(–Nov). Sandy loam; 60–1900 m; Ark., Calif., Ga., Tex.; c Mexico.

Oenothera hispida is native across the eastern half of Texas, south through Mexico to Oaxaca and Puebla; it is naturalized in Sevier County, Arkansas, coastal southern California, and Glynn County Georgia.

P. H. Raven and D. P. Gregory (1972[1973]) reported *Oenothera hispida* to be self-incompatible. It occasionally forms hybrids with *O. suffrutescens*.

17k.7. Oenothera Linnaeus (sect. Gaura) subsect. Gaura (Linnaeus) W. L. Wagner & Hoch, Syst. Bot. Monogr. 83: 171. 2007

Gaura Linnaeus, Sp. Pl. 1: 347. 1753; *Gaura* sect. *Pterogaura* P. H. Raven & D. P. Gregory; *Pleurostemon* Rafinesque

Herbs annual, biennial, or clumped perennial (*O. lindheimeri*); from robust taproot. **Stems** single or several from base, sometimes branched distally. **Inflorescences** erect. **Flowers** (3 or)4-merous, strongly zygomorphic, opening near sunset or sunrise; floral tube 3–14(–20) mm; petals white, slightly unequal; filaments with basal scales. **Capsules** erect, ellipsoid, ovoid, or narrowly obovoid, sharply (3- or)4-angled or narrowly (3- or)4-winged, abruptly constricted or tapered to base; sessile. **2*n*** = 14.

Species 13 (12 in the flora): North America, Mexico, Central America (Guatemala).

Subsection *Gaura* is the largest subsection of sect. *Gaura*. W. L. Wagner et al. (2007) combined the two sections (*Gaura* and *Pterogaura*) recognized by P. H. Raven and D. P. Gregory (1972[1973]) into a single subsection based on recent molecular studies by G. D. Hoggard et al. (2004) and R. A. Levin et al. (2004). The two groups are quite similar morphologically, differing primarily in the shape of the base of the fruit (abruptly constricted versus tapering). Subsequent analyses by Wagner et al. (2013; K. Krakos, unpubl.), which sampled all 13 taxa of subsect. *Gaura*, showed that none of the three species (*O. coloradensis*, *O. hexandra*, and *O. suffulta*) subdivided into two subspecies each by Raven and Gregory and maintained in *Oenothera* by Wagner et al. were monophyletic, and these taxa were recently recognized as six species with the additional ones being *O. dodgeniana*, *O. nealleyi*, and *O. podocarpa* (Wagner et al. 2013). A majority of the species of subsect. *Gaura* occur in the central and eastern United States, but the subsection extends west to Colorado and Wyoming (*O. coloradensis*), New Mexico (*O. dodgeniana*, *O. nealleyi*, *O. podocarpa*), and Arizona (*O. podocarpa*), southward into Mexico in the Sierra Madre Occidental and the Trans-Volcanic Belt, to Guatemala (*O. hexandra*). Five species (*O. demareei*, *O. filiformis*, *O. lindheimeri*, *O. nealleyi*, and *O. suffulta*) are self-incompatible with the other eight species self-compatible and exhibiting variable degrees of autogamy. Two of the self-compatible and autogamous species, *O. gaura* and *O. triangulata*, are PTH species and form a ring of 14 chromosomes in meiosis (Raven and Gregory).

Pleurandra Rafinesque (1817), not *Pleurandra* Labillardière (1806), is an illegitimate homonym that pertains here.

1. Flowers 3- or, rarely, 4-merous; capsules 3(or 4)-winged or 3(or 4)-angled.
    2. Plants usually unbranched proximally, 60–180 cm; capsules narrowly ellipsoid or ovoid . . . . . 52. *Oenothera simulans* (in part)
    2. Plants usually branched proximally, 15–65(–85) cm; capsules ellipsoid or narrowly obovoid.
        3. Sepals 10–15 mm; pollen 90–100% fertile . . . . . 46. *Oenothera patriciae* (in part)
        3. Sepals 4.5–6 mm; pollen 35–65% fertile . . . . . 48. *Oenothera triangulata*
1. Flowers 4-merous; capsules 4-winged or 4-angled.
    4. Flowers opening at sunrise; herbs perennial (clumped), usually branched from base, villous throughout, usually also glandular puberulent in distal parts . . . . . 56. *Oenothera lindheimeri*
    4. Flowers usually opening at sunset, rarely at sunrise (*O. demareei*); herbs usually annual or biennial, rarely monocarpic perennial, branched or unbranched, villous, strigillose, or short-hirtellous throughout, distal parts usually glandular puberulent, short-hirtellous, strigillose, or glabrate, sometimes villous.

[5. Shifted to left margin.—Ed.]

5. Capsules winged, furrowed between wings; stems 15–120 cm.
   6. Sepals 6–15 mm; floral tubes 6–12 mm.
      7. Sepals 6–12 mm; plants glabrate, glandular puberulent, and/or sparsely strigillose distally . . . . 45. *Oenothera podocarpa*
      7. Sepals 10–15 mm; plants usually glabrate, sometimes strigillose distally. . . . . 46. *Oenothera patriciae* (in part)
   6. Sepals 11–21 mm; floral tubes 6.5–20 mm.
      8. Plants glabrous distally, except sometimes proximal part of inflorescence, especially bracts, sparsely villous; capsules with stipes 0–1 mm . . . . 47. *Oenothera suffulta*
      8. Plants glandular puberulent distally; capsules with stipes 0.2–2 mm. . . . . 49. *Oenothera nealleyi*
5. Capsules angled, not winged; stems 50–400 cm.
   9. Sepals 2.5–8 mm; stems usually unbranched, sometimes with several branches from base. . . . . 52. *Oenothera simulans* (in part)
   9. Sepals (5–)9–20 mm; stems usually branched from base upwards.
      10. Flowers opening at sunrise; plants exclusively strigillose throughout . . . . 55. *Oenothera demareei*
      10. Flowers opening at sunset; plants with some combination of villous, glandular puberulent, short-hirtellous, or strigillose.
         11. Anthers: pollen 35–65% fertile; plants villous and glandular puberulent, rarely short-hirtellous, not strigillose . . . . 53. *Oenothera gaura*
         11. Anthers: pollen 90–100% fertile; plants strigillose, villous, short-hirtellous, glabrate, or glandular puberulent distally.
            12. Herbs annual; capsules 4.5–7 mm . . . . 51. *Oenothera filiformis*
            12. Herbs biennial or monocarpic perennial; capsules 6–11 mm.
               13. Herbs monocarpic perennial, strigillose proximally, becoming short-hirtellous and strigillose distally . . . . 50. *Oenothera coloradensis*
               13. Herbs biennial, villous and strigillose proximally, becoming also glandular puberulent distally and also sparsely villous . . . . . . 54. *Oenothera dodgeniana*

45. Oenothera podocarpa (Wooton & Standley) Krakos & W. L. Wagner, PhytoKeys 28: 68. 2013 [F]

Gaura podocarpa Wooton & Standley, Contr. U.S. Natl. Herb. 16: 154. 1913; *G. brassicacea* Wooton & Standley; *G. glandulosa* Wooton & Standley; *G. gracilis* Wooton & Standley; *G. hexandra* Ortega subsp. *gracilis* (Wooton & Standley) P. H. Raven & D. P. Gregory; *G. strigillosa* Wooton & Standley; *Oenothera hexandra* (Ortega) W. L. Wagner & Hoch subsp. *gracilis* (Wooton & Standley) W. L. Wagner & Hoch

Herbs annual, villous proximally, glabrate, strigillose and/or glandular puberulent distally, leaves glabrate to densely villous, glabrate in age; from stout taproot. **Stems** ascending to erect, unbranched or well-branched at base and distally, 15–100 cm. **Leaves** in a basal rosette and cauline, basal 3–15 × 0.5–1 cm, blade lyrate; cauline 1–9 × 0.1–0.8 cm, blade linear to very narrowly elliptic or narrowly lanceolate, margins sinuate-dentate to subentire. **Flowers** 4-merous, zygomorphic, opening at sunset; floral tube 6–10 mm; sepals 6–12 mm; petals white, fading pink to red, narrowly obovate, 5.5–9.5 mm, short-clawed; filaments 4–6 mm, anthers 2–3 mm, pollen 90–100% fertile; style 11–19 mm, stigma surrounded by anthers at anthesis. **Capsules** ellipsoid or narrowly obovoid, narrowly 4-winged, furrowed between wings, 6–8 × 2–3 mm, narrowed at base, stipe 0 mm; sessile. **Seeds** 4, yellowish to reddish brown, 2–3 × 1–1.5 mm. **2*n*** = 14.

Flowering (May–)Jun–Oct. Disturbed sites, sandy washes, slopes, grasslands, meadows, pinyon-juniper or ponderosa pine woodlands, on volcanic cinders; 700–2800 m; Ariz., N.Mex.; Mexico (Chihuahua, Durango, Sonora).

Oenothera podocarpa occurs in Arizona from eastern Mohave County south through the mountains of central Arizona to eastern Pima County and the southwestern quarter of New Mexico, and in Mexico southward in the Sierra Madre Occidental to eastern Sonora and throughout the western halves of Chihuahua and Durango. P. H. Raven and D. P. Gregory (1972[1973]) determined *O. podocarpa* to be self-compatible and primarily autogamous.

46. Oenothera patriciae W. L. Wagner & Hoch, Syst. Bot. Monogr. 83: 213. 2007 E F

Gaura brachycarpa Small, Fl. S.E. U.S., 848, 1335. 1903, not *Oenothera brachycarpa* A. Gray 1852; *G. hexandra* Ortega var. *coryi* (Munz) Munz; *G. tripetala* Cavanilles var. *coryi* Munz

Herbs annual, villous proximally, sparsely villous along leaf veins and on margins, usually glabrate or, sometimes, strigillose distally; from taproot. **Stems** usually well-branched from base and distally, rarely unbranched, 15–65(–85) cm. **Leaves** in a basal rosette and cauline, basal 6–9.5 × 1–2 cm, blade lyrate; cauline 1–7 × 0.1–2.3 cm, blade narrowly lanceolate to oblanceolate, margins shallowly sinuate-denticulate to subentire. **Flowers** (3- or)4-merous, zygomorphic, opening at sunset; floral tube 6–12 mm; sepals 10–15 mm; petals white, fading pink to purple, elliptic-obovate, 8–13 mm; filaments 5–8 mm, anthers 2–4 mm, pollen 90–100% fertile; style 15–24 mm, stigma exserted beyond anthers at anthesis. **Capsules** ellipsoid, narrowly (3- or)4-winged, deeply furrowed between wings, 6–10 × 2–3 mm, with or without prominent lower corners, narrowed to a stipe 0–1 mm; sessile. **Seeds** 3 or 4, yellowish to reddish brown, 2–3 × 1–1.5 mm. $2n = 14$.

Flowering Feb–Jun. Open, sandy sites; 0–300 m; La., Miss., Okla., Tex.

Oenothera patriciae is known from Acadia Parish, Louisiana, Amite County, Mississippi, Bryan and Love counties, Oklahoma, and eastern Texas.

Reports of *Oenothera patriciae* near Tulsa, Oklahoma, and at the single locations in Arkansas and Mississippi may represent introductions. P. H. Raven and D. P. Gregory (1972[1973]) determined *O. patriciae* to be self-compatible, but primarily outcrossing.

47. Oenothera suffulta (Engelmann ex A. Gray) W. L. Wagner & Hoch, Syst. Bot. Monogr. 83: 214. 2007 E

Gaura suffulta Engelmann ex A. Gray, Boston J. Nat. Hist. 6: 190. 1850

Herbs annual, sparsely villous proximally, leaves glabrate to sparsely villous along veins and on margins, usually glabrous distally, except sometimes proximal part of inflorescence, especially bracts, sparsely villous; from stout taproot. **Stems** usually well-branched, 25–120 cm. **Leaves** in a basal rosette and cauline, basal 7–11 × 0.1–2.3 cm, blade lyrate; cauline 1–9.5 × 0.1–2.3 cm, blade narrowly lanceolate to linear, margins sinuate-dentate, undulate. **Flowers** 4-merous, zygomorphic, opening at sunset; floral tube 6.5–14 mm; sepals 11–21 mm; petals white, fading pink to red or sometimes purple, elliptic-obovate, 10–15 mm; filaments 6–9 mm, anthers 2–6 mm, pollen 90–100% fertile; style 16–32 mm, stigma exserted beyond anthers at anthesis. **Capsules** ovoid, narrowly 4-winged, furrowed between angles, 4.5–8 × 2–5 mm, abruptly tapering to stipe 0–1 mm; sessile. **Seeds** (1 or)2–4(or 5), yellowish to light brown, 2–3 × 1 mm. $2n = 14$.

Flowering Apr–Aug. In open, sandy places; 10–1100 m; Okla., Tex.

Oenothera suffulta is more common in western Texas while uncommon elsewhere throughout Texas, and absent in the Trans-Pecos region. P. H. Raven and D. P. Gregory (1972[1973]) determined *O. suffulta* to be self-incompatible.

48. Oenothera triangulata (Buckley) W. L. Wagner & Hoch, Syst. Bot. Monogr. 83: 214. 2007 E

Gaura triangulata Buckley, Proc. Acad. Nat. Sci. Philadelphia 13: 454. 1862; *G. hexandra* Ortega var. *triangulata* (Buckley) Munz; *G. tripetala* Cavanilles var. *triangulata* (Buckley) Munz

Herbs annual, villous proximally, sparsely villous along veins and on margins, usually glabrate, sometimes strigillose distally; from taproot. **Stems** ascending, usually well-branched from base and distally, rarely unbranched, 15–60 cm. **Leaves** in a basal rosette and cauline, 1.5–8 × 0.2–0.6(–1.5) cm, blade very narrowly elliptic to oblanceolate or oblong-elliptic, margins entire or weakly sinuate-dentate. **Flowers** 3(or 4)-merous, zygomorphic, opening at sunset; floral tube 4–5.5 mm; sepals 4.5–6 mm; petals white, fading pink, elliptic-obovate, 3.5–5 mm; filaments 2–3.5 mm, anthers 1.5–3 mm, pollen 35–65% fertile; style 9–10 mm, stigma surrounded by anthers. **Capsules** narrowly obovoid, 3(or 4)-winged, furrowed between wings, 7–9 × 3–5 mm, narrowed at base; sessile. **Seeds** (1 or)2–5, yellowish to light brown, 1.5–3.5 × 1–1.5 mm. $2n = 14$.

Flowering Apr–Jul. Open, sandy sites; 200–600 m; Okla., Tex.

Oenothera triangulata is a PTH species and forms a ring of 14 chromosomes in meiosis. The species is self-compatible and autogamous (P. H. Raven and D. P. Gregory 1972[1973]). It may have been derived from hybridization between *O. patriciae* and *O. suffulta*. The species has a relatively narrow distribution across south-central Oklahoma and north-central Texas (Oklahoma in Cleveland, Comanche, Cotton, Grady, Oklahoma,

Rogers, Stephens, and Tulsa counties; Texas in Archer, Baylor, Callahan, Clay, Coleman, Crosby, Eastland, Erath, Jones, Montague, Taylor, Throckmorton, Tom Greene, Wichita, Wilbarger, and Young counties).

49. Oenothera nealleyi (J. M. Coulter) Krakos & W. L. Wagner, PhytoKeys 28: 64. 2013 [F]

Gaura nealleyi J. M. Coulter, Contr. U.S. Natl. Herb. 1: 38. 1890; *G. suffulta* Engelmann ex A. Gray subsp. *nealleyi* (J. M. Coulter) P. H. Raven & D. P. Gregory; *G. suffulta* var. *terrellensis* Munz; *Oenothera suffulta* (Engelmann ex A. Gray) W. L. Wagner & Hoch subsp. *nealleyi* (J. M. Coulter) W. L. Wagner & Hoch

Herbs annual, sparsely villous proximally, leaves glabrate to sparsely villous along veins and on margins, usually glandular puberulent in distal parts; from stout taproot. **Stems** usually well-branched, 20–70(–100) cm. **Leaves** in a basal rosette and cauline, basal 3.5–9 × 0.5–1.5 cm, blade lyrate; cauline 1.5–7 × 0.1–0.6 cm, blade narrowly lanceolate to linear, margins sinuate-dentate, undulate. **Flowers** 4-merous, zygomorphic, opening at sunset; floral tube 10–20 mm; sepals 11–21 mm; petals white, fading pink to red, elliptic to elliptic-obovate, 10–15 mm; filaments 8–13 mm, anthers 2–6 mm, pollen 90–100% fertile; style 22–36 mm, stigma exserted beyond anthers at anthesis. **Capsules** ellipsoid or ovoid, narrowly 4-winged, furrowed between angles, 4.5–8 × 2–5 mm, stipe 0.2–2.2 mm; sessile. **Seeds** 3 or 4 (or 5), yellowish to light brown, 2–3(–4) × 1 mm. **2*n*** = 14.

Flowering Apr–Oct. Washes, sandy places, grasslands, extending to pinyon-juniper woodlands; 1200–2200 m; N.Mex., Tex.; Mexico (Coahuila).

Oenothera nealleyi is restricted to an area from trans-Pecos Texas and northern Coahuila, Mexico, north to Bernalillo and Torrance counties, New Mexico. P. H. Raven and D. P. Gregory (1972[1973]) considered *O. nealleyi* to represent an unevenly intergrading entity with *O. suffulta* based on merging of distinguishing characteristics. The known intermediates occur in Terrell County, Texas, and were previously described as *Gaura suffulta* var. *terrellensis* Munz, but until new data on its status are available, we include this name with *O. nealleyi*. The molecular data (K. N. Krakos, unpubl.) suggest that *O. nealleyi* is not as closely related to *O. suffulta* as suggested by Raven and Gregory, given the placement in the phylogeny and the difference in scent profiles for these two taxa. *Oenothera suffulta* is a member of a strongly supported clade that also includes *O. patriciae* and *O. triangulata*, while *O. nealleyi* is a member of a polytomy that consists of other species of subsect. *Gaura*, with the *O. suffulta*—*O. triangulata*—*O. patriciae* clade sister to it (W. L. Wagner et al. 2013). *Oenothera nealleyi* has a strong sweet scent, whereas *O. suffulta* does not have a discernible scent (Wagner et al.). Raven and Gregory determined *O. nealleyi* to be self-incompatible.

50. Oenothera coloradensis (Rydberg) W. L. Wagner & Hoch, Syst. Bot. Monogr. 83: 211. 2007 [C] [E]

Gaura coloradensis Rydberg, Bull. Torrey Bot. Club 31: 572. 1904; *G. neomexicana* Wooton subsp. *coloradensis* (Rydberg) P. H. Raven & D. P. Gregory; *G. neomexicana* var. *coloradensis* (Rydberg) Munz

Herbs monocarpic perennial, strigillose proximally, short-hirtellous and strigillose distally, leaves sometimes glabrate; from stout, fleshy taproot. **Stems** 1–few-branched from base, 50–80(–100) cm. **Leaves** in a basal rosette and cauline, basal 4–18 × 1.5–4 cm, blade very narrowly elliptic, lanceolate, or oblanceolate; cauline 5–13 × 1–4 cm, blade narrowly elliptic, narrowly lanceolate, or narrowly oblanceolate, margins subentire or repand-denticulate. **Flowers** 4-merous, zygomorphic, opening at sunset; floral tube 8–12 mm; sepals 9.5–13 mm; petals white, fading pink, rhombic-obovate, 7–12 mm; filaments 6.5–9 mm, anthers 2.5–4 mm, pollen 90–100% fertile; style 19–25 mm, stigma exserted beyond anthers at anthesis. **Capsules** ellipsoid or ovoid, sharply 4-angled, with fairly deep furrows alternating with angles, 6–8.5 × 2–3 mm; sessile. **Seeds** 1–4, yellowish to light brown, 2–3 × 1 mm. **2*n*** = 14.

Flowering Jul–Sep. In wet meadow vegetation of North and South Platte River watersheds on high plains, sloping floodplains, drainage basins in heavy soil; of conservation concern; 1500–2000 m; Colo., Nebr., Wyo.

Oenothera coloradensis is currently known from fewer than two dozen populations from southern Laramie and Platte counties in Wyoming, northern Weld County, Colorado, formerly near Fort Collins, Larimer County, Colorado, and in western Kimball County, Nebraska. It is federally listed as a threatened species in the United States. The primary threats are agricultural use of habitat, herbicide spraying to control weed species, and livestock trampling and grazing (see W. L. Wagner et al. 2013). Recent study by K. N. Krakos (unpubl.) has determined this species to be self-compatible. P. H. Raven and D. P. Gregory (1972[1973]) described this species as glandular puberulent in inflorescence, which was repeated in the recent revised taxonomy (Wagner et al.); however, examination of specimens show that P. A. Munz (1965) was correct in describing the pubescence of the inflorescence as non-glandular.

51. Oenothera filiformis (Small) W. L. Wagner & Hoch, Syst. Bot. Monogr. 83: 212. 2007 [E]

Gaura filiformis Small, Bull. Torrey Bot. Club 25: 617. 1898; *G. biennis* Linnaeus var. *pitcheri* Torrey & A. Gray; *G. filiformis* var. *kearneyi* Munz; *G. longiflora* Spach 1836, not *Oenothera longiflora* Linnaeus 1753

Herbs usually robust **winter-annual, sometimes biennial,** moderately to densely strigillose, sometimes also glandular puberulent, villous and/or short-hirtellous; from fleshy taproot. **Stems** usually well-branched distal to base, (50–)100–400 cm. **Leaves** in a basal rosette and cauline, basal 8–15(–40) × 1.5–3.6 cm, blade lyrate, margins irregularly toothed to lobed; cauline 1.5–13 × 0.5–3 cm, blade narrowly elliptic to elliptic or lanceolate, margins subentire or shallowly undulate-denticulate. **Flowers** 4-merous, zygomorphic, opening at sunset; floral tube 4–13(–15) mm; sepals 7–18 mm; petals white, fading pink, elliptic to elliptic-obovate, 7–15 mm; filaments 5–13 mm, anthers 1.5–5 mm, pollen 90–100% fertile; style 12–34 mm, stigma exserted beyond anthers at anthesis. **Capsules** ellipsoid or ovoid, sharply 4-angled, 4.5–7 × 1.5–2.5 mm; sessile. **Seeds** 2–4, yellowish to reddish brown, 1.3–3 × 0.7–1.3 mm. **2*n*** = 14.

Flowering (Jun–)Jul–Oct(–Nov). Open woods, fields, along streams, sandy soil, disturbed sites, ditch banks, roadsides, railway embankments; 10–500 m; Ont.; Ala., Ark., Colo., Conn., Ill., Ind., Iowa, Kans., Ky., La., Md., Mass., Mich., Miss., Mo., Nebr., Ohio, Okla., Pa., Tenn., Tex., Wis.

P. H. Raven and D. P. Gregory (1972[1973]) found various levels of hybridization and intergradation between *Oenothera filiformis* and *O. lindheimeri* to occur where their ranges overlap. In the region of geographical overlap, most populations of *O. filiformis* are strigillose in the inflorescences and have evening-opening flowers, while *O. lindheimeri* is villous in the inflorescences, and has large morning-opening flowers. Moreover, *O. lindheimeri* occurs only on black clay prairie soil, while *O. filiformis* occurs in light, sandy soil, as it does throughout its range, and in more disturbed areas. Despite these differences, Raven and Gregory found hybridization between these species scattered across an area from eastern Texas across Louisiana to Alabama. At some locations there is apparently no hybridization, while at others hybrids were uncommon to relatively common. Intermediate morning-blooming plants in Alabama appear to represent evidence of past hybridization since *O. lindheimeri* does not occur there. Many of the individuals they tested had somewhat reduced pollen fertility (40–70% fertile). They suspected that habitat disturbance was primarily responsible in many cases, but they also detected what may have been intergradation resulting from past hybridization outside of the current distribution of *O. lindheimeri*. Many of these individual cases deserve further investigation to better understand the dynamics of the interactions between these species and if any of the interactions have led to stabilization of novel populations that might be recognized taxonomically.

Very few collections have been made from areas on the periphery of the range of *Oenothera filiformis* (southern Ontario, Connecticut, Maryland, Massachusetts, Pennsylvania), and populations in these areas probably represent recent human-based introductions. P. H. Raven and D. P. Gregory (1972[1973]) determined *O. filiformis* to be self-incompatible.

52. Oenothera simulans (Small) W. L. Wagner & Hoch, Syst. Bot. Monogr. 83: 213. 2007 [E]

Gaura simulans Small, Bull. New York Bot. Gard. 3: 432. 1905; *G. angustifolia* Michaux 1803, not *Oenothera angustifolia* Miller 1768; *G. angustifolia* var. *eatonii* (Small) Munz; *G. angustifolia* var. *simulans* (Small) Munz; *G. angustifolia* var. *strigosa* Munz; *G. eatonii* Small

Herbs annual, glabrate, strigillose, and/or hirtellous; from taproot. **Stems** usually unbranched, sometimes several-branched from base, 60–180 cm. **Leaves** in a basal rosette and cauline, 0.8–13 × 0.1–1.6 cm; blade often red-blotched, narrowly lanceolate to narrowly oblanceolate, margins slightly to conspicuously sinuate-dentate. **Flowers** 3 or 4-merous, often mixed on a single plant, zygomorphic, opening near sunset; floral tube 3–8 mm; sepals 2.5–8 mm; petals white, fading pink, narrowly elliptic-obovate, 4.5–8 mm; filaments 2.5–6 mm, anthers 0.5–2 mm, pollen 90–100% fertile; style 7.5–19 mm, stigma surrounded by or slightly exserted beyond anthers. **Capsules** ellipsoid or ovoid, 3-(or 4-) angled, 5–9 × 2–3 mm; sessile. **Seeds** 2–4, yellowish to light brown, 1.2–2.3 × 0.8–1.1 mm. **2*n*** = 14.

Flowering (Feb–)May–Sep(–Nov). Sandy soil in open woodlands, fields, roadsides, primarily in outer Coastal Plain; 0–10 m; Fla., Ga., N.C., S.C.

Oenothera simulans occurs along the Coastal Plain from Cape Hatteras, North Carolina, southward and throughout Florida. It is self-compatible and autogamous (P. H. Raven and D. P. Gregory 1972[1973]). The species occasionally persists through mild winters in the southern part of its range, appearing biennial.

Gaura fruticosa Jacquin 1786, not *G. fruticosa* Loefling 1758, is an illegitimate later homonym that pertains here.

53. Oenothera gaura W. L. Wagner & Hoch, Syst. Bot. Monogr. 83: 212. 2007 E W

Gaura biennis Linnaeus, Sp. Pl. 1: 347. 1753, not *Oenothera biennis* Linnaeus 1753

Herbs usually robust winter-annual, sometimes biennial, usually moderately to densely villous, rarely short-hirtellous, also glandular puberulent; from fleshy taproot. **Stems** usually well-branched distal to base, 50–180 cm. **Leaves** in a basal rosette and cauline, basal 8–20 × 1.5–3 cm, blade oblanceolate, margins irregularly toothed to lobed; cauline 1.5–12 × 0.5–3 cm, blade narrowly elliptic to elliptic or lanceolate, margins subentire or undulate-denticulate. **Flowers** 4-merous, zygomorphic, opening at sunset; floral tube 6–13 mm; sepals 5–13 mm; petals white, fading pink to red, narrowly elliptic-obovate, 6–12 mm; filaments 5–10 mm, anthers 2–4 mm, pollen 35–65% fertile; style 12–15 mm, stigma surrounded by anthers. **Capsules** ellipsoid, 4-angled, 5–9 × 2–3 mm; sessile. **Seeds** 3–6, light to reddish brown, 2–2.5 × 1–1.3 mm. **2*n*** = 14.

Flowering Jun–Oct. Open woods, fields, along streams, disturbed sites, ditch banks, roadsides, railway embankments; 100–600 m; Ont., Que.; Conn., Del., D.C., Ill., Ind., Iowa, Ky., Md., Mass., Mich., Minn., N.J., N.Y., N.C., Ohio, Pa., Tenn., Vt., Va., W.Va., Wis.

Oenothera gaura is a PTH species and forms a ring of 14 chromosomes in meiosis. It is self-compatible and autogamous (P. H. Raven and D. P. Gregory 1972[1973]), and may have been derived from *O. filiformis*.

54. Oenothera dodgeniana Krakos & W. L. Wagner, PhytoKeys 28: 66. 2013 E

Gaura neomexicana Wooton, Bull. Torrey Bot. Club 25: 307. 1898 (as neo-mexicana), not *Oenothera neomexicana* (Small) Munz 1931; *O. coloradensis* (Rydberg) W. L. Wagner & Hoch subsp. *neomexicana* (Wooton) W. L. Wagner & Hoch

Herbs biennial, villous and strigillose proximally, leaves glabrate or strigillose, also glandular puberulent distally, sometimes also sparsely villous; from stout, fleshy taproot. **Stems** 1 or few-branched from base, 50–120 cm. **Leaves** in a basal rosette and cauline, basal 6–20 × 1–3 cm, blade lanceolate to narrowly elliptic; cauline 5–10 × 1–2.5 cm, blade lanceolate to narrowly elliptic, margins subentire or repand-denticulate. **Flowers** 4-merous, zygomorphic, opening at sunset; floral tube 10–11 mm; sepals 11–15 mm; petals white, fading pink, rhombic-obovate, 11–14 mm; filaments 6.5–9 mm, anthers 2.5–4 mm, pollen 90–100% fertile; style 22–28 mm, stigma exserted beyond anthers. **Capsules** ellipsoid or ovoid, sharply 4-angled, with deep furrows alternating with angles for 2–3 mm from apex, ribbed from base of furrow to base of fruit, 9–11 × 3–5 mm; sessile. **Seeds** 2–4, yellowish to light brown, 2–3 mm. **2*n*** = 14.

Flowering Jun–Sep. Mountain meadow openings in coniferous forests; 1800–2700 m; Colo., N.Mex.

Oenothera dodgeniana occurs in two disjunct areas: the western foothills of the San Juan Mountains in Archuleta and Huerfano counties, Colorado, and Rio Arriba County, New Mexico; and Sierra Blanca and Sacramento Mountains in Lincoln and Otero counties, south-central New Mexico. The species was collected once at Durango, La Plata County, Colorado (P. H. Raven and D. P. Gregory 1972[1973]), but has not since been recollected. *Oenothera dodgeniana* and *O. coloradensis* were considered by Raven and Gregory to represent a relict species along the eastern flank of the Rocky Mountains that arose from more widespread species farther to the east, such as *O. filiformis*. *Oenothera dodgeniana* belongs to a subclade which is sister to that containing *O. coloradensis*, and within that subclade is sister to *O. demareei* and *O. lindheimeri* (W. L. Wagner et al. 2013). Although *O. dodgeniana* is fairly closely related to *O. coloradensis*, the two taxa seem to have had independent origins that have led to distributions along the eastern flank of the Rocky Mountains. *Oenothera dodgeniana* is self-compatible (Raven and Gregory).

55. Oenothera demareei (P. H. Raven & D. P. Gregory) W. L. Wagner & Hoch, Syst. Bot. Monogr. 83: 212. 2007 E

Gaura demareei P. H. Raven & D. P. Gregory, Mem. Torrey Bot. Club 23(1): 78, figs. 21, 54. 1973

Herbs usually robust winter-annual, sometimes biennial, densely strigillose throughout; from fleshy taproot. **Stems** usually well-branched distal to base, 50–400 cm. **Leaves** in a basal rosette and cauline, 3–7 × 0.2–0.8 cm; blade narrowly elliptic to elliptic or lanceolate, margins subentire or shallowly undulate-denticulate. **Flowers** 4-merous, zygomorphic, opening at sunrise; floral tube 4–13(–15) mm; sepals 13–20 mm; petals white, fading pink, rhombic-obovate, 10–17 mm; filaments 8–17 mm, anthers 3–7 mm, pollen 90–100% fertile; style 18–32 mm, stigma exserted beyond anthers at anthesis. **Capsules** ellipsoid or ovoid, sharply 4-angled, 4.5–7 × 1.5–2.5 mm; sessile. **Seeds** 2–4, yellowish to reddish brown, 1.2–3 × 0.7–1.3 mm. **2*n*** = 14.

Flowering Jul–Oct. Open meadows in sandy loam; 70–200 m; Ark.

Oenothera demareei is known only from Clark, Garland, Hempstead, Howard, Montgomery, Pike, Saline, and Sevier counties.

P. H. Raven and D. P. Gregory (1972[1973]) found *Oenothera demareei* to be self-incompatible.

56. Oenothera lindheimeri (Engelmann & A. Gray) W. L. Wagner & Hoch, Syst. Bot. Monogr. 83: 213. 2007 [E]

Gaura lindheimeri Engelmann & A. Gray, Boston J. Nat. Hist. 5: 217. 1845; *G. filiformis* Small var. *munzii* Cory

Herbs clumped perennial, villous, usually more densely so proximally, hairs erect or ± appressed on leaf blades, also glandular puberulent distally, rarely glabrate; from taproot. **Stems** many from base, ascending or erect, usually branched, 50–150 cm. **Leaves** in a basal rosette and cauline, 0.5–9 × 0. 1–1.3 cm; blade narrowly elliptic to narrowly oblanceolate, margins coarsely and remotely serrate. **Flowers** 4-merous, zygomorphic, opening at sunrise; floral tube 4–9 mm; sepals 9–17 mm; petals white, fading light or deep pink, rhombic-obovate to elliptic, 10–15 mm; filaments 7–12 mm, anthers 3.5–4.5 mm, pollen 90–100% fertile; style 16–27 mm, stigma exserted beyond anthers at anthesis. **Capsules** ellipsoid or ovoid, 4-angled, 6–9 × 2–3.5 mm; sessile. **Seeds** 1–4, yellowish to light brown, 2–3 × 1–1.5 mm. $2n$ = 14.

Flowering Apr–Jul(–Oct). Black soil in coastal prairies; 0–100 m; La., Tex.

Oenothera lindheimeri has a fairly narrow distribution and occurs only in Acadia, Allen, Beauregard, Calcasieu, Jefferson Davis, Lafayette, St. Mary, Tangipahoa, and Vermillion parishes in Louisiana, and Brazoria, Brazos, Chambers, Fort Bend, Galveston, Hardin, Harris, Jasper, Jefferson, Liberty, Orange, Victoria, and Victoria counties in Texas.

P. H. Raven and D. P. Gregory (1972[1973]) found *Oenothera lindheimeri* to be self-incompatible. It occasionally forms hybrids with *O. filiformis*. This species is widely cultivated and has many different cultivars.

17l. Oenothera Linnaeus sect. Eremia W. L. Wagner, Ann. Missouri Bot. Gard. 73: 477. 1986

Herbs winter-annual, caulescent to short-caulescent; from a weakly fleshy taproot. **Stems** (when present) ascending to erect, branched or unbranched. **Leaves** in a basal rosette, sometimes also cauline, (1.4–)6–15(–28) cm; blade margins sinuate-dentate to subentire. **Inflorescences** solitary flowers in axils of leaves. **Flowers** opening near sunset, with strong, sweet, lemony scent or pungent, spermaceous scent, to weakly scented in autogamous populations; bud apex curved downward by recurved floral tube, becoming erect before anthesis, quadrangular, without free tips; floral tube (20–)26–60(–72) mm; sepals separating individually or in pairs; petals deep yellow, fading reddish orange to purple, drying purple, obcordate to obovate; stigma deeply divided into 4 linear lobes. **Capsules** hard, woody in age, sigmoid or curved to nearly straight, lanceoloid to ovoid, 4-angled, apex gradually tapering to a sterile beak, dehiscent 1/4–2/3 their length; sessile. **Seeds** usually numerous, in 2 rows per locule, obovoid to oblanceoloid, surface papillose, coarsely rugose on distal 1/2 abaxially, adaxial face thickened with a cavity that externally appears as a pore and groove along raphae. $2n$ = 14.

Species 1: sw United States, n Mexico.

Section *Eremia* consists of a single species that occurs in the Chihuahuan, Mojave, and Sonoran deserts.

57. Oenothera primiveris A. Gray, Smithsonian Contr. Knowl. 5(6): 58. 1853

Lavauxia lobata A. Nelson; *L. primiveris* (A. Gray) Small; *Oenothera bufonis* M. E. Jones; *O. cespitosa* Nuttall var. *primiveris* (A. Gray) H. Léveillé; *O. johnsonii* Parry; *O. primiveris* subsp. *bufonis* (M. E. Jones) Munz; *O. primiveris* var. *bufonis* (M. E. Jones) Cronquist; *O. primiveris* subsp. *caulescens* (Munz) Munz; *O. primiveris* var. *caulescens* Munz; *Pachylophus johnsonii* (Parry) Rydberg

Herbs winter-annual, caulescent to short-caulescent, long-hirsute, hairs often with reddish purple pustulate bases, especially proximally, also moderately strigillose, and glandular puberulent distally, often on leaves; from a weakly fleshy taproot. **Stems** (when present) unbranched and erect or, sometimes, few branches from near base, in robust plants stems and caudex hollow and greatly enlarged, especially toward base, densely leafy, 5–35 cm. **Leaves** in a basal rosette, sometimes also cauline, (1.4–)6–15(–28) × (0.2–)1–3.5(–5.6) cm; petiole (0.9–)3.5–8(–14) cm; blade oblanceolate to linear-oblanceolate, pinnatifid or 2-pinnatifid to shallowly pinnately lobed, margins sinuate-dentate or subentire, apex obtuse. **Flowers** usually 1–4, rarely more, opening per day, 1–2 hours before sunset; sepals (7–)12–25(–30) mm; petals yellow, fading reddish orange to purple, obcordate to obovate, (6–)13–35(–40) mm; filaments 6–16 mm, anthers 3–10 mm; style (32–)40–90(–100) mm, stigma exserted beyond anthers or surrounded by them. **Capsules** woody in age, sigmoid or curved to nearly straight, lanceoloid to ovoid, 4-angled, 10–45 (–60) × 4–8 mm, beak 4–15 mm, dehiscent 1/4–2/3 their length; sessile. **Seeds** usually numerous, in 2 rows per locule, obovoid to oblanceoloid, 3–3.5 × 1–1.4 mm, surface thickened above raphe and at distal end into U-shaped structure. $2n$ = 14.

Flowering Feb–May(–Jun). Sandy soil on flats, low hills and margins of sand dunes, along arroyos, roadsides, in desert scrub, grasslands and oak-grasslands; 30–1600 m; Ariz., Calif., Nev., N.Mex., Tex., Utah; Mexico (Baja California, Chihuahua, Sonora).

Oenothera primiveris has a complex variation pattern (W. L. Wagner 2005). In the western part of the range from southeastern California across southern Nevada to southern Utah counties of Emery, Kane, and Washington, and northwestern Mohave County, Arizona, plants generally have a gray appearance, with dense pubescence and larger flowers with widespread self-compatibility, but with scattered populations retaining self-incompatibility. Populations from south of the Mogollon Plateau to southern New Mexico, western Texas, Chihuahua, Sonora, and Baja California, Mexico, are greener in appearance with smaller to much smaller flowers, and are all self-compatible with occasional outcrossing or complete autogamy. The transitions between these two extremes are so extensive and more or less gradual that it is not possible to subdivide into two subspecies as has been done previously (Wagner).

17m. Oenothera Linnaeus sect. Contortae W. L. Wagner, Ann. Missouri Bot. Gard. 73: 478. 1986 [E]

Herbs perennial, acaulescent; from a thick, fleshy taproot. **Leaves** in a basal rosette, 2.6–4(–6.2) cm; blade margins pinnately lobed, lateral lobes often greatly reduced. **Inflorescences** solitary flowers in axils of leaves. **Flowers** opening near sunset, with strong, sweet scent; buds erect, quadrangular, without free tips; floral tube 27–45(–55) mm; sepals separating individually or in pairs; petals intensely yellow, fading deep salmon red, obcordate; stigma deeply divided into 4 linear lobes. **Capsules** moderately thin and flexible, lanceoloid, often contorted and twisted, 4-angled, gradually tapering to a long, slender, sterile apex, dehiscent 2/3–3/4 their length; sessile. **Seeds** numerous, in 1 row per locule, often forming 2 rows near base of capsule, obovoid, surface coarsely rugose, with turgid and collapsed papillae. $2n$ = 14.

Species 1: w United States.

58. Oenothera xylocarpa Coville, Contr. U.S. Natl. Herb. 4: 105, plate 8. 1893 [E]

Anogra xylocarpa (Coville) Small

Herbs perennial, acaulescent, densely short-hirsute, also sometimes sparsely long-hirsute distally; from a thick, fleshy taproot. **Leaves** in a basal rosette, 2.6–4.2(–6.2) × 1.4–4.2 cm; petiole 2.5–9(–11.5) cm; blade usually oblanceolate to obovate, sometimes suborbiculate, margins dentate, pinnately lobed, lateral lobes oblong to lanceolate, often absent or reduced to only a few lobes toward terminal lobe, base rounded to cordate. **Flowers** usually 1–3, rarely more, opening per day near sunset; buds erect, quadrangular, without free tips; floral tube 27–45(–55) mm; sepals 25–30 mm; petals intensely yellow, fading deep salmon red, obcordate, 25–38 mm; filaments 17–23 mm, anthers 7–10 mm; style 44–65(–80) mm, stigma somewhat exserted beyond anthers at anthesis. **Capsules** moderately thin and flexible, lanceoloid, falcate or sigmoid, often contorted and twisted, 4-angled, 35–90 × 7–11 mm, gradually tapering to a long, slender, sterile apex, 10–30(–40) mm, valves conspicuously wrinkled, dehiscent ⅔–¾ their length; sessile. **Seeds** numerous, in 1 row per locule, often forming 2 rows near base of capsule, obovoid, 2.4–3.2 × 1.3–1.7 mm, surface coarsely rugose. **2*n*** = 14.

Flowering Jun–Jul(–Aug). Open meadows, flats or slopes on loose granitic gravel, sand, or pumice in *Pinus jeffreyi* forests with *Artemisia tridentata*, or in *Pinus contorta* subsp. *murrayana* and *Abies magnifica* forests; 2200–3100 m; Calif., Nev.

Oenothera xylocarpa is known from three disjunct areas in California and adjacent Nevada: Mount Rose, Washoe County, Nevada; southern Sierra Nevada, southwestern Mono County, California, from the vicinity of Crestview south to Casa Diablo; and the area in the southern Sierra Nevada bounded by Horseshoe and Big Whitney meadows to the east and north, and Casa Vieja and Volcano Meadows to the south and west, west-central Inyo and eastern Tulare counties, California.

17n. Oenothera Linnaeus sect. Kleinia Munz in N. L. Britton et al., N. Amer. Fl., ser. 2, 5: 110. 1965

Herbs winter-annual or perennial, caulescent; from a taproot or lateral roots producing adventitious shoots. **Stems** 1–several arising from rosette, decumbent to ascending or erect, unbranched or with short, lateral branches, epidermis white or pink, not exfoliating. **Leaves** in a basal rosette and cauline, cauline 1–10 cm; blade margins entire, with few coarse teeth, lobed or pinnatifid. **Inflorescences** solitary flowers in axils of distal leaves. **Flowers** opening near sunset, with a sweet scent or nearly unscented; buds nodding by recurved floral tube, weakly quadrangular, without free tips; floral tube 10–30 mm; sepals separating in pairs or individually; petals white, fading pink, obcordate to obovate; stigma deeply divided into 4 linear lobes. **Capsules** straight or sometimes curved upward, cylindrical or fusiform, obtusely 4-angled, tapering toward base and apex, dehiscent ½ their length; sessile. **Seeds** numerous, in 2 rows per locule, ellipsoid to subglobose, surface regularly pitted, pits in longitudinal lines. **2*n*** = 14, 28.

Species 2 (2 in the flora): w United States, n Mexico.

Section *Kleinia* consists of two species of usually open sandy or rocky sites from the Chihuahuan, Sonoran, and southern portions of the Great Basin deserts to the Great Plains, from southern Utah to southeastern Montana and western North Dakota, south to northern Mexico (Chihuahua and Sonora) at 1000 to 3000 m; *Oenothera coronopifolia* generally occurs at higher elevations than *O. albicaulis*. As summarized by W. L. Wagner et al. (2007; see also K. E. Theiss et al. 2010), *O. albicaulis* is diploid (2*n* = 14) and *O. coronopifolia* has both diploid and tetraploid (2*n* = 14, 28) populations. Both species are self-incompatible and the vespertine flowers are pollinated by hawkmoths, especially *Hyles* and *Manduca*. P. H. Raven (1979) reported that *O. coronopifolia* had both self-incompatible and self-compatible populations. Theiss et al. found only self-incompatible plants in one population examined.

Section *Kleinia* is included within a strongly supported clade with members of sect. *Anogra* in recent molecular studies (R. A. Levin et al. 2004; M. E. K. Evans et al. 2005, 2009); placement of the subclade of the two sect. *Kleinia* species within the overall clade is not strongly resolved. For the present, the classification, based on morphology, is maintained until further resolution can be obtained. Section *Kleinia* can be distinguished by a number of morphological characteristics, including capsule shape, seeds in two rows per locule, and seeds with anatomy similar to that found in sect. *Oenothera* subsect. *Raimannia* but unlike that in sect. *Anogra* (W. L. Wagner et al. 2007; H. Tobe et al. 1987). P. A. Munz (1965) described sect. *Kleinia* as part of his subg. *Raimannia*, thus including these two white-flowered species in an otherwise yellow-flowered group because of similarities of the capsules and seeds. This group has been viewed as intermediate between sect. *Oenothera* subsect. *Raimannia* and sect. *Anogra* (Wagner et al.). Although both sects. *Anogra* and *Kleinia* have morphological synapomorphies that define them, a section combining them would be united by nodding buds and white petals; both characters are not unique morphological synapomorphies.

1. Herbs annual, from a taproot; floral tube mouth glabrous. . . . . 59. *Oenothera albicaulis*
1. Herbs perennial, lateral roots producing adventitious shoots; floral tube mouth conspicuously pubescent . . . . 60. *Oenothera coronopifolia*

59. Oenothera albicaulis Pursh, Fl. Amer. Sept. 2: 733. 1813 [F] [W]

Anogra albicaulis (Pursh) Britton; *A. bradburiana* Rydberg; *A. buffumii* A. Nelson; *A. confusa* Rickett; *A. ctenophylla* Wooton & Standley; *A. perplexa* Rydberg; *Oenothera albicaulis* var. *xanthosperma* H. Léveillé; *O. ctenophylla* (Wooton & Standley) Tidestrom; *O. sinuata* Linnaeus var. *bicolor* H. Léveillé

Herbs winter-annual, densely strigillose, also sparsely villous; from a taproot. **Stems** ascending to decumbent, 1–several from base, sometimes unbranched, erect or ascending, 5–30 cm. **Leaves** in a basal rosette and cauline, 1.5–10 × 0.3–2.5 cm; blade oblanceolate to oblong, margins subentire or coarsely dentate or pinnatifid. **Flowers** 1–3 opening per day near sunset; buds nodding, weakly quadrangular, without free tips; floral tube 15–30 mm, mouth glabrous; sepals 15–30 mm; petals white, fading pink, usually obcordate, sometimes obovate, (15–)20–35(–40) mm; filaments 11–17 mm, anthers 6–10 mm; style 25–50 mm, stigma exserted beyond anthers at anthesis. **Capsules** ascending to erect, usually straight, sometimes curved, cylindrical, weakly 4-angled, 20–40 × 3–4 mm, dehiscent ½ their length; sessile. **Seeds** in 2 rows per locule, ellipsoid to subglobose, 0.8–1.5 × 0.5–0.9 mm, surface regularly pitted, pits in longitudinal lines. **2*n*** = 14.

Flowering (Feb–)Mar–Jun(–Dec). Dry, usually sandy flats and slopes; 1000–2300 m; Ariz., Colo., Idaho, Kans., Mont., Nebr., N.Mex., N.Dak., Okla., S.Dak., Tex., Utah, Wyo.; Mexico (Chihuahua, Sonora).

Oenothera albicaulis is self-incompatible (W. L. Wagner et al. 2007; K. E. Theiss et al. 2010). *Oenothera albicaulis* has been reported from southern Nevada, but documentation is needed of its occurrence there.

Anogra pinnatifida Spach, *Baumannia pinnatifida* Spach, *Oenothera pinnatifida* Nuttall, *O. purshiana* Steudel, and *O. purshii* G. Don are illegitimate names that pertain here.

60. Oenothera coronopifolia Torrey & A. Gray, Fl. N. Amer. 1: 495. 1840 [E]

Anogra coronopifolia (Torrey & A. Gray) Britton

Herbs perennial, strigillose, usually also hirsute; from a taproot, lateral roots producing adventitious shoots. **Stems** ascending to erect, 1–several from base, these unbranched to well-branched, 10–60 cm. **Leaves** in a weakly developed basal rosette and cauline, 2–7 × 0.2–1.5 cm, axillary fascicles of reduced leaves often present; blade oblanceolate to oblong, margins usually pinnatifid, sometimes proximal ones coarsely few-toothed. **Flowers** 1–3 opening per day near sunset; buds nodding, weakly quadrangular, without free tips; floral tube 10–25 mm, mouth conspicuously pubescent, closed with straight, white hairs, 1–2 mm; sepals 10–20 mm; petals white, fading pink, ovate or shallowly obcordate, 10–15(–20) mm; filaments 10–15 mm, anthers 4–7 mm; style 17–42 mm, stigma exserted beyond anthers at anthesis. **Capsules** ascending to erect, straight, fusiform, weakly 4-angled, 10–20 × 3–5 mm, dehiscent ½ their length; sessile. **Seeds** in 2 rows per locule, ellipsoid to subglobose, 1.5–2 × 1.2–1.5 mm, surface regularly pitted, pits in longitudinal lines. 2*n* = 14, 28.

Flowering (Mar–)Jun–Aug(–Sep). Dry, open sites, grassy meadows, slopes, along drainages, foothills and mountains; 1500–3000 m; Ariz., Colo., Idaho, Nebr., N.Mex., S.Dak., Utah, Wyo.

Oenothera coronopifolia apparently has both self-incompatible and self-compatible populations (P. H. Raven 1979; W. L. Wagner et al. 2007; K. E. Theiss et al. 2010).

17o. Oenothera Linnaeus sect. Anogra (Spach) W. L. Wagner & Hoch, Syst. Bot. Monogr. 83: 179. 2007

Anogra Spach, Ann. Sci. Nat., Bot., sér. 2, 4: 164. 1835, based on *Baumannia* Spach, Hist. Nat. Vég. 4: 351. 1835, not de Candolle 1834; *Oenothera* [unranked] *Anogra* (Spach) Endlicher; *Oenothera* subg. *Anogra* (Spach) Reichenbach

Herbs winter-annual or perennial, caulescent; from a taproot, sometimes lateral roots producing adventitious shoots. **Stems** decumbent to ascending or erect, unbranched or with short, lateral branches, epidermis white or pink, exfoliating proximally. **Leaves** basal and cauline, sometimes forming conspicuous basal rosette, sometimes this weakly developed or absent (at least during flowering), 1–13(–26) cm; blade margins sinuate-dentate to pinnatifid, denticulate, subentire, or entire. **Inflorescences** solitary flowers in axils of distal leaves. **Flowers** opening near sunset, with a sweet scent or nearly unscented; buds nodding by recurved floral tube, usually sharply or bluntly quadrangular in cross section (sometimes fluted in distal ½ in *O. deltoides*), without free tips or free tips short (sometimes to 9 mm in *O. deltoides*); floral tube 15–40(–50) mm; sepals separatingin pairs or individually; petals white, fading pink, obcordate to obovate; stigma deeply divided into 4 linear lobes. **Capsules** straight, curved upward, spreading, or contorted, sometimes woody in age, cylindrical, obtusely 4-angled, gradually tapering from base to apex, dehiscent ½ to nearly throughout; sessile. **Seeds** numerous, in 1 row per locule, obovoid, surface minutely alveolate, but appearing smooth. 2*n* = 14, 28.

Species 8 (7 in the flora): w, c North America, n Mexico.

Section *Anogra* consists of eight species (17 taxa) native to western North America including Mexico, found usually in dry, sandy soil in a wide variety of habitats in the Chihuahuan, Great Basin, Mojave, and Sonoran deserts, to grasslands and open sites in montane forest, -50 to 3300 m. Only one species, *Oenothera wigginsii* Klein, occurs entirely outside the United States, while four others occur within the flora area but extend into northern Mexico. Section *Anogra* is included within a strongly supported clade with the two species of sect. *Kleinia* in recent molecular studies (R. A. Levin et al. 2004; M. E. K. Evans et al. 2005, 2009). The support levels for the topology within this clade are generally very weak, with only a few taxa grouping into moderately to strongly supported groups (for example, members of *O. pallida* complex, *O. deltoides* + *O. wigginsii*, *O. californica* + *O. arizonica* and *O. neomexicana*).

Species of sect. *Anogra* have vespertine flowers that are outcrossed and pollinated by hawkmoths or have flowers that are partly autogamous (D. P. Gregory 1964; W. M. Klein 1964, 1970). In *Oenothera deltoides* the capsule valves split open widely and disperse seeds, while the entire plant forms a so-called tumbleweed. Other species in the section appear to have more passive seed dispersal; the capsules dehisce while the plant remains rooted. The basal rosette may not be evident at time of flowering or not developed. When this is the case, or when the dimensions of the basal leaves are very similar to the cauline ones, only one range for leaf dimensions is given.

SELECTED REFERENCES Klein, W. M. 1964. A Biosystematic Study of Four Species of *Oenothera* Subgenus *Anogra*. Ph.D. dissertation. Claremont Graduate School. Klein, W. M. 1970. The evolution of three diploid species of *Oenothera* subgenus *Anogra* (Onagraceae). Evolution 24: 578–597.

1. Herbs winter-annual or short-lived perennial from a taproot.
   2. Sepals conspicuously maroon-spotted . . . . .64. *Oenothera arizonica*
   2. Sepals without maroon spots.
      3. Plants villous throughout, also strigillose on leaves and distal parts; leaf blade margins coarsely repand-dentate or -pinnatifid . . . . 62. *Oenothera engelmannii*
      3. Plants villous, strigillose, or glabrous, sometimes more densely villous or strigillose distally; leaf blade margins subentire to sinuate-dentate or remotely denticulate, sometimes pinnatifid.
         4. Capsules 2.5–5 mm diam.; sepals (13–)15–35 mm . . . . 66. *Oenothera deltoides* (in part)
         4. Capsules 1.5–2.5 mm diam.; sepals 10–18 mm . . . . 67. *Oenothera pallida* (in part)
1. Herbs perennial, from a taproot, also with lateral roots producing adventitious shoots or with long, fleshy roots.
   5. Buds fluted in distal ½, with free tips 1–9 mm; plants from relatively long, fleshy roots . . . . .66. *Oenothera deltoides* (in part)
   5. Buds quadrangular in distal ½, with free tips 0–4 mm; plants from a taproot and with lateral roots producing adventitious shoots.
      6. Plants glabrous, sometimes strigillose on leaves and/or glandular puberulent distally, at least on floral tube; leaf blades 0.3–0.6(–1) cm wide, narrowly oblong to oblong-lanceolate, margins usually entire . . . . .61. *Oenothera nuttallii*
      6. Plants villous, strigillose, glabrate, or glabrous, not glandular puberulent; leaf blades (0.3–)1–2.5 cm wide, usually ovate, oblong to lanceolate or linear-lanceolate, oblanceolate, or spatulate, rarely rhombic-ovate, margins usually sinuate-dentate to pinnatifid or subentire, rarely entire.
         7. Capsules erect or strongly ascending . . . . .63. *Oenothera neomexicana*
         7. Capsules spreading to reflexed.
            8. Capsules 2–3.5 mm diam.; stems decumbent or ascending. . . . . . . 65. *Oenothera californica*
            8. Capsules 1.5–2.5 mm diam.; stems erect or ascending . . . . . . 67. *Oenothera pallida* (in part)

61. Oenothera nuttallii Sweet, Hort. Brit. ed. 2, 199. 1830 [E]

Oenothera albicaulis Nuttall, Gen. N. Amer. Pl. 1: 245. 1818, not Pursh 1813; *O. albicaulis* Pursh var. *nuttallii* (Sweet) Engelmann

Herbs perennial, mostly glabrous, sometimes strigillose on leaves and/or glandular puberulent on distal parts, at least on floral tube; from a taproot, lateral roots producing adventitious shoots. **Stems** erect, often branched, 30–100 cm. **Leaves** in a basal rosette and cauline, rosette weakly developed or absent, at least during flowering, 2–6(–10.5) × 0.3–0.6 (–1) cm; petiole 0–2 cm; blade narrowly oblong to oblong-lanceolate, margins usually entire, sometimes remotely denticulate or repand-denticulate. **Flowers** 1–several opening per day near sunset; buds nodding, weakly quadrangular, with free tips 1–2 mm; floral tube 15–40 mm; sepals 20–30 mm, not spotted; petals white, fading pink, broadly obovate or obcordate, 15–30 mm; filaments 15–18 mm, anthers 8–10 mm; style 35–45 mm, stigma exserted beyond anthers at anthesis. **Capsules** erect or ascending, woody in age, straight or slightly curved, cylindrical, obtusely 4-angled, especially toward base, tapering slightly from base to apex, 20–30 × 2–3 mm; sessile. **Seeds** numerous, in 1 row per locule, reddish dark brown, narrowly obovoid, 1.5–2.3 mm. $2n$ = 14, 28.

Flowering Jun–Sep. Dry, sandy or rocky prairies, open wooded hillsides, disturbed areas, roadsides; 500–2200(–2900) m; Alta., Man., Sask.; Colo., Ill., Mich., Minn., Mont., Nebr., N.Dak., S.Dak., Wis., Wyo.

Oenothera nuttallii had been assumed to be self-incompatible (W. L. Wagner et al. 2007), but K. E. Theiss et al. (2010) determined two plants to be self-compatible.

Anogra nuttalliana Spach and *Baumannia nuttalliana* Spach are illegitimate names that pertain here.

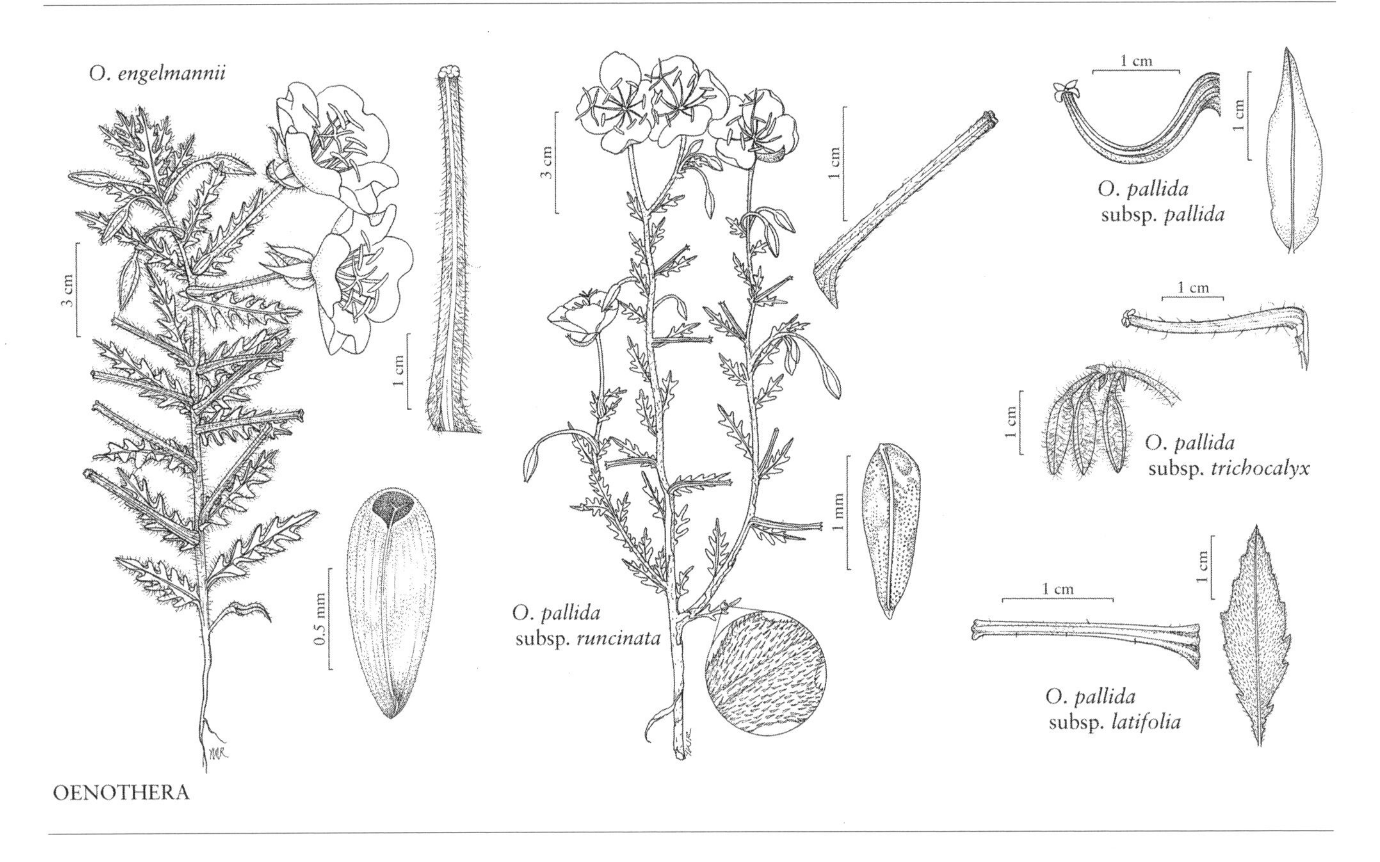

OENOTHERA

62. Oenothera engelmannii (Small) Munz, Amer. J. Bot. 18: 316. 1931 (as engelmanni) E F

Anogra pallida (Lindley) Britton var. *engelmannii* Small, Bull. Torrey Bot. Club 23: 176. 1896 (as engelmanni); *A. engelmannii* (Small) Wooton & Standley

Herbs winter-annual, conspicuously villous throughout, also strigillose on leaves and distal parts; from a taproot. **Stems** erect, unbranched or with few, spreading branches, 30–50(–80) cm. **Leaves** in a basal rosette and cauline, rosette weakly developed or absent, at least during flowering, (1–)2–6(–8) × 1–2(–3) cm; sessile; blade lanceolate to oblong lanceolate, proximal ones sometimes oblanceolate, margins coarsely repand-dentate or -pinnatifid. **Flowers** 1–several opening per day near sunset; buds nodding, weakly quadrangular, without free tips; floral tube 20–30 mm; sepals 13–21 mm, not spotted or with scattered small, maroon spots; petals white, fading pink, broadly obovate or obcordate, 15–30 mm; filaments 14–16 mm, anthers 6–8 mm; style 40–50 mm, stigma exserted beyond anthers at anthesis. **Capsules** widely spreading, woody in age, straight or slightly curved, cylindrical, obtusely 4-angled, especially toward base, tapering gradually from base to apex, 30–60 × 2–3 mm; sessile. **Seeds** numerous, in 1 row per locule, brown, narrowly obovoid, 1–1.5 mm. $2n = 14$.

Flowering Apr–Aug(–Sep). Sandy prairies, dunes, disturbed areas, roadsides; 500–1300 m; Colo., Kans., N.Mex., Okla., Tex.

Oenothera engelmannii is self-incompatible (W. L. Wagner et al. 2007; K. E. Theiss et al. 2010). It has a relatively narrow distribution in sandy areas of eastern New Mexico and western Texas, extending to southeastern Colorado, western Oklahoma, and southwestern Kansas. The flower size seems to vary, with larger flowers in eastern New Mexico and considerably smaller flowers in the eastern part of its range.

63. Oenothera neomexicana (Small) Munz, Amer. J. Bot. 18: 317. 1931 E

Anogra neomexicana Small, Bull. Torrey Bot. Club 23: 176. 1896 (as neo-mexicana)

Herbs perennial, glabrate proximally, strigillose and villous distally; from a taproot, also with lateral roots producing adventitious shoots. **Stems** erect or ascending, unbranched or branched, 30–60 cm. **Leaves** in a basal rosette and cauline, rosette weakly developed or absent, at least during flowering, 3–9 × (0.6–)1–2.5 cm; petiole 0–2 cm; blade oblong to lanceolate or narrowly ovate, margins irregularly sinuate-dentate. **Flowers** 1–several opening per day near sunset; buds nodding, weakly quadrangular, with free tips 0.5–4 mm; floral tube 30–50 mm; sepals 20–30 mm, not spotted; petals white, fading pink, broadly obovate, 20–30 mm; filaments 10–15 mm, anthers 8–15 mm; style 50–70 mm, stigma exserted beyond anthers at anthesis. **Capsules** erect or strongly ascending, not woody, straight or slightly curved, subcylindrical, obtusely 4-angled, tapering gradually from base to apex, 20–30 × 2–3 mm; sessile. **Seeds** numerous, in 1 row per locule, dark brown, narrowly obovoid, 1.5 mm. 2*n* = 14.

Flowering (Mar–)Jun–Jul(–Sep). Uncommon, in rocky or sandy clay or loamy soil in coniferous forest openings, stream valleys, roadsides; 1500–3300 m; Ariz., N.Mex.

Oenothera neomexicana is known from central to western New Mexico west of the Rio Grande Valley, except for the Organ Mountains, and eastern and central Arizona from the White Mountains south to Mount Graham and northwestward across the Mogollon Rim in Coconino and Yavapai counties. *Oenothera neomexicana* had been assumed to be self-incompatible (W. L. Wagner et al. 2007), but K. E. Theiss et al. (2010) determined one population sampled to be consistently self-compatible.

64. Oenothera arizonica (Munz) W. L. Wagner, Novon 8: 308. 1998

Oenothera deltoides Torrey & Frémont var. *arizonica* Munz, Amer. J. Bot. 18: 315. 1931; *O. avita* (W. M. Klein) W. M. Klein subsp. *arizonica* (Munz) W. M. Klein; *O. californica* (S. Watson) S. Watson subsp. *arizonica* (Munz) W. M. Klein

Herbs winter-annual, younger parts sparsely to densely strigillose and sparsely to densely hirsute, older stems glabrate; from a taproot. **Stems** ascending to erect, with decumbent branches, thickened at base, tapering toward apex, 10–35(–60) cm. **Leaves** in a basal rosette and cauline, basal 5–10(–26) × 0.6–1.5(–3.5) cm, cauline 5–8(–15.5) × 1–2 cm; petiole 0–12 cm; blade lanceolate to oblanceolate, margins pinnatifid or sometimes coarsely serrate. **Flowers** 1–several opening per day near sunset; buds nodding, weakly quadrangular, without free tips; floral tube 26–31 mm; sepals 19–26 mm, conspicuously maroon-spotted, each spot at base of a long hair; petals white, fading pink to deep pink, broadly obovate or obcordate, 16–26(–36) mm; filaments 9–15 mm, anthers 7–9 mm; style 45–50 mm, stigma exserted beyond anthers at anthesis. **Capsules** spreading, woody in age, curved upward, or distal end recurved, cylindrical, obtusely 4-angled, especially toward base, tapering gradually from base to apex, 30–80 × 2.5–3.5 mm; sessile. **Seeds** numerous, in 1 row per locule, light brown to yellowish brown with dark purple splotches, obovoid, 1.6–2 mm. 2*n* = 14.

Flowering (Oct–)Feb–May. Gravelly or sandy soil, along watercourses, disturbed sites; 200–1400 m; Ariz.; Mexico (Sonora).

Oenothera arizonica occurs in southern Arizona from Maricopa and Yuma counties to Cochise County, and from scattered localities in northern Sonora, Mexico, including Cerro Tepopa, Puerto Libertad, and Tastiota. The populations from southwestern Arizona (Yuma County) southward to Sonora often grow on low dunes.

Populations from sand dunes in Yuma County, Cabeza Prieta National Wildlife Refuge, appear to be a large phenotype of *Oenothera arizonica* that differ from all other specimens in the size of vegetative parts and flowers, and comprise all of the atypical measurements given in the description. *Oenothera arizonica* typically grows on dunes in Sonora, but rarely so in Arizona. Populations growing on dunes should be studied further and compared to non-dune populations in the northern and eastern portion of the range. *Oenothera arizonica* is self-compatible (W. L. Wagner et al. 2007; K. E. Theiss et al. 2010).

65. Oenothera californica (S. Watson) S. Watson in W. H. Brewer et al., Bot. California 1: 223. 1876

Oenothera albicaulis Pursh var. *californica* S. Watson, Proc. Amer. Acad. Arts 8: 582. 1873; *Anogra californica* (S. Watson) Small; *O. pallida* Lindley var. *californica* (S. Watson) Jepson

Herbs perennial, densely strigillose, sometimes also villous, or glabrous; from a taproot, lateral roots producing adventitious shoots, or rarely with fleshy underground horizontal rootstock (subsp. *eurekensis*). **Stems** ascending or decumbent, usually branched from near base, sometimes new rosettes forming at branch apex when buried in drifting sand, 10–60 cm. **Leaves** in a basal rosette and cauline, rosette sometimes weakly developed or absent, at least during flowering, 1–13 × 0.5–2 cm; petiole 0–2 (–4.5) cm; blade oblong to oblanceolate or spatulate, sometimes rhombic-ovate, margins entire or weakly to conspicuously dentate or pinnatifid. **Flowers** 1–several opening per day near sunset; buds nodding, weakly quadrangular, with free tips 0–0.8 mm; floral tube 20–40 mm; sepals 15–30 mm, not spotted; petals white, fading pink to deep pink, broadly obcordate, 15–35(–40) mm; filaments 10–17 mm, anthers 5–10 mm; style 30–60 mm, stigma exserted beyond anthers at anthesis. **Capsules** spreading to ascending, woody in age, often curved upward, cylindrical, obtusely 4-angled, tapering slightly from base to apex, 20–80 × 2–3.5 mm; sessile. **Seeds** numerous, in 1 row per locule, olive-brown or yellowish brown to black, sometimes with minute purple dots, obovoid, 1–2.5 mm.

Subspecies 3 (3 in the flora): w United States, nw Mexico.

Most populations of *Oenothera californica* are self-incompatible (W. L. Wagner et al. 2007; K. E. Theiss et al. 2010), but some populations of subsp. *californica* are self-compatible. All chromosome counts indicate that subspp. *avita* and *eurekensis* are diploid ($2n = 14$) and those of subsp. *californica* are tetraploid ($2n = 28$). *Oenothera californica* is polymorphic with subspp. *avita* and *californica* being very similar, and differing primarily in ecology, distribution, and relatively minor differences in leaf morphology and ploidy level, while the sand dune-restricted subsp. *eurekensis* is more distinctive in both morphology and habitat.

1. Plants with fleshy underground horizontal rootstocks; stems sometimes with new rosettes forming at stem apex when becoming buried in drifting sand; leaf blades rhombic-ovate to oblanceolate, margins entire or weakly dentate . . . . 65c. *Oenothera californica* subsp. *eurekensis*
1. Plants with a taproot and adventitious shoots from lateral roots; stems not with new rosettes forming at branch apex; leaf blades oblong to oblanceolate to spatulate, margins entire or weakly to conspicuously dentate or pinnatifid.
   2. Leaf blade margins usually entire or weakly dentate, sometimes more conspicuously dentate to pinnatifid; capsules 30–55 mm . . . . 65a. *Oenothera californica* subsp. *californica*
   2. Leaf blade margins conspicuously dentate to pinnatifid, rarely some or all of them entire or weakly dentate; capsules 20–80 mm . . . . 65b. *Oenothera californica* subsp. *avita*

65a. Oenothera californica (S. Watson) S. Watson subsp. **californica**

Oenothera albicaulis Pursh var. *melanosperma* H. Léveillé; *O. californica* var. *glabrata* Munz

Herbs perennial, densely strigillose and villous, sometimes glabrous; from a taproot, lateral roots producing adventitious shoots. **Stems** ascending to decumbent, unbranched or branched, new rosettes not forming at branch apex, 10–40 cm. **Leaves:** blade oblong to oblanceolate to spatulate, margins usually entire or weakly dentate, sometimes more conspicuously dentate to pinnatifid. **Flowers:** floral tube 20–40 mm; sepals 15–25 mm; petals 15–30 mm. **Capsules** 30–55 mm. $2n = 28$.

Flowering Apr–Jul(–Sep). Sandy or gravelly areas, open, coastal-sage scrub, chaparral, oak woodlands; 90–2000 m; Calif.; Mexico (Baja California).

Subspecies *californica* occurs in southwestern California from San Luis Obispo County south and into the Little San Bernardino Mountains to northern Baja California (Sierra de San Pedro Mártir).

65b. Oenothera californica (S. Watson) S. Watson subsp. **avita** W. M. Klein, Aliso 5: 179. 1962 [E]

Oenothera avita (W. M. Klein) W. M. Klein; *O. californica* var. *avita* (W. M. Klein) S. L. Welsh & N. D. Atwood

Herbs perennial, densely strigillose and villous; from a taproot, lateral roots producing adventitious shoots. **Stems** ascending to decumbent, unbranched or branched, new rosettes not forming at branch apex, 10–40 cm. **Leaves:** blade oblong to oblanceolate or spatulate, margins usually conspicuously dentate to pinnatifid, rarely some or all entire or weakly dentate. **Flowers:** floral tube 25–35 mm; sepals 15–30 mm; petals 25–35(–40) mm. **Capsules** 20–80 mm. $2n = 14$.

Flowering (Apr–)May–Jul. Sandy-gravelly flats, desert scrub, Joshua tree woodlands, oak woodlands, pinyon-juniper or pine woodlands; 800–2500 m; Ariz., Calif., Nev., Utah.

Subspecies *avita* occurs in southeastern California (south of areas just north of Bishop) mostly to the east of subsp. *californica*, eastward to northwestern Arizona, southern half of Nevada, and southwestern Utah. Some populations in the mountains of San Diego County, California, and northern Baja California (Sierra de San Pedro Mártir) appear to fit within subsp. *avita* (J. Rebman, pers. comm.).

65c. Oenothera californica (S. Watson) S. Watson subsp. **eurekensis** (Munz & J. C. Roos) W. M. Klein, Aliso 5: 179. 1962 • Eureka Dunes evening primrose [C] [E]

Oenothera deltoides Torrey & Frémont subsp. *eurekensis* Munz & J. C. Roos, Aliso 3: 118, fig. 7. 1955; *O. avita* (W. M. Klein) W. M. Klein subsp. *eurekensis* (Munz & J. C. Roos) W. M. Klein

Herbs perennial, densely strigillose and villous; with deep-seated, fleshy underground parts from underground horizontal rootstocks. **Stems** sprawling to decumbent, sometimes new rosettes forming at stem apex when becoming buried in drifting sand, 15–60 cm. **Leaves:** blade rhombic-ovate to oblanceolate, margins entire or weakly dentate. **Flowers:** floral tube 25–30 mm; sepals 15–25 mm; petals 20–30 mm. **Capsules** 30–70 mm. **2*n*** = 14.

Flowering (Mar)Jun–Jul(Sep). Sand dunes; of conservation concern; 900–1200 m; Calif.

Subspecies *eurekensis* is known from three main areas within the Eureka Dunes system, Inyo County. It is federally listed as endangered and is in the Center for Plant Conservation's National Collection of Endangered Plants.

66. Oenothera deltoides Torrey & Frémont in J. C. Frémont, Rep. Exped. Rocky Mts., 315. 1845

Anogra deltoides (Torrey & Frémont) Small

Herbs usually winter-annual, sometimes perennial, glabrous, glandular puberulent, strigillose, and/or villous, sometimes more villous distally, hairs sometimes very curly, especially on flower parts; from a taproot or relatively long, fleshy roots. **Stems:** central stem usually erect, usually thickened at base and spongy, branched or unbranched, branches few–several, slender, decumbent to ascending, from base, usually encircling central stem in older plants, 10–40(–100) cm. **Leaves** in a basal rosette and cauline, rosette usually well developed (except subsp. *howellii*), basal 5–25 × 1–5 cm, cauline 4–12(–18) × 0.5–4 cm; petiole 1.5–8 cm; blade rhombic-obovate, lanceolate, or oblanceolate, margins subentire, dentate, or pinnatifid. **Flowers** 1–several opening per day near sunset; buds nodding, weakly or strongly quadrangular or fluted in distal ½, with free tips 0–9 mm; floral tube 20–40 mm; sepals (13–)15–35 mm, not spotted; petals white, fading pink to deep pink, broadly obovate or obcordate, 15–44 mm; filaments 8–15 mm, anthers 5–14 mm; style 35–60 mm, stigma exserted beyond anthers at anthesis. **Capsules** spreading, straight to curved, becoming somewhat woody in age, cylindrical to slightly 4-angled, widest toward base, tapering from base to apex, (15–)30–80 × 1.5–5 mm; sessile. **Seeds** numerous, in 1 row per locule, buff with dark spots or black, narrowly obovoid, 1.5–2.8 mm.

Subspecies 5 (5 in the flora): w United States, nw Mexico.

Oenothera deltoides is self-incompatible or self-compatible (W. M. Klein 1964; W. L. Wagner et al. 2007; K. E. Theiss et al. 2010).

1. Herbs perennial, from long, fleshy roots; stems not thickened at base, branches erect or ascending, not encircling stems in older plants . . . . 66b. *Oenothera deltoides* subsp. *howellii*
1. Herbs annual or short-lived perennial from a taproot; stems usually thickened at base, branches ascending or decumbent, often encircling stems in older plants.
   2. Flower buds fluted or strongly quadrangular in distal ½, without free tips, villous with curly hairs, sometimes glabrous; leaf blade margins sinuate-dentate to pinnatifid; petals 15–25(–30) mm; capsules 15–25(–30) mm . . . . 66e. *Oenothera deltoides* subsp. *piperi*
   2. Flower buds weakly or strongly quadrangular in distal ½, with free tips 0–3 mm, strigillose, sparsely to moderately villous, or glabrous; leaf blade margins sinuate-dentate or subentire, rarely pinnatifid; petals 15–44 mm; capsules 40–80 mm.
      3. Flower buds with free tips 1–3 mm, quadrangular in distal ½; plants strigillose, especially distally . . . . .66d. *Oenothera deltoides* subsp. *ambigua*
      3. Flower buds with free tips 0–1.5 mm, weakly quadrangular in distal ½; plants glabrous, villous, or strigillose.
         4. Herbs annual, strigillose, sometimes also villous; capsules 2–3.5 mm diam.; flower buds with free tips 0–1.5 mm . . . 66a. *Oenothera deltoides* subsp. *deltoides*
         4. Herbs short-lived perennial or sometimes annual, glabrous or sparsely villous, rarely also strigillose; capsules 3–5 mm diam.; flower buds without free tips . . . . 66c. *Oenothera deltoides* subsp. *cognata*

66a. Oenothera deltoides Torrey & Frémont subsp. **deltoides**

Oenothera deltoides var. *cineracea* (Jepson) Munz; *O. kleinii* W. L. Wagner & S. W. Mill; *O. trichocalyx* Nuttall var. *cineracea* Jepson

Herbs annual, strigillose, sometimes also villous; from a taproot. **Stems:** central stem erect, usually thickened, with several leafy, ascending, slender branches from near base, encircling stems in older plants, 20–60(–100) cm. **Leaves** basal and cauline; blade rhombic-lanceolate, becoming oblanceolate to lanceolate distally, margins usually coarsely sinuate-dentate to subentire, rarely pinnatifid. **Flowers:** buds weakly quadrangular, with free tips 0–1.5 mm; sepals 20–35 mm; petals 18–44 mm. **Capsules** 40–80 × 2–3.5 mm. **2*n*** = 14.

Flowering (Mar–)Jun–Jul(–Sep). Sandy places, dunes, Mojave and Sonoran deserts; -50–1300 m; Ariz., Calif., Nev.; Mexico (Baja California, Sonora).

W. M. Klein (1964) determined most populations of subsp. *deltoides* that were studied to be self-incompatible, but found a few in Riverside County, California, to be self-compatible. *Oenothera kleinii* was described from a single, small roadside population near Wolf Creek Pass in Colorado. Subsequent study and discussion with W. A. Weber suggested that this population was not native to the location, but grew there accidently; it has not been collected again in Colorado. Further study of morphological diversity of *O. deltoides* has shown that the distinguishing characters mentioned when *O. kleinii* was described, such as longer hairs and larger capsules and seeds, do in fact occur in plants within the natural range of *O. deltoides*.

66b. Oenothera deltoides Torrey & Frémont subsp. **howellii** (Munz) W. M. Klein, Aliso 5: 180. 1962 C E

Oenothera deltoides var. *howellii* Munz, Aliso 2: 81. 1949

Herbs perennial, densely strigillose, also villous and glandular puberulent distally; with relatively long, fleshy roots, grayish. **Stems** erect or ascending, several from base, not thickened near base, with numerous shorter lateral branches, these not encircling stems in older plants, 40–80 cm. **Leaves** cauline, without clear basal rosette, at least at anthesis; blade lanceolate, margins runcinate-pinnatifid. **Flowers:** buds fluted in distal ½, with free tips 1–9 mm; sepals 20–30 mm; petals 20–40 mm. **Capsules** 45–60 × 3–4 mm. $2n = 14$.

Flowering (Mar–)Jun–Jul(–Sep). Sand dunes and bluffs; of conservation concern; 0–30 m; Calif.

W. M. Klein (1964) determined subsp. *howellii* to be self-incompatible.

Subspecies *howellii* is known from the Antioch Dunes in Contra Costa County. Subspecies *howellii* is federally listed as endangered and is in the Center for Plant Conservation's National Collection of Endangered Plants.

66c. Oenothera deltoides Torrey & Frémont subsp. **cognata** (Jepson) W. M. Klein, Aliso 5: 180. 1962 E

Oenothera trichocalyx Nuttall var. *cognata* Jepson, Man. Fl. Pl. Calif., 681. 1925; *O. deltoides* var. *cognata* (Jepson) Munz

Herbs short-lived perennial, sometimes annual, glabrous or sparsely villous, rarely also strigillose; from a taproot. **Stems:** central stem often thickened near base, branched from base, branches ascending, not encircling stems in older plants, 20–40 cm. **Leaves** basal and cauline; blade rhombic-lanceolate, becoming oblanceolate to lanceolate distally, margins usually coarsely sinuate-dentate to subentire, rarely pinnatifid. **Flowers:** buds weakly quadrangular, without free tips; sepals 20–35 mm; petals 25–40 mm. **Capsules** 40–60 × 3–5 mm. $2n = 14$.

Flowering (Mar–)Jun–Jul(–Sep). Sandy places; 30–500 m; Calif.

W. M. Klein (1964) determined subsp. *cognata* to be self-compatible. It is known from the San Joaquin Valley. Some specimens from Contra Costa County at the northern edge of the range of subsp. *cognata* appear to be somewhat intermediate toward subsp. *howellii* in having minute free sepal tips in bud and the plant is more pubescent.

66d. Oenothera deltoides Torrey & Frémont subsp. **ambigua** (Munz) W. M. Klein, Aliso 5: 179. 1962 E

Oenothera deltoides var. *ambigua* Munz, Amer. J. Bot. 18: 315. 1931, based on *O. ambigua* S. Watson, Proc. Amer. Acad. Arts 14: 293. 1879, not (Nuttall) Sprengel 1825; *Anogra simplex* Small; *O. albicaulis* Pursh var. *decumbens* S. Watson; *O. deltoides* var. *decumbens* (S. Watson) Munz; *O. simplex* (Small) Tidestrom

Herbs annual, strigillose, especially distally; from a taproot. **Stems:** central stem often thickened near base, branched from base, branches ascending to decumbent, not encircling stems in older plants, 5–25 cm. **Leaves** basal and cauline; blade rhombic-obovate to -lanceolate, becoming oblanceolate to lanceolate distally, margins subentire, sometimes coarsely sinuate-dentate distally. **Flowers:** buds quadrangular in distal ½, with free tips 1–3 mm; sepals 20–30 mm; petals 18–35 mm. **Capsules** 40–70 × 1.5–5 mm. $2n = 14$.

Flowering (Mar–)Jun–Jul(–Sep). Sandy soil in Mojave Desert with *Ambrosia*, *Cylindropuntia*, or *Larrea*; 200–1300 m; Ariz., Nev., Utah.

Subspecies *ambigua* is restricted to a relatively small area of Mojave Desert in southeastern Nevada and adjacent areas in southwestern Utah and Arizona. W. M. Klein (1964) determined subsp. *ambigua* to be self-incompatible.

66e. Oenothera deltoides Torrey & Frémont subsp. **piperi** (Munz) W. M. Klein, Aliso 5: 180. 1962 [E]

Oenothera deltoides var. *piperi* Munz, Amer. J. Bot. 18: 314. 1931; *O. trichocalyx* Nuttall var. *piperi* (Munz) Jepson

Herbs annual, usually villous, hairs relatively long, curly, especially distally and on buds, sometimes glabrous; from a taproot. **Stems:** central stem thickened proximally, unbranched or with several lateral, ascending to decumbent branches, 3–30(–40) cm. **Leaves** basal and cauline; blade rhombic, becoming lanceolate distally, margins deeply sinuate-dentate to pinnatifid. **Flowers:** buds fluted or strongly quadrangular in distal ½, without free tips; sepals 13–22(–27) mm; petals 15–25(–30) mm. **Capsules** 15–25(–30) × 3–5 mm. $2n = 14$.

Flowering (Mar–)Jun–Jul(–Sep). Sandy soil or dunes in Great Basin Desert with *Artemisia*, *Ericameria*, or *Sarcobatus*; 900–1900 m; Calif., Nev., Oreg.

W. M. Klein (1964) determined subsp. *piperi* to be self-incompatible. It occurs in the northern part of the range of *O. deltoides*, from northeastern California to southern Oregon and the western half of Nevada.

67. Oenothera pallida Lindley, Bot. Reg. 14: plate 1142. 1828 [F]

Anogra pallida (Lindley) Britton; *Oenothera albicaulis* Pursh var. *pallida* (Lindley) H. Léveillé

Herbs annual or perennial, glabrous, strigillose and/or villous, sometimes more villous distally, especially on flower parts; from a taproot, sometimes lateral roots producing adventitious shoots. **Stems** erect or ascending, single to several from base, unbranched or many-branched throughout, 10–50(–70) cm. **Leaves** cauline, rosette usually weakly developed or absent, at least during flowering, sometimes well developed, 1–5(–7.8) × 0.3–1(–1.5) cm; petiole 0–2(–4.5) cm; blade lanceolate, oblong, linear-lanceolate, or ovate, margins subentire or remotely denticulate, deeply sinuate-dentate, or pinnatifid, sometimes repand. **Flowers** 1–several opening per day near sunset; buds nodding, weakly quadrangular, with free tips 0–2 mm; floral tube 15–40 mm; sepals 10–30 mm, not spotted; petals white, fading pink to deep pink, broadly obovate or obcordate, (10–)15–25(–40) mm; filaments 9–15 mm, anthers 3–10 mm; style 25–55 mm, stigma exserted beyond anthers at anthesis. **Capsules** spreading to reflexed, straight to curved or contorted, cylindrical, obtusely 4-angled, tapering slightly from base to apex, 15–60 × 1.5–2.5 mm; sessile. **Seeds** numerous, in 1 row per locule, brownish with dark spots or black, narrowly obovoid, 1.5–2.2 mm.

Subspecies 4 (4 in the flora): w, c North America, n Mexico.

Oenothera pallida is a poorly understood species currently subdivided into four subspecies (W. L. Wagner et al. 2007) that differ largely in aspect, leaf division, capsule configuration, and pubescence. The variation pattern is rather complex with almost no diagnostic character uniformly distinguishing any one of the subspecies. Instead, each of the subspecies, which are mostly geographically separated although there is some level of overlap, have diagnostic suites of characters that maintain their linkage some of the time, but break down across the geographic area of each so that no single character uniquely identifies it. Each subspecies is characterized by leaf, pubescence, and, often, habit features. The issues with the integrity and intergradations of the subspecies are discussed below.

Oenothera pallida has been determined to be self-incompatible (W. L. Wagner et al. 2007), but K. E. Theiss et al. (2010) determined that although most populations of subsp. *pallida* are self-incompatible, one near Salt Lake City is self-compatible.

1. Herbs annual, sometimes perennial from a taproot, when perennial, sometimes with lateral roots producing adventitious shoots, strigillose throughout and villous distally, especially on flower parts . . . . 67d. *Oenothera pallida* subsp. *trichocalyx*
1. Herbs perennial from a taproot and with lateral roots producing adventitious shoots, glabrous, strigillose, or sparsely villous.
   2. Plants glabrous, sometimes strigillose, rarely sparsely villous; leaf blade margins usually subentire or remotely denticulate, rarely pinnatifid; capsules usually contorted to curved . . . . 67a. *Oenothera pallida* subsp. *pallida*
   2. Plants usually strigillose, rarely villous or glabrous; leaf blade margins shallowly sinuate-dentate or denticulate, or deeply sinuate-dentate to pinnatifid, rarely only dentate; capsules usually straight or curved, sometimes contorted.
      3. Leaf blades (0.4–)0.7–1.5 cm wide, margins shallowly sinuate-dentate or denticulate . . . . 67b. *Oenothera pallida* subsp. *latifolia*
      3. Leaf blades 0.4–1(–1.5) cm wide, margins usually deeply sinuate-dentate to pinnatifid, rarely dentate only . . . . 67c. *Oenothera pallida* subsp. *runcinata*

67a. Oenothera pallida Lindley subsp. **pallida** [E] [F]

Anogra leptophylla (Torrey & A. Gray) Rydberg; *O. pallida* var. *idahoensis* Munz; *O. pallida* var. *leptophylla* Torrey & A. Gray

Herbs perennial, usually glabrous, sometimes strigillose, rarely sparsely villous; from a taproot, lateral roots producing adventitious shoots. **Stems** usually branched throughout. **Leaves:** rosette not present at anthesis, 2–6 × 0.3–0.8(–1) cm; blade lanceolate to linear-lanceolate or oblong, margins usually subentire or remotely denticulate, rarely pinnatifid, usually repand. **Flowers:** buds with free tips 0.5–2 mm; floral tube 20–35 mm; sepals 12–18 mm; petals 12–25 mm. **Capsules** spreading, contorted to curved. $2n = 14$.

Flowering May–Sep. Sandy soil, dunes, disturbed areas, alkaline soil; 1100–2000 m; B.C.; Ariz., Colo., Idaho, Nev., N.Mex., Oreg., Utah, Wash., Wyo.

The distribution of subsp. *pallida* centers in the intermountain region from Oregon and Washington east of the mountains, adjacent southern British Columbia, south through southern Idaho, Wyoming, western half of Utah, southern Nevada, to northern Arizona, northwestern New Mexico, and adjacent southwestern Colorado. There are morphological intermediates with subspp. *runcinata* and *trichocalyx*. Densely strigillose plants occur within the range of subsp. *pallida*, especially near the St. Anthony Dunes in Idaho, and have been referred to as var. *idahoensis*.

Baumannia douglasiana Spach is an illegitimate name that pertains here.

67b. Oenothera pallida Lindley subsp. **latifolia** (Rydberg) Munz in N. L. Britton et al., N. Amer. Fl., ser. 2, 5: 119. 1965 [E] [F]

Oenothera pallida var. *latifolia* Rydberg, Contr. U.S. Natl. Herb. 3: 159. 1895; *Anogra cinerea* Rydberg; *A. latifolia* (Rydberg) Rydberg; *A. pallida* (Lindley) Britton var. *latifolia* (Rydberg) Small; *O. latifolia* (Rydberg) Munz

Herbs perennial, densely strigillose throughout; from a taproot, lateral roots producing adventitious shoots. **Stems** usually several, branched from base, sometimes unbranched. **Leaves:** rosette not present at anthesis, 1–5(–7) × (0.4–)0.7–1.5 cm; blade narrowly ovate to oblong-lanceolate or lanceolate, margins shallowly sinuate-dentate or denticulate. **Flowers:** buds with free tips 1–2 mm; floral tube 15–40 mm; sepals 12–30 mm; petals 15–40 mm. **Capsules** spreading, straight or curved, sometimes contorted. $2n = 14$.

Flowering Jun–Sep. Open sites, sandy soil, dunes, rocky sites in grasslands; 600–2000(–3100) m; Colo., Kans., Nebr., N.Mex., Okla., S.Dak., Utah, Wyo.

Some collections from mostly disturbed sites in northern Utah (Cache, Salt Lake, Summit, and Tooele counties) have been identified as subsp. *latifolia*; it is not clear if they represent a disjunct distribution area of this subspecies, naturalized populations, or if they are pubescent forms of subsp. *pallida*.

67c. Oenothera pallida Lindley subsp. **runcinata** (Engelmann) Munz & W. M. Klein in N. L. Britton et al., N. Amer. Fl., ser. 2, 5: 119. 1965 [F]

Oenothera albicaulis Pursh var. *runcinata* Engelmann, Amer. J. Sci. Arts, ser. 2, 34: 334. 1862; *Anogra gypsophila* (Eastwood) A. Heller; *A. leucotricha* Wooton & Standley; *A. pallida* (Lindley) Britton var. *runcinata* (Engelmann) Small; *A. runcinata* (Engelmann) Wooton & Standley; *O. albicaulis* var. *brevifolia* Engelmann; *O. albicaulis* var. *gypsophila* Eastwood; *O. pallida* subsp. *gypsophila* (Eastwood) Munz & W. M. Klein; *O. pallida* var. *runcinata* (Engelmann) Cronquist; *O. runcinata* (Engelmann) Munz; *O. runcinata* var. *brevifolia* (Engelmann) Munz; *O. runcinata* var. *gypsophila* (Eastwood) Munz; *O. runcinata* var. *leucotricha* (Wooton & Standley) Munz; *O. wislizeni* H. Léveillé

Herbs perennial, strigillose, sometimes also sparsely villous, or glabrous; from a taproot, lateral roots producing adventitious shoots. **Stems** usually branched throughout. **Leaves:** rosette not present at anthesis, 2–3.5(–5) × 0.4–1(–1.5) cm; blade oblong to narrowly lanceolate, margins usually deeply sinuate-dentate to pinnatifid, rarely dentate only. **Flowers:** buds with free tips 0–0.2 mm; floral tube 15–30 mm; sepals 10–25 mm; petals 10–30 mm. **Capsules** spreading to reflexed, straight or curved, sometimes contorted. $2n = 14$.

Flowering (Apr–)May–Sep. Sandy soil, dunes, disturbed areas, alkaline soil, pinyon-juniper woodlands, shrublands with *Artemisia*, *Ericameria*, or *Prosopis*, open ponderosa pine woodlands; 1100–2400 m; Ariz., Colo., N.Mex., Tex., Utah; Mexico (Chihuahua).

Subspecies *runcinata* replaces subsp. *pallida* in the southern part of the species range. In New Mexico, it is most morphologically diverse, with plants with subentire leaves densely strigillose recognized in the past as subsp. *gypsophila*. The distribution and characteristics of these plants should be studied more in relation to other types of populations included in subsp. *runcinata*. Glabrous plants in the southern portion of the range of subsp. *runcinata*, especially in New Mexico, adjacent Texas, and Chihuahua, Mexico, have been recognized as var. *brevifolia*.

67d. Oenothera pallida Lindley subsp. **trichocalyx** (Nuttall) Munz & W. M. Klein in N. L. Britton et al., N. Amer. Fl., ser. 2, 5: 119. 1965 E F

Oenothera trichocalyx Nuttall in J. Torrey and A. Gray, Fl. N. Amer. 1: 494. 1840; *Anogra rhizomata* A. Nelson; *A. trichocalyx* (Nuttall) Small; *A. violacea* A. Nelson; *A. vreelandii* Rydberg; *O. albicaulis* Pursh var. *trichocalyx* (Nuttall) Engelmann; *O. pallida* var. *trichocalyx* (Nuttall) Dorn

Herbs usually annual, sometimes perennial, strigillose throughout and villous distally, especially on flower parts; from a taproot, when perennial sometimes lateral roots producing adventitious shoots. **Stems** single to several from base, usually unbranched. **Leaves:** basal rosette usually present at anthesis, 3–5(–7.8) × 0.4–0.8 (–1.2) cm; blade narrowly lanceolate to oblong, margins pinnatifid or dentate. **Flowers:** buds with free tips 0–0.2 mm; floral tube 20–30 mm; sepals 10–18 mm; petals 10–20 mm. **Capsules** spreading to reflexed, straight or contorted. $2n = 14$.

Flowering (Apr–)May–Jun. Sandy, silty, or rocky soil in pinyon-juniper woodlands or shrublands, with *Artemisia* and *Ericameria*; 1100–2500 m; Ariz., Colo., N.Mex., Utah, Wyo.

Subspecies *trichocalyx* occurs across central to southern Wyoming, eastern Utah, western Colorado, northeastern Arizona, and northwestern New Mexico. Within its range it has slight overlap with subspp. *pallida* and *runcinata*. In its purest form, subsp. *trichocalyx* is the most distinctive phase of *Oenothera pallida*, but many of the populations have characteristics that approach other subspecies with perennial habit (versus annual) appearing occasionally. Plants that are glabrous, or nearly so, like subsp. *pallida*, but with apparent short-duration habit and divided leaves like subsp. *trichocalyx*, occur in southern Wyoming and in the Uinta Basin region of Utah; only more typical plants of subsp. *trichocalyx* otherwise occur in the region without any current evidence of the presence of subsp. *pallida*.

17p. OENOTHERA Linnaeus sect. OENOTHERA

Herbs annual, biennial, or perennial, caulescent; from a usually large taproot, sometimes developing adventitious shoots from lateral roots producing a fibrous root system. **Stems** erect to ascending or decumbent, branched or unbranched. **Leaves** in a basal rosette and cauline, cauline (1–)3–25 cm; blade margins pinnately lobed to sinuate-dentate, serrate to dentate or subentire. **Inflorescences** solitary flowers in axils of distal leaves, usually forming a dense or lax spike. **Flowers** opening near sunset, with a sweet scent or nearly unscented; buds erect or recurved, terete or weakly quadrangular, with free tips; floral tube 12–165(–190) mm; sepals separating in pairs and reflexed, or splitting along only 1 suture and reflexed to one side as a unit, or separate and reflexed individually; petals yellow, usually fading orange, reddish orange, yellow, or yellowish white, usually obcordate to obovate, sometimes rhombic, elliptic, or rhombic-ovate; stigma deeply divided into 4 linear lobes. **Capsules** straight, curved, or somewhat sigmoid, becoming somewhat woody in age, cylindrical to narrowly lanceoloid or ovoid, terete to weakly 4-angled dehiscent nearly throughout their length; sessile. **Seeds** numerous, in (1 or) 2 rows per locule, prismatic and angled, narrowly to broadly ellipsoid to subglobose, rarely obovoid and obtusely angled, surface reticulate and regularly or irregularly pitted, rarely flat. $2n = 14$.

Species 68 (26 in the flora): North America, Mexico, West Indies (Cuba), Bermuda, Central America, South America; introduced widely in temperate areas of the world.

Section *Oenothera* is the largest section in the genus, consisting of 68 species (79 taxa) divided into six subsections, one of which is further divided into three series. Section *Oenothera* has a wide geographic distribution from Canada south to Panama and throughout temperate South America, essentially encompassing the full natural distribution of the genus, although there is very sparse representation (only *O. elata*) from central Mexico south to Panama. Species of this section occur in a variety of habitats, often disturbed ones, from sea level to 5000 m. Most of the species are self-compatible, but with a few self-incompatible taxa and individual populations of others (W. Dietrich et al. 1997; W. L. Wagner et al. 2007). The flowers are vespertine, fading the following morning, and are pollinated by hawkmoths (in *O. versicolor* Lehmann perhaps by hummingbirds), or autogamous. In sect. *Oenothera*, as in several other sections of the genus, the diploid, bivalent-forming, usually outcrossing, species often have relatively narrow geographic and ecological ranges, whereas closely related PTH species derived from them are usually autogamous and have much wider ranges. There are 37 PTH species in sect. *Oenothera*, which includes the majority of species of angiosperms with this anomalous genetic system.

SELECTED REFERENCES Dietrich, W. and W. L. Wagner. 1988. Systematics of *Oenothera* section *Oenothera* subsection *Raimannia* and subsection *Nutantigemma* (Onagraceae). Syst. Bot. Monogr. 24: 1–91. Stubbe, W. and P. H. Raven. 1979. A genetic contribution to the taxonomy of *Oenothera* sect. *Oenothera* (including subsections *Euoenothera*, *Emersonia*, *Raimannia*, and *Munzia*). Pl. Syst. Evol. 133: 39–59.

1. Floral tubes 100–165(–190) mm; herbs perennial . . . . 17p.1. *Oenothera* subsect. *Emersonia*, p. 314
1. Floral tubes (5–)12–50(–160) mm; herbs annual, biennial, or short-lived perennial.
   2. Seeds prismatic and angled, surface irregularly pitted . . . . . 17p.2. *Oenothera* subsect. *Oenothera*, p. 316
   2. Seeds ellipsoid to subglobose, not angled, surface regularly pitted.
      3. Young flower buds nodding by recurved floral tube . . . . 17p.5. *Oenothera* subsect. *Nutantigemma*, p. 335
      3. Young flower buds with floral tube curved upward or straight.
         4. Petals rhombic to elliptic or rhombic-ovate . . . . 17p.3. *Oenothera* subsect. *Candela*, p. 327
         4. Petals shallowly or deeply obcordate.
            5. Young flower buds with floral tube curved upward . . . . 17p.4. *Oenothera* subsect. *Raimannia*, p. 331
            5. Young flower buds with floral tube straight. . . . . .17p.6. *Oenothera* subsect. *Munzia*, p. 335

17p.1. Oenothera Linnaeus (sect. Oenothera) subsect. Emersonia (Munz) W. Dietrich, P. H. Raven & W. L. Wagner, Syst. Bot. 10: 39. 1985

Oenothera sect. *Emersonia* Munz in N. L. Britton et al., N. Amer. Fl., ser. 2, 5: 105. 1965

Herbs perennial; from stout taproot, sometimes producing new shoots from lateral roots. **Stems** weakly erect to ascending [or decumbent], often rooting at nodes. **Inflorescences** solitary flowers in axils of distal leaves, sometimes forming a spike (*O. organensis*). **Flowers:** buds erect, weakly quadrangular, with free tips terminal, erect; floral tube straight, [55–]100–165(–190) mm; petals obovate to obcordate. **Capsules** straight, cylindrical to narrowly lanceoloid, subterete, [3–]4–5.5[–7] mm diam. **Seeds** in 2 rows per locule [or in 1 irregular row derived from 2 rows of ovules (*O. maysillesii* Munz)], dark reddish brown [brown], obovoid [oblong-ellipsoid], surface thickened at distal end and sometimes above raphe, [thickened area (*O. maysillesii*, *O. stubbei* W. Dietrich, P. H. Raven & W. L. Wagner) with an empty cavity]. $2n = 14$.

Species 4 (1 in the flora): New Mexico, n Mexico.

As delimited by W. Stubbe and P. H. Raven (1979) and W. Dietrich et al. (1985), subsect. *Emersonia* consists of four species of mesic or xeric habitats (*Oenothera stubbei*) in pine-oak forests, meadows, and canyons, or boggy sites in grasslands, from scattered areas ranging from the Organ Mountains in southern New Mexico to southeastern Chihuahua, southern Coahuila and Nuevo León, northern Zacatecas and San Luis Potosí, and southern Durango, Mexico, 1100 to 2600 m. Two species are self-incompatible and two are self-compatible.

SELECTED REFERENCE Dietrich, W., P. H. Raven, and W. L. Wagner. 1985. Revision of Oenothera sect. *Oenothera* subsect. *Emersonia* (Onagraceae). Syst. Bot. 10: 29–48.

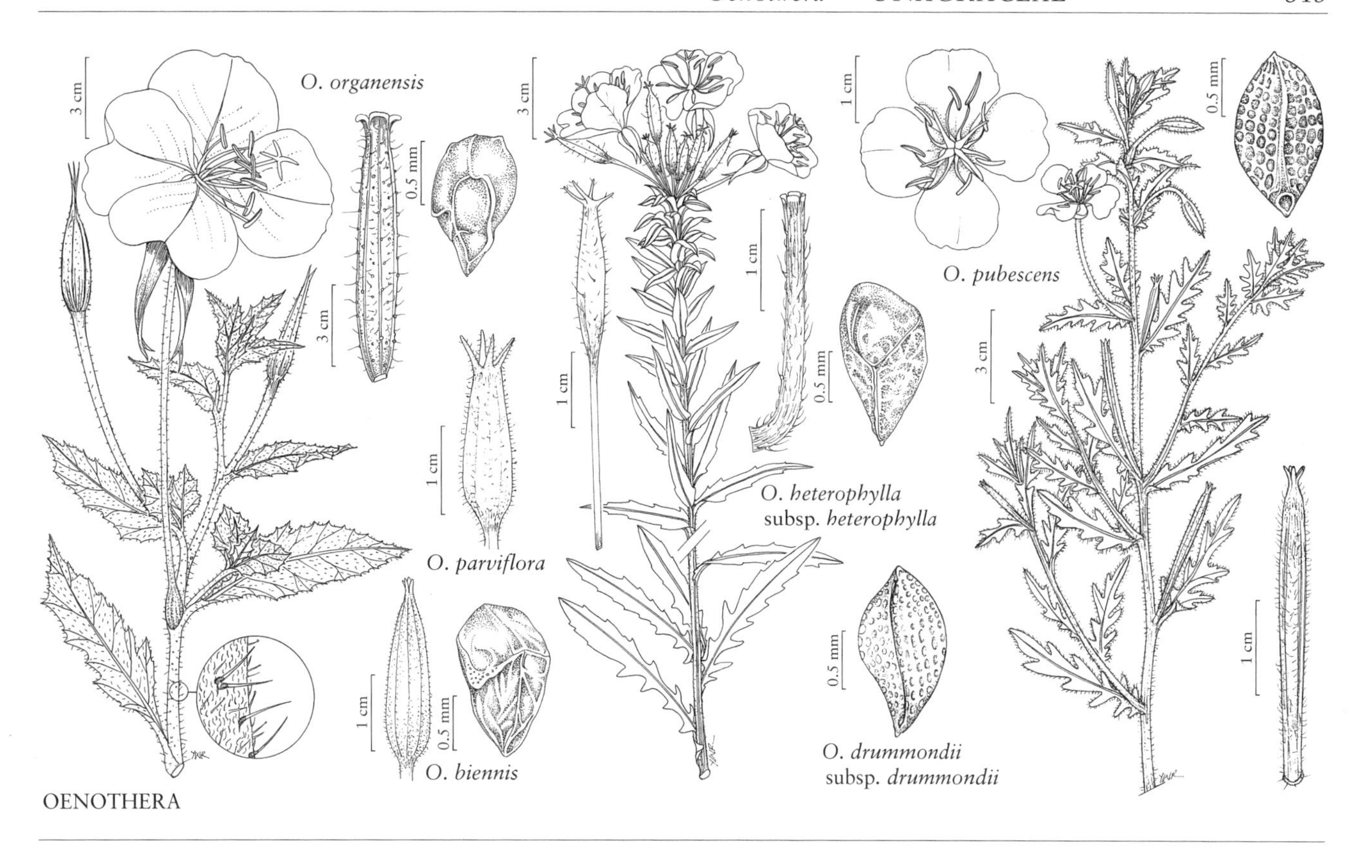

OENOTHERA

68. Oenothera organensis Munz ex S. Emerson, Genetics 23: 190. 1938 [C] [E] [F]

Oenothera macrosiphon Wooton & Standley, Contr. U.S. Natl. Herb. 16: 155. 1913, not Lehmann 1858

Herbs perennial, moderately hirsute (hairs often with reddish purple, pustulate bases), also strigillose and becoming glandular puberulent distally; initially from slender taproot with single rosette, later developing numerous adventitious shoots from taproot and lateral roots, root system then appearing fibrous. **Stems** weakly erect to ascending, many from base, forming clumps 1–1.5 m diam., often branched distally, 30–60 cm. **Leaves** in a weakly developed basal rosette and cauline, basal 9–23 × 1–2.5 cm, cauline 5–11 × 1.5–3.5 cm; petiole 0.5–1.5 cm; blade very narrowly oblanceolate to narrowly elliptic, margins undulate, remotely and bluntly dentate. **Flowers** opening near sunset, not strongly scented; buds with free tips terminal, erect, 3–10 mm; floral tube straight, 100–165 (–190) mm; sepals 25–50 mm; petals yellow, fading deep reddish orange, broadly obovate with truncate apex, or obcordate, 30–55 mm; filaments 18–35 mm, anthers 10–19 mm; style 140–235 mm, stigma exserted beyond anthers at anthesis. **Capsules** erect to slightly spreading at acute angle from stem, cylindrical, 25–35 × 4–5.5 mm, dehiscent at least ¾ their length. **Seeds** numerous, in 2 distinct rows per locule, dark reddish brown, sometimes with darker flecks, obovoid, asymmetrical, irregularly angled, 1.5–2.1 × 1–1.2 mm, surface irregularly pitted and with collapsed papillae. **2*n*** = 14.

Flowering Jun–Sep. In larger rhyolite canyons, along water courses, in eroded basins filled with gravel and rocks; of conservation concern; 1800–2300 m; N.Mex.

Oenothera organensis is known only from the Organ Mountains, Doña Ana County, especially on the east side. Various studies have been done to assess the status of this narrow endemic and estimate 2000 to 5000 individuals with wide fluctuations likely due to variation in rainfall (W. Dietrich et al. 1985). *Oenothera organensis* could decline if degradation of habitat increases in the Organ Mountains. The genetics and population biology of this taxon have been heavily studied in the past (summarized in Dietrich et al.). It is self-incompatible.

17p.2. OENOTHERA Linnaeus (sect. OENOTHERA) subsect. OENOTHERA

Oenothera sect. *Parviflorae* Rostański; *Oenothera* sect. *Strigosae* Rostański; *Oenothera* ser. *Devriesia* Rostański; *Oenothera* ser. *Linderia* Rostański; *Oenothera* ser. *Rugglesia* Rostański; *Oenothera* ser. *Stubbia* Rostański; *Onosuris* Rafinesque

Herbs usually facultatively biennial or short-lived perennial, rarely annual; from taproot. **Stems** usually erect or ascending, rarely decumbent. **Inflorescences** forming elongate, dense, erect or somewhat curved spike. **Flowers:** buds erect, weakly quadrangular, with free tips terminal or subterminal, erect or spreading; floral tube straight, (15–)22–135(–160) mm; petals obovate or obcordate. **Capsules** usually straight, rarely curved (*O. argillicola*), narrowly lanceoloid or ovoid, bluntly 4-angled, (3–)4–8[–12] mm diam. **Seeds** in 2 rows per locule, dark brown to almost black, prismatic and angled, surface reticulate and irregularly pitted. $2n = 14$.

Species 13 (12 in the flora): North America, Mexico, Central America; introduced in South America, Europe, Asia, Africa, Atlantic Islands, Pacific Islands (New Zealand), Australia.

Subsection *Oenothera* consists of 13 species (16 taxa) characterized by a weedy biennial, short-lived perennial, or rarely annual habit, stems mostly erect with many-flowered spikes, capsules lanceoloid or ovoid and bluntly 4-angled, and seeds black, prismatic, and angled, with the surface minutely pitted (W. Dietrich et al. 1997). All species are self-compatible, except rarely *O. grandiflora* is self-incompatible. The flowers are vespertine, with five species outcrossing and pollinated by hawkmoths (D. P. Gregory 1963, 1964), seven autogamous PTH species, or one regularly outcrossing PTH species (*O. glazioviana*) pollinated by hawkmoths (Dietrich et al.). The eight PTH species form a ring of 14 chromosomes or a ring of 12 and 1 bivalent in meiosis. This group of closely related species (in literature also as Euoenothera) has had a long history of scientific study resulting in many hundreds of research papers and several books, including the excellent summary by R. E. Cleland (1972) and more recent ones by W. Stubbe and P. H. Raven (1979), and C. Harte (1994), which recount nearly a century of experimental studies of the group. Among the first investigations of *Oenothera* were those of Hugo de Vries, which opened the modern era of study of mutation and its effect on evolution and speciation. Based on crossing studies of numerous individual types across the group, Stubbe (1964) formed a system of classification of plastome and genome types. His characterizations grouped the plastome types from hundreds of individual lines of species of subsect. *Oenothera* resulting in the recognition of five principal types, designated plastome I to V. These plastome types are genetically discernible by their compatibility or incompatibility with different nuclear genotypes that occur in homozygous (AA, BB, CC) and heterozygous (AB, BA, AC, BC) combinations (summarized by Stubbe and Raven; Dietrich et al.). Application of various molecular techniques has confirmed the genetic distinctness of each of the plastome types and has provided data relevant for the assessment of evolutionary relationships.

Species of subsect. *Oenothera* occur in open, often disturbed sites, in the drier parts of their range often in wet habitats. The indigenous range of the subsection extends in North America from southern Canada at sea level on both the Pacific and Atlantic coasts to elevations up to 3200 m in the Rocky Mountains southward through central Mexico, Guatemala, El Salvador, Costa Rica, and Panama. The range has been greatly extended by several of the PTH species (*O. biennis*, *O. oakesiana*, *O. parviflora*, and *O. villosa* subsp. *villosa*), which are widely naturalized in many parts of the world. One other species, the mostly outcrossing bivalent-forming *O. jamesii*, is sparingly naturalized in South Africa, the Canary Islands, and Japan. Two additional species, *O. glazioviana* and *O. stucchii* Soldano, apparently have arisen recently by stabilized hybridization and PTH formation; the former is now widely distributed around the world, and the latter occurs in Italy and Bouches-du-Rhône, France.

The revision presented by P. H. Raven et al. (1979) and detailed by W. Dietrich et al. (1997) accepts 13 species, eight of which are PTH. In that taxonomy, we recognize five mostly outcrossing, bivalent-forming, genomically homozygous species, and eight PTH species based on the combination of genomes and plastomes. Species were delimited in a broad sense, based on recognition of the three genomic types and the five plastome types, in conjunction with associated morphological characters, to provide a comprehensive taxonomic system that reflects the knowledge of the evolutionary history of the group and provides a reliable means for identification and for information synthesis and retrieval.

Several superfluous names, or names based on them, exist for this group: *Brunyera* Bubani; *Oenothera* sect. *Onagra* Seringe; *Oenothera* subg. *Onagra* (Seringe) Jepson; *Onagra* Miller; and *Usoricum* Lunell.

SELECTED REFERENCES Dietrich, W., W. L. Wagner, and P. H. Raven. 1997. Systematics of *Oenothera* section *Oenothera* subsection *Oenothera* (Onagraceae). Syst. Bot. Monogr. 50. Raven, P. H., W. Dietrich, and W. Stubbe. 1979. An outline of the systematics of *Oenothera* subsect. *Euoenothera* (Onagraceae). Syst. Bot. 4: 242–252.

1. Stigmas exserted beyond anthers at anthesis; petals (25–)30–65 mm; pollen 90–100% fertile (except 50% fertile in *O. glazioviana*).
    2. Floral tubes 60–135(–160) mm.
        3. Flower buds 5–9 mm diam.; capsules 4–9 mm diam., free tips of valves 1–2(–3) mm . . . . 70. *Oenothera longissima*
        3. Flower buds 7–12 mm diam.; capsules 6–12 mm diam., free tips of valves 2.5–5 mm . . . . 71. *Oenothera jamesii*
    2. Floral tubes (20–)30–55 mm.
        4. Inflorescences curved with ascending tip; flower buds with subterminal free tips, these divergent; cauline leaf blades 0.4–1 cm wide; capsules spreading at nearly a right angle to stem, apex long-attenuate, usually conspicuously curved . . . . 78. *Oenothera argillicola*
        4. Inflorescences erect; flower buds with terminal free tips, these usually erect; cauline leaf blades (1–)1.5–6.5 cm wide; capsules erect or slightly spreading, apex gradually attenuate, straight.
            5. Cauline leaf blades 2.5–4 cm wide, narrowly elliptic to lanceolate, margins usually strongly crinkled; pollen ca. 50% fertile; to 50% seeds abortive . . . . 77. *Oenothera glazioviana*
            5. Cauline leaf blades 1–4.5(–6.5) cm wide, narrowly oblanceolate, oblanceolate to narrowly obovate, or sometimes narrowly elliptic or elliptic, margins not strongly crinkled; flower pollen 90–100% fertile; few seeds abortive.
                6. Plant densely strigillose and either sparsely or moderately villous, with appressed or spreading hairs (some with red-pustulate bases), distally sometimes also glandular puberulent; bracts persistent . . . . 69. *Oenothera elata*
                6. Plant often appearing glabrous to naked eye, usually strigillose and sparsely to moderately villous with pustulate, translucent hairs proximal to inflorescence, pustules not red (in fresh material), inflorescence glabrous, glandular puberulent, or strigillose and glandular puberulent; bracts caducous . . . . 74. *Oenothera grandiflora*
1. Stigmas surrounded by anthers, sometimes (in *O. wolfii*) exserted beyond anthers and then petals conspicuously shorter than sepals; petals 7–25(–30) mm; pollen ca. 50% fertile.
    7. Plants predominately strigillose.
        8. Inflorescences with curved apices; flower buds with subterminal free tips, spreading to erect; dry capsules usually rusty brown. . . . . 80. *Oenothera oakesiana* (in part)
        8. Inflorescences with erect apices; flower buds with terminal free tips, erect; dry capsules grayish green or dull green.

9. Leaf blades dull green to grayish green; plants densely strigillose, sometimes also sparsely villous with appressed or subappressed hairs, these without, or sometimes with, red or green pustulate bases, rarely sparsely glandular puberulent distally . . . . 73. *Oenothera villosa* (in part)
9. Leaf blades green to pale green; plants sparsely to moderately strigillose and glandular puberulent, sometimes also scattered villous, hairs erect to appressed, and with pustulate bases . . . . 76. *Oenothera biennis* (in part)

[7. Shifted to left margin.—Ed.]

7. Plants conspicuously strigillose and villous with pustulate hairs, or appearing glabrate.
   10. Inflorescences with curved apices; flower buds with subterminal free tips.
      11. Plants predominantly villous or appearing glabrous to naked eye; leaf blades usually bright green; capsules usually greenish black when dry. . . . 79. *Oenothera parviflora* (in part)
      11. Plants, at least proximally, predominantly strigillose; leaf blades grayish green to dull green; capsules usually rusty brown when dry . . . . 80. *Oenothera oakesiana* (in part)
   10. Inflorescences with erect apices; flower buds with subterminal or terminal free tips.
      12. Inflorescences glabrous or appearing so to naked eye.
         13. Flower buds with terminal free tips; bracts caducous; petals 14–25(–30) mm, fading pale yellowish white; capsules dull green when dry . . . . 75. *Oenothera nutans*
         13. Flower buds with subterminal free tips; bracts persistent; petals 8–15 (–20) mm, fading pale yellow to pale yellowish orange; capsules usually greenish black when dry. . . . . 79. *Oenothera parviflora* (in part)
      12. Inflorescences conspicuously pubescent.
         14. Sepals 17–28 mm.
            15. Anthers 7–12 mm; petals conspicuously shorter than sepals . . . . 72. *Oenothera wolfii*
            15. Anthers 3–6(–9) mm; petals nearly equaling sepals . . . . 76. *Oenothera biennis* (in part)
         14. Sepals 8–18 mm.
            16. Ovary variously pubescent, never villous with pustulate-based hairs . . . . 76. *Oenothera biennis* (in part)
            16. Ovary villous with pustulate-based hairs, often also with other hair types.
               17. Inflorescences usually open (internodes in fruit usually as long as or longer than capsule), villous with appressed to erect hairs, with pustulate bases, also glandular puberulent; petals 7–20 mm; sepals 9–18 mm, usually red-striped or flushed red . . . . .73. *Oenothera villosa* (in part)
               17. Inflorescences dense (internodes in fruit usually shorter than capsule), sparsely to moderately strigillose and glandular puberulent, sometimes also scattered villous, the hairs erect to appressed, with or without pustulate bases; petals 10–25(–30) mm; sepals 12–22(–28) mm, usually green or yellowish, rarely red-striped or flushed red . . . . 76. *Oenothera biennis* (in part)

69. Oenothera elata Kunth in A. von Humboldt et al., Nov. Gen. Sp. 6(fol.): 72; 6(qto.): 90. 1823 W

Herbs biennial or short-lived perennial, densely strigillose and either sparsely or moderately villous, with appressed or spreading hairs (sometimes with red-pustulate bases), distally sometimes also glandular puberulent. **Stems** erect, green, flushed with red proximally or red throughout, unbranched or branches obliquely arising from rosette and secondary branches arising from main stem, 30–250 cm. **Leaves** in a basal rosette and cauline, basal 10–43 × 1.2–4(–6) cm, cauline 4–25 × 1–2.5(–4) cm; blade dull green to grayish green, rarely red, narrowly oblanceolate or oblanceolate to narrowly elliptic or narrowly lanceolate, margins usually flat, rarely undulate, bluntly dentate or subentire, teeth sometimes widely spaced, proximal blades sometimes sinuate-dentate toward base; bracts persistent. **Inflorescences** erect, unbranched. **Flowers** opening near sunset; buds erect, 6–10 mm diam., with free tips terminal, erect, 1–7 mm; floral tube (20–)30–45(–50) mm; sepals yellowish green, red-striped or strongly flushed with red, 27–50 mm; petals yellow to pale yellow, fading orange or pale yellow, very broadly obcordate, (25–)30–47(–55) mm; filaments 17–25 mm, anthers 8–23 mm, pollen 90–100% fertile; style 50–90 mm, stigma exserted beyond anthers at anthesis. **Capsules** erect or slightly spreading, dull green or gray-green when dry, narrowly lanceoloid, 20–65 × 4–7 mm, free tips of valves 0.5–2.5 mm. **Seeds** 1–1.9 × 0.6–1.2 mm.

Subspecies 3 (2 in the flora): c, w United States, Mexico, Central America.

Subspecies *elata* differs in anthers 7–12 mm, fewer or no pustulate-based hairs, and generally smaller flowers and habit. It ranges from the highlands of central Mexico, including Guanajuato, Hidalgo, México, Michoacán, Puebla, Querétaro, and Veracruz, south to Guatemala, El Salvador, Costa Rica, and Panama.

Oenothera elata has plastome I and a AA genome composition. *Onagra kunthiana* Spach is a superfluous name that pertains here.

1. Plants strigillose, usually also villous, with appressed or spreading hairs, sometimes these with red-pustulate bases, distally sometimes also glandular puberulent; buds green to yellowish green, red-striped, or sometimes red throughout; anthers 8–15(–20) mm . . . . 69a. *Oenothera elata* subsp. *hirsutissima*
1. Plants strigillose, also villous with appressed or spreading hairs, usually these with conspicuous red-pustulate bases, also glandular puberulent; buds flushed with red; anthers 12–23 mm. . . . . 69b. *Oenothera elata* subsp. *hookeri*

69a. Oenothera elata Kunth subsp. **hirsutissima** (A. Gray ex S. Watson) W. Dietrich, Ann. Missouri Bot. Gard. 70: 195. 1983

Oenothera biennis Linnaeus var. *hirsutissima* A. Gray ex S. Watson, Proc. Amer. Acad. Arts 8: 579, 603. 1873; *O. elata* var. *hirsutissima* (A. Gray ex S. Watson) Cronquist; *O. elata* subsp. *texensis* W. Dietrich & W. L. Wagner; *O. grisea* (Bartlett) Rostański; *O. hewettii* (Cockerell) Cockerell ex A. W. Hill; *O. hirsutissima* (A. Gray ex S. Watson) de Vries; *O. hookeri* Torrey & A. Gray subsp. *angustifolia* (R. R. Gates) Munz; *O. hookeri* var. *angustifolia* R. R. Gates; *O. hookeri* subsp. *grisea* (Bartlett) Munz; *O. hookeri* var. *grisea* (Bartlett) Munz; *O. hookeri* subsp. *hewettii* Cockerell; *O. hookeri* var. *hewettii* (Cockerell) R. R. Gates; *O. hookeri* subsp. *hirsutissima* (A. Gray ex S. Watson) Munz; *O. hookeri* var. *hirsutissima* (A. Gray ex S. Watson) Munz; *O. hookeri* var. *irrigua* (Wooton & Standley) R. R. Gates; *O. hookeri* subsp. *ornata* (A. Nelson) Munz; *O. hookeri* var. *ornata* (A. Nelson) Munz; *O. hookeri* var. *semiglabra* R. R. Gates; *O. hookeri* var. *simsiana* (Seringe) R. R. Gates; *O. hookeri* subsp. *venusta* (Bartlett) Munz; *O. hookeri* var. *venusta* (Bartlett) Munz; *O. irrigua* Wooton & Standley; *O. jepsonii* Greene; *O. macbrideae* (A. Nelson) R. R. Gates; *O. ornata* (A. Nelson) R. R. Gates; *O. simsiana* Seringe; *O. venusta* Bartlett; *O. venusta* var. *grisea* Bartlett; *Onagra macbrideae* A. Nelson; *O. ornata* A. Nelson

Herbs strigillose, usually also villous, with appressed or spreading hairs, sometimes these with red-pustulate bases, distally sometimes also glandular puberulent. **Flowers:** buds green to yellowish green, red-striped, or sometimes red throughout, with free tips 2–7 mm; petals 30–47(–55) mm; anthers 8–15(–22) mm. **2*n*** = 14.

Flowering (Apr–)Jul–Sep(–Oct). Montane sites along streams, mesic meadows, roadsides, near permanent or seasonally wet sites, ditch banks, riverbanks, flood plains, fallow agricultural land; 10–3000 m; Ariz., Calif., Colo., Idaho, Kans., Nev., N.Mex., Okla., Oreg., Tex., Utah, Wash.; Mexico (n Baja California, Chihuahua, Coahuila, Durango, Sinaloa, Sonora).

Subspecies *hirsutissima* occurs throughout much of the western United States, but with only scattered populations in Oklahoma (Custer, Logan, and McCurtain counties) and in eastern Texas (Anderson, Brazos, and Leon counties) and western Texas (Brewster, Culberson, Jeff Davis, and Presidio counties).

Onagra spectabilis Spach is an illegitimate name as is *Oenothera corymbosa* Sims 1818, not Lamarck 1798, and both pertain here.

69b. Oenothera elata Kunth subsp. **hookeri** (Torrey & A. Gray) W. Dietrich & W. L. Wagner, Ann. Missouri Bot. Gard. 74: 152. 1987 [E]

Oenothera hookeri Torrey & A. Gray, Fl. N. Amer. 1: 493. 1840; *O. biennis* Linnaeus var. *hookeri* (Torrey & A. Gray) B. Boivin; *O. communis* H. Léveillé var. *hookeri* (Torrey & A. Gray) H. Léveillé; *O. franciscana* Bartlett; *O. hookeri* var. *franciscana* (Bartlett) R. R. Gates; *O. hookeri* subsp. *montereyensis* Munz; *O. hookeri* var. *montereyensis* (Munz) Hoover; *O. montereyensis* (Munz) Rostański; *Onagra hookeri* (Torrey & A. Gray) Small

Herbs strigillose, also villous with appressed or spreading hairs, usually these with conspicuous red-pustulate bases, also glandular puberulent. **Flowers:** buds flushed with red, with free tips 1–5 mm; petals 25–40 mm; anthers 12–23 mm. **2*n*** = 14.

Flowering Aug–Oct. Moist coastal and slightly inland sandy and bluff habitats; 0–200 m; Calif.

Subspecies *hookeri* has a restricted range in California from around San Francisco Bay along the coast from the vicinity of Petaluma, Sonoma County, and Point Reyes south to Santa Barbara County, including Santa Cruz Island, and south to San Diego County. It intergrades along the inland margins of the range with subsp. *hirsutissima*.

70. Oenothera longissima Rydberg, Bull. Torrey Bot. Club 40: 65. 1913 [E]

Oenothera clutei A. Nelson; *O. longissima* subsp. *clutei* (A. Nelson) Munz; *O. longissima* var. *clutei* (A. Nelson) Munz

Herbs biennial or short-lived perennial, sparsely strigillose, sometimes also villous and with pustulate hairs near inflorescence, sometimes also glandular puberulent. **Stems** erect, usually flushed with red proximally or sometimes green, unbranched or with branches obliquely arising from base, secondary branches arising from main stem, 60–300 cm. **Leaves** in a basal rosette and cauline, basal 9–40 × 1.4–5 cm, cauline 5–22 × 0.8–2.5 cm; blade dull green, flat, narrowly oblanceolate, oblanceolate to narrowly elliptic, or narrowly lanceolate, margins bluntly dentate or subentire, teeth widely spaced; bracts persistent. **Inflorescences** open, erect, unbranched. **Flowers** opening near sunset; buds erect, 5–9 mm diam., with free tips terminal, erect, 2–6 mm; floral tube deciduous after anthesis, 60–135 mm; sepals yellowish green, flushed with some red or red to dark red throughout, 25–55 mm; petals yellow to pale yellow, fading orange or pale yellow, very broadly obcordate, 28–65 mm; filaments 20–40 mm, anthers 14–20 mm, pollen 90–100% fertile; style 90–180 mm, stigma exserted beyond anthers at anthesis. **Capsules** erect or slightly spreading, dull green or gray-green when dry, narrowly lanceoloid, 25–55 × 4–9 mm, free tips of valves 1–2(–3) mm. **Seeds** 1–1.9 × 0.6–1.2 mm. **2*n*** = 14.

Flowering Jul–Sep(–Oct). Seasonally moist sites, sandy or sandy-loam soil, sites with high alkalinity or associated with limestone, along desert washes, streams, seeps, roadsides; 800–2800 m; Ariz., Calif., Colo., Nev., N.Mex., Utah.

Oenothera longissima has plastome I and a AA genome composition.

Oenothera longissima is known from northern and western Arizona, Inyo, Los Angeles, and San Bernardino counties in California, Delta and Montezuma counties in Colorado, eastern Nevada, San Juan County in New Mexico, and southern Utah.

71. Oenothera jamesii Torrey & A. Gray, Fl. N. Amer. 1: 493. 1840

Onagra jamesii (Torrey & A. Gray) Small; *Oenothera communis* H. Léveillé var. *jamesii* (Torrey & A. Gray) H. Léveillé

Herbs biennial or winter-annual, usually predominately and densely strigillose, sometimes also villous with scattered, appressed hairs, rarely with a few pustulate hairs, inflorescence sometimes also glandular puberulent. **Stems** erect, usually green, rarely flushed with red, unbranched or with branches arising obliquely from rosette and secondary branches arising from main stem, 60–180 cm. **Leaves** in a basal rosette and cauline, basal 10–30 × 2.5–5 cm, cauline 4–20 × 1–5 cm; blade dull green, flat, narrowly oblanceolate, oblanceolate to narrowly elliptic, or narrowly lanceolate, margins bluntly dentate or subentire, teeth widely spaced; bracts persistent. **Inflorescences** erect, usually unbranched, rarely with few lateral branches. **Flowers** opening near sunset; buds erect, 7–12 mm diam., with free tips terminal, erect, 0.5–3 mm; floral tube persistent on ovary after anthesis, (60–)80–120(–160) mm; sepals yellowish green, red-striped to red throughout, 30–55 mm; petals yellow, fading orange or pale yellow, very broadly obcordate, 40–50 mm; filaments 23–30 mm, anthers 12–22 mm, pollen 90–100% fertile; style 90–170(–200) mm, stigma exserted beyond anthers at anthesis. **Capsules** erect or slightly spreading, dull green or gray-green when dry, narrowly lanceoloid, 20–50 × 6–12 mm, free tips of valves 2.5–5 mm. **Seeds** 1–1.2 × 0.7–1.3 mm. **2*n*** = 14.

Flowering Jul–Sep(–Oct). Sandy stream banks, ditches, moist areas, cultivated areas, disturbed roadsides; (30–) 300–1800 m; Kans., Okla., Tex.; Mexico (Coahuila, Nuevo León, Puebla); introduced in e Asia (Japan), s Africa, Atlantic Islands (Canary Islands).

Oenothera jamesii has plastome I and a AA genome composition; it is known in the flora area from southern Kansas (Clark County), central Oklahoma, and Texas.

72. Oenothera wolfii (Munz) P. H. Raven, W. Dietrich & Stubbe, Syst. Bot. 4: 244. 1980 [C] [E]

Oenothera hookeri Torrey & A. Gray subsp. *wolfii* Munz, Aliso 2: 16, plate 1, fig. B. 1949; *O. hookeri* var. *wolfii* Munz

Herbs biennial or short-lived perennial, densely strigillose, also villous with spreading to subappressed hairs, hairs sometimes pustulate, also glandular puberulent distally. **Stems** erect, green, flushed with red proximally or red throughout, unbranched or branches obliquely arising from rosette and sometimes secondary branches arising from main stem, 50–100 cm. **Leaves** in a basal rosette and cauline, basal 13–35 × 1.5–4(–5) cm, cauline 5–18 × 1–2.5(–4) cm; blade dull green, flat, oblanceolate to narrowly lanceolate or lanceolate to elliptic, margins bluntly dentate or subentire, teeth widely spaced or sinuate proximally; bracts persistent. **Inflorescences** erect, unbranched. **Flowers** opening near sunset; buds erect, 5–8 mm diam., with free tips terminal, erect, 1–3 mm; floral tube 30–46 mm; sepals yellowish green, also usually flushed with red or red-striped, 17–28 mm; petals yellow, fading orange, pale yellow and somewhat opaque, very broadly obcordate, 13–23 mm, conspicuously shorter than sepals; filaments 12–20 mm, anthers 7–12 mm, pollen ca. 50% fertile; style 43–58 mm, stigma usually slightly exserted beyond anthers or surrounded by them at anthesis. **Capsules** erect or slightly spreading, dark dull green and sometimes red-striped when dry, narrowly lanceoloid, 30–48 × 5–7 mm, free tips of valves 0.5–2.5mm. **Seeds** 0.9–2 × 0.9–1.3 mm. **2*n*** = 14.

Flowering Jun–Oct. Coarse-textured sandy or rocky sites, coastal dunes and bluffs, loose, sandy sites along roads, moist places; of conservation concern; 0–100 (–800) m; Calif., Oreg.

Oenothera wolfii is a PTH species and forms a ring of 14 chromosomes in meiosis, and is self-compatible and autogamous (W. Dietrich et al. 1997). It has plastome I and a AA genome composition. It is known only from the vicinity of Port Orford, Curry County, Oregon (currently apparently only as far north as Otter Rock), south in a scattered distribution through Del Norte County to the mouth of the Mattole River, Humboldt County, California. The distribution, at least in California, is closely associated with small patches of Cenozoic-age marine sediments, isolated from each other by Franciscan sedimentary and metamorphic rocks. Most populations appear to occur near river mouths or to the south of a headland. The largest populations center in the area about 11 km long in the vicinity of Crescent City in Del Norte County, between Point George and Enderts Beach in Redwood National Park. There are collections from two inland California localities, one at the eastern border of Humboldt County, California (Willow Creek, Trinity River Valley), and the other at Carville, Trinity County, that may be *O. wolfii*. If so, they would presumably represent recent introductions and should be studied further. As summarized by Dietrich et al., *O. wolfii* is a rare endemic of coastal habitats and known from about 20 different sites. The total number of individuals of *O. wolfii* apparently fluctuates, with perhaps no more than about 5000 individuals total. It is threatened by any potential development and alteration of its habitat, presently by road maintenance and foot traffic. Another possibly more serious threat comes from the recent spread of *O. glazioviana* to this area. *Oenothera glazioviana* could swamp populations through hybridization and, perhaps, by direct competition.

Oenothera wolfii is in the Center for Plant Conservation's National Collection of Endangered Plants.

73. Oenothera villosa Thunberg, Prodr. Pl. Cap., 75. 1794 [E] [W]

Herbs biennial, densely strigillose and either sparsely or moderately villous, with appressed or spreading hairs (sometimes with red-pustulate bases), distally sometimes also glandular puberulent. **Stems** erect, usually flushed with red proximally, sometimes green or red throughout, unbranched or with branches obliquely arising from rosette and secondary branches arising from main stem, 50–200 cm. **Leaves** in a basal rosette and cauline, basal 10–30 × 1.2–4(–5) cm, cauline 5–20 × 1–2.5(–4) cm; blade dull green or grayish green, narrowly oblanceolate, oblanceolate to narrowly elliptic, or narrowly lanceolate, margins flat or undulate, dentate to subentire, teeth sometimes widely spaced, sometimes sinuate-dentate proximally; bracts persistent. **Inflorescences** dense to open, erect, unbranched. **Flowers** opening near sunset; buds erect, 3–5 mm diam., with free tips terminal, erect, 0.5–3 mm; floral tube 23–44 mm; sepals green to yellowish green, red-striped, or flushed with red, 9–18 mm; petals yellow to pale yellow, fading orange or pale yellow, very broadly obcordate, 7–20 mm; filaments 7–15 mm, anthers 4–10 mm, pollen ca. 50% fertile; style 30–55 mm, stigma surrounded by anthers at anthesis. **Capsules** erect or slightly spreading, dull green or gray-green when dry, lanceoloid, 20–43 × 4–7 mm, free tips of valves 1–2 mm. **Seeds** 1–2 × 0.5–1.2 mm.

Subspecies 2 (2 in the flora): North America; introduced in s South America, Europe, Asia, s Africa.

Oenothera villosa is a PTH species and forms a ring of 14 chromosomes in meiosis, and is self-compatible and autogamous with plastome I and a AA genome composition (W. Dietrich et al. 1997). The original natural range of *O. villosa* was presumably from southern British Columbia south to California and east through the Rocky Mountain and the Great Plains regions. The wide occurrence east of this area in North America to eastern Quebec south throughout most of the eastern half of the United States, except for extreme southern and southeastern parts, is most likely the result of recent spread of this species, probably in the past several hundred years. *Oenothera villosa* is subdivided into two subspecies: subsp. *strigosa* occurs primarily in the Pacific Northwest southeast through the Rocky

Mountains; subsp. *villosa* is found primarily from the eastern foothills of the Rocky Mountains eastward throughout the Great Plains region. Both taxa occur sporadically beyond these regions, and subsp. *villosa* is naturalized in many parts of the world.

1. Inflorescences usually dense (internodes in fruit usually shorter than capsule); plants dull green to grayish green, densely strigillose and sometimes also sparsely villous with appressed or subappressed hairs, these without or with red or green pustulate bases, rarely sparsely glandular puberulent distally; sepals green to yellowish green; leaf blade margins conspicuously dentate, venation prominent, especially abaxially . . . . 73a. *Oenothera villosa* subsp. *villosa*
1. Inflorescences usually open (internodes in fruit usually as long as or longer than capsule); plants flushed with red at least proximally, often red throughout, strigillose, usually also villous with erect to sometimes appressed hairs with pustulate bases, pustules red, also glandular puberulent at least distally; sepals red-striped or flushed with red; leaf blade margins usually denticulate or subentire, sometimes moderately dentate, venation not prominent. . . . . 73b. *Oenothera villosa* subsp. *strigosa*

73a. Oenothera villosa Thunberg subsp. **villosa** [E] [W]

Oenothera albinervis R. R. Gates; *O. bauri* Boedijn; *O. biennis* Linnaeus var. *canescens* Torrey & A. Gray; *O. canovertex* Hudziok; *O. canovirens* E. S. Steele; *O. cockerellii* Bartlett ex de Vries; *O. depressa* Greene; *O. erosa* Lehmann; *O. hookeri* Torrey & A. Gray var. *parviflora* R. R. Gates; *O. hungarica* (Borbás) Borbás; *O. muricata* Linnaeus var. *canescens* (Torrey & A. Gray) B. L. Robinson; *O. muricata* subsp. *hungarica* (Borbás) Soó; *O. parviflora* Linnaeus var. *canescens* (Torrey & A. Gray) Farwell; *O. renneri* H. Scholz; *O. strigosa* (Rydberg) Mackenzie & Bush var. *albinervis* (R. R. Gates) R. R. Gates; *O. strigosa* subsp. *canovirens* (E. S. Steele) Munz; *O. strigosa* var. *cockerellii* (Bartlett ex de Vries) R. R. Gates; *O. strigosa* subsp. *hungarica* (Borbás) Á. Löve & D. Löve; *O. velutinifolia* Hudziok; *Onagra depressa* (Greene) Small; *O. hungarica* Borbás

Herbs dull green to grayish green, densely strigillose, sometimes also sparsely villous with appressed or subappressed hairs, these without or with red or green pustules, rarely glandular puberulent distally. **Leaves** grayish green to dull green, blade margins conspicuously dentate, sometimes sinuate-dentate proximally, venation prominent, especially abaxially, usually pale green, rarely red. **Inflorescences** relatively dense, apex truncate, internodes in fruit conspicuously shorter than capsules. **Flowers:** sepals green to yellowish green. **$2n$** = 14.

Flowering Jul–Aug(–Sep). Open, often wet sites, streamsides, fields, roadsides; 30–1500(–1700) m; Alta., B.C., Man., N.B., N.S., Ont., Que., Sask.; Ala., Ark., Colo., Conn., Ga., Ill., Ind., Iowa, Kans., Ky., Maine, Md., Mass., Mich., Minn., Miss., Mo., Mont., Nebr., N.H., N.J., N.Y., N.C., N.Dak., Ohio, Okla., Pa., S.Dak., Tenn., Tex., Vt., Va., W.Va., Wis., Wyo.; introduced in s South America, Europe, Asia, s Africa.

Subspecies *villosa* appears to be largely a taxon of the Great Plains that has subsequently, both naturally and with human assistance, spread to the north and east, primarily in historical times. It grows in a variety of habitats, primarily prairies, along streams or lakes, open woodlands, old fields, and other disturbed sites, and is widely naturalized in Asia, Europe, southern South America, and South Africa.

Oenothera strigosa (Rydberg) Mackenzie & Bush var. *depressa* (Greene) R. R. Gates is an illegitimate name, as is *O. salicifolia* Desfontaines ex G. Don 1837, not Lehmann 1824 and not Desfontaines ex Seringe 1825, and these names pertain here.

73b. Oenothera villosa Thunberg subsp. **strigosa** (Rydberg) W. Dietrich & P. H. Raven, Ann. Missouri Bot. Gard. 63: 383. 1977 [E]

Onagra strigosa Rydberg, Mem. New York Bot. Gard. 1: 278. 1900; *Oenothera biennis* Linnaeus var. *strigosa* (Rydberg) Cronquist; *O. cheradophila* Bartlett; *O. procera* Wooton & Standley; *O. strigosa* (Rydberg) Mackenzie & Bush subsp. *cheradophila* (Bartlett) Munz; *O. strigosa* var. *cheradophila* (Bartlett) R. R. Gates; *O. strigosa* var. *procera* (Wooton & Standley) R. R. Gates; *O. subulifera* Rydberg; *O. villosa* subsp. *cheradophila* (Bartlett) W. Dietrich & P. H. Raven; *O. villosa* var. *strigosa* (Rydberg) Dorn; *Onagra biennis* (Linnaeus) Scopoli var. *strigosa* (Rydberg) Piper; *O. strigosa* var. *subulata* Rydberg; *Usoricum strigosum* (Rydberg) Lunell

Herbs flushed with red, at least proximally, often red throughout, strigillose, rarely exclusively so, usually also villous with erect to ascending or subappressed red-pustulate hairs, and glandular puberulent, at least distally. **Leaves** green to dull green, blade margins usually denticulate or subentire, sometimes moderately dentate, venation not prominent. **Inflorescences** relatively open, apex obtuse, internodes in fruit usually equal to or longer than capsule. **Flowers:** sepals red-striped or flushed with red. **$2n$** = 14.

Flowering Jun–Sep. Open, often wet sites, streamsides, fields, roadsides; 30–3200 m; Alta., B.C., Man., N.W.T., Ont., Sask.; Ariz., Calif., Colo., Idaho, Mich., Minn., Mont., Nebr., Nev., N.Mex., N.Dak., Oreg., S.Dak., Utah, Wash., Wis., Wyo.

Subspecies *strigosa* occurs primarily in the Pacific Northwest southeast through the Rocky Mountains, and is found in mostly montane and foothill habitats. It has not spread much, if at all, outside of its native range.

Oenothera strigosa (Rydberg) Mackenzie & Bush is an illegitimate later homonym that pertains here; *O. rydbergii* House and *O. strigosa* (Rydberg) Mackenzie & Bush var. *subulifera* R. R. Gates also pertain here and are superfluous names.

74. Oenothera grandiflora L'Héritier in W. Aiton, Hort. Kew. 2: 2. 1789 [E]

Oenothera biennis Linnaeus var. *grandiflora* (L'Héritier) Lindley; *O. grandiflora* var. *glabra* Seringe; *O. grandiflora* var. *pubescens* Seringe; *O. lamarckiana* Seringe; *O. spectabilis* Lehmann

Herbs biennial, often appearing glabrous to naked eye, usually sparsely to moderately strigillose and villous with pustulate, translucent hairs proximal to inflorescence, pustules not red (in fresh material), inflorescence glabrous, glandular puberulent, or strigillose and glandular puberulent. **Stems** erect, red on proximal parts, usually green on distal ones, rarely red throughout, unbranched or with branches obliquely arising from rosette and secondary branches arising from main stem, 100–300(–400) cm. **Leaves** in a basal rosette and cauline, basal 18–32 × (2–)3–6.5 cm, cauline 6–20 × 1.5–6.5 cm; blade soft and thin, bright green, usually flat, rarely undulate, narrowly oblanceolate to narrowly obovate, or narrowly elliptic to elliptic, sometimes narrowly ovate distally, margins bluntly dentate or subentire, teeth widely spaced, sometimes sinuate-dentate proximally or lobed; bracts usually caducous. **Inflorescences** erect, often with secondary or tertiary branches just proximal to main one. **Flowers** opening near sunset; buds erect, 5–9 mm diam., with free tips terminal, erect, 2–9 mm; floral tube 35–55 mm; sepals yellowish green or flushed with red, 22–46 mm; petals yellow to pale yellow, fading pale yellowish white, very broadly obcordate or obovate, (25–)30–45 mm; filaments 18–27 mm, anthers 10–15 mm, pollen 90–100% fertile; style 57–90 mm, stigma exserted beyond anthers at anthesis. **Capsules** erect or slightly spreading, dull green when dry, narrowly lanceoloid to narrowly ovoid, 15–35 × 3.5–5.5 mm, free tips of valves 0.5–2.5 mm. **Seeds** 1–1.7 × 0.6–1.2 mm. $2n = 14$.

Flowering Jul–Aug(–Sep). Scattered, presumably relictual populations on chalky bluffs, loose sand over limestone, along streams, marshes, ditches, roadsides; 20–600 m; Ala., Fla., Miss., N.C., S.C., Tenn.

Oenothera grandiflora has a scattered distribution, from the eastern half of Mississippi and Alabama, east to Tennessee (Franklin and Marion counties), North Carolina (Cherokee, Macon, Martin, Moore, New Hanover, Sampson, and Swain counties), South Carolina (Oconee, Spartanburg, and Sumter counties), and Florida (Alachua, Escambia, Franklin, Lake, Leon, Polk, Putnam, and Santa Rosa counties). Collections from southern Canada, New York, Pennsylvania, Vermont, and West Virginia almost certainly represent cultivated plants, garden escapes, or adventive populations, and the single locality from central Kentucky also may be an introduction; it is sometimes a colonizer in disturbed sites such as along roads.

Oenothera grandiflora has plastome III and a BB genome composition. As summarized by W. Dietrich et al. (1997), some populations of *O. grandiflora* seem to be entirely or mostly composed of self-incompatible individuals, whereas others consist of self-compatible plants. This is an extremely uncommon phenomenon within a single species of *Oenothera*; the only other species known to exhibit mixed populations of self-incompatible and self-compatible individuals is *O. primiveris*.

Oenothera grandiflora Lamarck 1798, being a later homonym of *O. grandiflora* L'Héritier 1789, pertains here.

75. Oenothera nutans G. F. Atkinson & Bartlett, Rhodora 15: 83. 1913 [E]

Oenothera austromontana (Munz) P. H. Raven, W. Dietrich & Stubbe; *O. biennis* Linnaeus subsp. *austromontana* Munz; *O. biennis* var. *austromontana* (Munz) Cronquist; *O. biennis* var. *nutans* (G. F. Atkinson & Bartlett) Wiegand

Herbs biennial, often appearing glabrous to naked eye, usually strigillose and sparsely to moderately villous proximal to inflorescence, hairs translucent and with or without pustules, pustules not red (in fresh material), inflorescence glabrous, glandular puberulent, or strigillose and glandular puberulent. **Stems** erect, red on proximal parts, green on distal ones, rarely red throughout, unbranched or with branches obliquely arising from rosette and secondary branches arising from main stem, 30–200 cm. **Leaves** in a basal rosette and cauline, basal 10–32 × 3–7 cm, cauline 6–20 × 2–8 cm; blade green to pale green, narrowly oblanceolate to narrowly obovate, or narrowly elliptic, sometimes lanceolate distally, margins usually flat, rarely undulate, bluntly dentate, teeth widely spaced, sometimes sinuate-dentate proximally; bracts caducous. **Inflorescences** erect, unbranched or with secondary branches just proximal to main one. **Flowers** opening near sunset; buds erect, 4–6 mm diam., with free tips terminal, erect, 1.5–6 mm; floral tube 30–43 mm; sepals yellowish green or flushed with red, 10–23 mm; petals yellow to pale yellow, fading pale yellowish white, very broadly obcordate, 14–25(–30) mm; filaments 10–25 mm, anthers 4–10 mm, pollen ca. 50% fertile; style 35–63 mm, stigma surrounded by anthers at anthesis. **Capsules** erect or slightly spreading, dull green when dry, narrowly lanceoloid to narrowly ovoid, 12–36 × 3–6 mm, free tips of valves 1–1.5 mm. **Seeds** 1–1.9 × 0.6–0.9 mm. $2n$ = 14.

Flowering Jun–Aug(–Sep). Open, often disturbed sites, stream beds, flood plains, slopes, margins of mixed deciduous forests, roadsides, old fields; (200–) 400–1700 m; Ont.; Ala., Ark., Conn., Del., Fla., Ga., Ind., Ky., Maine, Md., Mass., Mich., Mo., N.H., N.J., N.Y., N.C., Ohio, Pa., S.C., Tenn., Vt., Va., W.Va.

Oenothera nutans is a PTH species and forms a ring of 14 chromosomes or a ring of 12 and 1 bivalent in meiosis, and is self-compatible and autogamous (W. Dietrich et al. 1997). It has plastome III and a BB genome composition. The disjunct occurrences of *O. nutans* in Arkansas and Missouri probably represent unintentional introductions by humans.

76. Oenothera biennis Linnaeus, Sp. Pl. 1: 346. 1753 E F W

Brunyera biennis (Linnaeus) Bubani; *Oenothera biennis* subsp. *caeciarum* Munz; *O. biennis* subsp. *centralis* Munz; *O. biennis* subsp. *chicaginensis* (de Vries ex Renner & Cleland) Á. Löve & D. Löve; *O. biennis* var. *leptomeres* Bartlett; *O. biennis* var. *muricata* (Linnaeus) Torrey & A. Gray; *O. biennis* var. *pycnocarpa* (G. F. Atkinson & Bartlett) Wiegand; *O. biennis* var. *rubricaulis* (Farwell) Farwell; *O. biennis* var. *vulgaris* Torrey & A. Gray; *O. brevicapsula* Bartlett; *O. chicaginensis* de Vries ex Renner & Cleland; *O. furca* Boedijn; *O. gauroides* Hornemann var. *brevicapsula* (Bartlett) R. R. Gates; *O. grandiflora* L'Héritier var. *tracyi* (Bartlett) R. R. Gates; *O. grandifolia* R. R. Gates; *O. muricata* Linnaeus; *O. muricata* var. *rubricaulis* Farwell; *O. novae-scotiae* R. R. Gates var. *serratifolia* R. R. Gates; *O. numismatica* Bartlett; *O. paralamarckiana* R. R. Gates; *O. parviflora* Linnaeus var. *muricata* (Linnaeus) Farwell; *O. pratincola* Bartlett; *O. pratincola* var. *numismatica* (Bartlett) R. R. Gates; *O. pycnocarpa* G. F. Atkinson & Bartlett; *O. pycnocarpa* var. *cleistogama* R. R. Gates; *O. pycnocarpa* var. *parviflora* R. R. Gates; *O. reynoldsii* Bartlett; *O. royfraseri* R. R. Gates; *O. rubricaulis* Klebahn; *O. ruderalis* Bartlett; *O. sabulosa* Farwell; *O. sackvillensis* R. R. Gates; *O. sackvillensis* var. *albiviridis* R. R. Gates; *O. sackvillensis* var. *royfraseri* (R. R. Gates) R. R. Gates; *O. shulliana* A. H. Sturtevant; *O. stenomeres* Bartlett; *O. suaveolens* Persoon; *O. tracyi* Bartlett; *O. victorinii* R. R. Gates & Catcheside; *O. victorinii* var. *intermedia* R. R. Gates; *O. victorinii* var. *parviflora* R. R. Gates; *O. victorinii* var. *undulata* R. R. Gates & Catcheside; *Onagra biennis* (Linnaeus) Scopoli; *O. chrysantha* Spach var. *grandiflora* Spach; *O. muricata* (Linnaeus) Moench

Herbs biennial, densely to sparsely strigillose and villous, with somewhat appressed to spreading hairs, those often pustulate, but sometimes predominantly strigillose, inflorescence sometimes also glandular puberulent. **Stems** erect, green or flushed with red on proximal parts, sometimes inflorescence axis red, unbranched or with side branches obliquely arising from rosette or stem, 30–200 cm. **Leaves** in a basal rosette and cauline, basal 10–30 × 2–5 cm, cauline 5–22 × (1–)1.5–5(–6) cm; blade usually green to pale green, narrowly oblanceolate to oblanceolate, sometimes narrowly elliptic to elliptic distally, margins usually flat, rarely undulate, bluntly dentate, teeth widely spaced, sometimes sinuate-dentate proximally or lobed; bracts persistent. **Inflorescences** erect, unbranched or with secondary branches just proximal to main one, internodes in fruit usually shorter than capsule. **Flowers** opening near sunset; buds erect, 3.5–6 mm diam., with free tips terminal, erect or spreading, 1.5–3 mm; floral tube (20–)25–40 mm; sepals yellowish green, rarely flushed with red or red-striped, 12–22(–28) mm; petals yellow to pale yellow, fading yellowish white and somewhat translucent, very broadly obcordate, 12–25(–30) mm; filaments 8–15 (–20) mm, anthers 3–6(–9) mm, pollen 50% fertile; style 30–55 mm, stigma surrounded by anthers at anthesis. **Capsules** erect or slightly spreading, dull green when dry, lanceoloid, 20–40 × 4–6 mm, free tips of valves 0.8–1.5 mm. **Seeds** 1–1.2 × 0.6–1.1 mm. $2n$ = 14.

Flowering Jul–Sep(–Oct). Open, disturbed sites; 10–700 m; Alta., B.C., Man., N.B., Nfld. and Labr. (Nfld.), N.S., Ont., P.E.I., Que., Sask.; Ala., Ark., Calif., Conn., Del., D.C., Fla., Ga., Idaho, Ill., Ind., Iowa, Kans., Ky., La., Maine, Md., Mass., Mich., Minn., Miss., Mo., Mont., Nebr., N.H., N.J., N.Mex., N.Y., N.C., N.Dak., Ohio, Okla., Oreg., Pa., R.I., S.C., S.Dak., Tenn., Tex., Utah, Vt., Va., Wash., W.Va., Wis.; introduced nearly worldwide in temperate and subtropical regions.

Oenothera biennis is a PTH species and usually forms a ring of 14 chromosomes or a ring of 12 and 1 bivalent in meiosis, and is self-compatible and autogamous (W. Dietrich et al. 1997). It has plastome II and a BA or AB genome composition across different populations.

W. Dietrich et al. (1997) found that in the western half of the United States, where *Oenothera biennis* is most likely introduced, there are distinctive series of forms that seemed similar to a phenotype of *O. biennis* that originated in Europe. The most distinctive morphological characters of this western form are the densely villous stems with pustulate hairs and the intense, often dark red color of the stems and sepals, characters which are not typical for the eastern North America forms of *O. biennis*. The pustulate pubescence of this form is also a characteristic feature of many *O. elata* and *O. villosa* subsp. *strigosa* forms, suggesting possible past hybridization with them. Crossing studies confirm that one of these taxa was most likely involved, and these studies showed that, like *O. biennis*, they are all AB and BA genomic combinations. One of the seemingly odd features of this form is its scattered distribution across a wide expanse of western states. A possible explanation is that perhaps *O. biennis* spread westward during glacial periods, hybridized with AA taxa (*O. elata* and *O. villosa* subsp. *strigosa*), followed by compression during warming periods since. A hypothesis of repeated recent hybridization does not seem likely since more typical eastern forms of *O. biennis* are not present throughout much of the western states. *Oenothera biennis* hybridizes with the other species of subsect. *Oenothera* with which it comes in contact, including *O. grandiflora*, *O. nutans*, *O. oakesiana*, *O. parviflora*, *O. villosa* subsp. *strigosa*, and *O. villosa* subsp. *villosa*.

Oenothera biennis subsp. *rubricaulis* (Klebahn) Stomps is a later homonym that pertains here.

77. Oenothera glazioviana Micheli in C. F. P. von Martius et al., Fl. Brasil. 13(2): 178. 1875 [I] [W]

Oenothera erythrosepala (Borbás) Borbás; *O. grandiflora* L'Héritier subsp. *erythrosepala* (Borbás) Á. Löve & D. Löve; *Onagra erythrosepala* Borbás

Herbs biennial, densely to sparsely strigillose and villous, with spreading, red-pustulate hairs, also glandular puberulent and with only a few appressed hairs near inflorescence. **Stems** erect, green or flushed with red on proximal parts, sometimes inflorescence axis red, usually with side branches obliquely arising from rosette and secondary branches from main stem, 50–150 cm. **Leaves** in a basal rosette and cauline, basal 13–30 × 3–5 cm, cauline 5–15 × 2.5–4 cm; blade dark to bright green, white- or red-veined, narrowly oblanceolate to oblanceolate, sometimes narrowly elliptic to lanceolate distally, margins usually conspicuously crinkled, sometimes undulate, bluntly dentate, teeth widely spaced, sometimes sinuate-dentate proximally or lobed; bracts persistent. **Inflorescences** erect, unbranched. **Flowers** opening near sunset; buds erect, 7–9 mm diam., with free tips terminal, erect to spreading, 5–8 mm; floral tube 35–50 mm; sepals yellowish green, usually flushed with red or red-striped, sometimes very dark red throughout, 28–45 mm; petals yellow to pale yellow, fading yellowish white and somewhat translucent, very broadly obcordate, 35–50 mm; filaments 17–25 mm, anthers 10–12 mm, pollen ca. 50% fertile; style 50–80 mm, stigma exserted beyond anthers at anthesis. **Capsules** erect or slightly spreading, dull green when dry, lanceoloid, 20–35 × 5–6 mm, free tips of valves 0.8–1.5 mm. **Seeds** 1.3–2 × 1–1.5 mm, ca. 50% abortive. $2n$ = 14.

Flowering (Jun–)Jul–Sep(–Oct). Open, disturbed sites; 20–600(–1400) m; introduced; B.C., Man., N.S., Ont., Que.; Ala., Ark., Calif., Conn., Ill., Ind., Maine, Mass., Mich., Mont., N.H., N.J., N.Y., N.C., Oreg., Pa., R.I., Vt., Wash., W.Va., Wis.; introduced nearly worldwide in temperate and subtropical regions.

Oenothera glazioviana originated by hybridization between two cultivated or naturalized species in Europe and was introduced into the horticultural trade by Carter and Company of England in 1860 (R. E. Cleland 1972; P. H. Raven et al. 1979). The oldest name applied to this entity was based on plants cultivated in Rio de Janeiro in 1868; clearly, *O. glazioviana* must have spread very rapidly.

Oenothera glazioviana is a PTH species and forms a ring of 12 chromosomes and 1 bivalent in meiosis, and is self-compatible and autogamous (W. Dietrich et al. 1997). It has plastome II or III and a AB genome composition.

78. Oenothera argillicola Mackenzie, Torreya 4: 56. 1904 [E]

Oenothera argillicola var. *pubescens* Core & H. A. Davis

Herbs biennial or short-lived perennial, strigillose and sparsely to moderately villous, hairs sometimes pustulate, pustules with green or red bases, inflorescence glabrous or sparsely glandular puberulent, sometimes also sparsely villous. **Stems** erect to ascending, green or red, unbranched or with branches obliquely arising from rosette or in distal ½ of main stem. **Leaves** in a basal rosette and cauline, basal 7–25 × 0.7–2 cm, cauline 6–13 × 0.4–1 cm; blade dark green, somewhat glossy, very narrowly oblanceolate to narrowly oblanceolate, linear-elliptic, lanceolate, or nearly linear, margins flat, entire or remotely and bluntly dentate, sometimes with larger teeth near base; bracts persistent. **Inflorescences** curved with ascending tip, unbranched. **Flowers** opening near sunset; buds erect, 4–8 mm diam., with free tips subterminal, divergent and hornlike, 3–9 mm; floral tube 32–52 mm; sepals yellowish green to yellow, sometimes flushed with red, especially at apex,

27–38 mm; petals yellow to pale yellow, fading pale yellow to pale yellowish orange, very broadly obcordate or obovate, 25–42 mm; filaments 20–27 mm, anthers 9–13 mm, pollen 90–100% fertile; style 60–85 mm, stigma exserted beyond anthers at anthesis. **Capsules** spreading at nearly a right angle to stem, curved upward, sometimes secund, dull green or rusty brown when dry, narrowly lanceoloid to lanceoloid, 20–40 × 4–6 mm, free tips of valves 1–2 mm. **Seeds** 1.3–1.9 × 0.7–1.1 mm. **2*n*** = 14.

Flowering (Jun–)Jul–Oct. Open sites on Devonian Brallier shale slopes, barrens, outcrops or adjacent roadsides in mid-Appalachian Allegheny Mountains; 150–700 m; Md., Pa., Va., W.Va.

Oenothera argillicola is one of eight angiosperm species restricted to the Devonian Brallier shale barrens, but among them only *O. argillicola* and *Trifolium virginicum* occur throughout the shale barren region. *Oenothera argillicola* has plastome V and a CC genome composition.

79. Oenothera parviflora Linnaeus, Syst. Nat. ed. 10, 2: 998. 1759 (as parviflor) [E] [F] [W]

Oenothera ammophiloides R. R. Gates & Catcheside var. *flecticaulis* (R. R. Gates) R. R. Gates; *O. ammophiloides* var. *parva* (R. R. Gates) R. R. Gates; *O. angustifolia* Miller; *O. angustissima* R. R. Gates; *O. angustissima* var. *quebecensis* R. R. Gates; *O. apicaborta* R. R. Gates; *O. atrovirens* Shull & Bartlett; *O. biennis* Linnaeus var. *cruciata* (Nuttall) Torrey & A. Gray; *O. biennis* var. *parviflora* (Linnaeus) Torrey & A. Gray; *O. biformiflora* R. R. Gates; *O. cleistantha* Shull & Bartlett; *O. comosa* R. R. Gates; *O. cruciata* Nuttall; *O. cruciata* var. *sabulonensis* Fernald; *O. deflexa* R. R. Gates; *O. flecticaulis* R. R. Gates; *O. hazeliae* R. R. Gates; *O. hazeliae* var. *parviflora* R. R. Gates; *O. hazeliae* var. *subterminalis* (R. R. Gates) R. R. Gates; *O. intermedia* R. R. Gates; *O. laevigata* Bartlett; *O. laevigata* var. *scitula* (Bartlett) R. R. Gates; *O. laevigata* var. *similis* R. R. Gates; *O. novae-scotiae* R. R. Gates; *O. novae-scotiae* var. *distantifolia* R. R. Gates; *O. novae-scotiae* var. *intermedia* (R. R. Gates) R. R. Gates; *O. parva* R. R. Gates; *O. parviflora* subsp. *angustissima* (R. R. Gates) Munz; *O. parviflora* var. *angustissima* (R. R. Gates) Wiegand; *O. robinsonii* Bartlett; *O. rubricapitata* R. R. Gates; *O. scitula* Bartlett; *O. subterminalis* R. R. Gates; *O. venosa* Shull & Bartlett; *Onagra biennis* (Linnaeus) Scopoli var. *cruciata* (Nuttall) Britton; *O. chrysantha* Spach var. *cruciata* (Nuttall) Spach; *O. cruciata* (Nuttall) Small; *O. parviflora* (Linnaeus) Moench

Herbs biennial, sparsely strigillose, glandular puberulent, and villous with pustulate or non-pustulate hairs, sometimes predominately strigillose proximally or predominately villous with pustulate or non-pustulate hairs distally, glabrous, or some mixture of strigillose, glandular puberulent, or sparsely villous distally, sometimes appearing glabrous to the naked eye. **Stems** erect, green or red on proximal parts or throughout, mostly branched from base or only in distal ½, 30–150 cm. **Leaves** in a basal rosette and cauline, basal 10–30 × 1–4 cm, cauline 4–18 × 1–3 cm; blade usually bright green, sometimes pale green distally, white- or red-veined, narrowly oblanceolate, narrowly elliptic to lanceolate, or oblong, margins usually flat, rarely undulate, regularly dentate to remotely denticulate, sometimes teeth widely spaced; bracts persistent. **Inflorescences** erect or ± curved, unbranched or with secondary branches just proximal to main one. **Flowers** opening near sunset; buds erect, 3–5 mm diam., with free tips subterminal, spreading to erect, 0.5–5 mm; floral tube 22–40 mm; sepals green to yellowish green or flushed with red or dark red, sometimes only red-flecked, 7–17 mm; petals yellow to pale yellow, fading pale yellow to pale yellowish orange, very broadly obcordate, 8–15(–20) mm; filaments 7–13 mm, anthers 3.5–6 mm, pollen ca. 50% fertile; style 25–50 mm, stigma surrounded by anthers at anthesis. **Capsules** erect or slightly spreading, usually greenish black when dry, narrowly lanceoloid to lanceoloid, 20–40 × 3.5–5 mm, free tips of valves 1–1.5 mm. **Seeds** 1.1–1.8 × 0.5–1 mm. **2*n*** = 14.

Flowering Jun–Aug. Open or disturbed, sandy or gravelly sites, roadsides, fallow fields, clearings, riverbanks, along water courses, salt marshes, coastal meadows; 0–1700 m; B.C., N.B., Nfld. and Labr. (Nfld.), N.S., Ont., P.E.I., Que.; Conn., Del., D.C., Ill., Ind., Iowa, Ky., Maine, Md., Mass., Mich., Minn., Mo., N.H., N.J., N.Y., N.C., N.Dak., Ohio, Pa., R.I., Tenn., Vt., Va., W.Va., Wis.; introduced in Europe, Asia (ne China, Japan), s Africa, Pacific Islands (New Zealand), Australia. It is introduced in British Columbia.

Oenothera parviflora is a PTH species and forms a ring of 14 chromosomes in meiosis, and is self-compatible and autogamous (W. Dietrich et al. 1997). It has plastome IV and a BC genome composition.

Onagra chrysantha Spach 1835, not Michaux 1803, is a superfluous name, as is *Onagra chrysantha* var. *parviflora* (Linnaeus) Spach, and both pertain here. *O. biformiflora* var. *cruciata* R. R. Gates is an invalid name that pertains here.

80. Oenothera oakesiana (A. Gray) J. W. Robbins ex S. Watson & J. M. Coulter in A. Gray et al., Manual ed. 6, 190. 1890 [E]

Oenothera biennis Linnaeus var. *oakesiana* A. Gray, Manual ed. 5, 178. 1867; *O. ammophila* Focke; *O. ammophiloides* R. R. Gates & Catcheside; *O. ammophiloides* var. *angustifolia* R. R. Gates; *O. ammophiloides* var. *laurensis* R. R. Gates; *O. atrovirens* Shull & Bartlett var. *ostreae* (A. H. Sturtevant) R. R. Gates; *O. canovirens* E. S. Steele var. *cymatilis* (Bartlett) R. R. Gates; *O. cruciata* Nuttall var. *stenopetala* (E. P. Bicknell) Fernald; *O. cymatilis* Bartlett; *O. deflexa* R. R. Gates var. *bracteata* R. R. Gates; *O. disjuncta* Boedijn; *O. eriensis* R. R. Gates; *O. eriensis* var. *niagarensis* (R. R. Gates) R. R. Gates; *O. eriensis* var. *repandodentata* (R. R. Gates) R. R. Gates; *O. germanica* Boedijn; *O. insignis* Bartlett; *O. laevigata* Bartlett var. *rubripunctata* R. R. Gates; *O. leucophylla* R. R. Gates; *O. litorea* Bartlett; *O. magdalena* R. R. Gates; *O. millersii* de Vries; *O. muricata* Linnaeus var. *parviflora* R. R. Gates; *O. niagarensis* R. R. Gates; *O. nobska* A. H. Sturtevant; *O. oakesiana* var. *nobska* (A. H. Sturtevant) R. R. Gates; *O. oakesiana* var. *tidestromii* (Bartlett) R. R. Gates; *O. ostreae* A. H. Sturtevant; *O. parviflora* var. *oakesiana* (A. Gray) Fernald; *O. perangusta* R. R. Gates; *O. perangusta* var. *rubricalyx* R. R. Gates; *O. repandodentata* R. R. Gates; *O. rubescens* Bartlett; *O. stenopetala* E. P. Bicknell; *O. tidestromii* Bartlett; *Onagra oakesiana* (A. Gray) Britton

Herbs biennial, densely silky-strigillose, at least proximally, also sparsely villous with long, appressed hairs, sometimes also villous with spreading, pustulate hairs and/or glandular puberulent distally. **Stems** erect to decumbent, green or flushed with red on proximal parts or throughout, unbranched or bushy and branched from base, with side branches arising obliquely or arcuately from rosette, 10–60 cm. **Leaves** in a basal rosette and cauline, basal 8–30 × 0.5–3 cm, cauline 3.5–20 × 0.5–2.7 cm; blade grayish green to dull green, very narrowly oblanceolate to narrowly oblanceolate or narrowly elliptic, margins flat, subentire or remotely dentate, teeth sometimes blunt, sometimes sinuate-dentate proximally; bracts persistent. **Inflorescences** usually recurved with ascending tip distally, rarely suberect, unbranched. **Flowers** opening near sunset; buds erect, 3–5 mm diam., with free tips subterminal, spreading to erect, 2.5–4 mm; floral tube 15–40 mm; sepals green to yellow, flushed with red and dark red flecked or red-striped, 9–17 mm; petals yellow to pale yellow, fading yellowish white to pale yellowish orange, very broadly obcordate, 7–20 mm; filaments 6–15 mm, anthers 3–7 mm, pollen ca. 50% fertile; style 20–45 mm, stigma surrounded by anthers at anthesis. **Capsules** erect or slightly spreading, usually rusty brown when dry, narrowly lanceoloid to lanceoloid, 15–40 × 4–8 mm, free tips of valves 0.5 mm. **Seeds** 1.1–1.2 × 0.8–1 mm. **2*n*** = 14.

Flowering Jun–Aug. Sandy coastal meadows and dunes, gravelly or rocky sites along rivers, disturbed sites, roadsides; 0–50(–500) m; Man., N.B., Nfld. and Labr. (Nfld.), N.S., Ont., P.E.I., Que.; Conn., Del., D.C., Ill., Ind., Maine, Md., Mass., Mich., Minn., N.H., N.J., N.Y., Ohio, Pa., R.I., Vt., Va., Wis.; introduced in Europe, Asia.

Oenothera oakesiana is a PTH species and forms a ring of 14 chromosomes or a ring of 12 and 1 bivalent in meiosis, and is self-compatible and autogamous (W. Dietrich et al. 1997). It has plastome IV and a AC genome composition.

17p.3. OENOTHERA Linnaeus (sect. OENOTHERA) subsect. CANDELA (W. Dietrich & W. L. Wagner) W. L. Wagner & Hoch, Syst. Bot. Monogr. 83: 187. 2007 [E]

Oenothera ser. *Candela* W. Dietrich & W. L. Wagner, Ann. Missouri Bot. Gard. 74: 147. 1987

Herbs annual, biennial, or short-lived perennial; from taproot. **Stems** erect. **Inflorescences** dense or lax, erect, terminal. **Flowers:** buds erect, terete to weakly quadrangular, with free tips terminal, erect or spreading; floral tube slightly curved upward or straight, 15–47 mm; petals rhombic to elliptic or rhombic-ovate. **Capsules** curved upward or straight, narrowly lanceoloid to lanceoloid, subterete, 2–4 mm diam. **Seeds** numerous, in 2 rows per locule, brown to dark brown, often flecked with darker spots, ellipsoid to broadly ellipsoid, surface reticulate and regularly pitted. **2*n*** = 14.

Species 5 (5 in the flora): c, e North America.

Subsection *Candela* consists of five closely related species found at low elevations in the central and eastern United States (W. Dietrich and W. L. Wagner 1988). They are self-incompatible

(*Oenothera cordata*, *O. rhombipetala*, and some populations of *O. heterophylla*) or self-compatible (*O. clelandii*, *O. curtissii*, and some populations of *O. heterophylla*). The flowers are vespertine, outcrossing and pollinated by hawkmoths, or autogamous in two PTH species (*O. clelandii* and *O. curtissii*) (Dietrich and Wagner). They were treated as part of a rather heterogeneous subg. *Raimannia* by P. A. Munz (1965). W. Stubbe and P. H. Raven (1979) placed these species in a reconfigured, narrower subsect. *Raimannia* based on crossing results; Dietrich and Wagner segregated them as a new series within subsect. *Raimannia* based on the earlier crossing results and morphology. The group is characterized by the usually dense flowering spikes with several flowers opening each evening, straight floral tubes, and petals acute to rounded at apex. R. A. Levin et al. (2004) did not test this group's monophyly, since they included only *O. heterophylla*, but the overall morphological similarity of these species and the clear synapomorphy of the petal shape support the monophyly of the subsection. Wagner et al. (2007) elevated the group to the status of a subsection, because it does not group with subsect. *Raimannia* in the molecular analyses; instead, it forms a weakly supported clade with subsect. *Oenothera*, with which it shares a similar habit, dense spikes, and lanceoloid capsules. Cytological analyses (Dietrich and Wagner), showed that species of subsect. *Candela* are all diploid ($2n = 14$) with *O. cordata*, *O. heterophylla*, and *O. rhombipetala* bivalent-forming species, and *O. clelandii* and *O. curtissii* PTH species forming a ring of 14 chromosomes in meiosis.

1. Petals 5–17 mm; stigmas surrounded by anthers at anthesis; pollen ca. 50% fertile.
   2. Inflorescences dense, 2+ flowers per spike opening per day . . . . 84. *Oenothera clelandii*
   2. Inflorescences lax, 1 or 2 flowers per spike opening per day . . . . 85. *Oenothera curtissii*
1. Petals 15–35 mm; stigmas exserted beyond anthers at anthesis; pollen 85–100% fertile.
   3. Mature flower buds not overtopping spike apex; distal parts of plant strigillose, sometimes also sparsely glandular puberulent . . . . 83. *Oenothera rhombipetala*
   3. Mature flower buds usually overtopping spike apex; distal parts of plant glandular puberulent, villous, strigillose, or hirsute with pustulate-based hairs, especially sepals and floral tube, or glabrous.
      4. Inflorescences dense; bracts longer than capsules they subtend, 1–3 cm . . . . 81. *Oenothera heterophylla*
      4. Inflorescences lax; bracts shorter than capsules they subtend, 0.5–1.7 cm. . . . . 82. *Oenothera cordata*

81. Oenothera heterophylla Spach, Nouv. Ann. Mus. Hist. Nat. 4: 348. 1836 [E] [F]

Raimannia heterophylla (Spach) Rose ex Sprague & L. Riley

Herbs annual or short-lived perennial, sparsely to densely strigillose, inflorescence sometimes also sparsely glandular puberulent, villous, or sparsely hirsute with spreading, pustulate-based hairs, or sometimes glabrate. **Stems** unbranched or branched mainly in distal part, 25–70 cm. **Leaves** in a basal rosette and cauline, basal 7–15 × 1–2.5 cm, cauline 3–13 × 0.4–2.3 cm; blade narrowly oblanceolate to oblanceolate, gradually narrowly lanceolate to lanceolate or elliptic distally, margins deeply lobed to remotely dentate or subentire; bracts longer than capsule they subtend, 1–3 cm. **Inflorescences** dense, often with several lateral branches, mature buds usually overtopping spike apex. **Flowers** 2–several per spike opening per day near sunset; buds erect, with free tips erect or spreading, 1–6 mm; floral tube nearly straight, 25–47 mm; sepals 15–30 mm; petals yellow, broadly elliptic to nearly rhombic, 18–35 mm; filaments 15–30 mm, anthers 3–8 mm, pollen 85–100% fertile; style 45–75 mm, stigma usually exserted beyond anthers at anthesis. **Capsules** lanceoloid, 13–25 × 2.5–4 mm. **Seeds** brown, often flecked with darker spots, ellipsoid to broadly ellipsoid, 1.1–1.8 × 0.4–0.8 mm.

Subspecies 2 (2 in the flora): s United States.

1. Herbs densely to sparsely strigillose, also at least parts of inflorescence sparsely hirsute with spreading, pustulate-based hairs, and often glandular puberulent and villous; flower buds with free tips spreading, 2–6 mm . . . . 81a. *Oenothera heterophylla* subsp. *heterophylla*
1. Herbs densely to sparsely strigillose, also at least part of inflorescence glabrate or glandular puberulent; flower buds with free tips erect, 1–3 mm . . . 81b. *Oenothera heterophylla* subsp. *orientalis*

81a. Oenothera heterophylla Spach subsp. **heterophylla** [E] [F]

Oenothera pyramidalis H. Léveillé var. *lindheimeri* H. Léveillé

Herbs densely to sparsely strigillose, also at least parts of inflorescence sparsely hirsute with spreading, pustulate-based hairs, and often glandular puberulent and villous. **Flowers:** buds with free tips spreading, 2–6 mm; floral tube 25–42 mm; sepals 15–28 mm; petals 18–35 mm. **2*n*** = 14.

Flowering Jun–Sep. Sandy to sandy-loam soil of open sites in woodlands, with *Persea borbonia*, *Pinus echinata*, *P. palustris*, *Quercus incana*, *Q. marilandica*, *Q. stellata*, and *Q. virginiana*; 0–200 m; La., Tex.

Populations of subsp. *heterophylla* were determined by W. Dietrich and W. L. Wagner (1988) to be self-incompatible. It occurs in a narrow range from eastern Texas (Austin, Bastrop, Brazos, Cass, Chambers, Cherokee, Dallas, Freestone, Gonzales, Gregg, Hardin, Harris, Henderson, Hopkins, Houston, Jasper, Lee, Leon, Liberty, Limestone, Nacogdoches, Newton, Robertson, Rusk, Sabine, San Augustine, San Jacinto, Smith, Sutton, Travis, Tyler, Upshur, Van Zandt, Victoria, Waller, and Wood counties) and southwestern Louisiana (Caddo, Calcasieu, Erwin, Natchitoches, and Winn parishes). It is known from several historical specimens in St. Louis, Missouri, as an adventive but is apparently no longer growing in that area.

Oenothera variifolia Steudel is a superfluous name that pertains here.

81b. Oenothera heterophylla Spach subsp. **orientalis** W. Dietrich, P. H. Raven & W. L. Wagner, Ann. Missouri Bot. Gard. 70: 196. 1983 [C] [E]

Herbs densely to sparsely strigillose, also at least parts of inflorescence glandular puberulent or glabrate. **Flowers:** buds with free tips usually erect, 1–3 mm; floral tube 30–47 mm; sepals 17–30 mm; petals 25–35 mm. **2*n*** = 14.

Flowering May–Jul. Sandy soil of open sites, old alluvium areas in woodlands; of conservation concern; 30–60 m; Ala., Ark.

Subspecies *orientalis* is known from two disjunct areas: Greene, Pickens, and Sumter counties, Alabama, and Calhoun, Nevada, and Ouachita counties, Arkansas. W. Dietrich and W. L. Wagner (1988) determined populations in Arkansas to be self-incompatible while those sampled in Alabama were self-compatible.

82. Oenothera cordata J. W. Loudon, Ladies' Flower-gard. Ornam. Perenn. 1: 167. 1843 [E]

Herbs annual or biennial, densely to sparsely strigillose, glandular puberulent or sometimes also sparsely villous distally. **Stems** unbranched or branched primarily distally, 25–70 cm. **Leaves** in a basal rosette and cauline, basal 6–12 × 0.7–2 cm, cauline 2–10 × 0.5–3 cm; subsessile; blade narrowly elliptic to oblanceolate, gradually narrowly lanceolate to lanceolate, elliptic, or ovate distally, margins lobed to remotely dentate or subentire; bracts shorter than capsule they subtend, 0.5–1.7 cm. **Inflorescences** open, lax, usually unbranched, mature buds usually overtopping spike apex. **Flowers** 1–few per spike opening per day near sunset; buds erect, with free tips erect, 1–3 mm; floral tube nearly straight, 20–40 mm; sepals 15–25 mm; petals yellow, broadly elliptic to rhombic-ovate, 20–30 mm; filaments 17–22 mm, anthers 4–7 mm, pollen 85–100% fertile; style 50–65 mm, stigma usually exserted beyond anthers at anthesis. **Capsules** narrowly lanceoloid, 15–33 × 2–3 mm. **Seeds** dark brown, ellipsoid, 1–1.4 ×0.4–0.6 mm. **2*n*** = 14.

Flowering Apr–Jul. Sandy, open places in oak woodlands; 30–200 m; Tex.

Oenothera cordata is self-incompatible. It occurs in a narrow range in eastern Texas (Austin, Bastrop, Colorado, Fayette, Guadalupe, Goliad, Matagorda, San Patricio, Victoria, Waller, and Wilson). It apparently occasionally hybridizes with *O. heterophylla* subsp. *heterophylla* where their ranges come together.

Oenothera bifrons D. Don 1838 (not Lindley 1831) pertains here.

83. Oenothera rhombipetala Nuttall ex Torrey & A. Gray, Fl. N. Amer. 1: 493. 1840 [E]

Oenothera heterophylla Spach var. *rhombipetala* (Nuttall ex Torrey & A. Gray) Fosberg; *O. leona* Buckley; *Raimannia rhombipetala* (Nuttall ex Torrey & A. Gray) Rose ex Britton & A. Brown

Herbs biennial, densely to sparsely strigillose, sometimes also sparsely glandular puberulent distally. **Stems** sometimes with lateral branches arising obliquely from rosette, 30–100(–150) cm. **Leaves** in a basal rosette and cauline, basal 6–20 × 0.6–2 cm, cauline 3–15 × 0.8–2.5 cm; blade narrowly oblanceolate, gradually narrowly elliptic to narrowly lanceolate, oblanceolate, or ovate distally, margins lobed to remotely dentate or subentire; bracts slightly longer than capsule they subtend. **Inflorescences** dense, usually without lateral branches, mature buds usually not overtopping spike apex. **Flowers** 2–several per spike opening per day near sunset; buds erect, with free tips erect, 0.5–3 mm; floral tube slightly curved upward to ± straight, 30–45 mm; sepals 15–30 mm; petals yellow, broadly elliptic to rhombic-elliptic, 15–35 mm; filaments 13–25 mm, anthers 3–8 mm, pollen 85–100% fertile; style 25–50 mm, stigma exserted beyond anthers at anthesis. **Capsules** narrowly lanceoloid, 13–25 × 2.5–3 mm. **Seeds** brown, sometimes flecked with dark red spots, ellipsoid, 1–1.7 × 0.4–0.7 mm. **2*n*** = 14.

Flowering May–Oct. Fields, prairies, sandy soil; 60–600(–1300) m; Ark., Colo., Ill., Kans., Mich., Minn., Mo., Nebr., N.Mex., Okla., S.Dak., Tex., Wis.

Oenothera rhombipetala is primarily a central plains species that has scattered localities in the Midwest to Illinois, Michigan, and Wisconsin, and barely entering the easternmost parts of Colorado and New Mexico.

Oenothera rhombipetala had a broader delimitation (P. A. Munz 1965) until W. Dietrich and W. L. Wagner (1988) divided it into three species (*O. clelandii*, *O. curtissii*, and *O. rhombipetala*), with both of the split-off species being PTH. Evidence gathered by Dietrich and Wagner showed that these PTH species are geographically separated populations of small-flowered plants, and although they are very close morphologically, their distributions and morphological differences suggest that they were each derived independently from *O. rhombipetala*. *Oenothera rhombipetala* is self-incompatible.

Oenothera pyramidalis H. Léveillé is a superfluous name and pertains here.

84. Oenothera clelandii W. Dietrich, P. H. Raven & W. L. Wagner, Ann. Missouri Bot. Gard. 70: 196. 1983 [E]

Herbs biennial, densely to sometimes sparsely strigillose, or also sparsely glandular puberulent distally. **Stems** sometimes with lateral branches arising obliquely from rosette, 20–70 (–100) cm. **Leaves** in a basal rosette and cauline, basal 5–16 × 0.5–1.5 cm, cauline 2–12 × 0.5–2 cm; blade narrowly oblanceolate, gradually narrowly elliptic to narrowly lanceolate distally, margins lobed to remotely dentate or subentire; bracts slightly longer than capsule they subtend. **Inflorescences** dense, without lateral branches, mature buds usually not overtopping spike apex. **Flowers** 2–several per spike opening per day near sunset; buds erect, with free tips erect, 0.5–2 mm; floral tube slightly curved upward to ± straight, 15–40 mm; sepals 6–13 mm; petals yellow, broadly elliptic to rhombic-ovate, 5–16 mm; filaments 4–18 mm, anthers 2–3.5 mm, pollen ca. 50% fertile; style 20–40 mm, stigma surrounded by anthers at anthesis. **Capsules** narrowly lanceoloid, 10–20 × 2–3 mm. **Seeds** brown, sometimes flecked with dark red spots, ellipsoid, 1–1.9 × 0.4–0.8 mm. **2*n*** = 14.

Flowering Jun–Aug(–Sep). Fields, prairies, sandy soil; 150–300 m; Ont.; Ark., Ill., Ind., Iowa, Ky., Mich., Minn., Mo., N.J., N.Y., Va., Wis.

Oenothera clelandii is a PTH species and forms a ring of 14 chromosomes in meiosis, and is self-compatible and autogamous (W. Dietrich and W. L. Wagner 1988). Some localities in the easternmost states may represent introductions, primarily occurring in disturbed areas along roads and railroad lines.

85. Oenothera curtissii Small, Fl. S.E. U.S. ed. 2, 1353. 1913 [E]

Oenothera heterophylla Spach var. *curtissii* (Small) Fosberg

Herbs biennial or short-lived perennial, densely to sparsely strigillose, sometimes also sparsely glandular puberulent distally. **Stems** sometimes with lateral branches arising obliquely from rosette, 30–80 cm. **Leaves** in a basal rosette and cauline, basal 7–17 × 0.5–1.5 cm, cauline 2–8 × 0.5–1.5 cm; blade narrowly oblanceolate, gradually narrowly elliptic to narrowly oblong distally, margins lobed to remotely dentate or subentire; bracts slightly longer than capsule they subtend. **Inflorescences** open, lax, without lateral branches, mature buds usually not overtopping spike apex. **Flowers** 1 or 2 per spike opening per day near sunset; buds erect, with free tips erect to spreading, 0.3–0.8 mm; floral tube slightly curved upward to straight, 23–37 mm; sepals 7–13 mm; petals yellow, broadly elliptic to rhombic-ovate, 8–17 mm; filaments 6–10 mm, anthers 1.5–4 mm, pollen ca. 50% fertile; style 30–45 mm, stigma surrounded by anthers at anthesis. **Capsules** narrowly lanceoloid, 10–25 × 2–3 mm. **Seeds** brown, sometimes flecked with dark red spots, ellipsoid, 1–1.3 × 0.5–0.7 mm. **2*n*** = 14.

Flowering Jun–Sep. Dry places, pine-oak woods, fields, roadsides, sandy soil; 0–60 m; Ala., Fla., Ga., S.C.

Oenothera curtissii is a PTH species and forms a ring of 14 chromosomes in meiosis, and is self-compatible and autogamous (W. Dietrich and W. L. Wagner 1988). It is known only from northern Florida, adjacent southern Georgia and southeastern Alabama, and one disjunct locality in South Carolina (Allendale County).

17p.4. Oenothera Linnaeus (sect. Oenothera) subsect. Raimannia (Rose ex Britton & A. Brown) W. Dietrich, Ann. Missouri Bot. Gard. 64: 612. 1978

Raimannia Rose ex Britton & A. Brown, Ill. Fl. N. U.S. ed. 2, 2: 596, figs. 3042–3044. 1913; *Oenothera* sect. *Raimannia* (Rose ex Britton & A. Brown) Munz; *Oenothera* ser. *Raimannia* (Rose ex Britton & A. Brown) W. Dietrich & W. L. Wagner; *Oenothera* subg. *Raimannia* (Rose ex Britton & A. Brown) Munz

Herbs annual or short-lived perennial; from taproot. **Stems** erect, ascending, or decumbent. **Flowers:** buds erect, terete to weakly quadrangular, with free tips terminal, erect or spreading; floral tube curved upward, 12–50 mm; petals broadly obcordate, sometimes shallowly so. **Capsules** usually straight, sometimes curved upward, cylindrical, sometimes slightly enlarged toward apex, subterete, 2–4(–5) mm diam. **Seeds** in 2 rows per locule, brown, ellipsoid to subglobose, surface reticulate and regularly pitted. **2*n*** = 14.

Species 6 (6 in the flora): c, s United States, n Mexico, West Indies (Cuba), Bermuda; introduced nearly worldwide in temperate and subtropical areas.

Subsection *Raimannia* consists of six diploid (2*n* = 14) species native to North America, primarily in the south-central United States, with some species extending into Mexico in Tamaulipas (*Oenothera grandis*), Campeche, and Baja California (*O. drummondii*) (W. Dietrich and W. L. Wagner 1988). Several species are widespread, including *O. laciniata* in most of eastern North America and naturalized in Australia, Europe, Hawaiian Islands, Japan, Paraguay, and South Africa; *O. drummondii* also is widely naturalized, in Australia, Europe, North Africa and the Middle East, China, South America, and South Africa. P. A. Munz (1965) included these species in his rather heterogeneous and broadly delimited subg. *Raimannia*, which consisted of species here assigned by Wagner et al. (2007) to sects. *Kleinia* and *Ravenia* W. L. Wagner, and *Oenothera* subsects. *Candela*, *Emersonia*, *Munzia*, and *Nutantigemma*, in addition to those retained here in subsect. *Raimannia*. Dietrich (1977) removed the South American species to a new subsect. *Munzia*. The current, narrower circumscription is based primarily on a wide series of crossing experiments (W. Stubbe and P. H. Raven 1979; Dietrich and Wagner). *Oenothera drummondii*, *O. falfurriae*, *O. grandis*, and *O. mexicana* are bivalent-forming species (2*n* = 14), whereas *O. humifusa* and *O. laciniata* are PTH. *Oenothera grandis* is self-incompatible, while the other species are self-compatible and largely autogamous, except *O. drummondii*, which, like *O. grandis*, is outcrossed by hawkmoths (Dietrich and Wagner).

1. Plants densely strigillose, sometimes also villous, becoming glandular puberulent distally; leaf blades grayish green, margins subentire or remotely, shallowly dentate; bracts flat; coastal sites.
   2. Sepals 20–30 mm; petals 20–45 mm; stigmas exserted beyond anthers at anthesis; pollen 85–100% fertile . . . . . 90. *Oenothera drummondii*
   2. Sepals 3–11 mm; petals 4.5–16 mm; stigmas surrounded by anthers at anthesis; pollen ca. 50% fertile. . . . . 91. *Oenothera humifusa*
1. Plants moderately to sparsely strigillose, usually also sparsely to sometimes densely villous, usually also glandular puberulent distally; leaf blades green, margins usually deeply lobed to dentate or pinnatifid, rarely subentire; bracts flat or, if grayish green, then margins revolute; inland sites, often in disturbed habitats.
   3. Petals 25–40 mm; styles 40–75 mm, stigmas exserted beyond anthers at anthesis; pollen 85–100% fertile . . . . . 86. *Oenothera grandis*
   3. Petals 5–25 mm; styles 20–50 mm, stigmas surrounded by, or slightly exserted beyond, anthers at anthesis; pollen 50–100% fertile.
      4. Bracts revolute, distalmost erect . . . . . 88. *Oenothera mexicana*
      4. Bracts flat, distalmost spreading.
         5. Petals 13–25 mm; stigmas slightly exserted beyond anthers at anthesis; pollen 85–100% fertile . . . . . 87. *Oenothera falfurriae*
         5. Petals 5–22 mm; stigmas surrounded by anthers at anthesis; pollen ca. 50% fertile . . . . . 89. *Oenothera laciniata*

86. Oenothera grandis Smyth, Trans. Kansas Acad. Sci. 16: 160. 1899

Oenothera laciniata Hill var. *grandiflora* (S. Watson) B. L. Robinson; *O. sinuata* Linnaeus var. *grandiflora* S. Watson; *Raimannia grandis* (Smyth) Rose ex Sprague & L. Riley

Herbs annual, strigillose and sparsely villous, also glandular puberulent distally. **Stems** erect to ascending, often with ascending lateral branches, 15–60(–100) cm. **Leaves** in a basal rosette and cauline, basal 5–13 × 1–3 cm, cauline 3–10 × 1.5–3.5 cm; blade green, narrowly oblanceolate to narrowly elliptic, margins lobed or dentate, lobes often dentate; bracts spreading, flat. **Flowers** 1–few opening per day near sunset; buds erect, with free tips terminal, erect or hornlike, 1.5–5 mm; floral tube 25–45 mm; sepals 15–30 mm; petals yellow, very broadly obovate or shallowly obcordate, 25–40 mm; filaments 12–22 mm, anthers 4–11 mm, pollen 85–100% fertile; style 40–75 mm, stigma exserted beyond anthers at anthesis. **Capsules** cylindrical, sometimes slightly enlarged toward apex, 25–50 × 2–3 mm. **Seeds** broadly ellipsoid to subglobose, 0.8–1.5 × 0.5–0.9 mm. $2n$ = 14.

Flowering Mar–Sep. Open, sandy sites; 0–1500 (–2200) m; Ala., Ark., Colo., Conn., Fla., Ill., Ind., Kans., La., Md., Mich., Mo., Nebr., N.J., N.Mex., N.C., Tex.; Mexico (Tamaulipas).

Oenothera grandis is probably native to eastern New Mexico and Colorado, Arkansas, Kansas, Louisiana, Missouri, Nebraska, Oklahoma, and Texas, and northeastern Tamaulipas, Mexico. Scattered collections made in other states probably represent introductions (W. Dietrich and W. L. Wagner 1988).

Oenothera grandis is self-incompatible (W. Dietrich and W. L. Wagner 1988).

Oenothera laciniata Hill var. *occidentalis* Small and *O. laciniata* var. *grandis* Britton are illegitimate superfluous names based on *O. sinuata* Linnaeus var. *grandiflora* S. Watson and pertain here.

87. Oenothera falfurriae W. Dietrich & W. L. Wagner, Ann. Missouri Bot. Gard. 74: 149. 1987 [E]

Herbs annual, moderately to sparsely strigillose and villous, sometimes glandular puberulent distally. **Stems** erect to ascending, usually unbranched, 10–40 cm. **Leaves** in a basal rosette and cauline, basal 5–12 × 1.3–3.5 cm, cauline 2–8.5 × 1–3 cm; blade green, narrowly oblanceolate to narrowly elliptic or narrowly lanceolate, margins usually dentate to pinnatifid, sometimes subentire; bracts spreading, flat. **Flowers** usually 1 opening per day near sunset; buds erect, with free tips erect, 0.5–2 mm; floral tube 25–40 mm; sepals 10–22 mm; petals yellow, fading orange or reddish tinged, broadly obovate or shallowly obcordate, 13–25 mm; filaments 10–17 mm, anthers 4–5 mm, pollen 85–100% fertile; style 35–50 mm, stigma slightly exserted beyond anthers at anthesis. **Capsules** cylindrical, sometimes slightly enlarged toward apex, 20–45 × 2–2.5 mm. **Seeds** ellipsoid, 0.8–1.4 × 0.3–0.6 mm. $2n$ = 14.

Flowering Apr–Aug. Open, sandy sites; 0–300 m; Tex.

Oenothera falfurriae is known only from southeastern Texas (Aransas, Brazos, Brooks, Cameron, Frio, Harris, Hidalgo, Jim Hogg, Jim Wells, Kenedy, Kleberg, Maverick, Nueces, Refugio, Starr, Val Verde, Victoria, Webb, Willacy, Wilson, and Zapata counties). It is self-compatible and autogamous, but not a PTH species.

88. Oenothera mexicana Spach, Nouv. Ann. Mus. Hist. Nat. 4: 347. 1836 [E]

Oenothera laciniata Hill var. *mexicana* (Spach) Small; *O. sinuata* Linnaeus var. *hirsuta* Torrey & A. Gray; *Raimannia mexicana* (Spach) Wooton & Standley

Herbs annual, moderately to sparsely strigillose and densely long-villous, sometimes also becoming glandular puberulent distally. **Stems** erect to ascending, usually unbranched, or with arcuate lateral branches arising from rosette, 15–40(–60) cm. **Leaves** in a basal rosette and cauline, basal 6–10 × 1–2.5 cm, cauline 3–7.5 × 0.8–2 cm; blade usually grayish green, narrowly oblanceolate to narrowly oblanceolate, margins deeply lobed, lobes usually dentate; bracts distalmost erect, revolute. **Flowers** usually 1 opening per day near sunset; buds erect, with free tips erect or appressed, 0.5–2.5 mm; floral tube 23–28 mm; sepals 5–12 mm; petals yellow, fading orange, broadly obovate or shallowly obcordate, 6–15 mm; filaments 4–12 mm, anthers 3–4 mm, pollen 85–100% fertile; style 27–40 mm, stigma surrounded by anthers at anthesis. **Capsules** cylindrical, sometimes slightly enlarged toward apex, 25–45 × 2.5–3 mm. **Seeds** ellipsoid to subglobose, 0.8–1.2 × 0.3–0.5 mm. $2n = 14$.

Flowering Apr–May. Open, sandy sites; 30–200 m; Tex.

Oenothera mexicana is known only from southeastern Texas (Atascosa, Aransas, Bexar, Brooks, Burleson, De Witt, Frio, Gonzales, Kenedy, Medina, Newton, Refugio, San Patricio, Waller, and Washington counties). It is self-compatible and autogamous, but not a PTH species.

89. Oenothera laciniata Hill, Veg. Syst. 12(app.): 64, plate 10. 1767 [E] [W]

Oenothera minima Pursh; *O. repanda* Medikus; *O. sinuata* Linnaeus; *O. sinuata* var. *minima* (Pursh) Nuttall; *Onagra sinuata* (Linnaeus) Moench; *Raimannia laciniata* (Hill) Rose ex Britton & A. Brown

Herbs annual, sparsely to moderately strigillose, sometimes also villous, sometimes also becoming glandular puberulent distally. **Stems** erect to ascending, unbranched to much branched, 5–50 cm. **Leaves** in a basal rosette and cauline, basal 4–15 × 1–3 cm, cauline 2–10 × 0.5–3.5 cm; blade green, narrowly oblanceolate to narrowly elliptic or narrowly oblong, margins usually dentate or deeply lobed; bracts spreading, flat. **Flowers** usually 1 opening per day near sunset; buds erect, with free tips erect, 0.3–3 mm; floral tube 12–35 mm; sepals 5–15 mm; petals yellow, fading orange or reddish tinged, broadly obovate or obcordate, 5–22 mm; filaments 3–14 mm, anthers 4–5 mm, pollen ca. 50% fertile; style 20–50 mm, stigma surrounded by anthers at anthesis. **Capsules** cylindrical, sometimes slightly enlarged toward apex, 20–50 × 2–4 mm. **Seeds** ellipsoid to subglobose, 0.9–1.8 × 0.4–0.9 mm. $2n = 14$.

Flowering (Feb–)Apr–Sep(–Oct). Open, usually sandy sites, disturbed habitats; 0–1000(–1300) m; Ala., Ark., Calif., Conn., Del., D.C., Fla., Ga., Ill., Ind., Iowa, Kans., Ky., La., Maine, Md., Mass., Mich., Minn., Miss., Mo., Nebr., N.J., N.Mex., N.Y., N.C., N.Dak., Ohio, Okla., Pa., R.I., S.C., S.Dak., Tenn., Tex., Vt., Va., W.Va., Wis., Wyo.; introduced nearly worldwide in temperate and subtropical areas.

Oenothera laciniata is a PTH species and forms a ring of 14 chromosomes in meiosis, and is self-compatible and autogamous (W. Dietrich and W. L. Wagner 1988).

Oenothera laciniata is known in New Mexico from Doña Ana and Roosevelt counties from non-montane habitats and thus do not appear to represent *O. pubescens*; however, a few collections from Brewster and Jeff Davis counties, Texas, reported by W. Dietrich and W. L. Wagner (1988) as *O. laciniata* appear to represent collections of *O. pubescens*. Dietrich and Wagner found that *O. laciniata* hybridizes not only with *O. grandis*, but also with *O. drummondii* subsp. *drummondii*, *O. humifusa*, and *O. mexicana*. It is naturalized nearly worldwide in temperate and subtropical areas.

90. Oenothera drummondii Hooker, Bot. Mag. 61: plate 3361. 1834 [F]

Raimannia drummondii (Hooker) Rose ex Sprague & L. Riley

Subspecies 2 (1 in the flora): s United States, n Mexico.

Subspecies *thalassaphila* (Brandegee) W. Dietrich & W. L. Wagner differs from subsp. *drummondii* in a number of modally distinctive morphological features, especially floral tubes 2–3.5 cm, sepal tips 0.3–1 mm, capsules 2–4 cm × 2.5–5 mm in diameter and those, coupled with the great disjunction from the Atlantic coast of the United States and Mexico to the southern tip of Baja California, make it worthy of recognition. *Oenothera drummondii* is self-compatible and outcrossing.

90a. Oenothera drummondii Hooker subsp. **drummondii** [F]

Oenothera drummondii var. *helleriana* H. Léveillé; *O. littoralis* Schlechtendal; *Raimannia littoralis* (Schlechtendal) Rose ex Sprague & L. Riley

Herbs annual, densely strigillose, sometimes also villous, also glandular puberulent distally. **Stems** erect to decumbent, with non-flowering lateral branches, these often with terminal rosette of crowded, small leaves, 10–50 cm. **Leaves** in a basal rosette and cauline, basal 5–14 × 1–2 cm, cauline 1–8 × 0.5–2.5 cm; blade grayish green, narrowly oblanceolate to elliptic, becoming elliptic to narrowly obovate to obovate distally, margins subentire or shallowly dentate; bracts spreading, flat. **Flowers** 1–few opening per day near sunset; buds erect, with free tips erect, 1–3 mm; floral tube 25–50 mm; sepals 20–30 mm; petals yellow, very broadly obovate or obcordate, 25–45 mm; filaments 10–23 mm, anthers 4–12 mm, pollen 85–100% fertile; style 35–75 mm, stigma exserted beyond anthers at anthesis. **Capsules** cylindrical, sometimes slightly enlarged toward apex, 25–55 × 2–3 mm. **Seeds** usually ellipsoid to broadly ellipsoid, rarely subglobose, 1–2 × 0.5–0.9 mm. $2n = 14$.

Flowering Jan–Dec. Along or near Atlantic coast on dunes and open sandy places; 0–10 m; Fla., La., N.C., S.C., Tex.; Mexico (Tamaulipas, Veracruz); introduced in South America, sw Europe, Asia (including Taiwan), Africa, Australia.

Collections of subsp. *drummondii* at inland localities in Bexar and Dallas counties, Texas, and Henderson County, North Carolina, presumably represent introductions; it is also widely naturalized and is known from Africa, Asia, Australia, southwestern Europe, South America, and Taiwan (W. Dietrich and W. L. Wagner 1988; Wagner et al. 2007).

91. Oenothera humifusa Nuttall, Gen. N. Amer. Pl. 1: 245. 1818

Oenothera niveifolia Gandoger; *O. sinuata* Linnaeus var. *humifusa* (Nuttall) Torrey & A. Gray; *Raimannia humifusa* (Nuttall) Rose ex Britton & A. Brown

Herbs annual or short-lived perennial, densely strigillose, sometimes also villous, also becoming glandular puberulent distally. **Stems** erect to decumbent, much branched, 10–50(–90) cm. **Leaves** in a basal rosette and cauline, basal 4–8 × 0.7–1 cm, cauline 1–7 × 0.3–1.5 cm; blade usually grayish green, narrowly oblong to narrowly elliptic or narrowly obovate, margins remotely shallowly dentate to subentire; bracts spreading, flat. **Flowers** usually 1 opening per day near sunset; buds erect, with free tips erect and appressed or slightly spreading, 0.5–2 mm; floral tube 15–35 mm; sepals 3–11 mm; petals yellow, very broadly obovate or obcordate, 4.5–16 mm; filaments 4–11 mm, anthers 2–5.5 mm, pollen ca. 50% fertile; style 23–45 mm, stigma surrounded by anthers at anthesis. **Capsules** cylindrical, sometimes slightly enlarged toward apex, 15–45 × 2–3 mm. **Seeds** usually ellipsoid to broadly ellipsoid, rarely subglobose, 1–2 × 0.5–0.9 mm. $2n = 14$.

Flowering Apr–Nov. Dunes, open sandy places along or near Atlantic coast; 0–10 m; Ala., Del., Fla., Ga., La., Md., Miss., N.J., N.C., Pa., S.C., Va.; West Indies (Cuba); Bermuda.

Oenothera humifusa is a PTH species and forms a ring of 14 chromosomes in meiosis, and is self-compatible and autogamous (W. Dietrich and W. L. Wagner 1988). The inland collection from Iredell County, North Carolina, presumably represents an introduction. There are two geographically separated morphological forms of *O. humifusa*. Plants of one form are somewhat decumbent, with subentire cauline leaves and bracts; this form occurs in the southern part of the range. The other form is more upright, with more deeply divided leaves; it occurs from North Carolina northward.

17p.5. OENOTHERA Linnaeus (sect. OENOTHERA) subsect. NUTANTIGEMMA W. Dietrich & W. L. Wagner, Ann. Missouri Bot. Gard. 74: 145. 1987

Herbs annual or biennial [perennial]; from taproot. **Stems** erect to ascending, [in *O. pennellii* very short and decumbent]. **Flowers:** buds nodding; floral tube erect, becoming recurved and nodding, then erect just before anthesis, 15–50 mm; petals broadly obovate to obcordate. **Capsules** usually straight, sometimes curved upward, cylindrical, sometimes slightly enlarged distally, subterete, 2–4[–5] mm diam. **Seeds** in 2 rows per locule, brown, sometimes dark-flecked, ellipsoid to subglobose, surface reticulate and regularly pitted. $2n = 14$.

Species 4 (1 in the flora): sw, sc United States, Mexico, Central America (Guatemala), South America (Colombia, Ecuador, Peru).

Subsection *Nutantigemma* consists of four species and has one of the widest distributions in the genus, from Arizona and New Mexico to western Texas, southward nearly throughout Mexico to Guatemala; it is also in Andean Colombia, Ecuador, and Peru from 1500 to 3900 m (W. Dietrich 1977; Dietrich and W. L. Wagner 1988). The four species are likely self-compatible, but only two have been examined; of the two studied, one is a PTH species and the other is not.

92. Oenothera pubescens Willdenow ex Sprengel, Syst. Veg. 2: 229. 1825 [F]

Anogra amplexicaulis Wooton & Standley; *Oenothera amplexicaulis* (Wooton & Standley) Tidestrom; *O. laciniata* Hill subsp. *pubescens* (Willdenow ex Sprengel) Munz; *O. laciniata* var. *pubescens* (Willdenow ex Sprengel) Munz; *O. nyctaginiifolia* Small; *O. stuebelii* Hieronymus; *Raimannia colimae* Rose ex Sprague & L. Riley; *R. confusa* Rose ex Sprague & L. Riley

Herbs annual or biennial, densely to sparsely strigillose, sometimes also villous and glandular puberulent distally; from a taproot. **Stems** unbranched or with branched central stem and ascending to decumbent lateral branches arising from rosette, 5–50(–80) cm. **Leaves** in a basal rosette and cauline, basal 5–14 × 0.5–2.5 cm, cauline 2–8 × 0.5–2.5 cm; blade narrowly oblanceolate to narrowly elliptic or narrowly oblong, margins usually dentate to deeply lobed; bracts spreading, flat. **Flowers** usually 1 opening per day near sunset; buds with free tips erect, 0.1–1 mm; floral tube erect, becoming recurved and nodding, then erect again just before anthesis, 15–50 mm; sepals 5–25 mm; petals yellow, fading reddish orange, broadly obovate to obcordate, 5–25(–35) mm; filaments 6–18 mm, anthers (2–)3–9 mm, pollen ca. 50% fertile; style 20–60 mm, stigma surrounded by or slightly exserted beyond anthers at anthesis. **Capsules** cylindrical, sometimes slightly enlarged distally, 20–45 × 2–4 mm. **Seeds** brown, sometimes dark-flecked, 0.9–1.5 × 0.6–1 mm. $2n = 14$.

Flowering (Feb–)Apr–Sep(–Oct). Open sites in montane habitats; (1300–)1500–2500(–3100) m; Ariz., N.Mex., Tex.; Mexico; West Indies; Central America (Guatemala); South America (Colombia, Ecuador, Peru).

Oenothera pubescens is a PTH species and forms a ring of 14 chromosomes in meiosis, and is self-compatible and autogamous (W. Dietrich and W. L. Wagner 1988).

Oenothera pubescens has been collected once in California in 1884 (Newberry Springs, San Bernardino County), where it was temporarily introduced or a natural occurrence that was extirpated. Collections from west Texas (Brewster, Jeff Davis, and Presidio counties) have been made since 1990 and a few others collected earlier were misidentified as *O. laciniata*.

17p.6. OENOTHERA Linnaeus (sect. OENOTHERA) subsect. MUNZIA W. Dietrich, Ann. Missouri Bot. Gard. 64: 443. 1978 [I]

Herbs annual or biennial; from taproot. **Stems** erect or ascending [prostrate]. **Flowers:** buds erect; floral tube straight, 5–130 mm; petals obcordate to obovate. **Capsules** straight, cylindrical or ovoid, narrowed at both ends or somewhat enlarged toward apex. **Seeds** numerous, in 2 rows per locule, pale to dark brown, ellipsoid to subglobose, surface reticulate and regularly pitted.

Species 36 (1 in the flora): introduced, California; South America.

Subsection *Munzia* is subdivided into three series (W. Dietrich 1977), only one of which occurs in the United States as a naturalized plant.

17p.6a. Oenothera Linnaeus (sect. Oenothera subsect. Munzia) ser. Allochroa (Fischer & C. A. Meyer) W. Dietrich, Ann. Missouri Bot. Gard. 64: 489. 1978 [I]

Oenothera sect. *Allochroa* Fischer & C. A. Meyer, Index Seminum (St. Petersburg) 2: 44. 1836; *Oenothera* subg. *Allochroa* (Fischer & C. A. Meyer) Reichenbach

Capsules narrowly fusiform, [2–]3–4[–5] mm diam., sometimes enlarged in distal ⅓.

Species 20 (1 in the flora): introduced, California; South America.

Series *Allochroa* consists of 20 species, which are all native to South America, 16 of which are PTH, forming a ring of 14 chromosomes in meiosis. Unlike other species of *Oenothera* that are PTH, those of subsect. *Munzia* exhibit pollen fertility of over 90% and are maintained by selective fertilization (W. Dietrich 1977). Species of subsect. *Munzia* ser. *Allochroa* are widespread in South America with four species naturalized outside of their native range, three of which have been recorded in the United States: *O. indecora* Cambessèdes, *O. mollissima* Linnaeus, and *O. stricta* subsp. *stricta*. None of these species seems to be naturalized and the only one collected several times, and also the most recently collected, is *O. stricta* subsp. *stricta*, treated here. *Oenothera mollissima* was collected once in 1866 in ballast in Camden, New Jersey, once about 1920 in a shipyard in Linnton, Oregon, and three times in 1885 and 1900 at Port Eads, Louisiana. Similarly, *O. indecora* was collected once in Wilmington, North Carolina, in 1890. One additional species, *O. villaricae* W. Dietrich of subsect. *Clelandia* W. Dietrich, native to southern South America, has been collected once (in 1994) at Eglin Air Force Base, Florida; its status is not known.

93. Oenothera stricta Ledebour ex Link, Enum. Hort. Berol. Alt. 1: 377. 1821 (as striata) [I]

Subspecies 2 (1 in the flora): introduced, California; South America.

Oenothera stricta is a PTH species and forms a ring of 14 chromosomes in meiosis, and is self-compatible and autogamous (W. Dietrich 1977).

Subspecies *stricta* is naturalized in many areas around the world and may be so in California. Subspecies *altissima* W. Dietrich occurs only in Argentina.

93a. Oenothera stricta Ledebour ex Link subsp. stricta [I]

Oenothera arguta Greene; *O. brachysepala* Spach; *O. bracteata* Philippi; *O. bracteata* var. *glabrescens* (Philippi) Reiche; *O. glabrescens* Philippi; *O. mollissima* Linnaeus subsp. *propinqua* (Spach) Thellung; *O. mollissima* var. *valdiviana* (Philippi) Reiche; *O. propinqua* Spach; *O. propinqua* var. *sparsiflora* Philippi; *O. stricta* var. *propinqua* (Spach) Reiche; *O. valdiviana* Philippi; *Onagra arguta* (Greene) Small

Herbs annual or biennial, strigillose, especially proximally, also villous and glandular puberulent. **Stems** erect or ascending, 25–100 cm. **Leaves** in a basal rosette and cauline, basal 10–15 × 0.8–1.3 cm, cauline 6–10 × 0.6–1 cm; blade very narrowly elliptic to oblanceolate or lanceolate, margins slightly wavy, serrate. **Flowers:** buds with free tips 1–3 mm; floral tube 20–45 mm; sepals 14–20 mm; petals yellow, fading reddish orange, 15–25(–35) mm; filaments 10–20 mm, anthers 7–11 mm; style 30–60 mm, stigma surrounded by anthers at anthesis. **Capsules** 30–40 mm, subtending bract not adnate to capsule base. **Seeds** 1.4–1.8 × 0.5–0.7 mm. $2n = 14$.

Flowering Apr–May. Open, disturbed sites; 10–500 m; introduced; Calif.; South America (Chile); introduced also widely in temperate, semiarid regions.

Subspecies *stricta* has been collected in several counties near the coast in California (Monterey, Santa Barbara, and Santa Cruz counties) a few times, but the most recent collections are from 1953 and 1995, and it may no longer be naturalized in California. It is naturalized widely in temperate, semiarid areas.

MYRTACEAE Jussieu

• Myrtle Family

Leslie R. Landrum

Trees, shrubs, or subshrubs, usually synoecious, terrestrial, unarmed, occasionally clonal by root sprouts; young growth usually glandular, usually aromatic when crushed; trunk often with smooth or scaly bark; hairs unicellular, simple or dibrachiate. **Leaves** usually persistent, opposite, alternate, or whorled, sometimes decussate, simple; without true stipules; usually petiolate; blade leathery, papery, or submembranous, margins entire, sometimes somewhat sinuate. **Inflorescences** usually axillary, sometimes terminal or pseudoterminal, solitary flowers, dichasia, racemes, panicles, or spikes; bracts often present; bracteoles usually present. **Flowers** usually bisexual, rarely unisexual, (0–)4 or 5(–7)-merous, actinomorphic; usually epigynous, rarely semiepigynous; hypanthium obconic, cylindric, or cup-shaped, sometimes prolonged beyond summit of ovary; calyx lobes distinct and, usually, imbricate, or fused in calyptra that falls as a unit or tears open; petals distinct and imbricate, or fused with calyx, rarely coherent, usually equaling calyx lobes (when lobes distinct); nectary glands, when present, produced on disc surrounding style; stamens 10–720; anthers basifixed or dorsifixed, usually dehiscing by slits; pistil 1; ovary inferior (partially so in *Melaleuca*), 1–6[–18]-locular and carpellate; placentation axile, subapical, or basal; style 1; stigma 1; ovules 2–300(–500), usually biseriate or multiseriate, bitegmic. **Fruits** berries, capsules with apical dehiscence, or nutlike. **Seeds** 1–100+; seed coat membranous, ± leathery, or hard and bony; embryo starchy or oily; endosperm scant or absent.

Genera ca. 130, species ca. 6000 (13 genera, 38 species in the flora): sw, s, se United States, Mexico, West Indies, Central America, South America, s, se Asia, Africa, Pacific Islands (Hawaii, New Guinea, Philippines), Australia; nearly worldwide in tropical, subtropical, and Mediterranean regions.

Myrtaceae are apparently of Gondwanan origin with centers of diversity in tropical America and Australasia and with fewer species in Africa and southern Asia. *Syzygium aromaticum* (Linnaeus) Merrill & L. M. Perry (clove) and *Pimenta dioica* (Linnaeus) Merrill (allspice) are economically important spices; *Psidium guajava* (guava) is a common tropical fruit; species of *Eucalyptus* are widely planted for fast growing timber and as ornamentals. *Melaleuca* (including *Callistemon*) and other genera are planted as ornamentals with *M. quinquenervia* having become an invasive pest in Florida.

Native and introduced genera of Myrtaceae in North America can be divided conveniently into two groups: those with dry fruit (*Chamelaucium*, *Eucalyptus*, *Leptospermum*, and *Melaleuca*) and those with fleshy fruit (*Calyptranthes*, *Eugenia*, *Luma*, *Mosiera*, *Myrcianthes*, *Myrtus*, *Psidium*, *Rhodomyrtus*, and *Syzygium*). This division based on fruit type has historically been the basis for recognizing subfamilies or tribes; molecular work shows that neither group is monophyletic. For the purposes of this treatment, the division based on fruit type will be retained, but without formal taxonomic standing. Among the fleshy-fruited genera, embryo structure has been taxonomically important. In the bony-seeded genera (*Mosiera*, *Myrtus*, *Psidium*, and *Rhodomyrtus*), the embryos are small and difficult to see. The C-shaped embryos in that group have small, leaflike or linear cotyledons equal to or shorter than the cylindrical hypocotyls. In other fleshy-fruited genera, it is usually possible to open the seed coat and see the embryo. In *Eugenia*, the embryo is mainly cotyledon tissue fused into a reniform to globose mass. In *Myrcianthes* and *Syzygium*, the cotyledons are similar to those of a bean and unfused and the hypocotyl is insignificant. In *Calyptranthes*, the cotyledons are leaflike and folded into a bundle and the equally long hypocotyl curls around the bundle. In *Luma*, the cotyledons are lenticular and pressed against each other and the hypocotyl about equals them in length.

Recently, *Pimenta dioica* has been shown to be naturalized near Miami, Florida. It is most similar to *Syzygium cumini* (both have berry fruits and many-flowered panicle inflorescences). The two species are compared under 2. *S. cumini.*

SELECTED REFERENCES Lucas, E. J. et al. 2005. Phylogenetic patterns in the fleshy-fruited Myrtaceae—Preliminary molecular evidence. Pl. Syst. Evol. 251: 35–51. Lucas, E. J. et al. 2007. Suprageneric phylogenetics of Myrteae, the generically richest tribe in Myrtaceae. Taxon 56: 1105–1128. McVaugh, R. 1956. Tropical American Myrtaceae. Fieldiana, Bot. 29: 145–228. McVaugh, R. 1963. Tropical American Myrtaceae, II. Fieldiana, Bot. 29: 395–532. McVaugh, R. 1968. The genera of American Myrtaceae—An interim report. Taxon 17: 354–418. McVaugh, R. 1969. The botany of the Guayana Highland—Part VIII. Myrtaceae. Mem. New York Bot. Gard. 18: 55–286. Wilson, P. G. et al. 2005. Relationships within Myrtaceae sensu lato based on a *mat*K phylogeny. Pl. Syst. Evol. 251: 3–19.

1. Fruits capsules or nutlike; leaves on mature stems mostly alternate (opposite in *Chamelaucium* and *Melaleuca linariifolia*); seeds usually 1–3 mm.
    2. Trees (sometimes large shrubs in *E. conferruminata*); perianth parts fused in calyptrae .... 2. *Eucalyptus*, p. 340
    2. Shrubs or trees; perianth parts distinct.
        3. Inflorescences dense cylindrical clusters of flowers surrounding stems apically or subapically, as a bottlebrush; stamens several times longer than perianth .... 4. *Melaleuca*, p. 347
        3. Inflorescences clustered or flowers solitary, not forming bottlebrush; stamens about as long as perianth.
            4. Leaves opposite (decussate); style with a ring of hairs just below stigma; ovaries 1-locular; fruits nutlike........ 1. *Chamelaucium*, p. 339
            4. Leaves alternate; style without a ring of hairs just below stigma; ovaries 6–12-locular; fruits capsules........ 3. *Leptospermum*, p. 346
1. Fruits berries; leaves on mature stems mostly opposite, rarely subopposite or whorled; seeds often 3+ mm.
    5. Calyces closed in bud or with porelike opening at apex (appearing closed in *Syzygium cumini*), persisting after anthesis as irregular parts or falling completely.
        6. Inflorescences 1-flowered, or 3-flowered dichasia; seed coats bony..... 11. *Psidium* (in part), p. 358
        6. Inflorescences 3–100-flowered, mostly panicles; seed coats membranous or papery.
            7. Petioles 2–8 mm; fruits spheroid to oblate (shorter than to as long as wide); calyptrae formed by connate calyx lobes; cotyledons thin, foliaceous (folded) .... 5. *Calyptranthes*, p. 349
            7. Petioles 10–20 mm; fruits ellipsoid or subglobose (longer than wide); calyptrae formed by coherent petals; cotyledons thick, plano-convex....... .... 13. *Syzygium* (in part), p. 362

[5. Shifted to left margin.—Ed.]

5. Calyces open in bud, lobes clearly distinguishable, persisting after anthesis as 4 or 5 distinct lobes.
    8. Flowers 5-merous (sometimes 7-merous in *Rhodomyrtus*).
        9. Petals pink or red; leaves with brochidodromous to acrodromous venation . . . . 12. *Rhodomyrtus*, p. 361
        9. Petals whitish; leaves usually with brochidodromous venation, less often partially eucamptodromous.
            10. Flower buds with calyx open before anthesis; calyx scarcely tearing, if at all, at anthesis; seed coats shiny, not notably dense, few cells thick, easily broken . . . . 10. *Myrtus*, p. 358
            10. Flower buds with calyx closed completely or open only by apical pore before anthesis; calyx tearing irregularly at anthesis; seed coats dull, dense, many cells thick, broken only with difficulty . . . . 11. *Psidium* (in part), p. 358
    8. Flowers [3 or]4-merous (except sometimes 5-merous in *Myrcianthes* and *Syzygium*).
        11. Inflorescences panicles . . . . 13. *Syzygium* (in part), p. 362
        11. Inflorescences solitary flowers, racemes, dichasia, fascicles, or umbel-like clusters.
            12. Seeds 1–27, 3–6 mm; embryos C-shaped, with reduced linear cotyledons, or lenticular with suborbicular, somewhat fleshy, cotyledons.
                13. Seed coats membranous; embryos lenticular; cotyledons about as long as hypocotyl; leaf blades apiculate to abruptly acuminate; California . . . . 7. *Luma*, p. 355
                13. Seed coats bony; embryos C-shaped; cotyledons much shorter than hypocotyl; leaf blades acute to rounded, mucronate, or emarginate; Florida. . . . . 8. *Mosiera*, p. 356
            12. Seeds usually 1(–4), usually 10–15 mm; embryos each a solid, globose or reniform mass, or with 2 distinct, plano-convex cotyledons (beanlike).
                14. Cotyledons connate, fused in a mass . . . . 6. *Eugenia*, p. 351
                14. Cotyledons distinct, each thick, plano-convex, beanlike.
                    15. Inflorescences solitary flowers or dichasia; berries 6–15 mm; hypanthia not prolonged beyond summit of ovary, base not attenuate . . . . 9. *Myrcianthes*, p. 357
                    15. Inflorescences racemes; berries 14–40 mm; hypanthia prolonged beyond summit of ovary, base often attenuate. . . . . 13. *Syzygium* (in part), p. 362

1. CHAMELAUCIUM Desfontaines, Mém. Mus. Hist. Nat. 5: 39, plate 3, fig. B, plate 4. 1819 • Waxflower [Derivation uncertain; possibly Greek *chamai*, dwarf, and *lauchis*, poplar, alluding to flower; or *kamelaukion*, headdress of medieval Popes, alluding to form of calyptra] [I]

Leslie R. Landrum

Shrubs, glabrous. **Leaves** decussate; blade venation inconspicuous. **Inflorescences** 1-flowered, axillary, flowers solitary in corymbose clusters. **Flowers** 5-merous, sessile or pedicellate; hypanthium obconic to cylindric-obconic, prolonged and bowl-like beyond ovary summit; perianth parts distinct; petals white, pink, or purple, [yellow]; stamens ca. 10, about as long as perianth, equal number of staminodes sometimes present; ovary 1-locular; style equal to or longer than stamens, with a ring of hairs just below stigma; ovules 4–8. **Fruits** nutlike. **Seed** 1, winged or not. $x = 11$.

Species ca. 13 (1 in the flora): introduced, California; Australia.

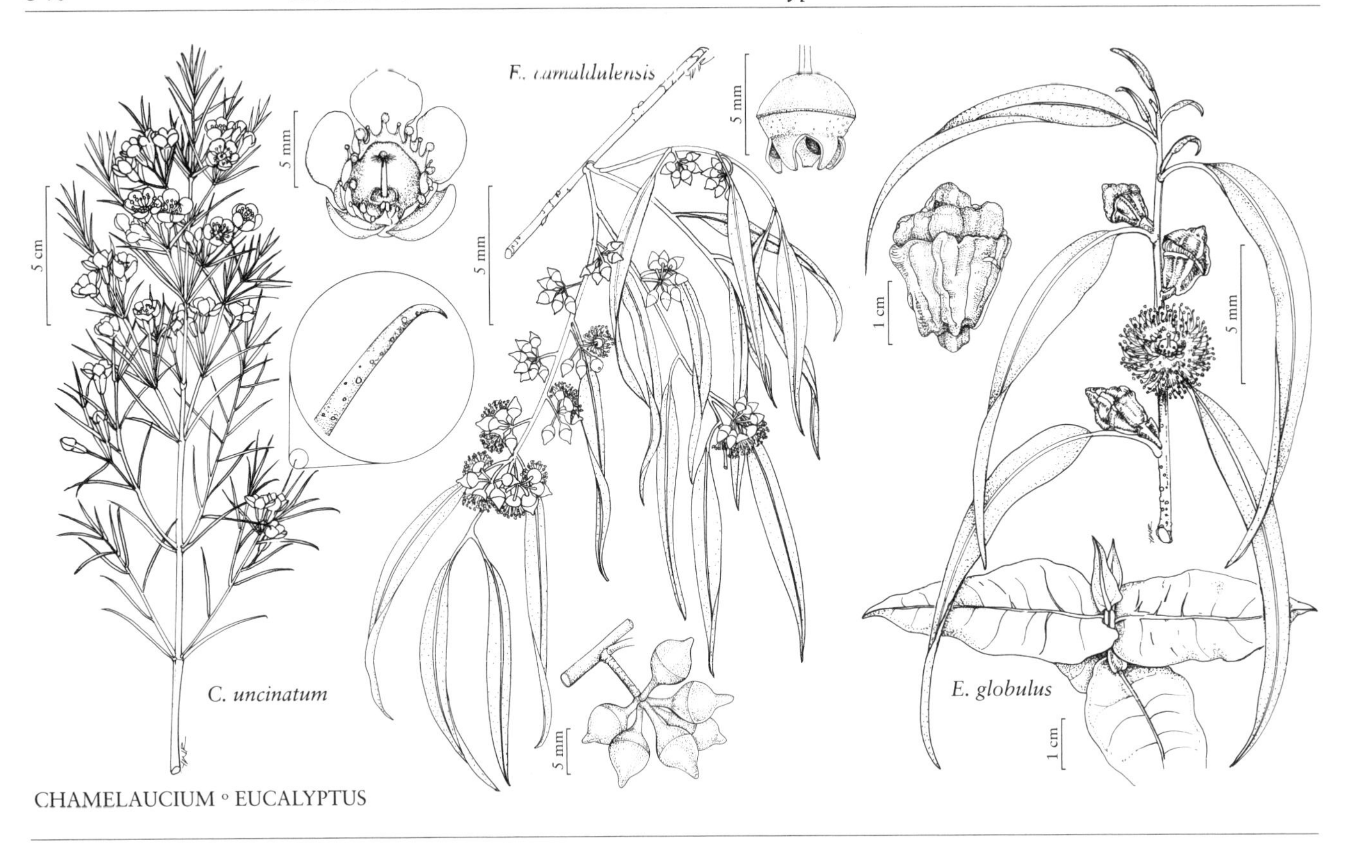

CHAMELAUCIUM ◦ EUCALYPTUS

1. **Chamelaucium uncinatum** Schauer in J. G. C. Lehmann, Pl. Preiss. 1: 97. 1844 (as Chamaelaucium) • Geraldton wax 

Shrubs to 3 m. **Leaves:** blade linear, 1.6–4 cm × 0.5–1 mm, about as thick as wide, midvein impressed adaxially, apex sharply pointed, often curved downward, surfaces conspicuously glandular. **Inflorescences** in axils of leaves or bracts, these often deciduous. **Peduncles** 4–12 mm. **Flowers** to 2 cm wide; calyx lobes broadly rounded, 0.5–1 mm; petals soon falling, spreading, suborbiculate, 4–5 mm; stamens 1–2 mm, alternate staminodes; filaments of stamens and staminodes fused in a short, tubular ring; style 4 mm.

Flowering spring. Disturbed areas; ca. 200 m; introduced; Calif.; w Australia.

In the flora area, *Chamelaucium uncinatum* is known from San Diego County.

2. EUCALYPTUS L'Heritier, Sert. Angl., 18. 1789; plate 20, 1792 • Eucalypt, gum tree [Greek *eu*, well, and *kalyptos*, covered, alluding to deciduous calyptra covering stamens in flower bud] [I]

Matt Ritter

Leslie R. Landrum

Trees or shrubs, usually erect, glabrous or pubescent, hairs simple; bark shedding, smooth, or occasionally persistent near base of trunk, or rough throughout. **Leaves** heterophyllous, juvenile usually opposite, horizontal, sessile, blade base ± cordate, surfaces glaucous, adult usually alternate, vertical, petiolate, blade surfaces often same color, glandular; blade venation usually pinnate, faint, lateral veins ascending, nearly straight, several. **Inflorescences** 1–19-flowered, flowers solitary in leaf axils, or in umbels or panicles of umbels and axillary or terminal. **Flowers** 4- or 5-merous, sessile or pedicellate; hypanthium hemispheric, cylindrical, urn-shaped, pyriform, ovoid, obconic, or campanulate; perianth parts fused in a calyptra (called an operculum or bud cap) that is shed at anthesis; stamens often 100+, often showy, usually fertile; ovary 3–6-locular; ovules 10–100+, sterile ones often present. **Fruits** capsules, brown to gray, hemispheric, obconic, ovoid, subpyriform, globose, cylindric, or urn-shaped, glaucous or not, thick-walled, woody, usually smooth, opening apically; valves exserted beyond apex or included (enclosed) below fruit apex. **Seeds** several–100, cuboid, usually 1–3 mm, wind dispersed. $x = 11$.

Species 700+ (15 in the flora): introduced; Australia; introduced also nearly worldwide.

Species of *Eucalyptus* are some of the world's largest flowering plants, some over 100 m; various species are most commonly planted as forestry and plantation trees for oil, timber, fuel, tannins, and paper pulp. Many are cultivated ornamentally in warm climates, with several naturalized in Arizona, California, and Florida, and, perhaps, other southern states.

Species of *Eucalyptus* in North America, where correlations with natural habitat do not exist, are often difficult to identify. Many species (over 200 in California alone) have been introduced into cultivation and more will surely be added.

Eucalyptus pulverulenta Sims has been reported to be naturalized in California but no supporting evidence has been found. Some putative hybrids have also been reported as naturalized.

In his treatment of eucalypts, M. I. H. Brooker (2000) included *Angophora* Cavanilles and *Corymbia* K. D. Hill & L. A. S. Johnson as subgenera of *Eucalyptus*, whereas other concurrent and more recent work has supported the status of *Angophora* and *Corymbia* as separate genera (P. Y. Ladiges et al. 1995; F. Udovicic and Ladiges 2000; D. A. Steane et al. 2002). Although the validity of the latter work is recognized herein, for the purpose of simplicity in treating a small group of naturalized species, *Eucalyptus* in the broad sense of Brooker is here adopted.

SELECTED REFERENCES Brooker, M. I. H. 2000. A new classification of the genus *Eucalyptus* L'Her. (Myrtaceae). Austral. Syst. Bot. 13: 79–148. Chippendale, G. M. 1988. Myrtaceae—*Eucalyptus*, *Angophora*. In: R. Robertson et al., eds. 1981+. Flora of Australia. 32+ vols. Canberra. Vol. 19.

1. Inflorescences umbels in panicles.
   2. Leaves lemon-scented; trunk smooth; bark shedding in irregular pieces . . . . 2. *Eucalyptus citriodora*
   2. Leaves medicinal or spicy scented (not lemon-scented); trunk smooth or rough; bark fissured, fibrous, or shedding in flakes or irregular strips.
      3. Capsules urn-shaped or truncate-globose, 8+ mm wide; juvenile leaves (common in mature crown) peltate, often reddish hirsute; Florida . . . . 14. *Eucalyptus torelliana*
      3. Capsules hemispheric, ovoid, or subpyriform, to 6 mm wide; leaves not peltate, not reddish hirsute; Arizona, California.
         4. Leaf blades narrowly lanceolate, 8–17 × 0.8–2.5 cm . . . . 5. *Eucalyptus coolabah*
         4. Leaf blades round, elliptic, or ovate, 5–10 × 1.5–5 cm . . . . .9. *Eucalyptus polyanthemos*
1. Inflorescences umbels not in panicles or flowers solitary.
   5. Flowers solitary in leaf axils, sessile or subsessile; capsules glaucous . . . . 6. *Eucalyptus globulus*
   5. Flowers 3–19 in umbels, pedicellate; capsules glaucous or not.
      6. Trunk rough, brown, reddish brown, dark brown, or black; bark deeply furrowed, persisting on trunks and limbs.
         7. Bark soft, spongy, fibrous; trunk reddish brown; leaves light green abaxially; anthers present on all stamens; peduncles flattened; fruit valves fused after dehiscence . . . . .11. *Eucalyptus robusta*
         7. Bark hard; trunk dark brown or ± black; leaves same color on both surfaces; anthers absent on outer stamens; peduncles subterete; fruit valves free. . . . . 12. *Eucalyptus sideroxylon*

[6. Shifted to left margin.—Ed]

6. Trunk smooth, gray, bluish gray, white, whitish gray, orange, tan, or mottled; bark shedding, occasionally rough near base (to ca. 1.5 m).
   8. Capsules connate, forming globose clusters. . . . . .4. *Eucalyptus conferruminata*
   8. Capsules not connate.
      9. Leaves lighter in color abaxially.
         10. Trunk mottled gray, orange, or tan; capsules ribbed; valves included; hypanthium length 3–4 times calyptra . . . . 3. *Eucalyptus cladocalyx*
         10. Trunk white; capsules smooth (not ribbed); valves exserted, incurved; hypanthium only slightly longer than calyptra . . . . .7. *Eucalyptus grandis*
      9. Leaves same color on both surfaces.
         11. Inflorescences usually 3-flowered; juvenile leaves opposite, sessile . . . . . 15. *Eucalyptus viminalis*
         11. Inflorescences 7- or 9+-flowered; juvenile leaves alternate, petiolate.
            12. Leaf blades linear, 0.1–0.5 cm wide; fruit valves ± level with apex or included. . . . . 10. *Eucalyptus pulchella*
            12. Leaf blades lanceolate, 1–3 cm wide; fruit valves exserted.
               13. Leaves usually green to bluish green, sometimes grayish green; trunk powdery to touch . . . . .8. *Eucalyptus mannifera*
               13. Leaves green; trunk not powdery to touch.
                  14. Calyptrae of flowers mostly hemispheric, often rostrate, rarely bluntly conic; hypanthium length ± equaling calyptra. . . . . .1. *Eucalyptus camaldulensis*
                  14. Calyptrae of flowers conic-acuminate or horn-shaped; hypanthium length 2–3 times shorter than calyptra . . . . . 13. *Eucalyptus tereticornis*

1. Eucalyptus camaldulensis Dehnhardt, Cat. Horti Camald. ed. 2, 20. 1832, name conserved • Red or river red gum [F] [I] [W]

Trees, to 25 m; trunk gray or tan, nearly straight, ± smooth; bark usually shed in irregular strips distally, sometimes persistent toward trunk base; branches often hanging in clusters. **Leaves** (juvenile alternate, petiolate); petiole 0.3–2 cm; blade green, lanceolate, 6–20 × 1.5–2.5 cm. **Peduncles** 0.8–1.2 cm. **Inflorescences** 7-flowered, umbels. **Flowers:** hypanthium hemispheric, 2–3 mm, length ± equaling calyptra; calyptra mostly hemispheric, often rostrate, rarely bluntly conic; stamens white. **Capsules** hemispheric, 5–9 mm, not glaucous; valves 3–5, exserted. $2n = 22$.

Flowering spring–early summer. Disturbed habitats, river bottoms; 0–1200 m; introduced; Ariz., Calif.; Australia; introduced also in South America.

Eucalyptus camaldulensis is the most widely cultivated eucalypt, with *E. globulus* and *E. grandis* being close second and third. Several infraspecific taxa have been recognized for this variable species and more than one of these may be naturalized in North America.

2. Eucalyptus citriodora Hooker in T. L. Mitchell, J. Exped. Trop. Australia, 235. 1848 • Lemon-scented gum [I]

Corymbia citriodora (Hooker) K. D. Hill & L. A. S. Johnson; *Eucalyptus maculata* Hooker var. *citriodora* (Hooker) L. H. Bailey

Trees, to 35 m; trunk golden becoming tan, straight, slender, graceful, smooth; bark shed in irregular pieces. **Leaves** lemon-scented; petiole 1–2 cm; blade green, lanceolate, often falcate, 10–20 × 1–2 cm. **Peduncles** terete, 1–1.5 cm. **Inflorescences** 3–5-flowered, umbels in panicles. **Flowers:** hypanthium hemispheric, 5–6 mm, length more than calyptra; calyptra mostly rostrate; stamens white. **Capsules** urn-shaped, to 15 mm, not glaucous; valves 3 or 4, included. $2n = 22, 44$.

Flowering winter–spring. Disturbed areas; 0–300 m; introduced; Calif.; e Australia.

Eucalyptus citriodora is often treated as *Corymbia citriodora*; it is found only in southern coastal and urban areas and is commonly cultivated.

3. Eucalyptus cladocalyx F. Mueller, Linnaea 25: 388. 1853 • Sugar gum [I]

Trees, 20 m; trunk white, often mottled gray, orange, or tan, mostly straight, graceful, ± smooth; bark shed in large, irregular patches. **Leaves:** petiole 0.1–0.2 cm; blade light green abaxially, ± widely lanceolate, 8–15 × 2–3 cm. **Peduncles** 1–3 cm. **Inflorescences** mostly 7–11-flowered, umbels, usually on leafless branches. **Flowers:** hypanthium cylindrical or urn-shaped, ± ribbed, less than 10 mm, length 3–4 times calyptra; calyptra cylindric to urn-shaped, abruptly pointed; stamens white. **Capsules** ± urn-shaped, ribbed, 10–15 mm, not glaucous; valves 3 or 4, included.

Flowering spring–summer. Disturbed areas; 0–200 m; introduced; Calif.; s Australia.

Eucalyptus cladocalyx is commonly cultivated in southern California.

Eucalyptus corynocalyx F. Mueller is an illegitimate name based on the same type as *E. cladocalyx*. Mueller may have thought *corynocalyx* (club-calyx) was a more appropriate name than *cladocalyx* (branch-calyx) and intended to change the name. The closed bud of this species is clublike.

4. Eucalyptus conferruminata D. J. Carr & S. G. M. Carr, Austral. J. Bot. 28: 535, figs. 2, 11, 17, 20, 27B, 28A, 30. 1980 • Spider gum, bushy yate [I]

Trees or shrubs, to 5 m; trunk light gray or tan, smooth; bark shed in strips and short ribbons. **Leaves:** petiole 0–1.2 cm; blade light green, elliptic to elongate-elliptic, 5–9 × 1–4 cm, surfaces glossy. **Peduncles** distinctly flattened, 3–7 × 1–3 cm. **Inflorescences** 7–19-flowered, umbels compact, globose. **Flowers:** hypanthium sessile, fused to adjacent hypanthia; calyptra horn- or finger-shaped; stamens yellowish green. **Capsules** connate; forming compact, globose cluster, 30–60 mm diam., not glaucous; valves 3, strongly exserted, with persistent style remnants.

Flowering spring–summer. Disturbed areas; 60–100 m; introduced; Calif.; sw Australia.

Eucalyptus conferruminata, commonly cultivated as a screen plant in southwestern coastal California, is often sold under the name *E. lehmannii* (Schauer) Bentham.

5. Eucalyptus coolabah Blakely & Jacobs in W. F. Blakely, Key Eucalypts, 245. 1934 • Coolibah [I] [W]

Trees, to 10 m; trunk gray to tan, rough; bark fissured. **Leaves:** petiole 1.4–2 cm; blade dull bluish gray-green, narrowly lanceolate, 8–17 × 0.8–2.5 cm, surfaces rarely glaucous. **Peduncles** 0.5–2.8 cm. **Inflorescences** 7-flowered, terminal, umbels in panicles. **Flowers:** hypanthium obconic, ± 2 mm, length ± equaling calyptra; calyptra obconic, apiculate; stamens white. **Capsules** hemispheric, 3–5 mm, to 6 mm wide; valves 3 or 4, exserted.

Flowering spring–summer. Urban ephemeral waterways; 300–400 m; introduced; Ariz.; n Australia.

Eucalyptus coolabah is an escape along the occasionally flooded Salt River and its tributaries near Phoenix. It is similar to *E. microtheca* F. Mueller and commonly sold under that name.

6. Eucalyptus globulus Labillardière, Voy. Rech. Pérouse 1: 153, plate 13. 1800 • Blue gum [F] [I] [W]

Trees, to 60 m; trunk bluish gray, straight, smooth; bark shed in irregular strips distally, sometimes persistent toward trunk base; twigs ± square or winged. **Leaves** mostly strongly aromatic; petiole 1.5–2.5 cm, flattened; blade green, usually narrowly lanceolate, often sickle-shaped, 10–30 × 2.5–4 cm. **Peduncles** 0.1–1 cm. **Inflorescences:** flowers solitary, sessile or subsessile. **Flowers:** hypanthium obconic, ± 4-ribbed, to 20 mm, glaucous; calyptra flattened-hemispheric, with central knob, warty, glaucous; stamens creamy white. **Capsules** hemispheric or obconic, ± 4-ribbed, 5–21 mm, glaucous, thickened, warty, rim wide; valves 3–5, ± level with apex or exserted. $2n = 22$.

Flowering fall–winter. Disturbed areas; 0–300 m; introduced; Calif.; se Australia; introduced also widely.

Eucalyptus globulus is known from the Outer North Coast Ranges, Great Central Valley, and central-western and southwestern California.

Eucalyptus globulus is commonly cultivated in warm regions of the world for its fast-growing timber and for paper pulp. The species is the tallest angiosperm in North America, easily recognized by the large, solitary flowers and fruit.

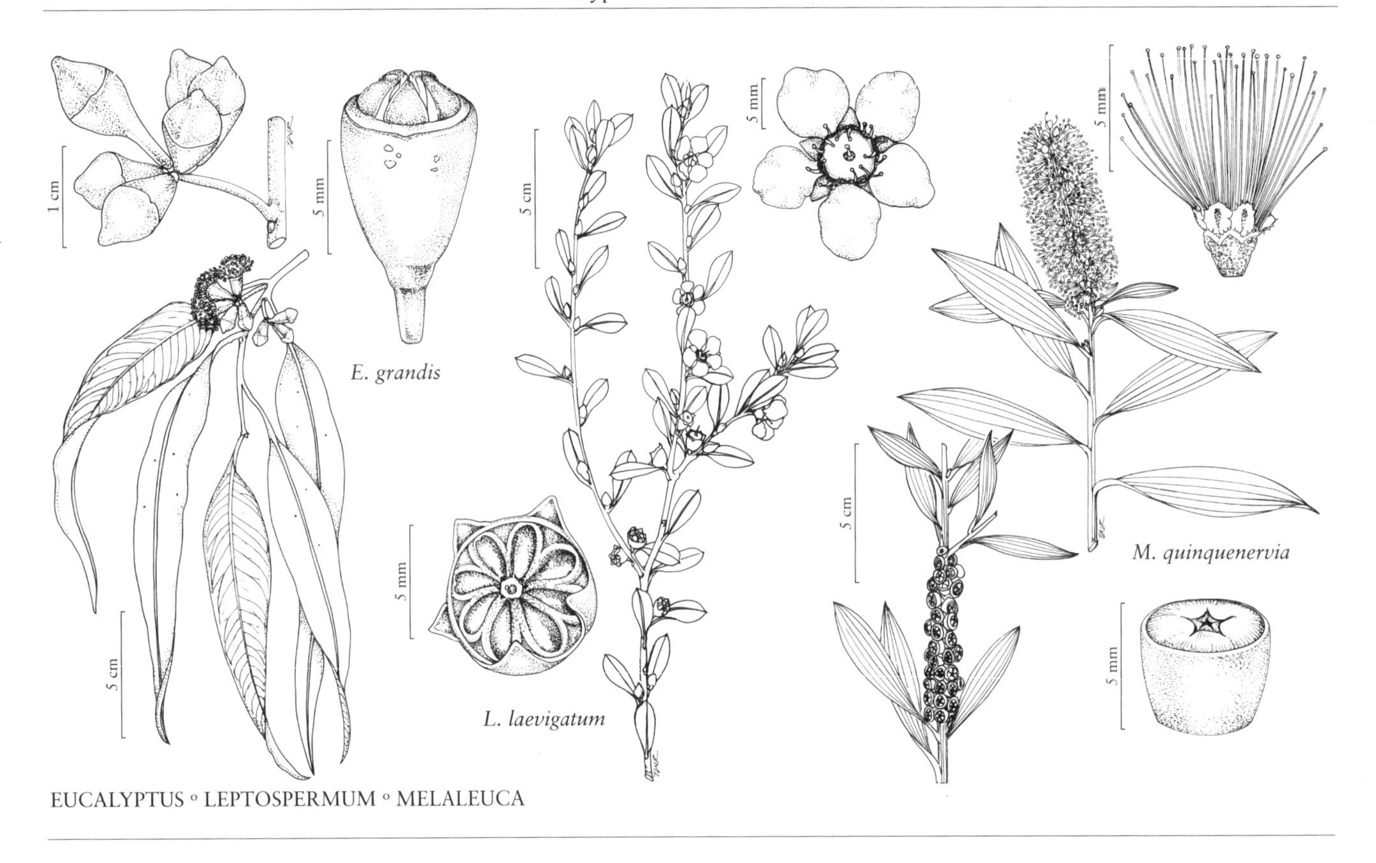

EUCALYPTUS ◦ LEPTOSPERMUM ◦ MELALEUCA

7. **Eucalyptus grandis** W. Mill ex Maiden, J. Proc. Roy. Soc. New S. Wales 52: 501. 1919 • Flooded or rose gum [F] [I]

Trees, to 55 m; trunk white, grayish white, or bluish gray, smooth; bark sparse, rough and flaky at trunk base. **Leaves:** petiole 1–2.2 cm; blade grayish green or yellow-green, lighter abaxially, lanceolate to elliptic, often falcate, 9.5–16 × 2–5 cm. **Peduncles** 1–1.5 cm, 3–4 mm wide apically. **Inflorescences** 7–11-flowered, umbels. **Flowers:** hypanthium obconic or campanulate, 3–4 mm, length only slightly greater than calyptra; calyptra conic or slightly rostrate, 3–4 mm; stamens white. **Capsules** subpyriform, 5–8 mm, not glaucous; valves 4 or 5, exserted, incurved. $2n = 22$.

Flowering fall. Disturbed areas; 0–200 m; introduced; Calif., Fla.; e Australia.

Eucalyptus grandis is known from Los Angeles, Monterey, Orange, San Diego, San Luis Obispo, Santa Barbara, Santa Cruz, and Ventura counties in California, and from Glades, Hendry, Palm Beach, and Pinellas counties in Florida.

8. **Eucalyptus mannifera** Mudie, Trans. Med.-Bot. Soc. London 1(3): 24. 1834 • Mottled or red spotted or brittle gum [I] [W]

Eucalyptus maculosa R. T. Baker

Trees, to 20 m; trunk white and grayish white, mottled, smooth, powdery to touch; bark shedding. **Leaves:** (juvenile alternate, petiolate); petiole 0.5–2.2 cm; blade green to bluish green or, occasionally, grayish green, narrowly lanceolate to lanceolate or falcate, 7–18.5 × 1–3 cm, surfaces dull. **Peduncles** 0.5–-1 cm. **Inflorescences** ca. 7-flowered; umbels. **Flowers:** hypanthium obconic, 2–4 mm, length ± equaling calyptra; calyptra conic to rounded, 2–4 mm; stamens white. **Capsules** hemispheric to obconic, 3–5 mm, not glaucous; valves 3, exserted.

Flowering spring–summer. Disturbed coastal and urban areas; 0–200 m; introduced; Calif.; se Australia.

Eucalyptus mannifera is known from the Central Coast, San Francisco Bay area, and Outer South Coast Ranges.

9. Eucalyptus polyanthemos Schauer in W. G. Walpers, Repert. Bot. Syst. 2: 924. 1843 • Silver dollar gum, red box [I]

Trees, to 25 m; trunk gray or tan, smooth or rough; bark rough, fibrous, and persistent, or smooth and shed in flakes or irregular strips. **Leaves:** petiole 1–2.5 cm; blade grayish green, silver, or bluish green, round, elliptic, or ovate, 5–10 × 1.5–5 cm, surfaces occasionally glaucous. **Peduncles** 1–4 cm. **Inflorescences** 5–7-flowered, terminal or axillary, umbels in panicles. **Flowers:** hypanthium ovoid to obconic, ca. 4 mm, length ca. 2 times calyptra; calyptra conic to hemispheric; stamens white; anthers rigid on filaments, adnate, absent on outer filaments. **Capsules** ovoid or subpyriform, 5–6 mm, to 6 mm wide, glaucous; valves 3 or 4, included.

Flowering winter–spring. Disturbed coastal urban areas; 0–200 m; introduced; Calif.; se Australia.

Eucalyptus polyanthemos is known from the San Joaquin Valley, Central Coast, San Francisco Bay Area, Outer South Coast Ranges, South Coast, Santa Catalina Islands, and Western Transverse Ranges.

Juvenile, adult, and transitional leaves are occasionally found in crowns of mature naturalized trees.

10. Eucalyptus pulchella Desfontaines, Tabl. École Bot. ed. 3, 284, 408. 1829 • White peppermint gum [I]

Trees, to 21 m; trunk white or bluish gray, mostly smooth, occasionally shaggy or rough near base; bark shed in relatively long strips. **Leaves:** (juvenile alternate, petiolate); with slight peppermint odor when crushed; petiole 0.1–0.6 cm; blade dark green, linear, 5–10 × 0.1–0.5 cm. **Peduncles** 0.1–0.3 cm. **Inflorescences** 9+-flowered, umbels. **Flowers:** hypanthium obconic, 2–3 mm, length ± equaling calyptra; calyptra hemispheric, rounded; stamens white. **Capsules** ovoid or subpyriform, 4–6 mm, not glaucous; valves 4, ± level with apex or included.

Flowering winter. Disturbed areas; 0–200 m; introduced; Calif.; Australia (Tasmania).

In the flora area, *Eucalyptus pulchella* is known only from Alameda County.

11. Eucalyptus robusta Smith, Spec. Bot. New Holland, 39, plate 13. 1795 • Swamp mahogany [I]

Trees, to 30 m; trunk reddish brown, rough; bark persistent, deeply furrowed, soft, spongy, fibrous, rough on small branches. **Leaves:** petiole 1.5–3 cm, slightly flattened; blade lighter green abaxially, broadly lanceolate, 8.5–17 × 2.5–7 cm, surfaces glossy. **Peduncles** broadly flattened, 1.5–3 cm. **Inflorescences** 9–15-flowered, umbels. **Flowers:** hypanthium obconic to pyriform, 6–7 mm, length ± equaling calyptra; calyptra conic to rostrate, 10–12 mm; stamens white. **Capsules** cylindric, 10–18 mm, not glaucous; valves 3 or 4, fused after dehiscence, ± level with apex or included and joined across orifice.

Flowering summer–fall. Disturbed areas; 0–200 m; introduced; Calif., Fla.; e Australia.

Eucalyptus robusta is known from Los Angeles, Monterey, Orange, San Diego, San Luis Obispo, Santa Barbara, Santa Cruz, and Ventura counties in California, and from Charlotte, Lee, Martin, and St. Lucie counties in Florida.

12. Eucalyptus sideroxylon A. Cunningham ex Woolls, Proc. Linn. Soc. New South Wales, ser. 2, 1: 859. 1887 • Red iron bark [I]

Trees, to 25 m; trunk dark brown or ± black, rough; bark persistent, deeply furrowed, hard. **Leaves:** petiole 1–2 cm; blade dull grayish green or blue-green, lanceolate, 6–14 × 1–2 cm. **Peduncles** subterete, 1–1.5 cm. **Inflorescences** 3–7-flowered, umbels, often pendent. **Flowers:** hypanthium ovoid to hemispheric, 4–6 mm, length greater than calyptra, occasionally glaucous; calyptra conic; stamens white, pink, or red; anthers absent on outer filaments. **Capsules** ovoid, 10 mm, glaucous or not; valves 4 or 5, included.

Flowering spring. Disturbed coastal, urban areas; 0–200 m; introduced; Calif.; se Australia.

In the flora area, *Eucalyptus sideroxylon* is known from southern California.

13. Eucalyptus tereticornis Smith, Spec. Bot. New Holland, 41. 1795 • Forest red gum [I]

Trees, to 40 m; trunk gray or tan, nearly straight, ± smooth; bark usually shed in irregular strips distally, sometimes persistent toward trunk base. **Leaves:** (juvenile alternate, petiolate); petiole 0.3–2 cm; blade green, lanceolate, 6–20 × 1.5–2.5 cm. **Peduncles** 0.8–1.2 cm. **Inflorescences** ca. 7-flowered, umbels. **Flowers:** hypanthium hemispheric, 2–3 mm, calyptra 2–3 times as long as hypanthium; calyptra mostly conic-acuminate or horn-shaped; stamens white. **Capsules** hemispheric, 5–9 mm, not glaucous; valves 3–5, exserted.

Flowering spring–summer. Disturbed coastal, urban areas; 0–200 m; introduced; Calif.; e Australia.

Eucalyptus tereticornis is known from the San Joaquin Valley, Outer South Coast Ranges, South Coast, northern Channel Islands, and Transverse and Peninsular ranges.

14. Eucalyptus torelliana F. Mueller, Fragm. 10: 106. 1877 • Cadaghi [I]

Corymbia torelliana (F. Mueller) K. D. Hill & L. A. S. Johnson

Trees, to 30 m; trunk slate green (young but reproductive trees) or gray or black (older trees), smooth (except on trunks of older trees); bark not shedding, new growth often reddish, hirsute or setose. **Leaves:** (juvenile alternate); petiole 2–2.5 cm; blade green, lighter abaxially, cordate to ovate or elliptic, peltate, or (adult) lanceolate to narrowly lanceolate, (juvenile) 8–15 × 5–11 cm or (adult) 10–14 × 1–4 cm, surfaces (juvenile) often reddish hirsute, (adult) glabrous. **Peduncles** ca. 4 cm. **Inflorescences** 3–7-flowered, umbels in panicles. **Flowers:** hypanthium ovoid, 6–8 mm, length greater than calyptra; calyptra rounded to conic to slightly rostrate; stamens white; anthers pivoting on filaments, versatile. **Capsules** urn-shaped or truncate-globose, 9–15 mm, 8+ mm wide, not glaucous; valves 3, deeply included.

Flowering spring. Disturbed areas; 0–50 m; introduced; Fla.; ne Australia.

Eucalyptus torelliana is known from Lee and Palm Beach counties. Juvenile leaves are often more common in the crown of reproducing trees.

Eucalyptus torelliana is often treated as *Corymbia torelliana* (K. D. Hill and L. A. S. Johnson 1995).

15. Eucalyptus viminalis Labillardière, Nov. Holl. Pl. 2: 12, plate 151. 1806 • Manna or ribbon gum [I]

Trees, to 50 m; trunk whitish, gray, or tan, straight, ± smooth; bark usually shed in relatively long, narrow, irregular strips distally, sometimes persistent toward trunk base. **Leaves:** petiole 1–1.5 cm; blade green, lanceolate to narrowly lanceolate, 10–15 × 1–2.5 cm. **Peduncles** to ca. 1 cm. **Inflorescences** usually 3-flowered, umbels. **Flowers:** hypanthium ovoid, 2–3 mm, length ± equaling calyptra; calyptra conic to rounded or, rarely, rostrate; stamens white. **Capsules** ± hemispheric, 5–10 mm, not glaucous; valves 3 or 4, exserted. $2n$ = 22, 90.

Flowering summer, fall. Disturbed urban areas; 0–100 m; introduced; Calif.; se Australia.

Eucalyptus viminalis is known from the Outer North Coast Ranges, Central Coast, Outer South Coast Ranges, South Coast, and Peninsular Ranges.

3. LEPTOSPERMUM J. R. Forster & G. Forster, Char. Gen. Pl. ed. 2, 71, plate 36. 1776, name conserved • Tea tree [Greek *leptos*, slender or small, and *sperma*, seed, alluding to form and size] [I]

Elizabeth McClintock†

Shrubs or small trees, glabrous or pubescent, hairs simple. **Leaves** alternate; blade venation parallelodromous, faint, often with visible midvein, sometimes also 2–4 veins arising from base, arching to apex. **Inflorescences** 1–3-flowered, axillary, flowers solitary or clustered. **Flowers** 5-merous, subsessile; hypanthium mostly widely cup-shaped; calyx lobes distinct; petals white [pink or red]; stamens 15–35, about as long as perianth; ovary [2–]6–12-locular; style equaling or shorter than stamens; ovules [6–]15[–28] per locule. **Fruits** capsules, brown or gray, obconic to broadly bowl-shaped, woody, opening apically. **Seeds** numerous, obovoid to irregular-linear, flattened, 1–2.5 mm; seed coat reticulate or striate.

Species ca. 70 (1 in the flora): introduced, California; Australasia.

1. **Leptospermum laevigatum** (Gaertner) F. Mueller, Rep. (Annual) Gov. Bot. Director Bot. Zool. Gard. 17: 22. 1858 [F] [I]

Fabricia laevigata Gaertner, Fruct. Sem. Pl. 1: 175. 1788

Shrubs or trees with spreading, twisted trunks; bark fibrous, shed in long strips. **Leaves:** blade obovate-oblong, 10–28 × 5–10 mm, principal veins inconspicuous, apex blunt or sharply apiculate. **Inflorescences** usually in pairs, of unequal age. **Flowers** 1.5–2 cm wide; calyx persisting until fruit maturity, lobes subtriangular 1–1.5 mm, internally sericeous; petals caducous, spreading; stamens 2 mm; style 1 mm, glabrous. **Capsules** 7–8 mm wide; valves 7–12.

Flowering spring. Disturbed coastal areas; 0–50 m; introduced; Calif.; se Australia.

Leptospermum laevigatum is known from the Central Coast. It is commonly cultivated along the coast, sometimes to stabilize moving sand.

4. MELALEUCA Linnaeus, Syst. Nat. ed. 12, 2: 507, 509. 1767; Mant. Pl. 1: 14, 105. 1767, name conserved • Paperbark, bottlebrush, tea tree [Greek *melas*, black, and *leukos*, white, alluding to colors of tree trunk and branches, respectively, in *M. leucadendron*, the type species] [I]

Lyn A. Craven†

Callistemon R. Brown

Shrubs or trees, glabrescent, hairs simple. **Leaves** usually alternate or opposite [ternate], sometimes decussate; blade venation pinnate to longitudinal. **Inflorescences** 4–80-flowered, pseudoterminal or axillary, usually spikes or clusters, sometimes flowers solitary (monad). **Flowers** usually 5-merous, sessile, in triplets (triads) or solitary (monads); hypanthium subglobose to subcylindrical, adnate to ovary proximally or to ¾ length of ovary; calyx lobes distinct, (0 or) 5; petals white; stamens 25–365, longer than perianth; filaments connate proximally into 5 bundles or, sometimes, distinct and not in bundles; ovary 3-locular; placenta peltate, axile-median to axile-basal; ovules 50–300 per locule. **Fruits** capsules, green, brown, or gray, subglobose to short-cylindrical, in spikes or clusters, a woody or subwoody hypanthium enclosing a capsule. **Seeds** 20+, usually obovoid-oblong to obovoid, not winged, with thin testa; cotyledons not or scarcely foliaceous, face-to-face, plano-convex or obvolute, wrapped around each other. $x = 11$.

Species ca. 300 (4 in the flora): introduced; Asia (Malesia), Pacific Islands (Lord Howe Island, New Caledonia), Australia; introduced also widely.

Melaleuca and *Callistemon* traditionally have been regarded as separate genera, distinguished by features of the androecium. The staminal filaments are usually distinct in *Callistemon* with the stamens arranged uniformly around the hypanthium rim. In some species of *Callistemon*, the stamens are connate towards the base and aggregated in five more or less distinct bundles. This condition is nearly universal in *Melaleuca*. In view of the similarities between the two genera in many features of the leaves, inflorescences, and fruits, they have been combined under *Melaleuca* (L. A. Craven 2006). Gill. K. Brown et al. (2001), in an analysis of 5S and ITS-1 rDNA data, found that *Callistemon* nested within *Melaleuca*, and a similar result was obtained by R. D. Edwards et al. (2010) on the basis of *ndh*F data.

Some species of *Melaleuca* are ornamental (notably *M. citrina* and *M. viminalis*) and are widely cultivated. Some of the taller paperbark species have potential for forestry use. Essential oils are produced commercially in Australia, Indonesia, Madagascar, and New Caledonia from some species, mainly the tea tree, *M. alternifolia* (Maiden & Betche) Cheel, cajuput, *M. cajuputi* Powell, tea tree or snow-in-summer, *M. linariifolia*, and punk tree, broad-leaved paperbark, or niauoli, *M. quinquenervia* (I. Southwell and R. Lowe 1999).

Melaleuca occurs in North America mostly as cultigens but one species, *Melaleuca quinquenervia*, has become a significant woody weed in Florida.

SELECTED REFERENCE Wrigley, J. W. and M. Fagg. 1993. Bottlebrushes, Paperbarks and Tea Trees. Sydney and Auckland.

1. Leaves opposite (decussate); flowers in monads . . . . 2. *Melaleuca linariifolia*
1. Leaves alternate; flowers in monads or triads.
    2. Flowers joined to inflorescence axes in clusters of 3 (triads), filaments white, cream, greenish white, green, creamy white, or creamy yellow . . . . 1. *Melaleuca quinquenervia*
    2. Flowers joined to inflorescence axes one-by-one (monads), filaments red, crimson, or mauve.
        3. Filaments distinct . . . . 3. *Melaleuca citrina*
        3. Filaments connate proximally, usually in 5 bundles (these obscure when bundle claws are short) . . . . 4. *Melaleuca viminalis*

1. Melaleuca quinquenervia (Cavanilles) S. T. Blake, Proc. Roy. Soc. Queensland 69: 76. 1958 • Punk tree, broad-leaved paperbark, niauoli [F] [I] [W]

Metrosideros quinquenervia Cavanilles, Icon. 4: 19, plate 333. 1797

Trees, 1–18 m; bark papery. **Leaves** alternate; blade usually narrowly elliptic to elliptic, rarely somewhat falcate, 5.5–12 × 1–3.1 cm, veins 5–7, longitudinal, surfaces glabrescent. **Inflorescences** 15–54-flowered, flowers in triads, pseudoterminal, sometimes also axillary distally, to 40 mm wide. **Flowers:** calyx lobes glabrous abaxially, margins scarious, 0.3–0.4 mm wide; petals deciduous, 2.5–3.5 mm; filaments connate in bundles of 5–10, white, cream, greenish white, green, creamy white, or creamy yellow, 10.5–20 mm, bundle claw 0.9–2.5 mm; style 11–18 mm; ovules ca. 50–65 per locule. **Capsules** 2.7–4 mm. **Cotyledons** obvolute. $2n = 22$.

Flowering year-round (commonly in fall); 0–30 m; introduced; Fla., La.; Asia (Malesia); Pacific Islands (New Caledonia); Australia; introduced also elsewhere in Pacific Islands (Hawaii), widely elsewhere.

Melaleuca quinquenervia is a serious woody weed of wetland habitats in Florida and Louisiana. Mechanical control has not been successful and research in recent years has been focused upon biological control.

2. Melaleuca linariifolia Smith, Trans. Linn. Soc. London 3: 278. 1797 • Tea tree, snow-in-summer [I]

Trees or shrubs, 2–10 m; bark papery. **Leaves** opposite (decussate); blade narrowly elliptic to linear-elliptic, 1.7–4.5 × 0.1–0.4 cm, veins 3, longitudinal, surfaces soon glabrescent. **Inflorescences** 4–20-flowered, flowers in monads, pseudoterminal, sometimes also axillary distally, to 40 mm wide. **Flowers:** calyx lobes glabrous abaxially, margins scarious, 0.1–0.2 mm wide; petals deciduous, 2.5–3.3 mm; filaments connate in bundles of 32–73, white or cream, 8.8–24 mm, bundle claw (5.5–)8–16 mm; style 3.5–5.2 mm; ovules 85–120 per locule. **Capsules** 2.5–4 mm. **Cotyledons** plano-convex.

Flowering late spring–early summer. Disturbed areas; 0–20 m; introduced; Fla.; Australia.

Melaleuca linariifolia is one of the sources of the essential oil called tea tree oil. Its massed flowers make it a striking garden plant in summer.

Melaleuca linariifolia has been erroneously called cajeput in North America (R. P. Wunderlin and B. F. Hansen 2011); cajeput or cajuput is *M. cajuputi*.

3. Melaleuca citrina (Curtis) Dumont de Courset, Bot. Cult. 3: 282. 1802 [I]

Metrosideros citrina Curtis, Bot. Mag. 8: plate 260. 1794; *Callistemon citrinus* (Curtis) Skeels

Shrubs, 1–5 m; bark fibrous or hard-papery. **Leaves** alternate; blade narrowly elliptic to elliptic or narrowly obovate, 2.6–9.9 × 0.5–2.5 cm, veins pinnate, surfaces glabrescent. **Inflorescences** (10–)20–80-flowered, flowers in monads, pseudoterminal, sometimes also axillary distally, 45–70 mm wide. **Flowers:** calyx lobes hairy abaxially, sometimes only marginally, margins scarious, 0.5–0.6 mm wide, or herbaceous; petals deciduous, 3.9–5.8 mm; filaments distinct, 30–45 per flower, red or mauve, 17–25 mm; style 23–31 mm; ovules 170–300 per locule. **Capsules** 4.4–7 mm. **Cotyledons** obvolute. $2n = 22$.

Flowering spring. Disturbed riparian areas; 0–50 m; introduced; Calif., La.; Australia.

Melaleuca citrina is widely cultivated for its showy flowers; it hybridizes with other species of bottlebrush and there are many named cultivars.

4. Melaleuca viminalis (Solander ex Gaertner) Byrnes, Austrobaileya 2: 75. 1984 • Weeping bottlebrush [I]

Metrosideros viminalis Solander ex Gaertner, Fruct. Sem. Pl. 1: 171, plate 34, fig. 4. 1788; *Callistemon viminalis* (Solander ex Gaertner) G. Don ex Loudon

Shrubs or trees, 1–35 m; bark fibrous, hard. **Leaves** alternate; blade narrowly elliptic to elliptic or narrowly obovate, 2.5–13.8 × 0.3–2.7 cm, veins pinnate, surfaces glabrescent. **Inflorescences** 15–50-flowered, flowers in monads, pseudoterminal or interstitial, 35–50 mm wide. **Flowers:** calyx lobes hairy or glabrescent abaxially, margins herbaceous; petals deciduous, 3.4–5.9 mm; filaments connate in bundles of 9–14, usually in 5 distinct bundles, these sometimes obscure, especially when claw is very short, red or crimson, 13–26 mm, bundle claw to 2.2 mm; style 16–29 mm; ovules ca. 100 per locule. **Capsules** 3.8–4.8 mm. **Cotyledons** obvolute.

Flowering spring. Disturbed riparian areas; 0–400 m; introduced; Calif., Fla.; Australia.

Melaleuca viminalis is well known in cultivation and is widely grown for its showy flowers. The species is an unusual member of the bottlebrush group of *Melaleuca* in that its staminal filaments are connate into five bundles; most other species of bottlebrush have distinct stamens.

5. CALYPTRANTHES Swartz, Prodr., 5, 79. 1788, name conserved • Spicewood, lid-flower [Greek *kalyptra*, cap or cover, and *anthos*, flower, alluding to calyx covering stamens in flower bud]

Bruce K. Holst

Chytraculia P. Browne, name rejected

Shrubs or trees, branching predominantly bifurcate, glabrous or pubescent, hairs simple or dibrachiate. **Young stems** terete, compressed, or quadrangular, often narrowly 2-keeled to 2-winged, the sharp keels or wings terminating distally between base of petioles. **Leaves** opposite; blade ± leathery, venation brochidodromous. **Inflorescences** 9–50+-flowered, subterminal, panicles [rarely spikes], often paired, opposite leaf axils at proximal node of normal or abortive branch, sometimes subtended by conspicuous, foliaceous bracts, axis usually compressed. **Flowers** 0–3(–5)-merous, usually sessile; hypanthium obconic to bowl-shaped, prolonged beyond summit of ovary; calyx closed in bud, forming calyptra, circumscissile, calyptra attached at 1 side in anthesis, usually deciduous; petals 0, 2, or 3(–5), inconspicuous, white, usually attached to calyptra and falling with it; stamens to ca. 200, filaments white; ovary 2-locular; style linear; stigma punctate; ovules 2 per locule. **Fruits** berries, bluish black or purplish black, spheroid to oblate, crowned by tubular portion of hypanthium or a circular scar, calyptra remnant occasionally persistent. **Seeds** 1 or 2, subglobose to oblate; seed coat papery; cotyledons foliaceous, thin, folded into bundle; hypocotyl elongate, as long as cotyledons and encircling them.

Species ca. 200 (2 in the flora): Florida, Mexico, West Indies, Bermuda, Central America, South America.

Calyptranthes is an easily identified genus throughout most of its range, though it tends to grade into *Marlierea* Cambessèdes and *Myrcia* de Candolle ex Guillemin in Amazonia. The combination of predominantly bifurcate branching and the calyx completely closed in bud and opening circumscissily characterize the genus. Some species have distinctive, narrowly winged or keeled twigs, with the wing/keel in the plane perpendicular to the insertion of the leaves. The inflorescence is usually of paired panicles, but these can be reduced to a few sessile flowers or a spike in some South American species.

Both species of *Calyptranthes* in Florida are handsome landscape shrubs or small trees for subtropical areas and provide abundant fruit for wildlife.

1. Inflorescences tomentulose; calyptrae obtuse to rounded; young stems narrowly 2-winged; midvein of leaves sulcate adaxially. . . . . 1. *Calyptranthes pallens*
1. Inflorescences glabrous; calyptrae apiculate; young stems terete to compressed (not winged); midvein of leaves convex adaxially . . . . 2. *Calyptranthes zuzygium*

1. Calyptranthes pallens Grisebach, Abh. Köngl. Ges. Wiss. Göttingen 7: 215. 1857 • Pale lid-flower [F]

Calyptranthes chytraculia (Linnaeus) Swartz var. *pauciflora* O. Berg, Linnaea 27: 27. 1855; *C. pallens* var. *mexicana* (Lundell) McVaugh; *C. pallens* var. *williamsii* (Standley) McVaugh

Shrubs or trees 5–10 m, with dense foliage and rounded crown, indumentum usually abundant on young stems and inflorescences, hairs light brown to yellowish white, dibrachiate. **Young stems** narrowly 2-winged; bark pale gray or nearly white, smooth or, ultimately, scaly. **Leaves:** petiole 4–8 mm, channeled, puberulous; blade elliptic to ovate, 3–6(–10) × 2–4(–6) cm, midvein sulcate adaxially, lateral veins 10–15 pairs, indistinct on both surfaces, marginal vein 1, 1–2 mm from margin, similar in prominence to lateral veins, slightly arched between lateral veins, base obtuse to cuneate, apex acute to acuminate, slightly recurved, surfaces usually appressed-pubescent abaxially, sometimes glabrescent, lustrous adaxially and appressed-puberulous or glabrous, glands slightly impressed adaxially when dried. **Peduncles** slender, 2–4 cm. **Inflorescences** (20–)30–50+-flowered, 7–15 cm, tomentulose; bracts deciduous, ovate, acuminate; bracteoles early deciduous, minute. **Pedicels** 1–2 mm. **Flowers** fragrant; bud narrowly obovoid, 2–3 mm, appressed-pubescent; hypanthium narrowly crateriform, cylindrical in fruit, 2 mm wide; calyptra 2 mm wide, obtuse to rounded; stamens 5 mm; style 4 mm. **Berries** lustrous dark purplish black, spheroid, 5–8 mm diam., crowned by cylindrical hypanthium when mature, glabrous or sparsely pubescent. **Seed** 4 mm diam.

Flowering late spring–mid summer. Subtropical hardwood hammocks; 0–10 m; Fla.; West Indies; Central America.

Calyptranthes pallens is known in the flora area from the southern peninsula coast, the Everglades, and the Florida Keys, in Collier, Miami-Dade, and Monroe counties.

Calyptranthes pallens is easily distinguished from *C. zuzygium* by the abundant, tightly appressed (versus glabrous) indumentum, the narrowly winged (versus unwinged) stems, the sulcate (versus convex) midvein on the upper leaf surface, and the obtuse to rounded (versus apiculate) calyptra. As circumscribed here to include a variety of forms from throughout its range, *C. pallens* is variable in leaf size and presence/absence of indumentum and stem wings. However, collections from Florida are uniform in these features with leaves on the smaller end of the size range given, the presence of pale brown or yellowish white indumentum on the young vegetative parts and inflorescence, and the presence of stem wings. *Calyptranthes pallens* is listed as threatened in Florida.

The fruits are edible and have an agreeable, spicy taste; ripening occurs unevenly over a period of weeks, giving the infructescence a distinctive appearance with green, red, orange, and purplish black fruit present at the same time. The brown heartwood is close-grained, heavy, and hard (J. K. Small 1933).

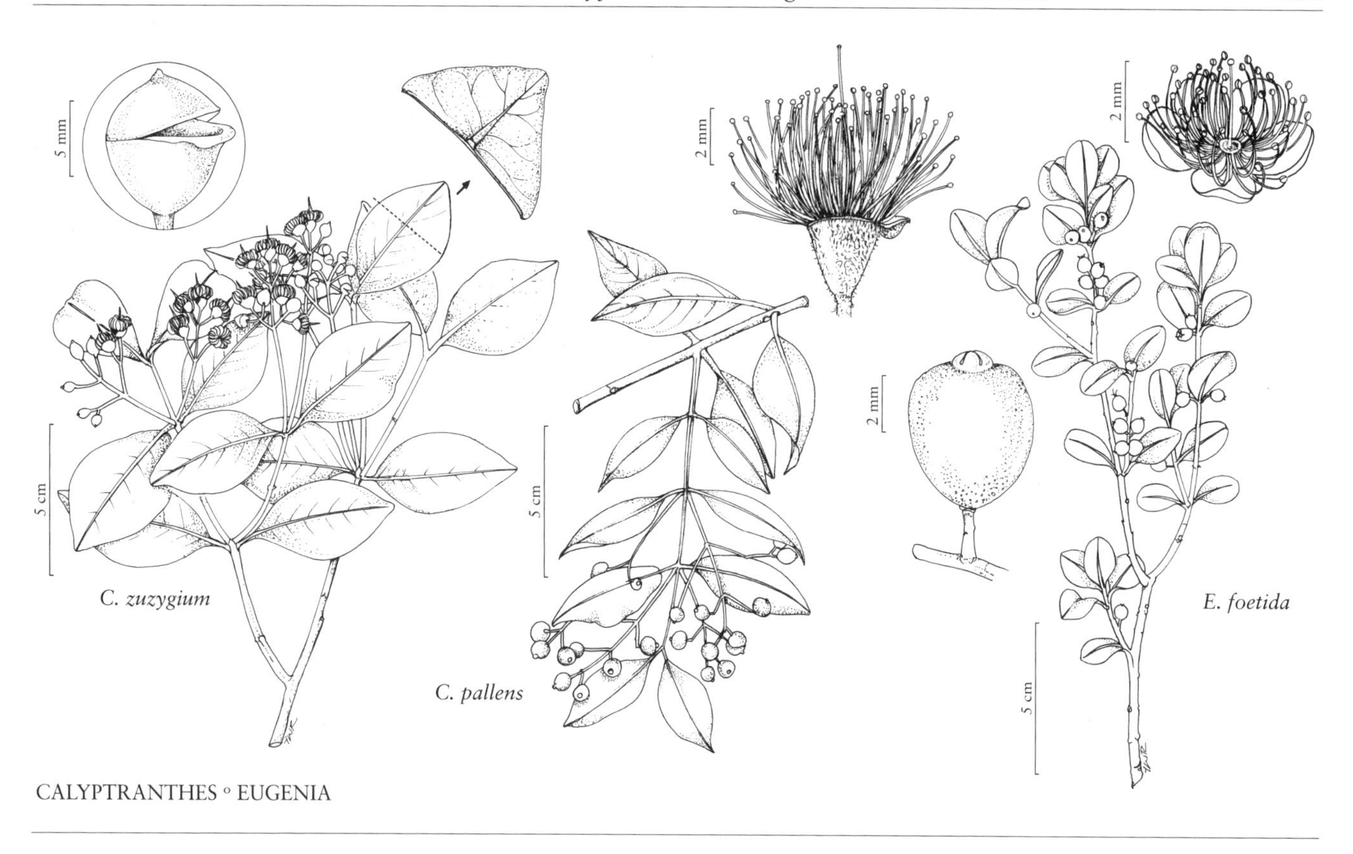

CALYPTRANTHES ◦ EUGENIA

2. **Calyptranthes zuzygium** (Linnaeus) Swartz, Prodr., 79. 1788 • Myrtle-of-the-river [F]

Myrtus zuzygium Linnaeus, Syst. Nat. ed. 10, 2: 1056. 1759 (as zuzygiu.)

Shrubs or trees to 12 m, glabrous throughout. **Young stems** terete to compressed; bark pale gray, smooth. **Leaves:** petiole stout, 2–4 mm, shallowly channeled; blade mostly elliptic to obovate, occasionally ovate, 4–6(–7) × 2–4 cm, midvein convex adaxially, distinctly broader proximally, lateral veins ca. 20 pairs, faint, ascending, marginal vein 1, 1–2 mm from margin, gently arching between lateral veins and margin, or with second faint marginal vein parallel to margin, base cuneate, apex obtuse to abruptly blunt-tipped, adaxial surface lustrous. **Peduncles** slender, 2.5–4 cm (longer than fertile portion of panicle). **Inflorescences** 9–20-flowered, flowers mostly in triads; panicles in pairs and laxly flowered, or sometimes 1 panicle present by abortion; bracts and bracteoles early deciduous. **Pedicels** (0–)1.5–5 mm. **Flowers** fragrant; bud obovoid, 3.5–4.5 mm; hypanthium broadly crateriform, shallowly cylindrical in fruit, 3.5–4.5 mm wide; calyptra 3.5–4.5 mm wide, apiculate; stamens 4 mm; style 5 mm. **Berries** bluish black, spheroid to oblate, 8–10 mm diam. **Seed** 5 mm diam.

Flowering late spring–early summer. Subtropical hardwood hammocks; 0–10 m; Fla.; West Indies.

Calyptranthes zuzygium is known from the Everglades and the Florida Keys, in Miami-Dade and Monroe counties; it is listed as endangered by the state of Florida.

6. EUGENIA Linnaeus, Sp. Pl. 1: 470. 1753; Gen. Pl. ed. 5, 211. 1754 • Stopper [For François-Eugène, Prince of Savoy, 1663–1736, Austrian General]

Fred R. Barrie

Shrubs or small trees, glabrous or pubescent, young growth glabrous or thinly to densely vested, hairs simple or dibrachiate. **Leaves** opposite; blade papery to leathery, venation brochidodromous, glands conspicuous to obscure on either or both surfaces. **Inflorescences**

1–8-flowered, axillary, racemes elongate or appearing fasciculate, or flowers solitary; bracteoles 2, caducous or persistent, distinct or connate basally, forming involucre beneath hypanthium. **Flowers** 4-merous, pedicellate; hypanthium obconic, campanulate, or bowl-shaped, not prolonged beyond summit of ovary; calyx lobes often persistent in fruit, in 2 opposing equal or markedly unequal pairs; petals white, conspicuous; stamens [20–]25–70[–600]; ovary 2-locular; ovules 12–25 per locule. **Fruits** berries, red, purple, black, or purplish black, globose, oblate, or obovoid, with fleshy pericarp. **Seeds** 1(or 2), subglobose to reniform; seed coat membranous or leathery; embryo a solid globose or reniform mass, mainly of cotyledonary tissue.

Species ca. 1000 (5 in the flora): Florida, Mexico, West Indies, Central America, South America, se Asia, Africa.

Eugenia is one of the largest genera of flowering plants; within the heart of its range, a significant number of species remain undescribed. The genus is characterized by the basically racemose inflorescences and flowers with four calyx lobes, four petals, 2-loculed ovaries with numerous ovules and an embryo with the cotyledons, radicle, and plumule fused into an undifferentiated mass. There is considerable overlap of characteristics among species, and often taxa can be differentiated only by the degree and type of vestiture or the persistence and degree of fusion of the bracteoles.

Eugenia is a common component of mesic and wet forests at all elevations in the New World tropics, where the vast majority of its species are found. About a dozen species are native to the Old World in tropical Africa and southeast Asia. Only four species range as far north as southern Florida, where they are restricted for the most part to coastal hammocks where winter temperatures are moderated by proximity to the ocean. One introduced species, *E. uniflora*, has naturalized locally in Florida.

The exceptionally hard wood of many species of *Eugenia* has been used for structural components in building, as well as for tools and cabinetry. The common name, stopper, apparently derived from the use of the fruit as a treatment for diarrhea (D. F. Austin 2004).

Although relatively safe in other parts of their range, all native species of *Eugenia* should be considered threatened in Florida due to habitat destruction.

SELECTED REFERENCE Barrie, F. R. 2009. *Eugenia*. In: G. Davidse et al., eds. 1994+. Flora Mesoamericana. 6+ vols. in parts. Mexico City, St. Louis, and London. Vol. 4(1), pp. 81–129.

1. Shrubs or trees mostly puberulent, hairs erect or recurved; leaf blades elliptic or obovate . . . . 3. *Eugenia foetida*
1. Shrubs or trees glabrous, glabrate, or pubescence limited to ciliate margins of bracts, bracteoles, and calyx lobes; leaf blades ovate to elliptic.
   2. Petioles 1–3 mm; leaf blades ovate, papery; hypanthia ribbed; berries costate, deep bright red . . . . 5. *Eugenia uniflora*
   2. Petioles 3–9 mm; leaf blades ovate to elliptic, leathery; hypanthia not ribbed; berries not costate, bright or dark red, purple, or purplish black.
      3. Petioles flattened or splayed; pedicels 1–3 mm; floral discs 0.7–1 mm diam.; berries purplish black . . . . 1. *Eugenia axillaris*
      3. Petioles channeled; pedicels 6–20(–30) mm; floral discs 2–3 mm diam.; berries bright or dark red or purple.
         4. Leaves with caudate-acuminate apex; berries bright red, 6–9 mm diam. . . . . . 2. *Eugenia confusa*
         4. Leaves with bluntly acute or acuminate apex; berries dark red or purple, 4–7 mm diam. . . . . 4. *Eugenia rhombea*

1. Eugenia axillaris (Swartz) Willdenow, Sp. Pl. 2: 960. 1799 • White stopper

Myrtus axillaris Swartz, Prodr., 78. 1788

Trees, to 10 m, slender, glabrous, except for a few simple, appressed, coppery hairs present on buds, and ciliate margins of bracts, bracteoles, and calyx lobes. **Twigs** terete or compressed at nodes; bark gray or brown. **Leaves** olive or tan abaxially, drying grayish green adaxially; petiole splayed or flattened, 3–8 mm; blade ovate or elliptic, 4–8 × 2–4 cm, leathery, base cuneate or oblique, margins decurrent into splayed distal edge of petiole, apex acute to rounded, surfaces with scattered glands abaxially, glands obscure adaxially. **Inflorescences** 4–8-flowered, racemes, solitary or 2 superposed; axis 3–6 mm, 4-angled; bud globose, 1.5 mm; bracteoles persistent, ovate, ca. 0.5 × 0.6 mm, base usually connate and involucrate, less commonly distinct, margins ciliate, apex rounded or truncate. **Pedicels** 1–3 mm (relatively equal). **Flowers:** hypanthium campanulate, 0.5–1 mm; calyx lobes elliptic, in unequal pairs, larger pair ca. 1 × 1 mm, margins ciliate, apex rounded; petals elliptic, 2.5–3 × 2–2.5 mm, apex rounded; disc 0.7–1 mm diam.; stamens 30–50, 2.2–3.5 mm; style 3–4.5 mm. **Berries** purplish black, globose or oblate, 5.5–9 × 5.5–7 mm; calyx persistent, not prominent.

Flowering and fruiting year-round. Coastal hammocks; 0–20 m; Fla.; Mexico; West Indies; Central America.

Eugenia axillaris is known in the flora area from the central and southern peninsula.

The Seminoles used *Eugenia axillaris* for making bows (D. F. Austin 2004).

2. Eugenia confusa de Candolle in A. P. de Candolle and A. L. P. P. de Candolle, Prodr. 3: 279. 1828 • Ironwood, redberry stopper

Eugenia garberi Sargent

Trees or shrubs, to ca. 6 m, glabrous or glabrate throughout. **Twigs** terete or weakly compressed; bark reddish brown, tan, or gray. **Leaves** drying green or olive; petiole channeled, 3–9 mm; blade ovate to elliptic-ovate, 2.5–6 × 1–3 cm, leathery, base rounded to cuneate, margins cartilaginous, apex caudate-acuminate, surfaces with numerous or obscure glands, smooth adaxially when fresh, often appearing wrinkled when dry. **Inflorescences** 1–8-flowered, racemes, solitary; axis 1–4 mm; bud obconic, ca. 2 mm; bracteoles caducous around anthesis, ovate, 0.5–1 mm, base distinct, apex obtuse to rounded. **Pedicels** 6–15 mm. **Flowers:** hypanthium obconic, 1.5–2 mm; calyx lobes in unequal pairs, larger pair 1–1.5 × 0.5–0.8 mm, apex rounded; petals obovate, 3–3.5 × 2.5–3.5 mm, apex rounded; disc 2 mm diam.; stamens ca. 40, 2–3.5 mm; style 2–5 mm. **Berries** bright red, globose to obovoid, 6–9 mm diam.; calyx persistent.

Flowering and fruiting year-round. Coastal hammocks; 0–20 m; Fla.; West Indies.

Eugenia confusa is known in the flora area from Martin and Miami-Dade counties, and the Keys in Monroe County.

3. Eugenia foetida Persoon, Syn. Pl. 2: 29. 1806 • Boxleaf or Spanish stopper [F]

Eugenia myrtoides Poiret

Shrubs or trees, to 10 m, puberulent on young stems, leaves, and inflorescences, hairs pale, recurved and 0.2–0.4 mm, or erect and 0.1 mm. **Twigs** compressed; bark tan or gray. **Leaves** drying green, olive, or tan; petiole terete or channeled, 2–5 mm; blade elliptic or obovate, 2.5–8 × 0.8–3.5 cm, papery, base cuneate, decurrent along petiole, margins revolute, apex rounded or bluntly acute, occasionally ultimately retuse, surfaces with numerous, tiny glands, hairs scattered, erect or recurved, more dense along midvein and margins, or glabrate. **Inflorescences** (1 or) 4–8-flowered, racemes, solitary or 2 or 3 superposed; axis 1–4 mm, puberulent, hairs straight and recurved; bud globose, 1 mm; bracteoles persistent, widely ovate, 0.5–0.7 × 0.7 mm, base distinct, margins scarious, apex acute, surfaces puberulent. **Pedicels** 1–3 mm. **Flowers:** hypanthium campanulate, 0.5–1 mm, pubescent; calyx lobes in about equal pairs, 0.5–1 × 0.5 –1 mm, margins ciliate, apex rounded, glabrous or with scattered hairs on outer surface; petals elliptic to widely ovate, 2–3.5 × 1.5–2.5 mm, margins ciliate, apex rounded; disc 1–1.5 mm diam.; stamens 25–30, 3–5 mm; style 4–6 mm. **Berries** black, globose, 4–6 mm diam.; calyx persistent, erect.

Flowering and fruiting year-round. Coastal hammocks; 0–20 m; Fla.; Mexico; West Indies; Central America (Belize, Guatemala).

Eugenia foetida is known in the flora area from Brevard and Manatee counties southward.

Eugenia buxifolia (Swartz) Willdenow is an illegitimate later homonym of *E. buxifolia* Lamarck and a synonym of *E. foetida*.

4. Eugenia rhombea (O. Berg) Krug & Urban, Bot. Jahrb. Syst. 19: 644. 1895 • Red stopper

Eugenia foetida Persoon var. *rhombea* O. Berg, Linnaea 27: 212. 1856

Shrubs or trees, 1–5(–10) m, glabrous except for scattered, simple hairs on bracts, bracteoles, calyces, and floral discs. **Twigs** compressed at nodes; bark gray. **Leaves** drying dull grayish green or olive, concolorous or abaxially paler; petiole channeled, 3–5 mm; blade narrowly ovate, ovate, or elliptic, 2–8 × 1–3.5 cm, rigidly leathery, base rounded or broadly cuneate, decurrent or merging into edge of petiole, margins cartilaginous, often white, apex bluntly acute or acuminate, surfaces with numerous glands, sometimes obscure adaxially on older leaves. **Inflorescences** 1–8-flowered, often short racemes, sometimes superficially fasciculate or a single flower; axis 0–2 mm; bud globose, 2–3 mm; bracteoles persistent, ovate to lanceolate, 0.5–0.7 × 0.3–0.5 mm, base distinct, margins scarious or with few hairs, hairs sometimes in apical tufts, apex acute. **Pedicels** gracile, 8–20(–30) mm, prominently glandular. **Flowers:** hypanthium globose or broadly campanulate, 1.5–2 mm, base rarely substipitate at anthesis, apex developing a narrow neck, separating developing fruit from calyx; calyx lobes reflexed at anthesis, concave, in unequal pairs, smaller pair 1.5–2 × 1.5–2, larger pair 2–4 × 2–3 mm, margins scarious, usually white or ciliolate, apex rounded; petals elliptic, 3 × 2 mm, margins ciliolate, apex rounded; disc 2–3 mm diam.; stamens ca. 60, 2.5–3.5 mm; style 3.5–4 mm. **Berries** dark red or purple, globose or oblate, 4–7 mm diam.; calyx persistent, lobes erect, paler.

Flowering mainly in summer; fruiting in fall. Coastal scrub; 0–10 m; Fla.; Mexico; West Indies; Central America.

Eugenia rhombea is known in the flora area from Miami-Dade County and the Keys in Monroe County.

5. Eugenia uniflora Linnaeus, Sp. Pl. 1: 470. 1753 • Surinam cherry, pitanga [I]

Shrubs or trees, to 10 m, glabrous except for few simple coppery hairs on buds, bracts, and bracteoles. **Twigs** slender, compressed distally; bark reddish, shredding, glandular. **Leaves** drying glossy pale green abaxially, darker adaxially; petiole channeled, 1–3 mm; blade ovate, 3–6 × 1.5–3 cm, papery, base rounded, margins merging abruptly into edge of petiole, apex acute to acuminate, surfaces with numerous, small, raised glands, becoming punctate adaxially on older leaves. **Inflorescences** (1 or) 2–6-flowered, short racemes, often appearing fasciculate, flowers rarely solitary; axis 1–2 mm; bud obovoid, 3–5 mm; bracteoles caducous, oblong-lanceolate, 1 × 0.5 mm, base distinct, margins ciliate, apex acute. **Pedicels** gracile, 15–25 mm. **Flowers:** hypanthium campanulate, 8-ribbed, 1–1.5 mm; calyx lobes oblong, subequal, 2.5–4 × 1.5–2 mm, margins ciliate, apex rounded or acute; petals obovate, 4–6 × 2.5–4 mm, margins ciliate, apex rounded; disc 2–2.5 mm diam.; stamens 40–70, 4–6 mm; style 4–7 mm. **Berries** deep bright red, globose, 12–15 mm diam., 8-costate; calyx persistent, erect.

Flowering and fruiting year-round. Hammocks, distrubed areas; 0–20 m; introduced; Fla.; South America.

Eugenia uniflora has escaped from cultivation in the flora area and is known from the central and southern parts of the peninsula.

Eugenia uniflora has been widely cultivated since pre-Columbian times. Its native range is unknown, but it is generally assumed to have originated in Brazil or, possibly, northern South America; R. McVaugh (1969) thought that southern Brazil was most likely. The species is prized for its fruit and is also grown as a specimen tree or trained as a formal hedge.

The Florida Exotic Pest Plant Council has listed *Eugenia uniflora* as a Class 1 invasive, a taxon that displaces native species or disrupts native habitats.

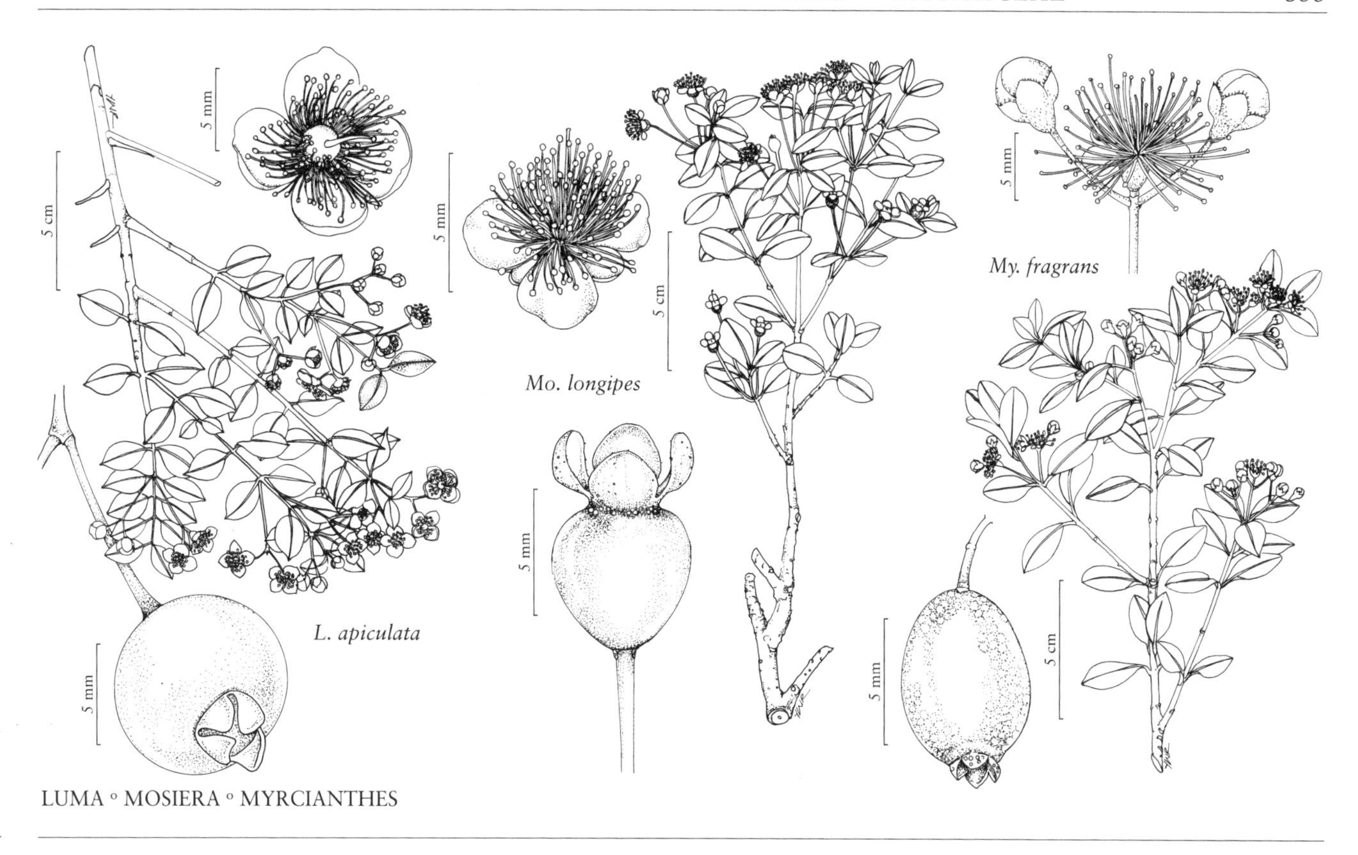

LUMA ◦ MOSIERA ◦ MYRCIANTHES

7. LUMA A. Gray, Proc. Amer. Acad. Arts 3: 52. 1853 • Arrayán [Chilean Native American (Mapuche) name for hardwood of *Amomyrtus luma*] [I]

Leslie R. Landrum

Shrubs or small trees, glabrous or pubescent, hairs simple. **Leaves** opposite; blade venation brochidodromous. **Inflorescences** 1 or 3-flowered, axillary, solitary flowers or dichasia. **Flowers** 4-merous, sessile or pedicellate; hypanthium obconic to campanulate; calyx green, lobes distinct; petals white; stamens 170–300; ovary 2-locular; ovules 6–14 per locule. **Fruits** berries, dark purple, subglobose, crowned by calyx lobes. **Seeds** 1–16, lenticular; seed coat membranous; embryo lenticular; cotyledons suborbicular, separate, thinly plano-convex; hypocotyls cylindrical, about as long as cotyledons.

Species 2 (1 in the flora): introduced, California; South America (Argentina, Chile, Peru); introduced also in temperate regions.

1. Luma apiculata (de Candolle) Burret, Notizbl. Bot. Gart. Berlin-Dahlem 15: 523. 1941 [F] [I]

Eugenia apiculata de Candolle in A. P. de Candolle and A. L. P. P. de Candolle, Prodr. 3: 276. 1828; *Myrceugenella apiculata* (de Candolle) Kausel

Shrubs or trees to 10 m; trunk smooth, often appearing somewhat twisted; bark grayish or bright orangish brown, often with lighter spots; young twigs puberulent to villous. **Leaves:** blade elliptic to suborbiculate, 1–4.5 × 0.5–3.5 cm, base cuneate to rounded, apex apiculate to abruptly acuminate. **Peduncles** 5–30 mm. **Flowers:** calyx open, lobes persistent, 2–3(–4) mm; petals 3–5 mm. **Berries** to ca. 10 mm. **Seeds** 3–6 mm.

Flowering mainly late summer; fruiting fall. Disturbed areas; 0–20 m; introduced; Calif.; South America (sw Argentina, s Chile).

In the flora area, *Luma apiculata* is known mainly from near San Francisco.

8. MOSIERA Small, Man. S.E. Fl., 936, 1506, fig. [p. 937]. 1933 • Stopper [For Charles A. Mosier, 1871–1936, first superintendent of Royal Palm State Park, Florida's first state park (now Everglades National Park)] [C]

Andrew Salywon

Shrubs or trees, glabrous or pubescent, hairs simple. **Leaves** usually subopposite or whorled, sometimes decussate; blade venation brochidodromous [hyphodromous]. **Inflorescences** 1- or 3(–5)-flowered, axillary, often solitary flowers, sometimes dichasia or racemes with 1–3(–5) decussate pairs of flowers. **Flowers** 4-merous, sessile or pedicellate; hypanthium obconic to campanulate; calyx lobes distinct in bud, usually erect in fruit; petals white; stamens [30–] 76–120[–250]; connective usually with 1 terminal oil gland; ovary [1 or]2–4-locular; placentation axile [to parietal], placenta not prominent; ovules [3–]9–40 per locule. **Fruits** berries, dark purple, red, or black, ellipsoid to globose. **Seeds** 2–27, subreniform; seed coat smooth or leathery, surface cells isodiametric and not overlapping; embryo whitish, C-shaped, oily; cotyledons linear, reflexed, less than 1/4 length of embryo; hypocotyl as wide as or wider than cotyledons.

Species ca. 20 (1 in the flora): Florida, Mexico, West Indies, Central America (Guatemala).

SELECTED REFERENCE Salywon, A. 2003. A Monograph of *Mosiera* (Myrtaceae). Ph.D. dissertation. Arizona State University.

1. Mosiera longipes (O. Berg) Small, Man. S.E. Fl., 937. 1933 • Mangroveberry, Bahama or long-stalked stopper [C] [F]

Eugenia longipes O. Berg, Linnaea 27: 150. 1856; *Anamomis bahamensis* (Kiaerskov) Britton ex Small; *A. longipes* (O. Berg) Britton ex Small; *E. bahamensis* Kiaerskov; *Myrtus bahamensis* (Kiaerskov) Urban; *M. verrucosa* O. Berg; *Psidium longipes* (O. Berg) McVaugh

Shrubs or trees to 4 m; older twigs gray, bronze, or reddish brown, bark smooth or peeling in small flakes, glabrous; young twigs reddish brown, gray, or yellowish green, terete, smooth, glandular, flattened near nodes, glabrous or, sometimes, sparsely to densely puberulent. **Leaves** fragrant when crushed; blade discolorous when fresh, dark green or yellowish green, elliptic to suborbiculate or ovate, (1.1–)1.8–4 (–5.2) × (0.2–)0.8–3(–3.8) cm, ± leathery, base cuneate to rounded, margins sometimes slightly crenulate, slightly revolute when dried, apex acute to rounded, or mucronate or emarginate, surfaces dull abaxially, shiny adaxially, densely glandular, glabrous or, rarely, sparsely pubescent. **Peduncles** (4–)12–40(–52) × 0.5–1.2 mm, flattened. **Inflorescences** in leafless nodes or in leaf axils on young shoots, often in opposite pairs on short, leafless shoots or on proximal 1/2 of new leafy shoots, sometimes solitary, glabrous or puberulent; bracteoles 2, caducous or present at anthesis, sometimes shortly petiolate, elliptic to orbiculate, 2–5 × 1–5 mm. **Flowers:** bud pyriform, ca. 4–5(–6) mm, globose portion 3–4 mm diam.; hypanthium campanulate, 2–3 mm, tube ca. 1 mm, tearing slightly between calyx lobes; calyx lobes ovate to hemiorbiculate, 2.5–3.5 × 3–3.5 mm, apex rounded, surfaces glandular, glabrous or margins ciliate; petals obovate or suborbiculate, 4–6 × 4–6 mm, margins ciliate; placentation axile; ovules multiseriate. **Berries** 7–10 × 6–8 mm, densely glandular, usually glabrous, rarely sparsely puberulent. **Seeds** ca. 2 × 1–1.5 mm; seed coat pale yellow, hard, lustrous; operculum present.

Flowering and fruiting year-round. Pine hammocks, dry coastal scrub, limestone substrates or dunes; of conservation concern; 0–50 m; Fla.; West Indies.

Mosiera longipes is listed as threatened in the Preservation of Native Flora of Florida Act, a result of the rapid habitat destruction in southern Florida.

9. MYRCIANTHES O. Berg, Linnaea 27: 315. 1856 • [Genus *Myrcia* and Greek *anthos*, flower, alluding to resemblance]

Fred R. Barrie

Trees or shrubs, glabrous or pubescent, hairs simple. **Leaves** opposite; blade venation brochidodromous. **Inflorescences** 1 or 3–7(–14)-flowered, axillary, solitary flowers or dichasia. **Flowers** 4- or 5-merous, sessile (terminal flowers) or pedicellate (lateral flowers); hypanthium obconic, not prolonged beyond summit of ovary, base not attenuate; calyx lobes distinct, about equal or 1 smaller; petals white, convex; stamens 100–150[–300]; ovary 2- or 3-locular; ovules 8–20. **Fruits** berries, purplish black, globose or ovoid, crowned by persistent, erect calyx. **Seeds** 1–4, reniform; seed coat thin, papery; embryo subreniform to subglobose; cotyledons distinct, plano-convex; hypocotyl terete, ca. ½ as long as cotyledons; plumule usually evident in mature seeds, much shorter than hypocotyl.

Species ca. 40 (1 in the flora): Florida, Mexico, West Indies, Bermuda, Central America, South America.

1. **Myrcianthes fragrans** (Swartz) McVaugh, Fieldiana, Bot. 29: 485. 1963 • Twinberry, Simpson's stopper [F]

Myrtus fragrans Swartz, Prodr., 79. 1788, name conserved; *Anamomis fragrans* (Swartz) Grisebach; *A. punctata* (Vahl) Grisebach; *A. simpsonii* Small; *Eugenia fragrans* (Swartz) Willdenow; *E. punctata* Vahl; *E. simpsonii* (Small) Sargent; *Myrcianthes fragrans* subsp. *simpsonii* (Small) A. E. Murray; *M. fragrans* var. *simpsonii* (Small) R. W. Long; *M. simpsonii* (Small) K. A. Wilson

Trees or shrubs to 20 m; bark reddish brown, smooth, exfoliating; twigs terete or compressed, becoming glabrate; young growth sparsely to densely appressed-pubescent, hairs cinereous or white, 0.2–0.4 mm. **Leaves:** petiole 2.5–10 mm, sericeous or glabrate; blade concolorous or paler abaxially, drying olive or tan, elliptic to obovate, 2–9 × 1.7–3 cm, leathery, base cuneate or narrowly so, decurrent into petiole, margins flat or revolute basally, apex acuminate to bluntly acute or rounded, or tip retuse, surfaces with numerous small glands, glabrate or hairs scattered, appressed, usually persistent along adaxial midvein. **Peduncles** 20–60 × 1–2 mm, compressed. **Inflorescences** sericeous to glabrate; bracts and bracteoles caducous, linear, 2–4 mm. **Pedicels** of lateral flowers 3–10 mm, compressed. **Flowers:** hypanthium 2–3 mm, coarsely sericeous; calyx lobes deltate to widely ovate, 1.5–2.2 × 1.3–2 mm, apex bluntly acute or rounded, surfaces becoming glabrate abaxially, persistently sericeous adaxially; petals oblong or obovate, 3.5–5 × 3.5–5 mm, convex, margins ciliate, apex rounded; disc round or quadrate, 3–4 mm diam., staminal ring usually pubescent; stamens 3–9 mm; style 4–8 mm, glabrous or sparsely pubescent. **Berries** 6–15 × 6–15 mm.

Flowering and fruiting year-round. Hammocks; 0 m; Fla.; Mexico; West Indies; Central America; South America (Colombia, Ecuador, Peru, Venezuela).

Myrcianthes fragrans is known in the flora area from St. Johns County and the central and southern peninsula.

Myrcianthes fragrans occurs in moist and wet forests at elevations (outside of the flora area) to 1500 m. Throughout its range, the species displays significant, but inconsistent, local variation in the degree of vestiture of the young twigs, inflorescences and flowers, and in the size and shape of the leaves. R. W. Long and O. Lakela (1971) recognized the more robust trees with buttressed base and more floriferous inflorescences (10–14 flowers) as var. *simpsonii*.

Myrcianthes fragrans is listed as threatened in Florida, due to habitat destruction.

10. MYRTUS Linnaeus, Sp. Pl. 1: 471. 1753; Gen. Pl. ed. 5, 212. 1754 • Myrtle [Classical name for a species of myrtle] [I]

Leslie R. Landrum

Shrubs or trees, usually glabrous or glabrate, hairs simple, whitish. **Leaves** usually opposite or whorled; blade venation brochidodromous, obscure. **Inflorescences** 1-flowered, axillary, flowers solitary. **Flowers** 5-merous, pedicellate; hypanthium obconic; calyx lobes persisting after anthesis, distinct, small tears sometimes forming between lobes; petals whitish; stamens 100–200; ovary 2- or 3-locular, septum often incomplete centrally to apically; placenta axile, not protruding, V- to O-shaped; ovules 22–34 per locule, 2-seriate. **Fruits** berries, bluish purple (pulp whitish), subglobose. **Seeds** 8–20, somewhat flattened, C-shaped to coiled; outer rim of seed coat hard and shiny, central portion often soft, external portion a few cells thick, not notably dense, easily broken; embryo C-shaped, cylindrical; cotyledons linear, ca. ½ as long as embryo. $x = 11$.

Species 1 or 2 (1 in the flora): introduced; s Europe, n Africa.

1. Myrtus communis Linnaeus, Sp. Pl. 1: 471. 1753 [F] [I]

Shrubs or trees to 4 m; trunk dark orangish brown, smooth or scaly, sometimes young twigs and floral disc puberulent. **Leaves:** blade elliptic to lanceolate, 1–3.5 × 0.3–1.8 cm, base cuneate to rounded, apex acute to acuminate, sometimes slightly mucronate. **Pedicels** 0.5–2.5 cm. **Flowers:** bud pyriform, 7–9 mm; calyx lobes ovate-triangular, 2–3 mm; petals 5–10 mm; bracteoles deciduous in very young bud, narrowly triangular, ca. 2 mm. **Berries** to ca. 10 mm. **Seeds** 3–4 mm. $2n = 22$.

Flowering spring; fruiting summer. Disturbed areas; 0–100 m; introduced; Calif., La., Tex.; s Europe; n Africa (Mediterranean region).

Myrtus communis is widely cultivated and has become naturalized in San Luis Obispo and Sonoma counties in California, Caddo Parish in Louisiana, and Brazoria and Hardin counties in Texas.

11. PSIDIUM Linnaeus, Sp. Pl. 1: 470. 1753; Gen. Pl. ed. 5, 211. 1754 • [Ancient Greek name *psidion* for *Punica*, alluding to supposed resemblance] [I]

Leslie R. Landrum

Shrubs or trees, glabrous or pubescent, hairs simple. **Leaves** sometimes drought deciduous, opposite; blade venation usually brochidodromous. **Inflorescences** 1- or 3-flowered, axillary, solitary flowers or dichasia; bracteoles caducous. **Flowers** usually 5-merous, sessile or pedicellate; hypanthium obconic; calyx lobes distinct or connate beyond summit of ovary to form calyx tube, sometimes forming calyptra (in closed flower bud, calyptra completely closed or open only as a terminal pore, tearing regularly into 5 lobes or irregularly); petals whitish; stamens [100–]280–720; ovary [2- or]3–6-locular; placenta bilamelate, often protruding as a peltate structure; ovules 12–180 per locule, biseriate or multiseriate. **Fruits** berries, green, yellow, or red, pyriform, globose, or subglobose. **Seeds** few–100+; seed coat dull, bony, densely woody, ca. 9–30 cells thick at narrowest point, covered with thin layer of pulpy tissue when wet, or glaze or crusty tissue when dry; embryo curved; cotyledons usually reflexed, linear to elliptic, shorter than hypocotyl.

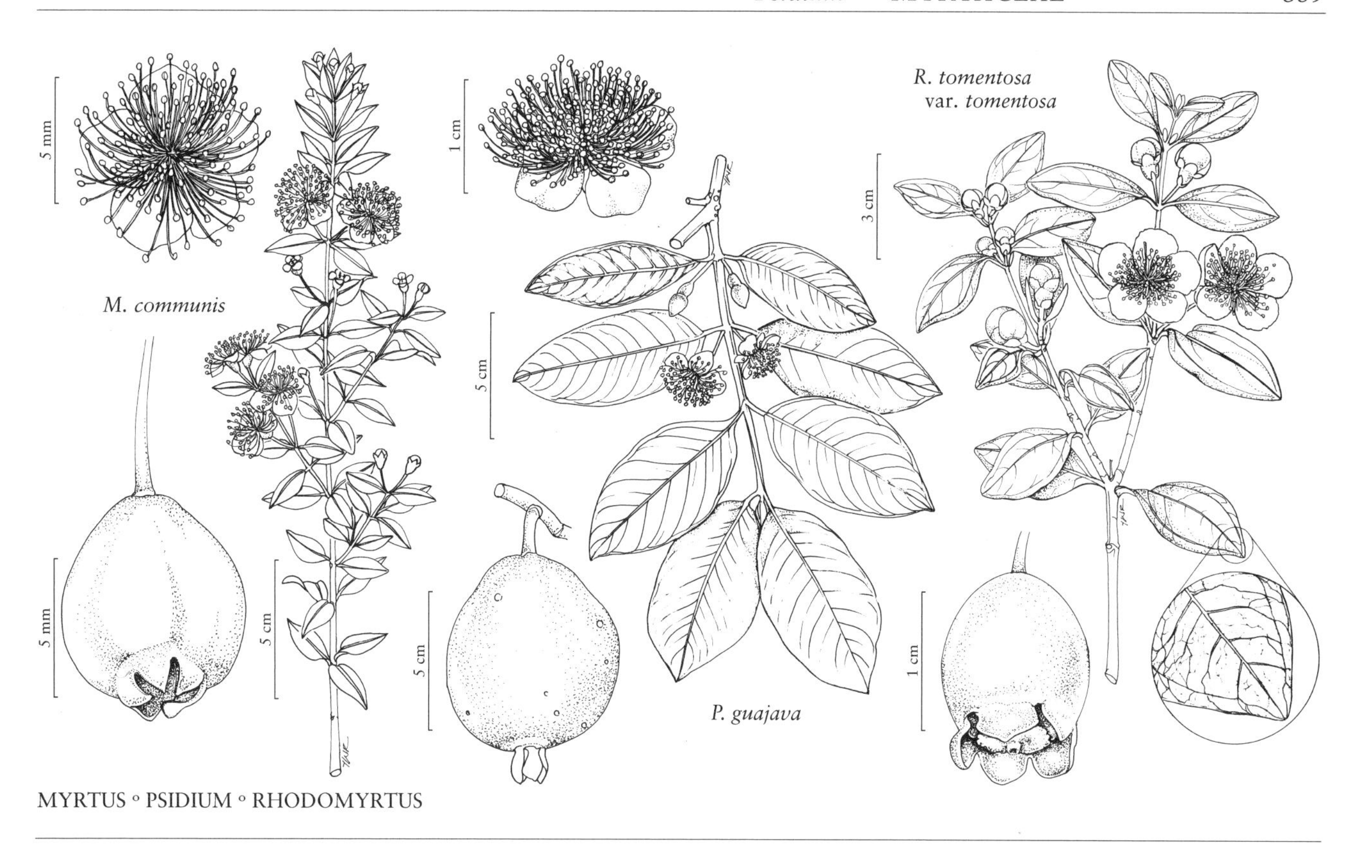

MYRTUS ◦ PSIDIUM ◦ RHODOMYRTUS

Species ca. 70 (2 in the flora): introduced, Florida; Mexico, West Indies, Central America, South America (except Chile).

1. Leaf blades glabrous, obovate, oblanceolate, or elliptic, lateral veins 8–13 pairs (weak to obscure); floral buds each usually with a terminal pore, apex rounded .... 1. *Psidium cattleyanum*
1. Leaf blades appressed-pubescent abaxially, elliptic, elliptic-oblanceolate, elliptic-obovate, lanceolate, or oblong, lateral veins 9–22 pairs (prominent); floral buds each without terminal pore, apex usually conic ....2. *Psidium guajava*

1. Psidium cattleyanum Sabine, Trans. Hort. Soc. London 4: [315–]317, plate 11. 1821 • Strawberry guava [I]

Psidium littorale Raddi; *P. variabile* O. Berg

Shrubs or trees to 8 m; trunk reddish brown, smooth to scaly; young twigs light reddish brown to light gray, flattened, becoming subterete, older twigs usually gray, remaining ± smooth; young growth glabrous or sparsely puberulent to strigose on some floral structures, hairs whitish, most less than 0.1 mm. **Leaves:** petiole channeled, 2–14 × 1–2 mm, glabrous; blade drying light or dark reddish brown or grayish green, nearly concolorous, obovate, oblanceolate, or elliptic, 5–10 × 2–5.8 cm, 1.5–2.6 times as long as wide, leathery (rubbery when fresh), midvein prominent abaxially, nearly flat to shallowly impressed adaxially, lateral veins 8–13 pairs, ascending, weak to obscure, alternating with weaker veins arising near margin and extending toward midvein, base usually attenuate to cuneate, rarely rounded, apex acute or acuminate to broadly rounded, surfaces glabrous. **Inflorescences** 1-flowered, borne in leaf axils, from leafless nodes, or in axils of leaflike or reduced bracts; bracteoles ovate, lanceolate, or oblong, 1–2 mm. **Flowers:** bud subpyriform, 6–13 mm, apex rounded; calyx tube extended 3–7 mm beyond ovary summit, terminating in sinuate-edged terminal pore (rarely completely closed), tearing irregularly at anthesis, tears cutting through staminal ring; hypanthium 3–5 mm (below calyx); petals suborbiculate to elliptic, 3–6 mm; disc within staminal ring ca. 4–6 mm across; stamens 280–400, 3–8 mm; anthers 0–1 mm; style 4–8 mm; stigma ca. 1 mm wide; ovary 3- or 4-locular; placenta reflexed; ovules ca. 12–25 per locule. **Berries** red or yellow, pyriform to subglobose, 15–30 mm. **Seeds** few–100, round to subreniform, ca. 5 mm, smooth.

Flowering spring. Disturbed areas; 0–15 m; introduced; Fla.; South America (Brazil); introduced also in Pacific Islands (Hawaii).

Psidium cattleyanum is known in the flora area from the central and southern peninsula and is commonly cultivated for its edible fruit.

2. Psidium guajava Linnaeus, Sp. Pl. 1: 470. 1753
• Guava, guayaba [F] [I]

Shrubs or trees to 8 m; trunk light brown, reddish brown, or light grayish green, mostly smooth, with large, flaky scales; young twigs green, quadrangular, slightly to strongly winged, often sulcate, at least when dry, older twigs reddish brown to grayish green, smooth or scaly; young growth glabrate to densely appressed-pubescent, hairs whitish, yellowish, or silvery, to ca. 0.7 mm. **Leaves:** petiole channeled, 2–5 × 1–2 mm, densely pubescent to glabrate; blade drying yellowish green, grayish green, or reddish brown, elliptic, oblong, elliptic-oblanceolate, elliptic-obovate, or lanceolate, 4.5–14 × 2.4–7.5 cm, 1.6–3.8 times as long as wide, leathery to submembranous, midvein prominent abaxially, impressed adaxially, lateral veins 9–22 pairs, prominent, ascending (at ca. 45°), nearly straight, curving upward near margin and connecting with next lateral vein, smaller veins connecting laterals in ladder-like to reticulate pattern, base rounded to slightly cordate, apex acute, acuminate, or rounded, surfaces densely to sparsely appressed-pubescent abaxially, glabrate adaxially (except midvein puberulent). **Peduncles** 1–3.5 cm × 1–1.5 mm, terete. **Inflorescences** 1- or 3-flowered, borne in leaf axils; bracteoles linear to narrowly triangular, 2–5 mm, sparsely pubescent. **Flowers:** bud subfusiform to pyriform, 10–17 mm, sometimes strongly constricted near midpoint, apex usually conic; hypanthium to summit of ovary obconic, ca. ½ as long as closed flower bud; calyx closed, conic in bud, tearing irregularly as bud opens, persisting or falling in ca. 3 parts; petals obovate to elliptic, 13–22 mm; disc 4–6 mm across; stamens 280–720, 7–15 mm; anthers 0.7–1 mm; style 10–15 mm; stigma ca. 0.5 mm wide; ovary 3–6-locular; ovules 90–180 per locule (multiseriate). **Berries** aromatic, green or yellow, with pink or white flesh inside, globose or pyriform, 20–60(–80) mm. **Seeds** usually 50+, subreniform, 3–4 mm, ± smooth.

Flowering spring. Roadsides, pastures, riparian areas; 0–100 m; introduced; Fla., La.; South America; introduced also in tropics and subtropics worldwide.

Psidium guajava is known in the flora area from the central and southern peninsula in Florida and Jefferson Parish in Louisiana.

Psidium guajava is commonly and widely cultivated for its edible fruit. It probably was originally cultivated in tropical South America. Archaeological evidence of guava cultivation has been reported for coastal Peru at about 4000 years ago (R. Shady-Solis et al. 2001) and even earlier in Rondônia, Brazil (J. Watling et al. 2018). In Central America and Mexico, the earliest archaeological find of *P. guajava* is about 2000 years old in the Tehuacán Valley of Mexico (C. E. Smith 1965). It reached the Caribbean Islands in pre-Columbian times (G. Fernández de Oviedo y Valdéz 1851). How much of the American distribution, which now extends from Mexico to Argentina, is due to the actions of humans is uncertain. In post-Columbian times it was rapidly spread to the tropical and subtropical regions around the world. Guava products are imported into the United States mainly from Brazil, Dominican Republic, Egypt, Mexico, Philippines, Taiwan, and Thailand. The leaves and bark are commonly used medicinally as a tea to remedy diarrhea.

Psidium guineense Swartz, common in tropical and subtropical America, is a similar weedy species that is often confused with *P. guajava*. One specimen collected at Bradenton, Florida, in 1916 has been seen; it may be expected in the southeastern United States. *Psidium guineense* differs from *P. guajava* in having leaves with fewer lateral veins, usually erect, reddish brown (not appressed and whitish) hairs on the abaxial surfaces, anthers 1–3 mm, and a calyx that tears in usually five (not three) segments.

12. RHODOMYRTUS (de Candolle) Reichenbach, Deut. Bot. Herb.-Buch, 177. 1841 • [Greek *rhodon*, rose or red, and genus *Myrtus*, alluding to flower color] [I]

Leslie R. Landrum

Myrtus Linnaeus sect. *Rhodomyrtus* de Candolle in A. P. de Candolle and A. L. P. P. de Candolle, Prodr. 3: 240. 1828

Shrubs or trees, pubescent, hairs simple. **Leaves** opposite; blade venation brochidodromous to acrodromous. **Inflorescences** 1- or 3[–7]-flowered, axillary, solitary flowers or dichasia. **Flowers** 4- or 5(–7)-merous, pedicellate; hypanthium campanulate; calyx lobes persistent, distinct; petals pink or red [whitish]; stamens ca. 150; ovary [1–]3 [or 4]-locular (false septae causing ovary to appear to have 2 times number of locules in fruit); placenta axile or parietal; ovules 16–20 per locule. **Fruits** berries, purplish black, subglobose, ellipsoidal, or elongate-cylindrical. **Seeds** usually 20+, reniform, in compact, stacklike rows, each seed surrounded by false, longitudinal and horizontal septae; seed coat hardened but porous; embryo C-shaped; cotyledons linear, shorter than hypocotyl.

Species ca. 11 (1 in the flora): introduced, Florida; s, se Asia (Borneo, Celebes Islands, India, Java, Lesser Sunda Islands, Malaya, Moluccas, New Guinea, Sri Lanka, Sumatra), Pacific Islands (New Caledonia, Philippines), e Australia.

SELECTED REFERENCE Scott, A. J. 1978c. A revision of *Rhodomyrtus* (Myrtaceae). Kew Bull. 33: 311–329.

1. Rhodomyrtus tomentosa (Aiton) Hasskarl, Flora 25(2,Beibl.): 35. 1842 • Rose myrtle [F] [I] [W]

Myrtus tomentosa Aiton, Hort. Kew 2: 159. 1789

Varieties 2 (1 in the flora): introduced, Florida; s, se Asia, Pacific Islands (Hawaii, Philippines).

A. J. Scott (1978c) assigned escaped Florida populations to var. *tomentosa* from southeast Asia to the Philippines; var. *parviflora* Craib occurs in India and Sri Lanka.

1a. Rhodomyrtus tomentosa (Aiton) Hasskarl var. tomentosa [F] [I] [W]

Shrubs or small trees, to 2 m, young growth densely tomentose, hairs white or yellowish, curled. **Leaves:** blade elliptic to oblong-elliptic, 3.5–7 × 2–4 cm, venation with 3 strong, subequal veins, midvein and 2 arcing veins that arise from midvein a few mm distal to base and unite with it near apex, base cuneate, apex rounded to obtuse. **Peduncles** 8–15 mm. **Inflorescences:** bracteoles persistent after anthesis, broadly ovate, 1–3 mm. **Flowers:** calyx lobes hemiorbiculate to ovate, unequal, to ca. 5 mm; petals elliptic-oblong, to ca. 2 cm; filaments pink or red; ovules biseriate, each ovule partially enclosed in a pocket of tissue that remains after ovule is extracted. **Berries** 10–15 mm. **Seeds** 20+, flattened horizontally.

Flowering spring–early summer (or longer). Pinelands, *Taxodium* swamps, disturbed forests; 0–30 m; introduced; Fla.; se Asia (Borneo, Celebes Islands, Java, Lesser Sunda Islands, Malaya, Moluccas, Sumatra); Pacific Islands (Hawaii, Philippines).

Variety *tomentosa* is known in the flora area from Collier County and the central peninsula.

13. SYZYGIUM P. Browne ex Gaertner, Fruct. Sem. Pl. 1: 166, plate 33, fig. 1. 1788, name conserved • [Greek *syzgios*, joined, alluding to paired leaves and branches] [I]

Fred R. Barrie

Leslie R. Landrum

Caryophyllus Linnaeus, name rejected; *Jambosa* Adanson, name rejected

Trees or shrubs, mostly glabrous. **Leaves** opposite; blade venation brochidodromous. **Inflorescences** (1 or)2–100-flowered, terminal or axillary, dichasia, panicles, or racemes. **Flowers** [3 or]4(or 5)-merous, sessile or pedicellate; bud turbinate, clavate, or obovoid; hypanthium forming a tube, prolonged well beyond summit of ovary, base often attenuate; calyx lobes usually distinct and well developed, in opposing subequal to equal pairs, rarely calyptrate and circumscissile at anthesis; petals white, distinct or coherent and calyptrate and falling as a unit at anthesis; stamens 50–300[–500], borne in a ring surmounting hypanthium; ovary 2(–4)-locular; style often persistent in developing fruit; ovules 2–90. **Fruits** berries, red, purple, purple-black, yellow, or reddish, globose, ellipsoid, or subglobose, usually excavated apically; calyx lobes persistent or caducous. **Seeds** usually 1[or 2, rarely 3–5], reniform to subglobose; seed coat membranous; embryo subglobose to reniform; cotyledons distinct, plano-convex, thick. $x = 11$.

Species ca. 1000 (3 in the flora): introduced; s, se Asia, Africa, Pacific Islands (New Guinea), Australia.

Many species of *Syzygium* are widely cultivated throughout the tropics as fruit or ornamental trees or as hedges. They are not tolerant of cold. Two species are naturalized in southern Florida and another in southern California.

1. Inflorescences 15–100-flowered, panicles; floral buds 4–5 mm; calyx lobes 0.5 × 0.5 mm, caducous; petals coherent . . . . 2. *Syzygium cumini*
1. Inflorescences (1 or) 2–8-flowered, dichasia or racemes; floral buds 6–30 mm; calyx lobes 2–8 × 2–10 mm, persistent; petals distinct.
    2. Leaf blades obovate or elliptic, 3–9 × 1.2–3.2 cm; petals 3–5 mm diam.; berries 1.4–2.3 cm; California . . . . 1. *Syzygium australe*
    2. Leaf blades narrowly elliptic or lanceolate, 12–24 × 3–5 cm; petals 10–15 mm diam.; berries 3–4 cm; Florida . . . . 3. *Syzygium jambos*

1. Syzygium australe (J. C. Wendland ex Link) B. Hyland, Austral. J. Bot., Suppl. Ser. 9: 55. 1983 [I]

Eugenia australis J. C. Wendland ex Link, Enum. Hort. Berol. Alt. 2: 28. 1822

Trees or shrubs usually to 3 m; older branches terete or nearly so; twigs weakly compressed, distally 4-winged or ribbed, wings merging in pairs, forming pocketlike structure just distal to many leaf nodes and decussate with petioles of that node; bark tan, flaky. **Leaves:** blade obovate or elliptic, 3–9 × 1.2–3.2 cm, base cuneate to narrowly so, apex acute or abruptly acuminate, mucronate, surfaces glandular or eglandular, glands sparse abaxially, small, obscure, or absent adaxially. **Inflorescences** 3–7-flowered, terminal, also axillary in distal leaf axils, dichasia; axis 10–15 mm; bracts deciduous well before anthesis (leaving prominent scar); bracteoles early deciduous. **Flowers** sessile or pedicellate; bud clavate, 6–10 mm; hypanthium narrowly obconic-campanulate; calyx lobes persistent, ovate, in subequal pairs, 2–3 × 2–4 mm, margins scarious, apex bluntly acute to rounded; petals distinct, orbiculate, 3–5 mm diam., margins scarious, apex rounded; stamens 100–150, ca. 10 mm; style 7–24 mm. **Berries** red or purple, globose or ellipsoid, 14–23 mm.

Flowering late summer–winter. Disturbed riparian areas; 0–50 m; introduced; Calif.; Australia.

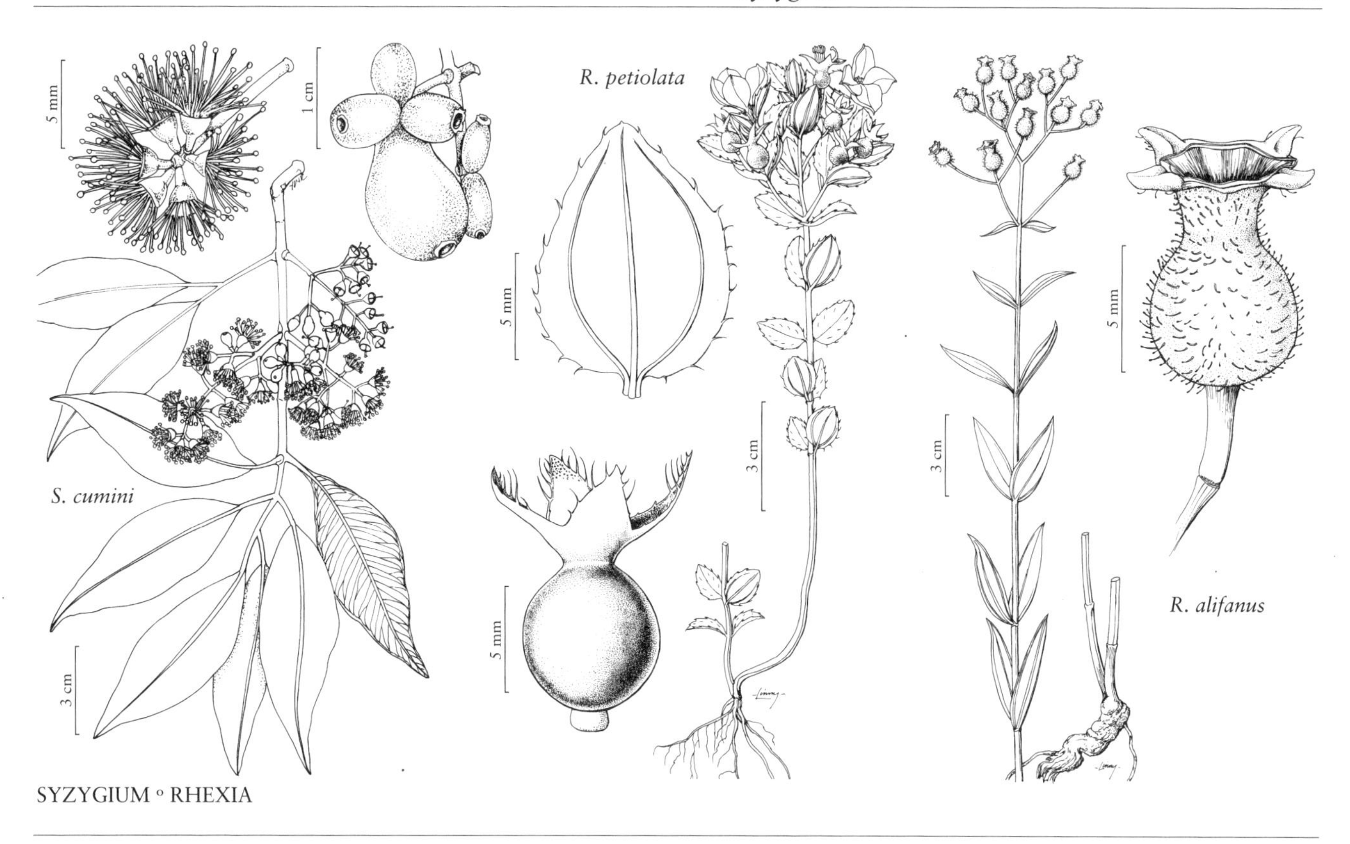

SYZYGIUM ◦ RHEXIA

Syzygium australe is known in the flora area from Los Angeles to San Diego in southern California,

Syzygium australe is sometimes confused with *S. paniculatum* Gaertner, which also is commonly cultivated in California and naturalized near San Diego. *Syzygium paniculatum* differs from *S. australe* in twigs not winged or ribbed and in not having a pocketlike structure just distal to leaf nodes, decussate with the petioles at that leaf node. The seeds of *S. paniculatum* are commonly polyembryonic.

2. **Syzygium cumini** (Linnaeus) Skeels, U.S.D.A. Bur. Pl. Industr. Bull. 248: 25. 1912 • Java plum [F] [I]

Myrtus cumini Linnaeus, Sp. Pl. 1: 471. 1753; *Calyptranthes oneillii* Lundell; *Eugenia cumini* (Linnaeus) Druce; *E. jambolana* Lamarck

Trees or shrubs 6–20+ m; twigs weakly compressed; bark white, smooth. **Leaves:** blade drying concolorous olive or tan, ovate, elliptic, or oblong, 8–17 × 3.5–7 cm, leathery, base cuneate, obtuse, or rounded, apex acuminate or obtuse, tip bluntly acute, surfaces glandular, glossy adaxially, glands small, often punctiform, numerous, sometimes more so abaxially. **Inflorescences** 15–100-flowered, axillary, panicles of dichasia, 1–3 times compound; axis 15–60 mm, lateral branches 5–20 mm, axis and branches compressed, glandular; bracts and bracteoles caducous. **Flowers** sessile at tips of lateral branches; bud pyriform, 4–5 mm; hypanthium obconic to narrowly campanulate, 3–5 mm; calyx lobes caducous, leaving round crateriform scar at ovary summit, equal, 0.5 × 0.5 mm; petals coherent, forming a calyptra, falling as a unit at anthesis; stamens 50–100, 3–5 mm; style 6–7 mm. **Berries** purple black, ellipsoid, 15–20 mm; calyx tube reduced to persistent, apical ring, 1–2 mm diam. $2n = 22$.

Flowering spring, summer. Disturbed areas, often near fresh water; 0–40 m; introduced; Fla.; se Asia (including India); introduced also elsewhere in tropics.

Syzygium cumini is known in the flora area from the central and southern peninsula.

Pimenta dioica has become established near Miami, Florida, and is perhaps most similar to *Syzygium cumini*. Of the berry fruited species, only these two have many-flowered panicles. *Pimenta dioica* is most easily distinguished from *S. cumini* by having pubescent (versus glabrous) flowering hypanthia, leaves with 10–15 prominent lateral veins (versus numerous weak lateral veins), embryos with a hypocotyl much longer than the cotyledons, and leaves with a strong spicy aroma when crushed.

3. **Syzygium jambos** (Linnaeus) Alston in H. Trimen et al., Handb. Fl. Ceylon 6: 115. 1931 • Malabar plum, rose-apple [I]

Eugenia jambos Linnaeus, Sp. Pl. 1: 470. 1753

Trees or shrubs to 10 m; twigs terete or quadrangular; bark reddish brown, flaky. **Leaves:** blade drying concolorous olive or adaxially dark green, narrowly elliptic or lanceolate, 12–24 × 3–5 cm, leathery, veins brown abaxially, base narrowly cuneate or gradually rounded, apex narrowly acuminate, surfaces glandular, glands numerous abaxially, obscure adaxially. **Inflorescences** usually 2–8-flowered, usually terminal, racemes; axis terete or quadrangular, 10–20 × 2–3 mm; bracts and bracteoles caducous. **Flowers** pedicellate (pedicels 7–15 mm); bud pyriform, 25–30 mm; hypanthium infundibular or obconic, 12–17 mm, tube 4–5 mm deep, 9–10 mm diam. at mouth, base abruptly contracted to pseudostalk, 3–5 mm; calyx lobes persistent, widely elliptic, in subequal pairs, 4–8 × 6–10 mm, convex, leathery, margins scarious, apex rounded; petals distinct, orbiculate, 10–15 mm diam., margins scarious, prominently glandular; stamens ca. 300, 20–40 mm; style 40–60 mm. **Berries** yellow or reddish, subglobose, 30–40 mm; calyx lobes persistent, erect in developing fruit. 2*n* = 22.

Flowering mainly in spring. Disturbed wooded areas; 0–50 m; introduced; Fla.; se Asia (Malaysia); introduced also widely elsewhere.

Syzygium jambos is known in the flora area from the central and southern peninsula.

Syzygium jambos is cultivated in tropical areas worldwide as an ornamental for its glossy, deep green leaves and showy flowers.

MELASTOMATACEAE Jussieu

• Melastome Family

Guy L. Nesom

Herbs, perennial, shrubs, or trees [lianas], usually synoecious, [rarely androdioecious], terrestrial, unarmed, not laticiferous, without colored juice, sometimes clonal. **Stems** erect, ascending-erect, or procumbent, often 4-angled when young; bud scales present. **Leaves** deciduous or persistent, usually opposite and decussate, 1 of each pair slightly smaller, rarely verticillate or alternate (by abortion of 1 of each pair), simple; stipules absent; petiolate, sessile, or subsessile; blade venation palmate or parallel, no dominant midrib, the several, strong veins diverging at base, converging at apex (acrodromous), usually strongly cross-veined, margins entire, subentire, serrate, serrulate, or crenulate. **Inflorescences** terminal or axillary, cymes or paniculate cymes (simple or compound dichasia), [umbels, corymbs, cincinni, or, rarely, fascicles, spikes, or flowers solitary]; bracts present, sometimes persistent and color conspicuous; bracteoles present, usually caducous, opposite. **Flowers** actinomorphic, androecium often slightly zygomorphic; subsessile or pedicellate; perianth and androecium perigynous, biseriate; hypanthium urceolate, campanulate, or globose-urceolate [funnelform, cyathiform]; calyx lobes [0, 3] 4 or 5 [or 6], usually as teeth or lobes on hypanthium rim, valvate (rarely connate), intersepalar segments present; petals (3 or)4 or 5[or 6], equal number to sepals, distinct, imbricate in bud; nectary glands absent [present]; stamens [4, 5] 8 or 10 [12–96], usually 2 times number of petals and in 2 whorls (except 1 whorl in *Tetrazygia*), or, rarely, equal to petals, isomorphic or dimorphic; filaments distinct, exserted, often geniculate, free of perianth lobes, equal or conspicuously unequal, often twisted, bringing anther to 1 side; anthers basifixed, not versatile, introrse, dehiscent by 1 or 2 apical pores [or by longitudinal slits]; staminodes (0 or)4 or 5; pistil 1, 4- or 5-carpellate; ovary inferior or semi-inferior, 3–5[–14]-locular; placentation axile [parietal or free-central]; style 1; stigma 1, capitate [truncate]; ovules (2–)6–50 per locule, anatropous or (*Rhexia*) orthotropous, bitegmic, crassinucellate. **Fruits** berries or capsules, dehiscent or indehiscent. **Seeds** 20–100, tawny to purple or black, cochleate or cuneate; endosperm absent. x = 7–18+.

Genera 150–170, species ca. 5000 (3 genera, 15 species in the flora): North America, Mexico, West Indies, Central America, South America, Asia, Indian Ocean Islands (Madagascar), Pacific Islands, Australia.

Species of Melastomataceae occur primarily in tropical and subtropical regions, especially in South America. *Miconia* Ruiz & Pavon, with ca. 1300 species, is the largest genus in the family; boundaries distinguishing potential generic-level taxa among these species are not clearly resolved. Most Melastomataceae can be recognized as members of the family by their opposite, decussate leaves with acrodromous venation (with three or more primary, arcuate longitudinal veins converging toward the apices and cross-veins at right angles to the primary veins), radially symmetric and diplostemonous flowers, apically dehiscent anthers, stamens often with enlarged and/or appendaged connectives, and relatively numerous small seeds. The species are notable for their diversity of hair types and modifications of the androecium. Molecular-based phylogenies suggest that berries have evolved from capsules at least four times within the family.

Species of some genera of Melastomataceae are grown as ornamentals in warm climates. *Tibouchina* Aublet (ca. 350 species) is particularly well represented among the cultivars in the United States.

Six genera with ca. 430 species have been treated as a tribe or subfamily of Melastomataceae, or as the separate family Memecylaceae (two of the genera, *Mouriri* Aublet and *Votomita* Aublet, are in South America, and four are in southeast Asia; none of them is in the flora area). These plants have primarily pinnate or brochidodromous venation, less commonly acrodromous, which is the ancestral state (R. D. Stone 2006). S. S. Renner (1993) treated them as Memecylaceae; molecular data (G. Clausing and Renner 2001) suggested that the group, along with the closely related *Pternandra* Jack of southeast Asia, is basal to the Melastomataceae. Angiosperm Phylogeny Group (2009) have placed this group within the Melastomataceae.

SELECTED REFERENCES Clausing, G. and S. S. Renner. 2001. Molecular phylogenetics of Melastomataceae and Memecylaceae: Implications for character evolution. Amer. J. Bot. 88: 486–498. Renner, S. S. 1993. Phylogeny and classification of the Melastomataceae and Memecylaceae. Nordic J. Bot. 13: 519–540. Renner, S. S., G. Clausing, and K. Meyer. 2001. Historical biogeography of Melastomataceae: The roles of Tertiary migration and long-distance dispersal. Amer. J. Bot. 88: 1290–1300. Stone, R. D. 2006. Phylogeny of major lineages in Melastomataceae, subfamily Olisbeoideae: Utility of nuclear glyceraldehyde 3-phosphate dehydrogenase (*GapC*) gene sequences. Syst. Bot. 31: 107–121. Wurdack, J. J. and R. Kral. 1982. The genera of Melastomataceae in the southeastern United States. J. Arnold Arbor. 63: 429–439.

1. Herbs; petals 4; fruits capsules; seeds usually cochleate . . . . 1. *Rhexia*, p. 366
1. Shrubs or small trees; petals 4 or 5; fruits berries or capsules; seeds cuneate or cochleate.
   2. Leaf surfaces with dense scales abaxially, glabrous adaxially; petals white or pink; fruits berries; seeds cuneate . . . . 2. *Tetrazygia*, p. 373
   2. Leaf surfaces usually strigose; petals light to dark pink, lavender, or purple; fruits fleshy and irregularly splitting-dehiscent; seeds cochleate . . . . 3. *Melastoma*, p. 375

1. RHEXIA Gronovius in C. Linnaeus, Sp. Pl. 1: 346. 1753; Gen. Pl. ed. 5, 163. 1754 • Meadow beauty, deergrass [Greek *rhexis*, rupture, alluding to reputed astringent property to cure wounds]

Herbs, sometimes suffrutescent; caudices relatively short and woody when present, sometimes becoming spongy-thickened when submerged; roots fibrous, root tubers produced by the roots in some species on the primary root at base of stem or at irregular positions on secondary roots. **Stems** erect to ascending-erect, ± 4-angled, faces subequal or unequal, if unequal then 1 opposing set convex, broader, the other concave, narrower, the different morphologies alternating at 90° at each node, usually hirsute to villous, hairs bristlelike, often gland-tipped; bark (if any) thin, exfoliating distally. **Leaves** petiolate, sessile, or subsessile; blade usually 3-veined (1-veined in *R. cubensis*), without strong cross veins, margins entire, subentire, serrate, serrulate, or crenulate, usually with bristle-tipped teeth, rarely gland-tipped, surfaces strigose, villous, hirsute, or glabrous. **Inflorescences** terminal, cymes or appearing as secund racemes (through abortion of inner branches); bracts often deciduous, subfoliaceous. **Flowers** subsessile or short-pedicellate; hypanthium urceolate to campanulate; calyx 4-lobed, lobes triangular; petals fugacious, 4, ascending or spreading, asymmetric, lavender, lavender-rose, lavender-purple, purple, pink, white, or yellow, short-clawed, midvein extending as a slender, multicellular hair; stamens 8, subequal, in 2 whorls, connective bases appendaged; anthers straight or downcurved, linear to linear-lanceolate, 4-locular (1-locular at anthesis through breakdown of septae), apically to subapically poricidal; ovary inferior, adnate to floral tube except apically, 4-locular; style exserted, curved, linear. **Fruits** capsules loculicidal, enclosed within hypanthium, dehiscent. **Seeds** cochleate (except cuneate-prismatic in *R. alifanus*). $x = 11$.

Species 13 (13 in the flora): North America, West Indies (Greater Antilles).

Hybrids are common between some of the species with elongate anthers (sect. *Rhexia*) and identifications sometimes are arbitrary. There is no evidence that the other species are involved in hybridization. Tuberous swellings on the roots are produced in some species of sect. *Rhexia*; presence of these root tubers appears to vary within and among populations.

Rhexia alifanus (sect. *Cymborhexia*) is the only species of the genus with glabrous, isofacial, subentire, and glaucous leaf blades, caducous bracts, and relatively large, oblong-cuneate, subprismatic seeds (versus small, cochleate seeds in the other species), but it shares with sect. *Rhexia* (perhaps as a plesiomorphic feature) elongate, curved, small-pored anthers.

Polyploidy occurs in many species of sect. *Rhexia*. In *R. cubensis* (2*x*, 4*x*, 6*x*), *R. nashii* (4*x*, 6*x*), and *R. virginica* (2*x*, 4*x*), conspecific plants of different ploidy apparently occur sympatrically and are completely reproductively isolated (no seeds formed in experimental crosses), but there are no obvious morphological differences among them (see chromosome counts, geography, and crossing data in R. Kral and P. E. Bostick 1969). The phylogenetic analysis by G. M. Ionta et al. (2007) indicates that the evolutionary origin of *R. cubensis*, *R. nashii*, *R. parviflora*, and *R. salicifolia* (sect. *Rhexia*), as well as *R. lutea*, probably was through hybridization.

The biological situation is different in *Rhexia mariana* (in the broad sense, as interpreted by R. Kral and P. E. Bostick 1969). Variety *mariana* is diploid and is completely reproductively isolated from the two tetraploid varieties (vars. *interior* and *ventricosa*); experimental crosses between vars. *interior* and *ventricosa* fail to produce viable seeds. Varieties *interior* and *ventricosa* are morphologically similar and completely allopatric; each differs from var. *mariana* in a prominent feature of stem morphology—typical *R. mariana* has unequal stem faces while each of the two tetraploids has equal faces. These two taxa are treated here at specific rank, apart from *R. mariana* in the strict sense, following C. W. James (1956).

Based on molecular evidence (G. Clausing and S. S. Renner 2001), *Rhexia* appears to be sister to *Arthrostemma* Pavon ex D. Don, a genus of perennial herbs native to Central America and northwestern South America. *Arthrostemma ciliatum* Pavon ex D. Don has become a damaging invader on the Hawaiian Islands.

The key to *Rhexia* species below is artificial, a nearly inevitable requirement to separate some of the species.

SELECTED REFERENCES Ionta, G. M. et al. 2007. Phylogenetic relationships in *Rhexia* (Melastomataceae): Evidence from DNA sequence data and morphology. Int. J. Pl. Sci. 168: 1055–1066. James, C. W. 1956. A revision of *Rhexia* (Melastomataceae). Brittonia 8: 201–230. Kral, R. and P. E. Bostick. 1969. The genus *Rhexia* (Melastomataceae). Sida 3: 387–440. Nesom, G. L. 2012. Infrageneric classification of *Rhexia* (Melastomataceae). Phytoneuron 2012-15: 1–9.

1. Petals golden-yellow . . . . 1. *Rhexia lutea*
1. Petals lavender, lavender-rose, lavender-purple, purple, pink, or white.
    2. Inflorescences condensed, mostly obscured by foliaceous bracts; leaf blades ovate to short-elliptic or suborbiculate, 1–2 cm; anthers 1.2–2 mm; roots relatively short, fibrous, lignescent and non-tuberiferous.
        3. Calyx lobes deltate, apices obtuse; hypanthia villous-hirsute, hairs gland-tipped; seeds irregularly ridged . . . . 2. *Rhexia nuttallii*
        3. Calyx lobes oblong-lanceolate, apices acute; hypanthia mostly glabrous except along calyx lobes; seeds pebbled or with ridges of domelike processes . . . . 3. *Rhexia petiolata*
    2. Inflorescences diffuse, not obscured by bracts; leaf blades lanceolate to elliptic, ovate, oblong, oblanceolate, linear, or spatulate, 1.5–7.5 cm; anthers 3–11 mm; roots relatively long, rhizomelike, sometimes tuberiferous, sometimes lignescent (except short and fibrous in *R. alifanus*).
        4. Stem internodes usually glabrous.
            5. Stem internodes and nodes glabrous; leaf blade margins entire or subentire, teeth remote, low, blunt not bristle-tipped, surfaces glabrous . . . . 4. *Rhexia alifanus*
            5. Stem internodes glabrous, nodes hirsute to villous; leaf blade margins serrate or crenulate, teeth bristle-tipped, surfaces glabrous, glabrate, strigose, hirsute, or villous.
                6. Anthers 3–3.5 mm . . . . 5. *Rhexia parviflora* (in part)
                6. Anthers 5–8 mm.
                    7. Leaf blades 3–9 mm wide, margins shallowly serrate to barely crenulate; hypanthia hispid-hirsute at neck, rim, and calyx lobes, hairs eglandular, yellowish . . . . 6. *Rhexia aristosa*
                    7. Leaf blades (5–)8–20(–35) mm wide, margins serrate or finely serrate; hypanthia glabrous, glabrate, hirsute-villous, or sparsely villous, hairs gland-tipped.
                        8. Leaf blades (7–)10–20(–35) mm wide; stem faces subequal, angles narrowly winged . . . . 8. *Rhexia virginica* (in part)
                        8. Leaf blades (5–)8–15(–20) mm wide; stem faces strongly unequal . . . . 11. *Rhexia mariana* (in part)
        4. Stem internodes hirsute to villous or hispid-villous.
            9. Stem faces unequal.
                10. Hypanthia glabrous or glabrate (except calyx rims and lobes); petals 2–2.5 cm . . . . 10. *Rhexia nashii*
                10. Hypanthia usually hirsute-villous or hirsute; petals 1.2–2 cm.
                    11. Hypanthia (10–)14–15(–16) mm . . . . 9. *Rhexia cubensis*
                    11. Hypanthia 6–10 mm . . . . 11. *Rhexia mariana* (in part)
            9. Stem faces subequal.
                12. Anthers 3–3.5 mm; petals white or pale lavender . . . . 5. *Rhexia parviflora* (in part)
                12. Anthers 5–8 mm; petals pink to lavender-rose, lavender-purple, or purple.
                    13. Leaf blades 1–5 mm wide, margins entire or minutely crenulate, ciliate, with gland-tipped hairs; petals 1.1–1.2 cm; hypanthia (4–)5–7(–8) mm . . . . 7. *Rhexia salicifolia*
                    13. Leaf blades (7–)10–20(–35) mm wide, margins serrate to serrulate, not ciliate; petals 1.2–2 cm; hypanthia 6–10 mm.
                        14. Stem angles narrowly winged; stems usually unbranched or few-branched proximally; stem internodes and hypanthia usually sparsely villous, sometimes glabrous . . . . 8. *Rhexia virginica* (in part)
                        14. Stem angles sharp, without wings or very narrowly winged; stems unbranched or few- to several-branched distally; stem internodes hirsute-villous, hypanthia hirsute-villous, glabrous, or glabrate.
                            15. Seed surfaces with papillae in concentric rows; Atlantic coast states . . . . 12. *Rhexia ventricosa*
                            15. Seed surfaces irregularly ridged in concentric rows or with laterally flattened, domelike processes; c, s United States . . . . 13. *Rhexia interior*

1. Rhexia lutea Walter, Fl. Carol., 130. 1788 • Yellow or golden meadow beauty [E]

Caudices developed; roots short, fibrous, lignescent, non-tuberiferous. **Stems** branched proximally, 10–40 cm, faces subequal, flat to convex, 4-angled distally from midstem, internodes and nodes hirsute, hairs eglandular. **Leaves** subsessile; blade spatulate to oblanceolate or elliptic, 2–3 cm × 2–8 mm, 2 lateral veins marginal on narrower leaves, margins subentire to shallowly serrate, surfaces loosely strigose, hairs yellowish. **Inflorescences** diffuse, not obscured by bracts. **Flowers:** hypanthium globose, much longer than the constricted neck, 6–7 mm, hirsute to villous, eglandular; calyx lobes triangular, apices aristate; petals ascending, golden-yellow, 1–1.5 cm; anthers straight, 2 mm. **Seeds** 0.7 mm, surfaces with few straight ridges of papillae along crest, sides with lower, more scattered papillae or ± smooth. **2*n*** = 44.

Flowering Apr–Jul. Wet pine flatwoods and savannas, slash pine scrub, cypress pond margins, seepage slopes, bogs, clearings, openings, sandy peat; 0–50 m; Ala., Fla., Ga., La., Miss., N.C., S.C., Tex.

2. Rhexia nuttallii C. W. James, Brittonia 8: 214. 1956 • Nuttall's meadow beauty [E]

Rhexia serrulata Nuttall, Gen. N. Amer. Pl. 1: 243. 1818, not Richard 1813

Caudices developed; roots short, fibrous, lignescent, non-tuberiferous. **Stems** branched proximally, 10–30(–40) cm, faces subequal, flat to convex, 4-angled, internodes glabrous, nodes sparsely hirsute, hairs eglandular. **Leaves:** petiole 0.5–1.5 mm; blade ovate to suborbiculate, 1–1.5 cm × 3–10 mm, margins serrate, surfaces glabrous abaxially, sparsely villous adaxially. **Inflorescences** condensed, mostly obscured by foliaceous bracts. **Flowers:** hypanthium globose, much longer than the constricted neck, 5–7 mm, villous-hirsute, hairs gland-tipped; calyx lobes deltate, apices obtuse, hirsute, hairs gland-tipped; petals ascending, lavender-rose, 1–1.2 cm; anthers straight, 2 mm. **Seeds** 0.6 mm, surfaces with relatively short, interrupted, irregularly spaced, ± longitudinal ridges. **2*n*** = 22.

Flowering Jun–Aug(–Sep). Pine flatwoods, bogs, seeps, pond margins, clearings, ditch banks, roadside depressions, sandy peat; 0–50 m; Fla., Ga.

3. Rhexia petiolata Walter, Fl. Carol., 130. 1788 • Fringed or short-stemmed meadow beauty [E] [F]

Rhexia ciliosa Michaux

Caudices developed; roots short, fibrous, lignescent, non-tuberiferous. **Stems** unbranched or few-branched, 10–50 cm, faces subequal, flat to convex, angles weakly ridged, internodes glabrous, nodes sparsely hirsute, hairs eglandular. **Leaves:** petiole 0.5–1.5 mm; blade ovate to short-elliptic or suborbiculate, 1–2 cm × 4–14 mm, margins serrate, surfaces glabrous abaxially, sparsely villous adaxially. **Inflorescences** condensed, mostly obscured by foliaceous bracts. **Flowers:** hypanthium globose, much longer than the constricted neck, 5–7(–9) mm, mostly glabrous except along calyx lobes; calyx lobes oblong-lanceolate, apices acute, spreading-ciliate, eglandular; petals ascending to divergent, lavender-rose, 1–2 cm; anthers straight, 1.2–1.8 mm. **Seeds** 0.6 mm, surfaces pebbled or with ridges of domelike processes. **2*n*** = 22.

Flowering Jun–Sep. Wet pine flatwoods and savannas, pine-cypress flats, cypress-pine-gum flats, cabbage palm hummocks, hillside bogs, swales, swamp and pocosin borders, borrow pits, ditches, roadsides, disturbed sites, sandy peat; 0–50 m; Ala., Fla., Ga., La., Md., Miss., N.C., S.C., Tex., Va.

4. Rhexia alifanus Walter, Fl. Carol., 130. 1788 • Savannah or smooth meadow beauty [E] [F]

Caudices developed, spongy; roots short, fibrous, lignescent, non-tuberiferous. **Stems** mostly unbranched, 6–11(–20) cm, faces equal, terete proximally, distal internodes somewhat flattened in a plane parallel to subtending leaf pair, longitudinally striate, narrower bands paler and aligned with leaf midribs, internodes and nodes glabrous. **Leaves** subsessile; blade lanceolate-ovate to lanceolate or elliptic, 3.5–7.5 cm × 4–15 mm, margins entire or subentire, teeth low, remote, blunt, not bristle-tipped, surfaces glabrous. **Inflorescences** diffuse, not obscured by bracts. **Flowers:** hypanthium subglobose to ovoid, longer than the constricted neck, 7.5–10 mm, hirsute, hairs gland-tipped; calyx lobes triangular, apices narrowed to linear-oblong extensions; petals spreading, lavender-rose, 2–2.5 cm; anthers curved, 7–8 mm. **Seeds** cuneate-prismatic, 1–2 mm, surfaces nearly smooth. **2*n*** = 22.

Flowering May–Sep. Pine flatwoods and savannas, pine-palmetto, pine-hardwood, pine-cypress, gum-maple flats, bogs, seeps, swamp and pocosin margins, ditches, roadsides, sandy peat; 0–50 m; Ala., Fla., Ga., La., Miss., N.C., S.C., Tex.

5. Rhexia parviflora Chapman, Fl. South. U.S. ed. 3, 156. 1897 • White or small-flowered white meadow beauty [E]

Caudices not developed; roots often long and rhizomelike, lignescent, non-tuberiferous. **Stems** mostly unbranched proximally, usually branched distally, 10–40 cm, faces subequal, angles weakly ridged, internodes glabrous or sparsely hirsute, hairs minutely gland-tipped, nodes hirsute, hairs eglandular. **Leaves:** petiole 1–5 mm; blade broadly ovate to elliptic, 1.5–3 cm × 4–8 mm, margins finely serrate, surfaces sparsely strigose. **Inflorescences** diffuse, not obscured by bracts. **Flowers:** hypanthium subglobose, much longer than the constricted neck, 5–7 mm, sparsely hirsute apically, hairs gland-tipped; calyx lobes triangular, apices short-acuminate; petals spreading, white or pale lavender, 0.8–1(–1.3) cm; anthers slightly curved, 3–3.5 mm. **Seeds** 0.6 mm, surfaces crested with irregular, roughly concentric, interrupted rows of laterally flattened tubercles, these in turn vertically grooved. $2n = 22$.

Flowering Jul–Aug. Swamp and lime sinkpond margins, gum swamps, cypress-titi swamps and depressions, ditches, roadsides; 0–50 m; Ala., Fla., Ga.

Rhexia parviflora is known from ten counties in the Florida panhandle and one adjacent county each in Alabama (Geneva County) and Georgia (Baker County).

6. Rhexia aristosa Britton, Bull. Torrey Bot. Club 17: 14, plate 99. 1890 • Awned or bristly meadow beauty [E]

Caudices usually not developed; roots often long and rhizomelike, lignescent, tuberiferous. **Stems** unbranched to several-branched, 40–70 cm, faces subequal, angles weakly ridged, sometimes narrowly winged, internodes glabrous, nodes sparsely hirsute to villous, hairs yellowish, eglandular. **Leaves** sessile to subsessile; blade usually lanceolate, 2–3 cm × 3–9 mm, margins shallowly serrate to barely crenulate, apex sometimes minutely apiculate, surfaces glabrous or sparsely long-strigose to ascending-villous. **Inflorescences** diffuse, not obscured by bracts. **Flowers:** hypanthium ovoid, about as long as the constricted neck, 7–10 mm, hispid-hirsute at neck, rim, and calyx lobes, hairs yellowish, eglandular; calyx lobes triangular, apices acuminate-aristate; petals spreading, dull lavender, 1–2 cm; anthers curved, 5–6 mm. **Seeds** ca 0.7 mm, surfaces with irregular, concentric ridges, with few isolated domes or papillae. $2n = 22$.

Flowering Jun–Aug(–Sep). Carolina bays, depression meadows, wetland margins, borrow pits, lime sinkponds, cypress flats and swamps, oak-pine-cypress, gum with slash and longleaf pines, sandy peat and clay; 0–50 m; Ala., Del., Ga., Md., N.J., N.C., S.C., Va.

7. Rhexia salicifolia Kral & Bostick, Sida 3: 402, fig. 4. 1969 • Panhandle or willowleaf meadow beauty [E]

Caudices not developed; roots often long and rhizomelike, lignescent, tuberiferous. **Stems** usually several-branched distally, 20–55 cm, faces subequal, angles narrowly winged, internodes and nodes hirsute-villous, hairs gland-tipped. **Leaves** sessile; blade narrowly elliptic or narrowly oblong to narrowly oblanceolate or linear, 1.5–4 cm × 1–5 mm, lateral veins relatively short or, in narrower leaves, absent, margins entire or minutely crenulate, ciliate, with gland-tipped hairs, apex sometimes apiculate, surfaces sparsely to moderately hirsute, hairs gland-tipped. **Inflorescences** diffuse, not obscured by bracts. **Flowers:** hypanthium globose, longer than the constricted neck, (4–)5–7(–8) mm, sparsely hirsute-villous, hairs gland-tipped; calyx lobes narrowly triangular, apices acute; petals spreading, pink to lavender-rose or purple, 1.1–1.2 cm; anthers curved, 4–5 mm. **Seeds** 0.7 mm, surfaces with 3–5 prominent, broad, symmetrical or tortuous longitudinal ridges or contiguous, domelike tubercles in rows. $2n = 22$.

Flowering Jun–Aug(–Sep). Inlet, pond, and lake shores, lime sinkpond margins, interdune swales, depressions, borrow pits, sandhills, longleaf pine savannas, slash pine flats, longleaf pine-turkey oak woods; 0–50 m; Ala., Fla.

8. Rhexia virginica Linnaeus, Sp. Pl. 1: 346. 1753 • Virginia meadow beauty, deergrass, handsome Harry [E]

Rhexia virginica var. *purshii* (Sprengel) C. W. James; *R. virginica* var. *septemnervia* (Walter) Pursh

Caudices barely developed; roots often long and rhizomelike, tuberiferous. **Stems** usually unbranched or few-branched proximally, 40–100 cm, faces subequal, angles narrowly winged, internodes and nodes usually sparsely villous, sometimes glabrous, hairs gland-tipped. **Leaves** sessile or subsessile; blade lanceolate to narrowly lanceolate, ovate, or elliptic,

3–5(–7) cm × (7–)10–20(–35) mm, margins finely serrate, surfaces glabrate abaxially, bristly villous adaxially. **Inflorescences** diffuse, not obscured by bracts. **Flowers:** hypanthium globose, longer than the constricted neck, (6–)7–10 mm, glabrous or sparsely villous, then hairs gland-tipped; calyx lobes narrowly triangular, apices acute to acuminate; petals spreading, lavender-rose to lavender-purple, 1.5–2 cm; anthers curved, 5–5.5 mm. **Seeds** 0.7 mm, surfaces low-muricate, papillose, or tuberculate in concentric rows, with sculpturing most prominent toward crest. **2*n*** = 22, 44.

Flowering (May–)Jun–Sep(–Oct). Pine flatwoods and savannas, pine-cypress savannas, bottomland hardwoods, turkey oak-pine flats, streamhead pocosins, hillside bogs, seepages, lake, pond, and stream edges, depressions, ditches, clearings, sandy fields, powerline rights-of-way, sand, sandy peat, sandy clay; 10–500 m; N.S., Ont.; Ala., Ark., Conn., Del., D.C., Fla., Ga., Ill., Ind., Iowa, Ky., La., Maine, Md., Mass., Mich., Miss., Mo., N.H., N.J., N.Y., N.C., Ohio, Okla., Pa., R.I., S.C., Tenn., Tex., Vt., Va., W.Va., Wis.

Plants that have been treated as var. *purshii* differ mainly in reduced vesture, shorter hypanthial neck, and erect-incurved (versus recurved) calyx lobes, but there is too much intergradation and geographic ambiguity to formally recognize variants. The geographic relationship of these plants to more typical ones is closely analogous to that between *R. mariana* vars. *exalbida* and *mariana*, with plants called var. *purshii* mostly in Florida, Georgia, and South Carolina. Polyploidy does not appear to be correlated with morphological variation.

Extremes of *Rhexia virginica* are similar to *R. interior* or *R. ventricosa*, both of which are sympatric with it. Both of the latter, however, are exclusively rhizomatous, have a tendency toward longer, hairier hypanthia, have stems usually without prominently winged angles, and never produce root tubers. Hybrids between the species appear to be common, although experimental crosses between *R. interior* or *R. ventricosa* (treated here as tetraploid *R. mariana*) and tetraploid *R. virginica* produced only non-germinable seeds (R. Kral and P. E. Bostick 1969).

9. Rhexia cubensis Grisebach, Cat. Pl. Cub., 104. 1866
• West Indies meadow beauty

Rhexia floridana Nash; *R. mariana* Linnaeus var. *portoricensis* Cogniaux

Caudices not developed; roots often long and rhizomelike, tuberiferous. **Stems** unbranched or few-branched proximally, 30–60 cm, faces strongly unequal, 1 pair of opposite faces rounded to convex, the other narrower, flat or concave, internodes and nodes sparsely villous, hairs gland-tipped. **Leaves** sessile; blade linear or linear-elliptic to oblong or narrowly spatulate, 2–4 cm × 1–4(–8) mm, apparently 1-veined, margins serrate, surfaces sparsely glandular-hirsute. **Inflorescences** diffuse, not obscured by bracts. **Flowers:** hypanthium ovoid to subglobose, about as long as the constricted neck, (10–)14–15 (–16) mm, sparsely hirsute, hairs gland-tipped; calyx lobes narrowly triangular to oblong-triangular, apices acute; petals spreading, bright to pale lavender-rose, 1.5–2 cm; anthers curved, 7–10 mm. **Seeds** 0.7 mm, surfaces concentrically, evenly ridged, especially along crest. **2*n*** = 22, 44, 66.

Flowering Jun–Aug(–Sep). Pine savannas and flatwoods, pine-cypress flats, lime sinkponds, longleaf pine hills, hillside bogs, seeps, sandy lakeshores and pond edges, swamp margins, salt marsh and canal banks, ditches, roadsides, sandy peat; 0–50 m; Ala., Fla., Ga., La., Miss., N.C., S.C.; West Indies (Cuba, Hispaniola, Puerto Rico).

10. Rhexia nashii Small, Fl. S.E. U.S., 824, 1335. 1903
• Maid Marian, hairy meadow beauty [E]

Rhexia mariana Linnaeus var. *purpurea* Michaux

Caudices not developed; roots often long and rhizomelike, tuberiferous. **Stems** mostly unbranched, 20–150 cm, faces strongly unequal, 1 pair of opposite faces rounded to convex, the other narrower, flat or concave, internodes and nodes hirsute to hirsute-villous or hispid-villous, hairs minutely gland-tipped. **Leaves:** petiole 0.5–1.5 mm; blade ovate to lanceolate-ovate, lanceolate, or elliptic, 3–7 cm × (5–)7–12(–15) mm, margins finely to coarsely serrate, surfaces hirsute. **Inflorescences** diffuse, not obscured by bracts. **Flowers:** hypanthium ovoid to subglobose, about as long as the constricted neck, 10–15(–20) mm, glabrous or glabrate, except calyx rims and lobes hirsute-villous, hairs minutely gland-tipped; calyx lobes narrowly triangular, apices acute to acuminate; petals spreading, dull lavender, 2–2.5 cm; anthers curved, 8–11 mm. **Seeds** 0.7 mm, surfaces with concentric rows of contiguous, dome-shaped, sometimes laterally flattened processes, with sculpturing most prominent toward crest. **2*n*** = 44, 66.

Flowering (May–)Jun–Sep(–Oct). Pine flatwoods and savannas, turkey oak-pine flats, sandhills, hardwood clearings, pond, creek, and swamp edges, marshes, hillside bogs, seepages, borrow pits, ditches, wet roadsides, powerline rights-of-way, sandy peat, sandy clay, sand; 0–50 m; Ala., Fla., Ga., La., Md., Miss., N.C., S.C., Va.

11. Rhexia mariana Linnaeus, Sp. Pl. 1: 346. 1753 • Maryland meadow beauty [E]

Caudices not developed; roots often long and rhizomelike, non-tuberiferous. **Stems** usually few- to several-branched distally, sometimes unbranched, 20–80 cm, faces strongly unequal, 1 pair of opposite faces rounded to convex, the other narrower, flat or concave, without wings or very narrowly winged, internodes and nodes usually hirsute-villous, rarely glabrous or glabrate, hairs gland-tipped. **Leaves** sessile, subsessile, or petioles 0.5–1.5 mm; blade linear, lanceolate, elliptic, or narrowly ovate, rarely linear-filiform, 2–4 cm × (5–)8–15(–20) mm, margins serrate, surfaces loosely strigose to strigose-hirsute or villous. **Inflorescences** diffuse, not obscured by bracts. **Flowers:** hypanthium ovoid to subglobose, about as long as the constricted neck, 6–10 mm, hirsute-villous, glabrous, or glabrate, hairs gland-tipped; calyx lobes narrowly triangular, apices acute to acuminate, narrowed to linear-oblong extensions; petals spreading, white or pale to dull lavender, 1.2–1.5 cm; anthers curved, 5–8 mm. **Seeds** 0.7 mm, surfaces longitudinally ridged with contiguous tubercles, papillae, or laterally flattened domes. $2n = 22$.

Varieties 2 (2 in the flora): sc, e United States.

As noted by R. Kral and P. E. Bostick (1969), *Rhexia mariana* is the most abundant and wide-ranging species of the genus. It is sympatric with every other species and hybridizes with, and often takes on characteristics of, other species.

Variety *exalbida* was formally recognized by C. W. James (1956) as distinct in its white petals and linear leaves and distribution mostly from southern Mississippi to Florida and along the coastal plain to the Carolinas; James noted that differences are quantitative and intergrading. R. Kral and P. E. Bostick (1969) observed that its distinct distribution might support its taxonomic recognition but that intergradation with var. *mariana*, especially in the Florida panhandle across the outer coastal plain to Texas, suggested that only a single entity should be recognized. The geography of chromosome counts reported by Kral and Bostick indicates that vars. *exalbida* and *mariana* are diploid. With the caveat that the decision is subjective, var. *exalbida* is treated here as distinct, emphasizing its geographic concentration in the southeastern corner of the species range.

In the concept of R. Kral and P. E. Bostick (1969), *Rhexia mariana* also includes vars. *interior* and *ventricosa*. The geographic ranges of each of these varieties lie almost completely within that of var. *mariana*; each of the varieties is tetraploid; var. *mariana* is diploid. The morphological differences that separate these entities are subtle but they appear to be consistent, and the ploidal differences probably act as isolating mechanisms. Of the alternatives, to treat all as a single species without subdivisions disregards the biology; to treat them as three varieties disregards the apparent isolation, which usually is a significant feature of a species concept.

1. Petals dull lavender to lavender-rose; leaf blades lanceolate to elliptic or narrowly ovate . . . . 11a. *Rhexia mariana* var. *mariana*
1. Petals white or pale lavender; leaf blades mostly linear. . . . . 11b. *Rhexia mariana* var. *exalbida*

11a. Rhexia mariana Linnaeus var. **mariana** • Pale meadow beauty [E]

Rhexia delicatula Small; *R. mariana* var. *leiosperma* Fernald & Griscom

Leaf blades lanceolate to elliptic or narrowly ovate. **Petals** dull lavender to lavender-rose. $2n = 22$.

Flowering May–Oct. Pine flatwoods, wet meadows, bog margins, ditches, wet roadsides, often weedy areas; 0–50 m; Ala., Ark., Del., D.C., Fla., Ga., Ill., Ind., Ky., La., Md., Mass., Mich., Miss., Mo., N.J., N.Y., N.C., Okla., Pa., S.C., Tenn., Tex., Va., W.Va.

11b. Rhexia mariana Linnaeus var. **exalbida** Michaux, Fl. Bor.-Amer. 1: 221. 1803 • White meadow beauty [E]

Rhexia lanceolata Walter, Fl. Carol., 129. 1788; *R. angustifolia* Nuttall; *R. filiformis* Small; *R. mariana* var. *lanceolata* (Walter) T. F. Wood & G. McCarthy

Leaf blades mostly linear. **Petals** white or pale lavender. $2n = 22$.

Flowering Jun–Sep. Wet pine flatwoods and savannas, wet meadows, pond edges, ditches, wet roadsides; 50–100 m; Ala., Fla., Ga., La., Miss., N.C., S.C., Tex.

12. Rhexia ventricosa Fernald & Griscom, Rhodora 37: 172, plate 346, figs. 1–4. 1935 • Swollen meadow beauty [E]

Rhexia mariana Linnaeus var. *ventricosa* (Fernald & Griscom) Kral & Bostick

Caudices not developed; roots often long and rhizomelike, lignescent, non-tuberiferous. **Stems** unbranched or few- to several-branched distally, 25–80 cm, faces subequal, angles sharp, without wings or very narrowly winged, internodes and nodes hirsute-villous, hairs gland-tipped. **Leaves** sessile, subsessile, or petiole 0.5–1.5 mm; blade broadly to narrowly elliptic to ovate-elliptic or lanceolate, 2–6 cm × 10–20 mm, margins serrate to serrulate, surfaces loosely strigose to strigose-hirsute or villous. **Inflorescences** diffuse, not obscured by bracts. **Flowers:** hypanthium ovoid to subglobose, about as long as the constricted neck, 6–10 mm, hirsute-villous, glabrous, or glabrate, hairs gland-tipped; calyx lobes narrowly triangular, apices acute to acuminate, narrowed to linear-oblong extensions; petals spreading, bright lavender-rose, 1.2–1.5 cm; anthers curved, 5–8 mm. **Seeds** 0.5–0.7 mm, surfaces with papillae in concentric rows. $2n = 44$.

Flowering Jun–Sep. Pine flatwoods and savannas, clearings in cypress-hardwood swamps, ditches, wet roadsides; 0–50 m; Del., Md., N.J., N.C., S.C., Va.

13. Rhexia interior Pennell, Bull. Torrey Bot. Club 45: 480. 1918 • Ozark meadow beauty [E]

Rhexia latifolia Bush, Rhodora 13: 167. 1911, not Aublet 1775; *R. mariana* Linnaeus var. *interior* (Pennell) Kral & Bostick

Caudices not developed; roots often long and rhizomelike, lignescent, non-tuberiferous. **Stems** unbranched or few- to several-branched distally, 40–60 cm, faces subequal, angles sharp, without wings or very narrowly winged, internodes and nodes hirsute-villous, hairs gland-tipped. **Leaves** sessile, subsessile, or petiole 0.5–1.5 mm; blade narrowly to broadly elliptic, 2–6(–7) cm × 10–25 mm, margins serrate to serrulate, surfaces loosely strigose to strigose-hirsute or villous. **Inflorescences** diffuse, not obscured by bracts. **Flowers:** hypanthium ovoid to subglobose, about as long as the constricted neck, 6–10 mm, hirsute-villous, glabrous, or glabrate, hairs gland-tipped; calyx lobes narrowly triangular, apices acute to acuminate, narrowed to linear-oblong extensions; petals spreading, bright lavender-rose, 1.2–1.5 cm; anthers curved, 5–8 mm. **Seeds** 0.5–0.7 mm, surfaces irregularly ridged in concentric rows or with laterally flattened, domelike processes. $2n = 44$.

Flowering Jun–Sep. Moist to wet areas, ditches, prairies; 0–300 m; Ala., Ark., Ill., Ind., Kans., Ky., La., Miss., Mo., Okla., Tenn., Tex.

2. TETRAZYGIA Richard ex de Candolle in A. P. de Candolle and A. L. P. P. de Candolle, Prodr. 3: 172. 1828 • Clover ash [Greek *tessares*, four, and *zygon* or *zygos*, yoke or crossbar, alluding to 4-merous flowers]

Shrubs or trees. Stems erect, terete, lepidote, scales dense, radiate, white or light brown and glabrescent; bark smooth to scaly or shallowly fissured on largest stems. **Leaves** petiolate, subsessile, or sessile; blade with 3 primary veins and prominent cross veins, margins entire, revolute, surfaces lepidote abaxially, glabrous adaxially. **Inflorescences** terminal, paniculate cymes; bracts usually persistent, triangular. **Flowers** pedicellate; hypanthium campanulate [urceolate]; calyx (0 or)5-lobed, lobes rounded to triangular; petals [4 or] 5 [or 6], spreading or slightly deflexed, symmetric, white or pink, often drying yellowish; stamens [8 or] 10 [or 12], subequal, in 1 whorl, connective bases unappendaged, relatively thick, oblong, shorter than thecae; anthers slightly downcurved, linear, 2-locular, apically poricidal; ovary inferior, adnate to floral tube except apically, 3-locular; style curved, linear, exserted; stigma minute, densely papillose. **Fruits** berries. **Seeds** cuneate. $x = 17$.

Species 20–25 (1 in the flora): Florida, West Indies; introduced in Pacific Islands.

Tetrazygia is an Antillean genus of ca. 20 (W. S. Judd and J. D. Skean 1991; Hno. Alain 1982+, vol. 9) or 25 species (R. A. Howard 1974–1989, vol. 5). The genus has been defined conventionally by a vesture of radiate scales, a terminal, paniculate inflorescence, the presence of a strong constriction in the hypanthium proximal to the calyx, and calyx lobes reduced or absent.

Hno. Alain (1957) considered *Tetrazygia* an artificial genus, perhaps better treated as a section of *Miconia* Ruiz & Pavon. Molecular-based phylogenetic analyses (F. A. Michelangeli et al. 2004, 2008) indicate that species of *Tetrazygia* are cladistically intermixed with members of other genera of Miconieae, especially *Pachyanthus* A. Richard and *Calycogonium* de Candolle, both predominantly Antillean. Species with 4-merous flowers may not be closely related to those with 5-merous or 6-merous flowers and conspicuously constricted hypanthia, such as *T. barbata* Borhidi, *T. bicolor*, *T. coriacea* Urban, and *T. lanceolata* Urban, which appear to be cladistically coherent. Other data (Michelangeli et al. 2004) suggest that *Charianthus* D. Don, and possibly *Calycogonium* de Candolle (in part), are derived from within *Tetrazygia*.

Species of *Tetrazygia* with conspicuous, elongated external calyx lobes were transferred by Borhidi to the new genus *Tetrazygiopsis*. In generic realignments of terminal-flowered taxa of Miconieae, W. S. Judd and J. D. Skean (1991) considered *Tetrazygiopsis* as a synonym of *Tetrazygia* and were followed in this assessment by Hno. Alain (1985–1997, vol. 4, Puerto Rico; 1982+, vol. 9, Hispaniola). Judd and Skean suggested that *Tetrazygia* eventually might be broadened to include some species now placed in *Miconia*, redefined by having a distinctive glabrous style, strongly curved apically and with a minute stigma, irrespective of the presence or absence of a constriction between calyx and hypanthium.

1. **Tetrazygia bicolor** (Miller) Cogniaux in A. L. P. P. de Candolle and C. de Candolle, Monogr. Phan. 7: 724. 1891 • Florida clover ash, Florida tetrazygia, West Indies lilac [F]

Melastoma bicolor Miller, Gard. Dict. ed. 8, Melastoma no. 6. 1768

Shrubs or trees 1–5(–10) m. **Stems:** twigs slender, tawny to gray-brown, lepidote; buds minute, globose; bark thin, gray-brown. **Leaves:** petiole 1–2.5 cm, lepidote; blade narrowly ovate-lanceolate to lanceolate or oblong-lanceolate, 5.5–12(–20) cm × 20–35 mm, base rounded to obtuse, apex acuminate, surfaces tawny abaxially, with dense, persistent scales, dark green adaxially. **Inflorescences** 4–10 cm, lepidote. **Pedicels** slender, 5–10 mm. **Flowers:** hypanthium urceolate (subglobose, with a short neck), 4–4.5 mm in flower, 4–5 mm at maturity, lepidote; calyx subtruncate at apex; petals obovate to obtriangular, 5–7 mm; anthers bright yellow, 4–5 mm. **Berries** persistent, purple to purplish black, subglobose (with persistent hypanthial rim), 7–9 mm diam.

Flowering Jun–Dec (year-round). Limestone sinks, mixed broadleaf hammocks, silver palm hammocks, slash pine woods with *Guettarda*, *Myrsine*, and *Serenoa*, disturbed areas, over limestone outcrops; 0–10 m; Fla.; West Indies (including Bahamas); introduced in Pacific Islands (Hawaii).

Tetrazygia bicolor is known in the flora area from Miami-Dade County. G. Nelson (1994) observed that in southern Florida it has two growth forms; in pinelands, plants mostly appear as relatively small, single-stemmed shrubs, but in adjacent hammocks, they may grow to 10 m with a trunk diameter of 6–10 cm.

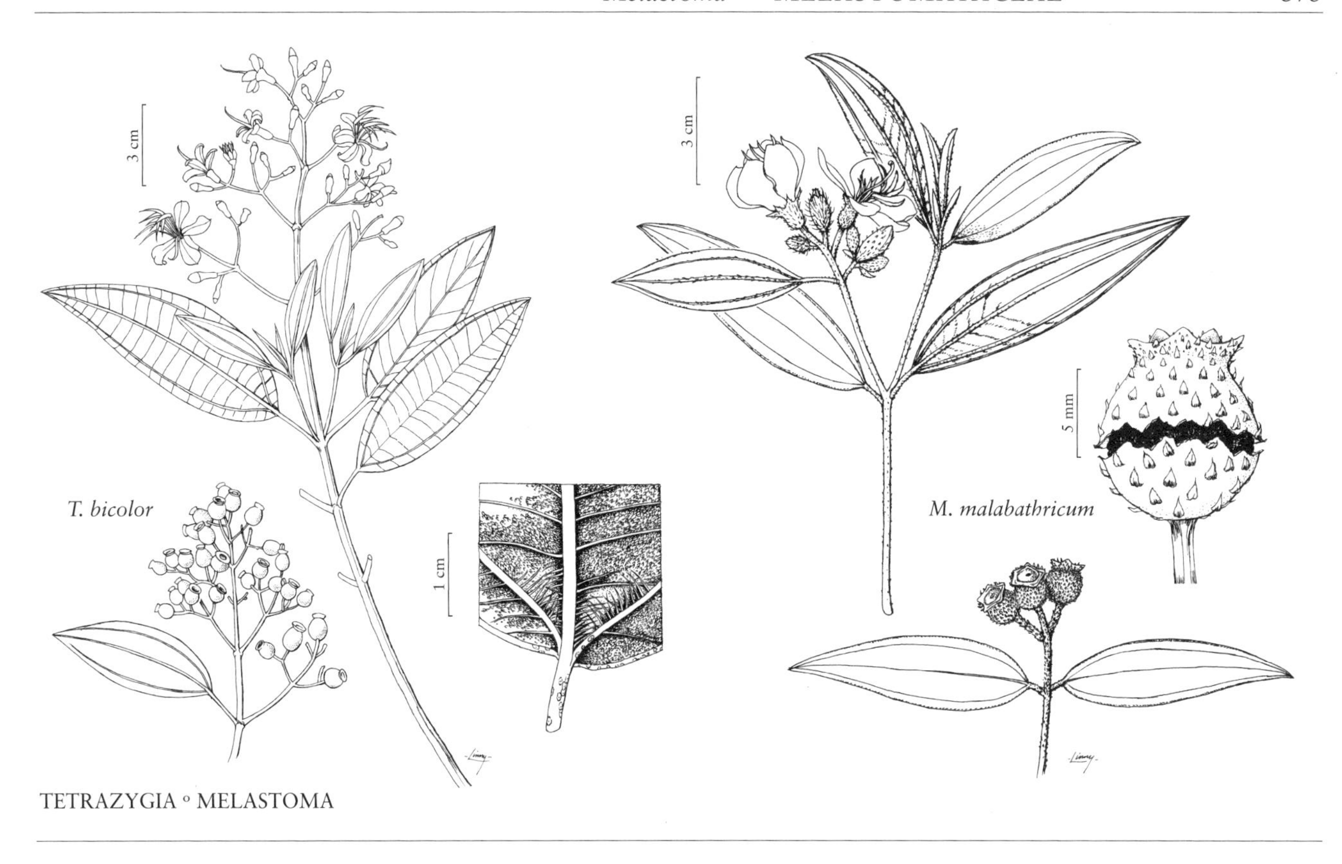

TETRAZYGIA ◦ MELASTOMA

3. MELASTOMA Linnaeus, Sp. Pl. 1: 389. 1753; Gen. Pl. ed. 5, 184. 1754 • Melastome [Greek *melas*, black, and *stoma*, opening, alluding to stained mouth, especially of children, when fruits of some species are eaten] [I]

Shrubs or trees. Stems erect or procumbent, 4-sided [subterete], often squamose-strigose; bark scaly. **Leaves** petiolate; blade with 1 or 2[–4] pairs of lateral primary veins, marginal pairs often inconspicuous, margins entire, surfaces usually strigose [subvillous to villous, rarely glabrate]. **Inflorescences** terminal or in distal foliar axils, usually cymes, rarely panicles or flowers solitary; bracts deciduous or persistent, leaflike, ovate, sometimes conspicuous. **Flowers** pedicellate; hypanthium campanulate to globose-urceolate; calyx deciduous, 5-lobed, lobes triangular to lanceolate or ovate; petals 5(–8), spreading, symmetric, light to dark pink, lavender, or purple; stamens 10, unequal, in 2 whorls, dimorphic, episepalous stamens with purple, upcurved anthers and long connectives, epipetalous stamens with yellow, straight anthers and shorter connectives, or stamens isomorphic and connectives slightly prolonged; anthers slightly downcurved, linear-oblong, 2-locular, dehiscent by 1 or 2 apical pores, or by short, longitudinal slits; ovary semi-inferior, adnate to floral tube, 5-locular; style straight, filiform, equal to petals. **Fruits** fleshy and irregularly splitting-dehiscent, [capsules and apically dehiscent, or berries and fleshy, indehiscent]. **Seeds** cochleate.

Species 22 (1 in the flora): introduced, Florida; Asia, Pacific Islands, Australia; introduced also in Mexico, elsewhere in Pacific Islands (New Zealand).

Most species of *Melastoma* have fleshy, irregularly dehiscent fruits; others have dry capsules [for example, *M. pellegrinianum* (H. Boissieu) Karsten Meyer].

SELECTED REFERENCE Meyer, K. 2001. Revision of the southeast Asian genus *Melastoma* (Melastomataceae). Blumea 46: 351–398.

1. Melastoma malabathricum Linnaeus, Sp. Pl. 1: 390. 1753 (as malabathrica) • Malabar melastome, straits or Singapore rhododendron, sendudok [F] [I]

Melastoma affine D. Don; *M. candidum* D. Don; *M. malabathricum* subsp. *normale* (D. Don) Karst. Meyer

Shrubs or trees 0.5–2(–3) m; branchlets with dense, short, appressed scales. **Leaves** persistent; petiole 0.5–1.9 cm; blade usually lanceolate to elliptic-lanceolate, rarely narrowly lanceolate to oblong, 4–15 cm × 20–50 mm, base rounded to subcordate, apex acuminate, surfaces densely strigose to strigose-sericeous. **Inflorescences:** (1 or)2–7-flowered, compact or loose; bracts 2, leaflike from base. **Pedicels** 2–8(–10) mm, strigose. **Flowers:** hypanthium 5–9 mm, densely appressed-scaly; calyx 5–13 mm; petals obovate, 25–35(–40) mm; ovary ovoid. **Fruits** broadly urceolate, 5–12 mm, succulent, pinkish tan, densely appressed-scaly, rupturing irregularly transversely at maturity. **Seeds** 1 mm diam., densely punctate, imbedded in purplish red, sweet, astringent pulp.

Flowering Feb–Apr. Wet flatwoods; 10 m; introduced; Fla.; Asia (Cambodia, India, Japan, Laos, Malaysia, Myanmar, Nepal, Sri Lanka, Thailand, Vietnam); Pacific Islands (Papua New Guinea, Philippines, Taiwan); Australia; introduced also in Mexico, West Indies (Jamaica), Indian Ocean Islands (Mauritius), Pacific Islands (New Zealand).

Melastoma malabathricum is known in the flora area from only a single report; a plant apparently naturalized in the middle of wet flatwoods in a park in Palm City (Martin County), collected in March 2000 (voucher USF). In a number of other states, it is explicitly regarded as a potential threat and listed as a noxious weed, even prohibited or quarantined in California, Massachusetts, Minnesota, and Oregon. All species of *Melastoma* are formally considered noxious in Hawaii. The seeds are bird-dispersed and *M. malabathricum* in southeast Asia sometimes forms dense thickets in forest plantations, orchards, pastures, rangelands, abandoned clearings, and disturbed sites.

Melastoma malabathricum was regarded by K. Meyer (2001) as the most widespread and morphologically variable species of the genus; he listed 65 synonyms for subsp. *malabathricum*. Meyer recognized only subsp. *normale* as distinct in its more densely pilose abaxial leaf surfaces and its pilose branches; intermediates occur and the geographic range of subsp. *normale* is completely within that of subsp. *malabathricum*.

SURIANACEAE Arnott

• Suriana Family

James S. Pringle

Shrubs or trees, hermaphroditic, terrestrial, unarmed, not clonal. **Leaves** persistent, alternate, simple [pinnately compound]; stipules absent [deciduous, small]; sessile [petiolate]; blade ± fleshy, margins entire. **Inflorescences** terminal or axillary, cymes or flowers solitary in axils [caulophyllous], irregular; bracts present. **Flowers** [some staminate only]; epicalyces absent; hypanthium absent or weakly developed and inconspicuous; sepals 5, proximally connate [distinct]; petals 5 [absent], distinct; intrastaminal nectary disc absent [present]; stamens 10, in 2 whorls of 5, distinct, free, inner whorl sometimes staminodial or vestigial; pistils (4 or)5[1–3], distinct; ovary 1-locular; placentation basal-marginal; style 1 per pistil, basally [ventrally] inserted; stigma 1; ovules 2[–5] per pistil, all but 1 abortive, basal, collateral, integument 1. **Fruits** drupelets [or nutlets]. **Seed** 1, reniform, embryo campylopterous [anatropous], curved, cotyledons 2, equal-sized, chlorophyllous.

Genera 5, species 8 (1 in the flora): Florida, Mexico, West Indies, Bermuda, Central America, South America, Asia, Africa, Indian Ocean Islands, Pacific Islands, Australia; tropical and subtropical coasts.

The generic count includes *Recchia* Moçiño & Sessé ex de Candolle, retained in Simaroubaceae by A. Cronquist (1981) but placed in Surianaceae by other authors, and *Stylobasium* Desfontaines, sometimes placed in a separate family. The other genera are *Cadellia* F. Mueller and *Guilfoylia* F. Mueller. *Recchia* has one Mexican species; the others are Australian.

A. Cronquist (1981) placed Surianaceae within Rosales. Formerly, this family, or at least the genus *Suriana*, was often included in the Simaroubaceae of the Rutales (J. Hutchinson 1964–1967, vol. 1; A. L. Takhtajan 1987). R. F. Thorne (1992b) allied Surianaceae closely with Fabaceae and Connaraceae. A phylogenetic analysis of molecular data by E. S. Fernando et al. (1993) supported the monophyly of Surianaceae (in the sense of Cronquist) and provided the first evidence for a close relationship to Fabaceae and Polygalaceae. Based on this, and evidence from other studies (P. A. Gadek et al. 1996; T. Kajita et al. 2001; F. Claxton et al. 2005), Surianaceae has been moved to Fabales (Angiosperm Phylogeny Group 2016).

SELECTED REFERENCES Fernando, E. S. et al. 1993. Rosid affinities of the Surianaceae: Molecular evidence. Molec. Phylogen. Evol. 2: 344–350. Fernando, E. S. and C. J. Quinn. 1992. Pericarp anatomy and systematics of the Simaroubaceae sensu lato. Austral. J. Bot. 40: 263–350. Gutzwiller, M.-A. 1961. Die phylogenetische Stellung von *Suriana maritima* L. Bot. Jahrb. Syst. 81: 1–49. Heo, K. and H. Tobe. 1994. Embryology and relationships of *Suriana maritima* L. (Surianaceae). J. Pl. Res. 107: 29–37. Wilson, P. 1911. Surianaceae. In: N. L. Britton et al., eds. 1905+. North American Flora.... 47+ vols. New York. Vol. 25, p. 225.

1. SURIANA Linnaeus, Sp. Pl. 1: 284. 1753; Gen. Pl. ed. 5, 107. 1754 • [For Joseph Donat Surian, d. 1691, French physician who collected plants in the West Indies]

Shrubs, sometimes trees; young stems, leaf surfaces, pedicels, and calyces densely puberulent, some hairs on herbage glandular; older stems with exfoliating bark. **Leaves** mostly crowded toward ends of twigs (older ones soon falling); blade narrowly oblanceolate, base tapering, apex obtuse to acute. **Inflorescences** usually scarcely exserted beyond leaves. **Flowers:** sepals persistent, narrowly ovate, apex short-acuminate; petals mostly fallen by midday, spreading, yellow, elliptic-obovate, base tapering to claw, apex rounded, erose-dentate; stamens: outer whorl fertile, filaments proximally pilose, inner whorl sterile, less than 1/2 as long as fertile, pilose throughout; anthers rudimentary or none; pistils short-stipitate, pilose; style slender, glabrous, much longer than ovaries. **Fruits** brown, round, becoming dry, achenelike.

Species 1: Florida, Mexico, West Indies, Bermuda, Central America, South America, Asia, Africa, Indian Ocean Islands, Pacific Islands, Australia.

1. Suriana maritima Linnaeus, Sp. Pl. 1: 284. 1753 • Bay-cedar, guitar 

Shrubs 0.7–3(–8) m, sometimes trees in sheltered sites, densely branched. **Leaves** 10–40 × 1–5(–7) mm. **Inflorescences** 1–23-flowered. **Flowers:** calyx 6–10 mm; petals ± 4 mm. **Fruits** 4–6 mm diam.

Flowering year-round. Sandy and coral-rock ocean shores, from just above high-water line to low dunes and hammocks; 0–10 m; Fla.; Mexico; West Indies; Bermuda; Central America; n South America; se Asia; e Africa; Indian Ocean Islands; Pacific Islands; n Australia.

Although *Suriana maritima* is widely distributed in the tropics, in the flora area it is native only to coastal Florida, west to the Dry Tortugas and north to Brevard and Pinellas counties, where much of its habitat has been eliminated by development.

A proportionately large air space within the drupelets enables them to float for long periods, thereby being dispersed by ocean currents. Because of its tolerance of salt and resistance to wind, *Suriana maritima* is valued as a landscape plant and for erosion control. In some parts of its range, the wood has a limited use for making small articles. Preparations from the leaves and bark have been used for medicinal purposes, including the treatment of fevers and intestinal problems (D. W. Nellis 1994).

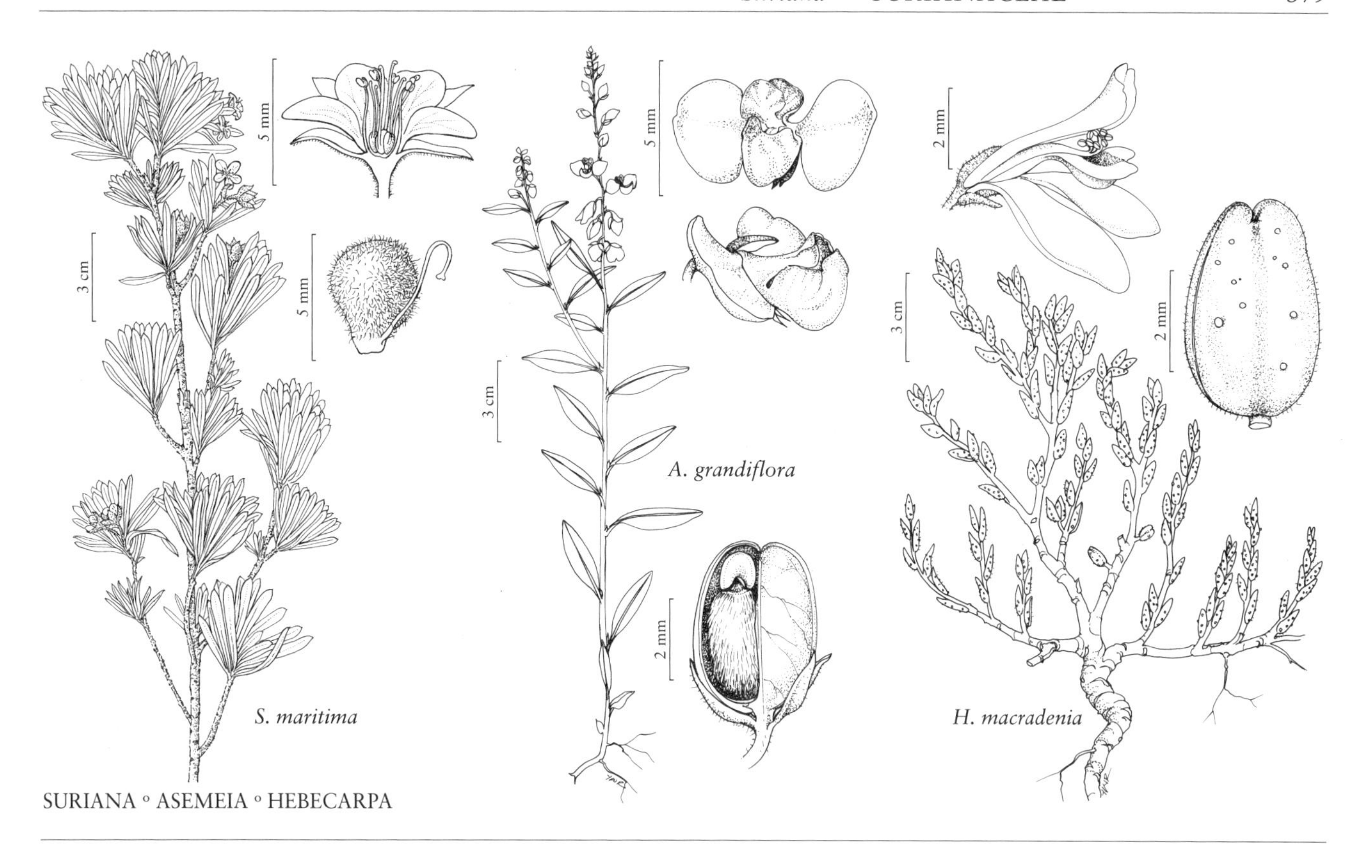

SURIANA ◦ ASEMEIA ◦ HEBECARPA

POLYGALACEAE Hoffmannsegg & Link • Milkwort Family

J. Richard Abbott

Herbs, annual, biennial, or perennial, subshrubs, shrubs, [trees, or lianas], synoecious, terrestrial, usually unarmed (sometimes thorns present as modified tips of racemes in *Rhinotropis*), sometimes clonal. **Rootstock** usually brownish to whitish, usually not exfoliating. **Stems** mostly erect, sometimes laxly so, or decumbent, prostrate, or creeping. **Leaves** usually alternate, sometimes opposite or whorled, simple, basal and/or cauline, basal usually not present at anthesis; stipules absent [present as glands]; petiolate or not; blade mostly ± uniform in shape and size, sometimes weakly to strongly dimorphic, pinnately veined, secondary venation usually obscure, usually not punctate, margins entire (rarely erose or appearing minutely serrulate). **Inflorescences** terminal, axillary, or leaf-opposed, usually racemes, sometimes panicles, spikes, corymbs, fascicles, or reduced to 1–5 flowers; bract 1; bracteoles present. **Flowers** bisexual, sometimes cleistogamous and subterranean or in proximal leaf axils, bilaterally [radially] symmetric, usually pedicellate, rarely sessile, pedicel usually not winged; perianth and androecium hypogynous [perigynous]; sepals persistent or deciduous, 5, imbricate in bud, unequal with 3 outer sepals, 1 upper and 2 lower (these 2 sometimes connate and appearing as 1 or 2-lobed), 2 lateral inner sepals (wings) expanded and petaloid [± similar to outer sepals], adnate proximally to petals and filaments; petals 3(or 5), upper 2 often apically reflexed at anthesis, lower 1 (keel) conduplicate and, sometimes, apically lobed and/or crested (crest typically fimbriate), rarely with 2 reduced lateral petals; stamens [2–](6 or 7)8(or 9)[10]; filaments connate [distinct]; anthers basifixed, dehiscent by 1 or 2 apical or subapical pores to very short slits [to introrsely longitudinally dehiscent]; pistil 2[–8]-carpellate, syncarpous; ovary superior, (1 or)2[–8]-locular; placentation axile; style 1; stigmas 2; ovule 1 per locule [2–40]. **Fruits** capsules or pseudomonomerous and samaroid [samaras, drupes, or berries], somewhat isodiametric to slightly oblong, pubescent or glabrous. **Seeds** 1 per locule, ± blackish at maturity in capsular species, ellipsoid, smooth (pitted in some *Polygala*), often whitish pubescent, usually arillate (the aril-like structure from integuments, not funiculus, thus called an arillode by some, a caruncle by others).

Genera 29, species ca. 1200 (6 genera, 53 species in the flora): North America, Mexico, West Indies, Central America, South America, Europe, Asia, Africa, Australia.

B. Eriksen and C. Persson (2007) provided a good synthesis of the taxonomic history of Polygalaceae, with modern phylogenetic hypotheses presented by Persson (2001), F. Forest et al. (2007), and J. R. Abbott (2009). All but one of the North American species have traditionally been treated within a broadly defined *Polygala*. At least six major clades within *Polygala* (in the traditional broad sense) are represented natively in North America (Abbott). These clades correspond to morphological differences and traditional infrageneric groups, in many cases correlated with geographic distribution. A seventh clade, represented by the type group and based on *P. vulgaris*, is native to Eurasia and is represented by one sparsely naturalized species. Multiple lines of robust evidence support that four of these native clades are so distant phylogenetically from the type group of *Polygala* that they can no longer be maintained within *Polygala*: the groups here recognized as the genera *Asemeia*, *Hebecarpa*, *Polygaloides*, and *Rhinotropis*, all of which correspond to traditional subgenera.

Much cytological and karyological variation has been documented in the small percentage of Polygalaceae that have been studied cytologically. Some species, including a few North American taxa, have different chromosome numbers, with discrepancies and uncertainties as to the base number for many groups. Even if some reports are in error, evidence of polyploid and aneuploid series is clear, as well as widespread hybridization in a few species, although few reports of hybridization are documented (T. L. Wendt 1978; A. J. Lack 1995). The taxa represented here generally do not intergrade.

The diversity of reported chromosome numbers for North American taxa shows several ploidy shifts, even within some taxa. The exact patterns cannot be fully tracked, given the number of taxa with undocumented chromosome numbers and the absence of solid hypotheses of sister taxa relationships. Despite the shortcomings of current cytological data, further investigations should be useful for illuminating relationships among North American Polygalaceae. There may still be diagnostic value, and even phylogenetic information, in the cytological and karyological data (T. L. Wendt 1978; J. A. R. Paiva 1998).

B. Eriksen and C. Persson (2007) noted Polygalaceae have little economic importance except in herbal medicine (for example, A. Freire-Fierro 1995). The North American *Polygala senega* is perhaps the most economically important taxon, with some commercial usage globally for various medicinal purposes (J. M. Gillett 1968b; C. J. Briggs 1988). Many species of Polygalaceae are showy in flower and are sometimes cultivated as ornamentals. Most commonly encountered in North America are *P. myrtifolia*, *P. virgata* Thunberg, *P.* ×*dalmasiana* L. H. Bailey (perhaps a form of *P. myrtifolia*), and *Polygaloides chamaebuxus* (Linnaeus) O. Schwarz.

Use of the term subshrub refers to a low, multi-stemmed, suffrutescent perennial herb; the herbaceous branches may be erect or decumbent, but they are woody above ground proximally. The term subsessile refers leaves with a petiole to 0.5 mm. Measurements of raceme length refer to the flowering and/or fruiting portion (including attachment scars), while width refers to the broadest portion with open (or nearly so) flowers. Raceme shape refers to the flowering and fruiting portion only; with age, there is often an elongated basal portion with scars left by the deciduous fruits. In descriptions of flower color, the outer sepals are not included, and it should be understood that the wings and petals are almost always paler in the proximal portion. The outer sepals are usually green, unless otherwise stated, although it is relatively common for some sepals to be whitish, pinkish, or purplish red marginally. It is also very common for the tips of the upper petals to be darker and for the apex of the keel to be greenish or yellowish. On drying, colors can become darker or paler, with whites often taking on a bluish or pinkish purple tinge, but, unless otherwise stated, the colors do not change dramatically. In key leads and descriptions for Polygalaceae, lower sepals refers to abaxial sepals and upper sepals refers to adaxial sepals; the two lateral (inner, often petaloid) sepals are referred to as wings. In reference to petals, the abaxial (lower) petal is referred to as the keel. Unless otherwise stated, if the sepals are deciduous, so are the wings and corolla; conversely, if the sepals are persistent (unless only the upper one is specified), then so are the wings and corolla. Descriptions of pubescence do not refer to marginal cilia; even a glabrous structure may be marginally ciliate. Measurements of seed length include aril and pubescence; the body itself is usually 0.3–0.7 mm shorter.

SELECTED REFERENCES Abbott, J. R. 2009. Revision of *Badiera* (Polygalaceae) and Phylogeny of the Polygaleae. Ph.D. dissertation. University of Florida. Banks, H. et al. 2008. Pollen morphology of the family Polygalaceae (Fabales). Bot. J. Linn. Soc. 156: 253–289. Bernardi, L. F. 2000. Consideraciones taxonómicas y fitogeográficas acerca de 101 *Polygalae* Americanas. Cavanillesia Altera 1: 1–456. Blake, S. F. 1924. Polygalaceae. In: N. L. Britton et al., eds. 1905+. North American Flora.... 47+ vols. New York. Vol. 25, pp. 305–379. Eriksen, B. 1993. Phylogeny of the Polygalaceae and its taxonomic implications. Pl. Syst. Evol. 186: 33–55. Eriksen, B. and C. Persson. 2007. Polygalaceae. In: K. Kubitzki et al., eds. 1990+. The Families and Genera of Vascular Plants. 15+ vols. Berlin etc. Vol. 9, pp. 345–363. Forest, F. et al. 2007. The role of biotic and abiotic factors in evolution of ant-dispersal in the milkwort family (Polygalaceae). Evolution 61: 1675–1694. Gillett, J. M. 1968b. The Milkworts of Canada. Ottawa. [Canada Dept. Agric. Monogr. 5.]. Holm, T. 1929b. Morphology of North American species of *Polygala*. Bot. Gaz. 88: 167–185. Miller, N. G. 1971. The Polygalaceae in the southeastern United States. J. Arnold Arbor. 52: 267–284. Persson, C. 2001. Phylogenetic relationships in the Polygalaceae based on plastid DNA sequences from the *trn*L-F region. Taxon 50: 763–779.

1. Ovaries 1-loculed; fruits indehiscent; keel petals entire, ± 3-lobed; seeds glabrous, not arillate; annuals . . . . 3. *Monnina*, p. 387
1. Ovaries 2-loculed (sometimes 1 abortive); fruits dehiscent; keel petals with crest or beak or entire; seeds usually pubescent, usually arillate; annuals, biennials, perennials, subshrubs, or shrubs.
   2. Keel petals with lobed crests, often fimbriate.
      3. Flowers and wing sepals 1–9(–10) mm; stems mostly erect (not creeping, rarely laxly decumbent or prostrate in a few species); leaves usually numerous, distributed evenly along stem or clustered in basal rosettes; all sepals usually persistent . . . . . . 4. *Polygala*, p. 388
      3. Flowers 15+ mm, wing sepals (10–)13–20 mm; stems creeping or decumbent with short, erect shoots; leaves few, well-developed ones ± clustered distally; all sepals deciduous . . . . 5. *Polygaloides*, p. 404
   2. Keel petals entire and blunt or beaked (with unlobed projection), without lobed crest.
      4. Lower 2 sepals connate ¾ their length; all sepals persistent . . . . 1. *Asemeia*, p. 382
      4. All sepals distinct (or appearing very slightly connate basally); all sepals usually deciduous, occasionally upper sepal persistent or rarely all sepals persistent.
         5. Shrubs or subshrubs; raceme rachis thorn-tipped . . . . 6. *Rhinotropis* (in part), p. 405
         5. Perennials, shrubs, or subshrubs; raceme rachis not thorn-tipped.
            6. Keel petals not beaked, sometimes bluntly 3-lobed; all sepals deciduous (sometimes upper 1 tardily so or subpersistent in *H. macradenia* and *H. punctata*) . . . . 2. *Hebecarpa*, p. 384
            6. Keel petals beaked, with unlobed projection, this sometimes reduced or obscure (rarely on all flowers of a given plant unless all flowers are cleistogamous); upper sepals persistent, others deciduous (except all sepals deciduous in *R. californica* and *R. cornuta*, and all sepals persistent in *R. rusbyi*) . . . . 6. *Rhinotropis* (in part), p. 405

1. ASEMEIA Rafinesque, Herb. Raf., 80. 1833 • [Greek *a*, absence, and *semion*, sign or flag, alluding to distinctness from *Polygala* in absence of vexillum]

Herbs, perennial, sometimes suffrutescent, single- to multi-stemmed. **Stems** usually erect, sometimes laxly spreading, pubescent or glabrous. **Leaves** alternate; petiolate; not strongly dimorphic; blade surfaces usually pubescent. **Inflorescences** terminal or axillary, racemes; peduncle present; bracts deciduous or persistent. **Pedicels** present. **Flowers** pink to purple (wings becoming green in fruit), chasmogamous, 5–9 mm; sepals persistent, lower 2 connate ¾ their length, distally 2-lobed, pubescent or subglabrous, sometimes marginally glandular-ciliate; wings persistent, (2.5–)4–7 mm, pubescent or subglabrous; keel entire, without crest or beak, often with lateral folds, pubescent; stamens 8, not grouped into bundles of 4; ovary 2-loculed. **Fruits** capsules, dehiscent, margins not or very narrowly winged, glabrous or sparsely pubescent. **Seeds** densely pubescent, arillate, aril lobes vestigial or present.

Species ca. 28 (1 in the flora): se United States, Mexico, West Indies, Central America, South America.

Rafinesque considered *Asemeia* to contain two species of *Polygala* combined by Elliott under the name *P. pubescens*. He characterized the genus by its absence of a crest, tuberculate keel petal, and 2-loculed, winged capsule with two seeds; the absence of a crest is known in other species of Polygalaceae. It is not clear what Rafinesque meant by a tuberculate keel petal (although it may be a reference to the lateral folds often found on the keel). Most members of the traditional *Polygala* in the broad sense have 2-loculed capsules, winged in many species, with two seeds.

1. **Asemeia grandiflora** (Walter) Small, Man. S.E. Fl., 766. 1933 • Showy milkwort [F]

Polygala grandiflora Walter, Fl. Carol., 179. 1788; *Asemeia cumulicola* (Small) Small; *A. leiodes* (S. F. Blake) Small; *A. miamiensis* (Small) Small; *P. corallicola* Small; *P. cumulicola* Small; *P. grandiflora* var. *angustifolia* Torrey & A. Gray; *P. grandiflora* var. *leiodes* S. F. Blake; *P. miamiensis* Small

Herbs (0.8–)2–5.5(–10) dm. **Stems** glabrous or hairs appressed or spreading. **Leaves:** petiole 1–4 mm; blade oblanceolate, linear-oblanceolate, elliptic, lanceolate, narrowly ovate, or linear, (6–)15–50(–64) × (0.3–)2–17 mm, base acute, apex acute, surfaces usually pubescent, at least on veins abaxially. **Racemes** usually loose, rarely very open, 5–10(–20) × 1–2.5 cm; peduncle 0.3–1.7 cm; bracts ovate to lanceolate. **Pedicels** (1–)2–4 mm, usually pubescent, sometimes glabrous. **Flowers:** sepals (1–)1.3–3.1 mm; wings orbiculate to reniform or obovate, often somewhat 4-angled, 2–6.8 mm wide; keel 2–5(–7) mm. **Capsules** ellipsoid, oblong, or ovoid, 3–5.3 × 2–3 mm. **Seeds** 1.1–2.5 mm; aril 1 mm, lobes helmetlike, to 1/4 length of seed. 2*n* = 28.

Flowering spring–summer (year-round). Open dry, sandy habitats, open pine and oak associations, dunes, roadsides, savannas, prairies; 0–300 m; Ala., Fla., Ga., La., Miss., N.C., S.C.; Mexico (Chiapas); West Indies; Central America.

Asemeia grandiflora is highly polymorphic, with numerous historically recognized segregate species or infraspecific taxa (C. E. Nauman 1981b; J. K. Small 1933). However, the characters artificially selected for differentiating the taxa, such as pubescence, presence of glandular trichomes, relative length of capsules and wing sepals, wing sepal coloration and shape, capsule shape, keel length, and leaf width, are polymorphic features without consistent geographical structure. Other features, such as being more strongly perennial (multi-branched and suffrutescent) farther south and more clearly herbaceous or, perhaps, even annual farther north, are clearly correlated with climate and not necessarily indicative of taxonomic distinctiveness. This situation is further complicated by the species being part of the even more widespread complex of *Asemeia violacea* (Aublet) J. F. B. Pastore & J. R. Abbott as circumscribed by M. C. M. Marques. More work is needed to determine whether *A. grandiflora* should be treated as conspecific with *A. violacea*, as many other related entities are still recognized and no phylogenetic or populational data have been presented to show that this entity is unworthy of recognition.

Asemeia grandiflora is the only species of Polygalaceae in eastern North America with an entire keel (lacks a beak or crest).

SELECTED REFERENCE Nauman, C. E. 1981b. *Polygala grandiflora* (Polygalaceae) Walter re-examined. Sida 9: 1–18.

2. HEBECARPA (Chodat) J. R. Abbott, J. Bot. Res. Inst. Texas 5: 134. 2011 • [Greek *hebe*, down of puberty or pubescent, and *karpos*, fruit, alluding to vestiture]

Polygala Linnaeus sect. *Hebecarpa* Chodat, Biblioth. Universelle Rev. Suisse, sér. 3, 25: 698. 1891

Herbs, perennial, or subshrubs, multi-stemmed [single]. **Stems** prostrate or decumbent to erect, usually densely pubescent, rarely glabrate. **Leaves** alternate; mostly sessile or subsessile (petiole to 2 mm in *H. ovatifolia*); sometimes dimorphic (proximal different from distal); blade surfaces not punctate (except translucent-punctate in *H. macradenia* and *H. punctata*), pubescent. **Inflorescences** terminal, axillary, or leaf-opposed, usually racemes, sometimes 1 or 2(–4)-flowered; peduncle present or absent; bracts deciduous. **Pedicels** present. **Flowers** usually purple, pink, yellow, greenish yellow, rarely whitish, wings sometimes cream, yellowish, or greenish, keel sometimes yellow or yellowish green distally, chasmogamous, 3–8 mm; sepals deciduous or upper 1 sometimes tardily so or subpersistent (in *H. macradenia* and *H. punctata*), distinct, pubescent; wings deciduous, 3.5–6(–7.5) mm, pubescent; keel not beaked or crested, sometimes bluntly 3-lobed, pubescent or glabrate (in *H. obscura*); stamens usually 8, sometimes 7 (in *H. macradenia*), not grouped; ovary 2-loculed. **Fruits** capsules, dehiscent, margins not winged (except sometimes narrowly winged in *H. barbeyana*), not punctate (except translucent-punctate in *H. macradenia* and *H. punctata*), pubescent to glabrate or glabrous and ciliolate. **Seeds** pubescent, arillate.

Species 40–70 (7 in the flora): sw, sc United States, Mexico, West Indies, Central America, South America.

Most species of *Hebecarpa* are Mexican, the center of diversity for this group. The seven in the flora area occur mostly in arid and semi-arid areas of the southwestern United States, representing contiguous northern expansions of their ranges in Mexico. A few species range into Central America and Andean South America, but they are not the basalmost species, suggesting that they have arrived in these regions secondarily.

Several species complexes in this group are in serious need of revision, with overly fine, historical splitting (often based on few specimens); thus the total number of species is probably fewer than 50.

1. Leaf blade surfaces and capsules densely translucent-punctate from internal cavities appearing as pustular glands 0.2–0.4 mm diam.
    2. Leaf blades linear-oblong to oblong-lanceolate, narrowly ovate, or linear . . . . . . 2. *Hebecarpa macradenia*
    2. Leaf blades usually obovate, rarely suborbiculate to broadly elliptic . . . . 6. *Hebecarpa punctata*
1. Leaf blade surfaces and capsules not translucent-punctate.
    3. Capsules densely pubescent to glabrate . . . . 3. *Hebecarpa obscura*
    3. Capsules glabrous, margins ciliolate.
        4. Stems and leaves with incurved hairs.
            5. Middle and distal leaf blades usually 1.5–4(–6) mm wide, elliptic, lanceolate, or linear; capsules 5–10(–11) mm. . . . . 1. *Hebecarpa barbeyana*
            5. Middle and distal leaf blades 4–7 mm wide, ovate, elliptic, or obovate; capsules 9–12 mm . . . . 5. *Hebecarpa palmeri*
        4. Stems and leaves with spreading hairs.
            6. Leaf blades ovate to ovate-oblong or elliptic; capsules 8.5–13 mm. . . . . . 4. *Hebecarpa ovatifolia*
            6. Leaf blades ovate, elliptic, spatulate, or linear proximally, elliptic, lanceolate, or linear distally (often longer than proximal ones); capsules 6–8 mm . . . . 7. *Hebecarpa rectipilis*

1. Hebecarpa barbeyana (Chodat) J. R. Abbott, J. Bot. Res. Inst. Texas 5: 134. 2011 • Blue milkwort

Polygala barbeyana Chodat, Mém. Soc. Phys. Genève 31(2): 16, plate 13, figs. 15–18. 1893; *P. longa* S. F. Blake; *P. racemosa* S. F. Blake; *P. reducta* S. F. Blake

Herbs, sometimes somewhat suffrutescent, (0.5–)0.8–6(10) dm. **Stems** usually erect, sometimes laxly so (rarely procumbent), hairs incurved (rarely glabrate). **Leaves:** blade ovate, elliptic, spatulate, or linear proximally, to elliptic, lanceolate, or linear distally (often longer than proximal ones), 4–45(–65) × 1.5–4(–6) mm, base usually acute, sometimes cuneate or obtuse, apex acute to acuminate or obtuse, surfaces not translucent-punctate, hairs incurved. **Inflorescences** terminal or leaf-opposed, racemes, loose, 2–12 × 0.9–1.4 cm; peduncle 0.5–1.5 cm; bracts linear to narrowly ovate. **Pedicels** 1–5 mm, pubescent. **Flowers** purplish, wings rarely yellow, keel distally yellow or yellowish green, 3.5–6 mm; sepals lanceolate, 2.2–3.5 mm; wings ovate to suborbiculate, 3.5–5.5 × 2.4–4 mm, sparsely pubescent; keel 4–5.7 mm, pubescent. **Capsules** oblong, oblong-ovoid, broadly ellipsoid, or subglobose, 5–10(–11) × 4–7 mm, not translucent-punctate, glabrous and ciliolate, margins sometimes narrowly winged. **Seeds** 4–4.5 mm; aril 0.6–1 mm, lobes vestigial or to 1/4 length of seed. 2*n* = 30.

Flowering spring–late fall. Gravelly or rocky soils, mostly limestone, or igneous or gypsum, arid or semi-arid scrub, open woodlands, chaparral; 400–2400 (–2800) m; Ariz., N.Mex., Tex.; Mexico (Coahuila, Durango, Guanajuato, Jalisco, Nuevo León, Querétaro, San Luis Potosí, Veracruz, Zacatecas).

Hebecarpa barbeyana is very similar to 7. *H. rectipilis*; see that species for comments.

2. Hebecarpa macradenia (A. Gray) J. R. Abbott, J. Bot. Res. Inst. Texas 5: 134. 2011 • Gland-leaf or purple milkwort [F]

Polygala macradenia A. Gray, Smithsonian Contr. Knowl. 3(5): 39. 1852

Herbs or subshrubs, 0.2–2.5 dm. **Stems** decumbent to laxly erect, forming low, compact clumps, hairs incurved or spreading. **Leaves:** blade linear-oblong to oblong-lanceolate, narrowly ovate, or linear, (2–)3–9 × 0.6–1.3 mm, base rounded, apex obtuse, surfaces densely translucent-punctate from internal cavities appearing as pustular glands 0.2–0.4 mm diam., hairs incurved or spreading. **Inflorescences** axillary, reduced to 1 or 2(–4) flowers, 0.4–1.5 × 0.4–1.8 cm; peduncle 0–0.3(–0.5) cm; bracts ovate. **Pedicels** (0.5–)1–2.5 mm, pubescent. **Flowers** usually purple or pink, rarely whitish, 3.5–6.5 mm; sepals: upper 1 sometimes tardily deciduous to subpersistent, oblong-ovate, 1.7–2.1 mm; wings obovate, 5–5.5 × 2.5 mm, sparsely pubescent; keel 4.5 mm, pubescent. **Capsules** oblong to oblong-ovoid, 4–6.5 × 2.5–4 mm, densely translucent-punctate from internal cavities appearing as pustular glands 0.2–0.4 mm diam., pubescent. **Seeds** 4 mm; aril 1–2 mm, lobes to 1/2 length of seed. 2*n* = 64, 68.

Flowering early spring–late fall. Open gravelly areas, usually loess-derived (or gypsum), scrub; 500–2400 m; Ariz., Nev., N.Mex., Tex.; Mexico (Baja California, Chihuahua, Coahuila, Durango, Guanajuato, Hidalgo, Nuevo León, Querétaro, San Luis Potosí, Sonora, Tamaulipas).

3. Hebecarpa obscura (Bentham) J. R. Abbott, J. Bot. Res. Inst. Texas 5: 134. 2011 • Velvet-seed milkwort

Polygala obscura Bentham, Pl. Hartw., 58. 1840; *P. orthotricha* S. F. Blake; *P. piliophora* S. F. Blake

Herbs or subshrubs, (0.7–) 1.2–4(–7.5) dm. **Stems** erect, sometimes laxly so, hairs incurved or spreading. **Leaves:** blade oblong to broadly elliptic proximally, to elliptic-ovate, lanceolate, or linear distally, often longer than proximal ones, (5–)12–42(–80) × 1.5–12 mm, base obtuse or acute, apex usually obtuse to acuminate, sometimes bluntly rounded, surfaces not translucent-punctate, hairs incurved or spreading. **Inflorescences** terminal or leaf-opposed, racemes, loose, 1.5–15 × 1.5–3 cm; peduncle 0.7–1.5 cm; bracts narrowly ovate or lanceolate-ovate. **Pedicels** 2–6 mm, pubescent. **Flowers** purplish, 4.5–7 mm; sepals lanceolate, 2.2–3.5 mm, pubescent; wings elliptic-obovate, elliptic, or suborbiculate, 4.5–5.8 × 2–3.5 mm, sparsely pubescent along middle and apically; keel 4.8–5 mm, subglabrous. **Capsules** broadly ellipsoid to subglobose, 7–10(–11) × 5–8 mm, not translucent-punctate, densely pubescent to glabrate. **Seeds** 4–5 mm; aril 1.3–3(–4) mm, covering less than 1/20 to 4/5 length of seed.

Flowering spring–fall. Sandy or rocky soils, mostly igneous, grasslands, open woodlands; 1200–2400 m; Ariz., Nev., N.Mex., Tex.; Mexico (Chihuahua, Guanajuato, Oaxaca, Puebla, Sonora).

The key feature provided by S. F. Blake for diagnosing *Polygala piliophora* [known only from the type collection (US) in the Huachuca Mountains of Cochise County, Arizona] as distinct from *Hebecarpa obscura* was having aril lobes subequal or smaller than the apical umbo portion of the aril (versus longer than the umbo). The veil-like distal portion of the aril (below the umbo region) is without distinct lobes in *H. obscura*, even when it covers nearly ⁴⁄₅ the seed body. Collections of *H. obscura* from scattered localities can have a reduced helmetlike aril, with the lobe region subequal or smaller than the apical umbo portion. The length of the aril lobe is not only variable within *H. obscura* but is a feature that varies within many species of *Hebecarpa*. Minor variations in pubescence and leaf and aril morphometrics within *H. obscura* do not seem worthy of taxonomic recognition, especially as the variants do not appear to be correlated with ecology or geography. *Hebecarpa obscura* is sometimes morphologically quite similar to *H. barbeyana*, despite their differences in ovary and fruit pubescence.

4. **Hebecarpa ovatifolia** (A. Gray) J. R. Abbott, J. Bot. Res. Inst. Texas 5: 134. 2011 • Egg-leaf milkwort

Polygala ovatifolia A. Gray, Smithsonian Contr. Knowl. 3(5): 39. 1852

Herbs or subshrubs, 0.5–3 dm. **Stems** erect to decumbent, hairs spreading. **Leaves:** petiole to 2 mm; blade ovate to ovate-oblong or elliptic, (5–)12–30 × 6–13 mm, base rounded to cuneate, apex acute to obtuse, surfaces not translucent-punctate, hairs spreading. **Inflorescences** terminal or leaf-opposed, racemes, loose, (1–)2–6.5(–8) × 1.5–2.5 cm; peduncle 0.5–2 cm; bracts ovate or lanceolate-ovate. **Pedicels** 2–3.5 mm, pubescent. **Flowers** yellow, wings greenish, 5–8 mm; sepals ovate-lanceolate, 3 mm; wings broadly elliptic to ovate, 4–5 × 3–4 mm, pubescent; keel 6–7.5 mm, pubescent. **Capsules** subglobose to broadly oblong, 8.5–13 × 8–12 mm, not translucent-punctate, glabrous and ciliolate. **Seeds** 4 mm; aril 1–1.5 mm, lobes to ⅓ length of seed. $2n$ = 56–60.

Flowering (early spring–)spring–summer(–late fall). Rocky slopes in dry scrub or grasslands; 500–1600 m; Tex.; Mexico (Coahuila, Nuevo León).

Hebecarpa ovatifolia is common in southwestern Texas in Tamaulipan thorn scrub and, rarely, oak-juniper woodlands of the western Edwards Plateau.

5. **Hebecarpa palmeri** (S. Watson) J. R. Abbott, J. Bot. Res. Inst. Texas 5: 134. 2011 • Palmer's milkwort

Polygala palmeri S. Watson, Proc. Amer. Acad. Arts 17: 325. 1882

Herbs, 0.6–3.5 dm. **Stems** erect to decumbent, hairs incurved. **Leaves:** blade ovate, elliptic, or obovate proximally, often narrower distally, 7–30 × 4–7 mm, base cuneate, apex obtuse to truncate and mucronate, surfaces not translucent-punctate, hairs incurved. **Inflorescences** terminal or leaf-opposed, racemes, loose, 2.5–7 × 1.5–2.5 cm; peduncle 0.5–2 cm; bracts ovate. **Pedicels** 2–4.5 mm, pubescent. **Flowers** pale greenish yellow and purplish, 5–7 mm; sepals lanceolate, 1–2 mm; wings obovate, 5.7–6 × 2.8 mm, pubescent; keel 5.7–6 mm, pubescent. **Capsules** broadly oblong-ovoid to subglobose, 9–12 × 7–11 mm, not translucent-punctate, glabrous and ciliolate. **Seeds** 6 mm; aril 1–1.4 mm, lobes to ¼ length of seed.

Flowering spring–summer. Open or brushy limestone slopes and plains in scrub; 400–1100 m; Tex.; Mexico (Coahuila, Nuevo León, Tamaulipas).

In the flora area, *Hebecarpa palmeri* is known only from Val Verde County.

6. **Hebecarpa punctata** (Humboldt & Bonpland ex Willdenow) J. R. Abbott & J. F. B. Pastore, Kew Bull. 70-39: 5. 2015 • Glandular milkwort

Viola punctata Humboldt & Bonpland ex Willdenow in J. J. Roemer et al., Syst. Veg. 5: 391. 1819; *Hebecarpa greggii* (S. Watson) J. R. Abbott; *Polygala greggii* S. Watson

Herbs or subshrubs, 0.5–2 dm. **Stems** prostrate to decumbent, hairs incurved. **Leaves:** blade usually obovate, rarely suborbiculate to broadly elliptic, 4–10 × 1–6 mm, base cuneate to rounded, apex rounded to subtruncate and mucronate, surfaces densely translucent-punctate from internal cavities appearing as pustular glands 0.2–0.4 mm diam., hairs incurved. **Inflorescences** axillary, reduced to 1 or 2 flowers, 0.3–0.4 × 0.6–0.8 cm; peduncle 0 cm; bracts ovate. **Pedicels** 0.5–1 mm, pubescent. **Flowers** purple, 3–4 mm; sepals with upper one sometimes tardily deciduous to subpersistent, oblong-ovate, 1.2–1.5 mm; wings spatulate-obovate, 3.5–4.5 × 1.5–2 mm, pubescent; keel 2.5(–4.5) mm, pubescent. **Capsules** ellipsoid, 1.5–3 × 2 mm, densely translucent-punctate from internal cavities appearing as pustular glands 0.2–0.4 mm diam., sparsely pubescent. **Seeds** 2.8–3 × 0.6–0.8 mm; aril 0.9–1.2 mm, lobes to ¼ length of seed.

Flowering spring–summer. Gravelly hills in scrub, chalky hills; 0–100 m; Tex.; Mexico (Nuevo León, Querétaro, Tamaulipas).

In the flora area, *Hebecarpa punctata* occurs in Duval, Hidalgo, Jim Wells, San Patricio, Starr, Webb, and Zapata counties. It is a common species in the southern Texas brushlands.

Polygala glandulosa Kunth is an illegitimate name that has been used in Texas for decades for *Hebecarpa punctata* and is in several on-line databases.

S. F. Blake (1924) reported the wings as 7.5 × 3.7 mm; none approaching that size was seen.

7. **Hebecarpa rectipilis** (S. F. Blake) J. R. Abbott, J. Bot. Res. Inst. Texas 5: 134. 2011 • New Mexico milkwort

Polygala rectipilis S. F. Blake, Contr. Gray Herb. 47: 27, plate 1, fig. 14. 1916

Herbs, 1.5–3.5 dm. **Stems** erect, sometimes laxly so, hairs spreading. **Leaves:** blade ovate, elliptic, spatulate, or linear proximally, to elliptic, lanceolate, or linear distally, often longer than proximal ones, 4–45(–65) × 1.5–4 mm, base often acute, sometimes cuneate or obtuse, apex acute to acuminate or obtuse, surfaces not translucent-punctate, hairs spreading. **Inflorescences** terminal or leaf-opposed, racemes, loose, 2–12 × 0.9–1.4 cm; peduncle 0.5–1.5 cm; bracts linear to narrowly ovate. **Pedicels** 1–5 mm, pubescent. **Flowers** purplish, wings usually cream or yellow, keel distally yellow or yellowish green, 3.5–6 mm; sepals lanceolate, 2.2–3.5 mm; wings ovate to suborbiculate, 3.5–5.5 × 2.4–4 mm, sparsely pubescent; keel 4–5.7 mm, pubescent. **Capsules** oblong or oblong-ovoid to broadly ellipsoid or subglobose, 6–8 × 4.5–7 mm, not translucent-punctate, glabrous and ciliolate. **Seeds** 4–4.5 mm; aril 0.6–1 mm, lobes vestigial to 1/4 length of seed.

Flowering (spring–)late summer. Rocky or gravelly soils in desert scrub; 1200–2000 m; N.Mex.; Mexico (Chihuahua, Coahuila, Durango, Zacatecas).

In the flora area, *Hebecarpa rectipilis* is known only from the type collection in Sierra County.

Hebecarpa rectipilis is very similar to, and may be conspecific with, *H. barbeyana*; it differs in the spreading hairs (versus incurved), wing sepals usually cream to yellow (versus usually purplish, rarely yellow), and fruits without the small and large size extremes (6–8 versus 5–10 mm). However, given the overlap in wing sepal color and fruit size, these taxa could represent phenotypic (and genotypic?) forms of a single polymorphic entity.

3. MONNINA Ruiz & Pavon, Syst. Veg. Fl. Peruv. Chil. 1: 169. 1798 • [For Josephus Monninus (José Moñino y Redondo), eighteenth-century Spanish Count of Florida-Blanca, administrator, and patron of botany]

Herbs, annual [perennial], [shrubs, small trees, rarely lianescent], single-stemmed. **Stems** erect, sparsely pubescent. **Leaves** alternate; petiolate; uniform or dimorphic; blade surfaces pubescent. **Inflorescences** terminal, racemes; peduncle present; bracts deciduous. **Pedicels** present. **Flowers** greenish or cream, to blue-tinged in age or on drying, wings bluish purple [flowers dark blue to purple and keel distally yellow], chasmogamous and, sometimes, cleistogamous, (2–)2.5–3.5 mm; sepals deciduous, rarely subpersistent, glabrous; wings deciduous, rarely subpersistent, 1.5–3 mm, glabrous or proximally sparsely ciliate; keel entire, ± 3-lobed, glabrous; stamens 6[–8] in chasmogamous flowers, divided into 2 groups by central tuft of hairs, fewer in cleistogamous flowers; ovary 1-loculed [2-loculed], or appearing so. **Fruits** samaroid [drupes], indehiscent, margins winged, sparsely pubescent or glabrate. **Seeds** glabrous, not arillate.

Species ca. 180 (1 in the flora): sw United States, Mexico, Central America, South America.

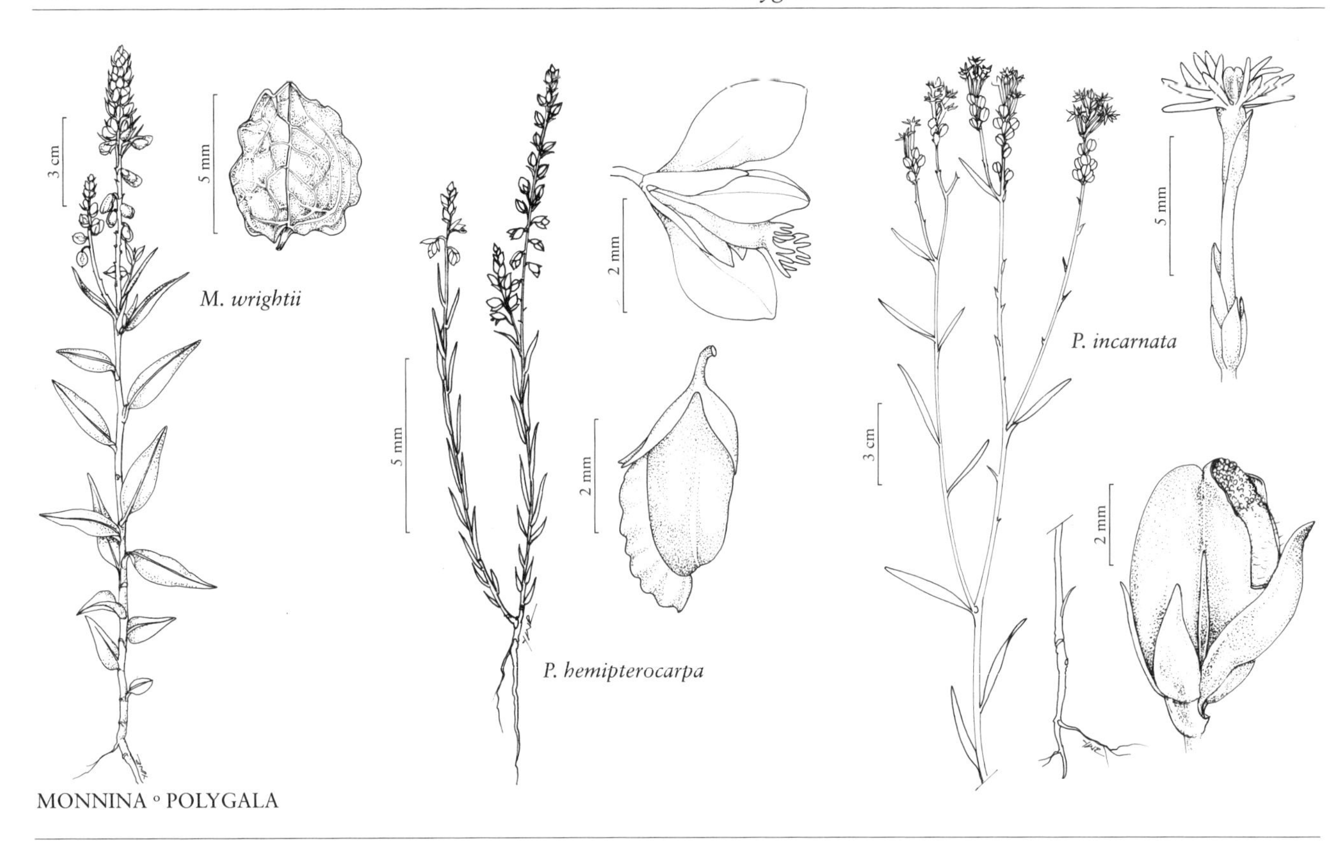

MONNINA ◦ POLYGALA

1. Monnina wrightii A. Gray, Smithsonian Contr. Knowl. 5(6): 31. 1853 • Blue pygmyflower [F]

Pteromonnina wrightii (A. Gray) B. Eriksen

Herbs (2–)3–6(–7) dm. **Stems** branched distally, hairs incurved. **Leaves:** petiole 1–2.5 mm; blade mostly lanceolate or narrowly ovate to elliptic, sometimes obovate proximally, 15–60 × 2–10 mm, base cuneate, apex usually acute to attenuate-acuminate, sometimes obtuse proximally. **Racemes** 2–20(–30) × 0.3–0.6 cm; peduncle to 6 cm; bracts narrowly lanceolate or ovate-subulate. **Pedicels** 0.4–0.5 mm, pubescent. **Cleistogamous flowers** usually present on much-reduced axillary branches, chasmogamous flowers sometimes absent, 1.5–2.1 mm. **Flowers:** non-wing sepals ovate to lanceolate, 1.2–1.8 mm; wings obovate, 1–2 mm wide; keel 1.5–3 mm. **Fruits** ovoid, subglobose, or subobovoid, 3–5 × 2.8–4 mm. **Seeds** 2–3 mm.

Flowering (mid–)late summer–fall. Igneous, limestone, gravelly, sandy, or silty loam soils, in dry to moist, open to shaded grasslands, shrublands, woodlands, disturbed areas; 1200–2600 m; Ariz., N.Mex.; Mexico (Chihuahua, Colima, Durango, Guanajuato, Jalisco, Michoacán, Nayarit, Sinaloa, Sonora, Zacatecas); South America.

Monnina wrightii is amphitropically disjunct in Argentina and Bolivia, with single collections each from Peru and Uruguay.

4. POLYGALA Linnaeus, Sp. Pl. 1: 701. 1753; Gen. Pl. ed. 5, 315. 1754 • [Greek *polys*, much, and *gala*, milk, alluding to supposed effect on foraging cattle]

Herbs, annual, biennial, or perennial, or subshrubs [lianas, shrubs, trees], single- to multi-stemmed. **Stems** mostly erect, sometimes lax to decumbent or prostrate [mat-forming], usually not glaucous, usually glabrous, sometimes pubescent or glabrate. **Leaves** basal and/or cauline, sometimes rosulate; usually alternate, sometimes opposite or whorled; sessile, subsessile, or petiolate; usually uniform; blade surfaces glabrous or pubescent. **Inflorescences**

terminal [axillary or leaf-opposed], racemes, sometimes in cymose panicles, or nearly spikes, [corymbs, fascicles, or 1–5-flowered], usually not interrupted; peduncle present or absent; bracts deciduous or persistent. **Pedicels** usually present, usually glabrous. **Flowers** white, cream, green or greenish, pink, purple, reddish purple, blue, yellow, or orange, chasmogamous (some cleistogamous in *P. crenata*, *P. lewtonii*, and *P. polygama*), 1–9(–10) mm; sepals persistent, glabrous (pubescent in *P. alba* and *P. vulgaris*), margins entire, sometimes ciliate or ciliolate; wings persistent, 1–9 mm, glabrous (pubescent in *P. alba*), margins entire, sometimes ciliolate; keel crested, crest 2-lobed, often fimbriate (lobes subdivided into fingerlike lobes), glabrous (pubescent in *P. alba*); stamens (6–)8 in chasmogamous flowers, fewer in cleistogamous flowers, not grouped; ovary 2-loculed (sometimes 1 abortive). **Fruits** capsules, dehiscent [indehiscent], margins winged or not, entire (crenately winged in *P. crenata*, erose-crenulate in *P. hemipterocarpa*), glabrous [pubescent]. **Seeds** coat without pits (except in *P. hemipterocarpa*, *P. scoparioides*, and *P. watsonii*), usually pubescent (hairs glochidiate in *P. glochidata*), usually arillate.

Species ca. 725 (31 in the flora): North America, Mexico, West Indies, Central America, South America, Europe, Asia, Africa, Australia.

Five non-native *Polygala* species have been reported for the flora area. Only *P. vulgaris* is truly naturalized; the other four, *P. longicaulis* Kunth, *P. myrtifolia* Linnaeus, *P. paniculata* Linnaeus, and *P. serpyllifolia* Hosé are historic waifs. *Polygala serpyllifolia* has been reported from Greenland (J. Devold and P. F. Scholander 1933) based on a single depauperate individual. Additional material has never been seen. *Polygala paniculata* was reported for Maryland by the USDA Plants database, but no specimens are known. It was reported also from Texas (D. S. Correll and M. C. Johnston 1970) but remains undocumented apart from a single specimen (*Eggleston 17398*, NY) collected in 1920 from Fort Davis in Jeff Davis County. *Polygala myrtifolia* is cultivated, especially in California and Florida, and is naturalized in Australia, Hawaii, and New Zealand. The USDA Plants website record for California is based on F. Hrusa et al. (2002), which was based on C. F. Smith (1976); however, the species was not listed by Smith (1998). *Polygala longicaulis* was collected only once, in 1917 from Jackson County, Alabama (ROCH). All four excluded, non-native species of *Polygala* are easily recognizable as part of the genus based on their fimbriate crests. They can be distinguished from the native *Polygala* species based on the following characters. *Polygala longicaulis* is separated by its obconic seeds densely covered in silky hairs, as well as the large, reddish to orangish resin dots medially on the sepals. *Polygala myrtifolia* would be the only shrubby *Polygala* in North America with a fimbriate crest. *Polygala paniculata* would look most similar to a robust *P. glochidata*, but lacks the glochidiate hairs on the seeds, usually has only the most proximal nodes with whorled leaves (often caducous by anthesis), and the stems are finely stipitate-glandular. *Polygala serpyllifolia* is very much like *P. vulgaris* but the proximal leaves are opposite or subopposite.

1. Capsule margins on abaxial locule not winged, adaxial locule longer and winged (sometimes not or obscurely winged in *P. watsonii*); seed coats with rows of pits (0.05 mm diam.); flowers white; racemes loosely flowered, cylindric or subcapitate; Arizona, New Mexico, Texas.
    2. Stems glabrous (rarely with a few scattered hairs) . . . . 12. *Polygala hemipterocarpa*
    2. Stems puberulent or glabrate.
        3. Wing sepals (2.4–)2.7–4.5 mm; racemes cylindric (flowering portion may appear subcapitate from deciduous fruit), (1–)1.5–7.8 × 0.4–0.8 cm . . . . 25. *Polygala scoparioides*
        3. Wing sepals (3.5–)4–6 mm; racemes subcapitate, 0.5–1.4(–2.5) × 1 cm . . . . 31. *Polygala watsonii*
1. Capsule margins not winged or narrowly and equally winged on both locules; seed coats without pits; flowers usually cream, green or greenish, pink, purple, reddish purple, blue, yellow, or orange, seldom white; racemes often densely flowered, sometimes spikelike or nearly so, capitate or cylindric; mostly e, n North America.

[4. Shifted to left margin.—Ed.]

4. Pedicels winged; flowers usually yellow, orange, green, or greenish, rarely white; racemes in cymose panicles or capitate.
   5. Racemes in cymose panicles, ± flat-topped; basal leaf blades obovate, spatulate, narrowly elliptic, linear-lanceolate, or linear, sometimes withered or nearly so at anthesis.
      6. Corollas pale yellowish apically, wing sepals white, cream, or, sometimes, greenish tinged; pedicels 0.6–1.2 mm . . . . 3. *Polygala balduinii*
      6. Corollas and wing sepals bright yellow; pedicels (1–)1.3–4 mm.
         7. Basal leaf blades linear-lanceolate to linear, forming large, persistent rosette; biennials, 4–12 dm . . . . 10. *Polygala cymosa*
         7. Basal leaf blades obovate or spatulate to narrowly elliptic, usually absent at anthesis; annuals or biennials, 1.5–4(–5) dm . . . . 22. *Polygala ramosa*
   5. Racemes solitary (not in panicles), capitate; basal leaf blades usually obovate, oblanceolate, linear-oblanceolate, or spatulate, rarely elliptic, often persistent.
      8. Flowers bright yellow to orange, drying pale yellow to dark greenish brown.
         9. Flowers bright orange, usually drying pale yellow; racemes (0.8–)1.2–2 cm diam. . . . . 17. *Polygala lutea*
         9. Flowers bright yellow, drying pale yellow to dark greenish brown; racemes 1.8–2.5 cm diam. . . . . 23. *Polygala rugelii*
      8. Flowers lemon yellow, greenish yellow, or green, sometimes drying green or yellowish green.
         10. Flowers lemon yellow to greenish yellow; wing sepals elliptic to oblong-lanceolate, tips 0.7–1.6 mm, apices involute; seeds 0.8–1.8 mm. . . . . 19. *Polygala nana*
         10. Flowers green or greenish yellow; wing sepals usually lanceolate, sometimes oblong-lanceolate, tips 0.5–0.9 mm, apices usually flat to slightly involute; seeds 1.9–2.3 mm. . . . . 28. *Polygala smallii*
4. Pedicels not winged; flowers white, cream, green or greenish, pink, purple, reddish purple, or blue, not bright yellow or orange; racemes solitary (not in cymose panicles), sometimes spikelike or nearly so, cylindric, cylindric-conic, or capitate.
   11. Seed hairs glochidiate; leaves whorled (at least proximally); Arizona . . . . 11. *Polygala glochidata*
   11. Seed hairs not glochidiate; leaves usually whorled or alternate, rarely opposite or subopposite; widespread.
      12. Leaves whorled, at least at proximalmost 1 or 2 nodes.
         13. Perennials, multi-stemmed, with thickened caudices; flowers white or greenish white.
            14. Capsules ovoid, ellipsoid, or oblong, 1.3–1.6 mm diam.; w North America . . . . 1. *Polygala alba*
            14. Capsules subglobose to broadly ellipsoid, (1.2–)2–3 mm diam.; se United States . . . . 4. *Polygala boykinii*
         13. Annuals, single-stemmed (usually branched distally), with slender taproots or fibrous roots; flowers mostly pink or purple, sometimes white, greenish, or greenish white.
            15. Racemes spikelike or nearly so, cylindric, cylindric-conic, or conic; bracts deciduous; pedicels 0–1(–2) mm.
               16. Seeds glabrous. . . . . 15. *Polygala leptostachys*
               16. Seeds pubescent. . . . . 29. *Polygala verticillata* (in part)
            15. Racemes mostly capitate or apically rounded, sometimes cylindric; bracts persistent; pedicels (0.7–)1–2.4 mm.
               17. Wing sepals ovate to deltate, apices acuminate, often strongly cuspidate . . . . 8. *Polygala cruciata*
               17. Wing sepals ovate to ovate-oblong or ovate-oblanceolate, apices acute, short-mucronate.
                  18. Wing sepals 1.5–2.5 mm wide; leaf blades 1–4(–7) mm wide . . . . 5. *Polygala brevifolia*
                  18. Wing sepals 1.3–2 mm wide; leaf blades 0.5–1 mm wide . . . . . 13. *Polygala hookeri*

[12. Shifted to left margin.—Ed.]

12. Leaves mostly alternate, proximal leaves sometimes opposite or whorled (in *P. incarnata*, *P. lewtonii*, and *P. polygama*).
  19. Cleistogamous flowers present, usually below, rarely just above, soil surface, sometimes in proximal leaf axils later in season; racemes loosely cylindric; short-lived perennials, biennials, or annuals, usually multi-stemmed, sometimes single-stemmed.
    20. Capsule margins winged . . . . 7. *Polygala crenata*
    20. Capsule margins not winged.
      21. Capsules ellipsoid or oblong, 2.5–3.5 × 1–1.5 mm, length 2–3 times width; flowers 6–8 mm; non-wing sepals 0.8–1.1 mm . . . . 16. *Polygala lewtonii*
      21. Capsules ellipsoid or globose-ovoid, 2–4 × 2–3 mm, length less than 2 times width; flowers 4–6 mm; non-wing sepals 1.3–2.5 mm . . . . 21. *Polygala polygama*
  19. Cleistogamous flowers absent; racemes sometimes spikelike or nearly so, often densely capitate or cylindric, sometimes cylindric-conic; mostly annuals, single-stemmed, sometimes perennials and single- or multi-stemmed.
    22. Leaf blades (1.5–)8–35 mm wide, margins often appearing serrulate (toothlike projections associated with marginal cilia . . . . 26. *Polygala senega*
    22. Leaf blades 0.3–2(–5.5) mm wide (rarely more than 4 mm); margins entire.
      23. Capsule margins narrowly winged; short-lived perennials, usually multi-stemmed (rarely 1 or few-stemmed) from small caudices; British Columbia, Michigan, Oregon . . . . 30. *Polygala vulgaris*
      23. Capsule margins not winged; annuals (sometimes perennial in *P. setacea*), single-stemmed, from taproots; s, e North America.
        24. Wing sepals less than ½ as long as keel petals; stems glaucous; arils scarcely lobed . . . . 14. *Polygala incarnata*
        24. Wing sepals usually (sub)equaling or exceeding keel petals, rarely slightly shorter; stems not or seldom glaucous; arils usually lobed, rarely vestigial.
          25. Leaf blades subulate scales, 0.5–1.6(–2) mm . . . . 27. *Polygala setacea*
          25. Leaf blades linear, linear-oblong, spatulate, oblanceolate, elliptic, linear-subulate, or subfiliform, 3–25(–40) mm.
            26. Wing sepals 1.5–2 times longer than keel petals; aril lobes (½–)⅔–nearly as long as seed . . . . 24. *Polygala sanguinea*
            26. Wing sepals equaling or slightly longer than keel petals (shorter in *P. chapmanii*, sometimes slightly shorter in *P. mariana* and *P. nuttallii*); aril lobes to ⅛–½ length of seed or vestigial.
              27. Bracts persistent (*P. chapmanii* often also with scattered deciduous ones).
                28. Racemes 0.4–0.7 cm diam.; wing sepals 2–3 mm; pedicels 0.5(–1) mm . . . . 20. *Polygala nuttallii*
                28. Racemes 0.6–1.3 cm diam.; wing sepals 2.5–5 mm; pedicels 0.7–2.8 mm.
                  29. Wing sepals 2.5–3.3(–3.8) mm; capsules 1.5–2.3 mm . . . . 6. *Polygala chapmanii*
                  29. Wing sepals 3–5 mm; capsules (1.7–)2.5–3 mm. . . . 9. *Polygala curtissii*
              27. Bracts deciduous (rarely a few persistent and scattered).
                30. Flowers white. . . . . 29. *Polygala verticillata* (in part)
                30. Flowers usually pink or purple, rarely white.
                  31. Racemes cylindric, 1.5–13 × 0.5–0.6 cm; flowers 1.6–2.2 mm; pedicels 0.8–1 mm . . . . 2. *Polygala appendiculata*
                  31. Racemes capitate, (0.5–)1–3.5 × 0.6–1.1 cm; flowers 4–5.7 mm; pedicels 1.5–3.5 mm . . . . 18. *Polygala mariana*

1. Polygala alba Nuttall, Gen. N. Amer. Pl. 2: 87. 1818 • White milkwort

Herbs perennial, multi-stemmed, (0.5–)2–4(–5) dm, infrequently branched; caudex thickened. **Stems** erect, sparsely pubescent or glabrate, hairs minute, incurved. **Leaves** whorled at 1 or 2 nodes proximally, rarely more; sessile or subsessile; blade narrowly oblanceolate or spatulate proximally, to narrowly oblong or linear distally, 4–18(–25) × 1.5–2.5 mm proximally, 8–25 × 0.5–1.2 (–2) mm distally, base cuneate, apex rounded to acute proximally, acute to acuminate distally, surfaces very sparsely pubescent or glabrate, hairs minute, incurved. **Racemes** to nearly spikes, cylindric-conic, (1–)2–8.5 (–12) × 0.4–0.8 cm, often interrupted proximally; peduncle (1–)2–13 cm; bracts deciduous, lanceolate. **Pedicels** 0.1–1 mm, pubescent to subglabrous. **Flowers** white, sepals, including wings, often with greenish or purple longitudinal stripe, keel distally green or yellow, crest often purplish, 2–3.5(–4) mm; sepals ovate to oblong, 1.3–1.8 mm, pubescent; wings elliptic or ovate, 2.2–4 × 1.3–1.5(–2) mm, apex usually obtuse, rarely acute to bluntly rounded, pubescent; keel 3 mm, crest 2-parted, with 4 lobes on each side, pubescent. **Capsules** ovoid, ellipsoid, or oblong, 1.5–3 × 1.3–1.6 mm, margins not winged. **Seeds** 2–2.5 mm, pubescent; aril 0.8–1.5 mm, lobes to (1/5–)1/3–3/4 length of seed. **2*n*** = 24, ca. 36, 52–54.

Flowering spring–summer. Grasslands, scrub, chaparral, open dry woods, open disturbed areas; 200–2000(–2600) m; Sask.; Ariz., Colo., Iowa, Kans., Mont., Nebr., N.Mex., N.Dak., Okla., S.Dak., Tex., Wyo.; Mexico (Baja California Sur, Chiapas, Chihuahua, Coahuila, Durango, Guanajuato, Jalisco, México, Nuevo León, Oaxaca, Querétaro, San Luis Potosí, Sinaloa, Sonora, Tamaulipas, Zacatecas).

Leaves of *Polygala alba* have abaxially embedded, linear to round, punctate glands, which are obscure in dried material but white and waxy when rehydrated.

2. Polygala appendiculata Vellozo, Fl. Flumin., 292. 1829; plate 66. 1831 • Swamp milkwort

Polygala leptocaulis Torrey & A. Gray

Herbs annual, single-stemmed, (1–)2–5(–6) dm, branched distally; from taproot, usually quickly becoming fibrous root cluster. **Stems** erect, glabrous. **Leaves** alternate; sessile or subsessile; blade linear to subfiliform, 8–25 × 0.5–1 mm, base obtuse to cuneate, apex acute to acuminate, surfaces glabrous. **Racemes** cylindric, 1.5–13 × 0.5–0.6 cm; peduncle 0.5–1 cm; bracts deciduous, narrowly lanceolate-ovate. **Pedicels** 0.8–1 mm, glabrous. **Flowers** usually purplish pink or lavender-pink, rarely white, 1.6–2.2 mm; sepals ovate to narrowly lanceolate-ovate, 0.6–1 mm; wings obovate or elliptic, 1.5–2 × 0.8–1 mm, apex obtuse to bluntly rounded; keel 1.5–2 mm, crest 2-parted, with 2–3 lobes on each side. **Capsules** oblong to ellipsoid, 1.4–2 × 0.7–1.1 mm, margins not winged. **Seeds** 1–1.2 mm, pubescent; aril 0.1(–0.2) mm, lobes less than 1/8 length of seed.

Flowering spring–early summer. Savannas, pastures, bogs, open pine woods, pond margins; 0–100 m; Fla., La., Miss., Tex.; Mexico (Campeche, México, Michoacán, Tabasco, Veracruz); West Indies; Central America; South America.

Polygala appendiculata is part of a widespread complex in Latin America, potentially representing more than one evolutionary lineage. Even if found to be a single lineage, it is one with a complicated nomenclatural history (J. F. B. Pastore 2013).

3. Polygala balduinii Nuttall, Gen. N. Amer. Pl. 2: 90. 1818 (as balduini) • Baldwin's milkwort

Pilostaxis balduinii (Nuttall) Small; *P. carteri* (Small) Small; *Polygala balduinii* var. *carteri* (Small) R. R. Smith & D. B. Ward; *P. carteri* Small

Herbs annual or biennial, single- to multi-stemmed, (1–)2–6.5 dm, rarely branched proximal to inflorescences; from fibrous root cluster. **Stems** erect, glabrous. **Leaves** with basal rosette sometimes present at anthesis, sometimes withered or nearly so; alternate; sessile or subsessile; basal blade obovate, to 12 mm wide, cauline blade obovate, elliptic, or narrowly spatulate, sometimes becoming linear distally, 3–25 × 1–5 mm, base cuneate or acute, apex usually rounded proximally, acute to obtuse distally, surfaces glabrous. **Racemes** in cymose panicles, ± flat-topped, 1.5–12 × 2.5–6 cm; each stem with (3–)5–20(–40) racemose branches, 0.4–6 × 0.7–1.5 cm; central one nearly sessile, lateral ones with peduncle to 2 cm; bracts persistent, lanceolate-ovate. **Pedicels** winged, 0.6–1.2 mm, glabrous. **Flowers** usually white or cream, sometimes greenish tinged, drying white to brownish green, corolla usually becoming pale yellowish apically, 3–6 mm; sepals decurrent on pedicel, ovate, lanceolate, or linear-lanceolate, 1.6–2.5 mm; wings narrowly ovate to slightly obovate or elliptic, 2.8–4.8 × 0.9–1.7 mm, apex narrowing into apical cusp, 0.6–0.8 mm; keel 2–4 mm, crest 2-parted, with 3 2-fid (rarely entire) lobes on each side. **Capsules** depressed-suborbicular, 0.6–1 × 0.8–1.2 mm, margins not winged. **Seeds** 0.5–1 mm, pubescent; aril 0–0.4 mm, lobes absent, reduced to minute scales, or to 1/3–2/5 length of seed. **2*n*** = 64, 68.

Flowering year-round. Bogs, marshes, prairies, wet flatwoods, coastal swales, open degraded areas; 0–100 m; Ala., Fla., Ga., Miss., Tex.; West Indies (Bahamas, w Cuba).

Plants from southern Florida have been recognized as var. *carteri*, based on their less robust stature and greener, less conspicuous inflorescences (R. R. Smith and D. B. Ward 1976). More specifically, extremes of var. *carteri* have more or less elongated racemes to 6 cm, bracts usually less than 2 mm, flowers cream to greenish white, seeds more than 0.6 mm, and arils a minute scale or absent, whereas extremes of var. *balduinii* have dense racemes to 3 cm, bracts usually more than 2 mm, flowers white, seeds less than 0.6 mm, and arils usually 0.2 mm (infrequently smaller). Such apparently distinctive features all intergrade and are not sharply geographically delineable; noting this, Smith and Ward stated that the northern limit of the variety was arbitrary.

Polygala ramosa is closely related to *P. balduinii*, despite the obvious contrast between the yellow-flowered (green when dry), loosely branched inflorescences of *P. ramosa* and the white or near-white, more compact inflorescences of *P. balduinii*. Hybrids occur and have been called *P. balduinii* var. *chlorogena* Torrey & A. Gray.

4. Polygala boykinii Nuttall, J. Acad. Nat. Sci. Philadelphia 7: 86. 1834 (as boykini) • Boykin's milkwort

Polygala boykinii var. *sparsifolia* Wheelock; *P. boykinii* var. *suborbicularis* R. W. Long; *P. flagellaris* Small; *P. praetervisa* Chodat

Herbs short-lived perennial, usually multi-stemmed, (1.5–)2–6 dm, infrequently branched distally; caudex thickened. **Stems** usually erect or decumbent, rarely prostrate, glabrous. **Leaves** whorled, at least proximally, to alternate distally; sessile or subsessile, rarely with narrowed petiolelike base to 1 mm; blade linear, elliptic, lanceolate, or obovate, 3–25 × 1–5 mm, base cuneate, margins sometimes erose, appearing toothed, apex often acute, sometimes obtuse or rounded proximally, surfaces glabrous. **Racemes** narrowly cylindric, 3–12(–15) × 0.4–0.8 cm; peduncle to 9 cm (elongate); bracts deciduous, lanceolate-ovate. **Pedicels** to 0.5 mm (to 1 mm in fruit), glabrous. **Flowers** white or greenish white, 2.5–3.2 mm; sepals ovate to elliptic, 1–1.3 mm; wings suborbiculate, broadly elliptic, or obovate, 2–3 × 1.5–2.5 mm, apex obtuse to bluntly rounded; keel 2–3 mm, crest 2-parted, with 2–4 lobes on each side. **Capsules** subglobose to broadly ellipsoid, 1.7–4 × (1.2–)2–3 mm, margins not winged. **Seeds** 1–2.5 mm, pubescent; aril 0.5–1.5 mm, lobes ⅓–⅔ length of seed. 2*n* = ca. 28, 96.

Flowering spring–summer. Open slopes, prairielike sites, cut-over oak-pine forests, limestone outcrops, flatwoods; 0–200 m; Ala., Fla., Ga., La., Miss., Tenn.; West Indies.

Polygala boykinii is highly variable, with numerous distinctive forms in southern Florida and the Caribbean, some of which may merit specific recognition.

5. Polygala brevifolia Nuttall, Gen. N. Amer. Pl. 2: 89. 1818 • Short-leaf milkwort [E]

Herbs annual, single-stemmed, 1–4(–6) dm, usually much-branched distally; from slender taproot, sometimes becoming fibrous root cluster. **Stems** erect, glabrous. **Leaves** whorled proximally, often alternate distally; sessile or subsessile; blade obovate, oblanceolate, narrowly elliptic, or linear, 10–50 × 1–4(–7) mm, base cuneate, apex acute, obtuse, or rounded, surfaces glabrous. **Racemes** capitate, 1–3.5 × 1–1.6 cm; peduncle 0.8–5 cm; bracts persistent, narrowly triangular-ovate. **Pedicels** 0.7–2 mm, glabrous. **Flowers** purple or pink, wings sometimes green-tinged, sepals often pink, 3.5–4.5 mm; sepals suborbiculate to ovate, 0.8–1.5 mm; wings ovate or ovate-oblong, 3–4.5 × 1.5–2.5 mm, apex acute, short-mucronate; keel 2.8–3.8 mm, crest 2-parted, with 3 or 4 lobes on each side, or lobes entire. **Capsules** globose or globose-reniform, 2 × 2 mm, margins not winged. **Seeds** 1.3–1.5 mm, sparsely pubescent; aril 0.8–1.3 mm, lobes ⅔+ length of seed.

Flowering summer–fall. Savannas, pocosins, sandy swamps, coastal swales; 0–50 m; Ala., Fla., Ga., La., Miss., N.J., N.Y., N.C.

Polygala brevifolia is very similar to *P. cruciata*, from which it is most readily distinguished by the usually much longer peduncle and by the wings acute or slightly acuminate (versus cuspidate). Ecologically, *P. brevifolia* tends to be more restricted to wetter areas, such as bogs and seeps. Although reported for Maryland, South Carolina, and Virginia, no vouchers are known.

6. Polygala chapmanii Torrey & A. Gray, Fl. N. Amer. 1: 131. 1838 • Chapman's milkwort [E]

Herbs annual, usually single-stemmed, rarely multi-stemmed, 2–5 dm, unbranched or few branched distally; from taproot, sometimes becoming fibrous root cluster. **Stems** erect, glabrous. **Leaves** alternate; sessile or subsessile; blade linear to narrowly elliptic proximally, or linear-subulate distally, 6–23 × (0.5–)1(–2) mm, base cuneate, apex acute to acuminate, surfaces glabrous.

Racemes capitate to densely cylindric, 0.7–3(–4) × 0.6–1.3 cm; peduncle 0.3–2 cm; bracts persistent, often with scattered deciduous ones as well, lanceolate-ovate. **Pedicels** 0.7–1.6 mm, glabrous. **Flowers** pink or pale purple, 2.8–4 mm; sepals ovate or lanceolate-ovate, 0.6–1.3 mm; wings ovate or elliptic, 2.5–3.3(–3.8) × 1.7–2 mm, apex obtuse to bluntly rounded; keel 2.8–4 mm, crest 2-parted, with 1 lobe on each side, other lobes reduced to low, blunt processes. **Capsules** subglobose, 1.5–2.3 × 1.5–2.3 mm, margins not winged. **Seeds** 1–1.5 mm, pubescent; aril 0.5–0.8 mm, lobes ⅓–½ length of seed. **2*n*** = 72.

Flowering spring–summer. Wet flatwoods, bogs; 0–100 m; Ala., Fla., Ga., La., Miss.

Polygala chapmanii occurs almost exclusively in the coastal counties (or one county removed) of the Gulf Coast, from eastern Louisiana to northwestern Florida, with the exception of an outlying occurrence in Georgia and a few counties in Alabama.

7. **Polygala crenata** C. W. James, Rhodora 59: 53. 1957 • Scalloped milkwort [E]

Polygala polygama Walter forma *obovata* S. F. Blake, Rhodora 17: 201. 1915

Herbs short-lived perennial (rarely biennial or annual), single- or multi-stemmed, (1.2–)2–3(–3.5) dm, usually unbranched, rarely branched distally; from taproot or fibrous root cluster. **Stems** erect, sometimes laxly so, to nearly decumbent, glabrous. **Leaves** alternate; sessile or subsessile, or with narrowed petiolelike base to 2 mm; blade obovate or elliptic, sometimes becoming scalelike proximally, (5–)8–23 × (2–)3–8(–9) mm, base cuneate, apex rounded, obtuse, or acute, surfaces glabrous. **Racemes** loosely cylindric, open, elongate, 7–10(–15) × 1–1.5 cm; peduncle 1–2(–2.5) cm; bracts deciduous, ovate. **Pedicels** 3–4(–5) mm, glabrous. **Cleistogamous flowers** present in racemes usually below, rarely just above, soil surface, sometimes in proximal leaf axils later in season. **Flowers** pink to pale purple, 4–6 mm; sepals ovate or elliptic, 1.1–2 mm; wings elliptic, ovate, or obovate, (3.3–)4–5 × (1.5–)2–2.7 mm, apex obtuse to bluntly rounded; keel 3–4 mm, crest 2-parted, with 2–4 entire or divided lobes on each side. **Capsules** broadly ellipsoid, 2–3.5 × 2–3 mm, margins narrowly winged, wings equal, crenate. **Seeds** 1.5–2 mm, pubescent; aril 1 mm, somewhat helmetlike, lobes to ¼ length of seed.

Flowering spring–summer. Wet flatwoods and savannas, sandy mesic hills, bogs, acidic swamps; 0–100 m; Ala., Fla., Ga., La., Miss., Tex.

Polygala crenata is known only from the East Gulf Coastal Plain.

8. **Polygala cruciata** Linnaeus, Sp. Pl. 2: 706. 1753 • Drumheads [E]

Polygala aquilonia (Fernald & B. G. Schubert) Sorrie & Weakley; *P. cruciata* subsp. *aquilonia* (Fernald & B. G. Schubert) A. Haines; *P. cruciata* var. *aquilonia* Fernald & B. G. Schubert; *P. ramosior* (Nash ex B. L. Robinson) Small

Herbs annual, single-stemmed, (0.5–)1–3(–5) dm, usually branched distally; from slender taproot. **Stems** erect, usually glabrous, rarely subglabrous. **Leaves** whorled, sometimes alternate distally; sessile or subsessile, or with narrow petiolelike base to 2 mm; blade mostly linear, to oblanceolate, spatulate, obovate, or narrowly elliptic, especially proximally, 8–35(–50) × 1–5(–7) mm, base cuneate to acute, apex rounded to obtuse or acute, surfaces glabrous. **Racemes** capitate to densely cylindric, 1–3.5(–6) × 1–1.7 cm; peduncle 0.5–3(–5) cm; bracts persistent, narrowly lanceolate-ovate. **Pedicels** 2–2.4 mm, glabrous. **Flowers** usually purple or pink, rarely white, wings and distal keel sometimes green-tinged, sepals often pink, 4–6 mm; sepals ovate, 0.8–1.5 mm, sometimes ciliolate; wings ovate to deltate, 3.5–6 × 2.7–4 mm, apex acuminate, often strongly cuspidate; keel 2.8–3.5 mm, crest 2-parted, with 2 or 3 entire or 2-fid lobes on each side. **Capsules** with winged, stipelike base, strongly oblique, subglobose, 2–2.5 × 1.8–2.1 mm, margins not winged. **Seeds** 1.1–1.5 mm, short-pubescent; aril 0.9–1.1 mm, lobes usually ⅔+ length of seed, rarely shorter or absent. **2*n*** = 36, 40.

Flowering spring–fall. Wet meadows, marshes, savannas, bogs, pocosins, sand dunes; 0–500 m; Ala., Ark., Conn., Del., D.C., Fla., Ga., Ill., Ind., Iowa, Ky., La., Maine, Md., Mass., Mich., Minn., Miss., N.H., N.J., N.Y., N.C., Ohio, Okla., Pa., R.I., S.C., Tenn., Tex., Va., W.Va., Wis.

Polygala cruciata var. *aquilonia* is usually a more northern form with broader leaf blades, shorter peduncles, narrower racemes, and more abruptly short-acuminate wing apices (versus strongly cuspidate and acuminate) than the more southern var. *cruciata*. Extreme forms can appear distinctive. However, in the absence of detailed populational study, these traditionally recognized varieties of *P. cruciata* do not merit taxonomical recognition at any rank. This species was known from Ontario but appears to be extirpated from Canada.

SELECTED REFERENCE Sorrie, B. A. and A. S. Weakley. 2017. Reassessment of variation within *Polygala cruciata* sensu lato (Polygalaceae). Phytoneuron 2017-37: 1–9.

9. Polygala curtissii A. Gray, Manual ed. 5, 121. 1867 • Curtiss's or Appalachian milkwort [E]

Herbs annual, single-stemmed, 1–4 dm, usually branched distally; from taproot, sometimes becoming fibrous root cluster. **Stems** erect, glabrous. **Leaves** alternate; sessile or subsessile; blade linear to linear-oblong or narrowly oblanceolate, 10–20 × 1–2 mm, base cuneate, apex acute, surfaces glabrous. **Racemes** capitate, 1–2 × 0.8–1.3 cm; peduncle to 5 cm (elongate); bracts persistent, lanceolate. **Pedicels** 1.3–2.8 mm, glabrous. **Flowers** usually pink or purple, rarely white, 3.2–5 mm; sepals ovate, 1.5–2 mm; wings elliptic to ovate-elliptic, 3–5 × 1–2 mm, apex acute, obtuse, or bluntly rounded; keel 2.5–2.7 mm, crest 2-lobed, with 2–4 lobes on each side. **Capsules** subglobose, (1.7–) 2.5–3 × 2.5–3 mm, margins not winged. **Seeds** 1–1.5 mm, pubescent; aril 1–1.2 mm, lobes to ⅓ length of seed. **2*n*** = 40.

Flowering summer–fall. Dry, sandy meadows, old fields, open woods; 0–1300 m; Ala., Del., D.C., Ga., Ky., Md., Miss., N.J., N.C., Ohio, Pa., S.C., Tenn., Va., W.Va.

Polygala curtissii is common in the southeastern United States, but rare in the northern part of its range in Ohio and Pennsylvania.

10. Polygala cymosa Walter, Fl. Carol., 179. 1788 • Tall pine-barren milkwort [E]

Pilostaxis cymosa (Walter) Small; *Polygala attenuata* Nuttall

Herbs biennial, usually single-stemmed, 4–12 dm, usually unbranched proximal to inflorescences; from fibrous root cluster. **Stems** erect, glabrous. **Leaves** with persistent, large basal rosette, sometimes withered at anthesis, brown leaves usually present; alternate; sessile or subsessile; basal blade linear-lanceolate to linear, (30–)35–70(–140) × 2–7 mm, cauline blade abruptly reduced to bracts of inflorescence, base obtuse, apex acute, obtuse, or rounded, surfaces glabrous. **Racemes** in cymose panicles, ± flat-topped, 5–11 × 6–20 cm; each stem with to 150 racemose branches, 2–5 × 1–1.5 cm; central ones nearly sessile, lateral ones with peduncle to 3 cm; bracts persistent, narrowly lanceolate-ovate. **Pedicels** winged, 2–4 mm, glabrous. **Flowers** bright yellow, becoming pale yellow to green on drying, 4.5–6.4 mm; sepals decurrent on pedicel, ovate to lanceolate-ovate, (1–)1.5–3.2 mm, sometimes sparsely ciliolate; wings elliptic to obovate or oblanceolate, (2–)4–6.2 × 1.4–2.2 mm, apex abruptly cuspidate, involute; keel 4.5–4.8 mm, crest 2-parted, with 2 or 3 entire or 2-fid lobes on each side. **Capsules** subglobose, 1–1.5 × 1–1.5 mm, margins not winged. **Seeds** 0.6–0.9 mm, glabrous; aril minute (usually appearing vestigial, usually less than 0.05 mm), unlobed. **2*n*** = 64, 68.

Flowering spring–summer. Wet depressions in savannas, pine flatwoods, bogs, edges of acidic swamps, cypress ponds, Carolina bays, marshes, emergent aquatics, especially in seasonally ponded Coastal Plain depressions; 0–100 m; Ala., Del., Fla., Ga., La., Miss., N.C., S.C.

Reports of *Polygala cymosa* from Maryland (M. L. Brown and R. G. Brown 1984, as cited in the USDA Plants database) and Texas (M. Pollard et al. 2009) are in error; no vouchers are known from Maryland and the specimen cited for Texas is *P. ramosa*.

11. Polygala glochidata Kunth in A. von Humboldt et al., Nov. Gen. Sp. 5(fol.): 313; 5(qto.): 400. 1823 • Glochidiate milkwort

Herbs annual, single-stemmed, 1–3 dm, unbranched or branched throughout; from taproot or fibrous root cluster. **Stems** erect, pubescent or subglabrous, hairs spreading. **Leaves** whorled, at least proximally; sessile or subsessile; blade narrowly oblanceolate-spatulate to linear or scalelike, 3–25 × 0.5–2(–5) mm, base cuneate, apex rounded or acute, surfaces pubescent. **Racemes** cylindric, 1–3 × 0.5 cm; peduncle to 1 cm; bracts deciduous, narrowly ovate. **Pedicels** 0.5–1 mm, glabrous. **Flowers** pink, purple, or white, 2.5–4.5 mm; sepals ovate to elliptic, 0.5–1 mm; wings mostly elliptic, to ovate or obovate, 3–5 × 2–3 mm, apex acute or obtuse; keel 2.7–3.8 mm, crest 2-parted with 2–4 entire or 2-fid lobes on each side. **Capsules** ellipsoid, 1.5–2 × 1 mm, margins not winged. **Seeds** 0.7–1.1 mm, pubescent, hairs glochidiate; aril vestigial or, rarely, to 0.1 mm, unlobed.

Flowering summer–late fall. Grasslands, montane slopes, near ephemeral springs or seeps; 1000–1800 m; Ariz.; Mexico (Aguascalientes, Baja California Sur, Chihuahua, Durango, Guanajuato, Jalisco, México, Michoacán, Sonora, Tabasco, Veracruz); West Indies; Central America; South America.

Polygala glochidata occurs in Cochise, Pima, and Santa Cruz counties.

12. Polygala hemipterocarpa A. Gray, Smithsonian Contr. Knowl. 5(6): 31. 1853 • Winged milkwort [F]

Herbs perennial, single- or multi-stemmed, 1.2–4(–5.6) [–7.5] dm, unbranched or branched distally; from thickened caudex. **Stems** erect, sometimes slightly glaucous, usually glabrous, rarely with few scattered, minute hairs, those appressed to incurved. **Leaves** alternate; sessile or subsessile; blade linear to linear-lanceolate, (3–)6–25 × 0.6–1 mm, base obtuse, apex acute to subacuminate, surfaces glabrous. **Racemes** loosely cylindric, 3–15(–21) × 0.4–0.8 cm; peduncle 0.5–2.5 cm; bracts deciduous, ovate to lanceolate-ovate. **Pedicels** 0.5–1 mm, usually glabrous, rarely puberulent. **Flowers** white, greenish veined, 3–4.5 mm; sepals ovate, 1.6–2.3 mm; wings obovate or elliptic-ovate, 3–4.3 × 1.6–1.8 mm, apex obtuse (rarely to bluntly rounded); keel 3.4 mm, crest 2-parted, with 3 lobes on each side, each 2–4-lobed. **Capsules** oblong, 4–6.3 × 2.3 mm, abaxial locule not winged, adaxial locule longer, winged, wing broadly scarious, erose-crenulate. **Seeds** 2.7 mm, pubescent, coat with rows of pits 0.05 mm wide; arils 2 mm, lobes ⅓–⅔ length of seed, vestigial, or to ⅓ length of seed in wingless abaxial locule.

Flowering spring–fall. Rocky slopes in grasslands, open woodlands; 1400–2600 m; Ariz., N.Mex., Tex.; Mexico (Chihuahua, Durango, eastern Sonora, Tamaulipas).

Polygala hemipterocarpa is known only from the Chihuahuan and eastern Sonoran deserts.

13. Polygala hookeri Torrey & A. Gray, Fl. N. Amer. 1: 671. 1840 • Hooker's milkwort [E]

Polygala attenuata Hooker, J. Bot. (Hooker) 1: 195. 1834, not Nuttall 1818

Herbs annual, single-stemmed, 1–2.5(–4) dm, usually sparsely branched distally; from slender taproot. **Stems** erect, sometimes laxly so or sprawling, glabrous or puberulent distally. **Leaves** whorled proximally, alternate distally; sessile or subsessile; blade linear to narrowly spatulate, sometimes scalelike proximally, 4–11 × 0.5–1 mm, base cuneate, apex acute, surfaces glabrous or puberulent, hairs glanduliform (as small glandlike projections less than 0.05 mm). **Racemes** loosely conic-cylindric to narrowly capitate, 0.7–4 × 0.6–0.9 cm; peduncle usually 3–7 cm; bracts persistent, triangular-ovate. **Pedicels** 1.5–2 mm, glabrous. **Flowers** pink or greenish, keel sometimes yellowish distally, 3–4 mm; sepals ovate, 1.1–1.5 mm; wings ovate, oblong-ovate, or ovate-lanceolate, 3–4 × 1.3–2 mm, apex acute, short-mucronate, ciliolate on distal margin, hairs sometimes resembling minute, stipitate glands; keel 2.8–3.8 mm, crest 2-parted, with 2 or 3 entire lobes on each side. **Capsules** with strongly winged stipelike base, globose, 1.4–2.2 × 1.4–2.2 mm, margins not winged. **Seeds** 1.2 mm, short-pubescent; aril 1.2 mm, lobes subequal to length of seed.

Flowering summer. Wet flatwoods, low pinelands, bogs, savannas; 0–100 m; Ala., Fla., La., Miss., N.C., S.C.

Reports of *Polygala hookeri* from Georgia (A. Weakley, pers. comm. as cited in the USDA Plants database) and Texas (S. L. Hatch et al. 1990) appear to be erroneous, as no vouchers are known.

14. Polygala incarnata Linnaeus, Sp. Pl. 2: 701. 1753 • Procession flower, slender milkwort [F]

Galypola incarnata (Linnaeus) Nieuwland

Herbs annual, single-stemmed, (1–)2–4(–7) dm, sparsely branched (rarely from near base); from taproot. **Stems** erect, glaucous, glabrous. **Leaves** mostly alternate, sometimes whorled or opposite at lowest 1–2 nodes proximally; sessile or subsessile; blade linear, (3–)4–12(–17) × 0.3–1 mm, base obtuse to acute or slightly cuneate, apex acuminate, surfaces glabrous. **Racemes** (rarely spikelike), densely cylindric, (0.6–)1–3.5(–4) × 1–1.3 cm; peduncle 2–4 cm; bracts deciduous, ovate to lanceolate-ovate. **Pedicels** (0–)0.5–1 mm, glabrous. **Flowers** usually pink to purple, rarely white, sepals pink, 5–8(–10) mm; sepals narrowly ovate to linear-oblong or lanceolate, 0.9–1.5 (–2.5) mm; wings narrowly ovate to linear-oblong or narrowly oblanceolate, 1.5–3(–4) × 0.6–1 mm, apex acute to obtuse; keel 5–8 mm, crest 2-parted, with 3 lobes on each side, each 2–4-lobed. **Capsules** ellipsoid or subglobose-ovoid, 2.4–3.5 × 1.5–3 mm, margins not winged. **Seeds** 1–2.2 mm, pubescent; aril to 1.1 mm, not obviously lobed, to ¼ length of seed.

Flowering spring–late summer (year-round). Prairies, meadows, savannas, bogs, old fields, open woodlands; 0–400 m; Ont.; Ala., Ark., Del., D.C., Fla., Ga., Ill., Ind., Iowa, Kans., Ky., La., Md., Mich., Miss., Mo., N.J., N.Y., N.C., Ohio, Okla., Pa., S.C., Tenn., Tex., Va., Wis.; Mexico (Veracruz); Central America (to Nicaragua).

Polygala incarnata is the only species in the flora area with the keel two (or rarely three) times the length of the sepal wings. It appears to be relatively rare over much of its range.

15. Polygala leptostachys Shuttleworth ex A. Gray, Smithsonian Contr. Knowl. 3(5): 41. 1852 • Threadleaf milkwort [E]

Herbs annual, single-stemmed, 1.5–3.5(–5) dm, unbranched or branched distally; from slender taproot, sometimes becoming fibrous root cluster. **Stems** erect, glabrous. **Leaves** whorled, at least proximally, to alternate distally; sessile or subsessile; blade narrowly obovate to linear-spatulate proximally, linear to filiform distally, 3–30 × 0.5–1(–2.5) mm, base cuneate, apex acute, surfaces glabrous. **Racemes** spikelike, narrowly cylindric, 0.5–6 × 0.2–0.4 cm; peduncle 0.7–3 cm; bracts deciduous, ovate to lanceolate-ovate. **Pedicels** 0–0.5 mm, glabrous. **Flowers** white or greenish white, 1.5–2.8 mm; sepals ovate to elliptic, 0.6–1.1 mm; wings elliptic or obovate, 1.5–2.6 × 0.8–1 mm, apex obtuse to bluntly rounded; keel 1.3–2.5 mm, crest 2-lobed, with 2 or 3 lobes on each side. **Capsules** ellipsoid to oblong, 1.7–2.2 × 1–1.5 mm, margins not winged. **Seeds** 1.5–2 mm, glabrous; aril 1 mm, lobes ½ length of seed.

Flowering spring–summer. Sandhills, dry to xeric pine-oak woodlands; 0–100 m; Ala., Fla., Ga., Miss.

Polygala leptostachys is most common in Florida and rare elsewhere in its range.

16. Polygala lewtonii Small, Bull. Torrey Bot. Club 25: 140. 1898 • Lewton's milkwort [C] [E]

Herbs short-lived perennial (rarely biennial or annual), usually multi-stemmed, 1–2.5 dm, unbranched (or rarely branched medially to distally); from taproot (or rarely fibrous root cluster). **Stems** erect, sometimes laxly so, to nearly decumbent, glabrous. **Leaves** usually alternate, sometimes subopposite or opposite when scalelike proximally (less than 2 mm); sessile or subsessile, or with narrow petiolelike portion to 2(–3) mm; blade spatulate to linear-oblong, sometimes scalelike proximally, 5–15(–30) × 1–3(–6) mm, base cuneate, apex obtuse to rounded or acute, surfaces glabrous. **Racemes** loosely cylindric (open, elongate), 1.3–6(–10) × 0.8–1.6 cm; peduncle 0.5–1(–2) cm; bracts deciduous, ovate to oblong-ovate or elliptic. **Pedicels** (1–)2(–3) mm, glabrous. **Cleistogamous flowers** present in racemes usually below, rarely just above, soil surface, sometimes in proximal leaf axils later in season. **Flowers** pink to pale purple, 6–8 mm; sepals ovate, 0.8–1.1 mm; wings elliptic or obovate, 3–6 × 1.5–2.7 mm, apex obtuse to bluntly rounded; keel (2–)3–5 mm, crest 2-parted, with 2 or 3 lobes on each side, each lobe subdivided into 2–4 lobes. **Capsules** ellipsoid or oblong, 2.5–3.5 × 1–1.5 mm, margins not winged. **Seeds** 2.3–3.3 mm, pubescent; aril 2–3 mm, lobes ¾ to equal length of seed.

Flowering spring–summer. Sandhills, scrub; of conservation concern; 0–100 m; Fla.

Polygala lewtonii occurs in Highlands, Lake, Marion, Orange, Osceola, and Polk counties, according to the U.S. Fish and Wildlife Service (2010).

Polygala lewtonii is in the Center for Plant Conservation's National Collection of Endangered Plants.

SELECTED REFERENCE U.S. Fish and Wildlife Service. 2010. Lewton's Polygala (*Polygala lewtonii*). Five Year Review: Summary and Evaluation. Vero Beach.

17. Polygala lutea Linnaeus, Sp. Pl. 2: 705. 1753 • Orange milkwort [E]

Pilostaxis lutea (Linnaeus) Small

Herbs biennial or short-lived perennial, single- or multi-stemmed, 0.6–5 dm, unbranched or branched distally; from taproot or fibrous root cluster. **Stems** erect, sometimes laxly so, to nearly decumbent, glabrous. **Leaves** usually with basal rosette; alternate; sessile or subsessile, or with narrow petiolelike region to 1–2 mm; basal blade obovate, oblanceolate, or spatulate, cauline becoming narrowly ovate or nearly linear distally, basal to 60 × 20 mm, cauline to 40 × 10 mm, succulent, base cuneate, apex bluntly rounded to obtuse or acute, especially distally, surfaces glabrous. **Racemes** capitate, 0.8–3.5(–4) × (0.8–)1.2–2 cm; peduncle 3–10 cm; bracts deciduous, narrowly lanceolate. **Pedicels** winged, 1.5–2.8 mm, glabrous. **Flowers** usually bright orange, rarely yellow-orange, usually drying pale yellow, 4.5–6 mm; sepals decurrent on pedicel, ovate, 1.2–2 mm, ciliolate; wings elliptic, 5–7.5 × 2.7–3.6 mm, apex acuminate to abruptly cuspidate, partially involute; keel 3.5–6 mm, crest 2-parted, with 2–4 lobes on each side, each lobe entire or divided. **Capsules** broadly ellipsoid to obovoid, 1.2–2.3 mm, margins not winged. **Seeds** 1–1.6 mm, pubescent; aril 0.5–1.6 mm, lobes ½ to subequal length of seed. $2n$ = 64, 68.

Flowering spring–fall (nearly year-round). Moist to wet soils (at least seasonally), open fields, savannas, pine flatwoods, sandy mixed pine-hardwoods, bogs, pocosins, pond margins; 0–200(–300) m; Ala., Del., Fla., Ga., La., Md., Miss., N.J., N.Y., N.C., Pa., S.C., Va.

A single lemon-yellow flowered plant of *Polygala lutea* has been reported from Brunswick County, North Carolina (R. R. Smith and D. B. Ward 1976); populations elsewhere may also produce yellow or yellow-orange flowers. Smith and Ward also reported that a possible hybrid with *P. rugelii* had over 65% apparently nonfunctional pollen grains. DNA analysis of the nrITS region (J. R. Abbott, unpubl.) found the hybrids to be polymorphic at all of the bases that differed between the parents; coupled with their rarity in the landscape despite common co-occurrence with the parents, this supports the hypothesis that they are F1 hybrids rather than established introgressives.

18. Polygala mariana Miller, Gard. Dict. ed. 8, Polygala no. 6. 1768 • Maryland milkwort [E]

Polygala harperi Small

Herbs annual, single-stemmed, 1.5–5 dm, unbranched or mostly branched distally (sometimes throughout); from taproot (or rarely fibrous root cluster). **Stems** erect, glabrous or sparsely pubescent distally, hairs incurved. **Leaves** alternate; sessile or petiolate, petiole to 1 mm; blade narrowly spatulate proximally to linear distally, (6–)10–20(–25) × (0.5–)1–2(–2.5) mm, base cuneate or acute, apex acute, surfaces glabrous or sparsely pubescent. **Racemes** capitate, (0.5–)1–3.5 × 0.6–1.1 cm; peduncle 0.2–2 cm; bracts usually deciduous, infrequently a few persistent, scattered, lanceolate-ovate. **Pedicels** 1.5–3.5 mm, glabrous. **Flowers** pink or purple, 4–5.7 mm; sepals elliptic to ovate-lanceolate, 0.8–1.8 mm; wings ovate, elliptic, or obovate, 2.5–4.5(–5.2) × 1.1–3(–3.6) mm, apex acute to obtuse, often minutely apiculate; keel 2.3–3.5(–4) mm, crest 2-parted, with 2–4 lobes on each side. **Capsules** subglobose or ellipsoid, 1.5–2.3 × 1–2 mm, margins not winged. **Seeds** 0.9–1.2 mm, pubescent; aril 0.4 mm, lobes nearly vestigial to ⅓ length of seed. **2*n*** = 34.

Flowering spring–fall. Dry to moist, sandy meadows, bogs, savannas, open wet areas, open mixed pine-hardwoods; 0–300 m; Ala., Ark., Del., D.C., Fla., Ga., Ky., La., Md., Miss., N.J., N.Y., N.C., S.C., Tenn., Tex., Va.

Polygala mariana is polymorphic (for example, flower color and inflorescence and flower size); separation into discrete taxa has been unsuccessful. Some specimens resemble *P. curtissii*, which has persistent bracts and usually deeper pink flowers. The two species are largely allopatric, with *P. mariana* predominantly on the coastal plain and *P. curtissii* more inland.

19. Polygala nana (Michaux) de Candolle in A. P. de Candolle and A. L. P. P. de Candolle, Prodr. 1: 328. 1824 • Dwarf milkwort [E]

Polygala lutea Linnaeus var. *nana* Michaux, Fl. Bor.-Amer. 2: 54. 1803; *Pilostaxis nana* (Michaux) Rafinesque

Herbs annual or biennial, single- or multi-stemmed, 0.3–1.8 dm, unbranched; from taproot (or rarely fibrous root cluster). **Stems** erect, glabrous. **Leaves** mostly basal, rarely cauline, with persistent rosette; alternate; usually with narrow petiolelike region to 15 mm, rarely (sub-)sessile; basal blade spatulate, usually oblanceolate or obovate, rarely elliptic, 11–55 × (1.5–)5–20 mm, succulent, base cuneate or acute, apex rounded to acute, occasionally apiculate, rarely acuminate, surfaces glabrous. **Racemes** capitate, 1–3.8 × 1–1.7 cm; peduncle 2.3–7.5 cm; bracts deciduous, often tardily so, or sometimes persistent, linear-subulate. **Pedicels** winged, 0.4–0.8(–1) mm, glabrous. **Flowers** lemon-yellow to greenish yellow, drying green or yellowish green, 5.5–8.2 mm; sepals decurrent on pedicel, lanceolate, 3–5.3 mm, sometimes ciliolate; wings elliptic to oblong-lanceolate, 5.5–8 × 1.2–2(–2.6) mm, apex long-acuminate to cuspidate, involute, tip 0.7–1.6 mm; keel 3.5–5.8 mm, crest 2-parted, with 3 entire or 2-fid, linear lobes on each side; stamens 6(–8). **Capsules** broadly ellipsoid to subglobose, 1.6–2 × 1.2–1.6 mm, margins not winged. **Seeds** 0.8–1.8 mm, pubescent; aril 0.7–1.1 mm, lobes ⅓ to equal length of seed. **2*n*** = 64, 68.

Flowering spring–fall (year-round). Savannas, sandy pine woods, low wet woods, seepage slopes, wet depressions, flatwoods, bogs, coastal swales; 0–300 m; Ala., Ark., Fla., Ga., La., Miss., N.C., S.C., Tenn., Tex.

Individuals of *Polygala nana* in scattered populations (especially in southern Florida) approach the habit of *P. smallii*, with the inflorescences scarcely exceeding the leaves; they can be distinguished using the differences discussed under 28. *P. smallii*. Herbarium specimens of robust individuals are sometimes confused with *P. lutea* and small, rosulate plants of *P. lutea* may be confused with *P. nana*. If fresh flower color (orange in *P. lutea*, yellow in *P. nana*) is not available, then the taxa can be distinguished readily by the pedicel length, 1.5–2.8 mm in *P. lutea* and less than 1 mm in *P. nana*, as well as the involute apical cusp on the sepal wings of *P. nana* usually ca. 1 mm (0.7–1.6 mm), versus sepal wings only partially involute apically and cusps (if present) less than 0.5 mm in *P. lutea*.

20. Polygala nuttallii Torrey & A. Gray, Fl. N. Amer. 1: 670. 1840 • Nuttall's milkwort [E]

Herbs annual, single-stemmed, 1–3 dm, usually branched; from taproot. **Stems** erect, appearing glabrous, but usually puberulent, at least proximally, hairs appressed (very rarely with a few sparse, incurved to spreading hairs). **Leaves** alternate; sessile or subsessile; blade oblanceolate proximally to linear distally, 3–16 × 0.5–1(–1.5) mm, base acute, apex acute or obtuse, surfaces glabrous or sparsely pubescent. **Racemes** densely cylindric, 0.5–4 × 0.4–0.7 cm; peduncle 0.5–2.5 cm; bracts persistent, linear-lanceolate. **Pedicels** 0.5(–1) mm, glabrous. **Flowers** purple, dull reddish pink, or greenish, 2.4–3.4 mm; sepals ovate, 0.5–0.8 mm; wings elliptic to elliptic-lanceolate, 2–3 × 0.5–0.7 mm, apex acute to obtuse, sometimes minutely apiculate; keel 2–3.2 mm, crest 2-parted, with 2 or 3 blunt lobes on each side. **Capsules** ovoid, ellipsoid, or subglobose, 1.5–2.4 × 1.1–1.8 mm, margins not winged. **Seeds** 0.9–1.4 mm, pubescent; aril 0.3–0.7 mm, lobes to ½ length of seed. **2*n*** = 46.

Flowering summer. Dry, sandy meadows, open moist disturbed areas, pocosin margins, pine woodlands, bogs, wet flatwoods, powerline cuts, roadsides; 0–300 m; Ala., Ark., Conn., Del., Fla., Ga., Ky., Md., Mass., Miss., N.J., N.Y., N.C., Pa., R.I., S.C., Tenn., Va.

Polygala nuttallii has been reported for Ontario, Canada; no vouchers are known (J. M. Gillett 1968b).

21. Polygala polygama Walter, Fl. Carol., 179. 1788 • Bitter or racemed milkwort, polygale polygame [E] [F]

Polygala aboriginum Small; *P. polygama* var. *obtusata* Chodat

Herbs short-lived perennial or biennial (rarely annual), single- or multi-stemmed, (1–)1.5–3 (–5) dm, mostly unbranched, or sparsely branched distally; from taproot or fibrous root cluster. **Stems** usually erect, rarely somewhat sprawling, glabrous. **Leaves** usually alternate, sometimes subopposite or opposite when leaves scalelike proximally (less than 1 mm); sessile or subsessile, sometimes with narrow petiolelike region to 2 mm; blade spatulate to obovate or, sometimes, scale-like proximally, linear-oblong or elliptic to oblanceolate distally, (8–)15–30(–40) × 2–6(–8) mm, base acute or cuneate, apex obtuse to rounded or, sometimes, acute proximally to mostly acute distally, surfaces glabrous. **Racemes** loosely cylindric, open, elongate, (2–)5–10 (–25) × 0.8–1.4 cm; peduncle 1–2 cm; bracts deciduous, ovate to oblong-ovate. **Pedicels** 1–4 mm, glabrous. **Cleistogamous flowers** present in racemes usually below, rarely just above, soil surface, sometimes in proximal leaf axils later in season. **Flowers** usually pink to pale purple, rarely whitish, outer sepals sometimes with pink or white margins, 4–6 mm; sepals ovate, 1.3–2.5 mm; wings elliptic or obovate, 3–6 × 1.6–2.7 mm, apex obtuse to bluntly rounded; keel (2–)3–5 mm, crest 2-parted, with 2 or 3 divided lobes on each side. **Capsules** broadly ellipsoid or globose-ovoid, 2–4 × 2–3 mm, margins not winged. **Seeds** 1.8–3 mm, usually densely short-pubescent to subglabrous; aril 0.8–2 mm, lobes ⅓–¾ length of seed. **2*n*** = 56.

Flowering spring–mid summer. Sandy meadows, savannas, bogs, sandhills, flatwoods, dry hammocks, floodplain swamps, open roadsides, clear-cuts, granite outcrops, coastal dunes; 0–300 m; N.S., Ont., Que.; Ala., Ark., Conn., Del., Fla., Ga., Ill., Ind., Iowa, Ky., La., Maine, Md., Mass., Mich., Minn., Miss., N.H., N.J., N.Y., N.C., Ohio, Okla., Pa., R.I., S.C., Tenn., Tex., Vt., Va., W.Va., Wis.

Variety *obtusata*, based primarily on having a slightly denser raceme, is not recognized here; that character is found throughout the range of the species, often in the same population as plants with more open inflorescences (for example, C. W. James 1957).

22. Polygala ramosa Elliott, Sketch Bot. S. Carolina 2: 186. 1822 • Low pine-barren milkwort [E]

Pilostaxis ramosa (Elliott) Small

Herbs annual or biennial, single- to multi-stemmed, 1.5–4 (–5) dm, unbranched or branched distally; from fibrous root cluster. **Stems** erect, glabrous. **Leaves** usually without basal rosette at anthesis; alternate; sessile or subsessile; basal blade obovate or spatulate to narrowly elliptic, cauline oblanceolate, spatulate-elliptic, or linear, (7–) 15–40(–70) × 1.5–8 mm, base often cuneately narrowed and petiolelike proximally, acute to obtuse distally, apex usually rounded proximally, acute to obtuse distally, surfaces glabrous. **Racemes** in cymose panicles, each stem with to 110 racemose branches, 1.8 × 0.7–1.1 cm, ± flat-topped, 2.5–15 × 3–15 cm; central ones nearly sessile or peduncle to 2 cm, lateral ones subsessile or peduncle to 2(–3) cm; bracts usually persistent, lanceolate-ovate. **Pedicels** narrowly winged, (1–)1.3–2.3 mm, glabrous. **Flowers** bright yellow, drying green or yellow-brown, 3–3.5 mm; sepals decurrent on pedicel, ovate to lanceolate-ovate, 1–3 mm; wings ovate, elliptic-obovate, obovate, or spatulate, 2.5–3.5 × 1–1.4 mm, apex acuminate to cuspidate, involute; keel 2–2.5 mm, crest 2-parted, with (1 or)2 or 3 lobes on each side. **Capsules** subglobose, 0.8–1 × 0.8–1 mm, margins not winged. **Seeds** 0.5–0.7 mm, densely pubescent; aril 0–0.2 mm. **2*n*** = 64, 68.

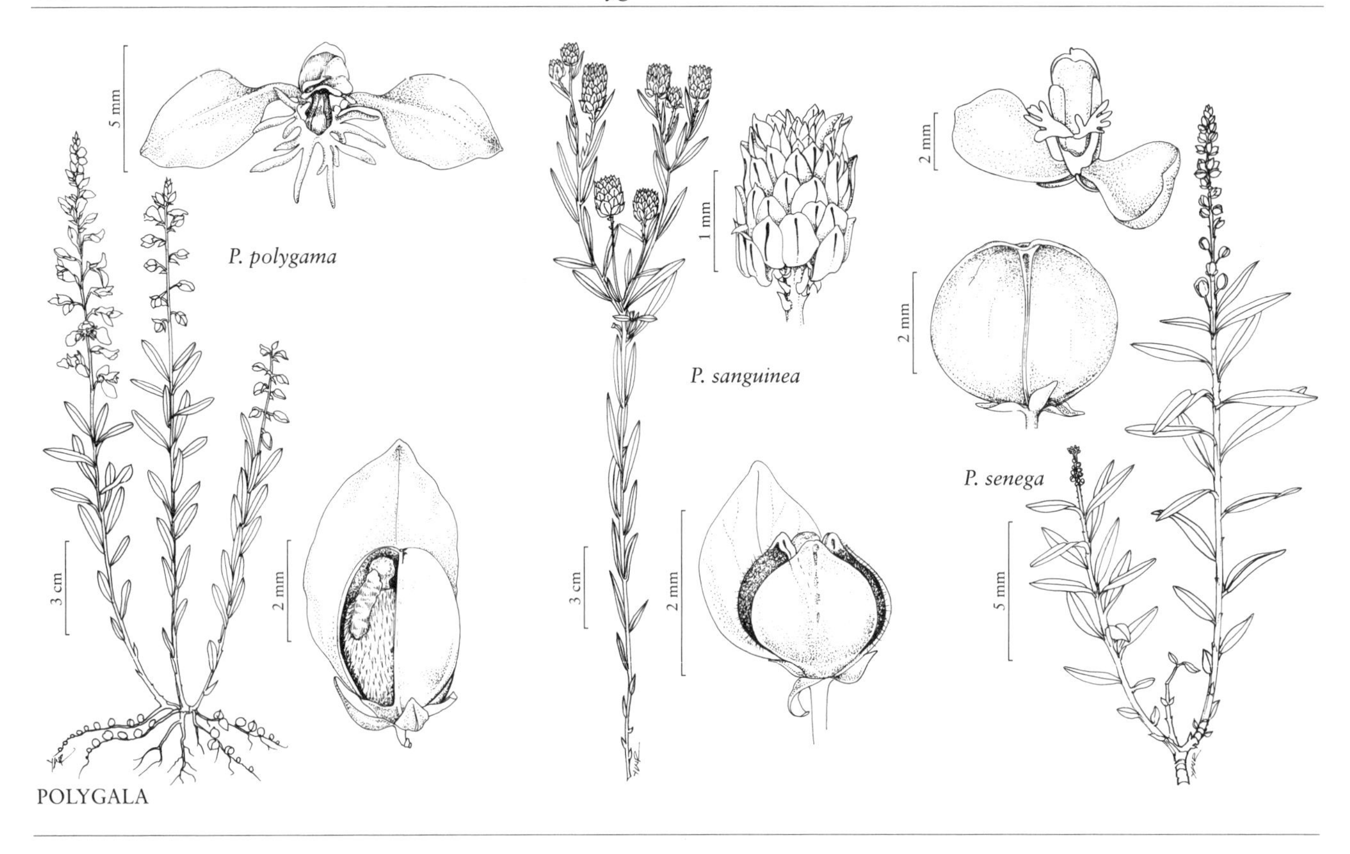

POLYGALA

Flowering spring–summer. Savannas, open pine woods, flatwoods, bogs, seepage slopes, coastal swales, exposed pond shores; 0–100 m; Ala., Del., Fla., Ga., La., Miss., N.J., N.C., S.C., Tex., Va.

Polygala ramosa has been reported from Maryland (M. L. Brown and R. G. Brown 1984, as cited in the USDA Plants database); no vouchers are known.

23. **Polygala rugelii** Shuttleworth ex A. Gray, Smithsonian Contr. Knowl. 3(5): 41. 1852 • Yellow milkwort [E]

Pilostaxis rugelii (Shuttleworth ex A. Gray) Small

Herbs annual or biennial, single- or multi-stemmed, 2–7.8 dm, usually branched distally or unbranched; usually from fibrous root cluster or taproot. **Stems** erect, glabrous. **Leaves** sometimes with basal rosette at anthesis; alternate; sessile or subsessile, proximal leaves sometimes with narrow, petiolelike base to 2 mm; basal blade spatulate, cauline obovate, oblanceolate, or spatulate proximally, becoming lanceolate distally, basal to 30–60 × 5–15 mm, cauline 8–45 × 2–8(–12) mm, succulent, base cuneate or acute, apex usually acute, sometimes obtuse or rounded proximally, surfaces glabrous. **Racemes** capitate, usually 2.5–3(–5) × 1.8–2.5 cm; peduncle 1–15 cm; bracts usually deciduous, narrowly ovate to linear-lanceolate. **Pedicels** winged, 2–2.5 mm, glabrous. **Flowers** bright yellow, drying pale yellow to dark greenish brown, 6–9 mm; sepals decurrent on pedicel, deltate-subulate to ovate, 1.5–2.5 mm, ciliolate; wings oblong, elliptic, or obovate, 4.5–9 × 2–4 mm, apex acuminate, acumen 0.5–0.8 mm, involute; keel 5–6 mm, crest 2-parted, with 2–4 2-fid lobes on each side. **Capsules** obovoid, ellipsoid or subglobose, 1.5–2 × 1.5–2 mm, margins not winged. **Seeds** 1.2–1.6 mm, pubescent; aril 1.1–1.6 mm, lobes subequal to or greater than length of seed. $2n = 64, 68$.

Flowering spring–late fall. Wet flatwoods, open areas (including prairielike or savannalike openings); 0–50 m; Fla.

Polygala rugelii occurs in all of the peninsular counties.

24. **Polygala sanguinea** Linnaeus, Sp. Pl. 2: 705. 1753 • Blood or purple or field milkwort, polygale sanguin 

Polygala viridescens Linnaeus

Herbs annual, single-stemmed, (0.5–)1–4 dm, usually branched distally; from taproot (or rarely fibrous root cluster). **Stems** erect, glabrous. **Leaves** alternate; sessile or subsessile; blade spatulate proximally to linear or narrowly elliptic distally, (5–)10–20(–40) × (0.5–)1–3(–5) mm, base acute or obtuse, apex acute to acuminate, surfaces glabrous. **Racemes** capitate to densely cylindric, (0.5–)1–2(–4) × 0.5–1.4 cm;

peduncle 0.3–2.5(–3) cm; bracts subpersistent to tardily deciduous, subulate. **Pedicels** 0.4–1.5 mm, glabrous. **Flowers** usually pink, purple, or reddish purple, rarely white, sometimes greenish tinged, sepals sometimes pink or white, 4–6 mm; sepals oval, elliptic-ovate, or lanceolate, 1–3 mm; wings ovate to broadly elliptic, (2.6–)4.5–6.3 × (1–)2.5–3.5 mm, apex obtuse to broadly rounded, sometimes minutely apiculate, rarely acute; keel 2.5–3 mm, crest 2-parted, with 2–4 lobes on each side. **Capsules** usually with flattened, sterile base, cuneate-subglobose, 2.5–3 × 2–2.5 mm, margins not winged (sometimes with raised rim). **Seeds** 1.3–1.7 mm, pubescent; aril 1–1.3 mm, lobes usually (½–)⅔ to ± length of seed, rarely minute.

Flowering spring–summer. Prairies, old fields, gravelly logging road margins, meadows, glades, bogs, flatwoods, open woods; 0–300 m; N.B., N.S., Ont., P.E.I., Que.; Ala., Ark., Conn., Del., D.C., Ga., Ill., Ind., Iowa, Kans., Ky., La., Maine, Md., Mass., Mich., Minn., Miss., Mo., Nebr., N.H., N.J., N.Mex., N.Y., N.C., Ohio, Okla., Pa., R.I., S.C., S.Dak., Tenn., Tex., Vt., Va., W.Va., Wis.

Polygala sanguinea is the only species of the genus in the flora area with the wings to twice the length of the keel. Late season flowers can have much smaller wings, some as small as 2.6 × 1 mm.

25. Polygala scoparioides Chodat, Mém. Soc. Phys. Genève 31(2): 284, plate 26, figs. 6, 7. 1893 • Broom milkwort

Herbs perennial, multi-stemmed, (0.5–)0.9–3(–5) dm, unbranched or sparsely branched distally; from thickened caudex. **Stems** erect, sometimes very slightly glaucous, puberulent, hairs incurved. **Leaves** alternate; sessile or subsessile; blade linear-lanceolate, linear, or linear-acicular, (3–)7–15 × 0.6–1.3 mm, base acute, apex acute to acuminate, surfaces subglabrous or glabrous. **Racemes** loosely cylindric (flowering portion may appear subcapitate from deciduous fruit), (1–)1.5–7.8 × 0.4–0.8 cm; peduncle 0.2–1 cm; bracts deciduous, ovate to lanceolate-ovate. **Pedicels** 0.5 mm, glabrous. **Flowers** white, greenish veined, wings with green or purple longitudinal stripe, keel green to purplish brown, (2.7–)2.9–4.7 mm; sepals oblong-ovate, 1.3 mm; wings spatulate-obovate, (2.4–)2.7–4.5 × 1.3–1.6 mm, apex acute to obtuse; keel 2.8 mm, crest 2-parted, with 3 usually divided lobes on each side. **Capsules** oblong-ellipsoid, 3–4.5 × 1.6–2 mm, abaxial locule not winged, adaxial locule slightly longer, winged, wing very narrow. **Seeds** 2.5–3 mm, puberulent, coat with rows of pits 0.05 mm wide; aril 1–1.9 mm, lobes (½–)⅔ to subequal length of seed. **2*n*** = 32, 34.

Flowering early spring–late fall. Various substrates, mostly on limestone, open rocky areas in scrub, grasslands, disturbed areas, chaparral, open woodlands; 500–1900 m; Ariz., N.Mex., Tex.; Mexico (Chihuahua, Coahuila, Nuevo León, San Luis Potosí, Zacatecas).

Polygala scoparioides is closely related to other Mexican taxa of the Monninopsis group, such as *P. mexicana* Moçiño ex Cavanilles, with which it intergrades morphologically.

26. Polygala senega Linnaeus, Sp. Pl. 2: 704. 1753 • Seneca snakeroot, polygale sénéca [E] [F]

Polygala lonchophylla Greene; *P. senega* var. *latifolia* Torrey & A. Gray

Herbs perennial, usually multi-stemmed, (1–)1.5–5 dm, unbranched or sparsely branched distally; from thickened caudex. **Stems** erect, puberulent or glabrous, hairs appressed, incurved, and spreading. **Leaves** alternate; subsessile or petiolate, petiole to 0.5–5 mm; blade scalelike proximally, linear-lanceolate to lanceolate-elliptic, lanceolate, or lanceolate-ovate distally, (13–)20–80 × (1.5–)8–35 mm, base acute, margins often appearing serrulate from toothlike projections associated with cilia, apex acute to acuminate, surfaces glabrous. **Racemes** conic or cylindric-conic, (1–)1.5–4 (–4.5) × 0.5–0.9 cm; peduncle 1–3 cm; bracts deciduous, ovate. **Pedicels** 0.5(–1) mm, glabrous. **Flowers** white, wings often with greenish veins, other sepals sometimes white, 2–4 mm; sepals ovate or lanceolate, 1–2 mm; wings suborbiculate to broadly elliptic or obovate, 2–3.7 × 2–3 mm, apex bluntly rounded (or rarely obtuse); keel 2–3.5 mm, crest 2-parted, with 2–4 lobes on each side. **Capsules** subglobose or ovoid, 2.5–4.5 × 3–4.3 mm, margins not winged. **Seeds** 2–3.5 mm, sparsely pubescent; aril 1.9–3.6 mm, lobes subequal to longer than length of seed. **2*n*** = 34.

Flowering spring–mid summer. Open woods, mesic forests, prairies, rocky creek bottoms, often on soils derived from limestone or mafic rocks, roadsides, clearings; 50–800 m; Alta., B.C., Man., N.B., Ont., Que., Sask.; Ark., Conn., Del., D.C., Ga., Ill., Ind., Iowa, Kans., Ky., Maine, Md., Mass., Mich., Minn., Mo., Nebr., N.J., N.Y., N.C., N.Dak., Ohio, Okla., Pa., R.I., S.C., S.Dak., Tenn., Vt., Va., W.Va., Wis., Wyo.

Two varieties have been recognized within *Polygala senega*: var. *latifolia* (= *P. lonchophylla*) with the distal leaf blades more than 1 cm wide (in correlation with an overall more robust habit and slightly larger size of most parts), and var. *senega*, with the distal leaf blades to 1 cm wide. A. E. Trauth-Nare and R. F. C. Naczi (1998) suggested that these entities may warrant specific recognition based on size and phenology differences, but in the absence of published details, the ranges and morphological features overlap too extensively to warrant taxonomic recognition.

Manitoba and Saskatchewan have been the major source of wild harvested roots of *Polygala senega* in North America, with up to several thousand kilograms being harvested annually (C. J. Briggs 1988). An increase in demand for *Polygala senega* has raised concerns about sustainable harvest (C. L. Turcotte 1997).

SELECTED REFERENCE Briggs, C. J. 1988. Senega snakeroot. A traditional Canadian herbal medicine. Canad. Pharm. J. 121: 199–201.

27. Polygala setacea Michaux, Fl. Bor.-Amer. 2: 52. 1803 • Coastal-plain or scale-leaf milkwort [E]

Herbs annual or short-lived perennial, usually single-stemmed, sometimes 2 or 3 stems near base, 1–5 dm, usually branched distally; from taproot or, when perennial, sometimes with slender taprootlike caudex with persistent stem base. **Stems** erect, glabrous. **Leaves** alternate; sessile; blade subulate, squamiform, 0.5–1.6(–2) × 0.3–0.7 mm, base obtuse, apex acute, surfaces glabrous. **Racemes** cylindric, 0.4–3.5 × 0.3–0.5 cm; peduncle 0.1–0.5 cm; bracts deciduous, lanceolate. **Pedicels** 0.2–0.5 mm, glabrous. **Flowers** usually white, sometimes pinkish tinged, 1.8–2.7 mm; sepals ovate to lanceolate-ovate, 0.6–1(–1.5) mm; wings elliptic to obovate, 1.5–2.5 × 0.6–1.1 mm, apex usually obtuse to bluntly rounded, rarely acute, often minutely apiculate or cuspidate; keel 1.5–2.2 mm, crest 2-parted, with 2 or 3 lobes on each side. **Capsules** ovoid to ellipsoid, 1.7–2.2 × 1.2–1.5 mm, margins not winged. **Seeds** 0.8–1.2 mm, pubescent; aril vestigial.

Flowering year-round. Moist to somewhat dry flatwoods, pine-palmetto woodlands, margins of seepage bogs; 0–100 m; Fla., Ga.

The protologue description of *Polygala setacea* with “in Carolina septentrionali” is the likely source of later reports of the species occurring in the Carolinas, (for example, J. K. Small 1933; R. W. Long and O. Lakela 1971; R. K. Godfrey and J. W. Wooten 1981). There are no known specimens from either of the Carolinas; the locality reported by Michaux may be erroneous. Small reported *Polygala setacea* also from Mississippi; no supporting specimens are known. The presence of this species in Mississippi or either of the Carolinas would represent a disjunction from the range documented by known vouchers.

28. Polygala smallii R. R. Smith & D. B. Ward, Sida 6: 307. 1976 • Tiny or Small’s milkwort [C] [E]

Polygala arenicola Small, Bull. New York Bot. Gard. 3: 426. 1905, not Gürke 1903; *Pilostaxis arenicola* Small

Herbs biennial, usually single-stemmed, rarely multi-stemmed, 0.2–0.5(–0.8) dm, unbranched; from taproot or fibrous root cluster. **Stems** erect, glabrous. **Leaves** with persistent basal rosette, clustered and irregular, crowded; alternate; sessile or subsessile with narrowed petiolelike region to 5(–10) mm (usually obscured by tightly clustered leaves); blade oblanceolate to linear-oblanceolate, 10–42 × 2–14 mm, succulent, base cuneate, apex usually rounded to obtuse, rarely acute, sometimes apiculate, surfaces glabrous. **Racemes** capitate, 0.4–3 × 0.5–1.8 cm; peduncle 0.5–5 cm; bracts deciduous, often tardily so, linear-subulate. **Pedicels** winged, 0.4–0.8 mm, glabrous. **Flowers** green or greenish yellow, 4.5–6(–8) mm; sepals decurrent on pedicel, narrowly ovate to lanceolate, (2–)3–5 mm, sometimes ciliolate; wings usually lanceolate, sometimes oblong-lanceolate, 4–6(–8) × 1–2 mm, apex long-acuminate, sometimes slightly involute, tip 0.5–0.9 mm; keel (3.5–)4(–5.5) mm, crest 2-parted, with 3 2-fid or entire lobes on each side; stamens 6–8. **Capsules** broadly ellipsoid to subglobose, 1.6–2 × 1.2–1.6 mm, margins not winged. **Seeds** 1.9–2.3 mm, pubescent; aril 0.9–1.6 mm, lobes ½ to ± equal length of seed. $2n$ = 64, 68.

Flowering spring (year-round). Pinelands; of conservation concern; 0–50 m; Fla.

Polygala smallii is known from Broward, Martin, Miami-Dade, Palm Beach, and St. Lucie counties.

Compared to *Polygala nana*, the most similar species, *P. smallii* is smaller and more compact (stems and inflorescences are shorter and almost always surpassed by the tightly clustered leaves), leaves are narrower with a less conspicuously narrowed petiolelike portion (less obviously spatulate), wing sepals are more strongly lanceolate (tapering more to the apex, whereas *P. nana* wings typically are more elliptic or oblong-lanceolate), flowers are much greener at anthesis, and seeds are longer.

Polygala smallii is in the Center for Plant Conservation’s National Collection of Endangered Plants.

29. Polygala verticillata Linnaeus, Sp. Pl. 2: 706. 1753 • Whorled milkwort, polygale verticillé [E]

Polygala ambigua Nuttall; *P. pretzii* Pennell; *P. verticillata* var. *ambigua* (Nuttall) Alph. Wood; *P. verticillata* var. *dolichoptera* Fernald; *P. verticillata* var. *isocycla* Fernald; *P. verticillata* var. *sphenostachya* Pennell

Herbs annual, single-stemmed, 0.5–3(–4) dm, usually branched distally; from slender taproot. **Stems** erect, sometimes slightly glaucous, glabrous. **Leaves** usually whorled proximally, sometimes whorled to inflorescence, opposite to alternate distally, rarely alternate throughout; sessile or subsessile, petiole rarely to 1 mm; blade sometimes linear-spatulate proximally, usually linear, linear-oblong, -elliptic, or -lanceolate distally, (5–)10–20(–30) × 0.5–5.5 mm, base cuneate to acute, apex acute to acuminate, surfaces glabrous. **Racemes** nearly spike-like, conic to cylindric-conic, 0.5–5 × 0.2–0.6 cm, sometimes interrupted proximally, with some proximal fruits persistent below gap; peduncle to 9 cm (usually elongate); bracts deciduous, subulate to lanceolate. **Pedicels** 0.2–1(–2) mm, glabrous. **Flowers** white or with greenish or pinkish tinge, sepals sometimes purplish or whitish, 1–2.2(–2.6) mm; sepals ovate or elliptic to lanceolate, 0.5–1.6 mm, ciliate; wings suborbiculate, ovate, elliptic, or obovate, 1–2.6 × 0.8–2.5 mm, apex obtuse to bluntly rounded; keel 1.2–1.8 mm, crest 2-parted, with 1 or 2 lobes on each side. **Capsules** subglobose or broadly ellipsoid to ovoid, 1.3–2.4 × 0.7–2.2 mm, margins not winged. **Seeds** 1.2–2.2 mm, pubescent; aril 0.5–1 mm, lobes ⅓–½ length of seed. **2*n*** = 34.

Flowering spring–fall. Meadows, prairies, open woodlands, sand dunes, old fields, open places (limestone glades, railroad rights-of-way, rock quarries); 0–2100 m; Man., N.B., Ont., Que., Sask.; Ala., Ark., Conn., Del., D.C., Fla., Ga., Ill., Ind., Iowa, Kans., Ky., La., Maine, Md., Mass., Mich., Minn., Miss., Mo., Mont., Nebr., N.H., N.J., N.Y., N.C., N.Dak., Ohio, Okla., Pa., R.I., S.C., S.Dak., Tenn., Tex., Utah, Vt., Va., W.Va., Wis., Wyo.

Varieties of *Polygala verticillata* have been recognized; they co-occur and intergrade, suggesting that they are polymorphisms not worthy of taxonomic recognition.

30. Polygala vulgaris Linnaeus, Sp. Pl. 2: 702. 1753 • Common or European milkwort [I]

Herbs short-lived perennial, usually multi-stemmed, rarely with 1 or few stems, 1.5–4 dm, unbranched or branched distally; from small, thickened, woody caudex. **Stems** erect to decumbent, pubescent or subglabrous, hairs incurved. **Leaves** alternate; sessile or subsessile; blade narrowly oblanceolate-spatulate to elliptic or lanceolate, 5–35 × 1–5 mm, base cuneate, apex rounded or acute, surfaces glabrous. **Racemes** broadly cylindric, usually elongate, 0.5–2.5(–9) × 0.5–2 cm; peduncle usually poorly developed, 0.2–1.5(–2.5) cm; peduncle and central axis pubescent; bracts deciduous, lanceolate. **Pedicels** 1.5–4(–5.8) mm, glabrous. **Flowers** blue, pink, or white, 5–8 mm; sepals ovate to elliptic, 1.6–3.5 mm, pubescent; wings obovate, 3–5 × 2–3 mm, apex obtuse to bluntly rounded; keel 5–8 mm, crest 2-parted, with 2–4 lobes on each side. **Capsules** broadly ellipsoid, ovoid, or slightly obovoid, (3–)4–6 × 2.5–4 mm, margins narrowly winged, rim 0.2–0.4 mm wide. **Seeds** 3.2–4.2 mm, pubescent; aril 1–1.7 mm, lobes ½ length of seed. **2*n*** = 24–32, 34, 38, ca. 56, 68, ca. 70.

Flowering spring–mid summer. Dry hillsides near dunes, meadows, grassy roadsides, disturbed areas; 0–400 m; introduced; B.C.; Mich., Oreg., Wash.; Europe; Asia.

A 1947 specimen of *Polygala vulgaris* from Vancouver Island (at V) reported this species as apparently naturalized (H. J. Scoggan 1978–1979, part 3). Two other collections (both at V) are known from nearby, at Comox in 1941 and on Texada Island in 1999, the latter with approximately 150 clumps, according to the collector. *Polygala vulgaris* appears to be naturalized, at least locally, along the Strait of Georgia in British Columbia. The Michigan collections (MICH) are from 1916 and 1974 from different counties, suggesting that *P. vulgaris* may be sparsely naturalized in Michigan or just a casual waif. In Oregon, there are six collections (OSC, WTU) from 1950 to 2018 from three different counties, documenting a long-term presence. There is a single collection (WTU) from 2014 from Clark County, Washington. This taxon seems naturalized in North America, at least in British Columbia, Oregon, and probably Washington.

31. Polygala watsonii Chodat, Mém. Soc. Phys. Genève 31(2): 285, plate 26, figs. 8, 9. 1893 (as watsoni) • Watson's milkwort

Subshrubs, multi-stemmed, 0.5–5 dm, branched throughout; from thickened caudex. **Stems** erect, sometimes laxly so, puberulent or subglabrous, hairs incurved. **Leaves** alternate; sessile or subsessile; blade usually linear-lanceolate to linear, sometimes falcate-linear with reflexed tip, 4–15 × 0.3–1 mm, base obtuse or acute, apex acute to acuminate, surfaces usually puberulent, rarely subglabrous, hairs incurved. **Racemes** subcapitate, 0.5–1.4(–2.5) × 1 cm; peduncle 0–0.5 cm; bracts deciduous, ovate to lanceolate-ovate. **Pedicels** 0.5 mm, glabrous. **Flowers** white, greenish or purplish veined, keel pink to purplish brown, often yellow-green distally, (3.7–)4–6 mm; sepals oblong-ovate, 1.3 mm; wings spatulate-obovate, (3.5–)4–6 × 1.3–1.6 mm, apex obtuse or acute, sometimes minutely apiculate; keel 2.8 mm, crest 2-parted, with 3 divided lobes on each side. **Capsules** oblong-ellipsoid to oblong, 2.3–3.5 × 1.6 mm, abaxial locule not winged, adaxial locule slightly longer, not or obscurely winged. **Seeds** 2.5–3 mm, puberulent, coat with rows of pits 0.05 mm wide; aril 1–1.9 mm, lobes ½ to subequal to length of seed. $2n = 16$.

Flowering early spring. Stony limestone slopes; 1800–1900 m; Tex.; Mexico (Chihuahua, Coahuila, Nuevo León).

Polygala watsonii is known in the flora area from a single collection from the Glass Mountains in Brewster County (T. L. Wendt 1979).

5. POLYGALOIDES Haller, Hist. Stirp. Helv. 1: 149. 1768 • [Genus *Polygala* and Latin -*oides*, resembling]

Chamaebuxus (de Candolle) Spach; *Polygala* Linnaeus sect. *Chamaebuxus* de Candolle; *Polygala* subg. *Chamaebuxus* (de Candolle) Duchesne

Herbs, perennial, usually multi-stemmed. **Stems** creeping, with short erect shoots, or decumbent, rhizomatous or stoloniferous, usually glabrous. **Leaves** alternate; petiolate or subsessile; dimorphic, bractlike proximally, well-developed, uniform and somewhat clustered distally; blade surfaces glabrous or pubescent. **Inflorescences** terminal, racemes or appearing corymblike or 1–4(or 5)-flowered (from poorly developed peduncle); peduncle usually present; bracts deciduous. **Pedicels** present. **Flowers** usually pink or rose-purple, rarely white, sepals pale pink or whitish, crest often yellowish, (2–)15–23 mm, cleistogamous sometimes present; sepals deciduous, glabrous; wings deciduous, (10–)13–20 mm, glabrous; keel crested, crest 2-parted, with 2–4 lobes on each side, glabrous; stamens 6(–8) in chasmogamous flowers, fewer in cleistogamous flowers, not grouped; ovary 2-loculed. **Fruits** capsules, dehiscent, margins narrowly winged apically, glabrous. **Seeds** pubescent, arillate.

Species 6 or 7 (1 in the flora): North America, Europe, n Africa.

1. Polygaloides paucifolia (Willdenow) J. R. Abbott, J. Bot. Res. Inst. Texas 5: 134. 2011 • Gaywings, flowering wintergreen, polygale paucifolié [E] [F]

Polygala paucifolia Willdenow, Sp. Pl. 3: 880. 1802; *Triclisperma paucifolia* (Willdenow) Nieuwland

Herbs 0.5–1.5(–2) dm. **Leaves:** petiole 4–10 mm (to almost sessile proximally); leaves proximally highly reduced, scalelike, 2–8 mm, distally with 3–6 well-developed leaves, somewhat clustered; blade elliptic to ovate, 15–40(–50) × 10–25 mm, base cuneate, apex rounded, obtuse, or acute, surfaces often glabrous, except pubescent adaxially on midrib and margins densely incurved-ciliolate, sometimes densely pubescent throughout. **Inflorescences:** peduncle (0–)1–2 cm; bracts ovate. **Pedicels** 10–20 mm, glabrous. **Cleistogamous flowers** usually present later in season, usually on short, erect, rarely leafy, bracteate stems proximally, 2 mm. **Chasmogamous flowers** 15–23 mm; sepals 4–6 mm; wings obovate or spatulate; keel 10–20 mm. **Capsules** obovoid or subglobose, 5–8 × 5–8 mm. **Seeds** 2.5–3.5 mm; aril 1.5–3.5 mm, lobes ½ to as long as seed. $2n = 34$.

Flowering spring–early summer(–mid summer). Sandy, limestone, or granite soils, coniferous and/or mixed forests on moist soils, wooded bogs; 50–700 m;

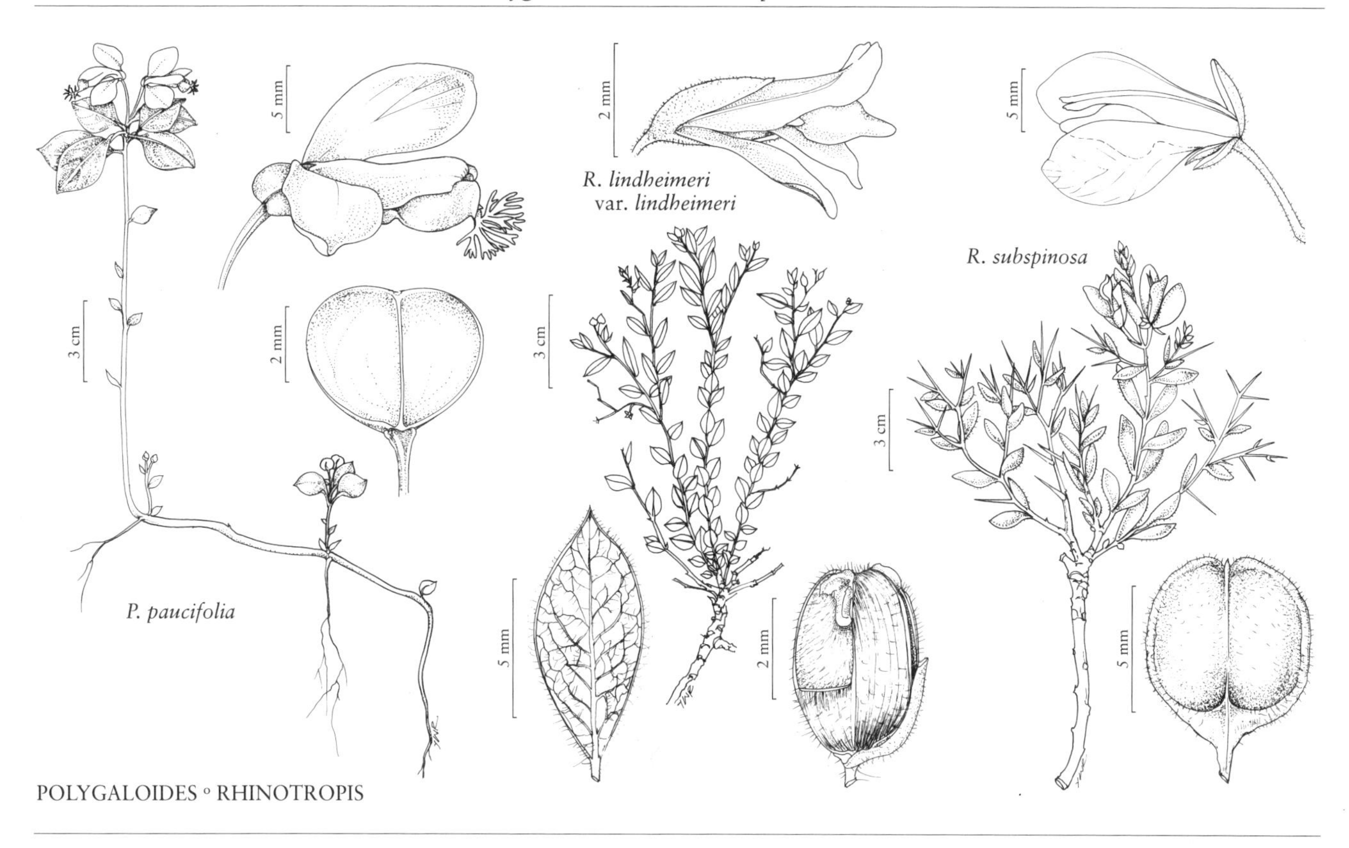

POLYGALOIDES ◦ RHINOTROPIS

Alta., Man., N.B., Ont., Que., Sask.; Conn., Del., Ga., Ill., Ind., Ky., Maine, Md., Mass., Mich., Minn., N.H., N.J., N.Y., N.C., Ohio, Pa., R.I., S.C., Tenn., Vt., Va., W.Va., Wis.

Polygaloides paucifolia is isolated and morphologically distinct from its closest relatives in Europe and northern Africa, perhaps being a vicariant relict of a historical circumboreal flora. Its creeping herbaceous habit with relatively thin leaves and large showy crest set *P. paucifolia* apart from the Old World taxa (subshrubs or shrubs to 1 m with subcoriaceous leaves and small, fairly inconspicuous crest); the form of the stigma, the glanduliform disc at the base of the flower, and the few-flowered inflorescences morphologically unite the taxa.

Polygaloides paucifolia is considered to be extirpated in Newfoundland.

6. RHINOTROPIS (S. F. Blake) J. R. Abbott, J. Bot. Res. Inst. Texas 5: 134. 2011 • [Greek *rhinos*, snout, and *tropis*, keel; alluding to beaked keel petal]

Polygala Linnaeus sect. *Rhinotropis* S. F. Blake, Contr. Gray Herb. 47: 70. 1916

Herbs, perennial, subshrubs, or shrubs, single- or multi-stemmed, with or without thorns, then as modified tips of racemes. **Stems** usually sprawling to erect, sometimes prostrate or decumbent, usually not glaucous, pubescent or glabrous. **Leaves** alternate; sessile, subsessile, or petiolate; usually not strongly dimorphic; blade surfaces pubescent or glabrous. **Inflorescences** terminal or leaf-opposed, sometimes appearing axillary if poorly developed, racemes, sometimes reduced and appearing fasciculate or aggregated into pseudopanicles; peduncle present or absent; bracts deciduous to subpersistent or persistent. **Pedicels** present. **Flowers** cream, yellowish green, yellow, white, pink, rose, or purple, cleistogamous usually absent, sometimes present (in *R. californica* and *R. lindheimeri*), (2.4–)3.5–14.5 mm; sepals deciduous or persistent (when persistent, usually only upper; all persistent in *R. rusbyi*), sometimes appearing very slightly connate basally, pubescent or glabrous; wings deciduous, 2.5–12.5 mm, glabrous or pubescent; keel usually beaked with unlobed projection, beak sometimes reduced or obscure (rarely on all flowers unless cleistogamous, and then inflorescence usually proximal), keel glabrous or pubescent; stamens usually 7 or 8, rarely 9 (in *R. acanthoclada*), in chasmogamous flowers, fewer in cleistogamous flowers, not grouped; ovary 2-loculed. **Fruits** capsules, dehiscent, margins winged or not, glabrous or pubescent. **Seeds** pubescent to subglabrous, arillate. $x = 9$.

Species 17 (12 in the flora): w, sc United States, Mexico, Central America (Guatemala).

Of the 17 species of *Rhinotropis* ranging from the southwestern United States and/or Mexico, only *R. purpusii* (Brandegee) J. R. Abbott extends into Guatemala. Of all the genera treated here, this is the only one that has been monographed within the last 100 years (T. L. Wendt 1978). *Rhinotropis* is probably sister to the Caribbean clade *Phlebotaenia* Grisebach, and appears to be fairly closely related also to the pantropical (although predominantly neotropical) genus *Securidaca* Linnaeus. *Rhinotropis* is largely endemic to arid regions but some species (*R. californica*) occur in mesic areas.

The flower beak is a cylindric, conic, or contorted non-fimbriate hollow projection from the lower (or central) apex of the keel region. It is highly reduced or absent in some species. The other diagnostic features of *Rhinotropis* are also not monothetic across all species. Many species have the upper sepal persistent in fruit and the other sepals, including the wings (and the corolla), deciduous. Unlike other North American Polygalaceae, species of *Rhinotropis* often have five petals; the lateral petals are much reduced, linear, and adnate for most of their length to the staminal column; additionally, several species are shrubs and a few have thorn-tipped inflorescence axes.

SELECTED REFERENCE Wendt, T. L. 1978. A Systematic Study of *Polygala* Section *Rhinotropis* (Polygalaceae). Ph.D. dissertation. University of Texas.

1. Shrubs or subshrubs; racemes thorn-tipped; sepals deciduous.
    2. Flowers (2.5–)3–5(–5.3) mm; sepal wings usually cream or greenish; keel petals without beaks or beaks to 0.5–0.7 mm; stems densely pubescent to glabrate, not glaucous.
        3. Leaves and stems densely pubescent, with short, spreading hairs; pedicels 1.5–4 (–5.8) mm, usually shorter than flowers, pubescent; non-wing sepals pubescent, hairs spreading . . . . 1. *Rhinotropis acanthoclada*
        3. Leaves densely pubescent with incurved hairs, stems densely pubescent or glabrate, hairs matted or shaggy, appressed, incurved or, sometimes, spreading; pedicels (2.5–) 3–7(–9) mm, usually longer than flowers, glabrous; non-wing sepals glabrous or subapically sparsely hairy . . . . 5. *Rhinotropis intermontana*
    2. Flowers (6–)7–14.5 mm; sepal wings usually pink, sometimes light green; keel petals with prominent beaks (1–4 mm); stems glabrous or pubescent, often glaucous, at least when young.
        4. Keel petal beaks with 1 or 2 prominent invaginations along abaxial side (formed by sinuate excess tissue); seeds most densely pubescent apically, proximal ½ sparsely and unevenly pubescent or glabrous . . . . 4. *Rhinotropis heterorhyncha*
        4. Keel petal beaks entire or slightly erose; seeds ± evenly and moderately densely pubescent, occasionally with glabrate patches . . . . 12. *Rhinotropis subspinosa*
1. Herbs, subshrubs, or shrubs; racemes usually not thorn-tipped (except usually weakly thorn-tipped in *R. rusbyi*, then all sepals persistent in fruit); upper sepal persistent in fruit or all deciduous.
    5. Flowers (2.5–)7–14.5 mm (smaller cleistogamous flowers sometimes present in *R. californica* proximally); sepals all deciduous or all persistent.
        6. Sepals persistent; racemes usually weakly thorn-tipped (often not clearly visible when young); Arizona . . . . 11. *Rhinotropis rusbyi*
        6. Sepals deciduous; racemes not thorn-tipped; California, Oregon.

7. Keel petal beaks oblong, usually notched or contorted, rarely subentire, 0.7–1 mm wide distally; cleistogamous flowers often present, 2.4–5 mm, chasmogamous flowers 9–14.5 mm; perennials, 0.5–3.5 dm; stems laxly erect, decumbent, or prostrate . . . . .2. *Rhinotropis californica*
7. Keel petal beaks linear, entire, 0.2 mm wide distally; flowers all chasmogamous, 7–14 mm; perennials, subshrubs, or shrubs, 1–25 dm; stems erect to sprawling or decumbent . . . . .3. *Rhinotropis cornuta*

[5. Shifted to left margin.—Ed.]

5. Flowers (2.4–)2.9–7.5(–7.7) mm; upper sepal persistent, others deciduous.
   8. Racemes usually leaf-opposed; keel petals usually with glabrous sacs, beaks (0–)0.5–2 mm; leaf blades usually elliptic, ovate, obovate, lanceolate, or linear, sometimes scalelike, (3–)4–35(–40) mm.
      9. Roots with brown, gray, or dull red-brown cortex, not exfoliating. . . . . . . 6. *Rhinotropis lindheimeri*
      9. Roots with prominent bright orange-red cortex, loosely exfoliating . . . . 8. *Rhinotropis nitida*
   8. Racemes terminal, sometimes appearing axillary, but not leaf-opposed; keel petals with incurved-puberulent sacs in distal ½, beaks small, occasionally to 1.1 mm; leaf blades usually scalelike, linear, linear-subulate, lanceolate, elliptic, obovate, or ovate, 1–5.9(–15) mm.
      10. Herbs often loosely mat-forming; stems prostrate to laxly erect, not glaucous, glabrous or very sparsely pubescent; leaf blades elliptic, obovate, ovate, or scalelike, (1.5–)2–5.9(–8) mm . . . . 10. *Rhinotropis rimulicola*
      10. Subshrubs broomlike, not mat-forming; stems usually erect, glaucous and glabrous, or not glaucous and sparsely puberulent, rarely glabrate; leaf blades often scalelike, sometimes elliptic, linear-subulate, lanceolate, or obovate, some early deciduous, appearing absent, 1–4(–15) mm.
         11. Flowers pink; capsule bases cuneate; seed bodies often more sparsely pubescent to subglabrous in distal ⅕–½; stems usually glaucous, at least proximally, and glabrous . . . . 7. *Rhinotropis maravillasensis*
         11. Flowers usually white or cream, often with purplish center stripe, rarely pink; capsule bases rounded or subtruncate; seed bodies usually evenly pubescent; stems not glaucous, usually sparsely pubescent, rarely glabrate . . . . 9. *Rhinotropis nudata*

1. Rhinotropis acanthoclada (A. Gray) J. R. Abbott, J. Bot. Res. Inst. Texas 5: 134. 2011 • Desert or thorny milkwort [E]

Polygala acanthoclada A. Gray, Proc. Amer. Acad. Arts 11: 73. 1876

Shrubs or subshrubs, single- to multi-stemmed, (1.5–)2–10(–12) dm. **Stems** sprawling to erect, densely pubescent to glabrate, hairs spreading and short. **Leaves** sessile or subsessile; blade oblanceolate, narrowly obovate, or narrowly elliptic, 3–25 × 1–5 mm, base long-cuneate, apex rounded or acute, surfaces usually densely pubescent, rarely subglabrous, hairs spreading and short. **Racemes** terminal, sometimes appearing axillary if poorly developed, sometimes aggregated into pseudopanicles or reduced and appearing fasciculate, 0.5–2.5 × 0.6–2 cm; rachis thorn-tipped; peduncle 0–0.2(–0.5) cm, often poorly developed; bracts deciduous, lanceolate. **Pedicels** 1.5–4(–5.8) mm, usually shorter than flowers, pubescent, sometimes sparsely so. **Flowers** cream to yellowish green, wings cream to greenish, upper petals often purple-tipped, distal keel often dark yellow to green, (3–)3.5–5(–5.3) mm; sepals deciduous, ovate to elliptic, 1.6–3.5 mm, spreading-pubescent, margins usually ciliate; wings obovate, 3–5 × 2–3 mm, glabrous or sparsely pubescent; keel 2.7–3.8 mm, sac glabrous, beak absent or obscure and bluntly rounded, to 0.7 × 0.5 mm, glabrous; stamens rarely 9. **Capsules** ellipsoid or slightly obovoid, (3–)4–6 × 2.5–4 mm, base rounded or, sometimes, cuneate, margins with very narrow and even wing, glabrous. **Seeds** 3.2–4.2 mm, pubescent; aril 1–1.7 mm, lobes to ⅓ length of seed. $2n$ = 18.

Flowering (early spring–)spring–summer(–late fall). Usually on loose silts or sands derived from limestone, granite, sandstone, or gypsum in open places or slopes in desert scrub or juniper woodlands; 800–1800 m; Ariz., Calif., Nev., Utah.

In northern Arizona and southern Utah, *Rhinotropis acanthoclada* overlaps geographically with *R. intermontana* and tetraploid hybrids are known (T. L. Wendt 1978, 1979).

2. Rhinotropis californica (Nuttall) J. R. Abbott, J. Bot. Res. Inst. Texas 5: 134. 2011 • California milkwort [E]

Polygala californica Nuttall in J. Torrey and A. Gray, Fl. N. Amer. 1: 671. 1840

Herbs, sometimes suffrutescent, multi-stemmed, often forming a ground cover, 0.5–3.5 dm. **Stems** laxly erect, decumbent, or prostrate, pubescent to subglabrous, hairs incurved. **Leaves** sessile or subsessile; blade ovate, elliptic, or obovate, 7–50(–60) × 3–20(–26) mm, base usually rounded to acute, sometimes cuneate, apex rounded to acute, surfaces pubescent, hairs incurved. **Racemes** terminal or leaf-opposed, 1–4(–5) × 1.8–3 cm; rachis not thorn-tipped; peduncle 0–1 cm; bracts early deciduous, linear to lanceolate. **Pedicels** (2.5–)3.5–8.5 mm, sparsely pubescent or glabrous. **Cleistogamous or semi-cleistogamous flowers** often present terminally, on much reduced scale-leaved lateral branches from proximal (or distal) leaf axils, or terminally on leafy branches that are often leaf-opposed. **Flowers** usually pink, rarely white, keel distally yellow (fading white), (2.5–) 9–14.5 mm, cleistogamous and semi-cleistogamous flowers mostly 2.5–5 mm, intergrading with chasmogamous flowers; sepals deciduous, elliptic, 4–6.5 mm, pubescent or glabrous; wings obovate, (7.5–)8–12 × 2.5–6 mm, glabrous or sparsely pubescent; keel (7–) 8–11 mm, sac glabrous (sometimes proximally ciliate), beak oblong, (1.2–)1.6–3 × 0.7–1 mm (mostly absent in cleistogamous flowers), usually notched or contorted abaxially, rarely subentire, pubescent. **Capsules** ellipsoid to ovoid, 7.3–10.5 × 4.5–7 mm, in cleistogamous and semi-cleistogamous flowers 4.5–8 mm, base obtuse, rounded, or subtruncate, margins with narrow, entire or slightly erose wing, glabrous, margins sometimes ciliolate. **Seeds** 3.5–6 mm, densely pubescent; aril 1.7–4 mm, less than ½ length of seed. $2n = 18$.

Flowering spring–summer. Rocky or clay soils, deep duff, rich soils, serpentine soils, slopes or drainages, full sun to deep shade, open habitat, chaparral, mixed evergreen forests, oak woodlands, coniferous forests; 10–1400 m; Calif., Oreg.

Rhinotropis californica occurs in western California and Oregon.

Cleistogamous and semi-cleistogamous flowers can appear earlier than chasmogamous flowers. Their flowers, fruits, and seeds are similar to those of chasmogamous flowers, but typically are smaller and without the keel beak.

3. Rhinotropis cornuta (Kellogg) J. R. Abbott, J. Bot. Res. Inst. Texas 5: 134. 2011 • Sierra milkwort

Polygala cornuta Kellogg, Proc. Calif. Acad. Sci. 1: 62. 1855

Herbs, shrubs, or subshrubs, single- or multi-stemmed, rhizomatous, sometimes spreading to form ground cover or thicket, 1–25 dm. **Stems** erect to sprawling or decumbent, sometimes glaucous, glabrous or sparsely pubescent, hairs incurved. **Leaves** petiolate, petiole 0.5–5 mm; blade linear to ovate, 10–50(–65) × (3–)4–11(–16) mm, base cuneate to rounded, apex acuminate to rounded, surfaces glabrous or pubescent at least along basal midvein abaxially, hairs incurved. **Racemes** terminal, sometimes appearing axillary, 1–8(–22) × 1.5–3 cm; rachis not thorn-tipped; peduncle 0.2–0.6 cm; bracts deciduous to subpersistent, lanceolate to ovate. **Pedicels** 2.5–11 mm, pubescent or glabrous. **Flowers** cream, greenish, or pink, keel distally rose, fading green, 7–14 mm; sepals deciduous, ovate, elliptic, or suborbiculate, 1.8–4.5(–5.2) mm, pubescent; wings obovate to elliptic, (6–)6.3–11(–12) × (3–)3.5–5.6(–6) mm, usually pubescent, rarely glabrous, margins ciliate; keel 7.5–10.2 mm, sac glabrous or sparsely pubescent, beak linear, entire, 0.5–2.5 × 0.2 mm, pubescent. **Capsules** ovoid to subglobose, 5.9–10.2 × 6–9.6 mm, base rounded to truncate or subcordate, margins with narrow, entire or slightly erose wing, glabrous, margins sometimes ciliolate. **Seeds** 4.5–7.3 mm, pubescent; aril 2–5.2 mm, lobes ¼–½ length of seed.

Varieties 2 (2 in the flora): California, nw Mexico.

1. Flowers (8.5–)9.5–14 mm; sepal wings abaxially densely short-pubescent; upper sepal apex usually acute to acuminate, rarely obtuse or rounded; petioles usually 2–5 mm . . . . 3a. *Rhinotropis cornuta* var. *cornuta*
1. Flowers 7–10.5(–11.2) mm; sepal wings abaxially glabrous or puberulent near apex; upper sepal apex usually rounded or obtuse, occasionally acute; petioles 0.5–3 mm. . . . . 3b. *Rhinotropis cornuta* var. *fishiae*

3a. Rhinotropis cornuta (Kellogg) J. R. Abbott var. **cornuta** [E]

Herbs or subshrubs, 1–6(–10) dm, often spreading rhizomatously, forming a ground cover. **Stems** sprawling to erect. **Leaves:** petiole usually 2–5 mm; blade width often more than ½ length. **Pedicels** 2.5–7 mm. **Flowers** (8.5–)9.5–14 mm; outer sepals and wings cream to greenish or pale pink in bud, occasionally deep pink, apex of upper sepal usually acute to acuminate, rarely obtuse or rounded; wings abaxially densely short-pubescent. **2*n*** = 18.

Flowering late spring–fall. Open to shaded rocky slopes, pine forests, chaparral; 100–2100 m; Calif.

Variety *cornuta* is found in northern and central California.

3b. Rhinotropis cornuta (Kellogg) J. R. Abbott var. **fishiae** (Parry) J. R. Abbott, J. Bot. Res. Inst. Texas 5: 134. 2011

Polygala fishiae Parry, Proc. Davenport Acad. Nat. Sci. 4: 39. 1884; *P. cornuta* Kellogg subsp. *fishiae* (Parry) Munz; *P. cornuta* var. *fishiae* (Parry) Jepson; *P. cornuta* subsp. *pollardii* (Munz) Munz; *P. cornuta* var. *pollardii* Munz

Shrubs, usually 6–25 dm, often forming rhizomatous thickets. **Stems** erect to decumbent. **Leaves:** petiole 0.5–3 mm; blade width less than ½ length. **Pedicels** 3–11 mm. **Flowers** 7–10.5 (–11.2) mm; outer sepals (at least distally) and wings dark pink in bud, occasionally lighter, apex of upper sepal rounded or obtuse, occasionally acute; wings abaxially glabrous or puberulent near apex. **2*n*** = 18.

Flowering (early spring–)spring–late summer. Shaded canyons to open rocky slopes in chaparral or oak woodlands; 100–1300 m; Calif.; Mexico (Baja California).

In the flora area, var. *fishiae* occurs in Los Angeles, Orange, Riverside, San Diego, Santa Barbara, and Ventura counties.

4. Rhinotropis heterorhyncha (Barneby) J. R. Abbott, J. Bot. Res. Inst. Texas 5: 135. 2011 • Beaked spiny polygala, desert or notch-beaked milkwort [E]

Polygala subspinosa S. Watson var. *heterorhyncha* Barneby, Leafl. W. Bot. 3: 194. 1943; *P. heterorhyncha* (Barneby) T. Wendt

Subshrubs, multi-stemmed, mat-forming, 1–2.5 dm. **Stems** prostrate to laxly erect, often glaucous, glabrous or pubescent, hairs spreading. **Leaves** sessile; blade ovate, elliptic, or obovate, 4–20 × 2–12 mm, base cuneate, rounded, or nearly clasping, apex acute or rounded, surfaces pubescent, hairs spreading. **Racemes** terminal, to 3.5(–5) × 1.5–3 cm; rachis thorn-tipped; peduncle 0.2–0.3 cm; bracts deciduous, ovate, elliptic, or linear. **Pedicels** (3–)4–8(–9.5) mm, glabrous or pubescent. **Flowers** pink, wings usually pink, keel distally yellow, (7.5–)9.5–13.5 mm; sepals deciduous, elliptic to ovate, lower sepals mostly obovate, (2–)2.5–6 mm, pubescent; wings obovate to elliptic-obovate, (6.5–)8–12.5 × (2.5–)3–5.5 mm, glabrous or sparsely pubescent; keel (6–)7.5–11.2 mm, sac glabrous, beak oblong, with 1 or 2 prominent invaginations along abaxial side formed by sinuate excess tissue, (1.4–)2–4 × (0.6–) 0.8–1.3 mm, glabrous. **Capsules** ellipsoid-ovoid to obovoid, 4.2–7.8 × 3.7–7 mm, base cuneate to rounded, margins with very narrow and even wing, pubescent or glabrous. **Seeds** 3–4.4 mm, most densely pubescent apically, proximal ½ sparsely and unevenly pubescent or glabrous; aril 1.3–2.6 mm, lobes ¼–½ length of seed. **2*n*** = 36(or 38).

Flowering spring–early summer. Sandy or gravelly open slopes and flats in desert scrub; 900–1600 m; Calif., Nev.

Rhinotropis heterorhyncha is known from the Funeral Mountains of Inyo County, California, in the Mojave Desert region, and from adjacent areas of southern Nevada.

5. Rhinotropis intermontana (T. Wendt) J. R. Abbott, J. Bot. Res. Inst. Texas 5: 135. 2011 • Intermountain milkwort [E]

Polygala intermontana T. Wendt, J. Arnold Arbor. 60: 505. 1979, based on *P. acanthoclada* A. Gray var. *intricata* Eastwood, Proc. Calif. Acad. Sci., ser. 2, 6: 283. 1896

Subshrubs or shrubs, multi-stemmed, sometimes mat-forming, 1.5–10 dm. **Stems** erect to sprawling, densely pubescent or glabrate, with dense, matted or shaggy tomentum, hairs appressed, incurved, or, occasionally, irregularly spreading. **Leaves** sessile or subsessile, rarely with narrow, petiolelike base to 1(–2) mm; blade linear to oblanceolate or obovate, (3–)4–20 (–25) × 0.8–3(–3.5) mm, base long-cuneate, apex rounded to acute, surfaces densely pubescent, hairs incurved. **Racemes** terminal, sometimes aggregated into pseudopanicles or reduced and appearing fasciculate, 1.5 × 0.7–1.3 cm; rachis thorn-tipped; peduncle 0–0.1 cm; bracts deciduous, lanceolate or ovate. **Pedicels** (2.5–)3–7(–9) mm, glabrous. **Flowers** cream or greenish, (2.5–)3–4.7(–5.2) mm; sepals deciduous, ovate or elliptic, 1.3–3.3 mm, glabrous or with few incurved hairs subapically, margins sparsely ciliate; wings obovate, 2.5–4.9 × 1.5–3 mm, glabrous or sparsely pubescent subapically; keel (2–)2.5–3.4 mm, sac glabrous or appressed-pubescent in upper part, beak mostly absent, when present, a bluntly rounded projection, 0(–0.5) × 0(–0.5) mm, glabrous or pubescent. **Capsules** broadly ellipsoid, ovoid, or subglobose, 3.5–5.8 × 3.3–4.6 mm, base truncate to rounded, margins with narrow and even wing, glabrous. **Seeds** 2.8–4.2 mm, sparsely pubescent to subglabrous; aril 1.2–2.3 mm, lobes to 1/3 length of seed. $2n$ = 18.

Flowering spring–early summer(–fall). Sandy, gravelly, or loose silt flats, slopes, dunes, ridges, and badlands of diverse parent materials in open desert scrub or mountain slopes in pinyon-juniper-sagebrush woodlands, sagebrush scrub; 600–3000 m; Ariz., Calif., Nev., Utah.

Rhinotropis intermontana is named for its distribution in the Intermountain region of the United States, which is bounded by the Rocky Mountains on the east, the Sierra Nevada and Cascade Range on the west, and the Mojave Desert to the south.

6. Rhinotropis lindheimeri (A. Gray) J. R. Abbott, J. Bot. Res. Inst. Texas 5: 135. 2011 • Shrubby milkwort [F]

Polygala lindheimeri A. Gray, Boston J. Nat. Hist. 6: 150. 1850

Herbs, multi-stemmed, 0.3–3(–3.5) dm (rarely straggling to 10 dm). **Stems** decumbent to erect, usually pubescent, rarely glabrous, hairs spreading or incurved. **Leaves** subsessile to petiolate, petiole to 1(–1.5) mm; blade elliptic to linear, lanceolate, ovate, obovate, or scalelike, (3–)4–41 × (0.5–)1–12(–18) mm, base rounded to cuneate, apex obtuse to rounded, surfaces pubescent or glabrous, hairs incurved or spreading. **Racemes** terminal, usually leaf-opposed, often also from near base of plant, these usually with chasmogamous flowers, occasionally bearing reduced, beakless cleistogamous or semi-cleistogamous flowers, rarely with cleistogamous or semi-cleistogamous flowers throughout, 1–12(–15) × 0.3–1.5 cm; rachis not thorn-tipped; peduncle 0–1 cm; bracts usually persistent, ovate, lanceolate, or elliptic. **Pedicels** 1–4.5 mm, pubescent. **Flowers** usually pink to purple, rarely white, keel yellowish distally, wings pink or rose, (3.7–)4–7.4(–7.7) mm; upper sepal persistent, other sepals deciduous, upper sepal ovate, 1.7–4.5(–5.2) mm, lower sepals lanceolate to obovate, (1.3–)1.6–3.5(–3.8) mm, pubescent or glabrous; wings obovate to oblong-obovate, 3–6.4(–7.2) × (1.2–)1.4–3.2 mm, glabrous or pubescent; keel (2.7–)3.1–6.2 mm, sac glabrous or with scattered hairs, beak linear (or bluntly rounded), (0–)0.5–2 × (0–)0.2–0.6 mm, glabrous or pubescent. **Capsules** ellipsoid, oblong, slightly ovoid, or obovoid, 3.3–6(–6.8) × 2–4 mm, base rounded to subtruncate, often oblique, margins with narrow wing or not winged, usually pubescent, rarely subglabrous. **Seeds** 2.8–4.3 mm, pubescent; aril 0.7–2.5 mm, lobes to 3/4 length of seed.

Varieties 3 (2 in the flora): sw, sc United States, n Mexico.

Variety *eucosma* (S. F. Blake) T. Wendt is known from northern Mexico.

1. Stems usually with spreading hairs, hairs rarely somewhat crisped, mostly 0.3–0.5 mm; leaf blades usually elliptic, ovate, or obovate proximally, distally becoming narrowly so, venation usually prominently reticulate, surfaces pubescent (not glabrous); keel sacs glabrous or with scattered, spreading hairs proximally, hairs not incurved in distal 1/2. . . . 6a. *Rhinotropis lindheimeri* var. *lindheimeri*
1. Stems usually with incurved hairs, hairs rarely irregularly spreading, 0.07–0.15 mm, rarely glabrous; leaf blades lanceolate, linear, or scalelike to elliptic, ovate, or obovate, venation usually not prominently reticulate (usually midvein prominent abaxially, occasionally reticulate), surfaces pubescent or glabrous; keel sacs glabrous or, rarely, with incurved hairs in distal 1/2 . . . . 6b. *Rhinotropis lindheimeri* var. *parvifolia*

6a. Rhinotropis lindheimeri (A. Gray) J. R. Abbott var. **lindheimeri** [F]

Stems usually pubescent, rarely glabrous, hairs strictly or irregularly spreading, rarely ± crisped, sparse to dense, mostly 0.3–0.5 mm. **Leaf blades** elliptic, ovate, or obovate proximally, becoming narrowly so to nearly lanceolate distally, mostly 7–20(–27) × 3–10(–18) mm, venation usually prominently and firmly reticulate, occasionally obscure, surfaces with pubescence similar to stems, or hairs slightly longer. **Keel sacs** glabrous or with scattered, spreading hairs proximally, hairs not incurved in distal 1/2. $2n$ = 18.

Flowering spring–fall (year-round). Limestone or caliche, sandstone, shale, infrequently on gypsum, granite, or igneous substrates on ridge tops, slopes, roadcuts, canyons in juniper-oak woodlands, grassland, thorn scrub, desert scrub, canyon brush; 90–1600 m; Tex.; Mexico (Chihuahua, Coahuila, Durango, Nuevo León, Tamaulipas).

Variety *lindheimeri* occurs in central, southern, and southwestern Texas in over 25 counties.

6b. Rhinotropis lindheimeri (A. Gray) J. R. Abbott var. **parvifolia** (Wheelock) J. R. Abbott, J. Bot. Res. Inst. Texas 5: 135. 2011

Polygala lindheimeri A. Gray var. *parvifolia* Wheelock, Mem. Torrey Bot. Club 2: 143. 1891; *P. tweedyi* Britton ex Wheelock

Stems usually pubescent, rarely glabrous, hairs usually closely incurved, occasionally loosely so, or rarely irregularly spreading, 0.07–0.15 mm. **Leaf blades** lanceolate, linear, scalelike to elliptic, ovate, or obovate, 3–41 × 0.5–10 mm, venation usually not prominently reticulate, except midvein abaxially, occasionally reticulate, surfaces with pubescence similar to stems or glabrous. **Keel sacs** glabrous or, rarely, with incurved hairs in distal ½. **2*n*** = 18.

Flowering early spring–fall (year-round). Rocky or clay soils of limestone or igneous origin, infrequently on gypseous substrates, occasionally in rock crevices of open slopes, ridge tops, canyons, savannas, desert grasslands, oak-pinyon woodlands, chaparral; 300–2400 m; Ariz., N.Mex., Okla., Tex.; Mexico (Chihuahua, Coahuila, Nuevo León, Zacatecas).

As discussed by T. L. Wendt (1978), var. *parvifolia*, the most widespread and variable variety, intergrades fairly extensively with the others, especially in the southern and western portions of its range, where it is relatively common to find morphological intermediates. Despite var. *parvifolia* having been treated as a distinct species by other workers, for example, S. F. Blake (1916, 1924), Wendt made a compelling case correlating morphology with geography, ecology, and karyology, for nomenclatural recognition at varietal rank.

7. Rhinotropis maravillasensis (Correll) J. R. Abbott, J. Bot. Res. Inst. Texas 5: 135. 2011 • Maravillas milkwort [C]

Polygala maravillasensis Correll, Wrightia 3: 131. 1965

Subshrubs, multi-stemmed, broomlike, 1.5–4 dm. **Stems** usually erect, usually stiff, sometimes lax or sprawling, usually glaucous, especially proximally, glabrous. **Leaves** early deciduous; usually sessile, rarely subsessile; blade scalelike, linear-subulate, lanceolate, or elliptic, 2(–3) × 0.5–1 mm, base and apex narrowly acute, surfaces pubescent, hairs incurved. **Racemes** terminal, often also appearing axillary (from branches proximal to racemes of major branches with vegetative portions highly reduced), 2–10(–15) × 0.8–1.9 cm; rachis not thorn-tipped; peduncle to 2 cm, sometimes vestigial, especially on reduced axillary racemes; bracts mostly deciduous, rarely persistent, lanceolate, narrowly ovate, or linear. **Pedicels** 1.5–3.2 (–3.6) mm, glabrous. **Flowers** pink, keel green to yellow distally, (3–)3.4–5 mm; upper sepal persistent, others deciduous, ovate to elliptic, lower sepals ovate or elliptic to narrowly obovate, 1.5–2.8 mm, glabrous, margins sparsely ciliate proximally; wings obovate, (2.7–) 3.5–4.7 × (1.5–)1.8–2.8 mm, glabrous, margins sometimes sparsely ciliate proximally; keel (2.5–)2.7–3.5 mm, sac incurved-puberulent in distal ½, beak bluntly rounded, 0.3–0.8 × 0.3–0.6 mm, pubescent. **Capsules** obovoid, usually narrowly so, (2.6–)3.3–4.4 × 1.8–2.6 mm, base cuneate, margins with very narrow and even wing, glabrous or sparsely pubescent apically. **Seeds** 2.3–2.9 mm, pubescent, usually more sparsely pubescent to often subglabrous in distal ⅕–½ (sometimes evenly pubescent throughout); aril 0.6–1.1 mm, lobes to ⅓ length of seed. **2*n*** = 18 (36).

Flowering spring–fall. Crevices of limestone rocks and cliffs in desert and semidesert canyons and hills; of conservation concern; 400–900 m; Tex.; Mexico (Coahuila).

In the flora area, *Rhinotropis maravillasensis* occurs along the Rio Grande in Brewster and Terrell counties.

8. Rhinotropis nitida (Brandegee) J. R. Abbott, J. Bot. Res. Inst. Texas 5: 135. 2011 • Shining milkwort

Polygala nitida Brandegee, Univ. Calif. Publ. Bot. 4: 272. 1912

Herbs, multi-stemmed, 0.5–2 dm; roots with bright red to orange cortex, loosely exfoliating in thin layers. **Stems** spreading to erect, pubescent, hairs incurved and appressed. **Leaves** subsessile or petiolate, petiole to 3 mm; blade ovate, lanceolate, elliptic, or linear, 5–35(–40) × 1–9 mm, base rounded to cuneate, apex acute to rounded or acuminate, surfaces subglabrous or pubescent, hairs incurved. **Racemes** terminal, usually leaf-opposed, sometimes near stem base, 0.8–5(–6) × 1–2.5 cm; rachis not thorn-tipped; peduncle 0–0.5 cm; bracts persistent, elliptic, ovate, or lanceolate. **Pedicels** 1.5–4 mm, pubescent or glabrous. **Flowers** pink to cream, wings sometimes greenish, (4.4–)5–7.5 mm; upper sepal persistent, others deciduous, ovate, lower sepals ovate to elliptic or obovate, (1.6–)1.9–4.1 mm, pubescent or glabrous; wings obovate to elliptic, 3–7 × 1.4–3.2 mm, pubescent or glabrous; keel 2.7–6.1 mm, sac usually glabrous, rarely sparsely incurved-pubescent distally, beak oblong or bluntly rounded, (0.2–)0.6–1.7 × 0.2–0.8 mm (rarely absent in var. *tamaulipana*), glabrous or pubescent. **Capsules** ellipsoid, sometimes broadly so, 3.7–6.2 × 2.4–4.3 mm, base subtruncate to acute, margins with very narrow wing or not winged, pubescent. **Seeds** 2.7–4.9 mm, densely pubescent; aril 1.3–2.6 mm, lobes to ½ length of seed.

Varieties 4 (2 in the flora): Texas, n Mexico.

Rhinotropis nitida is closely related to *R. lindheimeri*, with seven varieties recognized between the two. According to T. L. Wendt (1978), although the differences between the two species are fairly small, the recognition of a single species would obscure the differences in the evolution of several superficially similar taxa.

1. Stems spreading to erect, hairs mostly 0.15–0.3 mm, usually spreading to loosely incurved, tips not close to stem surface; leaf blades elliptic to ovate in proximal ⅓ of stem, length 1.5–3 times width, distal leaves similar or somewhat narrower, not lanceolate or linear; pedicels (2–)2.5–4 mm . . . . 8a. *Rhinotropis nitida* var. *goliadensis*
1. Stems erect, hairs mostly 0.07–0.1(–0.15) mm, closely incurved-appressed, very close to stem, tips touching stem surface; leaf blades narrowly elliptic to lanceolate or linear, including basal ones, length at least 5 times width, or when leaves in proximal ⅓ of stem are broader, then distal leaves lanceolate-elliptic, much narrower than proximal ones; pedicels 1.5–3 mm . . . . 8b. *Rhinotropis nitida* var. *tamaulipana*

8a. Rhinotropis nitida (Brandegee) J. R. Abbott var. **goliadensis** (T. Wendt) J. R. Abbott, J. Bot. Res. Inst. Texas 5: 135. 2011 [E]

Polygala nitida Brandegee var. *goliadensis* T. Wendt, J. Arnold Arbor. 60: 508. 1979

Stems spreading to erect, hairs usually spreading to loosely incurved, mostly 0.15–0.3 mm, tips not close to stem surface. **Leaf blades** elliptic to ovate on proximal ⅓ of stem, length 1.5–3 times width, distal leaves similar or somewhat narrower, not lanceolate or linear, 8–22(–25) × (2–)3–9 mm, apex acute or rounded. **Pedicels** (2–)2.5–4 mm. **Beaks** present. **2*n*** = 54.

Flowering spring–summer. Rocky, bare caliche roadcuts in thorn scrub; 50–150 m; Tex.

Variety *goliadensis* is known only from south-central Texas, and an apparently disjunct population in Blanco County in central Texas.

8b. Rhinotropis nitida (Brandegee) J. R. Abbott var. **tamaulipana** (T. Wendt) J. R. Abbott, J. Bot. Res. Inst. Texas 5: 135. 2011

Polygala nitida Brandegee var. *tamaulipana* T. Wendt, J. Arnold Arbor. 60: 508. 1979

Stems erect, hairs closely incurved-appressed, mostly 0.07–0.1(–0.15) mm, very close to stem, tips touching stem surface. **Leaf blades** narrowly elliptic to lanceolate or linear, including basal ones, length 5+ times width, or when leaves in proximal ⅓ of stem are broader, then distal leaves lanceolate-elliptic, much narrower than proximal leaves, 5–35(–40) × 1–3 mm, apex acute to acuminate. **Pedicels** 1.5–3 mm. **Beaks** rarely absent. **2*n*** = 36.

Flowering early spring–fall. Caliche or shale soils or sands in thorn scrub, oak-savannas, coastal prairies, or montane woodlands; 0–600 m; Tex.; Mexico (Nuevo León, San Luis Potosí, Tamaulipas).

Variety *tamaulipana* occurs from central Tamaulipas and eastern Nuevo León north into south-central and coastal Texas.

9. Rhinotropis nudata (Brandegee) J. R. Abbott, J. Bot. Res. Inst. Texas 5: 135. 2011 • Small-flower milkwort

Polygala nudata Brandegee, Univ. Calif. Publ. Bot. 4: 183. 1911

Subshrubs, multi-stemmed, broomlike, 1.5–4(–5) dm. **Stems** usually erect to decumbent or procumbent, usually sparsely pubescent, rarely glabrate, hairs incurved. **Leaves:** larger ones usually early deciduous; sessile or petiolate, petiole to 1 mm; blade linear, lanceolate, narrowly elliptic, obovate, or scalelike, 1–4(–15) × 0.5–1(–3.7) mm, base cuneate, apex acute, surfaces glabrous or sparsely pubescent, hairs incurved. **Racemes** terminal, often also appearing axillary (from branches proximal to racemes of major branches with vegetative portions highly reduced), 2–10(–15) × 0.8–1.5 cm; rachis not thorn-tipped; peduncle 0–2 cm; bracts persistent, ovate to lanceolate. **Pedicels** (1.3–)1.6–3.6 mm, usually pubescent, rarely subglabrous. **Flowers** usually white or cream, often with purplish center stripe, rarely pink, 3–5 mm; upper sepal persistent, others deciduous, ovate to elliptic or obovate, 1.4–3(–3.4) mm, pubescent; wings obovate to elliptic, (2.5–)3–4.6 × (1.2–)1.5–3 mm, usually glabrous, rarely sparsely pubescent; keel (2.2–)2.5–4 mm, sac incurved-puberulent in distal ½, beak bluntly rounded to oblong, (0.3–)0.4–1.1 × 0.2–0.5(–0.7) mm, pubescent. **Capsules** ovoid to broadly ellipsoid, 2.5–4(–4.5) × (2.1–)2.3–3.2(–3.4) mm, base rounded or subtruncate, margins with very narrow and even wing, usually pubescent, sometimes subglabrous proximally. **Seeds** 2.3–3(–3.4) mm, usually evenly pubescent; aril (0.8–) 1–1.6 mm, lobes to ½ length of seed. **2*n*** = 18.

Flowering spring–late fall. Mostly on limestone, also on gypsum, sandstone, or tuff, rocky slopes in desert scrub, chaparral, or mixed woodlands (pinyon, juniper, or oak); 1200–1700 m; Tex.; Mexico (Chihuahua, Coahuila, Nuevo León, Tamaulipas).

Rhinotropis nudata occurs in Brewster and Presidio counties.

The name *Polygala minutifolia* Rose was misapplied to this taxon by D. S. Correll and M. C. Johnston (1970).

10. Rhinotropis rimulicola (Steyermark) J. R. Abbott, J. Bot. Res. Inst. Texas 5: 135. 2011 [E]

Polygala rimulicola Steyermark, Ann. Missouri Bot. Gard. 19: 390. 1932

Herbs or subshrubs, multi-stemmed, often loosely mat-forming to 2.5 dm diam., 0.1–0.5 dm. **Stems** prostrate to laxly erect, glabrous or very sparsely pubescent, hairs incurved. **Leaves** sessile or subsessile; blade elliptic, obovate, ovate, or scalelike, (1.5–)2–5.9(–8) × 0.8–3.5 mm, base cuneate or rounded, apex acute, surfaces sparsely pubescent to subglabrous, hairs incurved. **Racemes** terminal, sometimes appearing axillary (from branches proximal to racemes of major branches with vegetative portions highly reduced), reduced to 1–3 (–5) flowers, 0.5–1 × 0.7–1.9 cm; peduncle 0–0.2 cm; rachis not thorn-tipped; bracts persistent, ovate, lanceolate, or linear. **Pedicels** (0.6–)1–3(–3.6) mm, usually glabrous, rarely sparsely pubescent. **Flowers** pink or purple with pale margins, keel sometimes cream, distally greenish yellow, (2.4–)2.9–5.1(–5.4) mm; sepals: lateral ones deciduous, elliptic, ovate, or obovate, (1.2–)1.6–3(–3.2) mm, upper sepal persistent, ovate, (1–)1.2–2.4 mm, glabrous, margins sometimes sparsely ciliate proximally; wings obovate, (2.2–)2.5–4.6(–5.1) × 1.8–2.8 mm, glabrous or sparsely pubescent proximally; keel (1.7–)2.3–3.6 mm, sac incurved-puberulent in distal ½, beak obscure, deltate, bluntly rounded, or linear to oblong, (0–)0.1–0.7 × (0–)0.1–0.5 mm, glabrous. **Capsules** ellipsoid-obovoid, sometimes broadly so, 1.9–3.6 × 1.6–2.9 mm, base cuneate to rounded, margins narrowly and evenly winged or slightly expanded apically, glabrous or sparsely pubescent apically. **Seeds** (1.7–)1.9–3.2 mm, body, excluding aril and pubescence, 1–1.9 mm, densely pubescent; aril (0.3–)0.4–0.7 mm, lobes often highly reduced, nearly absent to ⅙ length of seed.

Varieties 2 (2 in the flora): sw, sc United States.

1. Keel petals with deltate or rounded beaks (often obscure), (0–)0.1–0.3(–0.5) mm, base width equal to or greater than length; seeds (1.7–) 1.9–2.7 mm; sepals eciliate or, infrequently, with a few proximal cilia . . . . 10a. *Rhinotropis rimulicola* var. *rimulicola*
1. Keel petals with linear to oblong beaks, 0.3–0.7 mm, base width usually less than ⅔ length; seeds 2.8–3.2 mm; sepals sparsely ciliate in proximal ½. . . . . 10b. *Rhinotropis rimulicola* var. *mescalerorum*

10a. Rhinotropis rimulicola (Steyermark) J. R. Abbott var. **rimulicola** • Steyermark's milkwort [E]

Flowers (2.4–)3–5.1(–5.4) mm; sepals eciliate or infrequently with a few proximal cilia; keel beak deltate or rounded, often obscure, (0–)0.1–0.3(–0.5) mm, base width equal to or greater than length; staminal column glabrous adaxially, except marginally. **Seeds** (1.7–)1.9–2.7 mm, body (excluding aril and pubescence) (1–)1.1–1.5 (–1.7) mm. **2*n*** = 18.

Flowering summer. Crevices of limestone boulders and cliffs in mesic oak-maple canyons, chaparral, open coniferous forests; 1500–2400 m; N.Mex., Tex.

Variety *rimulicola* occurs in the Guadalupe Mountains in New Mexico (Eddy County) and Texas (Culberson County). It is also known from the Sierra Diablo Mountains of Texas (Culberson and Hudspeth counties).

10b. Rhinotropis rimulicola (Steyermark) J. R. Abbott var. **mescalerorum** (T. Wendt & Todsen) J. R. Abbott, J. Bot. Res. Inst. Texas 5: 135. 2011 • Mescalero milkwort [C] [E]

Polygala rimulicola Steyermark var. *mescalerorum* T. Wendt & Todsen, Madroño 29: 20, fig. 1. 1982

Flowers 2.9–4.5 mm; sepals sparsely ciliate in proximal ½; keel beak linear to oblong, 0.3–0.7 mm, length usually greater than 1.5 times width; staminal column pubescent adaxially. **Seeds** 2.8–3.2 mm, body (excluding aril and pubescence) (1.4–) 1.7–1.9 mm. **2*n*** = 18.

Flowering early–mid summer. Crevices of limestone cliffs; of conservation concern; 1500–1700 m; N.Mex.

Variety *mescalerorum* is known only from the type locality in the San Andres Mountains in Doña Ana County.

11. Rhinotropis rusbyi (Greene) J. R. Abbott, J. Bot. Res. Inst. Texas 5: 135. 2011 • Rusby's milkwort [E]

Polygala rusbyi Greene, Bull. Torrey Bot. Club 10: 125. 1883

Herbs or subshrubs, usually multi-stemmed, 0.2–1.2 dm. **Stems** decumbent to erect, densely pubescent, hairs spreading. **Leaves** subsessile; blade elliptic or obovate, 6–20(–26) × 3–10(–12) mm, base cuneate, apex acute or obtuse, surfaces pubescent, hairs spreading. **Racemes** terminal, sometimes leaf-opposed, 1–3.5 × 1.5–2.8 cm; rachis weakly thorn-tipped (often not clearly visible when young); peduncle 0–0.5 cm; bracts semipersistent, lanceolate to linear. **Pedicels** 3–7(–8.6) mm, pubescent. **Flowers** pink and white, sepals pink or brownish pink, margins white, keel beak yellow to yellow-green, (7.5–)8.5–13(–14) mm; sepals persistent, elliptic or ovate, (3–)3.5–6(–7) mm, glabrous or sparsely pubescent; wings narrowly elliptic to obovate, (7.2–)8–11.5(–12.3) × 3.2–5.6 mm, glabrous or sparsely pubescent; keel (6.5–)7.2–10.7 mm, sac spreading-pubescent in distal ½, sometimes also proximally, beak oblong, (1.6–)1.9–3.2 × (0.5–)0.7–1 mm, pubescent. **Capsules** ellipsoid, 4.8–8.5 × 4–5.7 mm, base obtuse or rounded, margins with very narrow wing, sparsely pubescent. **Seeds** 3.5–4.8 mm, pubescent; aril 0.8–2.6 mm, lobes 0–½ length of seed. 2*n* = 18.

Flowering spring–mid summer. Calcareous or gypseous soil in open pinyon-juniper woodlands or transition to desert scrub; 900–1800 m; Ariz.

Rhinotropis rusbyi occurs in Coconino, Maricopa, Mohave, and Yavapai counties.

12. Rhinotropis subspinosa (S. Watson) J. R. Abbott, J. Bot. Res. Inst. Texas 5: 135. 2011 • Spiny or cushion or showy milkwort [E] [F]

Polygala subspinosa S. Watson, Amer. Naturalist 7: 299. 1873

Subshrubs or shrubs, multi-stemmed, 0.5–2.5(–6) dm. **Stems** prostrate to erect, sometimes glaucous, at least when young, glabrous or pubescent, hairs spreading to slightly incurved. **Leaves** subsessile to petiolate, petiole to 1(–2) mm; blade obovate or elliptic, 4–31 × 0.8–11 mm, base cuneate, apex rounded or acute, surfaces densely to sparsely pubescent or subglabrous, hairs spreading to slightly incurved. **Racemes** terminal, sometimes reduced to (1 or)2–few flowers, 3–12.5 cm; rachis thorn-tipped; peduncle 0.1–0.5 cm; bracts usually deciduous, rarely persistent, elliptic, ovate, or lanceolate. **Pedicels** (1.5–)3.5–10(–20.5) mm, glabrous or pubescent. **Flowers** pink to rose, wings (and other sepals) sometimes light green, distal keel yellow or green, (6–)8–12(–13) mm; sepals deciduous, ovate, elliptic, or lanceolate, 2–7.2 mm, glabrous or pubescent; wings obovate to elliptic-obovate, (5–)7–11.5(–12.2) × (2.3–)3–5.2(–5.9) mm, glabrous or pubescent; keel (5.4–)6.2–10.5 mm, sac glabrous, beak oblong, 1–3 × 0.9–1.5 mm, glabrous. **Capsules** ellipsoid to obovoid, 4.3–8.8(–10) × 3.7–7.3 mm, base cuneate to rounded, margins with narrow, entire or slightly erose wing, glabrous or pubescent. **Seeds** 3.3–4.9 mm, ± evenly and moderately densely pubescent, occasionally with glabrate patches; aril 1.2–3.1 mm, lobes to ½ length of seed. 2*n* = 18, 36.

Flowering spring–mid summer. Gravelly soils derived from limestone, shale, lava, or tuff, or crevices of soft calcareous rocks on eroded hills, open slopes and flats, in desert scrub, open pinyon-juniper woodlands, mountain brush, ponderosa pine woodlands; 1300–2400 m; Ariz., Calif., Colo., Nev., N.Mex., Utah.

ELAEAGNACEAE Jussieu

• Oleaster Family

Leila M. Shultz
William A. Varga

Shrubs or trees, polygamous or dioecious, terrestrial, armed or unarmed, clonal or not. **Stems** scurfy-pubescent, glabrate, or glabrescent [glabrous]. **Leaves** deciduous or evergreen, opposite or alternate, simple; stipules absent; petiolate or sessile; blade membranous or leathery, venation pinnate, margins entire, surfaces pubescent, covered with silver, yellow, or rust scales, or stellate trichomes (sometimes glabrous adaxially in *Elaeagnus multiflora*). **Inflorescences** axillary, racemes, spikes, umbels, or flowers paired or solitary; bracts absent. **Pedicels** present or absent. **Flowers** bisexual or unisexual, actinomorphic; perianth in 1 series, hypogynous; hypanthium ± tubular, sometimes constricted, accrescent to pistil; sepals 2 or 4, appearing as lobes on hypanthium, valvate, connate; petals 0; nectary disc well-developed or rudimentary; stamens 4 or 8, filaments adnate to hypanthium, relatively short; anthers basifixed or dorsifixed, dehiscing laterally, pollen colporate; pistil 1-carpellate; ovary superior, 1-locular; placentation basal; style 1, apical, slender; stigma 1, capitate or linear; ovule 1, anatropous, bitegmic. **Fruits** achenes, covered by persistent and, sometimes, fleshy base of hypanthium, appearing drupe- or berrylike. **Seed** 1 per fruit, oblong, ovoid, or ellipsoid; embryo axile and centric, nearly filling testa; endosperm scanty or absent.

Genera 3, species ca. 45 (3 genera, 9 species in the flora): North America, Europe, Asia, Australia.

All species of Elaeagnaceae have root nodules with nitrogen-fixing bacteria (*Frankia*). The capacity to fix nitrogen is advantageous to species colonizing disturbed habitats and may account, in part, for the occurrence of Russian olive (*Elaeagnus angustifolia*) as an invasive plant in parts of North America.

Some species of Elaeagnaceae that have been introduced into the horticulture trade have become weedy or problem exotics; see discussion under 1. *Elaeagnus*. Some species treated here have been reported as naturalized and caution should be used in selecting plants for landscape use; most Elaeagnaceae species have the potential to become weedy.

Phylogenetic trees based on chloroplast *rbc*L sequences group Elaeagnaceae and Rhamnaceae in the same clade (M. Clawson et al. 1998); no proposal has been made to combine the families.

Some lines of evidence suggest a relationship with Rhamnaceae: wood anatomy and the presence of vestured pits (S. Jansen et al. 2000), DNA sequencing (J. E. Richardson et al. 2000), vegetative characteristics (R. F. Thorne 1992b), and the occurrence of nitrogen fixing symbioses in Elaeagnaceae and some Rhamnaceae, Rosaceae, and Ulmaceae (D. E. Soltis et al. 1995).

SELECTED REFERENCES Jansen, S., F. Piesschaert, and E. Smets. 2000. Wood anatomy of Elaeagnaceae, with comments on vestured pits, helical thickenings, and systematic relationships. Amer. J. Bot. 87: 20–28. Nelson, A. 1935. The Elaeagnaceae—A monogeneric family. Amer. J. Bot. 22: 681–683.

1. Flowers unisexual; plants dioecious; inflorescences appearing before leaves; leaves petiolate or sessile, alternate; calyces 2-lobed; hypanthia inconspicuous . . . . 3. *Hippophaë*, p. 420
1. Flowers bisexual or unisexual; plants polygamous or dioecious; inflorescences usually appearing after leaves (except *Shepherdia argentea*); leaves petiolate, alternate or opposite; calyces 4-lobed; hypanthia conspicuous.
   2. Leaves alternate; pedicels present; flowers bisexual; plants polygamous; stamens 4. . . 1. *Elaeagnus*, p. 416
   2. Leaves opposite; pedicels absent; flowers unisexual; plants dioecious; stamens 8 . . . 2. *Shepherdia*, p. 419

1. ELAEAGNUS Linnaeus, Sp. Pl. 1: 121. 1753; Gen. Pl. ed. 5, 57. 1754 • [Ancient name used by Theophrastus for a *Salix* taxon; derivation uncertain, probably Greek *elaia*, olive tree, and *agnos*, chaste tree, alluding to resemblance, or *helodes*, marsh, and *hagnos*, pure or sacred, alluding to habitat and fertility ritual of Thesmophoria]

Shrubs or trees, polygamous, armed or unarmed, clonal or not. **Stems** densely pubescent with scales and stellate hairs or glabrate. **Leaves** deciduous or evergreen, alternate; petiolate; blade ovate, ovate-oblong, elliptic, lanceolate, lanceolate-linear, or cuneate, base attenuate or blunt, apex rounded, surfaces with silvery scales and stellate hairs (scales sometimes brown abaxially in *E. commutata* and *E. multiflora*, sometimes glabrous adaxially in *E. multiflora*). **Inflorescences** umbellate, or flowers paired or solitary, appearing after leaves. **Pedicels** present. **Flowers** bisexual; hypanthium conspicuous; calyx lobes 5; nectary disc conspicuous or inconspicuous; stamens 4, alternate with calyx lobes; style linear, stigmatic on 1 side. **Fruits** drupelike, silver, pale green, red, reddish brown, or pink, fleshy or dry. **Seeds** striate. x = 6, 14.

Species ca. 45 (5 in the flora): North America, Eurasia, Australia.

The flowers of *Elaeagnus* are strongly sweet-scented with a fragrance that most people find pleasant; the fruits are generally edible and attractive to birds. Some Eurasian species have been introduced into the horticulture trade and are now naturalized in North America (M. A. Dirr 2009). Some of these species have the potential to hybridize with native species; *E. angustifolia*, *E. pungens*, and *E. umbellata* have become noxious weeds. New introductions should be carefully considered and monitored.

1. Shrubs or trees 2–10(–12) m; leaves deciduous, blade surfaces densely silver or silver-green, (sometimes less densely so or with scattered, brown scales abaxially).
   2. Leaf blades ovate-oblong or elliptic (lengths 2 times widths); nectary discs inconspicuous; stems unarmed, scales gray in age . . . . 1. *Elaeagnus commutata*
   2. Leaf blades lanceolate-linear to narrowly elliptic (lengths 3–8 times widths); nectary discs conspicuous; stems usually armed, with thornlike lateral branches, scales reddish brown in age . . . . 2. *Elaeagnus angustifolia*
1. Shrubs or trees 1–5 m; leaves deciduous or evergreen, blade surfaces silvery or silver-green abaxially, silver-green or green adaxially.
   3. Stems armed; leaves evergreen, blades leathery, margins wavy, surfaces with silver scales, more densely hairy and silvery-green abaxially, glabrous and lustrous dark green or dull silver-green adaxially; petioles woody . . . . 4. *Elaeagnus pungens*

[3. Shifted from left margin.—Ed.]

3. Stems usually unarmed (young ones thornlike in *E. umbellata*); leaves deciduous, blades not leathery, margins entire or ± wavy, surfaces silvery abaxially (sometimes with scattered, brown scales), green or dark green and sparsely pubescent or glabrous adaxially; petioles not woody.
    4. Flowers solitary or paired, hypanthium broadly flared, calyx lobes with brown scales outside; leaf blades 1–2.5(–5) cm, broadly lanceolate or cuneate. . . . . 3. *Elaeagnus multiflora*
    4. Flowers densely clustered, hypanthium narrowly funnelform, calyx lobes with silver scales outside; leaf blades (2–)3–8(–10) cm, elliptic or ovate . . . . 5. *Elaeagnus umbellata*

1. Elaeagnus commutata Bernhardi ex Rydberg, Fl. Rocky Mts., 582. 1917 • American silver-berry, wolf willow, chalef argenté [E] [F]

Elaeagnus argentea Pursh, Fl. Amer. Sept. 1: 114. 1813 (as Elaeagrus), not Moench 1794

Trees, 2–5 m, clonal. **Stems** unarmed, densely brownish scaly when young, scales fading gray in age. **Leaves** deciduous; blade elliptic or ovate-oblong, 2–7 × 1–3(–5) cm, length 2 times width, surfaces with dense, silvery scales and stellate hairs, sometimes with scattered, brown scales abaxially. **Flowers** usually in pairs; hypanthium broadly flared, 4–7 mm distal to constriction; calyx yellow or yellow-green, 2.5–4 mm, covered with silver scales; nectary disc inconspicuous. **Fruits** silver, orbicular, 5–15 mm, densely scaly. **2*n*** = 28.

Flowering Jun–Jul. Stream banks, moist, open slopes; 0–2500 m; Alta., B.C., Man., N.W.T., Ont., Que., Sask., Yukon; Alaska, Calif., Colo., Idaho, Minn., Mont., N.Dak., S.Dak., Tex., Utah, Wyo.

2. Elaeagnus angustifolia Linnaeus, Sp. Pl. 1: 121. 1753 • Russian olive, oleaster, olivier de Bohême [I] [W]

Shrubs or trees, 5–10(–12) m, not clonal. **Stems** usually armed, with thornlike lateral branches, densely silvery-scaly when young, scales reddish brown in age, glabrate. **Leaves** deciduous (often tardily); blade lanceolate-linear to narrowly elliptic, 3–8 (–10) × 0.5–1.5 cm, length 3–8 times width, surfaces silvery and densely stellate-hairy. **Flowers** solitary or 2 or 3 in clusters; hypanthium funnelform, 3.5–6 mm distal to constriction; calyx silver-green abaxially, yellow adaxially, 3–5 mm; nectary disc conspicuous, forming thick cylinders around styles. **Fruits** pale green, ovoid or ellipsoid, (8–)10–15(–20) mm, densely white-scaly and succulent, becoming dull orange-yellow and dry in age. **2*n*** = 24, 28.

Flowering May–Jul. Roadsides, along streams; 0–2000 m; introduced; Alta., B.C., Man., N.B., N.S., Ont., Que., Sask.; Ariz., Calif., Colo., Conn., Del., D.C., Idaho, Ill., Ind., Iowa, Kans., Ky., La., Maine, Md., Mass., Mich., Minn., Mo., Mont., Nebr., Nev., N.H., N.J., N.Mex., N.Y., N.C., N.Dak., Ohio, Okla., Oreg., Pa., R.I., S.Dak., Tenn., Tex., Utah, Vt., Va., Wash., W.Va., Wis., Wyo.; Eurasia.

Elaeagnus angustifolia was originally planted as an ornamental and as a windbreak and for erosion control; it has become weedy along waterways and in disturbed areas, especially in the western United States. The roots grow to great depths and, because the plants are heavy users of water, they are known to lower dramatically the water table. Russian olive is globally invasive and is spreading in arid regions. Where it lacks competition from other trees, it forms dense thickets that exclude most other vegetation.

The fruit is somewhat succulent when young but quickly becomes dry and mealy; it is sweet and edible, and is widely dispersed by birds. The plant is prone to diseases such as leaf spot, canker, rust, and *Verticillum* wilt in humid areas (M. A. Dirr 2009). Two varieties, var. *angustifolia* and var. *orientalis* (Linnaeus) Kuntze, have been recognized. The dried, powdered fruits are reportedly mixed with milk for the treatment of rheumatoid arthritis and joint pain.

3. Elaeagnus multiflora Thunberg in J. A. Murray, Syst. Veg. ed. 14, 163. 1784 • Cherry elaeagnus or silver-berry, gumi, natsugumi [I] [W]

Shrubs, 1–2.5(–3) m, not clonal. **Stems** unarmed, with gray or reddish gray scales. **Leaves** deciduous; blade broadly lanceolate or cuneate, 1–2.5(–5) × 1–1.5(–3) cm, surfaces silvery abaxially, also with scattered, brown scales, glabrous or sparsely stellate-hairy, green adaxially. **Flowers** solitary or paired; hypanthium broadly flared, 6–7 mm distal to constriction; calyx cream, 4.5–6 mm, lobes with brown scales outside, glabrous inside; nectary disc conspicuous. **Fruits** bright red with silver flecks, oblong, 10–15(–25) mm, sparsely pubescent.

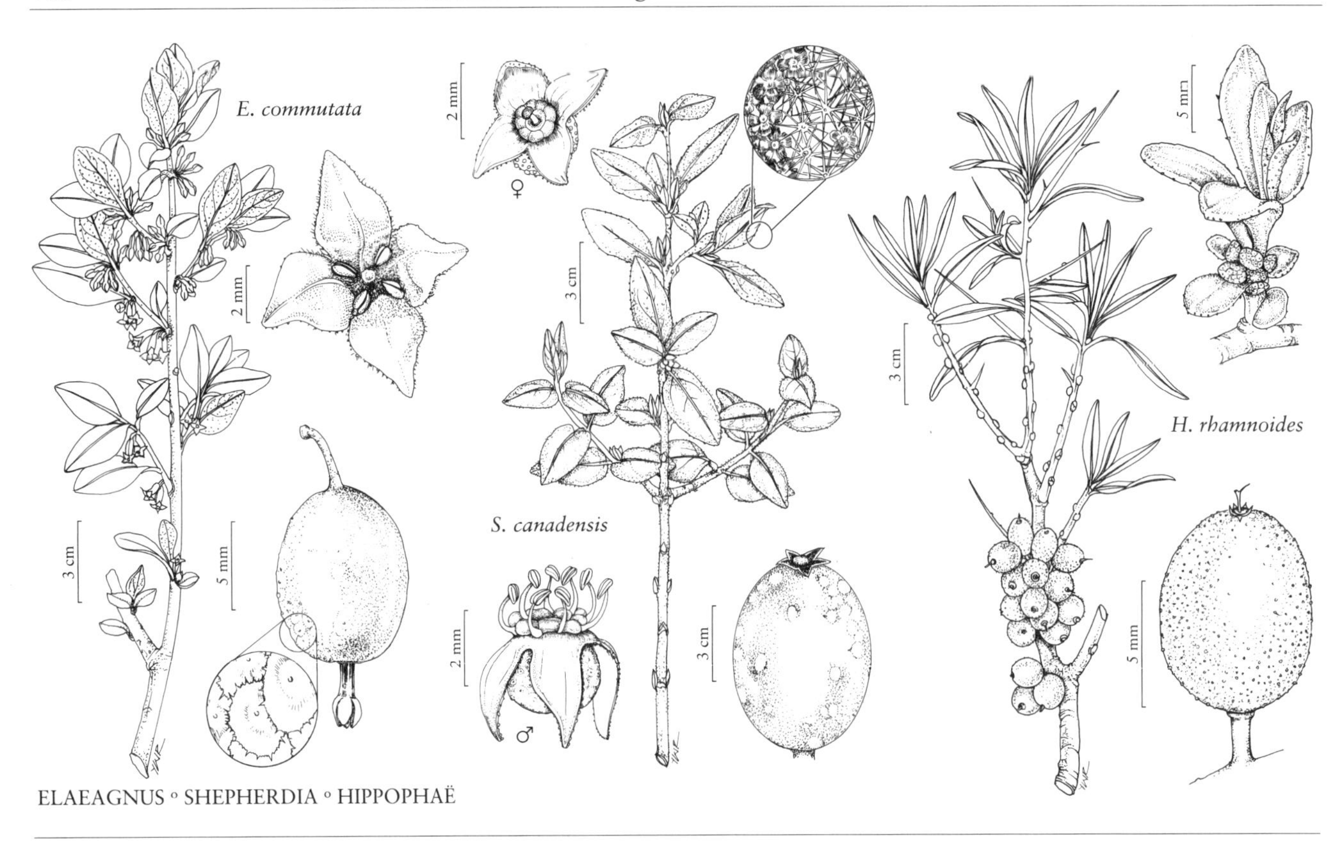

ELAEAGNUS ◦ SHEPHERDIA ◦ HIPPOPHAË

Flowering Apr–May. Sandy or clay soils, alkaline or saline soils; 100–400[–1800] m; introduced; Ala., Ga., Ill., Ky., Mass., Mich., Mo., N.Y., N.C., Ohio, Tenn., Va.; Asia (China, Japan).

Elaeagnus multiflora is similar to *E. umbellata* and may have been overlooked in some locations; it is distinguished by a calyx tube and limb that are more or less equal in length, and fruits on elongated pedicels. The species has been grown as an ornamental in Utah but, apparently, has not become naturalized in the western United States.

4. **Elaeagnus pungens** Thunberg in J. A. Murray, Syst. Veg. ed. 14, 164. 1784 • Silverthorn, spotted elaeagnus, thorny-olive [I]

Shrubs or trees, 2–5 m, clonal. **Stems** armed, dark gray-scaly or reddish. **Leaves** evergreen, leathery; petiole woody; blade broadly elliptic, (3.5–)4–8(–10) × 1.5–2.5 cm, margins wavy, surfaces with silvery scales, more densely pubescent and silver-green abaxially, glabrous and lustrous dark green or dull silver-green adaxially. **Flowers** 2–4 in clusters; hypanthium broadly flared, 2–3 mm distal to constriction; calyx white or cream, 6 mm, glabrous; nectary disc inconspicuous. **Fruits** reddish or red-brown, oblong, 8–12 mm, somewhat fleshy, sparsely lepidote, scales brown with silver margins.

Flowering Apr, Sep–Oct. Sandy soils; 0–500 m; introduced; Ala., D.C., Fla., Ga., Ky., La., Miss., N.C., S.C., Tenn., Va., W.Va.; Asia (China, Japan).

Flowers of *Elaeagnus pungens* have a sweet gardenia-like fragrance that attracts butterflies and its fruits are favored by birds. It forms clumps that are broader than tall, with canes that often grow into neighboring shrubs and vines. Plants grow prolifically and can be propagated by cuttings. It has become invasive in some areas. The Florida Exotic Pest Plant Council has placed it in the category of invasive exotics that have increased in abundance or frequency but have not yet altered Florida plant communities.

Cultivars vary in growth form and color: 'Fruitlandii' forms large bluish green mounds; 'Maculata' has large leaves with bright yellow variegations and blotches; and 'Marginata' has leaves with silvery-white margins. The leaves are distinctive with their leathery texture, wavy margins, and thick petioles. Most cultivars are thornless. Because these plants form large, vigorous clones, they are not recommended for small properties (M. A. Dirr 2009).

5. Elaeagnus umbellata Thunberg in J. A. Murray, Syst. Veg. ed. 14, 164. 1784 • Autumn-olive [I] [W]

Elaeagnus parvifolia Wallich ex Royle; *E. umbellata* var. *parvifolia* (Wallich ex Royle) C. K. Schneider

Shrubs or trees, to 5 m, clonal. **Stems** armed when young, unarmed when older, silvery-green becoming densely brown-scaly in age. **Leaves** deciduous; blade elliptic or ovate, (2–)3–8 (–10) × 1–2.5 cm, margins entire or ± wavy, surfaces silvery-scaly abaxially, sparsely pubescent, dark green, glossy adaxially. **Flowers** densely clustered, 3+, appearing to encircle stems; hypanthium narrowly funnelform, 7–8 mm distal to constriction; calyx yellow to cream-white, 3.5–4 mm, calyx lobes with silver scales outside, glabrous inside; nectary disc conspicuous. **Fruits** bright red or pink, ovoid, 6–8 mm, fleshy, lepidote.

Flowering Apr–May. Sandy soils, open areas, oak-hickory woodlands, mesic forests; 0–300 m; introduced; N.B., N.S., Ont.; Ala., Ark., Conn., Del., D.C., Fla., Ga., Ill., Ind., Iowa, Kans., Ky., La., Maine, Md., Mass., Mich., Miss., Mo., Mont., Nebr., N.J., N.Y., N.C., Ohio, Okla., Oreg., Pa., R.I., S.C., Tenn., Vt., Va., Wash., W.Va., Wis.; e Asia (China, Japan, Korea).

The flowers of *Elaeagnus umbellata* are more densely clustered and umbel-like than in other species of the genus in the flora area. Originally introduced for soil conservation and as food for wildlife, it is not considered a good plant for home landscapes because it has a tendency to become weedy (M. A. Dirr 2009). The species thrives in acidic or basic soils and birds spread the seeds.

2. SHEPHERDIA Nuttall, Gen. N. Amer. Pl. 2: 240. 1818, name conserved • Buffaloberry, shépherdie [For John Shepherd, 1764–1836, curator of the Liverpool Botanic Garden] [E]

Shrubs or trees, dioecious, armed or unarmed, clonal or not. **Stems** densely pubescent with brown or silver scales and stellate hairs or glabrescent. **Leaves** deciduous or evergreen, opposite; short-petiolate; blade elliptic, ovate, obovate, or subcordate, base attenuate or blunt, apex acute or blunt, surfaces sparsely pubescent or with silvery and rust scales and stellate hairs. **Inflorescences** usually umbellate, flowers rarely solitary, usually appearing after leaves (before in *S. argentea*). **Pedicels** absent. **Flowers** unisexual; hypanthium conspicuous elongated; calyx lobes 4; nectary disc conspicuous or inconspicuous; stamens 8, alternate and opposite calyx lobes; style linear, stigmatic on 1 side. **Fruits** berrylike, red, red-orange, yellow, or green, fleshy or dry. **Seeds** smooth. x = 11, 13.

Species 3 (3 in the flora): North America.

Shepherdia was considered part of *Elaeagnus* by A. Nelson (1935); it is easily distinguished by its opposite leaves and sessile, unisexual flowers. Species of *Shepherdia* are tolerant of alkaline soils and extremes of temperature. The first two species described here have been grown in arboreta; M. A. Dirr (2009) stated that neither is horticulturally attractive. The third species (*S. rotundifolia)* has horticultural potential for water-efficient landscapes in arid parts of the western United States.

1. Leaf blade surfaces bicolored, sparsely silvery-pubescent abaxially, with stellate hairs interspersed with rust-brown scales, glabrate and green adaxially . . . . 2. *Shepherdia canadensis*
1. Leaf blade surfaces not bicolored, silvery pubescent and without rust-brown scales.
   2. Leaves deciduous, blades elliptic or obovate, margins plane; plants clonal . . . . 1. *Shepherdia argentea*
   2. Leaves evergreen, blades broadly ovate, margins revolute; plants not clonal . . . 3. *Shepherdia rotundifolia*

1. Shepherdia argentea (Pursh) Nuttall, Gen. N. Amer. Pl. 2: 240. 1818 • Silver buffaloberry [E]

Hippophaë argentea Pursh, Fl. Amer. Sept. 1: 115. 1813; *Elaeagnus utilis* A. Nelson; *Lepargyrea argentea* (Pursh) Greene

Shrubs or trees, 1–5 m, densely clonal, from underground stems. **Stems** armed, spines 2–5 cm. **Leaves** deciduous; blade elliptic or obovate, 2–5 × 0.5–1.5(–2.5) cm, margins plane, surfaces silvery-pubescent. **Flowers:** sepals yellow, 2–3 mm on staminate flowers, 1–1.5 mm on pistillate flowers; nectary disc conspicuous. **Fruits** bright red-orange or yellow, globose, 6–9 mm, fleshy, sparsely lepidote. **Seeds** brown, 3–4 mm. $2n$ = 22, 26.

Flowering Apr–May. Moist habitats, canyon floors, meadows, open slopes, sometimes on alkaline soils; 300–2300 m; Alta., B.C., Man., Sask.; Ariz., Calif., Colo., Idaho, Iowa, Kans., Mich., Minn., Mont., Nebr., Nev., N.Mex., N.Dak., Oreg., S.Dak., Utah, Wis., Wyo.

Shepherdia argentea is possibly escaped in the eastern United States (reported from New York; possibly only in gardens).

2. Shepherdia canadensis (Linnaeus) Nuttall, Gen. N. Amer. Pl. 2: 241. 1818 • Canadian or russet buffaloberry, rabbitberry, shépherdie du Canada [E] [F] [W]

Hippophaë canadensis Linnaeus, Sp. Pl. 2: 1024. 1753; *Elaeagnus canadensis* (Linnaeus) A. Nelson; *Lepargyrea canadensis* (Linnaeus) Greene

Shrubs, 0.3–3 m, not clonal. **Stems** unarmed. **Leaves** deciduous; blade elliptic, ovate, or subcordate, 3–5(–7) × 1–3(–4) cm, margins plane, surfaces sparsely silvery stellate-hairy abaxially with interspersed rust-brown scales, glabrate and green adaxially. **Flowers:** sepals yellow, 1.5–3 mm on staminate flowers, 1–2 mm on pistillate flowers; nectary disc inconspicuous. **Fruits** bright red or yellow, ellipsoid, 6–9 mm, fleshy, sparsely lepidote. **Seeds** dark brown to black, 4–5 mm. $2n$ = 22.

Flowering Apr–Jun. Woods, open rocky slopes, sandy-gravelly shores, granitic sands; 0–3300 m; Alta., B.C., Man., N.B., Nfld. and Labr., N.W.T., N.S., Nunavut, Ont., Que., Sask., Yukon; Alaska, Ariz., Calif., Colo., Idaho, Ill., Ind., Maine, Mich., Minn., Mont., Nev., N.Mex., N.Y., N.Dak., Ohio, Oreg., Pa., S.Dak., Utah, Vt., Wash., Wis., Wyo.

3. Shepherdia rotundifolia Parry, Amer. Naturalist 9: 350. 1875 • Roundleaf buffaloberry, silver-scale [E]

Elaeagnus rotundifolia (Parry) A. Nelson; *Lepargyrea rotundifolia* (Parry) Greene

Shrubs, 0.5–2 m, not clonal. **Stems** unarmed. **Leaves** evergreen; blade broadly ovate, 1.5–3.5 × 1–3 cm, margins revolute, surfaces silvery-pubescent, hairs stellate. **Flowers:** sepals green, 2.5–4 mm on staminate flowers, 2–3 mm on pistillate flowers; nectary disc conspicuous. **Fruits** light green, ellipsoid, 6–8 mm, dry (not fleshy), densely silvery-scaly and stellate-pubescent. **Seeds** brown, 3–4 mm.

Flowering Mar–May. Dry, open, often rocky places, sandstone, sometimes on clay soils, pinyon-juniper zone; 1000–2600 m; Ariz., Utah.

Shepherdia rotundifolia is an attractive shrub of the southwestern deserts of North America; it grows on rock ledges or slick-rock sandstone habitats where the rounded growth form and bright silver indument are distinctive. Drought tolerance makes it a good candidate for gardens in arid regions; it has been planted in some botanical gardens.

3. HIPPOPHAË Linnaeus, Sp. Pl. 2: 1023. 1753; Gen. Pl. ed. 5, 449. 1754 (as Hippophae) • Seaberry [Greek *hippos*, of horse, and *phaeos*, splendor, probably alluding to ancient use of silvery leaves as horse fodder to supposedly make their coats shine or boost their energy] [I]

Trees or shrubs, dioecious, armed, clonal from root crowns. **Stems** glabrescent, trichomes gray. **Leaves** deciduous, alternate; petiolate or sessile; blade linear or linear-lanceolate, base attenuate or oblique, apex acute or rounded, surfaces covered with silver-green scales, silver-gray abaxially, dark gray-green adaxially, hairs sparsely interspersed with brown scales. **Inflorescences**

spikes (in staminate plants) or racemes (in pistillate plants), appearing before leaves. **Pedicels** present or absent. **Flowers** unisexual; hypanthium inconspicuous; calyx lobes 2; nectary disc inconspicuous; stamens 4, alternate and opposite calyx lobes; style inconspicuous; stigma sessile, capitate. **Fruits** drupelike, orange or yellow-orange, fleshy. **Seeds** smooth. $x = 12$.

Species ca. 4 (1 in the flora): introduced; Europe, Asia.

Unlike other genera of Elaeagnaceae, *Hippophaë* is wind-pollinated and the flowers are unscented. Flowers are conspicuous in spring because they develop before the leaves; in fall, the plants blaze with large clusters of bright orange fruits. The fruits are a rich source of vitamin C; the juice may protect against arsenic toxicity (R. Gupta and S. J. Flora 2005).

SELECTED REFERENCES Bartish, I. V. et al. 2000. Inter- and intraspecific genetic variation in *Hippophaë* (Elaeagnaceae) investigated by RAPD markers. Pl. Syst. Evol. 225: 85–101. Lian, Y., Chen X., and Lian H. 1998. Systematic classification of the genus *Hippophaë* L. Seabuckthorn Res. 1: 13–23. Rousi, A. 1971. The genus *Hippophaë* L. A taxonomic study. Ann. Bot. Fenn. 8: 177–227.

1. **Hippophaë rhamnoides** Linnaeus, Sp. Pl. 2: 1023. 1753 (as Hippophae) • Willow-leaved sea-buckthorn, argousier faux-nerprun [F] [I]

Elaeagnus rhamnoides (Linnaeus) A. Nelson

Trees or shrubs 3–6(–10) m, clonal, forming masses as broad as tall. **Stems** with terminal and axillary spines, spines 1–6 cm. **Leaf blades** 1–6 × 0.3–1 cm. **Pedicels** absent on staminate plants, present on pistillate plants. **Flowers:** sepals yellow, 3 mm, glabrous or sparsely pubescent. **Fruits** bright orange or yellow-orange, subglobose, 6–8 mm. **Seeds** orange-brown, oblong, 4–5 mm, shiny. $2n = 24$.

Flowering Apr–May; fruits often persisting through winter. Open areas, sandy loam or clay soils; 100–300 m; introduced; Alta., Ont., Que., Sask., Yukon; c, n Europe; Asia.

Hippophaë rhamnoides is planted for erosion control and as an ornamental; it does well in nutritionally poor soils and is potentially invasive. Because the species is dioecious, both staminate and pistillate plants must be planted for viable seeds to develop. Reports of the species in Wyoming are based on collections from gardens; it has not become naturalized in the area (R. Hartman, pers. comm.). M. A. Dirr (2009) called this one of the best plants available for winter fruit color, one that is also tolerant of salt-spray.

Literature Cited

Robert W. Kiger, Editor

This is a consolidated list of all works cited in volume 10, whether as selected references, in text, or in nomenclatural contexts. In citations of articles, both here and in the taxonomic treatments, and also in nomenclatural citations, the titles of serials are rendered in the forms recommended in G. D. R. Bridson and E. R. Smith (1991). When those forms are abbreviated, as most are, cross references to the corresponding full serial titles are interpolated here alphabetically by abbreviated form. In nomenclatural citations (only), book titles are rendered in the abbreviated forms recommended in F. A. Stafleu and R. S. Cowan (1976–1988) and Stafleu et al. (1992–2009). Here, those abbreviated forms are indicated parenthetically following the full citations of the corresponding works, and cross references to the full citations are interpolated in the list alphabetically by abbreviated form. Two or more works published in the same year by the same author or group of coauthors will be distinguished uniquely and consistently throughout all volumes of *Flora of North America* by lower-case letters (b, c, d, ...) suffixed to the date for the second and subsequent works in the set. The suffixes are assigned in order of editorial encounter and do not reflect chronological sequence of publication. The first work by any particular author or group from any given year carries the implicit date suffix "a"; thus, the sequence of explicit suffixes begins with "b". There may be citations in this list that have dates suffixed "b," "c," "d," etc. but that are not preceded by citations of "[a]," "b," and/or "c," etc. works for that year. In such cases, the missing "[a]," "b," and/or "c," etc. works are ones cited (and encountered first from) elsewhere in the *Flora* that are not pertinent in this volume.

Abbott, J. R. 2009. Revision of *Badiera* (Polygalaceae) and Phylogeny of the Polygaleae. Ph.D. dissertation. University of Florida.

Abh. Königl. Ges. Wiss. Göttingen = Abhandlungen der Königlichen Gesellschaft der Wissenschaften zu Göttingen.

Abh. Math.-Phys. Cl. Königl. Bayer. Akad. Wiss. = Abhandlungen der Mathematisch-physikalischen Classe der Königlich bayerischen Akademie der Wissenschaften.

Abrams, L. 1904. Flora of Los Angeles and Vicinity. Stanford. (Fl. Los Angeles)

Account Exped. Pittsburgh—See: E. James [1822]1823

Acta Bot. Yunnan. = Acta Botanica Yunnanica. [Yunnan Zhiwu Yanjiu.]

Adams, P. T. 1998. New plant distribution records. New Mexico Bot. Newslett. 7: 6.

Agric. Res. (Washington, DC) = Agricultural Research.

Aiken, S. G. 1976. Turion formation in watermilfoil, *Myriophyllum farwellii.* Michigan Bot. 15: 99–102.

Aiken, S. G. 1978. Pollen morphology in the genus *Myriophyllum* (Haloragaceae). Canad. J. Bot. 56: 976–982.

Aiken, S. G. 1981. A conspectus of *Myriophyllum* (Haloragaceae) in North America. Brittonia 33: 57–69.

Aiken, S. G. and A. Cronquist. 1988. Lectotypification of *Myriophyllum sibiricum* Komarov (Haloragaceae). Taxon 37: 958–966.

Aiken, S. G. and J. McNeill. 1980. The discovery of *Myriophyllum exalbescens* Fernald in Europe and the typification of *M. spicatum* Linnaeus and *M. verticillatum* Linnaeus. J. Linn. Soc., Bot. 80: 213–222.

Aiken, S. G., P. R. Newroth, and I. Wile. 1979. The biology of Canadian weeds. 34. *Myriophyllum spicatum* Linnaeus. Canad. J. Pl. Sci. 59: 201–215.

Aiken, S. G. and R. R. Picard. 1980. The influence of substrate on the growth and morphology of *Myriophyllum exalbescens* and *Myriophyllum spicatum*. Canad. J. Bot. 58: 1111–1118.

Aiton, W. 1789. Hortus Kewensis; or, a Catalogue of the Plants Cultivated in the Royal Botanic Garden at Kew. 3 vols. London. (Hort. Kew.)

Alain (Hno.). 1957. Flora de Cuba. Vol. 4. Havana.

Alain (Hno.). 1982+. La Flora de la Española. 9+ vols. San Pedro de Macorís.

Alain (Hno.). 1985–1997. Descriptive Flora of Puerto Rico and Adjacent Islands: Spermatophyta. 5 vols. Río Piedras.

Allen, G. A., V. S. Ford, and L. D. Gottlieb. 1990. A new subspecies of *Clarkia concinna* (Onagraceae) from Marin County, California Madroño 37: 305–310.

Allg. Naturgesch.—See: L. Oken 1833–1841

Amer. Bot. Fl.—See: A. Wood 1870

Amer. J. Bot. = American Journal of Botany.

Amer. J. Sci. Arts = American Journal of Science, and Arts.

Amer. Naturalist = American Naturalist....

Anales Real Acad. Ci. Méd. Fís. Nat. Habana Revista Ci. = Anales de la Real Academia de Ciencias Médicas, Físicas y Naturales de la Habana. Revista Científica.

Anales Soc. Ci. Argent. = Anales de la Sociedad Científica Argentina.

Anales Soc. Esp. Hist. Nat. = Anales de la Sociedad Española de Historia Natural.

Anderson, L. W. J., ed. 1984. Proceedings of the First International Symposium on Watermilfoil *(Myriophyllum spicatum)* and Related Haloragaceae Species. Vicksburg.

Anderson, N. O. and P. D. Ascher. 1993. Male and female fertility of loosestrife *(Lythrum)* cultivars. J. Amer. Soc. Hort. Sci. 118: 851–858.

Angiosperm Phylogeny Group. 2009. An update of the Angiosperm Phylogeny Group classification for the orders and families of flowering plants: APG III. Bot. J. Linn. Soc. 161: 105–121.

Angiosperm Phylogeny Group. 2016. An update of the Angiosperm Phylogeny Group classification for the orders and families of flowering plants: APG IV. Bot. J. Linn. Soc. 181: 1–20.

Ann. Bot. Fenn. = Annales Botanici Fennici.

Ann. Bot. (Oxford) = Annals of Botany. (Oxford.)

Ann. Missouri Bot. Gard. = Annals of the Missouri Botanical Garden.

Ann. Nat. = Annals of Nature; or, Annual Synopsis of New Genera and Species of Animals, Plants, &c. Discovered in North America.

Ann. Sci. Nat., Bot. = Annales des Sciences Naturelles. Botanique.

Ann. Sci. Nat. (Paris) = Annales des Sciences Naturelles. (Paris.)

Appl. Environm. Microbiol. = Applied and Environmental Microbiology.

Aquatic Bot. = Aquatic Botany; International Scientific Journal Dealing with Applied and Fundamental Research on Submerged, Floating and Emergent Plants in Marine and Freshwater Ecosystems.

Artz, D., C. Villagra, and R. A. Raguso. 2010. Spatiotemporal variation in the reproductive ecology of two parapatric subspecies of *Oenothera cespitosa* (Onagraceae). Amer. J. Bot. 97: 1498–1510.

Atlantic J. = Atlantic Journal, and Friend of Knowledge.

Austin, D. F. 2004. Florida Ethnobotany. Boca Raton.

Austral. J. Bot. = Australian Journal of Botany.

Austral. J. Bot., Suppl. Ser. = Australian Journal of Botany. Supplementary Series.

Austral. Syst. Bot. = Australian Systematic Botany.

Australas. J. Dermatol. = Australasian Journal of Dermatology.

Baillon, H. E. 1866–1895. Histoire des Plantes. 13 vols. Paris, London, and Leipzig. (Hist. Pl.)

Baldwin, B. G. et al., eds. 2012. The Jepson Manual: Vascular Plants of California, ed. 2. Berkeley.

Banks, H. et al. 2008. Pollen morphology of the family Polygalaceae (Fabales). Bot. J. Linn. Soc. 156: 253–289.

Barber, J. C., A. G. Ghebretinsae, and S. A. Graham. 2010. An expanded phylogeny of *Cuphea* (Lythraceae) and a North American monophyly. Pl. Syst. Evol. 289: 35–44.

Barrett, S. C. H., L. K. Jesson, and A. M. Baker. 2000. The evolution and function of stylar polymorphisms in flowering plants. Ann. Bot. (Oxford) 85(suppl. A): 253–265.

Barrie, F. R. 2009. *Eugenia*. In: G. Davidse et al., eds. 1994+. Flora Mesoamericana. 6+ vols. in parts. Mexico City, St. Louis, and London. Vol. 4(1), pp. 81–129.

Bartish, I. V. et al. 2000. Inter- and intraspecific genetic variation in *Hippophaë* (Elaeagnaceae) investigated by RAPD markers. Pl. Syst. Evol. 225: 85–101.

Bartonia = Bartonia; a Botanical Annual.

Baum, D. A., K. J. Sytsma, and P. C. Hoch. 1994. The phylogeny of *Epilobium* (Onagraceae) based on nuclear ribosomal DNA sequences. Syst. Bot. 19: 363–388.

Benoit, P. M. 1966. Synthesized *Circaea alpina* × *lutetiana*. Proc. Bot. Soc. Brit. Isles 6: 271.

Bentham, G. 1839[–1857]. Plantas Hartwegianas Imprimis Mexicanas.... London. [Issued by gatherings with consecutive signatures and pagination.] (Pl. Hartw.)

Bentham, G. 1863–1878. Flora Australiensis.... 7 vols. London. (Fl. Austral.)

Bernardi, L. F. 2000. Consideraciones taxonómicas y fitogeográficas acerca de 101 *Polygalae* Americanas. Cavanillesia Altera 1: 1–456.

Berry, P. E. 1989. A systematic revision of *Fuchsia* sect. *Quelusia* (Onagraceae). Ann. Missouri Bot. Gard. 76: 532–584.

Berry, P. E. et al. 2004. Phylogenetic relationships and biogeography of *Fuchsia* (Onagraceae) based on noncoding nuclear and chloroplast DNA data. Amer. J. Bot. 91: 601–614.

Biblioth. Universelle Rev. Suisse = Bibliothèque Universelle: Revue Suisse (et Étrangère); Archives des Sciences Physiques et Naturelles.

Biehl, R. and H. Kinzel. 1965. Blattbau und Salzhaushalt von *Laguncularia racemosa* (L.) Gaertn. f. und anderer Mangrovebaume auf Puerto Rico. Oesterr. Bot. Z. 112: 56–93.

Biehler, J. F. T. 1807. Plantarum Novarum ex Herbario Sprengelii Centuriam.... Halle. (Pl. Nov. Herb. Spreng.)

Bigelow, J. 1824. Florula Bostoniensis. A Collection of Plants of Boston and Its Vicinity..., ed. 2. Boston. (Fl. Boston. ed. 2)

Biodivers. & Conservation = Biodiversity and Conservation.

Biol. Invasions = Biological Invasions.

Blake, S. F. 1916. A revision of the genus *Polygala* in Mexico, Central America, and the West Indies. Contr. Gray Herb. 47: 1–122.

Blake, S. F. 1924. Polygalaceae. In: N. L. Britton et al., eds. 1905+. North American Flora.... 47+ vols. New York. Vol. 25, pp. 305–379.

Blossey, B., L. C. Skinner, and Janith Taylor. 2001. Impact and management of purple loosestrife *(Lythrum salicaria)* in North America. Biodivers. & Conservation 10: 1787–1807.

Blumea = Blumea; Tidjschrift voor die Systematiek en die Geografie der Planten (A Journal of Plant Taxonomy and Plant Geography).

Böcher, T. W. 1962. A cytological and morphological study of the species hybrid *Chamaenerion angustifolium* × *C. latifolium*. Bot. Tidsskr. 58: 1–34.

Boissiera = Boissiera; Mémoires des Conservatoire et de l'Institut de Botanique Systématique de l'Université de Genève (later: Mémoires des Conservatoire et Jardin Botaniques de la Ville de Genève). Supplement de Candollea.

Bol. Soc. Brot. = Boletim da Sociedade Broteriana.

Boston J. Nat. Hist. = Boston Journal of Natural History.

Bot. Beechey Voy.—See: W. J. Hooker and G. A. W. Arnott [1830–]1841

Bot. Bull. Acad. Sin. = Botanical Bulletin of Academia Sinica; a Quarterly Journal Containing Scientific Contributions from the Institute of Botany, Academia Sinica. [Kuo Li Chung Yang Yen Chiu Yuan Chih Wu Hsüeh Hui Pao.]

Bot. California—See: W. H. Brewer et al. 1876–1880

Bot. Cult.—See: G. L. M. Dumont de Courset 1802–1805

Bot. Gaz. = Botanical Gazette; Paper of Botanical Notes.

Bot. J. Linn. Soc. = Botanical Journal of the Linnean Society.

Bot. Jahrb. Syst. = Botanische Jahrbücher für Systematik, Pflanzengeschichte und Pflanzengeographie.

Bot. Mag. = Botanical Magazine; or, Flower-garden Displayed.... [Edited by Wm. Curtis.] = [With vol. 15, 1801, title became Curtis's Botanical Magazine; or....]

Bot. Misc. = Botanical Miscellany.

Bot. Not. = Botaniska Notiser.

Bot. Reg. = Botanical Register....

Bot. Rev. (Lancaster) = Botanical Review, Interpreting Botanical Progress.

Bot. Tidsskr. = Botanisk Tidsskrift.

Bot. Zeitung (Berlin) = Botanische Zeitung. (Berlin.)

Bot. Zhurn. S.S.S.R. = Botanicheskii Zhurnal S S S R.

Botany (Fortieth Parallel)—See: S. Watson 1871

Bothalia = Bothalia; a Record of Contributions from the National Herbarium, Union of South Africa.

Boufford, D. E. 1982b. The systematics and evolution of *Circaea* (Onagraceae). Ann. Missouri Bot. Gard. 69: 804–994.

Boufford, D. E. et al. 1990. A cladistic analysis of *Circaea* (Onagraceae). Cladistics 6: 171–182.

Boufford, D. E. and Xiang Q. Y. 1992. *Pachysandra* reexamined. Bot. Bull. Acad. Sin. 33: 201–207.

Bowman, R. N. and P. C. Hoch. 1979. A new combination in *Epilobium* (Onagraceae). Ann. Missouri Bot. Gard. 66: 897–898.

Brenan, J. P. M. 1953. Notes on African Onagraceae and Trapaceae. Kew Bull. 8: 163–172.

Brewer, W. H. et al. 1876–1880. Geological Survey of California.... Botany.... 2 vols. Cambridge, Mass. (Bot. California)

Bridson, G. D. R. 2004. BPH-2: Periodicals with Botanical Content. 2 vols. Pittsburgh.

Bridson, G. D. R. and E. R. Smith. 1991. B-P-H/S. Botanico-Periodicum-Huntianum/Supplementum. Pittsburgh.

Briggs, C. J. 1988. Senega snakeroot. A traditional Canadian herbal medicine. Canad. Pharm. J. 121: 199–201.

Britton, N. L. et al., eds. 1905+. North American Flora.... 47+ vols. New York. [Vols. 1–34, 1905–1957; ser. 2, parts 1–13+, 1954+.] (N. Amer. Fl.)

Britton, N. L. and A. Brown. 1913. An Illustrated Flora of the Northern United States, Canada and the British Possessions from Newfoundland to the Parallel of the Southern Boundary of Virginia, and from the Atlantic Ocean Westward to the 102d Meridian..., ed. 2. 3 vols. New York. (Ill. Fl. N. U.S. ed. 2)

Britton, N. L., E. E. Sterns, J. F. Poggenburg, et al. 1888. Preliminary Catalogue of Anthophyta and Pteridophyta Reported As Growing Spontaneously within One Hundred Miles of New York City. New York. [Authorship often attributed as B.S.P. in nomenclatural contexts.] (Prelim. Cat.)

Brittonia = Brittonia; a Journal of Systematic Botany....

Brooker, M. I. H. 2000. A new classification of the genus *Eucalyptus* L'Her. (Myrtaceae). Austral. Syst. Bot. 13: 79–148.

Brown, B. J., R. J. Mitchell, and S. A. Graham. 2002. Competition for pollination between an invasive species (purple loosestrife) and a native congener. Ecology 83: 2328–2336.

Brown, Gill. K., F. Udovicic, and P. Y. Ladiges. 2001. Molecular phylogeny and biogeography of *Melaleuca, Callistemon* and related genera (Myrtaceae). Austral. Syst. Bot. 14: 565–585.

Brown, M. L. and R. G. Brown. 1984. Herbaceous Plants of Maryland. College Park, Md.

Brown, R. 1830. Supplementum Primum Prodromi Florae Novae Hollandiae.... London. (Suppl. Prodr. Fl. Nov. Holl.)

Browne, P. 1756. The Civil and Natural History of Jamaica.... London. (Civ. Nat. Hist. Jamaica)

Brummitt, R. K. and C. E. Powell, eds. 1992. Authors of Plant Names. A List of Authors of Scientific Names of Plants, with Recommended Standard Forms of Their Names, Including Abbreviations. Kew.

Bull. Acad. Imp. Sci. Saint-Pétersbourg = Bulletin de l'Académie Impériale des Sciences de Saint Pétersbourg.

Bull. Auckland Inst. Mus. = Bulletin of the Auckland Institute and Museum.
Bull. Brit. Mus. (Nat. Hist.), Bot. = Bulletin of the British Museum (Natural History). Botany.
Bull. Calif. Acad. Sci. = Bulletin of the California Academy of Sciences.
Bull. Florida Mus. Nat. Hist., Biol. Sci. = Bulletin of the Florida Museum of Natural History. Biological Sciences.
Bull. New York Bot. Gard. = Bulletin of the New York Botanical Garden.
Bull. Soc. Bot. France = Bulletin de la Société Botanique de France.
Bull. Torrey Bot. Club = Bulletin of the Torrey Botanical Club.
Bult, C. J. and E. A. Zimmer. 1993. Nuclear ribosomal RNA sequences for inferring tribal relationships within Onagraceae. Syst. Bot. 18: 48–63.
Burns, G. P. 1904. Heterophylly in *Proserpinaca palustris.* Ann. Bot. (Oxford) 18: 579–589.
Calder, J. A. and R. L. Taylor. 1968. Flora of the Queen Charlotte Islands. 2 vols. Ottawa.
Callihan, R. H., S. L. Carson, and R. T. Dobbins. 1995. NAWEEDS, Computer-aided Weed Identification for North America. Illustrated User's Guide plus Computer Floppy Disk. Moscow, Idaho.
Canad. J. Bot. = Canadian Journal of Botany.
Canad. J. Pl. Sci. = Canadian Journal of Plant Science.
Canad. Pharm. J. = Canadian Pharmaceutical Journal.
Candolle, A. L. P. P. de and C. de Candolle, eds. 1878–1896. Monographiae Phanerogamarum.... 9 vols. Paris. (Monogr. Phan.)
Candolle, A. P. de. 1828b. Onagrariae. In: A. P. de Candolle and A. L. P. P. de Candolle, eds. 1823–1873. Prodromus Systematis Naturalis Regni Vegetabilis.... 17 vols. Paris etc. Vol. 3, pp. 35–64.
Candolle, A. P. de and A. L. P. P. de Candolle, eds. 1823–1873. Prodromus Systematis Naturalis Regni Vegetabilis.... 17 vols. Paris etc. [Vols. 1–7 edited by A. P. de Candolle, vols. 8–17 by A. L. P. P. de Candolle.] (Prodr.)
Candollea = Candollea; Organe du Conservatoire et du Jardin Botaniques de la Ville de Genève.
Carlquist, S. 1975. Wood anatomy of Onagraceae, with notes on alternative modes of photosynthate movement in dicotyledon woods. Ann. Missouri Bot. Gard. 62: 386–424.
Carlquist, S. 1977. Wood anatomy of Onagraceae: Additional species and concepts. Ann. Missouri Bot. Gard. 64: 627–637.
Carlquist, S. 1982b. Wood anatomy of Onagraceae: Further species; root anatomy; significance of vestured pits and allied structures in dicotyledons. Ann. Missouri Bot. Gard. 69: 755–769.
Carpenter, L., ed. 1992. Proceedings of a National Conference on Enhancing the States' Lake Management Programs. Monitoring and Lake Impact Assessment. Chicago.
Carpenter, R. J., R. S. Hill, and G. J. Jordan. 2005. Leaf cuticular morphology links Platanaceae and Proteaceae. Int. J. Pl. Sci. 166: 843–855.
Castanea = Castanea; Journal of the Southern Appalachian Botanical Club.
Cat. Horti Camald. ed. 2—See: F. Dehnhardt 1832
Cat. Pl. Cub.—See: A. H. R. Grisebach 1866
Cat. Pl. Upper Louisiana—See: T. Nuttall 1813
Catal. Bot.—See: A. W. Roth 1797–1806
Catling, P. M. 1998. A synopsis of the genus *Proserpinaca* in the southeastern United States. Castanea 63: 404–414.
Cavanilles, A. J. 1791–1801. Icones et Descriptiones Plantarum, Quae aut Sponte in Hispania Crescunt, aut in Hortis Hospitantur. 6 vols. Madrid. (Icon.)
Ceska, A. and O. Ceska. 1986. Notes on *Myriophyllum* (Haloragaceae) in the Far East: The identity of *Myriophyllum sibiricum* Komarov. Taxon 35: 95–100.
Ceska, O., A. Ceska, and P. D. Warrington. 1986. *Myriophyllum quitense* and *Myriophyllum ussuriense* (Haloragaceae) in British Columbia, Canada. Brittonia 38: 73–81.
Channell, R. B. and C. E. Wood Jr. 1987. The Buxaceae in the southeastern United States. J. Arnold Arbor. 68: 241–257.
Chapman, A. W. 1860. Flora of the Southern United States.... New York. (Fl. South. U.S.)
Chapman, A. W. 1883. Flora of the Southern United States..., ed. 2. New York. (Fl. South. U.S. ed. 2)
Chapman, A. W. 1892. Flora of the Southern United States..., ed. 2 reprint 2. New York, Cincinnati, and Chicago. (Fl. South. U.S. ed. 2 repr. 2)
Chapman, A. W. 1897. Flora of the Southern United States..., ed. 3. Cambridge, Mass. (Fl. South. U.S. ed. 3)
Char. Gen. Pl. ed. 2—See: J. R. Forster and G. Forster 1776
Chase, M. W. et al. 1993. Phylogenetics of seed plants: An analysis of nucleotide sequences from the plastid gene *rbc*L. Ann. Missouri Bot. Gard. 80: 528–580.
Chen, C. J., P. C. Hoch, and P. H. Raven. 1992. Systematics of *Epilobium* (Onagraceae) in China. Syst. Bot. Monogr. 34: 1–209.
Chippendale, G. M. 1988. Myrtaceae—*Eucalyptus, Angophora.* In: R. Robertson et al., eds. 1981+. Flora of Australia. 32+ vols. Canberra. Vol. 19.
Civ. Nat. Hist. Jamaica—See: P. Browne 1756
Cladistics = Cladistics; the International Journal of the Willi Hennig Society.
Clausen, J., D. D. Keck, and W. M. Hiesey. 1940. Experimental studies on the nature of species. I. Effect of varied environments on western North American plants. Publ. Carnegie Inst. Wash. 520.
Clausing, G. and S. S. Renner. 2001. Molecular phylogenetics of Melastomataceae and Memecylaceae: Implications for character evolution. Amer. J. Bot. 88: 486–498.
Clawson, M., M. Caru, and D. R. Benson. 1998. Diversity of *Frankia* strains in root nodules from plants from the families Elaeagnaceae and Rhamnaceae. Appl. Environm. Microbiol. 64: 3539–3543.
Claxton, F. et al. 2005. Pollen morphology of families Quillajaceae and Surianaceae (Fabales). Rev. Palaeobot. Palynol. 133: 221–233.
Cleland, R. E. 1972. *Oenothera:* Cytogenetics and Evolution. London.
Clinebell, R. R. et al. 2004. Pollination ecology of *Gaura* and *Calylophus* (Onagraceae, tribe Onagreae) in western Texas, U.S.A. Ann. Missouri Bot. Gard. 91: 369–400.
Commentat. Soc. Regiae Sci. Gott. = Commentationes Societatis Regiae Scientiarum Gottingensis.
Comun. Inst. Trop. Invest. Ci. Univ. El Salvador = Comunicaciones. Instituto Tropical de Investigaciones Científicas, Universidad de El Salvador.

Conti, E. et al. 1997. Interfamilial relationships in Myrtales: Molecular phylogeny and patterns of morphological evolution. Syst. Bot. 22: 629–647.

Conti, E., A. Litt, and K. J. Sytsma. 1996. Circumscription of Myrtales and their relationship to other rosids: Evidence from *rbc*L sequence data. Amer. J. Bot. 83: 221–233.

Contr. Gray Herb. = Contributions from the Gray Herbarium of Harvard University. [Some numbers reprinted from (or in?) other periodicals, e.g. Rhodora.]

Contr. U.S. Natl. Herb. = Contributions from the United States National Herbarium.

Cook, C. D. K. 1979. A revision of the genus *Rotala* (Lythraceae). Boissiera 29: 1–156.

Cooperrider, T. S. 1995. The Dicotyledoneae of Ohio: Linaceae through Campanulaceae. Columbus. [The Vascular Flora of Ohio, vol. 2(2).]

Cooperrider, T. S. and B. K. Andreas. 1991. Noteworthy collections. Ohio. *Epilobium parviflorum* Schreber (Onagraceae). Small-flowered hairy willow-herb. Michigan Bot. 30: 69–70.

Correll, D. S. and M. C. Johnston. 1970. Manual of the Vascular Plants of Texas. Renner, Tex.

Couch, R. and E. Nelson. 1985. *Myriophyllum spicatum* in North America. In: L. W. J. Anderson, ed. 1984. Proceedings of the First International Symposium on Watermilfoil *(Myriophyllum spicatum)* and Related Haloragaceae Species. Vicksburg. Pp. 8–18.

Couch, R. and E. Nelson. 1988. *Myriophyllum quitense* (Haloragaceae) in the United States. Brittonia 40: 85–88.

Couch, R. and E. Nelson. 1992. The exotic myriophyllums of North America. In: L. Carpenter, ed. 1992. Proceedings of a National Conference on Enhancing the States' Lake Management Programs. Monitoring and Lake Impact Assessment. Chicago. Pp. 5–11.

Couillault, J. 1973. Organisation de l'appareil conducteur de *Trapa natans* L. Bull. Soc. Bot. France 119: 177–198.

Craven, L. A. 2006. New combinations in *Melaleuca* for Australian species of *Callistemon* (Myrtaceae). Novon 16: 468–475.

Crawford, B. 2002. Loosestrife causes strife to young sheep on stubbles. Weedscene 13(2): 2.

Cronquist, A. 1968. The Evolution and Classification of Flowering Plants. Boston.

Cronquist, A. 1981. An Integrated System of Classification of Flowering Plants. New York.

Cronquist, A. et al. 1972–2017. Intermountain Flora. Vascular Plants of the Intermountain West, U.S.A. 7 vols. in 9. New York and London.

Cronquist, A., N. H. Holmgren, and P. K. Holmgren. 1997c. Onagraceae. In: A. Cronquist et al. 1972–2017. Intermountain Flora. Vascular Plants of the Intermountain West, U.S.A. 7 vols. in 9. New York and London. Vol. 3, part A, pp. 172–244.

Crous, P. W. et al. 2004. Cultivation and Diseases of Proteaceae: *Leucadendron, Leucospermum* and *Protea*. Utrecht.

Cult. Prot.—See: J. Knight 1809

Darwiniana = Darwiniana; Carpeta del "Darwinion."

Davidse, G., M. Sousa S., A. O. Chater, et al., eds. 1994+. Flora Mesoamericana. 6+ vols. in parts. Mexico City, St. Louis, and London.

Davis, G. J. 1967. *Proserpinaca:* Photoperiodic and chemical differentiation of leaf development and flowering. Pl. Physiol. (Lancaster) 42: 667–669.

DeFilipps, R. A. 1998. Useful Plants of the Commonwealth of Dominica, West Indies. Washington. (Useful Pl. Dominica)

Dehnhardt, F. 1832. Catalogus Plantarum Horti Camalduensis, ed. 2. Naples. (Cat. Horti Camald. ed. 2)

Derraik, J. G. B. and M. Rademaker. 2009. Allergic contact dermatitis from exposure to *Grevillea robusta* in New Zealand. Australas. J. Dermatol. 50: 125–128.

Desfontaines, R. L. 1829[–1832]. Tableau de l'École Botanique du Muséum d'Histoire Naturelle, ed. 3. 1 vol. + Add. Paris. [Title: Catalogus Plantarum Horti Regii Parisiensis....] (Tabl. École Bot. ed. 3)

Deut. Bot. Herb.-Buch—See: H. G. L. Reichenbach 1841

Deutsche Bot. Monatsschr. = Deutsche botanische Monatsschrift; zugleich Organ der Bot. Vereine in Hamburg und Nürnberg und der Thüring. bot. Gesellschaft "Irmischie" zu Arnstadt.

Devold, J. and P. F. Scholander. 1933. Flowering Plants and Ferns of Southeast Greenland. Oslo.

Dietrich, W. 1977. The South American species of *Oenothera* sect. *Oenothera* (*Raimannia, Renneria*; Onagraceae). Ann. Missouri Bot. Gard. 64: 425–626.

Dietrich, W., P. H. Raven, and W. L. Wagner. 1985. Revision of *Oenothera* sect. *Oenothera* subsect. *Emersonia* (Onagraceae). Syst. Bot. 10: 29–48.

Dietrich, W. and W. L. Wagner. 1988. Systematics of *Oenothera* section *Oenothera* subsection *Raimannia* and subsection *Nutantigemma* (Onagraceae). Syst. Bot. Monogr. 24: 1–91.

Dietrich, W., W. L. Wagner, and P. H. Raven. 1997. Systematics of *Oenothera* section *Oenothera* subsection *Oenothera*. Syst. Bot. Monogr. 50.

Dirr, M. A. 2009. Manual of Woody Landscape Plants..., ed. 6. Champaign.

Dorken, M. E. and C. G. Eckert. 2001. Severely reduced sexual reproduction in northern populations of a clonal plant, *Decodon verticillatus* (Lythraceae). J. Ecol. 89: 339–350.

Dumont de Courset, G. L. M. 1802–1805. Le Botaniste Cultivateur.... 5 vols. Paris. (Bot. Cult.)

Dumortier, B. C. J. 1827. Florula Belgica, Operis Majoris Prodromus.... Tournay. (Fl. Belg.)

Duncan, W. H. 1950. Stamen numbers in *Cuphea*. Rhodora 52: 185–188.

Durand, E. M. 1855. Plantae Prattenianae Californicae.... Philadelphia. [Preprinted from J. Acad. Nat. Sci. Philadelphia, n. s. 3: 79–104. 1855.] (Pl. Pratten. Calif.)

Eckert, C. G. and S. C. H. Barrett. 1994. Post-pollination mechanisms and the maintenance of outcrossing in self-compatible, tristylous, *Decodon verticillatus* (Lythraceae). Heredity 72: 396–411.

Eckhart, V. M., M. A. Geber, and C. McGuire. 2004. Experimental studies of selection and adaptation in *Clarkia xantiana* (Onagraceae). I. Sources of phenotypic variation across a subspecies border. Evolution 58: 59–70.

Ecology = Ecology, a Quarterly Journal Devoted to All Phases of Ecological Biology.

Econ. Bot. = Economic Botany; Devoted to Applied Botany and Plant Utilization.
Edwards, R. D. et al. 2010. *Melaleuca* revisited: cpDNA and morphological data confirm that *Melaleuca* L. (Myrtaceae) is not monophyletic. Taxon 59: 744–754.
Edwards's Bot. Reg. = Edwards's Botanical Register....
Elliott, S. [1816–]1821–1824. A Sketch of the Botany of South-Carolina and Georgia. 2 vols. in 13 parts. Charleston. (Sketch Bot. S. Carolina)
Encycl.—See: J. Lamarck et al. 1783–1817
Endlicher, S. L. 1830. Flora Posoniensis.... Poznan. (Fl. Poson.)
Engler, H. G. A., ed. 1900–1953. Das Pflanzenreich.... 107 vols. Berlin. [Sequence of vol. (Heft) numbers (order of publication) is independent of the sequence of series and family (Roman and Arabic) numbers (taxonomic order).]
Engler, H. G. A. and K. Prantl, eds. 1887–1915. Die natürlichen Pflanzenfamilien.... 254 fascs. Leipzig. [Sequence of fasc. (Lieferung) numbers (order of publication) is independent of the sequence of division (Teil) and subdivision (Abteilung) numbers (taxonomic order).] (Nat. Pflanzenfam.)
Enum. Hort. Berol. Alt.—See: J. H. F. Link 1821–1822
Enum. Syst. Pl.—See: N. J. Jacquin 1760
Eriksen, B. 1993. Phylogeny of the Polygalaceae and its taxonomic implications. Pl. Syst. Evol. 186: 33–55.
Eriksen, B. and C. Persson. 2007. Polygalaceae. In: K. Kubitzki et al., eds. 1990+. The Families and Genera of Vascular Plants. 15+ vols. Berlin etc. Vol. 9, pp. 345–363.
Ertter, B. and D. Gowen. 2019. Notes on *Lythrum* (Lythraceae) in California: *Lythrum junceum* as a new (but old) introduced species, typification of Greene's *Lythrum* types, and possible hybridization of *Lythrum* in California. Madroño 66: 97–102.
Erythea = Erythea; a Journal of Botany, West American and General.
Evans, M. E. K. et al. 2005. Climate and life-history evolution in evening primroses (*Oenothera,* Onagraceae): A phylogenetic comparative analysis. Evolution 59: 1914–1927.
Evans, M. E. K. et al. 2009. Climate, niche evolution, and diversification of the "bird-cage" evening primroses (*Oenothera,* sections *Anogra* and *Kleinia*). Amer. Naturalist 173: 225–240.
Evol. Applic. = Evolutionary Applications. [Electronic journal.]
Evolution = Evolution; International Journal of Organic Evolution.
Exell, A. W. 1931. The genera of Combretaceae. J. Bot. 69: 113–128.
Exell, A. W. and C. A. Stace. 1966. Revision of the Combretaceae. Bol. Soc. Brot., ser. 2, 40: 5–25.
Exell, A. W. and H. Wild, eds. 1960+. Flora Zambesiaca: Mozambique, Federation of Rhodesia and Nyasaland, Bechuanaland Protectorate. 12+ vols. in parts. London.
Eyde, R. H. 1977. Reproductive structures and evolution in *Ludwigia* (Onagraceae). I. Androecium, placentation, meristem. Ann. Missouri Bot. Gard. 64: 644–655.
Eyde, R. H. 1979. Reproductive structures and evolution in *Ludwigia* (Onagraceae). II. Fruit and seed. Ann. Missouri Bot. Gard. 66: 656–675.
Eyde, R. H. 1981. Reproductive structures and evolution in *Ludwigia* (Onagraceae). III. Vasculature, nectaries, conclusions. Ann. Missouri Bot. Gard. 68: 470–503.
Eyde, R. H. 1982. Evolution and systematics of the Onagraceae: Floral anatomy. Ann. Missouri Bot. Gard. 68: 470–503.
Fassett, N. C. 1953c. *Proserpinaca.* Comun. Inst. Trop. Invest. Ci. Univ. El Salvador 2: 139–162.
Feddes Repert. = Feddes Repertorium.
Ferguson, I. K. and J. Muller, eds. 1976. The Evolutionary Significance of the Exine. London.
Fernald, M. L. 1919c. Two new *Myriophyllum* and a species new to the United States. Rhodora 21: 121–124.
Fernald, M. L. 1944d. The identities of *Epilobium lineare, E. densum,* and *E. ciliatum.* Rhodora 46: 377–386.
Fernald, M. L. 1945d. Some inconvenient upheavals of familiar names and author-citations. Rhodora 47: 197–204.
Fernald, M. L. 1950. Gray's Manual of Botany, ed. 8. New York.
Fernández de Oviedo y Valdéz, G. 1851. Historia General y Natural de las Indias, Islas y Tierra-firme del Mar Oceano..., ed. D. José Amador de los Rios. Primera Parte. Madrid.
Fernando, E. S. et al. 1993. Rosid affinities of the Surianaceae: Molecular evidence. Molec. Phylogen. Evol. 2: 344–350.
Fernando, E. S. and C. J. Quinn. 1992. Pericarp anatomy and systematics of the Simaroubaceae sensu lato. Austral. J. Bot. 40: 263–350.
Field & Lab. = Field & Laboratory.
Fieldiana, Bot. = Fieldiana: Botany.
Fish Wildlife Res. = Fish and Wildlife Research.
Fisher, J. B. and H. Honda. 1979. Branch geometry and effective leaf area: A study of *Terminalia*-branching pattern. 1. Theoretical trees. Amer. J. Bot. 66: 633–644.
Fisher, J. B. and H. Honda. 1979b. Branch geometry and effective leaf area: A study of *Terminalia*-branching pattern. 2. Survey of real trees. Amer. J. Bot. 66: 645–655.
Fl. Amer. Sept.—See: J. R. Forster 1771; F. Pursh [1813]1814
Fl. Austral.—See: G. Bentham 1863–1878
Fl. Belg.—See: B. C. J. Dumortier 1827
Fl. Bor.-Amer.—See: A. Michaux 1803; W. J. Hooker [1829–]1833–1840
Fl. Boston. ed. 2—See: J. Bigelow 1824
Fl. Bras.—See: C. F. P. von Martius et al. 1840–1906
Fl. Calif.—See: W. L. Jepson 1909–1943
Fl. Carniol. ed. 2—See: J. A. Scopoli 1771–1772
Fl. Carol.—See: T. Walter 1788
Fl. Dan.—See: G. C. Oeder et al. [1761–]1764–1883
Fl. Flumin.—See: J. M. Vellozo 1825–1827[1829–1831]
Fl. Franç. ed. 3—See: J. Lamarck and A. P. de Candolle 1805–1815
Fl. Francisc.—See: E. L. Greene 1891–1897
Fl. Ind.—See: W. Roxburgh 1820–1824
Fl. Los Angeles—See: L. Abrams 1904
Fl. N. Amer.—See: J. Torrey and A. Gray 1838–1843
Fl. N.W. Coast—See: C. V. Piper and R. K. Beattie 1915
Fl. Poson.—See: S. L. Endlicher 1830

Fl. Rocky Mts.—See: P. A. Rydberg 1917
Fl. S.E. U.S.—See: J. K. Small 1903
Fl. S.E. U.S. ed. 2—See: J. K. Small 1913
Fl. South. U.S.—See: A. W. Chapman 1860
Fl. South. U.S. ed. 2—See: A. W. Chapman 1883
Fl. South. U.S. ed. 2 repr. 2—See: A. W. Chapman 1892
Fl. South. U.S. ed. 3—See: A. W. Chapman 1897
Fl. W. Calif.—See: W. L. Jepson 1901
Flexner, S. B. and L. C. Hauck, eds. 1987. The Random House Dictionary of the English Language, ed. 2 unabridged. New York.
Flora = Flora; oder (allgemeine) botanische Zeitung. [Vols. 1–16, 1818–1833, include "Beilage" and "Ergänzungsblätter"; vols. 17–25, 1834–1842, include "Beiblatt" and "Intelligenzblatt."]
Flora of North America Editorial Committee, eds. 1993+. Flora of North America North of Mexico. 21+ vols. New York and Oxford.
Folia Geobot. Phytotax. = Folia Geobotanica et Phytotaxonomica.
Ford, V. S. and L. D. Gottlieb. 2003. Reassessment of phylogenetic relationships in *Clarkia* sect. *Sympherica.* Amer. J. Bot. 90: 167–174.
Ford, V. S., and L. D. Gottlieb. 2007. Tribal relationships within Onagraceae inferred from *Pgi*C sequences. Syst. Bot. 32: 348–356.
Forde, M. B. 1964. *Haloragis erecta:* A species complex in evolution. New Zealand J. Bot. 2: 425–453.
Forest, F. et al. 2007. The role of biotic and abiotic factors in evolution of ant-dispersal in the milkwort family (Polygalaceae). Evolution 61: 1675–1694.
Forster, J. R. 1771. Flora Americae Septentrionalis.... London. (Fl. Amer. Sept.)
Forster, J. R. and G. Forster. 1776. Characteres Generum Plantarum, Quas in Itinere ad Insulas Maris Australis..., ed. 2. London. (Char. Gen. Pl. ed. 2)
Fourqurean, J. W. et al. 2010. Are mangroves in the tropical Atlantic ripe for invasion? Exotic mangrove trees in the forests of South Florida. Biol. Invasions 12: 2509–2522.
Fragm.—See: F. J. H. Mueller 1858–1882
Fredskild, B. 1984. Distribution and occurrence of Onagraceae in Greenland. Nordic J. Bot. 4: 475–480.
Freire-Fierro, A. 1995. Especies medicinales Sudamericanas de Polygalaceae. In: P. Naranjao and R. Escaleras, eds. 1995. La Medicina Tradicional en el Ecuador, Memorias de las Primeras Jornadas Ecuatorianas de Etnomedicina Andina. Quito. Pp. 103–105.
Frémont, J. C. 1843–1845. Report of the Exploring Expedition to the Rocky Mountains in the Year 1842, and to Oregon and North California in the Year 1843–44. 2 parts. Washington. [Parts paged consecutively.] (Rep. Exped. Rocky Mts.)
Fruct. Sem. Pl.—See: J. Gaertner 1788–1791[–1792]
Fuller, D. G. and L. J. Hickey. 2005. Systematics and leaf architecture of the Gunneraceae. Bot. Rev. (Lancaster) 71: 295–353.
Furtado, C. X. and S. Montien. 1969. A revision of *Lagerstroemia* Linnaeus (Lythraceae). Gard. Bull. Straits Settlem. 24: 185–334.
Gadek, P. A. et al. 1996. Sapindales: Molecular delimitation and infraordinal groups. Amer. J. Bot. 83: 802–811.
Gaertner, C. F. von. 1805–1807. Supplementum Carpologiae.... 2 fascs. Leipzig. [Fascs. paged consecutively.] (Suppl. Carp.)
Gaertner, J. 1788–1791[–1792]. De Fructibus et Seminibus Plantarum.... 2 vols. Stuttgart and Tübingen. [Vol. 1 in 1 part only, 1788. Vol. 2 in 4 parts paged consecutively: pp. 1–184, 1790; pp. 185–352, 353–504, 1791; pp. 505–520, 1792.] (Fruct. Sem. Pl.)
Gard. Bull. Straits Settlem. = Gardens' Bulletin. Straits Settlements.
Gard. Dict. ed. 8—See: P. Miller 1768
Garrett, A. O. 1927. Spring Flora of the Wasatch Region, ed. 4. Salt Lake City. (Spring Fl. Wasatch ed. 4)
Gen. N. Amer. Pl.—See: T. Nuttall 1818
Gen. Pl. ed. 5—See: C. Linnaeus 1754
Genetics = Genetics; a Periodical Record of Investigations Bearing on Heredity and Variation.
Ges. Naturf. Freunde Berlin Neue Schriften = Der Gesellschaft naturforschender Freunde zu Berlin, neue Schriften.
Gillett, J. M. 1968b. The Milkworts of Canada. Ottawa. [Canad. Dept. Agric. Monogr. 5.]
Gleason, H. A. 1952. The New Britton and Brown Illustrated Flora of the Northeastern United States and Adjacent Canada. 3 vols. New York.
Gmelin, J. F. 1791[–1792]. Caroli à Linné...Systema Naturae per Regna Tria Naturae.... Tomus II. Editio Decima Tertia, Aucta, Reformata. 2 parts. Leipzig. (Syst. Nat.)
Godfrey, R. K. and J. W. Wooten. 1981. Aquatic and Wetland Plants of Southeastern United States: Dicotyledons. Athens, Ga.
Gottlieb, L. D. and V. S. Ford. 1996. Phylogenetic relationships among the sections of *Clarkia* (Onagraceae) inferred from the nucleotide sequences of *Pgi*C. Syst. Bot. 21: 45–62.
Gottlieb, L. D. and S. K. Jain, eds. 1988. Plant Evolutionary Biology. A Symposium Honoring G. Ledyard Stebbins. London.
Gottlieb, L. D. and L. P. Janeway. 1995. The status of *Clarkia mosquinii* (Onagraceae). Madroño 42: 79–82.
Grace, J. B. and R. J. Wetzel. 1978. The production biology of Eurasian watermilfoil (*Myriophyllum spicatum* Linnaeus). A review. J. Aquatic Pl. Managem. 16: 1–11.
Graham, S. A. 1964. The genera of Rhizophoraceae and Combretaceae in the southeastern United States. J. Arnold Arbor. 45: 285–301.
Graham, S. A. 1975. Taxonomy of the Lythraceae in the southeastern United States. Sida 6: 80–103.
Graham, S. A. 1977. The American species of *Nesaea* (Lythraceae) and their relationship to *Heimia* and *Decodon.* Syst. Bot. 2: 61–71.
Graham, S. A. 1979. The origin of *Ammannia* ×*coccinea* Rottboell. Taxon 28: 169–178.
Graham, S. A. 1985. A revision of *Ammannia* (Lythraceae) in the Western Hemisphere. J. Arnold Arbor. 66: 395–420.
Graham, S. A. 1988. Revision of *Cuphea* section *Heterodon* (Lythraceae). Syst. Bot. Monogr. 20: 1–168.
Graham, S. A. 2007. Lythraceae. In: K. Kubitzki et al., eds. 1990+. The Families and Genera of Vascular Plants. 15+ vols. Berlin etc. Vol. 9, pp. 226–246.

Graham, S. A. 2013. Fossil records in the Lythraceae. Bot. Rev. (Lancaster) 79: 48–145.

Graham, S. A. et al. 2005. Phylogenetic analysis of the Lythraceae based on four gene regions and morphology. Int. J. Pl. Sci. 166: 995–1017.

Graham, S. A. et al. 2011. Relationships among the confounding genera *Ammannia, Hionanthera, Nesaea* and *Rotala* (Lythraceae). Bot. J. Linn. Soc. 166: 1–19.

Graham, S. A. and T. B. Cavalcanti. 2001. New chromosome counts in the Lythraceae and a review of chromosome numbers in the family. Syst. Bot. 26: 445–458.

Graham, S. A., J. V. Freudenstein, and M. Luker. 2006. A phylogenetic study of *Cuphea* (Lythraceae) based on morphology and nuclear rDNA ITS sequences. Syst. Bot. 31: 764–778.

Graham, S. A. and A. Graham. 2014. Ovary, fruit, and seed morphology of the Lythraceae. Int. J. Pl. Sci. 175: 202–240.

Grana = Grana; an International Journal of Palynology Including "World Pollen and Spore Flora."

Gray, A. 1867. A Manual of the Botany of the Northern United States..., ed. 5. New York and Chicago. [Pteridophytes by D. C. Eaton.] (Manual ed. 5)

Gray, A., S. Watson, and J. M. Coulter. 1890. A Manual of the Botany of the Northern United States..., ed. 6. New York and Chicago. (Manual ed. 6)

Greene, E. L. 1891–1897. Flora Franciscana. An Attempt to Classify and Describe the Vascular Plants of Middle California. 4 parts. San Francisco. [Parts paged consecutively.] (Fl. Francisc.)

Gregor, T. et al. 2013. *Epilobium brachycarpum:* A fast-spreading neophyte in Germany. Tuexenia 33: 259–283.

Gregory, D. P. 1963. Hawkmoth pollination in the genus *Oenothera.* Aliso 5: 375–384.

Gregory, D. P. 1964. Hawkmoth pollination in the genus *Oenothera.* Aliso 5: 385–419.

Greuter, W. and T. Raus. 1987. Med-checklist notulae, 14. Willdenowia 16: 430–452.

Grisebach, A. H. R. 1866. Catalogus Plantarum Cubensium Exhibens Collectionem Wrightianam Aliasque Minores ex Insula Cuba Missas. Leipzig. (Cat. Pl. Cub.)

Groth, A. T., L. Lovett-Doust, and J. Lovett-Doust. 1996. Population density and module demography in *Trapa natans* (Trapaceae), an annual, clonal aquatic macrophyte. Amer. J. Bot. 83: 1406–1415.

Gupta, R. and S. J. Flora. 2005. Therapeutic value of *Hippophaë rhamnoides* L. against subchronic arsenic toxicity in mice. J. Med. Food 8: 353–361.

Gutzwiller, M.-A. 1961. Die phylogenetische Stellung von *Suriana maritima* L. Bot. Jahrb. Syst. 81: 1–49.

Haller, A. von. 1768. Historia Stirpium Indigenarum Helvetiae.... 3 vols. Bern. (Hist. Stirp. Helv.)

Handb. Fl. Ceylon—See: H. Trimen et al. 1893–1931

Hara, H. 1953. *Ludwigia* versus *Jussiaea.* J. Jap. Bot. 28: 289–294.

Harte, C. 1994. *Oenothera:* Contributions of a Plant to Biology. Berlin.

Harvard Pap. Bot. = Harvard Papers in Botany.

Harwood, C. E. 1989. *Grevillea robusta:* An Annotated Bibliography. Nairobi.

Harwood, C. E., ed. 1992. *Grevillea robusta* in Agroforestry and Forestry. Nairobi.

Haston, E. M. et al. 2007. A linear sequence of Angiosperm Phylogeny Group II families. Taxon 56: 7–12.

Hatch, S. L., K. N. Gandhi, and L. E. Brown. 1990. Checklist of the Vascular Plants of Texas. College Station, Tex.

Haussknecht, C. 1884. Monographie der Gattung *Epilobium.* Jena. (Monogr. Epilobium)

Häussknechtia = Häussknechtia; Mitteilungen der Thüringischen botanischen Gesellschaft.

Hedrick, U. P., ed. 1919. Sturtevant's Notes on Edible Plants. Albany. [Reprinted 1972, New York.]

Henderson, G., P. G. Holland, and G. L. Werren. 1979. The natural history of a subarctic adventive: *Epilobium angustifolium* L. (Onagraceae) at Schefferville, Quebec. Naturaliste Canad. 106: 425–437.

Heo, K. and H. Tobe. 1994. Embryology and relationships of *Suriana maritima* L. (Surianaceae). J. Pl. Res. 107: 29–37.

Herb. Raf.—See: C. S. Rafinesque 1833

Heredity = Heredity; an International Journal of Genetics.

Hill, J. 1759–1775. The Vegetable System.... 26 vols. London. (Veg. Syst.)

Hill, K. D. and L. A. S. Johnson. 1995. Systematic studies in the ecalypts. 7. A revision of the bloodworms, genus *Corymbia* (Myrtaceae). Telopea 6: 173–505.

Hilu, K. W. 2003. Angiosperm phylogeny based on *mat*K sequence information. Amer. J. Bot. 90: 1758–1776.

Hist. Nat. Pl.—See: C. de Mirbel and N. Jolyclerc [1802]–1806

Hist. Nat. Vég.—See: É. Spach 1834–1848

Hist. Pl.—See: H. E. Baillon 1866–1895

Hist. Stirp. Helv.—See: A. von Haller 1768

Hoch, P. C. 2012. *Chamerion.* In: B. G. Baldwin et al., eds. 2012. The Jepson Manual: Vascular Plants of California, ed. 2. Berkeley. Pp. 930–931.

Hoch, P. C. et al. 1993. A cladistic analysis of the plant family Onagraceae. Syst. Bot. 18: 31–47.

Hoch, P. C. et al. 1995. Proposal to reject the name *Epilobium alpinum* L. (Onagraceae). Taxon 44: 237–239.

Hoch, P. C. and B. J. Grewell. 2012. *Ludwigia.* In: B. G. Baldwin et al., eds. 2012. The Jepson Manual: Vascular Plants of California, ed. 2. Berkeley. Pp. 948–949.

Hoch, P. C. and P. H. Raven. 1990. Perspectives on Monographie der Gattung *Epilobium* of Carl Haussknecht. Häussknechtia 5: 21–28.

Hoch, P. C. and P. H. Raven. 1992. *Boisduvalia,* a coma-less *Epilobium* (Onagraceae). Phytologia 73: 456–459.

Hoch, P. C. and P. H. Raven. 1999. Onagraceae. In: K. Iwatsuki et al., eds. 1993+. Flora of Japan. 3+ vols. in 6+. Tokyo. Vol. 2c, pp. 224–246.

Hoggard, G. D. et al. 2004. The phylogeny of *Gaura* (Onagraceae) based on ITS, ETS and *trn*L-F sequence data. Amer. J. Bot. 91: 139–148.

Holm, T. 1929b. Morphology of North American species of *Polygala.* Bot. Gaz. 88: 167–185.

Holsinger, K. E. 1985. Taxonomic and nomenclatural notes on *Clarkia* sect. *Phaeostoma* Lewis & Lewis (Onagraceae). Taxon 34: 704–706.

Holsinger, K. E. and N. C. Ellstrand. 1984. The evolution and ecology of permanent translocation heterozygotes. Amer. Naturalist 124: 48–71.

Holub, J. 1972b. Taxonomic and nomenclatural remarks on *Chamaenerion* auct. Folia Geobot. Phytotax. 7: 81–90.

Hooker, W. J. [1829–]1833–1840. Flora Boreali-Americana; or, the Botany of the Northern Parts of British America.... 2 vols. in 12 parts. London, Paris, and Strasbourg. (Fl. Bor.-Amer.)

Hooker, W. J. and G. A. W. Arnott. [1830–]1841. The Botany of Captain Beechey's Voyage; Comprising an Account of the Plants Collected by Messrs Lay and Collie, and Other Officers of the Expedition, during the Voyage to the Pacific and Bering's Strait, Performed in His Majesty's Ship Blossom, under the Command of Captain F. W. Beechey...in the Years 1825, 26, 27, and 28. 10 parts. London. [Parts paged and plates numbered consecutively.] (Bot. Beechey Voy.)

Hort. Berol.—See: C. L. Willdenow 1803–1816

Hort. Bot. Vindob.—See: N. J. Jacquin 1770–1776

Hort. Brit. ed. 2—See: R. Sweet 1830

Hort. Kew.—See: W. Aiton 1789

Howard, R. A. 1974–1989. Flora of the Lesser Antilles: Leeward and Windward Islands. 6 vols. Jamaica Plain.

Howell, J. T. 1985. The genus *Ammannia* (Lythraceae) in western United States and Canada. Wasmann J. Biol. 43: 72–74.

Howell, T. J. 1885–1896. Howell's Pacific Coast Plants. 9 fasc. Arthur, Oreg. [Pages unnumbered.] (Howell's Pacific Coast Pl.)

Howell's Pacific Coast Pl.—See: T. J. Howell 1885–1896

Hrusa, F. et al. 2002. Catalogue of non-native vascular plants occurring spontaneously in California beyond those addressed in The Jepson Manual—Part I. Madroño 49: 61–98.

Hultén, E. 1941–1950. Flora of Alaska and Yukon. 10 vols. Lund and Leipzig. [Vols. paged consecutively and designated as simultaneous numbers of Lunds Univ. Årsskr. (= Acta Univ. Lund.) and Kungl. Fysiogr. Sällsk. Handl.]

Humboldt, A. von, A. J. Bonpland, and C. S. Kunth. 1815[1816]–1825. Nova Genera et Species Plantarum Quas in Peregrinatione Orbis Novi Collegerunt, Descripserunt.... 7 vols. in 36 parts. Paris. (Nov. Gen. Sp.)

Husband, B. C. and H. A. Sabara. 2004. Reproductive isolation between autotetraploids and their diploid progenitors in fireweed, *Chamerion angustifolium*. New Phytol. 161: 703–713.

Husband, B. C. and D. W. Schemske. 1996. Evolution of the magnitude and timing of inbreeding depression in plants. Evolution 50: 54–70.

Hutchinson, G. E. 1975. Limnological Botany. New York.

Hutchinson, J. 1959. The Families of Flowering Plants, ed. 2. 2 vols. Oxford.

Hutchinson, J. 1964–1967. The Genera of Flowering Plants (Angiospermae).... 2 vols. Oxford.

Icon.—See: A. J. Cavanilles 1791–1801

Icon. Pl. Select.—See: J. H. F. Link and C. F. Otto 1820–1828

Iconogr. Bot. Pl. Crit.—See: H. G. L. Reichenbach 1823–1832

Ill. Fl. N. U.S. ed. 2—See: N. L. Britton and A. Brown 1913

Index Kew.—See: B. D. Jackson et al. [1893–]1895+

Index Seminum (Hamburg) = Semina in Horto Botanico Hamburgensi...Collecta Quae pro Mutua Commutatione Offeruntur. [1823–1840 for years 1822–1840. Title after 1829: Delectus Seminum Quae in Horto Hamburgensium Botanico e Collectioni Anni...Mutuae Commutatione Offeruntur.]

Index Seminum (St. Petersburg) = Index Seminum, Quae Hortus Botanicus Imperialis Petropolitanus pro Mutua Commutatione Offert.

Index Seminum (Zürich) = Verzeichniss im Tausch abgebbarer Samaereien und Fruchte des Botanischen Gartens der Universität Zürich.

Int. J. Pl. Sci. = International Journal of Plant Sciences.

Ionta, G. M. et al. 2007. Phylogenetic relationships in *Rhexia* (Melastomataceae): Evidence from DNA sequence data and morphology. Int. J. Pl. Sci. 168: 1055–1066.

Iter Hispan.—See: P. Loefling 1758

Iwatsuki, K. et al., eds. 1993+. Flora of Japan. 3+ vols. in 6+. Tokyo.

Iwatsuki, K. and P. H. Raven, eds. 1997. Evolution and Diversification of Land Plants. Tokyo and New York.

Izco, J. 1983. *Epilobium paniculatum* nueva adventicia para Europa. Candollea 38: 309–315.

J. Acad. Nat. Sci. Philadelphia = Journal of the Academy of Natural Sciences of Philadelphia.

J. Amer. Soc. Hort. Sci. = Journal of the American Society for Horticultural Science.

J. Aquatic Pl. Managem. = Journal of Aquatic Plant Management.

J. Arnold Arbor. = Journal of the Arnold Arboretum.

J. Biogeogr. = Journal of Biogeography.

J. Bot. = Journal of Botany, British and Foreign.

J. Bot. (Hooker) = Journal of Botany, (Being a Second Series of the Botanical Miscellany), Containing Figures and Descriptions....

J. Bot. Res. Inst. Texas = Journal of the Botanical Research Institute of Texas.

J. Ecol. = Journal of Ecology.

J. Ethnopharmacol. = Journal of Ethnopharmacology; Interdisciplinary Journal Devoted to Bioscientific Research on Indigenous Drugs.

J. Exped. Trop. Australia—See: T. L. Mitchell 1848

J. Jap. Bot. = Journal of Japanese Botany.

J. Linn. Soc., Bot. = Journal of the Linnean Society. Botany.

J. Med. Food = Journal of Medicinal Food.

J. Pl. Res. = Journal of Plant Research. [Shokubutsu-gaku zasshi.]

J. Proc. Linn. Soc., Bot. = Journal of the Proceedings of the Linnean Society. Botany.

J. Proc. Roy. Soc. New S. Wales = Journal of the Proceedings of the Royal Society of New South Wales.

J. Trop. Ecol. = Journal of Tropical Ecology.

Jackson, B. D. et al., comps. [1893–]1895+. Index Kewensis Plantarum Phanerogamarum.... 2 vols. + 21+ suppls. Oxford. (Index Kew.)

Jacono, C. C. and V. V. Vandiver. 2007. *Rotala rotundifolia*, purple loosestrife of the south? Aquatics 29: 4–9.

Jacquemont, V. [1835–]1841–1844. Voyage dans l'Inde.... 4 vols. Paris. (Voy. Inde)

Jacquin, N. J. 1760. Enumeratio Systematica Plantarum, Quas in Insulis Caribaeis Vicinaque Americes Continente Detexit Novas.... Leiden. (Enum. Syst. Pl.)

Jacquin, N. J. 1770–1776. Hortus Botanicus Vindobonensis.... 3 vols. Vienna. (Hort. Bot. Vindob.)

Jahrb. Gewächsk. = Jahrbücher der Gewächskunde.

James, C. W. 1956. A revision of *Rhexia* (Melastomataceae). Brittonia 8: 201–230.

James, C. W. 1957. Notes on the cleistogamous species of *Polygala* in southeastern United States. Rhodora 59: 51–56.

James, E. [1822]1823. Account of an Expedition from Pittsburgh to the Rocky Mountains, Performed in the Years 1819 and '20...under the Command of Major Stephen H. Long. 2 vols. + atlas. Philadelphia. (Account Exped. Pittsburgh)

Jansen, S., F. Piesschaert, and E. Smets. 2000. Wood anatomy of Elaeagnaceae, with comments on vestured pits, helical thickenings, and systematic relationships. Amer. J. Bot. 87: 20–28.

Jepson, W. L. 1901. A Flora of Western Middle California.... Berkeley. (Fl. W. Calif.)

Jepson, W. L. 1909–1943. A Flora of California.... 3 vols. in 12 parts. San Francisco etc. [Pagination consecutive within each vol.; vol. 1 page sequence independent of part number sequence (chronological); part 8 of vol. 1 (pp. 1–32, 579–index) never published.] (Fl. Calif.)

Jepson, W. L. [1923–1925.] A Manual of the Flowering Plants of California.... Berkeley. (Man. Fl. Pl. Calif.)

Johnson, L. A. S. and B. G. Briggs. 1984. Myrtales and Myrtaceae: A phylogenetic analysis. Ann. Missouri Bot. Gard. 71: 700–756.

Johnson, M. T. et al. 2011. Loss of sexual recombination and segregation is associated with increased diversification in evening primroses. Evolution 65: 3230–3240.

Johnson, M. T., S. D. Smith, and M. D. Rausher. 2009. Plant sex and the evolution of plant defenses against herbivores. Proc. Natl. Acad. Sci. U.S.A. 106: 18079–18084.

Judd, W. S. et al. 2008. Plant Systematics: A Phylogenetic Approach, ed. 3. Sunderland, Mass.

Judd, W. S. and J. D. Skean. 1991. Taxonomic studies in the Miconieae (Melastomataceae) IV. Generic realignments among terminal-flowered taxa. Bull. Florida Mus. Nat. Hist., Biol. Sci. 36: 25–84.

Kajita, T. et al. 2001. *rbc*L and legume phylogeny, with particular reference to the Phaseoleae, Milletieae, and allies. Syst. Bot. 26: 515–536.

Kane, M. E. and L. S. Albert. 1982. Environmental and growth regulator effects on heterophylly and growth of *Proserpinaca intermedia* (Haloragaceae). Aquatic Bot. 13: 73–85.

Kane, M. E. and L. S. Albert. 1989. Abscisic acid induction of aerial leaf development in *Myriophyllum* and *Proserpinaca* species cultured in vitro. J. Aquatic Pl. Managem. 27: 102–111.

Katinas, L. et al. 2004. Geographical diversification of tribes Epilobieae, Gongylocarpeae, and Onagreae (Onagraceae) in North America, based on parsimony analysis of endemicity and track compatibility analysis. Ann. Missouri Bot. Gard. 91: 159–185.

Keating, R. C. 1982. The evolution and systematics of Onagraceae: Leaf anatomy. Ann. Missouri Bot. Gard. 69: 770–803.

Keating, R. C., P. C. Hoch, and P. H. Raven. 1982. Perennation in *Epilobium* (Onagraceae) and its relation to classification and ecology. Syst. Bot. 7: 379–404.

Kew Bull. = Kew Bulletin.

Kiger, R. W. and D. M. Porter. 2001. Categorical Glossary for the Flora of North America Project. Pittsburgh.

Kim, S. C., S. A. Graham, and A. Graham. 1994. Palynology and pollen dimophism in the genus *Lagerstroemia* (Lythraceae). Grana 33: 1–20.

Klein, W. M. 1964. A Biosystematic Study of Four Species of *Oenothera* Subgenus *Anogra*. Ph.D. dissertation. Claremont Graduate School.

Klein, W. M. 1970. The evolution of three diploid species of *Oenothera* subgenus *Anogra* (Onagraceae. Evolution 24: 578–597.

Knight, J. 1809. On the Cultivation of the Plants Belonging to the Natural Order of Proteeae.... London. (Cult. Prot.)

Knuth, P. 1906–1909. Handbook of Flower Pollination..., transl. J. R. Ainsworth Davis. 3 vols. Oxford.

Koehne, E. 1903. Lythraceae. In: H. G. A. Engler, ed. 1900–1953. Das Pflanzenreich.... 107 vols. Berlin. Vol. 17[IV,216], pp. 1–326.

Komarov, V. L., B. K. Schischkin, and E. Bobrov, eds. 1934–1964. Flora URSS.... 30 vols. Leningrad.

Krakos, K. N., J. S. Reece, and P. H. Raven. 2014. Molecular phylogenetics and reproductive biology of *Oenothera* section *Kneiffia* (Onagraceae). Syst. Bot. 39: 523–532.

Kral, R. and P. E. Bostick. 1969. The genus *Rhexia* (Melastomataceae). Sida 3: 387–440.

Kubitzki, K. et al., eds. 1990+. The Families and Genera of Vascular Plants. 15+ vols. Berlin etc.

Kuntze, O. 1891–1898. Revisio Generum Plantarum Vascularium Omnium atque Cellularium Multarum.... 3 vols. Leipzig etc. (Revis. Gen. Pl.)

Kurabayashi, M., H. Lewis, and P. H. Raven. 1962. A comparative study of mitosis in the Onagraceae. Amer. J. Bot. 9: 1003–1026.

Kytövuori, I. 1972. The Alpinae group of the genus *Epilobium* in northernmost Fennoscandia. A morphological, taxonomical and ecological study. Ann. Bot. Fenn. 9: 163–203.

L'Héritier de Brutelle, C.-L. 1788[1789–1792]. Sertum Anglicum.... 4 fascs. Paris. [All text in fasc. 1; plates numbered consecutively.] (Sert. Angl.)

Labillardière, J. J. H. de. [1800.] Relation du Voyage à la Recherche de la Pérouse.... 2 vols. Paris. (Voy. Rech. Pérouse)

Labillardière, J. J. H. de. 1804–1806[1807]. Novae Hollandiae Plantarum Specimen.... 2 vols. in parts. Paris. [Parts paged consecutively within vols., plates numbered consecutively throughout.] (Nov. Holl. Pl.)

Lack, A. J. 1995. Relationships and hybridization between British species of *Polygala*—evidence from isozymes. New Phytol. 130: 217–223.

Ladies' Flower-gard. Ornam. Perenn.—See: J. W. Loudon 1843–1844

Ladiges, P. Y., F. Udovicic, and A. N. Drinnan. 1995. Eucalypt phylogeny: Molecules and morphology. Austral. Syst. Bot. 8: 483–497.

Lamarck, J. et al. 1783–1817. Encyclopédie Méthodique. Botanique.... 13 vols. Paris and Liège. [Vols. 1–8, suppls. 1–5.] (Encycl.)

Lamarck, J. and A. P. de Candolle. 1805–1815. Flore Française, ou Descriptions Succinctes de Toutes les Plantes Qui Croissent Naturellement en France..., ed. 3. 5 tomes in 6 vols. Paris. [Tomes 1–4(2), vols. 1–5, 1805; tome 5, vol. 6, 1815.] (Fl. Franç. ed. 3)

Lamarck, J. and J. Poiret. 1791–1823. Tableau Encyclopédique et Méthodique des Trois Règnes de la Nature. Botanique.... 6 vols. Paris. [Vols. 1–2 = tome 1; vols. 3–5 = tome 2; vol. [6] = tome 3. Vols. paged consecutively within tomes.] (Tabl. Encycl.)

Landry, C. L., B. J. Rathcke, and L. B. Kass. 2009. Distribution of androdioecious and hermaphroditic populations of the mangrove *Laguncularia racemosa* (Combretaceae) in Florida and the Bahamas. J. Trop. Ecol. 25: 75–83.

LaRue, E. A. et al. 2013. Hybrid watermilfoil lineages are more invasive and less sensitive to a commonly used herbicide than their exotic parent (Eurasian watermilfoil). Evol. Applic. 6: 462–471.

Leafl. W. Bot. = Leaflets of Western Botany.

Lehmann, J. G. C. 1828–1857. Novarum et Minus Cognitarum Stirpium Pugillus I–X.... 10 vols. Hamburg. (Nov. Stirp. Pug.)

Lehmann, J. G. C. 1844–1847[–1848]. Plantae Preissianae.... 2 vols. in parts. Hamburg. [Parts paged consecutively within vols.] (Pl. Preiss.)

Les, D. H. and L. J. Mehrhoff. 1999. Introduction of nonindigenous aquatic vascular plants in southern New England: A historical perspective. Biol. Invasions 1: 281–300.

Levin, R. A. et al. 2003. Family-level relationships of Onagraceae based on on chloroplast *rbc*L and *ndh*F data. Amer. J. Bot. 90: 107–115.

Levin, R. A. et al. 2004. Paraphyly in tribe Onagreae: Insights into phylogenetic relationships of Onagraceae based on nuclear and chloroplast sequence data. Syst. Bot. 29: 147–164.

Lewis, H. and M. E. Lewis. 1955. The genus *Clarkia*. Univ. Calif. Publ. Bot. 20: 241–392.

Lewis, H. and D. M. Moore. 1962. Natural hybridization between *Epilobium adenocaulon* and *E. brevistylum*. Bull. Torrey Bot. Club 89: 365–370.

Lewis, H. and P. H. Raven. 1961. Phylogeny of the Onagraceae. Recent Advances Bot. 2: 1466–1469.

Lewis, H. and J. Szweykowski. 1964. The genus *Gayophytum* (Onagraceae). Brittonia 16: 343–391.

Lewis, W. H. and M. P. F. Elvin-Lewis. 1977. Medical Botany: Plants Affecting Man's Health. New York.

Lian, Y., Chen X. and Lian H. 1998. Systematic classification of the genus *Hippophaë* L. Seabuckthorn Res. 1: 13–23.

Lievens, A. W. and P. C. Hoch. 1999. *Epilobium*. In: K. Iwatsuki et al., eds. 1993+. Flora of Japan. 3+ vols. in 6+. Tokyo. Vol. 2c, pp. 241–246.

Link, J. H. F. 1821–1822. Enumeratio Plantarum Horti Regii Berolinensis Altera.... 2 parts. Berlin. (Enum. Hort. Berol. Alt.)

Link, J. H. F. and C. F. Otto. 1820–1828. Icones Plantarum Selectarum Horti Regii Botanici Berolinensis.... 10 parts. Berlin. (Icon. Pl. Select.)

Linnaea = Linnaea; ein Journal für die Botanik in ihrem ganzen Umfange.

Linnaeus, C. 1753. Species Plantarum.... 2 vols. Stockholm. (Sp. Pl.)

Linnaeus, C. 1754. Genera Plantarum..., ed. 5. Stockholm. (Gen. Pl. ed. 5)

Linnaeus, C. 1758[–1759]. Systema Naturae per Regna Tria Naturae..., ed. 10. 2 vols. Stockholm. (Syst. Nat. ed. 10)

Linnaeus, C. 1762 1763. Species Plantarum..., ed. 2. 2 vols. Stockholm. (Sp. Pl. ed. 2)

Linnaeus, C. 1766–1768. Systema Naturae per Regna Tria Naturae..., ed. 12. 3 vols. Stockholm. (Syst. Nat. ed. 12)

Linnaeus, C. 1767[–1771]. Mantissa Plantarum. 2 parts. Stockholm. [Mantissa [1] and Mantissa [2] Altera paged consecutively.] (Mant. Pl.)

Linsley, E. G. et al. 1973. Comparative beharior of bees and Onagraceae. V. *Camissonia* and *Oenothera* bees of cismontane California and Baja California. Univ. Calif. Publ. Entomol. 71: 1–76.

Linsley, E. G., J. W. MacSwain, and P. H. Raven. 1963. Comparative behavior of bees and Onagraceae. I. *Oenothera* bees of the Colorado Desert. Univ. Calif. Publ. Entomol. 33: 1–24.

Linsley, E. G., J. W. MacSwain, and P. H. Raven. 1963b. Comparative behavior of bees and Onagraceae. II. *Oenothera* bees of the Great Basin. Univ. Calif. Publ. Entomol. 33: 25–58.

Linsley, E. G., J. W. MacSwain, and P. H. Raven. 1964. Comparative behavior of bees and Onagraceae. III. *Oenothera* bees of the Mojave Desert. Univ. Calif. Publ. Entomol. 33: 59–98.

Liu, S. H. et al. 2017. Multi-locus phylogeny of *Ludwigia* (Onagraceae): Insights on infrageneric relationships and the current classification of the genus. Taxon 66: 1112–1127.

Loefling, P. 1758. Iter Hispanicum, Eller Resa til Spanska Ländern uti Europa och America, Förrättad Iffrån År 1751 til År 1756 ... Utgifven Efter Dess Frånfälle af Carl Linnaeus. Stockholm. (Iter Hispan.)

Long, R. W. and O. Lakela. 1971. A Flora of Tropical Florida: A Manual of the Seed Plants and Ferns of Southern Peninsular Florida. Coral Gables. [Reprinted 1976, Miami.]

Loudon, J. W. 1843–1844. The Ladies' Flower-garden of Ornamental Perennials. 2 vols. London. (Ladies' Flower-gard. Ornam. Perenn.)

Löve, Á. 1961. Some notes on *Myriophyllum spicatum*. Rhodora 63: 139–145.

Lucas, E. J. et al. 2005. Phylogenetic patterns in the fleshy-fruited Myrtaceae—Preliminary molecular evidence. Pl. Syst. Evol. 251: 35–51.

Lucas, E. J. et al. 2007. Suprageneric phylogenetics of Myrteae, the generically richest tribe in Myrtaceae. Taxon 56: 1105–1128.

Madroño = Madroño; Journal of the California Botanical Society [from vol. 3: a West American Journal of Botany].

Mal, T. K. et al. 1992. The biology of Canadian weeds. 100. *Lythrum salicaria*. Canad. J. Pl. Sci. 72: 1305–1330.

Malone, M. H. and A. Rother. 1994. *Heimia salicifolia*: A phytochemical and phytopharmacologic review. J. Ethnopharmacol. 42: 135–159.

Man. Fl. Pl. Calif.—See: W. L. Jepson [1923–1925]

Man. S.E. Fl.—See: J. K. Small 1933

Mant. Pl.—See: C. Linnaeus 1767[–1771]

Manual ed. 5—See: A. Gray 1867

Manual ed. 6—See: A. Gray et al. 1890

Martius, C. F. P. von, A. W. Eichler, and I. Urban, eds. 1840–1906. Flora Brasiliensis. 15 vols. in 40 parts, 130 fascs. Munich, Vienna, and Leipzig. [Vols. and parts numbered in systematic sequence, fascs. numbered independently in chronological sequence.] (Fl. Bras.)

Maurin, O. et al. 2010. Phylogenetic relationships of Combretaceae inferred from nuclear and plastid DNA sequence data: Implications for generic classification. Bot. J. Linn. Soc. 262: 453–476.

Maurin, O. et al. 2017. The inclusion of *Anogeissus, Buchenavia* and *Pteleopsis* in *Terminalia* (Combretaceae: Terminaliinae). Bot. J. Linn. Soc. 184: 312–325.

McAlpine, D. F. et al. 2007. Andean watermilfoil, *Myriophyllum quitense* (Haloragaceae), in the Saint John River estuary system, New Brunswick, Canada. Rhodora 109: 101–107.

McCallum, W. B. 1902. On the nature of the stimulus causing the change of form and structure in *Prosperpinaca palustris*. Bot. Gaz. 34: 93–108.

McVaugh, R. 1956. Tropical American Myrtaceae. Fieldiana, Bot. 29: 145–228.

McVaugh, R. 1963. Tropical American Myrtaceae, II. Fieldiana, Bot. 29: 395–532.

McVaugh, R. 1968. The genera of American Myrtaceae—An interim report. Taxon 17: 354–418.

McVaugh, R. 1969. The botany of the Guyana Highland—Part VIII. Myrtaceae. Mem. New York Bot. Gard. 18: 55–286.

Med. Repos. = Medical Repository.

Mém. Acad. Imp. Sci. Saint Pétersbourg, Sér. 7 = Mémoires de l'Académie Impériale des Sciences de Saint Pétersbourg, Septième Série.

Mem. Amer. Acad. Arts = Memoirs of the American Academy of Arts and Science.

Mém. Mus. Hist. Nat. = Mémoires du Muséum d'Histoire Naturelle.

Mem. New York Bot. Gard. = Memoirs of the New York Botanical Garden.

Mém. Soc. Phys. Genève = Mémoires de la Société de Physique et d'Histoire Naturelle de Genève.

Mem. Torrey Bot. Club = Memoirs of the Torrey Botanical Club.

Merriam-Webster. 1988. Webster's New Geographical Dictionary. Springfield, Mass.

Meyer, K. 2001. Revision of the southeast Asian genus *Melastoma* (Melastomataceae). Blumea 46: 351–398.

Michaux, A. 1803. Flora Boreali-Americana.... 2 vols. Paris and Strasbourg. (Fl. Bor.-Amer.)

Michelangeli, F. A. et al. 2004. A preliminary phylogeny of the tribe Miconieae (Melastomataceae) based on nrITS sequence data and its implications on inflorescence position. Taxon 53: 279–290.

Michelangeli, F. A. et al. 2008. Multiple events of dispersal and radiation of the tribe Miconieae (Melastomataceae) in the Caribbean. Bot. Rev. (Lancaster) 74: 53–77.

Michigan Bot. = Michigan Botanist.

Miller, P. 1768. The Gardeners Dictionary..., ed. 8. London. (Gard. Dict. ed. 8)

Mirbel, C. de and N. Jolyclerc. [1802]–1806. Histoire Naturelle...des Plantes. 18 vols. Paris. (Hist. Nat. Pl.)

Mitchell, T. L. 1848. Journal of an Expedition into the Interior of Tropical Australia.... London. (J. Exped. Trop. Australia)

Mohr, C. T. 1901. Plant life of Alabama. Contr. U.S. Natl. Herb. 6.

Molec. Phylogen. Evol. = Molecular Phylogenetics and Evolution.

Molina, G. I. 1782. Saggio sulla Storia Naturale del Chili.... Bologna. (Sag. Stor. Nat. Chili)

Monogr. Epilobium—See: C. Haussknecht 1884

Monogr. Phan.—See: A. L. P. P. de Candolle and C. de Candolle 1878–1896

Monogr. Syst. Bot. Missouri Bot. Gard. = Monographs in Systematic Botany from the Missouri Botanical Garden.

Moody, M. L. et al. 2016. Unraveling the biogeographic origins of the Eurasian watermilfoil *(Myriophyllum spicatum)* invasion in North America. Amer. J. Bot. 103: 709–718.

Moody, M. L. and D. H. Les. 2002. Evidence of hybridity in invasive watermilfoil *(Myriophyllum)* populations. Proc. Natl. Acad. Sci. U.S.A. 99: 14867–14871.

Moody, M. L. and D. H. Les. 2007. Phylogenetic systematics and character evolution in the angiosperm family (Haloragaceae). Amer. J. Bot. 94: 2005–2025.

Moody, M. L. and D. H. Les. 2007b. Geographic distribution and genetic composition of invasive hybrid watermilfoil (*Myriophyllum spicatum* × *M. sibiricum*) populations in North America. Biol. Invasions 9: 559–570.

Moody, M. L. and D. H. Les. 2010. Systematics of the aquatic angiosperm genus *Myriophyllum* Haloragaceae. Syst. Bot. 35: 121–139.

Morgan, D. R. and D. E. Soltis. 1993. Phylogenetic relationships among members of Saxifragaceae sensu lato based on *rbc*L sequence data. Ann. Missouri Bot. Gard. 80: 631–660.

Morong, T. 1891. Notes on North American Haloragaceae. Bull. Torrey Bot. Club 18: 229–246.

Morris, J. A. 2007. A Molecular Phylogeny of the Lythraceae and Inference of the Evolution of Heterostyly. Ph.D. dissertation. Kent State University.

Mosquin, T. 1966. A new taxonomy for *Epilobium angustifolium* L. (Onagraceae). Brittonia 18: 678–682.

Mosquin, T. 1967. Evidence for autopolyploidy in *Epilobium angustifolium* (Onagraeae). Evolution 21: 713–719.

Mosquin, T. and E. Small. 1971. An example of parallel evolution in *Epilobium* (Onagraceae). Evolution 25: 678–682.

Moxley, G. L. 1920. A study in *Zauschneria*. S. W. Sci. Bull. 1: 13–29.

Mozingo, H. N. and Margaret Williams. 1980. Threatened and Endangered Plants of Nevada: An Illustrated Manual. [Washington.]

Mueller, F. J. H. 1858–1882. Fragmenta Phytographiae Australiae.... 12 vols. + suppl. Melbourne. (Fragm.)

Muhlenbergia = Muhlenbergia; a Journal of Botany.

Munz, P. A. 1928. Studies in Onagraceae I. A revision of the subgenus *Chylismia* of the genus *Oenothera*. Amer. J. Bot. 15: 223–240.

Munz, P. A. 1937. Studies in Onagraceae X. The subgenus *Kneiffia*. Bull. Torrey Bot. Club 64: 287–306.

Munz, P. A. 1941. A revision of the genus *Boisduvalia* (Onagraceae). Darwiniana 5: 124–153.

Munz, P. A. 1942. Studies in Onagraceae XII. A revision of the New World species of *Jussiaea*. Darwiniana 4: 179–284.

Munz, P. A. 1944. Studies in Onagraceae XIII. The American species of *Ludwigia.* Bull. Torrey Bot. Club 71: 152–165.

Munz, P. A. 1965. Onagraceae. In: N. L. Britton et al., eds. 1905+. North American Flora.... 47+ vols. New York. Ser. 2, part 5, pp. 1–278.

Murray, J. A. 1784. Caroli à Linné Equitis Systema Vegetabilium.... Editio Decima Quarta.... Göttingen. (Syst. Veg. ed. 14)

N. Amer. Fl.—See: N. L. Britton et al. 1905+

Naranjao, P. and R. Escaleras, eds. 1995. La Medicina Tradicional en el Ecuador, Memorias de las Primeras Jornadas Ecuatorianas de Etnomedicina Andina. Quito.

Nat. Hist. Aleppo ed. 2—See: A. Russell 1794

Nat. Pflanzenfam.—See: H. G. A. Engler and K. Prantl 1887–1915

Naturaliste Canad. = Naturaliste Canadien. Bulletin de Recherches, Observations et Découvertes se Rapportant à l'Histoire Naturelle du Canada.

Nauman, C. E. 1981b. *Polygala grandiflora* (Polygalaceae) Walter re-examined. Sida 9: 1–18.

Nellis, D. W. 1994. Seashore Plants of South Florida and the Caribbean: A Guide to Identification and Propagation of Xeriscape Plants. Sarasota.

Nelson, A. 1935. The Elaeagnaceae—A mono-generic family. Amer. J. Bot. 22: 681–683.

Nelson, G. 1994. The Trees of Florida: A Reference and Field Guide. Sarasota.

Nesom, G. L. 2012. Infrageneric classification of *Rhexia* (Melastomataceae). Phytoneuron 2012-15: 1–9.

Nesom, G. L. and J. T. Kartesz. 2000. Observations on the *Ludwigia uruguayensis* complex (Onagraceae) in the United States. Castanea 65: 123–125.

New Mexico Bot. Newslett. = The New Mexico Botanist Newsletter.

New Phytol. = New Phytologist; a British Botanical Journal.

New Zealand J. Bot. = New Zealand Journal of Botany.

Nichols, S. A. 1975. Identification and management of Eurasian watermilfoil in Wisconsin. Trans. Wisconsin Acad. Sci. 63: 116–128.

Nordic J. Bot. = Nordic Journal of Botany.

Notizbl. Bot. Gart. Berlin-Dahlem = Notizblatt des Botanischen Gartens und Museums zu Berlin-Dahlem.

Nouv. Ann. Mus. Hist. Nat. = Nouvelles Annales du Muséum d'Histoire Naturelle.

Nov. Gen. Sp.—See: A. von Humboldt et al. 1815 [1816]–1825

Nov. Holl. Pl.—See: J. J. H. de Labillardière 1804–1806 [1807]

Nov. Stirp. Pug.—See: J. G. C. Lehmann 1828–1857

Novi Provent.—See: K. Sprengel [1818]

Novon = Novon; a Journal for Botanical Nomenclature.

Nuttall, T. 1813. A Catalogue of New and Interesting Plants Collected in Upper Louisiana.... London. (Cat. Pl. Upper Louisiana)

Nuttall, T. 1818. The Genera of North American Plants, and Catalogue of the Species, to the Year 1817.... 2 vols. Philadelphia. (Gen. N. Amer. Pl.)

Oeder, G. C. et al., eds. [1761–]1764–1883. Icones Plantarum...Florae Danicae Nomine Inscriptum. 17 vols. in 51 fascs. Copenhagen. [Fascs. paged independently and numbered consecutively throughout vols.] (Fl. Dan.)

Oesterr. Bot. Z. = Oesterreichische botanische Zeitschrift. Gemeinütziges Organ für Botanik.

Ohri, D. 1996. Genome size and polyploidy variation in the tropical hardwood genus *Terminalia* (Combretaceae). Pl. Syst. Evol. 200: 225–232.

Oken, L. 1833–1841. Allgemeine Naturgeschichte.... 7 vols. in 12. Stuttgart. [Botanik: vol. 2, 1839; vol. 3 (in 3), 1841.] (Allg. Naturgesch.)

Orchard, A. E. 1975. Taxonomic revisions in the family Haloragaceae. I. The genera *Haloragis, Haloragodendron, Glischrocaryon, Meziella* and *Gonocarpus.* Bull. Auckland Inst. Mus. 10: 1–299.

Orchard, A. E. 1979. *Myriophyllum* (Haloragaceae) in Australasia. 1. New Zealand: A revision of the genus and a synopsis of the family. Brunonia 2: 247–287.

Orchard, A. E. 1981. A revision of South American *Myriophyllum* (Haloragaceae), and its repercussions on some Australian and North American species. Brunonia 4: 27–65.

Orchard, A. E. 1985. *Myriophyllum* (Haloragaceae) in Australasia. II. The Australian species. Brunonia 8: 173–291.

Organization for Flora Neotropica. 1968+. Flora Neotropica. 121+ nos. New York.

Osborne, B. et al. 1991. *Gunnera tinctoria:* An unusual nitrogen-fixing invader. BioScience 41: 224–234.

Owens, S. J. and P. J. Rudall, eds. 1998. Reproductive Biology in Systematics, Conservation and Economic Botany. Kew.

Pacif. Railr. Rep.—See: War Department 1855–1860

Paiva, J. A. R. 1998. Polygalarum Africanarum et Madagascariensium prodromus atque gerontogaei generis *Heterosamara* Kuntze, a genere *Polygala* L. segregati et a nobis denuo recepti, synopsis monographica. Fontqueria 50: 1–346.

Pastore, J. F. B. 2013. A review of Vellozo's names for Polygalaceae in his Flora Fluminensis. Phytotaxa 108: 41–48.

Patel, V. C., J. J. Skvarla, and P. H. Raven. 1984. Pollen characters in relation to the delimitation of the Myrtales. Ann. Missouri Bot. Gard. 71: 858–969.

Patten, B. C. 1954. The status of some American species of *Myriophyllum* as revealed by discovery of intergrade material between *M. exalbescens* Fern. and *M. spicatum* in New Jersey. Rhodora 56: 213–225.

Patterson, D. T. et al. 1989. Composite List of Weeds. Champaign.

Peng, C. I. 1988. The biosystematics of *Ludwigia* sect. *Microcarpium* (Onagraceae). Ann. Missouri Bot. Gard. 75: 970–1009.

Peng, C. I. 1989. The systematics and evolution of *Ludwigia* sect. *Microcarpium* (Onagraceae). Ann. Missouri Bot. Gard. 76: 221–302.

Peng, C. I. 1990. *Ludwigia* ×*taiwanensis* (Onagraceae), a new species from Taiwan, and its origin. Bot. Bull. Acad. Sin. 31: 343–349.

Peng, C. I. et al. 2005. Systematics and evolution of *Ludwigia* section *Dantia* (Onagraceae). Ann. Missouri Bot. Gard. 92: 307–359.

Peng, C. I. and H. Tobe. 1987. Capsule wall anatomy in relation to capsular dehiscence in *Ludwigia* sect. *Microcarpium* (Onagraceae). Amer. J. Bot. 74: 1102–1110.

Persoon, C. H. 1805–1807. Synopsis Plantarum.... 2 vols. Paris and Tubingen. (Syn. Pl.)
Persson, C. 2001. Phylogenetic relationships in the Polygalaceae based on plastid DNA sequences from the *trn*L-F region. Taxon 50: 763–779.
Phytologia = Phytologia; Designed to Expedite Botanical Publication.
Phytomorphology = Phytomorphology; an International Journal of Plant Morphology.
Piper, C. V. and R. K. Beattie. 1915. Flora of the Northwest Coast.... Lancaster, Pa. (Fl. N.W. Coast)
Pl. Hartw.—See: G. Bentham 1839[–1857]
Pl. Horti Univ. Rar. Progr.—See: C. F. Rottbøll 1773b
Pl. Nov. Herb. Spreng.—See: J. F. T. Biehler 1807
Pl. Physiol. (Lancaster) = Plant Physiology.
Pl. Pratten. Calif.—See: E. M. Durand 1855
Pl. Preiss.—See: J. G. C. Lehmann 1844–1847[–1848]
Pl. Syst. Evol. = Plant Systematics and Evolution.
Pl. Veron.—See: J. F. Séguier 1745–1754
Pollard, M., J. R. Singhurst, and W. C. Holmes. 2009. *Polygala cymosa* (Polygalaceae) new to Texas. J. Bot. Res. Inst. Texas 3: 969–970.
Pollen & Spores = Pollen et Spores.
Polska Akademia Nauk. 1919–1980. Flora Polska.... 14 vols. Cracow and Warsaw.
Praglowski, J. et al. 1983. Fuchsieae L. Jussiaeeae L. World Pollen Spore Fl. 12: 1–41.
Praglowski, J. et al. 1987. Onagraceae Jussieu: Onagreae P. Raimann pro parte. World Pollen Spore Fl. 15: 1–55.
Praglowski, J. et al. 1989. Onagraceae Jussieu: Onagreae R. Raimann pro parte and Lopezieae Spach. World Pollen Spore Fl. 16: 1–35.
Praglowski, J. et al. 1994. Onagraceae Jussieu: Circaeeae DC, Haureae Raimann, Epilobieae Spach. World Pollen Spore Fl. 19: 1–38.
Précis Découv. Somiol.—See: C. S. Rafinesque 1814
Prelim. Cat.—See: N. L. Britton et al. 1888
Presl, C. B. 1825–1835. Reliquiae Haenkeanae seu Descriptiones et Icones Plantarum, Quas in America Meridionali et Boreali, in Insulis Philippinis et Marianis Collegit Thaddeus Haenke.... 2 vols. in 7 parts. Prague. (Reliq. Haenk.)
Proc. Acad. Nat. Sci. Philadelphia = Proceedings of the Academy of Natural Sciences of Philadelphia.
Proc. Amer. Acad. Arts = Proceedings of the American Academy of Arts and Sciences.
Proc. Biol. Soc. Wash. = Proceedings of the Biological Society of Washington.
Proc. Boston Soc. Nat. Hist. = Proceedings of the Boston Society of Natural History.
Proc. Bot. Soc. Brit. Isles = Proceedings of the Botanical Society of the British Isles.
Proc. Calif. Acad. Sci. = Proceedings of the California Academy of Sciences.
Proc. Davenport Acad. Nat. Sci. = Proceedings of the Davenport Academy of Natural Sciences.
Proc. Linn. Soc. New South Wales = Proceedings of the Linnean Society of New South Wales.
Proc. Natl. Acad. Sci. U.S.A. = Proceedings of the National Academy of Sciences of the United States of America.
Proc. Roy. Soc. Queensland = Proceedings of the Royal Society of Queensland.
Prodr.—See: A. P. de Candolle and A. L. P. P. de Candolle 1823—1873; O. P. Swartz 1788
Prodr. Pl. Cap.—See: C. P. Thunberg 1794–1800
Publ. Carnegie Inst. Wash. = Publications of the Carnegie Institution of Washington.
Publ. Field Mus. Nat. Hist., Bot. Ser. = Publications of the Field Museum of Natural History. Botanical Series.
Purcell, N. J. 1978. *Epilobium parviflorum* Schreb. (Onagraceae) established in North America. Rhodora 78: 785–787.
Pursh, F. [1813]1814. Flora Americae Septentrionalis; or, a Systematic Arrangement and Description of the Plants of North America. 2 vols. London. (Fl. Amer. Sept.)
Quart. Rev. Biol. = Quarterly Review of Biology.
Rafinesque, C. S. 1814. Précis des Découvertes et Travaux Somiologiques.... Palermo. (Précis Découv. Somiol.)
Rafinesque, C. S. 1833. Herbarium Rafinesquianum. .... Herbals; or Botanical Collections of C. S. Rafinesque... the Labor of a Whole Life! 3 parts. Philadelphia. [Parts designated as extras of Atlantic J., paged consecutively.] (Herb. Raf.)
Raguso, R. A. et al. 2007. Floral biology of North American *Oenothera* sect. *Lavauxia:* Advertisements, rewards and extreme variation in the floral depth. Ann. Missouri Bot. Gard. 94: 236–257.
Raimann, R. 1893. Onagraceae. In: H. G. A. Engler and K. Prantl, eds. 1887–1915. Die natürlichen Pflanzenfamilien.... 254 fascs. Leipzig. Fasc. 96[III,7], pp. 199–223.
Ram, M. 1956. Floral morphology and embryology of *Trapa bispinosa* Roxb. with a discussion on the systematic position of the genus. Phytomorphology 6: 312–323.
Ramamoorthy, T. P. 1979. A sectional revision of *Ludwigia* sect. *Myrtocarpus* sensu lato (Onagraceae). Ann. Missouri Bot. Gard. 66: 893–896.
Ramamoorthy, T. P. and E. Zardini. 1987. The systematics and evolution of *Ludwigia* sect. *Myrtocarpus* sensu lato (Onagraceae). Monogr. Syst. Bot. Missouri Bot. Gard. 19: 1–120.
Raven, P. H. 1962. The systematics of *Oenothera* subgenus *Chylismia*. Univ. Calif. Publ. Bot. 34: 1–122.
Raven, P. H. 1962b. The genus *Epilobium* in the Himalayan region. Bull. Brit. Mus. (Nat. Hist.), Bot. 2: 327–382.
Raven, P. H. 1963c. Amphitropical relationships in the floras of North and South America. Quart. Rev. Biol. 38: 151–177.
Raven, P. H. [1963]1964. The Old World species of *Ludwigia* (including *Jussiaea*), with a synopsis of the genus (Onagraceae). Reinwardtia 6: 327–427.
Raven, P. H. 1964. The generic subdivision of Onagraceae, tribe Onagreae. Brittonia 16: 276–288.
Raven, P. H. 1967. A revision of the African species of *Epilobium* (Onagraceae). Bothalia 9: 309–333.
Raven, P. H. 1968. *Epilobium*. In: T. G. Tutin et al., eds. 1964–1980. Flora Europaea. 5 vols. Cambridge. Vol. 2, pp. 308–311.
Raven, P. H. 1969. A revision of the genus *Camissonia* (Onagraceae). Contr. U.S. Natl. Herb. 37: 161–396.
Raven, P. H. 1976. Generic and sectional delimitation in Onagraceae, tribe Epilobieae. Ann. Missouri Bot. Gard. 63: 326–340.

Raven, P. H. 1978. Onagraceae. In: A. W. Exell and H. Wild, eds. 1960+. Flora Zambesiaca: Mozambique, Federation of Rhodesia and Nyasaland, Bechuanaland Protectorate. 12+ vols. in parts. London. Vol. 4, pp. 329–346.

Raven, P. H. 1979. A survey of reproductive biology in Onagraceae. New Zealand J. Bot. 17: 575–593.

Raven, P. H. 1988. Onagraceae as a model of plant evolution In: L. D. Gottlieb and S. K. Jain, eds. 1988. Plant Evolutionary Biology. A Symposium Honoring G. Ledyard Stebbins. London. Pp. 85–107.

Raven, P. H. and D. I. Axelrod. 1974. Angiosperm biogeography and past continental movements. Ann. Missouri Bot. Gard. 61: 539–673.

Raven, P. H. and D. I. Axelrod. 1978. Origin and relationships of the California flora. Univ. Calif. Publ. Bot. 72: 1–134.

Raven, P. H., W. Dietrich, and W. Stubbe. 1979. An outline of the systematics of *Oenothera* subsect. *Euoenothera* (Onagraceae). Syst. Bot. 4: 242–252.

Raven, P. H. and D. P. Gregory. 1972[1973]. A revision of the genus *Gaura* (Onagraceae). Mem. Torrey Bot. Club 23(1): 1–96.

Raven, P. H. and D. M. Moore. 1965. A revision of *Boisduvalia* (Onagraceae). Brittonia 17: 238–254.

Raven, P. H. and D. R. Parnell. 1977. Reinterpretation of the type of *Godetia bottae* Spach (Onagraceae). Ann. Missouri Bot. Gard. 64: 642–643.

Raven, P. H. and T. E. Raven. 1976. The Genus *Epilobium* (Onagraceae) in Australasia: A Systematic and Evolutionary Study. Christchurch. [New Zealand Dept. Sci. Industr. Res. Bull. 216.]

Raven, P. H. and W. Tai. 1979. Observations of chromosomes in *Ludwigia* (Onagraceae). Ann. Missouri Bot. Gard. 66: 862–879.

Recent Advances Bot. = Recent Advances in Botany.

Reed, C. F. 1970. Selected Weeds of the United States. Washington. [Agric. Handb. 366.]

Reichenbach, H. G. L. 1823–1832. Iconographia Botanica seu Plantae Criticae. 10 vols. Leipzig. [Vols. 6 and 7 each published in two half-centuries paged independently.] (Iconogr. Bot. Pl. Crit.)

Reichenbach, H. G. L. 1841. Der deutsche Botaniker.... Erster Band. Das Herbarienbuch. Dresden and Leipzig. [Alt. title: Repertorium Herbarii....] (Deut. Bot. Herb.-Buch)

Reliq. Haenk.—See: C. B. Presl 1825–1835

Renner, S. S. 1993. Phylogeny and classification of the Melastomataceae and Memecylaceae. Nordic J. Bot. 13: 519–540.

Renner, S. S., G. Clausing, and K. Meyer. 2001. Historical biogeography of Melastomataceae: The roles of Tertiary migration and long-distance dispersal. Amer. J. Bot. 88: 1290–1300.

Rep. (Annual) Gov. Bot. Director Bot. Zool. Gard. = Annual Report of the Government Botanist and Director of the Botanical and Zoological Garden.

Rep. (Annual) Missouri Bot. Gard. = Report (Annual) of the Missouri Botanical Garden.

Rep. Exped. Rocky Mts.—See: J. C. Frémont 1843–1845

Repert. Bot. Syst.—See: W. G. Walpers 1842–1847

Repert. Spec. Nov. Regni Veg. = Repertorium Specierum Novarum Regni Vegetabilis.

Rev. Palaeobot. Palynol. = Review of Palaeobotany and Palynology; an International Journal.

Revis. Gen. Pl.—See: O. Kuntze 1891–1898

Rhodora = Rhodora; Journal of the New England Botanical Club.

Richardson, J. E. et al. 2000. A phylogenetic analysis of Rhamnaceae using *rbc*L and *trn*L-F plastid DNA sequences. Amer. J. Bot. 87: 1309–1324.

Ripley, S. D. 1975. Report on the Endangered and Threatened Plant Species of the United States. Washington.

Robbins, H. C. 1968. The genus *Pachysandra* (Buxaceae). Sida 3: 211–248.

Robertson, R. et al., eds. 1981+. Flora of Australia. 32+ vols. Canberra.

Roemer, J. J., J. A. Schultes, and J. H. Schultes. 1817[–1830]. Caroli a Linné...Systema Vegetabilium...Editione XV.... 7 vols. Stuttgart. (Syst. Veg.)

Roth, A. W. 1797–1806. Catalecta Botanica.... 3 parts. Leipzig. [Parts paged independently.] (Catal. Bot.)

Rottbøll, C. F. 1773b. Plantas Horti Universitatis Rariores Programmate.... Copenhagen. (Pl. Horti Univ. Rar. Progr.)

Rousi, A. 1971. The genus *Hippophaë* L. A taxonomic study. Ann. Bot. Fenn. 8: 177–227.

Roxburgh, W. 1820–1824. Flora Indica; or Descriptions of Indian Plants.... 2 vols. Serampore. (Fl. Ind.)

Ruiz López, H. and J. A. Pavon. 1798. Systema Vegetabilium Florae Peruvianae et Chilensis.... [Madrid.] (Syst. Veg. Fl. Peruv. Chil.)

Russell, A. 1794. The Natural History of Aleppo, ed. 2. 2 vols. London. (Nat. Hist. Aleppo ed. 2)

Rydberg, P. A. 1917. Flora of the Rocky Mountains and Adjacent Plains. New York. (Fl. Rocky Mts.)

S. W. Sci. Bull. = Southwest Science Bulletin.

Sabara, H. A., P. Kron, and B. C. Husband. 2013. Cytotype coexistence leads to triploid hybrid production in the diploid-tetraploid contact zone of *Chamerion angustifolium* (Onagraceae). Amer. J. Bot. 100: 962–970.

Sag. Stor. Nat. Chili—See: G. I. Molina 1782

Salywon, A. 2003. A Monograph of *Mosiera* (Myrtaceae). Ph.D. dissertation. Arizona State University.

Savolainen, V. et al. 2000. Phylogenetics of flowering plants based on combined analysis of plastid *atp*B and *rbc*L gene sequences. Syst. Biol. 49: 306–362.

Savolainen, V. et al. 2000b. Phylogeny of the eudicots: A nearly complete familial analysis based on *rbc*L gene sequences. Kew Bull. 55: 257–309.

Schmidt, B. L. and W. F. Millington. 1968. Regulation of leaf shape in *Proserpinaca palustris*. Bull. Torrey Bot. Club 95: 264–286.

Schreber, J. C. 1771. Spicilegium Florae Lipsicae.... Leipzig. (Spic. Fl. Lips.)

Science = Science; an Illustrated Journal [later: a Weekly Journal Devoted to the Advancement of Science].

Scoggan, H. J. 1978–1979. The Flora of Canada. 4 parts. Ottawa. [Natl. Mus. Nat. Sci. Publ. Bot. 7.]

Scopoli, J. A. 1771–1772. Flora Carniolica..., ed. 2. 2 vols. Vienna. (Fl. Carniol. ed. 2)

Scott, A. J. 1978c. A revision of *Rhodomyrtus* (Myrtaceae). Kew Bull. 33: 311–329.

Scribailo, R. W. and M. S. Alix. 2006. *Myriophyllum tenellum* (Haloragaceae): An addition to the aquatic plant flora of Indiana. Rhodora 108: 76–79.

Sculthorpe, C. D. 1967. The Biology of Vascular Plants. London.

Seabuckthorn Res. = Seabuckthorn Research.
Seavey, S. R. 1992. Experimental hybridization and chromosome homologies in *Boisduvalia* sect. *Boisduvalia* (Onagraceae). Syst. Bot. 17: 84–90.
Seavey, S. R. 1993. Natural and artificial hybrids involving *Epilobium luteum* (Onagraceae). Syst. Bot. 18: 218–228.
Seavey, S. R. et al. 1977. Evolution of seed size, shape, and surface architecture in the tribe Epilobieae (Onagraceae). Ann. Missouri Bot. Gard. 64: 18–47.
Seavey, S. R. et al. 1977b. A comparison of *Epilobium minutum* and *E. foliosum* (Onagraceae). Madroño 24: 6–12.
Seavey, S. R. and K. S. Bawa. 1986. Late-acting self-incompatibility in angiosperms. Bot. Rev. (Lancaster) 52: 195–219.
Seavey, S. R. and S. K. Carter. 1994. Self-sterility in *Epilobium obcordatum* (Onagraceae). Amer. J. Bot. 81: 331–338.
Seavey, S. R. and S. K. Carter. 1996. Ovule fates in *Epilobium obcordatum* (Onagraceae). Amer. J. Bot. 83: 316–325.
Seavey, S. R. and P. H. Raven. 1977. Chromosomal evolution in *Epilobium* sect. *Epilobium* (Onagraceae). Pl. Syst. Evol. 127: 107–119.
Seavey, S. R. and P. H. Raven. 1977b. Chromosomal evolution in *Epilobium* sect. *Epilobium* (Onagraceae), II. Pl. Syst. Evol. 128: 195–200.
Seavey, S. R. and P. H. Raven. 1977c. Experimental hybrids in *Epilobium* (including sect. *Zauschneria*) species with $n = 15$ (Onagraceae). Amer. J. Bot. 64: 439–442.
Seavey, S. R. and P. H. Raven. 1977d. Chromosomal differentiation and the sources of the South American species of *Epilobium* (Onagraceae). J. Biogeogr. 4: 55–59.
Seavey, S. R. and P. H. Raven. 1978. Chromosomal evolution in *Epilobium* sect. *Epilobium* (Onagraceae), III. Pl. Syst. Evol. 130: 79–83.
Séguier, J. F. 1745–1754. Plantae Veronenses, seu Stirpium Quae in Agro Veronensi Reperiuntur Methodica Synopsis.... 3 vols. Verona. (Pl. Veron.)
Sennikov, A. N. 2011. *Chamerion* or *Chamaenerion* (Onagraceae)? The old story in new words. Taxon 60: 1485–1488.
Sert. Angl.—See: C.-L. L'Héritier de Brutelle 1788 [1789–1792]
Shady-Solis, R., J. Haas, and W. Creamer. 2001. Dating Caral, a preceramic site in the Supe Valley on the central coast of Peru. Science, ser. 2, 292: 723–726.
Shinners, L. H. 1941. *Epilobium paniculatum* var. *subulatum* in Wisconsin. Rhodora 43: 335.
Shinners, L. H. 1953. Synopsis of the United States species of *Lythrum* (Lythraceae). Field & Lab. 21: 80–89.
Sida = Sida; Contributions to Botany.
Sketch Bot. S. Carolina—See: S. Elliott [1816–]1821–1824
Skvarla, J. J. et al. 1978. An ultrastructural study of viscin threads in Onagraceae pollen. Pollen & Spores 20: 5–143.
Skvarla, J. J., P. H. Raven, and J. Praglowski. 1975. The evolution of pollen tetrads in Onagraceae. Amer. J. Bot. 62: 6–35.
Skvarla, J. J., P. H. Raven, and J. Praglowski. 1976. Ultrastructural survey of Onagraceae pollen. In: I. K. Ferguson and J. Muller, eds. 1976. The Evolutionary Significance of the Exine. London. Pp. 447–479.
Small, E. 1968. The systematics of autopolyploidy in *Epilobium latifolium* (Onagraceae). Brittonia 20: 169–181.
Small, E. 1971. The systematics of *Clarkia*, section *Myxocarpa*. Canad. J. Bot. 49: 1211–1217.
Small, J. K. 1896. *Oenothera* and its segregates. Bull. Torrey Bot. Club 23: 167–194.
Small, J. K. 1903. Flora of the Southeastern United States.... New York. (Fl. S.E. U.S.)
Small, J. K. 1913. Flora of the Southeastern United States..., ed. 2. New York. (Fl. S.E. U.S. ed. 2)
Small, J. K. 1933. Manual of the Southeastern Flora, Being Descriptions of the Seed Plants Growing Naturally in Florida, Alabama, Mississippi, Eastern Louisiana, Tennessee, North Carolina, South Carolina and Georgia. New York. (Man. S.E. Fl.)
Smith, C. E. 1965. The archeological record of cultivated crops of New World origins. Econ. Bot. 19: 322–334.
Smith, C. F. 1976. A Flora of the Santa Barbara Region, California. Santa Barbara.
Smith, C. F. 1998. A Flora of the Santa Barbara Region, California, ed. 2. Santa Barbara.
Smith, J. E. 1793[–1795]. A Specimen of the Botany of New Holland.... 1 vol. in 4 parts. London. [Parts paged and plates numbered consecutively.] (Spec. Bot. New Holland)
Smith, N. P. et al., eds. 2004. Flowering Plants of the Neotropics. Princeton.
Smith, R. R. and D. B. Ward. 1976. Taxonomy of the genus *Polygala* series *Decurrentes* (Polygalaceae). Sida 6: 284–310.
Smithsonian Contr. Knowl. = Smithsonian Contributions to Knowledge.
Solomon, J. C. 1982. The systematics and evolution of *Epilobium* (Onagraceae) in South America. Ann. Missouri Bot. Gard. 69: 239–335.
Soltis, D. E. et al. 1997b. Molecular phylogenetic relationships among angiosperms: An overview based on *rbc*L and 18S rDNA sequences. In: K. Iwatsuki and P. H. Raven, eds. 1997. Evolution and Diversification of Land Plants. Tokyo and New York. Pp. 157–178.
Soltis, D. E. et al. 2000. Angiosperm phylogeny inferred from 18S rDNA, *rbc*L, and *atp*B sequences. Bot. J. Linn. Soc. 133: 381–461.
Soltis, D. E. et al. 2011. Angiosperm phylogeny: 17 genes, 640 taxa. Amer. J. Bot. 98: 704–730.
Sorrie, B. A. and A. S. Weakley. 2017. Reassessment of variation within *Polygala cruciata* sensu lato (Polygalaceae). Phytoneuron 2017-37: 1–9.
Southwell, I. and R. Lowe, eds. 1999. Tea Tree: The Genus *Melaleuca*. Amsterdam.
Sp. Pl.—See: C. Linnaeus 1753; C. L. Willdenow et al. 1797–1830
Sp. Pl. ed. 2—See: C. Linnaeus 1762–1763
Spach, É. 1834–1848. Histoire Naturelle des Végétaux. Phanérogames.... 14 vols., atlas. Paris. (Hist. Nat. Vég.)
Spec. Bot. New Holland—See: J. E. Smith 1793[–1795]
Spic. Fl. Lips.—See: J. C. Schreber 1771
Sprengel, K. [1818.] Novi Proventus Hortorum Academicorum Halensis et Berolinensis. Halle. (Novi Provent.)

Sprengel, K. [1824–]1825–1828. Caroli Linnaei...Systema Vegetabilium. Editio Decima Sexta.... 5 vols. Göttingen. [Vol. 4 in 2 parts paged independently; vol. 5 by A. Sprengel.] (Syst. Veg.)

Spring Fl. Wasatch ed. 4—See: A. O. Garrett 1927

Stace, C. A. 1965. The significance of the leaf epidermis in the taxonomy of the Combretaceae. Bot. J. Linn. Soc. 59: 229–252.

Stace, C. A. 2004. Combretaceae. In: N. P. Smith et al., eds. 2004. Flowering Plants of the Neotropics. Princeton. Pp. 110–111.

Stace, C. A. 2007. Combretaceae. In: K. Kubitzki et al., eds. 1990+. The Families and Genera of Vascular Plants. 15+ vols. Berlin etc. Vol. 9, pp. 67–82.

Stace, C. A. 2010. Combretaceae. In: Organization for Flora Neotropica. 1968+. Flora Neotropica. 121+ nos. New York. No. 107.

Stafleu, F. A. et al. 1992–2009. Taxonomic Literature: A Selective Guide to Botanical Publications and Collections with Dates, Commentaries and Types. Supplement. 8 vols. Königstein.

Stafleu, F. A. and R. S. Cowan. 1976–1988. Taxonomic Literature: A Selective Guide to Botanical Publications and Collections with Dates, Commentaries and Types, ed. 2. 7 vols. Utrecht etc.

Steane, D. A. et al. 2002. Higher-level relationships among the eucalypts are resolved by ITS-sequence data. Austral. Syst. Bot. 15: 49–62.

Stebbins, G. L. 1971. Chromosómal Evolution in Higher Plants. London.

Steinberg, E. I. 1949. Onagraceae. In: V. L. Komarov et al., eds. 1934–1964. Flora URSS.... 30 vols. Leningrad. Vol. 15, pp. 565–637.

Stone, R. D. 2006. Phylogeny of major lineages in Melastomataceae, subfamily Olisbeoideae: Utility of nuclear glyceraldehyde 3-phosphate dehydrogenase (*Gap*C) gene sequences. Syst. Bot. 31: 107–121.

Straley, G. B. 1977. Systematics of *Oenothera* sect. *Kneiffia* (Onagraceae). Ann. Missouri Bot. Gard. 64: 381–424.

Straley, G. B. 1982. Octoploid populations of *Oenothera fruticosa* L. (Onagraceae) from coastal North Carolina. Rhodora 84: 281–283.

Strefeler, M. S. et al. 1996. Isozyme characterization of genetic diversity in Minnesota populations of purple loosestrife, *Lythrum salicaria* (Lythraceae). Amer. J. Bot. 83: 265–273.

Stubbe, W. 1964. The role of the plastome in the evolution of the genus *Oenothera*. Genetica 35: 28–33.

Stubbe, W. and P. H. Raven. 1979. A genetic contribution to the taxonomy of *Oenothera* sect. *Oenothera* (including subsections *Euoenothers, Emersonia, Raimannia,* and *Munzia*). Pl. Syst. Evol. 133: 39–59.

Stuckey, R. L. 1970. Distributional history of *Epilobium hirsutum* (great hairy willow-herb) in North America. Rhodora 72: 164–181.

Stuckey, R. L. 1980. Distributional history of *Lythrum salicaria* (purple loosestrife) in North America. Bartonia 47: 3–20.

Sturtevant, A. P. et al. 2009. Molecular characterization of Eurasian watermilfoil, northern milfoil, and the invasive interspecific hybrid in Michigan lakes. J. Aquatic Pl. Managem. 47: 128–135.

Summers, H. E., S. M.Hartwick, and R. A. Raguso. 2015. Geographic variation in floral allometry suggests repeated transitions between selfing and outcrossing in a mixed mating plant. Amer. J. Bot. 102: 745–757.

Suppl. Carp.—See: C. F. von Gaertner 1805–1807

Suppl. Prodr. Fl. Nov. Holl.—See: R. Brown 1830

Swartz, O. P. 1788. Nova Genera & Species Plantarum seu Prodromus.... Stockholm, Uppsala, and Åbo. (Prodr.)

Sweet, R. 1830. Hortus Britannicus..., ed. 2. London. (Hort. Brit. ed. 2)

Syn. Pl.—See: C. H. Persoon 1805–1807

Syst. Biol. = Systematic Biology.

Syst. Bot. = Systematic Botany; Quarterly Journal of the American Society of Plant Taxonomists.

Syst. Bot. Monogr. = Systematic Botany Monographs; Monographic Series of the American Society of Plant Taxonomists.

Syst. Nat.—See: J. F. Gmelin 1791[–1792]

Syst. Nat. ed. 10—See: C. Linnaeus 1758[–1759]

Syst. Nat. ed. 12—See: C. Linnaeus 1766[–1768]

Syst. Veg.—See: J. J. Roemer et al. 1817[–1830]; K. Sprengel [1824–]1825–1828

Syst. Veg. ed. 14—See: J. A. Murray 1784

Syst. Veg. Fl. Peruv. Chil.—See: H. Ruiz López and J. A. Pavon 1798

Sytsma, K. J. et al. 2004. Clades, clocks, and continents: Historical and biogeographical analysis of Myrtaceae, Vochysiaceae and relatives in the Southern Hemisphere. Int. J. Pl. Sci. 165(4, suppl.): 85–105.

Sytsma, K. J. and L. D. Gottlieb. 1986. Chloroplast DNA, evolution and phylogenetic relationships in *Clarkia* sect. *Peripetasma* (Onagraceae). Evolution 40: 1248–1261.

Sytsma, K. J. and L. D. Gottlieb. 1986b. Chloroplast DNA evidence for the origin of the genus *Heterogaura* from a species of *Clarkia* (Onagraceae). Proc. Natl. Acad. Sci. U.S.A. 83: 5554–5557.

Sytsma, K. J., J. F. Smith, and L. D. Gottlieb. 1990. Phylogenetics in *Clarkia* (Onagraceae): Restriction site mapping of chloroplast DNA. Syst. Bot. 15: 280–295.

Tabl. École Bot. ed. 3—See: R. L. Desfontaines 1829[–1832]

Tabl. Encycl.—See: J. Lamarck and J. Poiret 1791–1823

Tacik, T. 1959. Onagraceae. In: Polska Akademia Nauk. 1919–1980. Flora Polska.... 14 vols. Cracow and Warsaw. Vol. 9, p. 254.

Takhtajan, A. L. 1969. Flowering Plants: Origin and Dispersal. Edinburgh.

Takhtajan, A. L. 1987. Systema Magnoliophytorum. [Sistema Magnoliofitov.] Leningrad.

Tan, F. X. et al. 2001. Analysis of nrDNA ITS sequences in the subfamily Combretoideae (Combretaceae) and its systematic significance. Acta Bot. Yunnan. 23: 239–242.

Tan, F. X. et al. 2002. Phylogenetic relationships of Combretoideae (Combretaceae) inferred from plastid, nuclear gene and spacer sequences. J. Pl. Res. 115: 475–481.

Taxon = Taxon; Journal of the International Association for Plant Taxonomy.

Theiss, K. E., K. E. Holsinger, and M. E. Evans. 2010. Breeding system variation in 10 evening primroses (*Oenothera* sections *Anogra* and *Kleinia*; Onagraceae). Amer. J. Bot. 97: 1031–1039.

Thien, L. B. 1969. Translocations in *Gayophytum* (Onagraceae). Evolution 23: 456–465.

Thompson, D. Q., R. L. Stuckey, and E. B. Thompson. 1987. Spread, impact, and control of purple loosestrife *(Lythrum salicaria)* in North American wetlands. Fish Wildlife Res. 2: i–v, 1–55.

Thorne, R. F. 1992b. Classification and geography of the flowering plants. Bot. Rev. (Lancaster) 58: 225–348.

Thorne, R. F. 1993d. Phytogeography. In: Flora of North America Editorial Committee, eds. 1993+. Flora of North America North of Mexico. 21+ vols. New York and Oxford. Vol. 1, pp. 132–153.

Thum, R. A. et al. 2011. Molecular markers reconstruct the invasion history of variable leaf watermilfoil *(Myriophyllum heterophyllum)* and distinguish it from closely related species. Biol. Invasions 13: 1687–1709.

Thunberg, C. P. 1794–1800. Prodromus Plantarum Capensium, Quas in Promontorio Bonae Spei Africes, Annis 1772–1775, Collegit.... 2 parts. Uppsala. [Parts paged consecutively.] (Prodr. Pl. Cap.)

Tidestrom, I. 1925. Flora of Utah and Nevada.... Contr. U.S. Natl. Herb. 25.

Tobe, H., S. A. Graham, and P. H. Raven. 1998. Floral morphology and evolution in Lythraceae sensu lato. In: S. J. Owens and P. J. Rudall, eds. 1998. Reproductive Biology in Systematics, Conservation and Economic Botany. Kew. Pp. 329–334.

Tobe, H. and P. H. Raven. 1983. An embryological analysis of Myrtales: Its definition and characteristics. Ann. Missouri Bot. Gard. 70: 71–94.

Tobe, H. and P. H. Raven. 1985. The histogenesis and evolution of integuments in Onagraceae. Ann. Missouri Bot. Gard. 72: 451–468.

Tobe, H. and P. H. Raven. 1986. Evolution of polysporangiate anthers in Onagraceae. Amer. J. Bot. 73: 475–488.

Tobe, H. and P. H. Raven. 1986b. A comparative study of the embryology of *Ludwigia* (Onagraceae): Characteristics, variation, and relationships. Ann. Missouri Bot. Gard. 73: 768–787.

Tobe, H. and P. H. Raven. 1996. Embryology of Onagraceae (Myrtales): Characteristics, variation and relationships. Telopea 6: 667–688.

Tobe, H., P. H. Raven, and C. I. Peng. 1988. Seed coat anatomy and relationships of *Ludwigia* sects. *Microcarpium, Dantia,* and *Miquelia* (Onagraceae) and notes on fossil seeds of *Ludwigia* from Europe. Bot. Gaz. 149: 450–457.

Tobe, H., W. L. Wagner, and Chin H. C. 1987. Systematic and evolutionary studies of *Oenothera* (Onagraceae): Seed-coat anatomy. Bot. Gaz. 148: 235–257.

Torrey, J. and A. Gray. 1838–1843. A Flora of North America.... 2 vols. in 7 parts. New York, London, and Paris. (Fl. N. Amer.)

Torreya = Torreya; a Monthly Journal of Botanical Notes and News.

Towner, H. F. 1977. The biosystematics of *Calylophus* (Onagraceae). Ann. Missouri Bot. Gard. 64: 48–120.

Trans. Hort. Soc. London = Transactions, of the Horticultural Society of London.

Trans. Kansas Acad. Sci. = Transactions of the Kansas Academy of Science.

Trans. Linn. Soc. London = Transactions of the Linnean Society of London.

Trans. Med.-Bot. Soc. London = Transactions of the Medico-Botanical Society of London.

Trans. New York Acad. Sci. = Transactions of the New York Academy of Sciences.

Trans. Wisconsin Acad. Sci. = Transactions of the Wisconsin Academy of Sciences, Arts and Letters.

Transylvania J. Med. Assoc. Sci. = The Transylvania Journal of Medicine and the Associate Sciences.

Trauth-Nare, A. E. and R. F. C. Naczi. 1998. Taxonomic status of the varieties of Seneca snakeroot, *Polygala senega* L. (Polygalaceae). [Abstract.] Amer. J. Bot. 85(6, suppl.): 163.

Trelease, W. 1891. A revision of the American species of *Epilobium* occurring north of Mexico. Rep. (Annual) Missouri Bot. Gard. 2: 69–117.

Trimen, H., J. D. Hooker, and A. H. G. Alston. 1893–1931. A Hand-book to the Flora of Ceylon.... 6 vols. London. (Handb. Fl. Ceylon)

Tucker, S. C. 1979. Ontogeny of the inflorescence of *Saururus cernuus* (Saururaceae). Amer. J. Bot. 66: 227–236.

Turcotte, C. L. 1997. Towards Sustainable Harvesting of Seneca Snakeroot *(Polygala senega)* on Manitoba Hydro Rights-of-way. M.S. thesis. University of Manitoba.

Turner, B. L. et al. 2003. Atlas of the Vascular Plants of Texas. 2 vols. Fort Worth. [Sida Bot. Misc. 24.]

Tutin, T. G. et al., eds. 1964–1980. Flora Europaea. 5 vols. Cambridge.

U.S. Fish and Wildlife Service. 2010. Lewton's Polygala *(Polygala lewtonii)*. Five Year Review: Summary and Evaluation. Vero Beach.

U.S.D.A. Bur. Pl. Industr. Bull. = U S Department of Agriculture. Bureau of Plant Industry. Bulletin.

Udovicic, F. and P. Y. Ladiges. 2000. Informativeness of nuclear and chloroplast DNA regions and the phylogeny of the eucalypts and related genera (Myrtaceae). Kew Bull. 55: 633–645.

Ueno, S. and Y. Kadono. 2001. Monoecious plants of *Myriophyllum ussuriense* (Regel) Maxim. in Japan. J. Pl. Res. 114: 375–376.

Univ. Calif. Publ. Bot. = University of California Publications in Botany.

Univ. Calif. Publ. Entomol. = University of California Publications in Entomology.

Univ. Missouri Stud., Sci. Ser = University of Missouri Studies. Science Series.

University of Chicago Press. 1993. The Chicago Manual of Style, ed. 14. Chicago.

Useful Pl. Dominica—See: R. A. DeFilipps 1998

Veg. Syst.—See: J. Hill 1759–1775

Vellozo, J. M. 1825–1827[1829–1831]. Florae Fluminensis, seu Descriptionum Plantarum Praefectura Fluminensi Sponte Nascentium.... 1 vol. text + 11 vols. plates + indexes. Rio de Janeiro. (Fl. Flumin.)

Vodolazskij, L. E. 1979. On variety in the adventitious buds and shoots of the root-sucker forming herbaceous perennial *Chamaenerion angustifolium* (L.) Scop. (Onagraceae). Bot. Zhurn. S.S.S.R. 64: 734–739.

Voss, E. G. 1972–1996. Michigan Flora.... 3 vols. Bloomfield Hills and Ann Arbor.

Voy. Inde—See: V. Jacquemont [1835–]1841–1844

Voy. Rech. Pérouse—See: J. J. H. de Labillardière [1800]

Wagner, W. L. 1984. Reconsideration of *Oenothera* subg. *Gauropsis* (Onagraceae). Ann. Missouri Bot. Gard. 71: 1114–1127.

Wagner, W. L. 2004. Resolving a nomenclatural and taxonomic problem in Mexican *Oenothera* sect. *Hartmannia* (tribe Onagreae, Onagraceae). Novon 14: 124–133.

Wagner, W. L. 2005. Systematics of *Oenothera* sections *Contortae, Eremia, Pachylophus,* and *Ravenia* (Onagraceae). Syst. Bot. 30: 332–355.

Wagner, W. L. and P. C. Hoch. 2009. Nomenclatural corrections in Onagraceae. Novon 19: 130–132.

Wagner, W. L., P. C. Hoch, and P. H. Raven. 2007. Revised classification of the Onagraceae. Syst. Bot. Monogr. 83: 1–240.

Wagner, W. L., K. N. Krakos, and P. C. Hoch. 2013. Taxonomic changes in *Oenothera* sections *Gaura* and *Calylophus* (Onagraceae). PhytoKeys 28: 61–72.

Wagner, W. L., R. Stockhouse, and W. M. Klein. 1985. A systematic and evolutionary study of the *Oenothera cespitosa* species complex (Onagraceae). Monogr. Syst. Bot. Missouri Bot. Gard. 12: 1–103.

Wallenstein, A. and L. S. Albert. 1963. Plant morphology: Its control in *Proserpinaca* by photoperiod, temperature and gibberellic acid. Science, ser. 2, 140: 990–1000.

Walpers, W. G. 1842–1847. Repertorium Botanices Systematicae.... 6 vols. Leipzig. (Repert. Bot. Syst.)

Walter, T. 1788. Flora Caroliniana, Secundum Systema Vegetabilium Perillustris Linnaei Digesta.... London. (Fl. Carol.)

Wang, H. et al. 2009. Rosid radiation and the rapid rise of angiosperm-dominated forests. Proc. Natl. Acad. Sci. U.S.A. 106: 3853–3858.

Wanntorp, L., H.-E. Wanntorp, and M. Källersjö. 2002. Phylogenetic relationships of *Gunnera* based on nuclear ribosomal DNA ITS region, *rbc*L and *rps*16 intron sequences. Syst. Bot. 27: 512–521.

Wanntorp, L., H.-E. Wanntorp, and R. Rutishauser. 2003. On the homology of the scales in *Gunnera* (Gunneraceae). Bot. J. Linn. Soc. 142: 303–308.

War Department [U.S.]. 1855–1860. Reports of Explorations and Surveys, to Ascertain the Most Practicable and Economical Route for a Railroad from the Mississippi River to the Pacific Ocean. Made under the Direction of the Secretary of War, in 1853[–1856].... 12 vols. in 13. Washington. (Pacif. Railr. Rep.)

Warnock, M. J. 1981. Biosystematics of the *Delphinium carolinianum* complex (Ranunculaceae). Syst. Bot. 6: 38–54.

Wasmann J. Biol. = Wasmann Journal of Biology.

Watling, J. et al. 2018. Direct archaeological evidence for southwestern Amazonia as an early plant domestication and food production center. PLOS ONE 13(7): e0199868 DOI: 10.1371/journal/pone.0199868.

Watson, S. 1871. United States Geological Expolration [sic] of the Fortieth Parallel. Clarence King, Geologist-in-charge. [Vol. 5] Botany. By Sereno Watson.... Washington. [Botanical portion of larger work by C. King.] [Botany (Fortieth Parallel)]

Webb, D. A. 1967. Generic limits in European Lythraceae. Feddes Repert. 74: 10–13.

Weber, J. A. and L. D. Noodén. 1974. Turion formation and germination in *Myriophyllum verticillatum:* Phenology and its interpretation. Michigan Bot. 13: 151–158.

Weber, J. A. and L. D. Noodén. 1976. Environmental and hormonal control of turion germination in *Myriophyllum verticillatum.* Amer. J. Bot. 63: 936–944.

Weedscene = Weedscene; the Newsletter of The Weed Society of Victoria.

Wendt, T. L. 1978. A Systematic Study of *Polygala* Section *Rhinotropis* (Polygalaceae). Ph.D. dissertation. University of Texas.

Wendt, T. L. 1979. Notes on the genus *Polygala* in the United States and Mexico. J. Arnold Arbor. 60: 504–514.

Wilkinson, H. P. and L. Wanntorp. 2007. Gunneraceae In: K. Kubitzki et al., eds. 1990+. The Families and Genera of Vascular Plants. 15+ vols. Berlin etc. Vol. 9, pp. 177–183.

Willdenow, C. L. 1803–1816. Hortus Berolinensis.... 2 vols. in 10 fascs. Berlin. [Fascs. and plates numbered consecutively.] (Hort. Berol.)

Willdenow, C. L., C. F. Schwägrichen, and J. H. F. Link. 1797–1830. Caroli a Linné Species Plantarum.... Editio Quarta.... 6 vols. Berlin. [Vols. 1–5(1), 1797–1810, by Willdenow; vol. 5(2), 1830, by Schwägrichen; vol. 6, 1824–1825, by Link.] (Sp. Pl.)

Williams, P. A. et al. 2005. Chilean Rhubarb *(Gunnera tinctoria):* Biology, Ecology and Conservation Impacts in New Zealand. Wellington. [DOC Res. Developm. Ser. 210.]

Wilson, P. 1911. Surianaceae. In: N. L. Britton et al., eds. 1905+. North American Flora.... 47+ vols. New York. Vol. 25, p. 225.

Wilson, P. G. et al. 2005. Relationships within Myrtaceae sensu lato based on a *mat*K phylogeny. Pl. Syst. Evol. 251: 3–19.

Wood, A. 1870. The American Botanist and Florist; Including Lessons in the Structure, Life, and Growth of Plants; Together with a Simple Analytical Flora, Descriptive of the Native and Cultivated Plants Growing in the Atlantic Division of the American Union.... New York and Chicago. (Amer. Bot. Fl.)

Wood, M. 2006. Squelching water primrose. Agric. Res. (Washington, DC) 54(5): 13.

World Pollen Spore Fl. = World Pollen and Spore Flora.

Wrightia = Wrightia; a Botanical Journal.

Wrigley, J. W. and M. Fagg. 1993. Bottlebrushes, Paperbarks and Tea Trees. Sydney and Auckland.

Wunderlin, R. P. and B. F. Hansen. 2011. Guide to the Vascular Plants of Florida, ed. 3. Gainesville.

Wurdack, J. J. and R. Kral. 1982. The genera of Melastomataceae in the southeastern United States. J. Arnold Arbor. 63: 429–439.

Xie, L. et al. 2009. Molecular phylogeny, divergence time estimates, and historical biogeography of *Circaea* (Onagraceae). Molec. Phylogen. Evol. 53: 995–1009.

Zardini, E., C. I. Peng, and P. C. Hoch. 1991. Chromosome numbers in *Ludwigia* sections *Oligospermum* and *Oöcarpon* (Onagraceae). Taxon 40: 221–230.

Zardini, E. and P. H. Raven. 1992. A new section of *Ludwigia* (Onagraceae) with a key to the sections of the genus. Syst. Bot. 17: 481–485.

Zoë = Zoë; a Biological Journal.

Index

Names in *italics* are synonyms, casually mentioned hybrids, or plants not established in the flora. Page numbers in **boldface** indicate the primary entry for a taxon. Page numbers in *italics* indicate an illustration. Roman type is used for all other entries, including author names, vernacular names, and accepted scientific names for plants treated as established members of the flora.

Flora of North America — Index to families/volumes of vascular plants, current as of August 2018. **Boldface** denotes published volume: page number.